RIVER, COASTAL AND ESTUARINE MORPHODYNAMICS: RCEM 2007

BALKEMA – Proceedings and Monographs
in Engineering, Water and Earth Sciences

PROCEEDINGS OF THE 5TH IAHR SYMPOSIUM ON RIVER, COASTAL AND ESTUARINE MORPHODYNAMICS, ENSCHEDE, THE NETHERLANDS, 17–21 SEPTEMBER 2007

# River, Coastal and Estuarine Morphodynamics: RCEM 2007

## VOLUME 2

*Editors*

C. Marjolein Dohmen-Janssen & Suzanne J.M.H. Hulscher
*University of Twente, Faculty of Engineering Technology, Department of Water Engineering and Management (WEM), Enschede, The Netherlands*

Taylor & Francis
Taylor & Francis Group

LONDON / LEIDEN / NEW YORK / PHILADELPHIA / SINGAPORE

*Taylor & Francis is an imprint of the Taylor & Francis Group, an informa business*

© 2008 Taylor & Francis Group, London, UK

Typeset by Charon Tec Ltd (A Macmillan Company), Chennai, India
Printed and bound in Great Britain by TJ International Ltd, Padstow Cornwall

Published by:   Taylor & Francis/Balkema
                P.O. Box 447, 2300 AK Leiden, The Netherlands
                e-mail: Pub.NL@tandf.co.uk
                www.taylorandfrancis.co.uk/engineering, www.crcpress.com

ISBN Set: 978-0-415-45363-9
ISBN Vol. 1: 978-0-415-44167-4
ISBN Vol. 2: 978-0-415-45471-1

*River, Coastal and Estuarine Morphodynamics: RCEM 2007 – Dohmen-Janssen & Hulscher (eds)*
*© 2008 Taylor & Francis Group, London, ISBN 978-0-415-45363-9*

# Table of contents

## Longterm morphodynamics of estuaries

## Longterm morphodynamics rivers

## General aspects of longterm morphodynamics

XIII

XIV

# Preface

Around the world, many people live, work and recreate in river, estuarine and coastal areas, systems which are also important wildlife habitats. It is imperative to understand the physics of such systems. A key element here is morphodynamics: the mutual interaction and adjustment of landform topography and fluid dynamics involving the motion of sediment. The numerous interacting processes involved, such as large- and small-scale hydrodynamics, sediment transport dynamics, growth and decay of bed perturbations or larger bed forms, biological processes and human interferences make morphodynamics a challenging scientific issue.

The 5th IAHR-Symposium on River, Coastal and Estuarine Morphodynamics, RCEM 2007, was organised from 17 to 21 September 2007 at the University of Twente, Enschede, The Netherlands. This conference formed the follow-up of the earlier, successful, biannual conferences that were organised in Genova, Italy (1999), Obihiro, Japan (2001), Barcelona, Spain (2003) and Urbana, USA (2005). These proceedings of RCEM 2007 contain about 150 scientific papers that were presented during the conference, either as oral presentation or as poster. The papers are written by scientists, engineering consultants and water managers of universities, research institutes, consultancies and governance institutes from more than 20 countries around the world. In addition, five key-note lectures introduced the five specific topics of RCEM 2007:

A. Longterm morphodynamics
B. Biogeomorphology
C. Small-scale processes and grain sorting
D. Morphodynamic free behaviour
E. Human interferences in morphodynamics

Topic A (Longterm morphodynamics) focuses on longterm morphodynamic evolution of tidal inlets, ebb-tidal deltas, estuaries, beaches and river channels. Several types of morphodynamic models are presented, varying from schematized conceptual and simple mathematical models to numerical process-based models, as well as field observations.

Topic B (Biogeomorphology) deals with the interaction between biology and the physical system, i.e. sediment transport and morphology. Examples are the effect of different species on mobility of sediment and erosion rates, the effect of sediment transport on growth of species, the effect of vegetation on flow characteristics and the interaction between these different processes. Most papers on biogeomorphology in these proceedings focus on estuaries, which are systems in which biogeomorphology is particularly important, due to the presence of fine sediments and many different species.

Topic C (Small-scale processes and grain sorting) focuses on various small-scale processes, such as incipient motion and mobility of grains, sediment transport processes (bed load and/or suspended load) due to currents and/or waves offshore, in the surf- and swash zone, in tidal inlets and in rivers. In addition, attention is paid to the evolution of small-scale bed forms, such as wave ripples, and to the interactions between small-scale sediment transport processes and large-scale bed forms, such as dunes, sand waves and sand banks. Moreover, specific processes that occur in non-uniform sediment, such as hiding and exposure, partial or selective transport, downstream fining and sorting over bed forms are considered explicitly.

Within topic D (Morphodynamic free behaviour) two scales can be distinguished: On the one hand, the very large scale of features such as deltas, tidal channels and tidal flats, and river channels and river meanders. On the other hand, the relatively smaller scale of different types of bed forms, such as offshore sand banks and sand waves, river dunes and anti-dunes. Different types of models, laboratory experiments and field observations are presented to describe, understand and ultimately predict the behaviour of these large and small-scale features.

Finally, Topic E (Human interferences in morphodynamics) focuses on the impact of artificial structures or other human interferences on the morphodynamics of shallow seas, coasts, estuaries and rivers. Examples are the impacts of offshore wind farms, beach and river nourishments, dredging activities, cutting off of river meanders, coastal and river groynes or artificial basins.

The organization of RCEM 2007 would not have been possible without the help of many companies, institutes and individuals. First of all, we are very grateful to the financial sponsors of RCEM 2007: *University of Twente,*

*WL\Delft Hydraulics*, *Rijkswaterstaat RIZA* (Directorate-General for Public Works and Water Management), *Impact* (Institute of Mechanics, Processes and Control – Twente), *KIVI NIRIA* (The Royal Institution of Engineers in the Netherlands), *NAM (Nederlandse Aardolie Maatschappij)*, *VBKO (Vereniging van Waterbouwers in Bagger-Kust- en Oeverwerken)*, *Gemeente Enschede* (Municipality of Enschede), *ARGOSS, Alkyon – Hydraulic Consultancy and Research, J.M. Burgerscentre, DHV, HKV Consultants, PAO Delft* (Postgraduate Education Delft) and *Royal Boskalis Westminster nv*.

The conference was organized by people from the *Department of Water Engineering and Management (WEM)* of the *University of Twente* and we thank all the people of the organization committee for their ideas and contributions in making this conference a success. At this place, we specifically want to mention René Buijsrogge and Arthur Kamst for their valuable help in all ICT-related matters, Coby van Houten-Vos, for all her work on the financial aspects of the conference, all the students of the WEM-department for their assistance during the conference and, finally, our three wonderful secretaries, Joke Meijer, Brigitte Leurink and Anke Wigger for their enormous job in preparing the conference, their careful attention to the countless important details and for making the conference run very smoothly.

Last but not least, we thank all the authors for their excellent contributions to these proceedings and all the participants for their contributions to this successful conference, either in the form of an oral presentation or poster or in the form of their attention to the presentations and contributions to the discussions.

We hope that everyone enjoys reading the proceedings and that the papers inspire you in your research or other professional activities. *See you at the 6th RCEM in 2009!*

C. Marjolein Dohmen-Janssen
Suzanne J.M.H. Hulscher
*Editors*

*Topic D: Morphodynamic free behaviour*
*Large-scale offshore morphodynamics*

*River, Coastal and Estuarine Morphodynamics: RCEM 2007 – Dohmen-Janssen & Hulscher (eds)*
*© 2008 Taylor & Francis Group, London, ISBN 978-0-415-45363-9*

# A numerical algorithm to compute the morphodynamics of shallow tidal seas

Paolo Blondeaux & Giovanna Vittori

*Department of Civil, Environmental and Architectural Engineering, University of Genoa, Genoa, Italy*

ABSTRACT: A second-order accurate, highly efficient method is proposed to simulate unsteady three-dimensional flows over a complex topography. Efficiency is achieved by using boundary body forces that allow the imposition of the no- slip boundary condition on a given surface not coinciding with the computational grid. Therefore, the governing equations are discretised and solved on a regular mesh thus retaining the advantages of using standard solution procedures. In particular, the solution is determined by means of a finite difference approach. Standard centered second-order finite difference approximations of the spatial derivatives are used and the time-advancement of momentum equations employs a fractional-step method. To enforce the no-slip boundary condition at the immersed surface, the approach of Fadlun et al. (2000) is used, since in morpho-dynamic applications the bed surface is largely aligned in the two horizontal directions. The code is then used to investigate the process which leads to the appearance and development of sand waves, i.e. large scale bedforms which are observed in shallow tidal seas like the North Sea and the Adriatic Sea, in Italy.

## 1 INTRODUCTION

As pointed out by Seminara & Blondeaux (2001), in the last decades morphodynamics has developed at a much faster rate than in the past. Starting from the status of an essentially descriptive, empirically based discipline, morphodynamics has progressively moved towards a more mature stage, attempting to become predictive and quantitative, i.e. to understand why, when, and how much.

On the theoretical side, the linear stability analyses of basic coastal morphologies, aimed at investigating the mechanisms of appearance of rhythmic patterns of the seabed and/or the shoreline, have been replaced by attempts to incorporate nonlinear effects. In particular the interaction of different components of a random perturbation has been investigated to explain the formation of complex topographies (Vittori & Blondeaux, 1992; Schuttelaars & De Swart, 1999) and the nonlinear coupling between hydrodynamic and morphodynamic modes has been considered to understand the formation of particular morphological patterns (Vittori et al., 1999; Coco et al., 2000). Moreover, attempts to explain the simultaneous appearance of bedforms of different length scales through a self-exciting mechanism have been made (Komarova & Newell, 2000). However, it is clear that further progresses can be made only by means of fully nonlinear models.

The continuous growth of computer power strongly encourages engineers to rely on computational fluid dynamics for the solution of nonlinear problems. Indeed numerical simulations allow the analysis of phenomena without resorting to difficult experimental measurements or expensive field surveys. On the other hand, while simple geometries can be discretised by means of regular grids and thus efficiently handled by currently available codes and hardware, flows in complex geometries require body-fitted curvilinear or unstructured meshes and pose challenging problems for actual computers. Moreover, in morphodynamic applications, geometrical complexity is combined with moving boundaries which considerably increase the computational difficulties since they require regeneration or deformation of the grid. As a result, computer simulations of morphodynamic problems are very expensive and time consuming. Despite these difficulties, some attempts have been made to predict the equilibrium configuration of bottom forms by means of the numerical solution of the fully nonlinear problem (Andersen, 1999; Calvete et al., 1999, 2001, 2002; Németh et al., 2006). However this line of research needs a significant progress in order to obtain results at low computational costs.

In view of the difficulties mentioned above, it is clear that a numerical procedure that can cope with flow complexity but at the same time retain the accuracy and high efficiency of the simulations performed on fixed regular grids would represent a significant advance in the study of morphodynamic problems. A second-order accurate, highly efficient method is presently proposed to simulate unsteady three-dimensional flows in a complex geometry. This is achieved by using boundary body forces that allow the imposition of the no-slip boundary condition on a given surface not coinciding with the computational grid. Therefore, the governing equations can be discretised and solved on a regular mesh thus retaining the advantages and the efficiency of the standard solution procedures. In this method, the solid boundaries are reconstructed by adding forcing terms to the flow field equations. The surface may split a computational cell removing the constraint of the near wall gridlines to be aligned with the surface. This feature greatly simplifies the grid generation process which is cumbersome and expensive. The method is ideally suited for Cartesian flow solvers. Indeed, the flow equations appear in a very simple form and several numerical algorithms can be used for an efficient solution of the equations. In the present work, the solution is determined with a finite difference approach. Standard centered second-order finite difference approximations of the spatial derivatives are used and the time-advancement of momentum equations employs a fractional-step method. To enforce the no-slip boundary condition at the immersed surface, the approach of Fadlun et al. (2000) is used, since in morphodynamic applications the bed surface is largely aligned in the two horizontal directions.

The present code is used to investigate the process which leads to the appearance and development of sand waves, i.e. large scale bedforms which are observed in shallow tidal seas like the North Sea and the Adriatic Sea, in Italy.

Almost all the models, which are presently available to study the dynamics of sand waves, are linear models and assume that the amplitude of the sand waves is much smaller than their wavelength. Such models can explain the formation of sand waves but are unable to predict the equilibrium amplitude of these bottom forms. As already pointed out, a first study of the non-linear evolution of sand waves has recently appeared (Németh et al., 2006). In Németh et al. (2006), the dynamics of the sand waves is described by means of the two-dimensional shallow water equations, i.e. by assuming a hydrostatic pressure distribution. Hence, the model is not able to describe flow separation and therefore it cannot be applied to study very large sand wave heights. On the contrary, in the present analysis, the fully three-dimensional momentum equations are considered. Moreover, in Németh et al. (2006) a coordinate transformation is introduced which allows for a smooth description of the seabed topography but it makes more difficult and time consuming the solution of the model equations, because of the presence of the terms which take into account the transformation from the physical plane to the computational one.

## 2 FORMULATION OF THE PROBLEM

The problem formulation is similar to that described in Blondeaux & Vittori (2005a, 2005b) and Besio et al. (2006). A shallow sea of small depth $h^*$ is considered and a cartesian coordinate system $(x^*, y^*, z^*)$ is introduced such that the $x^*$ and $z^*$ axes are in the horizontal directions and the $y^*$- axis is vertical, pointing upwards and such that $y^* = 0$ describes the still water surface. The seabed is supposed to be made of a cohesionless sediment of uniform size $d^*$ and density $\rho_s^*$ (hereinafter a star denotes dimensional quantities).

As pointed out in the Introduction, the aim of the work is to determine the time development of the bottom configuration $y^* = -h^*(x^*, z^*, t^*)$ forced by tidal currents.

As discussed in Besio et al. (2006), Coriolis effects related to the Earth rotation do affect the formation of sand banks but they can be neglected when studying the time development of sand waves (see also Gerkema, 2000). Hence, by assuming that sand waves are two-dimensional bottom forms with crests orthogonal to the $x^*$-axis, the problem can be simplified by considering a two- dimensional flow independent of $z^*$.

On defining the following dimensionless variables

$$(x, y) = \frac{(x^*, y^*)}{h_0^*} \ , \ t = t^* \omega^* \tag{1}$$

$$(u, v) = \frac{(u^*, v^*)}{U_0^*} \ , \ p = \frac{p^*}{\varrho^* \omega^* h_0^* U_0^*} \ , \ h = \frac{h^*}{h_0^*}$$

($\rho^*$ is the sea water density, $t^*$ is time, $(u^*, v^*)$ are the velocity components along the $x^*$ and $y^*$ axes, $p^*$ is pressure, $h_0^*$ is the average water depth, $\omega^*$ is the angular frequency of the tide, $U_0^*$ is the maximum value of the depth averaged velocity during the tidal cycle), the flow equations become:

$$\frac{\partial u}{\partial x} + \frac{\partial v}{\partial y} = 0 \tag{2}$$

$$\frac{\partial u}{\partial t} + r \left[ u \frac{\partial u}{\partial x} + v \frac{\partial u}{\partial y} \right] = -\frac{\partial p}{\partial x} \tag{3}$$

$$+ \delta^2 \left[ \nu_T \left( \frac{\partial^2 u}{\partial x^2} + \frac{\partial^2 u}{\partial y^2} \right) + 2 \frac{\partial \nu_T}{\partial x} \frac{\partial u}{\partial x} + \frac{\partial \nu_T}{\partial y} \left( \frac{\partial u}{\partial y} + \frac{\partial v}{\partial x} \right) \right]$$

$$\frac{\partial v}{\partial t} + r \left[ u \frac{\partial v}{\partial x} + v \frac{\partial v}{\partial y} \right] = -\frac{\partial p}{\partial y} \tag{4}$$

$$+ \delta^2 \left[ \nu_T \left( \frac{\partial^2 v}{\partial x^2} + \frac{\partial^2 v}{\partial y^2} \right) + 2 \frac{\partial \nu_T}{\partial y} \frac{\partial v}{\partial y} + \frac{\partial \nu_T}{\partial x} \left( \frac{\partial v}{\partial x} + \frac{\partial u}{\partial y} \right) \right]$$

In (3) and (4), the flow regime is assumed to be turbulent and viscous effects are neglected. As discussed by Soulsby (1983), who analysed a lot of field data, turbulence structure can be assumed isotropic. Hence, using the Boussinesq hypothesis to model Reynolds stresses, a scalar kinematic eddy viscosity $v_T^*$ is introduced. Then, the kinematic eddy viscosity $v_T^*$ is written as the product $v_{T0}^* v_T$. The constant $v_{T0}^*$ is dimensional and provides the order of magnitude of the eddy viscosity while $v_T = v_T(x, y, t)$ is a dimensionless function (of order 1) describing the spatial and temporal variations of the turbulence structure. In (3)–(4), two dimensionless parameters appear which are denoted by $r$ and $\delta$ respectively:

$$r = \frac{U_0^*}{\omega^* h_0^*} , \quad \delta = \frac{\sqrt{v_{T0}^*/\omega^*}}{h_0^*} . \tag{5}$$

The parameter $r$ is the ratio between the amplitude of horizontal fluid displacement oscillations and the local depth. Actual values of $r$ are of order $10^2$. The parameter $\delta$ is the ratio between the thickness of the bottom boundary layer and the local depth. A rough estimate of $\delta$ shows that $\delta$ is of order one.

The hydrodynamic problem is then closed by appropriate boundary conditions. As discussed by Besio et al. (2006), at the free surface, the rigid lid approximation can be applied. Hence

$$\frac{\partial u}{\partial y} = 0 , \quad v = 0 \quad \text{at } y = 0 \tag{6}$$

Finally, the velocity is forced to vanish at a distance from the seabed equal to a fraction of the dimensionless roughness $z_r = z_r^*/h_0^*$, $z_r^*$ being the size of the bottom roughness.

$$u = 0 , \quad v = 0 \quad \text{for } z = -h + (z_r/\text{constant}) \tag{7}$$

In (7) the constant has been chosen equal to 29.8 as suggested by Fredsøe & Deigaard (1992) on the basis of the analysis of data on steady velocity profiles.

The time development of the bottom configuration $y = -h$ is provided by the sediment continuity equation which in dimensionless form reads:

$$\frac{\partial h}{\partial T} = \frac{\partial Q}{\partial x} \tag{8}$$

where $Q^*$, $Q$ are the dimensional and dimensionless volumetric sediment transport rates in the $x$-direction per unit width, such that $Q = (Q^*/\sqrt{(\rho_s^*/\rho^* - 1) g^* (d^*)^3}$. Moreover, the slow time scale

$$T = \frac{td}{(1 - p_{or}) \sqrt{\Psi_d}} \tag{9}$$

is introduced. In (9) $p_{or}$ is the sediment porosity and $d$ is the dimensionless sediment size which, along

with the mobility number $\Psi_d$ characterize the sediment particles

$$d = \frac{d^*}{h_0^*} ; \quad \Psi_d = \frac{(U_0^*)^2}{(\rho_s^*/\rho^* - 1) g^* d^*} \tag{10}$$

The problem can be closed once a model for the eddy viscosity $v_T^*$ is given and a predictive approach for $Q^*$ is chosen. The eddy viscosity $v_T^*$ is assumed to be described by the relationship which is given in details in Besio et al. (2006) and is not written herein for the sake of space.

Moreover, the approach proposed by Fredsøe & Deigaard (1992) is used to evaluate the bed load while the suspended load is assumed to be negligible. To complete the description of the sediment transport which takes place close to the sea bed, it is necessary to account for the weak effects associated with a slow spatial variation of the bottom topography, which affects the bed load sediment transport. Assuming that the bottom slope $dh/dx$ is small, simple dimensional arguments coupled with linearization lead to the following form of $Q$

$$Q = 9.55 \left[ (\theta - \theta_c)(\sqrt{\theta} - 0.7\sqrt{\theta_c}) \right.$$

$$\left. - \frac{\theta_c}{\mu} \left( \frac{3\theta - \theta_c - 1.4\sqrt{\theta\theta_c}}{2\sqrt{\theta}} \right) \frac{\partial h}{\partial x} \right] \frac{\tau^*}{|\tau^*|}$$

In (11), $\theta$ is the Shields parameter due to the tidal current and defined as

$$\theta = \frac{|\tau^*|}{(\rho_s^* - \rho^*) g^* d^*} \tag{12}$$

where $|\tau^*|$ is the modulus of the dimensional shear stress, which can be easily evaluated by means of the constitutive law. Moreover, $\theta_c$ is the critical value of $\theta$ such that for $\theta < \theta_c$ no sediment moves. Needless to say that (11) provides the amount of sediment moved by the current only for $\theta > \theta_c$ while $Q$ vanishes for $\theta < \theta_c$. The dimensionless constant $\mu$ can be estimated on the basis of the experimental observations of various authors (a.o. Talmon et al., 1995). Seminara (1998) suggested a value equal to about 0.5 but later Parker et al. (2003) proposed a value equal to 0.3 which is presently used.

## 3 THE NUMERICAL APPROACH

The solution of the problem is determined numerically with a finite difference approach. Standard centered second-order finite difference approximations of the spatial derivatives are used. The time-advancement of Navier-Stokes equation employs a fractional-step method extensively described by Kim & Moin (1985), Orlandi (2000) and Rai & Moin (1991). The non-solenoidal intermediate velocity field is evaluated by means of the Adam-Bashforth scheme to discretize convective terms together with a Crank-Nicholson scheme for the diffusive terms. The implicit treatment of the viscous terms would require the inversion of large sparse matrices which are reduced to tridiagonal matrices by a factorization procedure with an error of order $(\Delta t)^3$ Beam & Warming (1976). Then, by forcing the continuity equation, a Poisson equation for the pressure field is obtained which is readily solved by taking advantage of the imposed periodicity along the $x$-direction.

The equations are solved in a computational domain the size of which is $L_x$ in the streamwise direction. $N_x$, $N_y$ denote the number of grid points in the streamwise and vertical directions respectively. The mesh is uniform in the $x$-direction while in the vertical one a non-uniform mesh is used to cluster the grid-points in the vicinity of the bed, where velocity gradients are expected to be stronger. Different values of $N_x$ and $N_y$ are used depending on the values of the parameters of the simulation. The results described in the following have been obtained with $N_x = 256$ and $N_y = 100$, even though some of the simulations have been repeated with larger numbers of the grid points, to ascertain that the results do not depend on $N_x$ and $N_y$.

## 4 DISCUSSION OF THE RESULTS

Because of the large number of parameters which control the phenomenon, an exhaustive investigation of the paramenter space is not possible. We start by describing the results of a simulation made for values of the parameters similar to those encountered in the field. As already pointed out, if semidiurnal tides and shallow seas are considered, it can be easily verified that $\delta$ is a parameter of order one while $r$ is of order $10^2$. Moreover, for fine sands, the sediment mobility number is of order $10^2$. Finally, since small scale bedforms (ripples) usually cover the sea bottom, the roughness size turns out to be of order $10^{-2}$. Therefore, we have fixed the following values.

$$r = 70 \quad \delta = 1 \quad \Psi = 90 \quad z_r = 0.002 \tag{13}$$

Figure 1 shows the $x$−component of the tidal velocity at different phases $\varphi$ of the tidal cycle for a wavy bottom configuration the amplitude of which is equal to 0.1 (for $\varphi = 0$, the forcing pressure gradient is maximum). The flow over the crests is accelerated, while a significant reduction of the velocity appears over the troughs. Moreover, because of continuity equation, a significant vertical velocity component is generated which is negative along the lee sides of the bottom forms and positive along the stoss sides (see figure 2).

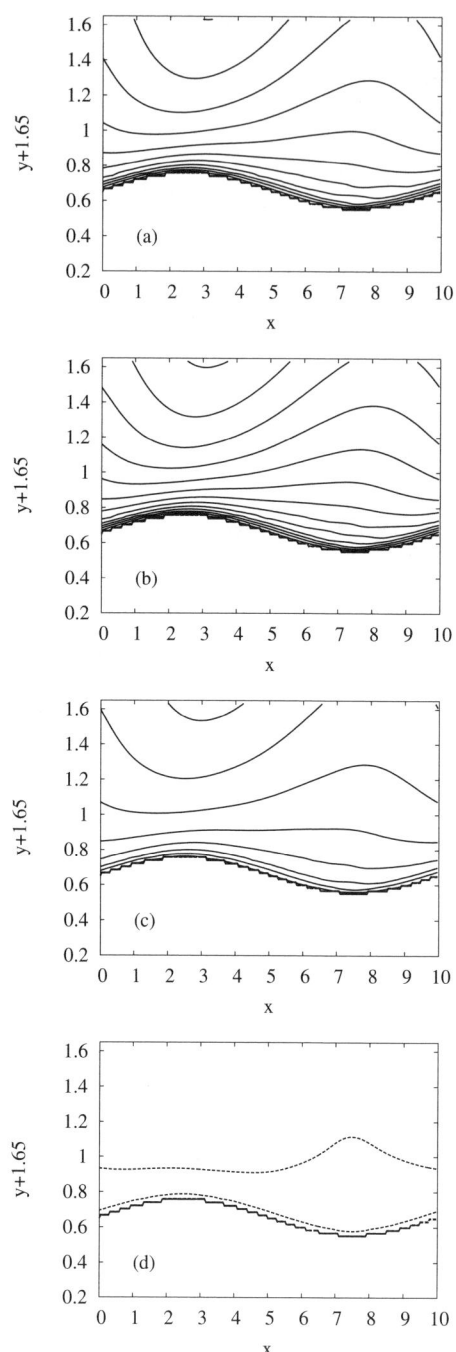

Figure 1. Contour lines of the streamwise velocity component for $r = 70$, $\delta = 1$, $\Psi = 90$, $z_r = 0.002$. (a) $\varphi = 0$, (b) $\varphi = \pi/4$, $\varphi = \pi/2$, $\varphi = 3\pi/4$. The dimensionless amplitude of the sand wave is equal to 0.1. Continuous lines = positive values, broken lines = negative values, $\Delta u = 0.05$.

676

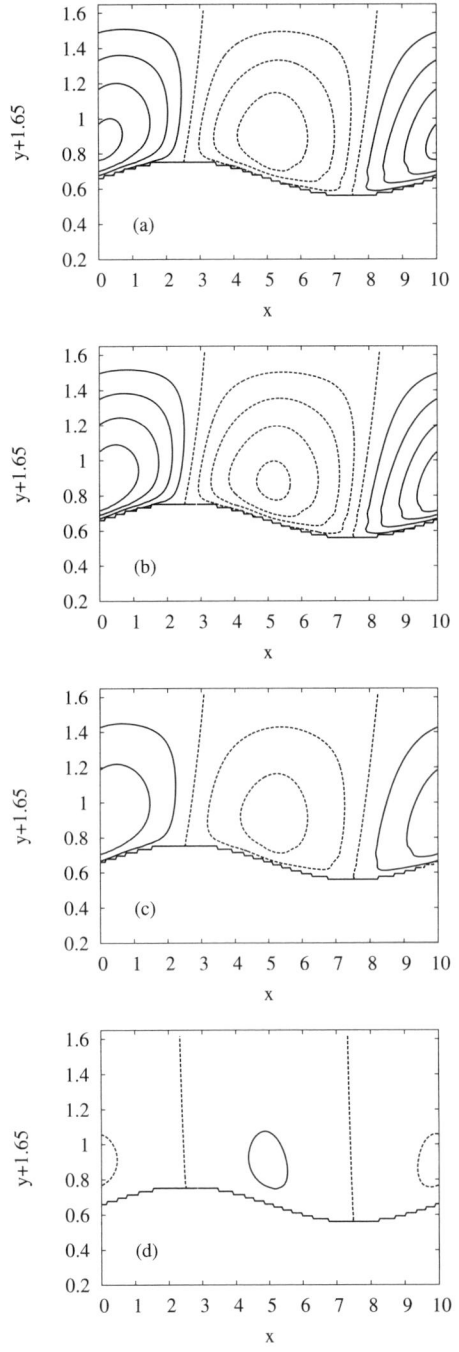

Figure 2. Contour lines of the vertical velocity component for $r = 70$, $\delta = 1$, $\Psi = 90$, $z_r = 0.002$. (a) $\varphi = 0$, (b) $\varphi = \pi/4$, $\varphi = \pi/2$, $\varphi = 3\pi/4$. The dimensionless amplitude of the sand wave is equal to 0.1. Continuous lines = positive values, broken lines = negative values, $\Delta v = 0.005$.

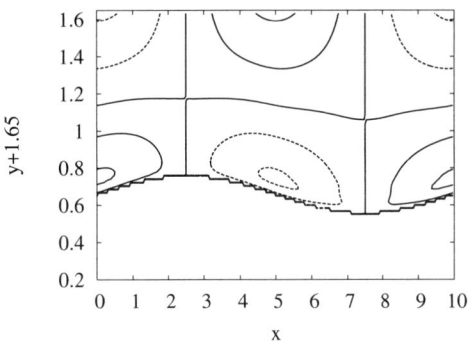

Figure 3. Contour lines of the steady steamwise velocity component for $r = 70$, $\delta = 1$, $\Psi = 90$, $z_r = 0.002$. (a) $\varphi = 0$, (b) $\varphi = \pi/4$, $\varphi = \pi/2$, $\varphi = 3\pi/4$. The dimensionless amplitude of the sand wave is equal to 0.1. Continuous lines = positive values, broken lines = negative values, $\Delta u = 0.05$.

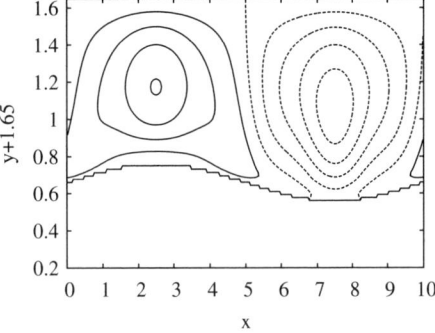

Figure 4. Contour lines of the vertical velocity component for $r = 70$, $\delta = 1$, $\Psi = 90$, $z_r = 0.002$. (a) $\varphi = 0$, (b) $\varphi = \pi/4$, $\varphi = \pi/2$, $\varphi = 3\pi/4$. The dimensionless amplitude of the sand wave is equal to 0.1. Continuous lines = positive values, broken lines = negative values, $\Delta v = 0.005$.

By averaging the flow over the forcing period, the steady streaming can be computed. Figures 3 and 4, where the obtained results for the horizontal and vertical velocity components are plotted respectively for an amplitude of the bottom waviness equal to 0.1, show that a steady velocity component is originated by the interaction of the oscillatory tidal current with the bottom waviness. The steady streaming is directed from the troughs towards the crests of the bottom waviness and tends to pile up the sediment at the crests of the bottom forms. The tendency of sediment to pile up near the crests is opposed by the gravity force acting down the slope. The growth or the decay of the bottom waviness is thus controlled by a balance between the above two effects. For small amplitudes of the bottom waviness, both the sediment transport induced by the steady streaming and that forced by gravity are linearly related to the amplitude of the bottom perturbation and

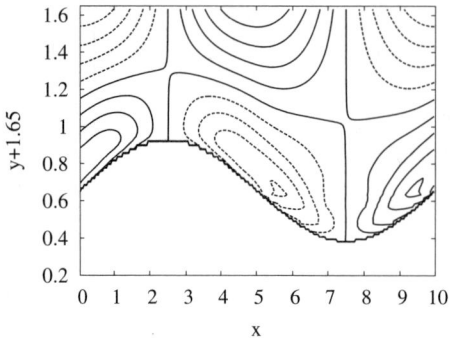

Figure 5. Contour lines of the steady steamwise velocity component for $r = 70$, $\delta = 1$, $\Psi = 90$, $z_r = 0.002$. (a) $\varphi = 0$, (b) $\varphi = \pi/4$, $\varphi = \pi/2$, $\varphi = 3\pi/4$. The dimensionless amplitude of the sand wave is equal to 0.27. Continuous lines = positive values, broken lines = negative values, $\Delta u = 0.05$.

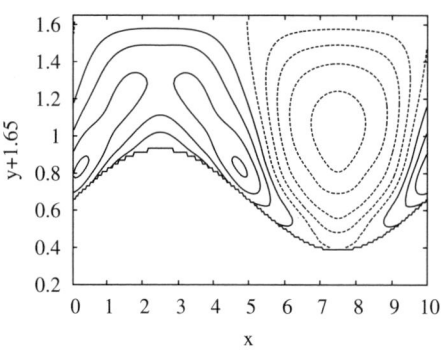

Figure 6. Contour lines of the vertical velocity component for $r = 70$, $\delta = 1$, $\Psi = 90$, $z_r = 0.002$. (a) $\varphi = 0$, (b) $\varphi = \pi/4$, $\varphi = \pi/2$, $\varphi = 3\pi/4$. The dimensionless amplitude of the sand wave is equal to 0.27. Continuous lines = positive values, broken lines = negative values, $\Delta v = 0.01$.

an exponential growth or decay of the bottom waviness is expected to take place. However, when the amplitude of the bottom form grows, nonlinear effects largely affect the strength of the steady streaming and a balance can be attained. The reader can compare figures 3 and 4 with figures 5 and 6 where the steady velocity components are plotted for an amplitude of the bottom waviness equal to 0.27.

In figure 7, the amplitude of the bottom profile is plotted versus time at the initial stage of the growth of the bottom form, to show the exponential growth which can be predicted also on the basis of a linear approach. The amplitude is obtained by making a Fourier analysis of the bottom profile and extracting the amplitude of the bottom waviness characterized by a wavelength equal to that of the initial perturbation. Moreover, in order to obtain the results displayed in figure 7 the morphodynamic phenomenon has been

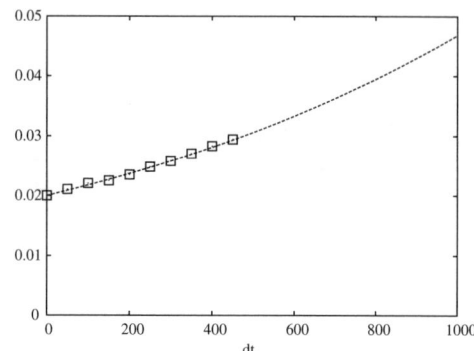

Figure 7. Amplitude of the simulated sand wave plotted versus time for $r = 70$, $\delta = 1$, $\Psi = 90$, $z_r = 0.002$.

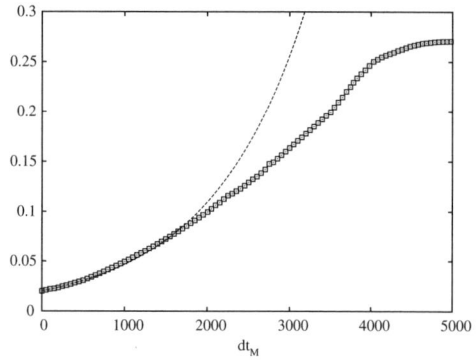

Figure 8. Amplitude of the simulated sand wave plotted versus time for $r = 70$, $\delta = 1$, $\Psi = 90$, $z_r = 0.002$.

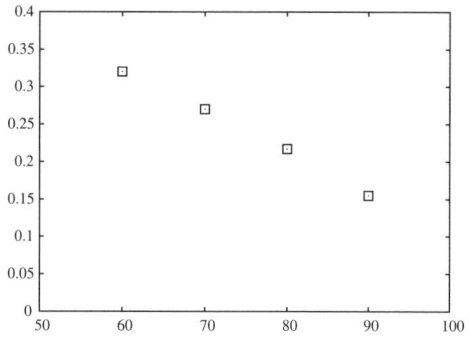

Figure 9. Equilibrium amplitude of the simulated sand wave plotted versus $r$ for $\delta = 1$, $\Psi = 90$, $z_r = 0.002$.

accelerated. In other words the flow during a tidal cycle has been computed explicitly, then, its effects on the bottom morphology have been multiplied by a factor N in such a way that the bottom changes become appreciable before computing the flow field over the updated bottom topography.

As already pointed out, even though the generation of the steady streaming takes place whatever amplitude of the bottom forms is present, the strength of the sediment transport induced by the steady velocity component increases at a different rate with respect to that induced by gravity effects till a balance is attained. The time development of the amplitude of the sand waves for large time is displayed in figure 8. After an initial exponential growth, during which nonlinear effects are negligible, significant values of the amplitude are attained and nonlinear effects make the bottom configuration to attain an equilibrium.

Of course the equilibrium amplitude depends on the parameters of the problem. Indeed if the Keulegan-Carpenter number of the phenomenon is varied and the other paramenters are kept fixed, the equilibrium amplitude changes. This can be clearly observed in figure 9, where the equilibrium amplitude of the sand waves is plotted versus $r$ for the simulated cases.

## 5   CONCLUSIONS

A numerical model able to describe the time development of the bottom of shallow tidal seas is described. The numerical model, which is based on a finite difference approach and uses the immersed boundary technique, turns out to be quite efficient since it uses a Cartesian coordinate system even for complex geometries of the bottom profile. Even though the results discussed in the paper describe the process which leads to the appearance of sand waves starting from a small amplitude bottom waviness, the model can be applied to investigate other phenomena like the time development of a sand pit or a large scale bottom form.

## ACKNOWLEDGEMENTS

This research has been supported by the Ministero dell'Universitá e della Ricerca under contract 'Cave sottomarine e ripascimenti: modellazione morfologica e applicazioni' n. 2005080197

## REFERENCES

Andersen, K.H 1999. The dynamics of ripples beneath surface waves and topics in shell models of turbulence. *Ph.D. Thesis, University of Copenhagen.*

Beam, R. M. & Warming R.F. 1976. An implicit finite-difference algorithm for hyperbolic system in conservation-law form. *J. Comput. Phys.* **22**, 87.

Besio, G., Blondeaux, P., & Vittori, G. 2006. On the formation of sand waves and sand banks. *J. Fluid Mech.* **557**, 1–27.

Blondeaux, P. & Vittori, G. 2005a. Flow and sediment transport induced by tide propagation. Part 1: the flat bottom case. *J. Geophys. Res.* **110**, C07020, doi:10.1029/2004JC002532

Blondeaux, P. & Vittori, G. 2005b. Flow and sediment transport induced by tide propagation. Part 2: the wavy bottom case . *J. Geophys. Res.* **110**, C08003, doi:10.1029/2004JC002545

Calvete, D., De Swart, H.E. & Falques, A. 2002. Effect of depth-dependent stirring on the final amplitude of shoreface-connected sand ridges. *Cont. Shelf Res.* **22**, 2763–2776.

Calvete, D., Falques, A., De Swart, H.E. & Dodd, D. 1999. Nonlinear modelling of shoreface-connected sand ridges. In *Coastal Sediments, A.S.C.E.*, June 20–24, 1123–1138.

Calvete, D., Falques, A., De Swart, H.E. & Walgreen, M. 2001. Modelling the formation of shoreface-connected sand ridges on storm-dominated inner shelves. *J. Fluid Mech.* 441,169–193.

Coco, G., Huntley, D.A. & O'Hare T.J. 2000. Investigation of a self-organized model for beach cusp formation and development. *J. Geophys. Res.* **C9, 105**, 21991–22002.

Fadlun, E.A. Verzicco, R., Orlandi, P. & Mohd-Yusof, J. 2000. Combined immersed-boundary/finite-difference methods for three-dimensional complex flow simulations *J. Comp. Phys.* **161**, 35.

Fredsøe, J. & Deigaard, R. 1992. *Mechanics of coastal sediment transport.* Advanced Series on Ocean Engineering. World Scientific, Singapore, xviii+369 p.

Gerkema, T. 2000. A linear stability analysis of tidally generated sand waves *J. Fluid Mech.* **417**, 303–322.

Kim, J. & Moin P. 1985. Application of a fractional-step method to incompressible Navier-Stokes equations. *J. Comput. Phys.* **59**, 308.

Komarova, N.L. & Newell A.C. 2000. Nonlinear dynamics of sand banks and sand waves. *J. Fluid Mech.* **415**, 285–312.

Németh, A.A., Hulscher, S.J.M.H. & van Damme, R.M.J. 2006. Simulating offshore sand waves. *Coastal Eng.* **53**, 265–275.

Orlandi, P. 2000. Fluid flow phenomena: A numerical toolkit, Dordrecht, Kluwer *Dordrechtm Kluver Cont.*.

Parker G., Seminara G. & Solari L. 2003. Bedload transport at low Shields stress on arbitrarily sloping beds: alternative entrainment formulation. *Water Resource Res.* **39, 7**, 1,183.

Rai M.M. & Moin P. 1991. Direct simulations of turbulent flow using finite-difference schemes. *J. Comput. Phys.* **96**, 15.

Schuttelaars, H.M. & De Swart H.E. 1999. Formation of channels and shoals in a short tidal embayment. *J. Fluid Mech.* **386**, 15–42.

Seminara, G. 1998. Stability and morphodynamics. *Meccanica.* **33**, 59–99.

Seminara, G. & Blondeaux, P. 2001. Perspectives in morphodynamics. In *River, Coastal and Estuarine Morphodynamics.* Ed. G. Seminara & P. Blondeaux. Springer

Soulsby, R.L. 1983. The bottom boundary layer of shelf seas. In *Physical Oceanography of Coastal and Shelf Seas.* Ed. B. Johns, Elsevier, Amsterdam, 189–266.

Talmon A.M., Struiksma N. & Van Mierlo M.C.L.M. 1995. Laboratory measurements of the direction of sediment transport on transverse alluvial-bed slopes. *J. Hydr. Res.* **33**, 495–517.

Vittori, G. & Blondeaux, P. 1992. Sea ripples under sea waves. Part 3. Brick-pattern ripple formation. *J. Fluid Mech.* **239**, 23–45.

Vittori, G., De Swart, H.E. & Blondeaux, P. 1999. Crescentic bedforms in the nearshore region. *J. Fluid Mech.* **381**, 271–303.

*River, Coastal and Estuarine Morphodynamics: RCEM 2007 – Dohmen-Janssen & Hulscher (eds)*
*© 2008 Taylor & Francis Group, London, ISBN 978-0-415-45363-9*

# Investigating plan-view asymmetry in wave-influenced deltas

A.D. Ashton & L. Giosan
*Woods Hole Oceanographic Institution, Woods Hole, Massachusetts, USA*

ABSTRACT: We investigate the asymmetrical development of deltas in terms of a high-angle-wave instability in the shape of a shoreline due to breaking-wave driven alongshore sediment transport. Demonstrating the strong wave-angle dependence of shoreline evolution, numerical modeling suggests that the characteristics of the wave climate can play an integral role in the morphological evolution of wave-influenced deltas. Systematic analysis demonstrates that delta development style, including asymmetrical evolution, depends on wave climate characteristics and the relative rate of sediment input. Although the 'river delta asymmetry index' presented by Bhattacharya and Giosan (2003) can predict asymmetrical behavior in natural cases, the simulated asymmetric development is poorly explained by this index, probably because the simplified modeling approach does not adequately resolve river mouth dynamics. However, the simulation results suggest that wave-angle distributions can constitute a first-order control on the plan-form expression of wave-influenced deltas.

## 1 INTRODUCTION

Deltas form where rivers deliver their sediment to open water; the competition between sediment delivery by the river and marine forces that rework this sediment determines the morphology of deltas. While many studies have explored how the balance of these forces affects the cross-shore evolution of deltas (e.g. Swenson et al 2005), there have been relatively few quantitative investigations of the plan-view evolution of deltas. Here, we apply numerical modeling and a new understanding of shoreline evolution due to alongshore sediment transport to study the plan-view evolution of deltas that are strongly influenced by sediment reworking by waves. When there is a regional net direction of alongshore sediment transport, this often leads to an asymmetrical development of a delta, where depositional styles differ on either side of a river mouth (Bhattacharya & Giosan 2003). Although the relative rates of sediment delivery to the coast versus alongshore sediment fluxes help determine the context for a delta's form, the results presented here suggest that the wave-angle distribution can play a first-order role in the plan-view morphology and asymmetrical development of wave-influenced deltas.

## 2 ASYMMETRICAL WAVE-INFLUENCED DELTAS

Classically, the plan-view expression of deltas has been understood via a tripartite model, whereby the relative strength of river, waves, and tides determines delta morphology (Galloway 1975). In this paper, we focus our attention on 'wave-influenced' deltas, where tidal forces are relatively weak and wave reworking of sediments at the coast controls the morphological evolution of the deltaic coast, as opposed to the 'wave-dominated' distinction meaning the delta does not protrude seawards. For example, Wright and Coleman (1973) investigated the transition from river-dominance to wave-dominance in terms of a balance between river forces and wave forces. In their classification, the Danube Delta would not be considered 'wave-dominated'. However, the St. George and Sulina lobes of the Danube delta exhibit a multitude of features, such as beach ridges and spits, indicating that wave reworking controls sediment movement at the shoreline (Giosan 1998, Giosan et al. 2005). Accordingly, we would consider these lobes of the Danube delta 'wave-influenced'.

Bhattacharya & Giosan (2003) presented new conceptual models of the asymmetric evolution of wave-influenced deltas, introducing a 'river delta asymmetry index', $A_{BG}$ (subscript added here for clarification purposes), the ratio between net alongshore transport at the river mouth (m³/yr) and river discharge ($Q_m$, $10^6$ m³/month). For a series of natural examples, this index performs well, asymmetrical delta development appears to require $A_{BG} > \sim 200$. However, this index does not appear to be able to distinguish between all observed behaviors; for instance it is unable to discern between 'asymmetric' and 'deflected' cases presented by the authors.

Numerical investigations by Ashton & Murray (2005) suggest instabilities associated with alongshore sediment transport could also play a significant role in the asymmetrical development of wave-influenced deltas. Building upon recent understanding of a fundamental instability in shoreline shape when waves approach with particularly oblique ('high') angles ($> \sim 43°$ in deep water) (Ashton et al. 2001), numerical simulations reproduce a range of delta morphologies as the characteristics of the wave-angle climate are varied. In this case, asymmetric behavior develops on the downdrift side of the river mouth as this portion of the delta coast becomes dominated by high-angle waves after the delta lobe's aspect ratio (regionally cross-shore extension versus alongshore width) increases. As a result, simulations show migrating fields of shoreline undulations (or 'alongshore sandwaves') and spits that extend downdrift of the river mouth.

Because these simulations neglect complex river-mouth dynamics, the striking asymmetrical behaviors arise only from presence of the instability in shoreline shape. Below, we take a more systematic approach to compare and contrast the asymmetrical behavior expressed in the quantitative model with that presented by Bhattacharya & Giosan's (2003) field-based model of asymmetrical model development.

## 3 NUMERICAL MODELING

Delta evolution is simulated using a numerical model which evolves the shoreline based upon gradients in alongshore sediment fluxes. This 'one-contour-line' approach assumes that the sandy shoreface sediment remains within the shoreface, and that cross-shore fluxes of sand beyond the shoreface depth, $D$, can be considered negligible compared to gradients in alongshore flux in the evolution of the coastline. The numerical model, summarized below, is described in greater detail by Ashton and Murray (2006a). The model discretizes equations for alongshore sediment flux and cross-shore sediment conservation similar to other one-contour-line models (e.g. Hanson & Kraus, 1989), with the unique ability to simulate a coast of arbitrary sinuosity and numerically accommodate unstable high-angle waves.

Within the model, the plan-view domain is discretized into cells filled with a fractional quantity of sediment, $F$, representing the plan-view excursion of the shore in each cell (Figure 1). Cells with $F = 0$ represent the open ocean, and cells with $F = 1$ are fully subaerial; the line of cells between the ocean and land with $0 < F < 1$ represents the shoreline. Deep-water waves with given height, $H$(m), and period, $T$(s), and approaching angle, $\phi_0$ (angle of wave crest) are refracted onshore using assumed shore-parallel contours until the waves break due to depth limitation. The CERC equation determines alongshore sediment fluxes ($Q_s$, m$^3$/s) between adjacent cells:

$$Q_s = KH_b^{5/2} \cos(\phi_b - \theta) \sin(\phi_b - \theta), \qquad (1)$$

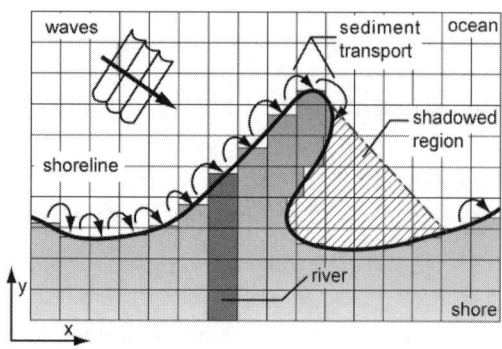

Figure 1. Plan-view schematic of model domain, showing the direction of sediment flux along the shoreline for a given angle of wave approach and the implementation of a fixed-location sediment source. Also shown is the region of the coast 'shadowed' by incoming waves. Figure after Ashton & Murray (2006a).

where $H_b$ is the breaking wave height (m), $\phi_b$ is the orientation of the breaking wave crests, $\theta$ is the shoreline orientation, and $K$ is an empirical constant dependent on sediment charcateristics (typically $\sim 0.4$ m$^{1/2}$/s for quartz density sand with a porosity of 0.6, although this value can vary greatly in nature) (Komar 1998, Rosati et al, 2002). The model has constant-flux boundary conditions, and allows large promontories to 'shadow' other regions of the coast from oblique waves (Figure 1). An additional process of barrier overwash maintains a minimum barrier width (Ashton & Murray 2006a). Shoreface depth is held at 10 m in all simulations.

Every simulated day, the deep-water angle of approaching waves, $\phi_0$, changes and can be any angle between $-90°$ and $90°$; this represents one of the main distinctions between this model and traditional one-contour-line approaches. Wave-approach angles are selected randomly from a probability distribution function defined by two variables, $U$, the fraction of high-angle waves, and $A_w$, the wave climate 'asymmetry' representing the fraction of waves approaching from the left, looking offshore. When $A_w = 0$, all waves approach from the left, when $A_w = 0.5$, waves approach equally from both directions. In all simulations, $U < 0.5$, representing a low-angle dominated climate; these simulations study the effect of active deposition at the shoreline, not the large-scale self-organization of the coast predicted when $U > 0.5$. Because an increased proportion of low-angle waves increases the effective 'diffusivity' of shoreline evolution (Ashton & Murray 2006a,b), increases in $U$ represent a reduction in the effective shoreline diffusivity, reducing the ability for waves to flatten a bump in the coast.

To simulate deltas, we represent the effects of riverine deposition in an extremely simplified manner. At a fixed point in the alongshore direction, an equal quantity of sediment, determined from the rate of sediment influx ($Q_b$, Mt/yr), is added each time step. $Q_b$ represents the rate of delivery of riverine bedload sediment, assuming that the finer-grained fraction typically held in suspension (mud) either bypasses the shoreface as it is advected by the river plume or is winnowed from the shoreface by wave action, eventually moving offshore.

The simplified approach here remains in keeping with our goal of simplifying the model dynamics to a point that the simulated behavior can be well-understood (e.g. Murray 2002). Simulations are run for a set of representative input variables, and are not calibrated to reproduce any one particular deltaic environment. As with a physical experiment, results could be rigorously scaled for direct comparison to a natural setting.

The ability to apply this model at deltas implicitly relies on several other assumptions. At natural deltas, deposition at the river mouth is typically complex, with other factors such as river avulsion, river mouth processes, multiple lobe interactions, and the seasonality of wave and river inputs playing important roles in the morphologic evolution (Wright & Coleman 1973, Giosan et al. 2005). Offshore deposition of fine-grained sediment must be sufficient to provide a platform over which the delta shoreface can prograde. However, if this deposition were too great, the prodelta could play a significant role in frictional wave attenuation, such as at the Mississippi River (Wright and Coleman 1973). If wave attenuation becomes dominant, the delta can no longer be considered wave-influenced, the model assumes prodelta deposition is neither too fast nor too slow.

## 4 RESULTS

As reported by Ashton & Murray (2005), the form of delta evolution changes as the inputs are varied. By systematically investigating the space occupied by the parameters $A_w$, $U$, and $Q_b$, we identified five 'prototype' forms of the simulated deltas (Figures 2 and 3) as the inputs are varied (Figure 4). Many natural deltas exhibit similar behaviors to the prototypes (Ashton & Murray 2005).

### 4.1 Prototype deltas

In the model's most expected behavior, the delta steadily progrades symmetrically about the river mouth, exhibiting a classic cuspate shape (Figure 2a). Similar behavior was modeled by Komar (1973), using waves approaching from one shore-normal (or one slightly oblique) wave approach angle. This behavior has long been the paradigm for delta evolution under the influence of waves. Although symmetrical wave climates favor this behavior, if sediment delivery

rates are low, this classic depositional style can arise even with a distinct asymmetry to the wave climate (Figure 4).

For generally symmetric wave climates, a higher rate of sediment input or an increased presence of high-angle waves (larger $U$) can result in more complex behavior where spits extend offshore of the coast near the river mouth in both directions, resulting in a discontinuous shore (Figure 2b). Increasingly complex interactions emerge as new spits are formed near the mouth and move downdrift towards the flanks, affecting previously created spits.

When sediment delivery is relatively high and with a pronounced asymmetry to the wave climate ($A_w > 0.6$), the dominance of high-angle waves is favored on the downdrift flank of the delta (Ashton and Murray 2005). As a result, high-wave-angle features such as migrating sandwaves and eventually offshore-extending spits begin to form along the downdrift coast (Figure 3a). Delta progradation increases the plan-view aspect ratio of the lobe, favoring the formation of spits, increasing the complexity of the downdrift coast over time.

For simulations with sediment fluxes between the extremes of the asymmetrical-spit delta and classic cuspate delta, an interesting behavior emerges. The delta develops a significant 'bend' in the downdrift coast, yet offshore-extending spits do not form (Figure 3b). This region may experience transient alongshore pulses of sediment in the form of alongshore sandwaves, but generally the delta grows maintaining this geometry. This behavior appears for a significant number of parameter combinations (Figure 4), so we do not consider it merely a continuous form between the Komar and spit-dominated morphologies.

The final type of behavior represents a far less realistic manifestation of the numerical model. Deposition at the 'river' far outpaces the ability for alongshore sediment transport to spread this sediment alongshore, forming a 'tree' shape (Figure 2c). In a natural delta, this rapid offshore extension would likely result in river avulsion towards the sides of the 'delta' or the development of many distributaries. The modeled behavior certainly does not resemble that of natural deltas, and we interpret these simulation results as an indication that the delta should be considered 'river-dominated'. Although waves could affect evolution of a delta with these input variables, it likely would display morphologies and behaviors reflecting river domination (Galloway 1975). Therefore, these simulations do provide some insight as they help suggest a quantitative upper threshold on relative sediment input rates for wave-influenced deltas.

### 4.2 Parameter space dependence

The wave climate parameters and sediment influx determine the basic model behavior (Figure 4).

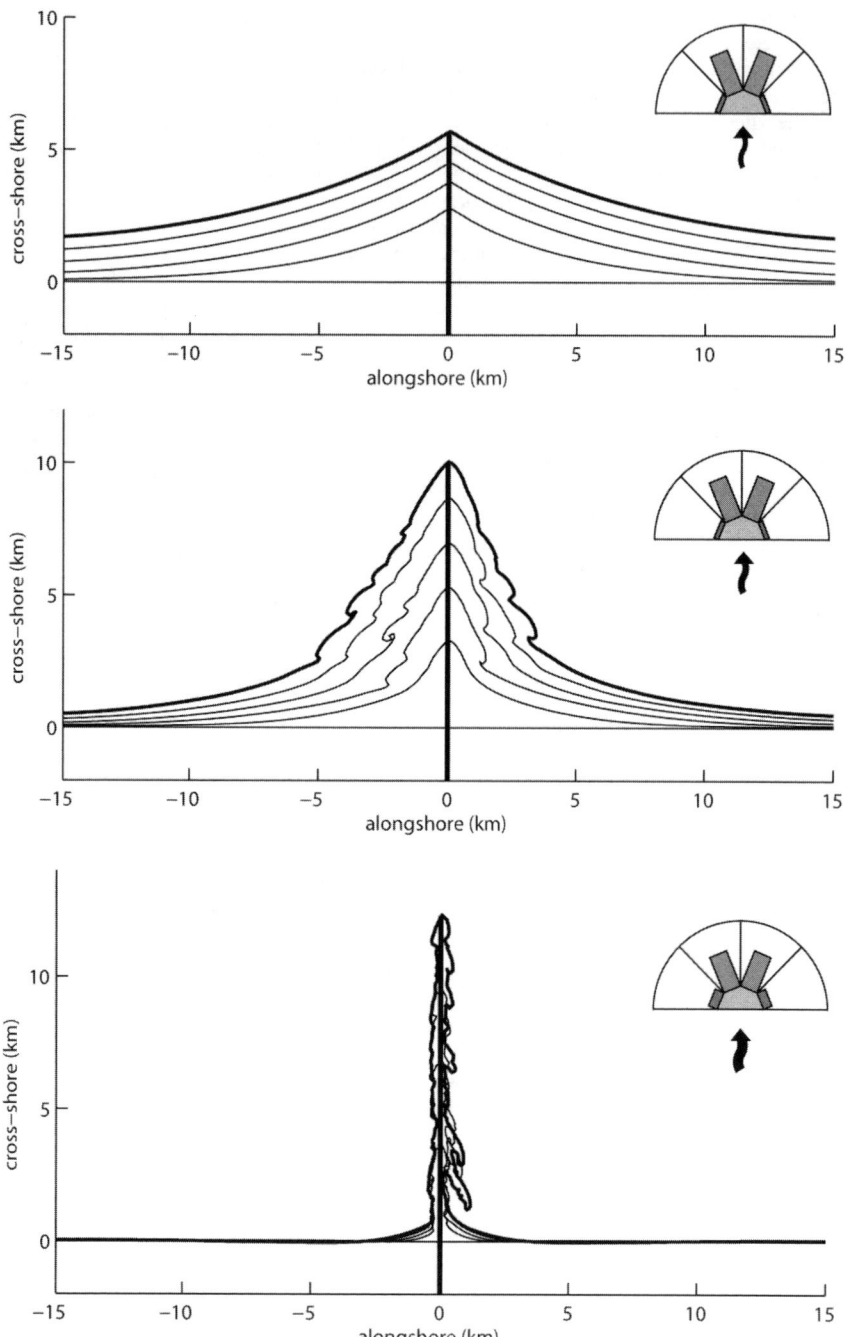

Figure 2. Plan-view simulated shorelines demonstrating symmetrical prototype model behaviors: (a) 'c', 'classic' cuspate, symmetrical delta ($A = 0.5$, $H = 0.1$, $R = 2.3$), (b) 'p', 'pointed' delta with shore-oblique spits ($A = 0.5$, $H = 0.1$, $R = 3.1$), and (c) 't', 'tree'-like river-dominated evolution ($A = 0.5$, $H = 0.2$, $R = 4.7$). Plots show shoreline configurations at equal time intervals, with final configurations after 440, 220, and 14 simulated years, respectively. Insets show roses of wave-approach direction and arrows representing the relative magnitude of $Q_b$.

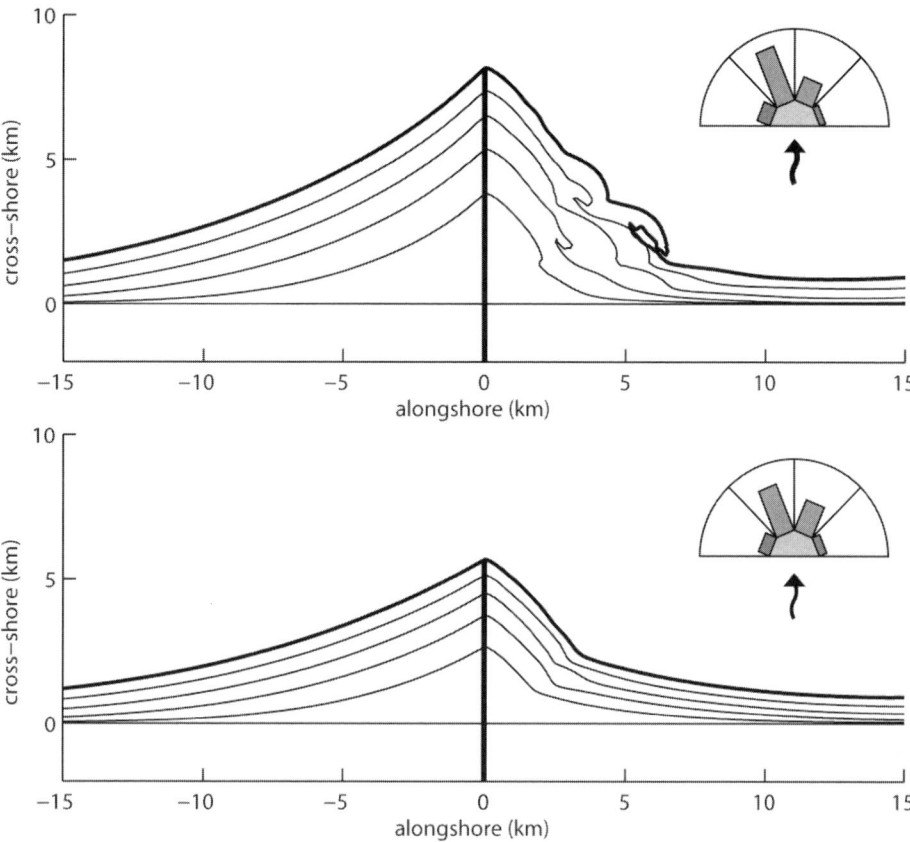

Figure 3. Plan-view simulated shorelines demonstrating asymmetrical prototype model behaviors: (a) 'a', 'asymmetrical' spit-forming ($A = 0.7, H = 0.2, R = 2.3$) and (b) 'b', asymmetrical delta with distinct 'bend' ($A = 0.6, H = 0.2, R = 1.6$). Plots show shoreline configurations at equal time intervals, with final configurations after 440 and 330 simulated years, respectively. Insets show roses of wave-approach direction and arrows representing the relative magnitude of $Q_b$.

Although most of the simulations are run with the same offshore wave height and period ($H = 1$ m and $T = 6$ s, respectively), simulations were run with different values to determine if modeled behavior were dependent on the particular input variables. To plot these different simulations in the same parameter space, we normalized the riverine sediment flux ($R$) by a maximum alongshore sediment flux. Because simulations use a full 'climate' of wave-approach directions, and not waves approaching from only one direction, determining this maximum flux is not straightforward. We computed the 'maximum alongshore sediment flux' using (1) and assuming a wave climate with $A_w = 1$ (fully asymmetric climate). Although variations in $U$ only have a slight effect on alongshore fluxes, simple computations reveal that alongshore fluxes are also maximized when $U = 1$.

$R$, the normalized riverine sediment flux, provides a rough, yet straightforward estimate of the river influx

versus the capability of waves through alongshore sediment transport to remove this sediment from the river mouth. As opposed to other approaches comparing variables such as the relative 'power' of waves versus the river (Wright & Coleman 1973), we believe that the focusing on the fluxes allows a direct comparison of the influence of the processes.

Simulation results collected for a range of $R$ are plotted together (Figure 4). In general, the 'classic' behavior is favored for small river inputs (small $R$), when waves are more low-angle (small $U$), and when wave climates are symmetric ($A_w \sim 0.5$). Increasing asymmetry and the proportion of high-angle waves ($U$) tends to favor complex behaviors. Model results suggest that river dominance becomes the norm after $R > 2.5$, a sensible result as the ability for alongshore sediment transport to remove sediment should begin to be limited beyond $R \sim 2$ (with sediment moved away from the river mouth in both directions).

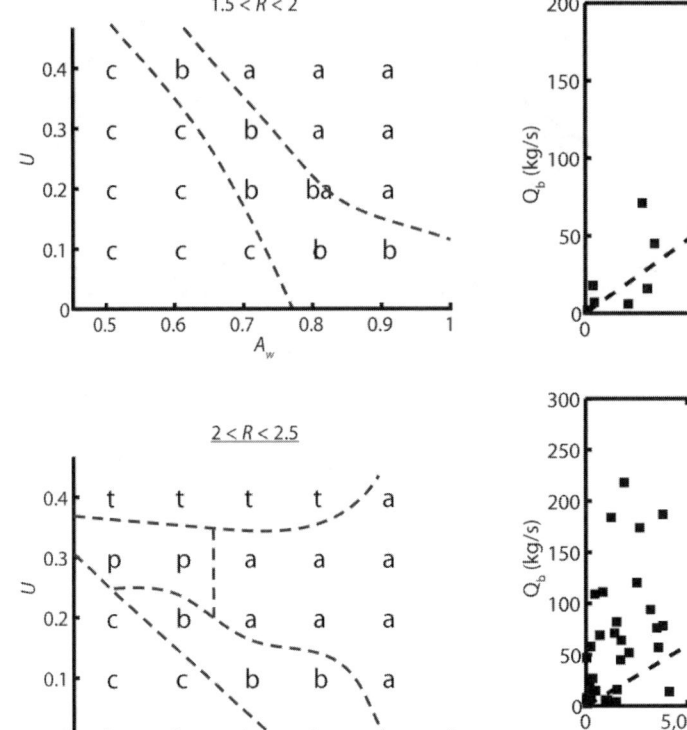

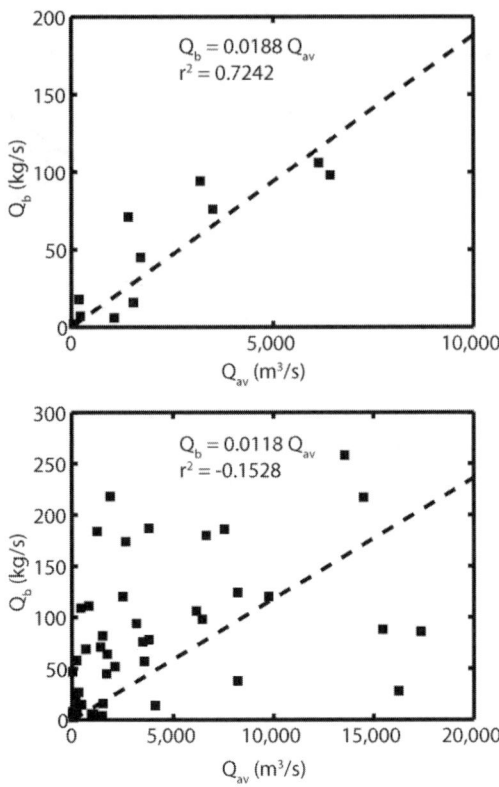

Figure 5. Bedload sediment transport ($Q_b$) vs. average water discharge ($Q_{av}$) (from Syvitski and Saito, in press) for (a) 11 wave-dominated deltas (Arno, Brazos, Ceyan, Danube, Ebro, Indus, Niger, Nile, Po, Rhone, and Vistula) and (b) 51-delta dataset (minus the 4 largest: Amazon, Ganges/Brahmaputra, Orinoco, and Yangtze). In both cases, linear regressions are forced through the origin.

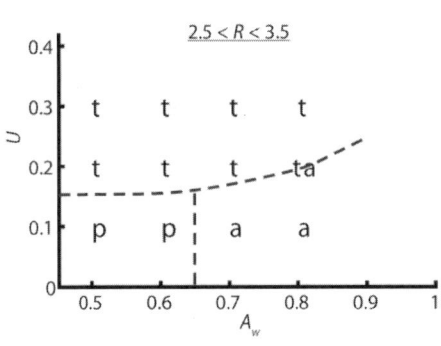

Figure 4. Parameter space plots of model behaviors for different relative river strengths (R). Prototype labels and examples found in Figures 2 and 3, and each plotted letter represents the results of (at least) one simulation.

### 4.3 Comparisons with delta asymmetry index

To compare the model results with the asymmetrical deltaic behavior described by Bhattacharya & Giosan (2003), the input variables need to be placed in terms of $A_{BG}$, which is computed in terms of a riverine water flux divided by a mass flux for alongshore sediment transport. To do so, the bedload river flux in the model needs to be converted to an average monthly water discharge ($Q_m$).

Although several rating relationships for suspended sediment transport have been presented (e.g. Syvitski et al, 2000), few relationships relate bedload and discharge. Using values from Syvitski & Saito (in press), a reasonable linear relationship exists between discharge and bedload transport for wave-influenced deltas (Figure 5a). Note that the values presented by Syvitski and Saito (in press) do not represent measurements; rather, they are computed values, essentially a function of water discharge and the delta gradient. Essentially, the linear trend in Figure 4a reflects that these river-influenced deltas have similar gradients. This linear relationship between bedload flux and water discharge is not apparent when all deltas are considered (Figure 5b), suggesting that there is some significance to the linear relationship found for wave-influenced deltas.

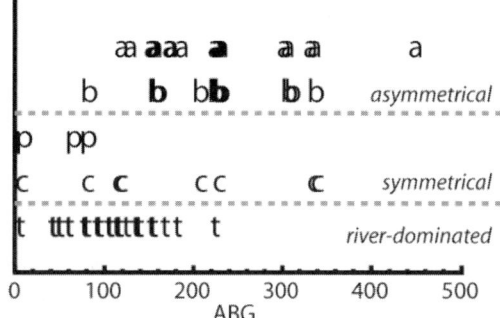

Figure 6. Model behaviors as the delta asymmetry index, $A_{BG}$, increases. Each letter represents (at least) one simulation. Vertical separation is for visualization purposes only.

Using this linear relationship and the known along-shore sediment fluxes, $A_{BG}$ can be computed for the simulations (Figure 6). The first interesting result is that despite the abstract, theoretical nature of the model and the crudeness of the relationship derived between discharge and bedload flux, the values for $A_{BG}$ in the model (up to ~450) fall within the same range as those reported for nature (up to ~350). Although the appearance of downdrift spits requires some degree of asymmetry, the variable $A_{BG}$ appears to have little ability to predict asymmetrical model behaviors.

## 5 DISCUSSION

Although it was unable to predict the modeled asymmetric behavior, $A_{BG}$ should not be dismissed, particularly as it performs well for natural examples. Rather, the results from Figure 6 suggest that some of the modeled behavior arises for reasons unexplained by this simple metric. The model is limited at the river mouth as it does not resolve processes that occur there that play vital roles in the asymmetrical behaviors described by Bhattacharya & Giosan (2003) and Giosan et al. (2005), including wave/river plume interactions and complex morphodynamic feedbacks that allow for the storage on the subaqueous delta. Furthermore, as the simulations assume a fixed river location, they could not generate 'deflected' river mouths seen in nature. They also assume that the river does not impede sediment from bypassing the mouth. Although seasonal and longer-term fluctuations in driving forces have been shown to affect the morphodynamic evolution of natural deltas (e.g. Fraticelli 2006, Giosan et al. 2005), the model also assumes constant sediment inputs and regular, uncorrelated inputs of wave energy. Future research will aim to better resolve and represent river mouth processes and episodicity in driving forces.

## 6 CONCLUSIONS

One-contour-line simulations of wave-influenced delta evolution demonstrate a range of symmetric and asymmetric behaviors, determined by the characteristics of the wave climate and the relative sediment influx. The details of the wave climate distribution can play a first-order control on delta evolution, showing that the influence of waves cannot be adequately represented by the average 'wave energy'. The simulated asymmetrical behavior does not arise from river mouth processes, these river mouth processes can result in asymmetrical development of natural deltas. However, as the asymmetric behavior from both the wave-climate and river mouth processes occur when waves approach obliquely, driving strong alongshore sediment transport, it may be likely that both types of asymmetrical behavior could occur concurrently.

## REFERENCES

Ashton, A., A. B. Murray & O. Arnoult 2001. Formation of coastline features by large-scale instabilities induced by high-angle waves. *Nature* 414: 296–300.
Ashton, A. D. & A. B. Murray 2005. *Delta simulations using a one-line model coupled with overwash*. Coastal Dynamics '05, Barcelona, Spain: 13 pp.
Ashton, A. & A. B. Murray 2006a. High-angle wave instability and emergent shoreline shapes: 1. Modeling of sand waves, flying spits, and capes. *Journal of Geophysical Research-Earth Surface* 111(F04011): doi:10.1029/2005JF000422.
Ashton, A. & A. B. Murray 2006b. High-angle wave instability and emergent shoreline shapes: 2. Wave climate analysis and comparisons to nature. *Journal of Geophysical Research-Earth Surface* 111(F04012): doi:10.1029/2005JF000423.
Bhattacharya, J. & L. Giosan 2003. Wave-influenced deltas: geomorphological implications for facies reconstruction. *Sedimentology* 50: 187–210.
Fraticelli, C. M. 2006. Climate Forcing in a Wave-Dominated Delta: The Effects of Drought-Flood Cycles on Delta Progradation. *Journal of Sedimentary Research*, 76: 1067–1076.
Galloway, W. E. 1975. Process framework for describing the morphologic and stratigraphic evolution of deltaic depositional systems. *Deltas, Models for Exploration*. M. L. Broussard. Houston, TX, Houston Geological Society: 87–89.
Giosan, L. 1998. Long term sediment dynamics on Danube delta coast. In J. Dronkers & M. Scheffers (eds), *Physics of Estuaries and Coastal Seas*. Rotterdam, Balkema: 365–376.
Giosan, L., J. P. Donnelly, et al. 2005. River delta morphodynamics: Examples from the Danube delta. *River Deltas: Concepts, Models, and Examples. SEPM Special Publication 83*. L. Giosan and J. Bhattacharya: 393–412.
Hanson, H. & N. C. Kraus 1989. *GENESIS: Generalized Model for Simulating Shoreline Change, Report 1: Technical Reference*. Vicksburg, MS, U.S. Army Eng. Waterways Experiment Station, Coastal Eng. Res. Cent.

Komar, P. D. 1973. Computer Models of Delta Growth due to Sediment Input from Rivers and Longshore Transport. *Geological Society of America Bulletin* 84: 2217–2226.

Komar, P. D. 1998. *Beach Processes and Sedimentation*. Upper Saddle River, New Jersey, Simon & Schuster.

Murray, A. B. 2002. Seeking Explanation Affects Numerical Modeling Strategies. *EOS, Transactions, American Geophysical Union* 83(38): 418,419.

Rosati, J. D., T. L. Walton & K. Bodge 2002. Longshore Sediment Transport. In D. B. King Ed.ˆEds, *Coastal Engineering Manual, Part II, Coastal Sediment Processes, Chapter III-2*. Washington, DC, U.S. Army Corps of Engineers. Engineer Manual 1110-2-1100.

Swenson, J. B. 2005. Fluviodeltaic response to sea level perturbations: Amplitude and timing of shoreline translation and coastal onlap. *J. Geophys. Res.* 110: F03007, doi:10.1029/2004JF000208.

Syvitski, J. P. M., M. D. Morehead, D. B. Bahr & T. Mulder 2000. Estimating fluvial sediment transport: The rating parameters. *Water Resources Research* 36(9): 2747–2760.

Syvitski, J. P. M. & Y. Saito, in press. Morphodynamics of deltas under the influence of humans. *Global and Planetary Change* In Press, Corrected Proof.

Wright, L. D. and J. M. Coleman 1973. Variations in morphology of major river deltas as functions of ocean wave and river discharge regimes. *AAPG Bulletin* 57: 370–398.

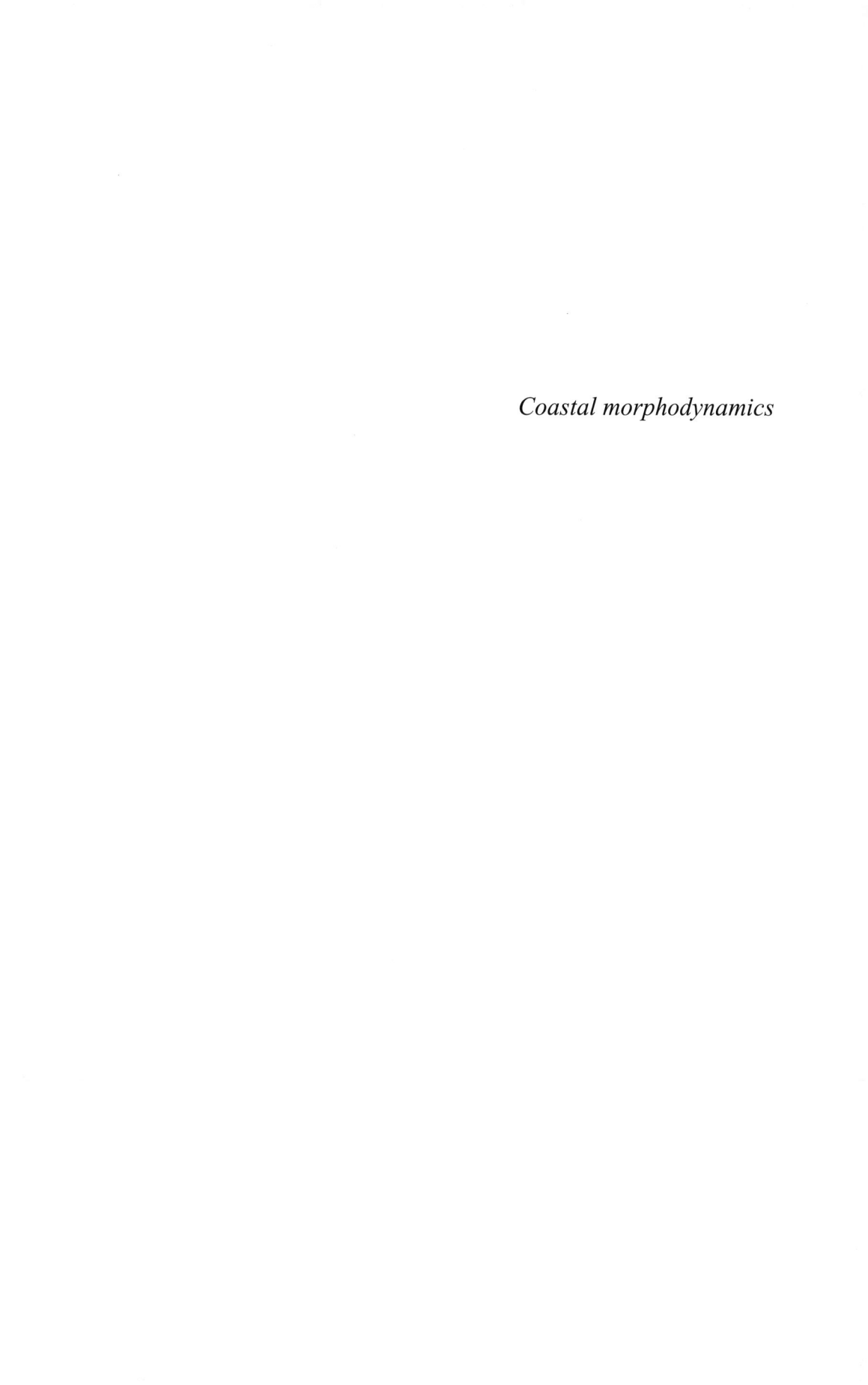

*Coastal morphodynamics*

*River, Coastal and Estuarine Morphodynamics: RCEM 2007 – Dohmen-Janssen & Hulscher (eds)*
*© 2008 Taylor & Francis Group, London, ISBN 978-0-415-45363-9*

# Finite amplitude dynamics of shoreface-connected ridges: Role of waves

N.C. Vis-Star & H.E. de Swart
*Institute for Marine and Atmospheric research Utrecht, Utrecht University, Utrecht, The Netherlands*

D. Calvete
*Dept. de Física Aplicada, Universitat Politécnica de Catalunya, Barcelona, Spain*

ABSTRACT: The long-term evolution of shoreface-connected sand ridges is investigated with a nonlinear morphodynamic spectral model. New is that wave properties are calculated with a shoaling-refraction model instead of using a crude parameterization. Model simulations show that the critical transverse bed slope of the inner shelf for which ridges start to grow is larger for larger offshore angles of wave incidence.Furthermore, the finite total height of the ridges decreases and the timescale on which the bedforms reach their final height increases if the waves approach the coast at a more oblique angle. An interesting phenomenon is the overshoot in bedform height just before the saturated state is reached. The competition between the bed slope-induced (diffusive) sediment transport and the current-induced (advective) sediment transport acts in favor of the latter when the waves are less oblique, thereby explaining the faster growth and larger total bedform height.

## 1 INTRODUCTION

Field data collected at various storm-dominated inner shelves of coastal seas (depths between 5–20 m) reveal the presence of patches of large-scale shoreface-connected sand ridges (Swift and Field 1981; Harrison et al. 2003). Typically, a patch consists of 4–8 ridges where the latter have heights of several meters and are spaced several kilometers apart. As the ridges seem to affect the stability of the beach (Van de Meene and Van Rijn 2000), gaining more fundamental understanding about their dynamics is relevant for coastal zone management purposes.

Several studies focused on understanding the initial growth of shoreface-connected ridges. Trowbridge (1995) demonstrated that ridges can grow due to feedbacks between the waves (which stir sediment from the bottom), the storm-driven current (which transports the sediment) and the sandy bottom. Calvete and de Swart (2003) investigated the long-term dynamics of shoreface-connected ridges. They showed that after the initial growth stage ridge profiles become asymmetric and reach a finite height.

A drawback of previous models is that the behavior of waves is described in a parametric manner. In the present contribution a new nonlinear model will be discussed in which wave properties follow from applying physical principles. Thus, processes like shoaling and refraction are explicitly accounted for. Recent studies

on the formation of bedforms by e.g. Caballeria et al. (2002) showed that this significantly affects the model results. The objective of the present work is to analyze the long-term behavior of ridges in dependence of the offshore angle of wave incidence and the transverse bottom slope of the inner shelf.

The paper is organized as follows. The model formulation is presented in section 2. The method of analysis follows in section 3 and continues in section 4 with a presentation of the results. The paper ends with a discussion and conclusions in section 5.

## 2 MODEL FORMULATION

### 2.1 *Geometry and hydrodynamics*

We consider an idealized inner shelf which, in the cross-shore direction $x$, is bounded by the shoreface at its shallow side ($x = 0$, depth $H_0$) and by the outer shelf at its deeper part ($x = L_s$, depth $H_s$). The bottom profile is uniform in the alongshore direction $y$ with a linear transverse bottom slope $\beta = (H_s - H_0)/L_s$. The geometry is similar as used by Trowbridge (1995) and shown in Figure 1.

Let $\vec{v}$ denote the depth- and wave-averaged flow velocity with components $u$ and $v$ in the $x$ and $y$ direction, respectively. The depth- and wave-averaged shallow water equations (representative for stormy

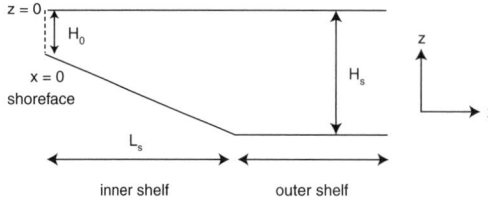

Figure 1. Side view of a typical longshore- and time-averaged bottom topography of the continental shelf, representing the inner and outer shelf, in the shore-normal direction. For explanation of the symbols, see the text.

conditions) are considered to govern the large-scale water motion, and read

$$\frac{\partial \vec{v}}{\partial t} + (\vec{v} \cdot \vec{\nabla})\vec{v} + f\vec{e}_z \times \vec{v} = -g\vec{\nabla}z_s + \frac{\vec{\tau}_s - \vec{\tau}_b}{\rho D}, \quad (1)$$

$$\frac{\partial D}{\partial t} + \vec{\nabla} \cdot (D\vec{v}) = 0. \quad (2)$$

Here, $f$ is the Coriolis parameter, $\vec{e}_z$ is a unit vector in the vertical direction, $\rho$ the water density, $D = z_s - z_b$ is the water depth and $\vec{\tau}_s$ and $\vec{\tau}_b$ represent the wind stress and bed shear-stress. In the definition for the water depth $z_s$ and $z_b$ are the free surface elevation and the bottom depth both measured with respect to the undisturbed water level $z = 0$. Finally, $t$ is time and $\vec{\nabla}$ is the horizontal nabla operator.

The mean alongshore wind stress $\tau_{sy}$, which is assumed to be constant, drives the longshore flow $v$. The bed shear-stress $\vec{\tau}_b$ is described by the linear friction law, as the amplitude of wave orbital motions is much larger than the magnitude of the storm-induced current during storms. Assuming waves and currents to be near-parallel it follows

$$\vec{\tau}_b = \rho r u_w \vec{v}, \quad (3)$$

where $r$ is a friction coefficient computed for random waves and $u_w$ is the root-mean-square amplitude of the near-bed wave orbital motion. The latter is calculated using linear wave theory. This involves solving the dispersion relation, generalized Snell law and wave energy balance for random linear gravity waves, following Mei et al. (2005), chapter 3. Refraction of waves by currents is not taken into account. Further details can be found in Vis-star et al(2007).

The boundary conditions are that the cross-shore flow component $u$ vanishes at $x = 0$ and far offshore. The bed level $z_b$ has a fixed level at these two positions. Offshore wave properties (height, period and angle of wave incidence) are imposed.

## 2.2 Bed evolution and sediment transport

Conservation of sediment mass yields the bed evolution equation

$$(1 - p)\frac{\partial z_b}{\partial t} + \vec{\nabla} \cdot \langle \vec{q}_b \rangle + \vec{\nabla} \cdot \langle \vec{q}_s \rangle = 0, \quad (4)$$

where $p$ is the porosity of the bed and $\langle \vec{q}_b \rangle$ and $\langle \vec{q}_s \rangle$ denote the wave-averaged sediment transport as bedload and suspended load, respectively. It is assumed that the sea bed consists of non-cohesive sediment of uniform size. The formulation of Bailard (1981) is used to derive an expression for the bedload transport of sediment:

$$\langle \vec{q}_b \rangle = \frac{3}{2}\nu_b u_w^2 \left( \vec{v} - \lambda_b u_w \vec{\nabla} z_b \right). \quad (5)$$

The coefficient $\nu_b$ depends on sediment properties and $\lambda_b$ is the bed slope parameter which is related to the angle of repose of the sediment. The first contribution to $\langle \vec{q}_b \rangle$ represents the net sediment transport due to a stirring of sediment by the waves and the subsequent transport by the net current. The second contribution accounts for gravitational effects on sediment grains in the bedload layer. Hereafter, the first contribution is called the current-induced sediment transport and the second the bed slope-induced sediment transport.

The wave-averaged suspended load transport reads

$$\langle \vec{q}_s \rangle = \mathcal{C}\vec{v} - \lambda_s u_w^5 \vec{\nabla} z_b. \quad (6)$$

Here, $\mathcal{C}$ is the depth-integrated volumetric concentration of sediment and $\lambda_s$ is the bed slope parameter for suspended sediment. To compute the current-induced flux of suspended sediment a concentration equation is used, which describes the depth-integrated concentration of sediment $\mathcal{C}$,

$$\frac{\partial \mathcal{C}}{\partial t} + \vec{\nabla} \cdot (\mathcal{C}\vec{v}) = \alpha u_w^3 - \gamma \frac{\mathcal{C}}{D}. \quad (7)$$

Here, $\alpha$ and $\gamma$ are known coefficients. The corresponding boundary condition is that the concentration vanishes far from the coast.

As the ridges evolve on a timescale which is long compared to that of hydrodynamic processes, the quasi-steady hypothesis is used: time derivatives in the hydrodynamic and concentration equations are neglected. Furthermore, since the Froude number is small, the rigid-lid approximation is used and hence $D \simeq -z_b$.

## 3 METHOD OF ANALYSIS

### 3.1 Basic state and expansion

Let $\Phi = (u, v, u_w, z_s, \mathcal{C}, z_b)$ denote the state of the system. The model allows for a basic state

$\Phi_b = (U, V, U_w, \zeta, C, -H)$ which is steady and long-shore uniform. The seabed of the basic state defines a morphodynamic equilibrium. Note that the wave orbital velocity $U_w$ is calculated using a shoaling-refraction model.

## 3.2 Stability analysis and linear theory

Subsequently, we investigate the dynamics of small perturbations evolving on the basic state by substituting $\Phi = \Phi_b + \phi$ into the equations of motion. The final result can be symbolically written as

$$S\frac{\partial\phi}{\partial t} = \mathcal{L}\phi + \mathcal{N}(\phi). \tag{8}$$

Here, $\phi = (u', v', u'_w, \eta', c', h')$ and the $6 \times 6$ matrix $S$ has one non-zero element: $S(6,6) = 1$. The linear matrix operator $\mathcal{L}$ contains all the linear terms and $\mathcal{N}$ is the nonlinear vector operator, which includes all nonlinear terms in the equations of motion for the perturbations.

Linearization of this system of equations implies that only terms that are proportional to the amplitude of the perturbations are retained and thus $\mathcal{N} = 0$. The linearized system sustains wave-like solutions which travel in the $y$ direction and of which the amplitude can grow (or decay) exponentially in time:

$$\phi(x, y, t) = \mathcal{R}e\left\{\tilde{\phi}(x)e^{iky+\sigma t}\right\}. \tag{9}$$

Here, $\mathcal{R}e$ denotes the real part of the solution, tildes denote the as yet unknown cross-shore structure of the solution, $k$ the longshore wavenumber (which can be assigned any value) and $\sigma$ the complex frequency. Substitution of expression (9) in the linearized equations of motion, together with the boundary conditions specified before, results in an eigenvalue problem

$$\sigma S\tilde{\phi} = \mathcal{L}_k\tilde{\phi}. \tag{10}$$

Note that $\sigma$ is the complex eigenvalue and the linear matrix operator $\mathcal{L}_k$ is obtained by substitution of $\frac{\partial}{\partial y}$ by $ik$ in $\mathcal{L}$. Solutions of the eigenvalue problem are the eigenmodes $\mathcal{R}e\{\tilde{\phi}_{kn_k}e^{iky}\}$ with corresponding eigenvalues $\sigma_{kn_k}$. Here, longshore wavenumber $k$ refers to the longshore structure and for each $k$, $n_k$ $(=1, 2, 3, \dots)$ gives the cross-shore structure of the eigenmodes. The growth rate and migration speed $V_m$ of the perturbation are given by $\mathcal{R}e(\sigma_{kn_k})$ and $-\mathcal{I}m(\sigma_{kn_k})/k$, respectively. Smaller $n_k$ values correspond with larger growth rates. Solutions of the eigenvalue problem are obtained numerically by applying a collocation method (Canuto et al. 1988).

## 3.3 Nonlinear theory: finite amplitude behavior

In order to investigate the long-term evolution of the shoreface-connected ridges the full set of nonlinear equations (8) has to be considered. First the perturbations are written as

$$\phi(x, y, t) = \overline{\phi}(x, t) + \phi'(x, y, t), \tag{11}$$

where the unknown contributions $\overline{\phi}$ have a longshore uniform structure and the contributions $\phi'$ are expanded into a truncated series of eigenmodes of the linear system:

$$\phi' = \mathcal{R}e\left\{\sum_{j=1}^{J}\sum_{n_j=1}^{N_j}\hat{\phi}_{jn_j}(t)\tilde{\phi}_{jn_j}(x)e^{ik_jy}\right\}. \tag{12}$$

Here, for each alongshore wavenumber $k_j$, $n_j$ refers to the cross-shore modenumber, $\hat{\phi}_{jn_j}(t)$ are the unknown modal amplitudes and $\tilde{\phi}_{jn_j}(x)$ the known cross-shore structures of the eigenfunctions of the linear problem. In the expansion only the most preferred mode with wavenumber $k_p$ (largest initial growth rate) and its superharmonics are used. Thus, the system is solved on a domain with finite longshore length $L_y = 2\pi/k_p$ and periodic boundary conditions are applied. According to equations (8) wavenumbers that fit in the domain are $k_j = 2\pi j/L_y$ $(j = 1, 2, 3, \dots)$.

Expansions (11) and (12) are substituted in the nonlinear equations of motion. After averaging over the alongshore direction equations are obtained for the longshore uniform flow modes and bottom mode, which are subsequently subtracted from the original equations. The results are projected onto the adjoint linear eigenmodes. The result is a set of nonlinear algebraic equations for the flow amplitudes and a set of nonlinear differential equations for the amplitudes of the bottom modes. For a more detailed explanation of the procedure followed and the time integration scheme applied we refer to Calvete and de Swart (2003).

## 4 RESULTS

Runs were performed with parameter values representative for the micro-tidal inner shelf of Long Island. At this location ($40°$ latitude) depth varies from $H_0 = 14$ m to $H_s = 20$ m over an inner shelf width of $L_s = 5.5$ km ($\beta = 1.1 \times 10^{-3}$). The wind stress $\tau_{sy} = -0.4$ N m$^{-2}$, offshore root-mean-square wave height $H_{rms,s} = 1.5$ m, wave period $T = 11$ s and offshore angle of wave incidence $\Theta_s = -20°$. Values of the other parameters are: $r = 2.0 \times 10^{-3}$, $v_b = 5.6 \times 10^{-5}$ s$^2$ m$^{-1}$, $\lambda_b = 0.65$, $\lambda_s = 7.5 \times 10^{-4}$ s$^4$ m$^{-3}$, $\alpha = 1.4 \times 10^{-5}$ s$^2$ m$^{-2}$, $\gamma = 0.20$ m s$^{-1}$ and $p = 0.4$.

In section 4.1 results of the basic state and linear analysis are presented for different offshore angles of wave incidence $\Theta_s$ and transverse bottom slope $\beta$. In section 4.2 the saturation behavior of the ridges is analyzed. The latter is possible up to a bottom slope which is about 60% of realistic values. In the simulations wave-topography interactions are neglected, which is done by excluding perturbations in the wave orbital motion: $u'_w = 0$, hence $u_w = U_w$. A motivation for this assumption is given in section 5.

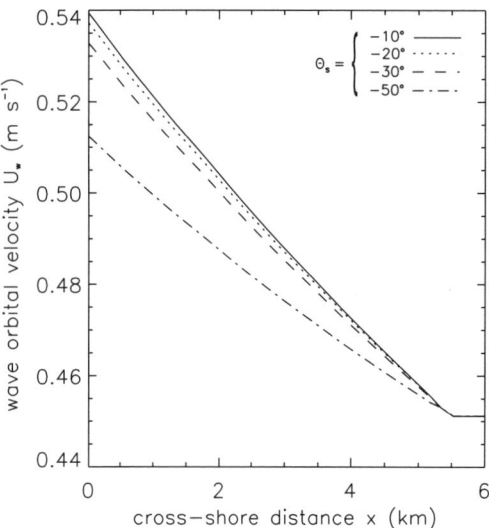

Figure 2. Cross-shore profiles of the wave orbital velocity amplitude in the basic state for different offshore angles of wave incidence for $\beta = 6.4 \times 10^{-4}$ (60% of the actual value).

### 4.1 Basic state and linear stability analysis

The cross-shore profile for the basic state near-bed wave orbital velocity $U_w$ is shown in Figure 2 for different offshore angles of wave incidence $\Theta_s$. The angle is measured with respect to the shore-normal, where positive (negative) $\Theta_s$ indicate a clockwise (anti-clockwise) deviation. The amplitude of the near-bed wave orbital velocity increases towards shallower depths, to become somewhat larger than $0.50\,\mathrm{m\,s^{-1}}$ at the shoreface. The increase in amplitude is weakened by wave refraction, which is stronger if incident waves become more oblique. The onshore increase of $U_w$ plays an important role in the growth process (see section 5). Besides, it causes an increase in bed friction, hence the longshore current decreases towards the shoreface (result not shown). The growth rate curve for linear perturbations evolving on the basic state of the model attains a maximum for a specific alongshore wavenumber $k_p$, the initially most preferred mode. The corresponding wavelength is $\lambda = 2\pi/k_p$. A contour plot of the growth rate of the initially most preferred mode as a function of the transverse bottom slope of the inner shelf $\beta$ and the offshore angle of wave incidence $\Theta_s$ is given in Figure 3. Here, a variation in $\beta$ corresponds to a variation in $H_s$. For realistic values of the inner shelf slope and offshore angle of wave incidence the growth rate is in the order of $2.5 \times 10^{-3}\,\mathrm{yr^{-1}}$. So, considering the fact that storms and fair weather alternate, the actual growth rate will be about a factor 10 smaller.

The critical inner shelf bottom slope $\beta_c$ which has to be exceeded before growing bottom perturbations are obtained is different for a change in the offshore angle of wave incidence. An increase from $\Theta_s = -2°$ to

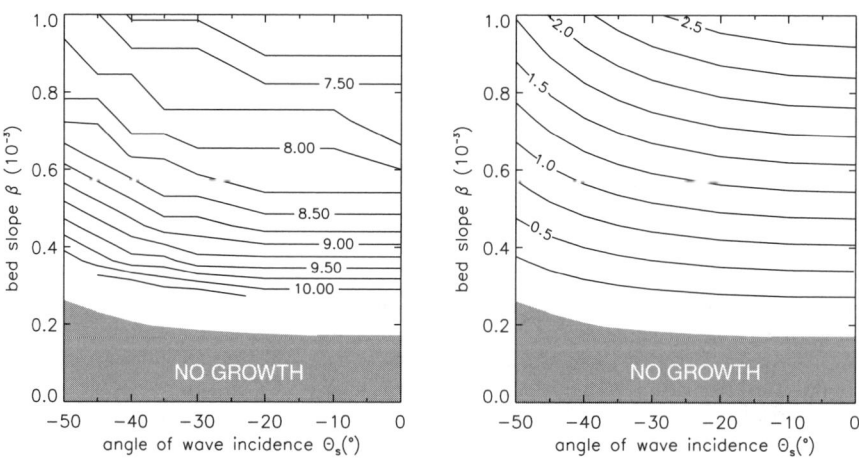

Figure 3. Contour plots of equal longshore spacing (km) and growth rate $(10^{-3}\,\mathrm{yr^{-1}})$ of the initially most preferred mode in the $\theta_s - \beta$ plane.

$\Theta_s = -50°$ results in an increase in $\beta_c$ from $1.8 \times 10^{-4}$ to $2.7 \times 10^{-4}$, respectively. Once bedforms start to form, increasing the slope of the inner shelf leads to a decrease in the alongshore spacing of the most preferred mode. The same trend is visible for a decrease in the offshore angle of wave incidence. The growth rate of the most preferred mode increases with increasing $\beta$ and with decreasing $\Theta_s$. Furthermore, migration speeds slightly decrease with increasing bottom slope (not shown), whereas no clear dependence is visible for a change in the offshore angle of wave incidence. The sensitivity of the migration speed for the transverse bottom slope $\beta$ can be approximated by $V_m/V^* = -1.0 + 0.34(\beta - \beta_c)$, where $\beta_c = 0.3 \times 10^{-3}$ and $V^* = 1.3 \, \mathrm{m \, yr^{-1}}$.

### 4.2 Nonlinear analysis

The nonlinear evolution was truncated at $J = 64$ and $N_j = 10$. In the time integration a time step of $10 \, \mathrm{yr}$ was used. Both adding more modes and decreasing the time step did not change or improve the solutions.

Figure 4 shows for an inner shelf slope of $\beta = 6.4 \times 10^{-4}$ the time evolution of the total height (crest to trough) of the bottom patterns as a function of the offshore angle of wave incidence. All bottom modes have small amplitudes at $t = 0$. The initial growth of the bottom patterns is exponential and therefore in accordance with linear theory. On the longer term growth becomes less than exponential as nonlinear processes become important. The total height of the bedforms tends to a constant finite value

which is larger for smaller offshore angles of wave incidence. Note that the change in finite total bedform height is stronger than linear. The saturation time is defined as the time needed to attain a height which is 98% of the finite total height. The saturation time is between 1000 and 2000 years, longer for larger angles of wave incidence. An interesting overshoot in bedform height is visible, just before the saturated state is reached, for waves which approach the coast under a relatively small angle.

Contour plots of the finite total bedform height and saturation time in the $\Theta_s - \beta$ plane are given in Figure 5 and Figure 6, respectively. As in Figure 3,

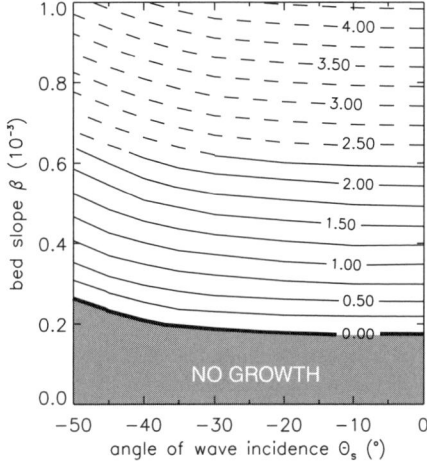

Figure 5. Contour plot of equal finite total height (m) of the bedforms in the $\Theta_s - \beta$ plane. Obtained with nonlinear model (solid lines) or extrapolated (dotted lines).

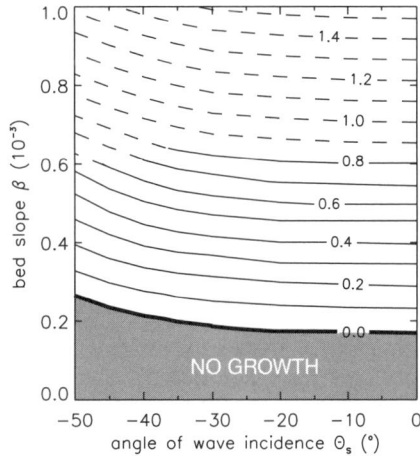

Figure 6. Contour plot of the inverse of time needed for saturation ($10^{-3} \, \mathrm{yr^{-1}}$) in the $\Theta_s - \beta$ plane. Obtained with nonlinear model (solid lines) or extrapolated (dotted lines).

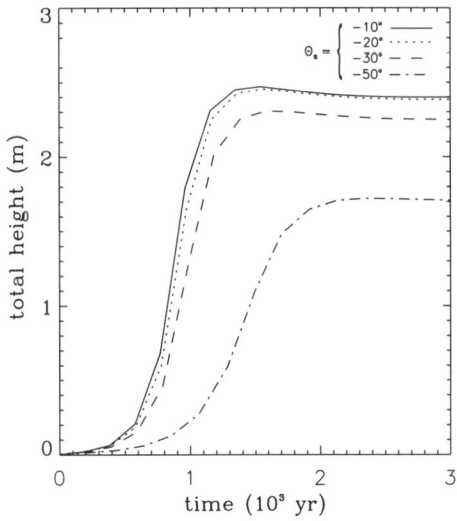

Figure 4. Time evolution of total height of the bedforms for different offshore angles of wave incidence for $\beta = 6.4 \times 10^{-4}$.

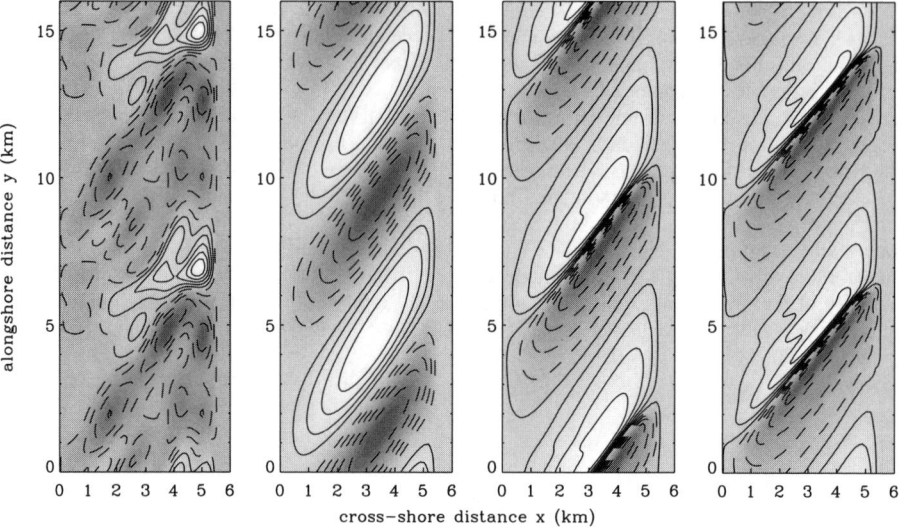

Figure 7. Bottom pattern (greyscale; light: bars, dark: troughs) at $t \sim 50, 500, 1000, 2000$ yr. Here, $\Theta_s = -20°$ and $\beta = 6.4 \times 10^{-4}$.

the critical transverse bottom slope $\beta_c$ increases from $1.8 \times 10^{-4}$ to $2.7 \times 10^{-4}$ for $\Theta_s = -2°$ to $\Theta_s = -50°$, respectively.

Numerically stable solutions are obtained up to about 60% of realistic values of the inner shelf slopes (solid lines) and are extrapolated into the regime of realistic bed slopes (dotted lines). Larger bottom slopes cause the growth to become faster and bedforms to attain a larger total height. For normal incident waves, evolving bedforms grow faster and they become higher than those formed under obliquely incident waves. Extrapolating to realistic values of the inner shelf slope and offshore angle of wave incidence, ridges would reach a finite amplitude somewhat larger than 4 m and saturation occurs in about 700 yr.

The bottom patterns at different stages in the long-term evolution are visible in Figure 7 for $\beta = 6.4 \times 10^{-4}$. After 50 years shoreface-connected ridges start to emerge. After 500 years nonlinear processes become important and the bottom pattern becomes more asymmetrical. Note the steepening of the seaward (downstream) flank with respect to the landward (upstream) flank. Both during the evolution and in the saturated state the ridges migrate in southward direction, which is in the direction of the mean storm-driven flow. Both the above-mentioned steepening and migration of the ridges are nicely illustrated in the longshore profiles of depth at two times during the evolution (see Fig. 8).

## 5 DISCUSSION AND CONCLUSIONS

In this paper the sensitivity of the long-term evolution of shoreface-connected ridges for both the transverse bottom slope of the inner shelf and the offshore angle of wave incidence have been investigated.

Normal incident waves do not experience wave refraction and therefore exhibit the largest near-bed wave orbital velocity amplitude on the inner shelf. Previous studies (Ribas et al. 2003, and references therein) indicated that shoreface-connected ridges grow if the spatial correlation between $u'$, $h'$ and the cross-shore gradient of the basic state depth-averaged volumetric concentration is negative. The latter quantity is defined as $\frac{d(C/H)}{dx}$, where $C$ is the solution of equation (7) for steady and longshore uniform conditions. Thus, $C$ is proportional to $U_w^3$. As both $U_w^3$ and $H$ decrease with increasing distance from the coast (see Fig. 2), $\frac{d(C/H)}{dx}$ is negative. Thus, in order to have growth, the current should exhibit an offshore deflection over the ridge. This only occurs if the ridges are up-current oriented. The ratio of the current-induced sediment transport over the bed slope-induced sediment transport becomes larger for stronger cross-shore gradients in the depth-averaged volumetric concentration. Therefore, the initial growth of bedforms for normal incident waves is faster. The latter results in a shorter saturation time on the longer term and higher finite amplitudes.

Numerical stable solutions could be obtained up to approximately 60% of the observed value of the slope

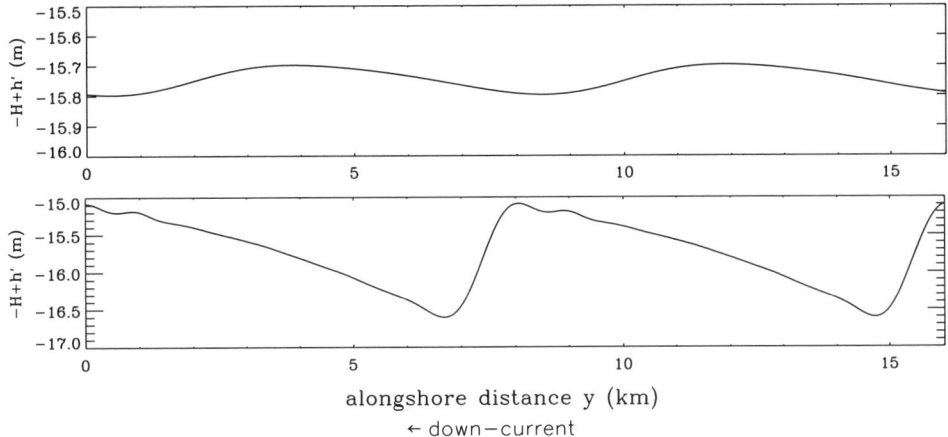

Figure 8. Longshore profiles of depth $-H + h'$ at a cross-shore location halfway the inner shelf. The profiles are taken at $t \sim 500$ yr (top) and $t \sim 1000$ yr (bottom). Note the difference in vertical scale. Here, $\Theta_s = -20°$ and $\beta = 6.4 \times 10^{-4}$.

of the inner shelf. For larger values of $\beta$ solutions become singular at some point during the evolution. Extrapolation to realistic values for the transverse slope of the inner shelf would yield bedforms with a finite amplitude of $\sim 4$ m and a saturation timescale of $\sim 700$ yr. Values fairly agree with field data shown in Duane et al (1972).

The results presented in this paper were obtained with a version of the model in which interactions between waves and growing bedforms were ignored. Vis-Star et al. (2007) showed that including these interactions is crucial as soon as waves are obliquely incident. However, nonlinear experiments including wave-topography interactions revealed spurious modes in the linear analysis which have to be filtered. When incorporated in the nonlinear model ridges continue to grow on the long term and numerical instabilities arise before the saturated state is reached. An explanation for the latter is that stirring of sediment by waves increases with increasing ridge height. This tendency will be counteracted by e.g. wave breaking, a process which is not yet implemented in the model. Caballeria et al. (2002) were able to obtain results with a nonlinear model including wave-topography interactions, but they employ Snell's law instead of the eikonal equation. The latter means that bedform-induced additional wave refraction is only partially accounted for.

For the future it would be interesting to investigate the sensitivity of the results for offshore root-mean-square wave height and wave period. Another interesting experiment is to consider subharmonic modes (i.e. increasing the finite longshore length $L_y$ of the domain). The study by Calvete and de Swart (2003) showed that adding subharmonic modes strongly influences the transient behavior of the ridges.

Furthermore, subharmonic modes can be excited in such a way that they dominate over the most preferred mode of the linear analysis in the saturated state.

ACKNOWLEDGEMENTS

The work of N.C. Vis-Star is supported by 'Stichting voor Fundamenteel Onderzoek der Materie' (FOM), which is supported by the 'Nederlandse Organisatie voor Wetenschappelijk Onderzoek' (NWO). The work of D. Calvete has been partially funded by the Ministerio de Ciencia y Tecnologia of Spain through the 'Ramón y Cajal' contract.

REFERENCES

Bailard, J. A. (1981). An energetics total load sediment transport model for a plane sloping beach. *J. Geophys. Res.* 86(C11), 10938–10954.

Caballeria, M., G. Coco, A. Falqués, and D. A. Huntley (2002). Self-organization mechanisms for the formation of nearshore crescentic and transverse sand bars. *J. Fluid Mech.* 465, 379–410.

Calvete, D. and H. E. De Swart (2003). A nonlinear model study on the long-term behavior of shore face-connected sand ridges. *J. Geophys. Res.* 108(C5), doi:10.1029/2001JC001091.

Canuto, C., M. Y. Hussaini, A. Quarteroni, and T. A. Zang (1988). *Spectral methods in fluid dynamics.* Springer-Verlag, New York.

Duane, D. B., M. E. Field, E. P. Miesberger, D. J. P. Swift, and S. J. Williams (1972). Linear shoals on the Atlantic continental shelf, Florida to Long Island. In D. J. P. Swift, D. B. Duane, and O. H. Pilkey (Eds.), *Shelf Sediment Transport: Process and Pattern,* pp. 447–498. Dowden, Hutchinson and Ross, Stroudsburg, Pa.

Harrison, S. E., S. D. Locker, A. C. Hine, J. H. Edwards, D. F. Naar, and D. C. Twichell (2003). Sediment-starved sand ridges on a mixed carbonate/siliciclastic inner shelf off west-central Florida. *Mar. Geol. 200*, 171–194.

Mei, C. C., M. Stiassnie, and D. K.-P. Yue (2005). *Theory and applications of ocean surface waves. Part 1: Linear aspects*. World Scientific, Singapore.

Ribas, F., A. Falqués, and A. Montoto (2003). Nearshore oblique sand bars. *J. Geophys. Res. 108*(C4), 1–17.

Swift, D. J. P. and M. E. Field (1981). Evolution of a classic sand ridge field: Maryland sector, North American inner shelf. *Sedimentology 28*, 461–482.

Trowbridge, J. H. (1995). A mechanism for the formation and maintenance of the shore oblique sand ridges on storm-dominated shelves. *J. Geophys. Res. 100*(C8), 16071–16086.

Van de Meene, J. W. H. and L. C. Van Rijn (2000). The shoreface-connected ridges along the central Dutch coast. part 1: field observations. *Cont. Shelf Res. 20*(17), 2295–2323.

Vis-Star, N. C., H. E. De Swart, and D. Calvete (2007). Effect of wave-topography interactions on the formation of sand ridges on the shelf. *J. Geophys. Res.*, submitted, *reprint can be obtained from first author.

*River, Coastal and Estuarine Morphodynamics: RCEM 2007 – Dohmen-Janssen & Hulscher (eds)*
*© 2008 Taylor & Francis Group, London, ISBN 978-0-415-45363-9*

# Observation of morphological development on the tsunami-affected coast of Banda Aceh using multi-temporal Digital Elevation Models

E. Meilianda, C.M. Dohmen-Janssen & S.J.M.H. Hulscher
*University of Twente, Enschede, The Netherlands*

B.H.P. Maathuis
*Departemen of Water Resources, International Institute for Geo-Information Science and Earth Observation (ITC), Enschede, The Netherlands*

ABSTRACT: In this study, a combination of different sources and multiple years is used of elevation data to study the coastal changes and development of the tsunami-affected area. The data which consists of contour lines and elevation points are used as the input to generate Digital Elevation Models (DEMs) by using Triangulated Irregular Network (TIN) method. The results exhibit 3 sub regions which have different patterns of morphological development in the early period (i.e. one year) after the tsunami. Meanwhile the development in the following year shows a tendency of smoothened coastal profiles (i.e cross-shore as well as alongshore) in the entire region, which could indicate a development towards a new coastal state.

## 1 INTRODUCTION

Indicator for coastal problems is often concentrated to shoreline positions and trends as well as the state of the coastal profiles, either by the scientists, coastal engineers or coastal managers. Monitoring morphological development in geospatial studies requires information on the horizontal and vertical positions of a coastal feature relative to the coordinate system and the elevation reference (datum) used.

To get an idea about the state of the coastal profile in time, a combination of different sources of data available to complement topographic and bathymetric point data for the studied area is required. Data sources representing this information in a tsunami affected area like Banda Aceh, Sumatra Island, Indonesia are scarce, both pre- or post-tsunami. This is particularly the case for elevation data, such as topographic and bathymetric elevation. Consequently, there is a gap of information in studying the morphological development in this particular area. One of the problems we may face is that for a specific area like the water-to-land boundary (foreshore zone), the interpolation between the scarce elevation point density may generate erroneous/unrealistic results. For example, the shoreline position may considerably deviate from the real position, or special features like shoals or trenches are not covered by the data. Therefore, the spatial resolution of the data should be increased.

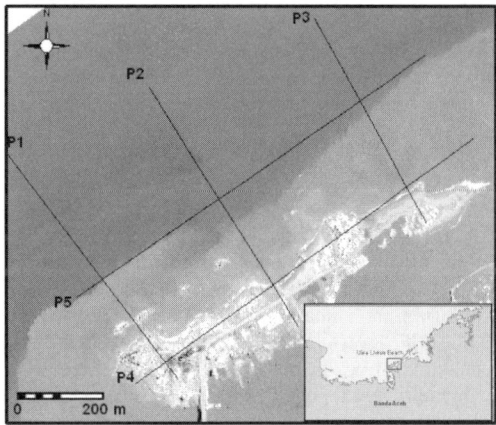

Figure 1. Location of sampled profiles on top of ortho-rectified aerial image of Ulee Lheue coast, northwest of Banda Aceh, Indonesia. (Images source: Bakosurtanal, 2006).

We study a small part of the northwest coast of Banda Aceh, namely Ulee Lheue (read: *Oe lé lé*) coast (Fig.1). This narrow coast stretched over 4.5 km length in between the two inlets; i.e. Krueng Cakra River on the southwest and Krueng Aceh River on the northeast. The coast also separated a coast-parallel lagoon system from the sea. Before the tsunami, almost the entire Ulee Lheue beach was protected by seawalls. Since the tsunami, the low-lying coast has suffered from permanent inundation due to the breaching of its narrow sandy beach. The breached beach was transformed into small barrier islands. The new Ulee Lheue coast is one of the largest barrier islands which length alongshore is about 800 m.

Elevation data for this location is considered to meet a minimum standard, in terms of bathymetric and topographic spatial resolutions. Bathymetric surveys conducted in the past could not reach through shallow water. Moreover, the topographic surveys had to deal with a difficult terrain, and had to be combined with the water depth measurement for the inner water body. Hence, there is a gap of data available. In the present study, this problem is solved by using the data from the other studies, such as water-line study by Meilianda et al. (2007) to fill this data gap.

A Triangulated Irregular Network (TIN) method is applied to generate a representative Digital Elevation Model (DEM) for different periods. Here, we use any possible data available to enforce the most realistic DEM results. Eventually, the objective of this study is to analyze the morphological changes due to the impact of the tsunami in Banda Aceh in December 2004, and how the coast has developed in the recent past as exhibited by the multi-temporal data.

The following chapter will describe the data used for the present study (2), the method to generate DEMs (3) and sampling of profiles derived from the multi-temporal DEMs (4). The analysis of the results will be discussed in the next chapter (5), and finally this paper will present the conclusion (6).

## 2   DATA

Different sources of data, in terms of topography and bathymetry are compiled as input for generating a DEM. We took the sample profiles along the region which morphologically similar; i.e. in this case the region which was protected by the seawalls in the pre-tsunami's period. In this study three multi-temporal bathymetry and topography data are used: i) the pre-tsunami's data were obtained from a survey, investigation and design project of coastal protection on Ulee Lheue coast conducted by a local surveyor in April 2001; ii) the one year post-tsunami data was obtained from the similar project conducted by national surveyor in January 2006; iii) the two year post-tsunami data was obtained from the similar survey conducted by the coastal defense consultant of Bureau of Reconstruction and Rehabilitation of tsunami disaster in Aceh and Nias (BRR Aceh-Nias) in January 2007. The first two data were measured using Mean Low Water Level (MLWL), while the third one was using Mean Sea Level (MSL) as the elevation reference.

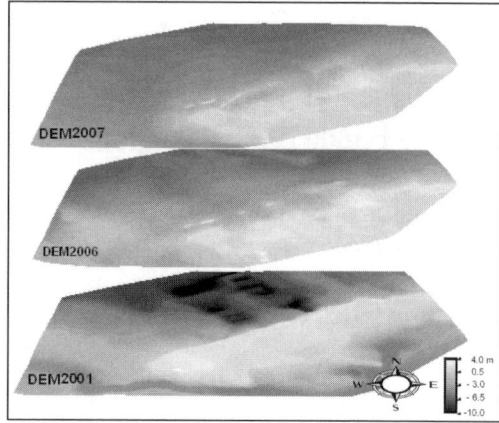

Figure 2.   Generated multi-temporal DEMs of Ulee Lheue coast.

In the present study the latter one was then converted into MLWL by subtracting each elevation data by 0.8 m value. This value was taken based on the accompanying metadata suggestion.

The fact that the data of 2006 and 2007 were measured in the same month of the year (January), the morphological changes in between the two periods can be straightforwardly compared; regardless the detail of the coastal processes leading to each current coastal state. The bathymetric and topographic data of the year 2001 were obtained as layers of contour maps. The data of the year 2006 and 2007 were obtained as elevation point maps. Topographic data of 2006 and 2007 were obtained as contour maps, which actually came from the same source; i.e. based on the photogrammetry of ortho-rectified aerial imagery taken in June 2005 (see Meilianda, et al. 2007). Here, we assume that the landward elevation for the period of 2005 till 2007 did not change significantly.

## 3   METHODS

We apply the TIN method which is commonly used in the GIS-based terrain generation. The advantage of using TIN method is that there is a possibility to combine any type of data, i.e. segment (contour) map and point elevation map as input layers. The disadvantage is that the further the distance between the data, the more generalized or erratic the interpolation result will be.

In this study, a nearshore is defined as a segment of a coastal profile between MLWL and the closure depth. A foreshore is defined as a segment from the Mean High Water Level (MHWL) to MLWL. As mentioned earlier, in the present study we have to face the fact that the bathymetric surveys conducted for all the three periods was hardly ever (yet available at some locations) reaching through the shallow water parts, and neither was it complemented by topographic surveys, which causing a gap of data such as the shallow part of a nearshore and a foreshore. Estimation of the properly assigned elevation values to integrate the bathymetry and the topography is therefore difficult, nevertheless, inevitable. To cope with this issue, some additional elevation points in combination with additional break lines have to be properly assigned. The points or break lines were estimated by using knowledge on the general morphological pattern of a foreshore zone. For instance, we use the foreshore contour lines generated from our previous study (see Meilianda, et al. 2007) as break lines.

The initial terrain generation by TIN in the present study was processed in ArcView$^{TM}$ software. As a result, generalized DEMs were generated for each period (2001, 2006 and 2007). Upon generating the first DEM by TIN method with the additional break lines, a contour map is subsequently converted from it, with an assigned contour interval of 0.5 m. The resulting contour map can be chaotic and unreliable at some locations.

An example of what may cause the erroneous result is described below: we take the case of generating the terrain in the vicinity of a seawall remnant which was located in the sea. This unique feature jumps up as a positive among the neighbouring negative elevation values (sea water). In most of the case, there were only a few elevation values existed; for instance, one existed on top of the seawall (which was positive elevation value) and the other was located at some distance further away from the toe of the seawall (negative value). The distance between this pair of values may generate a gentler slope than the actual breakwater's slope during the interpolation process. One way to fix this problem is by (often the case) manually digitizing the feature's contour lines (e.g. from top to toe of the seawall) as additional break lines. Herein, the dimension and the elevation of such specific coastal features are required.

Similar additional break lines should also be obtained at water-to-land boundary; otherwise the resulting coastline position will considerably deviate. For this matter, the generated foreshore contour lines which we have previously studied (see Meilianda, et al., 2007) were employed as an additional layer of input data for DEM generation. The contour lines are acting as break lines among the point data, which force the elevation interpolation between them and the surrounding points. Hence, the resulting DEMs will show more precise and realistic elevation values in the vicinity of the seawall remnant and along the land-to-water boundaries.

After the abovementioned adjustment process has been completed satisfactory, the final adjusted contour

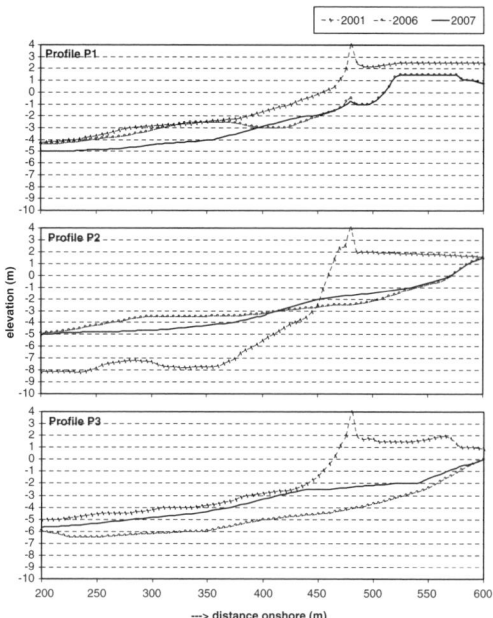

Figure 3. Three representative coastal profiles describing the morphological development in time along the coast. Profile P1, P2 and P3 represent the southwest, center and the northeast sub regions.

map is to be re-processed into DEM. The resulting DEM is converted into raster grids by assigning a proper spatial resolution. In this case we use a $0.5 \times 0.5$ m pixel size.

The subsequent process is to analyze the morphological changes by plotting some cross-shore profiles alongshore. From 18 sample profiles plotted along the coast with ca. 50 m intervals, we observed that there are three representative sub-regions which represent three different pattern of morphological changes and development. The location of representative profiles is depicted in Fig.1. The characteristics of the morphology of these representative profiles will be described later subtitle 4.

The profile plotting process was done in ILWIS$^{TM}$ which in this case gives a flexible adjustment on the distance interval of the overlaying multi-temporal data layers by using a script function. Here, we use a 5 m cross-shore interval. The resulting profiles are displayed in Fig.3 and Fig.4 for the cross-shore and alongshore directions, respectively.

## 4  ANALYSIS OF RESULTS

The three different periods of elevation information; i.e. the pre-tsunami's period (year 2001), one year after- (year 2006) and two years after the tsunami (year

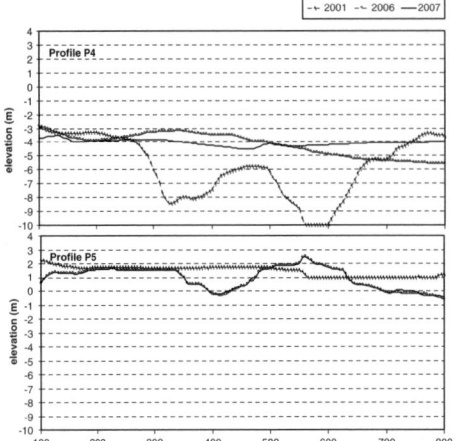

Figure 4. Alongshore profile representing the bathymetry (P4) and the backshore (P5).

2007) were simultaneously plotted for each profile. Observing these profiles we found 3 representative profiles which represent three sub regions. The three sub regions exhibited different morphological development. Analysis of the morphological changes and development along the coastal area are described based on the profile graphs as shown in Fig. 3.

## 4.1 Period 2001–2006 (before and 1 year after tsunami)

The profiles of the southwest sub-region exhibit severe erosion on the foreshore which was coincident with deposition offshore. The alongshore width of this sub region is about 150 m along the southwestern part of the studied area. The width of the eroded pre-tsunami's foreshore was rather narrow close to the inlet (Krueng Cakra Rivermouth) and was increasing towards the northeast.

At the southwest part of the coast the position of the 'turning points' between the eroding and the accreting profile segments was varying alongshore. Profile P1 in Fig. 3 represents the southwest sub-region. At the backshore (i.e. the land part of the coast) the elevation tends to lower in the order of 1 m from the pre-tsunami's elevation during the post-tsunami period, however was still about 1.5 m above MLWL which is more or less at the same level of high tide for this area. Therefore, the landward part is expected to have been less modified by the inter-tidal process. At present it is not certain whether the lowering effect of the backshore has something to do with the land subsidence since the evidence has only been observed in this particular region. The development of this sub-region until 2006 showed a tendency of erosion, which particularly occurred at

Table 1. Volume of changes.

| Profile | Periode | Change Volume (erosion) m³/m | Change Volume (deposition) m³/m |
|---------|---------|------------------------------|----------------------------------|
| 1 | 2001–2006 | 220 | |
|   | 2006–2007 | 225 | |
| 2 | 2001–2006 | | 802 |
|   | 2006–2007 | 75 | |
| 3 | 2001–2006 | 825 | |
|   | 2006–2007 | | 700 |

the foreshore. The average changes in volume are displayed in Table 1.

The center region is the second part of the studied area which shows specific morphological changes. It has an alongshore distance of about 350 m. The profile P2 in Fig. 3 represents the morphology of this sub region. An enormous deposition in 2006 occurred at the nearshore. It is revealed by the significant increase of the sample profile's elevation, which used to be much deeper in 2001 (ca. −10 to −16 m). In this region it can be observed that the sediment filling-up at the nearshore region was not entirely compensated by the erosion at the foreshore of the corresponding profile, even though the width of the eroded pre-tsunami's foreshore in this case also increased (ca. 115 m). This is because the amount of sediment deposition (filling-up) exceeds that of the foreshore erosion volume (see average change volume in table 1). Therefore, we presume that the deposited sediment to a certain portion must have come from other parts of the region by other processes than the normal cross shore sediment transport. It could be that the deposition occurred as an immediate impact of the multiple incoming tsunami wave action.

Indication of subsidence after the tsunami was fading along this sub region. This was because of the backshore was in the transition from the lower 2006's elevation at the adjacent southwest region to the higher 2006's elevation at the northeast by the fresh material deposition towards the northeast (both compared to the 2001's). This phenomenon can be better observed on the alongshore profile P4 in Fig. 4. At the northeastern part of this sub region, the backshore elevation of 2006 was increasing in the order of 1 m above the 2001's elevation which was built by the fresh material from the post-tsunami's process. The filling-up process was somehow terminated in this region; along with the abruptly increasing 2001's nearshore elevation towards the third sub region on the northeast.

In the third sub region (i.e. profile P3; the northeast region), again an erosion tendency occurred. Such erosion-deposition turning points which occurred in southwest sub region's profile only occurred here at the southwestern part, which is at the transition part between the center and the northeast sub region. The rest of the entire sub region exhibits that the nearshore and foreshore profiles were completely eroded and coincided with the deposition on the backshore.

The foreshore erosion distance was increasing towards the northeast (ca. 145 m). The 2001's backshore elevation was getting lower landward (profile P3 in Fig. 3). The pre-tsunami's backshore has turned into post-tsunami's foreshore and has been exposed to the sea ever since. Further to the northeast, the elevation of the backshore was getting lower. Immediately after the tsunami the backshore was almost completely submerged during the high tide (previous study of Meilianda, et al. 2007). After some months, the development of the backshore shows a tendency of deposition along this location, to form a sand-spit towards the northeast. The elevation of this newly deposited sediment was ca. +1.5 m above MLWL which correlate with the tidal range amplitude of this area. This indicates that the formation of the sediment deposition on the backshore was due to the normal inter-tidal processes. This deposition is contrary to what has happened on the southwest region (i.e. erosion on the backshore).

## 4.2 Period 2006–2007(development in 2 years after tsunami)

Two years aftermath, the coast is continuing its recovery by distributing the sediment to form smoothened cross-shore profiles. The overall sediment distribution tendency has shown that the alternating erosion coincides with accretion in most of the profiles, except at the northeast region, where a large amount of deposition on the nearshore and foreshore occurred.

The nearshore profile P1 (Fig. 3) in 2007 eroded. However, this particularly occurred at the nearby inlet profile where in 2006 a shoal appeared in the vicinity of the Krueng Cakra rivermouth. Further towards the center sub region, the alternating erosion on the nearshore and deposition on the foreshore coincided along the profile was common. However, due to the filling-up process that occurred in the previous development period, the nearshore profile of this region setup to a new coastal state. Contrary to this process, there is a sediment surplus along the entire part of the profile towards the northeast (Profile P3 in Fig 3). It is obvious that the nearshore profile in the northeast sub region is developing towards the pre-tsunami's elevation. The resume of the total change volumes is displayed in Table 1.

## 4.3 Discussions

Despite the fact that information of how the coast responded immediately after the tsunami is missing in the present study, a recovery process can be observed by how the coast re-organize its profiles after one and in between a couple of years. The overall tendency of the changes and development after one year (2006) is that in the early development after the tsunami the coastal profiles exhibits uncommon profile adjustments alongshore, where there were parts of the coast that were eroded (southwest and northeast sub region) and there were also parts in between where tremendous amount of deposition occurred (center sub region). Meanwhile, the width of the eroded foreshore region as a direct impact due to the tsunami was increasing towards the northeast. The deposition (filling-up) itself does not express a smooth transition, but more like an immediate filling-up of the trench by an enormous amount of supplied sediments (from elsewhere). This effect presumably occurred as the immediate response to the multiple (i.e. three consecutive waves according to Paris et al. (2007)) incoming tsunami wave action.

Upon the filling-up process, the normal coastal sediment transport process might have been taking control in the following development (i.e. in 2007). This can be explained by the adjustment of the coastal profile revealed by erosion which coincident with adjacent deposition across the coastal profiles which was actually a re-distribution of sediment towards a smoothened coastal profile. The coincidence of erosion at the upper part of the profile (foreshore) with deposition at the lower part (nearshore) can also be observed by another study by Zhang et al. (2005) on beach changes caused by hurricane Floyd in Florida. Such adjustment not only occurred along the cross-shore profiles, but also alongshore as depicted in Fig. 4. In the early development (2006) the alongshore profile tended to look bumpy with decreasing elevation towards the northeast. After two years (2007) it started to adjust into a smoothened profile due to erosion in the southwest and deposition in the northeast.

The large amount of deposition on the foreshore and towards the northeast as well as erosion on the southwest in one year after the tsunami (January 2006) may be affected by the alongshore current resulting from a pre-dominant rough wave climate which was coming from the southwest direction during the previous Southwest monsoon. However, in recent development the foreshore deposition has been coincident with a smoother nearshore profile that was evident at almost the entire profiles alongshore. The smoothened effect is also exhibited at the alongshore profile. This suggests that both longshore and cross-shore transport process have gradually organized the smoothened profiles during the transition into the calm Southeast monsoon. Eventually, the particular coastal morphology tends to develop into a new beach state with a new setting of foreshore and backshore, along with the level adjustment of the nearshore. The indication of land subsidence, however, remains weak in the present study.

The strong point of the applied method in this study is that there is a possibility to enhance the data quality of the initial scarce elevation data on the coastal zone by filling in the gaps of this data using any possible knowledge and additional data derived from remote-sensing-based studies. The basic idea is to complement information which was not available. The remaining uncertainty is how to accurately estimate the missing point values relative to the 'morphological judgments'. In addition to that, the order of accuracy from the combination of spatially and temporally different sources of data, as well as how the data was obtained from the measurements are difficult to determine. Overall, our close-to-realistic results from the generated DEM has been able to give insight of the changes and development of the coastal morphology through time, particularly with respect to the impact of the huge-scale natural event like the tsunami in December 2004.

## 5   CONCLUSION

In this study, combination of different sources and multiple years of topographic and bathymetric data were used to study the changes and development of the tsunami affected area. The data was used as the input to generate several Digital Elevation Models (DEMs). The Triangulated Irregular Network (TIN) method was applied to generate the DEM. In the period of 2001 to 2006, the results exhibited different morphological developments over 3 different sub regions in the early development after the tsunami impact. The southwest sub-region considerably eroded at the foreshore, but slightly eroded at the nearshore. The center sub-region experienced a filling-up process as the direct impact due to the tsunami. The northeast sub-region exhibited erosion for the entire profile. Meanwhile the development in the following one year (2006–2007) shows a tendency of smoothened coastal profiles on the cross-shore as well as the alongshore direction for the entire region. As a result, the coast experienced sediment deficit at the southwest and towards the center sub-region, and turned into sediment surplus at the northwest sub-region. Overall, the morphological development of the coastal area after two years indicates a development towards a new coastal state.

## ACKNOWLEDGEMENT

We address our acknowledgement to UP-PSDA Hydraulic Laboratory of Syiah Kuala University for providing the bathymetry and topography data of 2001, to Sea Defence Consortium – BRR Banda Aceh for providing the bathymetry, topography and other supporting data of 2006 and 2007. We also address our thanks to the Water Resource Department of ITC, the Netherlands, for providing image and DEM data processing facilities, as well as for providing supervision. This study is made possible by the fund of Asian Development Bank (ADB) through the TPSDP Project of the Higher Education Department of Republic of Indonesia, and the University of Twente, the Netherlands.

## REFERENCES

Meilianda, E., Dohmen-Janssen, C.M., Maathuis, B.H.P., Hulscher, S.J.M.H. and Mulder, J.P. 2007. Beach morphology at Banda Aceh, Indonesia in response to the tsunami on 26 December 2004, *to be published in Proceeding of Coastal Sediment Conference 2007 in New Orleans*, 14–18 May 2007

Paris, R., Lavigne, F., Wassmer, P. and Sartohadi, J. 2007. Coastal sedimentation associated with the December 26, 2004 tsunami in Lhok Nga, west Banda Aceh (Sumatra, Indonesia), *Marine Geology*, Vol. 238, pp. 93–106

Zhang, K., Whitman, D., Leatherman, S. and Robertson, W. 2005. Quantification of beach changes caused by hurricane Floyd along Florida's coast using Airborne laser surveys, *Journal of Coastal Research*, Vol. 21(1), pp. 123–134

*River, Coastal and Estuarine Morphodynamics: RCEM 2007 – Dohmen-Janssen & Hulscher (eds)*
*© 2008 Taylor & Francis Group, London, ISBN 978-0-415-45363-9*

# Predictability experiments of surf zone bathymetry using a process-based numerical model

B.G. Ruessink

*Department of Physical Geography, Faculty of Geosciences, IMAU, Utrecht University, Utrecht, Netherlands*

ABSTRACT: Surf zone sandbars constantly change their position in response to the time-varying forcing of off-shore surface gravity waves. Process-based predictions of cross-shore sandbar migration, relevant to understand the autonomous and humanly altered evolution of beaches, are intrinsically uncertain because of uncertainty in the model equations and, potentially, the sensitive dependence on the initial bathymetry. However, the magnitude of the resulting prediction limit and its dominant source are unknown. Here we show that cross-shore sandbar migration on the time scale of years is deterministic forced rather than deterministic chaotic and that the unpredictability of sandbar migration results primarily from model inadequacy during extreme wave events. Because the unpredictability of sandbar migration is related to the randomness of the forcing, the prediction limit is not a fixed value but depends on the timing of the extreme event. Our results demonstrate that cross-shore sandbar migration is deterministically predictable and imply that the efforts of detailed field and laboratory experiments to understand sandbar behaviour from the underlying first principles will eventually pay off to extend the prediction limit. We believe that other, in particular alongshore variable surf zone features are deterministic chaotic. The coexistence of deterministic forced and chaotic surf zone behaviour will provide an intriguing prediction problem.

## 1 INTRODUCTION

Surf zone sandbars (alongshore ridges of sand in 2–10 m water depth typical of microtidal, storm-dominated coasts) serve as a natural protection for beaches by causing waves to break away from the shoreline. Predictions of their behaviour – or, more generally, of surf zone bathymetry – using process-based, deterministic models are intrinsically uncertain. The underlying hydrodynamics and sediment transport processes are fundamentally non-linear and the models are likely to be chaotic (De Vriend 2001; Holman 2001; Reeve et al. 2004). Essentially, predictions are uncertain for three reasons: (1) uncertainty in the observations used to define the initial state; (2) uncertainty in the model equations and the numerical methods to solve these equations; and, (3) uncertainty in future external forcing. The last of these uncertainties relates to future time series of offshore wind-induced surface gravity waves, which are believed to be unpredictable more than a few days ahead. In this paper we will not consider the third source of uncertainty but rather concentrate on the first two. Both those sources impose a predictability limit beyond which the difference between the predicted and observed morphological state of the surf zone has become as large as the difference between two random states.

The analogy between the surf zone and the weather – both are natural dynamical systems dominated by non-linearity and strong dissipation (Lorenz 1963; Werner 1999) – sparked discussions (De Vriend 2001; Holman 2001; Reeve et al. 2004) on the (un)predictability of surf zone bathymetry. Surprisingly, these predominantly exploratory discussions were not followed by attempts to actually quantify surf zone prediction limits, even though theoretical models (Werner & Fink 1993; Calvete et al. 2007) and fractal analysis of bathymetric data sets (Southgate & Möller 2000) suggest chaotic behaviour is indeed possible. Quantitative estimates of the atmospheric predictability limit are mostly based on numerical integrations of a numerical weather prediction model by measuring how fast two or more similar states diverge (e.g. Thompson 1957; Charney et al. 1966; Smagorinsky 1969; Simmons et al. 1995). Because each initial state is integrated forward in time with the same model ('perfect-model scenario') under constant boundary forcing, the weather's unpredictability results completely from dynamical instabilities and their non-linear interaction (Lorenz 1969). The boundary forcing of the surf zone is, however, inevitably stochastic; we anticipate that the random character of the forcing may be more relevant to the divergence of initially similar morphological states than internal dynamics. Perfect-model estimates of predictability limits should be considered as an upper limit to predictability (Lorenz 1982), since perfect models do no exist. The skill of the model integrations verified against actual observations provides a lower, more realistic limit to predictability (Dalcher & Kalnay 1987).

In this paper we aim to determine the magnitude of the predictability limit and its dominant source – uncertain initial conditions vs. model inadequacy – for on/offshore sandbar behaviour. We use an 8-year data set of daily measured cross-shore depth profiles of the Hasaki coast, Japan and ensemble predictions generated with a waves-currents-bathymetric evolution model. First, we will discuss the model, data and methods; next, the temporal evolution of ensemble spread and model skill; and, finally, the implications of the results for surf zone research and modeling.

## 2 METHODOLOGY

### 2.1 Model

To simulate the temporal evolution of cross-shore bathymetry, the Unibest-TC model (Bosboom et al. 1997) was used, one of only few operational models that can accurately simulate onshore and offshore sandbar migration on time scale of days to weeks (Ruessink et al. sub judice). Unibest-TC uses an initial bed profile, median grain size of the bed material, and time series of offshore wave parameters (height, period, direction) and water levels to evolve the bed profile through commonly adopted coupled hydrodynamical (waves and currents) and sediment transport (bedload and suspended load) equations, for parameterization details see Ruessink et al. (sub judice). The wave-averaged nature of these equations ensures the applicability of the model on time scales up to years within reasonable computing time, while still incorporating many essential aspects of surf zone hydrodynamics and sediment transport. Alongshore variability in the bed profile, waves, and currents are neglected.

Unibest-TC incorporates the effect of wave skewness, infragravity waves bound to wave groups, gravity, near-bed streaming, and undertow in the sediment transport equations. Consistent with observations (Gallagher et al. 1998), the model predicts the sandbar to migrate offshore and to become more pronounced under breaking waves because of the feedback between these waves, undertow, suspended sediment transport, and the sandbar. Offshore migration continues until the storm subsides or the water depth above the sandbar becomes too large for the waves to break. When waves are energetic but non-breaking, the dominant feedback in the model involves the near-bed wave skewness, bedload transport, and the sandbar. This feedback causes the sandbar to migrate onshore and to reduce in height. Under small waves the model predicts the sandbar to decay in place slowly because of the small but persistent downslope, gravity-induced bedload. The numerical time step in the model

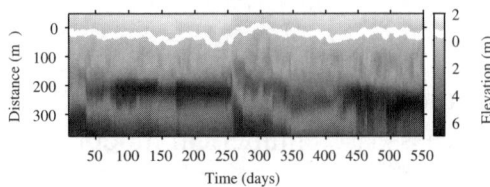

Figure 1. Time-space diagram of cross-shore bed profiles at Hasaki Beach, Japan. The white line is the $-0.5\,$m contour, separating the intertidal beach from the (subtidal) bar zone; its time-averaged position is about 30 m. Time $= 0$ corresponds to January 1, 1987; elevation $= 0$ equals mean sea level; and distance $= 375\,$m is the pier's tip. Occasional gaps with a 1–2 day duration were filled with linear interpolation.

is 1 hour, and the cross-shore grid size varies from 10–50 m well seaward of the breaker zone to 1 m on the beach. Parameters inherent to the formulation of subgrid processes, such as bed ripples, are set to values determined from earlier model calibrations (Ruessink et al. sub judice).

### 2.2 Data

The cross-shore bed profiles used to initialize and verify the model predictions are part of a data set of daily profiles collected along the pier of the Hazaki Oceanographic Research Station at Hasaki Beach, Kashima Coast, Japan between January 1987 and December 2004 (Kuriyama 2002). A subset of the data is displayed as a time-space diagram in Figure 1. The cross-shore mean difference $\Delta_{net}$ between two consecutive surveys, which indicates the overall loss or gain of sediment from a cross-shore profile (Gallagher et al. 1998), is typically a factor 2–4 less than the root-mean-square difference (Ruessink et al. sub judice). This implies that sediment was predominantly redistributed in the cross-shore direction, in line with the neglect in the model of alongshore variability in the bed profile. Occasional non-zero $\Delta_{net}$ indicate periods with a dominance of alongshore over cross-shore processes, potentially owing to the formation of crescentic bar patterns in the alongshore, or with a non-zero sediment flux past the offshore boundary. An example of the latter is visible in Figure 1 around time $t = 440$ days, when the outer bar migrated seaward of the pier's tip during a storm. Concurrent time series of daily averaged offshore root-mean-square wave height and peak period are available from the Kashima Port sensor located approximately 5 km offshore in 24-m depth. Wave heights peak at 3.5 to 4 m in autumn–spring during the passage of frontal systems and typhoons. Because of the non-directional nature of the Kashima sensor, the offshore wave direction was set to 30° from shore normal, a realistic value given the wind and observed wave climate at the site (Kuriyama et al. 2005). Water levels are available as hourly tidal values predicted for the tip of the pier. The median grain size is about 180 $\mu$m.

Figure 1 shows that the sandbar at Hasaki Beach migrates net offshore on the time scale of weeks to months, as the sandbar migrates more offshore during storms (up to 20 m/day) than onshore during inter-storm periods (up to 5 m/day). When the water depth above the sandbar crest has reached some 3.5–4 m, the sandbar starts to decay; during the next storm, a new sandbar is generated near the shore. This sandbar subsequently migrates offshore and takes its place as the new prominent sandbar (Fig. 1, $t = 250 - 270$ days). The time frame between successive sandbar decay is roughly 1 year (Kuriyama 2002). In the observations, bed variability induced by sandbar migration is restricted to $x = 30 - 375$ m (Fig. 1); this cross-shore range is henceforth referred to as the bar zone.

### 2.3 Ensemble approach

The sensitivity to initial conditions and the effect of model inadequacy on hindcasted sandbar evolution is examined from ensembles generated with lagged-average forecasting (Hoffman & Kalnay 1983). At each verification time $t_V$, an ensemble not only contains the prediction started from the initial conditions observed at some control time $t_C$ but also predictions for the same $t_V$ started one or more days earlier than $t_C$. Predictions initiated up to 7 days earlier than $t_C$ are included, implying an ensemble to consist of 8 members (Fig. 2). The measured and 7 predicted bed profiles at $t_C$ are viewed as differing but physically feasible initial conditions at that moment in time. All initial states from $t = 9 - 426$ days are integrated to $t = 556$ days (initial states for $t > \approx 430$ days are unreliable because the potential presence of a sandbar seaward of the pier's tip (Fig. 1) and are not used). This results in 411 ensembles ($t_C$ ranges from $t = 16$ to 426 days) and in simulation durations $\Delta t_s = t_V - t_C$ between 1 and 540 days ($t_V$ ranges from $t = t_C + 1$ to 556 days). The maximum simulation duration exceeds the dominant period in the forcing (days) by 2 orders of magnitude and is well beyond the typical simulation duration of model applications existing in the literature.

The divergence of the ensembles as they evolve from $t_C$ to $t_V$ is used to examine the limit on predictability because of a sensitive dependence on initial conditions. A convenient measure of the divergence is the standard deviation, or spread $s$, of the ensemble members about the ensemble mean, defined as the root-mean-square difference from the ensemble mean of all 8 members, averaged over the bar zone. Thus,

$$s(t_C, t_V) = \sqrt{\overline{\frac{1}{8} \sum_{i=1}^{8} (z_i(x, t_V) - \tilde{z}(x, t_V))^2}^x}, \quad (1)$$

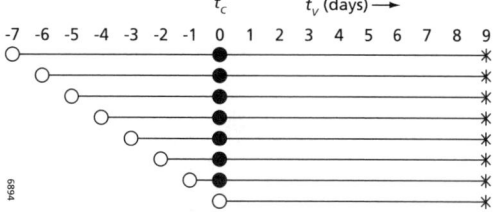

Figure 2. Schematic description of the ensemble method. The open circles denote the starting points of each simulation. From $t = t_C$ (filled circles and lowermost open circle) the 8 simulations act as ensemble, from which in this schematic spread and skill are computed (asterisks) at $t_V = 9$ days.

where $z_i(x, t_V)$ is the predicted bed elevation of the $i$th ensemble member at cross-shore location $x$ and time $t_V$, the overbar is the bar-zone average and the tilde represents the ensemble mean. The limit to predictability is reached when the ratio $\alpha(t_C, t_V)$ of $s(t_C, t_V)$ to the spread of 8 random states (=0.68 m based on 1000 draws of 8 random bed profiles from all available observations) exceeds some prechosen magnitude, which for practical purposes should exceed the magnitude of the normalized spread at $t_C$ (here, $\alpha(t_C, t_V) \approx 0.1 - 0.3$) but less than the magnitude of 8 random bed profiles ($\alpha(t_C, t_V) = 1$). Following Dalcher & Kalnay (1987), we examine when $\alpha(t_C, t_V)$ exceeds 0.5 and 0.95.

The effect of model inadequacy on forecast accuracy is examined with the temporal evolution of a skill score $SS$,

$$SS(t_C, t_V) = 1 - \frac{\overline{(\tilde{z}(x, t_V) - z_o(x, t_V))^2}^x}{\overline{\Delta z_o^2(\Delta t_s)}^x}, \quad (2)$$

where $z_o(x, t_V)$ is the observed cross-shore bed profile at $t_V$, and $\Delta z_o^2$ ($\Delta t_s$) is the squared difference between two random observed bed profiles separated by the simulation duration. Thus, the bar-zone averaged squared difference between the ensemble mean and the observed profile is compared to the bar-zone averaged squared difference of two random bed profiles measured $\Delta t_s$ from each other. We assume that the limit to predictability is reached when $SS(t_C, t_V)$ becomes negative. Values of $\overline{\Delta z_o^2}^x$ were computed for each $\Delta t_s$ by averaging $\overline{\Delta z_o^2}^x$ of 100 draws of two random bed profiles separated by $\overline{\Delta z_o^2}^x$ from all available observations. Here, $\overline{\Delta z_o^2}^x$ grows rapidly from $\approx 0.02$ m$^2$ at $\Delta t_s = 1$ day to $\approx 1.0$ m$^2$ at $\Delta t_s = 250$ days; for larger $\Delta t_s$, $\overline{\Delta z_o^2}^x$ fluctuates around 0.9 m$^2$ with minima at multiples of $\Delta t_s \approx 1$ year because of the sandbar cycle period.

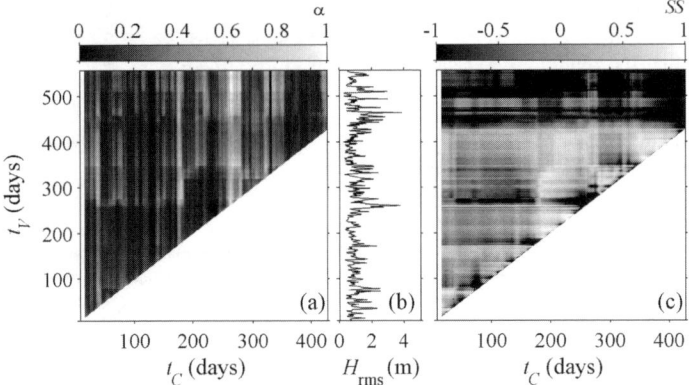

Figure 3. (a) Normalized spread $\alpha(t_C, t_V)$ and (c) model skill $SS(t_C, t_V)$ versus control time $t_C$. The root-mean-square wave height $H_{rms}$ at the model's seaward boundary is given for reference in (b).

## 3 RESULTS

### 3.1 Initial conditions

Figure 3a displays the $\alpha(t_C, t_V)$ array; each column represents the temporal evolution of $\alpha$ for a single ensemble, and each row shows the variability of $\alpha$ at the same $t_V$ as a function of $t_C$. If $\alpha(t_C, t_V)$ depended only upon the simulation duration, we would expect most variability in $\alpha(t_C, t_V)$ to line up with the line $t_C = t_V$. The most prominent property of Figure 3a, however, is the simultaneous increase and decrease in $\alpha(t_C, t_V)$ in all ensembles, which implies a relationship with the randomness of the boundary forcing. Abrupt $\alpha(t_C, t_V)$ increases at $t_V \approx 255$ and $450$ days coincide with major storms ($H_{rms} > \approx 3$ m in Figure 3b), when in most ensembles the model predicts the birth (near the shoreline) and subsequent offshore migration of a new sandbar. Moderate storms ($H_{rms} > \approx 2$ m, for example, $t_V \approx 70$, $350$ and $510$ days), resulting in the offshore migration of an existing sandbar, aggravate $\alpha(t_C, t_V)$ slightly. Gentle $\alpha(t_C, t_V)$ decreases concur with prolonged periods of relatively low-energy conditions ($H_{rms} < \approx 1 - 1.5$ m), when the model predicts an existing sandbar to migrate onshore and diminish in height. In contrast to suggestions in Stive & Reniers (2003) and Mariño-Tapia et al. (2007), the results demonstrate that sandbar behaviour during (major) storms is inherently less predictable than that during lower-energy conditions. We find no evidence from Figure 3a to suggest that internal dynamics is relevant to temporal changes in spread.

Another prominent feature of Figure 3a is the increased vertical striping after especially the first major storm ($t_V = 255$ days), which shows the great variability in $\alpha(t_C, t_V)$ growth in ensembles with comparable pre-storm spread. For example, some ensembles with a pre-storm ($t_V = 223$ days) spread in the $0.05$–$0.15$ range did not diverge at all, while others in

the same range diverged by a factor of four or more. The amplification is therefore not a 'surf zone' constant, but depends on the type of pre-storm uncertainty; some situations are more predictive than others.

To explore why storms impact spread differently than low-energy conditions and why some ensembles are more predictive than others, we computed various sandbar characteristics (position $x_b$, bar-trough relief $h_b$, and mean water depth $d_b$ above the bar crest) for each simulation suggested previously to result in divergent sandbar behaviour under constant forcing (Calvete et al. 2007; Mariño-Tapia et al. 2007). We found (Fig. 4) that the ensembles with subdued bar-trough morphology at $t_V = 223$ days (ensemble-average $d_b < 0.3$ m) were less predictive during the storm at $t_V = 255$ days than ensembles with pronounced bar-trough morphology (ensemble-average $d_b > 1.5$ m). In the latter ensembles the sandbar is sufficiently high to strongly localize wave breaking and sediment transport gradients; in fact, in these ensembles the sandbar does not decay and no new sandbar loosens itself from the shore. Further onshore, protected by the well-developed sandbar, waves, currents, and sediment transport are relatively low, causing minor (and similar) profile response. In contrast, a subdued sandbar barely affects the cross-shore evolution of hydrodynamics and sediment transport; the profile response is now concentrated further onshore, where small initial profile variability is amplified into variable response of the newly generated sandbar. These results show that the morphology response of major storms strongly depends on the antecedent morphology (predominantly, bar-trough relief) and may explain observations (Lippmann et al. 1993) at Duck, NC why some storms trigger a new bar cycle but other, similar storms don't.

We further found that, whereas all sandbars behave similarly during a moderate storm – they all migrate offshore –, the initial intra-ensemble variability in

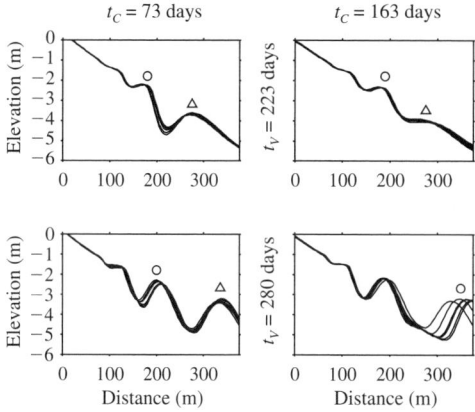

Figure 4. Elevation versus cross-shore distance at (upper row) $t_V = 223$ days and (lower row) $t_V = 280$ days for ensembles started at (left column) $t_C = 73$ days and (right column) $t_C = 163$ days. The symbols denote the corresponding bars. Normalized spread $\alpha$ increases from 0.11 to 0.16 in the $t_C = 73$ days ensemble and from 0.07 to 0.34 in the $t_C = 163$ days ensemble.

water depth above the sandbar crests causes a divergence in sandbar location and hence an increase in spread. With an increase in $d_b$, wave breaking lessens and, accordingly, the undertow, the current-induced suspended sediment transport and the offshore sandbar migration rate diminish. Ensembles with a larger pre-storm $d_b$ variability accordingly diverge more than ensembles with a smaller $d_b$ variability (Fig. 5). During low-energy conditions all sandbars reduce in bar-trough relief as they move onshore. Hence, all ensemble members are predicted to more and more resemble the same featureless cross-shore profile, which causes the ensemble spread to reduce.

Of all 411 ensembles, only 1% reached $\alpha \geq 0.95$; an additional 24.5% attained a normalized spread of at least 0.5. In 53 ensembles ($=12.9\%$) the spread actually diminished from the start and was never aggravated above its initial value. To examine the $\alpha$ evolution for longer simulation durations than considered so far, we integrated three selected ensembles with different maximum spread prior to $t_V = 556$ days for another 2 years. The resulting series (Fig. 6) fluctuate around a constant value not too different from their initial value, with, as already demonstrated from Figures 3a–b, sudden $\alpha$ growth during storms and gentle decline afterwards. There is no evidence that on/offshore sandbar migration is chaotic.

### 3.2 Model inadequacy

Figure 3c signifies that, similar to $\alpha(t_C, t_V)$, model skill $SS(t_C, t_V)$ is not simply a function of simulation duration. Instead, major changes in $SS(t_C, t_V)$ are

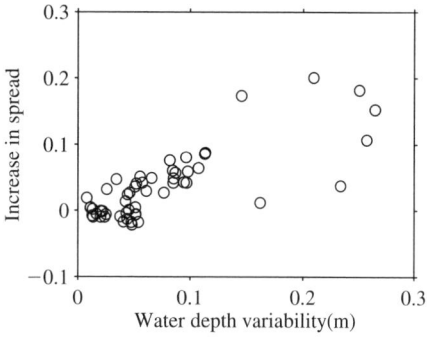

Figure 5. Increase in spread $\alpha$ from $t_V = 70$ to 79 days (a period encompassing 2 storms with offshore root-mean-square wave heights in excess of 2 m, see Fig. 3b) as a function of the intra-ensemble variability in water depth $d_b$ above the sandbar crests, quantified as the standard deviation in $d_b$ of the 8 ensemble members, at $t_V = 70$ days. The linear relation between the increase in $\alpha$ and $d_b$ is statistically significant ($r = 0.78$, $p \ll 0.01$).

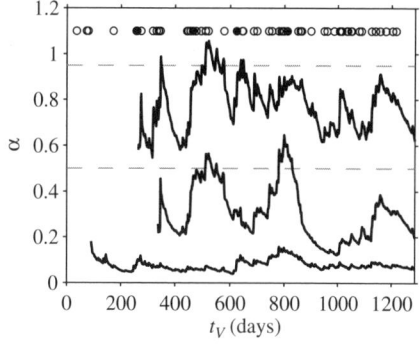

Figure 6. Normalized spread $\alpha(t_V)$ for $t_C = 90$, 265 and 339 days versus verification time $t_V$. Storms with offshore root-mean-square wave height exceeding 2 m (open circles) and 3 m (filled circles) are shown. The horizontal dashed lines are $\alpha = 0.5$ and $\alpha = 0.95$.

aligned horizontally, which, again, implies a relationship with the randomness of the external forcing. Intriguingly, several ensembles (e.g., $t_C \approx 9$–150 days) lose skill during storm events (e.g., $t_V \approx 80$, 175 and 260 days) to regain it slowly during the subsequent less energetic conditions. Our inspection of the magnitude of the individual terms in Eq. (2) revealed that the return of skill is not imposed artificially by the temporal dependence of $\overline{\Delta z_0^{2}}^x$, but results primarily from the reduction of the bar-zone averaged squared difference between the ensemble mean and the observed profile.

Although model skill is negative at $t_V = 260$ days for most $t_C$ (Fig. 3c), we find it encouraging that in most

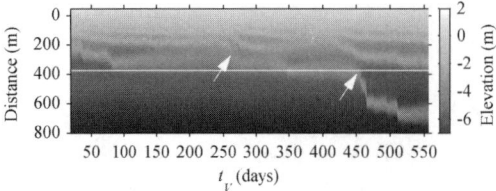

Figure 7. Time-space diagram of ensemble-mean ($t_C = 18$ days) cross-shore bed profiles at Hasaki Beach, Japan. The white line is the seaward extent of the pier. The white arrows indicate (left) the loosening of a new sandbar from the shore and (right) the migration of the sandbar seaward of the pier's tip.

ensembles the model actually predicts the loosening of a sandbar from the shore following outer-sandbar decay around this time. An example model output that demonstrates this behaviour is provided in Figure 7. To the best of our knowledge, no other existing process model is capable of simulating bar birth and subsequent offshore migration following outer-bar decay using realistic boundary forcing. Our model appears to be slightly off in the rate of the offshore migration of the new sandbar, compare Figure 7 to Figure 1. While the migration of the outer sandbar seaward of the pier's tip around $t_V = 440$ days (Fig. 7 and most other ensembles) is consistent with the observations (Fig. 1), the subsequent offshore migration of the next most seaward located sandbar was not observed. This inconsistency between model predictions and observations causes the skill of all ensembles (that is, independent of the variability in the initial conditions, the spread in the ensembles just before $t_V = 440$ days, and the simulation duration) to become negative (Fig. 3c).

To facilitate a more quantitative analysis of the temporal evolution of $SS(t_C, t_V)$, we first rewrite the denominator in Eq. (2) as (dropping the bar zone average) $((\bar{z}(x, t_V) - \overline{z_o}(x)) - (z_o(x, t_V) - \overline{z_o}(x)))^2$ or $(\tilde{z}'(x, t_V) - \tilde{z}'_o(x, t_V))$, where $\tilde{z}'(x, t_V) = \tilde{z}(x, t_V) - \overline{z_o}(x)$ and $\tilde{z}'_o(x, t_V) = z_o(x, t_V) - \overline{z_o}(x))$ are the anomalies in the model predictions and observations, respectively, relative to the observed cross-shore profile time-averaged over the entire data set, $\overline{z_o}(x)$. A linear regression model, in which the predicted anomalies are regressed on the observed anomalies, now provide a useful means to examine the reasons for non-perfect skill (Murphy & Epstein 1989). More specifically, the regression slope $b = (s_Y/s_X)r$, where $s_X$ and $s_Y$ are the standard deviation of the observed and predicted anomalies at a given $t_V$, respectively, and $r$ is the correlation coefficient, incorporates both a measure of amplitude error ($s_Y/s_X$ – the predicted and observed anomalies are of different magnitude when $s_Y/s_X \neq 1$) and phase error ($r$ – the predicted anomalies are in the wrong position when $r \neq 1$).

The graphical display of the ratio $s_Y/s_X$ (Fig. 8a) shows that the model overpredicts the reduction in bar-trough relief as the bar migrates onshore ($s_Y/s_X < 1$ for $t_V \approx 100$–260 days). The decrease in skill after the storms at $t_V \approx 80$, 175 and 260 days (Fig. 3c) results primarily from phase errors, as indicated by the sudden drops in $r$ (Fig. 8c). Our inspection of the ensembles revealed that the $r$ decrease during the first two storms results from an overestimate of the offshore sandbar migration. The subsequent returns of skill can be attributed to the gradual $r$ increases; because the model has just overpredicted offshore sandbar migration, the $r$ increases actually imply that the model also overpredicts onshore sandbar migration at these times. The $r$ decrease at $t_V = 260$ days confirms our earlier visual inspection (Fig. 7) that the model overestimates the rate of offshore sandbar migration following sandbar generation. In the weeks following $t_V = 260$ days the predicted sandbar is of the correct magnitude ($s_Y/s_X \approx 1$, Fig. 8a) but in the wrong position ($r < 1$, Fig. 8c). The loss of skill after $t_V = 440$ days is also induced primarily by phase errors (Fig. 8c); in many ensembles, especially after $t_C \approx 180$ days, $r$ is close to $-1$, implying the observed and predicted anomalies to mirror each other.

4  DISCUSSION AND CONCLUSIONS

To understand and predict surf zone sandbar behaviour it is vital to know whether we are dealing with a deterministic forced or deterministic chaotic response to the time-varying boundary forcing. With our numerical results we have established that initial variability in cross-shore depth profiles does not necessarily grow to the level that initial similar states become uncorrelated (Fig. 6). Thus, we conclude that on/offshore sandbar migration on the time scale of years is deterministic forced rather than deterministic chaotic, and hence is deterministically predictable. Our numerical results further show that the present-day unpredictability of on/offshore sandbar migration results primarily from model inadequacy during major storms (here, typhoons), with an overestimate of offshore sandbar migration. Since the loss in model skill is coupled to the randomness of the forcing, there is no 'absolute' value for the prediction limit; it depends on the timing of the extreme event. One of the implications of our results is that the enormous effort to measure and understand physical processes from detailed field (Gallagher et al. 1998; Hoefel & Elgar 2003) and laboratory (Roelvink & Stive 1989) experiments will eventually pay off to extend the prediction horizon of on/offshore sandbar migration.

Do our results imply that all surf zone morphological variability is deterministic forced? We answer this question negatively. Theoretical models suggest that alongshore variable morphology, such as beach

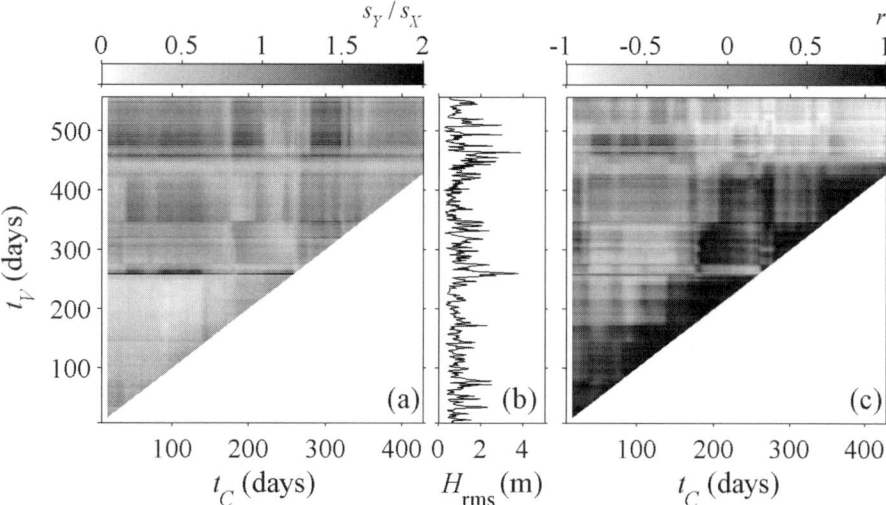

Figure 8. (a) Ratio of the standard deviations in the predicted and observed anomalies $s_Y/s_X$ and (c) correlation coefficient $r$ of the linear regression of the predicted on the observed anomalies versus control time $t_C$. As outlined in the text, $s_Y/s_X$ and $r$ represent an amplitude and phase error, respectively. The root-mean-square wave height $H_{rms}$ at the model's seaward boundary is given for reference in (b).

cusps, rip channels, oblique bars and crescentic bars, is emergent from the selective growth of small random perturbations in the sea bed or the water motion under constant boundary forcing (Werner & Fink 1993; Falqués et al. 2000; Garnier et al. 2006). These studies as well as field studies (Holman & Sallenger 1993; Ruessink et al. 1998) have falsified the earlier hypothesis of the deterministic forcing of these features by phase-locked edge waves trapped to the coast. We believe that deterministic forced and deterministic chaotic behaviour coexist in the surf zone. The extension of our simulations to include alongshore morphological variability will provide an intriguing prediction problem.

Although the similarity between the weather and the surf zone initiated discussions on the unpredictability of surf zone bathymetry, we did not find any evidence that the unresolved flow (i.e. fluid turbulence) and bed states (i.e. bed ripples) impose a predictability limit to the (resolved) cross-shore sandbar migration in the manner suggested by Lorenz (1969) for numerical weather prediction. One could argue that the limited range of scales considered here as well as the use of time-independent and cross-shore constant parameters in the subgrid parameterizations underestimate the ensemble divergence associated with internal dynamics. Now, as an example, suppose we would able to accurately predict temporal and spatial variability in bed ripple height and length, and their effect on near-bed flow, sediment transport and, hence, sandbar migration. It is conceivable that the improved model physics alleviates skill problems but, at the same time, triggers unpredictability because of the now potentially sensitive dependence of sandbar behaviour on initial ripple characteristics. A more complex model may be more realistic (at least, intuitively) but at the same time also more uncertain. We suggest that improved model physics should not only be judged on its ability to extend the model-inadequacy induced prediction limit but also on its potentially adverse affect on the sensitivity to initial conditions.

ACKNOWLEDGEMENTS

I gratefully acknowledge the many people who collected the Hasaki data set. In particular, Y. Kuriyama provided the bathymetric data; S. Yanagishima the water level data; and M. Mizuguchi the Kashima Port wave data. I further acknowledge J. A. Roelvink, A. J. H. M. Reniers and, in particular, D. J. R. Walstra for their help with the model, and H. N. Southgate for stimulating discussions. This work was supported by the Netherlands Organisation for Scientific Research Award 864.04.007.

REFERENCES

Bosboom, J., S. G. J. Aarninkhof, A. J. H. M. Reniers, J. A. Roelvink, & D. J. R. Walstra (1997). Unibest-TC 2.0. Overview of model formulations. WL|Delft Hydraulics Report H2305.42.

Calvete, D., G. Coco, A. Falqués, & N. Dodd (2007). (Un)predictability in rip channel systems. *Geophys. Res. Lett. 34*, L05605, doi:10.1029/2006GL028162.

Charney, J. G., R. G. Fleagle, H. Riehl, V. E. Lally, & D. Q. Wark (1966). The feasibility of a global observation and analysis experiment. *Bull. Am. Meteorol. Soc. 47*, 200–220.

Dalcher, A. & E. Kalnay (1987). Error growth and predictability in operational ECMWF forecasts. *Tellus 39A*, 474–491.

De Vriend, H. (2001). Long-term morphological prediction. In G. Seminara & P. Blondeaux (Eds.), *River, coastal and estuarine morphodynamics*, pp. 163–190. Berlin: Springer Verlag.

Falqués, A., G. Coco, & D. A. Huntley (2000). A mechanism for the generation of wave-driven rhythmic patterns in the surf zone. *J. Geophys. Res. 105*, 24071–24088.

Gallagher, E. L., S. Elgar, & R. T. Guza (1998). Observations of sand bar evolution on a natural beach. *J. Geophys. Res. 103*, 3203–3215.

Garnier, R., D. Calvete, A. Falqués, & M. Caballeria (2006). Generation and nonlinear evolution of shore-oblique/transverse sand bars. *J. Fluid Mech. 567*, 327–360.

Hoefel, F. & S. Elgar (2003). Wave-induced sediment transport and sandbar migration. *Science 299*, 1885–1887.

Hoffman, R. N. & E. Kalnay (1983). Lagged average forecasting, an alternative to Monte Carlo forecasting. *Tellus 35A*, 100–118.

Holman, R. (2001). Pattern formation in the nearshore. In G. Seminara & P. Blondeaux (Eds.), *River, coastal and estuarine morphodynamics*, pp. 141–162. Berlin: Springer Verlag.

Holman, R. A. & A. H. Sallenger (1993). Sand bar generation: a discussion of the Duck experiment series. *J. Coast. Res. SI 15*, 76–92.

Kuriyama, Y. (2002). Medium-term bar behavior and associated sediment transport at Hasaki, Japan. *J. Geophys. Res. 107*, doi: 10.1029/2001JC000899.

Kuriyama, Y., Y. Ito, & S. Yanagishima (2005). Field investigation on cross-shore distribution of predominant longshore current velocity. *JSCE J. Hydr. Coast. Env. Eng., Div. B. 803/II-73*, 145–153.

Lippmann, T. C., R. A. Holman, & K. K. Hathaway (1993). Episodic, nonstationary behavior of a two sand bar system at Duck, NC, USA. *J. Coast. Res. SI(15)*, 49–75.

Lorenz, E. N. (1963). Deterministic nonperiodic flow. *J. Atmos. Sci. 20*, 130–141.

Lorenz, E. N. (1969). The predictability of a flow which possess many scales of motion. *Tellus 21*, 289–307.

Lorenz, E. N. (1982). Atmospheric predictability experiments with a large numerical model. *Tellus 34*, 505–513.

Mariño-Tapia, I. J., P. E. Russell, T. J. O'Hare, M. A. Davidson, & D. A. Huntley (2007). Cross-shore sediment transport on natural beaches and its relation to sandbar migration patterns: 2. Application of the field transport parameterization. *J. Geophys. Res. 112*, C03002, doi:10.1029/2005JC002894.

Murphy, A. H. & E. S. Epstein (1989). Skill scores and correlation coefficients in model verification. *Mon. Wea. Rev. 117*, 572–581.

Reeve, D., A. Chadwick, & C. Fleming (2004). *Coastal Engineering - processes, theory and design practice*. London: Spon Press, Taylor & Francis Group.

Roelvink, J. A. & M. J. F. Stive (1989). Bar-generating cross-shore flow mechanisms on a beach. *J. Geophys. Res. 94*, 4785–4800.

Ruessink, B. G., M. G. Kleinhans, & P. G. L. Van den Beukel (1998). Observations of swash under highly dissipative conditions. *J. Geophys. Res. 103*, 3111–3118.

Ruessink, B. G., Y. Kuriyama, A. J. H. M. Reniers, J. A. Roelvink, & D. J. R. Walstra (sub judice). Modeling cross-shore sandbar behavior on the time scale of weeks. *J. Geophys. Res.*.

Simmons, A. J., R. Mureau, & T. Petroliagis (1995). Error growth and estimates of predictability from the ECMWF forecasting system. *Q. J. R. Meteorol. Soc. 121*, 1739–1771.

Smagorinsky, J. (1969). Problems and promises of deterministic extended range forecasting. *Bull. Am. Meteorol. Soc. 50*, 286–311.

Southgate, H. N. & I. Möller (2000). Fractal properties of coastal profile evolution at Duck, North Carolina. *J. Geophys. Res. 105*, 11489–11507.

Stive, M. J. F. & A. J. H. M. Reniers (2003). Sandbars in motion. *Science 299*, 1855–1856.

Thompson, P. D. (1957). Uncertainty of intial state as a factor in the predictability of large-scale atmospheric flow patterns. *Tellus 9*, 275–295.

Werner, B. T. (1999). Complexity in natural landform patterns. *Science 284*, 102–104.

Werner, B. T. & T. M. Fink (1993). Beach cusps as self-organized patterns. *Science 260*, 968–971.

River, Coastal and Estuarine Morphodynamics: RCEM 2007 – Dohmen-Janssen & Hulscher (eds)
© 2008 Taylor & Francis Group, London, ISBN 978-0-415-45363-9

# Comparing observed surfzone transverse finger bars with model results

F. Ribas
*Institut de Ciències del Mar, CSIC, Barcelona, Spain*

D. Calvete & A. Falqués
*Universitat Politècnica de Catalunya, Barcelona, Spain*

H.E. de Swart
*Utrecht University, Utrecht, The Netherlands*

A. Kroon
*Copenhagen University, Copenhagen, Denmark*

ABSTRACT: The results of a self-organization model for surfzone transverse bar formation are compared with field observations in Noordwijk beach (the Netherlands). Two events of bar formation and subsequent evolution in June-July 2000 and in August-September 2002 are studied. The wave length of the patches is 35 m and 46 m, respectively, and the mean celerity is 3–4 m/d in the direction of the longshore current. Bars have an oblique orientation, deviating some 30° from the shore-normal against the longshore current. Bars persist up to 2 months, coexisting with obliquely incident waves of intermediate heights. Application of the model to Noordwijk conditions yields wave lengths, crest orientations and growth rates that are in agreement with observations, but the model overestimates the migration rates. The clue to obtain up-current oriented bars is the assumption that the depth-integrated suspended sediment concentration is constant across the inner surf zone.

## 1 INTRODUCTION

Patches of several transverse finger sand bars have been observed in the surf zone of some beaches, spaced with a remarkable alongshore periodicity from 20 to 200 m. Transverse finger bars are thin and elongated accumulations of sand attached to the shoreline that extend inside the surf zone with a shore-normal or shore-oblique orientation. They emerge on gently sloping beaches (slope <0.03) both in microtidal sheltered areas (Gelfenbaum and Brooks 2003) and on mesotidal more energetic open coasts (Konicki and Holman 2000; Ribas and Kroon 2007).

The two latter studies used hourly video-images to describe the characteristics of transverse finger bars in Duck (USA) and in Noordwijk (the Netherlands), respectively (see Fig. 1). One to three shore-parallel subtidal bars are very often present in these beaches, sometimes showing a crescentic shape (van Enckevort et al. 2004). The detected finger bars were most often located inside the trough of the inner bar, attached to the low-tide shoreline. That is why Konicki and Holman (2000) named them 'trough bars'. Both the percentage of days with patches and the number of bars per patch were significantly larger in Noordwijk. The overall averaged wave length was 39 m in Noordwijk bars and 79 m in Duck 'trough bars'. Bar crests deviated from the shore-normal some 25° and bars migrated at rates up to a few tens of meters per day. Ribas and Kroon (2007) also correlated the characteristics of Noordwijk bars with the hourly wave conditions measured by an offshore buoy at 18 m water depth. Bar patches migrated as a whole in the direction of the longshore current and bar crests deviated from the shore-normal in the up-flow direction ('up-current orientation'). Wave conditions detected in Noordwijk during bar presence were characterized as intermediate waves ($\sim$ 1 m height) arriving with large angles of incidence with respect to the shore-normal ($\theta_{off} \simeq 50°$).

A viable explanation for nearshore bar formation is based on the concept of morphodynamic self-organization. Topographic perturbations superimposed on an alongshore uniform beach profile induce hydrodynamic perturbations, which can lead to convergence of sand transport over the bars, hence producing a positive feedback. Linear stability analysis (LSA) is a convenient tool to investigate this

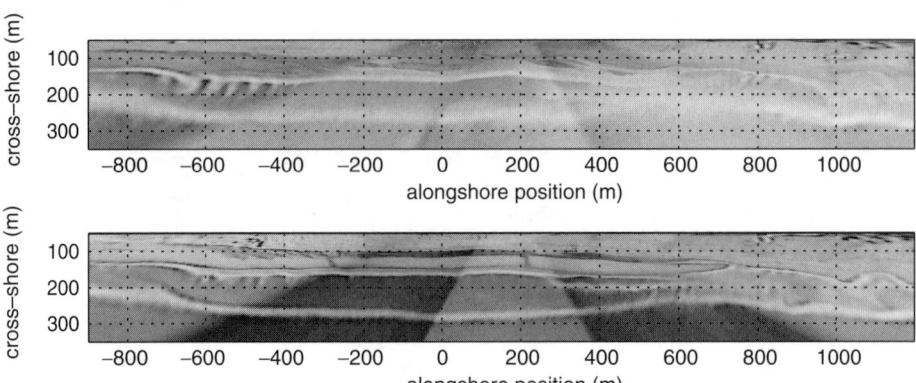

Figure 1. Time-exposure planview images of Noordwijk corresponding to the transverse bar event in August-September 2002. The horizontal axis is the alongshore coordinate, $y$, pointing southwards and the vertical axis is the cross-shore coordinate $x$, pointing seaward. A patch of surfzone transverse finger bars can be seen at $y = [-800, -400]$ m and $x = 180$ m, attached to the low-tide shoreline. Top panel: Image on September 22 at 11 h. Low panel: Image on October 4, 9 h.

possible feedback, yielding information about the shape, growth rate and migration speed of the initially emerging modes. It also allows for a systematic exploration of the sensitivity of bar characteristics to the beach conditions and to the model formulation of different physical processes. Nonlinear models are used to describe the finite-amplitude features and verify the results of LSA.

Several self-organization models for transverse bar formation have been developed in the last years (Ribas et al. 2003; Ribas et al. 2005; Garnier et al. 2006). They demonstrate that self-organization can explain bar formation since the computed topographic patterns resemble transverse bars in nature. However, the predicted shapes (orientation of oblique bars with respect to the longshore current) and the time scales for growth and migration strongly depend on the specific description of wave propagation and sediment transport. Performing a quantitative comparison with field observations is essential to test these models, verify the available predictions for bar characteristics and clarify the remaining open questions.

Ribas and Kroon (2007) used their observations to test qualitatively the predictions of some self-organization models. Considering that Noordwijk finger bars were 'up-current oriented' and that they only emerged for large $\theta$ only one of the existing physical mechanisms for transverse bar growth remained as a viable explanation: the 'bed-flow coupling'. As explained in Ribas et al. (2003), the 'bed-flow coupling' is dominant in case of $\theta_{off} > 40°$, the presence of strong longshore currents being essential. The growing bars locally modify the longshore current, which veers towards the direction of maximum topographic gradient due to mass conservation. Hence offshore

deflection occurs over up-current oriented bars. Positive feedback occurs if the depth-integrated sediment concentration is constant across the inner surf zone because this enhances the convergence of sediment flux in offshore-directed flows (since they slow down when moving into deeper water due to mass conservation). Ribas et al. (2003) presented a simplified model that included the bed-flow mechanism and predicted the formation of up-current oriented bars. It was based on LSA and a highly-idealized formulation for the beach geometry (constant sloping beach) and the wave transformation (regular wave height and no shoaling effects). A more accurate wave transformation was included in Ribas et al. (2005), with random wave heights and shoaling effects, and the up-current bars remained as a robust outcome. Finally, Garnier et al. (2006) studied the non-linear temporal evolution of up-current oriented bars. The wave transformation was similar to that of Ribas et al. (2005) and the initial beach profile was also constant sloping. The modelled bars emerged and reached saturation of their amplitude at values around 30 cm.

The present contribution aims at modelling the formation of up-current oriented bars on Noordwijk beach. For this, an up-to-date self-organization model based on LSA is applied to Noordwijk wave and bathymetric conditions. Whereas the previous models were applied to constant-sloping beach profiles, we will use the profiles with two shore-parallel bars measured at Noordwijk. Firstly, the characteristics of the bar patches detected in Noordwijk are summarized, focusing on two well-developed events of bar formation. Then the model equations are described together with the used methodology. Finally, the results of the model are compared with the field observations.

## 2  OBSERVATIONS

Noordwijk beach is located at the long and straight North Sea coast of the Netherlands. The median grain size is 0.20 mm and the profiles always display 2 to 3 shore-parallel bars over a gentle mean slope of 0.007. The averaged root mean square height is $H_{rms} = 0.76$ m, the mean peak period is $T_p = 5.7$ s and the average of the absolute value of the angle is $|\theta| = 43°$ (measured with respect to the shore-normal). The semidiurnal tide show a mean tidal range of 1.6 m.

Many events of formation, evolution and decay of bar patches were detected by Ribas and Kroon (2007) during the studied period (44 events in nearly 6 years). A bar patch consisted of 3 to 9 transverse finger bars spaced quasi-regular in the alongshore direction. Patches persisted during periods from 1 day to 2 months, coexisting with intermediate waves and large angles of incidence with respect to the shore-normal ($H_{rms} = 0.80$ m and $|\theta| = 49°$). The underlying bathymetry also affected bar growth: certain along-shore bar positions and surf zone slopes favoured their formation. Bar patch characteristics were anal-ysed using hourly longshore transects of image pixel intensity (located over the bars). The wave length of the transverse bar patches detected in Noordwijk ranged from 21 to 75 m. They migrated at rates up to 22 m/d in the direction of the longshore current, with a mean celerity of 3.8 m/d. Bars had an orientation either per-pendicular or oblique with respect to the shoreline. In the latter case, they deviated against the longshore current (up-current orientation) at angles up to 40° from the shore-normal. More details on these field observations can be found in Ribas and Kroon (2007).

Since bar growth was proved to be sensitive to the bathymetric conditions, we have selected two well-developed events that happened less than one week before (or after) a bathymetric survey was per-formed. A specially well-developed event occurred from August 7 until October 5, 2002 (Fig. 1). Waves during this period had an averaged $H_{rms} = 0.72$ m, $T_p = 5.6$ s and $|\theta| = 47°$, with a dominance of northern incident waves (which corresponds to positive values of $\theta$). Detected bar patches showed 5 to 8 finger bars, with an averaged wave length of $\lambda_e = 46$ m and a stan-dard deviation of $\sigma_{\lambda_e} = 12$ m. Bar crests made an angle of 30° with respect to the north ('up-current orienta-tion'). Migration rates varied from $-9$ to 22 m/d, where positive means migration to the south, with a mean value of 3.2 m/d. A bathymetric survey was performed on October 3, 2002, and the profile corresponding to $y = -500$ m yielded information about the large-scale bathymetric conditions that coexisted with this finger bar patch.

The second event occurred from June 14 until July 7, 2000, with waves conditions $H_{rms} = 0.76$ m, $T_p = 6.1$ s and $|\theta| = 49°$ (northern waves were again dominant).

A space-time diagram (time-stack) of the hourly inten-sity transects corresponding to this bar event is shown in Figure 2, together with the hourly offshore measured $H_{rms}$, $T_p$ and $\theta$ (positive values are waves from the north), the alongshore component of the wave radia-tion stress, $S_{xy}$, the daily-averaged wavelength, $\lambda_d$, and the daily migration rate, $c_m$. White in the time-stacks corresponds to high-intensity values (crests) while black represents low-intensity values (troughs). The horizontal alternations of white and black thus illus-trate the presence of transverse finger bars, whereas a general temporal shift in the bands (e.g. on June 24–27) reflects the alongshore migration of bars (driven by large positive values of $S_{xy}$). Grey hori-zontal bands correspond to missing data due to night hours, days with bad images or days with small waves (no wave breakers over the bars). The small horizon-tal lines in the panels for $\lambda_d$ and $c_m$ are equal to $\sigma_{\lambda_d}$ and to the ground accuracy, respectively (images had an alongshore accuracy of 5 m in this region). Patches showed 4 bars with an averaged wave length of $\lambda_e = 35$ m and $\sigma_{\lambda_e} = 11$ m. Bars were also up-current oriented at some 30° and the averaged migration rate was 4.1 m/d. A bathymetric survey performed in June 9, 2000, was used to describe the profile that corresponds to $y = -500$ m.

## 3  MODEL

The model used in this study describes the feedbacks between depth-uniform mean currents, waves and an erodible bed in a nearshore zone bounded by a straight coast. The $y$ (or $x_2$) axis is chosen to coincide with the rectilinear shoreline, the $x$ (or $x_1$) axis points in the seaward direction and the $z$ axis points upwards.

The large-scale fluid motions are governed by the wave- and depth-averaged momentum and mass conservation equations. The bed shear stresses, $\tau_{bi}$, are parameterized following the generalized equation developed by Feddersen et al. (2000), which we have extended to model the effect of a 2-dimensional flow,

$$\tau_{bi} = \rho c_D \frac{u_{rms}}{\sqrt{2}} v_i \left( v^2 + 2\frac{|\vec{v}|^2}{u_{rms}^2} \right)^{1/2} , \quad i = 1, 2 . \quad (1)$$

Here, $c_D$ is the drag coefficient, $u_{rms}$ is the root mean square wave orbital velocity amplitude, $\vec{v} = (v_1, v_2)$ is the depth-averaged fluid velocity, $v$ is a constant and $\rho$ is the water density. According to Feddersen et al. (2000) and Ruessink et al. (2001), this empir-ical parameterization adequately represents the shear stresses for the directionally spread random wave field at both Duck and Egmond beaches, respectively, when using the value $v = 1.16$. The drag coefficient $c_D$ is the dimensionless friction coefficient due to current

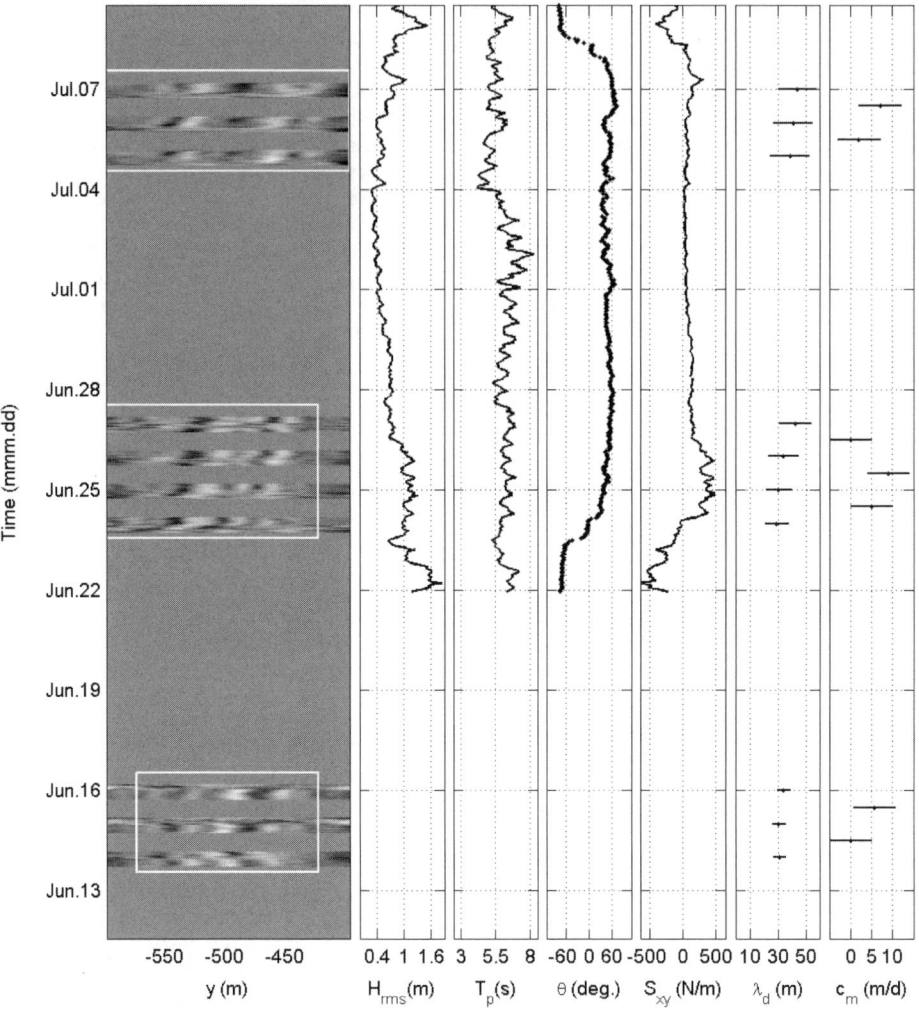

Figure 2. Time series corresponding to the bar event detected in June-July 2000. It shows the temporal evolution of (from left to right) the low-pass filtered intensity transects (time-stack), the offshore measured $H_{rms}$, $T_p$ and $\theta$, the alongshore component of the offshore wave radiation stresses, $S_{xy}$, the daily-averaged wavelength, $\lambda_d$, and the daily migration rate, $c_m$. In the time-stack, white corresponds to high-intensity values (crests) and black represents low-intensity values (troughs).

and waves. It is assumed to vary with depth following the Manning-Strickler law (Soulsby 1997). The bed roughness is assumed to be constant in time and space and a value of 0.022 m is used. This value is adopted from Ruessink et al. (2001), who compared modelled and observed longshore currents at the Dutch coast. The turbulent Reynolds stresses, $S_{ij}''$, are modeled with the standard eddy viscosity approach. The lateral turbulent mixing coefficient is directly linked to wave energy dissipation, $\nu_t = M(\mathcal{D}/\rho)^{\frac{1}{3}}$, where $M \simeq 1$ is a parameter and $\mathcal{D}$ is the energy dissipation rate due to wave breaking, which is described later on in this section. The fluid velocities are imposed to vanish at

both the coastline and the offshore boundary. Also, the free surface elevation must vanish far offshore.

Waves are assumed to have a narrow spectrum in frequency and angle. Their heights are supposed to be random and follow the Rayleigh distribution, characterized by $H_{rms}$ (with wave energy being $E = \rho g H_{rms}^2/8$). When waves approach the coast, their evolution is described using linear wave theory, which yields expressions for the radiation stresses, $\mathcal{S}_{ij}$, and the root mean square wave orbital velocity amplitude, $u_{rms}$. The dispersion relation for the intrinsic wave frequency is also computed with the standard linear wave theory. When introducing the Doppler shift to relate

716

the intrinsic frequency to the absolute frequency, $\hat{\omega}$, the following relation is obtained,

$$\hat{\omega} = \sqrt{g|\vec{K}|\tanh(|\vec{K}|D)} + v_i K_i, \quad i = 1, 2. \quad (2)$$

Here, $g$ is gravity, $\vec{K}$ is the wave number, $D = z_s - z_b$ is the water depth, where $z_s$ is the mean free surface elevation and $z_b$ is the sea bottom level. Steady conditions are assumed, $\hat{\omega} = $ constant. Equation 2 is finally rewritten in terms of the wave phase $\Phi$, from which $\hat{\omega}$, $\vec{K}$ and thereby $\theta$ can be computed. This equation describes the refraction of the waves due to both topography and currents. More complex processes in wave propagation, like wave diffraction, are not accounted for. Wave energy balance is described with a wave- and depth-averaged equation, where the energy dissipation rate due to wave breaking, $\mathcal{D}$, is parameterized using the formulation of Church and Thornton (1993). Wave conditions are prescribed far offshore, at 18 m depth, where Noordwijk conditions were measured ($H_{off}$, $\theta_{off}$ and $\hat{\omega} = 2\pi T_p^{-1}$). More details on the hydrodynamic equations can be found in Calvete et al. (2005).

Finally, conservation of sediment mass yields the bottom evolution equation,

$$(1-p)\frac{\partial z_b}{\partial t} + \frac{\partial q_j}{\partial x_j} = 0, \quad j = 1, 2, \quad (3)$$

with $p \simeq 0.4$ being the porosity of the bed and $q_1, q_2$, the two components of the depth- and wave-averaged volumetric sediment transport ($m^3 m^{-1} s^{-1}$). A widely accepted formulation for $q_i$ in the nearshore is the Soulsby and van Rijn formula (SvR-formula) given in Soulsby (1997). Their original expression has been extended to model the effect of a 2-dimensional flow and the preferred downslope transport of the sand,

$$q_i = \alpha \left( v_i - \Gamma \frac{\partial h}{\partial x_i} \right), \quad i = 1, 2. \quad (4)$$

The bedslope function $\Gamma$ is modelled following Calvete et al. (2005), $\Gamma = u_{rms}$. The corresponding term accounts for the tendency of the system to smooth out the sea bed disturbances, $h$, if the latter do not cause positive feedback into the flow.

The function $\alpha$ in Equation 4 describes the sediment stirring or depth-integrated sediment concentration and reads

$$\alpha = A_s(\hat{u}_s - u_{crit})^{2.4}. \quad (5)$$

Here, $\hat{u}_s$ is the stirring velocity, $u_{crit}$ is the threshold flow intensity for sediment transport and the parameter $A_s$ describes the sediment properties. The full expressions for the two latter quantities are given in Soulsby (1997). In the original SvR-formula, $\alpha$ was

assumed to be a result of the shear stresses produced in the bottom boundary layer of the wave orbital velocity and the depth-averaged currents hence the stirring velocity was

$$\hat{u}_s^{SvR} = \left( |\vec{v}|^2 + \frac{0.018}{c_D} u_{rms}^2 \right)^{1/2}, \quad (6)$$

where the Manning-Strickler law is again assumed for the drag coefficient $c_D$. The SvR-formula was tested to be accurate in the shoaling domain, at water depths of the order of 5 m Soulsby (1997). However, in the inner surf zone (depths $< 2$ m), where $u_{rms}$ and the longshore current decay, other processes like wave breaker turbulence or bore propagation could be dominant and produce significant sediment resuspension (Voulgaris and Collins 2000; butt et al. 2004). In the present study, the stirring velocity $\hat{u}_s$ in Equation 5 has been extended to allow inclusion of these other possible processes,

$$\hat{u}_s = \frac{1}{2} \left[ \hat{u}_s^{SvR} \left( 1 + \tanh\frac{(x - x_b)}{l_b} \right) + \right.$$

$$\left. + \hat{u}_s^b \left( 1 - \tanh\frac{(x - x_b)}{l_b} \right) \right]. \quad (7)$$

The first term corresponds to the original SvR-formula, $\hat{u}_s^{SvR}$ is given in Equation 6. The second term in Equation 7 describes the depth-integrated sediment concentration due to bores and wave breaker turbulence and $\hat{u}_s^b$ is considered to be constant. By varying $x_b$ and $l_b$ in Equation 7, we can change the cross-shore domain where the two types of stirring dominate. Three different sets of values for $x_b$ and $l_b$ are used in the present study: (a) $x_b \simeq 0$, $l_b \simeq 0$, i.e. the original SvR-formula is applied in the whole nearshore; (b) $x_b = x_{tro}$, $l_b = x_{tro}/3$, where $x_{tro}$ is the position of the inner trough in the profile, i.e. constant depth-integrated concentration is dominant in the inner surf zone whereas the stirring due to boundary layer processes prevail in the rest of the domain; and (c) $x_b = 3000$ m, $l_b = 1500$ m, constant depth-integrated concentration is applied in the whole domain. The latter case is equivalent to the one used by Ribas et al. (2003) and Garnier et al. (2006). In the two last cases, $\hat{u}_s^b$ is assumed to be 60% of the maximum of $\hat{u}_s^{SvR}$.

## 4 METHODOLOGY

The governing equations, together with the parameterizations used, define a closed dynamical system for the unknowns $\vec{v}$, $z_s$, $E$, $\Phi$ and $z_b$. The linear stability approach to the formation of bars by self-organization starts by defining a steady and alongshore uniform basic state (without longshore rhythmic topography).

In this study, we will use reference profiles, $z_b^o(x)$, measured in Noordwijk beach during the studied bar events. The superscript $^o$ refers to the basic state variables. The basic state is characterized by the presence of a longshore current, $\vec{v} = (0, V^o(x))$, and an elevation of the mean free surface, $z_s = z_s^o(x)$. This basic state only represents a morphodynamic equilibrium if the net cross-shore sediment flux vanishes.

Once the basic state has been computed, stability analysis can be applied in a standard way. A small perturbation, assumed to be periodic in time and in the alongshore coordinate, is added to this state,

$$(v_1, v_2, z_s, E, \Phi, z_b) = (0, V^o, z_s^o, E^o, \Phi^o, z_b^o) + \quad (8)$$

$$+ \Re\{e^{\omega t + i\kappa y} (u(x), v(x), \eta(x), e(x), \phi(x), h(x))\},$$

where $\kappa$ is the longshore wavenumber and $\omega = \omega_r + i\omega_i$ a complex growth rate. By inserting Equation 8 in the governing equations and linearizing with respect to the perturbations, we arrive at an eigenproblem. For each $\kappa$, different eigenvalues $\omega$ exist, which characterize the different growing modes, and the complex eigenfunctions are $(u(x), v(x), \eta(x), e(x), \phi(x), h(x))$. The growth rate of the emerging bars is given by $\omega_r$, so that $\omega_r > 0$ means growth. In case of an unstable basic state, some perturbations with $\omega_r > 0$ are found. The growth rate curves show these positive $\omega_r$ for different values of $\kappa$. Starting from arbitrary initial conditions, the dynamics after some time will be dominated by the mode with largest growth rate, which is called Fastest Growing Mode (FGM). Its $e$-folding growth time is given by $T_g = \omega_r^{-1}$ and the migration speed by $c_m = -\omega_i/\kappa$. The alongshore wavelength of the corresponding bar system is $\lambda = 2\pi/\kappa$ and their shape is given by $\Re\{e^{i\kappa y}h(x)\}$. The associated perturbations in the flow, the free surface elevation and the wave energy are obtained in a similar way from $u(x), v(x), \eta(x), e(x)$ and $\phi(x)$. Given the uncertainties in the available sediment transport formulations, the functions $\alpha$ and $\Gamma$ in Equation 4 are not perturbed.

# 5 MODEL RESULTS AND DATA COMPARISON

Firstly, the model was applied to the conditions measured in Noordwijk during the finger bar event in August-September 2002. The profile at $y = -500\,\text{m}$ of the bathymetric survey from October 3, 2002 was used to calculate the basic state. As input parameters for wave conditions, we used the event-averaged values: $H_{off} = 0.72\,\text{m}$, $T_p = 5.6\,\text{s}$ and $\theta = 47°$. The three different sets of values for $x_b$ and $l_b$ described at the end of Section 3 were used. Figure 3 shows the solution obtained in these three scenarios. The left panels display the basic state, the obtained $H^o$, $V^o$, $\theta^o$, $\alpha^o$ and the measured profile, $z_b^o$. The result of the LSA is

shown in the top right panels: the growth rate curve in the top middle and the migration rate curve in the top right. In the $\alpha^o$ panel and in the LSA curves, three lines have been plotted corresponding to the three scenarios given by the values of $x_b$ and $l_b$: case (a) is the original SvR-formula (dotted lines), case (b) is constant stirring only in the inner surf zone (solid lines) and case (c) is constant stirring in the whole domain (dashed lines).

When the original SvR-formula was used (case a), the only positive growth rates obtained corresponded to the transformation of the shore-parallel inner bar into a crescentic bar (maximum for $\kappa = 0.010\,\text{m}^{-1}$ in the dotted growth rate curve of Figure 3, which corresponds with the mode CB). This phenomenon occurs in Noordwijk beach (van Enckevort et al. 2004). However, the model could not predict formation of finger bar patches. When the sediment stirring was assumed to be constant in the inner surf zone (case (b), with $x_b = 80\,\text{m}$ as can be seen in the panel with $z_b^o$ of Figure 3), the model predicted the formation of both a crescentic bar and a system of 'up-current oriented bars'. The solid line in the growth rate curve showed a secondary maximum for $\kappa = 0.100\,\text{m}^{-1}$, which corresponded to a wave length $\lambda = 63\,\text{m}$, an $e$-folding growth time of $T_g = 47\,\text{h}$ and a migration rate $c_m = 38\,\text{m/d}$. The middle right panel in Figure 3 displays the shape of the topographic perturbation corresponding to the up-current bars, together with the current perturbations. In the topographic plots, white areas indicate crests and dark areas represent troughs. Waves approach the coast from the bottom left corner so the induced mean longshore current is directed from left to right. Small arrows indicate the main trend in the deviations of the longshore current due to the hydrodynamic circulation induced by the growing bars. As can be seen, the solution consisted of a patch of up-current oriented bars with current perturbations deflecting offshore over their crests. In order to visualize the final shape of the bottom, the reference profile, $z_b^o$, should be added. The same applies to the flow: the longshore current $V^o$ should be added to the perturbations of the velocity to obtain the total flow. The predicted wave length and crest orientation were in good agreement with the observed event-averaged $\lambda_e = 46\,\text{m}$ and the up-current orientation of 30° from the shore normal. The predicted migration rate was twice the maximum rate detected and one order of magnitude larger than the event-averaged rate (22 m/d and 3.8 m/d, respectively). Finally, the lower right panel in Figure 3 shows the crescentic bar obtained for $\kappa = 0.010\,\text{m}^{-1}$. When a constant stirring was applied to the whole domain (case c) only the 'up-current-oriented bars' were obtained (dashed growth rate curve of Figure 3).

The model was then applied to the conditions measured in June-July 2000 and the results were similar (see Fig. 4). The bathymetric survey used was performed in June 9, 2000, and the inner bar was located

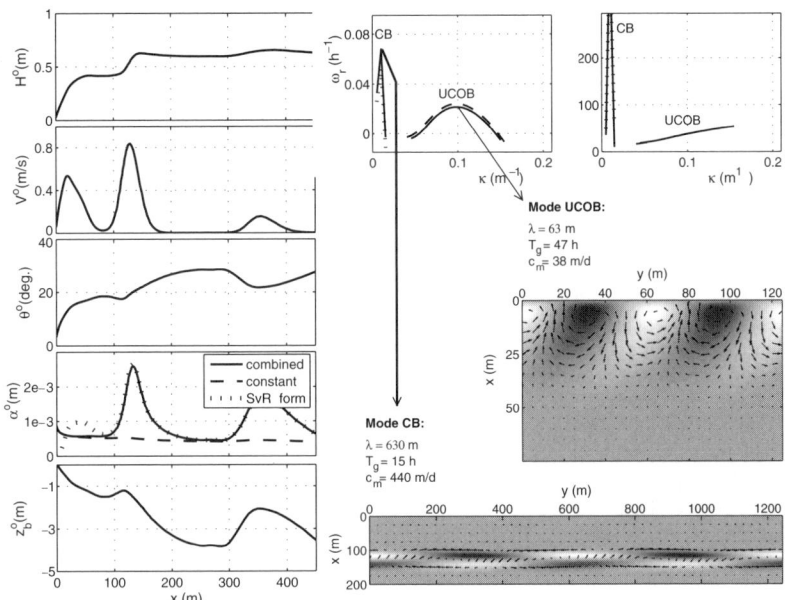

Figure 3. Model solution found for the conditions measured in Noordwijk in August-September 2002. Left panels: Basic state result for (from top to bottom) the root mean square wave height, $H°$, the longshore current, $V°$, the wave angle from the shore-normal, $\theta°$, the sediment stirring, $\alpha°$, and the bed level, $z_b°$. The horizontal axis is the cross-shore position. Right panels: LSA result for the growth rate curve (top middle), the migration rate curve (top right), the topographic and current perturbations corresponding to the up-current oriented bars (UCOB, middle right) and the the topographic and current perturbations for the crescentic bars (CB, bottom right). More explanation is given in the text.

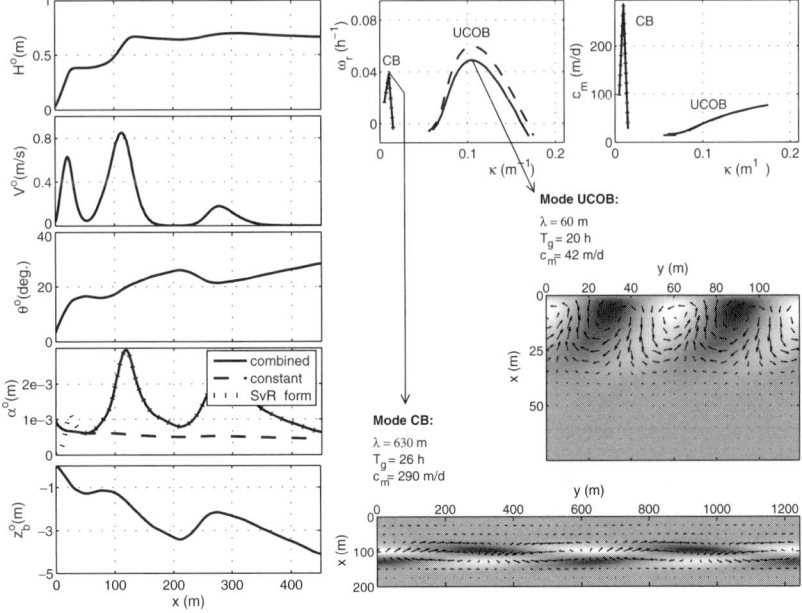

Figure 4. Model solution found for the conditions measured in Noordwijk in June-July 2000. See the caption of Figure 3 for an explanation of all the different subplots.

719

closer to shore, with a distance from the shoreline to the trough of 50 m. The event-averaged wave conditions were $H_{off} = 0.76$, $T_p = 6.1$ and $\theta = 49°$. Again, the formation of both a crescentic bar and a system of transverse finger bars could only be reproduced in case (b), now with $x_b = 50$ m. The maximum in the growth rate curve was in a wave number similar to the previous case, $\kappa = 0.105 \, \text{m}^{-1}$, corresponding to $\lambda = 60$ m. Bars were predicted to grow faster, with an $e$-folding growth time of $T_g = 14$ h and their migration rate was $c_m = 52$ m/d. The result for the topographic and current perturbations was also very similar to the previous event. In this case, the predicted wave length of the up-current bars was nearly twice the observed value, $\lambda_e = 35$ m. The model overpredicted the event-averaged migration rate by one order of magnitude.

The modelled $e$-folding growth times, ranging from one to two days, were coherent with the general observations described by Ribas and Kroon (2007). The wave conditions that were established to allow bar formation commonly persisted during at least one day before bar patches were detected for the first time. However, in a significant number of the events detected, wave conditions prior to bar formation could not be determined. For instance, in the event occurring in June-July 2000, there was no wave data available from May 29 till June 22, 2000, i.e. during bar formation (see Fig. 2). Regarding the event in August-September 2002, bar formation occurred during 1 day, from August 6 at 10 h till August 7 at 10 h.

## 6 CONCLUSIONS

A self-organization model for transverse bar formation, based on LSA, has been applied to the conditions prevailing in Noordwijk during two detected bar events. The predicted shape, the wave length and the growth rate of both up-current oriented bars and crescentic bars are in good agreement with observations. The model overestimates the migration rate by a factor ranging from 2 to 10. This study confirms that the 'bed-flow effect' can be responsible for the formation of transverse finger bars in Noordwijk.

The key aspect to obtain formation of up-current oriented bars, like those observed in the field, is the assumption that the depth-integrated suspended sediment concentration is constant across the inner surf zone. In that region, some studies indicate that vortices induced by breaking waves and bores are the main mechanism for sediment resuspension and create vertically well mixed suspended sediment concentration profiles. The effect of these vortices prevails over bottom boundary layer processes (Voulgaris and Collins 2000).

## ACKNOWLEDGEMENTS

The work of F. Ribas and D. Calvete is supported by the Spanish government through the programs 'Programa Juan de la Cierva' and 'Programa Ramon y Cajal', respectively. This research is part of the PUDEM project which is funded by the Spanish government under contract REN2003-06637-C02-01/MAR.

## REFERENCES

Butt, T., P. Russel, J. Puleo, J. Miles, and G. Masselink (2004). The influence of bore turbulence on sediment transport in the swash and inner surf zones. *Cont. Shelf Res. 24*, 757–771.

Calvete, D., N. Dodd, A. Falqués, and S. M. van Leeuwen (2005). Morphological development of rip channel systems: Normal and near normal wave incidence. *J. Geophys. Res. 110*(C10006), doi:10.1029/2004JC002803.

Church, J. C. and E. B. Thornton (1993). Effects of breaking wave induced turbulence within a longshore current model. *Coastal Eng. 20*, 1–28.

Feddersen, F., R. T. Guza, S. Elgar, and T. H. C. Herbers (2000). Velocity moments in alongshore bottom stress parameterizations. *J. Geophys. Res. 105*(C4), 8673–8686.

Garnier, R., D. Calvete, A. Falqués, and M. Caballeria (2006). Generation and nonlinear evolution of shore-oblique/transverse sand bars. *J. Fluid Mech. 567*, 327–360.

Gelfenbaum, G. and G. R. Brooks (2003). The morphology and migration of transverse bars off the west-central florida coast. *Mar. Geol. 200*, 273–289.

Konicki, K. M. and R. A. Holman (2000). The statistics and kinematics of transverse bars on an open coast. *Mar. Geol. 169*, 69–101.

Ribas, F., A. Falqués, and A. Montoto (2003). Nearshore oblique sand bars. *J. Geophys. Res. 108*(C43119), doi:10.1029/2001JC000985.

Ribas, F. and A. Kroon (2007). Characteristics and dynamics of transverse finger bars. *J. Geophys. Res.*, under revision.

Ribas, F., H. Swart, D.Calvete, A. Falqués, N. Dodd, and S. van Leeuwen (2005). Sensitivity of modeled nearshore morphology to wave and sediment transport formulations. In *Coastal Dynamics 2005*. Am. Soc. of Civ. Eng.

Ruessink, B. G., J. R. Miles, F. Feddersen, R. T. Guza, and S. Elgar (2001). Modeling the alongshore current on barred beaches. *J. Geophys. Res. 106* (C10), 22451–22463.

Soulsby, R. L. (1997). *Dynamics of Marine Sands*. London, U.K.: Thomas Telford.

van Enckevort, I. M. J., B. G. Ruessink, G. Coco, K. Suzuki, I. L. Turner, N. G. Plant, and R. A. Holman (2004). Observations of nearshore crescentic sandbars. *J. Geophys. Res. 109*(C06028), doi:10.1029/2003JC002214.

Voulgaris, G. and M. B. Collins (2000). Sediment resuspension on beaches: response to breaking waves. *Marine Geology 167*, 167–187.

*River, Coastal and Estuarine Morphodynamics: RCEM 2007 – Dohmen-Janssen & Hulscher (eds)*
*© 2008 Taylor & Francis Group, London, ISBN 978-0-415-45363-9*

# Morphodynamical modeling: Impact of the tide and seasonal conditions on characteristic bar systems of the Aquitanian Coast, France

N. Bruneau
*BRGM, Orléans, France*
*UMR CNRS 5805 EPOC, Bordeaux, France*

R. Pedreros & D. Idier
*BRGM, Orléans, France*

P. Bonneton
*UMR CNRS 5805 EPOC, Bordeaux, France*

F. Dumas
*IFREMER, Plouzané, France*

ABSTRACT: The morphological evolution of the beach and the bar formation are important research subjects in order to understand physical processes and to study the shoreline evolution. The purpose of this work is to analyze the effects of different physical phenomena on a crescentic subtidal bar which is a rhythmic characteristic pattern met along the Aquitanian French coast. This approach is based on a coupling between the spectral wave model SWAN, the shallow water MARS and a morphodynamic module. i) The sediment transport has been analyzed as a function of tide cycle types: a sinusoidal tide or a neap tide and spring tide. ii) Breaking roller contribution has been highlighted in this work for both the hydrodynamics and the morphodynamics. The bed evolution is faster with the roller effect. iii) Then, seasonal wave conditions like storms or wind sea have been simulated. The results show that during a storm period, the subtidal crescentic bar moves offshore whereas the wind sea involves low bottom evolution but seems to contribute to the beach nourishment.

## 1 INTRODUCTION

Crescentic subtidal bars and ridge and runnel systems are characteristic of the french Aquitanian coast (Lafon et al. 2004). These rythmic sediment patterns present along the whole coast (see Figure 1) have a significant impact on the wave-induced currents and on the beach stability. Currently, Integrated Coastal Zone Management (ICZM) knows an increasing interest for the international scientific community and for policymakers. Indeed some research codes have been developped these last years to model wave-induced currents and to understand the processes of bar formation such as Morpho55 (Garnier et al. 2006), MORPHODYN (Saint-Cast 2002) or NearCoM (Shi et al. 2005). These models can well-represent wave-induced currents and morphological evolution but they can not take into account tide and meteorological phenomena. In order

to model these processes, the spectral wave model SWAN (Booij et al. 2004) is coupled with the shallow water model MARS (Pérenne 2006). A sedimentary module based on MORPHODYN has been developed and included inside MARS to simulate bed evolution.

Following the hypothesis proposed by (Castelle and Bonneton 2003) that the wave refraction over a crescentic bar generates energy focalization areas, Bruneau et al. (2007) have shown the formation of ridge and runnel systems in the intertidal zone in the up-current direction. The aims of this work is to study, the formation of bar systems on a reference case. Then, to analyze the impact of the roller phenomenon which can not be negligible in the morphological evolution due to wave-induced currents. To finish, more real wave conditions and tide levels will be applied to the model to take into account: (1) swell, (2) neap and spring tides, (3) storms and (4) wind sea.

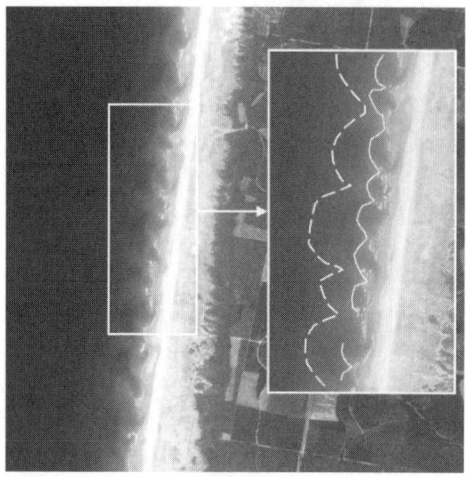

Figure 1. Rythmic systems of the Aquitanian coast $\approx$(1.24W,44.77N): crescentic subtidal bars (dotted line), inner bars (solid line) and ridge and runnel systems (along the shoreline). *Extracted from Google Earth.*

## 2 MODELS & COUPLING

In this section, the numerical models are presented briefly (for more details about the equations solved by the models, see Bruneau et al. (2007)). Only breaking roller contribution is detailed below.

### 2.1 Morphodynamical platform

The third-generation spectral wave driver SWAN (Simulating Waves Nearshore), developed at the Delft University of Technology (Booij et al. 2004), is used to compute the wave characteristics (significant wave height, mean wave direction, mean absolute wave period, etc.). It is coupled with the shallow water model MARS (Model for Applications at Regional Scale) developed at IFREMER (Pérenne 2006). The MARS model solves the unsteady shallow-water system of equations in two (depth-averaged, noted 2DH) or three dimensions. This code has proved his capacity to model the tide-induced and wind-induced currents on the whole french coast. In the present work, only the 2DH equations which are been modified to take into account radiation stresses (Bruneau et al. 2007), are solved.

The radiation stresses are computed according to the linear theory (Dingemans 1997) and the effect of the breaking roller is taken into account according to Dally (2001) (see next subsection). The gradients of radiation stresses $S_{ij}$ and roller contribution $R_{ij}$ are classicaly included in the momentum equations. Thus, defining $U_i$ as the depth-averaged current velocity in

the $i$-direction, the mean free surface elevation by $\bar{\zeta}$, the governing equations are in mathematical formulation:

$$
\begin{cases}
\dfrac{\partial U_i}{\partial t} + U_j \dfrac{\partial U_i}{\partial x_j} + g \dfrac{\partial \bar{\zeta}}{\partial x_i} = -\dfrac{1}{\rho \bar{h}} \dfrac{\partial \, (S_{ij} + R_{ij})}{\partial x_j} \\[2ex]
\qquad\qquad\qquad - \dfrac{\tau_i^b}{\rho \bar{h}} + \dfrac{\partial}{\partial x_j} \left( \nu_H \dfrac{\partial U_i}{\partial x_j} \right) \\[2ex]
\dfrac{\partial \bar{\zeta}}{\partial t} + \dfrac{\partial \bar{h} U_i}{\partial x_i} = 0
\end{cases}
\tag{1}
$$

with $\bar{h}$ the mean water depth, $\rho$ the mass density of sea water, $g$ the gravity, $\tau_i^b$ the bed shear stress and $\nu_H$ the horizontal eddy viscosity.

The resolution of the sediment conservation law is computed with a simple second order scheme and the transported sediment total fluxes obtained using the Bailard (1981) formulation. This formulation allows to take into account bed-load transport, suspended transport and the slope effects on both the types of transport. The flow velocity close to the bottom is a function of the mean velocity and the orbital velocity which depends of the time. But asymmetrical waves are not considered in the present work.

### 2.2 Breaking roller contribution

Roller contribution is a really significant phenomena in the breaking zone (Swendsen 1984) or (Goda 2003). It induces currents and free elevation surface changes and needs to be taken into account for a more realistic approach.

*Roller equation:*
In the present work, the energy balance presented by Dally (2001) is applied to model the roller. Thus, period-averaged mass flux in the roller, $\rho_r A_r/T$ is calculated starting from the equation (2) given below:

$$
\frac{\partial}{\partial x_i} \left[ c^2 \Big( \frac{\rho_r A_r}{T} \Big) \frac{k_i}{k} \right] + g \beta_D \frac{\rho_r A_r}{T} = D_E
\tag{2}
$$

where $c$ is the wave velocity, $\kappa$ the wave number, $T$ the wave period, $\rho_r$ the mass density in the roller with aeration, $A_r$ the cross-sectional area of the roller, $g$ gravity, $\beta_D$ is a model's calibration coefficient chosen equal to 0.2 according to Dally (2001) and $D_E$ is the energy dissipation of the waves (computed with the formulation of (Battjes and Janssen 1978)).

The roller terms of the equation system (1) are computed on the following way:

$$
R_{ij} = c \left( \frac{\rho_r A_r}{T} \right) \left( \frac{k_i k_j}{k^2} + \frac{\delta_{ij}}{2} \right)
\tag{3}
$$

with $\delta_{ij}$ the *Kronecker* symbol.

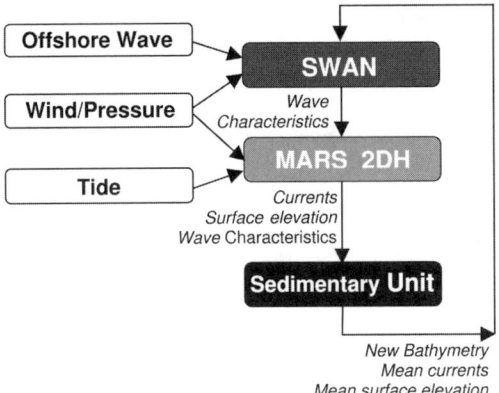

Figure 2. Global scheme of the morphological platform and differents forcing possibilities.

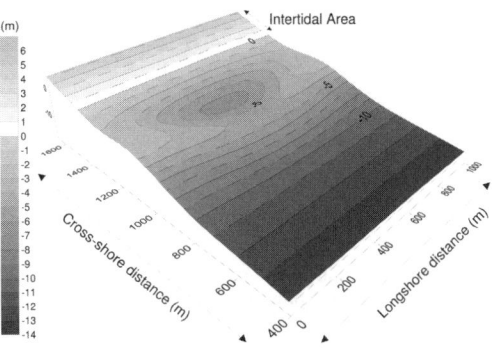

Figure 3. Ideal crescentic subtidal bar. Bathymetry build by Castelle (2004) from a synthesis of bathymetric surveys and SPOT images.

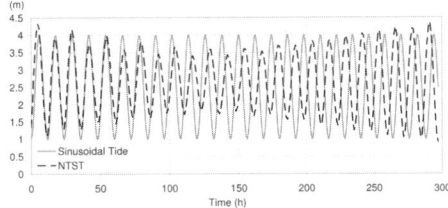

Figure 4. In plain line, the sinusoidal tide which represents a mean tide on the Aquitanian coast. In dash line, NTST tide elevation above the reference level.

*Undertow:*

The undertow is a relevant phenomenon for the hydro- and morphodynamics of the beach. Like it is not take into account in the 2DH equation system, it is added to the velocity which allows the computation of the trans- port sediment fluxes. According to Phillips (1977), the final mean depth-averaged velocity in the *i*-direction noted $U_i^{morph}$ is given by:

$$U_i^{morph} = U_i - \frac{Q_i^w}{h} - \frac{Q_i^r}{h} \qquad (4)$$

where $\bar{Q}_i^w = Ek_i/(\rho ck)$ is the volume flux associated with the organized wave motion ($E$ the wave energy) and $\bar{Q}_i^r = \rho_r A_r k_i/(\rho Tk)$ is the volume flux for roller.

### 2.3 Coupling scheme

The Figure 2 presents the scheme of the coupling between the different models previously described. The advantage of each of this two operational mod- els is the possibility to force them by measured datas or other numerical models (for wave conditions, tide, wind, etc). Both MARS and SWAN can compute on nested grids which allows a rapid computational time. Moreover this two models have been intensively tested.

### 2.4 Ideal complex bathymetry and set-up modeling

In the present study, the initial bathymetry represents a crescentic subtidal bar (see Figure 3 in 3D). It was created by Castelle (2004) using SPOT images and bathymetric surveys. The wavelength is equal to 1000 m, which is a little bit greater than the mean wave- length of this kind of systems observed on the coast (between 579 and 818 m, (Lafon et al. 2004)). A gently slope allows the connection between the crescent bar and the beach. The offshore depth is 19 m at low tide

to the shoreline. The mesh is made of a uniform 20 m grid in both crosshore and longshore directions with also periodic lateral conditions. The wave incidence can perturb the SWAN results close to the boundaries. So a three times larger domain in longshore direction is chosen to perform the SWAN computations.

The study proposes an analysis of the effects of the tide level on the bar formation. For theses rea- sons, two kinds of tide cycle have been tested. The first schematical cycle represents a simple sinusoidal cycle (Figure 4) with a tidal range of 3 m and a tidal cycle of 12 h (not a real but a schematical tide period). The second cycle is a real cycle with neap and spring tides (Figure 4), noted NTST; it represents the sea level above the reference level during May 2006 for the study area. In both cases, the tide is not com- puted in a continuous way but the level is actualized every hour. The new bathymetry and thus the new wave distribution are also computed every hour.

## 3 RESULTS

### 3.1 Reference case

To compare and to analyze the effects of the differ- ent physical processes, a reference case is used. A gaussian wave with a significant height $H_s = 1{:}5\,m$, a

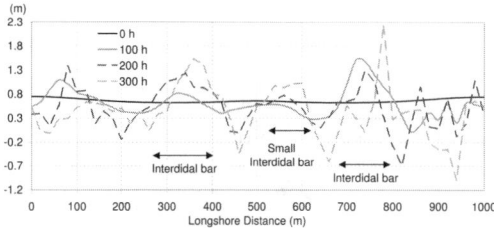

Figure 5. Evolution of a bottom level longshore profile ($y = 1480$ m, see on Figure 7) for the reference case: at initial stage, after 100 h, 200 h and 300 h of morphodynamical evolution.

mean period $T_m = 10$ s and a incidence $\theta = 10°$ which represents swell on the Aquitanian coast (Butel et al. 2002) is simulated without roller contribution. Figure 5 shows the evolution of a bottom level longshore profile function of the time. At time 0 and at low tide, the longshore profile $y = 1480$ m delimits the shoreline (see Figure 7).

On this figure (Fig. 5), we see in a first time the formation of erosion/accretion areas which correspond to intertidal bar systems like ridge and runnel systems. The two main intertidal patterns are spaced of about 450 m which are in according with Lafon et al. (2004) (the wavelengths oberved range between 370 m and 463 m). On the Aquitanian coast, the observations show that for a crescentic subtidal bar, two or three ridge and runnel systems are placed on the intertidal zone so the results are in agreement. Then along the longshore profile, the ridges and the runnels begin to move which shows the lengthening of the bars in up-current direction. Bruneau et al. (2007) have shown the formation of interdidal systems in up-current direction which is not observed on the Aquitanian coast (see on Figure 1, with waves mainly arriving from North-West). The up-current direction can be due to the transport formulation (Garnier et al. 2006) or maybe other phenomena not taking into account like the NTSP cycle or real wave conditions. So we can explain the bar formation in two main stages: i) the sediment accumulates on preferential places, ii) the bars widen in longshore direction (see boxes on Figure 6). We also observed on the Figure 7 the formation of inner bars (dark dashed line on the Figure 7) which connect the intertidal domain with the subtidal crescentic bar and which link the ridge and runnel systems together. Close to the shoreline, the wave-induced currents are in an opposite direction compared to the wave incidence.

## 3.2 Influence of the tide

This subsection deals with the importance of the tide level variations to form some bar systems in the intertidal domain. Indeed the modulation of the tide level allows a continuous evolution (erosion/accretion) of

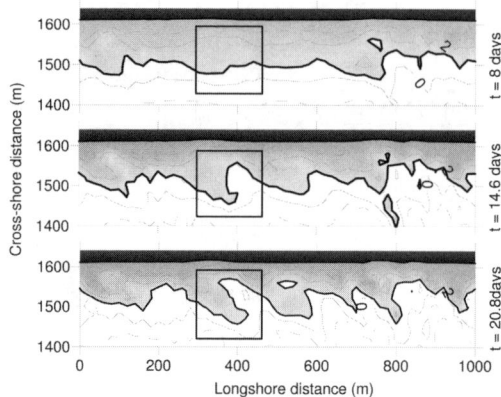

Figure 6. Intertidal 2D zone plots of the intertidal domain. From top to bottom: after 8, 16.6 and 25.5 days of morphodynamical evolution.

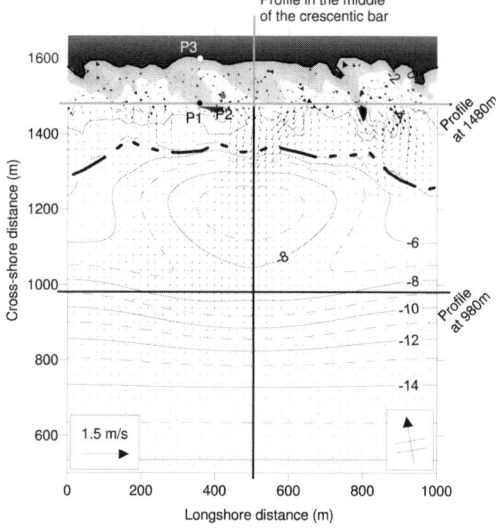

Figure 7. Wave-induced current vector map over the crescentic subtidal bar (bathymetry levels) after 19 days at middle tide ($h_t = 3{:}25$ m above the reference level). In black color, the beach and in grey color the intertidal area. Wave conditions: $H_s = 1.5$ m, $T_m = 10$ s and $\theta = 10°$. Dark dashed line shows the formation of inner crescentic bars. The grey lines are the different longshore and crosshore profiles cited in the paper. $P1$ is the probe point (360 m, 1480 m), $P2$ the point (420 m, 1500 m) and $P3$ the point (360 m, 1600 m).

the beach. It is interesting to note the fluctuations of the bottom level during the whole simulation period.

On Figure 8, two probe points have been chosen: one in an accretion zone ($P1$, Fig. 7) and the second in an erosion area ($P2$, Fig. 7) for the two tide cycles presented previously. The curves are not similar which proved that the tidal cycle has an influence on the bar

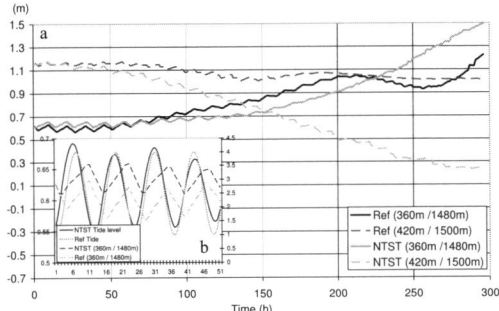

Figure 8. (a) Diagnostic points close to the shoreline at low tide. On a ridge (P1): 360 m; 1480 m – in a runnel (P2): 420 m; 1500 m for a sinusoidal cycle (dark line) and a NTST cycle (grey line). (b) Zoom.

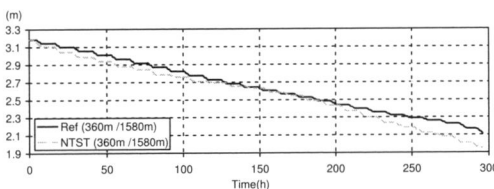

Figure 9. Probe points P3 close to the shoreline at high tide for a sinusoidal cycle (dark line) and a NTST cycle (grey line).

formation. The probe points show that the bottom level evolution oscillates with the tide level (see Figure 8b). For a simulation, the two signals (tide and accretion) have the same period but are not in phase. This phase changes with the position of the probe points. For neap tides, the fluctuations disappear but the accretion or erosion phenomena go on progressing. Figure 9 shows a point close to the shoreline at high tide (P3, Fig. 7). This point is always eroded when water is present. It is mainly due to the undertow effects which induce offshore currents which eroded the high shoreline.

### 3.3 Influence of the roller

Figure 10 shows, for the same simulation conditions, the free surface elevation and the velocity with or without inclusion of roller effects. To analyze the contribution of the roller in the inner surf zone, we compute $\Delta_r$ the maximum variation between the "roller" curve and the "without-roller" curve, divided by the maximum value:

$$\Delta_r = \frac{max\,|Val_{Roller} - Val_{no-Roller}|}{max(Val_{Roller}, Val_{no-Roller})} \qquad (5)$$

then, the value of $\Delta_r = 8\%$ is reached for the surface elevation and 15% for the velocity. In the surf zone, the set-up and set-down are greater (in absolute value)

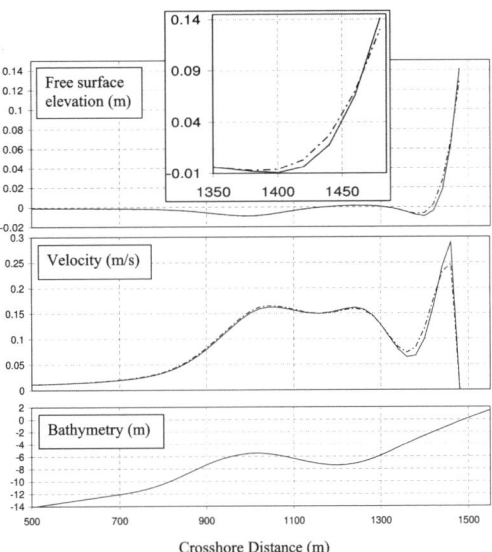

Figure 10. Crosshore profile in the middle of the crescentic bar. From top to the bottom, the graphs are respectively the free surface elevation, the current velocity ant the bathymetry profiles. Without roller contribution in dashed line and with roller in plain line.

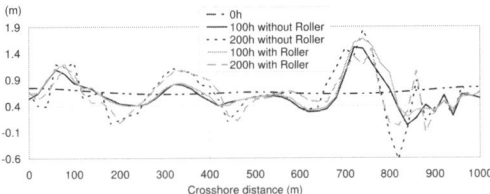

Figure 11. Longshore profile ($y = 1480$ m). Comparison of bottom level for 100 h and 200 h between simulations without (dark lines)/with (grey lines) breaking roller contribution.

with roller contribution than without. Figure 10 also shows:

- a first set-down when the waves arrive on the crescentic bar due to breaking of a unit of waves. Then the waves continue their propagation.
- at the end of the inner surf zone, the second set-down shows the final breaking following by an important set-up (14 cm for a wave of 1.5 m significant height).
- for the velocity, a peak is visible close to the shoreline due to important radiation stresses. Above the crescentic bar, the current velocity is also significant.

Figure 11 and Figure 12 give respectively the morphodynamical evolution of the bottom level, on a longshore profile in the intertidal area and for 2D plot of the inner surf zone. Like previously explained, the

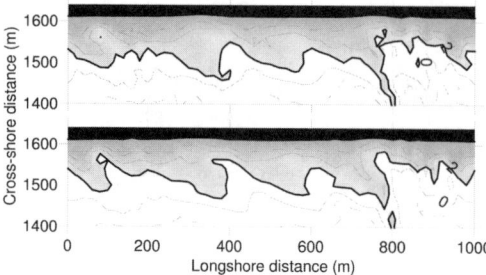

Figure 12. Intertidal 2D zone plots after 14 days morphological evolution. Top: without Roller; bottom: with roller contribution.

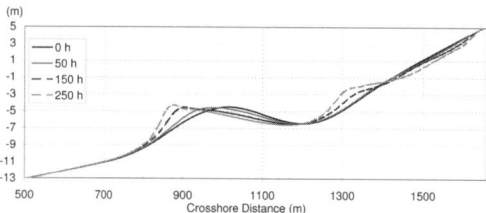

Figure 13. Crosshore profile of the bottom level in the middle of the crescentic bar. Initial level, after about 2, 6.25 and 10.5 days.

breaking roller contribution induces different free surface elevation and wave-induced current profiles. The more significant currents induce a faster evolution. Figure 12 shows that, at the same time, the intertidal bar systems are more developed with roller effects than without. But a more precisely study needs to be done to analyze in details the impacts on crosshore and longshore directions.

### 3.4 Seasonal wave forcing

Abadie et al. (2006) have shown the repartition of waves in three main classes: wind sea represents approximately 26% of the waves, swell 60% and storms only 14% (greater than 3 m). For these reasons, a storm and a wind sea are simulated.

### Extreme conditions: Storms

The modeling platform is tested on extreme wave conditions. In this case, the storm conditions are $H_s = 3$ m, $T_m = 10$ s and an incidence of $\theta = 10°$. A crosshore profile of the bed evolution in the middle of the crescent is plotted on Figure 13. The crescentic subtidal bar moves clearly in the offshore direction. Figure 13 highlights four main points:

- Erosion of the intertidal domain in a continuous and significant way,
- Formation of an inner bar at 3 m below the refernce water level,

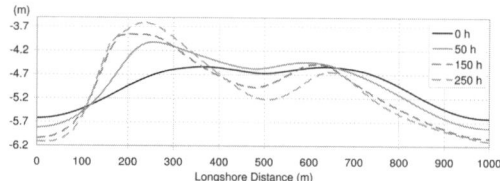

Figure 14. Longshore profile ($y = 980$ m) of the bottom level at the beginning of the offshore slope (see profile on Fig. 7). Initial level, after about 2, 6.25 and 10.5 days.

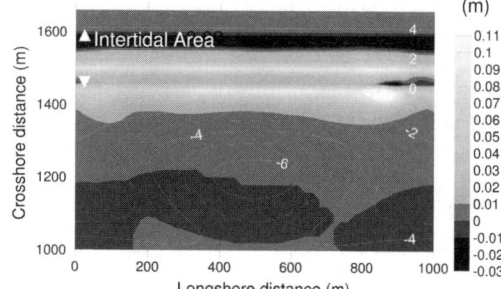

Figure 15. Difference between bottom level after 20 days of simulation and initial bathymetry superposed with the initial bathymetry levels above reference level.

- Accretion of sand in the hollow of the crescent,
- The crescentic bar lengthens in offshore direction and reaches a stability point due to the offshore slope.

The longshore profile, at the top of the crescentic bar (see Figure 14) gives the migration of the bar in the down-current direction. Indeed during swell, the waves are not enough significant to induce important currents above the crescentic bar (about 0.2 m/s at low tide) but during storms, with a wave height of 3 m, the wave induced currents reaches 0.7 m/s and can put in movement the sandy particules. All these characteristics induce erosion and migration of the subtidal bar and intertidal domain during storm events (which is observed on the fields).

### Wind sea

Then we simulated a wind sea with the following conditions: $H_s = 0.9$ m, $T_m = 3.5$ s and $\theta = 10°$. The wave-induced currents are really low (max current velocity reaches 0.25 m/s) and thus the bed evolution is not significant. We remark a constant longshore evolution which seems to prove that the wave refraction above crescentic bar is not relevant. Close to the shoreline, due to undertow effects, the model creates an erosion area. But in the whole remainder of the intertidal area, the accretion is predominant (see Figure 15) and seems to show that wind sea contribute to the beach accretion.

726

# 4 DISCUSSION

The platform has proved its capacity to model hydro- and morphodynamics in the surf zone. According to Castelle and Bonneton (2003), the wave refraction above the crescentic bar induces energy focalisation areas which tend to be one of the motor sources of sandy bar formation. However the formed patterns are directed in up-current which are not observed on the Aquitanian coast. Garnier et al. (2006) have shown that this direction is sensitive to the transport formulation. In the present work, two main stages during the bar formation are highlighted: the formation of the bar and its lengthening in up-current direction. Two kinds of tide cycles have been simulated here: a schematical sinusoidal tide cycle and a neap tide and spring tide cycle. In both cases, we see a narrow correspondence between the erosion/accretion cycles and the tide cycle. But taking into account NTST cycle can modify this link and changes final morphological evolution. Figure 9 shows one of the limits of the present coupling platform. Indeed, the undertow induces close to the shoreline, an erosion area. If we simulate a large period, the beach high level shoreline is eroded in a continuous way and a step appears at high tide level. In order to limit this shoreline erosion process, a swash zone module needs to be coupled (Puleo and Butt 2006) with the platform to take into account onshore sediment transport or/and the taking into account of the asymmetrical waves in the Bailard (1981) formulation. For the cresentic bar, simulations with or without roller effect give value approaching 15% of difference for velocity and 8% for free surface elevation which confirm the significant effect of this physical process. It induces larger undertow currents which modify the morphological evolution of the beach. To finish, storm and wind sea conditions have also be simulate. The model have shown the crosshore mobility of the crescentic subtidal bar due to significant wave-induced currents during storm episodes above the bar. The erosion of subtidal bar is observed along coast during significant wave conditions. We can also note if the duration of the storms increase in the future due to modifications of climate conditions, it is possible that the different sandy patterns have not enough time to be rebuilt before the next storm sequences. For the moment, we have not modelled the onshore migration of the cresentic bar for other kinds of wave conditions. The wind sea induces very low bed evolution but, except close to the high tide shoreline, the whole intertidal domain is in accretion. This phenomenon contributes to the re-nourishment of the beach. The future investigations need to be taking into account real wave (wave classes and their probability) and tide (NTST cycle) conditions in order to reproduce the morphological evolution.

# 5 CONCLUSIONS

The present paper has presented the impacts of forcing conditions in a morphological framework using the coupling between MARS and SWAN models. Four points have been studied: i) the tide cycles are a relevant phenomenon which induces various erosion zones function of time. ii) The addition of the breaking roller contribution for the computation of radiation stresses and undertow currents has been taken into account in the model. iii) The behaviors of the morphological model on seasonal wave conditions have been tested. Now further investigations are needed to well-estimated the bar formation and the long-run evolution of the beach like on asymmetrical waves influence or swash hydro- and morphodynamics influence for the sediment onshore transport. To conclude, the model can reproduce the bar formation but we need to validate the whole coupling for real beaches. The next stages of this approach are:

- Investigation of wave parameter effects on bar formation (formation velocity, bar wavelength, etc),
- Comparison with observations and field measurements.

## ACKNOWLEDGEMENT

This study is carried out within the framework of a collaboration between BRGM, UMR EPOC University Department and IFREMER. The authors would like to thank Bruno Castelle for his contribution.

## REFERENCES

Abadie, S., R. Butel, S. Mauriet, D. Morichon, and H. Dupuis (2006). Wave climate and longshore drift on the south aquitaine coast. *Continental Shelf Research 26*, 1924–1939.

Bailard, J. (1981). An energetic total load sediment transport model for a plane sloping beach. *Journal of Geophysical Research 86(C11)*, 10938–10954.

Battjes, J. and J. Janssen (1978). Energy loss and set-up due to breaking of random waves. In *Proceedings of the 16th International Conference on Coastal Engineering (Hamburg, Germany)*, pp. 569–587. American Society of Civil Engineers.

Booij, N., I. Haagsma, L. Holthuijsen, A. Kieftenburg, R. Ris, A. Van der Verthuysen, and M. Zijlema (2004). *SWAN User Manual SWAN Cycle III version 40.41*. Delft University of Technology.

Bruneau, N., P. Bonneton, R. Pedreros, F. Dumas, and D. Idier (2007). A new morphodynamic modeling platform: Application to characteristic sandy systems of the aquitanian coast, France. In *Proccedings of the 9th International Coastal Symposium (Gold Coast, Australia)*, Volume 50. Journal of Coastal Research SI.

Butel, R., H. Dupuis, and P. Bonneton (2002). Spatial variability of wave conditions on the french atlantic coast using in-situ data. In *Proccedings of the 7th International Coastal Symposium (Templepatrick, Northern Ireland)*, Volume 36, pp. 96–108. Journal of Coastal Research SI.

Castelle, B. (2004). *Modélisation de lhydrody-namique sédimentaire au-dessus des barres sableuses soumises laction de la houle: application la côte Aquitaine. Bordeaux, France*. Ph. D. thesis, University of Bordeaux I.

Castelle, B. and P. Bonneton (2003). Nearshore waves and currents over crescentic bars. In *Proccedings of the 8th International Coastal Symposium (Itajai, Brazil)*, Volume 39. Journal of Coastal Research SI.

Dally, W. (2001). Modeling nearshore currents on reef-fronted beaches. In *Proceedings of the Fourth Conference on Coastal Dynamics (Lund, Sweden)*. American Society of Civil Engineers.

Dingemans, M. (1997). *Water Wave Propagation Over Uneven Bottoms*, Volume 13. Advanced Series on Ocean Engineering.

Garnier, R., D. Calvete, A. Falques, and M. Ca-balleria (2006). Generation and nonlinear evolution of shore-oblique/transverse sand bars. *Journal of Fluid Mechanics 527*, 327–360.

Goda, Y. (2003). Examination of the influence of several factors on longshore current computation with random waves. *Coastal Engineering 53*, 157–170.

Lafon, V., D. De Melo Apoluceno, H. Dupuis, D. Michel, H. Howa, and J. Froidefond (2004). Morphodynamics of nearshore rhythmic sandbars in a mixed-energy environment (sw france): I. mapping beach changes using visible satellite imagery. *Estuarine, Costal and Shelf Science 61*, 289–299.

Pérenne, N. (2006). *MARS a Model for Applications at Regional Scale, Documentation scientifique version 0.1.* IFREMER/HOCER.

Phillips, O. (1977). *The dynamics of the upper ocean.* Cambridge University Press.

Puleo, J. and T. Butt (2006). Editorial : The first international workshop on swash-zone processes. In *International Workshop on swash-zone processes (Lisbon, Portugal)*, Volume 26, pp. 556–560. Continental Shelf Research SI.

Saint-Cast, F. (2002). *Modélisation de la morphodynamique des corps sableux en milieu littoral.* Ph. D. thesis, University of Bordeaux I.

Shi, F., J. Kirby, P. Newberger, and H. K. (2005). *NearCoM Master Program, Version 2005.4: User's manual and module integration.* University of Delaware.

Swendsen, I. (1984). Mass flux and undertow in a surf zone. *Coastal Engineering 8*, 347–365.

*River, Coastal and Estuarine Morphodynamics: RCEM 2007 – Dohmen-Janssen & Hulscher (eds)*
*© 2008 Taylor & Francis Group, London, ISBN 978-0-415-45363-9*

# Analysis of the morphodynamical development of crescentic bed patterns at Duck over two months in 1998

M.C.H.Tiessen
*University of Nottingham, School of Civil Engineering, Nottingham, UK*

S.M. van Leeuwen
*Centre for Environment, Fisheries and Aquaculture Science (CEFAS), Lowestoft, UK*

D. Calvete
*Department de Física Aplicada, Universitat Politécnica de Catalunya, Barcelona, Spain*

N. Dodd
*University of Nottingham, School of Civil Engineering, Nottingham, UK*

ABSTRACT:   A morphodynamical linear stability model is used to generate model predictions of the formation of crescentic bars. All wave data collected by one wave gauge at Duck (North Carolina, USA) for a two month period are used to conduct a linear stability analysis, using Morfo60 (Calvete et al. 2005). Model predictions of bed pattern development are similar to observations, especially for high wave energy periods. A program developed to distinguish significant events in the model predictions is able to identify these high wave periods.

## 1   INTRODUCTION

Beaches are constantly changing environments. Wave breaking induces currents, and forces sediments to be picked up by the flow, and be deposited at other locations. The erosion and sedimentation can vary for different locations and may cause the formation of bed patterns in the nearshore region.

A common bed pattern observed at sandy beaches are crescentic bars. Crescentic bars are lunate shaped rhythmic bed patterns of variable height, that occur along a beach at some distance offshore (Komar 1976).

Linear stability models can be applied to describe the formation of bed patterns. This type of model uses the full dynamical equations in linearised form, thus allowing a more rapid numerical solution at the cost of considering only initial developments (Falqués et al. 2000).

### 1.1   *Literature study*

Crescentic bars have been observed at many beaches around the world (Van Enckevort et al. 2004). They exist for mild wave conditions (Blondeaux 2001) and will disappear into a straight bar during storm surges (Falqués et al. 2006). In general they are found at locations where tidal variation and beach slope are small, together with situations where the alongshore current

is small or does not exist (Komar 1976). The forcing of this type of bed pattern was previously assumed to be due to edge waves (Komar 1976), but recent studies show that it can also be the result of positive feedback between the developing topography and the flow (Calvete et al. 2005). The spacing of crescentic bars ($\lambda$, the distance between subsequent crescentic tops in the alongshore direction) can vary, from 30 m to 3000 m (Blondeaux 2001) but are generally between 200 and 500 m (Komar 1976). This characteristic length scale is related to the distance from the shore to the mean alongshore bar crest ($X_b$). Initial estimates and observations of $\lambda/X_b$ ranged from 1 to 10. But recent observations showed a narrower band between 3 and 7 (Van Enckevort et al. 2004). Observed growth periods can vary from one day, up to several days. Tidal variation can slow down the development of crescentic bars significantly (Falqués et al. 2006). A small or moderate alongshore current can result in the migration of crescentic bars. Migration rates of this bed pattern vary due to wave conditions, from 0 to 60 m d$^{-1}$ on average (Van Enckevort et al. 2004), but can be as high as 180 m d$^{-1}$ (Falqués et al. 2006).

Recent studies have tried to capture the formation of crescentic bed patterns using stability models and examined the driving forces behind it. Deigaard et al. (1999) applied a linear stability analysis to describe the initial formation of these bed patterns from an uniform alongshore bar. They studied the sensitivity to key parameters on the formation of crescentic bars, and found that an increase in distance from the shore to the bar ($X_b$) resulted in increasing bed pattern lengths ($\lambda$) and decreasing growth rates. Deepening the alongshore bar trough resulted in faster growing crescentic bars with increased spacing. Finally, the incidence wave angle was varied. Wave angles up to 30° resulted in longer bed pattern lengths and bigger growth rates. Wave angles bigger than 30° resulted in rapidly decreasing growth rates and smaller pattern lengths.

Falqués et al. (2000) studied the dependence of bed pattern formation to model settings and physics, using a linear stability model. In the article, the "bed-surf" interaction was studied. The bed-surf interaction is the coupling of the bed patterns with the wave field. The basic settings were normal incidence waves on a plane beach. This research showed that the ratio the of stirring function ($\alpha$) and the water depth ($D$) is important in determining which bed pattern will develop. The stirring function describes the tendency of the waves to mobilise sediment. A stirring function that increases with increasing distance offshore would cause a crescentic bed pattern to develop.

Caballeria et al. (2002) also used a plane sloping beach, and normal incidence waves, but here a nonlinear model (Morfo55) was used. Similar conclusions were drawn with respect to the stirring function as Falqués et al. (2000). But here nearshore wave refraction was also included. The conclusion was that moderate waves and grain sizes resulted in the development of crescentic bars, while transverse bars would develop under low wave conditions and coarse grain sizes.

Damgaard et al. (2002) used a linear model in the situation of a barred beach and compared these results with a commercial (non-linear) model (2DH Coastal Area model). Comparison of both models showed similar spacings of crescentic bed patterns, and similar initial growth rates.

Calvete et al. (2005) used a linear model to describe the development of mainly crescentic bed patterns, projected on a alongshore bar. This model (Morfo60) was the first to use the full (linearised) shallow water equations. The sensitivity of the model results to several forcing parameters and other settings was described. Higher wave heights resulted in longer length scales of the bed pattern ($\lambda$), along with smaller growth rates. Longer wave periods resulted in longer bed patterns, with faster growing rates. Oblique incidence waves resulted in longer bed pattern lengths, and migration.

Van Leeuwen et al. (2006) used the same model as Calvete et al. (2005) to compare the effects of random waves and regular waves on the formation of bed patterns on a plane beach. The main conclusion was that random waves resulted in the development of transverse bars, where regular waves resulted in crescentic bar formation.

## 1.2 Present study

Here a linear stability model, Morfo60 (Calvete et al. 2005), is used to predict crescentic bed pattern formation at Duck, North Carolina, USA. Wave records collected at Duck from October 25th 1998 until December 31st 1998 are used to create model predictions of the bed evolution. These records contain wave characteristics at a three hour interval. For each of these wave record (528 in total) the model will predict the development of a certain bed pattern.

Observations suggest that after a storm surge, the seabed is cleared of bed patterns, and new bed patterns develop. A linear stability analysis can only predict initial developments of bed patterns. The main focus in this research is the Morfo60 predictions of bed pattern developments during and immediately after high wave conditions, since for these periods predictions are more likely to be realistic. Furthermore, we focus on developing an algorithm to allow us to identify significant developments in the model predictions.

## 2 MODEL FORMULATIONS

The Morfo60 model uses a comprehensive morphodynamic linear stability analysis to calculate preferred patterns that may arise as free instabilities of a coastal system. The model is based on the shallow water equations (see Calvete et al. (2005) for more details). The governing equations are:

$$\frac{\partial D}{\partial t} + \frac{\partial D v_i}{\partial x_i} = 0 \tag{1}$$

$$\frac{\partial v_i}{\partial t} + v_j \frac{\partial v_i}{\partial x_j} = -g \frac{\partial z_s}{\partial x_i}$$

$$-\frac{1}{\rho D} \frac{\partial}{\partial x_j} \left( S'_{ij} - S''_{ij} \right) - \frac{\tau_{bi}}{\rho D} \tag{2}$$

$$\frac{\partial E}{\partial t} + \frac{\partial}{\partial x_i}((v_i + c_{g_i})E) + S'_{ij} \frac{\partial v_j}{\partial x_i} = -d \tag{3}$$

$$\frac{\partial z_b}{\partial t} = -\frac{1}{1-\phi} \frac{\partial q_i}{\partial x_i}, \tag{4}$$

here $i,j = 1, 2$; $x_{1,2} = x, y$; $v_{1,2} = u, v$, where $x$ and $y$ are the cross shore and alongshore coordinates and $u$ and $v$ are the cross shore and alongshore depth averaged water motion, respectively. $z_s(x, y, t)$ is the mean sea level, $z_b(x, y, t)$ is the mean bed level and $D$ is the total mean depth ($D = z_s - z_b$) (see figure 1). $E(x, y, t)$ is the wave energy density, which can be expressed

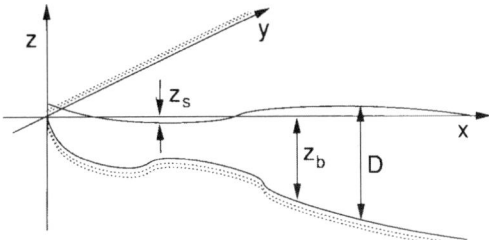

Figure 1. Coordinate system for the Morfo60 model (Calvete et al. 2005)

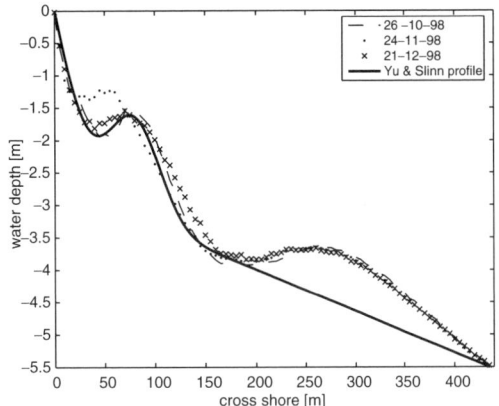

Figure 2. The cross shore beach profile at Duck: the measured profiles at several dates during the modelling period and the Yu and Slinn (2003) profile used in the Morfo60 runs.

in the wave height ($E = 1/8\rho g H^2$). $\tau_{bi}$ represents the linear bed shear stress, $g$ is the gravitational acceleration, and the sediment flux ($q_i$) is defined on the basis of the formula of Soulsby and Van Rijn (Soulsby 1997), resulting in a depth dependent stirring function. The bed porosity $\phi = 0.4$ and the seawater density $\rho = 1024\,\mathrm{kg\,m^{-3}}$. $S_{ij}'$ is the radiation stress term, and $S_{ij}''$ represents the Reynolds stresses (Calvete et al. 2005). $d$ is the wave energy dissipation due to breaking and bottom friction (Church and Thornton 1993).

The model comprises a basic state from wave records and an initial bed profile. We assume that the basic state represents an equilibrium state of the system. The model introduces perturbations into the basic state and determines the characteristics of perturbations of different length scales. Finally, it is assumed that the fastest growing perturbation length scale will dominate the development of other length scales, and the data of this fastest growing length scale is stored.

## 2.1 Basic state

An alongshore constant beach profile is applied, with an alongshore bar located 80 m offshore (Yu and Slinn 2003). This profile was developed according to the Duck beach profile in 1990 and is described by:

$$Z(x) = \left(a_1 - \frac{a_1}{\gamma_1}\right)\tanh\left(\frac{b_1 x}{a_1}\right) + \frac{b_1 x}{\gamma_1}$$

$$-a_2\exp\left[-5\left(\frac{x - X_b}{X_b}\right)^2\right] \quad (5)$$

where $X_b = 80\,\mathrm{m}$ is the location of the longshore bar and $\gamma_1 = \tan\beta_1/\beta_2$ with $\beta_1 = 0.075$ being the beach slope close to the shore, and $\beta_2 = 0.0064$ the slope offshore of the bar. $b_1 = \tan\beta_1$, $a_1 = 2.97\,\mathrm{m}$ and $a_2 = 1.5\,\mathrm{m}$. The Yu and Slinn (2003) profile extends to 4 km offshore, after which the profile becomes flat (constant depth 28.3 m).

The Yu and Slinn (2003) profile is similar to the observed beach profile of November 26th 1998, at the start of our modelling period. The main difference

between the observed profile and the modelled profile is the missing outer bar (see figure 2). More differences occur between the Yu and Slinn (2003) profile and the bathymetry measurements from later dates during the modelling period.

The wave records are collected at Duck, using 15 bottom mounted pressure gauges located 900 m offshore. The data is collected continuously and then averaged over three hours intervals. This way, every three hours the wave height ($H_{rms}$), the wave period ($T_p$) and the angle of the incoming waves ($\theta$) are recorded, resulting in 528 wave data records. The Morfo60 model applies wave conditions at the offshore boundary of the modelling grid. At this point, the water depth is fixed at 28.3 m. The recorded wave conditions are collected at 8 m water depth. The wave records are refracted back to simulate wave conditions at the offshore boundary.

## 2.2 Linear stability analysis

The cross shore current ($u$), the alongshore current ($v$), the free surface elevation ($z_s$), the bed level ($z_b$) and the energy density ($e$) are perturbed. The general form of the perturbation is given by:

$$F(x,y,t) = F_{bs}(x) + f(x,y,t) \quad (6)$$

$$f(x,y,t) = f'(x)e^{ik_y y + \omega t}, \quad (7)$$

where $F$ stands for the total value of any variable, $F_{bs}$ is the basic state value and $f$ is the perturbation. The growth rate is given by $\omega_r [d^{-1}]$ and the migration rate ($V_m$) is given by $\omega_i/k_y [md^{-1}]$. The perturbation length in the along shore direction ($\lambda\,[m]$) is given by

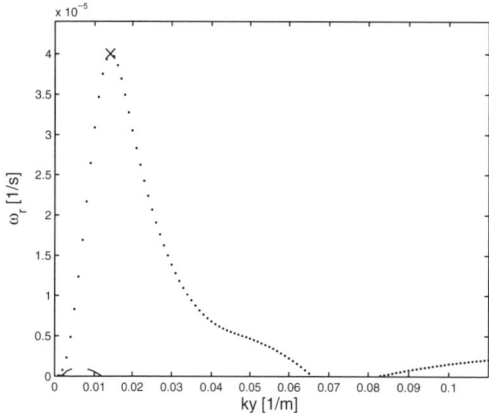

Figure 3. The growth rate curve as a function of the wave number, $H_{rms} = 2.2$ m, $T_p = 13.6$ s, $\theta = -8.7°$. The fastest growing length scale is represented by a cross.

$2\pi/k_y$, where $k_y$ is the alongshore wave number of the perturbation.

The height of the perturbations is assumed to be small, compared to the basic state values. Therefore, the system of equations can be linearised with respect to the basic state (Dodd et al. (2003), Calvete et al. (2005)). By examining the growth rate predictions of different $k_y$-values (see figure 3), a fastest growing perturbation length scale ($\lambda$) can be obtained. This fastest growing length scale is assumed to be dominant. Characteristics of this length scale, like the flow pattern, the energy density distribution and the bed pattern evolution, are stored (see figure 4). For each set of wave conditions, the Morfo60 model calculates a different fastest growing length scale and so for each set of wave conditions a different fastest growing crescentic bed pattern length scale and characteristics is obtained (Calvete et al. 2005).

## 3  SIGNIFICANT DEVELOPMENTS

A linear stability analysis only gives predictions of the initial development of morphological patterns. Initially a longshore uniform bar is assumed, and changing wave conditions will result in frequently changing model predictions. In reality, the development of crescentic bed patterns is a gradual process. To overcome the variability of the model predictions, a program is developed to find significant developments in the Morfo60 data.

The real redevelopment of crescentic bed patterns after storm surges is expected to show the most similarities with Morfo60 predictions. Therefore, high wave conditions are of most interest.

To observe significant model predictions, two parameters are important: the perturbation length ($\lambda$

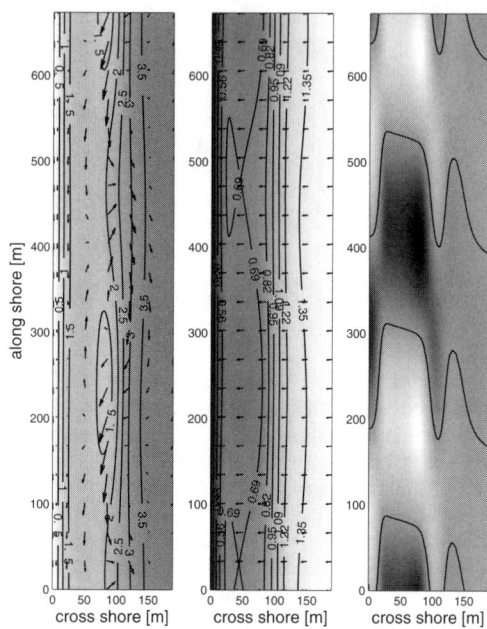

Figure 4. Morfo60 results for the fastest growing length scale during high wave conditions ($H_{rms} = 2.2$ m, $T_p = 13.6$ s, $\theta = -8.7°$, $\lambda = 448$ m, $\omega_r = 3.45$ d$^{-1}$, $V_m = -153$ m d$^{-1}$): (left) the bed evolution (contourlines and greyscale) and the flow pattern (arrows), (centre) the wave height reduction (due to wave breaking) while approaching the shore and (right) the perturbed free surface elevation.

[m]) and the growth rate ($\omega_r[d^{-1}]$). It is assumed that distinctive bed patterns only exist when two conditions are met: a certain bed pattern length is being predicted constantly for a period of time ($j = 1 \cdots N$) and at the same time the growth rate predictions are big.

Constant perturbation length predictions are assumed, if predicted length scales are within a bandwidth ($\Delta\lambda$ [m]) around a constant value ($\lambda_c$ [m]).

The period of time that is necessary for a specific bed pattern to become significant ($T_{sig}$ [d]) is dependent on the growth rate. The composite growth rate of the bed pattern lengths within the bandwidth should be bigger than a certain threshold, before this bed pattern might be observed in reality. We assume that a bed pattern should double in size ($h$), to count as a significant development (similar to the assumption made by Falqués (2005)).

$$\frac{h_{j=N}}{h_{j=1}} = \exp\left[\sum_{j=1}^{N} \omega_{r_j} \delta t\right] = 2, \qquad T_{sig} = N\delta t \qquad (8)$$

Finally, the growth of a bed pattern is only assumed to be significant if, during the doubling time, the number of bed pattern lengths outside the bandwidth is limited.

Table 1. The characteristics of the significant developments program for the three different cases.

| | $\lambda_c[m]$ | $\Delta\lambda[m]$ | Significance* |
|---|---|---|---|
| Case 1 | $\lambda_1$ | $\lambda_c \pm 100$ | $\sum_{j=1}^{N} j$ |
| Case 2 | $\dfrac{\sum_{j=1}^{N} \lambda_j}{N}$ | $\left(1 \pm \frac{1}{2}\right)\lambda_c$ | $\sum_{j=1}^{N} j$ |
| Case 3 | $\dfrac{\sum_{j=1}^{N} \omega_{r_j}\lambda_j}{\sum_{j=1}^{N} \omega_{r_j}}$ | $\dfrac{2\pi}{k_{yc} \pm 5\Delta k_y}$ | $\sum_{j=1}^{N} \omega_{r_j}$ |

* "Significance" stands for the determination of the maximum number of results outside the bandwidth, which is still allowed within a significant event.

Different versions of the program defining significant developments have been developed. Here, three different cases will be presented (see table 1).

## 3.1 Case 1

The most basic version of the algorithm for significant developments consists of the initial value ($j = 1$) of $\lambda$ to be used as the constant bed pattern length ($\lambda_c = \lambda_1$). The margin for constant wave length predictions ($\Delta\lambda$) is set at 100 m around $\lambda_c$. The number of bed pattern length predictions which is allowed outside this bandwidth is dependent on the number of bed pattern length predictions inside the bandwidth.

$$\left[\sum_{j=1}^{N} j\right]_{joutside} < 0.25 \left[\sum_{j=1}^{N} j\right]_{jinside} \quad (9)$$

## 3.2 Case 2

Instead of using a fixed value for the constant bed pattern length ($\lambda_c$), in the second case the value of $\lambda_c$ is determined as an average of the bed pattern lengths included. While the number of included model results ($N$) increases, the average will be composed of more bed pattern length predictions. Due to different spacings between adjacent length scales, the use of a constant value for the margin for constant wave length predictions ($\Delta\lambda$) is not optimal. In this case, $\Delta\lambda$ is a fraction of $\lambda_c$ (see table 1).

## 3.3 Case 3

In the final case it is assumed that faster growing lengths will dominate slower growing bed pattern lengths. To express this assumption in the significant developments program, the constant bed pattern length is determined as a weighted running average of the bed pattern lengths included. This means that all bed pattern lengths included are weighted by their growth rate.

$$\lambda_c = \frac{\sum_{j=1}^{N} \omega_{r_j}\lambda_j}{\sum_{j=1}^{N} \omega_{r_j}} \quad (10)$$

The bandwidth within a constant bed pattern lengths is assumed, is determined by the value of $k_y$.

$$\Delta\lambda = \frac{2\pi}{\pm 5\Delta k_y + k_{yc}}, \quad (11)$$

where $k_{yc}$ is the constant alongshore wave number ($2\pi/\lambda_c$). In the Morfo60 runs $\Delta k_y = 0.001$. By calculating the bandwidth for significant developments as a function of $k_y$, a fixed number of adjacent length scales is included.

The number of incorrect results that is allowed in a significant development is dependent of the growth rate. The composite growth rate of the bed patterns outside the bandwidth is supposed to be less than 25% of the composite growth rate of the bed patterns inside the bandwidth.

$$\left[\sum_{j=1}^{N} \omega_{r_j}\right]_{joutside} < 0.25 \left[\sum_{j=1}^{N} \omega_{r_j}\right]_{jinside} \quad (12)$$

## 4 RESULTS

Figure 5 shows the results of the Morfo60 experiments. The upper three graphs depict the forcing wave conditions (wave height ($H_{rms}$), wave period ($T_p$) and angle of the incoming waves ($\theta$)) as they were collected at the wave gauge in front of the Duck coast. The lower three graphs show the resulting bed pattern length ($\lambda$), growth rate ($\omega_r$) and migration rate ($V_m$). The x-axis represents time in all graphs, starting October 25th 1998 and finishing December 31st 1998.

The first graph ($H_{rms}$) shows four storm surges (around days 340, 344, 347 and 353), while minor high wave conditions occur in the period before day 340. The high wave conditions coincide with Morfo60 predictions of big growth rates. Bigger bed pattern length predictions also roughly coincide with high wave conditions, although fluctuations in bed pattern length predictions occur for low wave conditions due to changing wave angles and periods. Apart from a small number of very extreme bed pattern lengths (up to 2000 m), the length scales vary between 500 and 1000 m for high wave conditions. For low wave conditions, length scales less than 300 m are predicted, in general. Field observations suggest similar length scales directly after high wave conditions as those predicted by Morfo60 (Van Enckevort et al. 2004). The migration rate is generally less than 100 m d$^{-1}$, although more extreme results occur up to 250 m d$^{-1}$. Field observations of the migration of crescentic bars, generally suggest smaller migration rates (Van Enckevort et al. (2004), Falqués (2006).

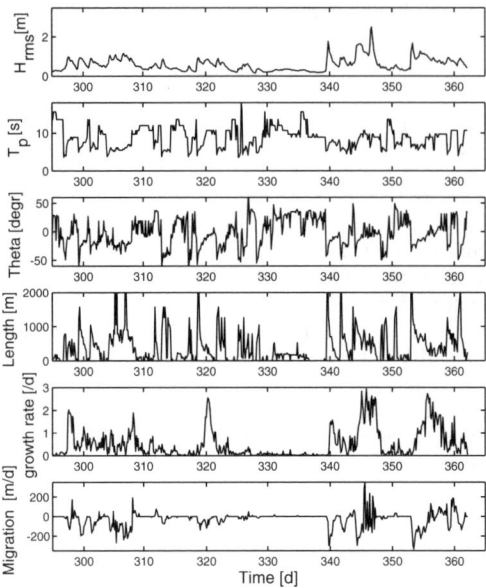

Figure 5. Morfo60 results: model input and output in time.

## 4.1 Case 1

The results of the first version of the significant developments program are shown in figure 6. Three plots are shown: the wave height is shown in the first graph and the length scale and growth rate plots are shown in the second and third graph, to examine the effectivity of the significant developments algorithm. The focus of this research lies with high wave conditions. The settings of the significant developments program do not specify these locations clearly. The developments seem to focus around low wave conditions that correspond with small bed pattern length predictions and growth rates. For low wave conditions, the bed pattern length predictions are rather constant. Due to the fixed value of $\Delta\lambda$, these slow growing bed patterns can become significant developments. In regions of longer bed pattern length predictions few significant developments are observed, due to the increased spacing between adjacent length scales.

## 4.2 Case 2

The significant developments determined for the second case (see figure 7), focus more around the high wave conditions. The significant developments program identifies big growth rate and bed pattern predictions. In this case, the value of $\Delta\lambda$ is dependent on $\lambda_c$, which results in the exclusion of most small bed pattern lengths as significant developments. Due to the wide margins for $\Delta\lambda$ $((1 \pm 0.5)\lambda_c)$, big variations in bed pattern length are not excluded as significant

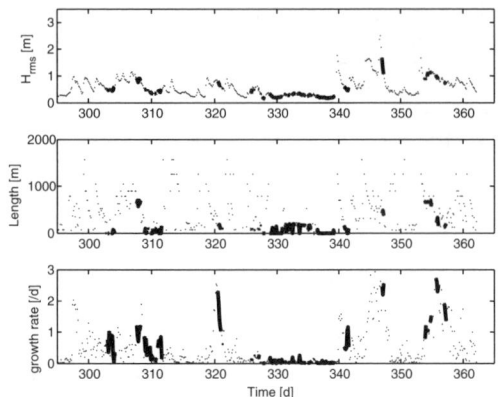

Figure 6. Significant developments for different parameters (Case 1). Insignificant results are shown as dots, significant results are presented as a black line.

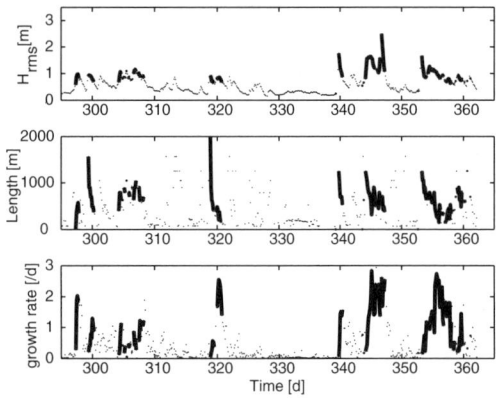

Figure 7. Significant developments for different parameters (Case 2). Insignificant results are shown as dots, significant results are presented as a black line.

developments and not only the fastest growing length scales are identified as significant developments.

## 4.3 Case 3

The use of a $\Delta\lambda$ which is dependent on the $k_{y_c}$, results in a bandwidth that is related to the distance between adjacent length scales. This means that for short length scales, the $\Delta\lambda$ is small, where for longer bed pattern lengths, this increases. The use of the growth rate $(\omega_r)$ for the determination whether a development is significant or not (12), results in more specific significant developments for big growth rates. The results for this case are shown in figure 8 indicating that the significant developments program has now focused only on high wave conditions with big growth rate predictions.

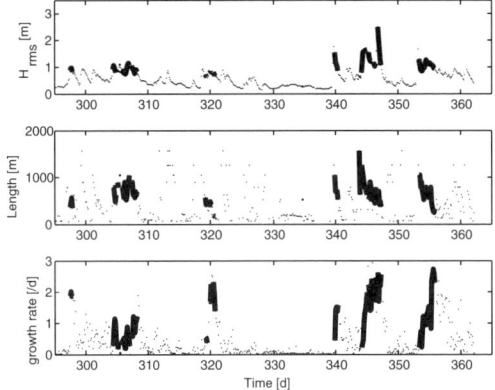

Figure 8. Significant developments for different parameters (Case 3). Insignificant results are shown as dots, significant results are presented as a black line.

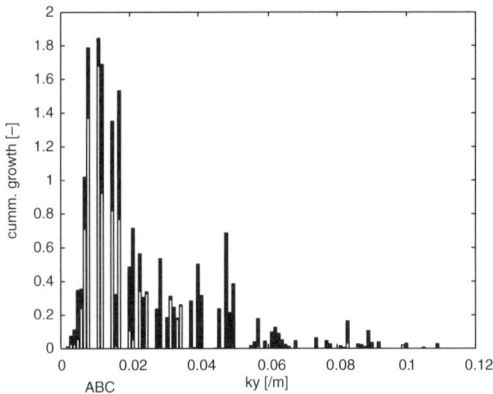

Figure 9. The cumulative growth rate for each wave number. The black bars represent the total cumulative growth, where the white part represents the part of the predictions that is supposed to be significant developments (Case 3).

Figure 9 shows the Morfo60 predictions of the cumulative growth for each length scale $(\lambda = 2\pi/k_y)$ for the full two month period, along with the significant cumulative growth as it is determined in the third case. Three fastest growing groups of length scales can be identified, "A" represents bed pattern lengths of about 900 m, while "B" represents lengths of 550 m, and "C" is about 400 m length. The significant developments program can find the fastest growing length scales, which correspond with length scales observed directly after high wave conditions (Van Enckevort et al. 2004). The cumulative growth of the bars around "C" are for less than 50 % captured by the significant developments program. This is unsatisfying because this length scale is generally the prevailing length scale (Van Enckevort et al. 2004)

## 5 DISCUSSION

The bed profile used in the Morfo60 runs shows much similarities with the beach profile at the beginning of the examined period, but it is different from the profiles measured later during this period. A different profile might improve model predictions significantly, as recent research shows that model predictions of for instance the bed pattern length and growth rate are sensitive to the dimensions of the beach profile (Calvete et al. 2006).

In the Morfo60 model runs, the tide is not taken into account. On average the tidal variation of the water level is about 1 m. This difference may cause the waves to break at a different location and this can result in the formation of different bed patterns.

In the second and third case, the significant developments program distinguishes the moments in time where the model predicts constant bed pattern lengths, together with big growth rates. The program locates most high wave conditions, which are in reality also the moments in time where most morphological action occurs. The assumed bandwidth for constant length scales $(\Delta\lambda)$ is very wide for longer crescentic pattern lengths.

A direct comparison of the model predictions with field observations is not presented in this paper. This is a main focus of this research and future research will include a comparison of field observations made by Van Enckevort (2004) with Morfo60 predictions of Duck.

## 6 CONCLUSION

A linear stability model (Morfo60) has been used to predict the development of crescentic bed patterns using a wide range of wave records. The wave records and the bed profile were collected at Duck, North Carolina, USA. High wave conditions result in larger growth rates and longer length scales predictions of the crescentic bed pattern. Model predictions vary due to changing wave conditions. An algorithm was developed to distinguish significant developments in the Morfo60 predictions. Model predictions during these significant developments are more likely to be observed in reality. Several versions of the algorithm were presented. In the final version, high wave conditions and the fastest growing crescentic bed pattern lengths can be identified.

REFERENCES

Blondeaux, P. (2001). Mechanics of coastal forms. *Annu. Rev. FluidMech 2001 33*, 339–370.
Caballeria, M., G. Coco, A. Falqués, and D. Huntley (2002). Self-organization mechanisms for the formation

of nearshore crescentic and transverse sand bars. *Journal of Fluid Mechanics 465*, 379–410.

Calvete, D., G. Coco, A. Falqués, and N. Dodd (2006). (un)predictability and (ir)regularity in rip channel systems. *To be published in Geophysical Research Letters*.

Calvete, D., N. Dodd, A. Falqués, and S. Van Leeuwen (2005). Morphological development of rip channel systems: Normal and near-normal wave incidence. *Journal of Geophysical Research 110*(C10006), 1–18.

Church, J. and E. Thornton (1993). Effects of breaking wave induced turbulence within a longshore current model. *Journal of Coastal Engineering 20*, 1–28.

Damgaard, J., N. Dodd, L. Hall, and T. Chesher (2002). Morphodynamic modelling of rip channel growth. *Journal of Coastal Engineering 45*, 199–221.

Deigaard, R., N. Drønen, J. Fredsøe, J. Hjelmager Jensen, and Jørgensen (1999). A morphological stability analysis for a long straight barred beach. *Journal of Coastal Engineering 36*, 171–195.

Dodd, N., P. Blondeaux, D. Calvete, H. d. Swart, A. Falqués, S. Hulscher, G. Różyński, and G. Vittori (2003). Understanding coastal morphodynamics using stability methods. *Journal of Coastal Research* 19(4), 849–865.

Falqués, A. (2005). Wave driven alongshore sediment transport and stability of the dutch coastline. *Journal of Coastal Engineering 53*, 243–254.

Falqués, A., G. Coco, and D. Huntley (2000). A mechanism for the generation of wave-driven rhythmic patterns in the surf zone. *Journal of Geophysical Research 105*(C10), 24071–24087.

Falqués, A., N. Dodd, R. Garnier, F. Ribas, L. MacHardy, P. Larroudé, D. Calvete, and F. Sancho (2006). Rhythmic surf zone bars and morphodynamic self-organisation. *To be published in Journal of Coastal Engineering*.

Komar, P. (1976). *Beach processes and sedimentation*. Englewood Cliffs, New Jersey: Prentice-Hall.

Soulsby, R. (1997). *Dynamics of Marine Sands*. London: Thomas Telford.

Van Enckevort, I., B. Ruessink, G. Coco, K. Suzuki, I. Turner, N. Plant, and R. Holman (2004). Observations of nearshore crescentic sandbars. *Journal of Geophysical Research 109*(C06028), 1–17.

Van Leeuwen, S., N. Dodd, D. Calvete, and A. Falqués (2006). Physics of nearshore bed patter formation under angular or random waves. *Journal of Geophysical Research 111*(F01023), 1–16.

Yu, J. and D. Slinn (2003). Effects of wave-current interaction on rip currents. *Journal of Geophysical Research 108*(C3), 1–19.

*River, Coastal and Estuarine Morphodynamics: RCEM 2007 – Dohmen-Janssen & Hulscher (eds)*
*© 2008 Taylor & Francis Group, London, ISBN 978-0-415-45363-9*

# Video-observations of shoreward propagating accretionary waves

K.M. Wijnberg
*University of Twente, Water Engineering & Management, Enschede, The Netherlands*

R.A. Holman
*Oregon State University, College of Oceanic and Atmospheric Sciences, Corvallis, OR, USA*

ABSTRACT: A 10 year database of daily time-exposure images of the nearshore zone, collected by an Argus video system installed at Duck, NC, revealed a new phenomenon in the evolution of nearshore topography. Under some conditions, a nearshore sand bar can shed a small bar-like feature from its shoreward facing side. This pinched off daughter bar subsequently transits the trough as an intact feature and merges with the beach. The average onshore propagation rate of the feature is about 3 m/day and, with an average size of about 130 m by 30 m and a height comparable to that of the inner bar, it represents a locally significant onshore sediment flux.

## 1 INTRODUCTION

Currently we are only partly aware of the range of morphologic behavior that can occur in the nearshore zone. Waves and currents have ample capacity to move sediment around in the nearshore zone. The nonlinearities in both the sediment transport processes and the surf zone hydrodynamics carry with them a large potential for generating unexpected gradients in sediment transport across the nearshore topography, hence producing unexpected bathymetric change. Documenting the natural range of morphologic behavior in the nearshore zone is therefore indispensable to focus our thinking about nearshore morphodynamics.

The present perception of breaker bar evolution is that bars generally reshape into linear forms in association with storm events Under subsequent lower energy conditions, these bars are usually observed to move shoreward and to develop three-dimensionality that may be rhythmic or irregular and complex (Lippmann & Holman 1990). It bas been commonly assumed that the response of a bar to changing wave conditions occurs as an intact, albeit evolving, sand bar form.

A large database of routinely collected time-exposure video images of the nearshore zone near Duck (North Carolina, USA) revealed that this is not always the case. Under some conditions, a sand bar can shed a small bar-like feature from its shoreward facing side. This pinched off daughter bar subsequently transits the trough as an intact feature and merges with the beach (Fig. 1). Tentatively, we have named this phenomenon a Shoreward Propagating Accretionary

Wave or, abbreviated, a SPAW. The term 'wave' was chosen to reflect the similarities between the observed phenomenon and a solitary wave in fluid dynamics. That is, both phenomena are single, isolated perturbations that maintain their shape as they propagate. In both cases, the latter involves a net displacement of material in the direction of propagation.

In this paper we will present a first documentation of this newly observed phenomenon and discuss its relevance for understanding the nearshore morphodynamic system.

## 2 DUCK FIELD SITE AND DATA BASE

The Duck study area is near the CERC Field Research Facility (FRF), which is located at about the middle of Currituck Spit, a 100 km long unbroken stretch of shoreline facing the Atlantic Ocean. The beach is generally fronted by one or two nearshore bars, with variable planforms through time (Lippmann & Holman 1989). The sediment is in the medium sand range with a mixture of coarser material on the beach (Larson & Kraus 1992). The mean annual wave height and period are about 1 meter and 8 second; the spring tide range is about 1.5 meter (Leffler et al. 1992).

The video camera that collected the images from the beach is mounted on a 43 m-high tower at the dune crest near the FRF-pier. The area of interest in the field of view of the camera extends about 800 m alongshore. In the cross-shore direction the study area extended from the inner nearshore bar to the shoreline.

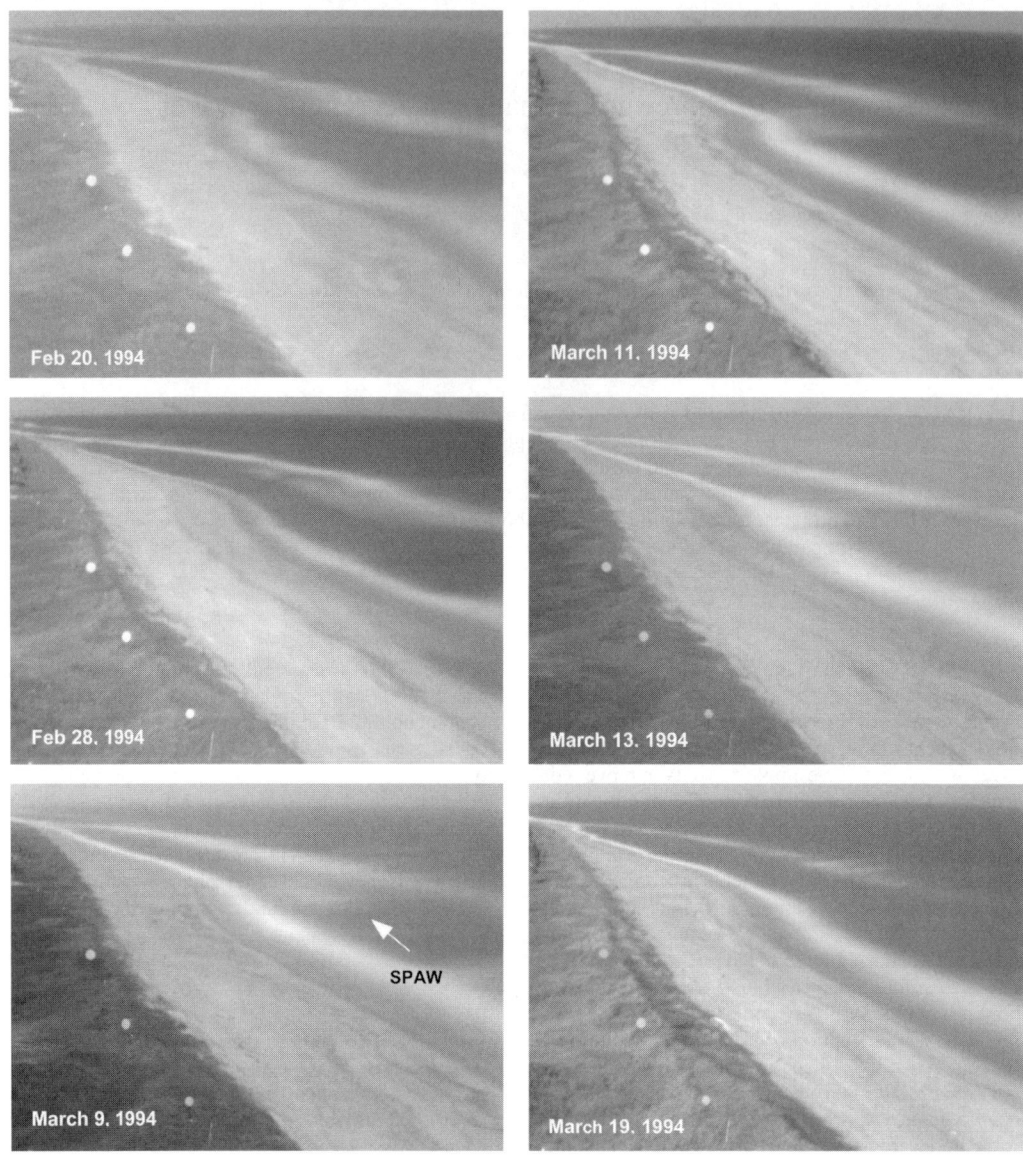

Figure 1.    Sequence of time exposure images near Duck illustrating a SPAW event.

The analyzed period spans from October 7, 1986 until December 1996. During this period one major data gap exists in the image time series, which extends from August 10, 1992, to January 28, 1993.

## 3   METHODOLOGY

SPAW events are identified by scanning long time series of time-exposure images by eye. The presence of an isolated, but coherent patch of foam in between a nearshore bar and the shoreline indicates the presence of a submerged mound of sand, the SPAW. Having recognized a SPAW feature, the dates of starting and ending of the event are determined to obtain statistics on the duration of a SPAW event.

The starting date of a SPAW is defined as the first day on which separation of the SPAW from the parent bar becomes apparent. The exact starting date may be obscured because of high wave conditions, which

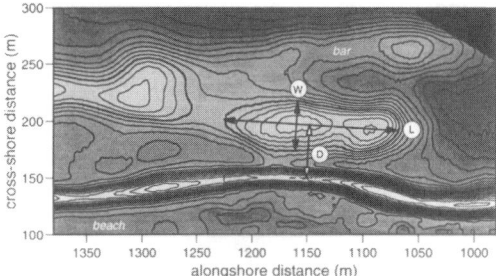

Figure 2. Definition sketch of morphometric measurements based on contoured time-exposure image (contours based on pixel intensity). W = SPAW width, L = SPAW length, D = SPAW initiation distance.

usually occur in conjunction with SPAW initiation. The residual foam that is generally present during those conditions may merge with the foam due to bathymetry-related wave breaking such that the bathymetric separation of the SPAW from the parent bar may not be visible from the earliest moment.

The ending date of a SPAW event is defined as the date on which no noticeable traces are left of its occurrence. Because the disappearance of the shoreline protrusion is generally gradual in nature, some arbitrariness exists in the choice of the ending date.

To account for the potential arbitrariness in the quantification of SPAW duration, three operators independently viewed the same time series of images to detect SPAWs and identify their starting and ending dates. Subsequently, the image time series was re-examined to reconcile the differences in interpretation of the individual operators. Dubious cases were omitted for the final statistics. Inter-operator variability will be presented in the results section.

In addition to assessing the frequency of occurrence and the duration of a SPAW event, we obtained several morphometric measures. These are: (1) the cross-shore position of the SPAW at initiation (which, divided by SPAW duration, provides an estimate of the propagation speed) and (2) the size of the SPAW as defined by the width and length of the foam patch, which is actually a proxy measure for the actual size of the submerged mound)

The surf zone time-exposure images discussed in this paper are obtained by land-based cameras. The oblique images of the beach produced by these cameras can be transformed into plan view images by standard photogrammatric relationships (Holland et al., 1997). These rectified images can be used to take undistorted measurements of the above indicated morphometric properties.

The cross-shore position (D) of the SPAW is measured from the time-exposure image by determining the distance between the crest of maximum intensity on the SPAW, as a proxy for the SPAW crest position,

and the maximum intensity at the shoreline (Fig. 2). The length (L) and width (W) of a SPAW just after its initiation are determined from a contour plot of the intensities on the time-exposure image (Fig. 2). After densely contouring the time-exposure image a single contour is picked to represent the SPAW. Either the outermost closed contour around the SPAW feature is picked or, in case of a less strongly developed separation from the parent bar, the outermost contour showing contractions around the SPAW feature is picked. The area where the contour contracts indicates the location where the daughter bar is separating from the parent bar.

It should be noted that the SPAW length scales as determined from the time exposure images are proxy measures, because only the shallower part of the SPAW where wave breaking occurs is visible on the image. In addition, image intensity of itself is not an exact measure for depth, so the equal intensity contour used to represent the circumference of a SPAW does not necessarily relate to a single depth contour around the SPAW. Nevertheless, using the same type of measure for all SPAWs gives us some handle on the average size of the phenomenon as well as its variability.

## 4 RESULTS

### 4.1 *Frequency and duration of SPAW events*

Near Duck, one can observe on average about two SPAW occurrences per year (19 events in 9.75 years covered by video-imagery, along 800 m of beach). However, the inter-annual variability is large (Fig. 3). The same holds for the duration of a SPAW event; on average it takes 17 days for a SPAW to transit the trough, merge with the beach, and be redistributed alongshore such that no noticeable traces are left of its occurrence. Some SPAWs, however, only need just over a week to complete this sequence, whereas one needed up to 7 weeks (Fig. 4); the standard deviation in SPAW duration was found to be 9 days.

The inter-operator variability in the identification of the SPAW events is presented in Table 1. Of the original 29 cases, 10 were omitted because we were not sufficiently confident that the features welding to the beach did truly separate from the parent bar. Of these 10 cases, 9 were identified by only one operator.

The estimates of the duration of individual SPAW events varied among the different operators. The magnitude of the inter-operator differences appeared to be proportional to the duration of the event. So, in terms of the number of days, the individual duration estimates for longer duration events tended to be wider apart than for shorter duration events. On average, the operator deviations from the final data set equaled 20% of the magnitude of the values used in that final,

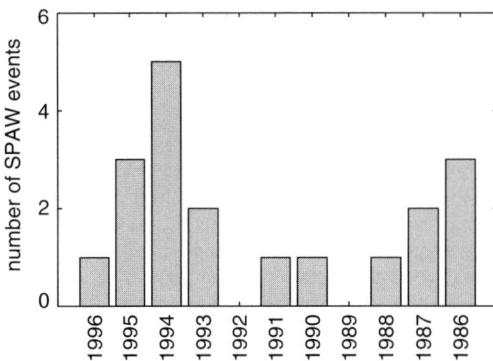

Figure 3. Number of SPAW events per year, as observed along 800 m of beach. Note that the value for 1986 is based on only 3 months of observations, and 1992 on only 7 months.

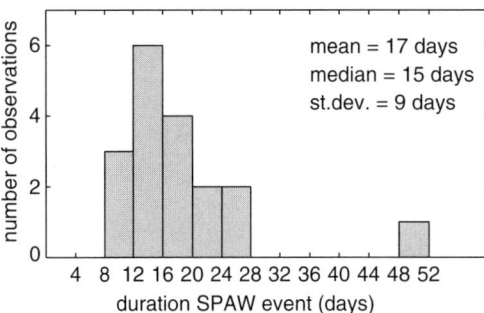

Figure 4. Histogram of observed durations of SPAW events (note Ntotal = 18, because for one SPAW event a 20 day data gap prevented observation of its evolution.)

Table 1. Inter-operator variability in SPAW identification.

| Number of operators $n$ identifying the same SPAW event | Number of SPAW events identified by $n$ operators | |
|---|---|---|
| | Original data set | Final, reconciled data set |
| 1 | 14 | 5 |
| 2 | 4 | 3 |
| 3 | 11 | 11 |
| ≥1 | 29 | 19 |

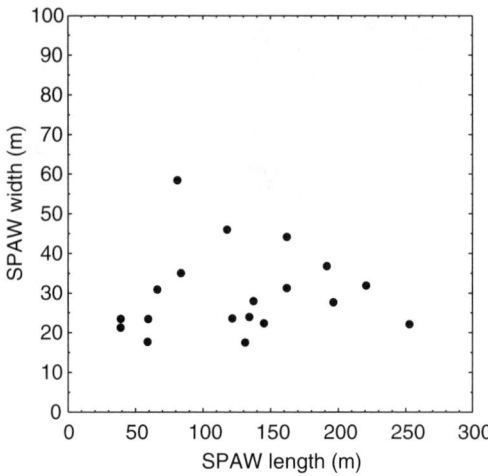

Figure 5. Plan shape geometry of SPAWs.

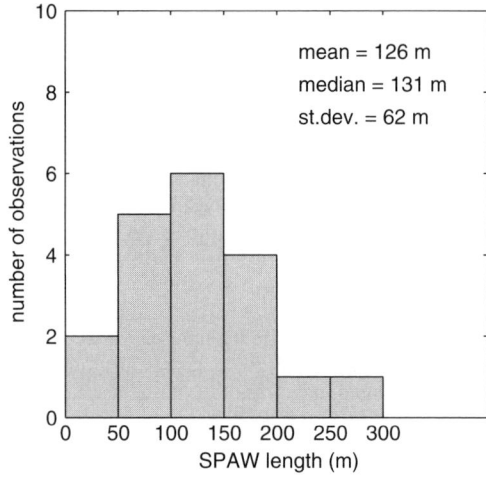

Figure 6. Histogram of observed SPAW lengths.

events occurring closely in time were interpreted as one single event by some of the operators. These interpretations were based on the oblique image time series. Rectified images of these cases helped to solve this ambiguity.

reconciled data set. Some of the larger deviations originated from differences in the interpretations of the start of an event. In three cases a bifurcation developed but it did not separate from the main bar for some time (1 to 2 weeks). Some interpreted the moment of initial bifurcation while others identified the day of separation as the start of the SPAW (the latter was used for the final data set). Further, on three occasions SPAW

4.2 SPAW size and shape

Generally, SPAWs are elongated features (Fig. 5). The length of the SPAWS, as determined from the white patches that disclose the subaqueous presence of SPAWs, varied between about 40 m and 255 m. The average length scale for SPAWs was found to be 126 m with a standard deviation of 60 m (Fig. 6). The width of these same patches varied between 18 m and 58 m,

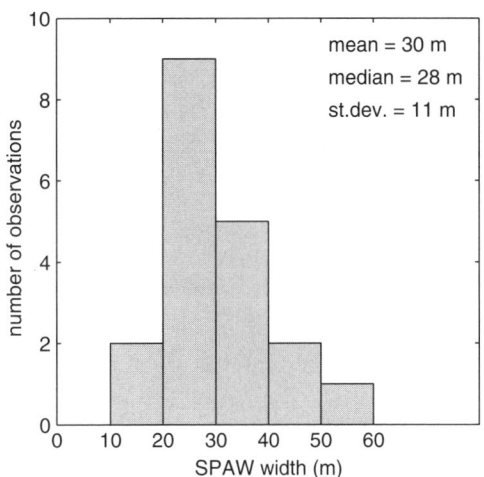

Figure 7.  Histogram of observed SPAW widths.

Figure 8.  Time exposure image of Duck beach, 6 Sep. 1994, showing the SPAW captured in the bathymetric survey shown in Figure 9.

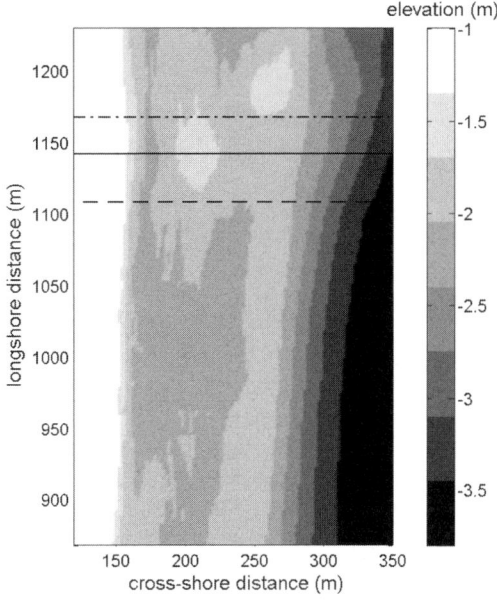

Figure 9.  Bathymetric survey, Duck, 7 Sep. 1994, capturing a SPAW. The three lines (dash-dotted, solid, dashed) indicate the location of transects shown in Figure 10.

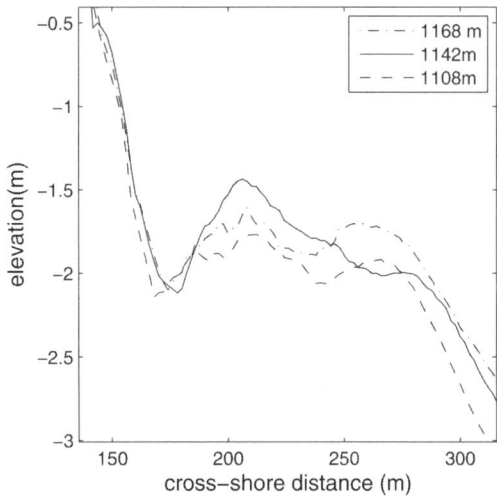

Figure 10.  Bathymetric transects crossing a SPAW near Duck (survey 7 Sep. 1994, see Figure 9 for plan view).

having an average of 30 m with a standard deviation of 10 m (Fig. 7). A measure for the vertical scale of a SPAW, such as the elevation difference between the crest of the SPAW and the trough landward of it, is hard to obtain from the video time-exposures. However, just after its initiation a SPAW will probably have a vertical scale comparable to that of the parent bar. The SPAW pinches off from the parent bar, so the trough level with which to compare the crest elevation is comparable for the SPAW and the parent bar. In addition, the crest elevation of the SPAW will be close to that of the parent bar, otherwise no waves would be seen breaking over the SPAW.

A bathymetric survey conducted during the Duck'94 experiment fortuitously captured a SPAW (Figs. 8, 9), which indeed shows that the vertical scale of the SPAW and parent bar are similar (Fig. 10). Figure 8 further shows that in the background of the surveyed SPAW a second SPAW event is occurring simultaneously (but was not covered by the

bathymetric survey). In the considered case, the SPAW crest-trough elevation difference is about 0.7 m. No bathymetric data were available to determine the amplitude evolution of the SPAW feature as it propagated onshore.

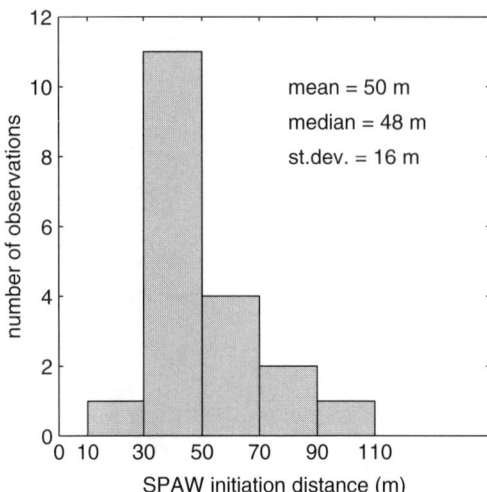

Figure 11. Histogram of observed SPAW initiation distances.

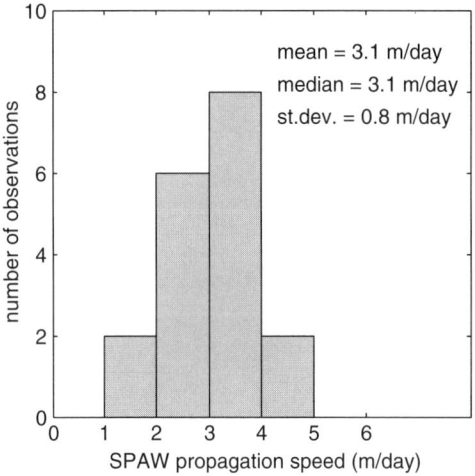

Figure 12. Histogram of observed SPAW propagation speeds.

### 4.3 SPAW propagation and beach accretion

Estimates of average onshore propagation speeds can be determined by taking the ratio of the distance to the shoreline at initiation and the duration of each individual SPAW event. On average, SPAWs are initiated at 50 m from the shoreline, but distances half and twice as large have been observed too (Fig. 11). Propagation speeds appeared to vary between 1.7 m/day and 4.8 m/day (Fig. 12). Averaged over all SPAW events, a mean onshore propagation speed of 3.1 m/day was found with a standard deviation of 0.8 m/day.

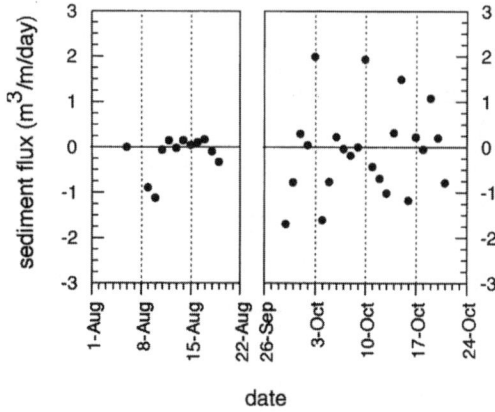

Figure 13. Daily mean sediment fluxes to the beach near Duck, based on beach surveys collected during Duck'94 experiment (courtesy: N.G. Plant).

The onshore sediment flux related to the SPAW propagation can be estimated from the above numbers to amount to about 1 to 2 $m^3$/m/day (assuming an approximate SPAW height of 0.5 m). This amount is a significant contribution to the daily cross-shore sediment flux in the inner nearshore zone. For comparison, net daily sediment fluxes to the beach, determined from densely sampled DGPS surveys of the beach elevation, are of the same order of magnitude (Fig. 13). The fluxes shown in Figure 13 represent the net cross-shore flux that occurred in the slice of beach between the 0.5 m and 2 m elevation contour, averaged over a 100 m longshore section of the beach (under the assumption that volume change in this reference box occurred due to cross-shore exchange with the nearshore).

## 5 DISCUSSION

Various models exist to simulate nearshore morphodynamics, but so far none of them has explicitly brought up the possibility of shedding isolated features like SPAWs. Many of these model are template models in which patterns in the nearshore flow field enforce similar patterns in the underlying sand bed (e.g. Dyhr-Nielsen & Sorensen 1970, Bowen & Inman 1971, Holman & Bowen 1982, Sallenger & Howd 1989, Howd et al. 1991). However, nearshore topographical change may also be explained by feedback models that, in contrast to template models, lack the requirement of pre-existing patterns in the flow field (cf. ripple formation under steady currents). Such types of models for the nearshore were already proposed three decades ago (Hino 1974) but also received attention more recently (e.g. Deigaard et al. 1999, Ribas et al. 2003).

The creation of an isolated mound of sediment in the trough of a sand bar that maintains its integrity as it transits the trough seems inconsistent with the concept of feature generation as a direct response to a fluid forcing template. Instead, we think that the creation and stability of SPAWs point out the probable importance of feedback between topographic features and their perturbations to overlying fluid motions.

## 6 CONCLUSIONS

A large database of video time-exposure images of the surf zone near Duck (NC, USA) revealed that a nearshore bar can occasionally become unstable and shed a small bar from its landward facing side. This daughter bar subsequently transits the trough as an intact feature and merges with the beach. These generally elongated features were found to have a length of $126 \pm 60$ m and a width of $30 \pm 10$ m. A typical duration of the event is $17 \pm 9$ days, and the average onshore propagation speed of a SPAW is $3.1 \pm 0.8$ m/day. The related onshore sediment flux was estimated to be about $1–2$ m$^3$/m/day, which is a locally significant contribution to the beach accretion rate that is normally observed along the studied beach.

None of the existing nearshore morphodynamic models has explicitly brought up the possibility of shedding isolated features like SPAWs so far. This may indicate a 'missing process' in the current models. Therefore, further study of SPAWs can contribute useful information to our general understanding of the nearshore processes, even though SPAWs have a relatively low frequency of occurrence and have only a local impact on the beach. For example, understanding the mechanism that allows the SPAW to maintain its integrity throughout its onshore propagation across the trough may help us better understand the cross-shore transport processes in the nearshore environment.

## ACKNOWLEDGEMENTS

We would like to acknowledge Amanda Barstow and Eric Gilbert for their conscientious efforts in viewing all the images for the occurrences of SPAWs which allowed us to establish the operator dependence of SPAW identification. Thanks also go to Nathaniel Plant for kindly providing the data on daily sediment fluxes to the beach of Duck. In addition we thank the CERC Field Research Facility for providing the bathymetric data. We gratefully acknowledge support from the Office of Naval Research, Coastal Geosciences program for helping make possible long-term Argus collections.

## REFERENCES

Bowen, A.J. & Inman, D.L. 1971. Edge waves and crescentic bars. *J. Geophys. Res.* 76: 8662–8671.

Deigaard, R., Drønen, N., Fredsøe, J., Jensen, J.H. & Jørgensen, M.P. 1999. A morphological stability analysis for a long straight barred coast, *Coastal Eng.* 36: 171–195.

Dyhr-Nielsen, M. & Sorensen, T. 1970. Some sand transport phenomena on coasts with bars. ASCE, *Proceedings 12th Int. Conf. on Coastal Eng.*: 855–865.

Hino, M. 1974. Theory on formation of rip-current and cuspidal coast. *Coastal Engineering in Japan* 17: 81–92.

Holland, K.T., Holman, R.A., Lippmann, T.C., Stanley, J. & Plant, N. 1997. IEEE *J. of Oceanic Engineering* 22(1): 81–92.

Holman, R.A. & Bowen, A.J. 1982. Bars, bumps, and holes: models for the generation of complex topography. *J. Geophys. Res.* 87: 457–468.

Howd, P., Bowen, T., Holman, R. & Oltman-Shay, J. 1991. Infragravity waves, longshore currents, and linear sand bar formation. *Proceedings Coastal Sediments '91*, ASCE, New York: 72–84.

Larson, M. & Kraus, N.C. 1992. Analysis of cross-shore movement of natural longshore bars and material placed to create longshore bars. CERC, US Army Engineers Waterways Experiment Station, Vicksburg, MS. *Techn. Report DRP-92-5*, 89 pages + Appendices.

Leffler, M.W., Baron, C.F., Scarborough, B.L. Hathaway, K.H. & Hayes, R.T. 1992. Annual Data Summary for 1990 CERC Field Research Facility, Vol. 1. CERC, US Army Engineers Waterways Experiment Station, Vicksburg, MS. *Techn. Report CERC-92-3.* 69 pages + Appendices.

Lippmann, T.C. & Holman, R.A. 1989. Quantification of sand bar morphology: a video technique based on wave dissipation. *J. Geophys. Res.* 94 (C1): 995–1011.

Lippmann, T.C. & Holman, R.A. 1990. The spatial and temporal variability of sand bar morphology. *J. Geophys. Res.* 95(C7): 11,575–11,590.

Ribas, F., Falqués, A. & Montoto, A. 2003. Nearshore oblique sand bars. *J. Geophys. Res.*, 108(C4): 3119, doi:10.1029/2001JC000985, 2003.

Sallenger, A.H. & Howd, P.A. 1989. Nearshore bars and the break-point hypothesis. *Coastal Eng.* 12: 301–313.

*River, Coastal and Estuarine Morphodynamics: RCEM 2007 – Dohmen-Janssen & Hulscher (eds)*
*© 2008 Taylor & Francis Group, London, ISBN 978-0-415-45363-9*

# The role of initial bathymetry on rip channel formation

Giovanni Coco
*National Institute of Water and Atmospheric Research, Hamilton, New Zealand*

Daniel Calvete & Albert Falqués
*Dept. de Física Aplicada, Universitat Politécnica de Catalunya, Barcelona, Spain*

Nicholas Dodd
*School of Civil Engineering, University of Nottingham, University Park, Nottingham, UK*

ABSTRACT: Rip channel systems are a common feature of nearshore system. They are often characterized by regular alongshore spacing. Field observations and computer simulations have focused on predicting the spacing and e-folding time of rip channels as a function of offshore hydrodynamics. Here, we show that the lack of predictability of rip channels is an inherent property of the system related to the high sensitivity to the bathymetry prior to pattern development. Sensitivity to the initial cross-shore profile appears to be as important as sensitivity to wave height. Because of this sensitivity, simple predictors of rip channel spacing or e-folding time have only limited success.

## 1 INTRODUCTION

Sandbars are typical features of the nearshore (Komar 1989; short 1999). They are often uniform in the alongshore direction, but sometimes regular alongshore undulations in the height and cross-shore position of the sandbar crest are evident (van Enchkevort et al. 2004). This type of nearshore feature is usually defined as "rip channel system" or "crescentic sandbar" (Figure 1). Mean current flows are onshore over the horns of the crescentic sandbar and offshore in the bays where they can be concentrated as rip currents. The spacing between horns can be regular and is usually of the order of hundreds of meters (van Enchkevort et al. 2004). If several sandbars are present in the cross-shore direction crescentic shapes can still develop, and are usually characterized by different spacings for inner and outer sandbars. Predicting the development of rip channel systems on beaches has been one of the key topics in nearshore research for decades.

Field observations and computer simulations have focused on predicting the spacing and growth time of rip channels as a function of wave characteristics (Damgaard et al. 2002, van Enchkevort et al. 2004; MacMahan et al. 2006) but satisfactory predictors of rip channel spacing and growth time have not yet been proposed. With respect to the physical

Figure 1. Rip channels at Pauanui Beach (North Island, New Zealand). The image results from the average of 600 individual snapshots. The white (high-intensity) band reflects the location of predominant wave breaking and is usually associated to the presence of a submerged sandbar. The location of the rip channels correspond to the areas of reduced wave breaking. Arrows represent typical flow pattern associated to rips.

mechanism leading to the growth of the pattern, linear stability analysis (Falqués et al. 2000) has been initially used to show the positive feedback leading to the development of crescentic shapes: at locations where water depth is smaller than average as a result of a seabed perturbation, wave breaking is enhanced resulting in an increased water level and an onshore flow. This implies the development of circulation cells with an offshore flow between locations of enhanced wave breaking (smaller water depth). Assuming a suspended sediment concentration that increases from the shoreline to the breaking location: a) where water depth is smaller than average and the flow is onshore-directed, sediment fluxes decrease in the flow direction (shoreward) leading to sediment deposition and a further decrease in water depth; b) where water depth is larger than average and the flow is offshore-directed, sediment fluxes increase in the flow direction leading to sediment erosion and a further increase in water depth. Assuming that suspended sediment concentrations decrease outside the surf zone, and that flow directions are unaltered there, opposite erosion/deposition patterns develop. The combination of the erosion/deposition patterns inside and outside the surf zone leads to the growth of crescentic sandbars.

This same mechanism is essentially at the core of all the linear stability analysis models presented in recent years. Fully nonlinear models have also been developed (Damgaard et al. 2002; Caballeria et al. 2002; Reniers et al. 2004) and simulate the development of crescentic sandbars through the same self-organization mechanism initially explored through linear stability analysis. Since the earliest numerical studies (Hino 1975), the spacing between rip channels has been suggested to be proportional to the width of the surfzone. This result has been essentially confirmed by other numerical studies (Deiggard et al. 1999; Falqués et al. 2000; Damgaard et al. 2002; Calvete et al. 2005) but has never been successfully tested through field observations (Huntley and short 1992; Van Enchkevort et al. 2004; Holman et al. 2006; Turner et al. 2007) that have in some cases also indicated lack of alongshore regularity. A recent study (Reniers et al. 2004) has provided insight on one possible mechanism, directional spreading of short waves, that can explain lack of regularity of the emerging rip channel pattern.

Here, we expand on recently published work (Calvete et al. 2007) and show that the irregularity and the lack of predictability of rip channel spacing can be related to the high sensitivity to the shape of the bathymetry prior to pattern development.

## 2  NUMERICAL MODEL

We assume an alongshore-uniform coast ($x_1$ the cross-shore, and $x_2$ the alongshore direction) and consider the depth- and time-averaged equations of continuity (1a) and momentum (1b), along with equations describing wave energy (1c) transformation and phase (1d):

$$\frac{\partial D}{\partial t} + \frac{\partial}{\partial x_j}(Du_j) = 0 \tag{1a}$$

$$\frac{\partial u_i}{\partial t} + u_j\frac{\partial u_i}{\partial x_j} = -g\frac{\partial z_s}{\partial x_i} - \frac{1}{\rho D}\frac{\partial}{\partial x_j}(S'_{ij} - S''_{ij}) - \frac{\tau_{bi}}{\rho D} \tag{1b}$$

$$\frac{\partial E}{\partial t} + \frac{\partial}{\partial x_j}((u_j + c_{gj})E) + S'_{ij}\frac{\partial u_j}{\partial x_i} = -\mathcal{D} \tag{1c}$$

$$\frac{\partial \Phi}{\partial t} + \varpi + u_j\frac{\partial \Phi}{\partial x_j} = 0 \tag{1d}$$

The above system of equations describes the hydrodynamics and is coupled to a sediment conservation equation(2):

$$\frac{\partial z_b}{\partial t} + \frac{1}{1-p}\frac{\partial q_j}{\partial x_j} = 0 \tag{2}$$

where $z_s(x_1,x_2,t)$ is the mean sea level, $z_b(x_1,x_2,t)$ is the mean bed level, $D(x_1,x_2,t)$ is the total mean depth $(D = z_s - z_b)$, $\vec{u}(x_1,x_2,t)$ is depth averaged current $(\vec{u} = (u_1,u_2))$ and $E(x_1,x_2,t)$ is the wave energy density.

We define wave phase $\Phi(x_1,x_2,t)$ through the wavenumber vector $\vec{K} = (K_1,K_2)$:

$$\vec{\nabla}\Phi = \vec{K}, \qquad \Phi_t = -\omega$$

where $\omega$ is the absolute frequency and

$$\varpi = \sqrt{gK\tanh KD}.$$

where $\varpi$ is the intrinsic frequency. We denote $|\vec{K}|$ as $K$, and the corresponding magnitude of the group velocity vector as $c_g$, where $c_g = (c/2)(1 + 2KD/\sinh(2KD))$ and $c_{gi} = (K_i/K)c_g$. Similarly the phase velocity $c$ is given by linear theory (Mei 1989).

The wave energy dissipation model (Thornton and Guza 1983) takes the form:

$$\mathcal{D} = \frac{3\sqrt{\pi}}{16}B^3 f_p\rho g\frac{H_{rms}^5}{\gamma_b^2 D^3}\left(1 - \frac{1}{(1 + (H_{rms}/\gamma_b D)^2)^{5/2}}\right),$$

where $f_p(=\varpi/2\pi)$ is the intrinsic peak frequency of the wave field, $B(=1.0)$ is a breaking related coefficient, $\gamma_b(=0.42)$ is the breaker index, $H_{rms}(E = \frac{1}{8}\rho gH_{rms}^2)$ is root mean square average of the wave height, $\rho$ is the water density and $g$ represents gravity. The waves drive the circulation through the radiation stress terms $S'_{ij}$ described as:

$$S'_{ij} = E\left(\frac{c_g}{c}\frac{K_iK_j}{K^2} + \left(\frac{c_g}{c} - \frac{1}{2}\right)\delta_{ij}\right)$$

746

where $\delta_{ij}$ is the Kronecker delta symbol. The Reynolds stresses ($S''$) are parameterized using a horizontal viscosity coefficient (Battjes 1975):

$$S''_{ij} = \rho \nu_t D \left( \frac{\partial u_i}{\partial x_j} + \frac{\partial u_j}{\partial x_i} \right)$$

where $\nu_t = M(\mathcal{D}/\rho)^{\frac{1}{3}} H_{rms}$ ($M = 1$). Bed shear stresses $\vec{\tau}_b$ are parameterized assuming linear friction:

$$\vec{\tau}_b = \rho \mu \vec{u} , \qquad \mu = \left( \frac{2}{\pi} \right) c_D u_{rms}$$

where $c_D = \left( \frac{0.40}{\ln(D/z_0) - 1} \right)^2$, $z_0 (= 0.01 \, m)$ is the roughness length and $u_{rms}$, the root-mean-square wave orbital velocity at the boundary layer edge $z_0$, is computed through linear wave theory, ie.

$$u_{rms} = \frac{H_{rms}}{2} \frac{gK}{\varpi} \frac{\cosh K z_0}{\cosh K D}.$$

We use a widely adopted formula (soulsby 1997) to describe sediment fluxes $\vec{q}$:

$$\vec{q}_{svr} = \alpha \left( \vec{u} - \gamma u_{rms} \vec{\nabla} h \right) \tag{3}$$

where the stirring $\alpha$ is described as:

$$\alpha = A_s \left[ \left( |\vec{u}|^2 + \frac{0.018}{c_D} u_{rms}^2 \right)^{1/2} - u_{crit} \right]^{2.4} ,$$

$$\text{if} \left( |\vec{u}|^2 + \frac{0.018}{c_D} u_{rms}^2 \right)^{1/2} > u_{crit} \tag{4}$$

$$= 0 , \quad \text{otherwise}$$

There are various parameters in this model. $A_s = A_{ss} + A_{sb}$ where $A_{ss}$ represents the suspended load and $A_{sb}$ the bedload; and $u_{crit}$ the threshold flow intensity for sediment transport, which depends on sediment properties and depth (soulsby 1997). Finally, $p (= 0.4)$ is the seabed porosity and $\gamma (=1.6)$ is a morphodynamic diffusion coefficient. The original equation (soulsby 1997) has been adapted for a 2-dimensional flow and to model a gravitational downslope transport proportional to the bottom wave orbital velocity.

The approach we adopt in this study is linear stability analysis: we first linearize with respect to a basic state and express the six dependent variables, $[\eta, h, u, v, e, \phi]^T$, at their basic state (equilibrium) plus a perturbation about that equilibrium. The perturbed quantities are assumed to be small (infinitesimal) with respect to the basic state, so that the resulting system

of equations can be linearized with respect to this state (Falqués et al. 2000). We also adopt the quasi-steady hypothesis, which implies that hydrodynamical instabilities are excluded by disregarding time derivatives of hydrodynamical quantities. Essentially, the flow field responds instantaneously to the bed evolution at the morphodynamic timescale. Since the coefficients of the governing equations do not depend on $y$ we can assume alongshore periodicity into our perturbations, so that

$$\Psi(x, y, t) = \psi(x) e^{\sigma t + iky} \tag{5}$$

where $\Psi$ and $\psi$ can represent any of our perturbation variables. Substituting (5) in (1a)–(2) and imposing the boundary conditions results in a system of linear equations that defines an eigenvalue problem. It is solved numerically by a spectral method (Falqués and Iranzo 1992), which allows computational nodes to be distributed efficiently where variations are most rapid (typically near the shoreline). Experiments used 300 collocation points in the cross-shore direction, with half of the points located in the first 150 m, to achieve numerical convergence (Falqués and Iranzo 1992;Falqués et al. 1996). The fortran code to solve for the basic state and its linear stability is referred to as MORFO60.

For each alongshore wavenumber $k$ as many eigenvalues $\sigma$ and associated eigenfunctions $\psi(x)$ as degrees of freedom of the discretization are obtained. Many of them are purely numerical, i.e., not describing any solution of the continuous system of stability equations (Falqués and Iranzo 1992;Falqués et al. 1996). Those which are physically sound and have $\text{Re}(\sigma) > 0$ represent a possible instability mode associated to the spatial pattern defined by $k$ and $\psi(x)$. For each mode, the e-folding time is $\tau = 1/\text{Re}(\sigma)$ and gives an indication of the characteristic formation time in nature. The alongshore migration speed is computed as $V_{mi} = -\text{Im}(\sigma)/k$. The growth rate curves show $\text{Re}(\sigma)$ as a function of $k$ and allow identification, for each mode, of the wavelength with the largest growth rate: the fastest growing mode (FGM) which is interpreted as the predominant alongshore lengthscale for that particular mode.

Growth rate curves have been evaluated for a number of cross-shore profiles (Figure 2) primarily differing in the slope offshore of the sandbar, $\beta_{off}$, the depth over the sandbar crest, $D_c$, and trough, $D_t$, and the crest position, $X_c$. Profiles are generated fixing the position and elevation of a number of points to describe the slope close to the shoreline and offshore of the sandbar as well as the depth over the sandbar trough and crest. These points are then interpolated using a cubic spline. The different profiles are generated changing the position and/or elevation of the points prior to the cubic spline interpolation.

747

## 3 RESULTS

Examples of some cross-shore profiles and the associated "basic state" variations in free surface elevations, wave energy and root mean square wave height are shown in Figure 3 and 4. To understand and maximize the effect of the initial bathymetry on rip channel development, we have kept offshore wave height, $H_{rms}$, constant and equal to 1.5 m.

For the same reason, waves have been assumed to approach the shore normally and with an incident wave period of 12 s. The fastest growing unstable mode obtained using MORFO60 displays a configuration (Calvete et al. 2005; van Leeuwen et al. 2006; Calvete et al. 2007) that closely resembles, in terms of both morphology (alongshore sequence of highs and lows)

and hydrodynamics (onshore flow over the highs in the surfzone and offshore flow concentrated in the channels) the features observed in the surfzone. Previous numerical modelling studies have attempted to relate the spacing of rip channels in sandbar systems to parameters related to the geometry of the system like the area of sandbar through $\sqrt{A_t}$ (Deigaard et al. 1999) or the distance between the sandbar crest and the shoreline $X_c$ (Damgaard et al. 2002). Our results (Figure 5) indicate that both parameters explain only part of the variability in rip channel spacing and have little explanatory power with respect to the e-folding time of the pattern. Overall, the distance between the sandbar crest and the shoreline appears to be a better predictor than the square root mean of the trough area. It is also interesting to notice that rip channel spacing and e-folding time appear to be entirely uncoupled (Figure 6).

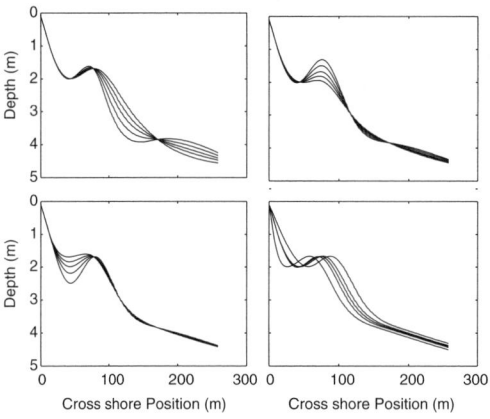

Figure 2. Geometry of initial cross-shore profiles used in the numerical simulations. Variations in the offshore beach slope $\beta_{off}$ (top-left panel), in the depth over the sandbar crest $D_c$ (top-right panel) and trough $D_t$ (bottom-left panel), and the distance between the sandbar crest and shoreline, $X_c$.

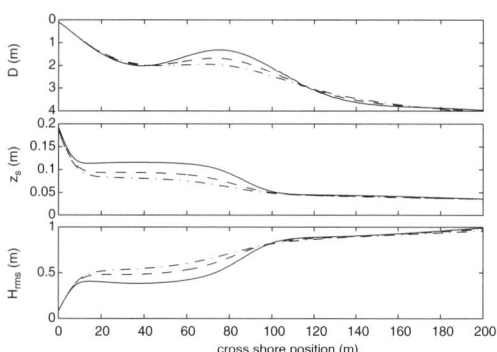

Figure 4. Basic state for three cross-shore profiles primarily differing for the depth over the sandbar: water depth (top), free surface elevation (middle), and wave height (bottom).

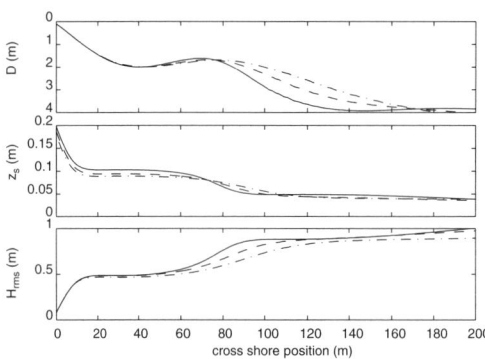

Figure 3. Basic state for three cross-shore profiles primarily differing for the offshore slope: water depth (top), free surface elevation (middle), and wave height (bottom).

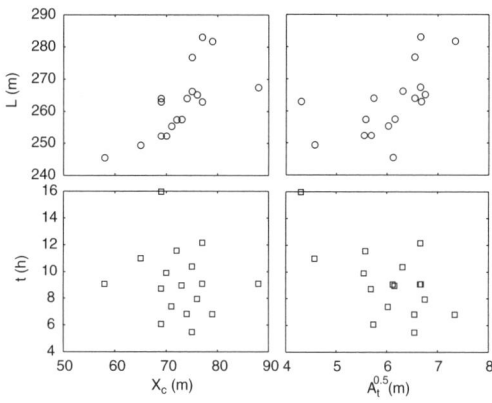

Figure 5. Alongshore spacing and e-folding time as a function of the area of the sandbar trough (top panels) and the distance between the sandbar crest and the shoreline (bottom panels).

More complicated parameters that attempt to account for the overall cross-shore profile geometry can explain a large part of the variability in e-folding time and spacing (Figure 7). In particular, with respect to the prediction of the e-folding time, it appears that predictors that account for the detailed shape of the cross-shore profile better explain e-folding variability.

Given that the predictors purely based on geometric considerations do not fully explain spacing and e-folding time variability, we have analyzed the system of equations focusing on its hydrodynamic components. Considering that the spacing of rip channels is somehow proportional to the velocity field, we have

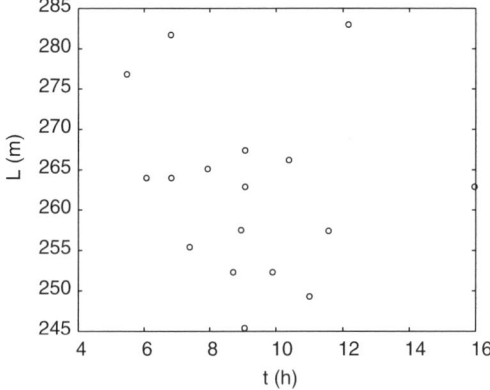

Figure 6.   Alongshore spacing versus e-folding time.

tried to understand if there is any correlation between the maximum velocity in the channel and the results obtained from the linear stability analysis. To this end, rather than correlating maximum velocity with alongshore spacing associated to the fastest growing mode, we have superimposed on each of the profiles an alongshore structure consisting of highs and lows resembling a rip channel pattern. This type of configuration has been superimposed with a different alongshore spacing and in each case we have evaluated the maximum velocity in the channel. The alongshore spacing corresponding to the maximum velocity in the channel, $\lambda_{VelMax}$, has been then compared to the spacing resulting from the linear stability analysis problem (the one characterized by the maximum growth rate), $\lambda_{LSA}$ (Figure 8). The limited agreement between the two ways of calculating the optimal spacing is likely to be related to the sediment transport parameterizations, which affect the instability and so the growth and geometry of the rip channels. If the term describing morphodynamic diffusion in (equation 3) is removed (essentially $\gamma$ is set to zero), the agreement between the spacing resulting from linear stability analysis, $\lambda_{LSA*}$, and the one related to the maximum velocity are better correlated (Figure 8). The same type of behavior is observed when comparing growth rates (not shown).

Recently published work (Calvete et al. 2007) has clearly indicated that the profile shape can have an effect on the spacing and growth of rip channels that is comparable to changes in wave height. Figure 9 shows how the same alongshore rip  channel spacing can

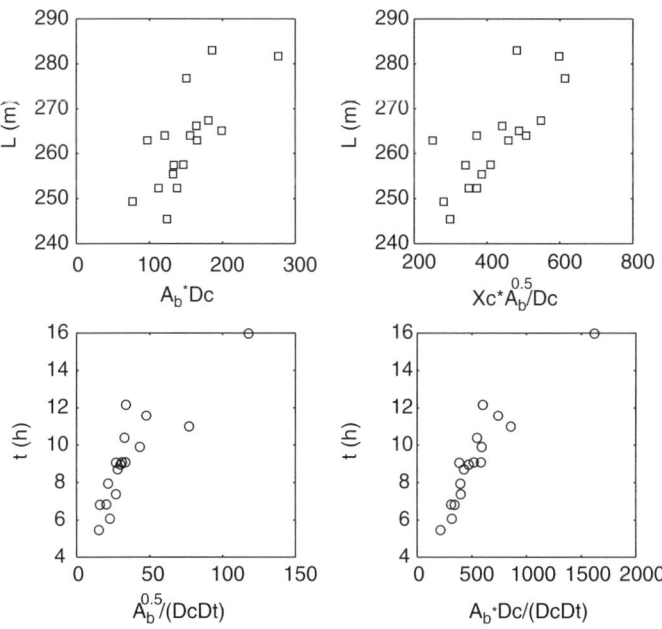

Figure 7.   Alongshore spacing (top panels) and e-folding time (bottom panels) as a function of geometric parameters.

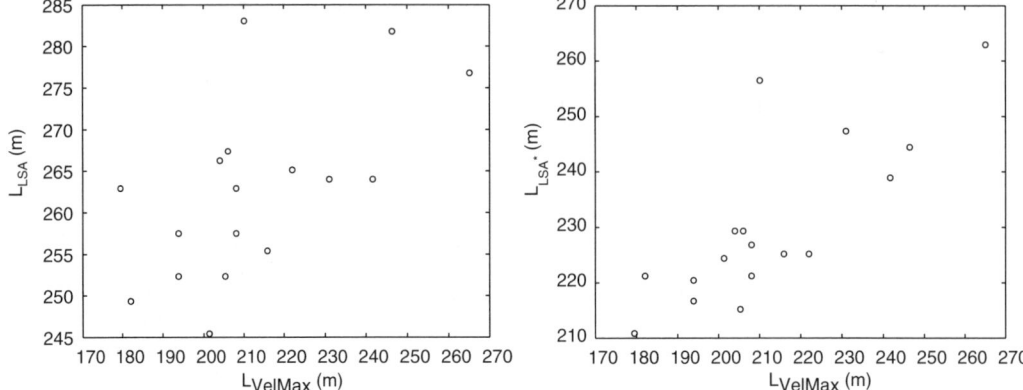

Figure 8. Comparison between the alongshore spacing corresponding to the maximum velocity in the channel, $\lambda_{VelMax}$, and the alongshore spacing associated to the largest growth rate with morphodynamic diffusion, $\lambda_{LSA}$ (left), and with no morphodynamic diffusion, $\lambda_{LSA*}$ (right).

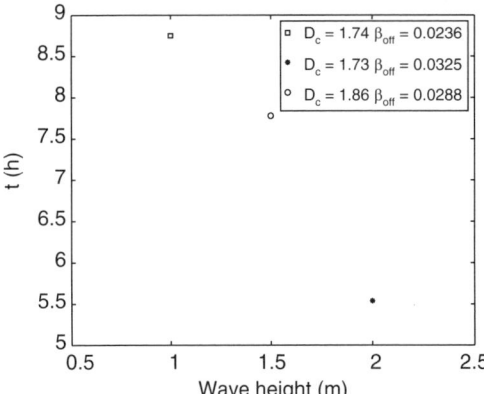

Figure 9. E-folding time and wave height combinations resulting in the same alongshore rip channel spacing. The legend indicates the differences in the geometric parameters between the three cross-shore profile considered. The alongshore spacing of the fastest growing rip channel configuration is the same for the three cases (251 m).

result from different combinations of wave height and cross-shore profile geometry (the geometric parameters primarily changing on each profile are indicated in the legend of the figure). At the same time, Figure 9 reiterates that substantially different (nearly a factor of two) e-folding times can be associated to the same alongshore rip channel spacing.

## 4 CONCLUSIONS

A numerical study based on linear stability analysis has been used to address the development of rip channels in the surfzone. Previous studies based on field observations tried to correlate rip channel spacing to wave characteristics (Huntley and short 1992) or indicators of surfzone width like the distance between the sandbar crest (as detected from video images) and the shoreline (Holman et al. 2006). The limited success of these studies is not surprising in light of our results indicating that the observed variability might not be related to uncertainties in the hydrodynamic forcing but in the shape of the cross-shore profile prior to rip channel development (Calvete et al. 2007). Our simulations confirm the link between wave height and rip channel spacing which in some situations might become the dominant source of spacing variability but also point at the so far neglected role of the pre-existing bathymetry.

## ACKNOWLEDGEMENTS

DC and AF acknowledge funding from the Ministerio de Ciencia y Tecnologia of Spain (PUDEM project, Contract REN2003-06637-C02–01/MAR) and the "Ramón y Cajal" contract of DC. GC acknowledges funding from the (New Zealand) Foundation for Research, Science and Technology (Contract C01X0401). ND gratefully acknowledges the financial support of EPSRC through the grant GR/S19172/0.

## REFERENCES

Battjes, J. A. (1975). A note on modeling of turbulence in the surf zone. In *Proc. Symp. Modeling Tech., San Francisco,* pp. 1050–1061. ASCE.

Caballeria, M., G. Coco, A. Falqués, and D. H. Huntley (2002). Self-organization mechanisms for the formation of nearshore crescentic and transverse sand bars. *J. Fluid Mech. 465*, 379–410.

Calvete, D., G. Coco, A. Falqués, and N. Dodd (2007). (Un)predictability in rip channel systems. *Geophys. Res. Lett. 34*(L05605), doi:10.1029/2006GL028162.

Calvete, D., N. Dodd, A. Falqués, and S. M. van Leeuwen (2005). Morphological development of rip channel systems: normal and near-normal wave incidence. *J. Geophys. Res. 110*(C10006), doi:10.1029/2004JC002803.

Damgaard, J., N. D. and L. Hall, and T. Chesher (2002). Morphodynamic modeling of rip channel growth. *Coast. Eng. 43*, 199–221.

Deigaard, R., N. Droonen, J. Fredsoe, J. H. Jensen, and M. Jorgesen (1999). A morphological stability analysis for a long straight barred coast. *Coastal Eng. 36*, 171–195.

Falqués, A., G. Coco, and D. A. Huntley (2000). A mechanism for the generation of wave-driven rhythmic patterns in the surf zone. *J. Geophys. Res. 105*(C10), 24071–24088.

Falqués, A. and V. Iranzo (1992). Edge waves on a longshore shear flow. *Phys. Fluids A 4*(10), 2169–2190.

Falqués, A., A. Montoto, and V. Iranzo (1996). Bed-flow instability of the longshore current. *Cont. Shelf Res. 16*(15), 1927–1964.

Hino, M. (1975). Theory formation of rip-current and cuspidal forms. In *Proc. 14th Int. Conf. Coastal Eng.*, pp. 901–919. ASCE, Reston.

Holman, R. A., G. Symonds, E. B. Thornton, and R. Ranasinghe (2006). Rip spacing and persistence on an embayed beach. *J. Geophys. Res. C1006.*, doi:10.1029/2005JC002965.

Huntley, D. A. and A. D. Short (1992). On the spacing between observed rip currents. *Coastal Engineering 17*, 211–225.

Komar, P. D. (1989). *Beach processes and sedimentaion*. Prentice Hall, New Jersey.

MacMahan, J. H., E. B. Thornton, and A. J. H. M. Reniers (2006). Rip current review. *Coastal Engineering 53*, 191–208.

Mei, C. C. (1989). *The applied dynamics of ocean surface waves*. World Scientific, Singapore.

Reniers, A. J. H. M., J. A. Roelvink, and E. Thornton (2004). Morphodynamic modeling of an embayed beach under wave group forcing. *J. Geophys. Res. 109*(C01030), doi:10.129/2002JC001586.

Short, A. D. (1999). *Handbook of beach and shoreface morphodynamics*. A. D. Short Ed., Wiley, Chichester.

Soulsby, R. L. (1997). *Dynamics of marine sands*. Thomas Telford, London, UK.

Thornton, E. B. and R. T. Guza (1983). Transformation of wave height distribution. *J. Geophys. Res. 88*, 5925–5938.

Turner, I. L., D. Whyte, B. G. Ruessink, and R. Ranasinghe (2007). Observations of rip spacing, persistence and mobility at a long, straight beach. *Marine Geology 236*, 209–221.

van Enckevort, I. M. J., B. G. Ruessink, G. Coco, K. Suzuki, I. L. Turner, N. G. Plant, and R. A. Holman (2004). Observations of nearshore crescentic sand-bars. *J. Geophys, Res. 109*(C06028), doi:10.1029/2003JC002214.

van Leeuwen, S. M., N. Dodd, D. Calvete, and A. Falqués (2006). Physics of nearshore bed pattern formation under regular or random waves. *J. Geophys. Res. 111*(F01023), doi:10.1029/2005JF000360.

751

*Estuarine morphodynamics*

*River, Coastal and Estuarine Morphodynamics: RCEM 2007 – Dohmen-Janssen & Hulscher (eds)*
*© 2008 Taylor & Francis Group, London, ISBN 978-0-415-45363-9*

# Simulation of morphology in the tidal environments

S. Masuya
*Docon Co., Ltd., River Eng. Dept., Japan*

Y. Shimizu & S. Giri
*Graduate School of Engineering. Hokkaido Univ., Japan*

ABSTRACT: This study aims to develop a physically-based morphodynamic model in order to reproduce incision of tidal creek networks as well as evolution of geomorphology in the tidal environments. The process of incision and long-term geomorphologic evolution on tidal flats and marsh is of practical significance from the environmental view point. Generally, it is postulated that tidal creeks are developed as a result of ecomorphologic processes of tidal system driven by tidal energy. Some previous investigations have provided significant insight to this problem. However, there is still lack of a physically-based modeling approach that can replicate the initiation and evolution process of this phenomenon. In this study, a two-dimensional morphodynamic model was developed and applied to compute morphologic change in the tidal environment. The proposed model was applied to a narrow area referred to as tidal flats in order to provide quantitative and qualitative evaluation of its reproducibility. The computation was carried out with no-flux boundary conditions in upstream end and sides of the calculation domain, whereas downstream end adjacent to the sea was opened with a flux condition of sinusoidal tidal cycle. Calculation was conducted for different time scale up to 350 days. Calculation results revealed that the incision of tidal network and its initial development process can satisfactorily be reproduced by a physically-based computational model.

## 1 INTRODUCTION

Notsuke Marsh in Hokkaido, Japan, is one of many existing tidal marsh throughout the world (Fig. 1). It creates feature of beautiful landscape with tidal creeks and vegetation. Landscape feature in the tidal environments such as tidal channel network or tidal creeks and their geomorphologic evolution is of significance form environmental point of view. On the other hand, tidal creeks control sediment transport and exchange processes within tidal flats. Tidal channels are closely associated with coastal hydraulics, hydrology (Rinaldo et al., 1999), geomorphology (Allen, 2000) and ecology. Since evolution of tidal creeks is a longstanding process, its incision process is almost unknown.

A number of authors have described the hydrodynamics of tidal channels and creeks, the consequences of tidal currents and asymmetries on sediment dynamics, and other morphological characteristics of tidal channels (e.g., Boon 1975; Speer & Aubrey 1985; Friedrichs & Aubrey 1988; Friedrichs 1995; Lanzoni & Seminara 2002). Morphometric analyses of tidal networks have been carried out by several researchers (e.g., Myrick & Leopold 1963; Steel & Pye 1997; Rinaldo et al. 1999a, b; Marani et al. 2002, 2003). Also,

Figure 1.    Notsuke Marsh in Hokkaido, Japan.

sedimentation and accretion patterns in salt marshes have been studied (e.g., Leonard & Luther 1995; Christiansen et al. 2000), together with ecological dynamics and patterns in salt marshes (e.g., Silvestri & Marani 2004). In addition, some simplified models have been proposed to simulate the morphological behavior of tidal basins (e.g., van Dongeren and de Vriend, 1994; Schuttelaars and de Swart, 1996).

755

Despite several studies that have noted the complexity of drainage channel systems in general (e.g. Knighton et al., 1992), very few provide quantitative evaluation of incision process and subsequent morphologic development of tidal creek networks. The factors controlling the origin and evolution of the complex patterns and dissection of tidal creeks on tidal flats and marshes are still more or less unidentified (Perillo and Iribarne, 2003). Consequently, understanding incision and evolution processes of tidal creek geomorphology is of importance. Recently, D'Alpaos et al. (2005) proposed a tidal ecomorphodynamic model to address the problem of initiation of creeks on tidal flats. The hydrodynamic component of their model is comprised of a simplified Poisson-like model- a simplified form of shallow water equations. Details of morphodynamic model can be found in their paper. This model seems to be a robust tool to simulate evolution of tidal channel networks on tidal flats; however proposed model is not rigorously physical.

Due to the reducing computational time and cost, application of physically-based computational modeling approach has become reliable. Recently, several morphodynamic models are being applied in river engineering issues. For an instance, a numerical model to calculate two-dimensional bed deformations was proposed by Shimizu & Itakura (1991).

In this study, a morphodynamic model has been proposed, the flow component of which is depth-averaged two-dimensional. The hydrodynamic model has been coupled with sediment continuity equation to calculate temporal change of bed morphology considering both bed load and suspended load transport. Within the scope of this study, some numerical experiments have been carried out in a square domain, considered to be tidal flats. The geomorphologic evolution of tidal flats in the tidal environments was calculated for moderate duration up to 350 days.

## 2  THE NUMERICAL MODEL

### 2.1  Hydrodynamic model

This study uses two-dimensional depth-averaged flow model to calculate flow-field. Continuity and momentum equations can be expressed as:

$$\frac{\partial h}{\partial t} + \frac{\partial (uh)}{\partial x} + \frac{\partial (vh)}{\partial y} = 0 \quad (1)$$

$$\frac{\partial (uh)}{\partial t} + \frac{\partial (u^2 h)}{\partial x} + \frac{\partial (uvh)}{\partial y} = -gh\frac{\partial H}{\partial x}$$
$$-\frac{\tau_x}{\rho} + \frac{\partial}{\partial x}\left[v_t\frac{\partial (uh)}{\partial x}\right] + \frac{\partial}{\partial y}\left[v_t\frac{\partial (uh)}{\partial y}\right] \quad (2)$$

$$\frac{\partial (vh)}{\partial t} + \frac{\partial (uv\,h)}{\partial x} + \frac{\partial (v^2 h)}{\partial y} = -gh\frac{\partial H}{\partial x}$$
$$-\frac{\tau_y}{\rho} + \frac{\partial}{\partial x}\left[v_t\frac{\partial (vh)}{\partial x}\right] + \frac{\partial}{\partial y}\left[v_t\frac{\partial (vh)}{\partial y}\right] \quad (3)$$

Where: $x$ and $y$ = the horizontal axis on the orthogonal coordinate system; $t$ = time; $u$ and $v$ = flow velocity in the $x$ and $y$ direction, respectively; $\rho$ = the density of water; $h$ = water depth; $H$ = water surface elevation; $g$ = the acceleration due to gravity; $v_t$ = depth-averaged eddy viscosity; $\tau_x$ and $\tau_y$ = bed shear stress in the $x$ and $y$ direction, respectively. The bed shear stress can be expressed as

$$\tau_x = \frac{\rho g n^2 u\sqrt{u^2 + v^2}}{h^{\frac{1}{3}}} \quad (4)$$

$$\tau_y = \frac{\rho g n^2 v\sqrt{u^2 + v^2}}{h^{\frac{1}{3}}} \quad (5)$$

Where: $n$ = coefficient of Manning's roughness.

A zero-equation turbulence closure has been used in this model.

### 2.2  Sediment transport model

Bed exchange on the tidal flats is governed by the following continuity equation of sediment transport considering bed load and suspended load.

$$\frac{\partial z}{\partial t} + \frac{1}{1-\lambda}\left(\frac{\partial q_{bx}}{\partial x} + \frac{\partial q_{by}}{\partial y} + q_{su} - w_f c_b\right) = 0 \quad (6)$$

where, $z$ = bed elevation; $\lambda$ = void ratio; $q_{bx}$ and $q_{by}$ = bed load in the $x$ and $y$ direction, respectively; $q_{su}$ = pickup rate of suspended sediment; $w_f$ = fall velocity; and $c_b$ = reference concentration.

Fall velocity can be obtained using well-known Rubey's formula. Bed load transport in the $x$ and $y$ direction are calculated by dividing it into both flow direction ($x$ and $y$ direction). Bed load transport rate ($q_b$) can be obtained from Ashida & Michiue's formula (Ashida & Michiue, 1972). Likewise, pickup rate of suspended sediment transport can be obtained using Itakura and Kishi's formula (Itakura and Kishi 1980). Reference concentration can be expressed as:

$$c_b = \frac{<c>\beta}{[1-\exp(-\beta)]} \quad (7)$$

Where: $<c>$ = depth-averaged suspended sediment concentration; $\beta = w_f h/\varepsilon$; $\varepsilon = \kappa u_* h/6$; $\kappa$ = Karman's constant.

Depth-averaged suspended sediment concentration is calculated by the following continuity equation of suspended sediment concentration:

$$\frac{\partial(<c>h)}{\partial t} + \frac{\partial(u<c>h)}{\partial x} \\ + \frac{\partial(v<c>h)}{\partial y} = q_{su} - w_f c_b$$

(8)

## 3 RESULTS OF NUMERICAL EXPERIMENT

Since evolution of tidal creeks is a long-term process, it is necessary to run the computation for sufficiently long duration so as to reproduce numerically the incision and evolution of tidal creeks in early stage. Consequently, computational time and stability are the important factors. In this study, the calculations have been performed for different time duration and only a couple of scenario have been selected for the simulation with a moderate duration (up to 350 days) so as to reproduce incision and early development of tidal channel networks.

### 3.1 Calculation conditions

Initially, some short-term computations have been conducted for a square area domain of 200 m × 200 m with the equal grid spacing in both direction ($\Delta x$ and $\Delta y$), namely 2.0 m to 10.0 m, to evaluate effect of grid spacing. Bed slope was set to be 1/1000; grain size of bed materials is set to be 0.01 mm. Coefficient of Manning's roughness can be obtained from the formula of Manning-Strickler. A random field of initial perturbation was added to flat-bed. Boundary at the upstream end and both sides have been set to be no-flux boundaries (for flow, bedload and suspended sediment concentration). The boundary at the downstream end of domain adjacent to sea level has been treated as an open boundary with a sinusoidal tidal cycle (see Figure 2). The boundary condition for bed load transport at the downstream end has been set to be free flux. The boundary condition of suspended sediment concentration at the downstream end is given 0.00005 m³. Proposed model is able to deal with repetitive wetting and drying processes of calculation domain.

### 3.2 Consideration of slope effect

Some trial calculations revealed that the computation result was unstable and unrealistic (as shown in Fig. 3), when used sediment continuity (Eg.6) without considering local slope effect. Therefore, an additional term has been added to this continuity equation in order to consider the local slope effect (see below Eq.9). Right hand side of Equation 9 denotes the sediment

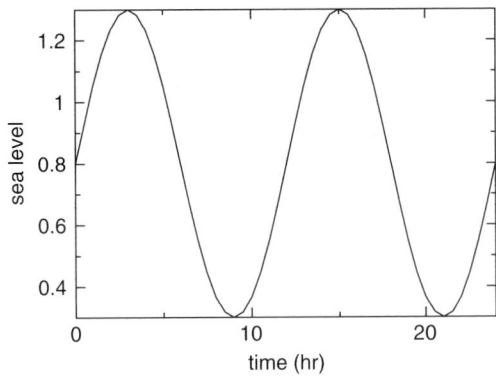

Figure 2. Boundary condition at downstream end (sea level).

flux in both directions caused by the slope effect and calculated using a correction coefficient.

$$\frac{\partial z}{\partial t} + \frac{1}{1-\lambda}\left(\frac{\partial q_{bx}}{\partial x} + \frac{\partial q_{by}}{\partial y} + q_{su} - w_f c_b\right) \\ = \frac{q_B}{1-\lambda}\sqrt{\frac{\tau_{*c}}{\mu_s \mu_k \tau_*}}\left(\frac{\partial^2 z}{\partial x^2} + \frac{\partial^2 z}{\partial y^2}\right)$$

(9)

where, $\tau_* =$ non-dimensional shear stress; $\tau_{*c} =$ non-dimensional critical shear stress; $\frac{\tau_{*c}}{\mu_s \mu_k \tau_*} = \gamma =$ correction coefficient; $\mu_s$ and $\mu_k =$ static and dynamic coefficient of Coulomb frictions, respectively.

It is evident that results, calculated by using Equation 9 (Figure 4), are more stable and reliable than those of using Equation (6). Therefore, Equation (9) has been applied to calculate bed deformation in this study.

### 3.3 Effect of grid spacing

Computational time is affected by the spatial grid size ($\Delta x$ and $\Delta y$), and the time step ($\Delta t$) as well. On the other hand, grid size has significant effect on calculation results. Therefore, some test runs have been conducted in order to examine the effect of grid size on calculation result before conducting long-term computation.

Figure 5 shows a comparison between calculation results obtained by using different spatial grid sizes, namely 2.0 m, 2.5 m, 5.0 m and 10.0 m in both directions. The model performance with grid sizes 2.0 m, 2.5 m and 5.0 m shows its ability to reproduce the channel network like tidal creeks. However, smaller the grid spacing, better the result, i.e. more detailed channel networks appear to be replicated by the model. Whereas, computation with 10.0 m grid size is unable to reproduce the channels. Furthermore, by analyzing

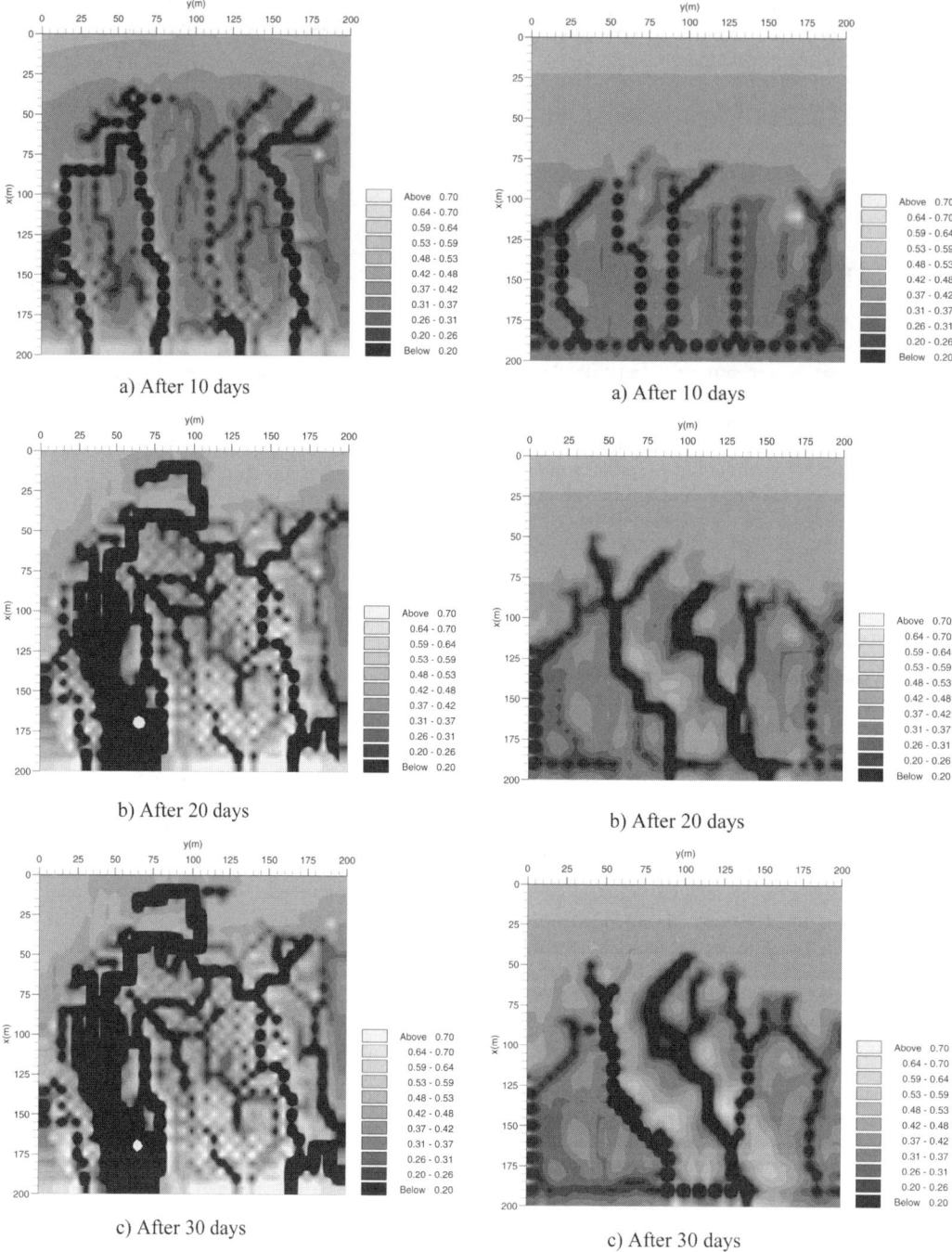

a) After 10 days

b) After 20 days

c) After 30 days

Figure 3.    Results calculated using Equation (6).

a) After 10 days

b) After 20 days

c) After 30 days

Figure 4.    Results calculated using Equation (9).

calculation result, it was found that the width of the reproduced channels was varied in size (within the range of 5.0 m to 10.0 m) depending on the grid spacing. The result of this comparison indicates that

the model with smaller grid size can reproduce channel networks with better accuracy and detail. However, with the small grid the computational time increases significantly. Based on the preliminary analysis, it

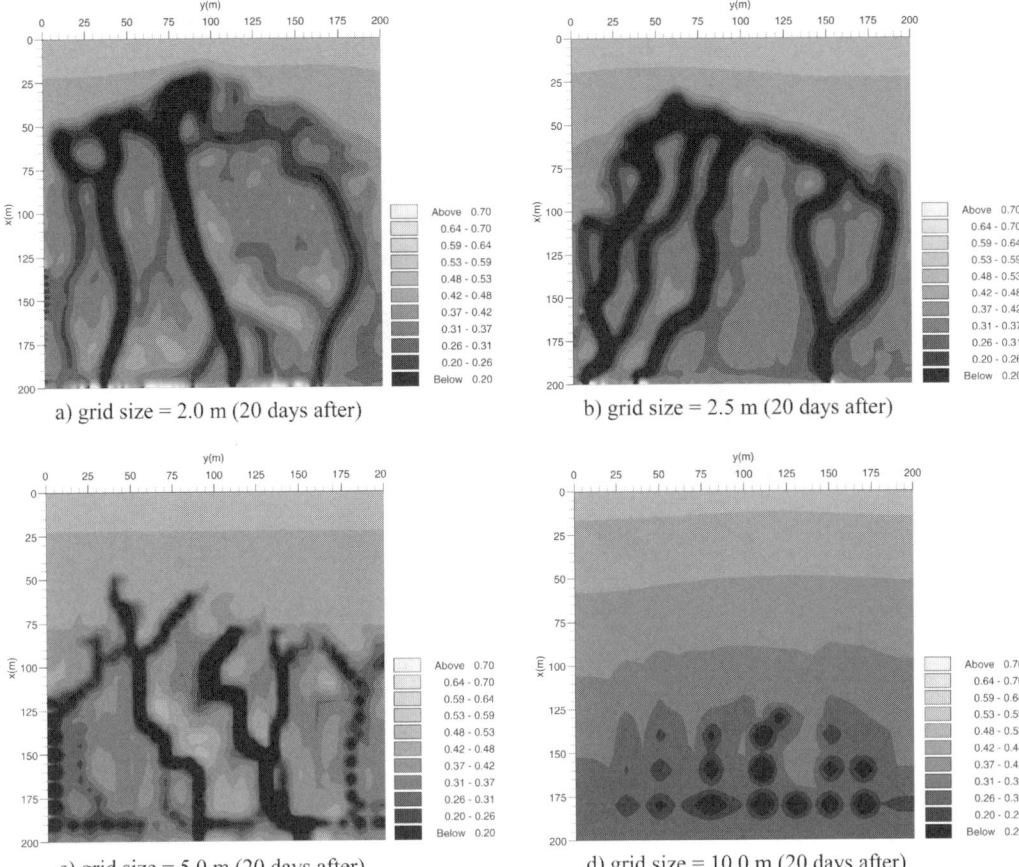

a) grid size = 2.0 m (20 days after)

b) grid size = 2.5 m (20 days after)

c) grid size = 5.0 m (20 days after)

d) grid size = 10.0 m (20 days after)

Figure 5.   Comparison of Calculation results.

was decided to use grid size equal to 5.0 m in both directions.

### 3.4  Long-term calculation of tidal morphology

For the long-term calculation, all the conditions are same as in the prior calculations (Section 3.2 & 3.3) except for the calculation area, which has been set to be 200 m × 400 m. Calculation results after 10 days, 200 days and 350 days has been depicted in figures 6, 7 and 8 respectively. The process of channel initiation and evolution in numerical computation can be described as follows: (i) At first, a large number of channels emerge due to the bed erosion as the boundary shear stress exceeds critical shear stress; (ii) Channels with low flows gradually become dead as a result of deposition around the initial channels due to the supply of sediment from downstream; (iii) Natural banks develop around the channels as a result of sediment accumulation, transported by the tidal cycle from the downstream end; (iv) Channels with natural

banks start to meander and form the feature of channel networks looks like tidal creeks. All these processes of initiation and early development of creeks are simulated in physically-based manner without any special treatments and manipulation.

## 4   CONCLUSIONS

The main objective of this study is to reproduce channel networks in tidal flats (tidal creeks) by using a two-dimensional morphodynamic model. It was revealed that a physically-based hydrodynamic approach, coupled with sediment continuity, can be used to describe the initiation and early development of geomorphologic processes in the tidal environment. The following conclusions can be drawn from this study: (i) Tidal creeks are appeared to be generated due to the tidal cycle and repetitive drying-wetting process accompanied by sediment erosion/deposition on tidal flats; (iii) Consideration of the slope effects on sediment

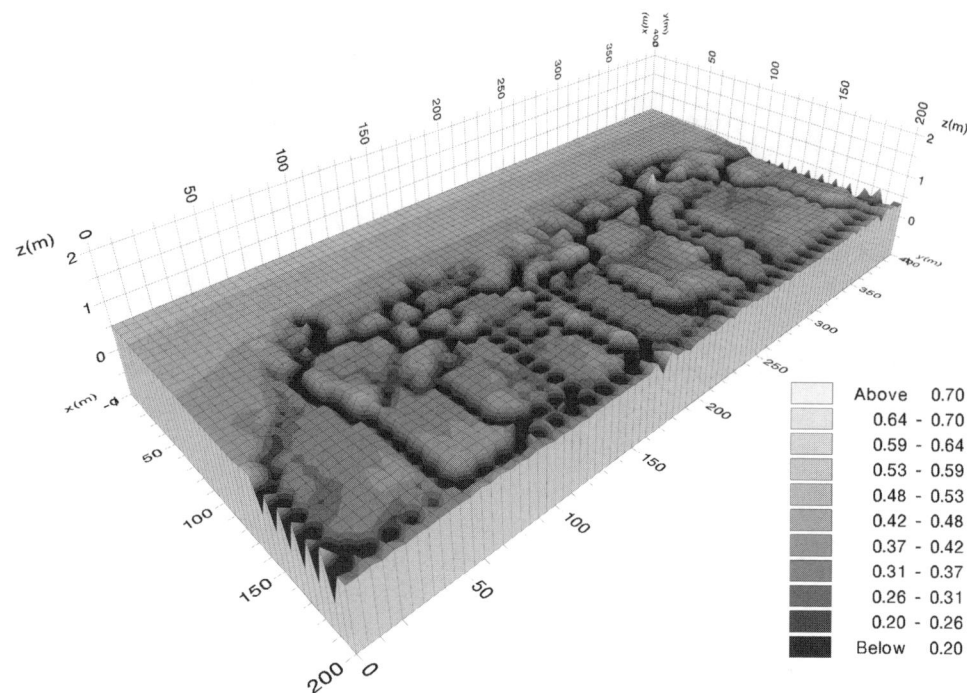

Figure 6.   Result of long term calculation (after 10 days).

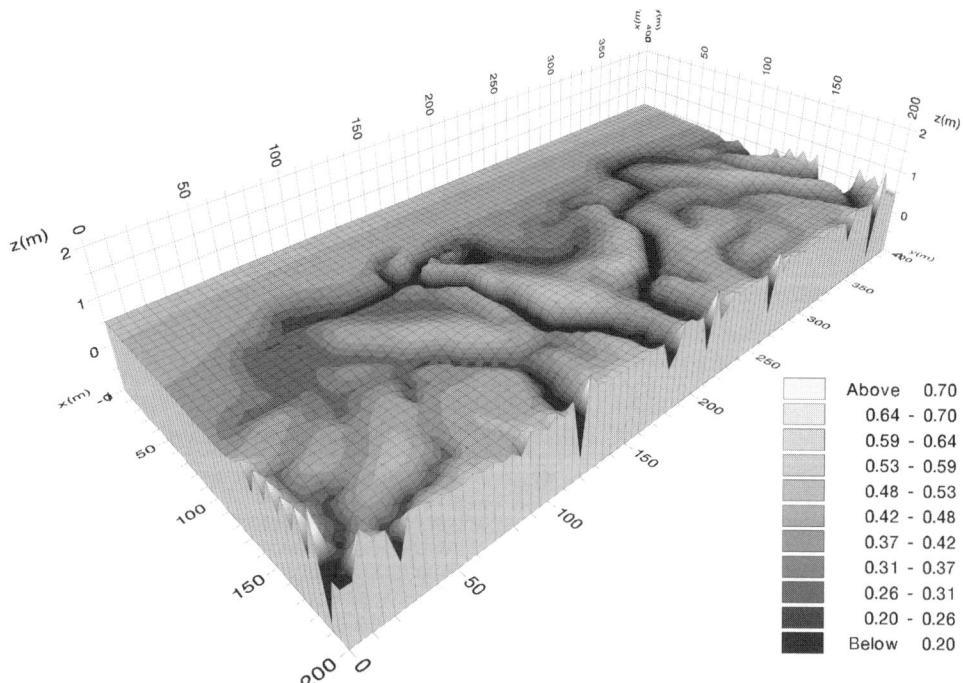

Figure 7.   Result of long term calculation (after 200 days).

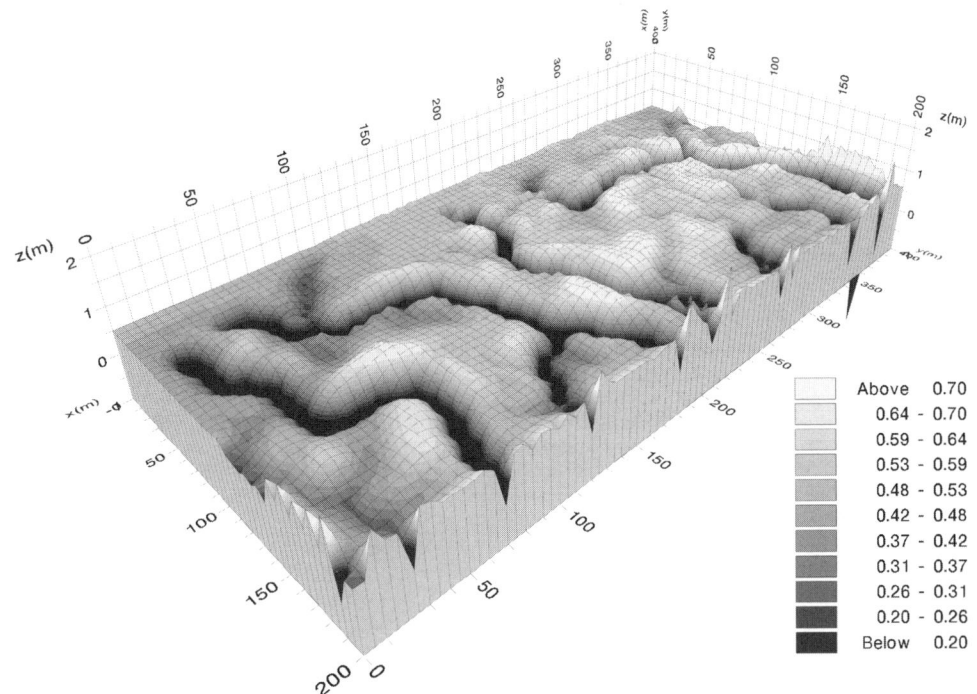

| | |
|---|---|
| Above | 0.70 |
| 0.64 - | 0.70 |
| 0.59 - | 0.64 |
| 0.53 - | 0.59 |
| 0.48 - | 0.53 |
| 0.42 - | 0.48 |
| 0.37 - | 0.42 |
| 0.31 - | 0.37 |
| 0.26 - | 0.31 |
| 0.20 - | 0.26 |
| Below | 0.20 |

Figure 8.   Result of long term calculation (after 350 days).

transport in both direction of tidal flats significantly improves computation result and stability; (iv) Natural banks are appeared to be developed around the channels as a result of sediment transport processes caused by the tidal cycle.

The model is capable to replicate incision and early development of tidal creeks in a physically-bases manner. The model performance seems to be rather promising. However, this study is still in its initial phase and further validation and detailed study, considering natural scale and much longer morphological processes, is supposed to be carried out in future. Furthermore, for such simulation, it is necessary to take into account the ecological dynamics in tidal flats and marsh. For an instance, sediment erosion and deposition may significantly be altered by the presence of the vegetation. This work is a first step towards the development of a fully physical modeling tool to be used for the mathematical description of the geomorphological evolution in the tidal environments.

REFERENCES

Allen, J.R.L. 2000. Morphodynamics of Holocene salt marshes: a review sketch from the Atlantic and Southern North Sea coasts of Europe, *Quaternary Science Reviews* 19: 1155–1231.

Ashida, K. & Michiue, M. 1972. Hydraulic resistance and bed transport rate in alluvial streams, *Proceedings of JSCE*, 201: 59–69 (in Japanese).

Boon, J.D. III. 1975. Tidal discharge asymmetry in a salt marsh drainage system. *Limnology & Oceanography* 20: 71–80.

Christiansen, T., P.L. Wiberg, & T.G. Milligan 2000. Flow and sediment transport on a tidal salt marsh surface. *Estuarine Coastal and Shelf Science* 50(3): 315–331.

D'Alpaos, A., S. Lanzoni, M. Marani, S. Fagherazzi, and A. Rinaldo 2005. Tidal network ontogeny: Channel initiation and early development, *J. Geophys. Res.*, 110, F02001, doi:10.1029/2004JF000182.

Friedrichs, C.T. 1995. Stability shear stress and cross-sectional geometry of sheltered tidal channels. *Journal of Coastal Research* 11(4): 1062–1074.

Friedrichs, C.T. & D.G. Aubrey 1998. Non-linear tidal distortion in shallow weel-mixed estuaries: a synthesis. *Estuarine Coastal Shelf Science* 27(5): 521–545.

Itakura, T. & Kishi, T. 1980. Open channel flow with suspended sediment, *Proceedings of ASCE*, No.106(HY8): 1325–1343.

Knighton, A.D., Woodroffe, C.D. and Mills, K. 1992. The evolution of tidal creek networks, Mary River, Northern Australia. *Earth Surface Processes and Landforms* 17: 167–190.

Lanzoni, S. & G. Seminara 2002. Long-term evolution and morphodynamic equilibrium of tidal channels, *Journal of Geophysical Research* 107(C1): 3001, doi:10.1029/2000JC000468.

Leonard, L.A. & M.E. Luther 1995. Flow hydrodynamics in tidal marsh canopies. *Limnology & Oceanography* 40(8): 1474–1484.

Myrick, R.M. & L.B. Leopold 1963. Hydraulics geometry of a small tidal estuary. *US Geol. Surv. Prof. Pap* 422-B: 18pp.

Marani, M., S. Lanzoni, D. Zandolin, G. Seminara, & A. Rinaldo 2002. Tidal meanders. *Water Resources Research* 38(11): 1225, doi:10.1029/2001WR000404.

Marani, M., E. Belluco, A. D'Alpaos, A. Defina, S. Lanzoni, & A. Rinaldo 2003. On the drainage density of tidal networks. *Water Resources Research* 39(2): 105–113.

Perillo, G.M.E. & Iribarne, O.O., 2003. Processes of tidal channel development in salt and freshwater marshes. *Earth Surface Processes and Landforms* 28: 1473–1482.

Rinaldo, A., S. Fagherazzi, S. Lanzoni, M. Marani, W.E. Dietrich 1999a. Tidal networks 2. Watershed delineation and comparative network morphology. *Water Resources Research* 35(12): 3905–3917.

Rinaldo, A., S. Fagherazzi, S. Lanzoni, M. Marani, W.E. Dietrich 1999b Tidal networks 3. Landscape-forming discharges and studies in empirical geomorphic relationships. *Water Resources Research* 35(12): 3919–3929.

Schuttelaars, H.M. & H.E. de Swart 1996. An idealized long-term morphodynamic model of a tidal embayment, *Eur. J. Mech. B Fluids*, 15: 55–80.

Shimizu, Y., & Itakura, T. 1991. Calculation of flow and bed deformation with a general non-orthogonal coordinate system. *Proceedings of XXIV IAHR Congress, Madrid, Spain*, C-2: 241–248.

Silvestri, S. & M. Marani 2004. Salt marsh vegetation and morphology: modeling and remote sensing observations. In Fagherazzi S., Marani M., & Blum L.K., (eds). *The Eco-geomorphology of Salt Marshes, Estuarine and Coastal Studies Series*, American Geophysical Union: 15–26.

Speer, P.E. & D.G. Aubrey 1985. A study on non-linear tidal propagation in shallow inlet/estuarine systems, part II, Theory. *Estuarine Coastal Shelf Science* 21: 206–240.

Steel, T.J. & K. Pye 1997. The development of salt marsh tidal creek networks: Evidence from the UK. *Proceedings of the 1997 Canadian Coastal Conference, Can. Coastal Sci. and Eng. Assoc., Guelph, Ont., May 21–24*.

van Dongeren, A.R. & H.J. de Vriend 1994. A model of morphological behaviour of tidal basins, *Coastal Eng.*, 22: 287–310.

*River, Coastal and Estuarine Morphodynamics: RCEM 2007 – Dohmen-Janssen & Hulscher (eds)*
*© 2008 Taylor & Francis Group, London, ISBN 978-0-415-45363-9*

# Meandering channel dynamics in highly cohesive sediment on an intertidal mud flat

Wiecher Bakx, Filip Schuurman, Maarten G. Kleinhans & Henk Markies
*Universiteit Utrecht, Fac. Geosciences, Dept. Physical Geography, Utrecht, The Netherlands*

ABSTRACT: On an intertidal mud flat in the Western Scheldt estuary (the Netherlands), small meandering channels (about 1 m wide) were studied with the aim to improve understanding of the effect of highly cohesive sediment on channel and meander geometry and dynamics. The morphology of several channels was mapped repeatedly using stereo photography. During a neap-spring tidal cycle the flow velocity and sediment concentration was measured in detail in one meandering channel. Grain size analysis showed that the mud consists of about equal portions of clay and silt plus fine sand. Laboratory flume experiments were done on a carefully installed bed of sediment from the Western Scheldt intertidal flat. Three processes dominate erosion and morphodynamics of the meandering channels: 1) splash-erosion by rain, 2) erosive steps under hydraulic jumps and 3) bank erosion in very sharp bends in addition to the usual common bank erosion by undercutting. The flow velocity during flooding is negligible compared to the ebb flow. Rain storms at low tide cause much larger morphological change and sediment concentrations than the diurnal tide-related change and concentrations, even though the tidal discharge volume is larger than the rain runoff. Also storms cause larger morphological change, probably because of sediment stirring by gravity waves. Laboratory experiments with rainfall simulations verified that splash erosion by rain-drops breaks down the cohesive structure of the consolidated clay, resulting in an easy erodible surface layer of fluid mud. We hypothesise that the smaller, upstream channels in the channel network on the intertidal flat are generated by high-concentration, channelising rainfall runoff. Vertical downward steps ranging 0.01–0.1 m in height were observed particularly in the sections with steeper slopes near the low-tide base level of the Western Scheldt main channel adjacent to the intertidal flat. At these times the flow discharge sources were rainfall runoff and base flow, causing critical flow conditions with discharges far below bankfull. The steps occur under highly erosive hydraulic jumps and migrate upstream, resulting in the excavation of the thalweg and undercutting of the banks as confirmed in preliminary laboratory flume experiments. Very sharp meander bends with $R/W < 2$ are common while meander cutoffs are rare. Dye tracer experiments in several meander bends at bankfull discharge revealed recirculating flows in two places: just downstream of the apex in the inner bend and just upstream of the apex in the outer bend. Thus, the main flow impinges on the bank of the outer bend, causing the bank erosion pattern that leads to the sharp meander in agreement with literature. We discuss a simple extension of a meander simulation model to include very sharp bends.

## 1 INTRODUCTION

### 1.1 *Scope and aim*

Intertidal mud flats commonly have small meandering streams (e.g Fagherazzi et al. 2004). Since meandering dynamics strongly depend on the strength of river banks, these meandering channels are interesting exemplars in the more cohesive part of the spectrum from noncohesive to very cohesive banks. The channels are so small that they could potentially be recreated in a laboratory flume. However, earlier attempts to do so proved that the meanders arrest after some time and become fixed in planform (e.g. Smithm, 1998). Dynamic meandering in cohesive

sediment has not been reproduced in the laboratory because the unscaled cohesion is too large compared to the downscaled laboratory channels. However, the similar-sized meandering channels on intertidal flats are dynamic (pers. comm. with locals at our field site, and Fagherazzi et al. (2004)).

The aim of this paper is to identify the main mechanisms for meander dynamics on a cohesive intertidal mud flat. We study natural meander geometry and dynamics during a neap-spring tide cycle, and study erosive processes in more detail on natural mud in a laboratory flume. We found three processes that dominate erosion and morphodynamics of the meandering channels: splash erosion, erosive backward migrating steps, and bank erosion in very sharp bends. These phenomena are reviewed first. Next, the field site, data collection and results are given, followed by the experimental methods and results. We then focus on modelling erosive steps with the model of Izumi and Parker (2000) and modelling sharp meander bend evolution with an adapted version of the model of Ikeda et al. (1981), followed by conclusions.

## 1.2 Review

Splash erosion is the process where individual raindrops break down the cohesive structure of the (consolidated) clay during low tide, resulting in an easy erodible surface layer of fluid mud. This potentially has a large influence on intertidal flat morphodynamics where the cohesion might otherwise be to high for morphological change to occur. This process is well known from studies on hill-slope erosion.

Erosive steps under hydraulic jumps migrate along the channels in an upstream direction. Izumi and Parker (2000); Parker and Izumi (2000) developed a downstream-driven model in which they assume a free overfall (hydraulic step) leading to a backwater curve as the water accelerated towards this overfall that results in the transition from subcritical flow to supercritical flow (Froude number towards unity) leading to intense erosion of the upstream side of the step. Based on the Froude number the model analyses the stability of the step in terms of migrational distance and velocity (in upstream direction). The stability of the channels on the intertidal flat and of the flume experiments was analysed using this model.

Bank erosion in very sharp bends shows a different pattern than commonly found in meander bends. In meander bends with $R/W > 2$, wherein $R =$ bend radius and $W =$ channel width, bank erosion and associated bend migration is coupled to the curvature of the bend. Meandering models (e.g. Ikeda et al., 1981) are only valid for those less sharp bends. In very sharp bends, horizontal recirculation cells form (Andrle, 1994; Hodskinson and Ferguson 1998: Ferguson and Parsons, 2003) resulting in flow separation and adjusted locations of bank erosion. These vortices form near the outer-bend bank just downstream of the bend apex, where the flow is perpendicular to the bank, and near the inner-bend bank just downstream of the bend. We will adapt the model of Ikeda et al. for very sharp meander bends of $R/W < 2$.

## 2 FIELD WORK

### 2.1 Site description and methods

The field site is located on the southern shore of the Western Scheldt estuary (the Netherlands), on an intertidal flat near the Paulinapolder (Temmerman et al., 2003). The estuary is embanked and the mud flat ranges from the dike to 220 m seawards.

The intertidal flat has a convex topography with a nearly horizontal section with a slope of 0.008 to about 150 m from the dike. Further towards the estuary the slope increases to a slope of 0.03 over another 70 m. Small meandering channels are nearly regularly spaced and have widths ranging between 0.2–1.3 m. All channels have at least some very sharp bends with $R/W < 2$.

We collected data for 6 weeks from neap tide to neap tide. Flow velocity, pressure and sediment concentration (at three heights above the bed) were measured at 2 Hz in one cross-section of a relative large meandering channel ($W \approx 1.2$ m and depth $h \approx 0.5$ m). To prevent disturbances of the mud at this channel, measurements were collected in several cross-sections along other channels of the flow velocity (using tracers), water depth, channel width, sinuosity and bend radius. For a detailed study of the morphology, three channels of different size (of which the channel with the high-frequency measurements was the largest) were mapped using stereo photography with a camera suspended from a fishing rod. Sediment samples were analyzed on grain size and showed equal portions of fine sand and mud (clay and silt).

### 2.2 Results

On the onshore half of the mud flat, visual observations strongly suggest that the surface runoff during rain storms results in larger morphological changes and higher sediment concentrations than due to the tide-related flows and concentrations. This is caused by rain-drop splash erosion which breaks down the cohesive structure of the sediment, which then forms a fluid mud layer of a few cm thickness. Rainfall runoff easily entrains the fluid mud so that small channels are formed. These small channels combine into fewer larger channels in the first onshore 100 m.

The flow velocity in the channels is at its maximum when the discharge is bankfull and the water level decreases (from flood to ebb in the estuary) (Fig. 1). After bankfull, the flow velocity rapidly decreases. The highest observed ebb flow velocity is −0.80 m/s while the maximum flood flow velocity is about 0.30 m/s. Contrary to the channels in significant reversing flow studied by Fagherazzi et al. (2004), the present study site is dominated by currents in one (ebb) direction only. The flow velocity is high only for about 30 minutes, namely when the mud flat is nearly empty (towards ebb) or only just submerged (towards flood). The sediment concentration is higher in ebb than in flood flow. This is caused by the upstream erosion of the tidal flat and the channel banks in the outflow, which results in larger concentrations than in the estuarine water.

764

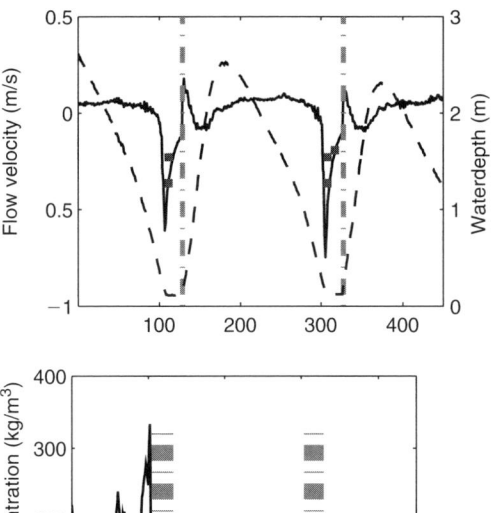

Figure 1. A. Flow velocity (full line), water depth (dashed line) and B. calibrated mud concentration (bottom) measured during two spring tidal cycles. Negative velocity is offshore. Squares indicate velocity measurements by tracers. Vertical (dash-dotted) lines indicate breaks in time when the flat was dry and data collection was stopped automatically.

A second important flow stage occurs for about an hour after the first 30 minutes of bankfull and waning flow. In this second stage the channels are fed by base flow from the mud flat, resulting in a waterdepth of 0.01–0.02 m with flow velocities (from tracers) of up to 0.2 m/s. Locally this resulted in supercritical flow where small-scale antidunes were observed. Most of the incision occurs in erosive, backward migrating steps under hydraulic jumps (Fig. 4), so that a secondary channel with vertical walls and a depth of a few cm is formed. This process of incision undercuts the primary channel banks, causing bank failure. The steps were only found in the downstream parts of the channels on the steeper part of the intertidal flat.

Meanders were found all over the intertidal area at a more or less regular spacing, but differed clearly in bend radii between the steep and gentle parts of the flat (Fig. 3). The sinuosity of the channels is 1.2 and 1.7 for the gentle and steep slope respectively. The meanders in the steep section of the tidal flat have relatively small width to depth ratios with the average $W/h = 3.6 \pm 1.6$, whereas the gently sloped meanders have $W/h = 38.1 \pm 20.6$. The most

well-developed and largest bends occur just downstream of the transition from gentle to steep slope, near the confluence Fig. 2 with vertical outer-bend banks and on average 45° inner-bend bank slope. The presence of the well developed meanders near the transition in bed slope is due to large amount of water that is stored on the tidal flat and that flows trough that section of the channel, caused by the convex topography of the tidal flat. Contrastingly, more downstream, the channel is much smaller, the meandering is less obviously and incision dominates compared to lateral erosion. However, during severe storms the entire system is probably reset (based on interviews with local fishermen).

In the sharp meander bends, horizontal circulation cells are present near the outer-bend bank just upstream of the apex and near the inner-bend bank just downstream of the apex (Fig. 5). The outer-bend vortex forces the thalweg towards the inner-bend bank upstream of the apex, where the flow velocity is the largest and bank erosion occurs. The deflected main flow can then no longer follow the curved channel and impinges on the outer-bend bank downstream of the apex, where upwelling and bank erosion also occurs. Hence, a positive feedback emerges once a sharper bend is formed: the sharper bend causes a vortex, which erodes the bend in two places so that the bend becomes sharper or maintains its shape. This mechanism potentially explains very sharp meanders commonly observed on cohesive tidal flats or in meandering rivers in cohesive sediments or when deeply incised.

## 3 EXPERIMENTS

### 3.1 Methods

Four processes were studied in the laboratory: splash erosion, incipient entrainment, backward step migration and meander bend erosion. Our laboratory flume of 0.4 m wide was carefully filled with sediment collected from the intertidal flat, which was consolidated for 1 week in stationary water.

Rainfall simulators were used to study the effect of splash erosion. These experiments were done with fresh and saline water to assess whether the splash erosion is of physical or chemical nature.

The incipient entrainment experiment was done in the flume before all other experiments so that the mud was as pristine as possible. The water depth was 0.27 m and the flow velocity was increased in steps from 0 to 0.8 m/s. The sediment concentration was measured at 3.0, 6.5 and 15.5 cm above the bed, using Optical Backscatter Sensors (OBS) at 2 Hz. Also, the water surface slope was measured with pitot tubes so that the shear stress could be calculated from the depth-slope product.

In the third set of experiments, straight initial channels of 0.06–0.08 m wide and 0.005–0.01 m deep

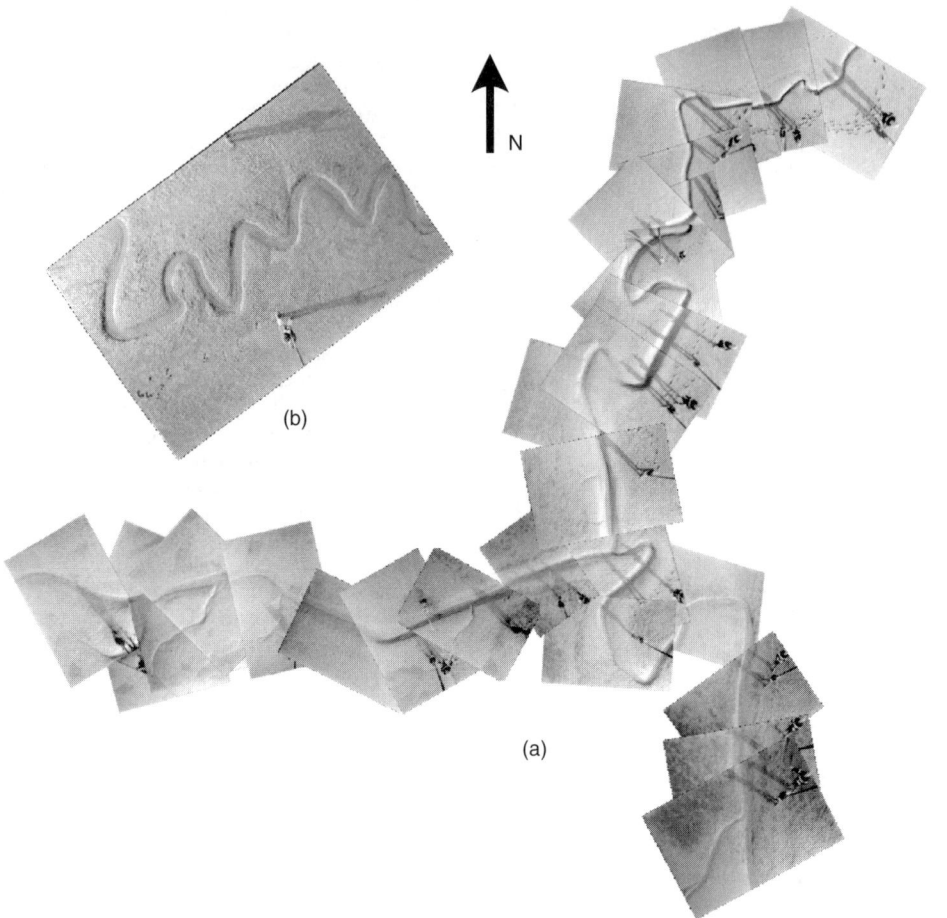

Figure 2. (A) Photo mosaic of the channel in which hydrodynamics were measured. The scale is about 1:480. Note the instrument bridge in the middle. The transition from gentle to steep slope occurs at the convergence of two main channels, where also the largest and sharpest meanders are observed. (B) Example of smaller channel on the steep slope (ebb current from left to upper right). The scale is about 1:150.

were carved in the sediment bed on an adjustable slope of 0.01–0.025 and a known discharge of 0.06–0.24 m$^3$/s. Morphodynamics were mapped with repeated digital stereo photography and time lapse photography. Flow velocity was measured using tracers and the depth was measured manually at several thalweg locations along the channel. The channels were monitored by time lapse photography with a 10 min interval from which the migration rate of the steps was estimated.

For the fourth set of experiments initial meander channels were carved with smooth banks and varying bend radii. The sharpest bends had $R/W = 0.5$ to study bend erosion in separated flow, while a straight section was carved to assess whether meander initiation occurred. Also, meanders of both upstream and downstream skewed bends were constructed. At the downstream boundary of the sediment bed incision was forced by a water level a few cm below the sediment bed. The initial incision formed a step that migrated upstream.

### 3.2 Results

Several experiments with drops at various rainfall intensities up to a constant jet of water were done. The impact of single drops was found to be important (Fig.6), whereas the rainfall intensity gave no different results. Saline and fresh water had equal impacts. The impact of single drops destroyed the cohesive clay structure and the surface layer turns into fluid mud of a few cm thickness in a matter of hours.

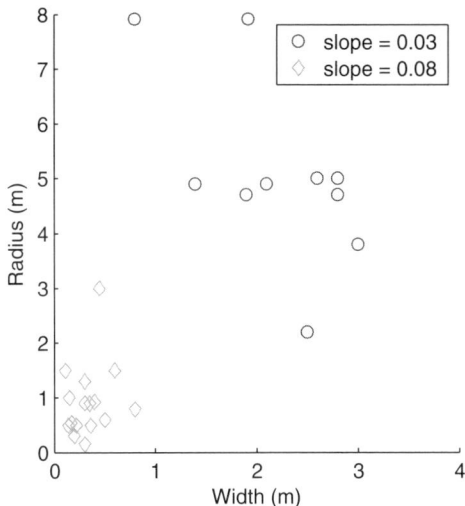

Figure 3. Measured channel width and bend radius on the steep and gentle slopes of the mud flat.

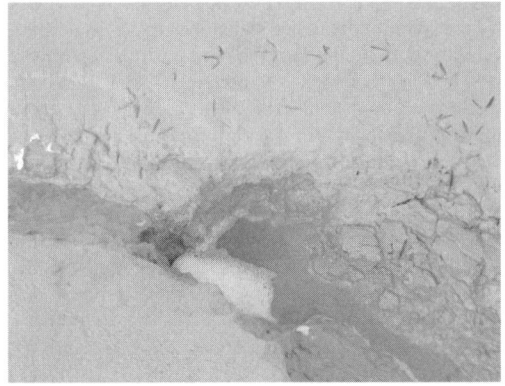

Figure 4. Erosive backward migrating step in a large meandering channel. The bird footprints are about 0.04 m wide. Note the remains of bank collapse by slumping.

The measured concentration increased rapidly after the shear stress was increased in each step, and then saturated around a maximum concentration for that step (Fig. 7). This behaviour is commonly observed for mud (e.g. Winterwerp and van Kesteren, 2004). The shear stress at which the sediment concentration starts to increase fast, is taken as the critical shear stress. The range for this value is rather large and arbitrary. A critical shear stress of 7 $N/m^2$ is assumed here in agreement with Winterwerp and van Kesteren, 2004).

In the experiments with initial channels steps initiated at the downstream end of the channel resulting in a hydraulic jump. The steps migrated in a constant speed in upstream direction following the straight or

Figure 5. Tracer in a sharp meander bend flowing to the left and top The vortex on the right bank downstream of the corner rotates clockwise; the vortex in the left outer-bank corner (note the bank erosion) rotates counterclockwise.

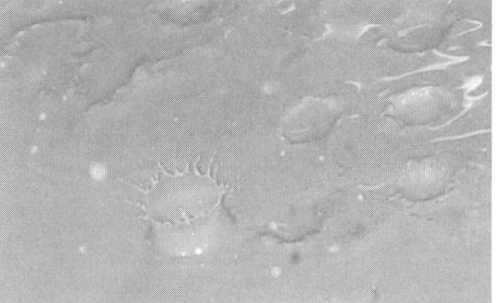

Figure 6. Rainfall experiment on a mud bed. A fluid mud layer is created by the impact of raindrops.

meandering pattern of the initial channel (Fig. 8). The erosion by the migrating step resulted in a deepening and widening of the meandering channel, and often removed the entire sediment layer from the flume. The velocity in which the step moved in an upstream direction depended mainly on the slope of the sediment bed and the flow velocity through the channel (presented in the modelling section).

The flow structure in the laboratory meandering channels is similar to that observed in the field. In the sharper bends both the outer- and inner-bend vortices formed, while only an inner-bend vortex was found in the less sharp bends. The outer-bend vortex occurred upstream of, and at the bend apex, and the inner-bend vortex occurred downstream of the apex. The latter extended up to 0.2 m downstream whereas the former are much shorter. Bank erosion was associated to the vortices on both sides of the channel. The thalweg upstream of the apex is forced towards the inner-bend by the outer-bend vortices so that the flow impinges on the outer bank. Bank erosion occurred only in very

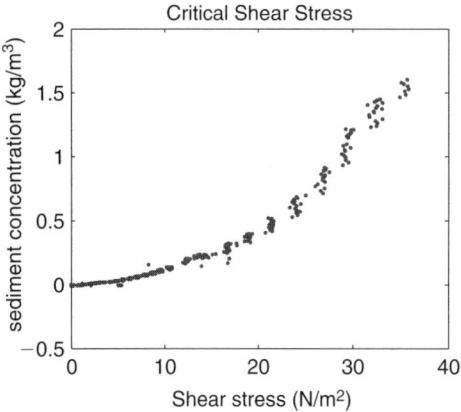

Figure 7. Sediment concentration versus applied shear stress.

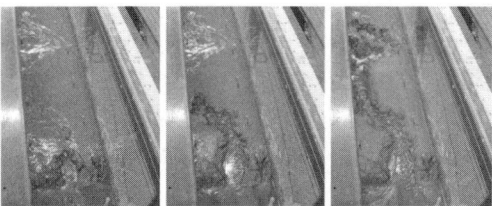

Figure 8. Time series of a migrating step in an initially meandering channel. The times are 10, 110 and 550 minutes from left to right during which the step migrated 1.5 m resulting in a step celerity of $4.5 \times 10^{-5}$ m/s.

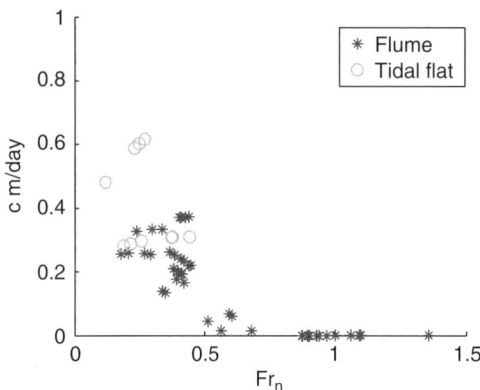

Figure 9. Step celerity plotted against the Froude number of normal flow far upstream of the step for the field and laboratory data.

few places and did not lead to meander migration. The meandering experiment shows that the erosive power of the flow is too small in comparison to the cohesion of the sediment to initiate bend migration, so that the meandering is arrested.

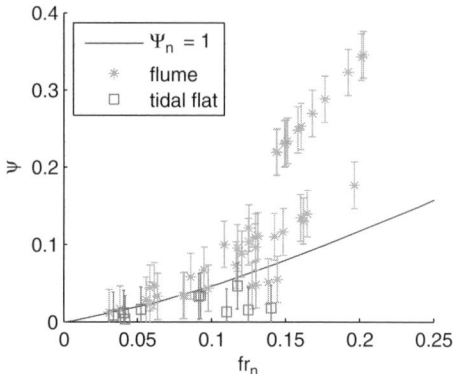

Figure 10. Mobility parameter against Froude number. Steps above the line migrate infinitely far upstream and arrest below the line.

## 4 MODELLING

### 4.1 Step migration

Izumi and Parker (2000) describe an analytical model for the stability of erosive steps. Starting from normal, subcritical flow, the flow becomes critical above the step leading to intense erosion at the step. Based on the backwater formulation, a shock at the step and a formulation for the erosion of cohesive sediment or bedrock, Izumi and Parker find solutions for the celerity of the step and a stability criterion determining whether the step migrates upstream infinitely or only for a limited distance. The Froude number $Fr_n$ for the normal flow far upstream of the step is a function of the nondimensional normal flow velocity $U_n$ far upstream of the step:

$$U_n = Fr_n^{2/3} \tag{1}$$

where $U_n = u_{x=\infty}/U_c$ with $u_{x=\infty}$ = velocity infinitely far upstream, $U_c = (qg)^{1/3}$ with $q$ = specific flow discharge and $g = 9.81$ m/s$^2$ is the gravitational acceleration. They define a mobility parameter $psi_n$ as the ratio of critical and actual bed shear stress:

$$\psi_n = \frac{\tau_c}{\tau_n} \tag{2}$$

The dimensional version is $\psi = U_n^2 \psi_n$. Herein

$$\tau_n = \rho C_f U_n^2 \tag{3}$$

in which $\rho$ = density of the flow and the friction coefficient $C_f$ is given as:

$$C_f = \frac{gh_n i}{U_n^2} \tag{4}$$

where $h_n$ = normal flow depth and $i$ = upstream slope.

From the stability analysis Izumi and Parker (2000) find that the step will migrate infinitely far upstream for $\psi_n > 1$ and present a stability diagram in which we plot our field and laboratory data (Fig. 10). The channels from the tidal flat are plotted below and to the left of the line which is in agreement with the observation in the field were erosional steps where only found in the downstream part of the channels. The channels from the flume experiments almost all plot above the line, meaning they will reach infinite far upstream which was also observed during the flume experiments.

The nondimensional step celerity in the upstream direction is calculated by (Izumi and Parker 2000):

$$c_* = \frac{(1 - \psi)^\gamma - \left(Fr_n^{4/3} - \psi\right)^\gamma}{1 - Fr_n^2} \quad (5)$$

for $Fr_n^{4/3} - \psi > 0$ and

$$c_* = \frac{(1 - \psi)^\gamma}{1 - Fr_n^2} \quad (6)$$

for $Fr_n^{4/3} - \psi \leq 0$, where $\gamma = 1.5$ is the power of the assumed bed erosion equation. The dimensional celerity is calculated from $c = c_* U_c$.

The calculated dimensional step celerity is $O(10^{-6})$ m/s, which is a factor of ten smaller than the measured celerity in the flume experiments (Fig. 9). Given the large uncertainty in the erodibility coefficient and measured roughness we consider this a reasonable agreement.

### 4.2  Sharp meander bends

The meander migration model of Ikeda et al. (1981), including the momentum distribution $A'$ introduced by Johannesson and Parker (1989) is based on the shallow flow equation:

$$U\frac{\partial u_b}{\partial s} + 2\frac{U}{h}K_f u_b = \quad (7)$$

$$W\left[-U^2\frac{\partial C}{\partial s} + K_f C\left[\frac{U^4}{gh^2} + \left(A + A' - 1\right)\frac{U^2}{h}\right]\right] \quad (8)$$

where $U$ = reach-averaged flow velocity, $u_b$ = flow velocity at the eroding bank, $K_f$ = roughness coefficient, $C$ = local curvature calculated as $C = \partial\phi/\partial s$ where $\phi$ = angle and $s$ = streamline coordinate and $A$ = parameter for transverse bed slope.

The model ad hoc is adjusted to generate sharp bends as follows. The relative high erosion rate at the outer banks downstream of the apex and the erosion

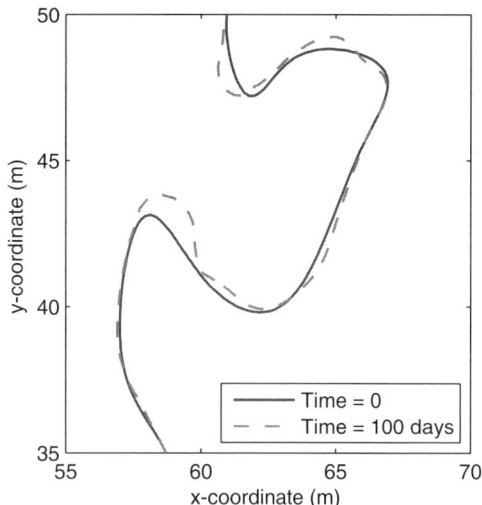

Figure 11.  Initial and final step of the meander model, which was run for a period of about a month (the same as between field surveys).

at the inner bank upstream of the apex is reached by introducing a delay of the curvature. For the calculation of the bank flow velocity at a node, the curvature is taken at 6 nodes upstream. The consequences of this are that the location of maximum erosion is further downstream and the erosion is more concentrated to the location of maximum erosion. The second adjustment is that the curvature is calculated for a larger $\partial s$ of $1.5W$ (6 nodes in this case), so that sharp bends are recognised as such independent of the $s$. This adjusted curvature is only used in the term $\partial C/\partial s$ in the equation above.

The model is tested on a tidal channel at the Paulinapolder, for which the coordinates of the nodes were extracted from a cubic spline function, fitted trough the planform (Fig. 2). The initial distance $ds$ between the nodes is 0.2 m, but varies trough time in the model. The values of the constants are: $b = 0.8$ m; $h = 0.7$ m; $U = 0.7$ m/s, $A = 6$, $A' = 16.7$ and $K_f = 0.0981$.

With the adjusted model the bends remain quite sharp (Fig. 11), in contrast to the original model (Crosato, 2007). Bank erosion and bend migration occurs only in the sharpest bends in agreement with the observations. Although the results of the adjusted model are more satisfying than the original model, the ad hoc nature of the model adaptation does not yet have a physical basis. We are working on improving this by schematising the vortex formation.

We hypothesise that the meandering channels are initially formed in the fluid mud layer after the entire mud flat has been reset (e.g. during a large storm). Deepening of the channels proceeds by backward migrating steps over hydraulic jumps during the final ebb phase.

## 5 CONCLUSIONS

In highly cohesive sediment meandering is less dynamic. Bend erosion occurs only in the sharpest bends where the flow is directed towards the bank. Meander simulation models are not yet able to cover this phenomenon satisfactoraly. Moreover, the channels are mostly erosive in nature, and the erosion takes place in backward cutting steps under hydraulic jumps. These could easily be reproduced in a laboratory flume, while the celerity was not very well predicted with an existing theory due to the uncertainty of the erodibility coefficient. Rain splash impact destroys the cohesive structure of the mud and forms a fluid mud layer of a few cm thick, in which incipient channels may form. In short, meandering channels on a tidal flat resemble larger meandering rivers with cohesive banks in their sharp meander bend dynamics, but not in the channel bed process which is alluvial in larger rivers and like bed rock erosion with hydraulic jumps over steps in the smaller tidal meanders.

## ACKNOWLEDGEMENTS

MGK is supported by the Netherlands Earth and Life sciences Foundation (ALW) with financial aid from the Netherlands Organisation for Scientific Research (NWO) (grant ALW-VENI-863.04.016).

## REFERENCES

Andrle, R. (1994). Flow structure and development of circular meander pools. *Geomorphology 9*, 261–270.

Crosato, A. (2007). Effects of smoothing and regridding in numerical meander migration models. *Water Resources Research 43*, W01401.

Fagherazzi, S., E. Gabet, and D. Furbish (2004). The effect of bidirectional flow on tidal channel plan-forms. *Earth Surface Processes and Land forms 29*, 295–309.

Ferguson, R. and D. Parsons (2003). Flow in meander bends with recirculation at the inner bank. *Water Resources Research 39*, 1322–1333.

Hodskinson, A. and R. Ferguson (1998). Numerical modelling of separated flow in river bends: Model testing and experimental investigation of geometric controls on the extent of flow separation at the concave bank. *Hydrological Processes 12*, 1323–1338.

Ikeda, S., G. Parker, and K. Sawai (1981). Bend theory of river meanders. Part 1. Linear development. *J. of Fluid Mechanics 112*, 363–377.

Izumi, N. and G. Parker (2000). Linear stability analysis of channel inception: downstream-driven theory. *J. of Fluid Mechanics 419*, 239–262.

Johannesson, H. and G. Parker (1989). *Linear Theory of river meanders, Volume 12 of Water Res. Monogr. Ser.*, pp. 181–213. Washington D.C.: American Geophysical Union.

Parker, G. and N. Izumi (2000). Purely erosional cyclic and solitary steps created by flow over a cohesive bed. *J. of Fluid Mechanics 419*, 203–238.

Smith, C. (1998). Modeling high sinuosity meanders in a small flume. *Geomorphology 25*, 19–30.

Temmerman, S., G. Govers, S. Wartel, and P. Meire (2003). Spatial and temporal factors controlling short-term sedimentation in a salt and freshwater tidal marsh, Scheldt estuary, Belgium, SW Netherlands. *Earth Surface Processes and Land forms 28*, 739–755.

Winterwerp, J. and W. van Kesteren (2004). *Introduction to the physics of cohesive sediment in the marine environment*. Number 56 in Developments in Sedimentology. Amsterdam, The Netherlands: Elsevier.

*River, Coastal and Estuarine Morphodynamics: RCEM 2007 – Dohmen-Janssen & Hulscher (eds)*
*© 2008 Taylor & Francis Group, London, ISBN 978-0-415-45363-9*

# Morphological characteristics of laboratory generated tidal networks

G. Tesser, A. D'Alpaos & S. Lanzoni
*Department IMAGE, University of Padova, Padova, Italy*

ABSTRACT: In this paper we present the first results of a series of laboratory experiments carried out in a large experimental apparatus, aimed at reproducing a typical lagoonal environment subject to tidal forcings. We observed the growth and development of a tidal network and analyzed its most relevant features, taking into account the role played by the characteristics of the tidal forcings in driving the development of channelized patterns. Such experiments were designed in order to improve our understanding of the main processes responsible for channel network ontogeny and evolution. Mathematical and theoretical analyses of network configurations were also carried out through the use of simplified and complete morphodynamic models. In particular, we analyzed the evolution in time of the morphometric characteristics of the developed networks, and studied the hydrodynamics and sediment transport processes related to different channel configurations. The evolution of channel cross-sectional areas, width-to-depth ratios, and unchanneled flow lengths shows that relevant features of actual tidal networks are reproduced by the tidal patterns developed within our laboratory experiments.

## 1 INTRODUCTION

Tidal channel networks exert a strong control on hydrodynamics, sediment and nutrient dynamics within tidal environments. Improving our understanding of their origins and evolution is of critical importance when addressing issues of conservation of tidal systems.

A wide literature exists, developed especially in the last three decades, describing the hydrodynamics of tidal systems and their morphodynamic evolution (see e.g. Allen 2000, Friedrichs & Perry 2001, for thorough reviews). In spite of their fundamental role in driving the morphological evolution of tidal basins, only in the last few years mathematical and numerical models analyzing the morphogenesis and long-term morphodynamic evolution of tidal channels have been proposed (Schuttelaars & de Swart 2000, Lanzoni & Seminara 2002, Fagherazzi & Furbish 2001, Fagherazzi & Sun 2004; D'Alpaos et al. 2005, 2007). Moreover, even though the development of tidal channel networks has been analyzed both through field observations and conceptual models (Pestrong 1965, French & Stoddart 1992) the description of the processes leading to the initiation and early development of tidal networks still lacks a proper delineation. In particular, attempts to investigate such processes on the basis of controlled laboratory experiments have not been pursued, except for those recently carried out by Tambroni et al. (2005), who investigated the morphodynamic evolution of the bottom of a single straight tidal channel closed at one end and connected at the other end

to a rectangular basin representing the sea. Towards the goal of gaining fundamental knowledge into the description of the main physical processes responsible for tidal network ontogeny, we carried out a series of laboratory experiments in a large experimental apparatus schematizing a typical lagoonal environment, subject to tidal forcings.

We furthermore used simplified and complete hydrodynamic models to carry out numerical and theoretical analyses of the experimental network configurations and to compare the relevant features of experimental and observed morphologies.

The paper is organized as follows. Section 2 describes the experimental apparatus. Section 3 reports on the experiments and related results. In Section 4 we use simplified and complete hydrodynamic models to compare relevant morphometric features of the synthetic and observed networks. Finally, Section 5 draws a set of conclusions and some remarks on forthcoming developments.

## 2 EXPERIMENTAL APPARATUS

The experimental apparatus, schematically depicted in Figure 1, consists of two adjoining basins reproducing schematically the sea and the lagoon. The lagoon basin is $5.3 \times 4.0$ m wide, while the much deeper adjacent sea basin is $1.6 \times 4.0$ m wide. The bed of the lagoon was uniformly covered with a 30 cm-thick layer of sediments during the experiments. The sea is separated

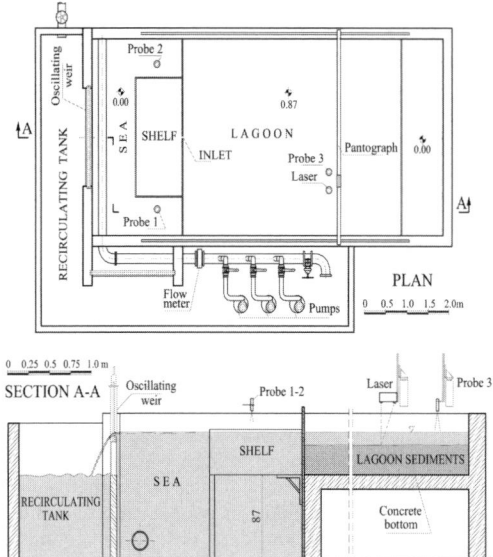

Figure 1.  Sketch of the experimental apparatus.

from the lagoon by a barrier of wooden panels; the lagoon inlet (whose shape and width may be varied) is located in the middle of this barrier while in front of the inlet a shelf enables to reproduce the gentle slope of the sea bed (Fig. 1).

The tide is generated at the sea by a vertical steel sharp-edge weir, oscillating vertically. The water continuously flowing over the weir is collected to a separated tank, where a set of pumps recirculates the flow.

An *ad hoc* software has been implemented to drive the weir, allowing us to reproduce a sinusoidal tide of fixed amplitude and period, oscillating around a prescribed average level. The software continuously corrects the motion of the weir on the basis of a feedback instantaneously controlled by water-level measurements at the sea, carried out through ultrasonic probes.

A computer-driven pantograph is used to survey bottom elevations within the lagoon. The apparatus consists of a laser system (300 μm resolution) which measures bottom elevation coupled to a ultrasonic probe which simultaneously gauges the associated water level. The latter measurement is used to determine the local flow depth and to correct laser measurements from refraction effects induced by the presence of water. The bathymetric survey of the lagoonal bottom, in fact, is carried out without stopping the experiment and drying the sediment surface, thus avoiding undesired perturbations of bed topography.

The sediments used in the experiments are cohesionless plastic grains, with density of 1041 kg/m$^3$ and median grain size, $d_{50}$, of 0.8 mm.

## 3  EXPERIMENTS

The experiments carried out so far mainly aimed at understanding under which conditions a channel network develops. To this end various tides (i.e., characterized by different amplitude, period and mean level) and different shapes and dimensions of the tidal inlet have been considered.

Each experiment started forcing an initially flat bed topography with a given tidal wave. A tidal network was observed to form only for small enough values of the tidal amplitude (1.0–2.0 cm) and of the flow depth (1.0–2.0 cm) and for a mean water level allowing the drying of the sediment surface during the ebb phase. A tidal period of 8–12 minutes was chosen in order to avoid perturbing vortexes and ensure the long-wave character of the tide.

In the presence of a too high tidal amplitude the sediments tended to be transported as suspended load, large dunes formed, and the growth of the channel network was inhibited. In any case, a wide scour, covered by dunes, formed in correspondence of the inlet. The channel network was eventually observed to originate only from the landward border of such a scoured region. It is worthwhile to observe that bed load sediment transport was active during the whole experiment, while ebb and flood peak velocities promoted suspended load along the channels. Finally, in all the experiments the lagoon experienced a progressive net erosion, with the consequent reduction of its mean bottom level.

In the following we focus our attention on two typical experiments, denoted as Run 1 and Run 2.

### 3.1  *Run 1*

Let us here briefly describe the evolution of the lagoon bed which, starting from the initial flat configuration, was observed in Run 1.

The forcing tide with amplitude of 2 cm and period of 8 minutes, was oscillating around a mean level equal to the initial bed elevation. A 0.50 m wide rectangular-shaped inlet was located at the center of the sea-lagoon boundary.

As pointed out before, a wide scour region rapidly formed in front of the inlet. After 220 tidal cycles a few isolated channel started to form at the landward edges of this scour region, beginning to cut the portions of the lagoon subject to draining. These channels progressively experienced a head cut-growth during the ebb phase and, after 500 cycles, some little ramifications begun to form. After 1200 cycles, four main, nearly straight channels were present (Fig. 2a). The ramifications of these channels in some cases were unstable: some smaller creeks were observed to abandon a main channel to join another one. In some cases these smaller creeks migrated laterally, forming little

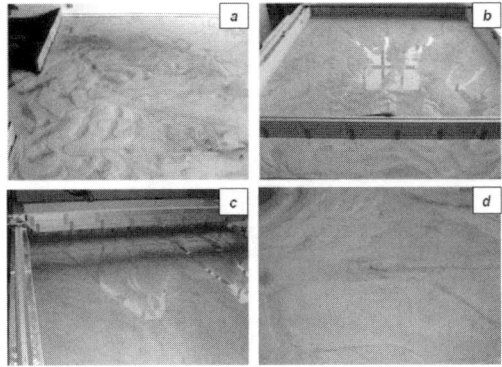

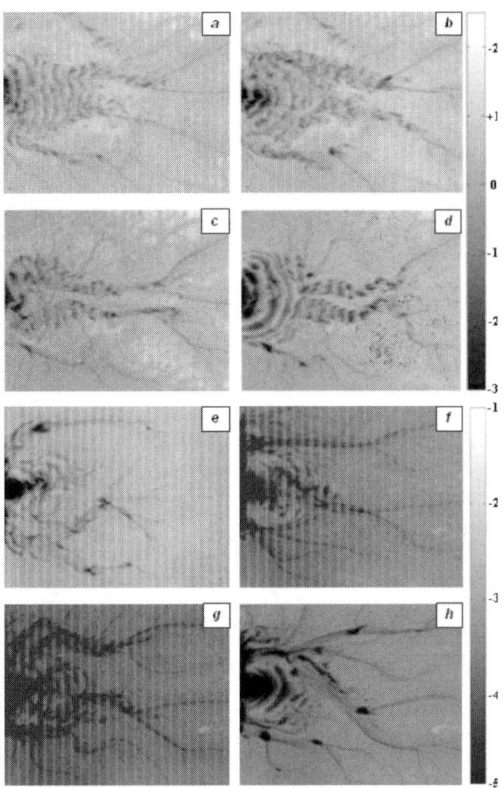

Figure 2. Lagoon topography observed in Run 1 after: a) 1200 cycles b) 6000 cycles and in Run 2 after: c) 6000 cycles; d) 7710 cycles.

bends. After 2300 cycles, the main channels lengthened significantly and at 2800 cycles three pronounced bends were observed to develop along one of these channels. However, the small scale bed forms (mainly dunes), initially covering the deep scour facing the inlet, progressively extended throughout the lagoon, tending to destroy the channel network. In order to smooth out these undesired bed forms and enhance the growth of a well defined channel network, we slightly decreased the mean sea level (about 0.5 cm). Figure 2b shows the bed configuration obtained after 6000 cycles, when the experiment was stopped.

## 3.2  Run 2

This experiment was characterized by a forcing tide with amplitude of 2 cm, period of 10 minutes, oscillating around a mean level 1.5 cm higher than the bed elevation. The width of the trapezoidal-shaped inlet varied from 0.05 m at the lagoon concrete bottom to 0.20 m at the sediment surface.

At the beginning of the experiment the mean sea level submerged the sediments during the whole tidal cycle. A wide scour region, covered by dunes, rapidly formed near the inlet. The scour was deeper than in Run 1 because of the larger values attained by the velocity (and hence by the bed shear stress) as a consequence of the narrower inlet. After 800 cycles, some isolated and unstable channel started to form at the landward boundary of the inlet scour. At 1300 cycles we reduced the mean sea level to allow draining of the sediment surface during the ebb phase and, soon after, a more defined tidal network began to form, characterized by three main channels that lengthened significantly during the following tide cycles. Some bends were also observed to form. The well developed small scale bed forms, previously observed to form in the main channels, tended to reduce sensibly their

Figure 3. Distribution of bottom elevations measured within the lagoon in Run 1 after: a) 1650 cycles; b) 2900 cycles; c) 3910 cycles; d) 5950 cycles; in Run 2 after: e) 2200 cycles; f)6270 cycles; g) 6845 cycles; h) 7700 cycles. Vertical bands are the consequences of changing tidal levels during the acquisition of the bathymetry. Elevations are referred to the initial uniform bottom elevation.

dimensions. Moreover, pronounced localized scours formed at the main junctions (Fig. 2d). The evolution was anyhow very slow and only in the last 1000 cycles the channels deepened appreciably, forming a more complex network, with some bends forming also along the main channels (Fig. 2c). The run was stopped after 7710 cycles.

## 4  ANALYSIS OF EXPERIMENTAL DATA

Bed elevations acquired through the laser system and suitably corrected to account for refraction effects were used to produce topographic maps of the lagoon bed at various instants. Figure 3 shows the distribution of bed elevations at four different instants of Run 1 and Run 2: the temporal evolution of channel networks clearly emerges.

In order to classify small scale bed forms, which were observed to form during both the experiments, we used the bottom topographies measured at the end of Run 1 and Run 2 to carry out some fixed bed numerical simulations. The 2D shallow water equations were solved numerically by using a semi-implicit staggered finite element model, based on Galerkin's approach. The equations were modified to deal with wetting and drying processes in irregular domains. We refer the reader to D'Alpaos & Defina (1995), Defina (2000), D'Alpaos & Defina (2006) for a detailed description of the modified shallow water equations, the numerical techniques adopted, and for model validation.

The numerical results, obtained setting to 30 m$^{1/3}$s$^{-1}$ the Gauckler-Strickler friction coefficient, $K_s$, indicate that in Run 1 the higher values of both flow velocity U ($\sim 3.7$ cm/s), and bed shear stress $\tau$ ($\sim 0.091$ Pa), were attained during the ebb phase, in correspondence of the deep scour which formed in front of the inlet and in the first reaches of the channels departing from it. Secondary channel ramifications exhibited lower values of U ($\sim 2$ cm/s) and $\tau$ ($\sim 0.0002$ Pa).

A similar picture emerges from the simulation of Run 2 but with higher values of both the velocity ($\sim 5.8$ cm/s) and the bed shear stress ($\sim 0.24$ Pa) in the scour facing the inlet and in the reaches of the two main channels developing from it. Conversely, secondary channel ramifications exhibited lower values of U ($\sim 1.5$ cm/s) and $\tau$ ($\sim 0.0015$ Pa).

The bed shear stress resulting from numerical calculation was used to determine the dimensionless Shields stress, $\tau^*$:

$$\tau^* = \frac{\tau_0}{g \cdot (\rho_s - \rho) \cdot d} = \frac{u^{*2}}{\Delta \cdot g \cdot d} \qquad (1)$$

where: $d$ is the representative sediment grain size, g is the gravity constant, $\rho$ and $\rho_s$ are water and sediment specific weight, respectively, $u^* = (\tau^*/\rho)^{1/2}$ is the friction velocity, $v$ is the kinematic water viscosity, and $\Delta = (\rho_s - \rho)/\rho$ is the relative density of submerged sediment.

Figures 4a,b show the spatial distribution of the excess Shield stress, $\tau^* - \tau_c^*$, during the ebb phases of Run 1 and Run 2, i.e., when bottom shear stress attains its maximum. The critical value $\tau_c^*$ for incipient sediment motion has been evaluated using the analytical relationship proposed by Brownlie (1981):

$$\tau_c^* = 0.22\, R_p^{-0.6} + 0.06 \exp(-17.77\, R_p^{-0.6}) \dots\dots \qquad (2)$$

where $R_p$ is particle Reynolds number:

$$R_p = \frac{\sqrt{\Delta \cdot g \cdot d^3}}{v} \qquad (3)$$

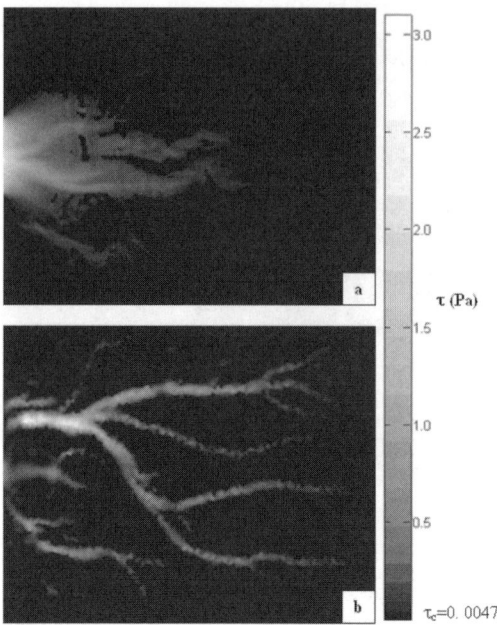

Figure 4. Spatial distribution of the excess Shield stress, $\tau^* - \tau_{cr}^*$, computed: (a) or the final configuration of Run 1 and (b) of Run 2. Both $\tau^*$ and $\tau_{cr}^*$ have been calculated using $d_{50} = 0.8$ mm, as representative grain size of the adopted sediments.

It clearly emerges that in both runs a large portion of the lagoon is interested by sediment transport. Additional analyses showed that choosing a different value of the representative grain diameter (e.g., $d_{90}$) does not significantly modify the percentage of lagoonal surface interested by sediment transport.

The computed values of the Shields stress, $\tau^*$, and of the particle Reynolds number, $R_p$, have been used to determine the portions of the lagoon in which, according to the criterion proposed by Simon & Richardson (1966), small scale bed forms are likely to develop. Figure 5 reports a comparison between the computed and experimental spatial distribution of bed forms for Run 1 and Run 2. The agreement between computed and observed results is pretty satisfactory.

Indeed, dunes are observed to form in the scoured zone facing the inlet and in the first reaches of the main channels. Dune wavelength ($\lambda_d \approx 20$ cm) appears to be in accordance with predictions given by the empirical relationships proposed by Van Rjin ($\lambda_d \approx 6$ D) and Yalin ($\lambda_d \approx 7.3$ D) (Van Rjin 1984c, Yalin 1977, Yalin & Ferreira 2001). In fact, the mean flow depth, D, in the portions of the lagoon covered by dunes was of about 3 cm.

The final configuration of Run 2 was characterized by less developed bed forms, except for the dunes ($\lambda_d \approx 7$–8 cm) covering the 0.8–1 cm deep scour close

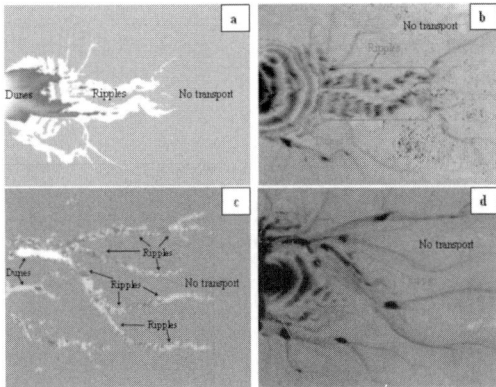

Figure 5. Spatial distribution of bed forms: (a, c) predicted through the criterion of Simons & Richardson (1966) applied to the computed values of $R_p$ and $\tau^*$ for Run 1 (a) and for Run 2 (c); (b, d) arising from experimental observations for Run 1 (b) and for Run 2 (d).

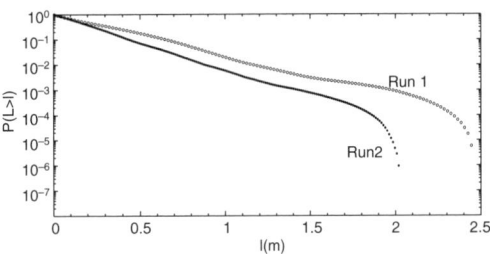

Figure 6. Probability distribution of unchanneled lengths for the channel network observed at the end of Run 1 and Run 2.

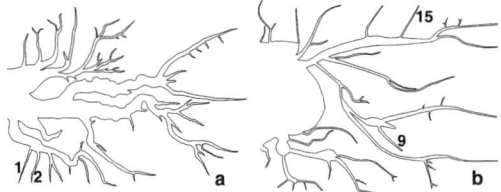

Figure 7. Border of final network configurations obtained in Run 1 (a) and Run 2 (b). Channels denoted by numbers 1, 2, 9 and 15 are those considered in the plots of Figs. 8, 9, 10.

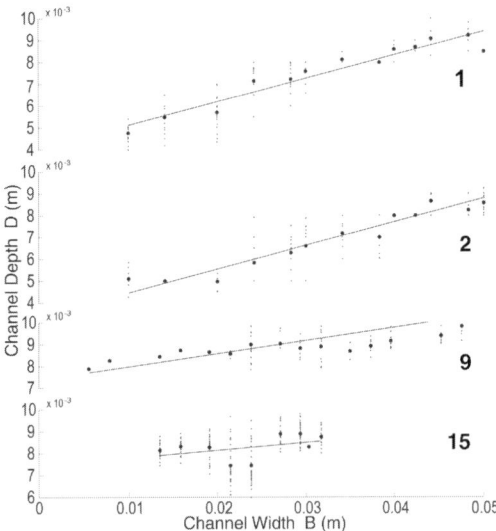

Figure 8. Channel depth versus width for the cross sections of channels indicated in Fig. 7.

to the inlet. Very small ripples affected the larger reaches of main channels.

We also carried out morphometric analyses of the experimental networks by using a simplified hydrodynamic model (Rinaldo et al. 1999a, b), successfully applied to the study of various tidal environments (Marani et al. 2003). The hydrodynamic flow field obtained through this simplified model makes it possible to determine the unchanneled hydrodynamic flow path connecting any unchanneled site to the nearest tidal channel and to compute its length. The probability distributions of unchanneled flow lengths, $\ell$, exhibits a linear trend in a semi-log plot (Fig. 6), suggesting the same type of exponential decay determined by Marani et al. (2003) in different areas of the Venice lagoon.

Finally, in Figure 7 we report the two digitalized final network configurations of Run 1 and Run 2, whereas in Figures 8 and 9 we report some relevant features of channel network cross-sectional geometry, namely channel width, B, channel depth, D (measured

along channel axis), width-to-depth ratio, $\beta = B/D$, and cross-sectional area, $\Omega$.

We also reported in a semi-log plot channel width, B, versus the along channel intrinsic coordinate, s (Fig. 10).

In both runs, B and D attained relatively small values, falling in the range 1–2 cm and 8–10 cm, respectively. The results shown in Figure 8 suggest that, for a given channel, a nearly linear relationship exists between D and B.

The aspect ratio, $\beta$, displayed in Fig. 9, was found to vary between 2, for smaller channels, and 20, for the main channels, with an upper limit of 50. These values are in accordance with field surveys carried out in the Venice Lagoon (Marani et al. 2002). Finally, Figure 10 emphasizes the progressive exponential widening experienced seaward by channels, which is typical of tidal environments (Lanzoni & Seminara 1998, Marani et al. 2002).

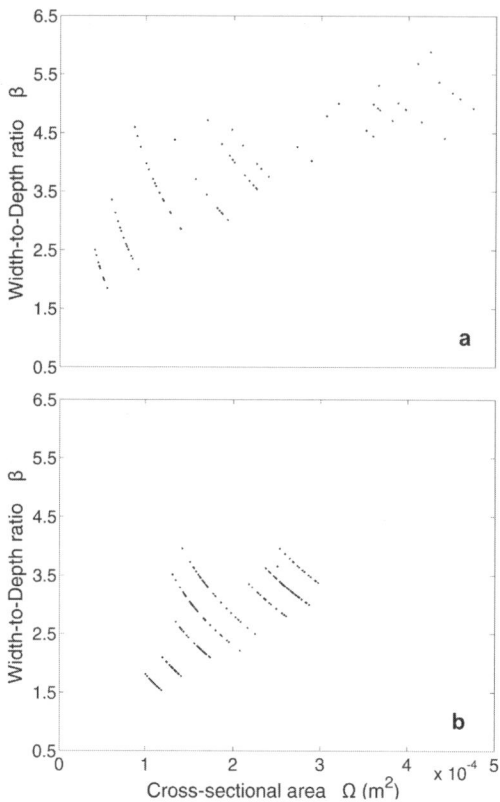

Figure 9. Comparison between the curve of width-to-depth ratio $\beta$-cross sectional area $\Omega$ of two straight channels: a) channel 1 of Run 1; b) channel 15 of Run 2.

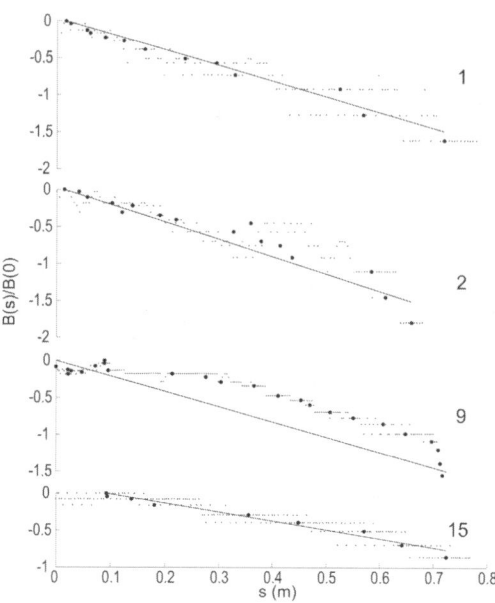

Figure 10. Logarithm of the ratio B(s)/B(0), with versus the intrinsic coordinate, s, for the channels indicated in Figure 7. B(s) is channel width at s, B(0) is width at channel inlet.

## 5 CONCLUSIONS

The morphometric analyses of the laboratory generated tidal channels described in the present contribution suggest a close analogy with real networks. The similarity between field and laboratory probability distributions of unchanneled flow lengths, as well as the range of variation of the aspect ratio, suggest that present experiments can be used to get a better understanding of the morphodynamic processes responsible for the initial growth and subsequent development of channel networks within tidal basins.

Laboratory channel networks were observed to form, starting from an initially flat and horizontal bed, only for: i) low enough values of the flow depth; ii) a mean sea level allowing the drying of the bottom during the ebb phase; iii) a tidal period of about 8–12 minutes. Higher flow depths enhance the formation of bed forms (ripples and dunes) and prevent the formation of well defined channels.

Channel networks were observed to form mainly through headward growth.

The rate of network growth was very slow at the beginning of the experiment whereas as soon as a well developed network began to form, the mean rate of channel elongation was approximately of 1.5–2 cm every 150 tidal cycles. After a quite rapid growth phase the tidal network appears to be subject only to small adjustments, thus supporting the usually adopted hypothesis that a time in the life of a tidal network exists during which it quickly cuts down the intertidal areas giving them a permanent imprinting. Such a process is later followed by slower elaborations including meandering and network contractions/expansions.

Clearly, a more systematic series of experiments is needed to further support the above findings.

Other points also need to be addressed in future developments of the research. In particular, the effects of the initial configuration (plane in the present experiments) on channel network structure has to be analysed, possibly accounting for the random irregularities that likely characterize the bottom of real tidal flats. The analysis of sediment cohesion, typical of lagoonal environments is deemed of importance, as well. Finally, the study of the influence exerted by different boundary conditions (e.g., the role of different lagoon inlet locations, and of external inputs of sediments) deserves further attention.

# REFERENCES

Allen, J. R. L. 2000. Morphodynamics of Holocene salt marshes: a review sketch from the Atlantic and Southern North Sea coasts of Europe. *Quat. Sci. Rev.* 19 (17–18), 1155–1231.

D'Alpaos, A., S. Lanzoni, M. Marani, S. Fagherazzi & A. Rinaldo 2005. Tidal network ontogeny: channel initation and early development. *J. Geophys. Res.* 110, F02001, doi:10.1029/2004JF000182.

D'Alpaos, A., S. Lanzoni, M. Marani, & A. Rinaldo 2007. Landscape evolution in tidal embayments: Modeling the interplay of erosion, sedimentation and vegetation dynamics. *J. Geophys. Res.* 110, doi:10.1029/2006JF 000537.

D'Alpaos, L. & A. Defina 1993. Venice lagoon hydrodynamics simulation by coupling 2D and 1D finite element models. *Proceedings of the 8th Conference on Finite Elements in Fluids. New trends and Applications.* Barcelona (Spain), 20–24 September 1993, p. 917–926.

D'Alpaos, L. & A. Defina 1995. Modellazione matematica del comportamento idrodinamico delle zone di barena solcate da una rete di canali minori. *Istituto Veneto di SS.LL.AA., Rapporti e studi,* Vol. XII, 353–372.

D'Alpaos, L. & A. Defina, 2006. Mathematical modeling of tidal hydrodynamics in shallow lagoons: A review of open issues and applications to the Venice lagoon. *Computer & Geoscience.* doi:10.1016/j.cageo.2006.07.009.

Defina, A. 2000. Two Dimensional Shallow Flow Equations for Partially Dry Areas. *Water Resources Research*, vol.36, 11, 3251–3264.

Fagherazzi, S. & D. J. Furbish 2001. On the shape and widening of salt-marsh creeks. *J. Geophys. Res.*, 106(C1), 991–1003.

Fagherazzi, S. & T. Sun 2004. A stochastic model for the formation of channel networks in tidal marshes. *Geophys. Res. Lett.* 31, doi:10.1029/2004GL020965.

French, J.R. & Stoddart D.R. 1992. Hydrodynamics of salt-marsh creek systems: Implications for marsh morphological development and material exchange. *Earth Surf. Proc. Landforms* 17: 235–252.

Friedrichs, C. T. & J. E. Perry 2001. Tidal salt marsh morphodynamics. *J. Coastal Res.* 11(4), 1062–1074.

Lanzoni, S. & G. Seminara 1998. On tide propagation in convergent estuaries. *J. Geophys. Res.* 103, 30, 793–30, 812.

Lanzoni, S. & G. Seminara 2002. Long-term evolution and morphodynamic equilibrium of tidal channels. *J. Geophys. Res.* 107(C1), 3001, doi:10.1029/2000JC000468.

Marani, M., S. Lanzoni, D. Zandolin, G. Seminara & A. Rinaldo 2002. Tidal Meanders. *Water Resources Research* 38 (11), 1225–1239.

Marani, M., S. Lanzoni, E. Belluco, A. D'Alpaos, A. Defina & A. Rinaldo 2003. On the drainage density of tidal networks. *Water Resources Research* 39 (2), 105–113.

Pestrong, R. 1965. The development of drainage patterns on tidal marshes. *Stanford Univ. Publ. Geol. Sci.,* Tech. Rep. 10, 87pp.

Rinaldo, A., S. Fagherazzi, S. Lanzoni, M. Marani, & W.E. Dietrich 1999a. Tidal networks 2. Watershed delineation and comparative network morphology. *Water Resources Research* 35(12), 3905–3917.

Rinaldo, A., S. Fagherazzi, S. Lanzoni, M. Marani, & W.E. Dietrich 1999b. Tidal networks 3. Landscape-forming discharges and studies in empirical geomorphic relationships. *Water Resources Research* 35(12), 3919–3929.

Schuttelaars, H. M. & H. E. de Swart 2000. Multiple morphodynamic equilibria in tidal embayments, *J. Geophys. Res.* 105, 24, 10524, 118.

Tambroni, N., M. Bolla Pittaluga & G. Seminara 2005. Laboratory observations of the morphodynamic evolution of tidal channels and tidal inlets. *J. Geophys. Res.* 110, F04009, doi:10.1029/2004JF000243.

Van Rijn, L. C. 1984a. Sediment transport. Part I: bed load transport. *Journal of Hydraulic Engineering 110 (11)*, 1613–1641.

Van Rijn, L. C. 1984b. Sediment transport. Part II: suspended load transport. *Journal of Hydraulic Engineering 110 (11)*, 1613–1641.

Van Rijn, L. C. 1984c. Sediment transport. Part III: bed forms and alluvial roughness. *Journal of Hydraulic Engineering 110 (11)*, 1613–1641.

Yalin, M. S. 1977. Mechanics of sediment transport. *Pergamon Press, Oxford.*

Yalin, M. S. & Ferreira da Silva, A. M. 2001. Fluvial processes, *Queen's University, Kingston, Canada.*

*River, Coastal and Estuarine Morphodynamics: RCEM 2007 – Dohmen-Janssen & Hulscher (eds)*
*© 2008 Taylor & Francis Group, London, ISBN 978-0-415-45363-9*

# Flood/ebb tidal dominance in an estuary: Sediment transport and morphology

J.M. Brown & A.G. Davies

*School of Ocean Sciences, College of Natural Science, University of Wales (Bangor), Anglesey, Wales*

ABSTRACT: Flood/ebb tidal dominance plays a pivotal role in estuarine sediment transport and morphodynamics. The Dyfi Estuary, UK, is used as a case study to illustrate some key processes giving rise to flood/ebb dominance. Observations indicate that the system is ebb-dominant and predictions made across the mouth of the estuary using the Telemac Modelling System suggest that net sand transport is out of the estuary. In contrast, due to the varying distribution of channels and flats within the estuary, the tidal flow in the upper estuary is flood-dominant and net transport is up-estuary. Flood/ebb dominance is attributable to the relative extent of the channels and sandbanks/flats. The implication is explored that, once an estuary has infilled from its initial ice-age state to present day conditions, it then oscillates around a dynamic equilibrium.

## 1 INTRODUCTION

In an estuary the tidal dynamics are modified by frictional influence if $\zeta >> h/10$, where $\zeta =$ the tidal elevation amplitude at the mouth and $h =$ mean depth (Prandle, 2003). The frictional drag, $C_D$, decreases as $z_0/h$ increases, where $z_0 =$ the bed roughness length. The greater depth at high water (HW) compared to low water (LW) serves to reduce the duration between LW and the following HW and increases it between HW and the subsequent LW. The result is commonly a short flood tidal phase and long ebb tidal phase. To conserve water flux the magnitude of the peak flood exceeds that of the peak ebb (Fig. 1). Although the ebb duration is considerable the velocities may only exceed the critical sediment threshold velocity for a minimal time period. During the flood phase the fast flow is predominantly over the critical velocity. This leads to a net sediment transport in the flood tide direction. The

estuary becomes a sediment sink leading to the build up of inter-tidal banks. The tidal asymmetry can be further complicated by the channel-sandbank system. A flood-dominant channel and an ebb-dominant channel may co-exist causing varying net transport directions. If the banks constrict the water volume during peak ebb, more so than at peak flood, then the magnitude of the ebb tide will increase to conserve water flux. The enhanced ebb velocity combined with the long ebb duration may then produce ebb-dominant net transport, as found in Southampton Water (Townend & Wells, 2003) and the Dyfi Estuary (investigated in this paper).

The Dyfi Estuary, mid-Wales, UK, is a large sandy tidal estuary containing a vast expanse of inter-tidal sand flats. In the lower estuary two channels are present a northern ebb-dominated channel and a southern flood-dominated channel. In the upper estuary a single river channel is present. At the mouth of the Dyfi and in the main northern channel in the lower estuary the peak ebb velocity has been found to exceed the peak flood velocity (Fig. 2). In contrast, the upper reaches of the estuary and the southern channel in the lower estuary exhibit typical flood-asymmetry. As a result the estuary is losing sediment through the mouth, while inter-tidal sediment is being redistributed internally to maintain the upper estuary morphology where the tidal force is weaker.

## 2 TIDAL ASYMMETRY

After the Holocene, rapid sea level rise drowned river valleys creating deep wide estuaries. Shallow water

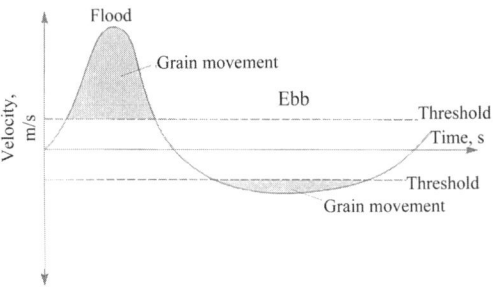

Figure 1. Typical shallow water tidal asymmetry.

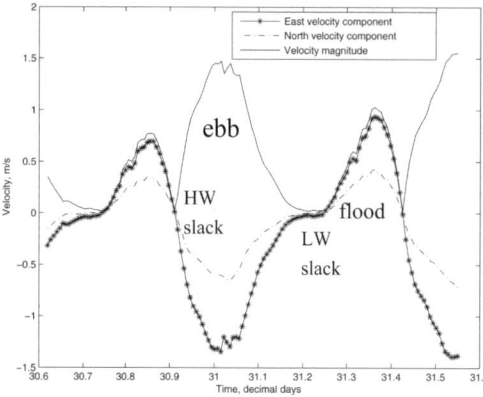

Figure 2. Tidal velocity measured in the northern East-West oriented channel near the mouth of the Dyfi Estuary (2nd dashed line Fig. 4). Positive eastward velocity occurs on the flood.

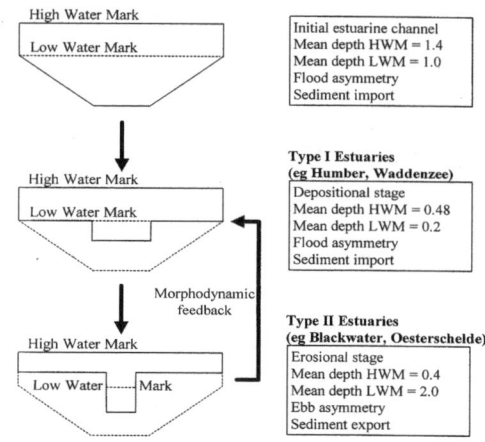

Figure 3. Typical stages of estuarine development, (Pethick, 1994).

tidal asymmetry produced flood-dominance in the sediment transport. The sediment brought into the system was deposited in the inter-tidal zone reducing the mean cross-section depth. Eventually a central deep channel bound by high inter-tidal banks was established, with river flow modifying the position of the channel in the system or producing multiple channels. The mean LW depth (with the flow constrained in the channel) can exceed mean HW depth (due to the inundation of the extensive shallow flats), and then reversed asymmetry (ebb-asymmetry) occurs, involving ebb-dominant sediment transport. Sediment loss from the inter-tidal banks increases the average depth, eventually restoring flood-dominance as the cross-section returns to its former configuration. Over long periods, perhaps 100 years, an estuary is thought to alternate between erosion and deposition phases. The estuary size and sediment supply controls whether an estuary is presently mature enough to have reached a stable oscillatory state. In the future sea level rise is likely to increase the average estuarine depth forcing flood-dominance until a new oscillating equilibrium is achieved (Pethick, 1994).

Estuaries can be classified in relation to their tidal asymmetry. Type I estuaries have been classified by Dronkers (1986) to consist of a wide, deep rectangular shaped channel. The inter-tidal flats are low, generally below mean sea level allowing flood-asymmetry. Most deposition of fine sediment fractions occurs at slack HW on the inter-tidal banks. Deposition during slack LW, when water is restricted to the main channels, will be re-suspended later by the peak flood flow. The inter-tidal flats will therefore build up, changing the initially wide deep channel into a central 'slot' shaped channel within relatively high bounding banks, forming a Type II estuary. The entire estuary displays a reduction in mean (width-average) depth under the

flood tide as the water flows over the large areas of highly elevated banks. Ebb-asymmetry is then experienced and a net export of sediment from the system occurs. These two estuary types represent the successive temporal stages in estuarine development. Erosion of the inter-tidal flats in a Type II estuary causes the estuary to revert back to a Type I estuary. Such morphological feedback (Fig. 3) keeps the estuary in a dynamic equilibrium oscillating between Type I and II characteristics (Pethick, 1994). From this it can be concluded that the Dyfi Estuary is currently a Type II estuary consisting of vast inter-tidal areas. Although reversed ebb-asymmetry has not been achieved (short fast ebb, long slow flood), the banks constrain the tidal discharge during low water elevations enhancing the ebb velocities (Fig. 2). This comes about due to reduced frictional influence on the ebb compared to the flood due to the greater average depth of the flow now restricted to the channels. In the Dyfi it is not only the magnitude of the ebb tide but also the duration of the ebb tide that leads to ebb-dominated net sediment transport. As a result a net sediment loss from the estuary occurs.

It is not necessarily the flood/ebb-dominance only that causes net erosion or deposition, but the asymmetry in slack water duration that may control the evolution of the sand flats. If HW slack is longer than that at LW then greater sediment accumulation may occur on the upper parts of the flats than on the lower levels, resulting in a sediment sink and vice versa if LW slack is longer. The long LW slack relative to the HW slack in the Dyfi (Fig. 2) causes sediment movement from the inter-tidal to the sub-tidal. Bathymetric surveys in 2002 and 2006 imply that the northern channel in the Dyfi (Fig. 4) is starting to become wider and shallower as a result, implying that the estuary

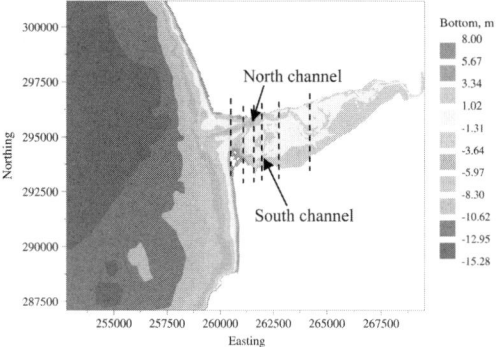

Figure 4. The actual estuary bathymetry and analyzed sections.

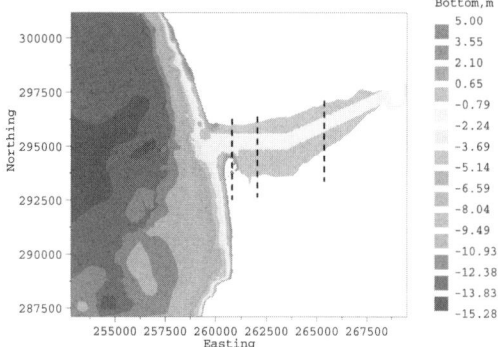

Figure 5. The imposed wide channel bathymetry with analyzed sections.

may be approaching the limit of a Type II estuary and will revert to a Type I estuary. Slack water asymmetry results in tidal flat build up due to fine sediment accumulation (Dronkers, 2005). Sediment is not always lost from or gained by the estuary system during the transitional stages between estuary types. The sediment may be redistributed from the inter-tidal banks to the sub-tidal channels

Dronkers (1998) developed a parameter, the asymmetry ratio $\gamma$, to determine the flood/ebb duration asymmetry of an estuarine system (Equation 1). It is assumed that if the flood-tide duration is shortened then the peak velocities will increase giving flood-asymmetry. This gives the potential to assess the net sediment transport.

$$\gamma = \left(\frac{h+a}{h-a}\right)^2 \frac{s_{lw}}{s_{hw}} \qquad (1)$$

where $h$ = mean hydraulic depth of the estuary system $(a + v_{lw}/s_{lw}, v_{lw} = $ low water volume$)$, $a$ = tidal amplitude, $s_{lw}$ = surface area at LW and $s_{hw}$ = surface area at HW. If $\gamma \approx 1$ then the tide is symmetrical, $\gamma > 1$ results in flood-asymmetry (long slow ebb short fast flood) and $\gamma < 1$ results in ebb-asymmetry (long slow flood short fast ebb). The relationship equates to the statement that channel width and depth are mutually dependent (Dronkers, 1998).

In this paper the dominant net transport through an estuarine cross-section is investigated to account for changing transport dominance over different estuary domains. The Dyfi as it is and with a reconfigured idealized channel/flat system has been used along with an idealized trumpet shape estuary with the aim to develop a simple parameter to determine the net sand transport over the estuarine system. A variety of channel widths and depths were studied following Pethick's (1994) idea to obtain different net transport regimes. Dronkers' (1998) duration ratio combined

with assumed flow patterns consistent with tidal asymmetry (phase with the short period would have faster flow) has been extended to account for enhanced flows and sediment thresholds, allowing more accurate prediction of the net sediment transport.

## 3 NUMERICAL INVESTIGATION OF FLOOD/EBB DOMINANCE

The role of the channel-sandbank system in producing flood/ebb-dominant net sediment transport has been investigated. The Telemac Modelling System (Hervouet & Bates, 2000) has been used to predict the tidal field and resulting sediment transport patterns in a set of estuaries. Telemac2D simulated the tide by forcing the outer boundary of the domain. Depth-averaged velocity components and elevation, obtained from the POLCOMS model of the Irish Sea, were imposed over the offshore bathymetry to simulate the tides internally within the model domain. Wetting and drying of banks was accounted for to accurately simulate the flow within the estuary. Sisyphe, the morphological module of Telemac, predicted the sediment transport as a result of the flow patterns and updated the bed morphology accordingly. The Dyfi Estuary (Fig. 4) has been used as a field case study. The estuary is 9 km long and the mouth is restricted by a spit extending from the southern bank. A morphological tide (average tide) representative of the spring-neap cycle in the Dyfi Estuary was applied using Latteux's (1995) method. The sediment transport was predicted using the UWB 1DV model (Davies & Li, 1997). This model was parameterized for a range of hydrodynamic and sediment conditions in the Dyfi Estuary and coded into Sisyphe. A realistic equal sediment mix of two grains $(D = 0.24, 0.3 \text{ mm})$ was applied. The same tide and grain sizes were applied in the idealized Dyfi case. The bed roughness was predicted by the Wiberg and Harris (1994) ripple prediction procedure. No waves

781

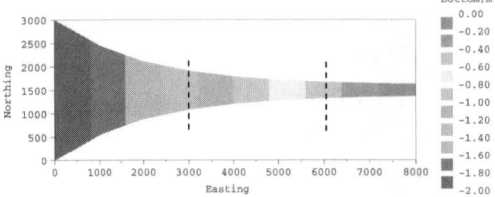

Figure 6. The trumpet shape estuary bathymetry and analyzed cross-sections.

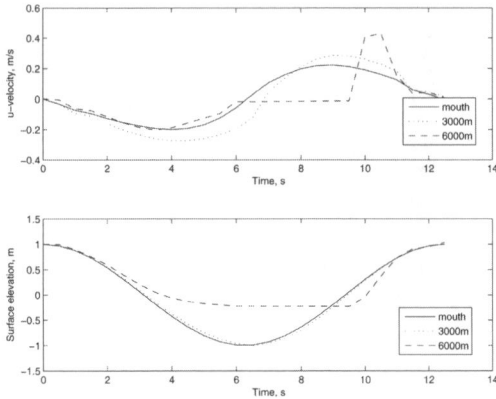

Figure 7. The stages in the development of tidal asymmetry.

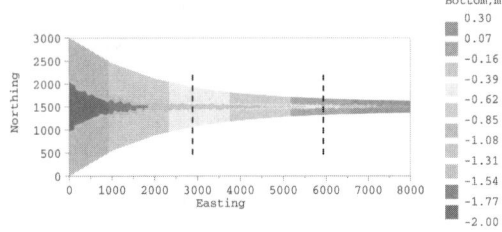

Figure 8. The trumpet shape estuary bathymetry with high wide sand flats imposed with investigated cross-sections.

or river inputs were included in the presented simulations. In the lower Dyfi the river has minimal influence since it is insignificant compared to the tidal currents and the waves only act to enhance the transport at the estuary mouth. Internally any waves are insignificant as they are forced to break on the sandbanks at the estuary entrance.

Four scenario cases were run using the Dyfi's geomorphologic shape, but with an imposed single (uniform) channel-bank system (Fig. 5). This generated model data for an estuary with a slightly restricted mouth. The channel was imposed from the estuary mouth to the upper limit with a depth of $-1$ m (ODN), and widths of 775 m and 387.5 m to represent a wide and narrow channel system. Banks were imposed at the mean LW level (-0.61 m (ODN)) and 0.5 m above mean LW level (0.1 m (ODN)). These scenarios represented bank levels and channel widths suitable for testing Pethick's (1994) classification.

A simple trumpet shape estuary (Fig. 6) with a gentle bed slope was also tested to analyze a typical estuary geomorphology. The trumpet estuary was 3 km wide at its mouth and 8 km long. The estuary mouth was forced by a sinusoidal tide with 1 m amplitude around a 0 m. The mean depth at the estuary mouth was $-2$ m. This allowed friction to modify the tidal wave as it propagated into the estuary. A uniform grain size of 0.2 mm was applied across the domain. This smaller grain size allowed a larger range of sand sizes to be investigated, although the later results will be biased by the Dyfi sand size due to a greater number of model results. This simple shape allowed typical shallow water effects to be modelled (Fig. 7). As the tide entered the mouth it was quite symmetrical. Then with distance into the estuary the period between LW to HW reduces resulting in 'saw tooth' then asymmetrical velocity curves. At the upper limit of the estuary a period of drying occurs during low water levels.

A narrow and a wide channel were next imposed in the trumpet estuary. The flats imposed either side of the channel increased in height with distance into the estuary (Fig. 8). This simulation attempted to recreate the typical estuarine infill patterns. The banks were imposed to a height above and below mean LW depth. This allowed wetting and drying over the banks to

occur differently in each scenario over the estuary domain.

### 3.1 The effect of not having any sandbanks

Three estuary bathymetries were imposed: i) the present day channel-sandbank system (Fig. 4), ii) a constant depth within the interior of the estuary of $-5$ m ODN (Fig. 9) and iii) a constant depth of $-1$ m ODN. The net flux through the mouth over a tidal cycle was out of the estuary when the channel-sandbank system was present (Fig. 10). This shows that the Dyfi Estuary has reached its equilibrium volume and is no longer a sediment sink fed by the offshore domain. The average flux through the mouth over an annual cycle with wave and river activity was $-0.073$ m$^3$/s. Approximating the estuary to a triangle produced an estuarine area of $9.54 \times 10^7$ m$^2$. The estuary therefore loses sediment at roughly the rate of 2.4 cm/year.

A constant depth of $-5$ m throughout the estuary (Fig. 9) was taken here as an approximation to the initial estuarine bathymetry before any infill occurred ($-5$ m being the depth in the channel through the estuary mouth leading offshore). In geological terms the 'base depth' of the estuary if it were emptied of sediment would be deeper than this. The net flux through

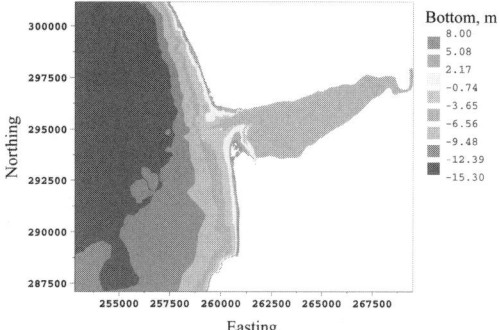

Figure 9. The flattened −5 m (ODN) bathymetry.

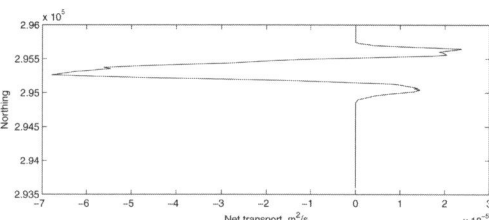

Figure 10. Net Flux through the estuary mouth with present day bathymetry.

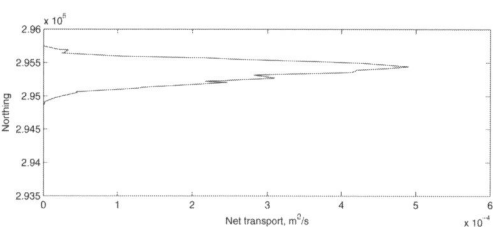

Figure 11. Net Flux through the estuary mouth with constant 5 m depth.

the mouth over a tidal cycle was into the estuary demonstrating that the Dyfi was initially a sink for offshore sediment (Fig. 11). The magnitude of the predicted net transport was 0.164 m$^3$/s, double that of the present day.

A constant depth of −1 m was an approximation to the estuarine bathymetry today, assuming that the sand was redistributed to remove the channel-sandbank system. The channels are generally −3 m and the banks 2 m above ODN. The average depth of −1 m was therefore imposed. The net flux through the mouth over a tidal cycle was into the estuary demonstrating that the Dyfi is still a sink for offshore sediment when it is 1 m deep. The net flux was 0.124 m$^3$/s, a factor of 1.7 times greater than the flux lost today. Although there are more banks than channels the banks are generally lower than this maximum depth and form inter-tidal

areas. Prandle (2003) indicates that the mean depth of the Dyfi Estuary in 1996 was 2.60 m. The tidal regime study carried out by Pethick (1996) shows that the estuary was ebb-dominant in the net sediment transport at this time. Therefore the estuary is unlikely to have become much shallower since no major windblown sand events have occurred since late 1995. The present day average depth is therefore between −5 m and −1 m so these simulations give an idea of how the estuary would behave if there was no channel-sandbank system.

The reduced magnitude of sediment input with reduced constant depth suggests that the estuary may be approaching a 'saturation state'. These results imply that the nature of the channel-sandbank system determines whether the estuary has reached its equilibrium volume. Estuaries with constant depth throughout will continually infill until the tide can no longer flood the domain. This results because the constant depth allows flood-dominant transport throughout the estuary.

### 3.2 The effect of the sandbank-channel system

Pethick (1994) has qualitatively described the oscillatory equilibrium state of an estuary. Here a quantitative description has been sought after. Cross-sections over the estuary domains were used to obtain the values of the maximum, average and minimum channel depths in the cross-sections. Seven cross-sections were used over the actual Dyfi Estuary (Fig. 4), three in the scenario Dyfi estuaries (Fig. 5) and two in the scenario trumpet estuaries (Figs. 6 & 8). The approach was applied to the full estuary width to avoid problems when multiple channels were present in the cross-section. Mean water level (mwl) at the mouth was applied as the reference level to maintain a constant level over the tidal cycle and estuary domain. The cross-section was defined as the area between the estuary boundaries at which the bathymetry cuts mwl. Three parameters were used to define the cross-section profile: $h_{min}$, the minimum bank depth below mwl, $h_{max}$, the maximum channel depth below mwl and $h_{ave}$, the average cross-section depth below mwl. In each simulation the net transport over the estuary cross-section was denoted as flood-dominant transport if it occurred into the estuary and ebb-dominant transport if it was towards the estuary mouth. No conclusive parameter related to the depths within the cross-section determining the net sediment transport was found. It was felt that using a depth only approach was too simple.

Applying Dronkers duration ratio (Equation 1) to the Dyfi Estuary, where $a \approx 2$ m, $h \approx 2.5$ m, $s_{lw} \approx 1.35 \times 10^7$ m$^2$ and $s_{hw} \approx 9.54 \times 10^7$ m$^2$ over the estuarine domain, gave $\gamma \approx 11.46$. Since the ratio is greater than 1 the estuary system should experience flood-asymmetry. Using a range of $a$ and $h$ values

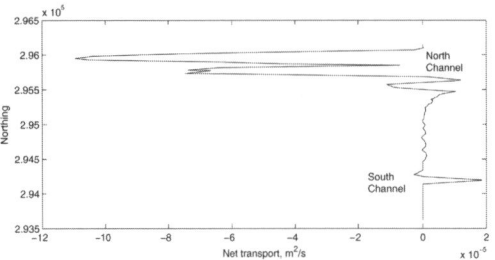

Figure 12. Net sediment transport through the 2nd cross-section Fig. 4 over the lower Dyfi Estuary width.

for the estuary conditions found $\gamma$ ranged from 0.45–11.46, implying different parts of the estuary would experience different asymmetry through the spring-neap cycle. ADCP data and model results confirmed a greater flood duration occurred over the estuary domain both at neap and spring tide. Dronkers (1998) implied that associated with reduced flood duration relative to the ebb duration should be an increase in peak flood velocity compared to the ebb velocity. In the Dyfi this is not the case over the full estuary domain. The ratio is therefore accurate when applied to tidal duration but does not predict which tidal phase will dominant in magnitude. Therefore it appears that it cannot be used to assess the net sediment transport in the estuary. This parameter assesses the tidal asymmetry of the estuarine system. The peak velocities indicate the net transport direction for the coarse sediment. A more precise direction is obtained when the duration of the velocities over the threshold velocity is accounted for. This indicates the net transport at a point in an estuary cross-section, the net transport over the full width is not captured. The water level and stream cross-section are greater at peak flood than peak ebb and bank exposure will reduce the net transport during the ebb phase.

The balance between magnitude and duration of the tidal phases leads to a net transport of sediment. It was found flood-dominant transport occurred over the inter-tidal zone when the height of the banks exceed low water level, due to exposure of the banks during the ebb phase. The width of the inter-tidal compared to the sub-tidal channel through the cross-section determines whether net flood- or ebb-dominance occurs in the transport averaged over the full cross-section width. For example a shallow flood-dominant channel and a deep ebb-dominant channel co-exist separated by inter-tidal banks in the lower Dyfi (Fig. 12); it is the balance between the width of the channels and flats that control the net transport of the cross-section. The flood-dominant transport over the inter-tidal (Fig. 12) is minimal due to the height of the banks leading to exposure for a significant part of the flood tide.

To determine the net flood/ebb dominance in the net transport through an estuary cross-section a different approach depending on the duration of each tidal phase and peak velocity was applied. Two ratios were used to describe the dominant direction of net sediment transport over an estuary cross-section of width, $w$. The first (Equation 2) represents the dominant velocity in the asymmetric tide. If the ratio of the peak flood velocity, $\hat{u}_f$, to the peak ebb velocity, $\hat{u}_e$, is greater than 1 then the flood tide is faster as found in flood-asymmetry. If a value less than 1 is achieved then the ebb has faster velocity either as a result of enhanced flow or ebb-asymmetry. The second ratio (Equation 3) compares the period that the flood tide is above the threshold velocity for sediment movement, $T_{f>th}$, to the period the ebb tide is above threshold velocity, $T_{e>th}$. This is a measure of the duration difference of the actual transport period of the tidal flow due to asymmetry. A value less than 1 results when the flood transport is shorter than the ebb transport duration and a value greater than 1 when the ebb transport is shorter than the flood transport duration. The durations will depend on the threshold velocity of the sediment. The period over a threshold is used to allow the effect of wetting and drying banks to be captured as well as the frictional influence of the banks reducing flow speed. $T_d$ uses velocity as a proxy to measure the channel depth bank height (faster flows occurring in deeper channels). By obtaining the cross-section averaged value of these ratios the width of the banks to the channel is accounted for and whether the banks or the channel dominate to produce long flows over threshold. This also allows the net transport to be found when ebb and flood dominant channels co-exist in an estuary cross-section. Using the cross-section averaged ratios of the peak flow, $u_p$, and the duration of transporting capability, $T_d$, captures the tidal asymmetry and any flow enhancement effects. When the values equal 1 then the tide is symmetrical and no net transport will result. The results for the trumpet estuary, simplified Dyfi Estuary and the real Dyfi Estuary are presented in Figure 13.

$$u_p = \frac{1}{w}\int_0^w \frac{\hat{u}_f}{\hat{u}_e}\, dw \qquad (2)$$

$$T_d = \frac{1}{w}\int_0^w \frac{T_{f>th}}{T_{e>th}}\, dw \qquad (3)$$

$$\text{ebb} - \text{dominant t ransport} : u_p < 1.15 \,\&\, T_d < 1 \\ u_p < 0.87 \,\&\, T_d > 1 \qquad (4)$$

The results were obtained for flood-asymmetry flow conditions and show that ebb-dominant transport occurs when the peak ebb flow is at least 87% of the flood-flow (Equation 4, the horizontal full line

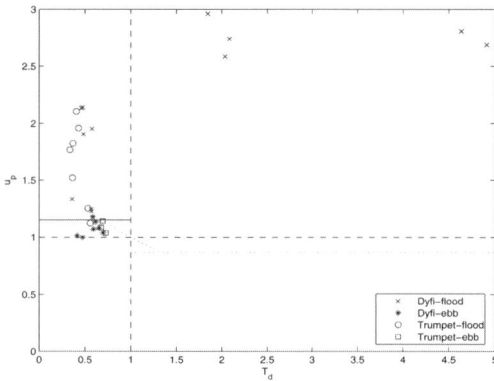

Figure 13. The net transport in the respective flood and ebb directions along with the proposed and assumed criteria for ebb-dominant sediment transport.

Fig. 13). This demonstrates that flood tide asymmetry produces flood-dominant transport unless the ebb flow is enhanced to a similar magnitude as the peak flood flow, in which case the duration of the ebb above the threshold velocity forces ebb-dominant transport. There is some scatter about the proposed limit. There is one flood-dominant point for the trumpet estuary, which falls below the line. This point has a small net transport compared to the peak gross transport. This implies cross-sections with weak net transports will fall close to the line. Adding more data for estuaries with known significant transport rates would consolidate the criteria proposed here (Equation 4).

### 3.3 Discussion

The diagram presented (Fig. 13) was developed for shallow estuaries, compared to the tidal range at the mouth. The case studies consisted of large sandy estuaries with significant inter-tidal zones to represent present day estuarine conditions in Wales, UK. Both a natural estuary morphology with a restricted mouth and an idealized trumpet shape estuary with an open mouth have been investigated. Further data for a variety of estuary shapes, depths and sediment size would develop the idea presented further and improve the robustness of the criterion (Equation 4). The diagram is an extension of Pethick's (1994) conceptual model. The ratios presented also account for wetting and drying of banks and frictional influence of the banks. Therefore extra interpretation for modified flows has been added to tidal asymmetry theory. This approach is applicable to an estuary cross-section to determine transport patterns over an estuary domain, whereas Pethick's (1994) and Dronkers (1998) methods can only be applied to the estuary as a whole, since they are determining whether the tide is flood- or

ebb-asymmetric due to the presence of banks/flats within the estuary.

The diagram presented (Fig. 13) is likely to contain blank data zones along the $T_d = 1$ line since the chance of having a symmetrical tidal period above threshold and different peak velocities is low. Symmetrical flows ($u_p = 1$) with asymmetric periods are more likely due to flow enhancement. Data in the region $T_d < 1$ and $u_p < 1$ is most likely to fall close to the $u_p = 1$ line. A long ebb even with enhanced flow is unlikely to have a peak flow velocity much larger than the peak flood. The quadrant $T_d < 1$ and $u_p > 1$ should be most densely populated. Flood-asymmetry is more likely to occur in an estuary and ebb flow enhancement will rarely result in a flow greater than the flood flow. Some data points will occur (as shown) in the quadrant $T_d > 1$ and $u_p > 1$ with $u_p >> 1$ (significantly faster flood flows). This is due to shallow water depth forcing extreme flood-asymmetry so that the weak ebb only exceeds a threshold for a duration less than that of the flood; or due to high banks being exposed during the ebb tide in a flood-asymmetry situation reducing the cross-section averaged ebb duration above threshold over the cross-section. The region $T_d > 1$ will be least densely populated since ebb-asymmetry is less likely to occur in shallow estuaries. Enhanced ebb flow under flood-asymmetry conditions due to bank growth is expected to reduce the contrast in channel to bank depth restoring typical flood-asymmetry preventing ebb-asymmetry occurring. Data representing ebb-asymmetry will most likely fall quite significantly below the $u_p < 1$ line. The $T_d > 1$ region will not have points close to $u_p = 1$ since it is unlikely the flood flow will be enhanced during ebb-asymmetry since the flood will not be restricted in deep channels. If the flood flow was enhanced during ebb-asymmetry then the flood-dominant transport would occur. This therefore leads to an additional assumed limit on the diagram (dotted line in Fig. 13). The proposed and assumed lines are discontinuous at $T_d = 1$ i.e. when the tide becomes symmetric and $u_p = 1$. Further data would be required to determine how the criteria changed as $T_d$ tended to 1. The lines might be linked, by a decay curve or line, with a decay rate dependent on the estuary morphology. It is felt the line would probably decay through the point $T_d = 1, u_p = 1$ (dashed line Fig. 13) rather than discontinue. This would imply less enhancement of the slightly weak, longer tidal flow would be required as the durations of the flows became symmetric.

The criterion presented (Fig. 13, Equation 4) was developed for sandy, shallow, well mixed estuaries. A stratified estuary would experience different sediment transport patterns due to a gravitational cycle (intrusion of the salt wedge). Different grain size would modify the threshold velocity and the amount of flow enhancement required in Equation 4. For the data presented inclusion of a river would enhance the ebb flow leading to a downward shift in the data points. For $T_d < 1$ this would increase the likelihood of ebb-dominant sediment transport ($u_p < 1.15$). In the Dyfi case study waves at the mouth enhance the transport patterns of the tide, but have no effect on the net direction of the transport only increasing its magnitude. In other estuaries waves may affect more than just the estuary mouth and interact differently with the tide/banks modifying the transport patterns. The idea presented is therefore valid when wave activity is minimal. An increase in sea level would increase the depth reducing ebb flow enhancement by frictional effects and/or flow constraint effects. Such a change would force an upward shift in the data points reducing the likelihood of ebb-dominant transport. This supports the suggestion that a rise in sea level will increase estuarine infill. In extreme cases of increased depth the data may be shifted towards $T_d = 1$ and $u_p = 1$ as the asymmetry is reduced. But this is unlikely since such an increase in depth would flood the surrounding area extending the estuary cross-section over shallow marsh lands.

## 4 CONCLUDING REMARKS

Two simple parameters have been derived to represent the flood/ebb-dominance of sand transport through an estuary cross-section (Equation 4). It is thought that this parameter could also be applied to a single channel within the cross-section to determine the transport patterns within the channel system. Changes in an estuary from sediment sink to source are assumed to be related to the channel cross-section (Pethick, 1994). From the results presented it was found that a narrow deep channel system through the estuary domain resulted in ebb-dominant sediment transport at the mouth, but not across the entire estuary domain. Similarly a wide shallow internal channel system resulted in flood-dominant sediment transport at the mouth but not always over the entire domain. The parameter presented therefore captures Pethick's (1994) classification when applying his description to the estuary system rather than an individual cross-section.

If an estuary is found to be losing sediment it can be concluded to have reached its oscillatory equilibrium state. Previous sediment infill must have occurred to build up the inter-tidal banks to force ebb-dominance. Using an 'average tide' will give an idea of the net transport dominance of the system over the spring-neap cycle. During the spring tide the ebb- or flood-dominance in net transport is likely to become enhanced; the larger tidal prism will allow the banks to play a more significant role. During the neap tide it will weaken or may even reverse, a result of the smaller tidal prism having less interaction with the banks. Sediment transport is nonlinear and will be greater at spring tide. The net transport in a spring-neap cycle will be related to the patterns during mean to spring tidal ranges, thus, applying this parameter to an 'average tide' will determine the overall dominance through a spring-neap cycle.

Not only does the net transport dominance depend on the cross-sectional profile (channel/bank width and relative depth) and the wetting and drying of banks (if only low water levels or low and intermediate water levels result in exposure) it also depends on whether the tidal wave has been significantly deformed at that point in the estuary, i.e. whether a 'saw tooth', 'asymmetrical' or 'symmetrical' wave form is present. It can be concluded that on inter-tidal zones flood-dominance occurs due to exposure during critical stages of the ebb phase. The strength of the flood-dominance will reduce if inter-tidal zone increases in height, drying out during critical parts of the flood tide. If an ebb-dominant channel exists in an inter-tidal zone the net transport will be controlled by the width and height of the inter-tidal flats. A wide inter-tidal zone with banks only affecting lower tidal elevations will favour flood-dominance. If the inter-tidal dries out for the majority of the tide or the channel width is significant then ebb-dominance will be favoured. Higher banks constrain both phases of the tide enhancing the peak flows but the ebb phase will be affected most.

From this study is seems unlikely that reversed ebb-asymmetry will occur in a shallow estuary. It seems more plausible that an enhanced ebb velocity combined with a long duration will result in ebb-dominance causing sediment loss from an estuary. In reality it is unlikely that ebb-dominance due to enhance velocities would occur throughout an estuary system. In general it will be the lower estuary that experiences ebb-dominance; this may or may not be accompanied by a net sediment loss through the mouth. Sediment may be internally redistributed once an equilibrium tidal volume is achieved to control the channel-sandbank cross-section profile over the estuary.

## REFERENCES

Davies, A.G. & Li, Z. 1997. Modelling sediment transport b neath regular symmetrical and asymmetrical waves above a plane bed. *Continental Shelf Research*, **17**(5): 555–582.

Dronkers, J. 1998. Morphodynamics of the Dutch Delta. Morphodynamics of the Dutch Delta in Physics of Estuaries and Coastal Seas, J. Dronkers & M.B.A.M. Scheffers (Ed). Rotterdam: A.A. Balkema.

Dronkers, J.J. 1986. Tidal asymmetry and estuarine morphology. *Journal of Sea Research*, **20**(2/3): 117–131.

Dronkers, J.J. 2005. Dynamics of Coastal Systems. Advanced Series on Ocean Engineering. **25**. World Scientific.

Hervouet, J-M. & Bates, P. 2000. The Telemac Modelling System. Special issue of Hydrological Processes. **14**(13): 2207–2364.

.Latteux, B. 1995. Techniques for long-term morphological simulation under tidal action. *Marine Geology*, **126**(1–4): 129–141.

Pethick, J. 1996. Shoreline intervention proposals: Aforn Dysynni to Aberdyfi. The Dyfi Estuary and Aberdyfi Coast. Prepared for Gwynedd Council. 11pp.

Pethick, J.S. 1994. Estuaries and wetlands: function and form. Wetland Management. London: Thompson Telford.

Prandle, D. 2003. Relationships between tidal dynamics and bathymetry in strongly convergent estuaries. *Journal of Physical Oceanography*, **33**(12): 2738–2750.

Townend, I.H. & Wells, T.J. 2003. Southampton Water case study. ABP Marine Environmental Research, Southampton.

Wiberg, P.L. & Harris, C.K. 1994. Ripple geometry in wave dominated environments. *Journal of Geophysical Research*, **99**(C1): 775–790.

*River, Coastal and Estuarine Morphodynamics: RCEM 2007 – Dohmen-Janssen & Hulscher (eds)*
*© 2008 Taylor & Francis Group, London, ISBN 978-0-415-45363-9*

# On the formation and migration of free bars in finite tidal channels

A. Giacchino, N. Tambroni & G. Seminara
*DICAT University of Genoa, Genoa, Italy*

ABSTRACT: This contribution investigates the formation and migration of free bars in tidal channels of finite length. We employ a 2-D numerical model based on the solution of the classical shallow water equations for the fluid phase coupled with the Exner equation and some semi empirical closure for the sediment transport rates. The model is applied to the case of a straight rectangular channel closed at the landward end and characterized by an initial plane bed, perturbed by randomly distributed disturbances of small amplitude. Numerical simulations have been performed for fixed values of the aspect ratio relatively close to the critical conditions of the linear theory. Preliminary results of our analysis show that bottom instability originates in the seaward part of the channel, giving rise to the formation of free alternate bars, with amplitudes decaying landward and wavelengths shorter than those predicted by the linear theory. Moreover, such features grow and migrate landward in time.

## 1 INTRODUCTION

The bed topography of tidal channels is characterized by the presence of a wide variety of bedforms of spatial scales falling in the range of a few centimeters to some kilometers. Tidal bars are sediment waves with wavelengths scaling with channel width. Dalrymple & Rhodes (1995) classified estuarine bars into *'repetitive barforms'* (alternate, point and braid bars), *'elongate tidal bars'* (typically observed at locations where tidal flow is strong and rectilinear), *'delta-like bodies'* (located at points of flow expansion, typically at channel outlets). Our attention is focused here on the first class of estuarine bars, in particular we investigate the formation and migration of free alternate bars in tidal channels of finite length.

The basic mechanism whereby bars form and migrate in tidal channels and estuaries has been theoretically investigated by Seminara & Tubino (2001) under the assumption of infinitely long channels, by means of a three-dimensional hydrostatic model. Results show that bars arise from a mechanism of instability of the erodible bed subject to the propagation of a tidal wave. In particular, instability occurs for large enough values of the mean aspect ratio of the channel, defined as half width to depth ratio, for given mean values of the Shields parameter and of the relative grain roughness. Furthermore, unlike fluvial bars, tidal bars forming in an infinitely long channel under a symmetrical small amplitude tide are non migrating features (in the mean). More precisely, they migrate alternatively forward and backward in a symmetric fashion, displaying a vanishing net migration in a tidal cycle.

The recent work by Garotta et al. (2006), shows that linear bars are found to exhibit a net migration over a tidal cycle only in the presence of overtides.

However, in the above works, the channel was assumed to be long enough for the effects of the end conditions on the process of bar formation to be negligible. As numerically investigated by Lanzoni & Seminara (2002) and later experimentally confirmed by Tambroni et al. (2005), in the long term, tidal channels evolve towards an equilibrium configuration which is typically characterized by a finite length, fixed by the longitudinal extension of the very shallow area which tends to form in the landward portion of the estuary.

The aim of this work is then to remove the hypothesis of infinitely-long channel and ascertain how the finite-length of the channel can modify the conditions of formation and migration of tidal free bars. In order to achieve this goal we employ a 2-D numerical model based on the solution of the classical shallow water equations for the fluid phase coupled with the Exner equation and some semi empirical closure for the sediment transport rate. We choose a numerical approach because it is able to capture all the physical processes held responsible for the formation of sediment patterns included in the idealized models and allows us to perform a fully non-linear analysis of this phenomenon.

The model is at first applied to the case of a straight rectangular channel characterized by an initial plane bed, perturbed by randomly distributed disturbances of small amplitude. The channel is closed at the landward end while at the seaward boundary an oscillation of free surface elevation is imposed; moreover, at the inlet, the bed is free to erode, allowing bar migration through this section. In this respect, the present analysis differs from that of Schuttelaars & de Swart (1999), who investigated theoretically the development of bedforms in a rectangular channel with finite length adopting the boundary condition of in-erodible bed at the bay entrance.

Numerical simulations have been performed for fixed values of the aspect ratio relatively close to the critical conditions, in order to investigate the formation of simple patterns, characterized by the presence of alternate bars. For very large values of the aspect ratio Hibma et al. (2003) have already shown how a simple and regular pattern of initial perturbation merge into complex larger-scale channel/shoal patterns. For the lower aspect ratios considered herein, results of our analysis show that bottom instability originates in the seaward part of the channel, giving rise to the formation of free alternate bars. Moreover, such features grow and migrate landward in time, at least in the initial phase of the process.

The paper is organized as follows. The next section is devoted to the theoretical formulation of the model. Section 3 describes some preliminary morphodynamic results, finally Section 4 concludes the paper with some discussion and final remarks.

## 2  FORMULATION OF THE PROBLEM

Let us consider a straight tidal channel of length $L_c$ and width $B_c$, closed at one end and connected at the other end with a tidal sea. We assume that the channel is not convergent and it has a rectangular cross-section with erodible bottom. Cohesionless sediments are taken to have uniform size and the channel is laterally bounded by non erodible banks. The mathematical problem can be formulated taking advantage of the fact that the phenomenon to be investigated has horizontal scales much larger than the vertical depth-limited scale, a feature which suggests that the shallow water equations can be safely employed to describe the hydrodynamics. Moreover, the morphological time scale is typically much larger than the flow time scale, which implies that a quasi steady approach can be used for the flow field. Recalling that the evolution of the bed interface is governed by the 2D form of Exner (1925) equation, we end up with the following differential problem:

$$\frac{\partial d}{\partial t} + \frac{\partial (du)}{\partial x} + \frac{\partial (dv)}{\partial y} = 0, \tag{1}$$

$$\frac{\partial (du)}{\partial t} + \frac{\partial (du^2)}{\partial x} + \frac{\partial (duv)}{\partial y} = -gd\frac{\partial (\eta + d)}{\partial x} +$$
$$-\frac{\tau_{bx}}{\rho} + \frac{\partial}{\partial x}\left(2\nu_e d\frac{\partial u}{\partial x}\right) + \frac{\partial}{\partial y}\left(\nu_e d\left(\frac{\partial u}{\partial y} + \frac{\partial v}{\partial x}\right)\right), \tag{2}$$

$$\frac{\partial (dv)}{\partial t} + \frac{\partial (duv)}{\partial x} + \frac{\partial (dv^2)}{\partial y} = -gd\frac{\partial (\eta + d)}{\partial y} +$$
$$\frac{\tau_{by}}{\rho} + \frac{\partial}{\partial y}\left(2\nu_e d\frac{\partial v}{\partial y}\right) + \frac{\partial}{\partial x}\left(\nu_e d\left(\frac{\partial v}{\partial x} + \frac{\partial u}{\partial y}\right)\right), \tag{3}$$

$$\frac{\partial \eta}{\partial t} = -\frac{1}{1-p}\left(\frac{\partial q_{sx}}{\partial x} + \frac{\partial q_{sy}}{\partial y}\right), \tag{4}$$

where $t$ denotes time, $x$ is a landward oriented longitudinal axis, $y$ is a transverse axis, $d$ is local water depth, $\eta$ is bed elevation, $u$ and $v$ are local depth-averaged components of velocity in the $x$ and $y$ directions respectively, $q_{sx}$ and $q_{sy}$ are the components of the depth averaged sediment flux accounting for both bed load and suspended load, $\rho$ is water density and $\nu_e$ is the eddy viscosity obtained from the turbulence model discussed below. The bottom stresses $\tau_{bx}$ and $\tau_{by}$ are evaluated based on the following classical relationship:

$$(\tau_{bx}, \tau_{by}) = (u, v)\rho\frac{\sqrt{u^2 + v^2}}{C^2}, \tag{5}$$

where $C$ is flow conductance which, for the case of plane bed, may be given the classical logarithmic form:

$$C = 6 + 2.5\ln\left(\frac{d}{2.5d_s}\right). \tag{6}$$

Here $d_s$ is the grain size and $2.5\,d_s$ is a roughness height after Engelund & Hansen (1967). The above formulation requires that the governing equations be supplemented with closure relationships for the eddy viscosity and the sediment flux. As regards the closure relationship for the eddy viscosity, we follow Stansby (2006), who has recently proposed the following two-dimensional formula for $\nu_e$:

$$\nu_e =$$

$$\sqrt{l_h^4\left[2\left(\frac{\partial u}{\partial x}\right)^2 + 2\left(\frac{\partial v}{\partial y}\right)^2 + \left(\frac{\partial v}{\partial x} + \frac{\partial u}{\partial y}\right)^2\right] + (\gamma u^* d)^2}, \tag{7}$$

where $l_h$ is a horizontal length scale, $u^*$ is the friction velocity, and $\gamma$ is a constant. The term involving $\gamma$ accounts for vertical mixing and, in parallel flow, Stansby (2003) finds $\gamma = 0.0067$. However, here it also accounts for dispersion and the effect of the horizontal length scale on vertical mixing which in turn affects the bed shear stress (and dispersion). There are thus three disposable parameters, $l_h$, $\gamma$ and $C$, which describe complex coupled physical effects and may be tuned for a particular application. In this case we set $l_h$ equal to 0.14 times the channel width (a typical value for a two-dimensional jet) while $\gamma$ is set equal to 0.2. Finally for $C$ we use the closure relationship (6). The sediment flux still remains to be defined. For the bedload contribution we assume:

$$(q_{sx}^b, q_{sy}^b) = (cos\psi, sin\psi)\Phi\sqrt{(s-1)gd_s^3}, \qquad (8)$$

where $s$ is the relative density of sediments, $g$ is gravity, and $\psi$ is the deviation angle of the direction of the bed load flux relative to the longitudinal direction due to gravitational effects (Seminara 1995). Neglecting the effect of the longitudinal bed slope (a restriction which can be easily removed) the deviation angle may be given the following form:

$$(cos\psi, sin\psi) = \left( \frac{u}{\sqrt{u^2 + v^2}}, \frac{v}{\sqrt{u^2 + v^2}} - \frac{r}{\sqrt{\theta}}\frac{\partial \eta}{\partial y} \right), \qquad (9)$$

where $r$ is an empirical parameter ranging about 0.5–0.6. Moreover, $\Phi$ is the intensity of bed load, a monotonically increasing function of the excess Shields stress which can be given the classical Meyer-Peter & Muller (1948) form:

$$\Phi = 8(\theta - \theta_c)^{1.5}, \qquad (10)$$

where $\theta$ is the Shields stress defined as

$$\theta = \frac{u_*^2}{(s-1)gd_s}, \qquad (11)$$

and $\theta_c$ is its critical value, below which no sediment moves. The flux of suspended sediment is calculated assuming local equilibrium, i.e. employing the classical Rouse form of the vertical distribution of sediment concentration associated with the local and instantaneous characteristics of the flow field. The resulting expression for $q_{sx}^s$ and $q_{sy}^s$, reads:

$$(q_{sx}^s, q_{sy}^s) = (u, v)\Psi d, \qquad (12)$$

with

$$\Psi = \frac{C_e}{kC}(I_2 + K_1 I_1). \qquad (13)$$

Here $k$ is von Karman constant, $C_e$ is the reference concentration for which various semi-empirical expressions have been proposed in the literature (e.g. van Rijn (1984), while $K_1$, $I_1$ and $I_2$ are integral functions depending on two parameters, namely: the Rouse number $R = w_s/(ku_*)$ and $\zeta_R$ the conventional dimensionless value of the reference elevation where the boundary condition is imposed under uniform condition.

Clearly the equations (1–4) require appropriate boundary conditions. As regards the hydrodynamics, at the seaward boundary of the channel the free surface elevation is imposed, according to the following law:

$$h(0, y, t) = d_0 + a_0 cos\left(\frac{2\pi t}{T}\right), \qquad (14)$$

where $d_0$ is the initial mean flow depth, $a_0$ is the amplitude of the tidal wave and $T$ is the tidal period. Note that we have concentrated our attention on the hydrodynamical response of the tidal system to an $M_2$ forcing oscillation ignoring, for the moment, the effect of higher harmonics. Vanishing normal component of velocity and free slip (no stress) for the longitudinal component of velocity are imposed at the side walls. The formulation of the morphodynamic problem is completed by imposing that the instantaneous sediment flux through the closed boundaries of the system must vanish. Note that this condition is automatically satisfied having imposed that the instantaneous normal velocity must also vanish at the walls. Finally, at the inlet of the basin, we assume that the sediment flux during both the ebb and flood phases is controlled by the local transport capacity of the flow in the channel.

The solution of the complete unsteady equations for the liquid (1–3) and solid phases (4) are obtained numerically. The numerical approach will only be summarised here as it is similar to that proposed for 3-D shallow water flows in Stansby & Lloyd (2001) (see also Stansby (2003)). The initial conditions for time stepping have a horizontal water surface in still water. Furthermore, at the beginning of each numerical simulations, we impose a horizontal bed, perturbed by randomly distributed disturbances of small amplitude. A staggered mesh is used to avoid saw-tooth oscillations. The advection terms are discretised using the QUICK upwind scheme. All other spatial discretisation is central difference, giving second-order accuracy. Temporally the advection and diffusion terms are treated explicitly using 2nd order Adams Bashforth, the bed shear stress semi-implicitly to 1st order and the depth gradient semi-implicitly using 2nd order Crank-Nicholson. To solve the system, expressions for $du$ and $dv$, determined from the momentum equations, are substituted into the mass conservation equation, forming a pentadiagonal equation set for $d$ which is solved by an efficient conjugate gradient method. Back substitution into the momentum equations then gives $du$ and $dv$. The scheme has negligible wave damping and numerical diffusion. The bed level is then advanced at the end of each time step. Finally, in order to accelerate computational times, we took advantage of the fact that the characteristic timescale for the bed evolution is much greater than the timescale for the flow (the tidal period). In particular, after the first cycles, it is possible to assume that the morphodynamics evolution is slowly varying and that the bed variations occurring at the time scale of the period are too small to affect the hydrodynamics significantly. Hence, the bottom is updated multiplying the instantaneous sediment discharge at each time step by a coefficient linearly increasing from zero to 20 in the first 20 cycle and then kept constant for the rest of the simulation.

Note that a validation of this approach has been provided by a comparison of the results with those obtained advancing the bed in the classical way.

## 3 RESULTS

We have applied our model to the case of a straight rectangular tidal channel, 30 km long and 120 m wide. The grid has a mesh size of 10 m × 120 m in the transverse and longitudinal directions, respectively. The initial depth is 5 m throughout the entire channel, hence the corresponding value of the aspect ratio of the channel turns out to be equal to 12. Note that the latter value is greater than the critical value for bar inception proposed by the linear theory of Garotta et al. (2006) which is approximately equal to 7. The lateral and landward boundaries of the channel are fixed and impermeable to the liquid and solid phases while, at the inlet, a periodic boundary for the water elevation is imposed to simulate the $M_2$ tidal component, with an amplitude of 1.5 m. The bed material consists of uniform sediments with $d_s = 0.1$ mm. Finally, the initial bed-level is plane and perturbed by randomly distributed disturbances of small amplitude, which do not exceed ±0.5% of the initial water depth.

At the present stage, our simulation covered a period of 4 *years*. Results show that bottom instabilities originate in the seaward reach of the channel, giving rise to the formation of a pattern of free alternate bars. Furthermore, numerical results show that bars grow in time and, at the end of each tidal cycle, exhibit a net migration which is directed landward, according with the flood-dominated character of the flow field.

The temporal evolution of the cross-sectionally averaged bed profile is shown in Figure 1. We can observe that sediments are scoured in the seaward part of the channel, driven landward and deposited in the inner part of the channel, giving rise to the formation of a sharp front. This mechanism is associated with the flood dominant character of the flow field. In fact, the temporal distribution of the cross-sectionally averaged flow speed is increasingly asymmetric proceeding from the inlet landward, with the ebb phase longer than the flood phase and the flood maximum larger than the ebb maximum in the entire channel, as shown in Figure 2.

A detailed view of the pattern of bed topography in the seaward half of the channel, at different times, is shown in Figure 3. Note that the latter plot has been obtained subtracting the average bed profile to the

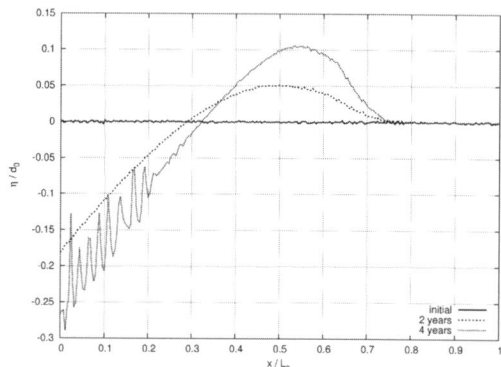

Figure 1. Temporal evolution of the cross-sectionally averaged bed profile. Disturbances superimposed on the average bed profile after 4 years (2920 cycles) are associated with the presence of bedforms in the seaward part of the channel.

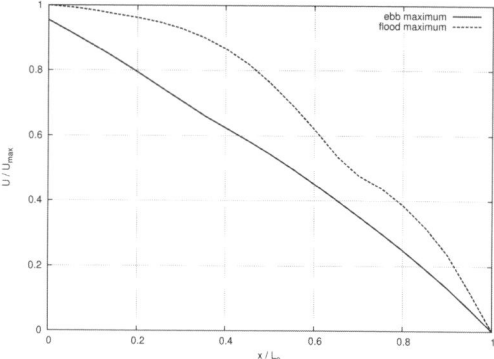

Figure 2. Cross-sectionally averaged maximum flood and ebb velocity over a tidal cycle at the initial stage of bar formation (after 1460 tidal cycles).

local bed elevations. In the initial phase of bar formation, bottom instability is confined in the seaward reach of the channel, where we observe the occurrence of structures having the same order of magnitude as the initial random disturbances. The first pattern of alternate bars becomes visible after about 1460 cycles corresponding to 2 years. At this time bars roughly cover about the 20% of the entire channel length and their amplitudes are approximately 0.1 the mean initial water depth, while their wavelength in the longitudinal direction is about 10 times the channel width. Proceeding with the simulation, bars exhibit a strong amplitude growth and a tendency to migrate landward, while their wavelengths remain nearly constants. After 4 years, a pattern of alternate bars with heights up to 0.8 the initial water depth covers about the 40% of the total channel length. However, at this time equilibrium is not yet reached and the bar amplitudes are still growing.

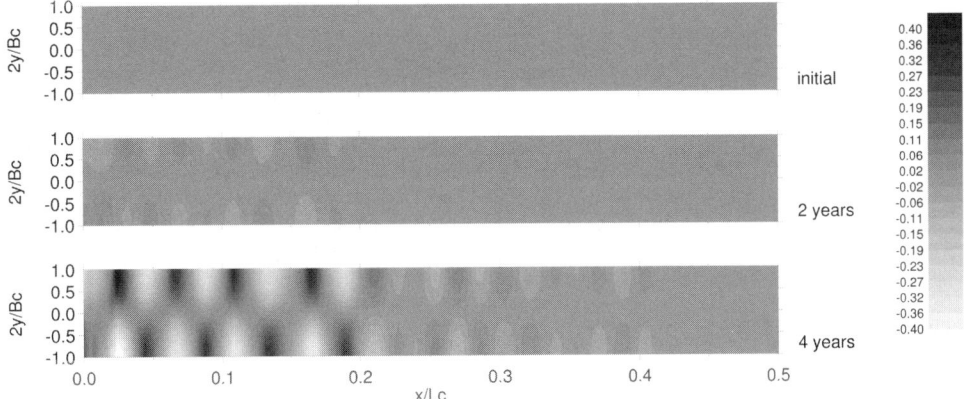

Figure 3. Pattern of bed topography in the seaward portion of the channel at different times, showing the presence of alternate bars. The cross sectionally averaged bed profile has been filtered out. Bottom elevation is scaled by the initial mean flow depth (darker zone corresponds to deposition and lighter zone to erosion).

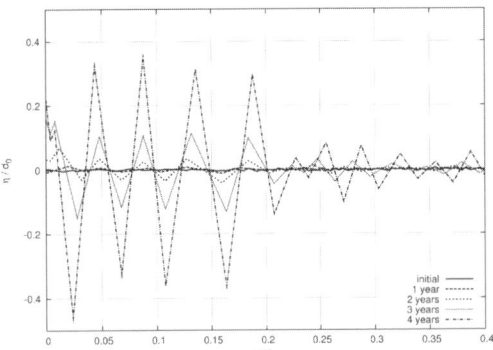

Figure 4. Differences between local bed elevations and the cross-sectionally averaged bed profile along the right bank of the channel at different times.

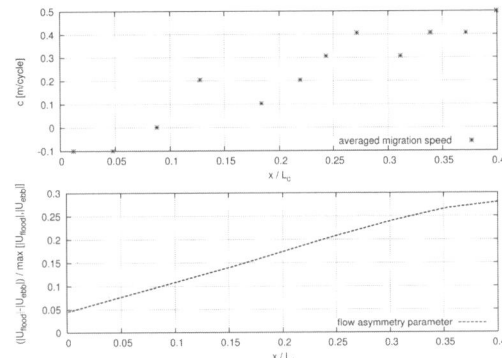

Figure 5. The bar migration speed averaged over the last 1460 cycles of simulations (2 years) and the local flow asymmetry parameter, defined as the dimensionless difference between the ebb and flood peak velocities over the 1460th tidal cycle, are plotted as a function of the longitudinal coordinate $x$ along the channel axis.

The figure 4 shows the temporal evolution of the longitudinal bed profile along the right bank of the seaward part of the channel. In the plot the cross sectionally averaged bed profile has been subtracted from the local bed elevation. Furthermore, in order to highlight the presence of alternate bars, we have plotted only the maximum and minimum elevations of each bed form. Results show the formation of bars which undergo temporal amplification and migrate landward according with the flood-dominant character of the flow field. Moreover, inner bars show a larger migration speed than those located close to the inlet. This feature is confirmed in Figure 5. Here we plot the local bar migration speed averaged over the last 1460 cycles (2 years) and the local flow asymmetry (measured by the dimensionless difference between the maximum ebb and flood flow velocities at the 1460th tidal cycle), as functions of the longitudinal coordinate $x$ along the channel axis. From a glance at Figure 5, it clearly

emerges that the landward increasing bar migration speed is correlated with the increasing flood-dominant character of the flow field in this part of the channel.

## 4 CONCLUSIONS

Numerical simulations show that in finite-length tidal channels bed instability can give rise to a bottom pattern characterized by the presence of alternate bars. Given an initial plane bed perturbed by randomly distributed disturbances of small amplitude, bars initially form in the seaward zone and then grow in time. Moreover, such features show a tendency to migrate landward, according with the flood-dominant character of the flow field.

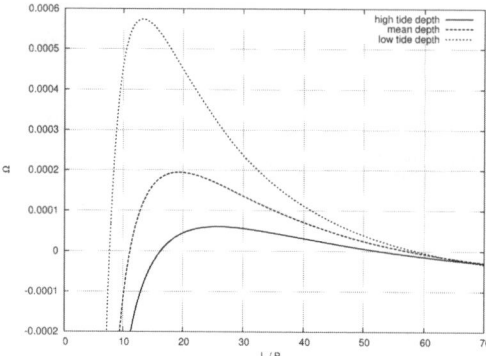

Figure 6. The bar growth rate $\Omega$ predicted by the linear theory for infinitely long channels Garatto et al.(2006) is plotted as a function of the bar wavelength scaled by channel width. The curves are calculated for the set of hydrodynamic and sediment parameters employed in the numerical calculations, the temporal distribution of the cross sectionally averaged longitudinal velocity numerically computed at half the channel length after 1460 cycles and three different values of depth equal to the local mean depth, low tide depth and hight tide dept, respectively.

At this stage it may be of interest to compare the numerical results with the predictions of the linear theory for infinitely long channels performed by Garotta et al. (2006). In particular, the wavelengths of the computed alternate bars turn out to range about 10 times the channel width. For the same set of flow and sediment parameters employed in the numerical simulation, the linear theory predicts positive values for the growth rate $\Omega$ of the generic small amplitude bottom perturbation, when it is associated with bottom waviness with wavelengths ranging between 7 and 60 times the channel width (see Figure 6). Note that the range of the unstable wavelengths predicted by the linear theory varies within each tidal cycle as, in the numerical simulation, the local depth varies significantly throughout each tidal cycle. In general, the results of computations seem to suggest the formation of bedforms shorter than those predicted by the linear theory. However, at low tide the wavelength corresponding to the most unstable mode, i.e. associated with the maximum value of the growth rate, is almost 12 times the channel width, hence close to the range of the computed bar wavelengths.

Furthermore, numerical results show a net landward displacement of bars over a tidal cycle which averages from zero, near the inlet zone, to about 0.4 $m$ in the central part of the channel. In this case we find a satisfactory agreement between linear theory and numerical results: the net displacement of bars provided by the linear theory at the end of each tidal cycle is directed landward, although it is slightly smaller (about 0.1 $m/cycle$).

These differences may be subject to various possible interpretations, which obviously will need to be substantiated by further investigations in the future. It is possible that the finite length of the channel is able to force the development of shorter bars with respect to those occurring in infinitely long channels. Furthermore, it is well known that the validity of the linear theory is restricted to cases characterised by small ratio of the amplitude of the free surface oscillation to the mean water depth ($\epsilon \ll 1$), while in our simulation the initial value of the dimensionless amplitude forcing at the inlet was $\epsilon = 0.3$, implying that non-linear effects are likely to be significant.

Non linearity implies that the flood dominant character of the tidal wave is felt in terms of landward phase of migration prevailing on the seaward phase.

The development of this study will be pursued by carrying on the calculation till equilibrium. This will allow us to estimate the bar equilibrium amplitude and wavelength, and verify their dependence on the distance from the inlet.

Moreover, further numerical simulations characterized by different values of the flow and sediment parameters will be performed with the aim to compare the results provided by the fully non-linear numerical approach with the predictions of the laboratory observation of Tambroni et al. (2005).

Some further results will be reported at the meeting.

REFERENCES

Dalrymple, R. W. & Rhodes, R. M. (1995). Estuarine dunes and bars. *Geomorph. and sediment. of Estuaries (ed. G.M.E. Percillo). Developments in sedimentology 53*, 359–422 Elsevier.

Engelund, F. & Hansen, E. (1967). A monograph on sediment transport in alluvial streams. *Copenhagen: Danish Technical Press.*

Exner, F. M. (1925). Uber die wechselwirkung zwishen wasser und geschiebe in flussen. *Sitzer Akad. Wiss. Wien*, 165–180.

Garotta, V., Pittaluga, M. B., & Seminara, G. (2006). On the migration of tidal free bars. *Physics of Fluids 18*, doi: 10.1063/1.2221346.

Hibma, A., de Vriend, H. J., & Stive, M. J. F. (2003). Numerical modeling of shoal pattern formation in well-mixed elongated estuaries. *Est., Coast. and Shelf Science 57*, 981–991.

Lanzoni, S. & Seminara, G. (2002). Long-term evolution and morphodynamic equilibrium of tidal channels. *J. Geophys. Res. 107(C1)*, doi:10.1029/2000JC000468.

Meyer-Peter, E. & Muller, R. (1948). Formulas for bedload transport. *Conf. of Internat. Ass. of Hydraul. Res. Stockholm, Sweden.*

Schuttelaars, H.M. & de Swart, H. E. (1999) Initial formation of channels and shoals in a short tidal embayment. *J. Fluid Mech. 386*, 15–42.

Seminara, G. (1995). Effect of grain sorting on formation of bedrooms. *Applied Mechanics Review 48(9)*, 549–563.

Seminara, G. & Tubino, M. (2001). Sand bars in tidal channels, part 1: Free bars. *J. Fluid Mech. 440*, 49–74.

Stansby, P. K. (2003). A mixing length model for shallow turbulent wakes. *J. Fluid Mech. 495*, 369–384.

Stansby, P. K. (2006). Limitations of depth-averaged modeling of shallow wakes. *Journal of Hydraulic Engineering 137(7)*, 737–740.

Stansby, P. K. & Lloyd, P. M. (2001). Wakes formulation around islands in oscillatory laminar shallow-water flows: Part ii boundary layer modeling. *J. Fluid Mech. 429*, 239–254.

Tambroni, N., Pittaluga, M. B. & Seminara, G. (2005). Laboratory observations of the morphodynamic evolution of tidal channels and tidal inlets. *J. Geophys. Res. 110(F04009)*, doi:10.1029/2004JF000243.

van Rijn, L. C. (1984). Sediment transport, part ii: Suspended load transport. *J. Hydraul. Eng., ASCE 110(1)*, 1616–1641.

*River morphodynamics*

*River, Coastal and Estuarine Morphodynamics: RCEM 2007 – Dohmen-Janssen & Hulscher (eds)*
*© 2008 Taylor & Francis Group, London, ISBN 978-0-415-45363-9*

# Process of channel formation at low flow on a bar created at high flow

T. Nogami
*Hokkaido Development Bureau, Ministry of Land, Infrastructure and Transport, Hokkaido, Japan*

Y. Watanabe
*Civil Engineering Research Institute for Cold Region, Hokkaido, Japan*

Y. Shimizu & Y. Kondo
*Hokkaido University, Hokkaido, Japan*

ABSTRACT: In various regions of Japan, local torrential rainfall often causes flooding on river plains. Clarifying land-forming processes and mechanism has become essential in proposing flood control and disaster mitigation measures for valley flats. To understand how channels form, this research studies the Appetsu River (Hasegawa 2004) in Hokkaido, which flooded in 2003. We conducted experiments on sandbar formation under three cases: i) high discharge until double-row bars transformed to alternate bars, at which time the discharge was reduced to low, ii) high discharge until double-row bars formed, at which time the discharge was reduced to low, and iii) low discharge only. The topography of the Appetsu River was compared to the topographies of the three experimental cases to determine which of the experimental topographies most closely corresponded to that observed in the Appetsu River. From this correspondence, the hydraulic conditions that produced the Appetsu River topography are hypothesized.

## 1 HYDRAULIC EXPERIMENT: CONDITIONS AND RESULTS

### 1.1 Previous studies

Many studies have addressed the formation of sandbars and meanders in rivers under constant discharges. However, sandbars in rivers form under conditions of fluctuant flow and changes in discharge, and the sandbars that form determine the channels that form after them. Few experiments have studied how river topography forms.

Sandbar formation under flood discharge has been studied experimentally by Uchijima et al. (1987) and Watanabe et al. (2005). These have identified changes in sandbars with time. Furthermore, through experiments or numerical analyses, Miwa et al. (2005), Takebayashi et al. (2001) and Teramoto et al. (2005) have studied changes in riverbeds according to long-term changes in discharge, and found that riverbed topography greatly differs by hydraulic conditions.

Experiments were performed to simulate how sandbars that form at high discharge subsequently become flow paths, including meandering channels, at low discharge. A steady-flow experiment was conducted for comparison with those results. The topographies produced from three cases of experiments were then compared with the topography of the Appetsu River

to determine which of the three cases most closely corresponds with that of the Appetsu River.

### 1.2 Conditions of the experiments

Hydraulic experiments were performed with the aim of studying how sandbars that had formed at high discharge ($Q_a$) change form at low discharge ($Q_b$). In the experiments, discharge was first high and then low.

The high discharge ($Q_a$) was used to produce the sandbars that occur during floods in flat riverbeds, and the discharge was such that the double-row bars that form at the initial stage of flow would transform into alternate bars with time. In Case 1, $Q_a$ of long duration was followed by low discharge ($Q_b$), after it was confirmed that the initially formed double-row bars had changed to alternate bars. In Case 2, $Q_a$ of shorter duration was followed by $Q_b$. Double-row bars formed during the period of $Q_a$. In Case 3 (steady-discharge experiment), the discharge was kept at $Q_b$, for comparison with Cases 1 and 2. The width of the water surface at $Q_b$ for Cases 1 and 2 was about 0.15 to 0.4 m, with an average of about 0.3 m. Therefore, in Case 3, the initial width of the channel, which was dug 1 cm deep, was set at 0.3 m.

The discharge $Q_b$ was set at less than the critical shear stress at the channel width of 0.9 m of flat bed. $T_a$ is the duration of $Q_a$, and $T_b$ is that of $Q_b$ (Table 1).

Table 1. Conditions of the experiment.

| Case nos. | $B_a$: Initial channel width (m) | $Q_a$ $Q_a$ ($\ell$/s) | High $\tau_*$ | Discharge $T_a$: Flow duration (min) | $Q_b$ $Q_b$ ($\ell$/s) | Low $\tau_*$ | discharge $T_b$ (min) |
|---|---|---|---|---|---|---|---|
| Case1 | 0.9 | 5.3 | 0.137 | 125 | 0.5 | 0.033 | 120 |
| Case2 | 0.9 | 5.3 | 0.137 | 25 | 0.5 | 0.033 | 120 |
| Case3 | 0.3 | – | – | – | 0.5 | 0.069 | 120 |

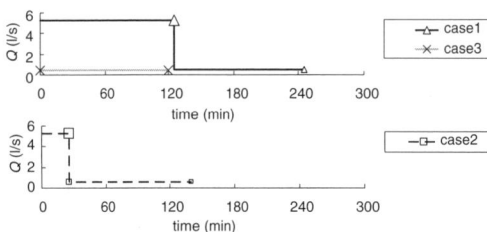

Figure 1. Discharge hydrograph.

The experiment used a straight channel 50 m long and 0.9 m wide. A meander channel will form only if the bed material is fine-grained. In the experiment, a mix of No. 4 silica sand ($d = 0.764$ mm) and No. 7 silica sand ($d = 0.154$ mm) was used (mix ration of 6:1) to produce an average grain size of 0.66 mm, which was used for the riverbed. The initial riverbed was set at a slope of 1/80, and at the upstream and downstream ends of the channel, weirs were fixed at the initial riverbed height. The bed load that flowed through the downstream weir during the experiment was measured, and that amount of sand was replaced at the upstream end.

After the prescribed flow duration of $T_a + T_b$, the shape of the riverbed surface for an approximately 15-m section from a point 26.25 m upstream of the channel end to a point 10.5 m upstream of the channel end was measured with a laser 3D scanner. Hydraulic conditions were given the subscript $a$ for the period of high flow and $b$ for the low flow.

### 1.3 Changes in sandbar shape

#### a) Case 1

After about 25 minutes of discharge $Q_a$, double-row bars with a wavelength ($L_o$) of 2.8 m and a wave height ($z$) of 1 cm had formed. At 1 hour of discharge $Q_a$, the double-row bars started to transform into alternate bars ($L_o = 6$ to 9 m, $z = 1$ to 2 cm). After the transition to alternate bars, the discharge was reduced to $Q_b$. In the beginning, the flow spread over the entire riverbed, but after about 1 hour, a flow path began to form. After 2 hours, a single meandering flow path had formed (Figure 2(c)).

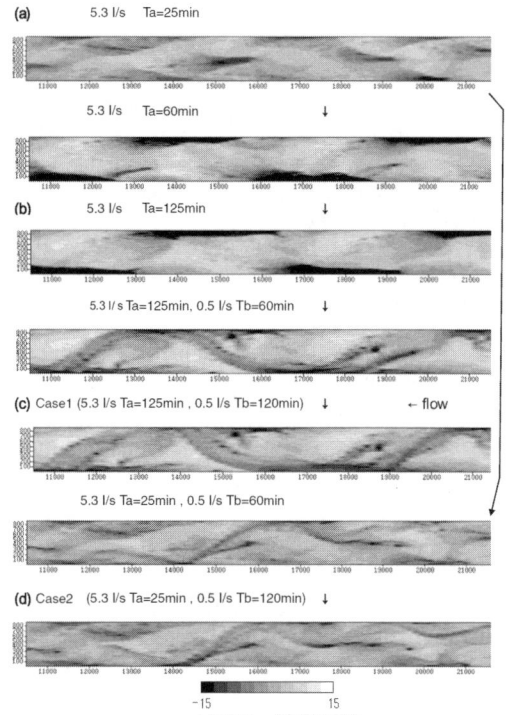

Figure 2. Sandbar shape (Cases 1 and 2).

#### b) Case 2

After about 25 minutes of discharge $Q_a$, double-row bars formed (Figure 2(a)). Observation was continued after reducing the discharge to $Q_b$. After the flow was reduced, a braided stream formed on the side of the sandbar where the riverbed was lower. Two hours into the experiment, two irregular meanders had completed their formation (Figure 2(d)).

#### c) Case 3

In Case 3 (steady-discharge experiment), discharge was constant at $Q_b$. The topographies were compared with those of Cases 1 and 2. The initial condition was $B/h = 50$, which is a condition in between the conditions under which alternate bars and double-row bars

800

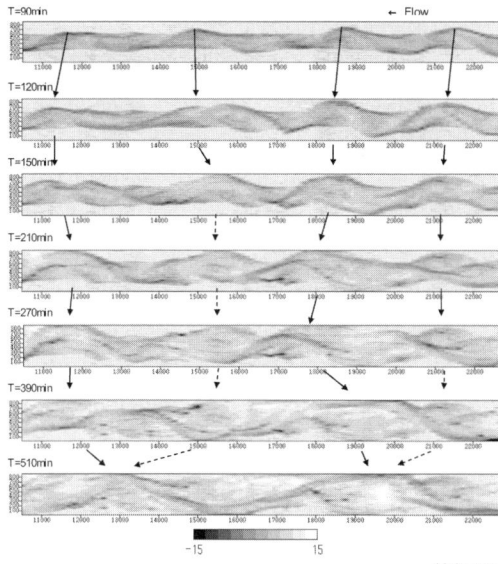

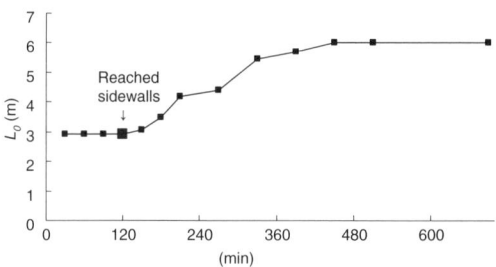

Unit: mm.

Arrows indicate migration of the deeply scoured por-tions at the right bank.

Figure 3.   Meandering shape (Case 3).

Figure 4.   Meander wavelength (Case3, $Q = 0.5$ l/s).

are formed. The process of change is shown in Figure 3. Arrows in the figure show the migration of the deeply scoured portions at the right bank. At 30 minutes into the experiment, five alternate bars ($L_o = 3$ m) had formed in the 15-m section. With the formation of the alternate bar, the banks at the deeply scoured portions began to erode and the flow changed to a meandering watercourse. At 120 minutes into the experiment, the meanders had reached the channel sidewalls. As shown in Figure 4, the meanders had a wavelength of about 3 m until 120 minutes into the experiment, after which the meanders reached the sidewalls of the channel and the wavelength changed to long. At 390 minutes into the experiment, the meander wavelength had increased to $L_o = 5$ m. The channel shape in Case 3 120 minutes into the experiment was compared with those in Cases 1 and 2. This time period was chosen for comparison because within this time period the meander in Case 3 remained free without reaching the sidewalls.

## 2   DOUBLE FOURIER ANALYSIS OF THE BED CONFIGURATION

The riverbed shape is expressed by Equation (1), and $\alpha_{ij}$ is obtained through double Fourier analysis (Hasegawa et al. 2004).

$$z = \sum_{i=0}^{I} \sum_{j=0}^{J} \alpha_{ij} \sin(i\frac{2\pi}{2B_v}n - \frac{\pi}{2}\frac{1+(-1)^i}{2})\cos(\frac{2\pi}{L}s - \sigma_{ij}) \quad (1)$$

where $B_v$ is the width of the channel; $i$ is the mode of sandbars; $j$ is the longitudinal wave number based the wave length of alternate bar; $\alpha_{ij}$ is the wave amplitude when the mode is $i$ and the wave number is $j$; $\sigma_{ij}$ is the phase of the wave ($i = 1$ and $j = 1$); and $L$ is the basic wavelength set as meandering wave length.

The results are shown in Figure 5. The wavelengths observed in the steady-flow experiment and when the double-row bars formed were 3 m. In contrast, the wavelengths of the alternate bars that formed during high discharge and those of the subsequently formed meanders were about 6 m. Therefore, the study was performed with the basic wavelength fixed at 6 m. Wavelengths in different experiments were compared. Furthermore, since the average water depth in each case was about 1 cm, and time-series comparisons were necessary, the riverbed data were used without dividing water depth by average water depth. Wave amplitude was expressed in centimeters.

In Case 2, waves with $i = 2, j = 2$ predominated, and this shows that double-row bars had formed at the $Q_a$ term. On the other hand in Case 1, the waves with $i = 1$, $j = 1$ predominated. This is in keeping with the char-acteristics of the alternate bar that was formed after the transition from double-row bars to alternate bars.

The above results show that when the riverbed is subject to low discharge after the riverbed has been shaped by high discharge, the riverbed shape that was formed during high flow is retained.

In Case 3, the predominant waves changed from those with $i = 1$, $j = 2$ (wavelength: 3 m) to those with $i = 1, j = 1$ (wavelength: 6 m). These waves that formed in Case 3 have the characteristics of alternate bars.

## 3   FLOOD ON THE APPETSU RIVER, AND TERRAIN CHARACTERISTICS

The previous sections examined alternate bars and double-row bars and their changes in form with time. These results were used in studying the flood on the Appetsu River and the terrain characteristics.

### 3.1   Outline of the river

The Appetsu River in Hokkaido has a basin area of 290.7 km$^2$ and length of 45 km. Typhoon Etau hit in

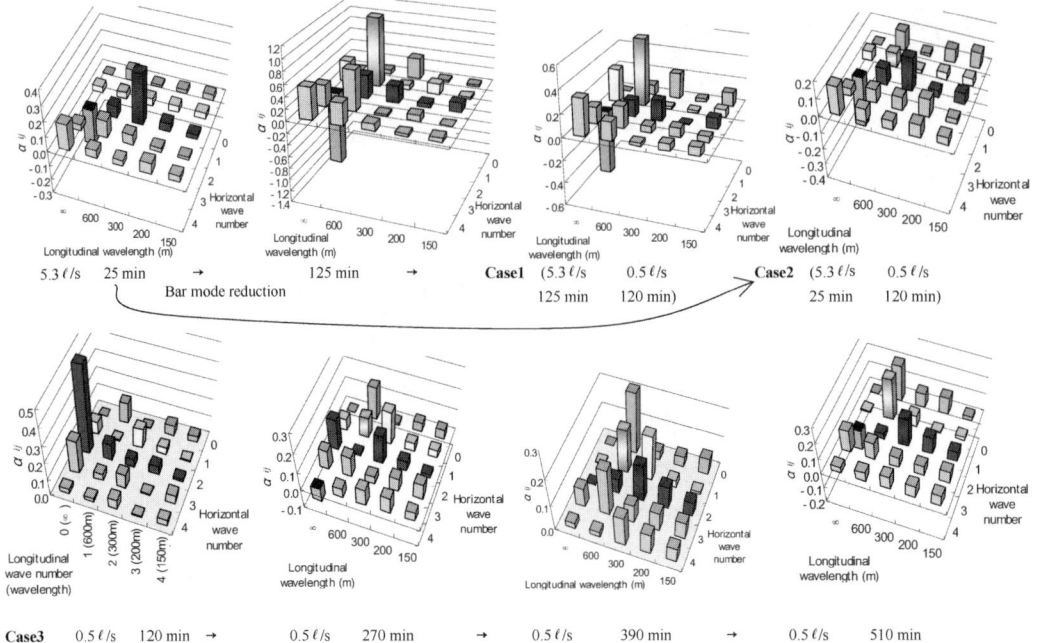

5.3 $\ell$/s   25 min   →    125 min   →   **Case1**   (5.3 $\ell$/s   0.5 $\ell$/s     **Case2** (5.3 $\ell$/s   0.5 $\ell$/s

Bar mode reduction            125 min    120 min)      25 min    120 min)

**Case3**   0.5 $\ell$/s   120 min   →    0.5 $\ell$/s   270 min    →    0.5 $\ell$/s   390 min   →    0.5 $\ell$/s   510 min

Figure 5.   Double Fourier analysis.

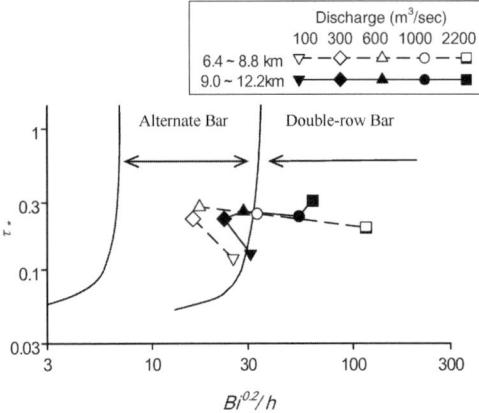

Figure 6.   Discharge and bed configuration.

Hokkaido, Japan, on 9 August 2003. A flood with an unprecedented discharge of about 2,200 m³/sec, which was far beyond the design discharge of 1,000 m³/sec, caused the entire valley flats of Appetsu River to flood (Hasegawa et al. 2004).

The meso-scale riverbed configuration (Kuroki & Kishi 1984) of the valley, including land protected by embankments, is broken down in Figure 6 by flow volume. From this figure, it was judged that discharges of up to about 1,000 m³/sec are in the domain of alternate bars, and discharges of above that are in the domain of

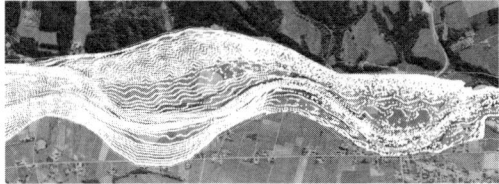

Figure 7.   Two-dimensional flood calculation of the downstream section (from the river mouth to 4 km upstream) on the Appetsu River (Hasegawa & Shimizu 2005).

double-row bars. At the peak of the flood, the discharge was 2,200 m³/sec, which is in domain of double-row bars.

Figure 7 shows two-dimensional flood calculations (Hasegawa & Shimizu 2005) from the river mouth to 4 km upstream on the Appetsu River. This figure also shows that flow line on the double-row bars occurred.

### 3.2   Shape analysis of the valley

Double Fourier analysis of the valley shape was performed, as in the previous section, under the assumption that the terrain was formed by repeated flooding. The topography of the Appetsu River was compared to the topographies produced in the three experimental cases. The sidewalls of the test channel in the experiment correspond to valleys in the actual river.

Figure 8 is a three-dimensional representation of data used for analysis. In the transverse direction,

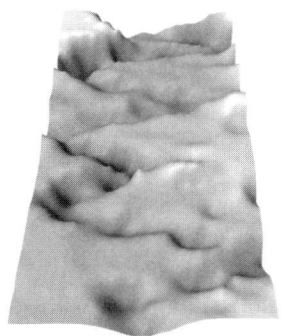

Figure 8.   Topographical data used in the analysis.

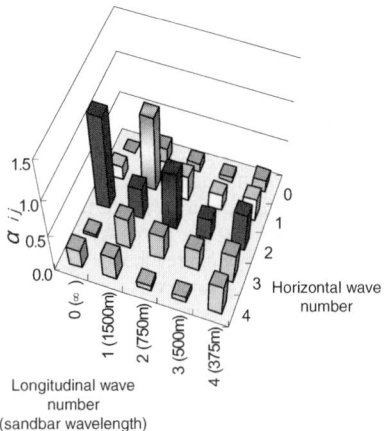

Figure 9.   Double Fourier analysis (Appetsu River; 10 km from the river mouth).

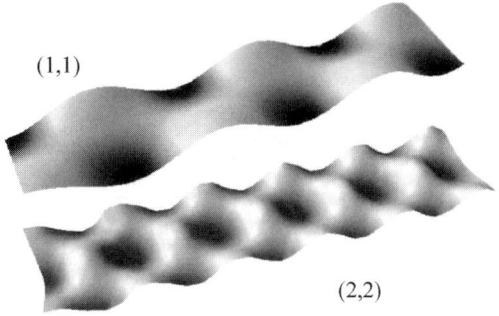

Figure 10.   Wave of $(i=1, j=1)$ and $(i=2, j=2)$.

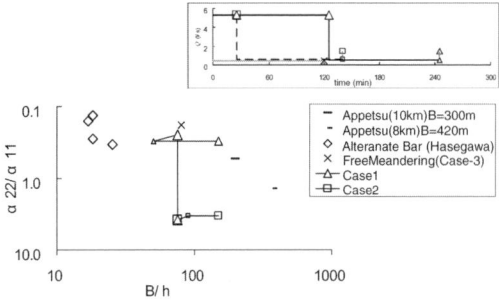

Figure 11.   $B/h$ vs. $\alpha_{22}/\alpha_{11}$ in the range of experimental discharges (bottom) and temporal change in discharge (top).

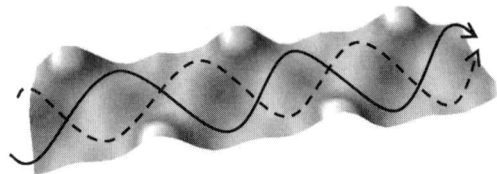

Figure 12.   Numerically modeled terrain of the Appetsu River.

the width of each 400- to 480-meter-long section is presented as 1.

The elevation is made dimensionless by dividing the difference in elevation from the mean valley flat (0) by the mean water depth.

The results of the analysis are shown in Figure 9.

On the Appetsu River, floods did not greatly change the form of the channel. The water surface width and depth in the experiment when discharge was 1.5 l/sec (equivalent to discharge under flood in the river) were used for comparison with the river.

Figure 11 shows the relationship between $\alpha_{22}/\alpha_{11}$ and $B/h$ from the experiment results and from the Appetsu River. The $\alpha_{22}/\alpha_{11}$ is the ratio of wave amplitude when $i=2$ and $j=2$ to that when $i=1$ and $j=1$, and $B/h$ is the ratio of channel width to water depth.

Figure 11 shows that the ratio of $\alpha_{22}/\alpha_{11}$ ranges from 0.1 to 0.4 in Cases1 and 3 and in the experiment conducted by Hasegawa (1982). However, that ratio is 3.4 in Case 2. The ratio of $\alpha_{22}/\alpha_{11}$ for the Appetsu River ranges between that at which alternate bars form

(Cases 1, 3 and Hasegawa) and that at which double-row bars form (Case 2).

Figure 12 shows the terrain recreated using an $\alpha_{22}/\alpha_{11}$ value of 0.75 (Figure 9). This figure is made from the wave $(i=1$ and $j=1)$ and the wave $(i=2$ and $j=2)$. This figure shows that the main flow runs like a figure of eight in the case of large discharge. However, in the case of small discharge, the flow course follows a meandering pattern along the lowest part of wave $(i=1$ and $j=1)$.

## 4   COMPARING THE CALCULATED WAVELENGTHS

This chapter compares the calculated wavelength of the Appetsu River with those of other rivers. By making a comparison firstly with the calculated double-row bar wavelength and latterly with the meander

803

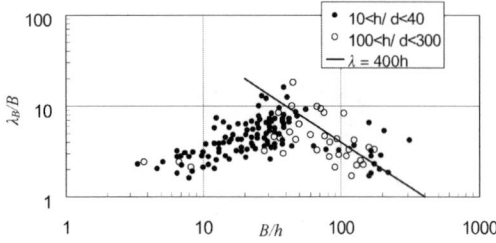

Figure 13. Water depth vs. bar wavelength ($\lambda_B = L/2$) (Muramoto, Y. & Fujita, Y. 1978).

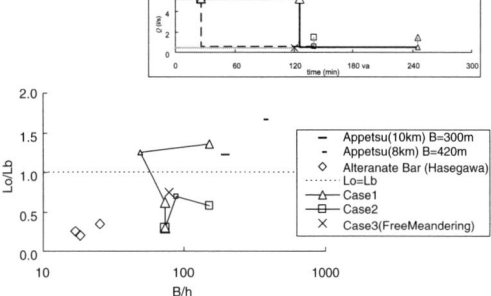

Figure 14. Calculated double-row bar wavelengths ($L_b$) and observed ($L_o$) in the range of experimental discharges (bottom), and temporal changes in discharge (top).

wavelength, the wavelengths of both the double- and alternate bars were analyzed based on the valley width flow (In the Appetsu River, $Q = 2200\,\text{m}^3/\text{s}$, $B_v = 300\text{–}420\,\text{m}$) and the mean annual maximum flow ($Q = 300\,\text{m}^3/\text{s}$, $B = 70\text{–}80\,\text{m}$).

### 4.1 Comparing the double-row bar wavelength

Figure 13 shows the relation between water depth and bar wavelength. When $B/h$ is small, bar wavelength has an inverse relationship to water depth. When $B/h$ is large, the double-row bar wavelength ($L_b$) has a proportional relationship to water depth ($L_b = 2\lambda_{Bb} = 800h$) (Muramoto, Y. & Fujita, Y. 1978).

Figure 14 compares observed bar wavelengths ($L_o$) to calculated double-row bar wavelengths. $L_b$ was calculated using Equation (2).

$$L_b = 800h \tag{2}$$

(the double-row bar wavelength)

The values for $L_o$, $B$, $h$ observed at 8 to 10 km from the river mouth in the Appetsu River and in Case 1 (between the discharge of 0.5 l/s and 1.5 l/s) are greater than average ($L_o/L_b > 1.0$).

Figure 15 compares $L_o$ to $L_b$ at discharges ranging from 100 to 2,200 m³/s in the Appetsu River. The $L_o/L_b$ was 1.2 to 1.5 in the discharge range of 300 to 2,200 m³/s.

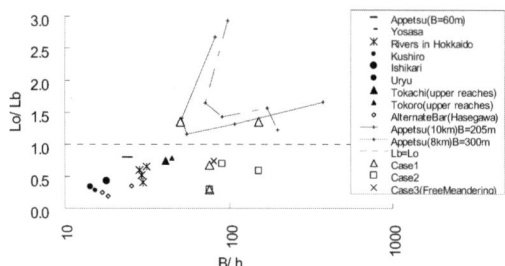

Figure 15. Calculated double-row bar wavelengths ($L_b$) and observed ($L_o$) at the range of discharges in the Appetsu River.

The shapes in Figure 15 shows $L_o/L_b$ vs. $B/h$ for various rivers in Hokkaido. They were obtained under a condition of constant water surface width ($B$), with Manning roughness coefficient fixed at $n = 0.03$, discharge fixed at average annual maximum discharge ($Q$), and observed wavelength ($L_o$), and uniform flow depth ($h$). $B$ and $L_o$ were obtained in a previous study (Hasegawa et al.1982).

In rivers with a gentle gradient, such as the Uryu and Ishikari rivers, and in meandering rivers, such as the Kushiro River, and experiments of alternate bars, the $L_o/L_b$ value was 0.1 to 0.5. In river sections with steep gradients, such as the upper reaches of the Tokachi and Tokoro rivers, the $L_o/L_b$ value was about 0.7. However, river sections on valley flats, such as the Appetsu river, showed long meander wavelengths with $L_o/L_b$ values of about 0.8 to 3.0. In this experiment, $L_o/L_b$ was 1.2 in Case 1, but 0.7 in Cases 2 and 3.

### 4.2 Comparing the meander wavelength

Figure 16 compares observed wavelengths ($L_o$) to calculated alternate bar wavelengths ($L_c = 2\lambda_{Bc}$) and calculated meander wavelengths ($L_d = 2\lambda_{Bd}$). $L_c$ was calculated using Equation (3), a theoretical equation developed by Ikeda and Parker (1981) that is based on the linear stability analysis of alternate bars and that is used to calculate the wavelength of alternate bars; $L_d$ was calculated using Equation (4) by Ikeda and Parker(1981), a theoretical equation that is based on linear theory for meander development and that is drawn from the meander equation.

$$L_c = 5v\sqrt{\frac{B}{gi}} = \frac{5Q}{h\sqrt{Bgi}} = \frac{5B^{1/10}}{n^{3/5}\sqrt{g}}\left(\frac{Q^2}{i}\right)^{1/5} \tag{3}$$

the alternate bar wavelength

$$L_d = \frac{4\pi}{1.5}\frac{h}{C_f} = \frac{8.4v^2}{gi} = \frac{0.85}{i}\frac{Q^2}{B^2h^2} = \frac{0.85}{n^{6/5}}\left(\frac{Q^2}{B^2i}\right)^{2/5} \tag{4}$$

the meander wavelength
where $Q$ is discharge, $h$ is water depth, $B$ is water surface width and $i$ is bed slope. $L_c$ and $L_d$ were calculated using Equations (3) and (4), and $Q$, $h$, $B$ were

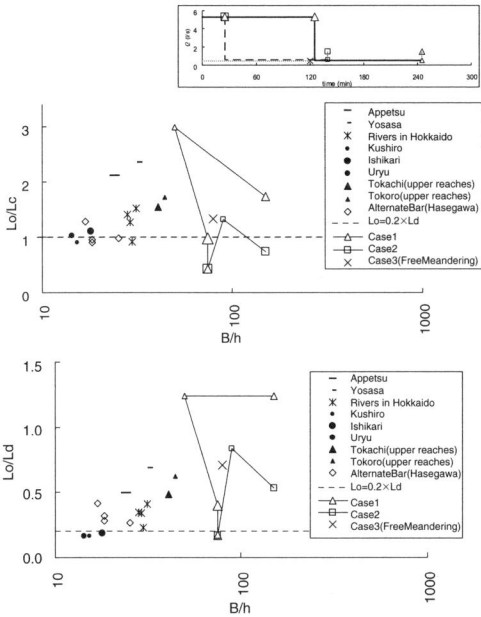

Figure 16. Calculated Alternate bar wavelengths ($L_c$), meander wavelengths ($L_d$) and observed ($L_o$).

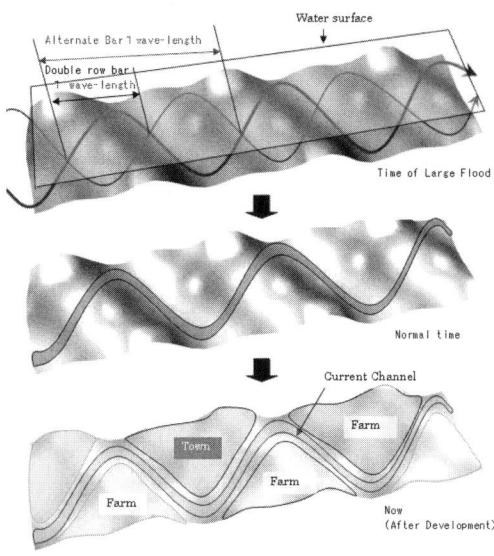

Figure 17. Process of topographic formation.

obtained from experiments. $L_c$ & $L_d$ for the rivers were calculated using $Q$, $i$, $B$.

In rivers with a gentle gradient, and in meandering rivers, the $L_o/L_c$ value was about 1.0. In river sections with steep gradients, the $L_o/L_c$ value was about 1.5. However, river sections on valley flats, such as the Appetsu river, showed long meander wavelengths with $L_o/L_c$ values of about 2.0 to 2.5. In this experiment, $L_o/L_c$ was 3.0 in Case 1, but 1.0 in Cases 2 and 3. Therefore, the findings suggest that in the Appetsu River, the meander wavelength at low discharge derives from the bar wavelength that was formed under the high discharge (Case 1).

In rivers with a gentle gradient, and experiments of alternate bars conducted by Hasegawa (1982), the $L_o/L_d$ values were 0.1 to 0.4, but in the Appetsu River, the Yosasa River, and at the upper reaches of the Tokachi River, the $L_o/L_d$ values were 0.4 to 0.6.

The $L_o/L_d$ value calculated for Case 3 was 0.7. In Cases 2 and 3, its value deviates greatly from $L_d$. This suggests that Equation (4) can only be used for the steady-flow experiment. The above finding could have resulted from discrepancies in riverbank conditions and variations in discharge used in calculations.

But when using the average annual maximum flow volume, a value of about $L_o/L_d$ of 0.2 should be considered appropriate. Rivers with $L_o/L_d$ values greater than approximately 0.2 are not meandering rivers; therefore, they are out of the scope of consideration.

The results calculated using Equation (3), which obtains the alternate bar wavelength, are similar to the experiment results and to the observed wavelengths in the rivers. Therefore, $L_o/L_c$ and $L_o/L_d$ might be usable as an index of river morphology.

## 5 STUDY OF TOPOGRAPHIC FORMATION

The studies above have clarified that meandering at low discharge is dictated by the sandbar shape that had formed by high discharge, and that this finding does not contradict current hydraulic equations. From the findings, the topographic formation of the valley flats on the Appetsu River was studied.

It became clear that the meander wavelength of the river's normal course and the wavelength of the alternate bars in valley flats were in rough agreement, and that the valley flat topography was made up of two kinds of sandbars: alternate and double-row. It was judged that the rim double-row bars on the Appetsu River derive from alternate bars that formed during large-scale flooding that spanned the width of the valley.

Figure 17, which is based on Figure 12, is a schematic of the process of topographic formation. The top figure shows sandbars that formed under large-scale flooding. The second figure from the top shows the same river channel at subsequent normal flow. The bottom figure shows the present channel, whose wavelength is determined by the wavelength of the sandbars that form during high discharge.

805

The findings made clear that it is possible to roughly estimate the behavior of flood flow at valley flats from the previously obtained behavior of flow over sandbars.

## 6 CONCLUSION

This research found the following:

(1) The experiment showed that the shape of riverbeds that were formed at high flow are retained after formation of the channel. Therefore, the morphological analysis showed that channels formed with the same meander wavelengths as those of sandbars formed at high flow.
(2) The data from flooding on the Appetsu River and the morphological analysis showed two coexisting sand wave components: double-row bars and alternate bars. The alternate bars have a wavelength double that of the double-row bars. Furthermore, the wavelength fell between those from the experiment results of Case 1 and 2.
(3) The channel formation process was analyzed by comparing the meander wavelength measured for the Appetsu River on the valley flat with those obtained from the hydraulic experiments. The ratio of the observed wavelength of the channel to the calculated meander wavelength was greater for the Appetsu River than for alluvial rivers. It is thought that the meander was caused by double-row bars which formed from large-scale flooding (i.e., those occurring less than once a year). The meander wavelength was similar to that in the channel formed at low discharge after high flow.
(4) In light of these findings, it is possible that the entire valley flats was formed by repeated large-scale floods that covered the entire valley. These findings show that hydraulic research can be used to estimate how valley terrain forms, for forecasting the likelihood of flooding caused by torrential rainfall and for disaster prevention.

ACKNOWLEDGEMENTS

This work was supported in part by the Hokkaido Regional Development Bureau of the Ministry of Land, Infrastructure and Transport, and the Foundation of River and Watershed Environment Management.

REFERENCES

Fujita, T., Tatsuzawa, H. & Hasegawa, K. 1999. Experimental reproduction and analyses of medium-scale bedforms in mountain rivers, IAHR Symposium on River, Coastal and Estuarine Morphodynamics: 273–282.

Hasegawa, K. 2004. Changes in the channel of the Appetsu River, Report of disasters caused by torrential rainfall from Typhoon Etau in Hokkaido in 2003. Hydraulic Engineering Committee of the Japan Society of Civil Engineers: 142–148.

Hasegawa, K. & Yamaoka, T. 1982. Analysis of alternating bars, Japan Society of Civil Engineers, Proceedings of Committee on Hydroscience and Hydraulic Engineering, Vol.26: 31–38.

Hasegawa, Y. & Shimizu, Y. 2005. Flood simulation for the Appetsu River in Hokkaido, Japan, during Typhoon Etau of 2003. Proceedings of the 4th IAHR Symposium on River, Coastal and Estuarine Morphodynamics: 301–307.

Ikeda, S., Parker, G. & Sawai, K. 1981. Bend theory of river meanders. Part 1.Linear development, Journal of Fluid Mech.112: 363–377.

Miwa, M., Daido, A. & Yokokawa, J. 2005. Effects of sediment supply and grain sorting on low-watercourse formation in channel with alternate bars, Annual Journal of Hydraulic Engineering of JSCE 49: 949–954 (in Japanese with English abstract).

Muramoto, Y. & Fujita, Y. 1978. The classification of meso-scale river bed configuration and the criterion of its formation, Annual Journal of Hydraulic Engineering of JSCE 22: 275–282 (in Japanese).

Nogami, T., Watanabe, Y., Hiroyasu, Y. & Hasegawa, K. 2007. Analysis of stream channel geometry and physiographic formative process in valley flats, Annual Journal of Hydraulic Engineering of JSCE 51: 991–996 (in Japanese with English abstract).

Takebayashi, H., Egashira, S. & Okabe, T. 2001. Stream formation process between confining banks of straight wide channels. Proceedings of the 2nd IAHR Symposium on River, Coastal and Estuarine Morphodynamics: 575–584.

Teramoto, A. & Tsujimoto, T. 2005. Effects of size heterogeneity of bed materials on mechanism to determine bar mode. Proceedings of the 4th IAHR Symposium on River, Coastal and Estuarine Morphodynamics: 433–444.

Uchijima, K. & Hayakawa, H. 1987. Characteristics Deformation of Alternate Bars due to Low Flow, Annual Journal of Hydraulic Engineering of JSCE 31: 683–688 (in Japanese).

Watanabe, Y., Hiroyasu, Y. & Shimada, T. 2007. Hydraulic experiments on changing process of bar shape in small discharge period, Annual Journal of Hydraulic Engineering of JSCE 51: 1039–1044 (in Japanese with English abstract).

Watanabe, Y. & Kuwamura, T. 2005. Mode-decrease process of double-row bars. Proceedings of the 4th IAHR Symposium on River, Coastal and Estuarine Morphodynamics: 445–453.

River, Coastal and Estuarine Morphodynamics: RCEM 2007 – Dohmen-Janssen & Hulscher (eds)
© 2008 Taylor & Francis Group, London, ISBN 978-0-415-45363-9

# Comparisons of morphology and flow structure at two braid-bar confluences in a large river

R.N. Szupiany & M.L. Amsler
*Consejo Nacional de Investigaciones Científicas y Técnicas (CONICET), Argentina Facultad de Ingeniería y Ciencias Hídricas, UNL, Argentina*

D.R. Parsons
*Earth and Biosphere Institute, School of Earth and Environment, University of Leeds, UK*

J.L. Best & R. Haydel
*Department of Civil and Environmental Engineering and Department of Geology, University of Illinois at Urbana-Champaign, USA*

ABSTRACT: This paper details the morphology and flow structure (primary and secondary currents) at two separate asymmetrical bar-confluences of the Paraná River, Argentina. The two sites occur over channel reaches ~3.8 km long and were surveyed in June 2006 with a single beam echo-sounder and an acoustic Doppler profiler (aDp). The width:depth ratios (W/D) are relatively high, providing opportunity to investigate similarities and differences between small and large scale fluvial confluences. The findings highlight that transversal distributions of the flow is critical as they combine within the confluence zone, essentially controlling the position of the scour and thus the flow structures in the downstream channel. In both confluences, and similar to studies of small confluence flows, counter rotating surface convergent flows are identified. However, the spatial extent of these secondary cells is limited to a low proportion of the total channel width. Moreover the cells are rapidly dissipated in the downstream direction.

## 1 INTRODUCTION

Mid-channel bars and islands in braided rivers are connected by zones of flow convergence, constriction and flow expansion. The dynamics of these channel-bar units are key in controlling the hydrological and sedimentological behaviour of such braided river systems. The dominant feature of flow within such morphological units are regions of flow convergence, the morphology and flow structure of which have been described extensively in literature. For example, the position and size of scour and deposition in relation to both the flow structure and sediment transport rates have been elucidated by field and laboratory research (e.g. Komura 1973; Mosley 1975, 1982; Ashmore & Parker 1983; Best 1986, 1987; Rhoads 2005). These investigations have documented the existence of several typical flow features at confluence zones (Best 1988), and have also found that the principal variables controlling the flow structure and channel morphology are the confluence angle and its planform asymmetry (Ashmore & Parker 1983; Best 1987, 1988), the flow and sediment discharge (or momentum) ratios

between the two confluent channels (Best 1987, 1988; Rhoads 1996, 2005), and the degree of bed concordance between the two confluent streams (Best & Roy 1991; Biron et al. 1996a,b). The role of bed discordance of the tributaries on the shear layer stability and the scour hole position as alternative origins of back-to-back secondary cells, has been studied by several authors, e.g. Rhoads & Kenworthy 1998; Biron et al. 2002. Understanding the morphology and hydrodynamic conditions within these morphodynamic zones thus permits insight into controls on the transfer and distribution of flow and sediment through the bar units, and the river channel as a whole (e.g. Ashmore & Parker 1983; Best 1988; Rhoads 1996; Rhoads & Kenworthy 1998).

However, despite a number of studies reporting measurements of the characteristics of such sites, the vast majority of studies have reported on findings from relatively small-scale field sites (Best 1988; Mclelland et al. 1996; Rhoads & Sukhodolov 2001; Rhoads 2005), detailed measurements made in laboratory flumes (Best & Roy 1991; Biron et al. 2002), or numerical models with boundary conditions based on these small-scale field and laboratory cases (e.g. Bradbrook et al. 2000; Bradbrook et al. 2001). There is thus a significant gap in our current understanding of how the findings from these small scales studies, which tend to have relatively low width:depth (W/D) ratios, scale to much larger river systems. We currently possess neither the data nor the understanding of how our findings from such small scale investigations relate to the processes occurring within much larger rivers, whose channels tend to have much higher W:D ratios. Recent developments in technology however, and in particular advances in global positioning systems and the advent of acoustic Doppler velocimetry, have begun to facilitate investigations of large river system dynamics (e.g. Richardson & Thorne 1998, 2001; Parsons et al. 2004; Parsons et al. 2005; Szupiany & Amsler 2005; Szupiany et al. 2005, 2006a,b,c). Parsons et al. (2007) show preliminary findings that question whether large-scale helical motions are present in large river channel confluences, and suggest that the role of form roughness at high W:D ratios might suppress the development of such flow structures.

The present paper details the morphology and flow structure, in both primary and secondary currents, measured at two separate asymmetrical bar-confluence units on the Paraná River, Argentina, where the W:D ratio is very large and typically above 100. The results expand the present database on bar-confluences in large river channels and provide an opportunity to begin to identify and analyse the similarities and differences between confluences of different spatial scales.

## 2 STUDY SITES AND FIELD MEASUREMENTS

The study areas on the Paraná River are located in its lower reach near the cities of San Martin (A) and Rosario (B), Argentina (Fig. 1). The two sites investigated consist of two large, asymmetrical, bar-confluences. Site A, at San Martin, was surveyed on 13th of June 2006 and site B, at Rosario, on 7th June 2006 (Fig. 1).

The Paraná River is one of the largest in the world (Schumm & Winkley 1994), with a drainage basin of $2.3 \times 10^6$ km$^2$. Downstream of the major confluence with the Paraguay River (Fig. 1), the mean discharge of the Paraná River is 19,500 m$^3$ s$^{-1}$, and the water surface slope is in the order of $10^{-5}$. The channel bed is composed largely of fine and medium sands (Drago & Amsler 1998), and the channel planform pattern has been classified as braided with a meandering thalweg (e.g. Ramonell et al. 2002). From a plan view, the river has a succession of wider and narrower nodal sections, with the mean widths and depths ranging between 600 to 2,500 meters and 5 to 16 meters, respectively.

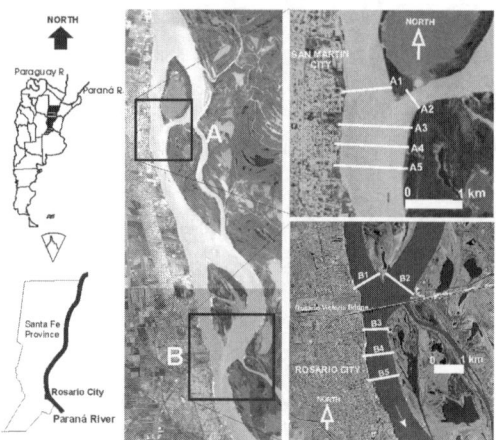

Figure 1. Study sites and measured cross sections within the two confluences.

At the two bar-confluences studied herein, the river bed morphology was surveyed using a Raytheon single beam echo-sounder which was coupled to a global positioning system (GPS) that was deployed on a small survey vessel. The GPS system provided horizontal positions to an accuracy of ±1 m at approximately 1 Hz. Morphological measurements were made along a series of cross sections through the bar-confluences, with each line separated by approximately 100 meters. The point x,y,z morphology data were interpolated onto a regular grid to create bathymetric maps of the two bar-confluence areas.

Once the bathymetry had been obtained, the three-dimensional (3-D) flow velocity was measured with a 1000 kHz Sontek acoustic Doppler profiler (aDp). At each of the two sites, flow measurements were made at 5 individual cross sections that were positioned in each of the confluences so as to provide details of the incoming flow fields and post-confluence flow structures (Fig. 1). Since the aDp was used from a moving vessel, it was linked to the GPS to provide both position and boat velocity. The aDp's own bottom tracking function was not used for boat motion due to the measurement errors that bed-load transport can introduce into the results obtained. The boat velocity and track position of the survey lines were monitored online and held as constant as possible during surveying, with a boat velocity of 1.5 m s$^{-1}$ (see Szupiany et al. 2006a for details). The aDp has 3 divergent monostatic transducers (beams), oriented 25° to the vertical and spaced at 120° from each other (Sontek 2000). The cell size (CS) and the averaging interval (AI), or ensemble, used herein were 0.75 meters and 10 seconds, respectively. In moving-vessel measurements, the velocity values represent the flow at a position halfway between two ensembles since during the AI the boat moves at a

808

Table 1. Main geometric and hydraulic characteristics of confluences A and B.

| Characteristic parameters | A | | | B | | |
|---|---|---|---|---|---|---|
| | LB | RB | A3 | LB | RB | B3 |
| Mean flow velocity (m s$^{-1}$) | 0.81 | 1.08 | 1.03 | 0.81 | 0.94 | 1.0 |
| Maximum flow velocity (m s$^{-1}$) | 1.25 | 1.63 | 1.73 | 1.09 | 1.14 | 1.26 |
| Width (m) | 310 | 700 | 1000 | 565 | 935 | 900 |
| Discharge (m$^3$ s$^{-1}$) | 2420 | 12250 | 14670 | 4800 | 9538 | 14338 |
| Mean depth (m) | 7.95 | 16.2 | 14.3 | 10.4 | 10.6 | 10.6 |
| Maximum depth (m) | 12.4 | 25.6 | 18 | 13.9 | 18 | 23.4 |
| Junction angle (degrees) | | 77 | | | 70 | |
| $\alpha_L^{(*)}$ | | 77 | | | 70 | |
| $\alpha_R^{(*)}$ | | 0 | | | 0 | |
| Flow momentum ratio, $M^{**}$ | | 0.15 | | | 0.43 | |
| Discharge ratio | | 0.2 | | | 0.5 | |

(*)$\alpha$ is the angle of deviation between the major tributary (defined on the basis of its discharge) and the downstream channel. (**)$M = (\rho Q_L V_L) / (\rho Q_R V_R)$, where $\rho$, water density (kg m$^{-3}$), Q, total discharge (m$^3$ s-1), V, cross-section average velocity (m s$^{-1}$) and the subscripts L and R denote left and right branches.

certain speed (i.e. the width of the measured water column will depend on the AI and the boat speed). Due to the beam separation, the water velocities are also spatially-averaged in a cone volume of 0.93*D diameter, where D is distance from the aDp transducer. These factors thus imply an assumption of temporal and spatial flow homogeneity in the measured volumes, which become particularly large at greater depths. This assumption may be valid depending on the flow characteristics. Considering an AI of 10 seconds and a boat speed of 1.5 m s$^{-1}$ (of the order of the flow velocity), the measured velocity would represent the flow conditions of an average volume that is 15 m in length and 0.93*D in width. In the Rio Paraná, with channel widths ranging between ~600 m to 2,500 m, the flow homogeneity assumption would thus be properly satisfied at this scale. The spatial and temporal averaging of the acoustic profiles obtained from a single transect with a moving-vessel yielded a large scatter in the measured velocity values that were deemed unacceptable for analysis of the detailed flow structure. Thus in order to obtain representative values of the time-averaged velocities, a series of 5 or more repeat transects were averaged (see Szupiany et al. 2006a for details).

## 3 SECONDARY CURRENTS IN LARGE RIVERS

The velocity vectors at confluences tend to cross obliquely relative to any applied cross-section reference system. Under such conditions the magnitude of cross-stream velocity components due to flow skewing obliquely through the section predominates and obscures the identification of any underlying coherent secondary circulation within the flow. Therefore a reliable procedure should be adopted if separation

of secondary velocities from cross-stream components is intended. The correct and robust definition of secondary currents within river channels has been a source of recent debate (Rhoads & Kenworthy 1999; Lane et al. 1999). Lane et al. (2000) summarize the calculation methods and classification of secondary flows into 4 definitions. However, discussion is still ongoing as to which of these methods yield the most realistic secondary velocity field. In the Paraná River, with channel widths varying between 600 and 2,500 m, constraints exist on the methods of data collection. These constraints result in both the centreline definition and the discharge continuity procedures being problematic to apply, given the problems of rapidly changing channel geometry, flow division and combination, and the requirement of closely spaced cross-sections respectively. Indeed, large river cross sections tend to have significant transverse variations in the main flow direction, which often complicates identification of the primary flow direction and thus definition of a robust frame of reference. Thus, Szupiany et al. (2006c) concluded that the Rozovskii and zero net-cross stream discharge definitions are the most effective methods for deriving secondary flows, yielding comparable and realistic results. Furthermore, Szupiany et al. (2006c) conducted a series of tests to show that the Rozovskii method offered the most effective method within very large, wide rivers, such as the Paraná River.

The Rozovskii method defines the secondary circulation on the basis of individual vertical profiles. The primary velocity direction is obtained for each profile as the depth-integrated flow vector and the secondary currents are identified by the differences from this average vector within the profile. This procedure effectively identifies individual secondary planes at each vertical profile across a given section, thus permitting

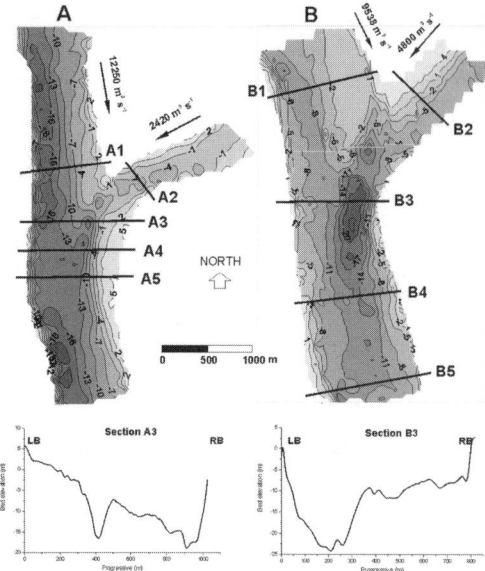

Figure 2. Bed contour maps of confluences A and B (Depths referred to the 0 m level in the Rosario Port gage). Survey date: 13/06/06 (A) and 07/06/06 (B). Water level: 2.74 m and 2.25 m respectively.

variations in the primary flow direction within a section, without distorting the secondary flow results (see Rhoads & Kenworthy 1998 for details).

## 4 RESULTS AND DISCUSSION

### 4.1 *General hydraulic and morphologic characteristics*

The gross geometric and hydraulic conditions of confluences A and B during the measurements are summarized in Table 1. The junction angle at each of the sites was measured using the average thalweg path along both branches and the angles were very similar, being 77 degrees and 70 degrees in A and B respectively. River stage was approximately equal during the two surveys with the stage records at Rosario Port being 2.3 m on June 7th and 2.7 m on June 13th. At both sites the right branch channel dominates in its flow discharge contribution to the confluence. However, due to differences in channel configuration, the discharge and momentum ratios are different in A and B (see Table 1).

The channel bed topography is shown in Figure 2. At B the incoming channels are of a similar depth with no distinct discordance in bed elevation. However, a deep scour region is present within the immediate confluence zone located at section B3. The scour is positioned towards the left bank, where the two flows

combine, and extends to a depth of approximately 23 meters and is around ~1,000 meters in length. The position and scale of the scour resembles those scours found and reported in studies of smaller confluences (e.g Ashmore & Parker 1983; Best 1988; Rhoads 1996; Bradbrook et al. 2000; Rhoads & Sukhodolov 2001). In the case of confluence A, there is a pronounced discordance in the bed height of the confluent channels, with the true left channel being much shallower (~8 m) than the true right channel (~16 m). The true right channel possesses the deepest flow and this continues thorough the confluence zone. The confluence scour is not as well defined as in B, although there is a small area of central scour approximately 15 m depth that is downstream of where the flows combine and extends into a deeper, larger scour that is a continuation of the deeper true right channel. Bed discordance and avalanche faces at the mouth of each tributary would present different conditions when small and large confluences are compared. At large scales like are the cases herein the avalanche faces slope are very small with angles lower than 4–3°. Unlike the case in small confluences with discordant beds (Biron et al. 2002) flow separation is not possible with that slopes thus loosing the avalanche faces their identity.

Horizontal (planform) flow separation and recirculation exits below the downstream junction corner on the true left banks were observed in field at both confluences and a large separation zone bar has formed at the downstream junction corner of A. However, such a bar is very much less pronounced in confluence B and separation was not measured at this site. Separation zone bars have been noted in numerous studies performed at small river confluences (Best 1988; Rhoads & Kenworthy 1995; Rhoads 2005). Reasons for the difference in the nature of the bed morphology at these two sites seem to be related to the relative positions of the scour and the rapidity of flow deflection as the flows combine. In A, the scour is closer to the true right bank and is thus much further from the left bank than in B, where the scour is close to the true left bank which thus suppresses the extent of the separation zone and the separation zone bar, particularly at higher flows.

### 4.2 *Primary velocity currents*

Figures 3 and 4 show the primary ($v_P$) and secondary components of velocity ($v_S$) at the 5 cross sections at confluences A and B respectively (A1 to A5 at A, and B1 to B5 at B). In both confluences, the first two sections were placed upstream of the junction within the confluent channels (A1 and A2 in A, and B1 and B2 in B). The third sections, A3 and B3, were positioned such that they incorporated the maximum depth of the scour hole. The fourth and fifth sections (A4 and A5 in A and B4 and B5 in B) were placed downstream to

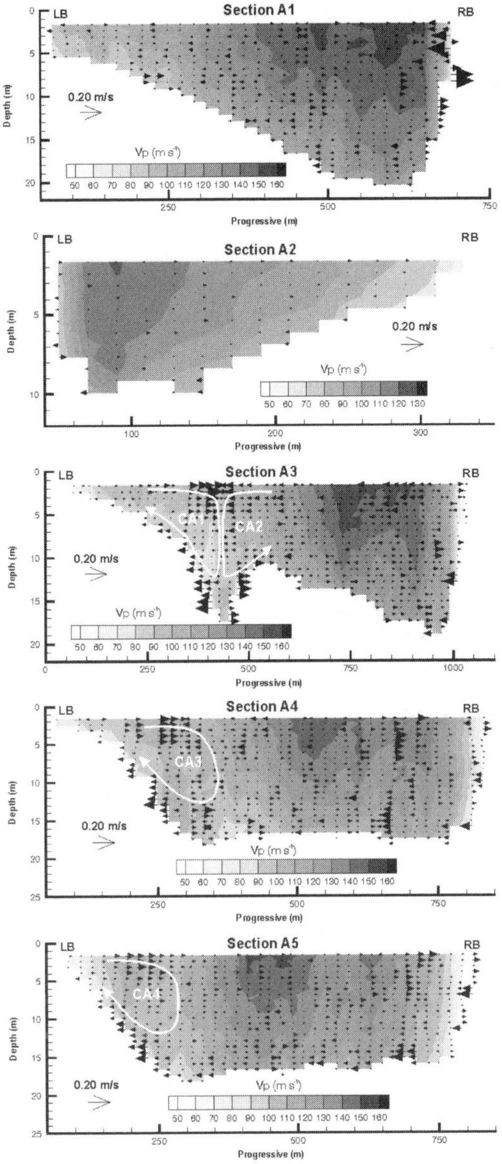

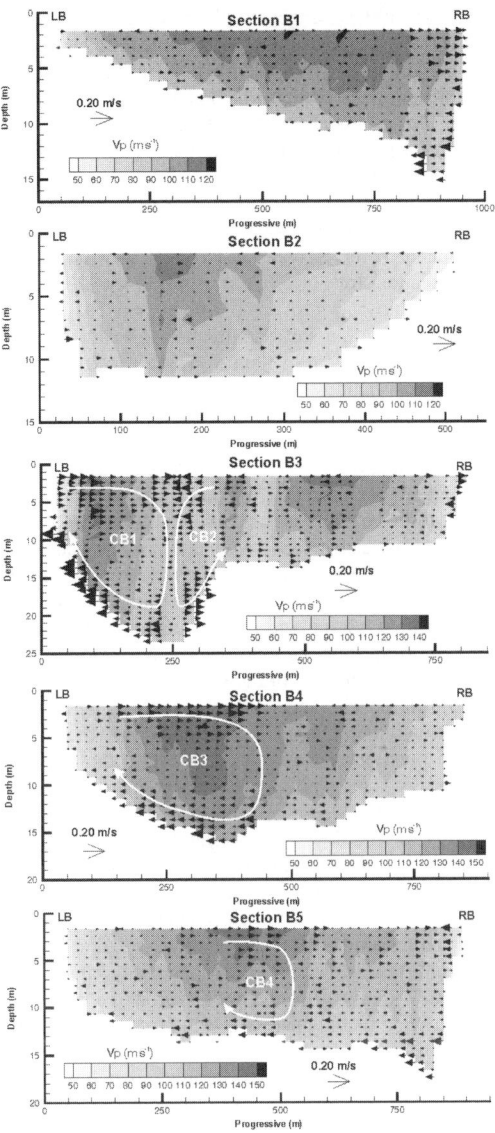

Figure 3. Primary and secondary velocity fields in the surveyed sections at confluence A.

Figure 4. Primary and secondary velocity fields in the surveyed sections at confluence B.

investigate the structure of flow as the flows recovered in post-confluence channel (Fig. 1).

Within both of the sections placed across the maximum scour depth within the two confluences (A3 and B3, Fig. 1), the two cores of maximum primary velocities from the upstream branches remain distinct (Figs. 3 and 4). A zone of lower primary velocity exists within the central area as the flows combine. At site B, between 400 to 450 m across section B3 there is an

area of slower flow that is interpreted as an effect of one of the central piers of the Rosario-Victoria Bridge located just upstream (see Fig. 1). Nevertheless, the location of the cores of primary velocity in the confluent channels, and in particular their position within the dominant branch channel (the right branch in both cases), appears to exert an important control on the structure of flow as the flows combine. This influence extends to the behaviour of the incoming flow from the smaller branch channel. In confluence A, the main flow in the main branch channel is positioned between the centre of the channel and the true right bank (A1), and remains close to the right bank through sections A3 and A4 (Fig. 3). In confluence B, the main flow in the dominant branch channel is more uniformly distributed across channel (section B1) and maintains that position downstream through the main confluence zone (B3, Fig. 4).

A distinct difference is observed when comparing the discharge distribution at sections A3 and B3. In confluence A the left branch is ∼ 310 m wide with a cross section area of 2761 m². In A3, the discharge flows within a width of ∼ 400 m and an area of 3245 m², extending between the true left bank and the region of slow primary velocities where the flows combine. However, the left branch discharge at B3 (flowing though an area of 6170 m² with a width of ∼ 560 m at B2) is distributed across a smaller width with a cross-sectional area of ∼ 260 m and 5350 m², respectively. This finding suggests that through confluence B the flow within the left branch channel in B2 is constricted in the downstream direction, resulting in flow acceleration between the mixing interface and the true left bank at B3. This may thus prevent formation of an extensive separation zone and result in erosion of the left bank. At A, despite the lower momentum ratio (Table 1), flow from the left branch channel is not so constricted and accelerated in a similar manner, permitting the formation of a larger separation zone at the downstream junction corner and producing sedimentation within this zone. This is because the primary velocity core in A1 is well displaced towards the right bank through A3 and does not produce large flow accelerations when mixing with the discharge coming from A2. Such interaction of the combining flows and lack of large accelerations would also explain the existence of a smaller scour hole in A than that found in B.

Recovery of the flow in B is relatively rapid, with only one primary velocity core discernable in sections B4 and B5, the intensity of which declines downstream. However, in the case of A, the two cores with marked different velocity magnitudes are still discernable in both sections A4 and A5, possibly due to these sections being nearer the junction. This maintenance of the difference in velocity between the two cores is despite the fact that downstream of A3 the maximum depths corresponding to the scour hole disappear and the sections are nearly uniform in terms of their spanwise geometry.

### 4.3 Secondary velocity currents

In the central portions of both confluences (sections A3 and B3), two counter-rotating surface-convergent secondary cells (CA1 and CA2 at A3 and CB1 and CB2 at B3, Figs. 3 and 4) are clearly visible across the scour holes where the two flows converge, with different directions between near bed and near surface velocity vectors (Figure 5). These velocities are coherent and

Table 2. Widths and maximum velocities of secondary cells.

|  | A (section A3) | | B (section B3) | |
|---|---|---|---|---|
| Secondary cell | left | right | left | right |
| Width (m) | 190 | 100 | 220 | 80 |
| Maximum velocity (m s-1) | 0.16 | 0.10 | 0.15 | 0.10 |

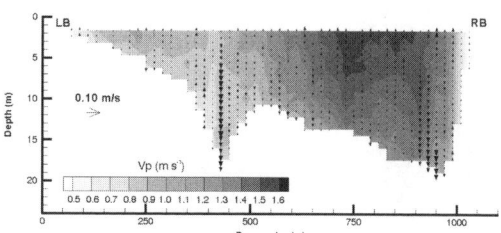

Figure 5. Vertical and primary components of flow velocity at section A3 (confluence A).

consistently above the measurement errors produced by aDp (see Szupiany et al. 2006a). The size of the two secondary flow cells seem depth limited, while their widths and maximum velocities are shown in Table 2. In both confluences, the secondary cells are of a similar spatial extent, flow intensity and pattern, with the cells being much larger and more intense in the left sections of the channel at both sites. However, given the widths of the channels, the spatial extent of these secondary cells is limited to a very small area of the channel, covering less than 25 % of the channel width. Such a finding is very different to that which has been observed for smaller channel confluences where the secondary cells cover up to 80–90% of the channel widths (Rhoads & Kenworthy 1995; Rhoads 2005). This has important implications for the dynamics of such channels and on the mixing rates as the flows combine (e.g. Parsons et al. 2007).

Meaningful vertical velocities were only captured in the shear layer at section A3 (Fig. 5) with values in the order of $0.05$ m s$^{-1}$, being above those comparable to the errors associated with the measurement equipment and are thus indiscernible from noise. The pattern of vertical velocity through the section indicates that the secondary flow cells identified above are reasonable, with downwelling over the central scour.

In sections A4 and B4, the larger secondary cells on the left side of the channels (CA3 and CB3, Figs. 3 and 4) that are generated upstream at sections A3 and B3 respectively, maintain their coherence. The cells retain approximately the same width in A4 and in B4 slightly increase in size (∼320 m). The magnitude of the secondary velocities remains the same in sections A4 and A5 (CA3 and CA4) in confluence A but decrease downstream in the case of B, although a small region of secondary flow components remain in B5 (CB4). However, the distance to B5 is further downstream than A5 (~600 m separating A3 from A5 and ~2000 m, B3 from B5). In both confluences the right-hand secondary cells lose their coherence much more rapidly than the secondary cells on the left of the channel (Figs. 3 and 4).

## 5 CONCLUDING REMARKS

The present paper provides a description and comparison of the main characteristics of bed morphology and flow structure at two large, asymmetrical, bar-confluence units in the Paraná River.

A number of similar features to those reported from past work detailing the confluence of far smaller streams were recorded: a zone of flow stagnation near the upstream junction corner; shear or mixing layers where the two flows combine; two counter-rotating surface-convergent secondary cells when the flows converge within the central scour hole, and a region of progressive flow recovery downstream from the confluence.

A zone of separated flow and sedimentation at the downstream junction corner, although not present in B, is evident in A. In the case of B, the lack of a significant separation zone seems related to the position of the scour and acceleration of the left branch channel through this left bank area, producing a zone of bank erosion. These findings also suggest that the $Q_L/Q_R$ may not the most important variable controlling the scour hole position on large rivers, and the position and distribution of the cores of primary velocity in the confluent channels may have a significant influence. Thus in confluence B, the position of the primary flow from the right branch channel is closer to the centre of the channel as the flows combine, and displaces and accelerates the flow from the left confluent channel through the confluence zone close to the true left bank. The possibility of a large asymmetry in the spanwise distribution of flow increases in large rivers, where W/D ratios are larger, and thus may exert an increasing influence on the confluence flow dynamics. An approach that assesses the 'local' momentum ratio between the cores of high velocity fluid, rather than the momentum ratio across the entire channel, may thus be a better predictor of confluence morphology in such wide channels. Additionally, this also suggests the major influence in the dynamics of such large confluences of the flow structure inherited from upstream and the role of upstream diffluence morphology and dynamics (Parsons et al. 2007).

In both confluences, two counter rotating, surface convergent secondary cells were identified. The secondary cells were generally small compared to the channel width. For example, widths were found to be ~ 250 m (left cell) and ~ 100 m (right cell) at B3 and ~ 200 m (left cell) and ~ 100 m (right cell) in

A3. The lack of spatial extent of these cells has profound implications in the dynamics and flow mixing in large rivers. Parsons et al. 2007 suggested that in large rivers with high width/depth ratios, the influence of form roughness dominates and suppresses the development of channel scale secondary currents that are a feature of much smaller rivers. These findings have considerable implications specifically in relation to the morphologic and hydraulic changes recorded at these units over time.

This research has gathered a unique data set concerning flow structure and morphology measured at two asymmetric disconcordant braid-bar confluences. The distribution of primary and secondary flows and morphology in large rivers confluences detailed herein are rarely collected in these types of large rivers and have been found crucial to adequately describe the characteristics of the flow to future research of the behavior of sediment transport in these zones.

## ACKNOWLEDGEMENTS

The authors are grateful for the field surveys conducted by Eng. Jose Huespe, who is responsible of the hydrographic measurements at the FICH. This study was made within the framework of the project: "Morphology and Hydraulic Features of Nodal Points in the Paraná River main channel" granted by the Universidad Nacional del Litoral of Santa Fe, Argentina. RNS thanks the Earth and Biosphere Institute at the University of Leeds for funding a 3-month research fellowship, which enabled the writing of this paper. HR thanks the STC International Research Experience Program of the NSF through the Nanobiotechnology Center under Agreement No. ECS-9876771. DP thanks the UK Natural Environment Research Council for his Fellowship funding (NE/C002636/1).

## REFERENCES

Ashmore, P.E., & Parker, G. 1983. Confluence scour in coarse braided streams. *Water Resources Research*, 19, 392–402.
Best, J.L. 1986. The morphology of river channel confluences. *Progress in Physical Geography*, 10, 157–174.
Best, J.L. 1987. Flow dynamics at river channel confluences: implications for sediment transport and bed morphology. *The Society of Economic Paleontologists and Mineralogists*.
Best, J.L. 1988. Sediment transport and bed morphology at river channel confluences, *Sedimentology*, 35, 481–498.
Best, J.L. & Roy, A.G. 1991. Mixing-layer distortion at the confluence of channels of different depth. *Nature*, 350, 6317, 411–413.
Biron P., Best J. L., & Roy A.G. 1996a. Effects of bed discordance on flow dynamics at river channel confluences. *Journal of Hydraulic Engineering*, ASCE 122(12), 676–682.

813

Biron P., Roy A.G. & Best J. L. 1996b. Turbulent flow structure at concordant and discordant open-channel confluences. *Experiments in Fluids*, 21, 437–446.

Biron P. M., Richer A., Kirkbride A. D., Roy A. G., & Han S. 2002. Spatial patterns of water surface topography at a river confluence. *Earth Surface Processes and Landforms*, 27, 913–928.

Bradbrook, K. F, Lane, S. N. & Richards K. S. 2000. Numerical simulation of three-dimensional, time-averaged flow structure at river channel confluences. *Water Resources Research*, 36, 9, 2731–2746.

Bradbrook, K. F, Lane S. N., Richards K. S., Biron P. M., & Roy A.G. 2001. Role of bed discordance at asymmetrical river confluences. *Journal of Hydraulic Engineering*, may, 351–368.

Drago, E. & Amsler M. L. 1998. Bed sediment characteristics in the Parana and Paraguay rivers. *Water Internat*. 23, 174–183.

Komura, S. 1973. River-bed variations at confluences. Proceeding of the Symposium on River Mechanics, Bangkok, 9–12 January, Paper A66, 773–784.

Mclelland, S.J., Ashworth P.J., & Best J.L. 1996. The Origin and Downstream Development of Coherent Flow Structures at Channel Junction. *In Coherent flow structures in open channels*, Ashworth, Bennett, Best and McLelland (eds), Wiley and Sons, 491–519.

Mosley, M. P. 1975. An experimental study of channel confluences. PhD. Thesis, Colorado State University, 216.

Mosley, M. P. 1982. Scour depths in branch channel confluences: Ohau River, Otago, New Zealand. Proceedings of New Zealand Institute of Professional Engineers, 9, 17–24.

Lane, S. N., Bradbrook, K. F., Richards, K. F., Biron, P. M. & Roy A. G. 1999. Time-Average flow structure in the central region of a stream comfluence: a discussion. *Earth Surface Processes and Landforms*, 24: 361–367.

Lane, S. N., Bradbrook, K. F., Richards, K. F., Biron, P. M., & Roy A. G. 2000. Secondary circulation cells in river channel confluence: measurement artifacts or coherent flow structures?. *Hydrological Processes*, 14, 2047–2071.

Parsons D. R., Lane S. N., Hardy R. J., Orfeo O., & Kostaschuk R. 2004. The Morphology, 3D flow structure and sediment dynamics of a large river confluence: the Río Parana and Río Paraguay, NE Argentina. Proc. Of the Sec. Int. Conference on Fluvial Hydr. Eds. M. Greco, A. Canaveta & R. Della Morte. Vol. 1. River Flow 2004. Napoly, Italy.

Parsons D. R., Best J., Lane S., Kostaschuk R., Orfeo, O., & Hardy R. 2005. Flow structure and morphology of a confluence-diffluence: Río Parana, Argentina. 4th IAHR Symposium on River, Coastal and Estuarine Morphodynamics, Urbana, Illinois, USA, October, 2005.

Parsons D.R., Best J.L., Lane S.N., Orfeo O., Hardy R.J & Kostaschuk R. 2007. Form roughness and the absence of secondary flow in a large confluence-diffluence, Río Paraná, Argentina. *Earth Surface Processes and Landforms*. Esex Commentary. DOI: 10.1002/esp. 1457.

Ramonell, C.G., Amsler M.L., & Toniolo H. 2002. Shifting modes of the Paraná River thalweg in its middle/lower reach. *Zeitschrift fur Geomorphologie*. Suppl.-Bd. 129, 129–142.

Rhoads, B.L., & Kenworthy S.T. 1999. On secondary circulation, helical motion and Rozovskii-Based analysis

of time-averaged two-dimensional velocity fields at confluences. *Earth Surface Processes and Landforms*, 24, 369–375.

Rhoads, B.L. 1996. Mean structure of transport-effective flows at an asymmetrical confluence when the main stream is dominant. In Coherent flows structure in Open Channel, Ashworth P.J., Bennett S.J., Best J.L., McLelland S.J. (eds). Wiley and Sons; 491–517.

Rhoads, B.L. 2005. Scaling of confluence dynamics in river systems: some general considerations. 4TH IAHR Symposium on River, Coastal and Estuarine Morphodynamics, Urbana, Illinois, USA, October, 2005.

Rhoads, B.L., & Kenworthy S.T. 1995. Flow structure in asymmetrical stream confluence. *Geomorphology*, 11: 273–293.

Rhoads, B.L., & Kenworthy S.T. 1998. Time-average flow structure in the central region of a stream confluence. *Earth Surface Processes and Landforms*, 23: 171–191.

Rhoads, B.L. & Kenworthy S.T. 1999. On secondary circulation, helical motion and Rozovskii-based analysis of time-averaged two-dimensional velocity fields at confluence. *Earth Surface Processes and Landforms*, 24: 369–375.

Rhoads, B. L. & Sukhodolov A. N. 2001. Field investigation of three-dimensional flow structure at stream confluences: 1. Thermal mixing and time-averaged velocities. *Water Resouces Research*, 37, 9, 2393–2410.

Richardson, W. R. & Thorne C. R. 1998. Secondary Currents around a Braid Bar in the Brahmaputra River, Bangladesh. *Journal of Hydraulic Engineering*, 124: 3, 325–328.

Richardson, W. R. & Thorne C. R. 2001. Multiple thread flow and channel bifurcation in a braided river: Brahmaputra-Jamuna River, Bangladesh. *Geomorphology*, 38, 185–196.

Schumm S.A., & Winkley B. R. 1994. *The character of large alluvial rivers*. p. 1–9. In: Schumm S. A. & Winkley, B. R. (eds.): The variability of large alluvial rivers, ASCE, 467 pp.

Sontek 2000. ADP Acoustic Doppler Profiler. Technical Documentation.

Szupiany, R. N., & Amsler M. L. 2005. Estrategia de medición del campo de velocidades en un gran río con la tecnología acústica Doppler. XX° Congreso Nacional del Agua, Mendoza, Argentina.

Szupiany R. N., Amsler M. L., & Fedele J. J. 2005. Secondary flow at a scour hole downstream a bar-confluence (Paraná River, Argentina). 4TH IAHR Symposium on River, Coastal and Estuarine Morphodynamics, Urbana, Illinois, USA.

Szupiany, R. N., Amsler M. L., Best J.L. & Parsons D.R. 2006a. Comparison of Fixed- and Moving Vessel Measurements with an ADP in a Large River. *Journal of Hydraulic Engineering*, in press.

Szupiany, R. N., Parsons D.R.; Best J.L.; Amsler M. L. & Orfeo O. 2006b. Morfología y corrientes secundarias en una confluencia con cauces discordantes, Río Paraná, Argentina. XXII Congreso Latinoamericano de Hidráulica. Ciudad Guayana, Venezuela.

Szupiany, R. N., Parsons D.R.; Best J.L. & Amsler M. L. 2006c. Secondary currents at a large river: definition, methods and problems. (in prep).

*River, Coastal and Estuarine Morphodynamics: RCEM 2007 – Dohmen-Janssen & Hulscher (eds)*
© *2008 Taylor & Francis Group, London, ISBN 978-0-415-45363-9*

# Evolution of a new tidal river bifurcation: Numerical modelling of an avulsion after a catastrophic storm surge

Maarten G. Kleinhans
*Universiteit Utrecht, Fac. Geosciences, Dept. Physical Geography, Utrecht, The Netherlands*

Henk T. Weerts
*TNO Built Environment & Geosciences, Geological Survey of the Netherlands, Utrecht, The Netherlands*

Kim Cohen
*Universiteit Utrecht, Fac. Geosciences, Dept. Physical Geography, Utrecht, The Netherlands*

ABSTRACT: In 1421, the St. Elisabeth storm surge catastrophy transformed medieval reclaimed land into a wide tidal embayment known now as the Biesbosch area. The main delta distributary of the Rhine (river Merwede) diverted into the Biesbosch area estuary, effectively shortening its course to the North Sea by 10–15 km. A shallow delta, about $10 \times 20$ km large and 2 m thick formed in about 200 years. This study aims to investigate 1) the evolution of delta formation into the Biesbosch area and simultaneously 2) the evolution of the river channel upstream of the Merwede diversion site, using insights from numerical modelling (Delft2D). The initial and boundary conditions for the model scenarios are derived from historical maps since 1550 AD, geological cross-sections, modern tidal ranges and river discharge. The model simulates delta formation at progradation rates that are in agreement with the historical data. Basin depth and tidal boundary conditions were varied in 20 scenarios. Geological reconstructions for these properties are relatively uncertain. The scenario with 2 m depth and 2 m tides best resembles measured delta progradation rates. The pre-storm surge river channel, downstream of the diversion site, does not silt up because the partly subaerial levee directs water into it. In the river upstream of the diversion site, no significant incision nor aggradation is simulated for the post-storm surge condition for any scenario. 17th century river engineers suggested that the developments in the Biesbosch area explain a change of discharge division at the bifurcation at the delta apex some 100 km upstream (Lobith), which is believed until today. However, our results do not support such a causal link between the bifurcation events in the Biesbosch area and in the bifurcation at the delta apex. This is of importance for the development of a larger model, covering the complete Rhine delta and simulating their development for the period 800–1800 AD.

## 1 INTRODUCTION

Rivers depositing their sediment in deltas and coastal areas frequently change course, a process that is called avulsion. As a part of the process, a splay develops where the river diverts out of the old course, and in favourable conditions a new river develops from the channels feeding the splay (e.g. Smith et al., 1989; Stouthamer and Berendsen, 2000; Slingerland and Smith, 2004). The avulsion is (at least temporarily) a river channel bifurcation (Kleinhans et al., 2006). Its evolution is partly determined by downstream boundary conditions and partly by morphodynamics at the node. We study a deltaic avulsion site along the river Merwede (The Netherlands) that was initiated about 600 years ago by a series of catastrophic storm surges and river floods. The avulsion splay formed a shallow

water delta in an area now known as the Biesbosch (Fig. 1) (Weerts et al., 2005).

When there is a slope advantage (e.g. shortened distance to the sea) for the new course compared to the old course, avulsions are widely believed to have an increased chance to succeed (see Slingerland and Smith, 2004, for review). The transport capacity increases in the new and in the upstream channels because of the slope increase, and erosion is expected to propagate into the upstream river (e.g. Kriele et al., 1998). The upstream extent of this influence is unknown. Considering the many avulsion sites within the Rhine delta, establishing this distance is important to decide which of the many avulsion/bifurcation events should be considered independent. The Biesbosch avulsion is located about 110 km downstream of the largest Rhine bifurcation (Lobith). Historically

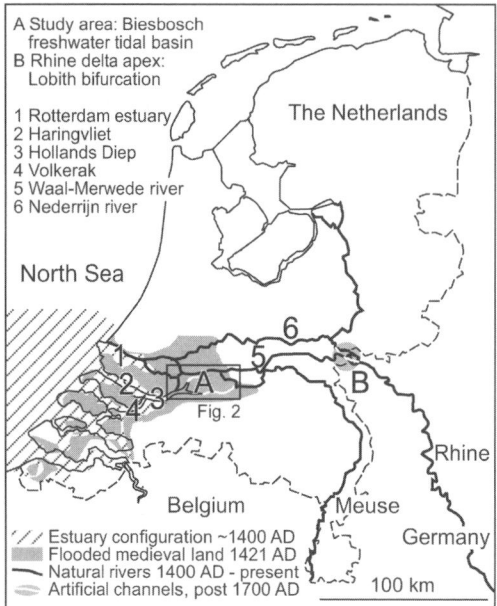

A Study area: Biesbosch
  freshwater tidal basin
B Rhine delta apex:
  Lobith bifurcation

1 Rotterdam estuary
2 Haringvliet
3 Hollands Diep
4 Volkerak
5 Waal-Merwede river
6 Nederrijn river

The Netherlands

North Sea

Fig. 2

Belgium

Rhine

Meuse

Germany

/// Estuary configuration ~1400 AD
▓ Flooded medieval land 1421 AD
➤ Natural rivers 1400 AD - present
➥ Artificial channels, post 1700 AD

100 km

Figure 1.  Study site and context.

it has been suggested that the Biesbosch avulsion in the 15th century AD affected the discharge and sediment distribution at the upstream Rhine bifurcation, accelerating the near-closure of the northern branch (Nederrijn) (see 1658 and 1749 references in Gottschalk, 1975; van de Ven, 1976).

The aim of this paper is to simultaneously assess the effects of morphodynamic developments downstream and upstream of the Merwede diversion into the Biesbosch area. The sequence of events that led to the formation of the Biesbosch tidal basin is briefly described here to serve as the background for modelling, including an estimate of the involved sediment budgets. After a short description of the 2DH model system, the chosen initial and boundary conditions are presented. Given the poorly constrained initial and boundary conditions, results of several simplified modelling scenarios are presented to assess the evolution and the sensitivity to the assumptions. This notwithstanding the delta and upstream channel evolution are robust and reproduce the observations, so that general conclusions can be drawn.

## 2  RECONSTRUCTION OF THE FORMATION OF THE BIESBOSCH

### 2.1  Catastrophies and delta formation

Zonneveld (1960) reconstructed the palaeogeography of the Biesbosch delta from historical maps and

geological data (Fig. 2) and Gottschalk (1975) added historical facts. Two smaller NW branches and one large SW branch connect the basin to estuaries. The 1550 AD map features a small delta at the mouth of the truncated river Merwede in the NE of the basin. The breached levee of the abandoned downstream part of the Merwede can be seen on the map, merged with the delta. In 1620 and 1685 AD, this delta had grown considerably larger.

The formation of the Biesbosch tidal basin was the result of a series of coincidences (Gottschalk, 1975). On 19 November 1421 a storm surge, the St. Elisabeth Flood, led to substantial loss of medieval embanked reclaimed land ('polder') known as the Grote Waard. In the preceding period, political struggles and increasing bureaucracy had caused neglection of dyke maintenance (Zonneveld, 1960; Gottschalk, 1975). In addition, the Grote Waard was created by artificially draining (digging ditches, debouching at low tide) and peat was mined for fuel and salt. Consequently, the land surface in the embanked polder was lowering towards ebb tide level. Several storm surges and river floods in the second half of the 14th century repeatedly breached the dykes, which deteriorated despite repairs. Hence, the St. Elisabeth flood dike breaches, which happened during the neap tide, was a disaster waiting to happen. To make things worse, a river flood reinundated the flooded area six weeks later. The damage had only partly been repaired in 1423, when yet another river flood breached the dykes. In 1424 a second storm surge followed, undoing all repairs and extending the damaged area to about $10 \times 30$ km.

By 1461 AD, parts of the flooded area were reclaimed through restoration of dykes (Land van Heusden en Altena E of the Biesbosch area, and land in the vicinity of the city of Dordrecht to the NW). Meanwhile a civil war flared up and the remainder of the lost land, about $10 \times 20$ km, was left unreclaimed and became the Biesbosch (Gottschalk, 1975). The area is a fresh water intertidal basin, mainly because the river Merwede diverted into the basin (Gottschalk, 1975). Thereby Rhine discharge newly connected to an estuary in the SW (Hollandsch Diep, Haringvliet) while, since about 1450, a delta formed of mud and sand (Zonneveld, 1960; Weerts et al., 2005) (Fig. 3). The tidal currents, meanwhile, eroded the southwestern part of the basin and perhaps imported estuarine mud into the area.

A Late Holocene channel belt of the river Meuse ran E-W through the study area, forming a modest ridge (Weerts and Berendsen, 1995) that probably acted as a divide between mud and sand deposition. The river deposited its sand to the north of the ridge while the easily suspended mud was also transported over the ridge (Weerts et al., 2005). The tidal currents on the other hand imported mud which mostly settled to the

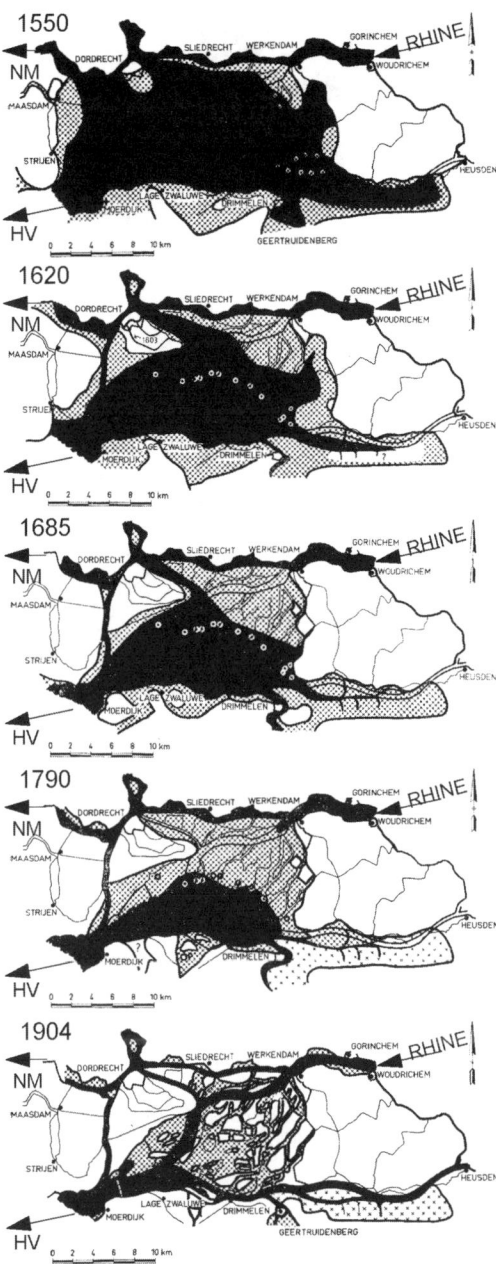

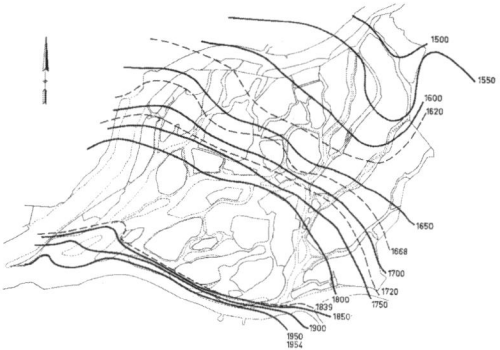

Figure 3. Progradation of the Biesbosch delta (after Zonneveld, 1960).

Figure 2. Palaeomaps of the Biesbosch delta (after Zonneveld, 1960). Black: open water, dark dotted: unembanked accreting areas, light dotted: embanked areas flooded during storm surges only, white: well embanked. Dots are beacons emplaced in 1560 on the old Meuse channel ridge. NM: Nieuwe Maas estuary (or Rotterdam estuary), HV: Haringvliet estuary. In 1904, the Nieuwe Merwede canal was dug in the northern part of the delta, and the Bergse Maas to connect the Meuse to the HV.

south of the ridge (Zonneveld, 1960), gradually filling the entire basin (Fig. 3).

After 1700 AD, the growing delta hindered the flood discharge of the river which caused high water levels in the upstream areas, and storm surges remained a threat. Therefore, a channel was designed (Nieuwe Merwede, dredged between 1864–1884) to reconnect the Merwede to the open estuaries in the south-west. For similar reasons of water management, a second channel was designed to independently connect the Meuse river to SW estuary (Bergse Maas, dug between 1886–1908), by connecting to the Amer tidal channel in the very south of the Biesbosch area. Until the 1960 s, the tidal range at the downstream boundary of the basin was about 2.1 m and dominated by M2 and O1 tides (Weerts et al., 2005). Major flood protection works in the SW Netherlands (Deltaplan following 1953 storm surge; Volkerak closure in 1968, Haringvliet closure in 1970) resulted in a reduction of the tidal range to 0.3 m leading to further mud deposition.

The period between 1461 and the early 18*th* century is the most interesting to model because the Biesbosch area developed with more or less constant boundary conditions, relative free of major human interference.

### 2.2 *Deposits and morphology*

Morphologically, the delta is characterised by channels that dissect sandy bars topped by silt and clay. Channels are filled by thin alternating clay and sand layers indicating a late-stage tidal origin. Some tidal-river channels scoured and reworked older sediments up to 10 m deep. Lithological cross-sections show that the delta has a variable thickness, typically 2–3 m. The variable thickness does not directly represent the depth of the basin: underlying unconsolidated Holocene sediments and peat were subject to compaction due to load of the delta (Weerts et al., 2005). The contribution of compaction to the total thickness is unknown. This is the main source of uncertainty in reconstructing 15th century topography/bathymetry and a reason to define a range of scenario's in the modelling exercise.

We use two reconstructions of depositional environments: the lithological profiles in Zonneveld (1960) which are based on his shallow soil augering and deeper mechanical corings by Rijkswaterstaat, and Weerts et al. (2005) based on the 'DINO' coring database of TNO.

In the NE of the tidal basin, only fresh-water macrobenthos is encountered, showing that the delta is the product of fresh-water river discharge predominantly (Weerts et al., 2005). Here, the first phase of delta deposition is marked by splays of relative coarse fluvial sand (origin testified by mineralogy by Zonneveld, 1960). The deposits locally buried the medieval soil horizon of the former embanked land. These bar deposits occur as a belt of splays south of the Merwede channel, continuing into the 1461-AD reclaimed part of the former Grote Waard polder immediately upstream of the Biesbosch area. This fits a 15th century age for these deposits.

Deposits attributed to the St. Elisabeth storm surge itself are limited to a few centimeters thickness and absent/unrecognizable in most places. In the W and S of the Biesbosch, the event-layer typically contains many juvenile *Cerastoderma* bivalves, indicative for short-lived brackish conditions. The event layer marks the thickness of the post-15th century AD deposition (Weerts et al., 2005).

The corings and Fig. 3 can be used to estimate the volume of accommodated sediment over time. Over an area of $10 \times 10 \, \text{km}^2$ (NE corner) the basin received a fill of about 2 m thickness between 1450–1650. Infilled channels cover about 10% of the area. These reworked a mixed substratum (clay, peat, locally sand) to about 10 m depth. This results in a total sediment volume of $280 \times 10^6 \, \text{m}^3$ trapped within the NE half of the basin. Some areas adjacent to the basin also silted up. Weerts et al. (2005) include these and the entire Biesbosch basin so that the displaced volume approximates $800 \times 10^6 \, \text{m}^3$), about half of which is fluvial sand and the remainder is mud. We estimate that at least $180 \times 10^6 \, \text{m}^3$ is sand imported from the river in 200 years which represents a robust test for the scenario modelling.

## 3  SCENARIO MODELLING

### 3.1  *Scenarios*

Several possible ways to schematise the area into a grid were tested. Curvilinear grids which expanded strongly into the area violated the requirements of orthogonality which predestined the model outcomes. A rectangular grid is the most appropriate with a

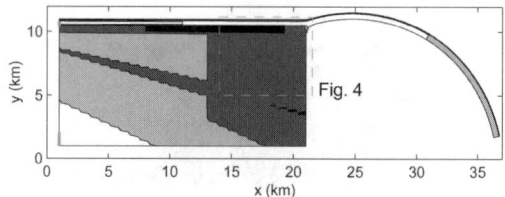

Figure 4.  Initial bathymetry between −6–2 m in 2 m classes. Bold lines are tidal water level boundaries. Dashed line indicates area shown in Fig. 5.

curvilinear-gridded river channel flowing into the northeast corner and two tidal boundaries in the western corners (bold lines in Fig. 4). The grid size in the basin is $100 \times 50$ cells of $200 \times 200$ m and in the river is $50 \times 6$ cells of $400 \times 100$ m.

We ran the model with this grid for about 20 scenarios: 1) the initial bathymetry of the basin was taken planar at the general gradient of the area, and alternatively included the northern channel with its southern levee (schematised), the Meuse channel belt in the middle of the basin and a deepening south-western corner to represent the Amer. The basin depth in the NE was varied between 4 m (called 'deep' hereafter) and 2 m ('shallow') below MSL, and control runs without the Biesbosch (called 'no basin' hereafter). 2) The tides were varied between amplitudes of 0–2.1 m with and without small phase differences between the two tidal outlets Weerts et al., (2005). 3) The sediment transport predictor was varied. The Engelund-Hansen predictor for total load was used. Also an updated version of the Van Rijn sediment transport predictors were used (see manual of Delft3D for formulations). The sand input at the upstream boundary is kept at capacity of the flow.

The model simplifies reality in several ways, most importantly the lack of mud deposition, which is imported from both the river and the sea. Large parts of the Biesbosch delta are covered in mud. However, most of the islands and bars in the northeastern corner consist of sand, and the mud deposited by tides provided only a late-stage cover on the islands and channel fills. The models in this paper only refer to the early stage of delta development in the northeastern corner, which is dominantly sandy.

### 3.2  *Delft2D model description*

The Delft2D morphodynamic model system was used (version FLOW3.54.23.00, 5 December 2006; no differences in a reference bifurcation run compared to the earlier version in Kleinhans et al. (2006)). The model solves the non-linear shallow-water equations and a parameterisation for spiral flow. The imposed boundary conditions are upstream flow discharge and downstream water levels. The discharge was constant at a value ($Q = 1600 \, \text{m}^3/\text{s}$) at which the same amount of sediment is transported annually in the upstream channel as on average for the complete discharge record of the past century. The downstream water levels were specified as tidal components M2 and O1 with varying amplitudes and in some model runs a phase difference. The flow and morphology are calculated according to Lesser et al. (2004). The roughness formulation is Colebrook-White with a constant Chézy value of $C = 45$. The initial bed of the upstream channel was plane and had a gradient of $S = 1 \times 10^{-4}$ m/m. The sediment size $D_{50} = 0.25$ mm. The time step of the flow was 60 seconds to ensure numerical stability, and an initial period without morphological updating was allowed to stabilize the flow. Assuming that the flow is not appreciably affected by erosion and sedimentation during a time step, the morphological change in each time step can be multiplied by a large multiplication factor to predict the morphological evolution. The chosen factor was 10 which gave no significant differences with a factor of 1 (no acceleration). The flow was calculated on a staggered grid by a second order ADI-scheme based on the dissipative reduced phase error scheme. Advection of turbulent quantities was computed using a third order upwind ADI scheme in horizontal directions and second order central in vertical direction. Total load transport was computed at cell centres; a first-order upwind Lax scheme was used to determine the bed level changes Lesser et al. (2004). Grid cells were converted to permanently dry cells for water depths $h < 0.1$ m. The grids were curvilinear, and were generated automatically as concatenated straight and curved sections to ensure repeatability. The grids were orthogonalised as much as possible in an automated procedure.

## 4  MODELLING RESULTS

In all model runs a delta formed (see example in Fig. 5), but the morphology and progradation rate varied between scenarios. The progradation rate depends on the depth of the tidal basin and on the sediment input from the river. From geometry it follows that, for a constant sediment input, the delta volume increase is constant, which for a radial feature leads delta front progradation related to the square-root of time (Fig. 6) in agreement with the data (Fig. 3, between 1650–1800).

For the deep basins, the original northern channel silted up in the first decade so that the entire discharge went through the Biesbosch area. For the shallow basins the channel deepened, because its levee was above the average sea level and acted as a longitudinal dam, and alternating bars developed (Fig. 6). The resistance of the levee against erosion, which is not modelled in Delft2D, will therefore have had a large effect of the bifurcation evolution and potential

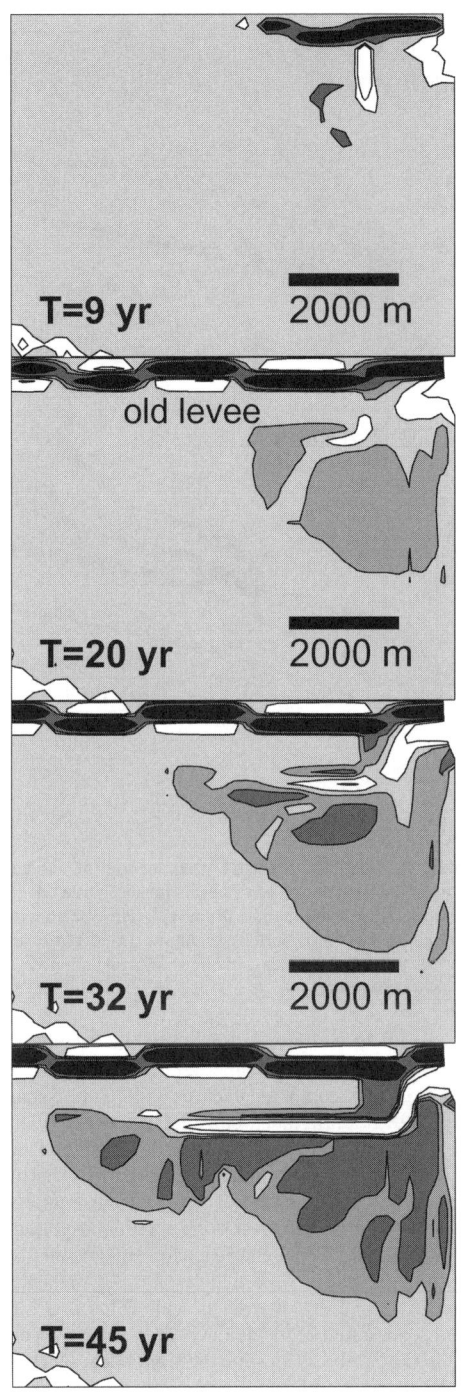

Figure 5.  Classified bed levels at 9, 20, 34 and 48 years minus the initial bed level, illustrating the channel and emergent bar evolution. White: <1 m erosion, light grey: no change, dark gray: >1 m deposition, black: >2 m deposition.

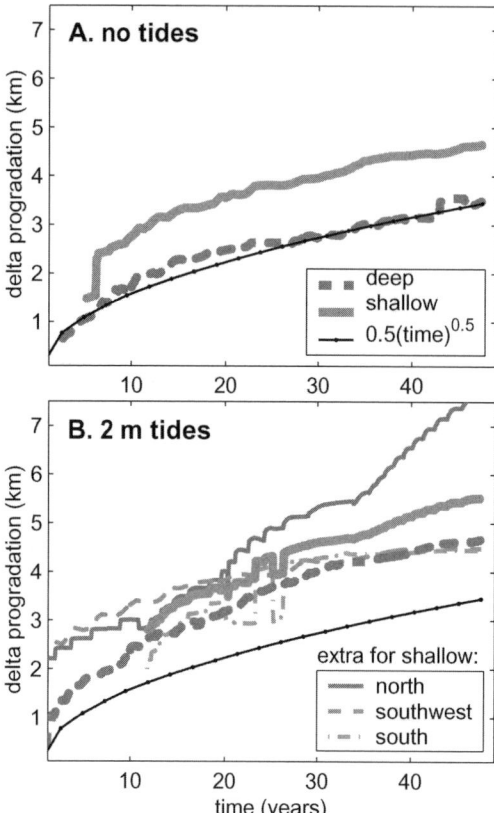

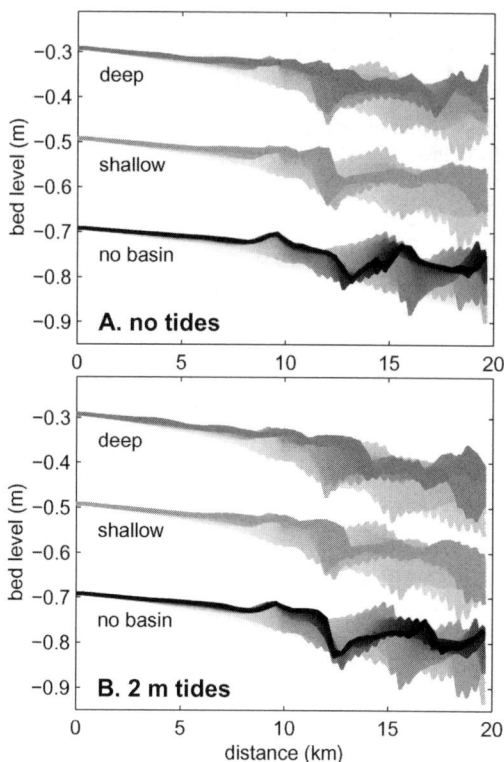

Figure 6. Spatially averaged progradation of modelled deltas in the deep and shallow basins for discharge only (A) and tides (B). For the shallow basin with tides the progradations along east-west, northeast-southwest and north-south lines is given as well to illustrate delta asymmetry and avulsive behaviour.

Figure 7. Bed levels of the channel upstream of the delta for discharge only (A) and 2 m tides (B), averaged over the width. The 'shallow' and no 'basin' have extra offsets of −0.2 and −0.4 m for visibility. The most saturated colour is the final model time step.

closure of the northern bifurcate, which would be a complete avulsion stouthamer and Berendsen, 2000).

In tidal flow compared to cases with river discharge only, the delta prograded considerably faster (Fig. 6) due to the larger ebb-velocities and was higher because the water levels were higher during part of the tidal cycle (Fig. 6). The effect of net tidal currents combined with the east-to-west ridge and tidal asymmetry in the 'shallow' basin scenario led to a significant faster progradation of the northern delta rim (Fig. 6B), which agrees to some extent with the data (Figs 2, 3). The progradation rate of the deltas is of the same order as that in the Biesbosch despite the simplifications in the model scenario and the uncertain initial conditions.

The width-averaged bed level of the upstream channel varied strongly in the longitudinal direction due to formation of bars (Fig. 7). The 'deep' and 'shallow' model results show no obvious bed degradation compared to the 'no basin' model results. This strongly

suggests that the erosion in the Merwede due to the St. Elisabeth Flood was insignificant. The bed level change averaged over the entire length and width of the channel (Fig. 8) shows a rapid initial sedimentation in the first 5–10 years because the initial depth apparently was chosen a few cm too large. Thereafter the bed adapts much slower with only 0.02 m in 50 years. The runs with tides show less sedimentation, which agrees with a faster delta migration (Fig. 6) for which the sediment was derived from the channel (Fig. 8; compare A and B).

The channels in the delta avulse frequently, leading to significant differences in migration along the delta front (Fig. 6B). The active channels are dominated by river and ebb-tidal flow whereas just abandoned channels are dominated by flood-tidal flow and rapidly silt up. In the runs with the Engelund-Hansen transport predictor the channels prograded much further than the main delta body before avulsion took place, and levees were formed, maintained and prograded that were emergent for at least part of the tidal cycle.

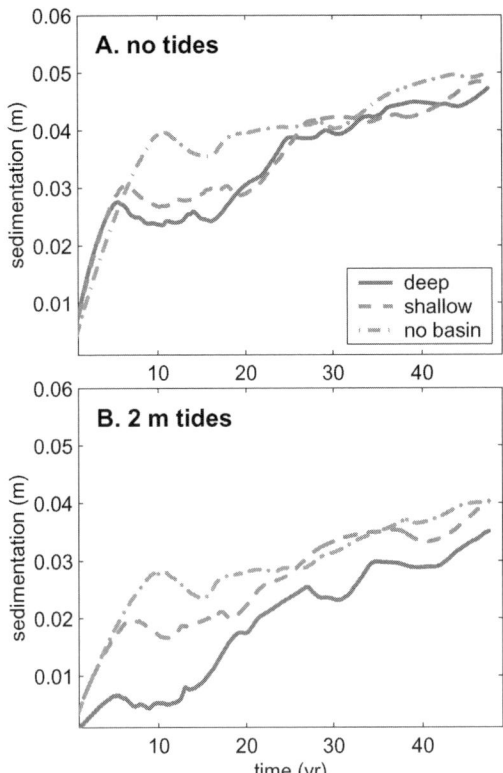

Figure 8. Bed level change averaged over the entire upstream channel for discharge only (A) and 2 m tides (B).

# 5 DISCUSSION

## 5.1 *Sediment budget*

The modelled annual delta growth averaged over 50 years is $5 \times 10^5$ m³/yr for the 'shallow' basin and $6–7 \times 10^5$ m³/yr for the deep basin without and with tides; all without mud (Fig. 8). The annual sedimentation for the shallow basin is smaller than for the deeper basin because the northern pre-flood channel remains active. Based on the soil augers, the sediment input is estimated at $900000$ m³/yr (accurate within about 20% and perhaps including some mud). These numbers are both of the same order of magnitude as the present-day annual sediment transport rate of the Rhine near the border with Germany, which is about $560000$ m³/yr.

The modelled migration rate is larger than the observed rate. The Biesbosch area prograded about 3 km between 1550 and 1620 AD (Fig. 2), while the deep delta without tides migrates the same distance in 50 years. Note that the migration is the fastest in the very beginning, so this disagreement could very well be due to mapping and interpretation uncertainties in the historical maps and the presence of a scour hole where the river breached through the levees. The soil augers in Zonneveld (1960) certainly suggest a scour hole that is several meters deeper than the tidal basin which extends over at least a kilometer. If the initial five model years were disregarded as initial effects different from those in reality (e.g. scour hole), then the modelled migration is one meter less. A basin with a depth between the 'shallow' and 'deep' scenarios will fit the observed delta migration rate and morphology best. The morphology of the Biesbosch delta indicates a small preference for migration to the west, which is not present in the 'deep' model scenario and exaggerated in the 'shallow' model, so the model scenarios bracket the natural conditions.

## 5.2 *Upstream effect of the Biesbosch avulsion*

The upstream channel silts up in most model runs including the control runs of the situation without a tidal basin (Fig. 8). Only the 'deep' scenario with tides has about 0.02 m erosion in the upstream channel, which is insignificant compared to the large bars and bed level fluctuations. Moreover, the slow sedimentation observed in all runs (disregarding the rapid adaptation to equilibrium depth in the first decade or less) can entirely be explained by the progradation of the delta, which causes some aggradation in the upstream channel (Kriele et al., 1998).

Kriele et al. (1998) studied the effect of cyclic avulsion in the Yellow River delta on the upstream channel. The active river mouth prograde rapidly, and given an equilibrium slope necessary to transport the entire upstream sediment load, the river channel must aggrade as well. At some point the superelevation of the channel leads to such large transverse slopes that the channel avulses into another direction. As a result, the distance to the sea is much shorter and the slope between avulsion site and the sea is much larger. The entire river discharge is then transferred to the new channel immediately and the river bed at the avulsion site (where the break in slope is situated) erodes rapidly. Meanwhile the new channel prograde by sedimentation at the river mouth, extending the channel again. Hence, the upstream migration of the eroded area is more limited by more rapid delta progradation, or, in other words, the relaxation length for the Yellow River is about 200 km Kriele et al., (1998). Kriele et al. (1998) presented a simplified parabolic model (in addition to their numerical model) for the relaxation length of the upstream erosion due to sudden avulsion. The relaxation length is:

$$\lambda = \sqrt{\frac{KT}{\pi}} \tag{1}$$

where $T =$ period since the avulsion in which the coastline has prograded to the original position, and $K =$ sediment diffusion coefficient:

$$K = \frac{nq_s}{3S} \qquad (2)$$

with $n = 4$ is the power in a sediment transport relation $q_s = mu^n$ for mixed bed load and suspended load, where sediment transport $q_s \approx 3 \times 10^{-5}$ for the Merwede. Hence, $K = 5 \times 3 \times 10^{-5}/(3 \times 1 \times 10^{-4}) = 0.5 \, \text{m}^2/\text{s}$. For a 'cycle' of delta progradation of $T = 200$ years in the Biesbosch, $\lambda \approx 30$ km upstream of the tidal basin. In the first decades, the erosion wave will not have migrated far upstream while the amplitude is the largest. That is, the erosion wave remained within the model in the first 50 years. The model results strongly suggest that no significant erosion took place. In addition, the relaxation distance is only half the length of the Merwede-Waal river length between the upstream bifurcation and the Biesbosch.

The evolution of the channel upstream of the Biesbosch tidal basin remained largely unaffected by the St. Elisabeth Flood, in contradiction to hypotheses voiced since the 1850s that the Biesbosch avulsion affected the discharge and sediment distribution at the upstream Rhine bifurcation, accelerating the near-closure of the northern branch (see historical references in Gottschalk, 1975; van de Ven, 1976). The reason for this behavioural difference with the Yellow river is likely that the tidal Biesbosch basin is much shallower than the upstream channel so that friction effects dominate and the river is not effectively closer to the sea, but just wider and shallower and without a slope advantage.

# 6 CONCLUSIONS

A combination of storm surges, river floods and poor dike maintenance led to the inundation of a medieval polder, which subsequently became the Biesbosch tidal basin connected to the North Sea through two estuaries and connected to the river Rhine (Merwede) in the east. Initially a sandy delta formed, and later most of the tidal basin was filled by this delta and by mud imported from the sea as well as from the river. Soil augers and modelling show that the delta captured nearly all sediment transported into the basin by the Rhine, leading to a progradation rate that decreased quadratically over time as the delta widened. Despite the shorter distance to the sea for the Rhine, model results strongly suggest that no river bed degradation took place following the catastrophe. This contradicts a well-established historical hypothesis that upstream channel erosion altered the discharge distribution at a bifurcation 100 km upstream.

ACKNOWLEDGEMENTS

MGK is supported by the Netherlands Earth and Life sciences Foundation (ALW) with financial aid from the Netherlands Organisation for Scientific Research (NWO) (grant ALW-VENI-863.04.016). Aat Barendrecht and Janrik van den Berg are cordially thanked for discussion and literature suggestions.

REFERENCES

Gottschalk, M. (1975). *Storm surges and river floods in the Netherlands I (the period 1400–1600*. Assen, The Netherlands: Van Gorcum.

Kleinhans, M., B. Jagers, E. Mosselman, and K. Sloff (2006). Effect of upstream meanders on bifurcation stability and sediment division in 1d, 2d and 3d models. In R. Ferreira, E. Alves, J. Leal, and A. Cardoso (Eds.), *River Flow 2006*, London, UK, pp. 1355–1362. International Conference on Fluvial Hydraulics, Lisbon, Portugal, Taylor and Francis/Balkema.

Kriele, H., Z. Wang, and M. de Vries (1998). Morphological interaction between the Yellow River and its estuary. In J. Dronkers and M. Scheffers (Eds.), *Physics of estuaries and coastal seas*, London, UK, pp. 287–295. 8th International Biennial Conference on Physics of Estuaries and Coastal Seas, The Hague, The Netherlands, 1996, Taylor and Francis/Balkema.

Lesser, G. R., J. A. Roelvink, J. A. T. M. van Kester, and G. Stelling (2004). Development and validation of a three-dimensional morphological model. *Journal of Coastal Engineering 51*, 883–915.

Slingerland, R. and N. Smith (2004). River avulsions and their deposits. *Annual Reviews of Earth and Planetary Science 32*, 257–285.

Smith, N., T. Cross, J. Dufficy, and S. Clough (1989). Anatomy of an avulsion. *Sedimentology 36*, 1–23.

Stouthamer, E. and H. J. A. Berendsen (2000). Factors Controlling the Holocene Avulsion History of the Rhine-Meuse Delta (The Netherlands). *J. of Sedimentary Research 70(5)*, 1051–1064.

van de Ven, G. (1976). *Aan de wieg van Rijkswaterstaat-wordingsgeschiedenis van het Pannerdens Kanaal (in Dutch)*. Zutphen, The Netherlands: De Walburg Pers.

Weerts, H. and H. Berendsen (1995). Late Weichselian and Holocene fluvial palaeogeography of the southern Rhine-Meuse delta (The Netherlands). *Geologie en Mijnbouw 74*, 199–212.

Weerts, H., P. Cleveringa, J. Hendriks, C. de Bont, G. Maas, T. Meijer, and B. Smit (2005). The last avulsion of the Rhine (1421). Delft, The Netherlands. International Conference for Fluvial Sedimentology 8.

Zonneveld, I. (1960). *De Brabantse Biesbosch, a study of soil and vegetation of a freshwater tidal delta*, Volume A,B,C. Wageningen, The Netherlands: PUDOC Centrum voor Landbouwpublicaties. Published PhD-thesis.

*River, Coastal and Estuarine Morphodynamics: RCEM 2007 – Dohmen-Janssen & Hulscher (eds)*
*© 2008 Taylor & Francis Group, London, ISBN 978-0-415-45363-9*

# Autogenic cycles of sheet and channelised flow on fluvial fan-deltas

M. van Dijk & G. Postma
*Sedimentology Group, Department of Earth Sciences, Faculty of Geosciences, Utrecht University, Utrecht, The Netherlands*

M.G. Kleinhans
*Fluvial Research Group, Department of Physical Geography, Faculty of Geosciences, Utrecht University, Utrecht, The Netherlands*

ABSTRACT: In the development of fluvial fan-deltas incised channels are usually regarded as signatures of changes in climate, tectonics or base level of the receiving basin. The role of autogenic mechanisms influencing the evolution of these deltas is poorly known. Experiments were performed to study the intrinsic evolution of fluvial fan deltas without external variations of base level, climate and/or tectonics. Water and sediment were led onto a submerged shelf, where flow deceleration caused the formation of fluvial deltas. The deltas evolved by sheet flow, incidentally alternating with channelised flow. The transition from sheet flow to channelised flow was initiated by an increased surface gradient caused by aggradation of the delta apex. This led to incision, headward erosion and ultimately to channelisation. Renewal of the sheet flow occurred after the incised channel was back filled. The back filling was initiated by bar deposition, caused by flow deceleration. The flow bifurcated around this bar, causing bank erosion and additional flow deceleration. The process of autogenic incision and back filling contrasts with commonly recognised autogenic processes; lateral shift of channels and active parts of fan lobes. The autogenic incisions can be easily confused with climate, tectonic and sea-level driven erosional events. Our flume models suggest that autogenic erosion might just as well be part of the intrinsic evolution of the system and need to be distinguished from allogenic signatures in modern and ancient fluvial fan-delta systems.

## 1 INTRODUCTION

Alluvial fans and fan deltas are common sedimentary environments along basin margins. They often constitute thick sediment wedges that form in geologically short periods of time and contain valuable, high resolution records of tectonic, climate and sea-level change. Alluvial fans are formed of coarse sediments, created where high-bed-load streams enter zones of reduced stream power, and deposit their coarser fraction of their loads (Bull 1977). Fan deltas are considered their coastal prism when alluvial fans prograde in a standing body of water (Nemec and Steel 1988). Moscariello (2005) distinguished two broad fan categories, which are referred to as 1) alluvial fans, deposited mainly by debris flows, and 2) fluvial fans, deposited mainly by sediment fluid flows. Research has been carried out to unravel the effects of catchments type (e.g. Blair 1999), climate changes (e.g. Postma 2001, Nemec and Kacanzi 1999), tectonic movements (e.g. Blair 2000, Gawthorpe and Collela 1990, Garcia-Garcia et al. 2006) and base-level variations (e.g. Nichols 2005, Gutsell et al. 2004) on the morphology, sedimentology and architecture of both modern and ancient fan deltas and alluvial fans.

Recently, the interest for the intrinsic aspects of fluvial systems has grown, since Miall (1996) named them autogenic processes to distinguish them from allogenic processes such as climate change as described above. During the past decade, Muto and co-workers have developed an entirely new concept on delta-scale systems from experimental and numerical modelling studies; autostratigrapy (Muto 2001, Muto and Steel 1992, 2001, 2004, Muto and Swenson 2006, Muto et al. 2007). It investigates the behaviour of deltas during constant sea-level fall and rise, while variables like discharge etc. are kept constant. They showed that the experimental deltas showed predictable and quantifyable responses ('autoretreat' and 'autobreak') over a large range of different allogenic conditions (Muto & Steel 2001, 2004). As the experiments were performed with constant sea-level rise and fall, the investigated deltas were still influenced and modified by the changing base level. To investigate the behaviour of deltas with the absence of all external forcings, the experiments reported herein were carried out without any variations in base level.

A disadvantage in the study of autogenic behaviour of fluvial systems is the lack of proper field study objects. During the Holocene the allogenic conditions

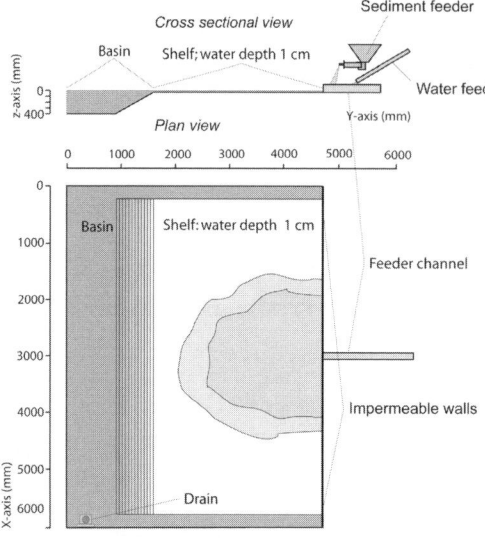

Figure 1.  Schematic drawing of the experimental setup.

have not been constant, and in fact they are still changing. Therefore, additional knowledge must come from experimental research where allogenic conditions can be kept constant. We aim to quantify the vertical and longitudinal changes in fluvial fan-delta deposition related the different flow mechanisms through time with the following objectives: 1) determine autogenically controlled morphodynamics of fluvial fan deltas; 2) relate observed behaviour to observed flow mechanisms; and 3) discuss implications for real world systems.

## 2  EXPERIMENTAL DESIGN

### 2.1  Setup

The experimental setup consisted of a shelf and a rectangular duct acting as a feeder channel (Fig. 1). The horizontal shelf measured 4.5 by 3.5 m. The bed of both the shelf and the channel consisted of sand. The shelf was covered by 1 cm of water. An adjustable drain located in the basin ensured a constant water level. The channel was 5 cm wide and 1 meter long and consisted of smooth impermeable walls. The outlet of the duct was located in the centre of one of the edges of the shelf, consisting of the same impermeable walls to ensure that no water or sediment was lost. The duct was positioned horizontally, and its width was chosen carefully to ensure that no alternating bars would occur in the duct at the range of discharges used to avoid variations in the outflow direction at the outlet of the duct. The sediment was mixed with the water flow by means of a sediment feeder. The sediment feeder delivered a

constant sediment volume by means of a worm gear. Deviations of the delivery by the feeder were measured to be less than 1% over 15 minutes. The grain size of the sediment used in the experiments was chosen such that sediment transport would occur as bed load, with Froude numbers around unity.

### 2.2  Measurements

Every four hours the experiment was halted and the surface topography was measured in detail by means of photogrammetry (a way of measuring topography based on stereoscopy). An automated movable platform contained a set of cameras, which took photographs of the surface from different angles simultaneously. With help of specially designed software (SANDPHOX™), the images are processed into a Digital Elevation Model (DEM) of the sand surface. The DEMs are here presented as shaded relief maps (Figs 3–4), processed in the commercially available program Surfer. The vertical accuracy of the photogrammetry is 250 μm and the horizontal accuracy 100 μm at the very least. To prevent reflection by the running water, we stopped the experiment and drained the water carefully before each photogrammetry session.

## 3  RESULTS

### 3.1  Autogenic behaviour

#### 3.1.1  Stage 1: sheet flow and fractionation
From the start of the experiment the delta was being congregated by sheet flow, i.e. a continuous shallow sheet of water flowing over the delta. As a result, the delta was smooth, and almost circular in shape (see Fig. 2). After 12¼ hours the sheet flow started to fractionate. As a result of enlargement of the surface area of the delta plain, it could no longer sustain a single continuous sheet flow, and the flow started to break up in smaller, but still shallow and wide flows. This has been named 'fractionated flow' by Whipple et al. (1998). After fractionation occurred, the delta continued to grow in diameter, but the smoothness of the shoreline of the delta was lost, as shown by Figure 3A. An important feature of the delta at this point is the zonation of flow mechanisms (Fig. 3A). On the apex sheet flow continues as a single sheet, but around that a zone of smaller and shallow flows exists. These 'fractionated sheet flows' then developed into distinct small-scale channels, which left a rill-like mark on the delta plain. These small-scale channels migrated laterally across the delta plain together with the upstream fractionated sheet flows, leading the more irregularly shaped shoreline (Fig. 3A). Flow migration meant that at a given location, water flow was active only part of the time, leading to lower sediment transport capacity

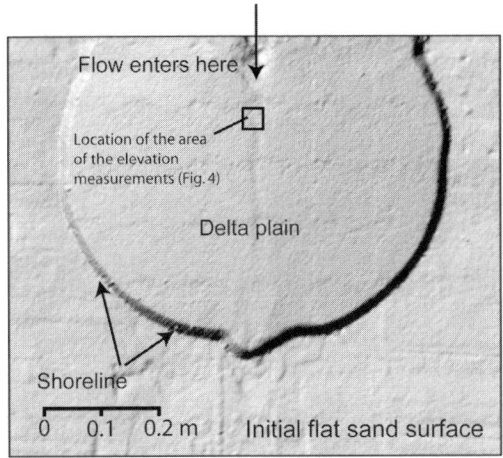

Figure 2.   Shaded relief map of the delta after 12 hours of evolving. Note the smooth, circular shoreline of the delta, which is basically a slip face with a height of one cm (the imposed water depth) and the angle of repose as gradient. The small square indicates the area of which the average elevation was measured, shown in Figure 4.

*averaged over longer periods* compared to the sheet flows on the apex. These lower transport capacity just downstream of the apex formed an impediment for the sediment on the apex, and as result the sediment was deposited on the apex, leading to aggradation and over-steepening of the gradient from the apex towards the shoreline.

### 3.1.2   *Stage 2: transition to channelised flow*
After the delta grew and aggraded for almost 40 hours (see Figs 3A and 4), a scour occurred on two third of the distance from apex to the delta shoreline, in front of the entering flow (Fig. 3B). This scour was initiated by the increasing gradient from apex to delta shoreline as the apex aggraded. The increasing gradient caused flows to accelerate, and led to incision. Once developed, the scour attracted increasingly more water as it deepened, and the water jetted from the scour towards the shoreline. This concentration of flow caused deposition of a lobe of sediment in front of the delta. As even more water was converging into the scour, headward erosion caused upstream migration. Within 30 minutes it reached the feeder channel, and a fully confined channel occurred (Fig. 3C), producing an erosional surface extending to the delta apex.

### 3.1.3   *Stage 3: Channelisation and channel fill-up*
After the flow became fully confined within the eroded channel walls, part of the delta plain was deprived from its supply of water and sediment, and was abandoned (Fig. 3C). As a result, part of the shoreline of the delta

was starved of its supply of sand, and all the material was deposited at the channel mouth. This area of the delta therefore showed high progradation rates as a delta lobe was formed (see also Fig. 4, change in length). As progradation continued, the lengthening of the channel caused the overall gradient to decrease, leading to loss of momentum. The decreasing capacity of the flow initiated deposition of the sediment it carried, and a bar was formed in the middle of the channel (Fig. 3D). The remaining flow bifurcated around the bar while at the same time two scours occurred upstream and to the sides of the newly formed bar (Fig. 3D). These scours started to erode the walls of the channel, and caused it to widen (compare the width of the channel in Figs 3C and D). This led to shallowing of the flow, more flow deceleration and more sediment was deposited adjacent to the mid-channel bar. Eventually, the bar migrated upstream, and the entire channel was backfilled. The zones of sheet flow, fractionated sheet flow and small channels were reestablished, and continued until a new cycle of incision, channelised flow and channel filling was reinitiated.

### 3.2   *Cyclic behaviour*
Four complete cycles of channel incision and fill-up have been observed within the experiment, and a fifth phase of incision was just initiated when the experiment was stopped. The delta behaviour throughout the experiment is shown in Figure 4. The periods of incision and subsequent channelised flow are indicated by the light-grey bars, whereas the white parts depict phases of sheet flow. Furthermore, the height of the apex of the delta is shown by the thick black line (see the axis on the left), while the other two thinner lines express the change in length (measured from apex to the delta edge, parallel to the inflow) and width (measured perpendicular to the inflow) of the delta (see axis on the right).

   The pattern occurring shows erosional events at quite regular intervals, and have comparable durations. Therefore this type of autogenic mechanism can be named 'autocyclic'. The erosion occurring during channelised flow is clearly shown by the elevation of the apex of the delta, which decreases by at least 10 mm during every incisional event. The height of the apex continues to grow, despite the incisions. The rate of growth, however, is declining. This can be explained by the continuously increasing lateral size of the delta through time. Furthermore, the start of a period of channelisation is accompanied by an increase in the change in length (see Fig. 4, at 40 and 120 hours into the experiment), which represents the effect of concentration of the water flow due to channelisation and corresponding delta progradation. In contrast, the width of the delta increases just before periods of channel initiation. This relates to the delta aggrading its apex to be able to initiate incision.

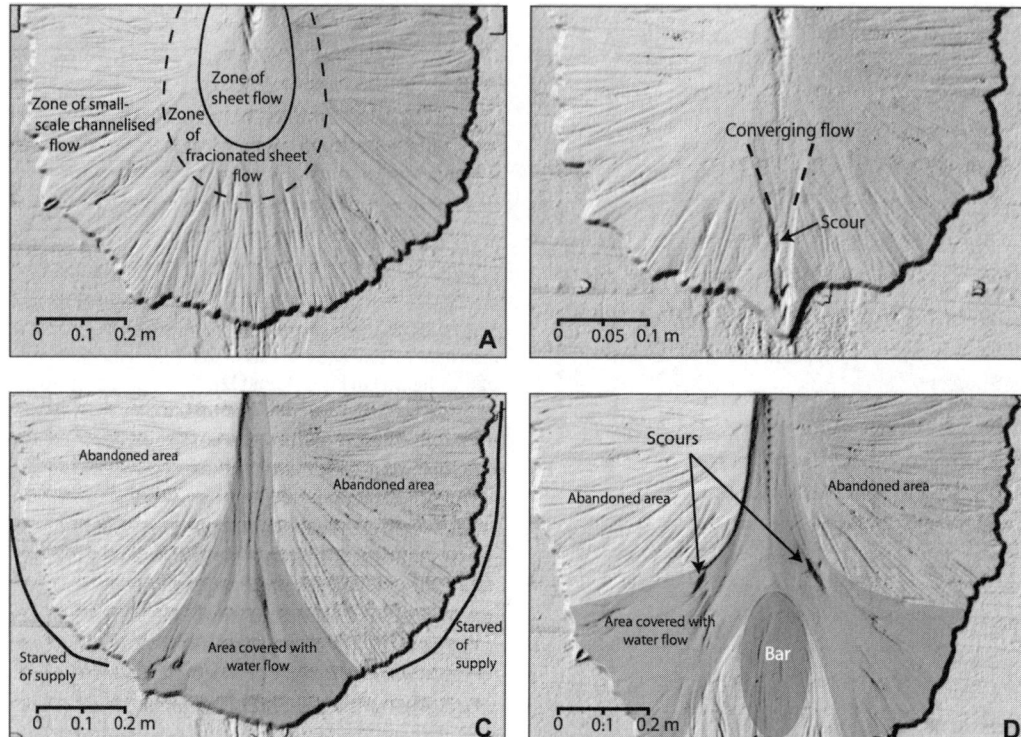

Figure 3. Shaded relief maps of the delta through a cycle of built up, incision, channelised flow, and channel fill. (A) The delta after 70 hours, showing zonation of the flow and small channels towards the delta shoreline. Note that although no 'area covered with water' is indicated, in Figures 4B and B the entire delta plain is covered with water. (B) The delta after 38 hours, showing a scour hole which caused the flow to converge and eventually led to channelisation. (C) The delta after 81 hours, with fully channelised flow. Note the abandoned areas on the delta plain and starving of the adjacent delta shoreline. (D) The delta after 85 hours. The channel widened compared to Figure 4C, and bar formation commenced.

## 4 IMPLICATIONS

These experiments show that the behaviour of a fluvial fan delta under constant conditions is more complex than previously thought. The alternation of periods of incision and channelised flow with periods of sheet flow dominance is very similar to the results reported by Kim et al. (2007). They propose a variation in slope related to storage and release of sediment to account for shoreline migrations. Essentially, the periods of sheet flow dominance in our model represent storage of sediment, the delta building up its slope for incision to be initiated, whereas during channelised flow sediment is released, leading to increased progradation. As Figure 4 shows, the change of the length and width of the delta is far from constant. They both show periods of increased and decreased growth, representing the storage and release of sediment.

The fact that subsequent periods of incision do not cut down to the level of the previous periods of channelised flow, has important implications. It means that the erosional surfaces are preserved within the architecture of the deltaic body. These erosional surfaces will look very similar to erosional surfaces formed by allogenic causes, such as climate changes or tectonic movements. In fact, it might well be impossible to distinguish between allogenic and autogenic formed erosional surfaces. This means that incisions previously interpreted as caused by allogenic processes, might also have an entirely autogenic origin.

## 5 CONCLUSIONS

Autocyclic aspects of fluvial fan delta evolution have been investigated by means of laboratory experiments. An experiment was performed, in which all external (allogenic) variables (water and sediment input, and sea level) were kept constant. As a result the processes observed had to be of an autogenic nature. Measurements done during the experiments consisted

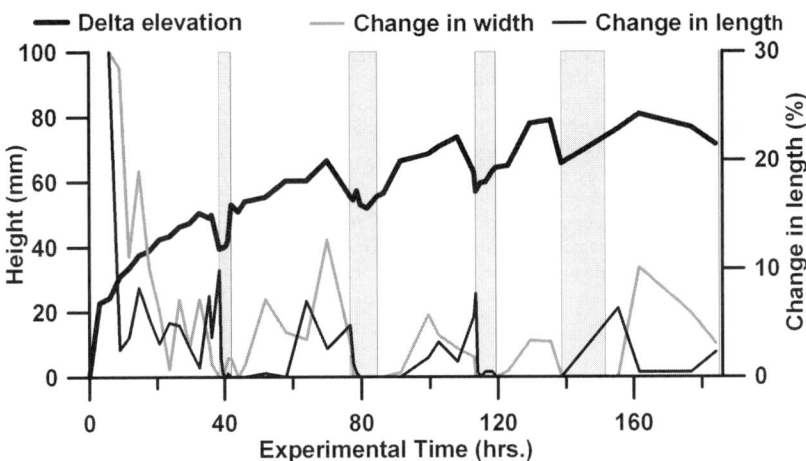

Figure 4. Behaviour of the delta throughout the experiment. The light grey blocks represent periods of channelised flow, whereas the white blocks represent sheet flow dominance. In total 5 incision events have been observed. However, the final event does not show up as a wide bar, as the experiment was stopped just after incision commenced. The fat black line indicates the average height of the delta apex (measurement location shown in Figure 2), while the other two lines show the change in length and width of the delta. Note that the delta elevation is measured in mm (left axis), and the changes in length and width in % (right axis).

of detailed topographic scanning (DEM's) and video images. Our main conclusions are:

– Autocyclic behaviour on fluvial fan deltas consists not only of sheet flow and flow fractionation and lateral migration of channels as commonly suggested, but also by cycles of fan incision that lead to channelisation.
– The transition from sheet to channelised flow was observed to be initiated by aggradation controlled oversteepening of the apex of the delta, caused by zonation of sheet flow, fractionated sheet flow and small-scale channelised flow. The transition back to sheet flow was initiated by flow deceleration and bar formation and subsequent backfilling of the channel.
– Incisions on alluvial fans or fan deltas which have been attributed to allogenic forces (climate, tectonics and base level), might actually have been caused by the intrinsic behaviour of the system itself.

REFERENCES

Blair, T.C. 1999. Sedimentology of the debris-flow-dominated Warm Spring Canyon alluvial fan, Death Valley, California. *Sedimentology* 46: 941–965.
Blair, T.C. 2000. Sedimentology and progressive unconformities of the sheetflood-dominated Hell's Gate alluvial fan, Death Valley, California. *Sedimentary Geology* 132: 233–262.
Bull, W.B. 1977. The Alluvial Fan environment. Progress in Physical Geography 1: 222–270.

Garcia-Garcia, F., Fernandez, J., Viseras, U. & Soria, J.M. 2006. Architecture and sedimentary facies evolution in a delta stack controlled by fault growth (Betic Cordillera, southern Spain, late Tortonian). *Sedimentary Geology* 185: 79–92.
Gawthorpe, R.L. & Collela A. 1990. Tectonic controls on coarse-grained delta depositional systems in rift basins. In: A. Collela and D.B. Prior (eds), *Coarse-grained deltas, Spec. Publ. Int. Ass. Sediment.* 10: 113–127. Oxford, Blackwell Int.
Gutsell, J.E., Clague, J.J., Best, M.E., Bobrowsky, P.T. & Hutchinson, I. 2004. *Journal of Quaternary Science* 19 (5): 497–511.
Kim, W., Paola, C., Swenson, J.B. & Voller, V.R. 2006. Shoreline response to autogenic processes of sediment storage and release in the fluvial system. *Journal of Geophysical Research* 111: F04013, doi:10.1029/2006J F000470.
Miall, A.D. 1996. *The Geology of Fluvial Deposits: Sedimentary Facies, Basin Analysis, and Petroleum Geology, and Petroleum Geology*, New York, Springer.
Moscariello, A. 2005. Exploration potential of the mature Southern North Sea basin margins: some unconventional plays based on alluvial fan sedimentation models. In: A.G. Doré & B.A. Vining (eds), *Northwest Europe and Global perspectives – Proc. 6th Petr. Geol.Conf.*: 595–605. Geological Society of London.
Muto, T. 2001. Shoreline autoretreat substantiated in flume experiments. *Journal of Sedimentary Research* 71 (2): 246–254.
Muto, T. & Steel, R.J. 1992. Retreat of the front of a prograding delta. *Geology* 20: 967–970.
Muto, T. & Steel, R.J. 2001. Autostepping during the transgressive growth of deltas: Results from flume experiments. *Geology* 29: 771–774.

Muto, T. & Steel, R.J. 2004. Autogenic response of fluvial deltas to steady sea-level fall: Implications from flume-tank experiments. *Geology* 32: 401–404.

Muto, T., Steel, R.J. & Swenson, J.B. 2007. Autostratigraphy: a framework norm for genetic stratigraphy. *Journal of Sedimentary Research* 77: 2–12.

Muto, T. & Swenson, J.B. 2006. Autogenic attainment of large-scale alluvial grade with steady sea-level fall: An analog tank-flume experiment. *Geology* 34: 161–164.

Nemec, W. & Kazanci, N. 1999. Quaternary colluvium in west-central Anatolia/: sedimentary facies and palaeoclimatic significance. *Sedimentology* 46 (1): 139–170.

Nemec, W. & Steel, R.J. 1988. What is a fan delta and how do we recognize it? In: W. Nemec & R.J. Steel (eds), *Fan deltas, sedimentology and tectonic setting*: 3–13. Glasgow, Blackie & Son.

Nichols, G. 2005. Tertiary alluvial fans at the northern margin of the Ebro Basin: a review. In: Harvey, A.M., Mather, A.E. & Stokes, M. (eds), *Alluvial fans: Geomorphology, Sedimentology, Dynamics Spec. Publ. Geol. Soc. London 251*: 187–206, London, Geol. Soc. London.

Postma, G. 2001. Physical climate signatures in shallow- and deep-water deltas. *Global and Planetary Change* 28: 93–106.

Whipple, K.X., Parker, G., Paola, C. & Mohrig, D. 1998. Channel dynamics, sediment transport, and the slope of alluvial fans: Experimental Study. *Journal of Geology* 106: 677–693.

*River, Coastal and Estuarine Morphodynamics: RCEM 2007 – Dohmen-Janssen & Hulscher (eds)*
*© 2008 Taylor & Francis Group, London, ISBN 978-0-415-45363-9*

# Flume experiment of debris flow confluence formed alluvial fan in the main channel

Su-Chin Chen & Shiuan-Pei An

*Department of Soil and Water Conservation, National Chung-Hsing University,*
*Taichung, Taiwan, R.O.C.*

ABSTRACT:  This study simulated the processes of debris flow forms an alluvial fan at a confluence. Two different experimental flumes were carried out. Additionally, this study used the theory and experiment to analyze river flow in a main channel and a tributary influences alluvial fan's longitudinal and horizontal ratio. The angle of the confluences with main flow was 90 degrees. Several experiments under different testing conditions were applied. Based on the experiments results, the process of an alluvial fan formed by debris flow could be divided into two stages. First stage was sediments diffusions and deposition that the process was short. Second stage was pushing expansion. In the second stage, the mass ratio between sediment transporting in main channel and supplied by debris flow was used to classify the alluvial fan forms in four types, such as no alluvial fan, transitional alluvial fan, slight retrogressive deposition alluvial fan and heavy one. The accumulated debris volume at the alluvial fan was observed and the corresponding length factors were discussed using the delay of retrogressive depositing.

## 1 INTRODUCTION

Taiwan, being located at the subtropical zone in western Pacific Ocean, is characterized by a prevalence of mountainous terrains, steep topography, complicated geological structures and concentrated rainfall. As mentioned above, it is quite often that floods, which are due to intense rainfall and carry sediment, cause debris flows at torrents in the mountain area. At the confluence, characters of river flow and slope in a main river are different with a tributary, so sediment deposits forms an alluvial fan in the main river. The flow conditions in main river and tributary are controlled by the alluvial fan, but at the same time they also influence the alluvial fan types. Therefore, the forming process of an alluvial fan is very complex.

Chumm (1977) and Bliar and McPherson (1994) divided the types of alluvial fans into channelized fluvial flow fans, sheet flow fans and debris flow fans. The water and sediment conditions of the first two types were more uniform than the debris flow fans, so the governing equation of their forming processes were much simpler, and Paker (1998) already provided the analytical solutions of the channelized fluvial flow and sheet flow fans. Unfortunately, due to the complex of forming process, there were no analytical solutions for the debris flow fans. Kuang (1995) proposed the formulas to predicting the travelling locus of the front for tributary debris flow and in main channel by the simplified governing equations. The paper in

Chen (2004) provided the regularity of confluence between a debris flow and the main river by model similarity rules and a series of model experiments. Chen (2002) proposed a numerical scheme to simulate the confluence phenomenon, and developed the technique of surveying confluence behavior by applying digital image processing (Chen, 2003). Others studies, such as the experiments of alluvial fans formed by hyperconcentrated tributaries (Chen, 2004), debris flow fan blocking the river (Chen, 2006) and hazardous degree of debris flow, have been done over the years. However, there have been few attempts to establish a direct relation between forms of alluvial fans and the conditions of water and sediments at confluence. To overcome these shortcomings, this paper focused on the confluence behavior. This study simulated the processes of debris flows which come out of tributary at junction area by experimental flumes, and lead to an understanding of influence between debris flows and alluvial fans.

## 2 METHOD

This study set the experimental conditions of a base flow in main river and a debris flow in tributary, and simulated the progressive alluvial fan at the confluence. We tried to understand the relation between a debris flow alluvial fan and inflow and sediment supply. For a large time scale, such as duration between

Figure 1. Alluvial fan cut by main river.

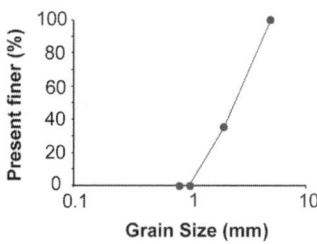

Figure 3. Grain size distribution.

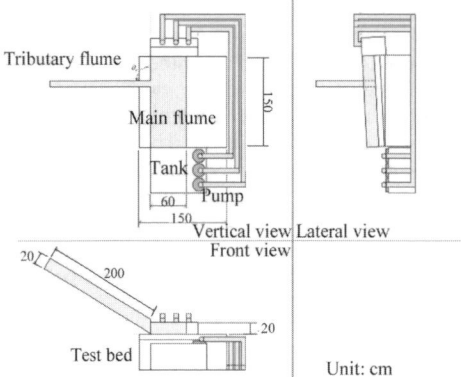

Figure 2. Experiment flume design chart.

several storm flood events, alluvial fan is an unstable structure in a river reach, and it would be scoured gradually by the flood flows. Figure 1 shows that the alluvial fan from Abang Creek was cut by Tachia Creek and had vanished. But if we take a short duration within two flood events, it can find complete alluvial fans after the storm floods from the field observation. Additionally, these alluvial fans will maintain their shape until next flood event. In order to conform to the field observation, the experiment duration was set in debris flow left tributary, and terminated when debris flow stopped. Furthermore, the side of the main river could be influenced the form of alluvial fan, so the main channel width of our experimental flume is 60 cm to avoid the effect from the side of main channel.

## 2.1 Experiment flume design

In this experiment (Figure 2), we consisted of a main flume, a tributary flume, a sediment supply system, and channel flow supply and recycle systems. The bed of the main flume was steel, and the side wall was constructed by acrylic board. The main flume had a longitudinal length of 150 cm, a cross sectional area with a height of 20 cm, a width of 60 cm, and a slope that changes according to the test bed angel. The acrylic tributary flume which is 200 cm long, 10 cm

wide and 20 cm high, was constructed with an angle of 90 degrees at the confluence. Besides, there was a circulatory system in main flume. We used three pumps, which can be shut or opened to determine the discharge of inflow into the main channel, to draw out the water in the downstream tank into the upstream. In tributary sediment supplying system, the experimental sand was characterized by $SG = 2.59$, $d_{50} = 2.46$ mm and $d_m = 2.38$ mm. 9 kg dry sand was filled in a container, and a hole in the tank bottom was cut so that dry sand could drop below the tank into the tributary upstream end. Then, the mixture of dry sand and inflow caused a debris flow in tributary. The grain size distribution is presented in Figure 3.

## 2.2 Experiment conditions

The experiment composed of 54 experimental conditions taken from 3 main flume discharges, 6 main flume slopes and 3 tributary debris flow concentrations. The experiment data of experimental case, main flume discharge $Q_m$, debris flow volume concentration $C_v$, debris flow volume discharge $Q_t$ and main flume slope are in Table 1.

## 3 RESULT

### 3.1 Experiment phenomenon

At the instant of debris flows leave tributary at confluence, behaviors of sediments transport are associated with the flow condition in main flume. Sediment could diffuse to upstream at the confluence with a strong flow condition in comparison to weak flow in main flume. In the reason of friction between sediment and flume bed, the first wave of sediments transport will stop and obstruct other sediments; hence, lots of sediments deposited at the confluence and form an alluvial fan gradually. If the intensity of main flume flow is large enough, however, the sediments kinetic energies which are taken from main flow will be larger than the frictional loss. Few sediments deposits in the main flume yield none of alluvial fan appearing at the confluence in this situation.

Table 1. Experiment condition.

| Case | $Q_m$ cms | $C_v$ | $Q_t$ cm³/sec | $\alpha_m$ degree |
|------|-----------|-------|---------------|-------------------|
| AA1 | 0.0044 | 0.35 | 604.435 | 1 |
| AA2 | 0.0044 | 0.35 | 604.435 | 2 |
| AA3 | 0.0044 | 0.35 | 604.435 | 3 |
| AA4 | 0.0044 | 0.35 | 604.435 | 4 |
| AA5 | 0.0044 | 0.35 | 604.435 | 5 |
| AA6 | 0.0044 | 0.35 | 657.820 | 6 |
| AB1 | 0.0044 | 0.40 | 657.820 | 1 |
| AB2 | 0.0044 | 0.40 | 657.820 | 2 |
| AB3 | 0.0044 | 0.40 | 657.820 | 3 |
| AB4 | 0.0044 | 0.40 | 657.820 | 4 |
| AB5 | 0.0044 | 0.40 | 657.820 | 5 |
| AB6 | 0.0044 | 0.40 | 657.820 | 6 |
| AC1 | 0.0044 | 0.50 | 790.469 | 1 |
| AC2 | 0.0044 | 0.50 | 790.469 | 2 |
| AC3 | 0.0044 | 0.50 | 790.469 | 3 |
| AC4 | 0.0044 | 0.50 | 790.469 | 4 |
| AC5 | 0.0044 | 0.50 | 790.469 | 5 |
| AC6 | 0.0044 | 0.50 | 790.469 | 6 |
| BA1 | 0.0081 | 0.35 | 604.435 | 1 |
| BA2 | 0.0081 | 0.35 | 604.435 | 2 |
| BA3 | 0.0081 | 0.35 | 604.435 | 3 |
| BA4 | 0.0081 | 0.35 | 604.435 | 4 |
| BA5 | 0.0081 | 0.35 | 604.435 | 5 |
| BA6 | 0.0081 | 0.35 | 657.820 | 6 |
| BB1 | 0.0081 | 0.40 | 657.820 | 1 |
| BB2 | 0.0081 | 0.40 | 657.820 | 2 |
| BB3 | 0.0081 | 0.40 | 657.820 | 3 |
| BB4 | 0.0081 | 0.40 | 657.820 | 4 |
| BB5 | 0.0081 | 0.40 | 657.820 | 5 |
| BB6 | 0.0081 | 0.40 | 657.820 | 6 |
| BC1 | 0.0081 | 0.50 | 790.469 | 1 |
| BC2 | 0.0081 | 0.50 | 790.469 | 2 |
| BC3 | 0.0081 | 0.50 | 790.469 | 3 |
| BC4 | 0.0081 | 0.50 | 790.469 | 4 |
| BC5 | 0.0081 | 0.50 | 790.469 | 5 |
| BC6 | 0.0081 | 0.50 | 790.469 | 6 |
| CA1 | 0.0124 | 0.35 | 604.435 | 1 |
| CA2 | 0.0124 | 0.35 | 604.435 | 2 |
| CA3 | 0.0124 | 0.35 | 604.435 | 3 |
| CA4 | 0.0124 | 0.35 | 604.435 | 4 |
| CA5 | 0.0124 | 0.35 | 604.435 | 5 |
| CA6 | 0.0124 | 0.35 | 657.820 | 6 |
| CB1 | 0.0124 | 0.40 | 657.820 | 1 |
| CB2 | 0.0124 | 0.40 | 657.820 | 2 |
| CB3 | 0.0124 | 0.40 | 657.820 | 3 |
| CB4 | 0.0124 | 0.40 | 657.820 | 4 |
| CB5 | 0.0124 | 0.40 | 657.820 | 5 |
| CB6 | 0.0124 | 0.40 | 657.820 | 6 |
| CC1 | 0.0124 | 0.50 | 790.469 | 1 |
| CC2 | 0.0124 | 0.50 | 790.469 | 2 |
| CC3 | 0.0124 | 0.50 | 790.469 | 3 |
| CC4 | 0.0124 | 0.50 | 790.469 | 4 |
| CC5 | 0.0124 | 0.50 | 790.469 | 5 |
| CC6 | 0.0124 | 0.50 | 790.469 | 6 |

Based on the experiment image, two steps of alluvial fan forming process can be distinguished by the model of sediments transport. The first stage of the process is sediments diffusions step. The step duration is short and sediments has not deposit yet. The sediments transport type is similar to concentration diffusions such as the point discharge of the wastewater, so the first stage of the process is called sediments diffusions step. In this step, sediments could diffuse to upstream and downstream at the same time, but it may only transports to downstream.

Since the first wave of sediments deposit and hence the termination of sediment diffusion step. In the second step, sediments come out tributary flume continually, then deposit and cover the first wave of sediments at the confluence. It leads to an increase in width and height of an alluvial fan. The growth of alluvial fan obstructs the flux of debris flow, so part sediments of debris flow is deposited and causes a phenomenon of retrogressive deposition in tributary flume. In the end of tributary near the confluence, sediments accumulate a mass of retrogressive deposit; because of this, the block of retrogressive deposition slides into confluence to push and increase the volume of alluvial fan at once. Therefore, the second stage of the forming process is called sediments deposition and pushing expansion step. The reason for the short duration of first step, this study will focus on the second stage of alluvial fan forming process.

### 3.2 Sediments deposition and pushing expansion step

Based on the observation of experiment image, this research differentiated alluvial fan forming processes into four types, no alluvial fan (Type 1), transitional alluvial fan (Type 2), slight retrogressive deposition alluvial fan (Type 3) and heavy retrogressive deposition alluvial fan (Type 4). Following were descriptions of four types of alluvial fans. The intensity of the main flume flow of type 1 was comparatively large, so sediments couldn't deposit in main flume. It was responsible none of alluvial fan in the confluence. The main flow intensity in type 3 and type 4 was smaller than that of type 1. Therefore, sediments could stop and deposit downstream the confluence, and caused the upper edge of alluvial fan moving closely to the end of tributary. Because of the upper edge of alluvial fan obstructed the tributary flume when arriving at confluence, debris flow deposited and yielded the phenomenon of retrogressive deposition. The alluvial fan with retrogressive deposition could be distinguished into slight and heavy retrogressive deposition alluvial fan by the quantity of retrogression in the tributary. In type 4, the main flow condition was between type 1 and type 3. There could be an alluvial fan at the confluence, but the fan was not large enough to obstruct the tributary flume. Therefore the kind of alluvial fans were called transitional alluvial fan. Figures 4–7 show four types of alluvial fans at confluence and retrogressive deposition in tributary

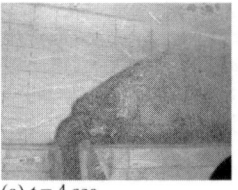

(a) $t = 4$ sec

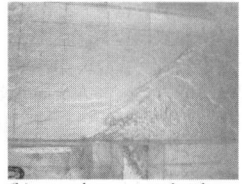

(b) experiment termination

Figure 4.  Flow condition with no alluvial fan, BA5.

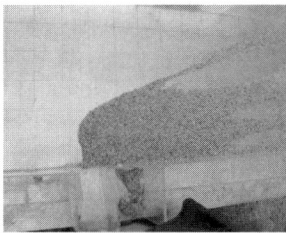

Figure 5.  Form of transition alluvial fan, AB4.

(a) main flume

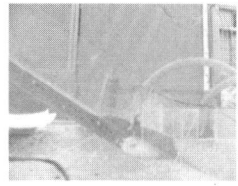

(b) tributary flume

Figure 6.  Slight retrogressive deposition alluvial fan, BA1.

(a) main flume

(b) tributary flume

Figure 7.  Heavy retrogressive deposition alluvial fan, AC1.

flume. Table 2 provides the list of 54 experiment conditions that were used for presenting the results in this study, grouped into four types.

This study had three parts of results to discuss: (a) the conditions of four alluvial fan types; (b) the delay of retrogressive depositing; (c) the corresponding width and length factors of alluvial fans.

Table 2.  Summary of alluvial fan forms vs. experiment conditions.

| 1° | | | | | 2° | | | |
|---|---|---|---|---|---|---|---|---|
| $Q$ | $Cv$ | | | | $Q$ | $Cv$ | | |
| cms | 0.35 | 0.4 | 0.5 | | cms | 0.35 | 0.4 | 0.5 |
| 0.0044 | AA1 | AB1 | AC1 | | 0.0044 | AA2 | AB2 | AC2 |
| 0.0081 | BA1 | BB1 | BC1 | | 0.0081 | BA2 | BB2 | BC2 |
| 0.0124 | CA1 | CB1 | CC1 | | 0.0124 | CA2 | CB2 | CC2 |

| 3° | | | | | 4° | | | | |
|---|---|---|---|---|---|---|---|---|---|
| $Q$ | $Cv$ | | | | $Q$ | $Cv$ | | | Type 1 |
| cms | 0.35 | 0.4 | 0.5 | | cms | 0.35 | 0.4 | 0.5 | |
| 0.0044 | AA3 | AB3 | AC3 | | 0.0044 | AA4 | AB4 | AC4 | |
| 0.0081 | BA3 | BB3 | BC3 | | 0.0081 | BA4 | BB4 | BC4 | Type 2 |
| 0.0124 | CA3 | CB3 | CC3 | | 0.0124 | CA4 | CB4 | CC4 | |

| 5° | | | | | 6° | | | | |
|---|---|---|---|---|---|---|---|---|---|
| $Q$ | $Cv$ | | | | $Q$ | $Cv$ | | | Type 3 |
| cms | 0.35 | 0.4 | 0.5 | | cms | 0.35 | 0.4 | 0.5 | |
| 0.0044 | AA5 | AB5 | AC5 | | 0.0044 | AA6 | AB6 | AC6 | |
| 0.0081 | BA5 | BB5 | BC5 | | 0.0081 | BA6 | BB6 | BC6 | |
| 0.0124 | CA5 | CB5 | CC5 | | 0.0124 | CA6 | CB6 | CC6 | Type 4 |

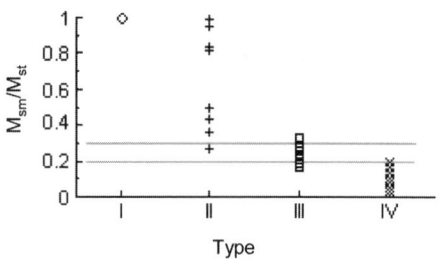

Figure 8.  Ratio of sediment mass transport in main and tributary vs. type of alluvial fans.

### 3.3  Conditions of four types of alluvial fan

Figure 8 presented the mass ratio between sediments transport in main flume and sediments supply in tributary, $M_{sm}/M_{st}$. Where $M_{sm}$ was the mass of sediments which left the end of main flume during an experiment case, and $M_{st}$ was mass of sediment supply in tributary. If the sediment transport capacity in main flume was larger than that of the tributary, there was none sediments deposits at the confluence, and $M_{sm}/M_{st} = 1$. It resulted in none alluvial fans developments in the main flume. On the other hand, part of sediments from debris flow could stay in main flume when $M_{sm}/M_{st} < 1$, as a result, an alluvial fan could develop in main flume. In addition, For smaller value of $M_{sm}/M_{st}$, sediments could stay in the tributary.

The relation between $M_{sm}/M_{st}$ and type of alluvial fans were given in Figure 8. The result reflected in this figure indicated that there were none alluvial fans when $M_{sm}/M_{st} = 1$, and $M_{sm}/M_{st}$ for transitional alluvial fans (type 2) were between 1 and 0.3. Furthermore, alluvial fans of type 3 had a range for $0.3 \leq M_{sm}/M_{st} \leq 0.2$, and $M_{sm}/M_{st} \leq 0.2$ for alluvial fans of type 4.

### 3.4  The delay of retrogressive deposition

Because retrogressive deposition didn't arise in the experiment conditions in type 1 and type 2, we used

Table 3. Experiment data of retrogressive deposition analysis.

| Type | Case | $C_v$ | $Q_m$ cms | $\alpha_m$ degree | $t_r$ sec | $Q_{sm}/Q_t$ | $M_r/M_{st}$ |
|---|---|---|---|---|---|---|---|
| III | AA2 | 0.35 | 0.0044 | 2 | 7.900 | 0.074 | 0.484 |
| III | BA1 | 0.35 | 0.0081 | 1 | 6.467 | 0.068 | 0.396 |
| III | AB3 | 0.40 | 0.0044 | 3 | 11.500 | 0.231 | 0.881 |
| III | BB2 | 0.40 | 0.0081 | 2 | 4.933 | 0.087 | 0.378 |
| III | CB1 | 0.40 | 0.0124 | 1 | 5.500 | 0.078 | 0.421 |
| III | AC4 | 0.50 | 0.0044 | 4 | 5.600 | 0.111 | 0.643 |
| III | BC3 | 0.50 | 0.0081 | 3 | 7.267 | 0.153 | 0.834 |
| III | CC2 | 0.50 | 0.0124 | 2 | 4.400 | 0.124 | 0.505 |
| IV | AA1 | 0.35 | 0.0044 | 1 | 3.633 | 0.029 | 0.223 |
| IV | AB2 | 0.40 | 0.0044 | 2 | 4.100 | 0.074 | 0.314 |
| IV | BB1 | 0.40 | 0.0081 | 1 | 4.800 | 0.048 | 0.368 |
| IV | AC3 | 0.50 | 0.0044 | 3 | 3.833 | 0.059 | 0.440 |
| IV | BC2 | 0.50 | 0.0081 | 2 | 2.567 | 0.072 | 0.295 |
| IV | CC1 | 0.50 | 0.0124 | 1 | 3.100 | 0.060 | 0.356 |
| IV | AB1 | 0.40 | 0.0044 | 1 | 4.400 | 0.043 | 0.337 |
| IV | AC2 | 0.50 | 0.0044 | 2 | 2.333 | 0.037 | 0.268 |
| IV | BC1 | 0.50 | 0.0081 | 1 | 3.633 | 0.041 | 0.417 |
| IV | AC1 | 0.50 | 0.0044 | 1 | 2.433 | 0.023 | 0.279 |

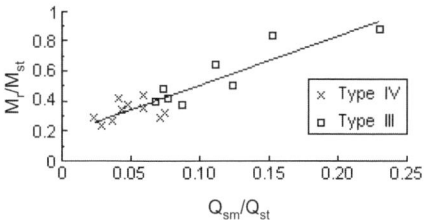

Figure 9. Time of retrogressive deposition started vs. ratio of sediment transport chart.

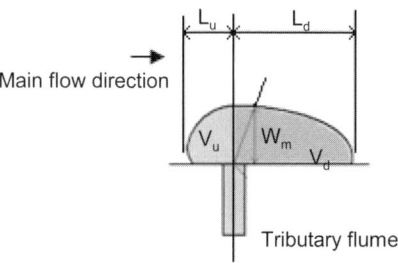

Figure 10. Geometry factor of alluvial fan.

the data which were taken form the experiment cases of type 3 and type 4 to analyze the delay of retrogressive depositing in the tributary. The experiment data of debris flow volume concentration $C_v$, main flume discharge $Q_m$, main flume slope $\alpha_m$, $Q_{sm}/Q_t$ and $M_r/M_{st}$ are in Table 3. Where $Q_{sm}$ is the volume discharge of dry sands in main flume, $Q_t$ is the volume discharge of debris flow in tributary, $M_r$ is the mass of sediments which flow into the main flume before the retrogressive deposition arising, $M_{st}$ is the mass of sediment supplying in tributary during a case.

It is worth noting that the discharge of sediments, which is entered the confluence from the tributary, is uniform and steady before the retrogressive deposition arising. Therefore, the ratio between $M_r$ and $M_{st}$ could be seen to relevant to the delay of retrogressive deposition reasonably. Otherwise, the ratio of $Q_{sm}/Q_t$ is related to the capacity of sediment transport in main and tributary flume respectively. Based on the observation of experiment, the delay of retrogressive deposition was associated with the capacity of sediment transport. The relation between $Q_{sm}/Q_t$ and $M_r/M_{st}$ was shown in Fig 9, and their linear model was given by

$$\frac{M_r}{M_{st}} = 0.18 + 3.255 \frac{Q_{sm}}{Q_{st}}$$ (1)

It could be found that the delay of retrogressive deposition was closely pertinent to the ratio of sediment transport capacity $Q_{sm}/Q_t$ in Figure 9. A larger value of $Q_{sm}/Q_t$ (i.e. large sediment transport capacity in

main flume) resulted in a more delay of retrogressive deposition in tributary. It was because that the high flow condition in main flume led to a long depositing distant for sediment. It took a long time for the upper edge of the alluvial fan to move towards the confluence; hence, the delay of retrogressive deposition was increased. Besides, there was a strong statistical evidence of the intercept in this model. Therefore, although there was none of flow in main flume ($Q_{sm}/Q_t = 0$), the delay of retrogressive deposition still existed in the tributary.

### 3.5 The corresponding width and length factors of alluvial fans

The geometry features of an alluvial fan are shown in Figure 10. Where $V_u$ is the volume of alluvial fan above the centerline of tributary flume, $V_d$ is the volume of alluvial fan below the centerline, $L_u$ and $L_d$ are the length of the alluvial fan above and below the centerline respectively, $W_m$ is the max width of alluvial fan. This study offers three the corresponding width and length factors of alluvial fans. The factors are $V_d/V_u$, $L_d/L_u$ and $L_d/W_m$. The first two terms could represent the downstream slant of an alluvial fan, and the other one describes the shape of an alluvial fan.

Both the values of $V_d/V_u$ and $L_d/L_u$ are described the slant of an alluvial fan, but the meaning of slant differs in different definition. The factor $L_d/L_u$ is described the length slant of alluvial fan appearance, and $L_d/L_u$ can be explained the volume slant of sediment which contained in the alluvial fan.

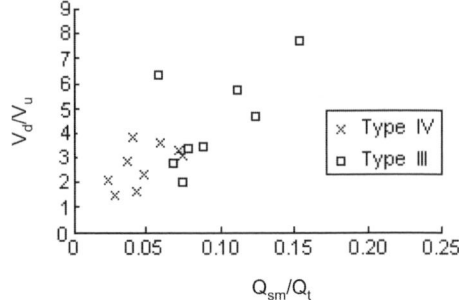

Figure 11. Sediment transport ratio vs. volume factor of alluvial fan.

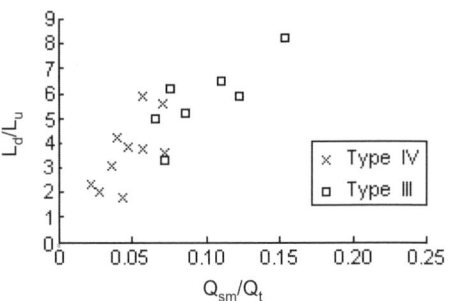

Figure 12. Sediment transport ratio vs. length factor of alluvial fan.

The relations between $Q_{sm}/Q_t$ and factors of $V_d/V_u$ and $L_d/L_u$ are compared here (see Figure 11 and Figure 12). The results showed that the increasing in $Q_{sm}/Q_t$ (i.e. large sediments transport capacity in main flume or small sediment supplying in tributary) led to a large slant in alluvial fan. However, Figure 11 showed that the relation between $Q_{sm}/Q_t$ and $V_d/V_u$ was worse than $Q_{sm}/Q_t$ and $L_d/L_u$. It could be explained below. The block of sediments which deposit in tributary after the retrogressive deposition arising could slide into the confluence and push the alluvial fan to expend upstream and downstream respectively. The volume of expansion of the upstream alluvial fan was equal to the downstream alluvial fan; consequently, it led to a small value of $V_d/V_u$. Compare with Figure 11, the value $Q_{sm}/Q_t$ and $L_d/L_u$ appeared to be highly related in the sense that showed in Figure 12. Sediments which carried by main flume flow could transport a long distance. Although the amount of sediment transports was a small proportion of the alluvial fan, it made a significant downstream increasing of the alluvial fan.

A debris flow alluvial fan in Tachia Creek at Taichung was show in Figure 13. The alluvial fan had a large downstream slant, that was, the value of $L_d/L_u$ of this alluvial fan was very significant. It could say that the discharge of Tachia Creek was larger

Figure 13. Example of length factor of alluvial fan.

than the tributary during the event of debris flow in tributary.

### 3.6 Factors of $L_d/W_m$

Besides longitudinal length of alluvial in main flume, the flow conditions in the main and tributary flume could effect upon the width of an alluvial fan according to the experiment result. For example, the increasing of discharge in main flume leads to a long and narrow alluvial fan with a large value of $L_d/W_m$. Because the flow and sediment conditions at the confluence are complex, we use a dimensional analysis to discuss the association between $L_d/W_m$ and flow conditions of main and tributary flume.

The first step in the planning of an experiment to study this problem was to decide on the factors that will have an effect on the $L_d/W_m$. This study expected the list to include the discharge in main flume, $Q_m$, the dimensionless effective shear stress of main flume flow, $\tau^* - \tau_c^*$, and the volume discharge of debris flow, $Q_t$. thus, we can be express this relation as

$$L_d = f_1\left(Q_t, \quad Q_m, \quad \tau^* - \tau_c^*, \quad W_m\right) \tag{2}$$

According to the experiment data, we could find the relation between $Q_m/Q_t$, $\tau^* - \tau_c^*$ and $L_d/W_m$. Besides, their multiple linear model was given by

$$\frac{L_d}{W_m} = \left(\frac{Q_m}{Q_t}\right)^{0.489} \left(\tau^* - \tau_c^*\right)^{0.146} \tag{3}$$

A large value of $Q_m/Q_t$ meant that the kinetic energy of flow in main flume was larger than that of the flow in tributary. Sediments which transported in main flume could take more energy form the flow in this situation; as a result, it took a long transport distance for sediments to deposit. Finally, the shape of an alluvial fan became long and narrow. In addition, large discharge of debris flow meant high concentration in our experiment. Debris flow with high concentration was hard to flow. Because sediments of debris flow deposits nearly the confluence, the shape of an alluvial fan became short and vast. As mentioned above, the increasing of the value of $Q_m/Q_t$ yielded a narrow alluvial fan, and it conformed to the result of equation (1). Besides, a steep slope of main flume led to a decreasing in effective shear stress. Sediments were easy to transport in

Figure 14. Example for length-width ratio of alluvial fans.

main flume, so the length of alluvial fan increased significantly. As had been discussed, our model was in complete agreement with the physical phenomenon in the confluence. In addition, this model passed F-test and t-test, and was found to be statistically significant at 0.05.

Figure 14 showed two case of debris flow alluvial fan in Wu Creek at Taichung. The main flow direction was right to left, and the angles between the main river and two tributaries were 90 degrees. The upstream alluvial fan was close to another, so the main flow conditions of two fans could be seen similarly here. Therefore, it could be said that the different shapes of these two alluvial fans were due to the discrepancy in debris flow discharge in tributary. The downstream alluvial fan was narrower than the upstream one as showing in Figure 14; consequently, we knew that the volume discharge or concentration of upstream tributary was larger than downstream one according to the equation 3.

4   CONCLUSIONS

In this paper, we presented the results of the types of alluvial fans in different main and tributary flow conditions and the effect of retrogressive deposition to the alluvial fan. We differentiated the alluvial fan forming process into sediments diffusions step and sediments deposition and pushing expansion step. Furthermore, four types of alluvial fan could be distinguished according to the existence of alluvial fan and retrogressive deposition in the second stage of forming process. Four types were no alluvial fan, transitional alluvial fan, slight retrogressive deposition alluvial fan and heavy retrogressive deposition alluvial fan.

This paper also defined the factor of $M_r/M_{st}$ as the delay of retrogressive deposition in the tributary. The results showed a strong relation between the factors of $M_r/M_{st}$ and $Q_{sm}/Q_t$.

The factors of $V_d/V_u$ and $L_d/L_u$ were offered and could be explained the downstream slant of an alluvial fan. It was shown that the factor of $Q_{sm}/Q_t$ was positive related to the $V_d/V_u$ and $L_d/L_u$. However, the relation between $Q_{sm}/Q_t$ and $L_d/L_u$ was greater than $Q_{sm}/Q_t$ and $V_d/V_u$. The reasons for the lower correlation between $Q_{sm}/Q_t$ and $V_d/V_u$ could be the equally downstream and upstream expansion which was due to the push of retrogressive deposition. The shape of an alluvial fan could be described by the factor of $L_d/W_m$. The large value of $L_d/W_m$ led to a long and narrow alluvial fan. We also offered the model which composes of $Q_m/Q_t$, $\tau^* - \tau_c^*$ and $L_d/W_m$, and it was found to be statistically significant.

REFERENCES

Schumm, S. A. 1977. *The Fluvial System*. John Wiley & Sons, Inc., New York, N.Y.
Blair, T. C. & McPherson, J. G. 1994. Alluvial Fans and Their Natural Distinction from Rivers Based on Morphology, Hydraulic Processes, Sedimentary Processes and Facies Assemblages. *J. Sed. Res.* A64(3): 450–489.
Parker, G. 1998. Alluvial Fans Formed by Channelized Fluvial and Sheet Flow. I: Theory. *Journal of Hydraulic Engineering* 124(10): 985–995.
Kuanh, S.F. 1995. Study on Behaviors and Deposit Processes of Debris Flow at the confluence. *Journal of Sediment Research* 3: 1–15.
Chen, S.C. & Peng, S.H. 2002. Numerical Modeling of Dam-Break Rapidly Varied Flow. *Journal of Soil and Water Conservation* 33(4): 293–303.
Chen, S.C. & Peng, S.H. 2003. Experiment of Confluence between Main River and Tributary by Applying Digital Image Processing. *Journal of Chinese Soil and Water Conservation* 34(2): 195–205.
Chen, C.G., Yao, L.K. & Yang, Q.H. 2004. Experimental Research on Confluence between Debris Flow and Main River. *Journal of Southwest Jiaotong University* 39(1): 10–14.
Chen, S.C. & Peng, S.H. 2004. Experiments of Alluvial Fans Formed by Hyperconcentrated Tributaries. *Researches on Mountain Disasters and Environmental Protection across Taiwan Strait* 4: 259–265.
Chen, S.C. & Li, C.C 2006. The Fluvial Processes of Main Channel Induced by the Tributary of Debris Flow. *Journal of Chinese Soil and Water Conservation*. 37(1): 9–22.

*River meandering*

*River, Coastal and Estuarine Morphodynamics: RCEM 2007 – Dohmen-Janssen & Hulscher (eds)*
*© 2008 Taylor & Francis Group, London, ISBN 978-0-415-45363-9*

# Long-term behaviour of meandering rivers

Alessandro Frascati & Stefano Lanzoni

*Dept. of Hydraulic, Maritime, Environmental and Geotechnical Engineering, University of Padova, Italy*

ABSTRACT: Natural rivers are self-formed features whose shapes are the result of interaction between erosion, deposition and transport of sediments. The study of their morphodynamics and the characterization of related sedimentary processes are of great interest not only to environmental engineers but also to hydrology and historical geology, contributing to the interpretation of stratigraphic records. In the present contribution we focus our attention on the long-term behaviour of meandering rivers, a very common pattern in nature, which belongs to a class of dynamical systems occurring at the spatial scale of the channel width and driven by the coexistence of complex linear and non-linear processes. On the short term time scale, the formation of meandering patterns can be suitably explained as an instability process, driven by bank erosion (bend instability). The planar development of the river is described by a non-linear integro-differential bend evolution equation, complemented with a suitable model for flow and bed topography in sinuous channels with cohesionless bed. On the long-term timescale, a further highly non-linear process must be accounted for, namely channel shortening via cutoff processes. Depending on the description adopted for the flow field, various mathematical models allowing the description of the temporal evolution of the channel axis can be developed. The problem then arise to compare the morphologic characteristics of the planimetric configurations obtained using the different flow field models as well as the differences/analogies between calculated patterns and those observed in the field. Usually, the comparison is pursued by considering a few typical variables (e.g., cartesian and intrinsic wavelengths, sinuosity, curvature) which, however, if separately investigated cannot provide an objective and discriminant description of the complex morphologic features exhibited by either calculated or observed planimetric configurations. In order to evaluate the differences/similarities of the patterns calculated considering different flow field models and to test the related long-term prediction capabilities, a significant sample of both computed planar configurations and of planimetric patterns extracted from Landsat mosaic images is analyzed through a more complete statistical method of characterization of channel axis configuration, based on a wide set of morphological variables.

## 1 INTRODUCTION

Modelling hydrodynamic and bed topography in natural rivers has long attracted attention of scientific community in the fields of hydraulic engineering and fluvial geomorphology. In the present contribution we focus our attention on meandering rivers whose planimetric migration is essentially governed by erosion at outer bank and deposition at inner bank of channel bends (Figure 1). Mathematically, the problem of channel axis migration can be described by a non linear integro-differential evolution equation which must be supplemented by a suitable relationship relating the lateral bank migration rate to the flow field in sinuous erodible channels.

Usually, bank erosion is an intermittent and heterogeneous process, whose mechanism is rather complex (Darby & Thorne 1996, Darby 1998 and Darby et al. 2002). However, on the very slow time scale associated with channel planimetric development, erosion and

Figure 1. Meandering river and point bar deposits (Alatna river, Alaska).

deposition processes can be considered as contemporaneous events and, consequently, bank erosion may be modelled as a continuous process which keeps nearly constant the average width of the river. Coupling this simplified treatment of bank erosion with a suitably linearized flow field model leads to the mathematical description of river meandering on short time scales. On longer time scales bend cutoff processes introduce a further highly non linear mechanism which must be properly accounted for. As a consequence of the non linear character of physical processes governing the planimetric evolution of meandering rivers, the planimetric patterns resulting from mathematical models, as well as the observed ones, are rather complex. The problem then arise of comparing the complex long-term morphologic characteristics of observed and synthetic meandering patterns.

In this contribute, in order to provide an objective and discriminant analysis, the comparison is pursued performing a Principal Component Analysis (PCA) on a wide set of morphological variables. The investigation is carried out on significant sample of planar configurations, computed using two different linearized flow field (Ikeda et al. 1981, Zolezzi & Seminara 2001), along with some real planimetric patterns, extracted from Landsat7 ETM+ mosaic images in different environmental contexts (Alaska, Canada, USA, Brazil, Papua-New Guinea, Peru).

The paper is organized as follow. Section 2 is devoted to summarize briefly the mathematical formulation of bend instability in erodible channels. In section 3, we emphasize the differences/similarities exhibited by the two linearized flow field solutions and describe the numerical procedure adopted to simulate the long-term behavior of meandering rivers. In section 4, we compare simulated planforms and real meandering patterns. Finally, section 5 concludes the paper with some discussion of open issues which will deserve attention in the near future.

## 2 MATHEMATICAL FORMULATION OF BEND INSTABILITY PROCESS

The problem governing the planform evolution of meandering channels can be essentially reduced to that of describing the motion of their centreline, lying on a plane, defining that each point of this line moving in the normal direction with some lateral migration speed $\zeta^*$ driven by bank erosion process. Let us consider the permanent flow in a wide meandering channel with constant width $2B_0^*$ (hereinafter a star superscript will denote dimensional quantities), uniform and cohesionless sediments, and gently sloping erodible banks. Moreover, let $D_0^*$ and $U_0^*$ the reach averaged flow depth and depth averaged longitudinal velocity. We refer the flow field and bed topography to the orthogonal intrinsic reference system $(s^*, n^*)$ illustrated in Figure 2, where $s^*$ is the longitudinal coordinate locally coinciding with channel axis tangent, and $n^*$ is the transverse coordinate, with origin at the channel axis.

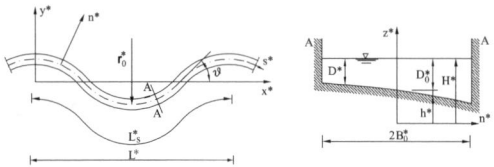

Figure 2. Sketch of a meandering channel and notation. A meander is defined as segment of stream lying among three inflection points.

Observing that, at a given time $t^*$, the channel axis configuration is described by the distribution $\theta(s^*, t^*)$ of the angle that the local tangent to the channel axis forms with the direction of a Cartesian axis $x^*$ (Fig. 2), the following non linear integro-differential planimetric evolution equation can be derived (Seminara et al. 1994, Seminara et al. 2001):

$$\frac{\partial \zeta}{\partial s} = \frac{\partial \theta}{\partial t} - \frac{\partial \theta}{\partial s} \int_0^s \zeta \frac{\partial \theta}{\partial s} ds \tag{1}$$

where:

$$(s^*, n^*) = B_0^*(s, n), \quad \zeta^* = U_0^* \zeta, \quad t^* = U_0^* t / B_0^* \tag{2}$$

The mathematical formulation of bend instability next requires an erosion law relating the lateral migration speed of the channel $\zeta$ to the stream hydrodynamics, which depends on channel curvature. Following the bend-erosion theory of Ikeda et al. (1981), the bank erosion process is modeled using a simple law relating linearly $\zeta$ to flow perturbations at the banks $\Delta U$. Hence we write:

$$\zeta = E\Delta U = E\left(U|_{n=1} - U|_{n=-1}\right) \tag{3}$$

where the depth averaged velocity is scaled by $U_0^*$ and $E$ is a dimensionless long-term erosion coefficient. In order to complete the formulation of the problem we finally need a hydrodynamic model of flow field in sinuous channel with cohesionless bed and an arbitrary distribution of channel curvature. In the present paper we will refer to two linear model which have been widely employed in the scientific community. The first model was proposed by Ikeda et al. (1981) (hereinafter referred to as IPS) for sequences of sine generate meanders, the second model (hereinafter referred to as ZS) was proposed for sequences of sine generate meanders by Blondeaux & Seminara (1985) and later on extended to rivers with an arbitrary distribution of channel curvature by Zolezzi & Seminara (2001). In this latter case the longitudinal velocity reads

$$U = 1 + \nu_0 \sum_{m=0}^{\infty} \left\{ \sin \frac{(2m+1)\pi n}{2} \sum_{j=1}^{4} \left[ A_m g_{j0} \cdot \right. \right.$$

$$\int_{s_0}^{s} \mathcal{C} e^{\lambda_{mj}(s-\xi)} d\xi + c_{mj} e^{\lambda_{mj}(s-s_0)} + A_m g_{j1} \mathcal{C} \bigg] \bigg\} \quad (4)$$

while the the near-bank velocity increment $u_b$ prescribed by the uncoupled IPS model is

$$u_b = \nu_0 \left[ -\mathcal{C} + \beta C_{f0}(A + F_0^2) \int_0^s e^{-2\beta C_{f0}(s-t)} \mathcal{C} dt \right] (5)$$

Here $\nu_0$ is the curvature ratio, defined as the ratio of $B_0^*$ to some typical value of radius of curvature $r_0^*$ (say its minimum value in the meandering reach, see Figure 2), $\mathcal{C}$ is the local value of the dimensionless channel curvature, $C_{f0}$ and $F_0$ are the reach averaged friction coefficient and the Froude number, respectively. Moreover, $g_{jk}$ ($j = 1, 4; k = 0, 1$) are coefficients depending on the relevant physical parameters ($\beta$, $C_{f0}$, and $\tau_*$), $c_{mj}$ are integration constants to be specified on the basis of the boundary conditions at the channel ends, while $\lambda_{mj}$ ($m = 0, \infty; j = 1, 4$) are characteristic exponents and $A$ is a positive constant, evaluated theoretically trough the linear relationship proposed by Engelund (1964) and afterward corrected by Johannesson & Parker (1989), which takes into account the effects of secondary flow in bends and the resulting variation in bed elevation.

In the ZS model the two-dimensional linearized flow-field is strictly coupled with the Exner sediment balance equation. It turns out that meanders behave as linear oscillators which resonate at specific values $\lambda_r$ and $\beta_r$ of their intrinsic dimensionless wavenumber $\lambda$ ($= \lambda^* B_0^*$) and aspect ratio $\beta$ ($= B_0^*/D_0^*$), depending on the intensity of sediment transport and friction. Conversely, in the linearized IPS model the near-bank velocity increment is evaluated by decoupling the hydrodynamic and the sediment conservation equations. The main consequence of this simplification is that the resonance phenomenon cannot occur and, therefore, the uncoupled model can provide only a part of the complete solution of the problem.

Resonance is crucial for discriminating the dynamics of meandering rivers, at least on short time scales. Following a classical normal mode approach, it can be demonstrated that the growth rate of small amplitude planimetric perturbations (given by a regular sequence of sine generated meanders) attains a maximum in the range 0.1–0.4 of the dimensionless wavenumber (corresponding to cartesian wavelengths of about 40–10 channel widths, respectively). This maximum tends to infinity when the dimensionless wavenumber and the aspect ratio take the resonant values $(\lambda_r, \beta_r)$ pointed out by Blondeaux & Seminara (1985). Moreover, the meandering pattern migrates while amplifying, owing to the occurrence of a phase lag between the erosion and curvature peaks. Crossing resonance (i.e., when $\beta$ increase beyond $\beta_r$ for given meander wavenumber, $\lambda$ or, vice versa, when $\lambda$ increases beyond the $\lambda_r$ for a

given aspect ratio) implies that the location of velocity and hence erosion) peak crosses the bend apex. As a result, in the sub-resonant ($\beta < \beta_r$) regime meanders are skewed upstream and migrate downstream. An opposite scenario arises in the super-resonant ($\beta > \beta_r$) regime: meander are skewed downstream and migrate upstream.

As thoroughly discussed in Lanzoni and Seminara (2006) and in Lanzoni et al. (2006), the ability of a model to discriminate between sub- or super-resonant regime has important implications on the issues of upstream/downstream morphodynamic and hydrodynamic influence and on the appropriate choice of the boundary conditions to be adopted in numerical modeling of the planimetric evolution of meandering rivers. Clearly, the uncoupled IPS model, being unable to predict the resonance phenomenon, cannot describe the super-resonant regime and the related dominant upstream morphodynamic influence. On short timescales, the coupling between hydrodynamics and morphodynamics is thus crucial in reproducing correctly the meandering dynamics. The question we address in the following is whether the appropriate choice of the flow field modeling is also discriminant on longer time scales.

## 3 NUMERICAL MODELING

The formation of sequences of periodic meanders usually characterizes only relatively short reaches of natural rivers. On long time scales typical of river meander evolution besides the non-linearities associated to equation (1), the geometry of planimetric patterns is further complicated by the strongly non linear interactions arising from abrupt channel shortening via cutoff processes and spatial-temporal heterogeneity of flood plain and bank erodibility.

It is thus apparent that the analysis of long-term behavior of meandering rivers requires a numerical approach. To this end, we discretize the channel axis through a sequence of equally spaced nodes $P_i = (x_i, y_i)$. At every time step $\Delta t$, the planimetric evolution of the channel axis is calculated by displacing each node orthogonally to the channel centerline by an amount $\zeta_i \Delta t$. On the very slow time scale associated with the planimetric development of the channel, bank erosion is modeled as a continuous process and described in term of the slow temporal variable $\tau = Et$. Although the long-term evolution of meandering rivers produces a rearrangement of the floodplain, in this contribution a constant erodibility coefficient is assumed (say $10^{-8}$). In other words, we neglect here the different erodibility exhibited by the undisturbed floodplain, the point bar deposits (see Figure 1) and the oxbow lakes deposits (see Figure 3).

a)

b)

Figure 3. Examples of two planimetric patterns showing some examples of neck cutoffs leading to the formation of oxbow lakes. The images refer to a) Mara River (Kenya) and b) Alatna River (Alaska).

The longitudinal velocity given by ZS and IPS linearized flow field models are alternatively used to predict, at each time step, the local value of the channel axis migration rate $\zeta$. The local value of dimensionless channel curvature $\mathcal{C}$ appearing in the velocity relationships is determined approximating through a centered finite difference scheme the geometrical relationship:

$$\nu_0 \mathcal{C}(s) = -\frac{d\theta}{ds} \qquad (6)$$

where the spatial distribution of the angle $\theta$ is determined by averaging back and forth

$$\theta_i = \frac{1}{2}\left(\arctan\frac{y_{i+1} - y_i}{x_{i+1} - x_i} + \arctan\frac{y_i - y_{i-1}}{x_i - x_{i-1}}\right) \qquad (7)$$

The computation of flow-field require also the computation of convolution integrals which appear in (5) and in (4). In order to significantly reduce the computational effort, the integration is performed using a semi-analytical approach, based on the assumption that the curvature varies linearly between two consecutive nodes. Moreover, taking advantage of the fact that the function to be integrated decays exponentially, the integration is truncated when the function is smaller than a given tolerance (say 0.0001), with a further significant reduction of the computational time.

In order to improve the accuracy and efficiency of computation, a time marching predictor-corrector method (Crosato 1990) is used. The forward time step is performed using for each node the normalized migration rate obtained by averaging its values at the previous and present time steps. The step size $\Delta\tau$ is controlled by requiring that

$$\Delta\tau \leq \epsilon \frac{\Delta s}{\Delta U_{max}} \qquad (8)$$

where $\Delta s$ is the distance between two consecutive nodes, and $\epsilon$ is an empirical parameter defining the threshold between stable and unstable computations. The choice of the value of the parameter $\epsilon$ (=0.005), was made through suitable simulation tests. As the channel migrates, the distance $\Delta s$ between individual nodes may increase or decrease. A standard cubic spline interpolation was used to remesh the points uniformly after each time step. Moreover, new nodes are periodically added to maintain the size of the spatial step in the range 0.8–0.9.

In order to study the long-term morphological evolution of a meandering rivers the highly non-linear process of channel shortening via cutoff processes must be accounted for. These processes provide local mechanisms to straighten the channel axis, thus limiting the growth of channel sinuosity $\sigma$ (defined as the ratio between meander intrinsic wavelength and meander cartesian wavelength; i.e $\sigma = L_s^*/L^*$, see Figure 2). Two different types of cutoff are usually recognized in natural channels, chute and neck cutoffs. Chute cutoffs are relatively long flow diversions which occur when a meander loop is bypassed through a new channel which forms across the bar enclosed by the loop. This process occurs most frequently in wide channels with large curvature bends, high discharges, poorly cohesive, weakly vegetated banks, and high gradients (Howard & Knutson 1984). Neck cutoffs occur when the local sinuosity becomes so large that adjacent loops intersect each other, leading to the formation of an abandoned loop (oxbow lake) when sedimentation closes the ends of the former loop (see Figure 3).

In the present simulations we focus our attention only on neck cutoff processes, which characterize the long-term development of highly sinuous meandering rivers. Following Howard & Knutson (1984) and Sun et al. (1996), the presence of potential neck cutoffs was detected by controlling whether the distance between the $i$-th node $P_i$ and a given node $P_{i+k}$ (not immediately upstream or downstream of $P_i$) approaches some critical distance (say $2.2B_0^*$). When such a distance is reached, all the points $P_{i+j}, j = 1, k$, representing the abandoned channel loop, are removed. To recognize the nodes closer than the selected critical distance we used a matrix algorithm proposed by Camporeale et al. (2005) with a significant reduction of the computational effort. Furthermore, in order to avoid the

Table 1. Morphodynamic parameters used in the simulations.

| Runs | $\beta$ | $d_s$ | $\tau_*$ | bed | influence |
|------|------|--------|------|-------|---------------|
| 1 | 13 | 0.0001 | 0.3 | dune | sub-resonant |
| 2 | 11 | 0.001 | 0.5 | dune | sub-resonant |
| 3 | 9 | 0.005 | 0.5 | dune | sub-resonant |
| 4 | 17 | 0.005 | 0.4 | dune | super-resonant |
| 5 | 18 | 0.001 | 0.15 | dune | super-resonant |
| 6 | 13 | 0.005 | 0.1 | plane | sub-resonant |
| 7 | 12 | 0.01 | 0.08 | plane | sub-resonant |
| 8 | 14 | 0.025 | 0.08 | plane | super-resonant |
| 9 | 18 | 0.01 | 0.1 | plane | super-resonant |
| 10 | 8 | 0.005 | 0.03 | plane | sub-resonant |

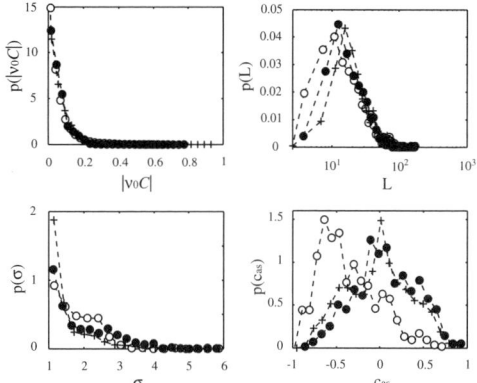

Figure 5. Probability density function of: a) absolute channel axis curvature $|C|$; b) cartesian wavelength $L$; c) sinuosity $\sigma$; d) asymmetry coefficient $c_{as}$ for the planimetric patterns observed in the field ($+$) and calculated adopting the ZS ($\bullet$) and IPS ($\circ$) models.

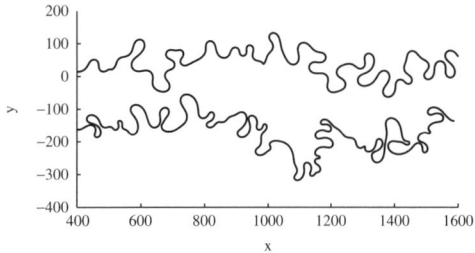

Figure 4. Planimetric configurations obtained for ZS model (top) and IPS model (bottom), run 7.

presence of physically nonrealistic cusplike regions with very high curvature at a reconnection, we chose to remove also three nodes upstream of $P_i$ and three points downstream of $P_{i+k}$. Such a somehow artificial procedure is justified by the fact that, in meandering rivers, the rapid smoothing of sharp bends is a well-known process.

## 4 ANALYSIS AND RESULTS

Starting from a initial straight planimetric configuration, slightly and randomly perturbed, we have simulated the long term planform evolution of meandering rivers using both the ZS and the IPS linearized flow field models.

Table 1 reports the sets (ten) of hydraulic parameters that have been used in the simulations. Plane and dune covered bed conditions, as well as sub-resonant and super-resonant regimes have been investigated. In all cases, no constraints have been imposed at both the channel ends (for a thorough discussion of suitable boundary conditions see Lanzoni & Seminara 2006). Two typical examples of the resulting planimetric configurations are shown in Figure 4. In order to emphasize the difference/similarities between the

complex meandering patterns which arise from simulations and those extracted from satellite images, following Howard & Hemberger (1991) we investigate the statistics of absolute channel axis curvature $|C|$, of half meander sinuosity $\sigma_i = L_{si}/L_i$, of cartesian half-meander wavelength $L_i$, of tortuosity $T$, defined as $T = \sum_{i=1}^{N} L_{si}/L_0$, and of the asymmetry coefficient (Howard & Hemberger, 1991):

$$c_{as} = \frac{L_{su} - L_{sd}}{L_s} \qquad (9)$$

where as shown in Figure 6, $L_0$ denotes the overall cartesian length of the meandering reach; $L_{si}$ and $L_i$ ($i = 1, 2, \cdots$) are the half-meander intrinsic and cartesian length, respectively; $L_{su}$ and $L_{sd}$ are the intrinsic lengths of the channel axis upstream and downstream the locus of maximum curvature $C_{max}$ in the half-meander.

An overall view of the probability density functions (pdf) of absolute curvature, cartesian wavelength, sinuosity and asymmetry coefficient obtained for both synthetic and natural patterns is reported in Figure 5. The pdf of $|C|$ and $L$ are pretty similar for all the three groups of planimetric configurations investigated. Some slight differences emerge in the plot of the $\sigma$ pdf. On the contrary, the $c_{as}$ pdf resulting from the analysis of IPS patterns is significantly different from the trend common to both ZS and natural configurations. On the other hand, natural rivers typically exhibit values of the tortuosity smaller than those characterizing ZS and IPS patterns, the respective average values of $T$ being 1.83, 3.07 and 3.72.

A quantitative comparison between the statistical properties intrinsic of the the various pdf patterns can

843

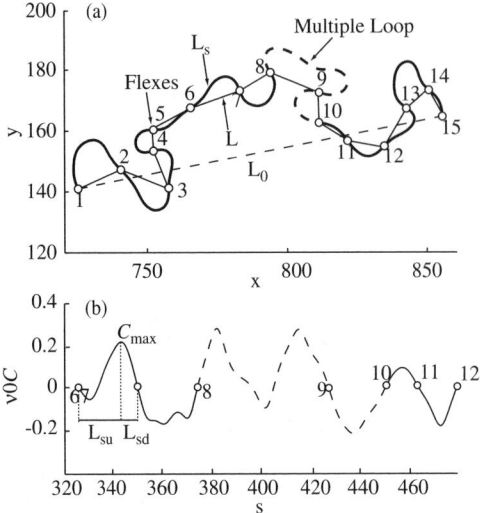

Figure 6. (a) Planimetric pattern of channel centerline showing: the inflection points (solid dots), the half-meanders intrinsic ($L_s$) and cartesian ($L$) wavelengths, and the straight distance, $L_0$, between the initial and final inflection point of the investigated reach. The dashed line portion of the channel axis shows a typical example of multiple loop. (b) Correspondent behavior of the local curvature $v_0 C$.

be easily pursued introducing a suitable set of statistic variables, grouped in two categories (Howard & Hemberger 1991). The first includes quantities related to the entire reach of the pattern under investigation, namely the mean ($|C|_{avg}$), the variance ($|C|_{var}$), the skewness ($|C|_{sk}$), and the kurtosis ($|C|_{kr}$) of the absolute curvature $|C|$ of channel axis, and the tortuosity $T$. The second category contains statistics of half-meander variables, i.e. measured on each portion of channel between two successive inflection points. The statistics include: the mean ($\sigma_{avg}$) and the variance ($\sigma_{var}$) half-meander sinuosity $\sigma_i = L_{si}/L_i$, the mean ($L_{avg}$), the variance ($L_{var}$), the skewness ($L_{sk}$), and the kurtosis ($L_{kr}$) of the cartesian half-meander wavelength $L_i$, the mean of the asymmetry coefficient $c_{as}$.

We now apply to the above suite of (twelve) statistical variables the so called Principal Component Analysis (PCA), a decomposition technique commonly adopted for recognizing patterns in data sets, and expressing the data in such a way as to highlight their similarities and differences.

The PCA methodology can be briefly summarized as follows. Let $\mathbf{X} = [\mathbf{x}_1 \mathbf{x}_2 \dots \mathbf{x}_M]$ the $N \times M$ data matrix, where the components of vector $\mathbf{x}_j$ (j = 1,M) represents the sampled values of the N (= 12) statistical variables associated to the j-th planform configuration. We first create a new data set matrix $\mathbf{Y} = [\mathbf{y}_1 \mathbf{y}_2 \dots \mathbf{y}_M]$

whose elements are obtained from $\mathbf{X}$ elements by the relationship:

$$y_{ij} = \frac{x_{ij} - \mu_i}{\sigma_{Di}} \qquad (i = 1, N; j = 1, M) \qquad (10)$$

where $\mu_i$ and $\sigma_i$ are the mean and the standard deviation of the i-th row of $\mathbf{X}$, respectively. The operation (10) produce a standardized data set with zero mean and unit standard deviation. We next compute the square sample covariance matrix $\mathbf{Z}$ and its eigenvalues ($\lambda_1 > \lambda_2 > \dots > \lambda_N$) and eigenvectors ($\mathbf{e}_1, \mathbf{e}_2, \dots, \mathbf{e}_N$). The eigenvector $\mathbf{e}_1$ with the highest eigenvalue ($\lambda_1$) represents the first principal component of the data set, the eigenvector $\mathbf{e}_2$ associated with the second highest eigenvalue is the second principal component, and so on. Hence, the full set of principal components is as large as the original set of variables. Since $\mathbf{Z}$ is symmetric all the principal components are orthogonal to each others and any vector column $\mathbf{y}_j$ can be written as a linear combination:

$$\mathbf{y}_j = \sum_{i=1}^{N} a_{ij} \mathbf{e}_i \qquad (j = 1, M) \qquad (11)$$

However, observing that the percent of total variability associated to each principal component decreases progressively from the first to the N-th principal component, it is possible to neglect the components of lesser significance (i.e., associated with smaller eigenvalues). Keeping only the term corresponding to the $K$ largest eigenvalue we finally obtain the reduced matrix $\hat{\mathbf{Y}}$ with components:

$$\hat{\mathbf{y}}_j = \sum_{i=1}^{K} a_{ij} \mathbf{e}_i \qquad (K << N; j = 1, M) \qquad (12)$$

where $\mathbf{a}_j = (a_{1j} a_{2j} \dots a_{kj})^T$ is the representation of $\hat{\mathbf{y}}_j$ in the principal component basis ($\mathbf{e}_1, \mathbf{e}_2, \dots, \mathbf{e}_K$).

In the specific case treated here the first two principal components account for about 60% of the total variance of the original data. Figure 7, displaying the scatter plot of the principal component coefficients $a_{1j}$ and $a_{2j}$, suggests a certain degree of similarity between natural streams and synthetic patterns obtained adopting the ZS model. Conversely, the group of points representing real streams and that synthetically generated though the IPS model tends to lie on opposite $a_{2j}$-half planes.

## 5 CONCLUSIONS

The fully nonlinear simulation of the lateral migration of meandering channels, carried out adopting the analytical flow field solutions provided by linearized models, provides a powerful and yet computationally

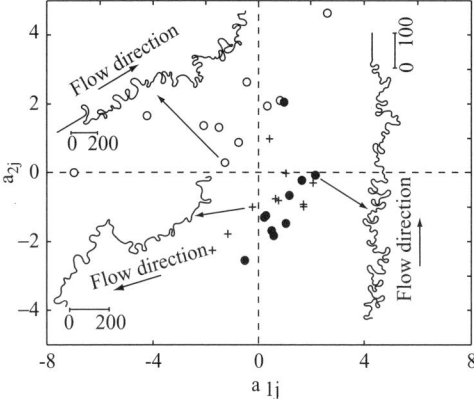

Figure 7. Scatter plot of the principal component coefficients $a_{1j}$ and $a_{2j}$. Symbols are as follows: o IPS model; • ZS model; + natural rivers. The typical planform configurations also reported in the figure refer to a natural river located in Papua-New Guinea, and to two synthetic patterns simulated considering the IPS model (run 7), and the ZS model (run 4). The planimetric dimensionless scale is expressed in half-width units.

accessible tool to investigate both short and long term evolution of alluvial rivers. A general consent exists (Lanzoni et al. 2006, Camporeale et al. 2007) that, on short time scales typical of the evolution of single meanders before cutoff, the computed planimetric forms crucially depend on the model chosen to specify the excess flow velocity at channel banks. On longer timescales, the repeated occurrence of highly nonlinear cut-off events leads to very complex planimetric patterns. Whether meander shapes computed on long term time scales resemble or not those observed in nature, is a question that can be investigated only from a statistical point of view, as pursued in the present contribution. Our results indicate that, although various morphological aspects of observed meandering rivers (e.g., the pdf of absolute curvature) are captured by strongly simplified flow field models (e.g, the IPS model), nevertheless a closer similarity with natural meandering shapes is achieved only though flow field models (e.g., the ZS model) which take into account more completely morphodynamic mechanisms, as resonance. Obviously, wider sets of both synthetical and observed meandering patterns are required to further validate the above results. Moreover, the statistical effects exerted on meander shapes by spatial variations of floodplain erodibility associated to sedimentation processes (e.g. oxbow lake formation), geological constraints and vegetation dynamics needs to be addressed. Finally, the role of cutoff processes, in particular chute cutoff (not accounted for here), will require attention in the near future.

ACKNOWLEDGEMENTS

This work was supported by the MoDiTe project, "Modelli di Generazione, Propagazione e del Trasporto per la Difesa del Territorio", funded by Fondazione Cassa di Risparmio di Verona, Vicenza, Belluno ed Ancona.

REFERENCES

Blondeaux, P., and G. Seminara (1985), A unified bar-bend theory of river meanders, J. Fluid Mech., 157, 449–470.

Chitale, S.V. (1970), River channel patterns, J. Hydr. Div. ASCE, 96(HY1), 201–221.

Camporeale, C., P. Perona, A. Porporato and L. Ridolfi (2005), On the long-term behavior of meandering rivers, Water Resour. Res., 41, doi:10.1029/2005WR004109.

Camporeale, C., P. Perona, A. Porporato, and L. Ridolfi (2007), Hierarchy of models for meandering rivers and related morphodynamics processes, Rev. Geophys., 45, RG1001, doi:10.1029/2005RG000185.

Colombini, M., M. Tubino, P. Whiting (1992), Topographic expressions of bars in meandering channels. In Dynamics of Gravel-Bed Rivers, edited by P. Billi, R.D. Hey, C.R. Thorne and P. Tacconi, pp. 457–474, John Wiley, Cichester.

Crosato, A. (1990) Simulation of meandering river processes, Communication on Hydraulic and Geotechnical Engineering, Report 90-3, Delft University of Technology, Delft, NL.

Darby, S.E. (1998), Modelling width adjustment in straight alluvial channels, Hydrol. Processes, 12, 1299–1321.

Darby, S.E., A.M. Alabyan, and M. Van de Wiel (2002), Numerical simulation of bank erosion and channel migration in meandering rivers, Water Resour. Res., 38(9), 1163, doi:10.1029/2001WR000602.

Federici, B., and G. Seminara (2003), On the convective nature of bar instability, J. Fluid Mech., 487, 125–145.

Hey, R.D., and C.R. Thorne (1986), Stable channels with mobile gravel beds, J. Hydr. Eng., 112(8), 671–689.

Howard, A. (1992), Modelling channel evolution and floodplain morphology. In Floodplain Processes, edited by M.G. Anderson, D.E. Walling and P.D. Bates, pp. 15–62, John Wiley, New York.

Howard, A. (1996), Modelling channel migration and floodplain development in meandering streams. In Lowland Flood-plain Rivers: Geomorphological Perspectives, edited by P.A. Carling and G.E. Petts, pp. 2–41, John Wiley, New York.

Howard, A., and T.R. Knutson (1984), Sufficient conditions for river meandering: A simulation approach, Water Resour. Res., 20(11), 1659–1667.

Howard, A., and A.T. Hemberger (1991), Multivariate characterization of meandering, Geomorphology, 4, 161–186.

Ikeda, S., G. Parker, G., and K. Sawai (1981), Bend theory of river meanders. Part 1. Linear development, J. Fluid Mech., 112, 363–377.

Johanneson, H., and G. Parker (1989), Linear theory of river meanders. In River Meandering, edited by S. Ikeda and G. Parker, Water Resour. Monograph 12, pp. 181–214, AGU, Washington D.C.

845

Lanzoni, S., and G. Seminara (2006), On the nature of meander instability, *J. Geophys. Res.*, 111, F04006, doi:10.1029/2005JF000416.

Lanzoni, S., A. Sivglia, A.Frascati, and G. Seminara (2006), Long waves in erodible channels and morphodynamic influence, *Water Resour. Res.*, 42, W06D17, doi:10.1029/2006WR04916.

Leopold, L.B., M.G. Wolman, and J.P. Miller (1964), *Fluvial processes in Geomorphology*, W.H. Freeman, New York.

Parker, G., and H. Johanneson, (1989), Observations on several recent theories of resonance and overdeepening in meandering channels. In *River Meandering*, edited by S. Ikeda and G. Parker, Water Resour. Monograph 12, pp. 379–415, AGU, Washington D.C.

Seminara, G. (2006), Meanders, *J. Fluid Mech.*, 554, 271–297.

Seminara, G., and M. Tubino (1989), Alternate bars and meandering: free, forced and mixed interactions, In *River Meandering*, edited by S. Ikeda and G. Parker, Water Resour. Monograph 12, pp. 267–320, AGU, Washington D.C.

Seminara, G., and M. Tubino (1992), Weakly nonlinear theory of regular meanders, *J. Fluid Mech.*, 244, 257–288.

Seminara, G., G. Zolezzi, M. Tubino, and D. Zardi (2001), Downstream and upstream influence in river meandering. Part 2. Planimetric development, *J. Fluid Mech.*, 438, 213–230.

Sun, T., P. Meaking, T. Jossang, and K. Schwarz (1996), A simulation model for meandering rivers, *Water Resour. Res.*, 32(9), 2937–2954.

Sun, T., P. Meaking, and T. Jossang (2001), A computer model for meandering rivers with multiple bed load sediment sizes. 2. Computer simulations *Water Resour. Res.*, 37(8), 2243–2258.

Tubino, M., and G. Seminara (1990), Free-forced interactions in developing meanders and suppression of free bars, *J. Fluid Mech.*, 214, 131–159.

Zolezzi, G., and G. Seminara (2001), Downstream and upstream influence in river meandering. Part 1. General theory and application to overdeepening, *J. Fluid Mech.*, 438, 183–211.

*River, Coastal and Estuarine Morphodynamics: RCEM 2007 – Dohmen-Janssen & Hulscher (eds)*
*© 2008 Taylor & Francis Group, London, ISBN 978-0-415-45363-9*

# A non linear model for river meandering

G. Nobile, M. Bolla Pittaluga & G. Seminara

*Department of Civil, Environmental and Architectural Engineering*

ABSTRACT: An analytical three dimensional model for flow and bed topography in alluvial meandering channels is proposed. The non linear model extends the analysis developed by Seminara and Solari (1998) on finite bed deformations in constant curvature channel with constant width. The former constraint is relaxed here accounting for arbitrary variations of channel curvature. Flow and bed topography are assumed to be slowly varying functions in both longitudinal and lateral directions.

Moreover, the model is also used to formulate a non-linear version of bend instability theory. The wavelengths selected in the meandering process turns out to depend on the amplitude of the initial perturbation and agree with values typically observed in nature, smaller than the ones obtained within a linear context.

The computational effort needed in the present model is many orders of magnitude smaller than the one which would be required in the context of a conventional 3D numerical model. This will hopefully allow us to investigate the planimetric evolution of meandering patterns accounting for the so far neglected role of flow non linearities.

## 1 INTRODUCTION

In the present contribution we extend the ideas developed in a previous work by Seminara and Solari(1998) where the classical assumption of linearity employed to investigate bed deformations in constant curvature channels with constant width was relaxed. We further extend the analysis to the general case of rivers with arbitrary distribution of channel curvature, the only constraint being that flow and bottom topography must be 'slowly varying' in both longitudinal and lateral directions and channel curvature must be 'sufficiently small': the former assumption requires the channel to be 'wide' with width and channel alignment varying on a longitudinal scale much larger than channel width, while the latter assumption is satisfied provided the radius of curvature of channel axis is large compared with channel width. Both conditions are typically met in actual rivers but, in spite of the popularity enjoyed by linear models, neither of them implies that perturbations of bottom topography are necessarily small. Taking advantage of the slowly varying assumption, we are able to develop a theory accounting for finite amplitude perturbations of flow and bed topography. The recent analysis of Solari and Seminara (2005) concerning the equilibrium width of meandering rivers is here extended and corrected by including convective effects through an appropriate extension of the perturbation expansion.

The paper is organized as follows. In chapter 2 we formulate the 3D problem of flow in sinuous channels with a cohesionless bed. In the analysis, the direct effect of secondary flow on the transverse distribution of the main flow is accounted for through transverse convection of longitudinal momentum. This effect, which has been argued to be important by many authors (e.g. Nelson and Smith (1989), Johannesson and Parker (1989), Imran and Parker (1999)) appears at the second order of approximation in the present scheme along with further convective terms appearing in the equation of motion. Chapter 3 is then devoted to the formulation of the bend theory by coupling the morphodynamic model with a bank erosion law, expressing the dependence of erosion intensity on the flow. We are then able to predict the wavelength selected by bend instability as well as meander wave speed in a non-linear context. The model is then applied to a small reach of the Cecina River, Italy (Chapter 4) to predict both the equilibrium configuration and the wavenumber selected in the meandering process. Finally, some discussion concludes the paper.

## 2 THE ANALYTICAL MODEL

### 2.1 *Formulation of the problem*

Let us then consider a sinuous channel with a cohesionless bed and refer it to intrinsic coordinates ($s^*, n^*$ and $z^*$ representing the longitudinal, lateral and vertical coordinates, respectively). In the case of channels with constant width, say $2B_u^*$, the appropriate scaling for the intrinsic coordinates, the local mean velocity averaged over turbulence $\mathbf{u}^* = (u^*, v^*, w^*)^T$, the flow depth $D^*$, the free surface elevation $h^*$, the eddy viscosity $\nu_T^*$ and the sediment flux per unit width $(q_s^*, q_n^*)^T$ reads:

$$(s^*, n^*) = B_u^*(s, n), \quad (z^*, D^*, h^*) = D_u^*(z, D, F_r^2 h),$$

$$(u^*, v^*, w^*) = U_u^*(u, v, \frac{w}{\beta_u}), \quad \nu_T^* = (\sqrt{C_{fu}} U_u^* D_u^*) \nu_T$$

$$(q_s^*, q_n^*) = \sqrt{(s_p - 1)gd^{*3}}(q_s, q_n), \qquad (1a\text{-}g)$$

where a star denotes dimensional quantities. Moreover, $s_p$ is the relative particle density ($= \rho_s/\rho$, with $\rho$ and $\rho_s$ water and particle density respectively), $d^*$ is the particle diameter (taken to be uniform), $C_{fu}$ is the friction coefficient, $\beta_u$ is the aspect ratio of the channel, $F_r$ is the the Froude number and the index $_u$ refers to properties of uniform flow in a straight channel with the same flow discharge and the average channel slope $S$. Two parameters arise, namely:

$$\beta_u = \frac{B_u^*}{D_u^*}, \qquad F_r^2 = \frac{U_u^{*2}}{gD_u^*}. \qquad (2a, b)$$

We then take advantage of the hydrostatic approximation, which applies when the spatial scale of the relevant hydrodynamic processes largely exceeds the flow depth. The steady turbulent flow of water in a channel characterized by a slowly varying spatial distribution of channel curvature $c^*(s)$, is then governed by the longitudinal and lateral components of the Reynolds equations, along with the continuity equations for the fluid and solid phases. In dimensionless form, they read:

$$\nabla \cdot \mathbf{u} = -h_s^{-1} \nu_0 c(s) v, \qquad (3)$$

$$(\mathbf{u} \cdot \nabla) u + h_s^{-1} h_{,s} - \beta_u \sqrt{C_{fu}} (\nu_T u_{,z})_{,z} =$$

$$= h_s^{-1} [\beta_u C_{fu} - \nu_0 c(s) uv], \qquad (4)$$

$$(\mathbf{u} \cdot \nabla) v + h_{,n} - \beta_u \sqrt{C_{fu}} (\nu_T v_{,z})_{,z} =$$

$$= h_s^{-1} \nu_0 c(s) u^2, \qquad (5)$$

$$h_s^{-1} q_{s,s} + q_{n,n} = -h_s^{-1} \nu_0 c(s) q_n, \qquad (7)$$

where $\nabla$ is defined by $(h_s^{-1} \partial/\partial s, \partial/\partial n, \partial/\partial z)$. Moreover $\nu_0$ is a curvature parameter, $c(s)$ is dimensionless curvature and $h_s$ is a metric coefficient, such that:

$$\nu_0 = \frac{B_u^*}{r_0^*}, \quad c(s) = r_0^* c^*(s), \quad h_s = [1 + \nu_0 c(s)]^{-1},$$

$$(8a, b, c)$$

where $r_0^*$ is some typical radius of curvature of the channel axis. The equations (3–7) must be supplemented with boundary conditions which may be written in the dimensionless form:

$$u = v = w = 0 \qquad z = z_0, \qquad (9a, b, c)$$

$$u_{,z} = v_{,z} = w - h_s^{-1} u F_r^2 h_{,s} - v F_r^2 h_{,n} = 0 \quad z = F_r^2 h,$$
$$(10a, b, c)$$

$$\int_{z_0}^{F_r^2 h} v \, dz = q_n = 0 \qquad n = \pm 1. \qquad (11)$$

The equations (9) impose no slip at the conventional reference level $z_0$; the equations (10) impose the conditions of no stress at the free surface and the requirement that the latter must be a material surface; finally, the condition (11) imposes the constraint that both the water and the sediment flux must vanish at the banks.

Closure relationships are then needed for the sediment flux per unit width $\mathbf{q}$ and for the eddy viscosity $\nu_T$.

We take advantage of the slowly varying character of flow field and bed topography to assume that the turbulent structure is in quasi equilibrium with the local conditions, only slightly perturbed by weak curvature effects. Hence we write:

$$\nu_T = \left( \frac{|\boldsymbol{\tau}^*|}{\rho C_{fu} U_u^{*2}} \right)^{1/2} D(n, s) N(\xi), \qquad (12)$$

where $\tau^*$ is the local value of the bottom stress, $D(n, s)$ is the local dimensionless value of the flow depth and $N(\xi)$ is the vertical distribution of the eddy viscosity in a plane uniform free surface flow. Note that $\xi$ is a normalized vertical coordinate which reads:

$$\xi = \frac{z - [F_r^2 h(n, s) - D(n, s)]}{D(n, s)}. \qquad (13)$$

Hence, $\xi$ attains values in the range $\xi_0 \leq \xi \leq 1$, with $\xi_0$ normalized reference level, a weakly dependent function of the longitudinal and lateral coordinates, here assumed to be constant. The distribution $N(\xi)$ is taken to coincide with the classical parabolic distribution characteristic of uniform flows corrected by Dean's wake function (1974):

$$N(\xi) = \frac{k\xi(1 - \xi)}{1 + 2A\xi^2 + 3B\xi^3}. \qquad (14)$$

The closure for the sediment flux per unit width $\mathbf{q}$ derives from a well established approach of semi empirical nature. In uniform open channel flow over a homogeneous cohesionless plane bed no significant sediment transport occurs below some *critical value* $\theta_c$

of a dimensionless form $\theta$ of the average shear stress $\tau^*$ acting on the bed, depending on the *particle Reynolds number* $R_p$. With $\nu_f$ kinematic viscosity of the fluid, the Shields stress (Shields 1936)) and $R_p$ read:

$$\theta = \frac{|\tau^*|}{(\varrho_s - \varrho)gd^*} \,, \qquad R_p = \frac{\sqrt{(s_p - 1)gd^{*3}}}{\nu_f} \,. \tag{15a, b}$$

For values of $\theta$ exceeding $\theta_c$ but lower than a second threshold value $\theta_s$, particles are transported as bedload with a distinct dynamics driven by, but different from, the dynamics of fluid particles. Under these conditions, on pure dimensional ground, the average bedload flux per unit width on a weakly sloping bottom may be given the general form:

$$\mathbf{q} = \Phi(\theta - \theta_c; R_p) \left( \frac{\tau^*}{|\tau^*|} + \mathbf{G} \cdot \nabla_h \eta \right) \,, \tag{16}$$

where $\eta (= F_r^2 h - D)$ is the dimensionless bed elevation, $\nabla_h$ is $(h_s^{-1}\partial/\partial s, \partial/\partial n)$, $\Phi$ is a monotonically increasing function of the excess Shields stress $(\theta - \theta_c)$ for given particle Reynolds number $R_p$, while $\mathbf{G}$ is a $(2 \times 2)$ matrix dependent on $\theta$, $\theta_c$ and the angle of repose of the sediment. The function $\Phi$ can be estimated through well known empirical of semi empirical relationships: in the following we use the relation proposed by Parker (1990). Moreover we only account for the prevailing lateral effect of gravity on the particle motion and write (Parker 1984):

$$G_{ss} = G_{sn} = G_{ns} = 0 \,, \qquad G_{nn} = -R \,, \tag{17a, b}$$

with $R$ a typically small parameter which reads:

$$R = \frac{r_c}{\beta_u \sqrt{\theta}} \,, \tag{18}$$

$r_c$ being an empirical constant ranging about 0.56 (Talmon et al. 1995).

At last, the problem formulated above is subject to two integral constraints stipulating that flow and sediment supply must be constant at any cross section, hence:

$$\int_{-1}^{+1} D \, dn \int_{\xi_0}^{+1} u(\xi; n, s) \, d\xi = constant \,, \tag{19}$$

$$\int_{-1}^{+1} \Phi \left[ \theta(n, s) \right] \, dn = constant \,. \tag{20}$$

### 2.2 Solution for channels with slowly varying distribution of curvature

Let us consider a sinuous channel characterized by a slowly varying distribution of curvature of the channel axis. Flow and bottom topography are then assumed to be 'slowly varying' in both longitudinal and lateral directions. The above assumptions do not imply that perturbations of flow and bottom topography are necessarily small. It is then appropriate to rescale the longitudinal coordinate $s$ as follows:

$$\sigma = \frac{s^*}{r_0^*} = \nu_0 s \,. \tag{21}$$

We may then expand the unknown functions in a neighborhood of the solution for uniform flow in a *straight channel with an unknown shape of the cross section*, described by a slowly varying function $D(n, \sigma)$ of both the longitudinal and lateral coordinates. We can then expand the solution in the form:

$$(u, v, w, h, D) = [u_0(\xi; n), 0, 0, h_0(\sigma)/\delta, D_0(n)] +$$

$$+ \sum_{1}^{\infty} \left( u_m, v_m, \frac{w_m}{\beta_u}, h_m, D_m \right) (\delta)^m \,, \tag{22}$$

where $\delta$ is the small parameter $(\nu_0/\beta_u \sqrt{C_{fu}})$. Note that the free surface elevation at leading order is taken to be a function of $\sigma$, in order to account for the small variations of the longitudinal free surface slope associated with channel curvature.

The latter expansion is then substituted into the governing differential problem (3-7), conveniently rewritten in terms of the transformed variables $\sigma, n$ and $\xi$ using the chain rules:

$$\frac{\partial}{\partial z} \to \frac{1}{D} \frac{\partial}{\partial \xi}, \quad \frac{\partial}{\partial n} \to \frac{\partial}{\partial n} + \left[ \frac{(1 - \xi)D_{,n} - F_r^2 h_{,n}}{D} \right] \frac{\partial}{\partial \xi}, \tag{23a, b}$$

$$\frac{\partial}{\partial s} \to \nu_0 \frac{\partial}{\partial \sigma} + \nu_0 \left[ \frac{(1 - \xi)D_{,\sigma} - F_r^2 h_{,\sigma}}{D} \right] \frac{\partial}{\partial \xi} \,. \tag{24}$$

We then equate likewise powers of $\delta$ to obtain a sequence of differential problems, to be solved in terms of the unknown functions $D$ and $h_{,\sigma}$.

$$\boxed{O(\delta^0)}$$

At the leading order of approximation, the longitudinal component of the Reynolds equations reduces to a uniform balance between gravity and friction in a channel with unknown distribution of flow depth $D_0(n, \sigma)$ and free surface slope $(-h_{0,\sigma}(\sigma))$. After setting:

$$u_0 = D_0^{1/2}(n, \sigma) R_0^{1/2}(\sigma) F_0(\xi; n, \sigma) \,, \tag{25}$$

$$\nu_{T0} = D_0^{3/2}(n, \sigma) R_0^{1/2}(\sigma) N(\xi) \,, \tag{26}$$

with $R_0 = 1 - h_{0,\sigma}/\sqrt{C_{fu}}$, one finds

$$[N(\xi)F_{0,\xi}]_{,\xi} = -\sqrt{C_{fu}},$$

$$F_0|_{\xi=\xi_0} = F_{0,\xi}|_{\xi=1} = 0. \qquad (27a,b,c)$$

The solution for $F_0$ is the classical logarithmic distribution corrected by a wake function

$$F_0(\xi) = \frac{\sqrt{C_{fu}}}{k}\left[\ln\frac{\xi}{\xi_0} + A(\xi^2 - \xi_0^2) + B(\xi^3 - \xi_0^3)\right], \qquad (28)$$

where $\xi_0$ is the normalized conventional reference level, here assumed to be constant.

$$\boxed{O(\delta^1)}$$

At first order, the lateral component of the Reynolds equations reduces to a balance between lateral component of gravity, centripetal inertia and lateral friction in a channel with unknown, yet slowly varying, distributions of flow depth $D_0(n,\sigma)$ and free surface slope $(-h_{0,\sigma})$ as well as given slowly varying distribution of channel curvature $c(\sigma)$. We find:

$$\frac{1}{D_0^2}[\nu_{T0}v_{1,\xi}]_{,\xi} = \frac{h_{1,n}}{\beta_u\sqrt{C_{fu}}} - u_0^2(\xi;n,\sigma)c(\sigma),$$

$$v_1|_{\xi=\xi_0} = v_{1,\xi}|_{\xi=1} = 0. \qquad (29a,b,c)$$

We then set

$$v_1 = D_0^{3/2}(n,\sigma)R_0(\sigma)G_1(\xi;n,\sigma)c(\sigma), \qquad (30)$$

$$\frac{\partial h_1}{\partial y} = \beta_u\sqrt{C_{fu}}a_1(n,\sigma)D_0(n,\sigma)R_0(\sigma)c(\sigma), \qquad (31)$$

where $G_1$ is the solution of the following ordinary differential problem:

$$[N(\xi)G_{1,\xi}]_{,\xi} = a_1(n,\sigma) - F_0^2(\xi),$$

$$G_1|_{\xi=\xi_0} = G_{1,\xi}|_{\xi=1} = 0. \qquad (32a,b,c)$$

Let us write the solution for $G_1$ in the form:

$$G_1 = a_1(n,\sigma)G_{11}(\xi) + G_{12}(\xi), \qquad (33)$$

where

$$G_{1j} = g_1(\xi) - \frac{g'_j|_{\xi=1}}{g'_0|_{\xi=1}}g_0(\xi), \qquad (j=1,2), \qquad (34)$$

and $g_j(j=0,1,2)$ solutions of ordinary differential problems identical with those solved by Seminara and Solari (1998) (equations 26 and 27 of that paper).

We may then proceed to determine the function $a_1(n,\sigma)$ firstly integrating the continuity equation (3) over the flow depth, at $O(\delta)$, with the use of the dynamic and kinematic boundary conditions, (9c) and (10c) respectively, and secondly integrating over the cross section, with the use of the boundary condition (11) at the sidewall.

Let us finally come to the sediment continuity equation (7). At $O(\delta)$, in the rescaled coordinates, it reads:

$$\beta_u\sqrt{C_{fu}}q_{s0,\sigma} + q_{n1,n} = 0. \qquad (35)$$

With the help of the closure relationship for $\mathbf{q}$ (equations $16-18$) rewritten in terms of the rescaled coordinates and expanded in powers of $\delta$, the equation (35) can be reduced to a non linear partial differential equation for the unknown functions $D_0(n;\sigma)$ and $h_{0,\sigma}(\sigma)$. We find:

$$D_{0,n} = -\frac{\sqrt{D_0R_0}}{R/\delta}\left[\frac{v_{1,\xi}|_{\xi_0}}{u_{0,\xi}|_{\xi_0}} + \frac{\beta_u\sqrt{C_{fu}}}{q_{s0}}\frac{\partial}{\partial\sigma}\int_{-1}^{n}q_{s0}\,dn\right] \qquad (36)$$

The equation (36) is to be solved with boundary conditions forcing the normal component of sediment flux to vanish at the side walls and integral constraints (19) whereby the flow and sediment discharges must keep constant in the longitudinal direction. The above differential problem can be solved numerically, marching in $n$ for every single cross section. This allows us to determine the unknown functions $D_0(n,\sigma)$ and $h_{0,\sigma}(\sigma)$ by a trial and error procedure.

Proceeding at the second order of approximation, convective terms are also considered in the equations of motion (4 and 6). Following a procedure similar to the one reported for the previous order, not reported here for the sake of brevity, we end up with the following differential equation:

$$D_{1,n} = -\frac{\sqrt{D_0R_0}}{R/\delta}\left\{\frac{v_{2,\xi}|_{\xi_0}}{u_{0,\xi}|_{\xi_0}} + \frac{\beta_u\sqrt{C_{fu}}}{q_{s0}}\frac{\partial}{\partial\sigma}\int_{-1}^{n}q_{s1}\,dn+\right.$$

$$\left.+\frac{q_{n1}}{q_{s0}}\left[\beta_u\sqrt{C_{fu}}nc(\sigma) + \frac{\phi_1}{q_{s0}} - \frac{u_{1,\xi}|_{,\xi_0}}{u_{0,\xi}|_{,\xi_0}}\right] + F_r^2 h_{1,n}\right\}, \qquad (37)$$

to be solved with similar boundary conditions as for equation (36). Again, the above differential problem is solved numerically, marching in $n$ for every cross section and it allows us to determine the unknown functions $D_1(n,\sigma)$ and $h_{1,\sigma}(n,\sigma)$ by a trial and error procedure.

## 3 NON LINEAR BEND THEORY FOR RIVER MEANDERS

The model presented here is also suitable for the formulation of a bend instability theory by associating a bank erosion equation with the governing equation for flow and bed topography. Many attempts to describe in some detail the mechanics of bank erosion have been proposed in the literature. Here, following the approach proposed by Ikeda et al. (1981), we assume the local lateral migration speed $\zeta(s)$ being a function of the longitudinal excess velocity near the bank. Hence the erosion law is:

$$\zeta(s) = E\left(U|_{n=1} - U|_{n=-1}\right) \tag{38}$$

where E is a dimensionless long-term erosion coefficient and the lateral migration speed $\zeta(s)$ is scaled by $U_u^*$. We then define the mean lateral migration speed $(\bar{\zeta}_x, \bar{\zeta}_y)$ integrating the local values of equation 38 along the intrinsic coordinate $s$, between two consecutive flex points. The components of the vector $(\bar{\zeta}_x, \bar{\zeta}_y)$ will then represent measures of meander wave speed and meander growth rate, respectively. A positive (negative) value of $\bar{\zeta}_x$ will correspond to downstream (upstream) meander migration. Similarly, meander amplification (attenuation) will correspond to positive (negative) values of $\bar{\zeta}_y$.

Moreover, we assume that the channel axis follows a sinusoidal curve in the $(x, y)$ plane, characterized by a cartesian wavenumber $k$ and amplitude $\epsilon$, having normalized both quantities by the half-width of the channel $B_u^*$.

In figure 1a we show the values of the meander amplification speed $\bar{\zeta}_y$ divided by the erosion coefficient $E$ as a function of meander wavenumber $\lambda$. Note that, for a given value of $\beta_u$, the curve is characterized by a peak which corresponds to the value of the critical wavenumber selected in the meandering process. All the curves are interrupted when the value of the shields stress falls below a certain threshold anywhere throughout the meander. The location of the peak, i.e. the wavenumber selected, increases monotonically as $\beta_u$ increases.

The wavenumbers corresponding to the maximum bend amplification are plotted in figure 1b versus $\beta_u$ for different values of the amplitude $\epsilon$ are reported . Note that the selected wavenumber depends on the initial perturbation and in particular longer wavelengths (smaller wavenumbers) are associated with greater amplitudes $\epsilon$. The wavenumbers selected according to the linear theory of Blondeaux and Seminara (1985) are also reported in the figure and show a peak at the resonant values $(\beta_R, \lambda_R)$. Note also that the perturbative parameter $\delta$ in this case attains values that are always smaller than 0.18.

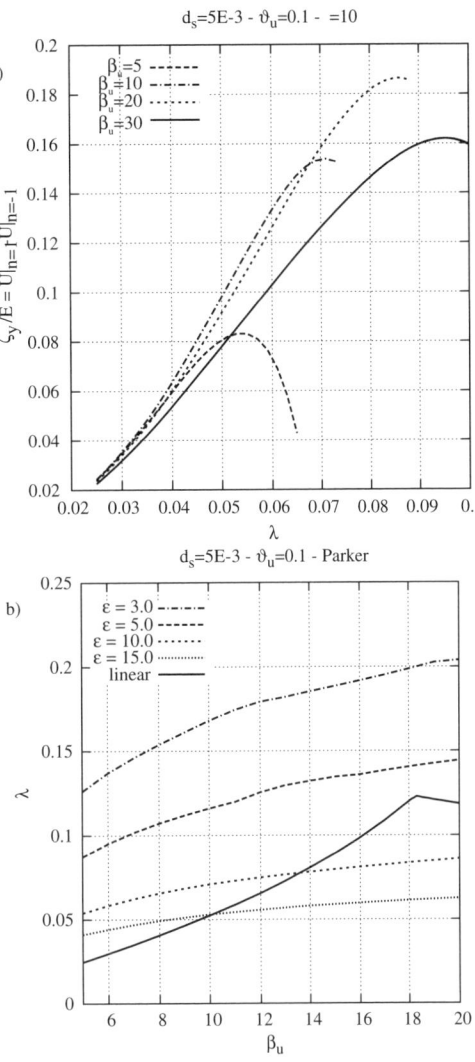

Figure 1. (a) Meander amplification $\bar{\zeta}_y$ divided by the erosion coefficient $E$ versus the wavenumber is reported for different values of $\beta_u$ in the case of a meander following a sinusoidal planimetric pattern with amplitude $\epsilon = 10$. (b) The selected wavenumbers for bend instability are reported for different values of the amplitude $\epsilon$ and compared with the linear theory. ($d_s = 5 \cdot 10^{-3}$, $\vartheta_u = 0.1$)

Figure 2 reports the values of meander wave speed $\bar{\zeta}_x$ divided by the erosion coefficient $E$ corresponding to the selected wavenumbers for bend instability versus the aspect ratio $\beta_u$, for different values of the amplitude $\epsilon$. For small values of the aspect ratio $\beta_u$ the meander wave speed is positive and it grows as the meander amplitude decreases. Every curve then shows a threshold value of the aspect ratio $\beta_u$ above which the meander exhibits an upstream migration. This finding

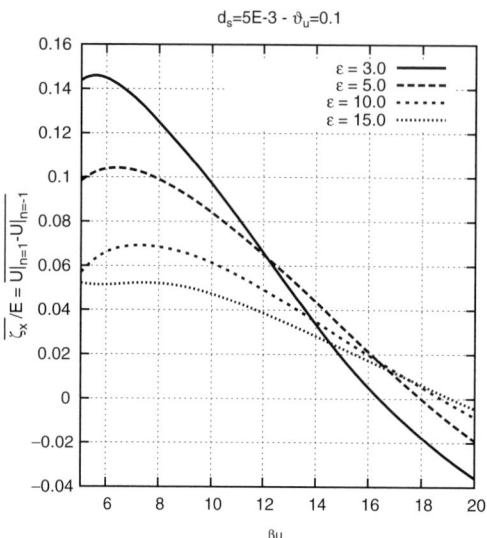

Figure 2. The values of meander wave speed $\bar{\zeta}_x$ divided by the erosion coefficient $E$ corresponding to the selected wavenumbers for bend instability versus the aspect ratio $\beta_u$ are reported for different values of the amplitude $\epsilon$. ($d_s = 5 \cdot 10^{-3}$, $\vartheta_u = 0.1$).

confirms the possibility of uptream migration already pointed out by Zolezzi and Seminara (2001).

## 4 APPLICATION TO THE CECINA RIVER

The model is then applied to a short reach of the Cecina River (Tuscany, Italy), a gravel bed river with actively migrating outer banks (Romanelli et al. 2004).

The Cecina River Basin is located in central Italy and comprises a basin surface area of approximately 900 km² and a total length of about 79 km. The study site is located a few kilometers upstream the confluence between the tributary Sterza and the Cecina. The main criteria for selecting this site were the availability of aerial photographs taken at different years (1954, 1978, 1986, 2004) showing the formation of a meander from a nearly straight reach (Fig. 3). River flow discharge data were also available from a gauging station, situated just downstream the study site, at Ponte di Monterufoli. Grain size analyses were also performed in the study site and made available to the authors (M. Rinaldi, personal communication). The study reach is about 1000 m long and is characterized by an average gradient of about 0.002.

The first set of simulations was performed extrapolating from the aerial picture of 1978 the planimetric development of the channel axis, which turned out to follow closely a sine generated curve (Langbein and Leopold 1966), characterized by a minimum

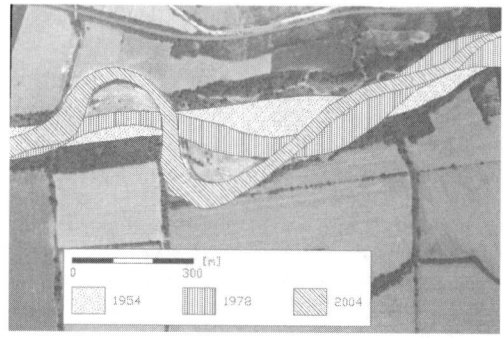

Figure 3. The reach of the Cecina river showing the formation of a meander from a nearly straight configuration. Flow is from right to left.

radius of curvature $r_0^* \simeq 325$ m, an intrinsic wavelength $L_s^* \simeq 970$ m and a channel width $2B_u^* \simeq 40$ m, taken as constant along the reach.

The value of bankfull water discharge $Q_u^*$ ($\simeq 110$ m³/s) was used in the simulations, corresponding to uniform flow depth of $D_u^* \simeq 1.3$ m. Sediments were considered uniform and characterized by $d^* = 7.4$ mm.

The values of the dimensionless relevant parameters were then calculated to give:

$$\beta_u \simeq 15 \quad \vartheta_u \simeq 0.210 \quad d_s = \frac{d^*}{D_u^*} \simeq 0.005$$

$$\nu_0 \simeq 0.062 \quad \lambda \simeq 0.129 . \qquad (39)$$

With the latter values, the model was then ran to predict the equilibrium bed pattern and the associated flow field. In figure 4a we report the equilibrium configuration of the bed topography obtained through our simulations which show that the value of the maximum scour depth with respect to the mean bed level is slightly greater than the uniform flow depth and is roughly located at the bend apex. On the contrary, the position of the forced bars is upstream the bend apex and shows a value of bed elevation which does not lead to bar emergence. The corresponding values of the vertically averaged longitudinal velocity predicted in this configuration are reported in figure 4b. Note that the high velocity core shifts from one side to the other side of the channel with distance through the meander, displaying a peak just downstream the bend apex. The value of the maximum velocity is slightly smaller than 4 m/s.

A second simulation was performed in order to investigate whether the model is able to predict the wavelength selected in the meandering process. We then applied the model to the same reach of the Cecina river (Fig. 3). From the first aerial picture available (year 1954) it can be easily observed

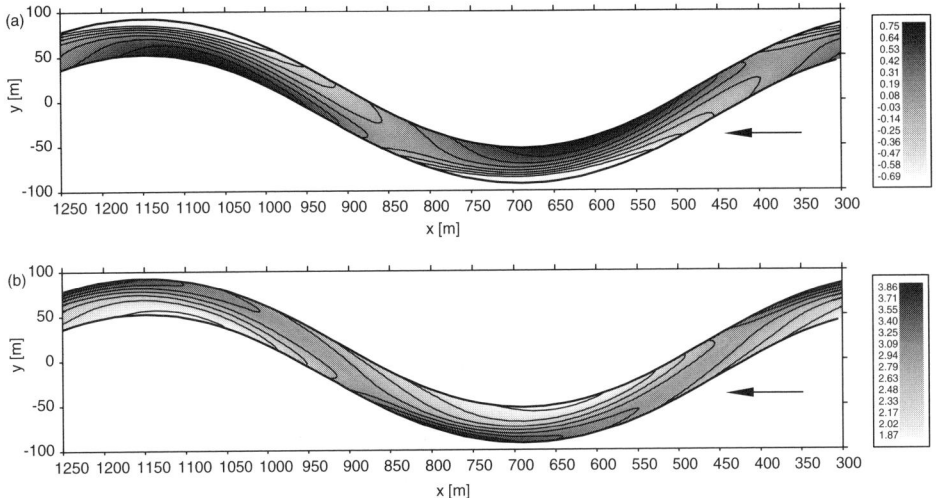

Figure 4. The equilibrium configuration of the bed topography simulated in the Cecina River. The bed elevations (in meters) represent the deviation from the mean longitudinal slope. (b) The pattern of longitudinal velocity corresponding to the equilibrium configuration simulated in the Cecina River. Velocities are expressed in m/s. Arrow represent flow direction. The values of the relevant dimensionless parameter are $\nu_0 \simeq 0.062$; $\lambda \simeq 0.129$; $\beta_u \simeq 15$, $\vartheta_u \simeq 0.210$, $d_s \simeq 0.005$.

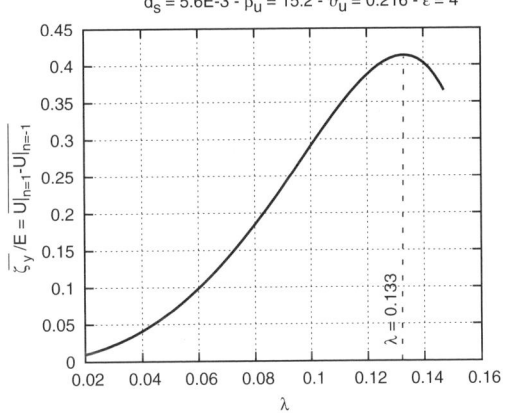

Figure 5. The meander amplification $\bar{\zeta}_y$ divided by the erosion coefficient $E$ is plotted versus the wavenumber for the values of the dimensionless parameters corresponding to the reach of the Cecina river. The selected wavenumber for bend instability turns out to be $\lambda \simeq 0.133$. ($d_s \simeq 5 \cdot 10^{-3}$, $\beta_u \simeq 15.2$, $\vartheta_u \simeq 0.210$, $\epsilon = 4$).

that two parallel straight reaches, probably rectified by human intervention, are joined together by an oblique stretch. The distance between the two parallel straight reaches is roughly equal to two channel widths. The aerial picture of 1978 already shows the evidence of a meandering process taking place downstream the connection between the oblique and straight reaches, characterized by and intrinsic wavelength $L_s^* \simeq 970$ m (corresponding to an intrinsic dimensionless wavenumber $\lambda \simeq 0.129$). The most recent pictures then clearly reveal the processes of meander amplification and downstream migration occurred in the following years.

Figure 5 reports the meander amplification $\bar{\zeta}_y$ divided by the erosion coefficient $E$ versus the wavenumber for the values of the dimensionless parameters corresponding to the reach of the Cecina river. The selected wavenumber for bend instability turns out to be $\lambda \simeq 0.133$ and corresponds to an intrinsic wavelength $L_s^* \simeq 945$ m, very close to that observed ($L_s^* \simeq 970$ m) from the aerial pictures. Assuming a typical value of the erosion coefficient $E$ ($=10^{-7}$), the values of the lateral migration speed and wave speed corresponding to the wavenumber selected are 3.5 m/year and 0.6 m/year, respectively.

## 5 CONCLUSIONS

A non linear analytic approach able to account for arbitrary, yet slow, variations of channel curvature has been developed. The theory is able to describe finite amplitude deformations with a computational effort much smaller than that needed in the context of numerical solutions of the full 3-D or of the shallow water equations (e.g. Mosselman (1991), Shimizu (2002) among others). A non-linear formulation of bend theory was also presented and is able to predict the wavelength selected in the meandering process.

The model was applied to a reach of the Cecina river to simulate the equilibrium configuration of bed and flow field and compare the prediction of the wavelength selected in the meandering process with the one observed. Results are in agreement with observations.

The extension of the model to predict the planimetric development of meandering rivers driven by both geometric and flow non linearities will follow in the near future.

## ACKNOWLEDGEMENTS

M. Rinaldi is kindly acknowledged for providing the aerial picture and the data of the Cecina river. The present work has been funded by Cariverona (Progetto MODITE). Partial support has also come in the framework of the National Project cofunded by the Italian Ministry of University and of Scientific and Technological Research 'Evoluzione morfodinamica di ambienti lagunari' (PRIN 2006).

## REFERENCES

Blondeaux, P. and G. Seminara (1985). A unified bar-bend theory of river meanders. *J. Fluid Mech. 157*, 449–470.

Ikeda, S., G. Parker, and K. Sawai (1981). Bend theory of river meanders. part 1. linear development. *J. Fluid Mech. 112*, 363–377.

Imran, J. and G. Parker (1999). A nonlinear model of flow in meandering submarine and subaerial channels. *J. Fluid Mech. 400*, 295–331.

Johannesson, H. and G. Parker (1989). Linear theory of river meandering. In S. Ikeda and G. Parker (Eds.), *River Meandering*, Washington, D.C., pp. 181–213. AGU, Water Resources Monograph 12.

Langbein, W. and L. Leopold (1966). River meanders: theory of minimum variance. *U.S. Geol. Survey 422-H*. Prof. Paper.

Mosselman, E. (1991). Modelling of river morphology with non-orthogonal horizontal curvilinear coordinates. *Communications on Hydraulic and Geotechnical Engineering 91*. Delft University of Technology, Delft, NL.

Nelson, J. and J. Smith (1989). Evolution and stability of erodible channel beds. In S. Ikeda and G. Parker (Eds.), *River Meandering*, Washington, D.C., pp. 321–377. AGU, Water Resources Monograph 12.

Parker, G. (1984). Lateral bed load transport on side slopes. *J. Hydraul. Engng. ASCE 110(HY2)*, 197–199.

Parker, G. (1990). Surface-based bedload transport relation for gravel rivers. *J. Hydraul. Res. 28*(4), 417–436.

Romanelli, L. R., M. Rinaldi, S. Darby, L. Luppi, and L. Nardi (2004). Monitoring and modelling river bank processes: a new methodological approach. In M. Greco, A. Carravetta, and R. D. Morte (Eds.), *River Flow 2004*, Volume 2, pp. 993–998. A.A. Balkema.

Seminara, G. and L. Solari (1998). Finite amplitude bed deformations in totally and partially transporting wide chanel bends. *Water Resour. Res. 34*(6), 1585–1598.

Shields, I. A. (1936). Anwendung der ahnlichkeit-mechanik und der turbulenzforschung auf die gescheibebewegung. *Mitt. Preuss Ver.-Anst., Berlin, Germany 26*.

Shimizu, Y. (2002). A method for simultaneous computation of bed and bank deformation of a river. In *River Flow 2002*, Volume 2, pp. 793–801.

Solari, L. and G. Seminara (2005). On width variations in meandering rivers. In G. Parker and M. Garcia (Eds.), *River, Coastal and Estuarine Morphodynamics, RCEM 2005*, Volume 2, Urbana, Illinois, USA, pp. 745–751.

Talmon, A. M., N. Struiksma, and M. V. Mierlo (1995). Laboratory measurements of the sediment transport on transverse alluvial-bed slopes. *J. Hydraul. Res. 33*, 495–517.

Zolezzi, G. and G. Seminara (2001). Downstream and upstream inuence in river meandering. Part 1. General theory and application to overdeepening. *J. Fluid Mech. 438*, 183–211.

*River, Coastal and Estuarine Morphodynamics: RCEM 2007 – Dohmen-Janssen & Hulscher (eds)*
*© 2008 Taylor & Francis Group, London, ISBN 978-0-415-45363-9*

# Bend theory of river meanders with channel width variations

Rossella Luchi, Guido Zolezzi & Marco Tubino

*Department of Civil and Environmental Engineering, University of Trento, Trento, Italy*

ABSTRACT: The assumption of a uniform channel width has been used in most existing theoretical and computer models for meandering rivers. However width variations, as well as channel curvature, are a major planimetric forcing for both single thread meandering streams and individual branches of braided rivers, especially at their state prior to bifurcation. How does their interaction affect meander dynamics? In the present work we have made a first attempt to understand how regular width oscillations may condition the well known process of bend instability. For these purpose we have relaxed the assumption of constant channel width in a 2D model able to describe the morphodynamics of channels characterized by both width variations and curvilinear axis. Centrifugal effects are related to the curvature of streamlines instead to that of the channel axis, in order to properly cope with both planimetric forcings. A modified bend theory has been carried out accounting for the presence of regular width variations. The effect of oscillating width can be reproduced at the first nonlinear interaction of the two-parameters perturbation expansion due to the peculiar wavelength ratio between the two planimetric forcing typically observed in natural streams. Results allow to detect the conditions that may lead to modified stability properties with respect to classical linear meander models, and to understand the effect of multiple resonance conditions, associated with different transverse modes, on the complete solution.

## 1 INTRODUCTION

Width variations in sinuous channels can be observed in both single thread, meandering streams and in individual branches of multiple thread rivers. Ashmore (1982) and Ashmore (1991) among others, has shown experimentally how the planimetric evolution of a weakly meandering channel with width variations of progressively increasing amplitude is the dynamics that triggers bifurcations of individual channels with cohesionless bed and banks. Bertoldi and Tubino (2005) have further demonstrated that incipient chute-cutoff bifurcations, responsible for initiating a braided pattern, occur when the amplitude of width variations peaks in time; Repetto and Tubino (1999) and Repetto et al. (2002) have modelled the morphodynamic response of a straight channel subject to regular, periodic oscillations of channel width, demonstrating their tendency to promote symmetrical, central bar altimetric patterns, whose amplitude may be large enough to suppress the migration of alternate bars.

Width variations in curved channels seem also relevant for simulation models of meandering processes, which commonly assume channel width to be constant (e.g. Seminara 2006). However, field evidence from different geographical and geomorphic settings suggests that this assumption may be incorrect, and that many meandering rivers exist with fairly

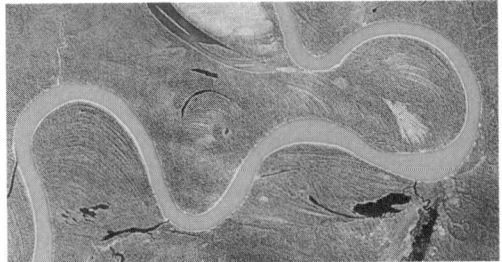

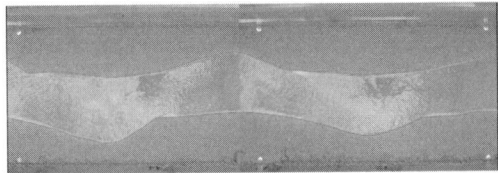

Figure 1. Examples of meandering channels with width variations. (a) River in the Amazon floodplain, Brasil; (b) laboratory experiments of Bertoldi and Tubino (2005) (courtesy of W. Bertoldi).

regular spatial oscillations of channel width (Figure 1). In spite of the absence of quantitative field studies on channel width fluctuations in meandering channels, however it is straightforward, by simply looking at

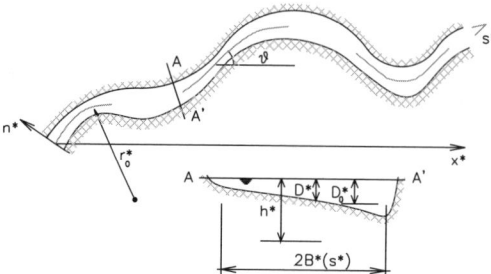

Figure 2. Meandering channel with variable width: notations.

aerial images, to detect at least one ubiquitous feature: the spatial frequency of width oscillations is roughly twice that of channel curvature variations. Frequently indeed, although not invariably, channel width peaks close to bend apexes, regardless of bend orientation: in other words, while the curvature of channel axis intrinsically represents an asymmetrical, external forcing, width variations are by nature a symmetrical disturbance to channel morphodynamics. Several studies have attempted to model the presence of width variations on the morphodynamic response of meandering rivers, through both semi-analytical (Chen and Duan 2006) and numerical models (Rüther and Olsen 2007, Duan and Julien 2005). Other work (Solari and Seminara 2005) has explicitly focussed on determining the theoretical amplitude of width variations and their phase lag with respect to channel curvature fluctuations, based on a 3D nonlinear approach.

In the present work we make a first attempt to gain a deeper insight on how meander evolution is affected by the presence of width variations explicitly accounting for the mutual nonlinear interactions between the two planimetric forcings and exploring the dependence of channel response on the governing parameters. Namely we focus on how the well known bend instability mechanism (e.g. Ikeda et al. 1981) is affected by periodic oscillations of channel width whose wavelength is twice that of channel curvature variations. To this aim, after reviewing the main properties of the two basic morphodynamic problems (channel width and curvature variations separately), we solve a suitable depth averaged morphodynamic model able to describe channel response to both width and curvature variations (Zolezzi et al. 2005) through a two parameters perturbation approach.

## 2 MATHEMATICAL FORMULATION

We intend to determine the flow structure and bed topography of sinuous erodible channels with variable width $2B^*$ (Figure 2). The width of the channel is assumed large enough for the shallow water

approximation to hold. We refer to an orthogonal curvilinear reference system $(s^*, n^*, z^*)$ where $s^*$ is the longitudinal coordinate of the channel axis, $n^*$ is the transverse coordinate lying in the plane to which $z^*$ is orthogonal. An asterisk denotes a dimensional quantity. Let us define dimensionless variables in the following form:

$$(s^*, n^*) = B_0^*(s, n'); \quad (U^*, V^*) = U_0^*(U, V);$$

$$(h^*, D^*) = D_0^*(F_0^2 h, D);$$

$$(q_s^*, q_n^*) = (q_s, q_n)\sqrt{g\frac{\rho_s - \rho}{\rho}d_s^{*3}}; \qquad (1)$$

here $B_0^*$ is the average channel width, $\mathbf{U}^* = (U^*, V^*)$ is the velocity vector, $D^*$ is local depth, $h^*$ is water surface elevation, $F_0$ is the Froude number, $\mathbf{q}^* = (q_s^*, q_n^*)$ is the sediment rate per unit width, $\rho$ and $\rho_s$ are water and sediment density respectively, $d_s^*$ the mean sediment diameter and $g$ the gravity acceleration. The subscript "0" refers to the reference conditions, taken to coincide with a uniform flow in a straight channel with the same average longitudinal slope, water discharge and width. The transverse variable $n'$ is scaled with the dimensionless local half channel width $B(s) = B^*(s)/B_0^*$ in order to stretch the physical domain into a rectangle. The depth-averaged, steady momentum and continuity equations for flow and sediments can be derived from a 3D shallow water scheme following a classical approach originally proposed by Kalkwijk and Vriend (1980) and further extended by Seminara and Tubino (1989); it takes the following form:

$$NL_1\left(DU^2\right) + L_2\left(DUV + D^2U\epsilon\phi\right) +$$

$$+\nu NC_a\left(DUV + D^2U\epsilon\phi\right) + NDL_1(h) + \beta\tau_s = 0$$

$$NL_1\left(DUV + D^2U\epsilon\phi\right) + L_2(DV^2 + 2D^2V\epsilon\phi +$$

$$+D^3\epsilon^2\psi) - \nu NC_aDU^2 + \frac{D}{b}\frac{\partial h}{\partial n} + \beta\tau_n = 0$$

$$NL_1(UD) + L_2(VD) = 0$$

$$NL_1(q_s) + L_2(q_n) = 0. \qquad (2)$$

In the above expression $L_1$ and $L_2$ are differential operators defined as:

$$(L_1; L_2) = \left(\frac{\partial}{\partial s} - nb_s\frac{\partial}{\partial n}; \quad \frac{1}{b}\frac{\partial}{\partial n} + \nu NC_a\right). \qquad (3)$$

The dimensionless geometry of the system sketched in Figure 2 is expressed by the following relationships:

$$\frac{d\vartheta}{ds} = -\nu C_a(s); \quad B(s) = 1 + \delta \mathcal{B}(s); \qquad (4)$$

where the functions $C_a$ and $\mathcal{B}$ measure the longitudinal variability of the planimetric disturbances while $\nu$ and $\delta$ quantify their intensity, being $\vartheta$ the meander inflection angle. The following metrics and dimensionless parameters arise from the analysis:

$$N = (1 + \nu n C_a)^{-1}; \quad b_s = \mathcal{B}^{-1}\frac{d\mathcal{B}}{ds};$$

$$d_s = \frac{d_s^*}{D_0^*}; \quad \beta = \frac{B_0^*}{D_0^*}; \quad \tau_* = \frac{\mathbf{t}^*}{(\rho_s - \rho)gd_s^*}; \quad (5a-e)$$

where $d_s$ is the relative roughness, $\beta$ the width ratio, $\tau_*$ the Shields parameter and $\mathbf{t}^* = (\tau_s^*, \tau_n^*)$ the bed shear stress. In (2) the formulation of the terms $\epsilon\phi$ and $\epsilon^2\psi$ derive from the parametrization scheme for secondary flow (Zolezzi et al. 2005), which is based on stream-lines curvature rather than on the curvature of the channel axis to evaluate centrifugal effects responsible for secondary circulations with vanishing depth average. The deviation of the streamlines from the channel axis can be considerable and, besides large scale vorticity, it can be due to both altimetric and planimetric effects. Note that this formulation can also take into account secondary circulations that generate in a straight channel with variable width (Repetto et al. 2002).

The curvature of streamlines $C(s,n)$ can be expressed as follows (Zolezzi et al. 2005):

$$\epsilon C = \nu C_a - \frac{U\frac{\partial V}{\partial s} - V\frac{\partial U}{\partial s}}{U^2 + V^2}, \qquad (6)$$

having defined $\epsilon = B_0^*/R_{sl}^*$ a dimensionless curvature ratio built with a reference value of the radius of streamline curvature $R_{sl}^*$. The dimensionless lateral shear stress can be expressed as the sum of two main contributions, due to secondary flows ($\tau_H$) and to the depth-averaged lateral velocity $V$:

$$(\tau_s, \tau_n) = (C_{f0}U|\mathbf{U}|, C_{f0}V|\mathbf{U}| + \tau_H). \qquad (7)$$

In order to close the governing system (2) we introduce a closure relationship for the sediment discharge: we have employed the bedload formula of Parker (1990).

Finally boundary conditions impose the walls of the channel be impermeable both to flow and sediments: defining $\hat{\mathbf{n}}_b$ as the unit vector normal to the bank lines yields:

$$(\mathbf{U} \cdot \hat{\mathbf{n}}_b) = 0; (\mathbf{q} \cdot \hat{\mathbf{n}}_b) = 0 \, (n = \pm 1). \qquad (8)$$

# 3 STABILITY AND RESONANCE OF BASIC PLANIMETRIC DISTURBANCES

We review the main features of the two morphodynamic problems that are taken as basis for the present study: the width-oscillating straight channel (Repetto et al. 2002) and the meander with constant width (Zolezzi and Seminara 2001), which will be referred to as the *width* and the *curvature* morphodynamic problems. For the sake of simplicity we will focus on regular, periodic variations of both channel width and axis curvature, such that:

$$[\mathcal{C}_a(s), \mathcal{B}(s)] = \exp i \, [\lambda_m, \lambda_b] \, s + c.c. \qquad (9)$$

with $\lambda_m$ and $\lambda_b$ the dimensionless wavenumbers of curvature and width variations respectively and c.c. denoting the complex conjugate. This assumption has often been taken as an effective, although approximate, representation of natural conditions, at least when focussing on the description of physical processes. Moreover $\nu$ and $\delta$ are commonly small numbers, allowing for linearization of the system (2).

Channel curvature is typically an asymmetric forcing: its longitudinal variations promote a sequence of scour – deposition patterns which take the form of alternate "point" bars whereby erosion and deposition are alternately spaced across the lateral direction, with the same bank experiencing scour close to a given bend and then deposition at the following downstream bend. This altimetric pattern is often associated with a perturbation of the uniform velocity field whereby deeper pools correspond to faster flow, thus creating an excess longitudinal velocity at one bank with respect to the other. This asymmetry has been shown to be a major driver of channel planform deformation and of bend instability: the rate of channel shift $\zeta$ is typically linearly related to the bank excess depth averaged longitudinal velocity (Hasegawa 1989) through an empirical erosion coefficient $E$:

$$\zeta(s) = E \, [u(s, n = 1) - u(s, n = -1)], \qquad (10)$$

with $u(s, n)$ excess velocity relative to the centerline value. Such asymmetry is related to the impermeability of lateral walls to both water and sediments (8), that determine non symmetrical boundary conditions at the banks for the longitudinal velocity as well as for the water depth.

The definition of bend stability implies that unstable bends tend to amplify during time, while stable meanders tend to rectify; in periodic systems bend stability is therefore related to the position of the $u(s, n = 1)$ peak with respect to the bend apex or, more generally, to the relative longitudinal lag between $u(s, n = 1)$ and $C_a(s)$.

An opposite behavior applies when channel width regularly oscillates in a river channel characterized by

857

a straight axis: typically a central bar pattern is induced, with central deposits that tend more often, although not invariably, to appear in the widest sections with laterally symmetrical scour holes close to the narrowest reaches. Boundary conditions in this case impose a symmetrical transverse distribution of the excess longitudinal velocity relative to the centerline value. Moreover, according to Repetto et al. (2002), the longitudinal lag between the distributions of $u(s, n = \pm 1)$ and of $B(s)$ determines the stability of the river planform, i.e. its tendency to enhance or reduce the width oscillations.

The transverse structure of the excess longitudinal velocity $u(s, n)$ can be therefore taken as a crucial control on the properties of channel stability in both the curvature and the width morphodynamic problems. It is possible to express this dependence through a peculiar Fourier expansion (e.g. Zolezzi and Seminara, 2001 in the curvature case) having the same formal structure for both planimetric configurations:

$$u(s, n) = \sum_{m=0}^{\infty} u_m(s) \sin\left(\frac{\pi}{2}mn\right),$$
(11)

where only odd transverse modes ($m = 1, 3, \ldots$) appear in the curvature case while only even modes ($m = 2, 4, \ldots$) contribute to the width case.

How can these two basic linear solutions affect the mixed curvature-width problem? To this purpose it is instructive to jointly review their resonance and stability properties taking advantage of the considerations exposed above.

Both linear meanders and the oscillating-width-channel have the typical features of driven-damped linear oscillators (e.g. Blondeaux and Seminara 1985) where resonance occurs for specific values of the longitudinal frequencies of the planimetric forcings determined by flow conditions ($\tau_*, d_s$) and related to the behavior of the homogeneous component of the governing differential system. Resonance is responsible for a theoretically infinite amplitude of the deformation of the flow-bed topography field and, given ($\tau_*, d_s$), it occurs for different combinations ($\lambda_R^{(m)}, \beta_R^{(m)}$) such that (Seminara and Tubino, 1992):

$$\left[\lambda_R^{(m)}, \beta_R^{(m)}\right] = m(\lambda_R^{(1)}, \beta_R^{(1)}).$$
(12)

This behavior is summarized in Figure 3, where a measure of the bank excess longitudinal velocity corresponding to both planimetric configurations is plotted under the same flow conditions ($\tau_* = 0.08$, $d_s = 0.05$) as function of the wavenumber $\lambda$ of the planimetric disturbances. The two thick lines refer to the curvature problem for two different resonant values of the width ratio $\beta$ corresponding to the first two odd transverse modes ($m = 1$ and $m = 3$), while thin

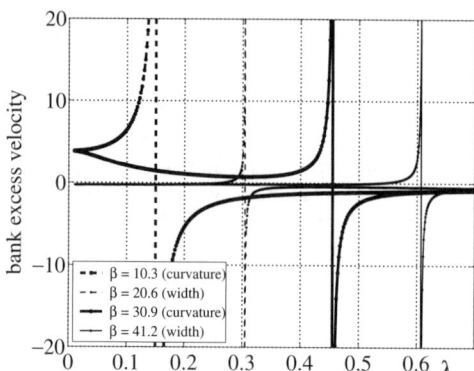

Figure 3. Occurrence of resonance in the linear responses of the curvature and width problems: real part of the excess longitudinal velocity.

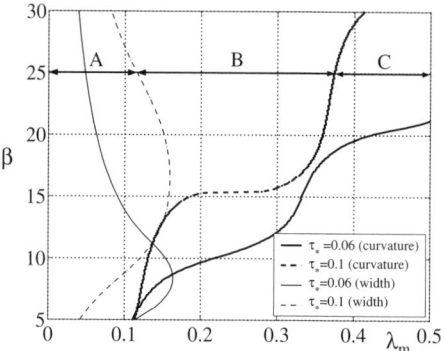

Figure 4. Neutral stability curves for bend and width variations stability ($d_s = 0.1$).

lines represent the same quantity in the width morphodynamic problem, which resonates for $\beta = 2\beta_R^{(1)}$ and for $\beta = 4\beta_R^{(1)}$. Note how the resonant wavenumber range is narrower in the width problem. The contemporary effect of curvature and width variations on channel response is therefore expected to be affected by the presence of these multiple resonance conditions, in particular when a peculiar ratio is imposed between the two wavelengths, such that $\lambda_b = 2\lambda_m$ as we assume in the present paper based on simple field observations.

Under this assumption we finally examine the linear stability properties of both curvature and width linear problems. Figure 4 shows neutral stability curves for two different values of the Shields parameter $\tau_*$: long enough planimetric wavelengths correspond to the instability of both the meandering and of the oscillating-width channel planform (region A, for

$\tau_* = 0.1$), with the bend instability range extending towards shorter wavelengths, especially for channels with high values of the width ratio. Region B in Figure 4 is characterized by only unstable bends, while any planimetrical disturbance whose wavelength falls in region C will tend to a straight, constant width channel configuration. The unstable regions of the curvature and of the width problem tend to match more closely when increasing the Shields parameter $\tau_*$ or reducing the relative roughness $d_s$.

The above results reinforce the need to investigate the mutual interaction between the curvature and the width morphodynamic effects, since unstable meander bends may often correspond to channel width oscillations (at double spatial frequency) that also tend to enhance their amplitude, and are therefore likely to affect the properties of linear bend stability and, in general, of meander evolution.

## 4 PERTURBATION SOLUTION AT DIFFERENT ORDERS OF APPROXIMATION

Recalling that $\lambda_b = 2\lambda_m$ we employ a two-parameters $(\nu, \delta)$ perturbation approach whereby each variable is expanded according to a common structure, that, for the longitudinal velocity $U$ takes the form:

$$
\begin{aligned}
U &= 1 + u(s,n) = \\
&= 1 + \delta\left[U_{01}\exp(2i\lambda_m s) + c.c.\right] + \\
&+ \nu\left[(U_{10} + \delta U_{11})\exp(i\lambda_m s) + c.c.\right] + \\
&+ \text{h.o.t.}
\end{aligned}
\tag{13}
$$

This approach allows us to predict the mutual interaction between the two planimetric forcings; namely the $\mathcal{O}(\nu\delta)$ term on the right hand side of (13) quantifies the correction of the meander response due to the presence of width variations of amplitude $\delta$. Indeed this order of approximation is the lowest at which the nonlinear interaction between curvature and width forced solutions reproduces the longitudinal structure of the fundamental perturbation ($\exp i\lambda_m s$). This is related to the assumption $\lambda_b = 2\lambda_m$ and represents the key specificity of the adopted perturbation scheme.

On substituting (13) into (2) we obtain separately $\mathcal{O}(\nu)$, $\mathcal{O}(\delta)$ and $\mathcal{O}(\nu\delta)$ solutions by solving three linear ordinary differential systems that can be written according to the same following structure:

$$
\mathbf{L}(\mathbf{V}) = \mathbf{b}(n)
\tag{14}
$$

where $\mathbf{V} = (U_{ij}, V_{ij}, h_{ij}, D_{ij})$ is the vector of the unknowns at the $\mathcal{O}(\nu^i \delta^j)$, while $\mathbf{L}$ and the vector $\mathbf{b}(n)$ read:

$$
\mathbf{L} = \begin{pmatrix}
a_{1ij} & a_{2ij}\frac{d}{dn} & a_{3ij} & a_{4ij} \\
0 & a_{5ij} & a_{6ij}\frac{d}{dn} & 0 \\
a_{7ij} & a_{8ij}\frac{d}{dn} & 0 & a_{9ij} \\
a_{10ij} & a_{11ij}\frac{d}{dn} & a_{12ij}\frac{d^2}{dn^2} & a_{13ij} + a_{14ij}\frac{d^2}{dn^2}
\end{pmatrix};
$$

$$
\mathbf{b} = \begin{pmatrix}
b_{1ij}(n) \\
b_{2ij}(n) \\
b_{3ij}(n) \\
b_{4ij}(n)
\end{pmatrix}.
$$

Accordingly, boundary conditions take the form:

$$
\left(V_{ij}, \frac{d(F_0^2 h_{ij} - D_{ij})}{dn}\right) = \pm(b_{5ij}, b_{6ij}) \quad (n = \pm 1)
$$

The systems (14) with the above boundary conditions are reduced to three fourth-order linear ordinary problems for the lateral velocity $V_{ij}$ at the $\mathcal{O}(\nu^i \delta^j)$:

$$
\frac{d^4 V_{ij}}{dn^4} + \Gamma_{1ij}\frac{d^2 V_{ij}}{dn^2} + \Gamma_{2ij}V_{ij} = \Gamma_{0ij}
$$

$$
\left(V_{ij}, \frac{d^2 V_{ij}}{dn^2}\right) = \pm(b_{5ij}, \Gamma_{3ij}) \quad (n = \pm 1)
\tag{15}
$$

The complete expression of the coefficients $a_{1ij} - b_{6ij}$ and $\Gamma_{0ij} - \Gamma_{3ij}$ has not been reported herein for the sake of brevity and is available from the authors upon request. We now summarize the main properties of the solution at the different orders of approximation.

$\mathcal{O}(\nu)$
Here we find the classical linear meander solution (Blondeaux and Seminara 1985) where the coefficients of $\mathbf{L}$ are modified by the parameterization of the secondary flow based on streamlines curvature. Due to the symmetric character of the non homogeneous terms $\Gamma_{010}$, $\Gamma_{310}$, the solution is symmetric for $V_{10}$ and antisymmetric for $(U_{10}, D_{10}, h_{10})$.

$\mathcal{O}(\delta)$
We obtain the solution of Repetto et al. (2002) for a straight channel with small amplitude sinusoidal variations of the width. In this case the governing fourth order differential equation is homogeneous ($\Gamma_{001} = 0$) and the effects of the forcing term associated with width variations is felt in the boundary conditions due to the curvilinear character of the banklines ($b_{501}, \Gamma_{301} \neq 0$). The antisymmetrical boundary conditions are satisfied through the antisymmetrical character of $V_{01}$ and symmetrical for $(U_{01}, D_{01}, h_{01})$. Note that the $\mathcal{O}(\delta)$ solution $U_{01}$ does not produce any contribution to the bend stability

859

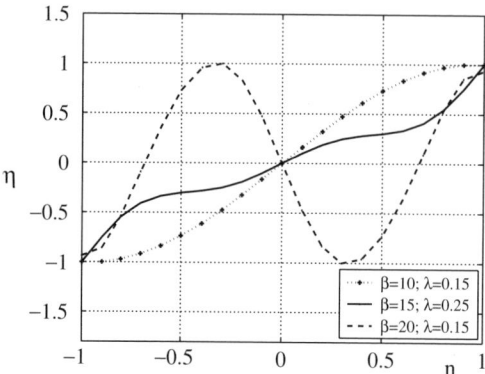

Figure 5. Structure of the eigenfunctions for bed elevation $\eta$ at the $\mathcal{O}(\nu\delta)$.

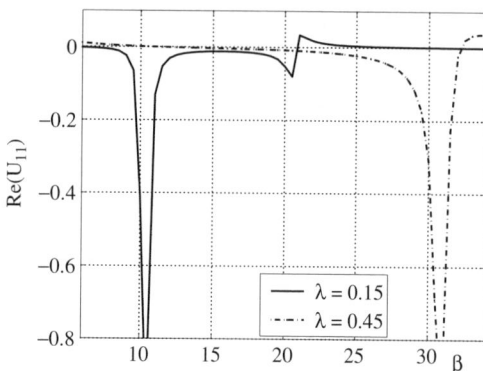

Figure 6. The real part of $U_{11}$, normalized with peak values, as function of the width ratio $\beta$ for the resonant value $\lambda_R$ of the $\mathcal{O}(\nu, \nu\delta)$ solutions.

mechanism, since its symmetrical transverse structure determines a vanishing effect according to (10).

$\mathcal{O}(\nu\delta)$

Here the solution for $V_{11}$ is symmetrical, while it is antisymmetrical for $U_{11}, D_{11}, h_{11}$. Indeed the homogeneous part of the governing equation coincides with that at $O(\nu)$ while the forcing term is a linear combination of the solutions at the previous orders $(\nu, \delta)$ so as to present a symmetric character. Figure 5 shows a typical form of the eigenfunctions corresponding to the $O(\nu\delta)$ contribution to bed deformation $\eta_{11} = F_0^2 h_{11} - D_{11}$. It can be clearly detected how increasing the channel width ratio $\beta$ gradually promotes the development of higher transverse modes (from $m=1$ to $m=3$).

## 5 RESULTS

We now examine the contribution of width variations to the process of bend stability related to the $\mathcal{O}(\nu\delta)$ solution. Recalling (9), (10) and (13), this is controlled by the real part of $U_{11}$, while its imaginary part is related to bend migration.

Figure 6 shows the dependence of the real part of $U_{11}$ (normalized with peak values) on the channel width ratio $\beta$ at different values of the longitudinal meander wavenumber $\lambda$. Note that $(\lambda, \beta) = (0.15, 10.3)$ correspond to resonance for the alternate $(m=1)$ bar mode at the given flow conditions $(\tau_* = 0.08, d_s = 0.05)$. Several interesting features arise. Since the homogeneous part of the governing system at the $\mathcal{O}(\nu\delta)$ coincides with that at the $\mathcal{O}(\nu)$, the real part of $U_{11}$ resonates exactly for $(\lambda_R^{(1)}, \beta_R^{(1)})$, and also for $(\lambda_R^{(3)}, \beta_R^{(3)}) = (0.45, 30.9)$ (third transverse mode). It can also be noted from Figure 6 that a third peak in the $\mathcal{O}(\nu\delta)$ solution corresponds to the resonance conditions for the second mode still for $\lambda = \lambda_R^{(1)} = 0.15$: indeed this condition corresponds to $\lambda_b = 0.3$, which is the resonant wavenumber of width

variations for $\beta = \beta_R^{(2)} = 20.6$. In this latter case the peak of the mixed response is not associated with resonance at the $\mathcal{O}(\nu\delta)$ but with a high amplitude forcing term that reflects resonance of the $\mathcal{O}(\delta)$ solution.

Moreover, for narrow channels ($\beta < 20$ at $\lambda = 0.15$) the mixed interaction always seems to stabilize linearly unstable bends, while width variations become destabilizing at higher values of $\beta$.

The effect of width variations on bend stability is more carefully examined in Figure 7. The bend amplification coefficient is plotted versus $\lambda$ for non-resonant $\beta$ values. The $\mathcal{O}(\nu\delta)$ correction stabilizes linearly unstable meander wavenumbers, with the stabilizing effect extending progressively towards larger values of $\lambda$ as $\beta$ grows (from $\beta = 8$ to $\beta = 12$ in Figure 7a). For given $(\tau_*, d_s)$ a threshold value of the aspect ratio ($\simeq 14$ in the present case) fixes a change in this dynamics: for higher aspect ratios ($\beta = 15, 25$) indeed width variations enhance the instability of longer bends. Figure 7b shows the behavior of the linear meander growth rate $Re(U_{10})$: the destabilizing $\mathcal{O}(\nu\delta)$ effect at $\lambda \simeq 0.12 \div 0.38$ ($\beta > 14$) further excites linearly unstable meanders, while progressively extending the instability range towards shorter bends (larger $\lambda$) when $\beta$ increases.

The effect of width variations on the migration of the meandering planform is shown in Figure 8. It appears (Figure 8a) that width variations almost invariably oppose to the direction of linear meander migration, which can be detected from Figure 8b.

Note that the above results are indicators of the qualitative role played by width variations on meander stability and migration: the actual effect in specific cases will ultimately depend on the absolute value of $\nu$ and $\delta$, which are assigned as parameters in the present analysis. In particular, the magnitude of $\delta$ is likely to increase within the range of planimetric instability of the width problem (Figure 4, region A).

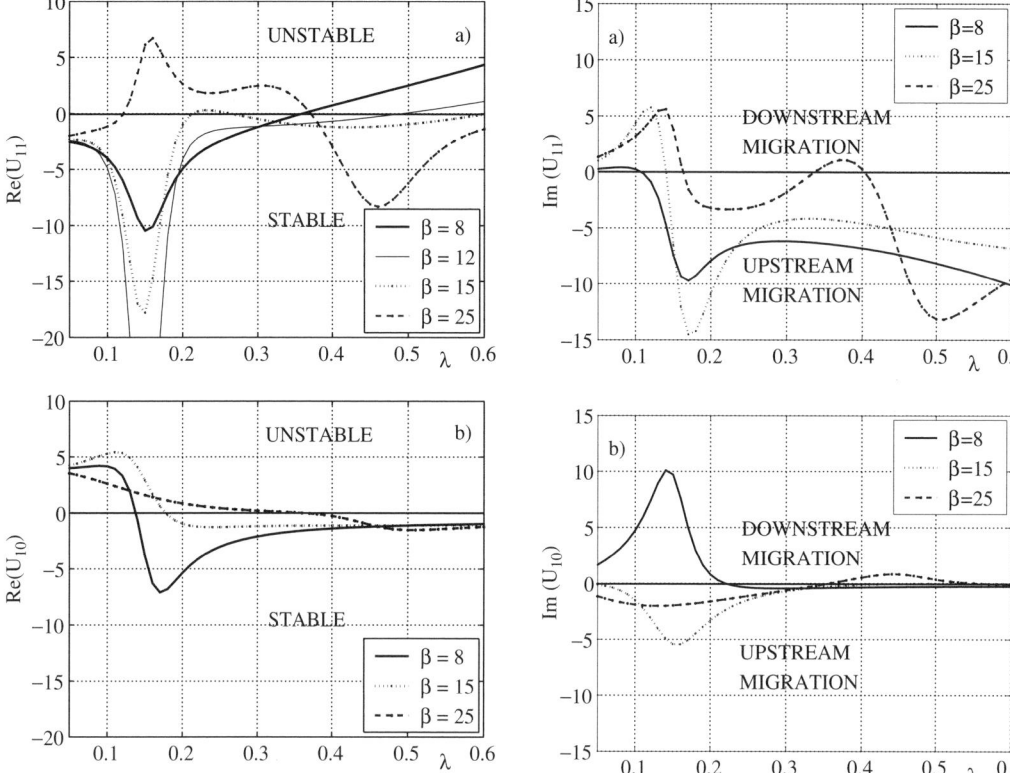

Figure 7. The real part of $U_{11}$ (a), and of $U_{10}$ (b) as functions of the width ratio $\beta$ for different values of $\beta$ (far from resonance).

Figure 8. The imaginary part of $U_{11}$ (a), and of $U_{10}$ (b) as functions of the width ratio $\lambda_R$ for different values of $\beta$ (far from resonance).

## 6 CONCLUSIONS

A bend stability analysis accounting for channel width variations has been performed by means of a two parameters $(\nu, \delta)$ perturbation expansion. This allows a first investigation of the role of longitudinal width variations on the classical, linear bend stability process originally solved by Ikeda et al. (1981).

The peculiar ratio between the wavenumbers of channel curvature and of width oscillations $(\lambda_b = 2\lambda_m)$ that can be typically observed in natural meanders, implies that the $\mathcal{O}(\delta)$ correction to the linear meander amplification rate can be reproduced by one simple interaction between the "curvature" and the "width" linear perturbations. Despite some degree of variability of the mixed $\mathcal{O}(\nu\delta)$ response with the governing dimensionless parameters, however we can state that width variations tend to enhance bend instability for large values of the channel width ratio, which may coincide with super-resonant meanders (Zolezzi and Seminara 2001). Moreover width oscillations tend to stabilize long, linearly unstable meander bends for both wide and narrow streams; at low width ratios their

destabilizing effect is confined to shorter bends that are unlikely to be observed in nature.

The actual degree to which linear bend stability is modified will eventually depend on the amplitude $\delta$ on width variations, which is playing as a parameter in the present analysis. Moreover the interaction between the two forcings is affected by multiple resonant regions where the linear solution is not valid any longer, although still qualitatively capturing major morphodynamic features (Seminara and Tubino, 1992, Zolezzi and Seminara, 2001). A nonlinear analysis might be called for, for instance on the line proposed by Nobile et al. (2007).

Two main research needs arise from the present study. First, quantitative field observations shall be collected to substantiate the present results. Second, a thorough understanding of the mixed curvature-width interactions requires to explore the complementary problem related to the effect of channel curvature variations on the planimetric stability of a width-oscillating river stream.

ACKNOWLEDGEMENTS

The present project has been carried out under the framework of the Grant "IMAIPO - CRS" funded by the University of Trento. The authors positively acknowledge the gradual birth of a national, informal scientific network devoted to facilitate knowledge sharing on meandering river processes, which will hopefully grow also in the future.

REFERENCES

Ashmore, P. (1982). Laboratory modelling of gravel braided stream morphology. *Earth Surface Processes and Landforms 7*, 201–225.

Ashmore, P. (1991). How do gravel-bed rivers braid? *Canadian Journal of Earth Sciences 28*, 326–341.

Bertoldi, W. and M. Tubino (2005). Bed and bank evolution of bifurcating channels. *Water Resources Research 41*, W07001, doi:10.1029/2004WR003333.

Blondeaux, P. and G. Seminara (1985). A unified bar-bend theory of river meanders. *Journal of Fluid Mechanics 112*, 363–377.

Chen, D. and J. Duan (2006). Modeling width adjustment in meandering channels. *Journal of Hydrology 321*, 59–76.

Duan, J. and P. Julien (2005). Numerical simulation of the inception of channel meandering. *Earth Surf. Process. Landforms 30*, 10931110. DOI: 10.1002/esp.1264.

Hasegawa, K. (1989). Studies on qualitative and quantitative prediction of meander channel shift. S. Ikeda & G. Parker (Eds.), River Meandering, Washington DC, Water Res. Monograph **12**, Amer. Geoph. Union.

Ikeda, S., G. Parker, and K. Sawai (1981). Bend theory of river meanders. Part 1 Linear development. *Journal of Fluid Mechanics 112*, 363– 377.

Kalkwijk, J. and H. D. Vriend (1980). Computation of the flow in shallow river bends. *Journal of Hydraulic Research 18 (4)*, 327–342.

Nobile, G., M. Bolla Pittaluga, and G. Seminara (2007). A nonlinear model for river meandering. In *Submitted to RCEM2007 Conference, Enschede, The Netherlands, September 17–21.* Balkema.

Parker, G. (1990). Surfacebased bedload transport relation for gravel rivers. *Journal of Hydraulic Research 28*, 417–436.

Repetto, R. and M. Tubino (1999). Transition from migrating alternate bars to steady central bars in channels with variable width. In *Proceedings of International Symposium on River, Coastal and Estuarine Morphodynamics, Genova, Italy, 6–10 September.*

Repetto, R., M. Tubino, and C. Paola (2002). Planimetric instability of channels with variable width. *Journal of Fluid Mechanics 457*, 79–109.

Rüther, N. and N. Olsen (2007). Modelling free-forming meander evolution in a laboratory channel using three-dimensional computational fluid dynamics. *Geomorphology*. doi: 10.1016/j.geomorph.2006.12.009.

Seminara, G. (2006). Meanders. *Paper invited for the 50th Anniversary issue of the J. Fluid. Mech. 554*, 271–297. doi: 10.1017/S0022112006008925.

Seminara, G. and M. Tubino (1989). *Alternate bars and meandering: Free, forced and mixed interactions*, pp. 267–320. in River Meandering, S. Ikeda and G. Parker (Eds.), Water Resources Monographies, 12.

Seminara, G. and M. Tubino (1992). Weakly nonlinear theory of regular meanders. *Journal of Fluid Mechanics 244*, 257–288.

Solari, L. and G. Seminara (2005). On width variations in river meanders. In *Proceedings of RCEM2005 Conference, Urbana, Illinois, USA, 4–7 October.* Balkema.

Zolezzi, G., R. Andreatta, and M. Tubino (2005). Streamline-based parametrization of centrifugally induced secondary flows in natural streams. In *Proceedings of RCEM2005 Conference, Urbana, Illinois, USA, 4–7 October*, pp. 783–791. London:Taylor & Francis Group.

Zolezzi, G. and G. Seminara (2001). Downstream and upstream influence in river meandering. Part 1. general theory and application of overdeepening. *Journal of Fluid Mechanics 438*, 183–211.

*River, Coastal and Estuarine Morphodynamics: RCEM 2007 – Dohmen-Janssen & Hulscher (eds)*
*© 2008 Taylor & Francis Group, London, ISBN 978-0-415-45363-9*

# Numerical analysis of meandering channel with a new boundary fitting coordinate system

H. Yasuda
*Civil Engineering Research Institute for Cold Region, Sapporo, Japan*

Y. Shimizu
*Hokkaido University, Sapporo, Japan*

ABSTRACT: Boundary-fitted calculation methods have traditionally used the generalized coordinate system which is required the condition that the coordinate axes intersect. When meandering formation extremely develop, the channel eventually is short-circuited. Because of this, general coordinate system should not be applied to evaluate that problem. This study proposes a new boundary-fitted calculation method that combines a co-orthogonal coordinate system and a boundary-fitted cell system. To validate the accuracy of this method, it was applied to reproduce the flow regime of a meandering channel measured in a previous flume experiment. The calculation method was generally successful at reproducing the flow velocity distribution.

## 1 INTRODUCTION

Rivers tend to form complicated morphology. One might even say that the mission of river engineering is to predict the changes in flow regime and channel morphology. However, accurate calculation of channel morphology based on the co-orthogonal coordinate system, the basic method of numerical calculation, requires very fine grids, which make the calculation load extremely high.

Shimizu(1991) developed a boundary fitted calculation method that uses a generalized coordinate system. In his pioneering work on calculating the flow regime of river channels that allows the flexible incorporation of channel morphology. This method allows the flow to be calculated for different channel morphologies; however, it requires that the computational grid be set such that the coordinate axes intersect. This method cannot deal with the channel shortening (cut-off) that is observed in meandering channels, one of the most important problems in river engineering.

Yasuda(2005) applied morphology-fitted cells, which flexibly and efficiently express morphology by combining triangles and polygons, to areas where high resolution is required while applying the co-orthogonal coordinate system to other areas, successfully combining the two methods.

It may be possible to develop Yasuda's combined calculation method into a new boundary-fitted calculation method in which morphology-fitted cells are applied to boundary areas such as complicated curve forms, which the co-orthogonal coordinate system has difficulty expressing without using fine grids, and the co-orthogonal coordinate system is applied to all other areas. Such a calculation method would allow the calculation area to be expressed without depending on coordinate axes, as the rectangular grid is used for most of the calculation area and the morphology-fitted cells are used only for the areas that require boundary fitting. An additional advantage is that it would make it very easy to generate grids. This method could be using for analyzing meander channel development and shortening, and calculating flow when there are river structures such as bridge piers.

As the first step toward the development of a new boundary-fitted calculation method and to evaluate the accuracy of a calculation method that combines the boundary-fitted-cell method and the co-orthogonal coordinate system method, this study compares the results of flow calculations using this method with those using the generalized coordinate system and with those measured in a previous flume experiment.

## 2 MATHEMATICAL MODEL

### 2.1 *Governing equation of the combined calculation method*

#### 2.1.1 *Calculation procedures*
The objective of the combined calculation method proposed in this paper is to accurately express the flow

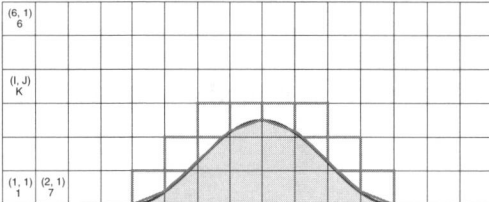

Figure 1.  The new combined calculation method. At the outer bank of a curve (blue) in a straight channel, boundary fitting is performed only for the red grids.

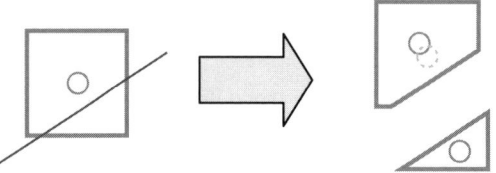

Figure 2.  Division of a rectangular grid that contains boundaries between riverbank and river channel, and the redefinition of water level calculation points at the center of gravity of each new cells.

regime in a complicated form channel. The calculation method employes two coordinate system in same calculation region at same time, one is co-orthogonal coordinate system, and other one is boundary fitted coordinate system. The cells in boundary fitted coordinate system are deformed rectangular grids that are difficult to express with rectangular grid in co-orthogonal coordinate system. For example, at the outer bank of a curve (blue) in a straight channel (Fig–1), the channel morphology is incorporated into the calculation by locally applying the boundary-fitted cells (red) to the grids through which the curve passes.

As shown in Fig–2, the grids of the co-orthogonal coordinate system that contain the boundary between riverbank and river channel (blue) are divided by that boundary, and water-level calculation points are redefined at the center of gravity of each new cells.

The allocation of calculation points in the combined calculation method is shown in Fig–3. Staggered grids are used to allocate the calculation points for the co-orthogonal coordinate system area. For the fitted coordinate system area, the calculation points are allocated by naturally expanding the staggered grids.

The connecting of two coordinate systems is done via the velocity calculation points (green triangles in Fig–3). The velocity at the calculation points on connecting point are within the co-orthogonal coordinate system area, including at the green triangles, and is calculated using the conventional two-dimensional shallow water theory described in the next section. The velocity at the calculation points within the fitted coordinate system area (red triangles) is calculated using

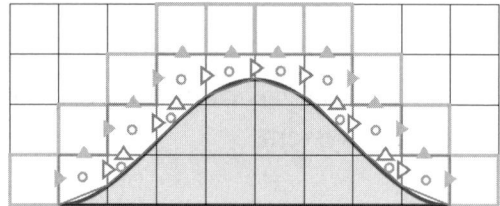

Figure 3.  Definition of calculation points and combining of grids.

an equation of motion obtained by naturally expanding the two-dimensional shallow water theory. The water level at calculation points within the fitted coordinate system area (red circles) is calculated using the green and red triangles.

### 2.1.2  Co–orthogonal coordinate system area

The discharge flux in the $x$ and $y$ directions and the water level of the co-orthogonal coordinate system area are calculated using Equations (1), (2) and (3):

$$\frac{\partial H}{\partial t} + \frac{\partial q_x}{\partial x} + \frac{\partial q_x}{\partial y} = 0 \tag{1}$$

$$\frac{\partial q_x}{\partial t} + \frac{\partial}{\partial x}\left(\frac{q_x^2}{h}\right) + \frac{\partial}{\partial y}\left(\frac{q_x q_y}{h}\right) + gh\frac{\partial H}{\partial x} =$$
$$-\frac{gn^2 q_x}{h^{7/3}}\sqrt{q_x^2 + q_y^2} \tag{2}$$

$$\frac{\partial q_y}{\partial t} + \frac{\partial}{\partial x}\left(\frac{q_x q_y}{h}\right) + \frac{\partial}{\partial y}\left(\frac{q_y^2}{h}\right) + gh\frac{\partial H}{\partial y} =$$
$$-\frac{gn^2 q_y}{h^{7/3}}\sqrt{q_x^2 + q_y^2} \tag{3}$$

Where, $q_x$ and $q_y$ are the discharge flux in the $x$ and $y$ direction, respectively, $t$ is the time coordinate, $x$ and $y$ are the plane coordinates, $h$ is the water depth, $g$ is the gravitational acceleration and $H$ is the water level.

Staggered grids are applied to the numerical calculation of these equations, except for the advective term of the equation of motion, to which upwind differences are applied.

### 2.1.3  Boundary-fitted area

The governing equation for the boundary-fitted area is expanded so that calculations can be performed with polygons (cells) in addition to rectangular grids based on the two-dimensional shallow water theory equation.

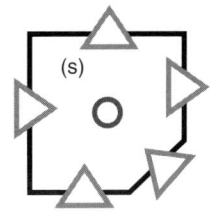

(a) Equation of continuity

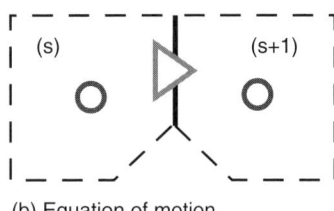

(b) Equation of motion

Figure 4. Calculation points within the boundary-fitted area.

First, as easily inferred from Figure 4(a), the equation of continuity is rewritten as follows:

$$\frac{\partial \eta}{\partial t} - \frac{1}{A}\sum_{i=1}^{k} Q_i = 0 \qquad (4)$$

Next, the equation of motion, with the advective terms omitted, is expressed as follow:

$$\frac{1}{l}\frac{\partial Q}{\partial t} + gD\frac{\partial \eta}{\partial s} = -\frac{gn^2 Q|Q|}{D^{7/3}} \qquad (5)$$

Where, $l$ is length of the side of cell, $Q$ is the discharge on the side of the cell, $A$ is the area of the cell for which the water level is obtained, and $k$ is the number of sides of each cell.

With this calculation method, calculations that ignore the advective terms are done only for the calculation points showed as red triangles in Fig–3.

Numerical calculations for the two equations are the same as the calculation method described in the previous section.

### 2.2 Governing equation of the generalized coordinate system

The shallow water theory equations, when transformed to the generalized coordinate system, are as follows:
Equation of continuity:

$$\frac{\partial}{\partial t}\left(\frac{h}{J}\right) + \frac{\partial}{\partial \xi}\left(\frac{hu^\xi}{J}\right) + \frac{\partial}{\partial \eta}\left(\frac{hu^\eta}{J}\right) = 0 \qquad (6)$$

Equation of motion for direction $\xi$:

$$\frac{\partial u^\xi}{\partial t} + u^\xi\frac{\partial u^\xi}{\partial \xi} + u^\eta\frac{\partial u^\xi}{\partial \eta} + \alpha_1 u^\xi u^\xi + \alpha_2 u^\xi u^\eta + \alpha_3 u^\eta u^\eta$$
$$= -g\left[(\xi_x^2 + \xi_y^2)\frac{\partial H}{\partial \xi} + (\xi_x\eta_x + \xi_y\eta_y)\frac{\partial H}{\partial \eta}\right]$$
$$- \frac{C_d u^\xi}{hJ}\sqrt{K1^2 + K2^2} + D^\xi \qquad (7)$$

Equation of motion for direction $\eta$:

$$\frac{\partial u^\eta}{\partial t} + u^\xi\frac{\partial u^\eta}{\partial \xi} + u^\eta\frac{\partial u^\eta}{\partial \eta} + \alpha_4 u^\xi u^\xi + \alpha_5 u^\xi u^\eta + \alpha_6 u^\eta u^\eta$$
$$= -g\left[(\eta_x\xi_x + \eta_x\xi_y)\frac{\partial H}{\partial \xi} + (\eta_x^2 + \eta_y^2)\frac{\partial H}{\partial \eta}\right]$$
$$- \frac{C_d u^\eta}{hJ}\sqrt{K1^2 + K2^2} + D^\eta \qquad (8)$$

Where, $u^\xi$ is the directional component $\xi$ of velocity, $u^\eta$: the directional component $\eta$ of velocity, $J$ is the the Jacobian of coordinate transformation, $1/(x_\xi y_\eta - x_\eta y_\xi)$, $K1 = \eta_y u^\xi - \xi_y u^\eta$, $K2 = -\eta_x u^\xi + \xi_x u^\eta$, $\alpha_1 = \xi_x\frac{\partial^2 x}{\partial\xi^2} + \xi_y\frac{\partial^2 y}{\partial\xi^2}$, $\alpha_2 = 2\left(\xi_x\frac{\partial^2 x}{\partial\xi\partial\eta} + \xi_y\frac{\partial^2 y}{\partial\xi\partial\eta}\right)$, $\alpha_3 = \xi_x\frac{\partial^2 x}{\partial\eta^2} + \xi_y\frac{\partial^2 y}{\partial\eta^2}$, $\alpha_4 = \eta_x\frac{\partial^2 x}{\partial\xi^2} + \eta_y\frac{\partial^2 y}{\partial\xi^2}$, $\alpha_5 = 2\left(\eta_x\frac{\partial^2 x}{\partial\xi\partial\eta} + \eta_y\frac{\partial^2 y}{\partial\xi\partial\eta}\right)$, $\alpha_6 = \eta_x\frac{\partial^2 x}{\partial\eta^2} + \eta_y\frac{\partial^2 y}{\partial\eta^2}$.

Also, where $D^\xi$, $D^\eta$ is diffusion terms:

$$D^\xi = \frac{\partial}{\partial \xi}\left(\nu_t\xi_r^2\frac{\partial u^\xi}{\partial \xi}\right) + \frac{\partial}{\partial \eta}\left(\nu_t\eta_r^2\frac{\partial u^\xi}{\partial \eta}\right) \qquad (9)$$

$$D^\eta = \frac{\partial}{\partial \xi}\left(\nu_t\xi_r^2\frac{\partial u^\eta}{\partial \xi}\right) + \frac{\partial}{\partial \eta}\left(\nu_t\eta_r^2\frac{\partial u^\eta}{\partial \eta}\right) \qquad (10)$$

For these equations, the decoupled solution was used for the advective, diffusion and other terms of the equation of motion, the CIP method was used to calculate the advective terms, and the center-difference method was used to calculate the diffusion terms.

## 3 COMPARISON WITH FLUME EXPERIMENT

### 3.1 Outline of the experiment with meander channel

Hasegawa et al.(1982, 1984, 1991) conducted a meandering channel experiment with a movable bed to understand how water flows in such a channel. This experiment was conducted at a channel that follows the sine-generated curve expressed by Equation (11).

$$\theta = -\theta_0 \sin\frac{2\pi}{L}s \qquad (11)$$

Table 1. Conditions of the hydraulic experiment.

| $\tilde{L}$ | $\tilde{B}$ | $\theta_\circ$ | Bed gradient | $q$ |
|---|---|---|---|---|
| 220 cm | 30 cm | 30 | 0.00333 | 1.87l/s |

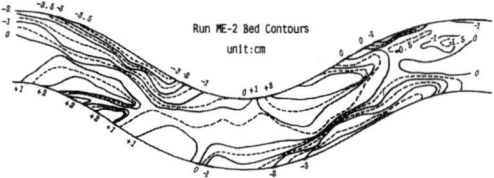

Figure 5. Riverbed form at steady state.

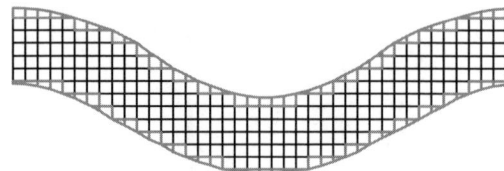

(a) Combined calculation method

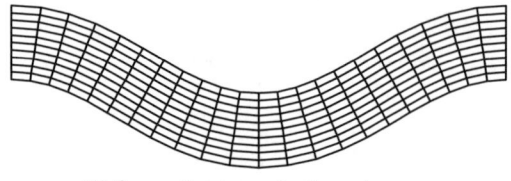

(b) Generalized coordinate system

Figure 6. Computational grids.

The scale and hydraulic conditions of the channel are shown in Table–1. Under the conditions shown in Table–1, steady state was reached 240 minutes after the beginning of the experiment. The riverbed morphology at steady state is shown in Fig–5.

This study compared the flow regime in a meandering channel that was measured in a flume experiment conducted by Hasegawa et al., in order to validate the accuracy of a calculation method that combines the boundary-fitted-cell method and the co-orthogonal coordinate system method. The riverbed morphology at steady state is used for calculations by the combined calculation method and the generalized coordinate system method as fixed bed.

### 3.2 Computational grid

Computational grids (Figures 6 a), b)) were set for the channel shown in Figure 5 for the numerical calculations. Figure 6 (a) shows computational grids and cells for the combined calculation method and Figure 6 (b) shows computational grids for the generalized coordinate system method.

The computational grid for the combined calculation method (Figure 6 (a)) has square coordinate cells (5 cm × 5 cm). The red cells, which contain the sidewall, are expressed by the boundary-fitted cells. There are 83 such cells in the calculation area. Rectangular grids number 40 in the longitudinal direction and 5 in the transverse direction.

Computational cells in the computational grid set by the generalized coordinate system for comparison (Fig. 6 (b)) number 22 in the longitudinal direction and 10 in the transverse direction. The number of computational cells is similar for the two methods.

### 3.3 Initial and boundary conditions

In this study, the Manning equation was used to evaluate the shear force at the riverbed, and the Manning's roughness coefficient (0.025) was back-calculated from the average velocity and water depth obtained in the experiment.

As initial conditions, the uniform flow depth calculated based on this roughness coefficient was applied to each water-level calculation point.

A boundary condition at the upstream end was given as velocity. The transverse distribution of velocity found in the experiment was applied to the computational grid. For calculations under the generalized coordinate system, the experimental values were transformed to the directional velocity components $\xi$ and $\eta$.

For the combined calculation method, however, the calculation points differed from the velocity measurement points in the experiment; therefore, interpolated values of velocities at the calculation points were calculated and were used as a boundary condition at the upstream end.

The water level, defined by the aforementioned uniform flow depth and riverbed height, was used as a boundary condition at the downstream end.

### 3.4 Results of calculation

Figures 7 and 9 compare the results of the two calculation methods with the flume experimental values. Figure 7 shows the two-dimensional distribution of velocity, and Figure 9 compares the three velocities (the results of the two calculation methods and the experimental values) in the longitudinal and transverse directions on two lines shown in Figure 8 ($j = 3$ and 8). Because the calculation points differ between the combined calculation method and the generalized coordinate system method, the results of the combined calculation method were interpolated with values at the red points in Figure 8 for comparison.

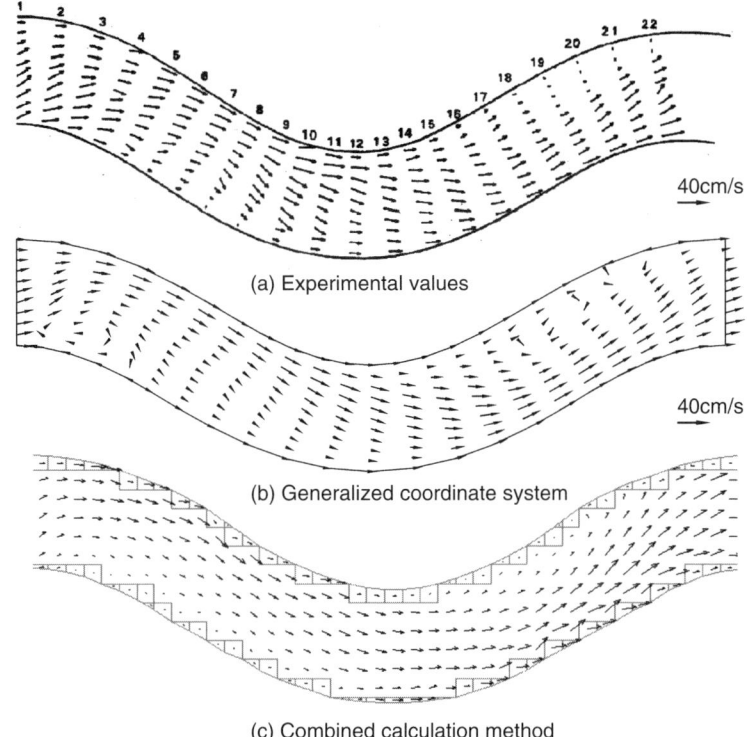

40cm/s

(a) Experimental values

40cm/s

(b) Generalized coordinate system

(c) Combined calculation method

Figure 7.   Comparison of velocity vectors.

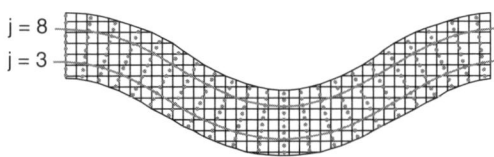

j = 8
j = 3

Figure 8.   Interpolation of calculation value in the combined method.

As shown in Figure 7, the combined calculation method and the generalized coordinate system method tend to accurately reproduce the experimental values: The mainstreams obtained by the two methods are gently meandering, as was found in the experiment. However, the flow directions near the sidewalls obtained by the combined calculation method differed from those measured in the experiment.

As for the flow velocities on the two lines ($j = 3$ and 8) in Figure 9, the velocity vectors determined from the $x$- and $y$-direction components obtained by the two methods for the line $j = 3$ accurately reproduced the experimental values. For the line $j = 8$, the $y$-direction velocity obtained by the two methods accurately reproduced the experimental values, but

the combined calculation method underestimated the velocity vectors and $x$-direction velocities relative to the experimental values at the upstream region.

Investigation into this problem is difficult, as the data are limited to what we have presented here. To improve the accuracy of the new combined calculation method, it will be necessary to identify its problems and their causes by applying it to various cases.

4   CONCLUSIONS

To validate the accuracy of the combined calculation method, the results obtained by the method were compared with those obtained by an flume experiment and the generalized coordinate system method. The comparison has demonstrated that the combined calculation method is useful as a new boundary-fitted calculation method, and that the coordinate-axis-independent method can be applied to calculations of the development of meandering and shortening of a meandering channel. There is, however, room for improving the calculation accuracy. Further investigation into the cause of the underestimation is currently difficult, as the data are limited to those presented here.

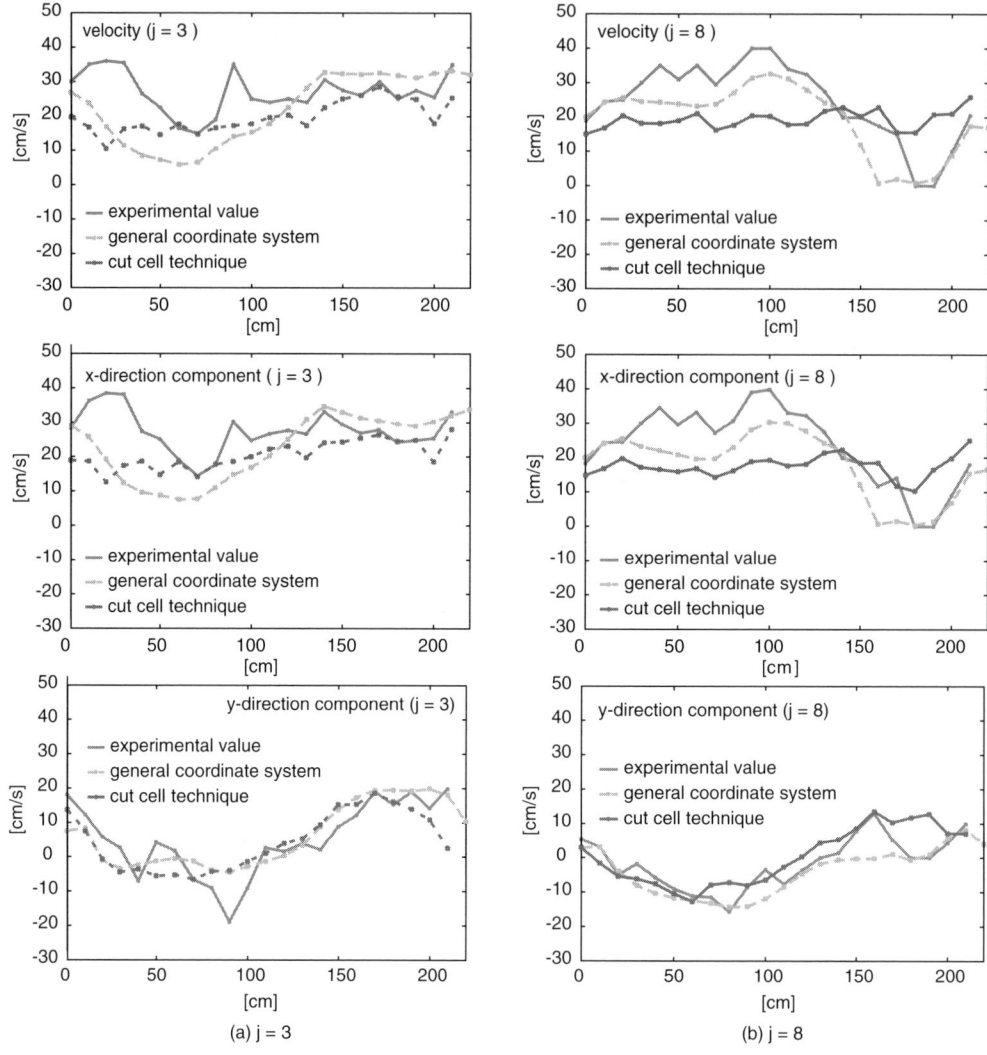

Figure 9.   Comparison of velocity.

To improve the accuracy of the technique, it will be necessary to identify its problems and their causes by applying it to various cases.

## ACKNOWLEDGEMENTS

The author would like to thank the Foundation of Hokkaido River Disaster Prevention Research Center for supporting this study.

## REFERENCES

Hasegawa, K. (1984). Hydraulic study on two-dimensional surface and bed shapes of alluvial meandering channels. *Ph.D. Thesis, Hokkaido University*.

Hasegawa, K., Yamaoka, I. and Suzuki, Y. (1982). Flow over sand waves in a meandering channel. *Proceedings for the Hokkaido branch of the JSCE 38*.

Hasegawa, K., Yamaoka, I. and Takana, N. (1991). Shape characteristics of sand waves influenced by channel meandering. *Proceedings for the Hokkaido branch of the JSCE 38*.

Shimizu, Y. (1991). Calculation of twodimensional flow and bed variation using the generalized coordinate system. *Proceedings of the Annual Academic Lectures of the JSCE 46*, pp.634–635.

Yasuda, H. (2005). Numerical simulation of tsunami runup onto a complex beach with a boundary-fitting cell system. *Proceedings for the Hokkaido branch of the JSCE 61*, CD–ROM.

*River, Coastal and Estuarine Morphodynamics: RCEM 2007 – Dohmen-Janssen & Hulscher (eds)*
*© 2008 Taylor & Francis Group, London, ISBN 978-0-415-45363-9*

# Bed morphology in Kinoshita meandering channels: Experiments and numerical simulations

Jorge D. Abad & Marcelo H. Garcia

*Ven Te Chow Hydrosystems Laboratory, Dept. of Civil and Environmental Engineering, University of Illinois at Urbana-Champaign, Urbana, USA*

ABSTRACT: Meandering rivers evolve and interact with floodplains constantly. During this evolution, several hydrodynamic (primary and secondary velocities and turbulence flow structure distribution) and morphodynamic features are presented (double heading bends, symmetric and asymmetric bends, upstream- and downstream-valler oriented bends). The detailed study of several of these stages is necessary to understand how rivers migrate. In the past, several experimental studies and analytical formulations were performed mostly in symmetric low-sinuosity meandering channels. In this study, a high-sinuosity asymmetric meandering channel, "the Kinoshita channel" was built at the Ven Te Chow Hydrosystems Laboratory. The channel was designed to keep high harmonic modes on the planform configuration, which differentiates it from a purely sine-generated meandering channel. In this regard, the channel could have the meander bends oriented upstream- and downstream-valley (skewed) by switching the water and sediment flow directions. Abad et al.(2007) perfomed several flat bed smooth experiments under both bend orientations, showing that the 3D mean and turbulence flow structure are unique for each bend orientation; therefore, inferring that hydrodynamics will influence the morphological patterns. In this study, mobile bed conditions are studied by performing experiments in the "Kinoshita channel" and prediction of morphology is done by using a 2D depth averaged model previously validated (Abad et al. 2007).

## 1 INTRODUCTION

The interaction between meandering rivers and their floodplains occur over several hundreds or even thousand of years. Rivers migrate laterally and downstream by modifying their banks by fluvial erosion combined with retreat of the banks as observed in Figure 1. The process of bank erosion is more complicated when interactions with groundwater, vegetation, geological characteristics are present. Bed alluvial bathymetry is the result of the interaction of free and forced bars where altimetric and planimetric instabilities interact. Several conditions for free bars suppression have been described in the past (Colombini et al. 1987, Whiting and Dietrich 1993a, Garcia and Nino 1993a, Bittner 1994). Whiting and Dietrich (1993b) and Whiting and Dietrich (1993c) observed that free bars interact with forced bars in high-amplitude symmetric meandering channels. The morphology of these macroforms may change depending on flow and sediment discharges, and the hydrodynamics of the meandering channel itself is particularly important for the development of complex migrating sediment patterns. Abad and Garcia (2005) have shown that by changing the channel sinuosity while maintaining other hydraulic parameters, the core of maximum velocity and turbulent kinetic energy shifts from the inner bank to the outer bank, allowing the shear stress to be modified. For the case of low sinuosity meandering channels, it was shown that the core of maximum velocity is located near the inner bank upstream of the apex and the shear stress is mainly distributed along the width of the channel. However, for high channel sinuosity, even when the core of maximum velocity is concentrated near the inner bank upstream of the apex, this core of maximum velocity shifts to the outer bank downstream of the bend apex. This shifting of the core of maximum velocity produces preferential concentrations of water velocities, shear velocities, and turbulent kinetic energy distributions along the cross section. Langbein and Leopold 1966) and Leopold and Wolman (1960) have presented sinusoidal formulations for meandering rivers. However, in natural rivers, more complicated features

Figure 1. River complex patterns, http://wwww.trekearth.com/gallery/Asia/Russia.

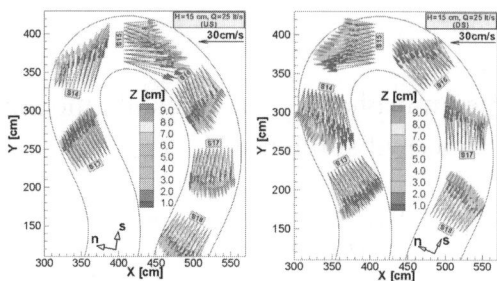

Figure 2. Plan view of velocity vectors. Left: upstream-valley skewed. Right: downstream-valley skewed.

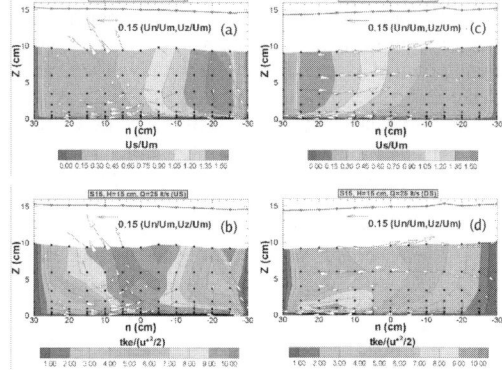

Figure 3. $H = 15$ cm, $Q = 25$ lt/s. Upstream-valley skewed: (a) Normalized velocity magnitudes, (b) Normalized TKE. Downstream-valley skewed: (c) Normalized velocity magnitudes, (d): Normalized TKE.

are found Seminara 2006). High-order (incorporation of skewness and flatness) idealized equations are the well-known Kinoshita curves (Kinoshita and Miwa 1974, Parker et al. 1983, Seminara 1998). Blondeaux and Seminara (19895) described a resonant region (defined by $\beta_R$:half-width-to-depth ratio and $\lambda_R$:resonance wavenumber) where bend instabilities select the wavenumber of spatial free modes, in which the bed erosion is amplified. Seminara et al. (2001) have presented that for sub-resonant morphodynamic conditions ($\beta < \beta_R$), meandering rivers tend to have bends oriented upstream-valley, and for the super-resonant morphodynamic conditions ($\beta < \beta_R$), the bends are oriented downstream-valley. Abad et al. (2007) have analyzed the effect of bend orientation on the hydrodynamics of the "Kinoshita channel" by means of experimental and numerical work under flat bed smooth conditions. Abad et al. (2007) described that the location of the core of maximum depth-averaged velocities is located near the inner bank for both conditions (see Figure 2). However, the secondary flow distribution for the upstream-valley bend orientations (flow from east to west tank) describes the interaction of at least two recirculating cells. One cell, which was produced by the previous bend and the one generated by the local curvature, therefore, at the middle cross section (half length of the "Kinoshita channel": Cross section 15), the secondary flow seems not well developed. On the other hand, for the case of bends oriented downstream valley (flow from west to east tank), the effect of previous bends does not affect the development of the local secondary flow, therefore at the same cross section (middle of the channel: Cross section 15), the secondary flow presents very well defined recirculation areas as commonly observed in well developed bends (see Figure 3). When comparing the turbulence flow structure (see Figure 3), it seems that flow separation (at inner bank upstream of bend

apex) in upstream-valley oriented bends produces high turbulence intensities near the central region around the bend apex, which differs from a well distributed turbulent intensity in the case of downstream-valley oriented bends. Therefore, without even performing morphological experiments, it is obvious that the variation in hydrodynamic parameters (mean velocities, shear stresses and turbulent kinetic energy) due to bend orientation will induce different morphodynamic patterns along the bends and therefore, in a system where bank migration is allowed, it will produce more complex migration patterns such as those described in Figure 1. This research is dedicated to study the hydrodynamics and morphodynamics of the "Kinoshita channel". Several hydraulic conditions under both bend orientations and under sub-resonance and super-resonance conditions will be performed. However, in this paper, only preliminary results are presented based on experiments and 2D depth averaged numerical simulations.

## 2 EXPERIMENTAL SETUP

A water and sediment recirculating meandering channel was built at the "Ven Te Chow" Hydrosystems Laboratory, University of Illinois. Abad and Garcia (2005) performed preliminary three-dimensional CFD modeling of periodic meander bends, which indicated that three consecutive bends were necessary to have both a fully developed turbulent flow and a fully developed secondary flow around the middle bend (between CS10 and CS20 in Figure 4), thus, inlet and outlet boundary conditions effects are minimized, if not eliminated. The designed channel consists of three high-amplitude meander bends defined by the Kinoshita curves (Kinoshita 1961, Parker et al. 1983, Parker and Andrews 1986, Abad and Garica 2005). The arc-wavelenth along the centerline of the channel is around 30.0 m (10.0 m per bend), as well as 1-m straight reach upstream and downstream of the meandering reach. The width and depth of the channel is 0.60 m and 0.40 m respectively. The Kinoshita curves are defined by the instrinsic-type equation as follows:

$$\theta = \theta_0 Sin\left(\frac{2\pi s}{\lambda}\right)$$

$$+\theta_0^3\left(J_s Cos\left(3\frac{2\pi s}{\lambda}\right) - J_f Sin\left(3\frac{2\pi s}{\lambda}\right)\right) \quad (1)$$

Where $J_s = 1/32$ and $J_f = 1/192$ are the skewness and flatness coefficients respectively, $\theta_0 = 110°$ is the maximum angular amplitude. Notice that by reducing $\theta_0$, the Kinoshita equation reduces to the well-known sine-generated symmetric curves (Langbein and Leopold 1966), which were widely used in physical models. A Micro Acoustic Doppler Velocimeter (16 MHz) is used to measure the three components of instantaneous velocities and a sand detector instrument is used to measure the bed morphology during the evolution and during bed equilibrium conditions.

## 3 NUMERICAL SETUP

The hydrodynamic, Finite Volume (FV), depth-averaged model *STREMR* was developed by Robert S. Bernard at the Waterways Experiment Station (WES) of the US Army Corps of Engineers (Bernard 1993). We have used *STREMR* in a natural bend where it was shown that mean velocity modeling results are in agreement with field measurements performed by using Acoustic Doppler Velocimeters (Rrodriguez et al. 2004). The acceptable hydrodynamic model predictions is associated to the fact that *STREMR* has a three-dimensional (3D) secondary flow correction based on a formulation proposed by Bernard (1993). Based on the depth-averaged hydrodynamic model,

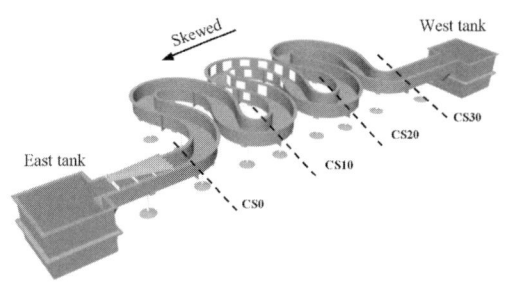

Figure 4. Experimental setup: "The Kinoshita channel". Upstream-valley skewed: water flows from East to West tank. Downstream-valley skewed: water flows from West to East tank.

modules for suspended and bed-load transport have been derived and incorporated into *STREMR*. Herein, only a brief description of the hydrodynamic and sediment transport modules will be presented; however, more detailes could be found at Bernard (1993) and Abad et al. (2007).

### 3.1 Hydrodynamic model

The depth-averaged flow equations are transformed from Cartesian (x,y) to curvilinear coordinates ($\xi = \xi[x,y]$, $\eta = \eta[x,y]$) (see Figure 5). This allows that the spatial discretization in the Cartesian coordinates (x,y) is set arbitrarily while the discretization in the computational plane (i,j) has a unit spacing ($\Delta i = \Delta j = 1$).

#### 3.1.1 Governing equations
The continuity and momentum equations are given by Equations (2) and (3) respectively.

$$\frac{\partial(h\bar{u}_i)}{\partial x_i} = 0 \quad (2)$$

$$\frac{\partial \bar{u}_i}{\partial t} + \bar{u}_j \frac{\partial \bar{u}_i}{\partial x_j} = -\frac{1}{\rho}\frac{\partial P}{\partial x_i} + T_i - X_i + S_i \quad (3)$$

where $\bar{u}_i$ is the depth-averaged velocity component in the i-th direction (i: 1, 2, with components $\bar{u}$ and $\bar{v}$), $t$ is the time, $x_i$ is the spatial coordinate in the i-th direction (namely $x$ and $y$), $\rho$ is the fluid density, $P$ is the pressure, and $h$ is the water depth. In turn, $T_i$, $X_i$, and $S_i$ refer to the viscous, friction and secondary-flow forces respectively in the i-th direction. For more details about the Viscous and friction forces, please read Abad et al. (2007).

The secondary flow correction is given by:

$$S_i \approx \frac{1}{\rho}\frac{\bar{u}_i}{|\bar{u}_i|}\left[h^{-1}\underline{n}\cdot\nabla(h\tau_s) + 2\frac{\tau_s}{r_s}\right] \quad (4)$$

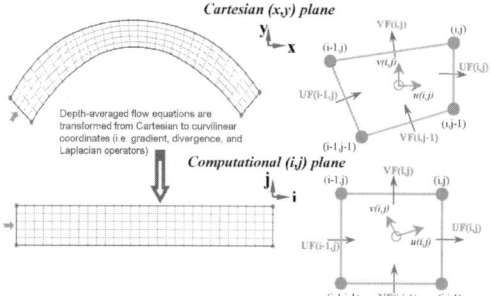

Figure 5. Transformation of coordinates (Cartesian to computational). UF and VF are the advective face-centered fluxes.

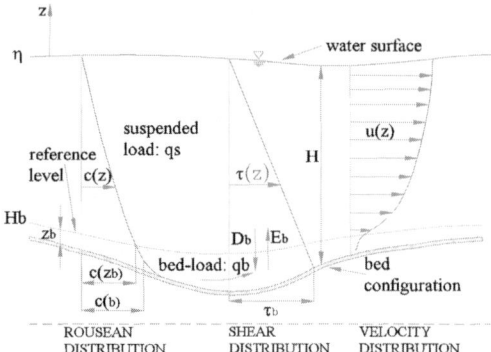

Figure 6. Sediment transport model configuration.

where $\tau_s = \rho h \Omega \sqrt{C_f} \, |\bar{u}_i|$ is the depth-averaged shear stress produced by the secondary circulation , $\underline{n}$ is the unit vector normal to the velocity vector, $r_s$ is the streamline curvature radius, and $\Omega$ is a vorticity-type variable for which a conservation equation can be prescribed (Abad et al.2007). STREMR uses the standard $k - \epsilon$ turbulence model (Launder and spalding 1974). The turbulent kinetic energy ($k$) and the rate of dissipation ($\epsilon$) of turbulent kinetic energy are given by Equations (5) and (6) respectively. Thus, the turbulent eddy viscosity is given by $\bar{\nu}_t = C_\nu f(R_k) \frac{k^2}{\epsilon}$ where $f(R_k)$ is the correction due to recirculating flows.

The turbulent kinetic energy ($k$) and dissipation of $k$ are given by:

$$\frac{\partial \bar{k}}{\partial t} + \bar{u}_j \frac{\partial \bar{k}}{\partial x_j} = \bar{\nu}_t \left( \frac{\partial \bar{u}_i}{\partial x_j} + \frac{\partial \bar{u}_j}{\partial x_i} \right) \frac{\partial \bar{u}_i}{\partial x_j}$$

$$+ \frac{1}{\sigma_k} \left[ h^{-1} \bar{\nu}_t \nabla \cdot \left( h \nabla \bar{k} \right) + \nabla \bar{\nu}_t \cdot \nabla \bar{k} \right] - \bar{\epsilon} + P_{kb} \quad (5)$$

$$\frac{\partial \bar{\epsilon}}{\partial t} + \bar{u}_j \frac{\partial \bar{\epsilon}}{\partial x_j} = c_1 \frac{\bar{\epsilon}}{k} \bar{\nu}_t \left( \frac{\partial \bar{u}_i}{\partial x_j} + \frac{\partial \bar{u}_j}{\partial x_i} \right) \frac{\partial \bar{u}_i}{\partial x_j}$$

$$+ \frac{1}{\sigma_\epsilon} \left[ h^{-1} \bar{\nu}_t \nabla \cdot (h \nabla \bar{\epsilon}) + \nabla \bar{\nu}_t \cdot \nabla \bar{\epsilon} \right] - C_{2\epsilon} \frac{\bar{\epsilon}^2}{k} + P_{\epsilon b} \quad (6)$$

Where,

$$P_{kb} = C_{kb} \frac{u_*^3}{h} \qquad C_{kb} = C_f^{-1/2}$$

$$P_{\epsilon b} = C_{\epsilon b} \frac{u_*^4}{h^2} \qquad C_{\epsilon b} = E_* \frac{C_{2\epsilon} C_\nu^{1/2}}{C_f^{3/4}} \quad (7)$$

Here $C_\nu = 0.09$, $C_1 = 1.44$, $C_2 = 1.92$, $\sigma_k = 1.0$, $\sigma_\epsilon = 1.3$ and $E_* = 3.6$.

### 3.2 Sediment transport model

Figure 6 shows the sediment transport configuration where the bed-load and suspended transport are separated by the reference level ($H_b$).

#### 3.2.1 Suspended transport module
The vertical coordinate $z$ is oriented downward from the water surface elevation. Abad et al. (2007) limited the derivation to the dilute multiphase case. The derivation starts by considering the 3D Reynolds-Averaged Navier-Stokes (RANS) equations for dilute multiphase flow as follows,

$$\frac{\partial C}{\partial t} + \nabla (J_c) = 0 \quad (8)$$

where $C$ is the concentration of suspended sediment, $J_c$ is the sediment flux defined as $J_c = U_c C - \overline{\nu_c} \nabla C$, $\overline{\nu_c} = \frac{\nu_t}{\sigma_c}$ is the sediment eddy diffusivity and $\sigma_c$ is the turbulent Schmidt number. Equation (8) is integrated from $-H_b$ (reference level) to $\eta$ (free surface) to get the depth-averaged suspended transport equation. After performing the integration, using the Leibniz rule, defining $h_b = H_b + \eta$, $\frac{\partial}{\partial t} \int_{-H_b}^{\eta} c\,dz = \frac{\partial}{\partial t}(h_b \overline{C})$, $\int_{-H_b}^{\eta} J_{cH}\,dz = h_b \overline{J_{cH}}$, using appropiate boundary condition for the total intake of sediment at $I( - H_b)$ and at the free surface, and neglecting unsteady terms, the depth averaged suspended sediment equations read as:

$$\frac{\partial}{\partial t}(h_b \overline{C}) + \nabla_H (h_b \overline{J}_{cH}) = I(-H_b) \quad (9)$$

Where $I( - H_b)$ is the sediment entrainment and deposition at the reference layer, which can be calculated as $I(-H_b) = w_s E_s - w_s C(-H_b)$, where $w_s$, $E_s$, and $C(-H_b)$ are the settling particle velocity, dimensionless rate of entrainment into suspension and the concentration at the reference layer. By using a Rousean distribution profile $C(z) = C(z_b) \left[ \frac{(H-z)/z}{(H-z_b)/z_b} \right]^{Z_R}$, the sediment entrainment and deposition rate could be expressed as $I(-H_b) = w_s E_s - \frac{w_s \overline{C}}{INT_1(\delta_b, Z_R)}$. Abad and Garcia (2006) presented a practical expression for $INT_1$. In the above equations, $Z_R$ represents the Rouse number equal to $\frac{w_s}{\kappa u*}$, where $u*$ is the shear velocity.

### 3.2.2 Bed-load transport module
In the present study, a general formulation based on excess shear stress is implemented ($q_b* = \alpha$ $(\phi_s \tau * - \tau_c*)^{n_e}$). Two effects can cause deviation between the depth-averaged water velocities and the direction of bed-load transport: (1) secondary flow, (2) downward acceleration along transverse bed slopes due to gravity (Mosselman 2005). To describe the effect of transverse slope on the bed-load transport, Struiksma et al. (1985)'s formulation is used herein. Thus, the total bed-load can be decomposed into $x-$ and $y-$direction components as $q_{bx} = q_b \cos\alpha$ and $q_{by} = q_b \sin\alpha$, where $\alpha$ is expressed as:

$$tan\alpha = \frac{sin\delta - \frac{1}{f_s \tau *} \frac{\partial \eta_b}{\partial n}}{cos\delta - \frac{1}{f_s \tau *} \frac{\partial \eta_b}{\partial s}} \qquad (10)$$

Where $\delta = tan^{-1}\left(\frac{v}{u}\right) - tan^{-1}\left(\frac{A}{r_s}h\right) = $ direction of the bed shear stress, $\partial \eta_b/\partial s$ and $\partial \eta_b/\partial n$ are the bed slopes in the streamwise and transverse directions respectively (intrinsic coordinates). For expressions of $A$ and $r_s$, the reader is invited to read Abad et al. (2007).

### 3.3 Bed evolution model – exner equation

Equation (8) is integrated between -H and $-H_b$. Following the same derivation as for the suspended sediment equation, resulting in:

$$\frac{\partial}{\partial t}(z_b \overline{C}_b) + \nabla_H(q_b) = I_b(-H) + I_b(-H_b)$$

$$-C(-H_b)\frac{\partial H_b}{\partial t} + C(-H)\frac{\partial H}{\partial t} \qquad (11)$$

where the total flux $I_b(-H_b)$ equals $-I(-H_b)$ because of continuity at $z = -H_b$ and because it is assumed that the concentration field is continuous there. Some terms on RHS of Equation 11 are eliminated (for more details, see Abad et al. 2007). Thus, the Exner equation is obtained as:

$$C_B \frac{\partial H}{\partial t} = \nabla_H(q_b) + I(-H_b) \qquad (12)$$

Commonly $C_B$ can be calculated as $(1-\lambda_p)$, where $\lambda_p$ is the porosity at the bed.

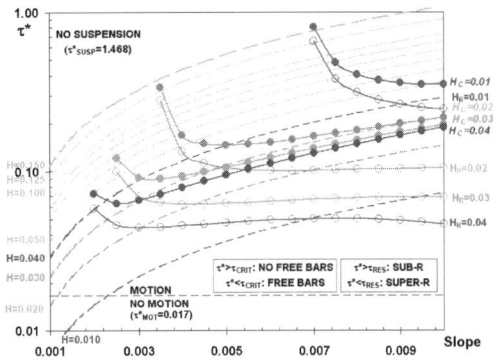

Figure 7. Experimental conditions in the Kinoshita channel. Using Wong and Parker (2006) bedload formulation. The mean particle size is equal to 0.832 mm.

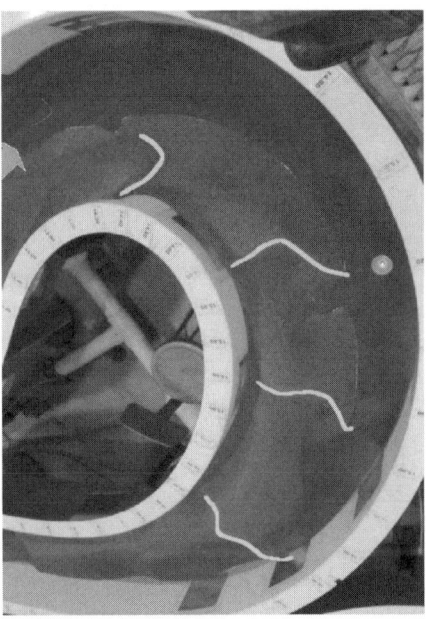

Figure 8. Preliminary bed-equilibrium conditions for H = 15 cm and Q = 25 lt/s. Notice that several bars are found close to the inner-bank.

## 4 PRELIMINARY RESULTS

As explained previously, several morphodynamic regimes (sub-resonant and super-resonant) were defined by using the resonance condition. By following the stability analysis performed by Blondeaux and Seminara (1985), a phase-type diagram for the Kinoshita is built as presented in Figure 7. Notice that the critical condition for sediment motion is ensured and the present experimental design is dedicated to study only bedload (no suspension). However, the

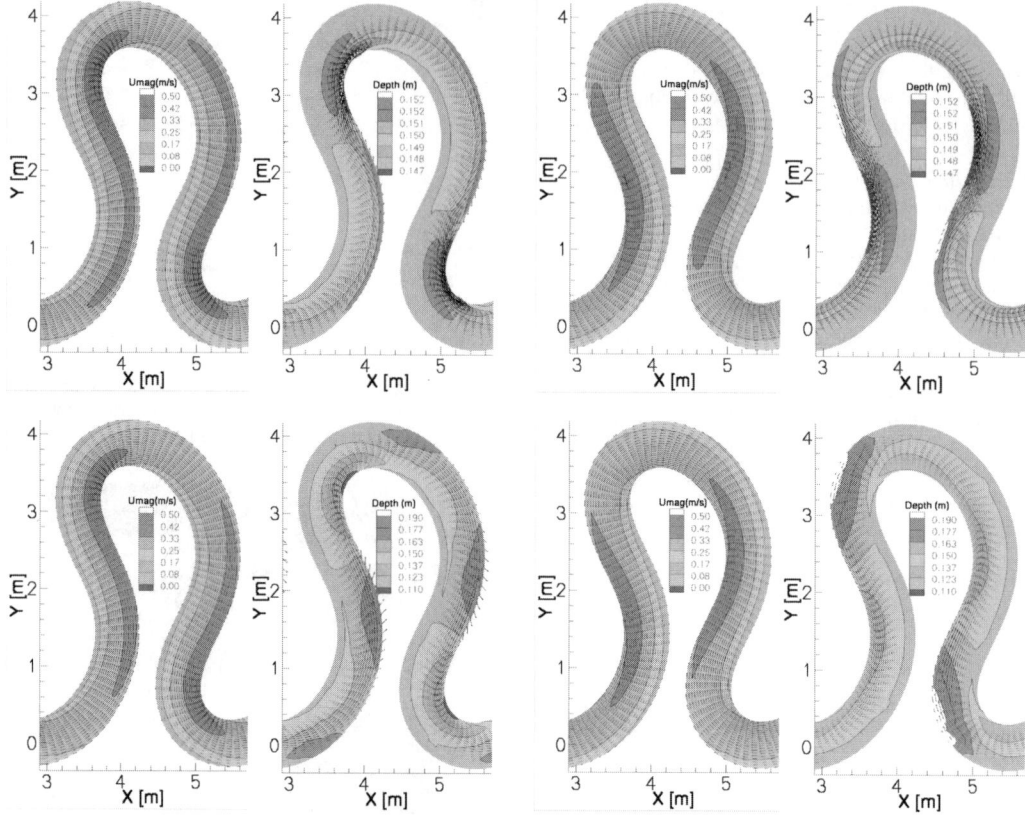

Figure 9. Upstream-valley bend orientation. Middle bend results for H = 15 cm and Q = 25 lt/s. Top: T = 22s, bottom: T = 8000s. The left figures show velocity contours and vectors. The right figures show water depth contours and bedload vectors.

Figure 10. Downstream-valley bend orientation. Middle bend results for H = 15 cm and Q = 25 lt/s. Top: T = 22s, bottom: T = 8000s. The left figures show velocity contours and vectors. The right figures show water depth contours and bedload vectors.

phase-type graph could be easily adapted to account for suspended sediment. The dashed lines are the experimental conditions to be used in the Kinoshita channel. The lines with filled circles are the critical conditions for the presence of free bars, and the lines with empty circles are the resonance conditions. The presence of free bars is defined by $\tau *_{exp} < \tau_{crit}$. The experimental condition will be in the sub-resonant condition if $\tau *_{exp} > \tau_{res}$, while having a super-resonant condition if $\tau *_{exp} < \tau_{res}$. Notice that for high flow depths (narrow channels), the subresonant condition is predominant. Currently, several experiments are undergoing, however, for this paper, the case of a sub-resonant condition (H = 15 cm, Q = 25 lt/s) for upstream-valley oriented bends (similar to the one used in the flat-bed experiments) is presented in Figure 8.

Figures 9 and 10 show the numerical simulation (no suspended sediment) for the case of upstream- and

downstream-valley oriented bends. Notice the different patterns of erosion, obtained by just modifying the water and sediment direction, and by allowing the banks to migrate, this process will produce more complicated planform migration patterns.

## 5 CONCLUSIONS

Bend orientation and therefore effective curvature seems to control the hydrodynamics (mean and turbulence flow structure) and morphodynamics (interaction of free and forced bars) in high-amplitude meandering channels. Several hydraulic and morphodynamic conditions are currently studied in the "Kinoshita channel" as well as 2D and 3D numerical simulations to understand the complex morphology in this channel. The results presented in this paper are preliminar, since currently bedload sediment transport rates are computed in order to re-examine the morphodynamic conditions (stability phase diagram) and calibrate the numerical simulations.

ACKNOWLEDGEMENTS

Profs. Gustavo Buscaglia, Gary Parker, Jim Best and Bruce Rhoads are thanked for providing valuable discussion in this research. The help of Javier Ancalle and Andy Waratuke during the experiments is deeply appreciated.

REFERENCES

Abad, J. D., J. Ancalle, G. Buscaglia, and M. H. Garcia (2007). Influence of bend orientation on the hydrodynamics of meandering channels: Experimental and numerical work. *XXXII International Association of Hydraulic Engineering and Research (IAHR) Congress, July 1–6, Venice, Italy,* 10 pp.

Abad, J. D., G. C. Buscaglia, and M. H. Garcia (2007). 2d stream hydrodynamic, sediment transport and bed morphology for engineering applications. *In press, Hydrological Processes.*

Abad, J. D. and M. H. Garcia (2005). Hydrodynamics in kinoshita-generated meandering bends: importance for river-planform evolution. *In proceedings of the 4th IAHR Symposium on River, Coastal and Estuarine Morphodynamics RCEM, edited by G. Parker and M. H. Garcia, Taylor and Francis, Balkema, Urbana,Il, October 4–7.*

Abad, J. D. and M. H. Garcia (2006). Discussion of "efficient algorithm for computing einstein integrals by junke guo and pierre y. julien". *Journal of Hydraulic Engineering 132*(3), 332–334.

Bernard, R. S. (1993). Stremr: Numerical model for depth-averaged incompressible flow. *Technical Report REMR-HY-11, US Army Corps of Engineers.*

Bittner, L. (1994). River bed response to channel width variation. *Master thesis, University of Illinois.*

Blondeaux, P. and G. Seminara (1985). A unified bar-theory of river meanders. *Journal of Fluid Mechanics 157,* 449–470.

Colombini, M., G. Seminara, and M. Tubino (1987). Finite-amplitude alternate bars. *Journal of Fluid Mechanics 181,* 213–232.

Garcia, M. H. and Y. Nino (1993). Dynamics of sediment bars in straight and meandering channels: experiments on the resonance phenomenon. *Journal of Hydraulic Research 31*(6), 739–761.

Kinoshita, R. (1961). Investigation of channel deformation in ishikari river. *Report of the Bureau of Resources,* 1–174.

Kinoshita, R. and H. Miwa (1974). River channel formation which prevents downstream translation of transverse bars. *ShinSabo (Translation)*(94), 12–17.

Langbein, W. D. and L. B. Leopold (1966). River meanders, a theory of minimum variance. *U.S. Geol. Surv. Prof. Pap. 422-H.*

Launder, B. E. and D. B. Spalding (1974). The numerical computation of turbulent flows. *Computer Methods in Applied Mechanics and Engineering 3,* 269–289.

Leopold, L. B. and M. G. Wolman (1960). River meanders. *71,* 769–794.

Mosselman, E. (2005). Basic equations for sediment transport in cfd for fluvial morphodynamics.*Chapter 4, Computational Fluid Dynamics, Applications in Environmental Hydraulics. P. D. Bates, S. N. Lane and R. I. Ferguson (Eds), John Wiley and Sons, Ltd.,* 71–89.

Parker, G. and E. Andrews (1986). On the time development of meander bends. *Journal of Fluid Mechanics 162,* 139–156.

Parker, G., P. Diplas, and J. Akiyama (1983). Meander bends of high amplitude. *Journal of Hydraulic Engineering 109*(10), 1323–1337.

Rodriguez, J. F., F. A. Bombardelli, M. Garcia, K. Frothingham, B. L. Rhoads, and J. D. Abad (2004). High-resolution numerical simulation of flow through a highly sinuous river reach. *Water Resources Managment 18,* 177–199.

Seminara, G. (1998). Stability and morphodynamics. *Meccanica 33,* 59–99.

Seminara, G. (2006). Meanders. *Journal of Fluid Mechanics 554,* 271–297.

Seminara, G., G. Zolezzi, M. Tubino, and D. Zardi (2001). Downstream and upstream influence in river meandering. part 2. planimetric development. *Journal of Fluid Mechanics 438,* 213–230.

Struiksma, N., K. W. Olesen, C. Flokstra, and H. De Vriend (1985). Bed deformation in curved alluvial channels. *Journal of Hydraulic Research 23*(1), 57–79.

Whitting, P. J. and W. E. Dietrich (1993a). Experimental constrains on bar migration through bends: Implications for meander wavelength selection. *Water Resources Research 29*(4), 1091–1102.

Whitting, P. J. and W. E. Dietrich (1993b). Experimental studies of bed topography and flow patterns in large-amplitude meanders 1. observations. *Water Resources Research 29*(11), 3605–3614.

Whitting, P. J. and W. E. Dietrich (1993c). Experimental studies of bed topography and flow patterns in large-amplitude meanders 2. mechanisms. *Water Resources Research 29*(11), 3615–3622.

Wong, M. and G. Parker (2006). Reanalysis and correction of bed-load relation of meyer-peter and muller using their own database. *Journal of Hydraulic Engineering 132*(11), 1159–1168.

*General aspects of morphodynamics*

*River, Coastal and Estuarine Morphodynamics: RCEM 2007 – Dohmen-Janssen & Hulscher (eds)*
*© 2008 Taylor & Francis Group, London, ISBN 978-0-415-45363-9*

# Estimation of nature systems morphodynamics

V.I. Klenov

*Personal, Profsouznaya, Moscow, Russia*

ABSTRACT: The objective of River Basins and Coastal Zone management requires for efficient technology, which couples 1) satellite data of the Digital Earth Technology; 2) Virtual Nature Systems (VNS), and 3) quick computing of water/mass/energy flows through the Virtual Nature Systems. Join of the three components is the Moving Digital Earth (MDE). The MDE reflects the Morphodynamics by regional monitoring and corresponding computer mapping.

## 1 INTRODUCTION

The MDE is in progress. Discussed below are components of the MDE, which are Virtual Nature Systems (VNS). The VNS are computer doubles of the actual ones. These VNS are the Coastal Zone and the River Basin(s). The expanded VNS joins the Coastal Zone, Delta/Estuary, and River Basins, being titled as the RIDEC (Klenov, 1999). The providing of Virtual Systems by the real-time data and quick reforming of the data to water/mass/energy flows is the Moving Digital Earth (MDE). The corresponding computer mapping of the area supports monitoring of the Morphodynamics and other tasks.

These VNS differ in governing processes, but they have common water related processes. The RIDEC have dissimilar processes in separate VNS, but joins them by that flows through the Coast Line cause these active interaction.

Governing processes in the VNS – River Basin are water flows and water related processes as follows: erosion, sedimentation, and others. Governing processes in the VNS Coastal zone are under influence of a wind-wave energy and currents, causing bottom and coastal abrasion and sedimentation. The Virtual System RIDEC responds also on any sea level changes, on external impacts in any separate system in any spatial-temporal combinations. All the VNS generate own Morphodynamics.

The Virtual Nature System calculates interactions between all cells of the Matrix, including a whole interested area. The case studies were gulfs of Japan Sea, Asov Sea, several river basins, and the join of the Rhine Basin, Delta, and Coastal zone of the North Sea. Monitoring and estimation of varied scenarios were computed by continual influence of meteorological external powers (precipitation, temperature, wind,

tides, and others), by tectonics (sea level changes, earth crush distorts), and by human activity. The repeated corresponding computer mapping demonstrates the Morphodynamics of the VNS. The case studies resulted in satisfactory decisions for a practice (submarine agriculture, pollution spread over a sea, land, river net, beach abrasion, basins engineering, and others).

Now it is the necessity to join the Virtual Systems with regular inflows of remote sensing data - with the Moving Digital Earth technology (MDE) for running of Virtual Doubles of the Nature Systems for regional monitoring, management, and other aims. The discussed below results are based on a field and computer modeling experience.

## 2 PROPERTIES FOR NATURE SYSTEMS MORPHODYNAMICS

River basins and Coastal zones belong to the Open Non-Equilibrium Systems (ONES) due their properties for response on incessant exterior influence, being both exogenous and endogenous. The morphodynamics of River Basins generates oscillations of flows, generates thresholds, and streams branching, meandering. Oscillations of a surface and of streams with a widely varied frequency result in dynamics of sedimentation and erosion, in forming of river and sea terraces and terrace ranks. The Morphodynamics of the Nature Systems results in corresponding self-organization of water, mass, and energy flows to stable structures in river basins and in coastal zone.

The Morphodynamics of the Virtual Nature Systems has the properties of the ONES, mentioned above. The every external influence and following system's response changes a free behavior of the system, which is irreversible to non disturbed regime. History of a system is 'written' in the only phase trajectory (Prigogine et al., 1984) so, that any external (and human) influence is never lost in the system's memory.

The influence of the ONES Morphodynamics on restoration of its history outcome in two seemingly opposite laws: the Information Loss Law (ILL) in natural records and forms (sediments layers and terrace ranks), and the Information Storage Law (ISL) for the Nature System.

The ILL property is the creation an illusion of trends in natural records and corresponding history, and the wiping up for most part of the records. Otherwise, the ISL, in accordance with the theory on Non-Equilibrium Systems provides a single history realization. By the ISL, a history is never losses, but only becomes 'hidden', what makes an opportunity to 'turn back' the Time to restore it through step by step restoration of the past powers (Klenov, 1987), of past processes, and past relief. The restoration has a single decision. The history is wiped from natural records due the Information Loss Law (ILL), and the true history is dissipated over a system due the Information Storage Law (ISL).

## 3  MORPHODYNAMICS OF RIVER BASINS

A number of Virtual Nature Systems–River Basins (VNS-RB) were worked out for the scales from local (Small tributary of the Moscow River), for regional scale (Upstream of the Moscow River), and for sub-continental (the Rhine Basin). River Basin is continually under continual pressure and impacts of long- and short-term external influence by the Meteorology and Geophysics. The VNS reforms external power to water/mass/energy flows due exchanges between all neighbor cells of the multi-layer Matrix. The multiply repeated interaction during the time results in environmental flows through a system. The continual and changeable external influences provide non-stabilities in the both actual and virtual systems because of their belonging to the Open Non-Equilibrium Systems, and because of their property for self-organization and evolution by the stochastic Morphodynamics. Non-stable flows of streams and sediments outcome in meandering, branches, cones, but in a stable river nets. The Multi-layer Matrix includes 15–20 layers of variables and parameters being as follows: elevation, soil/rock resistance, pollution, surface water, underground water, infiltration, evaporation, among others. The high spatial resolution of the Matrix is wanted due the strong necessity to recognize major geomorphology elements of a relief.

By presence records of precipitation and air temperature, of high resolution elevation grids and of other relevant information the dynamics of a river basin was computed with a satisfactory coincidence with independently observed hydrology data. The VNS calculates processes of floods and debris flows, processes of surface and underground pollution, dynamics of soil erosion and sedimentation, and the Morphodynamics.

The major advantage of the VNS is it ability to estimate hazardous and catastrophic processes even before their initiation in actual nature systems, because of nature systems always delay on exterior influences. For the efficient use of this effect it is necessary to join three following conditions: a) the VNS and it's property for efficient quick Digital System Analysis; b) the incessant input of meteorology and other necessary 2D data by satellite observation; c) the quick reforming of the data to recognize and to follow up environmental flows through river basins of any scale and complexity. The join of the three component turns the Virtual Systems into the Information Time Machine, being titled as the Moving Digital Earth (MDE) with a proficiency for outstripping monitoring of common and disastrous processes by usage for own energetic resources (volume and distribution of a free water in the area), and without use of meteorological and geophysics prognoses, and without of statistic extrapolation of most nature processes (which are almost a 'white noise'). The outstripping time changes from a few hours (debris flows) to some weeks (floods in large rivers).

The method of continual repeated recurrent estimation of all flow through the VNS was worked out as the Method of the Genetic String Coding (GSC) or the Genetic Matrixes (GM) (Klenov, 1987, 1999, 2003). The GSC for each time step writes the VNS in the multi-layer string. The GM uses repeated scanning of the Matrix and provides the VNS morphodynamics and evolution by the Matrix' memory. The simultaneous computed mapping (Figure 1) provides high-regularity computer mapping of the chosen layers with a smooth replacement of previous map by current map.

Finally, the VNS-RB, being coupled with the Digital Earth technology (Fukui, 2003), turns into the Moving Digital Earth (MDE) (Klenov, 2005, 2006). The skill for the estimation of the nearest future only from the Past (without any prognosis) data depends on value of the system's active energy (free water, et al.). The exactness of the view is provided only for the nearest time steps. Further this becomes 'foggy' due unknown future external influences. The similar outstripping effect exists after geodynamics shocks (earthquakes), sharply activating flows and debris-flows because of spatial and linear destruction of a rock resistance.

## 4  MORPHODYNAMICS OF THE COASTAL ZONE

The problem is to determine most safety areas for a submarine agriculture and to provide a beach protection. The main drivers for grid based models are wind power and direction. The VNS CZ uses records,

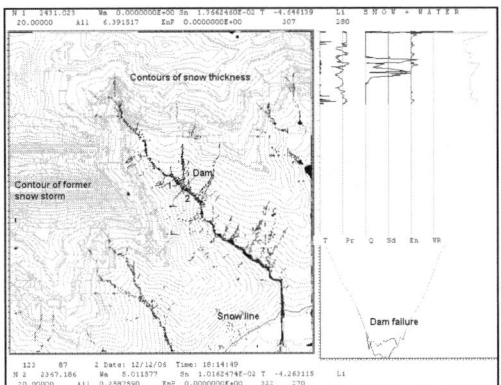

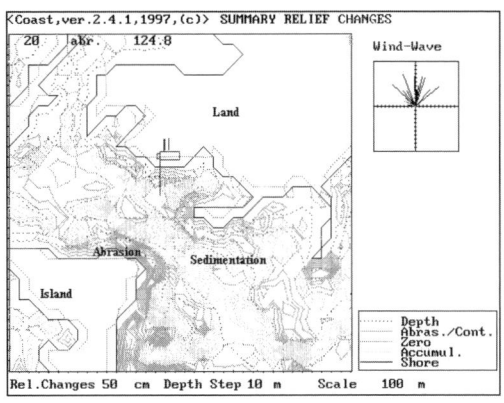

Figure 1. An ongoing computing and mapping of the basin, where: on the left – regular computer map of the basin; right bottom – dam failure due overfilling; right top – graphics of the following: T-air temperature, Pr–precipitation, Q–water depth, Sd–sediments thickness, En-Current Active energy, Wr-summary water resources. On the top and on the bottom – current data by the Gauges. Spatial resolution-10 m.

Figure 2. Continual computing by the VNS – 'Coastal Zone' of a gulf in the Japan Sea: areas and values of bottom/sedimentation under wind pressure (top right) with 100 m resolution.

bathymetry, and bottom grunt resistance for evaluation of bottom and shore abrasion and sedimentation pattern.

The influence of a wind with direction 0°–360° on the bottom depends on the initial influence by exterior waves from the open sea. In a case of a wind from a nearby land it is estimated gradual increasing of wind-wave energy across the Coastal Zone. The wave energy decreases in depth. The empirical deepness of wave energy penetration through a water is about 50 m, for the 5 m height of waves.

During a moving to shore wind-wave energy losses by any touching with a bottom. The remained energy attacks the shore and causes it abrasion and replace of removed matter on the nearest down-slope bottom. The matter of a bottom under pressure of wind energy periodically moves in direction of wind and in direction of a bottom incline. For the purpose was worked out a relevant method of a grid scanning. It is scanned across direction of wind as the 'oblique' scanning. The direction of scanning is determined by direction of wind-waves, and is changed in accordance with it. This principles was efficient for calculation of bottom sediments flows both in two directions. The wind 'feels' nearby islands and sandbanks and turned around them in accordance with empiric parameters. The received map of bottom abrasion/sedimentation was verified by independently observed data. The rose of winds, coastline contours determined most safety areas of marine and submarine agriculture.

The value a beached destroys is foundation for insurance. The other task is estimation of bottom contamination spread from sites of wastes and pipelines. Destructive influence on submarine agriculture also

was assessed. Moving of bottom matter along coastline is continually computed in dependence wind direction, but absence of a data on the sediment balance on both sides of coastal zone and sediments inflow by rivers were absent. Long time modeling under conditions of sea level tides or trends is also foreseen. More, it was computed scenarios of a shore and beach protection against storms by various set of dams with assigned parameters of its resistance. The coastal abrasion is estimated in all installations of the VNS (Figure 2). Special objects were as follow: part of a Coastal zone in the Asov Sea, and part of the Laptev Sea coastal zone. These case studies were specified by a small depth (the Asov Sea) and by Ice Coastline (the Laptev Sea). In the first case under-level channels in the Asov Sea are quickly filled by sediments. The peculiarity of the last case is that a thermo-abrasion is the governing process.

The Virtual Nature System Coastal Zone is the simplified computer double of any Coastal Zone's area. It may be enliven for any part of the area by regular wind-wave data. The sample case (Figure 2) uses real data of a depth and grunt resistance and artificial data of wind-wave regime. The Wave energy is a loader also for solid pollutants, what is estimated in separate layer of the Matrix. The separate interacting layers are as follow: bathymetry (depth of a sea/reservoir up to about 50 m), bottom grunt resistance, contamination, current and summary wind- wave energy in contact with a bottom and shore, and others. The VNS was applied also for estimation for a Coastal Zone placers, by mechanics of concentration of a hard fraction, for assessment of ice shore thermo-abrasion in the Arctic seas, and for coastal engineering under sea level changes. The problem for wide imply of the VNS is that exactness of the VNS exceeds existence and availability of data both for estimation of the morphodynamics and validation for results.

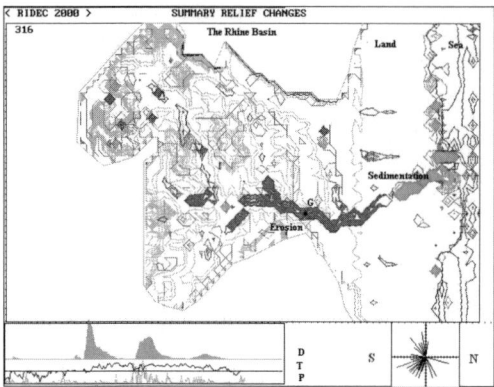

Figure 3. Continual modeling of the coupled system: the Rhine basin – Delta – Coastal zone, where: D – Water depth (computed), T- Air temperature (observed), P- Precipitation (observed), G – Gauge station, S–N – wind direction and power. Resolution – 9 km, time – 316th day.

## 5 MORPHODYNAMICS OF THE COUPLED SYSTEM: RIVER BASINS-DELTA-COASTAL ZONE

The join any river basins with a nearby coastal zone is the Complex Nature System. This causes interaction of processes of water and water related flows over the land with processes on a sea induced by a wind-wave energy, under influence of a sea level changes.

This process is governing in the Virtual (and Actual) Nature System RIDEC of a sub-continental scale (Klenov, 1999), which couples the whole Rhine basin, Delta, and the nearby part of the North Sea (Figure 3). This large area was set up in a rectangular grid with a spatial resolution 8–10 km. The other neighboring area is not belongs to the RIDEC and was automatically cut off for economy of the computing time. Due deficiency of independent hydrology records it was available only visual verification of the computed morphodynamics by corresponding computer mapping of governing processes. The sample meteorological record of precipitation an air temperature was continually reformed by the RIDEC to complex of water flows and water related processes (snow fall and melt, sedimentation and erosion, pollution spread, soil erosion, and others) through the area during at last one annual cycle. Many scenarios of local and regional precipitation, dragging, setting of dams, trends of the sea level, pollution, and others in various combinations of them were computed with computer mapping for thematic layers.

The peculiarity of the RIDEC is that the Coastline becomes a powerful factor of the Morphodynamics, which is the basis of erosion for all river basins. The RIDEC automatically determines zones of land and sea processes and offers their overlapping in the delta. The coast line is not a sharp line because overlapping of

erosion processes with processes under wind pressure in a zone of water level incline and the enough depth of a river.

In the Coastal zone the morphodynamics is offered by a wind-wave energy of changeable power and direction. It provides sediments' flows along the coastline, abrasion of the Coastline, and a pattern of abrasion and sedimentation in the sea bottom.

The Delta is under influence by both upstream water flows until water surface incline is meaningful, by influence of a wind-wave energy outside parts the delta and during tides and others sea level changes. The morphodynamics resulted in a forming of delta, in river branches, in forming of under-sea terraces, and in other effects.

The RIDEC includes layers of elevation/depth, precipitation, snow thickness, pollution, sedimentation, and others. The most of them is offered for computer mapping, by user's choice. At last, the feeling of the morphodynamics effects directly depends on the Matrix resolution, what is required to be increased as possible by a computing power and time. The 10 m resolution is may be satisfactory in view to sense geomorphologic forms and processes.

## 6 THE MORPHODYNAMICS AS RESPONSE ON GEODYNAMICS

Of course, morphodynamics is induced not only by exogenous (meteorology) influence, but also by endogenous (geodynamics) power. The digital assessment of tectonic deforms of the earth surface is foreseen in the all VNSs. For the VNS–CZ it is trend of a sea level, for the RIDEC it is also large scale tectonics deforms in a basin, for mountainous basins it is deforms of earth crust by earthquakes and by active tectonic zones. Meaningful are all former linear deforms, which decreased soil/rock resistance to erosion, and soft deformations of earth surface. All these types may be detected by remote sensing (Moutaz, 2005) to be computed by the VNS.

The effect of the earth surface deforming by earthquakes is in activation of erosion and gravitation processes as follows: landslides, rock collapses, dam's failure, and debris flows activation (Figures 4 and 5).

Noteworthy, that the division of exogenous influences from the simultaneous endogenous ones is often indivisible by observation, but becomes possible by the Virtual Nature Systems, by scenarios of exterior influence on Nature Systems.

The variability of regional, local, and linear soil resistance due former history is the factor of dissimilarity in morphodynamics over an area.

Areas out a site of local storm are potentially hazardous, because of these 'wait' the next storm(s) to be activated. The VNS takes into account both

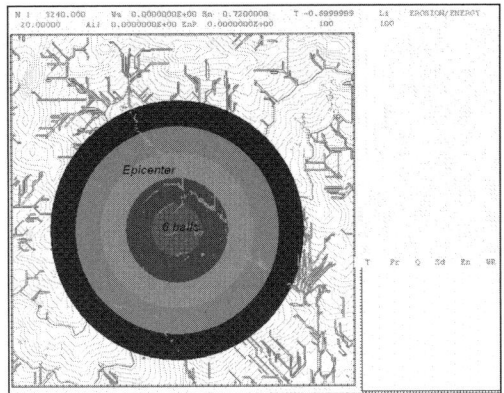

Figure 4. Location of the earthquake with gradations of the power in balls (scenario). The signs are the same as on Figure 1.

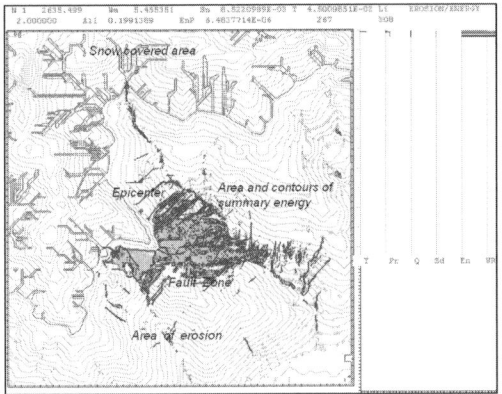

Figure 5. The area of hazardous flows after the earthquake (Fig. 4), and with the cross area fault zone, being covered by zone of precipitation. The signs are the same as on Figure 1.

endogenous and exogenous impacts which becomes hazardous for a long time forever. It was also possible and done artificial estimation of morphodynamics in centennial and geological time, and under human activity. The active geodynamic decreases soil resistance in fault zones and in epicenters of earthquakes due destroys of soil resistance by shocks. It strongly increases transport capacity of water flows. Non-stability of flows abruptly increases in sites of local extreme precipitation over fault zones. (Figure 5). This is a foundation for an insurance (by estimation of strong historical storms over an area). The artificial experience with the VNS is a tool for insurance evaluation.

Earthquakes being happened under a sea bottom cause strong hydraulic shock, which spread with catastrophic velocity over the sea/ocean and results in tsunami in coastal zones. This process should be computed by the expanded Virtual Nature System CZ in view to assess a power of the Wave in various sites of the Coastline due Wave energy lost in the way. Instead of the 'oblique' scanning there is used the algorithm of 'circle' scanning around epicenter, with estimation of Wave's energy loss above shoals.

Commonly, because of earthquakes are now non-predicable by the coupled components Time – Place – Power simultaneously, then response on the events must be based on a retard of nature system's response, for the following automatic and human respond.

## 7 CONCLUSIONS

The first group of conclusions respects to the 'Applied Morphodynamics'. The MDE has a skill to estimate forthcoming disasters and to their mapping at each time step, and before the end of a current time step. It last case the time for response becomes less for a discuss about decision. It seems, that decisions are should be estimated and done by the MDE independently, and a crisis response should be immediate too. It is the real problem, problem for reliable and safety automatic response by the MDE. For example, it is automatic corrections for stairs of dams filling; the other example is automatic warning of people through communication nets.

The elaboration of the VNS in not in equal state for different nature systems. The most advanced is the VNS-River Basin. For the case may be applied a system of insurance, based on skill of the VNS to estimate the nearest future. The outstripping monitoring offers forecasting for tomorrow events and automatic mapping of hazardous sites and tracks of hazardous events (floods and debris flows). For the long time insurance were computed maps of summary active energy of hazardous flows. Being enhanced, the map of the Summary Active Energy (SAE) should be calibrated in gradations of the insurance costs. The source step by step mapping of the Current Active Energy (CAE). The CAE recognizes upcoming hazards immediately in coupling of the Time – Place – Power.

The second group of conclusions respects to the 'Theoretical Morphodynamics'. It includes the following:

1) The outcome of the VNS morphodynamics is in the Information Loss Law (ILL) in natural records, which restricts a saving of history and distorts the saved history. It assesses not only a loss of the Past, but assesses a distort the view in the Past, making widespread illusions of trends. The ILL is valid to rivers, lakes, sea, and ice sediments layers, and to geomorphology forms (river terraces and terrace ranks). Observation and statistic modeling of the erosion and erosion-sedimentation processes

(Klenov, 1980) was resulted in that natural records can't be any foundation for restoration of a history of the past Morphodynamics and of the Past exterior powers. The Information Loss Law (ILL) confirms that: restoration of the Past by natural records does not have single decisions. The natural records reflect only the major property of the Nature Systems Morphodynamics, by that most records of past events are self-destroyed by the System, by oscillations of a river bed width and cutting, meandering, and others. It was received by statistics modeling of sedimentation/erosion process and river ranks forming over a river valley's cross-sections and by field observations in intermountain depressions. The computed records do not contradict to observed data (Klenov, 1980, 1981). The later high resolution Virtual Systems also confirm it (Klenov, 2003).

There are two types of processes oscillations: High Frequency and Low Frequency. The Low Frequency oscillations contain vast volumes of sediments, which slowly 'move' downstream. They were computed and observed by research of river beds widening (Klenov, 1999b). The High frequency morphodynamics forms micro-relief of streams, slopes, and sediment layers.

2) The Information Storage Law (ISL) confirms: All information about the Past of a Nature System is dissipated over a whole system. The Nature System remembers all it's Past, which is 'hidden' in it surfaces and theoretically may be restored by a computing back in the Time by the VNS (Klenov, 1987).

The ISL does not disagree with belonging to Open Non Equilibrium Systems (Prigogine et al., 1984). The ONES are under continual pressure of exterior power, which cause variable flows through nature systems. Increasing of external power increases the system non-stability by generation of oscillations and thresholds, and by catastrophic flows. Stochastic morphodynamics wipe out own former layers and dissipates own memory over the system's structure, with the single way to restore the history by moving back in the Time along own phase trajectory. This is common property of the ONES.

3) The concrete Nature Systems are changeable. The VNS – River Basin skills for self-organization and has a peculiarity by steadiness of river net structure, in spite (and by) oscillations and thresholds, what makes possible to evaluate any stochastic spatial-temporal external influences and to determine all sites of hazardous flows. The Coastal Zone is non-stable due changeable distribution and power of wind-wave energy and currents, and does not make structures like a river net, but makes specific structure across coastal zone (terraces along coast line).

For the purpose of computing the VNS Morphodynamics were worked out relevant methods as follow: the String Genetic Coding, and the Genetic Matrixes method for the River basins; for the VNS – Coastal Zone it was a join of the Genetic Matrixes and the Oblique Scanning; for the tsunami it is the Circle Scanning of the Genetic Matrixes.

REFERENCES

Hiromochi Fukui. 2003. From Digital Earth to Digital Asia. *GIS Next. Digital Earth, its' ideal, situation and future,* 9, pp 11–13.

John Mitchell, Hans Joachim Schellnhuber, Chris Rapley. 2001 Simulating and Observing the Earth System. *Challenges of a Changing Earth.* Springer, the Netherlands, pp. 144–159.

Klenov V.I. 1980. Some bylaws of terrace ranks forming. Vestnik MGU, Geography, 4, 53–58 (in Russian).

Klenov V.I. 1981. Modeling of correlations between structure and history of river valleys. Vestnik MGU, Geography, 5, 84–88 (in Russian).

Klenov V.I. 1987. On the methods of automatic reconstruction of the relief history. Geomorphologia, 1, 39–43 (in Russian).

Klenov V.I. 1997. Simulation of coastal zone morphodynamics and of oil slicks. *Third LOICZ Open Science meeting. Global Change Science in the Coastal Zone,* Noordwijkerhout, The Netherlands.

Klenov V.I. 1999a. River Basin Simulation for flood forecasting and management. *Coping with Floods: Lessons Learned from Recent Experience. Proceedings of NATO Advanced Research Workshop.* Malenovice, Chech Republic,

Klenov V.I. 1999b. Simulation of River Basin and Coastal Zone morphodynamics and evolution. *Proceedings on IAHR Symposium on River, Coastal and Estuarine Morphodynamics, (on CD-ROM),* Genova, Italy, 8 p.

Klenov V.I. 2003. Debris-flow recognition using an extended version of the river basin simulation model. *Debris – Flow hazards mitigation: Mechanics, Prediction, and assessment,* Millpress, Rotterdam, 139–145.

Klenov Valeriy. 2005. Estimation of Future Disasters. *The First International Symposium on Geoinformation for Disaster Management, Late Papers.* Delft University of Technology, the Netherlands, pp. 71–74.

Klenov Valeriy. 2006. The Moving Digital Earth Technology (MDE) for monitoring of Forthcoming Disasters, *Proceedings of the 3rd International ISCRAM Conference,* Newark, USA, pp. 17–23.

Moutaz Dalati 2005. Applications of Remote Sensing to Geological Hazards: Case Study Detecting the Active Faulting Zones NW of Damascus, Syria. *Geo-information for Disaster Management.* Delft University of Technology, the Netherlands, pp. 59–64.

Prigogine I., Stengers I. 1984. Order out of chaos, Heinemann, London.

Timothy B. Love. 2005. The Use of GIS Technologies within the NOAA Climate Predictions Center's FEWS-NET Program. *Geo-information for Disaster Management. Late papers,* Delft University of Technology, the Netherlands, pp. 365–378.

*River, Coastal and Estuarine Morphodynamics: RCEM 2007 – Dohmen-Janssen & Hulscher (eds)*
*© 2008 Taylor & Francis Group, London, ISBN 978-0-415-45363-9*

# Formation process of depositional landform over a permeable flat plain

M. Ogasawara & M. Sekine

*Department of Civil and Environmental Engineering, Waseda University, Tokyo, Japan*

ABSTRACT: The formation process of depositional landform over a one-dimensional permeable flat plain was investigated numerically in this paper. In such a process, a water exchange between a surface flow and a subsurface flow is important to be taken into account. The governing equations of the flows are one-dimensional shallow water equations for surface flow, and two-dimensional Richards' equation for subsurface flow. As a result of this study, the following two processes were confirmed: (a) a front of surface flow migrates a little in the downstream direction due to the loss of surface-flow discharge, and the thickness of the depositional landform increases; (b) the front migrates with relatively high speed due to the recovery of surface-flow discharge, and the depositional landform spread in the downstream direction.

## 1 INTRODUCTION

If the water which carries a sand or a gravel flows on a wide plain or a permeable plain, depositional landform is developed. Alluvial fan is the typical example of this landform. In the formation process of it, the water exchange through a ground surface between a surface flow and a subsurface flow is important and it has to be considered when we understand this process or mechanism behind this process.

Studies on the formation process of alluvial fans have been conducted experimentally over four decades by some researchers (for example, Hooke 1967, Schumm 1977). In these studies, depositional landforms in a laboratory scale were studied and a valuable piece of information was reported although some discussion about the scale effect was required. Second author of this paper also studied experimentally the formation process of depositional landforms on a permeable flat plain (Sekine et al 1998). It was reported by them that the pattern of sediment deposition and the expansion of the landform has the feature like a chaotic behavior due to the loss of water discharge.

The purpose of this study is to develop a numerical model which enables us to simulate the formation process of depositional landform. To understand the mechanism of this process more precisely is another purpose of this study. One-dimensional numerical model was constructed here as a first attempt. In this computation, the target to be analyzed here is a phenomenon which occurs in the flume that the bottom of the upstream half reach is impermeable and that of the downstream half is permeable. In the latter reach, there exists a relatively deep "permeable layer" just like an aquifer. This permeable layer is constituted of gravels.

And a "sand layer" with a small thickness is set in the entire reach, that is, on the permeable layer as well as on the bottom of the upstream reach. The water which is supplied at the upstream end flows downstream on the surface of this sand layer. As time goes on, the front of the flow migrates downstream, and it reaches the downstream half reach. After that, some amount of water inflows into the permeable layer and the sediment which is carried by the flow deposits. Finally, depositional landform develops. Sekine et al. (1998) conducted such an experiment although it was a two-dimensional experiment instead of a one-dimensional one. The computation here was the first attempt to simulate the phenomenon which was observed by them.

## 2 NUMERICAL SIMULATION MODEL

### 2.1 Summary of the computation

In this study, the computational condition was set by referring to the experiment by Sekine et al. (1998). Figure 1 shows an illustration of computational

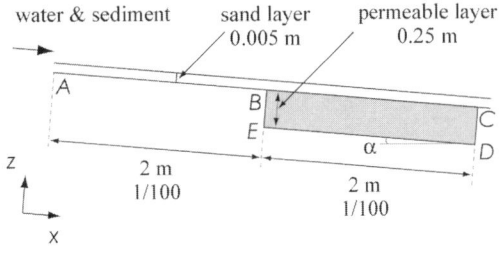

water & sediment      sand layer      permeable layer
                      0.005 m          0.25 m

Figure 1. Illustration of computational domain (Initial condition).

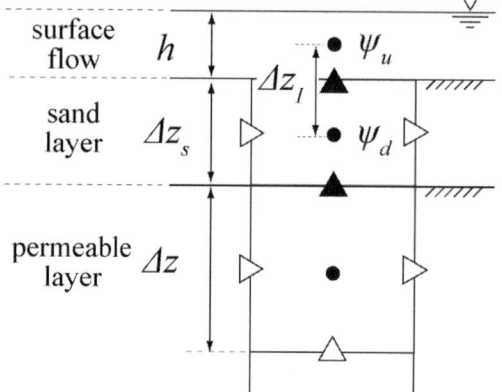

Figure 2.   Grid system of flow field.

domain. The area of BCDE in Figure 1 is defined as "permeable layer". As is shown in Figure 1, the length and the thickness of this layer are 2 m and 0.25 m, and the gradient of it is 0.01. This layer is constituted of gravel whose grain size is 5 mm. An impermeable flume is connected to the permeable layer at the point of B. The length of it is 2 m and the gradient is same as that of permeable layer. So a permeable layer is set between two points at x = 2 and 4 (m). The impermeable flume and the permeable layer are covered with the sand whose grain size is 0.48 mm. This layer is called here a "sand layer" whose thickness was 0.005 m in the initial condition. Under the condition of this numerical computation, this sand bed is movable and sand can be transported as bed load. The gravel which constitutes a permeable layer is not exposed to a surface flow and does not move at all.

Boundary conditions are as follows. No inflow or outflow occurs through the lines of AB, BE, ED and DC in Figure 2. At the upstream end of this computational domain, a constant discharge of water was supplied, and the unit discharge per width was 0.002 m$^3$/s/m. Sand was also supplied there by the amount which was estimated by substituting a dimensionless tractive force at the upstream end into a bedload function.

Initial Conditions are as follows. Dry bed condition was given for surface flow. The free surface in permeable layer was made horizontal and the elevation of it was set 0.05 m at x = 4 (m) (see Figure 3(a)). And the initial velocity of subsurface flow is zero.

Table 1 shows hydraulic parameters of a permeable layer and a sand layer. The exponent $m$ in Equation 5 is 3.0, which was estimated theoretically by Irmay (1954). And the value of Manning's coefficient was 0.03, and the value of dimensionless critical tractive force for sand was 0.032.

Table 1.   Hydraulic parameters of permeable layer and sand layer.

| | Grain size (mm) | $\theta s$ | $\theta r$ | $\psi_0$ (m) | $Ks$ (cm/s) | $m$ | $Ss$ (m$^{-1}$) |
|---|---|---|---|---|---|---|---|
| Permeable layer | 5.0 | 0.40 | 0.001 | −0.05 | 5.0 | 3.0 | 10$^{-6}$ |
| Sand layer | 0.48 | | | −0.15 | 0.1 | | 10$^{-5}$ |

### 2.2  Surface flow

In this study, a depth of surface flow is so small (less than 13 mm) that so-called "shallow water assumption" is effective. Therefore governing equations of surface flow are one-dimensional shallow water equations;

$$\frac{\partial h}{\partial t} + \frac{\partial hu}{\partial x} = w_I \tag{1}$$

$$\frac{\partial u}{\partial t} + u\frac{\partial u}{\partial x} + \frac{uw_I}{h} = g\sin\alpha - g\cos\alpha\left(\frac{\partial h}{\partial x} + \frac{\partial \eta}{\partial x}\right)$$
$$-\frac{C_f}{h}u^2 + \frac{\partial}{\partial x}\left(v_t\frac{\partial u}{\partial x}\right) \tag{2}$$

where $h$ is a flow depth, $u$ is a depth averaged velocity, $g$ is a gravitational acceleration, $\alpha$ is the angle between the x coordinate and a horizontal line, $\eta$ is a bed elevation, $v_t$ is a turbulent diffusion coefficient, respectively. $w_I$ is a flux (volumetric flow rate per unit area) through an interface between a surface flow and a subsurface flow. If the inflow from surface flow to subsurface flow occurs, the $w_I$ takes a minus value. If the outflow from subsurface flow to surface flow occurs, on the other hand, $w_I$ takes a plus value. $C_f$ is a frictional coefficient and was evaluated by Manning's law in this study.

### 2.3  Subsurface flow

A water flow in unsaturated porous media is governed by a continuity equation and the equations of flow velocity which is a modified Darcy's law. Substituting the equations of flow velocity into a continuity equation, Richards' equation is obtained. We can evaluate the flow in both unsaturated porous media and saturated ones by Richards' equation. In this study, the phenomenon in a vertical x-z plain was to be analyzed and a following Richards' equation was solved numerically;

$$\left(\frac{\partial \theta}{\partial \psi} + \beta S_s\right)\frac{\partial \psi}{\partial t}$$

$$= \frac{\partial}{\partial x}\left\{K\left(\frac{\partial \psi}{\partial x} - \sin\alpha\right)\right\} + \frac{\partial}{\partial z}\left\{K\left(\frac{\partial \psi}{\partial z} + \cos\alpha\right)\right\} \tag{3}$$

where $\theta$ is volumetric water content, $\psi$ is a pressure head, $S_s$ is a specific storage coefficient, $K$ is a hydraulic conductivity, respectively. The x and z coordinates are taken as is seen in Figure 1. $\beta$ is the coefficient which is defined as follows; when the value of $\psi$ is plus or zero, $\beta = 1$; when the value of $\psi$ is minus, $\beta = 0$.

In order to solve the Richards' equation, we need some relationship among the parameters $\psi, \theta$ and $K$. In this study, the relationship by Tani (1982) between $\psi$ and $\theta$ was adopted, and the same relationship between $\theta$ and $K$ as Tani (1982) used in his analysis was adopted.

$$\theta = \left(\theta_s - \theta_r\right)\left(\frac{\psi}{\psi_0} + 1\right)\exp\left(-\frac{\psi}{\psi_0}\right) + \theta_r \qquad (4)$$

$$K = K_s\left(\frac{\theta - \theta_r}{\theta_s - \theta_r}\right)^m \qquad (5)$$

where $\theta_s$ is saturated water content, $\theta_r$ is residual water content, $\psi_0$ is a pressure head which gives the maximum value of $\partial\theta/\partial\psi$, $K_S$ is a saturated hydraulic conductivity, $m$ is an exponent in Equation (5).

### 2.4 Modeling of a water exchange

In this computation, a following vertical structure is assumed; (a) a permeable layer which is constituted by gravel, (b) a sand layer on it, and (c) the space for surface flow above a sand layer. Subsurface flow was calculated in the two layer of (a) and (b). Water exchange through one of two interfaces, one of which is between (a) and (b), and the other is between (b) and (c) was evaluated in a following manner.

The flux through the upper surface of a sand layer, which corresponds to the one between (b) and (c), is defined as $w_I$ in Equations (1) and (2). This flux is evaluated on the basis of so-called Darcy's law and is computed by

$$w_I = \begin{cases} -K_d\left(\dfrac{\psi_u - \psi_d}{\Delta z_I} + \cos\alpha\right); & if \quad h > 0 \\ 0; & if \quad h = 0 \end{cases} \qquad (6)$$

where $K_d$ is a saturated hydraulic conductivity of a sand layer. In this computation, $\psi_u$ or $\psi_d$ is defined at the point of half depth or the midpoint of a sand layer. And $\Delta z_I$ is the vertical distance between these two points (see in Figure 2). $\psi_u$ and $\Delta z_I$ are estimated as follows;

$$\psi_u = 0.5h\cos\alpha, \qquad \Delta z_I = 0.5\Delta z_s + 0.5h$$
$$for \quad 0 < h < \Delta z \qquad (7)$$

$$\psi_u = (h - 0.5\Delta z)\cos\alpha, \quad \Delta z_I = 0.5\Delta z_s + 0.5\Delta z$$
$$for \quad h \geq \Delta z \qquad (8)$$

where h is a flow depth, $\Delta z_s$ is the thickness of sand layer which is not a constant and can vary due to the deposition or erosion of sand.

The flux through the interface between a sand layer and a permeable layer was evaluated in the same manner as explained above. The hydraulic conductivity at the interface is evaluated to be a smaller value of the two, that is, the value of sand layer or the one of permeable layer.

### 2.5 Sediment transport

Bed load is only considered here as the type of sediment transport. Bed load transport rate was evaluated by Meyer-Peter and Muller's formula. "Slope collapse model" by Sekine (2004) was introduced to evaluate an additional sediment transport rate. Local angle of a depositional landform grows larger as it develops, and then the angle reaches tentatively an angle of repose of sediment in this computation. But actually slope collapse occurs before the above condition is satisfied. We can consider such an event reasonably by this slope collapse model. Local angle has to be maintained in order not to exceed the angle of repose. And the sediment volume yielded by a collapse is treated as an additional sediment transport. The bed elevation at each grid point is calculated by solving the Exner's equation, and the additional sediment transport as well as that by Meyer-Peter and Muller's formula is considered. For further details, the reader may refer to the original paper by Sekine (2004).

### 2.6 Summary of computation

The procedure of this computation is as follows; (1) a subsurface flow was analyzed first, (2) a surface flow is analyzed by considering the value of $w_I$ which is estimated by using a latest value of pressure head of subsurface flow, (3) after such computations of the flow, the deformation of bed elevation is evaluated. These are the computation in one time step. The time interval $\Delta t$ is set to be 0.0005 s in this analysis.

In solving governing equations (1) and (2) for surface flow, the following individual solution techniques were adopted: (a) the staggered grid system, (b) explicit scheme, (c) the CIP scheme for convective term. In solving equations (3)–(5) for subsurface flow, on the other hand, the fully implicit scheme and the successive over relaxation (SOR) method were adopted.

## 3 FORMATION PROCESS OF DEPOSITIONAL LANDFORM

The formation process of depositional landform is discussed in this section referring to the numerical results which are seen in Figure 3 and Figure 4.

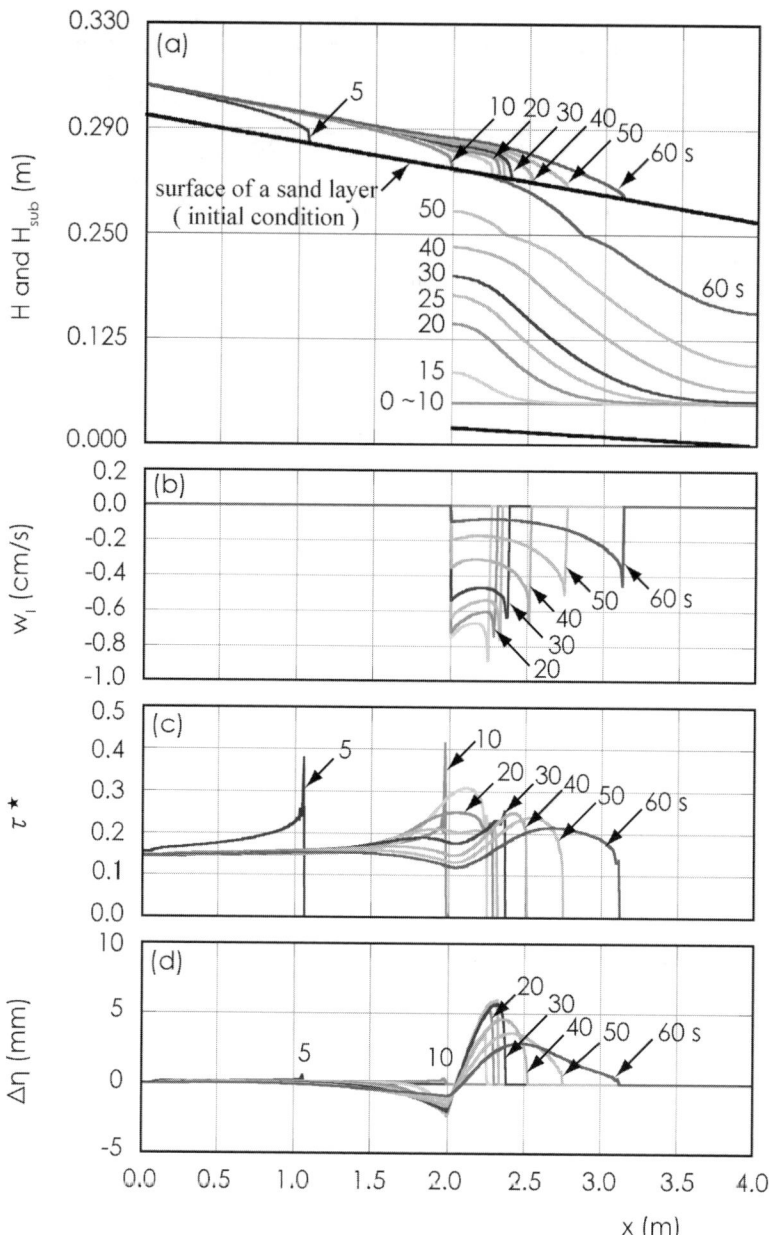

Figure 3.  Numerical results of the formation of depositional landform: (a) water surface elevation of surface flow $H$ and free surface elevation of subsurface flow $H_{sub}$, (b) the flux between surface flow and subsurface flow $w_I$, (c) dimensionless tractive force, (d) the amount of bed elevation change from initial bed. The number in each figure denotes a time from the beginning of computation.

In Figure 3(a), a time series of water surface profiles are seen. The upper lines correspond to the profiles of surface flow and the lower lines correspond to the free surface profiles of subsurface flow. We can see from this figure that a water front of surface flow migrates downstream but the movement is not continuous. The pattern of surface flow can be summarized as follows: (1) A front of surface flow migrated with almost constant speed until the front reaches the point $x = 2$ (m) at which there exists an upstream boundary of the

888

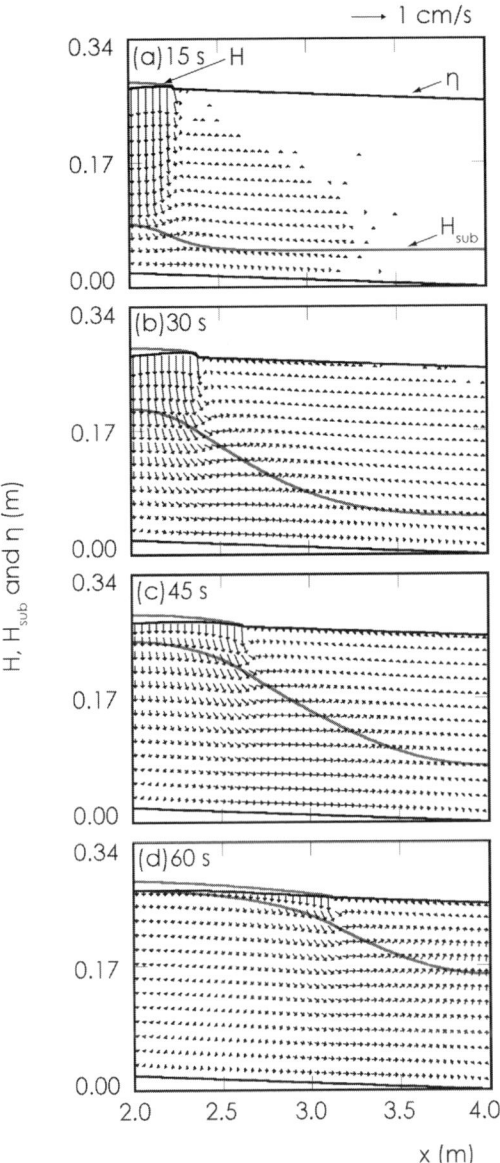

→ 1 cm/s

Figure 4. Velocity vectors of subsurface flow: (a) 15 s, (b) 30 s, (c) 45 s, and (d) 60 s from the beginning of computation.

permeable layer. The reason is that a water discharge keeps constant and no net loss of water discharge occurs due to the impermeability of the approaching channel bed. (2) The front reached the point $x = 2$ (m) at about 10 s, and then it migrated slightly in downstream direction till 25 s approximately. It looks like stopping during this period. (3) After that, the front started to migrate again with a relatively high speed.

Figure 3(b) denotes a temporal and a spatial variation of the flux $w_I$ through the interface between a surface flow and a subsurface flow. Figure 4 shows the velocity vector maps of subsurface flow. It is obvious from these two figures that the influx from a surface flow to a subsurface flow is dominant and causes the increase of moisture in the permeable layer. This results in the rise of free surface of water in this layer. Discharge of surface flow, on the other hand, decreases in downstream direction over a permeable layer due to this influx. But it is also recognized from these two figures that the flux $w_I$ decreases and the discharge loss of surface flow grows smaller as time goes on.

In Figure 3(c) and Figure 3(d), the variations of a dimensionless tractive force $\tau^*$ and the amount of bed deformation $\Delta\eta$ are seen. Now a bed is equivalent to the surface of landform here. The profiles of $\tau^*$ coincides with the pattern of the flux $w_I$, as was explained above. Figure 3(d) shows that depositional landform develops on the permeable bed after the front of surface flow reached the upstream end of the bed ($x = 2$(m)). Sediment deposition occurs in the reach where $\tau^*$ decreases in downstream direction, and the erosion, on the other hand, occurs in the reach where $\tau^*$ increases. The pattern of development can be summarized as follows; (1) the deposited area extends further in downstream direction and the landform migrates downstream, (2) the landform in the upstream reach is re-eroded and the thickness of it grows smaller. Such a phenomenon became definite after 25 s approximately when the water front of surface flow started to migrate again.

As is seen from the results explained above, there are two stage of development of depositional landform especially in the one-dimensional case. One is the initial stage, and a water front of surface flow almost stays still and does not migrate considerably in the downstream direction due to the discharge loss of surface-flow. In this stage, depositional landform develops in a restricted area, but the thickness of it grows larger as time goes on. The other is the second stage where the front migrates downstream. In this computation, this stage begins at about 25 s. The discharge of surface flow is much larger than that of inflow into a permeable layer in this stage, and the depositional landform continues to expand in such a manner as already explained.

Under the condition of this study, the above two sub-processes including a transition from one to the other were shown. But it was found from the results of more computations under different condition that only one of the two sub-processes appeared. For example, if the vertical walls at upstream and downstream end of the permeable layer (lines BE and CD in Figure 2) were changed into permeable one, a free surface of subsurface flow rose so little that the former process only appeared. If the free surface in initial state was close to the upper end of permeable layer, on the other hand, the discharge loss of surface flow was so little that the second stage only appeared and the depositional landform did not develop significantly. Under this condition, some other aspects were observed. For example, "return flow" occurs from a certain point toward a downstream direction. So-called "wadi" or "dried-up river" also appears in the upstream reach from this point.

## 4 CONCLUSION

The formation process of depositional landform over a permeable flat plain was investigated in this study. A numerical simulation model was developed to simulate the process over a one-dimensional plain as a first attempt. In conclusion, it was verified that the numerical model worked well, and the mechanism of the process was understood to some extent and discussed precisely. Following two sub-processes were confirmed:

(a) In the initial stage, a front of surface flow migrates a little in the downstream direction due to the discharge loss of surface-flow, and the thickness of depositional landform increases.

(b) After that, the front migrates with relatively high speed due to the recovery of surface flow discharge, and the depositional landform spread in the downstream direction.

In the next step, the numerical model will be extended to a two-dimensional one in order to simulate an alluvial fan formation, just like the experiment by Sekine et al. (1998).

## REFERENCES

Hooke, R.L. 1967. Process on arid-region alluvial fans. Journal of Geology Vol.75: 438–460.

Irmay, S. 1954. On the hydraulic conductivity of unsaturated soils. Trans. AGU. Vol.35: 463–467.

Schumm, S.A. 1977. The Fluvial System. New York, Wiley: 338.

Sekine, M., Arai, T., and Kubota, Y. 1998. Fluvial Depositional Landforms on Permeable Plain. Annual Journal of Hydraulic Engineering, No.42: 1087–1092. (in Japanese).

Sekine, M. 2004. Numerical simulation of braided stream formation on the basis of slope-collapse model. Journal of Hydroscience and Hydraulic Engineering. vol.22, No.2: 1–10.

Tani, M. 1982. The properties of a water-table rise produced by a one-dimensional, vertical, unsaturated flow. Journal of Japanese Forest Research. Vol.64: 409–418 (in Japanese).

*River, Coastal and Estuarine Morphodynamics: RCEM 2007 – Dohmen-Janssen & Hulscher (eds)*
*© 2008 Taylor & Francis Group, London, ISBN 978-0-415-45363-9*

# Correlation decay and dynamic equilibria in sandy transport systems

B. McElroy & D. Mohrig
*Jackson School of Geosciences, University of Texas, Austin, Texas, USA*

ABSTRACT: Active deformation of sandy river bottoms during periods of steady discharge represents a microscale morphodynamic free behavior. Coupled spatial variation of sediment transport and topography results in bedforms that continuously deform as they migrate, resulting in a statistical steady-state bed elevation field. We propose that the sediment-fluid interface is evolved by two processes, translation and deformation. Total evolution is captured by correlations between successive profiles. Cross-correlation distances and associated evolution durations serve to estimate rates of mean bed translation. The magnitude of cross-correlation serves as a measure of bed memory during its evolution and deformation. The correlation coefficient initially decays with an exponential form. This suggests a natural length scale, the interface half-life, which describes the decoupling of the bed from its previous states. In our study this length was less than half of the characteristic length of the bedforms. Overall, this analysis reveals the relevant length and time scales of bed deformation in sandy systems.

## 1 MOTIVATION

Anyone who has ever sat and watched bedforms migrate in a lab, on a beach, or anywhere else with a consistent flow of water over a sandy substrate has noticed that each crest and trough combination is at least slightly different in size and shape than all its neighbors. With a little patience they also notice that the bedforms change their shape and arrangement as they move along. This active deformation of sandy river bottoms during periods of steady boundary conditions represents a microscale morphodynamic free behavior. Even though the sizes or scales of ripples and dunes are largely constrained by the flow conditions (Southard & Boguchwal, 1990), the exact details of sand bed topography are independent of the mean flow properties (Jerolmack & Mohrig, 2005a) as well as being path dependent (Allen 1973; Jerolmack & Mohrig, 2005a). In this condition dynamic equilibrium is evident from a statistical steady-state of the bed elevation field. Its relevant measures are a characteristic height and length (McElroy et al., *in review*; Jerolmack & Mohrig, 2005b; Nordin, 1971). However, coupled spatial variations of transport and topography result in continuous net translation and deformation of bed topographic elements.

We therefore seek a method to resolve the amount of information preserved during the evolution of a bed (translation), the complement to which is the amount of information added during evolution (deformation). Many have observed that bedforms migrating under constant forcing tend to maintain their dimensions,

i.e. equilibrium conditions. (Allen, 1970; Southard, 1975; McLean, 1989) Our data (Fig. 1A) from the North Loup River near Taylor, Nebraska (after Mohrig, 1994) also show an equilibrium condition, and additionally, they exhibit deformation in such a manner that maintains the inherent statistical properties of the bed within a narrow range (Fig. 2). This is quantified by a coefficient of variation, $c_v$, which is the sample deviation normalized by the mean. (Davis, 1986) In this case for the characteristic height and length, $c_v = 0.08$, and $c_v = 0.06$, respectively. These imply that there is little variation in the characteristic height and length of bed topography between the sequentially collected profiles even though the range of heights and lengths for individual bedforms is large.

While the evidence for statistical steady-state described above accounts for the overall character, it does not address the internal deformation of a bed explicitly. We posit that over very short durations, the amount of topographic deformation is quite small and that an evolved bed can look qualitatively like a replicate of its precursor bed. In contrast, evolution over longer durations results in a divergence of form while completely preserving the salient statistical properties. This is the process we intend to quantify.

## 2 DATA

The data used in the analyses presented here were collected with low-altitude aerial photography over the North Loup River in Nebraska (Mohrig, 1994). The

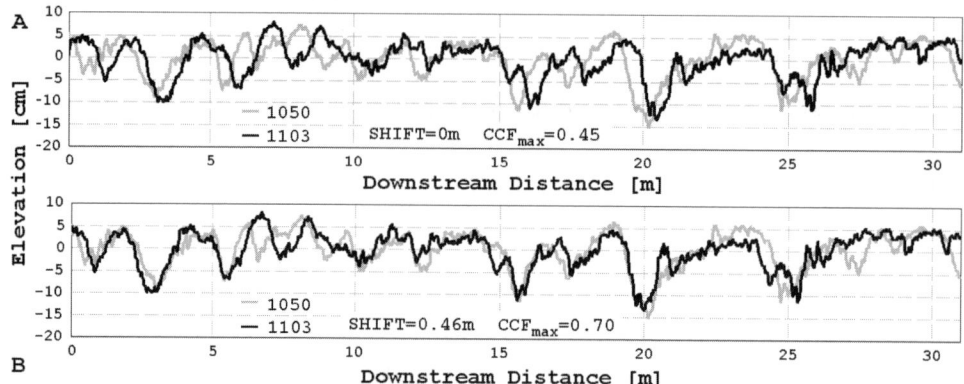

Figure 1. A) Two detrended bed-elevation profiles collected at the same location, one 13 minutes before the other. Detrending removed the local mean elevation and a slight downstream shallowing. B) The same two profiles as in A. The evolved profile, 1103, is shifted relative to the initial profile, 1050, by the distance that results in the maximization of the cross-correlation function (CCF). Note that some regions show large fractions of direct overlap (translation dominant) while other regions exhibit very little overlap (deformation dominant). Flow is from left to right.

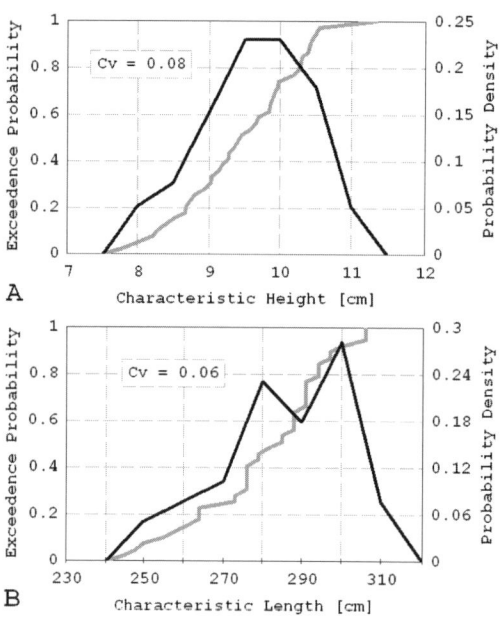

Figure 2. Characteristic height and length for bedform fields in the N. Loup R. collected from 40 detrended elevation profiles. A) Distribution of characteristic heights and B) Distribution of characteristic lengths. The coefficients of variation (Cv) are small, suggesting that these distributions are drawn from a single stationary population of bed configurations.

40 minute sequence of images obtained at 1 minute intervals has recently been rectified and registered to a coordinate system created from a survey grid constructed in the river. Bed elevation fields were calculated from the light intensity of the river bottom images by taking advantage of the attenuation of light trough the water column. An outline of the processing method is found in Jerolmack & Mohrig (2005a) and complete details as well as the data will be published shortly.

The resulting data is a complete assessment of the bed in four dimensions: x, y, z, & t, streamwise, cross-stream, vertical, and time, respectively. The horizontal resolution is 2 cm. The vertical resolution is 1 mm, with a nominal accuracy of $+/-2$ mm, and the temporal resolution is 1 minute. The lengths along x & y are 31 and 15.5 meters respectively. For our purposes here, we are working with a single profile drawn from the middle of the bedform field. This profile has been detrended to remove the local mean bed elevation as well as a slight downstream shallowing. The data are therefore centered around zero but still express the original range of the bed elevation field (Fig. 1). The whole profile comprises about 10 dune scale bedforms and many smaller bedforms. The characteristic height and length are 11 cm and 2.9 m, respectively. (McElroy ct al., *in review*)

## 3 METHODS

We are interested in quantifying the evolution of the bed topography and thereby assessing the amount of memory in a bed and its dependence on previous states. To achieve this, profiles of the bed that have evolved through a time, $\Delta t$, are compared. (Fig. 1) The result is two major metrics: a profile averaged translational velocity and a profile averaged deformational rate.

A fundamental formulation for comparison is Pearson's product-moment coefficient, $R$, (Eq. 1; Davis, 1986) where $\eta$ is the bed elevation, and $\sigma$ is

$$R = \frac{Cov(\eta_1, \eta_2)}{\sigma_1 \sigma_2} \qquad (1)$$

the standard deviation of elevations. It is the covariance of the two datasets normalized by the product of their deviations. For the case such as ours in which the mean of the set has been subtracted from the data, equation 1 can be reduced to equation 2.

$$R = \frac{\sum \eta_1(x) \times \eta_2(x)}{\sigma_1 \sigma_2} \qquad (2)$$

Correlation provides a measure of similarity or dissimilarity for two data series by summing the normalized product of their values over their shared domain. For very similar series the $R$ value approaches unity and for diametrically dissimilar sets it tends toward zero.

We use correlation to quantify the similarity between an evolving set of profiles. If for instance a bed evolves by translating, then it becomes less correlated with itself as the peaks and troughs move away from their original positions. For this case the R statistic would decrease and even become negative when troughs of the evolved profile began to coincide with peaks of the original profile. However, if the coordinates' origin translates with the profile, then no decrease in correlation would be observed. This can be thought of as shifting the profile back in space to its original location before it began to evolve. To accomplish this mathematically, a parameter that acts as a phase shift for periodic systems is introduced to the correlation function. The result is the cross-correlation function (Eq. 3; Davis, 1986). Where

$$CCF(L) = \sum_{x=0}^{D} \frac{\eta_1(x) \times \eta_2(x+L)}{\sigma_1 \sigma_2} \qquad (3)$$

$D$ is the domain of the bed profiles and L is the length by which the evolved profile is shifted relative to its original position. This method and similar ones have been used previously by Nikora et al. (1997) and Rubin (1992) in the study of sandy bedforms. By performing a cross-correlation and searching for the shift length, $L$, at which the $CCF(L) = 1$, the translational distance of the bed profile is found.

The CCF is also quite useful in characterizing a sandy bed that undergoes both deformation and translation throughout its evolution. In this case the CCF value is still constrained $CCF(L) \in \{1, -1\}$, however the maximum value after bed evolution is less than one

because the topography deforms as it translates. Deformation is the result of bed material transfer between bedforms and to a minor extent within bedforms. It is manifest as a change in the shapes and relative spacing of bed topographic elements. As long as the amount of bed deformation is relatively minor, the translational length can still be estimated robustly as the shift length, $L$, that yields the maximum value for the $CCF(L)$. At small translation lengths, deformation will similarly be slight, and the CCF will have a maximum value very near unity. At longer durations the CCF maxima decrease and broaden as a result of an increasing degree of bed deformation. (Fig. 3)

Finding the CCF maxima for an evolutionary series of bed topographies provides a method for relating the secular variation in position and shape of a bed profile. Associated with the maximum of the CCF for each profile pair is an interval of time and the length of translation. (Fig. 3) As long as the relation between the translation length and the duration is linear, a characteristic velocity for the bed profile is straightforward to calculate (Fig. 4). Its general formulation is a linear least-squares regression of the covarying translation lengths and times.

In addition, the relationship between the maximum CCF value and time interval can be investigated in order to determine the amount of memory associated with the bed topography itself. As the bed translates and deforms, some amount of its state can be predicted from it previous state; the remainder is ascribed to a new condition imparted by the transport process. That process is manifest as a secular decorrelation of the bed topographies (Fig 3).

## 4  DISCUSSION

Linear regression of the relation between bed translation length and time interval yields a characteristic velocity, $V_c$ (Eq. 4). In our case from the

$$L_T = V_c \Delta t \qquad (4)$$

N. Loup, $V_c = 3.7$ cm/min for the maxima of 39 CCFs that represent 40 minutes of bed evolution. That relation explains some 99% of the covariance of translation length, $L_T$, and evolution duration, $\Delta t$. The very high degree of goodness-of-fit for the characteristic velocity is due to the bed topography sample size and its resolution. Sample size is measured as the total length of the bed profile sample relative to the characteristic length of the bed, $L_c$, and the resolution is the distance between measurements relative to the characteristic length. For our measurements the total profile length is $D = 31$ m, and with $L_c = 2.9$ m, the sample size is approximate 10 characteristic bed lengths. Similarly, the elevation measurement spacing is $\Delta x = 2$ cm and

893

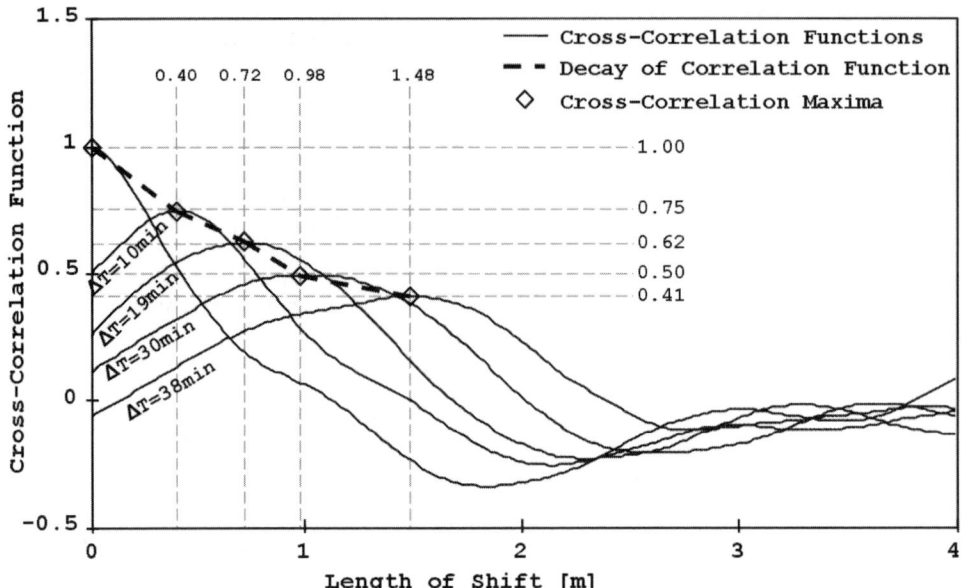

Figure 3. Cross-correlation functions (CCF) for a series of profile sets; the elapsed time is labeled as ΔT. The unlabeled CCF is the initial profile correlated with itself (autocorrelation). The maxima for the CCFs are associated with two other pieces of information: elapsed time between profiles and the shift lengths that result in maximization. Note that the CCF maxima decrease in magnitude but broaden in width during bed evolution.

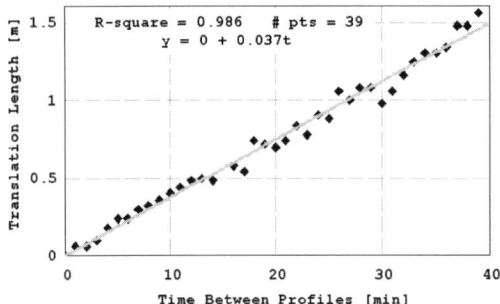

Figure 4. Plot of bed profile translation length as a function of bed evolution duration. Regression yields a characteristic velocity for the bed of 3.7 cm/min and intercepts the origin.

therefore the resolution is ~150 measurements per characteristic length. We surmise that greater resolution and greater sample size both correspond to a more robust characteristic velocity estimate.

Inspection of Figure 1 demonstrates two salient features relevant to the evolution of sand bed topographies. First, that there are subregions within the domain that could be locally characterized by greater or lesser velocities than the profile as a whole. Second, that deformation occurs in minor amounts everywhere

but also to quite significant extents locally. The first type of behavior is a product of the relative translation rates of individual topographic elements. In Figure 1 the bedform at ~24 m downstream has stalled relative to the rest of the profile. The result is a change within the overall arrangement of the bed and a decrease of the CCF maximum for the bed as a whole. This is visible in Figure 1B as ~0.5 m overlap between the bedform peak and trough. The second behavior, located around 6–7 m downstream, is localized deformation that is evidenced by a major reorganization of the bedforms. Over the 13 minutes of evolution, that particular section of the bed merged three large-scale topographic elements into two. The major result is a decrease in correlation; again the CCF maximum decreases because peaks and troughs have greater overlap. This is all in addition to the relatively diminutive elevation changes, centimeter-scale bed forms, trough scours, etc., that occur along the whole profile and also contribute to decorrelation.

The agglomerate of all the kinds of deformation has the effect of reducing the CCF maxima, and the size of that reduction can be ascribed to secular and spatial variations in topographic change. Using the characteristic velocity as a transformation from evolution time to translation length, the decorrelation can be investigated as a function of distance traversed by the bed

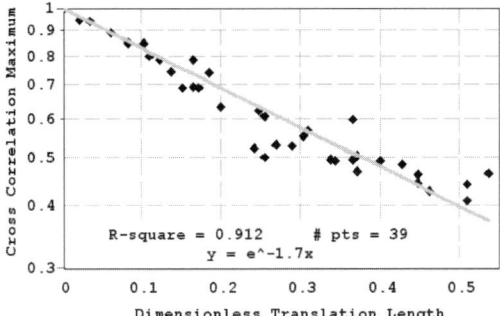

Figure 5. Log-linear plot of CCF maxima as a function of dimensionless bed translation length (Eq. 5). This data confirms an exponential decay of correlation and demonstrates that the bed decouples from its previous states rapidly relative to translation.

topography. An appropriate dimensionless translation length, $L_T^*$, is calculated by normalizing the translation length by the characteristic length of bed topography (Eq. 5).

$$L_T^* = \frac{V_c \Delta t}{L_c} \qquad (5)$$

CCF maxima, $CCF_{max}$, decrease exponentially as a function of $L_T^*$ (Eq. 6; Fig. 5). For the same

$$CCF_{max} = e^{-1.8 L_T^*} \qquad (6)$$

39 CCFs that represent translation of just over half a characteristic length, regression produces an exponential function with a decay constant, $d$, greater than unity. This relation explains over 90% of the covariance between the $CCF_{max}$ and $L_T^*$.

The significance of the magnitude of the decay constant is in its relevance to the rate at which the bed topography becomes decoupled from previous conditions. For small steps $L_T^*$ the relation can be discretized through $d$ with only small errors. (e.g. for $L_T^* = 0.1$, $CCF_{max} = 1 - (1.8 \times 0.1) = 0.83$) In the spirit of decay processes of radionuclides, an interfacial half-life, $\lambda$, can be defined which has no units but is written in terms of the dimensionless translation length (Eq. 7). In this case, by the time

$$\lambda = \frac{\ln(0.5)}{-1.8} = 0.39 \qquad (7)$$

the bed has translated 40% of its characteristic length, it is only 50% correlated with its original state.

Over longer evolution durations, the magnitude of correlation decay cannot continue. This is due to an

effective minimum similarity between two profiles that have the same statistical character. This behavior is not evident in our data because it represents a rather short normalized translation length. However, we think that decorrelation would become significantly slowed from its initial, exponential form before a bed translates by one characteristic length. This is related to the broadening of the width of the CCF maxima associated with the decrease in their magnitude (Fig. 3).

The statistical interpretation of the important process here is in the exponential nature of the correlation decay. As the bed evolves from the transport of material, the topography demonstrates two major behaviors, translation and deformation. Each translative step is accompanied by a deformative change, and that change can be cast as being an injection of new information into the bed that had not been present before. At each time step the bed is constituted by some percent of its previous state and by some percent of that new information. As long as flow conditions are consistent, then the relative dependence upon the previous state, memory, is constant. The memory, $m$, is defined as the compliment of the decay after being discretized and normalized as above (i.e. $m = 83\%$ per 0.1 $L_T^*$). As in the example $m$ is a function of the dimensionless translation length used for discretizing.

The physical interpretation is that the transfer of sediment along the bed comprises two major modes. Part is from the stoss sides to the lee sides of individual bedforms that are of the scale of the characteristic length. The remainder of the transport occurs between those largest (dominant) scale bedforms, from one to the next, etc. All other styles of bed material transport are subsumed within this description. Indeed it is not exclusive of them; rather it is intended only to allow for the mathematical methodology provided by the bed-load equation (Simons, et al., 1971). Investigations into the details of this framework and its quantification are ongoing. Within the context of dynamic equilibria, the physical and statistical interpretations are consistent with notion that the mean bed properties of height and length are conservative while the bed is continually evolving.

## 5 CONCLUSIONS

Cross-correlations of evolving sand bed topography results in the association of three key variables that are synoptic of bed behavior: evolution duration, translation distance, and correlation magnitude. From those can be deduced a characteristic bed velocity as well as a deformation rate, or memory. The characteristic velocity linearly relates the time interval of bed evolution to translation lengths. The relation of CCF maxima and translation length is exponential in form, and the exponential decay constant defines the interfacial half-life as well as the memory of the bed. Further, the dimensionless translation length is the appropriate metric for investigating the rate of bed deformation across scales in sandy transport systems.

In the case of the N. Loup R. specifically, sandy bottom topography is significantly decoupled from itself by deformation of bedform shape and arrangement before the bed translates by one characteristic distance. We speculate that this is also a rather general result and that values of bed memory do not vary between systems by more than an order of magnitude. The rapid decoupling is of import to studies of sediment transport because it suggests a component of bed-material load that is not captured by the bedload equation; one that does not participate in the mean translation of the bed but rather in its deformation. The rapid decoupling is also significant in setting the downstream persistence of stratification and deposit properties affiliated with this surface topography migrating across an aggrading bed. Quantification of these effects is underway.

## REFERENCES

Allen, J.R.L. 1970. *Physical Processes in Sedimentation.* 248pp.

Allen, J.R.L. 1973. Phase differences between bed configuration and flow in natural environments, and their geological relevance, *Sedimentology*, v. 20, p.323–329.

Davis, J. 1986. *Statistics and Data Analysis in Geology.* Wiley & Sons, New York, 646pp.

Jerolmack, D., & Mohrig, D. 2005a. A unified model for subaqueous bed form dynamics, *Water Resources Research*, v. 41, W12421.

Jerolmack, D., & Mohrig, D. 2005b. Interactions between bed forms: Topography, turbulence and transport, *Journal of Geophysical Research-Earth Surface*, v. 110, F02014.

McElroy, B., Mohrig, D., Jerolmack, D. *in review.* Quantifying topographic roughness in sandy transport systems, *Geophysical Research Letters.*

McLean, S. 1989. The stability of ripples and dunes, *Earth-Science Reviews*, v. 29, p131–144.

Mohrig, D. 1994. Spatial evolution of dunes in a sandy river, doctoral dissertation, University of Washington, 119pp.

Nikora, V., Sukhodolov, A., Rowinski, P. 1997. Statistical sand wave dynamics in one-directional water flows, *Journal of Fluid Mechanics*, v. 351, p. 17–39.

Nordin, C. 1971. Statistical properties of dune profiles, *United States Geologic Survey Professional Paper*, 562-F, 41pp.

Rubin, D. 1992. Use of forecasting signatures to help distinguish periodicity, randomness, and chaos in ripples and other spatial patterns, *Chaos*, v. 2, p. 525–535.

Simons, D, Richardson, E., Nordin, C. Bedload equation for ripples and dunes, *United States Geologic Survey Professional Paper*, 462-H, 9pp.

Southard, J. 1975. Bed configurations, in Eds. Harms, J, Southard, J, Spearing, D., & Walker, R., *Depositional environments as interpreted from primary structures and stratification sequences*, SEPM short course 2, p. 5–43.

Southard, J., & Boguchwal, L. 1990. Bed configurations in steady unidirectional water flows. Part 2. Synthesis of flume data, *Journal of Sedimentary Petrology*, v. 60, p. 658–679.

*River, Coastal and Estuarine Morphodynamics: RCEM 2007 – Dohmen-Janssen & Hulscher (eds)*
*© 2008 Taylor & Francis Group, London, ISBN 978-0-415-45363-9*

# Quasi-two-dimensional enhancement of the De Saint Venant-Exner coupled model for unsteady simulations in natural channels

Annunziato Siviglia & Marco Toffolon
*Department of Civil and Environmental Engineering, University of Trento, Trento, Italy*

ABSTRACT: Sediment routing simulation models are designed to reproduce the hydro-morphodynamic behaviour of natural rivers as completely as possible. Three- and two-dimensional models are accurate but time-consuming. Hence, they usually are not suitable for long-term simulations. Such purpose currently requires one-dimensional models, which ignore the lateral variations of the scour/deposition processes within the cross-section. This paper aims at including those effects for reliable predictions of lateral channel deformation. This is accomplished by solving in a coupled way the De Saint Venant-Exner equations. The resulting exchanged sediment volumes (scour/deposition) are then partitioned along the cross-section by using a two-dimensional version of the mass conservation laws (for both water and sediments), thus extending the one-dimensional model to a quasi-two-dimensional model. The proposed formulation is tested against the results of a fully three-dimensional model in some idealized benchmark problems. Then, the influence of the quasi-two-dimensional reconstruction is evaluated in some one-dimensional problems. The differences between formulations currently used in literature and the proposed model become more clear as the role of the cross-sectional shape gains relevance.

## 1 INTRODUCTION

One dimensional (1D) sediment routing models (SRM) are widely applied to practical engineering and management problems in the fields of water resource exploitation, environment protection and ecology management. Almost all SRMs are 1D in the sense that they treat flow and sediment transport on a width-averaged basis. This simplifies the calculations, allows better temporal and spatial resolution with available computing power, and minimizes input data requirements. However, this approach ignores the lateral variations of the scour/deposition processes within the cross-section, which is an important element in predicting morphodynamics evolution in natural rivers. In fact, such evolution is governed by non linear effects included both in the friction term and in the solid transport. Some efforts have been made in order to include those effects. A well known model of this type, which is largely used in practice engineering, is GSTARS (Molinas and Yang 1986) which is a quasi-steady semi two-dimensional (2D) model with the use of stream tube concept to simulate the bed evolution process. An improved model, which aims to remedy the shortcoming of GSTARS, is due to Lee et al. (1997), in which the model is extended to solve unsteady flows, but using the steady concept of stream tubes.

In this paper a new methodology able to describe the lateral distribution of erosion/deposition for any arbitrary cross section is presented. The exchanged sediment volumes (scour/deposition), resulting from solving numerically the De Saint-Venant-Exner model, is partitioned along the cross section, according to a quasi two-dimensional flow field, obtained by using a two-dimensional version of the mass conservation laws (for both water and sediments).

## 2 FORMULATION OF THE PROBLEM

One-dimensional mobile bed models are based on a systems of differential equations which involves the De Saint Venant momentum and continuity equation and the Exner sediment continuity equation,

$$\frac{\partial Q}{\partial t} + \frac{\partial}{\partial x}\left(\beta\frac{Q^2}{A} + gI_1\right) - gA\frac{\partial \eta}{\partial x} = gI_2 - gAS_f \tag{1}$$

$$\frac{\partial A}{\partial t} + \frac{\partial Q}{\partial x} = 0, \tag{2}$$

$$(1-p)\frac{\partial A_s}{\partial t} + \frac{\partial Q_s}{\partial x} = 0, \tag{3}$$

where $t$ is time, $x$ is the longitudinal direction, $A$ is the transversal wetted area, $Q$ is the liquid discharge, $Q_s$ is the solid discharge, $p$ is the porosity, $\eta$ is the elevation of the deepest point (thalweg) within the cross-section,

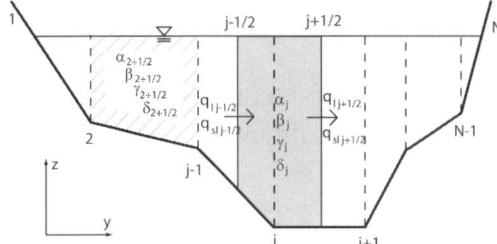

Figure 1. Sketch of the cross-section.

$I_1$ represents the hydrostatic pressure force term while $I_2$ accounts for the pressure forces in a volume of constant depth due to longitudinal variations. $A_s$ is the solid area of sediment within the cross- section, and it is computed as:

$$A_s = \int_{B_0} z_b(y)dy,\qquad(4)$$

where $B_0$ is the total width of the cross-section, $z_b$ is the bottom elevation.

The solution of the above system requires the closure relationships for the friction term $S_f$ and the quantification of $Q_s$ as a function of the Shields parameter $\theta$:

$$S_f = \frac{Q^2}{g(A^2C^2R)}, \qquad \theta = \frac{Q^2}{gA^2C^2\Delta d_s},\qquad(5)$$

where $C$ is the global conductivity in the cross-section, $R$ is the hydraulic radius, $\Delta$ the relative sediment density and $d_s$ a characteristic grain size.

In morphodynamics problems, non-linear effects are crucial. Some of them are related to the form of the section and should be included into one-dimensional models. They affect the friction term $S_f$ which in turn could affect sedimentation varying the solid discharge $Q_s(\theta)$. Usually, the cross section is defined by $N$ points ($j = 1..N$). Thus it can be subdivided in $N - 1$ cells (between the points), or in $N$ cells (around the points), as described in Figure 1. For each interval, it is possible to evaluate the local parameters (partial area, water discharge and solid discharge, bottom shear stress, etc.), according to the procedure proposed by Engelund (1964). Hence, the overall behavior of the cross-section, with respect to one-dimensional simulations, can be obtained by suitable sums of the fraction associated with its cells:

$$gA^2C^2R = \left\{ \int_B k_s(y)\,[h - z_b(y)]^{2/3}\,dy \right\}^2\qquad(6)$$

where $B$ is the width of the wetted cross-section, $h$ is the water level and $k_s$ is the local value of the Gauckler-Strickler coefficient.

## 2.1 Modification of river cross-section

In spite of the solution strategy (coupled or uncoupled) or numerical techniques (finite difference, finite volume, etc.) adopted for solving the governing equations (1), (2), (3), at the end of each time step all 1D mobile bed models, for each cross section $i$, give a sediment volume $\Delta A_{si}$ which should be distributed along the cross section. To ascertain change in cross section is always difficult in 1D computation because it actually goes beyond the conventional capacity of such computations. A reasonably accurate prediction of a river cross-section is nevertheless quite important. For instance, if navigation had to be considered, one would need to know the maximum depth after deposition. Even in the ecohydraulics field, change in the cross section may modify the suitability of habitat.

Some methods, empirically based, are available in literature and each of them is developed according to mass conservation. In the simplest method (CLASSIC) the amount of change $\Delta A_s$ is distributed along the wetted perimeter, increasing or decreasing each point of the same quantity

$$\Delta z_{bj} = \frac{\Delta A_s}{B}.\qquad(7)$$

A more complex method assumes the allocation of scour and fill across a section to be a power law of the local value of the Shields parameter as follow:

$$\Delta z_{bj} = \frac{\theta_j^m}{\sum_{j=1}^N \theta_j^m \Delta y_j}\Delta A_s.\qquad(8)$$

The coefficient $m$ is empirically determined and it is generally between 0 and 1; it affects the pattern of scour and fill allocation. A small value of $m$, say 0.1, would mean a fairly uniform distribution of $\Delta z_{bj}$ across the section; a larger value, say 1, will give a less uniform distribution of $\Delta z_{bj}$.

## 3 A QUASI TWO-DIMENSIONAL MODEL

In this section we propose a model that aims at obtaining two-dimensional information on the basis of the geometrical features of the cross-section. Thus we consider a generalized version of the one-dimensional equations (2)–(3) for each cell within the cross-section. Non vanishing liquid and solid transversal discharges represent a coupling between the different cells and allow for a quasi-two-dimensional (Q2D) description of the flow.

The continuity equation for the liquid phase in the cell associated to the point $j$ (the shaded region in Figure 1) reads

$$\frac{\partial A_j}{\partial t} + \frac{\partial Q_j}{\partial x} + q_{l\,j+\frac{1}{2}} - q_{l\,j-\frac{1}{2}} = 0,\qquad(9)$$

where $q_{lj+\frac{1}{2}}$ is the discharge per unit length exchanged between the regions associated to the points $j$ and $j+1$. Analogously, the sediment continuity can be written as

$$(1-p)\frac{\partial A_{sj}}{\partial t} + \frac{\partial Q_{sj}}{\partial x} + q_{sl\,j+\frac{1}{2}} - q_{sl\,j-\frac{1}{2}} = 0. \quad (10)$$

The main point of the model is the reconstruction of the lateral distribution of the one-dimensional variables, in particular of the solid area variation $\partial A_{sj}/\partial t$.

As a first step, we approximate the distribution of main quantities along the cross-section (assuming a laterally constant free surface elevation) by means of the following dimensionless parameters:

$$\alpha_j = \frac{\hat{Q}_j}{\hat{Q}}, \quad \beta_j = \frac{\hat{Q}_{sj}}{\hat{Q}_s}, \quad \gamma_j = \frac{A_j}{A}, \quad (11)$$

where $\hat{Q}_j$ and $\hat{Q}_{sj}$ are the liquid and solid discharges evaluated by means of a suitable subdivision model for each cell (with transversal area $A_j$ and width $B_j$), while $\hat{Q}$ and $\hat{Q}_s$ are their cumulate values summed for the whole cross-section (with total area $A$ and width $B$). The obvious fact that the sum of $\alpha_j$, $\beta_j$ and $\gamma_j$ over the whole cross-section is unitary will be important in the following considerations.

The sediment transport capacity can be estimated locally through a simplified relationship $Q_{sj} \propto B_j \theta_j^m$ (thus neglecting any threshold). With this choice, the ratio $\beta_j$ depends only on the geometrical characteristics of the section. In the following examples, we have chosen $m = 3/2$.

Secondly, we introduce the following assumptions:

$$\frac{\partial Q_j}{\partial x} \simeq \alpha_j \frac{\partial Q}{\partial x}, \quad (12)$$

$$\frac{\partial Q_{sj}}{\partial x} \simeq \beta_j \frac{\partial Q_s}{\partial x}, \quad (13)$$

which correspond to assume that the river reach is locally cylindrical and there are no effects of the water surface lateral variation. In fact, considering for instance the liquid discharge $Q_j$, its derivative should be expressed as

$$\frac{\partial Q_j}{\partial x} = \frac{\partial Q}{\partial x} \left[ \alpha_j + \frac{A}{nB} \frac{\partial \alpha_j}{\partial h} \Big|_x \right] + Q \frac{\partial \alpha_j}{\partial x} \Big|_h \quad (14)$$

where $n$ is the exponent of a simplified discharge power law $Q \propto A^n$. Neglecting the last two terms of (14) as in (12), we are not considering the differences in the shape of two adjacent cross-sections and the variations of the lateral distribution of discharge with changing water levels. An analogous assumption, though not strictly necessary, is introduced for the temporal variation of the area of the cell:

$$\frac{\partial A_j}{\partial t} \simeq \gamma_j \frac{\partial A}{\partial t}. \quad (15)$$

It is important to note that non-cylindrical terms can be important in real applications. We are currently working on the inclusion of non-cylindrical terms in our model, since they can be taken into account in a relatively straightforward manner. Nevertheless, in this paper we present some preliminary results based on the simplified assumptions (12), (13) and (15), which show the capability of the model to reproduce important features of the complete phenomenon.

At this point, the transversal liquid discharge can be estimated from the water continuity (9), introducing the distributions (11), the assumptions (12), (15), and the one-dimensional equation (2), as follows:

$$q_{lj+\frac{1}{2}} = q_{lj-\frac{1}{2}} + \frac{\partial A}{\partial t}(\alpha_j - \gamma_j), \quad (16)$$

where at the left boundary is imposed that $q_{l\frac{1}{2}} = 0$ (impermeable bank). The opposite boundary condition (e.g. $q_{lN+\frac{1}{2}} = 0$ at the other bank) is automatically satisfied; in fact, summing over the number of cells, it can be easily obtained

$$q_{lN+\frac{1}{2}} = q_{l\frac{1}{2}} + \frac{\partial A}{\partial t}\sum_{j=1}^{N}(\alpha_j - \gamma_j) = 0. \quad (17)$$

Note that the equation (16) implies that the lateral discharges disappear in steady conditions ($\partial A/\partial t = 0$).

Then, the lateral sediment transport between adjacent regions can be obtained as the sum of convective transport and gravitational effects, which can be estimated following Ikeda (1982) (for a discussion of a more refined model, see Parker et al. (2003)). Herein we adopt the simplified version

$$q_{sl} = q_s\left[\frac{\tau_y}{\tau} - \frac{r}{\sqrt{\theta}}\frac{\partial z_b}{\partial y}\right], \quad (18)$$

where the local values (e.g. at the point $j + 1/2$) of the variables are considered. The relationship (18) is based on the assumption that the direction of the sediment transport is given by the direction of the shear stress ($\tau_y/\tau \simeq v/\sqrt{u^2 + v^2}$, where $v = q_l/D$ and $u = q/D$, being $q = \alpha_{j+1/2}Q/B_{j+1/2}$ the specific longitudinal discharge) with a correction due to the gravitational effect. Considering, for instance, the right side of the cell containing the point $j$, the local values in (18) can be estimated as follows:

$$q_{sj+\frac{1}{2}} = \frac{\beta_{j+\frac{1}{2}}Q_s}{B_{j+\frac{1}{2}}}, \quad v_{j+\frac{1}{2}} = \frac{q_{lj+\frac{1}{2}}}{A_{j+\frac{1}{2}}/B_{j+\frac{1}{2}}}, \quad (19)$$

with a slope $\partial z_b/\partial y = (z_{bj+1} - z_{bj})/(y_{j+1} - y_j)$. In the following, when considering the effect of gravity, we have chosen a value $r = 0.3$ (Talmon et al. 1995).

899

Finally, the variation of the solid area in the cell can be obtained from the sediment continuity equation (10) by using the distributions (11), the assumption (13), and the one-dimensional equation (3), as

$$\frac{\partial A_{sj}}{\partial t} = \frac{\partial A_s}{\partial t}\beta_j + \frac{q_{sl\,j-\frac{1}{2}} - q_{sl\,j+\frac{1}{2}}}{1-p}. \quad (20)$$

When the lateral solid discharges are negligible (e.g. in the case of stationary conditions and no gravitational effects), the above relationship suggests that the solid area variations (erosion/deposition) are distributed along the cross section according to the local intensity of sediment transport $\beta_j$. Hence, if we use a power law sediment transport relation $q_s \propto \theta^m$, the erosion/deposition distribution is proportional to the local value of Shields number elevated to a power $m$ (in the range $1.5 \div 2.5$, significantly larger than the value proposed by previous models).

The temporal evolution of the elevation of the points must respect the sediment mass balance. Thus the following condition must be satisfied:

$$\Delta A_s = \sum_{j=1}^{N-1} \frac{(\Delta z_{bj} + \Delta z_{bj+1})(y_{j+1} - y_j)}{2}, \quad (21)$$

where $\Delta A_s$ is the total variation of the solid area calculated by the one-dimensional model. Since the sum (21) can also be written as

$$\Delta A_s = \sum_{j=1}^{N} \frac{\Delta z_{bj}(y_{j+1} - y_{j-1})}{2}, \quad (22)$$

it is easy to verify that the sediment balance is respected if we impose a local bottom variation in the form

$$\Delta z_{bj} = \frac{\Delta A_{sj}}{(y_{j+1} - y_{j-1})/2}. \quad (23)$$

Thus, the variations (23) can also be recast in terms of the temporal derivative as

$$\frac{\partial z_{bj}}{\partial t} = \frac{2}{(y_{j+1} - y_{j-1})}\left[\frac{\partial A_s}{\partial t}\beta_j + \frac{q_{sl\,j-\frac{1}{2}} - q_{sl\,j+\frac{1}{2}}}{1-p}\right]. \quad (24)$$

## 4 NUMERICAL RESULTS AND DISCUSSION

The performance of the model are compared with the simulations obtained by using a fully three-dimensional (3D) model (Vignoli and Tubino 2002; Francalanci et al. 2006). The comparison is made in the following way: at each time step, the results of the 3D

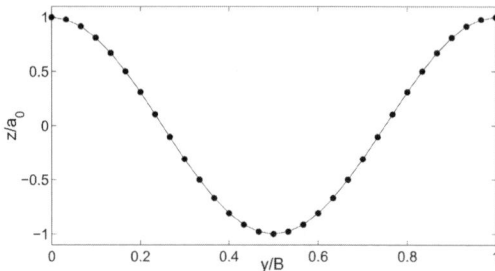

Figure 2. Test section: $z = a_0 \cos(2\pi y/B_0)$.

model (concerning both hydrodynamics and bottom evolution) are integrated over the cross-section in order to give the 1D input to our Q2D model, which independently modifies a parallel bottom configuration (starting from the same initial condition).

Moreover, an analysis of the influence of the lateral distribution of erosion and deposition on one-dimensional results is pursued by comparing different models of cross-section reconstruction.

The cross-section used hereafter is shown in Figure 2. For most examples, the following parameters have been chosen: $a_0 = 2\,m$, $B_0 = 100\,m$. For the case in Figures 8–10, $B_0 = 10\,m$.

### 4.1 Comparisons with a 3D model

In this section we compare the results of our model (Q2D, Quasi-Two-Dimensional) and those of the three-dimensional (3D) model in some significant cases. For sake of simplicity, when not specified, we refer to a base flow which is characterized as follows: discharge $Q = 1510\,m^3/s$, maximum water depth $D = 5\,m$, slope $S = 0.006$, length $L = 2000\,m$, sediment diameter $d_s = 0.05\,m$, Gauckler-Strickler coefficient $k_s = 30\,m^{1/3}/s$; the resulting Shields number is $\theta = 0.22$.

As a first test, we aim at verifying that the liquid and solid transversal discharge are reproduced correctly in the case of cylindrical reaches. For this purpose, we consider the propagation of a very rapid and intense flood wave ($\Delta h \simeq 15\,m$, $\Delta Q \simeq 2 \cdot 10^4\,m^3/s$ in a period $\Delta t = 1800\,s$, with a maximum rate of variation $dA/dt \simeq 3\,m^2/s$). For this test, the effect of gravity on lateral sediment transport is neglected, i.e. we set $r = 0$ in (18). In this case, the modifications of the cross-section are not relevant: the variation of the mean bottom level is negligible (few centimeters for a maximum depth of $5\,m$). Moreover, the transversal fluxes are several orders of magnitude smaller than the longitudinal ones ($q \sim 200\,m^2/s$, $q_l \sim 0.05\,m^2/s$, $q_s \sim 0.7\,m^2/s$, $q_{sl} \sim 1.5 \cdot 10^{-4}\,m^2/s$). The comparison between 3D and Q2D results is shown in Figure 3, where the lateral distributions of $q_l(y)$ and $q_{sl}(y)$ and the temporal variation of their maximum and

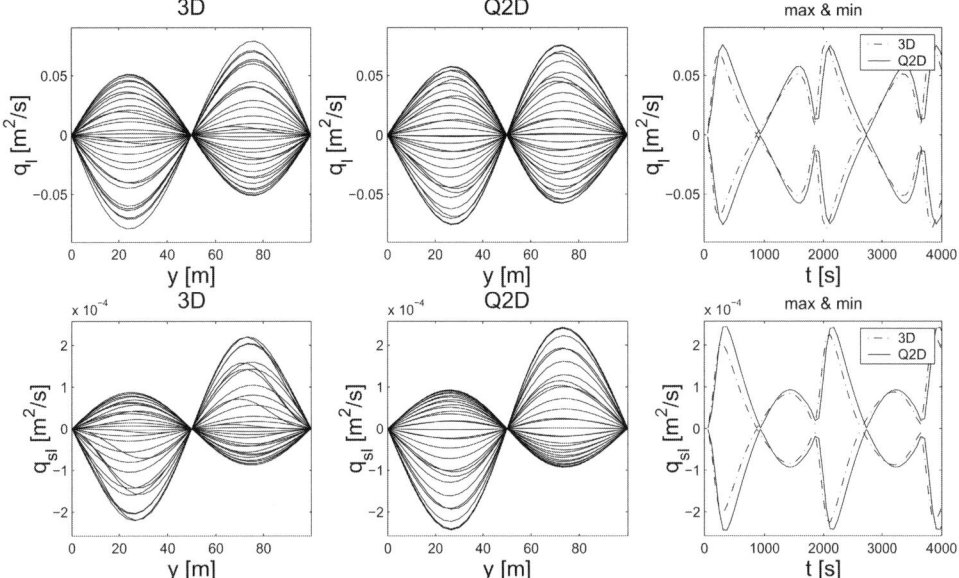

Figure 3. Transversal liquid ($q_l$) and solid ($q_{sl}$) discharges: comparison between Quasi-Two-Dimensional model (Q2D) and Three-Dimensional model (3D) at different times at section $x = 500\,m$.

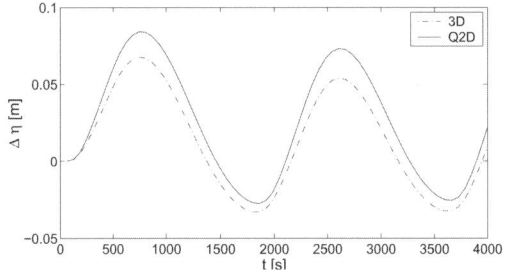

Figure 4. Temporal variation of thalweg elevation for the case of Figure 3.

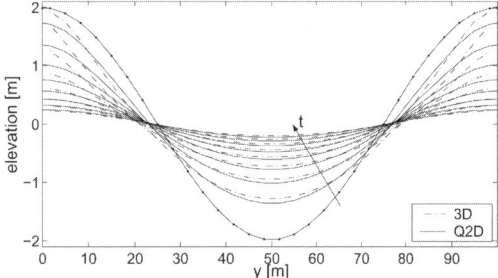

Figure 5. Cross-section evolution considering gravity and a constant liquid discharge: comparison between Q2D and 3D results. The transversal profiles are plotted every $3103\,s$.

minimum values along the cross-section are plotted. The performance of Q2D model are good, even if the role of transversal fluxes is likely to be negligible.

As a result, the elevation of the deepest point is mainly driven by the total erosion/deposition process. The lateral distribution of the solid area variation is governed by the transversal distribution of $\beta_j$ along the cross- section. The predictions of the altimetric variations of the thalweg made by 3D and Q2D model show a satisfactory agreement (see Figure 4); note that the variation of the mean bed elevation is the same for both model due to the specific formulation of Q2D model.

The second test aims at showing the effect of gravity in the case of cohesionless sediments. In the case of constant liquid discharge, gravitational correction of the sediment transport (18) is responsible for the

lateral flattening of the cross-section altimetry, which asymptotically tends to a rectangular section where lateral slope vanishes. The process can be quite fast if cohesive resistance is not taken into account, as shown in Figure 5, where the cross-section evolution at different times is plotted as calculated by the 3D and Q2D models: our model seems to capture the main features of the evolution in a correct manner. In natural rivers the process can be significantly different, because of the different material of the banks (which, for instance, usually have a slope that is much larger than the angle of repose).

Finally, it is interesting to verify the Q2D approximation in two opposite cases: net deposition and net erosion. The former case is reproduce by modifying

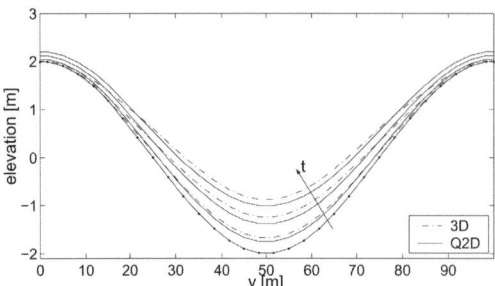

Figure 6. Deposition process, without gravity: cross-section evolution 3D-Q2D at different times ($t = 1550\,s$, $t = 3100\,s$, $t = 6200\,s$). Hydrodynamic conditions: discharge $Q = 1600\,m^3/s$; slope $S = 0.006$; length $L = 2000\,m$; initial maximum depth: $D = 5\,m$ (upstream), $D = 9\,m$ (upstream); the section is at $x = 1200\,m$; gravity is neglected.

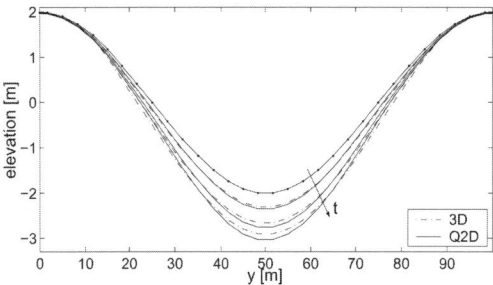

Figure 7. Erosion process, without gravity: cross-section evolution 3D-Q2D at different times ($t = 1900\,s$, $t = 5700\,s$, $t = 9500\,s$). Hydrodynamic conditions: discharge $Q = 1350\,m^3/s$; upstream: $S = 0.004$; $L = 3000\,m$; downstream: $S = 0.007$, $L = 1000\,m$; the section is at $x = 3000\,m$; gravity is neglected.

a uniform flow by imposing a large increase in the water elevation downstream (Figure 6). The latter case is given by a change of slope between a subcritical flow upstream and a supercritical flow downstream, which induces erosion in the transition (Figure 7). In both cases our model behaves in a satisfactory way.

### 4.2 Influence of Q2D reconstruction on 1D simulations

The aim of these simulations is twofold: first, to compare different models for computing the lateral evolution of the cross sections, and second, to investigate the role played by such models on the overall 1D morphodynamic evolution. The 1D SRM model used for obtaining the results presented in this section is the one developed by Siviglia et al. (2006), which has been coupled with 4 different cross section evolution models, namely: the CLASSIC and the SHIELDS (we set $m = 1$ in equation 8) models, described in section

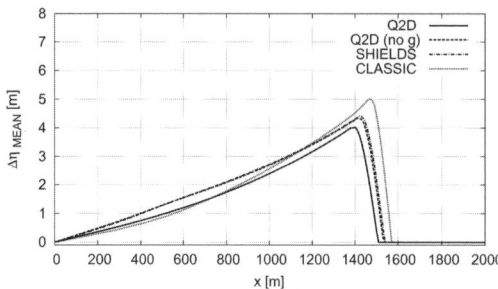

Figure 8. Deposition process: 1D longitudinal evolution of the mean bottom elevation. Hydrodynamic conditions: discharge $Q = 40\,m^3/s$; $S = 0.012$, $L = 2000\,m$; $d_s = 0.005\,m$ the section is at $x = 1500\,m$.

2.1, and the proposed model in two different configurations, accounting (Q2D model) and neglecting the effects of gravity (Q2D no g) (imposing $r = 0$ in 18).

*Deposition*
In this section we investigate the 1D evolution of a deposit generated under an hydraulic jump obtained by imposing downstream a large increase in the water elevation to a uniform supercritical flow. The bottom is considered cohesionless. In order to increase effects due to non linearity we have considered a narrower cross-section respect to the 3D simulations, i.e. the one sketched in Figure 2 characterized by a width $B_0 = 10\,m$.

In Figure 8, we have plotted the difference of the mean value of bottom profile between the final ($t = 37\,h$) and initial configurations ($t = 0$) along the longitudinal axes. There are some differences among the results of the compared models. The CLASSIC propagates the front of the deposition faster than the others models, while the Q2D is characterized by the slowest propagation. The fronts of the deposit obtained by using the Q2D (no g) and the SHIELDS models are located in between of these two. It is worth noting that such differences increase during the time evolution, possibly making long term morphodynamic simulations meaningless.

In Figure 9 the time evolution of the friction term $gAS_f$ at section $x = 1250\,m$ is given. Application of the four updating models gives rise to different behavior of such a term. During the first time lag $t_1$, the front of the deposition has not reached the coordinate $x = 1250\,m$, thus the value of the friction term predicted by all the models is constant and equal to the initial value. During the phase lag $t_2$ the front reaches the cross section under investigation, resulting in different behaviours for the value of the friction term. The Q2D (no g) and SHIELDS models start to change the shape of the cross-section almost immediately, thus resulting in larger value of the friction term, while the gravity effects, included in the Q2D model seem to induce a

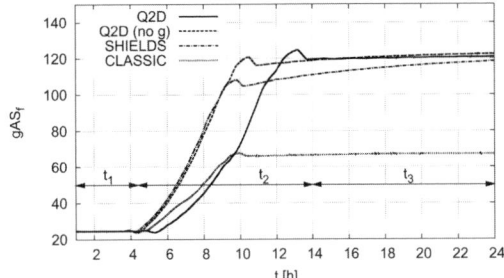

Figure 9. Depositional process: comparison of cross-section evolution at $t = 37\,h$ (Classic, Shields, Q2D, Q2D neglecting gravity). Hydrodynamic conditions: discharge $Q = 40\,m^3/s$; $S = 0.012, L = 2000\,m$; $d_s = 0.005\,m$ the section is at $x = 1250\,m$.

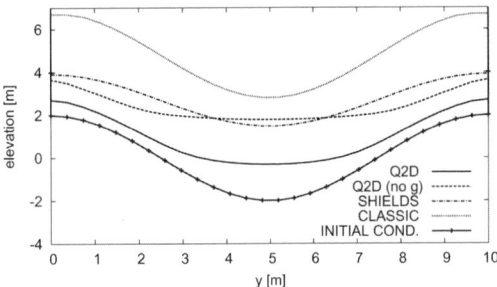

Figure 10. Depositional process: comparison of crosssection evolution at $t = 37\,h$ (Classic, Shields, Q2D, Q2D neglecting gravity). Hydrodynamic conditions: discharge $Q = 40\,m^3/s$; upstream: $S = 0.012$, $L = 2000\,m$; the section is at $x = 1500\,m$.

delay in such growth. When the front of the deposition has passed $x = 1250\,m$, the friction term reaches a constant value determined by both the shape of the cross-section and the water depth. The two proposed models Q2d and Q2D (no g) and the SHIELDS model converge to similar constant values while the CLASSIC method tends to converge to a smaller value. In this case, in which considerable non linear effects have been included, these values differ of a factor 2. These numerical results demonstrate a remarkable tradeoff: such a considerable variation can greatly affect the time evolution of hydraulic phenomena governed by gravitational effects, e.g. flood propagation.

The modification of the cross-section located at $x = 1500\,m$ at time $t = 37\,h$, due to the depositional process, is shown in Figure 10. At first glance it appears that the mobilized sediment volume are very different for each configuration. This can be explained by analyzing Figure 8; in fact, for the Q2D model, the front of deposition has not passed the longitudinal coordinate $x = 1500\,m$ yet, while it has moved on in the CLASSIC model. The other two models lie in between. Since all points are wetted, the application

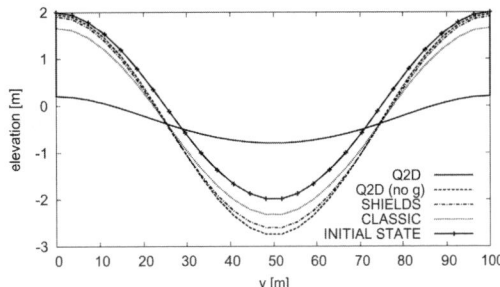

Figure 11. Erosional process: comparison of cross-section evolution at $t = 4\,h$ (Classic, Shields, Q2D, Q2D neglecting gravity). Hydrodynamic conditions: discharge $Q = 1350\,m^3/s$; upstream: $S = 0.004$, $L = 3000\,m$; downstream: $S = 0.007$, $L = 1000\,m$; the section is at $x = 2750\,m$.

of the CLASSIC method does not change the shape of the cross-section, which moves up unchanged. Application of the Q2D (no g) and SHIELDS models tends to flatten the cross-section, filling the deepest regions, and moving towards a fairly uniform deposit distribution. This effect is accelerated by gravity in the applications of the Q2D model. As a result, the lateral cross section evolution is characterized by very different shapes with high differences in the thalweg elevation.

*Scour*
In this section we investigate the 1D evolution of an erosion process generated by a change of slope (upstream: $S = 0.004$ and $L = 3000\,m$, downstream: $S = 0.007$ and $L = 1000\,m$). The water discharge is constant ($Q = 1350\,m^3/s$) and the bottom is considered cohesionless. The geometry is characterized by the cross-section depicted in Figure 2 with width $B_0 = 100\,m$. The modification of the cross-section located at $x = 2750\,m$ at time $t = 4\,h$ due to the erosional process is shown in Figure 11. CLASSIC, Q2D (no g) and SHIELDS models tend to erode the cross-section, scouring the deepest regions more intensively than the shallow one. It results in a decrease of the thalweg elevation. The results of such numerical experiments demonstrate a remarkable side effect: the thalweg can become deeper and deeper, and it can reach meaningless values, which, in some cases, can crash the numerical model. Such side effect is naturally solved in the Q2D model; in fact, inclusion of gravitational effects lead to a smoother cross-section, as it is shown in Figure 11. The elevation of the thalweg rises during time evolution while the value of the mean bottom elevation gradually decreases.

## 5 CONCLUSIONS

The quasi-two-dimensional model presented in this paper allows for reconstructing the transversal distribution of scour and deposition along the cross-section. It emerges that the local bottom variation (24) depends on two main factors: the distribution of the sediment transport, described by the dimensionless parameter $\beta_j$, and the lateral fluxes of sediments, $q_{sl}$. A preliminary analysis, concerning channels without strong non-cylindrical variations, show that the former factor is predominant in those cases for which the gravitational effects on transversal sediment transport can be neglected. In fact, even in the case of rapid flood waves, the liquid transversal fluxes $q_l$ are some orders of magnitude smaller than the longitudinal fluxes. Thus, the only relevant effect on bed load direction is given by the local transversal slope.

In the case of $\partial A/\partial t = 0$ and no gravitational effect, the solid area variation (20) is simply

$$\frac{\partial A_{sj}}{\partial t} = \frac{\partial A_s}{\partial t}\beta_j. \tag{25}$$

Given that $\beta_j$ is a non-linear function of the local depth through the value of the Shields parameter $\theta$ (i.e. $\beta_j \propto \theta_j^m$), we find a conceptual basis for a class of available models that assume a dependence on Shields as in (8). One of the relevant outcomes of the present analysis is that an exponent $m \simeq 3/2$ seems to be a good choice on the basis of the comparison with 3D results, whereas empirical relationships assume $m \in (0, 1)$.

The relevance of considering a correct shape of the cross-section in 1D models has been investigated in the case of the propagation of a sediment front. In particular, it is shown how the nonlinear terms $gAS_f$, from which the propagation velocity of liquid phase waves depends, and the Shields parameter $\theta$, which quantifies the solid transport, depend crucially from the model adopted for lateral evolution.

Finally, we want to point out the main strengths and weaknesses of the present model. The main limitation is that, in the current formulation, the model is not suitable to deal with strong longitudinal variation of the cross-section because of the assumptions (12) and (13), but its extension is feasible.

On the other hand, unlike other models which are based on the concept of stream tubes (which can be defined in a consistent manner only in a steady state context), the formulation of the present model concerns only a single cross-section at a time, avoiding the use of spatial derivatives. In this way, it is intrinsically consistent with whatever 1D model, since it does not require any assumption on the 1D numerical scheme. This fact, in addition to the simplicity of the formulation and the possibility to include gravitational effects on the lateral transport, gives the possibility to easily include it in existing models. Furthermore, the physically-based formulation represents an advantage over other semi-empirical models of lateral distribution of scour and deposition.

It is also important to note that, in order to apply the model to real cases, further considerations are required about the role of gravity in the regions of the river banks.

## ACKNOWLEDGEMENTS

The Authors thank Marco Tubino for the initial discussions about the problem of quasi-two-dimensional reconstruction and Gianluca Vignoli for the development of the three-dimensional model.

## REFERENCES

Engelund, F. (1964). Book of abstracts. Technical Report 6, University of Denmark.

Francalanci, S., L. Solari, and G. Vignoli (2006). Gravitational effects on river morphodynamics. In *Proceedings, River Flow 2006*, Volume 2, pp. 1129–1136. Lisbon, Portugal.

Ikeda, S. (1982). Incipient motion of sand particles on side slopes. *J. Hydr. Div. 108*(1), 95–114.

Lee, H., H. Hsieh, J. Yang, and C. Yang (1997). Quasi-two dimensional simulation of scour and deposition in alluvial channels. *Journal of Hydraulic Engineering, ASCE 123*(7), 600–609.

Molinas, A. and C. Yang (1986). *Computer program user's manual for GSTARS*. Department of Interior Bureau of Reclamatio Engrg. and Res. Ctr. Denver, Colorado.

Parker, G., G. Seminara, and L. Solari (2003). Bedload at low Shields stress on arbitrarily sloping beds: alternative entrainment formulation. *Water Resour. Res. 39*(7), doi:10.1029/2001WR001253.

Siviglia, A., G. Nobile, and M. Colombini (2006). The role of quasi-conservative form for morphodynamic modelling in river flow computations. In *Proceedings, River Flow 2006*, Volume 2, pp. 1453–1462. Lisbon, Portugal.

Talmon, A. M., N. Struiksma, and M. C. L. M. van Mierlo (1995). Laboratory measurements of the direction of sediment transport on transverse alluvial-bed slopes. *J. of Hydraulic Res. 33*(4), 495–517.

Vignoli, G. and M. Tubino (2002). A numerical model for sand bar stability. In *Proceedings, River Flow 2002*, Volume 2, pp. 833–841. Louvain-La-Neuve, Belgium.

*Bed forms – general*

*River, Coastal and Estuarine Morphodynamics: RCEM 2007 – Dohmen-Janssen & Hulscher (eds)*
*© 2008 Taylor & Francis Group, London, ISBN 978-0-415-45363-9*

# Spectral analysis of compound dunes

C. Winter & V.B. Ernstsen

*MARUM, Bremen University, Germany*

ABSTRACT:   Successive multibeam echo sounder surveys in a tidal channel on the North Sea coast reveal the dynamics of subaquatic compound dunes. Mainly driven by tidal currents, dune structures show complex migration patterns. Common methods for the analysis of bedform migration are based on the description of average characteristics as dune length, height and celerity. Their application to superimposed structures is dissatisfying as the recognition of dunes is subjective and work intense. Thus the bathymetric signal of a cross-section of compound subaquatic dunes has been approximated by the sum of a set of harmonic functions, derived by Fourier transformation. Dune migration has been analysed and quantified by taking into account the phase differences of individual harmonic constituents. The separate re-composition of harmonic constituents with zero or low phase shifts sums up to what can be regarded as the stable part of the original signal. On the other hand the summation of constituents with high phase differences forms the purely kinematic signal.

The further analysis of the kinematic signal revealed spectral properties of the concerning components, which lead to the formulation of a simple phenomenological model. Although the model is by no matter meant to represent the natural system of bed-form dynamics and super-position, it was shown to be capable of reproducing some morphological characteristics of compound dunes as e.g. their asymmetric shape. Model shortcomings have been identified as the lack of a wiping-out function and the necessity of including extreme events in the model forcing.

## 1 INTRODUCTION

### 1.1 *Compound dunes in the Gradyb tidal channel*

Tidal inlets are typically covered with complex bed-form patterns of various size and morphology as a reaction of the sandy beds to the forcing tidal- and wind driven currents and waves (Flemming & Davis, 1992; Hennings et al., 2004; Carling et al., 2000; Hoekstra et al., 2004; Flemming, 1988, Diesing et al., 2006).

Compound bedforms are large subaquatic sand bodies composed by dunes of different dimensions. Typically large dunes (in the order of 100 m length) are superimposed by smaller bedforms (in the order of a few meters in length). Compound dunes indicate a complex relationship between hydrodynamic forcing and morphological response. Common explanations for compound dunes differentiate between equilibrium and disequilibrium superposition (Dalrymple & Rhodes, 1995). The former describes the formation of smaller dunes within an internal boundary layer on the stoss side of the larger dunes (Smith & McLean, 1977, Rubin & McCulloch 1980, Dalrymple, 1984). Disequilibrium superposition is explained by temporal changes in water depth and currents which allow large dunes to form first. These large bedforms remain because of their long lag time, while smaller dunes form on their back in shallower water. (Allen,

1968). The observed superimposed dune structures show complex migration patterns in various temporal scales.

Common approaches for the description and analysis of dune morphology and migration mainly consider triangular shaped elements, which are far from similar to the complex bathymetric patterns found in tidal domains. Thus the application to compound dunes must consider some shortcomings: The manual recognition of dunes is subjective and very work intense. Additionally a thorough description of dune dynamics is complicated as the migration has to be described by several parameters involving the re-location of the toe, the stoss side, the crest, the lee side and the trough.

Bartholdy et al. (2002), Bartholomä et al. (2004), and Ernstsen et al. (2005, 2006) have described large compound dunes in the Grådyb tidal channel in the Danish Wadden Sea. The morphological characteristics of these highly dynamic structures were found to be related to the variability in sediment composition and hydrodynamic forcing. However it remains unclear how the compound morphology of these features can be explained.

### 1.2 *Spectral analysis*

Here a mathematical procedure is applied, that has been a standard for the analysis of water waves and tides for long times. We use the spectral decomposition of measured bathymetrical transects into a set of harmonic constituents using Fourier transformation. This approach had already been followed by Nordin & Algert (1966) who used autocovariance and spectral density analysis to overcome the simplistic description of bedforms as triangular shaped bodies. Also Robert & Richards (1988) applied semi-variograms as a statistical method for the investigation of roughness properties of bed profiles obtained from field work and laboratory experiments. More recent studies concentrate on wave generated ripples as Walker et al. (2003) and Davis et al. (2004) who used a spectral method to describe changes in the ripple height, length, and shape considered as an energy transfer process. However, this approach has been restricted to bedforms under currents and waves in laboratory scale.

The high accuracy of modern instrumentation nowadays allows the application of spectral methods also to field data on large scale dune bathymetries: In this study a spectral decomposition of compound dunes in a tidal inlet of the Danish Wadden Sea is applied, which enables the detailed description of dune patterns and their migration, a classification and analysis and serves as the basis of an phenomenological equilibrium model for the simulation of dune migration.

## 2   METHODS

### 2.1   Multi-beam sonar data processing

Bathymetric surveys were conducted in the Grådyb tidal inlet, located on the Danish West coast connecting the northernmost tidal basin of the Wadden Sea with the North Sea at about N55.46/E8.33. For this study a 1600 m section of the channel measured on 10 September 2002 and 11 July 2003 is taken into account. The bathymetric data were recorded from RV Senckenberg using a Seabat 8125 TM (RESON) multibeam echo sounder (MBES) system operating at 455 kHz (Ernstsen et al., 2005). The MBES system was coupled with an AQUARIUS 5002 (THALES) dual frequency (L1/L2) Long Range Kinematic (LRK) Global Positioning System (GPS). Corrections for ship movements were applied using an Octans Surface gyrocompass and motion sensor. Repeated recordings revealed a horizontal and vertical precision of the MBES system of ±20 and ±2 cm, respectively, at a 95% confidence level (Ernstsen et al., 2006). Data was interpolated by linear triangulation to three lines along the main current direction.

### 2.2   Fourier Transformation

A large number of scientific problems are tackled by Fourier transform methods which decompose

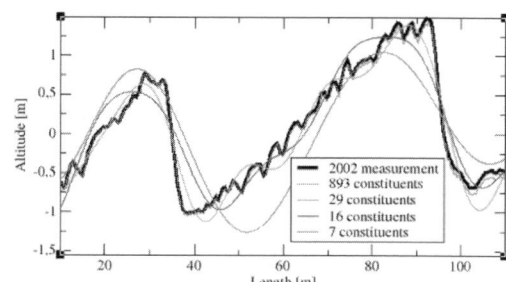

Figure 1.   Re-composition of a measured compound dune in Grådyb tidal channel (Denmark) by different numbers of harmonic constituents. An exact reconstruction is possible when using the full set of constituents.

functions into spectra of their frequency components.

The common transformation of data from the time domain, where the process is a function of time, h(t) to the frequency domain, where the process is a function of frequency, H(1/t) allows a decomposition of complex waveforms into a number of sinusoid harmonics of different amplitude, frequency and phase, which then are to be analysed adequately. The processing is recursive as the summation of function values of the single harmonics will return the original signal. The Fourier equations describing the transformation work well with different units. Thus a Fourier decomposition of a bed profile results in:

$$z(x) = \sum_i a_i \cos(k_i x + \varphi_i) \tag{1}$$

In which the sum of $i$ harmonic constituents (sinusoids) of amplitude $a_i$, wavenumber $k_i$ and phase $\varphi_i$ along $x$ equals the measured bed profile $z(x)$ (Figure 1).

As Fourier transform methods are described in detail in mathematical textbooks and excellent introductions to computational methods, codes and software can be found in textbooks and literature (e.g. Press et al., 1992) further mathematical details are omitted here. For this study the Discrete Fourier transform (DFT) method as implemented in GRACE has been applied (Grace, 2007).

## 3   RESULTS AND DISCUSSION

### 3.1   Morphology

Figure 2b show longitudinal transects of an exemplary compound dune of the Grådyb tidal channel of about 100 m length and 3.5 m height from trough to crest. This large dune is covered by small dunes of a few meters in length and decimeters in height. Successive measurements at high (black line) and low tide (red

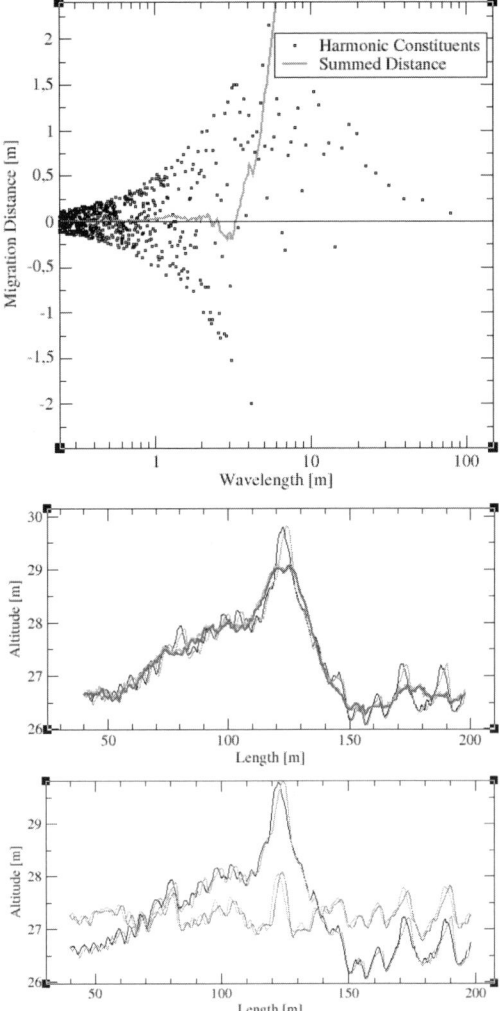

Figure 2. a) De-composition of bed profiles reveal migration distances of all harmonic waveforms between high-water and low-water. An integration helps in the detection of significant forms. b) Re-composition of waveforms considered stable (blue is dataset high-water, green is dataset low-water). c) Re-composition of waveforms considered kinematic.

line) exemplarily reveal the morphological change during the ebb tide period. It is observed that in this time mainly the upper parts of the compound dune change position, whereas the deeper troughs seem stable. Difficulties arise in the recognition of the super-imposed small dunes and thus the allocation of dynamic properties to specific elements. Thus for a further description and analysis of dune dynamics a separation of the stable and dynamic components of the compound dune has been determined.

## 3.2 Spectral characteristics

A spectral decomposition of the two profiles results in two sets of harmonic waveforms individually described by their amplitudes, wavelengths and phases. A comparison of individual waveforms for the two profiles thus reveals their specific evolution in time. The change in size can be determined by the difference in amplitude, while the migration distance can be calculated from the difference in phase. The changes in amplitude remain small, close to the instrument accuracy and thus are neglected for further analysis.

Figure 2a shows the derived migration distances of all waveforms. Numerical noise is shown by positive and negative numbers for small wavelengths. The summed distances of wavelengths smaller than two meters thus oscillate near zero. For the following it is assumed that these waveforms of indifferent migration and those featuring a migration distance of less than 0.25 m (instrument accuracy) are considered as stable. By contrast waveforms that show larger migration distances are considered kinematic. The re-composition of the classified harmonic constituents then is given for the part considered stable in Figure 2b and the kinematic part in Figure 2c. While the stable constituents sum up to a roughly triangular shaped body which is considered not to move during the ebb time period, the re-composition of the kinematic constituents exposes the migrating forms of approximately ten meters length and up to one meter height.

A further inspection of the kinematic signal constituents reveals that it can be closely approximated by the sum of harmonically related sinusoids:

$$z(x) = \sum_n \frac{\cos(nkx + \varphi_n)}{n} \qquad (2)$$

In which n are integers. Equation (2) which results in a sawtooth like signal if the phase of all harmonics is zero ($\varphi_n = 0$), and a cnoidal shaped signal if all constituents are shifted a quarter of a wavelength ($\varphi_n = \pi/2$). Thus the exposed kinematic signal can be understood as an intermediate state between these two extremes.

The fact that there is no continuous spectrum in wavelengths is also known from the evolution of regular bedforms under changing hydrodynamic forcing. According to Flemming (2004) the transition between size characteristics of bedforms is not continuous but determined by rather discrete "amalgamation" and "bifurcation" steps.

In the following a simple spectral model is formulated that accounts for the disequilibrium superposition of bedforms. It is by no matter meant to represent the physical system of bedform superposition, but is to test the changing morphology resulting from the migration of single harmonic constituents.

# 4 MODEL

## 4.1 Model formulation

As mentioned above the disequilibrium superposition of dunes of various sizes has been explained by temporal changes in water depth and currents which allow large dunes to form in higher currents and smaller dunes on the back of the first at lower energetic conditions while the large forms remain (Allen, 1968). To test the resulting shape of super-imposed harmonics a simple dynamic phenomenological model is formulated that takes into account the above spectral findings and the following model assumptions:

For each discrete time step $\Delta t$ a momentary hydrodynamic condition (water level $h$, mean current velocity $u$) determines the activation and migration of a coherent harmonic waveform:

1. The amplitude $a$ of this active equilibrium harmonic is determined by the Van Rijn (1984) formula for bedform height:

$$a = \tfrac{1}{2} 0.11 h (d_{50}/h)^{0.3} (1 - \exp(-0.5T))(25 - T) \quad (3)$$

   In which $h$ is the water depth, $d_{50}$ is the median grain size, and $T = ((\tau - \tau_{cr})/\tau_{cr})$ is a dimensionless parameter relating the excess of local bed shear stress $\tau$ to the critical Shields shear stress $\tau_{cr}$. The associated wavelength is calculated by the global relationship of bedform length and height compiled by Flemming (1988):

$$1/k = (a/0.0677)^{1/0.8098} \quad (4)$$

   In which $1/k = L$ is the length of the active bedform.
3. The derived wavelength (period) is classified into an array of harmonics, the bin size following $2^n d_{50}$, with integer $n = 1, \ldots 25$.
4. Exner's (1920) bed form celerity formula, calibrated by Carling (2000) for dunes of similar geometry but somewhat coarser material in river Rhine, Germany, is used to calculate the concerning migration distance $\Delta x$.

$$\Delta x = \Delta t \, 5.67 u^{2.04} \quad (5)$$

This distance is transformed to a distinct shift of phase and added to the phase of the concerning harmonic. The shifted harmonic then is updated in the array of harmonics. All steps are repeated until the end time is reached.

## 4.2 Model results

Exemplarily the model has been applied for a two months period using a mean water depth $h = 10$ m, a median grain size $d_{50} = 300.\text{e-}6$ m, and tidal current velocities using the Bartholdy (2006) model. The resulting bathymetrical profile is plotted in Figure 3.

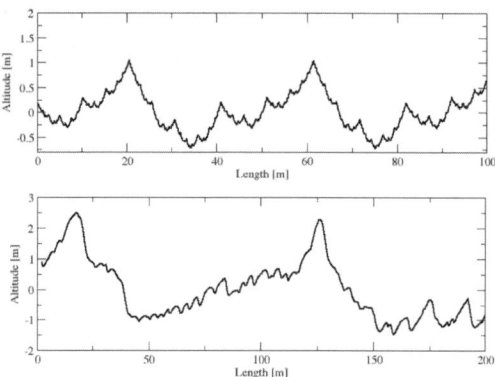

Figure 3. Upper graph: Result of a model simulation: Bathymetrical transect after two months of calculation. Lower graph: measured bathymetrical profile in the Gradyb tidal channel.

The calculated profile does exhibit some similar characteristics to the measured profile. The general asymmetric shape with a gently sloping stoss side and a steeper lee side due to the tidal asymmetry is captured as well as the overall appearance of smaller dunes super-imposing larger ones. However, some distinct differences reveal model and application shortcomings: Obviously the dimensions of the calculated dunes are much smaller than what was measured in nature. This difference is explained by the forcing of the model. The simulation was driven by the horizontal astronomical tide for a rather short period of two months compared to the geological age of the natural bedforms. Thus extreme hydrodynamic conditions which would have caused the activation of larger harmonic constituents were not covered in this simulation. However, if compared to the analysed kinematic signal of a single ebb tide (Figure 2c), the calculated transect exhibits dune properties of similar dimensions.

An obvious model shortcoming is the lack of a wipe-out criterion. The very sharp edges of the simulated compound bedforms result from the fact that all waveforms are kept in full shape for the model simulation time instead of being degraded.

## 5 CONCLUSIONS

In this study a spectral decomposition of compound dunes has resulted in some analytical findings and a phenomenological model of dune super-position.

It was shown that spectral decomposition overcomes the problem of subjective and manual work in the analysis of bathymetrical transects. Thus the distinction between stable and kinematic components of compound dunes when comparing successive measurements is possible. The further analysis of the kinematic signal revealed spectral properties of the concerning components, which lead to the formulation of a simple model. Although this model is by no matter meant to represent the natural system of bedform dynamics and super-position, it is capable of reproducing some morphological characteristics of compound dunes. Model shortcomings have been identified as the lack of a wiping out function and the necessity of including extreme events in the model forcing.

ACKNOWLEDGEMENTS

This study was supported by the German Science Foundation (DFG) as part of the DFG Research Center Ocean Margins (RCOM MARUM) at the University of Bremen, Germany and the Senckenberg Institute. Comments on an earlier version of the text by M.S. Yalin are highly appreciated.

REFERENCES

Allen, J.R.L. (1968). *Current Ripples*. Amsterdam, North Holland Publishing Company.

Bartholdy, J. (2006) A simple model for estimating current velocity in tidal inlets: example from Grådyb in the Danish Wadden Sea. *Geo-Marine Letters* 26(3): 133–140.

Bartholdy, J., Bartholomae, A., Flemming, B.W. 2002 Grain-size control of large compound flow-transverse bedforms in a tidal inlet of the Danish Wadden Sea. *Marine Geology* 188:391–413.

Bartholomä, A., Ernstsen, V.B., Flemming, B.W., Bartholdy, J. 2004. Bedform dynamics and net sediment transport paths over a flood-ebb tidal cycle in the Grådyb channel (Denmark), determined by high-resolution multi-beam echosounding. *Danish Journal of Geography* 104(1) 45–55.

Carling, P.A., Gölz, E., Orr, H.G., Radecki-Pawlik, A. 2000 The morphodynamics of fluvial sand dunes in the River Rhine, near Mainz, Germany. *Sedimentology* 47:227–252.

Dalrymple, R.W. 1984. Morphology and internal structure of sandwaves in the Bay of Fundy. *Sedimentology* 31: 365–382.

Dalrymple, R.W., Rhodes, R.N. 1995 Estuarine dunes and bars. In: Perillo (ed) Geomorphology and Sedimentology of Estuaries. *Developments in Sedimentology* 53. Elsevier Science, 359–422.

Diesing, M, Winter, C, Kubicki, A, Schwarzer, K. (2006) Decadal scale stability of sorted bedforms, German Bight, south-eastern North Sea. *Continental Shelf Research* 26: 902–916.

Ernstsen, V.B., Noormets, R., Winter, C., Hebbeln, D., Bartholomä, A., Flemming, B.W., Bartholdy, J. 2005: Development of subaqueous barchanoid-shaped dunes due to lateral grain size variability in a tidal inlet channel of the Danish Wadden Sea. *Journal of Geophysical Research*, 110(F04S08).

Ernstsen, V.B., R. Noormets, C. Winter, D. Hebbeln, A. Bartholomä, B.W. Flemming and J. Bartholdy (2006): Quantification of dune dynamics during a tidal cycle in an inlet channel of the Danish Wadden Sea. *Geo-Marine Letters*, 26(3): 151–163.

Ernstsen, V.B., Noormets, R., Hebbeln, D., Bartholomä, A., Flemming, B.W. 2006. Precision of high-resolution multi-beam echo sounding coupled with high-accuracy positioning in a shallow water coastal environment. *Geo-Marine Letters* 26:141–149.

Ernstsen, V., Becker, M., Winter, C., Bartholomä, A., Flemming, B.W., Bartholdy, J. (submitted to RCEM proceedings) Bedload transport in an inlet channel during a tidal cycle.

Exner, F.M. (1920) Zur Physik der Dunen. *Sitzber. Akad. Wiss Wien*, Part IIa, Bd. 129, 929–952.

Flemming, B.W. 1988. Zur Klassifikation subaquatischer, strömungstransversaler Transportkörper. *Bochumer geologische und geotechnische Arbeiten* 29:44–47.

Flemming, B.W., Davis, R.A. 1992 Dimensional adjustment of subaqueous dunes in the course of a spring-neap semi-cycle in a mesotidal backbarrier channel environment. In: Flemming BW (Ed) *Proceedings of the 3rd International Research Symposium on Modern and Ancient Clastic Tidal Deposits*. Senckenberg Institute, Wilhelmshaven, Germany, 28–30.

Flemming, B.W. (2000) The role of grain size, water depth and flow velocity as scaling factors controlling the size of subaqueous dunes, in Proceedings of the *First International Workshop on Marine Sandwave Dynamics*, edited by A. Trentesaux and T. Garlan, pp. 55–60, Univ. of Lille, Lille, France.

Grace Team, The (2007) GRACE, GNU licensed software online http://plasma-gate.weizmann.ac.il/Grace/.

Hennings, I., Dherbers, D., Prinz, K., Ziemer, F. 2004 *On waterspouts related to marine sandwaves*. In: Hulscher S, Garlan T, Idier D (eds) Proceedings of Marine Sandwave and River Dune Dynamics II, International Workshop, April 1–2 2004, University of Twente, The Netherlands,. 88–95.

Hoekstra, P, Bell, P, van Santen, P, Roode, N, Levoy, F, Whitehouse, R (2004) Bedform migration and bedload transport on an intertidal shoal. *Continental Shelf Research* 24(11):1249–126.

Nordin, C.F. & Algert, J.H. 1966. Spectral Analysis of sandwaves. *J. of the Hydr. Division*, ASCE 92(HY5) 95–114.

Press, W., Flannery, B., Teukolsky, S., Vetterling, W. 1992 *Numeric Recipes in C-The Art of Scientific Computing*. Cambridge, UR: Cambridge University Press.

Robert, A., Richards, K.S. 1988 On the modelling of sand bedforms using the semivariogram. *Earth Surface Processes and Landforms*.

Rubin, D.M. & McCulloch, D.S. 1980. Single and superimposed bedforms: a synthesis of San Francisco Bay and flume observations. *Sedimentary Geology* 26: 207–231.

Walker, D. 2003 Observations of the growth of wave-induced ripples using a spectral approach. *IAHR Symp. On River, Coastal and Estuarine Morphodynamics*. 341–351.

*River, Coastal and Estuarine Morphodynamics: RCEM 2007 – Dohmen-Janssen & Hulscher (eds)*
© *2008 Taylor & Francis Group, London, ISBN 978-0-415-45363-9*

# Evaluation of statistical properties of dune profiles

H. Friedrich
*The University of Auckland, Auckland, New Zealand*

A.J. Paarlberg & J. Lansink
*University of Twente, Enschede, The Netherlands*

ABSTRACT: A set experiments for dunes developing from a flattened sand bed was obtained in a narrow 0.44-m-wide and 12-m-long glass-sided open channel. The sand in use was a coarse uniform sand of $D_{50} = 0.85$-mm and was exposed to a series of steady and uniform flow conditions. The chosen flow depths generated practically 2D dunes in flow direction over the length of the channel. Spatial sand-bed-elevation profiles were recorded on the centreline of the flume over a distance of 6-m, roughly every 23-sec over the time of development. The recorded dune geometries were evaluated with both the discrete and the continuous approach. For the discrete approach, discrete height and length values are obtained for dunes. For the continuous approach, the second-order distribution moments, standard deviations, were used to obtain characteristic height and length values of the dune field. It is shown that a lack of clear definitions for the discrete approach results in a wide range of averaged dune geometries, depending on how thresholds were set during the analysis process. The continuous approach provides more objective results, but interpreting the results of the analysis requires careful consideration. For this paper, the analytical results of applying both approaches are compared. A preliminary correlation between both approaches is discussed. The physical relationship between the flow field and characteristic height and length values, as obtained through the continuous approach, is not yet clear.

## 1 INTRODUCTIONS

### 1.1 Background

Dunes are sediment transport objects found commonly on river beds and in coastal areas (Bridge, 2003). The objective of this study is to evaluate in detail two different methods of obtaining statistical properties of dune profiles in laboratory channels. Determining the topographical characteristics of the bed surface is important for forecasting the growth of dunes under different flow conditions and modelling alluvial channel processes. As dunes strongly influence the hydraulic roughness and the sediment transport, a quantitative description of dunes is vital in providing relationships between the dune parameters and the hydraulic conditions.

### 1.2 Dune bed forms

Dunes, also called large-scale bed forms, have discrete lengths generally larger than 0.6-m and influence and interact with the water flow (Ashley, 1990). A significant portion of the flow depth is occupied by dunes. Therefore, dune topographies affect the flow with accelerations and decelerations of the flow field

all the way to the free surface (Maddux, 2002). Dune wavelength $l$, and dune height $h$, are related to the flow depth $H$. In laboratory channels the discrete average length of developed dunes can be generally estimated as $6H$ (Yalin and da Silva, 2001). The averaged dune height can be estimated as $H/3$. The generated flow field over dunes is vastly different from the uniform flow field found over a flat bed (Best, 2005).

### 1.3 Analytical methods

Current methods for quantifying hydraulic roughness of river dunes generated in laboratory environments can be distinguished by two groups: a) discrete approach (also called direct approach); b) continuous approach (also called indirect or random-field approach). Commonly, the discrete approach is applied in determining geometric characteristics such as height and length of dune profiles. When applying the continuous approach, dune profiles are treated as a random field of sand-bed elevations.

### 1.4 Discrete approach methodology

Dunes are conventionally considered as periodic features with certain length, height and steepness

Figure 1. Roughly 2D dunes as generated in a narrow 0.44-m wide flume. Flow is from left to right. $D_{50} = 0.85$-mm.

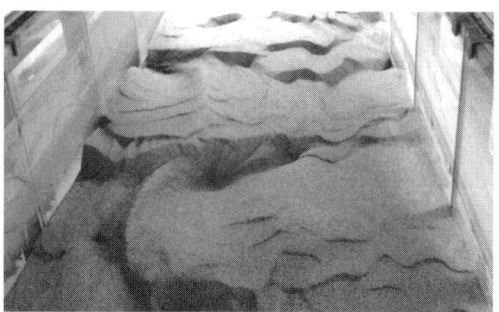

Figure 2. Irregular 3D dunes as generated in a wide 1.5-m wide flume. Flow is from top to bottom. Superposed sand sheets are clearly visible. $D_{50} = 0.85$-mm.

$(h/l)$ values. Average height and length parameters, obtained through discrete analysis, describe 2D dunes quite well. However, river dunes are highly irregular features, often not exposing such a periodic regularity in their natural environment. In reality the idealized triangular shapes of dunes (Figure 1), which form the basis of the discrete approach, do not exist (Figure 2). Additionally, the discrete approach has the drawback of being exposed to subjective identification (Van der Mark et al., 2005).

Commonly the height of a dune is the vertical distance between crest and trough. The first problem arises when dunes grow on cross-sets, where the vertical distance between crest and subsequent trough is different to the vertical distance of trough and subsequent crest. Furthermore the question arises, what constitutes a dune? Often sand sheets (Whiting et al., 1988; Venditti, 2003) migrate over the stoss side of dunes and can be identified in longitudinal records as individual dunes. If so, this so called superposition of dunes can reduce the statistical mean character of dune geometries significantly and averaged discrete height and length values will not represent the bed profile adequately.

When defining crest and trough elevations for dunes, one can determine local minima and maxima

and use them for the growth analysis. One also can determine the top and bottom positions of the lee-side and use them for the growth analysis. Additionally, Van der Mark et al. (2005) point out that some researchers take into account smaller-scale dunes whilst some do not. For instance Allen (1982) defines dunes as bed forms longer than 600-mm and higher than 40-mm. The introduction of such threshold values can influence the histograms and the fitted Probability Density Functions (PDFs) considerably.

Similar to the lack of agreement about how to define bed-form height, there exists a lack of agreement about how to define bed-form length. Van der Mark et al. (2005) summarizes that bed-form length can be defined as the horizontal distance between two successive mean bed level upcrossings. A mean bed level upcrossing is defined as the point where the upward going bed elevation profile crosses the mean bed level. Again it is questionable if all mean bed level upcrossings of the bed elevation should be treated as distinguished individual dunes or if some dunes should be excluded. Dune length can also be defined as the horizontal distance between two successive mean bed level downcrossings or between two successive troughs or between two successive crests.

Another problem arises during bed-form initiation. Depth-sounding recording devices are not sensitive enough to let researchers analyse early discrete bed-form geometries when generated from a flat bed. Research showed (Coleman and Melville, 1996) that a wavelength does not start at zero; instead bed-form initiation growth depends on the grain size. Measurement device inaccuracies and different spatial or temporal resolutions of the device between subsequent data points can also result in different inaccuracies when determining bed-form length and height characteristics.

Generally, an experimental data set of developing bed forms is analysed with the same criteria for the whole bed-form development. A criteria adjustment, depending at which stage of growth development the bed-form analysis is carried out, could result in a more objective statistical evaluation of bed-form growth.

### 1.5 Continuous approach methodology

Compared to the empirical distribution functions, which are obtained through discrete analysis, the continuous approach focuses on the characterization of the stochastic process, which generates the bed forms (Moll et al., 1987). Nordin and Algert (1966), Hino (1968) and Jain and Kennedy (1974) treated bed-form profiles as continuous longitudinal profiles and applied spectral analysis. Nikora et al. (1997) treated sand-bed elevation fields as random fields rather than discrete sand waves, using longitudinal and transverse spectra, correlation and structure functions to describe statistical sand wave dynamics.

The continuous approach is widely accepted when studying turbulent structures in water flow and is an important tool to identify coherent structures in the turbulent flow field. Scaling relationships are used to facilitate understanding of the flow dynamics. Researchers re-introduced the continuous approach to bed-form studies during the last decade (Nikora and Hicks, 1997; Nikora and Goring, 2000; Nikora et al., 1997). The continuous approach is based on the premise that random fields can be described completely by the $n$−dimensional probability density when $n \rightarrow \infty$. A complete quantitative description using the $n$-dimensional probability functions is theoretically possible, but practically beyond reach (Nikora et al., 1997). Therefore selected applications of higher-order distribution moments (mean, standard deviation, skewness and kurtosis) and moment functions (correlations, spectra, structure functions, etc.) are evaluated, allowing for a quantitative evaluation of sand-bed elevations. In the following, the focus is on analysing the change in standard deviation of bottom elevations of developing dune profiles. The framework of the moving-bottom elevation field model (Nikora and Hicks, 1997) is described in more detail in Section 3.2.

### 1.6 Outline

For this paper, the direct and the continuous approaches as applied for this study are introduced. The experimental setup is described. The results are compared and the relationship between the two approaches is discussed.

## 2 EXPERIMENTAL SETUP

A new data set of developing dune bed forms from a flattened sand bed was obtained at the Fluid Mechanics Laboratory at the University of Auckland. The experiments were conducted in a 12-m-long, 0.38-m-deep and 0.44-m-wide glass-sided open-channel flume. Water is recirculated, as well as the sediment. The sediment in use is a coarse ($D_{50} = 0.85$-mm) filter sand and was carefully flattened along the whole length of the flume. Once flattened, the flume was carefully filled with water, in order to not change the flat sand bed structure. At a given time $t = 0$ the pump was turned on and the spatial recording of the sand bed proceeded simultaneously.

Bed profiles were measured with the help of a moving carriage. On the rails of the flume, the moving carriage traversed the flume back and forth. A depth sounding probe was installed on the carriage and recorded the sand-bed elevation field for every downstream and upstream traverse. The initial position

of the moving carriage was located 4-m downstream of the inlet. The recordings covered a distance of 6-m, before the carriage stopped and returned to its initial position. A slight slippage of the chain-and-sprocket driven programmable-speed motorized carriage occurred, which was corrected for during the bed-profile analysis.

A depth-sounding system was used to record the profiles of the developing bed. The bed-profile measurement system is based on Coleman (1997), and has since been further improved. According to Coleman (1997), the accuracy of bed-elevation measurement is ±0.4-mm. The distance along the flume was measured using a potentiometer mounted on a 0.199-m diameter wheel running down the outside of the flume side-wall. The streamwise position was recorded at 2.45-mm distance increments, accuracy of measurement being ±1.23-mm.

Before each run, the slope of the flume ($S_b$) and water surface, as well as the temperature of the water were measured and recorded.

For each experiment, measurements were taken on the centreline of the flume. Therefore, a two-dimensional field of bottom elevations $z(x, t)$ was obtained for different flow fields. Bed profiles were recorded roughly every 23-sec for each experiment, lasting between 2.5 and 10 hours.

The mean depth-averaged flow velocity for each experiment was estimated by utilizing ADV flow measurements in the upper part of the water column, at the start of each experiment (with flat bed). The average flow velocity $U$ was determined by fitting a logarithmic velocity profile through the data, and assuming average velocity at a height of 1/e of the water depth above the sand bed. Two shear velocities were calculated for each flow setting. $u_*^{(1)}$ was calculated as $\sqrt{gHS_b}$ (see explanation of symbols in Table 1), whereas $u_*^{(2)}$ was determined based on the ADV measurements.

All together a set of 24 experiments were recorded, of which 14 experiments started with a flattened bed and were exposed to steady uniform flow conditions. The remaining 10 experiments were part of a flood-wave research project and a PIV flow-field study and will not be discussed herein. The experimental conditions for the experiments discussed are given in Table 1.

## 3 ANALYSIS

### 3.1 Discrete approach methodology

The discrete approach, based on the presentation of sand-bed elevations as individual discrete bed forms, is commonly used to characterize dune geometry and that information can be used for incorporation in hydraulic formulas.

Table 1. Experimental conditions

| Flow Conditions | Experiment Names | Flow parameters | | | | | | | | | | |
|---|---|---|---|---|---|---|---|---|---|---|---|---|
| | | B [m] | Q [l/s] | q [m2/s] | H [m] | B/H [-] | Sb [-] | U [m/s] | Fr [-] | Re [-] | $u_*^{(1)}$ [m/s] | $u_*^{(2)}$ [m/s] |
| I | IT3,T8 | 0.44 | 29.42 | 0.0669 | 0.1250 | 3.52 | 0.0010 | 0.53 | 0.48 | 66788 | 0.035 | 0.032 |
| II | T6,T23 | 0.44 | 35.94 | 0.0817 | 0.1250 | 3.52 | 0.0015 | 0.62 | 0.56 | 77604 | 0.043 | 0.036 |
| III | T4,T7,T11 | 0.44 | 38.56 | 0.0876 | 0.1250 | 3.52 | 0.0020 | 0.70 | 0.63 | 87633 | 0.050 | 0.041 |
| IV | T5,T24 | 0.44 | 27.19 | 0.0618 | 0.1000 | 4.40 | 0.0015 | 0.58 | 0.58 | 57709 | 0.038 | 0.035 |
| V | T13 | 0.44 | 20.46 | 0.0465 | 0.1000 | 4.40 | 0.0010 | 0.47 | 0.47 | 46508 | 0.031 | 0.028 |
| VI | T14 | 0.44 | 31.47 | 0.0715 | 0.1125 | 3.91 | 0.0015 | 0.62 | 0.59 | 69975 | 0.041 | 0.037 |
| VII | T15,T22 | 0.44 | 43.74 | 0.0994 | 0.1500 | 2.93 | 0.0015 | 0.65 | 0.54 | 97463 | 0.047 | 0.037 |

B = flume width, Q = discharge, q = specific discharge, H = water depth, $S_b$ = initial water surface and bottom slope, U = average flow velocity, Fr = Froudenumber = $U/sqrt(gH)$, Re = Reynoldsnumber = $UH/v$, Kinematic viscosity $v = 0.000001$-m2/s, $u_*$ = shear velocity, with $u_*^{(}1) = sqrt(gHSb)$, and $u_*^{(}2) =$ based on ADV measurements. Note, for experiment T22,T23,T24 the downstream water depth was changed during the experiment, such that the water-surface slope was approximately equal to the bed-surface slope; for the other experiments the downstream water level was kept constant to the initial value.

As introduced in Section 1.4, different algorithms can be used to determine discrete dune height h, dune length l and steepness of bed forms h/l. The applied algorithm for the recorded profiles is based on a routine which determines crest and trough positions based on identification of the lee face. This method requires a sufficiently small sampling resolution in the longitudinal direction. The following input threshold values are required:

- minimum dune height (default value: 0.005-m),
- minimum allowable horizontal separation distance between identified lee-face points (default value: 0.045-m),
- minimum number of points taken into account to constitute a lee face (default value: 6),
- minimum angle of the lee-face slope (default value: 12-degrees).

Figure 3 shows the results of applying the routine with different input parameters for each of the thresholds. Only one input threshold is changed in each case, the other three being set to the default values.

Figure 3a shows that the threshold for the dune height can influence the discrete height and length values for dune geometries significantly. For instance, a minimum dune height threshold of 1.5-cm results in smaller averaged discrete dune heights than a minimum threshold of 2.5-cm. Consistently, the number of discrete dunes which are identified is much larger for a smaller threshold than for a larger threshold.

Figures 3b and 3c show the change of statistical dune geometry for changing input thresholds for allowable horizontal separation distance between identified lee-face points and for the number of points which must constitute a lee face, respectively. One can clearly see that the geometric values of the developing dune geometries do not vary significantly with changing input thresholds for these discrete dune characteristics.

Figure 3d shows the influence of input thresholds for the minimum angle of the lee face. The statistical data here are more scattered than those in Figures 3b and 3c but still more uniform than the data displayed in Figure 3a.

In conclusion, it can be shown that the threshold for minimum dune heights as set during the analysis can skew the averaged discrete dune geometries significantly. A lower threshold results in depicting dune bed forms from early on during dune development. Additionally, sand sheets superposed on top of dune stoss sides, which develop later during an experiment, will be identified as independent bed forms and therefore reduce the average dune dimensions significantly. Alternatively, if a higher threshold for minimum dune heights is used for the analysis, dunes which develop early during bed-form development are not accounted for. Additionally, the averaged dune dimensions are substantially larger than for a smaller minimum dune height threshold.

### 3.2 Continuous approach methodology

The continuous approach, based on the presentation of sand-bed elevations as a 2D random field, is seen as an alternative to the conventional discrete approach. For the presented data set, the continuous approach considers the sand-bed surface as a random field $z(x)$ of bed elevations, where $x$ is the longitudinal (the main flow direction) coordinate. Depending on the measurement frequency, $z$ is also a function of time $t$. In order to compare statistical results relating to dune geometry, the detail of the continuous approach for this paper is restricted to the analysis of the change of the

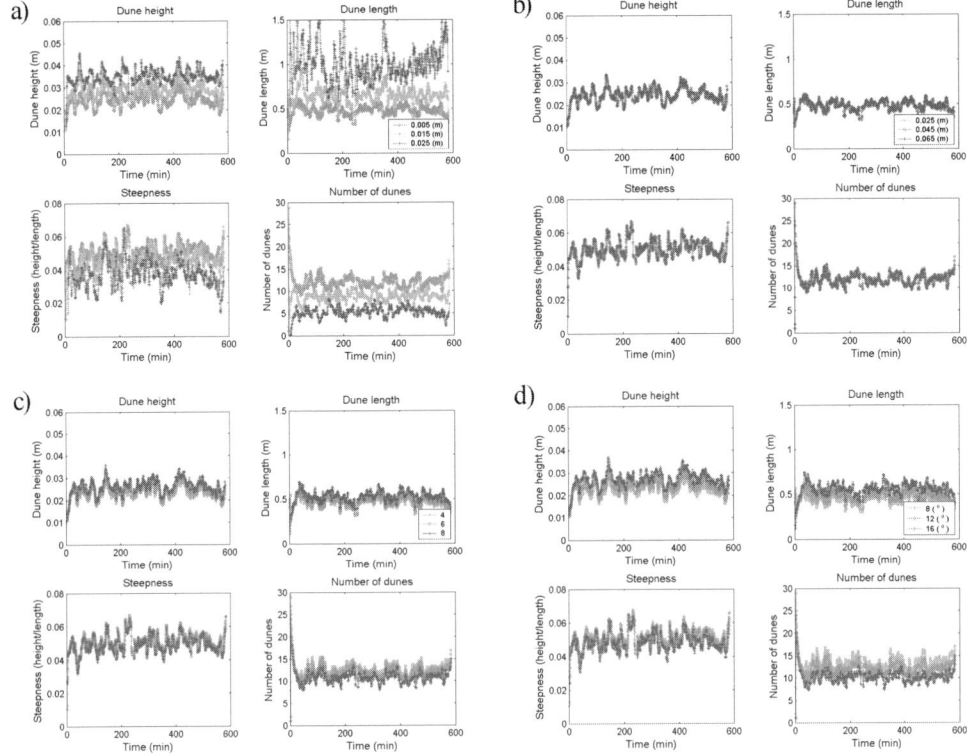

Figure 3.  Discrete bed-form statistics of experiment T14. a) Change in threshold for dune height; b) change in threshold for allowable horizontal separation distance between identified lee-face points; c) change in threshold for number of sample points of lee face; d) change in threshold for angle of lee face.

second-order distribution moments, i.e. standard deviations of bottom elevations, during dune development.

Nikora and Hicks (1997) introduce characteristic height and length values for sand-wave fields that are considered as a random field of bottom elevations. The root-mean-square deviation $\sigma_z(\delta x, t)$ over a certain distance $\delta x$ (ranges from the smallest possible sub distance available for the spatial profile up to the whole length of the spatial profile) along the flume, at a certain time $t$ during development, can be determined for every spatial profile during dune development starting from a flattened bed (Figure 4). When $\delta x \geq \delta x_0$ and $t \geq t_e$ ($t_e$ is the development time necessary for establishment of equilibrium conditions), $\sigma_{z0} = \sigma_z(\delta x, t)$. When $\sigma_z(\delta x, t_e)$ becomes saturated for equilibrium conditions, $\delta x_0 = \delta x$ (see Nikora and Hicks (1997) for more details).

By analogy to height and length parameters being inter-changeable growth parameters when applying the power law (Nikora and Hicks, 1997) for discrete sand waves, there must exist a similarity relationship between $\sigma_z(\delta x)$ and $\delta x$ for random fields. For self-similar features, which longitudinal spatial sand-wave profiles are considered to be, $\sigma_z(\delta x) \propto \delta x$ (Figure 5).

Therefore $\delta x$ is a characteristic scaling feature of the developing dune bed, similar to the wave length feature of discrete sand wave analysis. $\delta x$ will grow similar to a discrete wave length during the development of the dune bed and will reach an equilibrium stage once saturation is reached.

## 4  RESULTS AND DISCUSSION

Dune development was recorded for seven different flow conditions. Table 2 shows the equilibrium values of characteristic height and length obtained through the continuous analysis as well the averaged dune height and length as obtained through the discrete analysis. Additionally bed-form geometry predictions after van Rijn (1984) and Julien and Klaassen (1995) are shown. The bed-form predictors generally overestimate the discrete dune dimensions, as they are based on mostly field data. For our laboratory data, the flow field was practically two-dimensional. The minimum width/ depth ratio is just under 3. Therefore side walls are expected to influence the dune formations in close proximity to the walls, but the centreline spatial dune

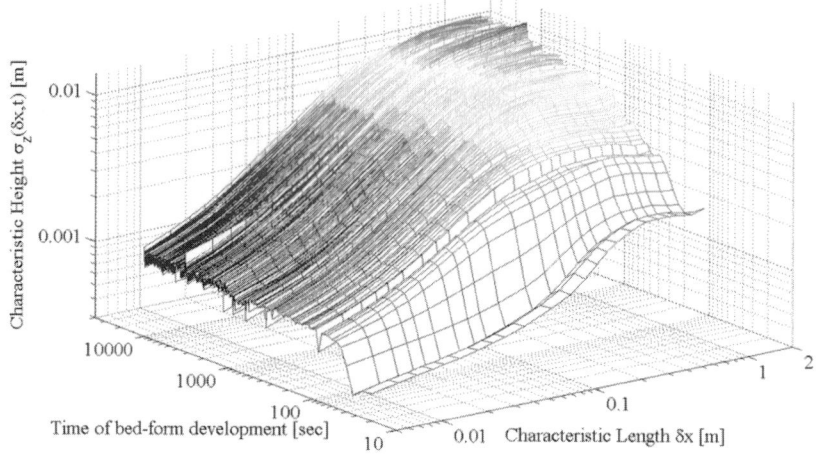

Figure 4. Development of characteristic height $\sigma_z$ and characteristic length $\delta x$ during dune development (here shown for experiment T13). It shows the initial fast development of the characteristic height $\sigma_z$, as well as the asymptotic approach of the characteristic length $\delta x$ towards a saturated value during development of an equilibrium dune bed.

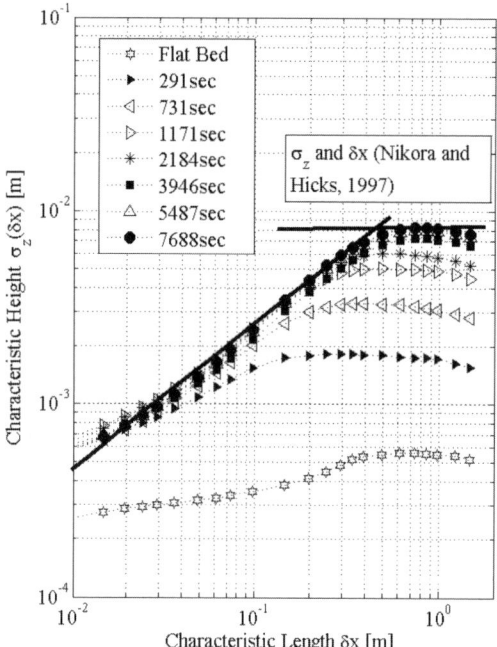

Figure 5. Exemplary functions $\sigma_z(\delta x, t)$ for experiment T3 during dune development starting from a flat bed. The self-similarity of dunes is characterized by the similar slopes in the scaling region after initial dune development.

profiles can be regarded as recording regular 2D dunes formed under interaction with 2D flow.

Figure 6 shows a typical comparison between dimensionless (in regards to the flow depth $H$)

characteristic dune statistics and averaged discrete dune statistics, as obtained for experiment T3. The general trend as shown in Figure 6 is valid for all analysed experiments.

From Figure 6 it can be seen that averaged discrete height and length show a slightly jagged growth, compared to a statistically smoother growth for characteristic dune geometries. Besides the discussed variations of dune geometries (caused by applying different thresholds during the discrete analysis), other factors can be associated with the more jagged statistical growth behaviour of the discrete parameters, such as the limited spatial length of the profiles. The measurement section is only 6-m in length. This implies that average discrete dune dimensions are strongly influenced by dunes of different sizes migrating in and out of the recording length, resulting in the more jagged statistical growth.

The statistical growth as obtained with the help of the standard deviation of bottom elevations, however, results in a smoother growth, indicating that this analysis provides a more objective description of the actual averaged dune dimensions.

Both approaches, the direct and the continuous analysis, are related. Nikora et al. (1997) state that the average height $h$ of bed forms relates to the standard deviation of bottom elevations $\sigma_z$ in the following way:

$$h = m\sigma_z \qquad (1)$$

where $m$ is in the range 1.7 to 2 for rivers. Table 2 shows values of $m$ for our laboratory data for equilibrium conditions, which are all significantly larger than 2. Figure 7 shows the development of $m$ during dune development for experiment T3. The general

Table 2.  Dune Parameters.

| Experiment Name | Flow conditions | Rec. Profiles [-] | Development Time [hrs] | Bed-form geometry (continuous) Char. Height [cm] | Char. Length [cm] | Steepness [-] | Bed-form geometry (discrete) Height [cm] | Length [cm] | Steepness [-] | m (Eq.1) [-] | Bed-form estimators (Van Rijn, 1984) Height [cm] | Length [cm] | Steepness [-] | Bed-form estimators (Julien and Klalaassen 1995) Height [cm] | Length [cm] | Steepness [-] | H/3 Height [cm] | 6H Length [cm] |
|---|---|---|---|---|---|---|---|---|---|---|---|---|---|---|---|---|---|---|
| T3 | I | 563 | 3.50 | 0.71 | 112 | 0.0064 | 2.5 | 67 | 0.038 | 3.5 | 4.17 | 91 | 0.046 | 6.87 | 78.54 | 0.087 | 4.2 | 75 |
| T4 | III | 412 | 2.53 | 0.83 | 99 | 0.0084 | 2.9 | 54 | 0.054 | 3.5 | 5.55 | 91 | 0.061 | 6.87 | 78.54 | 0.087 | 4.2 | 75 |
| T5 | IV | 391 | 2.44 | 0.53 | 90 | 0.0059 | 2.6 | 60 | 0.043 | 4.8 | 4.06 | 73 | 0.056 | 5.87 | 62.83 | 0.093 | 3.3 | 60 |
| T6 | II | 482 | 3.03 | 0.90 | 130 | 0.0070 | 2.9 | 50 | 0.057 | 3.2 | 5.26 | 91 | 0.058 | 6.87 | 78.54 | 0.087 | 4.2 | 75 |
| T7 | III | 325 | 3.02 | 0.64 | 100 | 0.0064 | 3.2 | 54 | 0.059 | 5.0 | 5.55 | 91 | 0.061 | 6.87 | 78.54 | 0.087 | 4.2 | 75 |
| T8 | I | 524 | 3.60 | 0.79 | 111 | 0.0071 | 2.8 | 75 | 0.038 | 3.6 | 4.17 | 91 | 0.046 | 6.87 | 78.54 | 0.087 | 4.2 | 75 |
| T11 | III | 458 | 2.92 | 0.91 | 159 | 0.0057 | 3.2 | 63 | 0.051 | 5.5 | 5.55 | 91 | 0.061 | 6.87 | 78.54 | 0.087 | 4.2 | 75 |
| T13 | V | 1137 | 8.57 | 0.67 | 144 | 0.0046 | 2.3 | 65 | 0.036 | 3.5 | 2.85 | 73 | 0.039 | 5.87 | 62.83 | 0.093 | 3.3 | 60 |
| T14 | VI | 1496 | 9.72 | 0.77 | 96 | 0.0080 | 2.6 | 58 | 0.044 | 3.4 | 4.69 | 82 | 0.057 | 6.38 | 70.69 | 0.090 | 3.7 | 67.5 |
| T15 | VII | 676 | 4.39 | 1.12 | 158 | 0.0071 | 3.4 | 62 | 0.055 | 3.1 | 6.24 | 110 | 0.057 | 7.80 | 94.25 | 0.083 | 5.0 | 90 |
| T22 | VII | 455 | 2.90 | 1.48 | 161 | 0.0092 | 4.1 | 67 | 0.061 | 2.8 | 6.24 | 110 | 0.057 | 7.80 | 94.25 | 0.083 | 5.0 | 90 |
| T23 | II | 503 | 3.26 | 1.33 | 126 | 0.0106 | 3.6 | 69 | 0.052 | 2.7 | 5.26 | 91 | 0.058 | 6.87 | 78.54 | 0.087 | 4.2 | 75 |
| T24 | IV | 575 | 3.74 | 1.07 | 149 | 0.0072 | 3.2 | 60 | 0.052 | 3.0 | 4.06 | 73 | 0.056 | 5.87 | 62.83 | 0.093 | 3.3 | 60 |

trend shown in Figure 7 is valid for all experiments, with $m$ decreasing during dune development and fluctuating around a certain mean value (between 2.7 and 5; see Table 2) once the dune bed is in equilibrium. Most of the fluctuation is attributed to the migration of dunes in and out of the recording region, which influences the discrete averaged dune heights significantly.

5 CONCLUSION

The present study was conducted in order to investigate two different deterministic and statistical analysis tools for dunes and to compare the results. A flattened sand bed of coarse sand was exposed to different flow strengths in a narrow 0.44-m-wide flume and the spatial development of dunes was recorded on the centreline of the flume.

Sand-bed elevation fields are treated in two different ways: as discrete bed forms and as a random field of bed-form elevations.

The discrete analysis utilizes a routine which can filter discrete dunes of less than a certain geometric threshold criterion, and exclude them from computing average dune properties. Slight variations of the input thresholds for detectable minimum dune height varies the statistical dune geometries significantly, although the general trend of exponential growth from initial bed features to equilibrium dunes is visible for all experiments.

In contrast, the continuous analysis results in unambiguous statistical characterisation of the growth of dunes during development from a flattened sand bed.

For total dune development, the value of $m = h/\sigma_z$ (Eq.1) is significantly larger than 2, the value associated with river data.

This paper shows that characteristic height and length values are independent of subjective thresholds during statistical analysis and describe the complete growth of average dune dimensions without discontinuity. This approach provides a substantial advantage compared to the traditional discrete approach.

More analysis and comparison of additional experimental data sets are required in order to improve the suggested relationship between these characteristic dune statistics and discrete dune geometries. Such a relationship can then be utilized for inclusion in flow resistance formulas and to help better understand the feedback mechanism between coherent flow structures and river morphology.

ACKNOWLEDGEMENTS

The project is supported by the Technology Foundation STW, the applied science division of NWO and the technology programme of the Ministry of Economic Affairs (Project No. TCB.6222). The authors wish to

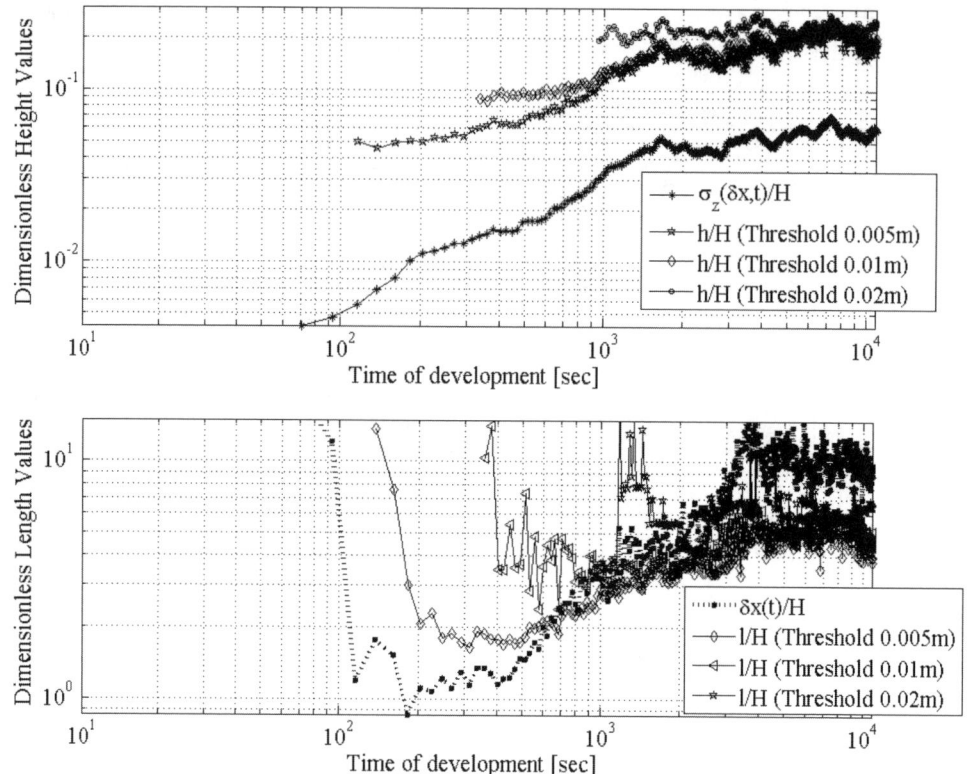

Figure 6. Relationship of averaged dimensionless discrete height $h/H$ and length $l/H$ values (different minimum dune height thresholds are applied) and dimensionless characteristic height $\sigma_z(\delta x, t)/H$ and length $\delta x(t)/H$ values during dune development. $m$ (Eq. 1) does change significantly during dune development (see also Figure 7). The development is shown for the first 3-hrs for experiment T3.

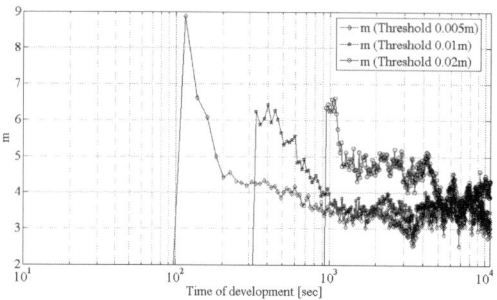

Figure 7. Development of $m$ during dune development. The development is shown for experiment T3.

acknowledge the support by the technical staff, Geoff Kirby and Jim Luo. We are furthermore grateful to Assoc. Prof. S. Coleman and T. Clunie for their input during the experimental stage. Useful discussions with Prof. B. Melville were very helpful and improved the quality of the paper.

REFERENCES

Allen, J. R. L. 1982. Sedimentary structures – Their character and physical basis, Elsevier Scientific Publishing Company.

Ashley, G. M. 1990. Classification of large-scale subaqueous bedforms: a new look at an old problem. Journal of Sedimentary Petrology, 60(1), 160–172.

Best, J. 2005. The fluid dynamics of river dunes: A review and some future research directions. Journal of Geophysical Research. 110, F04S02.

Bridge, J. S. 2003. Rivers and floodplains: forms, processes, and sedimentary record, Malden, MA: Blackwell Pub., Malden, MA.

Coleman, S.E., and Melville, B.W. 1996. Initiation of Bed Forms on a Flat Sand Bed. Journal of Hydraulic Engineering, 122(6), 301–309.

Coleman, S. E. 1997. Ultrasonic Measurement of Sediment Bed Profiles. 27th Congress of the International Association for Hydraulic Research, San Francisco, California, U.S.A., B221-B226.

Hino, M. 1968. Equilibrium-range spectra of sand waves formed by flowing water. Journal of Fluid Mechanics, 34(3), 565–573.

Jain, S. C., and Kennedy, J. F. 1974. The spectral evolution of sedimentary bed forms. Journal of Fluid Mechanics, 63(2), 301–314.

Julien, P. Y. and Klaassen, G. J. 1995. Sand-dune geometry of large rivers during floods. Journal of Hydraulic Engineering, 121 (9), 657–663.

Maddux, T. B. 2002. Turbulent Open Channel Flow over Fixed Three-Dimensional Dune Shapes, Doctoral Thesis, University of California, California, U.S.A., Santa Barbara.

Moll, J. R., Schilperoort, T., and De Leeuw, A. J. 1987. Stochastic Analysis of Bedform Dimensions. Journal of Hydraulic Research, 25(4), 465–478.

Nikora, V., and Hicks, D. M. 1997. Scaling Relationships for Sand Wave Development in Unidirectional Flow. Journal of Hydraulic Engineering, 123(12), 1152–1156.

Nikora, V. I., and Goring, D. G. 2000. Sand waves in unidirectional flows: Scaling and intermittency. Physics of Fluids, 12(3), 703–706.

Nikora, V. I., Sukhodolov, A. N., and Rowinski, P. M. 1997. Statistical sand wave dynamics in one-directional water flows. Journal of Fluid Mechanics, 351, 17–39.

Nordin, C. F., and Algert, J. H. 1966. Spectral Analysis of Sand Waves. Journal of Hydraulic Division, 92(HY5), 95–114.

Van der Mark, C. F., Blom, A., Hulscher, S. J. M. H., Leclair, S. F., and Mohrig, D. 2005. On modeling the variability of bedform dimensions. In River, Coastal and Estuarine Morphodynamics. RCEM 2005, 831–841.

Van Rijn, L. C. 1984. Sediment transport, part III: Bed forms and alluvial roughness. Journal of Hydraulic Engineering, 110 (12), 1733–1754.

Venditti, J. G. 2003. The Initiation and Development of Sand Dunes in River Channels, Doctoral Thesis, University of British Columbia, Vancouver.

Whiting, P. J., Dietrich, W. E., Leopold, L. B., Drake, T. G., and Shreve, R. L. 1988. Bedload sheets in heterogeneous sediment. Geology, 16, 105–108.

Yalin, M. S., and Ferreira Da Silva, A. M. 2001. Fluvial Processes, Balkema, Delft, The Netherlands.

*River, Coastal and Estuarine Morphodynamics: RCEM 2007 – Dohmen-Janssen & Hulscher (eds)*
*© 2008 Taylor & Francis Group, London, ISBN 978-0-415-45363-9*

# Variability in bedform characteristics using flume and river data

Caroline F. van der Mark
*University of Twente, Water Engineering & Management, Enschede, The Netherlands*

Astrid Blom
*Delft University of Technology, Environmental Fluid Mechanics, Delft, The Netherlands*

Suzanne J.M.H. Hulscher
*University of Twente, Water Engineering & Management, Enschede, The Netherlands*

ABSTRACT: Measured bed elevation profiles show that bedforms such as river dunes and sand waves are far from regular. Even under controlled steady flow conditions in laboratory flumes bedforms are irregular in size, shape and spacing, also in case of well sorted sediment. In this paper we study the variability in geometric characteristics of bedforms as measured in laboratory and field studies. We have developed a bedform tracking tool to determine, as objectively as possible, geometric properties of bedforms from measured bed elevation profiles. For each flume and field data set we analyzed variability in (1) trough elevation, (2) crest elevation, (3) bedform height, (4) bedform length, (5) slope of the bedform lee face. We find that the variability in bedform characteristics in field measurements is comparable to flume experiments, except for the lee face slope. This means that the stochastics of geometric characteristics of bedforms as found in the flume are representative for the stochastics found in the field.

## 1 INTRODUCTION

Bedforms such as river dunes or marine sand waves are rhythmic bed features which develop by the interaction between water flow and sediment transport. Often river dunes are schematized as a train of regular triangular features and sand waves as a sinusoidal wave train. Measured bed elevation profiles in a laboratory flume or in the field show that bedforms are not regular, even under steady conditions and for well-sorted sediment (e.g., Nordin 1971; Paola and Borgman 1991). Bedforms are usually irregular in size, shape and spacing (Figure 1).

For several economic activities water managers do not only require knowledge on mean values of geometric properties, but also on the more extreme values. Navigational channels in seas and rivers are dredged to keep them navigable. It is then desirable to have information on the highest bedforms or highest crest elevations. Pipelines and cables buried in the river or sea bed should not be exposed to the flow as the result of a migrating deep bedform trough. Safety against uplifting of a tunnel constructed underneath the river bed is no longer guaranteed if a deep trough of a large dune migrates above the tunnel (Amsler and García 1997). Knowledge on the deepest troughs is therefore important.

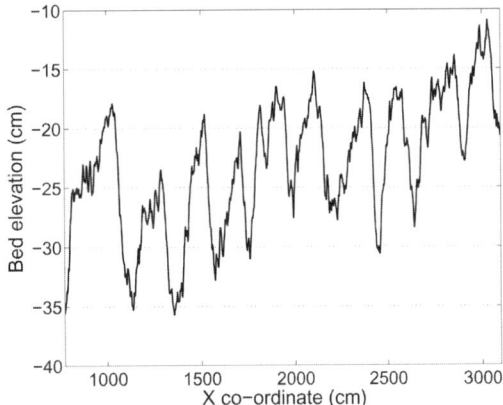

Figure 1. Bedforms are irregular in size, shape and spacing. This example shows a measured bed elevation profile of the North Loup River (Mohrig 1994; Mohrig and Smith 1996; Jerolmack and Mohrig 2005).

Variability in geometric properties of bedforms also needs to be taken into account for some modeling purposes. We illustrate this with three examples. First, we need information on the variability in trough elevations to reconstruct the original heights of bedforms from the thickness of cross-strata in preserved deposits. The variability in trough elevations mainly determines the cross-set thickness of preserved bedforms (e.g., Paola and Borgman 1991; Leclair 2002). Secondly, we also require information on the variability in trough elevations for modeling vertical sorting within bedforms. A sub-model for variability in trough elevations is required as a sub-model for a stochastic model for mass conservation of sediment mixtures (Blom et al. 2006). The third example considers the focus of our research, namely the effect of the variability in geometric properties of bedforms upon form roughness. Form drag due to the presence of bedforms results in a component of flow resistance that is often called form roughness. As form roughness depends on the size, shape and spacing of the bedforms (Allen 1983; Nelson et al. 1993), we hypothesize that the variability of geometric properties of individual bedforms within a reach affects the reach-averaged form roughness. We ground this hypothesis by making an analogy between grain roughness and form roughness. Van Rijn (1982) and others have found that mainly a relatively large grain size is representative in its effect on grain roughness, as it protrudes more into the flow. As such, often the 84% grain size ($D_{84}$) or the 90% grain size ($D_{90}$) is used as a representative diameter of the grains in predicting the grain roughness height. Analogously, form roughness may also be determined by the highest, longest or steepest bedforms.

In this paper we compare flume experiments to field measurements with respect to the variability in bedform characteristics.

## 2 DATA

### 2.1 Flume Experiments

We use laboratory flume data of Driegen (1986), Klaassen (1990), Leclair (2002) and Blom et al. (2003) to study variability in bedform characteristics. The experiments of Driegen (1986), Klaassen (1990) and Blom et al. (2003) were conducted in the Sand Flume of WL Delft Hydraulics in the Netherlands. Leclair (2002) performed a series of runs under varying flow conditions at Binghamton University (BU), New York, USA. The present paper is based on the BU runs in which no net aggradation occurred. The original bed elevation profiles of all flume experiments are available. Except for the experiments of Leclair (2002) (where bed elevation profiles were measured in the center), bed elevation profiles were measured in the center of the flume, as well as left and right from the center. We only consider measured data from the flume region that was not influenced by the entrance and exit of the flume. All measurements were taken under equilibrium (i.e. steady and uniform) conditions, which means that bedform characteristics, flow and sediment transport rate varied around steady mean

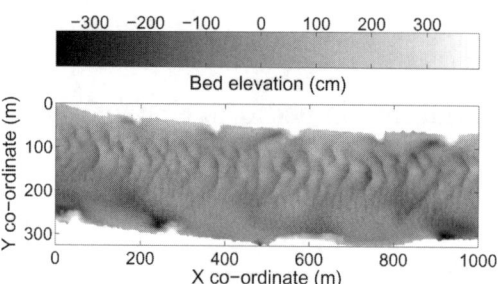

Figure 2. Bed elevation measurements of a part of the Waal River. The section shown here has a length of 1000 m. The width of the river is approximately 250 m. The flow is from right to left.

values, and that there was no net degradation or aggradation.

### 2.2 Field measurements

We study field data from a part of a branch of the Rhine River in the Netherlands, i.e. the Waal River, as well as field data from the North Loup River, Nebraska, USA (Mohrig 1994; Mohrig and Smith 1996; Jerolmack and Mohrig 2005).

The measurements made within the main channel of the Rhine River branch are multi-beam echo sounder measurements taken by the Dutch Ministry of Transport, Public Works and Water Management (*Rijkswaterstaat*) in April 2006. The considered reach is 6 km long and 250 m wide. The considered reach of the Waal River has a sandy bed ($D_{10} \approx 0.4$ mm, $D_{50} \approx 0.8$ mm, $D_{90} \approx 2$ mm).

The topographic data of the braided North Loup River are derived from low-altitude aerial photography. The river has a bed consisting of well-sorted medium sand (median grain diameter $D_{50} = 0.31$ mm) (Jerolmack and Mohrig 2005). Here we consider observations taken on 2 days (13 and 22 July 1990), taken with an interval of 2 minutes and 1 minute, respectively, for a period of 2 hours and 40 minutes, respectively. The observed section of the river was 30 m long and 15 m wide. Approximately constant river stage ensured that flow was essentially steady over the observation period (Jerolmack and Mohrig 2005).

## 3 DATA PROCESSING

### 3.1 Longitudinal bed elevation profiles

We use original bed elevation profiles (BEPs) in analyzing variability in bedform characteristics. BEPs from the flume were measured in longitudinal direction, but for the field measurements we convert the original bed elevations given in X and Y co-ordinates to bed elevations in the downstream flow direction. In order to study bedform characteristics of a run during steady state, the considered BEPs grouped together need to be statistically homogeneous in space and time. In other words, individual bedforms in a bed elevation profile (BEP) may differ as bedforms continuously merge and split, but the statistics of BEPs must be homogeneous (Paola and Borgman 1991). We verify whether BEPs measured at a given location at various moments in time are statistically homogeneous using a spatial scaling technique (Jerolmack and Mohrig 2005; Nikora and 1997).

Here we provide a short description of the spatial scaling technique. A measure of roughness is the root mean square of bed elevations, sometimes referred to as the interface width, $\omega$ (Barabási and Stanley 1995; Jerolmack and Mohrig 2005). The scaling of interface width with observed length or window size, $l$, provides, among other things, the characteristic bedform length. We plot interface width against window size for BEPs taken at different times but at the same location. It appears that BEPs measured at the same location show equal spatial scaling of interface width. The slope, characterizing the scaling of elevation fluctuations (Barabási and Stanley 1995; Dodds and Rothman 2000), and the location of the inflection point, characterizing the characteristic bedform length (Jerolmack and Mohrig 2005) are the same. This suggests that the scaling of roughness elements is stationary and that the river bed maintains a statistical steady state in terms of roughness (Jerolmack and Mohrig 2005). We refer to Jerolmack and Mohrig (2005) for a more detailed description of the scaling technique.

We also use the spatial scaling technique to divide the Waal River into statistically homogeneous parts. Figure 2 shows that bedform characteristics at the left bank differ from those at the right bank; bedform length at the right bank is larger than at the left bank. This can also be seen in the spatial scaling plot (Figure 3). The inflection point of the BEP at the right bank is located at a larger window size than the inflection points of the BEPs at the center and at the left bank. This indicates that characteristic bedform length at the right bank is larger than at the center and at the left bank. We use the technique to divide the Waal River into smaller parts, so that in each part the BEPs are statistically homogeneous.

### 3.2 Determination of bedform characteristics

We use the longitudinal BEPs to determine the geometric characteristics of bedforms. There exist several methods to find crest and trough locations and determine the geometric characteristics of individual bedforms (e.g., Prent 1998). Possibilities are (1) to select crests and troughs manually, (2) to find local

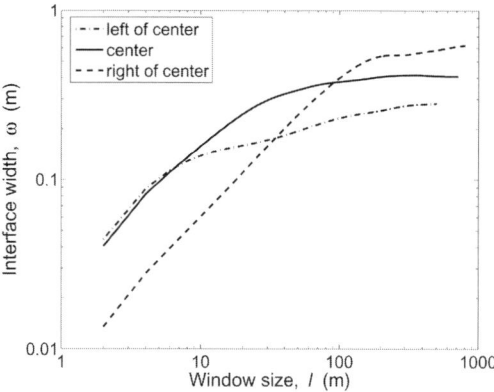

Figure 3. Example of the Waal River spatial scaling of roughness for three longitudinal bed elevation profiles. The bed elevation profiles are located at the center, at the left bank, and at the right bank of the river, respectively. The inflection points for the three bed elevation profiles differ, indicating that characteristic bedform lengths differ for the three bed elevation profiles.

extremes and select bedform heights and bedform lengths by introducing threshold values, (3) to find crests and troughs between zero upcrossings and zero downcrossings.

Other considerations in the determination of the geometric characteristics of the bedforms are how to detrend the BEPs (e.g., by fitting a linear line or by applying a moving average), and how to define the geometric characteristics of bedforms. For instance, bedform length can be defined as the distance between two consecutive crests or as the distance between two consecutive troughs. The chosen method to find crest and trough locations and the above considerations may influence the resulting bedform characteristics (Prent 1998) and are usually made subjectively on the basis of the whole bed configuration (Crickmore 1970). In order to compare various sets of experiments, we need (a) to use the same method for finding crests and troughs and (b) to use the same definitions for bedform characteristics for each data set. Therefore, it is generally not desirable to compare bedform data of different researchers if the original BEPs are lacking (Crickmore 1970).

We developed a bedform tracking tool (BTT), which determines the bedform characteristics of each individual bedform from the original BEPs as objectively as possible (Van der Mark and Blom 2007). Roughly, the procedure of the BTT is as follows:

1. For each BEP we find and replace outliers.
2. For each BEP we determine the trend line. For flume experiments under steady and uniform conditions (i.e. no spatial variations), this trend line is a linear fit to the measured bed elevations. For field

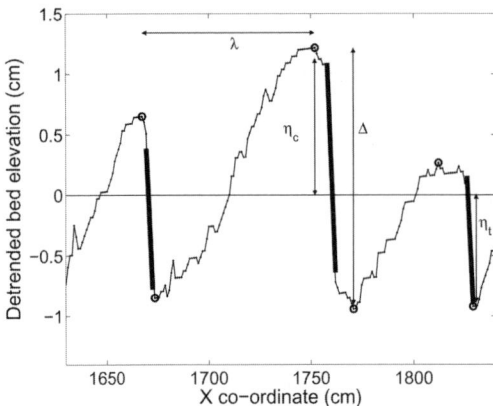

Figure 4. Definitions of the determined bedform characteristics in a detrended BEP: $\lambda$ denotes bedform length, $\Delta$ denotes bedform height, $\eta_c$ and $\eta_t$ denote crest elevation and trough elevation, respectively. The lee face slope is indicated with the bold lines. Crests and troughs are indicated with circles. Flow is from left to right.

measurements, we determine the trend line using a weighted moving average procedure.

3. We detrend the BEP using the trend line.
4. We apply a weighted moving average filter which yields a filtered BEP. The filtered BEP is only used to avoid the effect of small fluctuations in the BEP around the zero line on the resulting zero up- and downcrossings.
5. We determine zero upcrossings and zero downcrossings in the filtered BEP.
6. We determine crests and troughs. A crest is located at the maximum value between a zero up- and zero downcrossing. A trough is located at the minimum value between a zero down- and zero upcrossing.
7. We determine the geometric characteristics of individual bedforms (crest locations $\eta_c$, trough locations $\eta_t$, bedform heights $\Delta$, bedform lengths $\lambda$, lee face slope $S$).

Figure 4 shows how geometric characteristics are defined in the detrended BEP. For determining the lee face slope, we do not consider the whole lee face region between crest and its subsequent trough. We exclude a distance of one sixth of the bedform height below the crest and a distance of one sixth of the bedform height above the trough as these regions are usually transitional areas.

In the development of the BTT, we have tried to avoid subjective decisions as much as possible. We have written the numerical code such that it can easily be applied to each data set, without the necessity to 'tune' the code to the data set. At the moment, we have applied the code successfully to flume data, river data, and marine sand wave data (De Koning 2007).

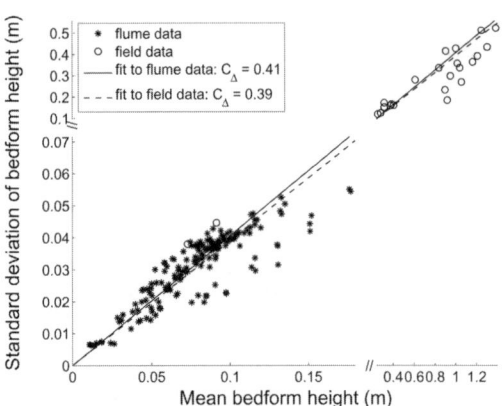

Figure 5. Standard deviation of bedform height versus mean bedform height. $C_\Delta$ denotes the coefficient of variation of bedform height.

## 4 RESULTS

We study the stochastics of geometric properties of bedforms by analyzing the coefficient of variation (Section 4.1), the probability density function (Section 4.2) and the 95% value of the bedform characteristics (Section 4.3).

### 4.1 Coefficient of variation

We study the variability in bedform height by determining for each experiment (1) the bedform height averaged over the statistically homogeneous BEPs, (2) the standard deviation in bedform height and (3) the coefficient of variation $C$, which is defined as the standard deviation divided by the mean value. Likewise, we study the coefficient of variation of bedform length, crest elevation, trough elevation and lee face slope. We compare the results of field data and flume data.

Figure 5 shows the standard deviation as a function of the mean value for bedform height. Each data point in the figure (192 in total) represents a statistically homogeneous flume bed (168 data points) or river bed (24 data points). Figure 5 shows that a linear relationship exists between the standard deviation of bedform height and the mean bedform height. We determine the mean coefficient of variation of bedform height $C_\Delta$ by taking the average of the coefficients of variation for the individual data points. Figure 5 distinguishes between flume experiments and field measurements, and shows that the difference between the mean coefficient of variation of bedform height for flume experiments ($C_\Delta = 0.41$) and the one for field experiments ($C_\Delta = 0.39$) is negligible. We thus propose a generic value for the coefficient of variation of bedform height, $C_\Delta = 0.40$.

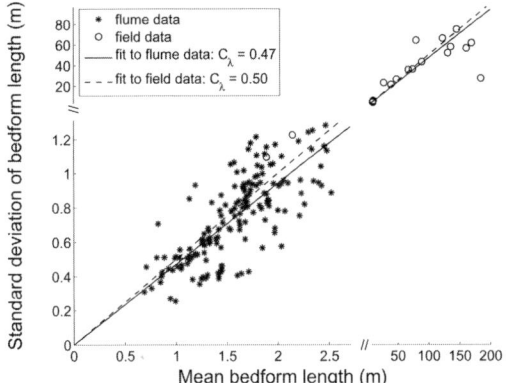

Figure 6. Standard deviation of bedform length versus mean bedform length. $C_\lambda$ denotes the coefficient of variation of bedform length.

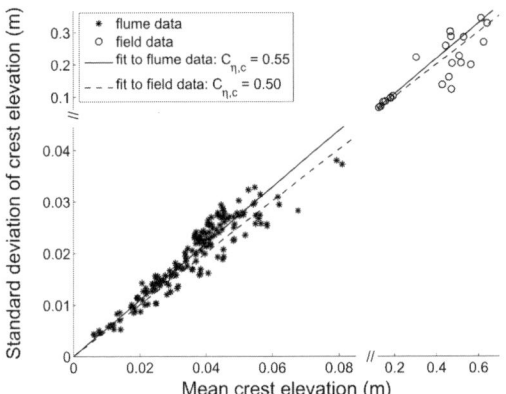

Figure 7. Standard deviation of crest elevation versus mean crest elevation. $C_{\eta,c}$ denotes the coefficient of variation of crest elevation.

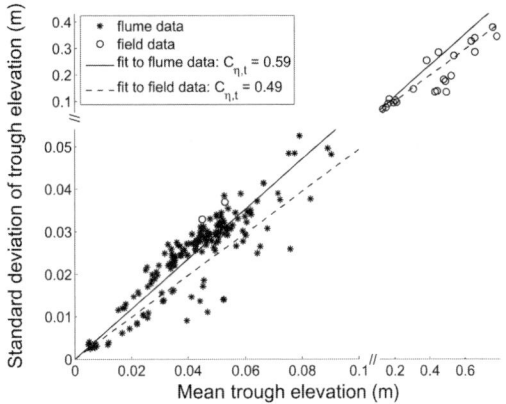

Figure 8. Standard deviation of trough elevation versus mean trough elevation. $C_{\eta,t}$ denotes the coefficient of variation of trough elevation.

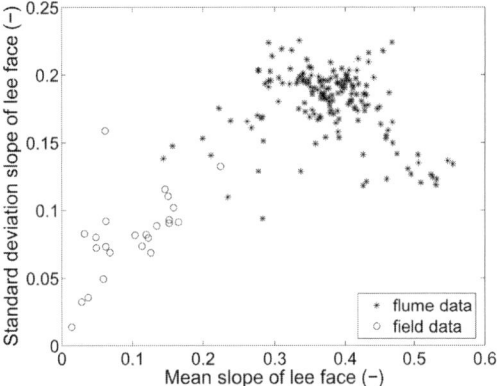

Figure 9. Standard deviation of lee face slope versus mean lee face slope.

We show similar plots of the standard deviation against the mean value for bedform length (Figure 6), crest elevation (Figure 7), trough elevation (Figure 8), and slope of the bedform lee face (Figure 9). We see a linear relationship between standard deviation and mean value for bedform length, crest elevation and trough elevation. Differences between the mean coefficients of variation for flume experiments and these for field experiments are small. We therefore propose the following general values: the coefficient of variation of bedform length $C_\lambda = 0.50$, the coefficient of variation of crest elevation $C_{\eta,c} = 0.52$, and the coefficient of variation of trough elevation $C_{\eta,t} = 0.55$.

For the Calamus River, Nebraska, USA, Gabel (1993) found values of $C_\Delta$ varying between 0.36 and 0.53. For dune length Gabel (1993) reports values of $C_\lambda$ between 0.30 and 0.55. The coefficients of variation of bedform height and bedform length found here are

in agreement with the values found by Gabel (1993). The values of the coefficients of variation also agree with the ones in the flume experiments of Wang and Shen (1980) and Leclair et al. (1997).

A coefficient of variation of bedform height for flume experiments that is representative to the coefficient of variation of bedform height for field measurements means that we can use the flume experiments for studying the variability in bedform heights. The same holds for bedform length, crest elevation and trough elevation. This is advantageous, as flume measurements are usually more accurate than field experiments and can be conducted under pre-defined flow and sediment conditions. External forces occurring in the field (e.g. waves or wind) are not present in a flume. As such, flume studies are valuable to get detailed insight in how the stochastics of geometric properties of bedforms are related to flow and sediment conditions.

Figure 9 shows the standard deviation against the mean value for the lee face slope. Figure 9 shows that the scatter is large and that no clear linear relationship exists between the standard deviation and the mean value for the lee face slope. Very roughly, we may see a linear trend in the field data ($C_{S,field} \approx 1$); the flume data seems to have a standard deviation roughly varying between 0.15 and 0.21, independent of the mean lee face slope. We can see that lee faces in the flume are generally somewhat steeper than those in the field. This agrees with field observations by, for instance, Roden (1998) and Carling et al. (2000) (see Best and Kostaschuk 2002). In studying the variability in lee face slopes, we recommend to keep in mind that flume and field measurements do not show similar trends.

## 4.2 Probability density functions

Previous researchers have assigned several types of probability density functions to bedform heights and bedform lengths as found from BEPs of flume and field experiments. Types that are identified are Beta, Exponential, Gamma, Gaussian, Rayleigh and Weibull (e.g., Ashida and Tanaka 1967; Nordin 1971; Annambhotla et al. 1972; Cheong and shen 1976; Wang and Shen 1980; Prent and Hickin (2001). The reasons for these discrepancies are not clear, but may be influenced by data processing and filtering (Annambhotla et al. 1972). Furthermore, the applied method for determining crests and troughs from BEPs may influence the shape of a probability density function (PDF). For instance, a sudden transition in the PDF may occur if bedforms with a height below a certain threshold value are not considered to be bedforms of interest. The PDF will not show such an abrupt step if a bedform is defined as the bedform between two consecutive upcrossings of the mean BEP. Also the chosen definition of bedform characteristics may influence the shape of the PDF. Some authors define bedform length as the distance between two successive bedform troughs (e.g., Wang and shen 1980), others use the distance between two successive zero upcrossings (e.g., Annambhotla et al. 1972) or the distance between two crests. Van der Mark and Blom (2007) show that the standard deviation in bedform length defined as the distance between two troughs may differ from the standard deviation defined as the distance between two crests. It is therefore difficult to compare the results of different authors if different data processing and definitions are used.

For each bedform characteristic in a data set (bedform height, bedform length, crest elevation, trough elevation, lee face slope) we now fit the following eight probability density functions using a maximum likelihood estimation: Beta, Exponential, Extreme value, Gamma, Gaussian, Log-normal, Rayleigh and Weibull. Figure 10 shows an example of fitted PDFs

for bedform heights measured during one of the flume experiments. For each experiment we then determine the best fit using a least squares goodness of fit evaluation. We determine the average sum of squares due to error (SSE) by averaging over all 192 experiments. The fit of the probability density function with the minimum SSE corresponds to the best fit. We express the SSE of a PDF fit as a percentage of the total error of all the eight PDF fits. Often, the best fit and the second best fit appear to be close to each other in terms of sum of squares. For bedform height the Weibull ($SSE_{\Delta,W} = 0.9\%$) and Beta ($SSE_{\Delta,B} = 1.3\%$) distributions provide the best fit. Figure 10 shows that the Exponential distribution provides the worst fit. We find $SSE_{\Delta,E} = 67.3\%$ for the Exponential distribution. Bedform length follows the Log-normal ($SSE_{\lambda,L} = 1.1\%$) or Gamma ($SSE_{\lambda,G} = 1.2\%$) distribution. For crest elevation, trough elevation and the lee face slope we find that the Beta distribution yields the best fit ($SSE_{\eta,c,B} = 1.7\%$, $SSE_{\eta,t,B} = 1.8\%$, $SSE_{S,B} = 2.8\%$, respectively), and the Weibull distribution the second best ($SSE_{\eta,c,W} = 2.9\%$, $SSE_{\eta,t,W} = 2.0\%$, $SSE_{S,W} = 5.0\%$, respectively).

We use the Kolmogorov-Smirnov hypothesis test to test the goodness of the distribution fits. The test indicates whether the null hypothesis that a certain data set comes from a pre-defined distribution can be rejected. The null hypothesis that a set follows the best fit distribution cannot be rejected in roughly 75% of the considered 192 experiments.

The best fit distribution types all appear to have a more or less similar and asymmetric shape with a longer tail for the higher values. We may conclude that bedform length can best be described by a Log-normal or Gamma distribution. Bedform height, crest elevation, trough elevation and lee face slope can best be described by a Beta or Weibull distribution.

## 4.3 Extreme values

We now analyze the 95% bedform height, indicated as $\Delta_{95}$, the 95% bedform length, indicated as $\lambda_{95}$, the 95% crest elevation, indicated as $\eta_{c,95}$, and the 95% trough elevation, indicated as $\eta_{t,95}$, in order to get insight in the extreme values of the distribution. Like the standard deviation of bedform height, we find that the 95% bedform height scales with the mean value. This also holds for bedform length, crest elevation and trough elevation.

We also analyze the relationship between extreme value (95%) and standard deviation. Figure 11 shows the 95% highest bedform minus the mean bedform height as a function of the standard deviation in bedform height. For bedform height, we find that

$$C_{\Delta,95} = \frac{\Delta_{95} - \mu_{\Delta}}{\sigma_{\Delta}} = 1.7 \tag{1}$$

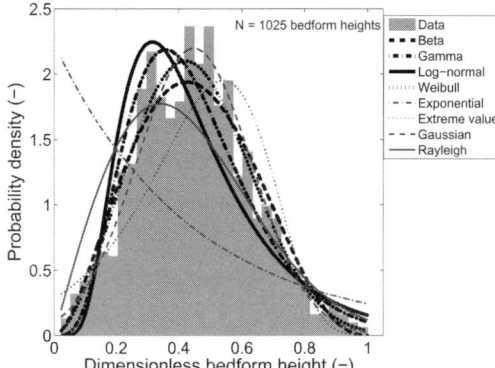

Figure 10. Probability density function of dimensionless bedform height (bedform heights divided by maximum bedform height) for one of the flume experiments. The lines indicate the fitted probability distributions.

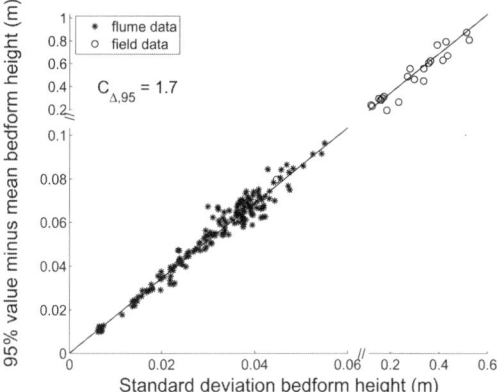

Figure 11. 95% bedform height minus mean bedform height versus standard deviation of bedform height.

(see Figure 11), where the constant $C_{\Delta,95}$, is a measure of how the 95% bedform height relates to the standard deviation in bedform height. Similar plots for crest elevation and trough elevation show that $C_{\eta,c,95}$ and $C_{\eta,t,95}$ equal 1.7, and that for bedform length $C_{\lambda,95}$ equals 1.9.

Theoretically it can be shown that $C_{X,95} = 1.64$ if the stochastic variable $X$ follows a Gaussian distribution. A value of $C_{95}$ that is larger than the value corresponding to a Gaussian distribution (1.64) indicates that the extreme value is more remote from the mean value than in the case of a Gaussian distribution. This confirms the finding in Section 4.2 that bedform characteristics are distributed according to a type of distribution that has a longer tail for the higher values.

## 5 CONCLUSIONS

Even under steady flow conditions and well-sorted sediment, bedforms are irregular in height, length

and shape. The stochastics of geometric properties of bedforms are studied for flume and field data. We find regularity in the stochastics of geometric properties of bedforms. A linear relationship exists between standard deviation and mean value for bedform characteristics. This appears to be valid for bedform height, bedform length, crest elevation and trough elevation.

Variability in bedform height, bedform length, crest elevation and trough elevation in the field is comparable to variability of bedform height, bedform length, crest elevation and trough elevation in flume experiments. This means that the stochastics of these bedform characteristics can be studied using flume experiments.

The standard deviation of the slope of the bedform lee face does not show a linear relationship with the mean slope of the bedform lee face. Also the lee face slope as measured in the flume is steeper than the lee face slope as measured in the field.

Bedform height, bedform length, crest elevation and trough elevation are best described by an asymmetric type of probability distribution with a longer tail for higher values (Beta, Gamma, Log-normal, Weibull). The extreme value analysis, in which we study the 95% value of bedform height, bedform length, crest and trough elevation, confirms the longer tail for higher values.

## ACKNOWLEDGEMENTS

This research project, which is part of the VICI project ROUGH WATER (project number TCB.6231), is supported by the Technology Foundation STW, applied science division of the Netherlands Organization for Scientific Research (NWO) and the technology programme of the Ministry of Economic Affairs.

The authors would like to acknowledge the Institute for Inland Water Management and Waste Water Treatment (Dr Arjan Sieben and Adri Wagener) for providing Waal River data and Dr Doug Jerolmack and Dr David Mohrig for providing North Loup data. We thank Dr Suzanne Leclair for providing flume data.

## REFERENCES

Allen, J. R. L. (1983). River bedforms: progress and problems. In J. D. Collinson and J. Lewin (Eds.), *Modern and ancient fluvial systems*, Volume 6, Blackwell, Boston, pp. 19–33. Int. Assoc. Sedimentol. Spec. Publ.

Amsler, M. L. and M. H. García (1997). Discussion: Sand dune geometry of large rivers during floods. *Journal of Hydraulic Engineering 123*, 582–585.

Annambhotla, V. S. S., W. W. Sayre, and R. H. Livesey (1972). Statistical properties of Missouri River bedforms. *Journal of Waterways, Harbors Coastal Eng. Div. ASCE 98*(WW4), 489–510.

Ashida, K. and Y. Tanaka (1967). A statistical study of sand waves. In *Proceedings of the 12th IAHR congress*, Fort Collins, Colorado, pp. 103–110.

Barabási, A. L. and H. E. Stanley (1995). *Fractal Concepts in Surface Growth*. New York: Cambridge University Press.

Best, J. and R. Kostaschuk (2002). An experimental study of turbulent flow over a low-angle dune. *Journal of Geophysical Research 107 (C9)* (3135, doi:10.1029/2000JC000294), 1118–1129.

Blom, A., G. Parker, J. S. Ribberink, and H. J. De Vriend (2006). Vertical sorting and the morphodynamics of bed-form-dominated rivers: An equilibrium sorting model. *Journal of Geophysical Research 111* (F01006, doi:10.1029/2004JF000175).

Blom, A., J. S. Ribberink, and H. J. De Vriend (2003). Vertical sorting in bed forms: Flume experiments with a natural and a trimodal sediment mixture. *Water Resources Research 39 (2)* (1025, doi:10.1029/2001WR001088).

Carling, P. A., E. Gölz, H. G. Orr, and A. Radecki-Pawlick (2000). The morphodynamics of fluvial sand dunes in the River Rhine, near Mainz, Germany. I. Sedimentology and morphology. *Sedimentology 47*, 227–252.

Cheong, H. F. and H. W. Shen (1976). The intervals between the successive zero crossings of sand bed profiles. In *Proc. 2nd International IAHR Symposium on Stochastic Hydraulics*, Lund, Sweden, pp. 245–265.

Crickmore, M. J. (1970, February). Effect of flume width on bedform characteristics. *Journal of the Hydraulics Division, ASCE 96*(HY2), 473–496.

De Koning, M. F. (2007). The stochastic characteristics of geometric properties of sand waves in the North Sea. M.Sc. thesis, University of Twente. In press.

Dodds, P. S. and D. H. Rothman (2000). Scaling, universality and geomorphology. *Annual Review of Earth and Planetary Sciences 28*, 571–610.

Driegen, J. (1986). Flume experiments on dunes under steady flow conditions (uniform sand, Dm = 0.77 mm). Description of bed forms. TOW Rivieren R 657 - XXVII / M 1314 part XV, WL | Delft Hydraulics, Delft, The Netherlands.

Gabel, S. L. (1993). Geometry and kinematics of dunes during steady and unsteady flows in the Calamus River, Nebraska, USA. *Sedimentology 40*, 237–269.

Jerolmack, D. J. and D. Mohrig (2005). A unified model for subaqueous bed form dynamics. *Water Resources Research 41* (W12421, doi:10.1029/2005WR004329).

Klaassen, G. J. (1990). Experiments with graded sediments in a straight flume. Vol. A (Text) and Vol. B (Tables and Figures). Technical Report Q788, Delft Hydraulics.

Leclair, S. F. (2002). Preservation of cross-strata due to the migration of subaqueous dunes: an experimental investigation. *Sedimentology 49*, 1157–1180.

Leclair, S. F., J. S. Bridge, and F. Wang (1997). Preservation of cross-strata due to migration of subaqueous dunes over aggrading and non-aggrading beds: comparison of experimental data with theory. *Geoscience Canada 24*(1), 55–66.

Mohrig, D. (1994). *Spatial evolution of dunes in a sandy river*. Ph.D. thesis, University of Washington, Seattle.

Mohrig, D. and J. D. Smith (1996, October). Predicting the migration rates of subaqueous dunes. *Water Resources Research 32*(10), 3207–3217.

Nelson, J. M., S. R. McLean, and S. R. Wolfe (1993, December). Mean flow and turbulence fields over 2-dimensional bed forms. *Water Resources Research 29*(12), 3935–3953.

Nikora, V. I. and D. M. Hicks (1997, December). Scaling relationships for sand wave development in unidirectional flow. *Journal of Hydraulic Engineering 123*(12), 1152–1156.

Nordin, C. F. (1971). Statistical properties of dune profiles. Sediment transport in alluvial channels. *U.S. Geological Survey Professional paper 562-F. US government Printing Office, Washington*.

Paola, C. and L. Borgman (1991). Reconstructing random topography from preserved stratification. *Sedimentology 38*, 553–565.

Prent, M. T. H. (1998). Seasonal regime of bedform and hydraulic geometry, Lillooet River, Pemberton, BC. M.Sc. thesis, Simon Fraser University.

Prent, M. T. H. and E. J. Hickin (2001). Annual regime of bedforms, roughness and flow resistance, Lillooet River, British Columbia, BC. *Geomorphology 41*, 369–390.

Roden, J. E. (1998). *The sedimentology and dynamics of mega-dunes, Jamuna River, Bangladesh*. Ph.D. thesis, Dep. of Earth Sci. and School of Geogr., Univ. of Leeds, Leeds, UK.

Van der Mark, C. F. and A. Blom (2007). A new and widely applicable tool for determining the geometric properties of bedforms. Technical report, University of Twente, Enschede, the Netherlands. In press.

Van Rijn, L. C. (1982, October). Equivalent roughness of alluvial bed. *Journal of the Hydraulics Division, ASCE 108*(HY10), 1215–1218.

Wang, W. C. and H. W. Shen (1980). Statistical properties of alluvial bed forms. In *Proceedings 3rd International Symposium on Stochastic Hydraulics*, Tokyo, Japan, pp. 371–389.

River, Coastal and Estuarine Morphodynamics: RCEM 2007 – Dohmen-Janssen & Hulscher (eds)
© 2008 Taylor & Francis Group, London, ISBN 978-0-415-45363-9

# Modelling the interaction between transverse and crescentic bar systems

R. Garnier
*School of Civil Engineering, University of Nottingham, Nottingham, UK*

D. Calvete
*Applied Physics Department, Universitat Politècnica de Catalunya, Barcelona, Spain*

N. Dodd
*School of Civil Engineering, University of Nottingham, Nottingham, UK*

A. Falqués
*Applied Physics Department, Universitat Politècnica de Catalunya, Barcelona, Spain*

ABSTRACT: It is nowadays increasingly recognized that different types of surf zone sand bars and morphological patterns can emerge from free instabilities of the coupling between topography and water motions. Among them, the surf zone exhibits rhythmic features as (1) transverse bars attached to the shore and (2) crescentic bars farther off-shore which develop from the deformation of a longshore bar. By using the 2DH morphodynamical numerical models, (1) linear (MORFO60) and (2) nonlinear (MORFO55), the stability analysis of a planar and a barred beach is given. This allows an understanding of the possible interaction between transverse and crescentic bars.

## 1 INTRODUCTION

The nearshore zone is an open system exposed to an external forcing but, at the same time and since it is highly nonlinear, it has a very rich internal or self-organized behavior. This is nowadays increasingly recognized as different types of surf zone sand bars and morphological patterns have been seen to emerge from free instabilities of the coupling between topography and water motions. Among them, the surf zone exhibits rhythmic features as (1) transverse bars attached to the shore and (2) crescentic bars farther off-shore (Short 2006; Konicki and Holman 2000).

A number of numerical modeling studies have explained the emergence of these patterns (Deigaard et al. 1999; Falqués et al. 2000; Caballeria et al. 2002; Ribas et al. 2003; Reniers et al. 2004). In the case of a planar beach (i.e., unbarred), the nonlinear study of Garnier et al. (2006) describes the evolution of transverse bars from their formation until their dynamical equilibrium for constant wave conditions. With a similar overall beach profile but featuring a shore parallel bar, the linear stability analysis of Calvete et al. (2005) predicts the development of a crescentic shape on the bar and the formation of transverse bars close to the shore as two separate modes.

This finding could suggest that, even with the presence of a longshore bar on the intial beach, transverse

bars can form. However, i) does the longshore bar have an effect on the properties of transverse bars ? On the other hand, crescentic bars might act on transverse bars simply because they may induce oscillations on the coastline. Therefore, some other crucial questions remain open. ii$_{1,2}$) Is the initial formation of one of the two rhythmic systems altered by the other one ? iii) Is there some finite amplitude interaction between them ? iv) Is there any equilibrium state composed of the superposition of the two modes ?

To shed light into these questions, the nonlinear numerical model MORFO55 (Garnier et al. 2006; Garnier 2007) and the linear numerical model MORFO60 (Calvete et al. 2005) are used. They are based on a wave and depth averaged shallow water equations solver with wave driver, sediment transport and bed updating and are used to further investigate the morphodynamical instability of a planar beach and a single barred beach.

## 2 MODELLING

The linear (MORFO60) and nonlinear numerical (MORFO55) models are based on a wave and depth averaged shallow water equations solver with wave driver, sediment transport and bed updating. They are governed by the same kind of equations. MORFO60

has been presented in Calvete et al. (2005) while
MORFO55 in Garnier et al. (2006).

## 2.1 Governing equations

The governing equations are given by the Mei's theory
(Mei 1989), and read:

$$\frac{\partial D}{\partial t} + \frac{\partial}{\partial x_j}(D v_j) = 0 \ , \tag{1}$$

$$\frac{\partial v_i}{\partial t} + v_j \frac{\partial v_i}{\partial x_j} = -g \frac{\partial z_s}{\partial x_i} - \frac{1}{\rho D} \frac{\partial}{\partial x_j}(S'_{ij} - S''_{ij}) - \frac{\tau_{bi}}{\rho D} \ , \tag{2}$$

$$\frac{\partial E}{\partial t} + \frac{\partial}{\partial x_j}((v_j + c_{gj})E) + S'_{ij}\frac{\partial v_j}{\partial x_i} = -\varepsilon \ , \tag{3}$$

$$k \sin\theta = k^0 \sin\theta^0 \ , \tag{4}$$

$$\frac{\partial z_b}{\partial t} + \frac{1}{1-p}\frac{\partial q_j}{\partial x_j} = 0 \ . \tag{5}$$

The five time- and depth-averaged dynamical
unknows are: the sea level $z_s(x_1,x_2,t)$, the horizontal velocity vector $\vec{v}(x_1,x_2,t)$, the wave energy density
$E(x_1,x_2,t)$, the wave angle $\theta(x_1,x_2,t)$ and the bed level
$z_b(x_1,x_2,t)$.

The other variables are defined as follow. $D$ is the
total mean depth ($D=z_s-z_b$). $\bar{S}'$ is the wave radiation
stress tensor. $\bar{S}''$ is the turbulent Reynolds stress tensor.
$\vec{\tau}_b$ is the bed shear stress vector. $g$ is the acceleration
due to gravity. $\rho$ the water density. $\vec{c}_g$ is the group
velocity vector. $\varepsilon$ is the dissipation rate due to wave
breaking and bottom friction. $\vec{k}$ is the wave vector. $\vec{q}$ is
the horizontal sediment flux vector, determined by the
Soulsby and Van Rijn formula (Soulsby 1997; Garnier
et al. 2006). More details on the parameterization are
given in Garnier et al. (2006).

On the one hand, these equations have been linearized with respect to a basic state. The fortran code
to solve for the basic state and its linear stablity is
referred to as MORFO60. The detailed method is introduced in Calvete et al. (2005). On the other hand, the
fully nonlinear equations have been solved by using
the finite difference model MORFO55 described in
Garnier et al. (2006).

## 2.2 Initial topographies

These two models have been used to study the morphodynamical instability of (1) a planar beach (Fig. 1a) and
(2) a single barred beach (Fig. 1b). MORFO55 uses a
finite mesh on a rectangular domain fixed to 250 m in
the cross-shore direction and 500 m in the long-shore

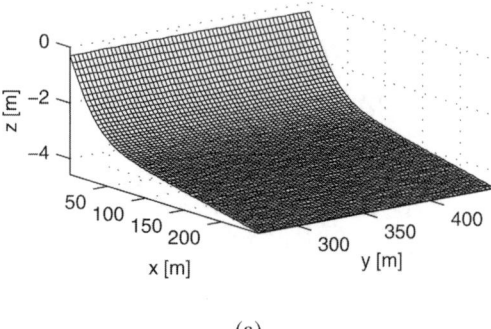

(a)

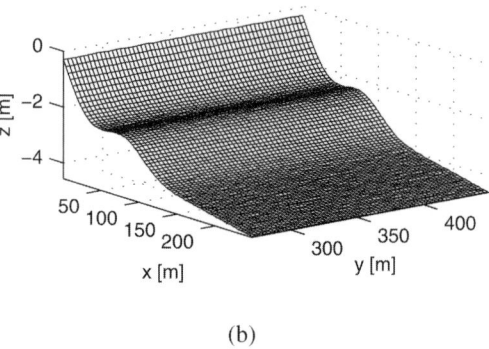

(b)

Figure 1. MORFO55 results. 3D view of a part of the topography at the initial time. (a) Planar beach. (b) Barred beach.
The x, y and z axis stand for the cross-shore direction, the
longshore direction and the bottom level.

direction. In case (2), the crest of the bar is at 80 m offshore. Experiments have been done for normal waves
of 1 m height, 6 s period at the finite longshore boundary ($x = 500$ m). The wave height as been calibrated in
this sense, giving the input wave height of MORFO60
of 1.5 m at the infinity.

## 3 RESULTS

### 3.1 Basic states

Figure 2 shows the cross-shore profile of the wave
height, the sea level and the bed level at the basic state
in the two experiments. This result has been obtained
by using MORFO55, in the case of non-perturbed initial topography. In this case, the morphodynamical
instablities are not excited. Because of the absence
of cross-shore sediment transport, the initial topography is stable and a hydrodynamical equilibrium state
is reached. In the case of a barred beach, the waves
strongly dissipate at the cross-shore location of the
longshore bar. Thus, the longshore bar imposes a

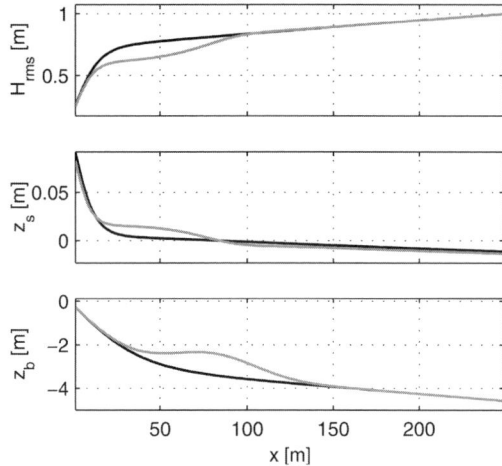

Figure 2. MORFO55 results. Basic States for the planar beach (black lines) and the barred beach (gray lines) experiments. From up to down, cross-shore profiles at the basic states of the root-mean-square wave height, the sea level and the bed level.

strong reduction of the wave height incoming in the inner surf zone. For instance, at $x = 50$ m, the presence of the longshore bar imposes a reduction in wave height of 20% compared to the case of a planar beach. As shown by Garnier et al. (2006), such a difference in wave height in the inner surf zone can have a strong influence on the formation of transverse bars.

### 3.2 Linear results

This latter supposition is confirmed by the linear stability analysis. Figure 3 shows the instability curves for (1) the planar beach (Fig. 3a) and (2) the barred beach (Fig. 3b). The stability analysis of the planar beach reveals a unique mode of transverse bars while for the barred beach, two modes are present: a transverse bar mode and a crescentic bar mode, the latter is characterized by a larger wave length and an outer cross-shore position. While the presence of the longshore bar does not remove the transverse bar mode which almost keeps the same wave length, it has an influence on its growth rate, which is reduced by 30% for the barred beach (answer i).

Moreover, because transverse and crescentic bars appear as two distinct modes, and because the growth rate of transverse bars is higher than the growth rate of crescentic bars, the linear theory suggests that the initial formation of transverse bars is not altered by the crescentic bars (answer $ii_1$). Indead, as shows Figure 4, a transverse bar system is formed at day 5 both for the two initial topographies, with the same wave length. In the case of barred beach, the crescentic bar system is not already formed. The difference in growth

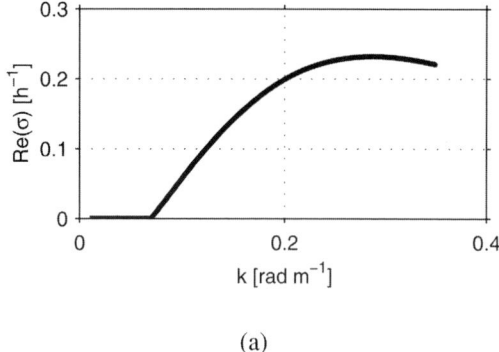

(a)

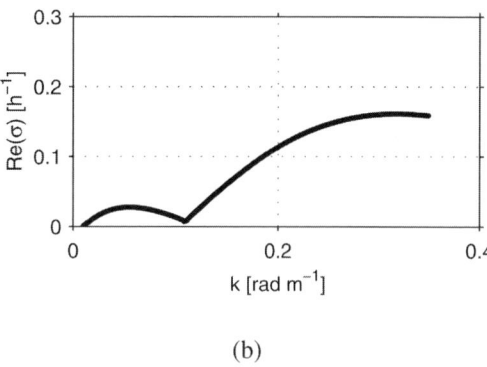

(b)

Figure 3. MORFO60 results. Instability curves. (a) Planar beach. (b) Barred beach. The vertical axis is the growth rate; the horizontal is the wave number.

rate between (1) and (2) is accompanied by a higher amplitude of the transverse bar system in case (1) than in case (2).

The reciprocity ($ii_2$), i.e. 'is the initial formation of crescentic bar system altered by the transverse bar system ?', is not trivial, however, because these two systems have a different scale and cross-shore location, it can be investigated. A comparison of the initial state of the crescentic bar system obtained by using MORFO60 (Fig. 5a) and MORFO55 (Fig. 5b) illustrates this. Although Figure 5b displays a transverse bar system close to the shoreline, the crescentic bar system seems independently to develop and is similar to the corresponding mode obtained by linear stability analysis (Fig. 5a).

### 3.3 Nonlinear results

Figure 6 shows the total topography after 25 days of morphological beach evolution. The transverse bar system appearing on planar beach (Fig. 6a) resembles the results of Garnier et al. (2006). The spacing

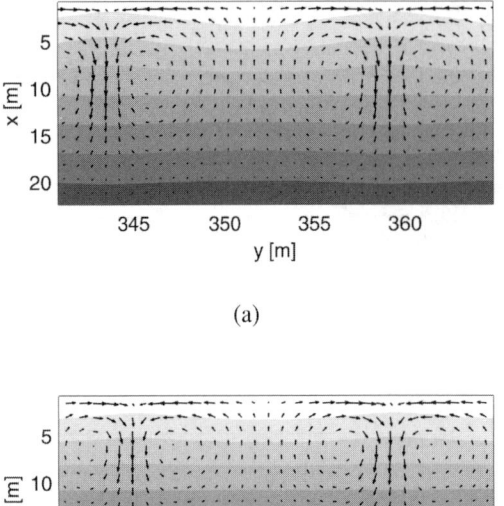

(a)

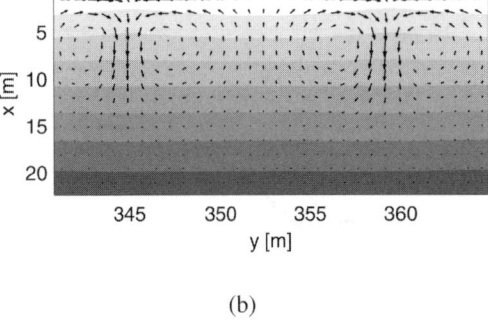

(b)

Figure 4. MORFO55 results. Plane view of the topography at day 5, colors represent the bottom level (deeper areas are darker), vectors represent the current. (a) Planar beach. (b) Barred beach. The x and y axis stand for the cross-shore direction and the longshore direction.

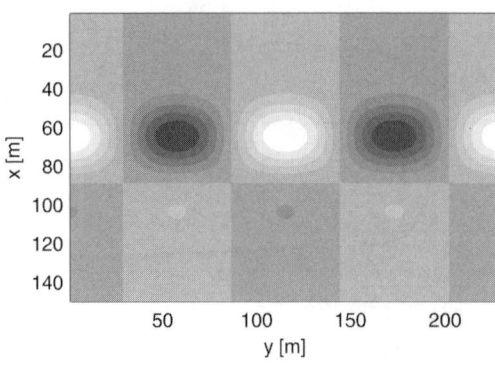

(a)

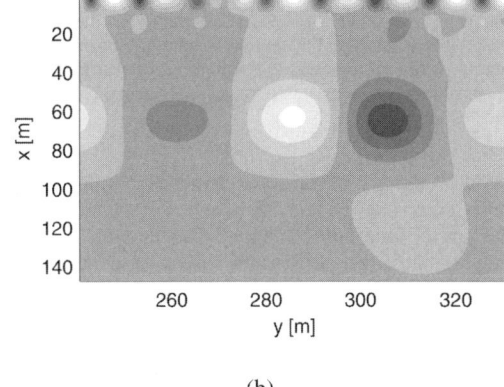

(b)

Figure 5. Plane view of the bottom perturbation h at the initial state of the crescentic bar system (barred beach). The x and y axis stand for the cross-shore direction and the longshore direction. (a) MORFO60. (b) MORFO55.

between two successive bars is $\lambda = 30$ m. The bars are attached to the shoreline and have a cross-shore span of 20 m. Figure 8a shows the corresponding rip current system: the current going offshore in the troughs and onshore over the shoals. The maximum magnitude of the current is 0.3 m/s. Figure 7a shows the time evolution of the bed level along the longshore section $x = 15$ m and the mean spacing between bars obtained by Fourier analysis. At the initial state, a random perturbation has been applied to the longitudinally uniform bathymetry, so a combination of a lot of modes is present. At day 10, the mode corresponding to $\lambda = 30$ m becomes clearly dominant. From day 10 to 15, the system has reached its equilibrium state with only one mode present.

For the barred beach (case 2), Figure 7b shows that the small scale transverse bars start to form while the shore parallel bar remains almost inactive. In this sense the morphodynamic evolution until day 7 is similar to case (1). After day 7, the longshore bar becomes crescentic with a spacing of $\lambda = 100$ m and both modes are present. The crescentic morphology forces in turn an

oscillation near the coastline consisting of large scale transverse bars in phase with the onshore shoals of the crescentic system. At day 25 (Figs 6b, 8b) the smaller scale transverse bars have disappeared in the inner zone corresponding to the troughs of the crescentic bar system. They are only present on the shoals of the large scale transverse bars.

This clearly suggests the occurence of non linear interactions between transverse bars and crescentic bars, at least between transverse bars and the large scale transverse bars induced by the crescentic bars (answer iii). The time evolution of the system wave length (Fig. 7) shows that larger scale system seems to 'eat' the smaller scale system. In fact, small scale transverse bars are still present, but only in the shoal of the large scale system. The superposition of the shoals of the small scale system on a shoal of the larger ones

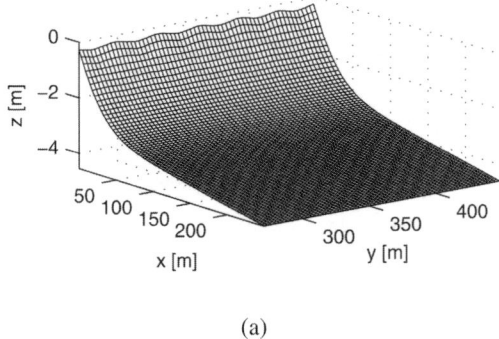

(a)

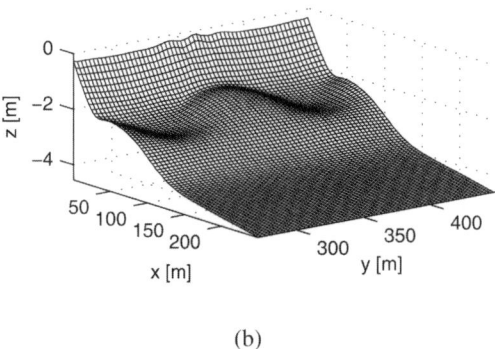

(b)

Figure 6. MORFO55 results. 3D view of a part of the topography at day 25. (a) Planar beach. (b) Barred beach. The x, y and z axis stand for the cross-shore direction, the longshore direction and the bottom level.

leads to a very shallow area, the bars tending to emerge up to the sea level. This leads to model numerical overflow, and no equilibrium state is reached (answer iv). On the other hand, in the troughs of the large scale system, a deep area leads to the disappearance of small scale transverse bars.

### 3.4 Model limitations

The sensitivity of the transverse bars' presence on the depth at the coastline suggests limitations of the present model. Firstly, the abscence of small scale transverse bars in the large scale troughs does not mean that the transverse bar can not appear farther onshore. The current model, discretized on a rectilinear shore boundary, is not able to predict it. A curvilinear coastline should be considered. Secondly, the overflow occuring when the depth is too small, reveals a limitation of the model in very shallow water. We determined that simulations are numerically stable if the ratio of maximum bar amplitude to total mean water depth at the initial state does not exceed the

thereshold value of 0.6. An increase of this value would allow a larger simulation domain i.e. the model could be used in shallower areas. This could be improved by including other processes such as 3D processes or swash zone processes, however, even if the model does not resolve the individual waves, the threshold value of 0.6 corresponds to the situation where the water depth at the troughs of the waves would be roughly zero. Thus, for larger values of this ratio, a first approximation could be to allow features to emerge up to sea level, so that it corresponds to a part of the dry beach. In order to take into account both the dry and wet part of the beach, the incorporation of a moving shoreline in the MORFO55 model, based on the algorithm of Falqués et al. (2007), is under development.

## 4 CONCLUSIONS AND FURTHER WORK

While the small scale transverse bar system ($\lambda = 30$ m) is an equilibrium state of a planar beach, the nonlinear model MORFO55 predicts that it is only a transient state of an initially longshore barred beach. Indeed, in the case of a barred beach, small scale transverse bars develop before the crescentic bars ($\lambda = 100$ m). At the emergence of the crescentic bars, both modes interact, the small scale transverse bars only persist in the inner part of each onshore bar of the crescentic system. The crescentic bar system forces the development of large scale transverse bars which are in phase with the onshore part of the crescentic system. This causes an undulation of the coastline, where, in the deepest areas, the small scale transverse bars are damped while in the shallowest areas, they tend to emerge up to sea level and model overflow occurs. This suggests the importance of considering a moving shoreline in order to integrate dry areas of the beach. At the same time the hypothesis of a rectilinear coastline should be relaxed in order to avoid large depths at the shore boundary, which cause the damping of smaller scale features. These results may also have relevance to bed-forms on planar beach subject to narrow-banded swell, as it is the strong peak in the dissipation that appears to lead to crescentic bedforms (see van Leeuwen et al. 2006).

ACKNOWLEDGEMENTS

This research is part of the PUDEM project, funded by the Spanish Government under contract REN2003-06637-C02-01/MAR and of the CTM2006-08875/MAR project. It is also funded through the 'Ramón y Cajal' contract of D. Calvete. The work of R. Garnier was supported by the University of Nottingham. Their support is gratefully acknowledged.

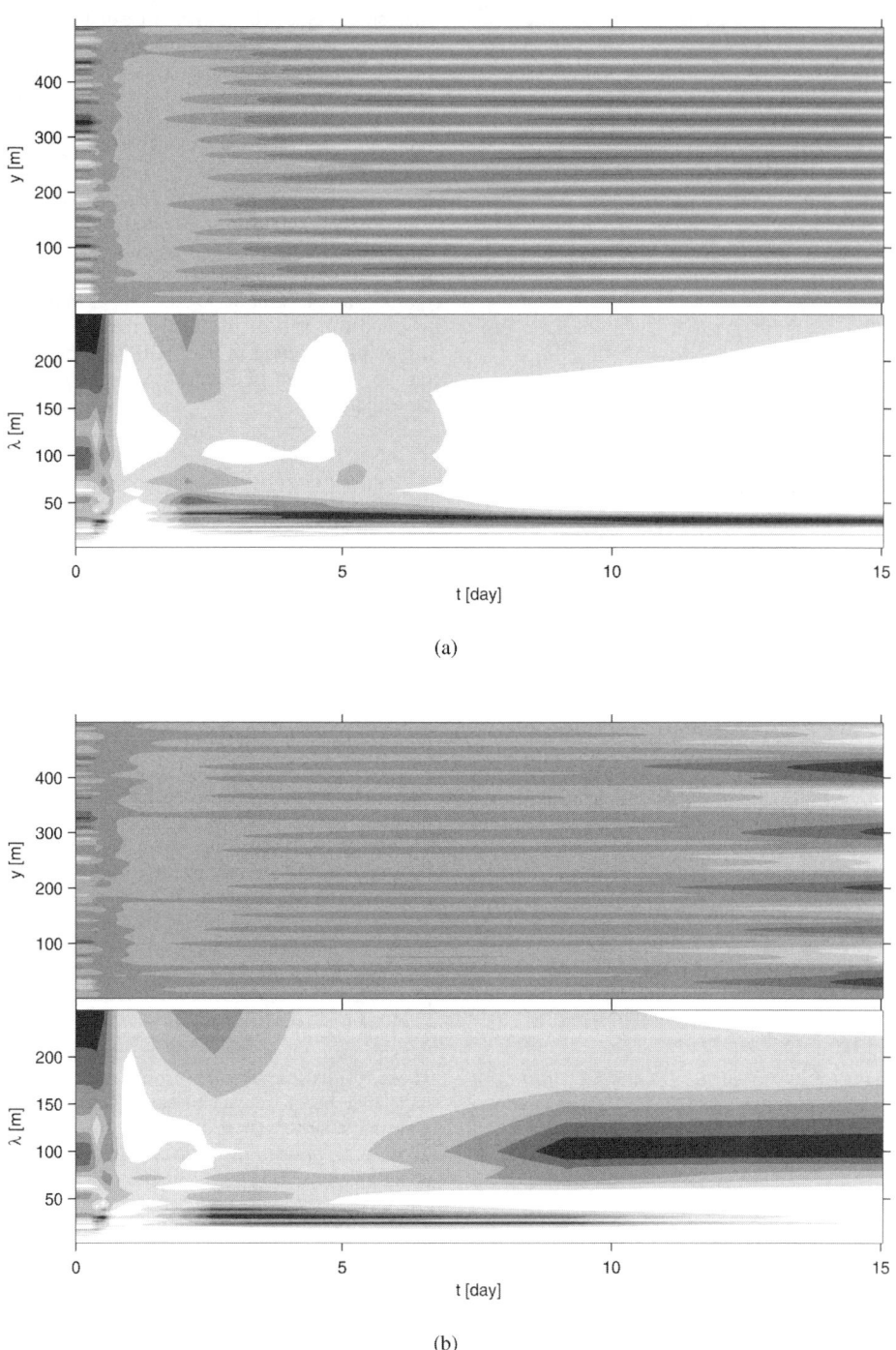

(a)

(b)

Figure 7. MORFO55 results. Time serie of (up) $h(x = 15\,\mathrm{m}, y, t)$, the bed level along the longshore section $x = 15\,\mathrm{m}$, the darker colors represent the deeper areas, (down) $\mathcal{F}(x = 15\,\mathrm{m}, \lambda, t)$, its Fourier analysis, the darker colors represent the more instable modes. (a) Planar beach. (b) Barred beach. The $y$, $\lambda$ and t axis stand for the longshore location, the bar spacing and the time.

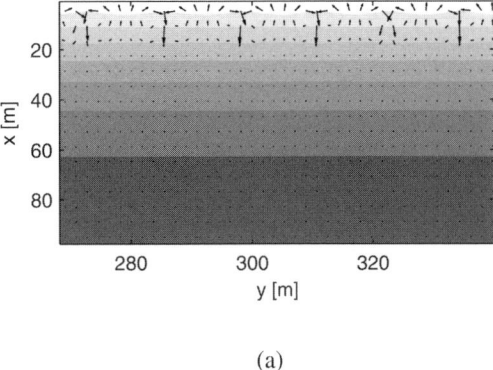

(a)

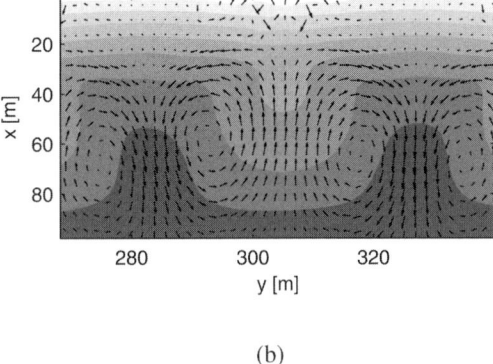

(b)

Figure 8. MORFO55 results. Plane view of the topography at day 25, colors represent the bottom level (deeper areas are darker), vectors represent the current. (a) Planar beach. (b) Barred beach. The x and y axis stand for the cross-shore direction and the longshore direction.

## REFERENCES

Caballeria, M., G. Coco, A. Falqués, and D. A. Huntley (2002). Self-organization mechanisms for the formation of nearshore crescentic and transverse sand bars. *J. Fluid Mech. 465*, 379–410.

Calvete, D., N. Dodd, A. Falqués, and S. M. van Leeuwen (2005). Morphological development of rip channel systems: Normal and near normal wave incidence. *J. Geophys. Res. 110*(C10006). doi: 10.1029/2004JC002803.

Deigaard, R., N. Drønen, J. Fredsoe, J. H. Jensen, and M. P. Jørgesen (1999). A morphological stability analysis for a long straight barred coast. *Coastal Eng. 36*(3), 171–195.

Falqués, A., G. Coco, and D. A. Huntley (2000). A mechanism for the generation of wave-driven rhythmic patterns in the surf zone. *J. Geophys. Res. 105*(C10), 24071–24088.

Falaqués, A., R. Garnier, E. Ojeda, F. Ribas, and J. Guillén (2007). Q2d-morfo: a medium to long term model for beach morphodynamics. In *Proc. 5th IAHR symposium on River, Coastal and Estuarine Morphodynamics*, Enschede, The Netherlands. International Association for Hydraulic Research. in preparation.

Garnier, R. (2007). *Nonlinear modelling of surf zone morphodynamical instabilities*. Ph. D. thesis, Appl. Physics Dept., Univ. Politècnica de Catalunya, Barcelona, Spain.

Garnier, R., D. Calvete. A. Falqués, and M. Caballeria (2006). Generation and nonlinear evolution of shoreoblique/transverse sand bars. *J. Fluid Mech. 567*, 327–360.

Konicki, K. M. and R. A. Holman (2000). The statistics and kinematics of transverse bars on an open coast. *Mar. Geol. 169*, 69–101.

Mei, C. C. (1989). *The Applied Dynamics of Ocean Surface Waves*, Volume 1 of *Advanced Series on Ocean Engineering*. Singapore: World Scientific.

Reniers, A. J. H. M., J. A. Roelvink, and E. B. Thornton (2004). Morphodynamic modeling of an embayed beach under wave group forcing. *J. Geophys. Res. 109*(C01030). doi:10.1029/2002JC001586.

Ribas, F., A. Falqués, and A. Montoto (2003). Nearshore oblique sand bars. *J. Geophys. Res. 108*(C43119). doi: 10.1029/2001JC000985.

Short, A. D. (2006). Australian beach systems – nature and distribution. *J. Coastal Res. 22*(1), 11–27.

Soulsby, R. L. (1997). *Dynamics of Marine Sands*. London, U.K.: Thomas Telford.

van Leeuwen, S. M., N. Dodd, D. Calvete, and A. Falqués (2006). Physics of nearshore bed pattern formation under regular or random waves. *J. Geophys. Res. 111*(F01023). doi:10.1029/2005JF000360.

*River, Coastal and Estuarine Morphodynamics: RCEM 2007 – Dohmen-Janssen & Hulscher (eds)*
*© 2008 Taylor & Francis Group, London, ISBN 978-0-415-45363-9*

# Simulating temporal response of bedform characteristics to varying flows

S. Giri, S. Yamaguchi & Y. Shimizu
*Hokkaido University, Japan*

J. Nelson
*USGS-Geomorphology & Sediment Transport Laboratory, USA*

ABSTRACT:   This work presents recent advances on morphodynamic modeling of bedforms under unsteady discharge. This paper includes further enhancement of a morphodynamic model, proposed earlier by Giri & Shimizu. The model reproduces the temporal evolution of bedform characteristics for temporally varying flows and is capable to accurately replicate the physical properties associated with bedform evolution under such flows. Based on comparison to previous observations, the model results appear to provide accurate predictions of the form drag over bedforms for both simple steady flows and temporally varying flows. Accurate predictions of form drag and total drag are key to the prediction of local and spatially averaged value of skin friction stress, as used in many sediment transport relations. Proposed model is able to replicate bed shear stress variation in accordance with the variation of form drag produced by the temporal growth or decay of bedforms resulting in a hysteresis loop of stage-discharge relationship. The results for strongly temporally varying flows show strong hysteresis in stage-discharge relationship; this is in good agreement with observation but has been treated in the past only using empirical methods. Proposed numerical model demonstrates its ability to solve an important practical problem associated with bedform evolution and flow resistance in varying flows.

## 1   BACKGROUND

Prediction of stage-discharge relation for alluvial rivers is a longstanding problem as it is associated with the hydraulic resistance exerted by bedforms. The total flow resistance for flows over bedforms is associated with skin friction of sediment particles and form drag exerted by bedforms on the flow. In the case of a flat-bed, the effective shear stress is equivalent to the grain shear stress, but the contribution of form drag becomes significant with the presence of bedforms due to the flow separation and spatial pressure variation. An adequate determination of bedform-induced resistance to flow, including its role in flows with temporal variation, is essential from a practical engineering point of view, because the relation of stage to discharge in temporally varying flows with berforms depends critically on the total drag. Furthermore, accurate predictions of form drag and total drag are key to the prediction of local and spatially averaged value of skin friction stress, as used in many sediment transport relations.

Some attempts have been made to analyze hysteresis characteristics observed during bedform transition in temporally varying flows, i.e. in rising and falling stages. In some past investigation, a strong hysteresis

between time varying flow discharge and bed resistance was observed. The relationship is usually found to be in the form of a loop (Izumi et.al, 2003). This phenomenon is attributed to the distinctive characteristics of bedform evolution/transition during rising and falling stages of flow. Yamaguchi & Izumi (2002, 2003) provided a physical explanation of such a hysteresis using weakly non-linear stability analysis. They described that the hysteresis is characterized by the subcritical bifurcation. Likewise, Izumi et.al (2003) conducted a laboratory observation on the transition process of the bedform configuration, in which they reproduced the hysteresis of stage- discharge relationship. The investigations for strongly temporally varying flows show strong hysteresis in form drag and sediment transport; but has been treated in the past only using empirical methods (Engelund & Fredsoe, 1982). Fedele et al. (2000) proposed a methodology to compute the components of the total shear stress, namely grain shear stress and form drag, for a steady, two-dimensional flow over fully developed dunes. The method is based on a simple energy balance with an analysis of spatially-averaged shear stress profile.

Some field and laboratory studies have been performed to analyze dynamic behavior of sediment motion under unsteady flow. Itakura et al. (1986) carried out some observations to characterize bed evolution in Ishikari River during the flood event. They identified the distinctive bed configuration depending upon the hydraulic properties of the flow, such as flat-bed to dunes and again flat-bet transition at the maximum flow rate and reappearance of dunes during falling stage. Kuhnle (1992) investigated bedload transport during rising and falling stages in two natural streams. Julien et al. (2002) also conducted detailed field measurement in Rhine River during flood and found noticeable hysteresis of the bedform height and discharge. Sutter et al. (2001) and Lee et al. (2004) performed some laboratory experiments on bedload transport process under unsteady flow condition and also evinced the hysteresis of the bedload transport and water level. However, the nature of the hysteresis curve in these two works is absolutely opposite in nature. The difference in bedload transport during rising and falling stages is found to be depending on the flow condition. For high flow condition, sediment flux is greater during rising stage, whereas for low flow condition, it is greater during falling stage of varying flows.

The interaction between flow-field, bed geometry and the sediment transport is quite complex to be quantified. The bedforms are created and altered by the flow and, conversely, the flow is acted upon by the bedforms through the production of form drag and significant variation in local mean flow and turbulent fields (Nelson et.al, 1993). Recently, a numerical model has been proposed by Giri & Shimizu (2006) to reproduce fluid and bedform dynamics under arbitrary steady or unsteady flow condition. The flow model component of the coupled morphodynamic model, which is two-dimensional in the streamwise and vertical directions, explicitly treats unsteadiness and nonhydrostatic effects. This model is able to capture most of the flow characteristics and morphodynamic feature of bedforms in a physically based manner. Likewise, model possesses ability to replicate temporal variation of physical characteristics associated with bedform evolution, such as temporal variation of form drag and thus total shear stress and associated flow-depth.

In present work, the aforesaid model has been enhanced that enables it to replicate bed shear stress variation in accordance with the variation of form drag produced by the temporal growth or decay of bedforms under temporally varying flows. Furthermore, model shows its capability to reproduce dune evolution, transition to flat-bed and reappearance of bedforms in falling stage of varying flow. In order to simulate this phenomenon, an assumption should be imposed on the variability of an empirical parameter used in pick up-deposition sediment transport formulation, namely mean step-length (which is usually treated as a constant parameter). It is evinced that mean step-length is supposed to be a varied parameter, particularly for the case of temporally varying flows. The bedform evolution process under unsteady discharge has been found to be depending on the varied condition of step-length parameter.

## 2 BRIEF DESCRIPTION OF HYDRODYNAMIC MODEL

A three-dimensional model with direct numerical simulation as proposed by Shimizu et al.(2001) seems to be more complete approach for the computation of flow and turbulence over bedforms. However, implication of this flow solver with a sediment transport model for morphodynamic simulation may inhibit it from being applied efficiently because of extremely high computational effort. Consequently, aforementioned three-dimensional hydrodynamic model has been simplified to a vertical two-dimensional and enhanced by imposing non-hydrostatic, free surface flow condition and, subsequently, coupled with a sediment transport module.

Computation of time-dependent water surface change is of importance for realistic reproduction of free surface flow over migrating bedforms. The kinematic condition, which constraints fluid particles to remain on the water surface at any time following the local flow velocity, is imposed along the water surface in order to compute water surface variation.

A no-slip boundary condition at the bed was employed; particularly a logarithmic expression for near-bed region was adopted. The periodic boundary condition was used in this computation. Some test computations with different length of calculation domain was performed in order to assess the sensitivity of periodic boundary condition on domain length.

The equations were transformed into a boundary fitted coordinate system. Transformed equations were numerically solved by splitting them into non-advection and pure advection phase. Non-advection phase was computed using central difference method. The pressure term was resolved using SOR method. Advection phase was calculated using a high-order Godunov scheme known as CIP method (Giri & Shimizu, 2006).

So far as turbulence model is concerned, a nonlinear k-$\varepsilon$ turbulence closure was employed that enables the anisotropy of Reynolds stresses to be considered to some extent. In a conventional k-$\varepsilon$ model turbulence stress tensors are evaluated using linear relationship. In order to reproduce turbulence characteristics more precisely in shear flows with separation zone, a nonlinear term is added to the standard k-$\varepsilon$ model. Kimura & Hosoda (2003) made detailed analysis and comparison of a nonlinear k-$\varepsilon$ model with other turbulence closures and found it more efficient than RSM or LES model in terms of CPU time. Later, we tested the performance of a standard k-ε model and also a zero-equation model so as to assess the significance of turbulence closure in context of morphodynamic simulation (Giri & Shimizu, 2007). In present study, the capability of proposed model is enhanced so as to simulate bedform-induced stage to discharge relationship as well as evolution of effective shear stress in relation to grain shear stress under varying flows.

## 3   SEDIMENT TRANSPORT APPROACH

### 3.1   Model description

With regard to sediment transport model, a pick up-deposition model for non-equilibrium sediment transport proposed by Nakagawa & Tsujimoto(1980) was employed.

The dimensionless pick up rate is expressed as follows:

$$p_s\sqrt{d/(\rho_s/\rho-1)g} = 0.03\tau_*(1-0.035/\tau_*)^3$$

where $p_s$ = sediment pick up rate, $\rho$ and $\rho_s$ = fluid and sediment density respectively and $\tau_*$ = dimensionless local bed shear stress.

The sediment deposition rate reads as:

$$p_d = p_s f_s(s)$$

where $p_d$ = sediment deposition rate and $f_s(s)$ = distribution function of step length.

Distribution function of mean step-length is found to be exponential as follows:

$$f_s(s) = \frac{1}{\Lambda}\exp\left(-\frac{s}{\Lambda}\right)$$

where $\Lambda$ = the mean step length and $s$ = the distance of sediment motion from pick up point.

Likewise, a suspended sediment pick up model, proposed by Itakura & Kishi (1980), was also incorporated in this model.

## 4   RESULT AND DISCUSSION

In this work, we have conducted some numerical tests in order to assess the capability of proposed numerical model to simulate bedform evolution process under unsteady discharge. Also, we attempted to simulate the stage-discharge relationship induced by bedforms. We performed basic analysis of the results of numerical experiments, which were carried out for different scenarios of temporally varying flows, i.e. different pattern of unit hydrograph (Fig.1).

Computations were carried out for a numerical flume with a slope of 0.002 and sediment diameter

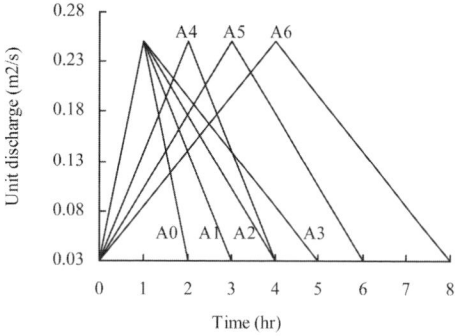

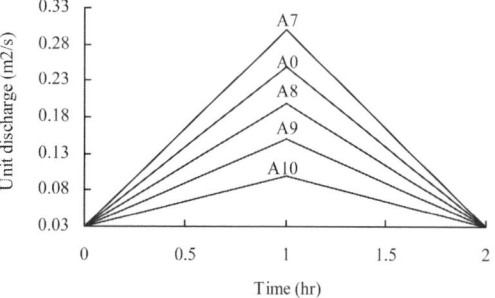

Figure 1.   Variations of unit hydrograph used in numerical experiments.

0.28 mm, for all calculation scenarios. Since a periodic boundary condition was used, a short domain-length (1.6 m) was used for these numerical experiments to reduce the computational time. In our recent study (Giri & Shimizu, 2007), we found that simulation result was insensitive to domain-length. In addition, a random field of initial perturbation was imposed on initial bed.

### 4.1   Sensitivity to mean step-length

The mean step length is conventionally proposed to be calculated as $\Lambda = \alpha d$, in which $\alpha$ is an empirical constant and $d$ is sediment diameter (Einstein, 1950). However, parameter mean step-length was found to be sensitive to the bedform evolution process, particularly for the temporally varying flow condition. In our recent study (Giri & Shimizu, 2007; Toyama et al., submitted) we have tested sensitivity of this parameter for some calculation cases with constant discharge. It was revealed that bedform length was sensitive to the empirical coefficient $\alpha$ for some of the flow condition even in steady flows. A linear stability analysis (Yamaguchi & Izumi, 2007) also confirmed similar trend. A basic attempt was made to explore this situation using some experimental analysis so as to calibrate a reliable value of coefficient $\alpha$ for steady flow condition based

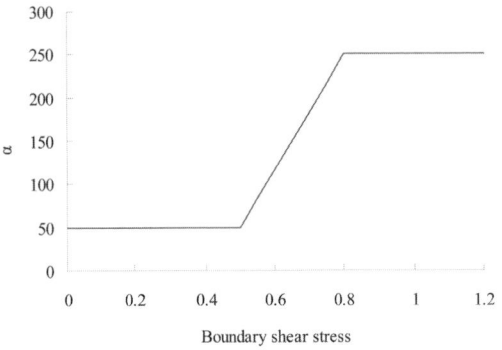

Figure 2.  Adapted relation between boundary shear stress and coefficient $\alpha$.

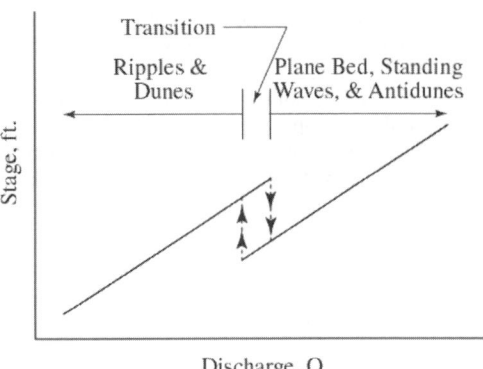

Figure 3.  Hysteresis of stage to discharge curve (After Simons & Richardson, 1961).

on the comparison of numerical result with experimental case. An adequate value of coefficient $\alpha$ was proposed to be about 50 (Toyama et al., submitted). However, for some of the cases, a conventional value, namely 100, is also appeared to be acceptable. Particularly in case of varying flows, this coefficient seems to be rather sensitive. It is revealed that mean step-length should not be treated as a constant parameter depending merely on particle size. Physically, mean step-length is supposed to be variable with respect to the local flow variability and temporal bed evolution. Nonetheless, no any investigation can be found regarding this issue.

In present study, we advanced a hypothesis that the empirical coefficient $\alpha$ is a function of boundary shear stress. As a first approximation, we simply propose a linear relationship between $\alpha$ and boundary shear stress as shown in Figure 2. Since, boundary shear stress varies temporally in our computation with varying flows, this coefficient also varies. We made some attempt to find out the effect of variation of this relation and selected the most reliable one.

### 4.2 Simulating bedform characteristics under varying flow

As mentioned above, we tested several combinations of upper and lower values of $\alpha$ and its relationship with boundary shear stress. We found the reliable variation of $\alpha$ is in the range of 50 to 250. Nakagawa and Tsujimoto (1980) also proposed the similar range of this coefficient. A typical example of numerical simulation of bedform evolution and transition can be seen in some selected instantaneous features depicted in Figure 4. In this figure, the dots in bottom plot (left) of unit hydrograph correspond to the instantaneous bed configuration depicted in upper plot of the same figure. From the result, it can be inferred that proposed model is capable to simulate dune evolution process, transition to flat-bed in rising stage and reappearance

of dune in falling stage. Despite the hypothesis we employed in relation to coefficient $\alpha$, the model capability seems to be quite promising to simulate such a complicated and practically significant phenomenon.

### 4.3 Simulating boundary shear stress variation under varying flows

The total flow resistance for flows over bedforms is associated with skin friction of sediment particles and form drag exerted by bedforms on the flow. In the case of a flat-bed, the effective shear stress is equivalent to the grain shear stress, but the contribution of form drag becomes significant with the presence of bedforms. Proposed model is able to reproduce the lag in total shear stress during rising and falling stage of temporally varying flows. The bottom plot of Figure 4 (right) shows the relation between grain shear stress and total shear stress for one of the calculation scenarios (case A0). In this case, a hysteresis can be seen, which is in agreement with previous physical investigations. For example, Kishi & Kuroki (1973) obtained the relationship between total shear stress and grain shear stress, which also comprises a hysteresis loop.

### 4.4 Hysteresis of stage to discharge relations

Flow in alluvial channels may be classified as lower and upper flow regime with a transitional region between. Dunes are found to appear in lower flow regime, whereas transition regime is characterized by bedforms ranges from dunes to plane bed or standing waves (Simons & Richardson, 1961). Likewise, plane bed, standing waves and antidunes may also appear in upper flow regime.

Simons and Richardson (1961) presented a typical quantitative stage to discharge relation for alluvial channel (Fig.3). There is usually considerable scatter in stage-discharge curve with ripples and dunes due to

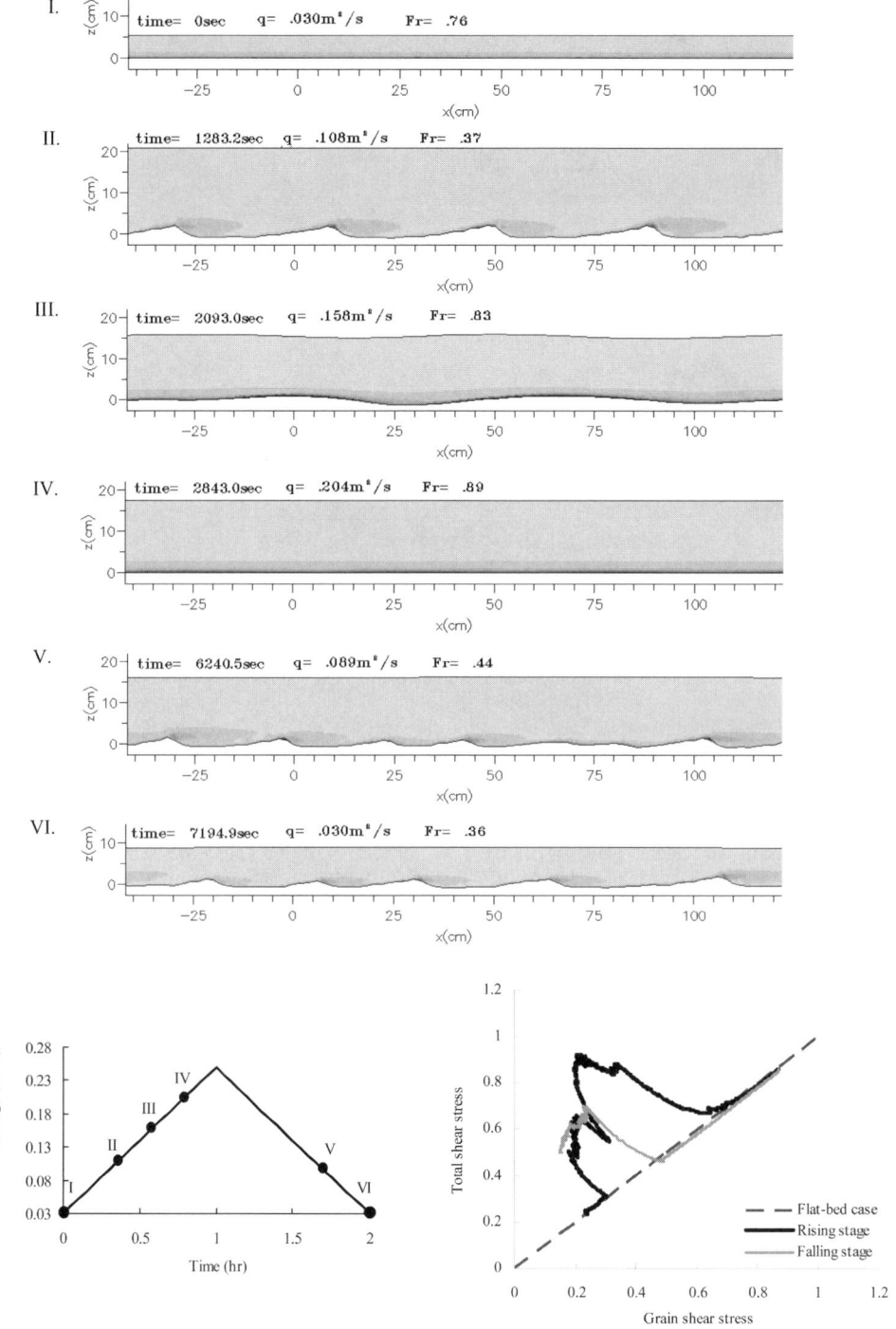

Figure 4.   Bedform evolution (top plots) for a typical condition of temporally varying flow (bottom plot, left) and associated evolution of boundary shear stress (bottom plot, right). The dots on unit hydrograph correspond to depicted bedform pattern.

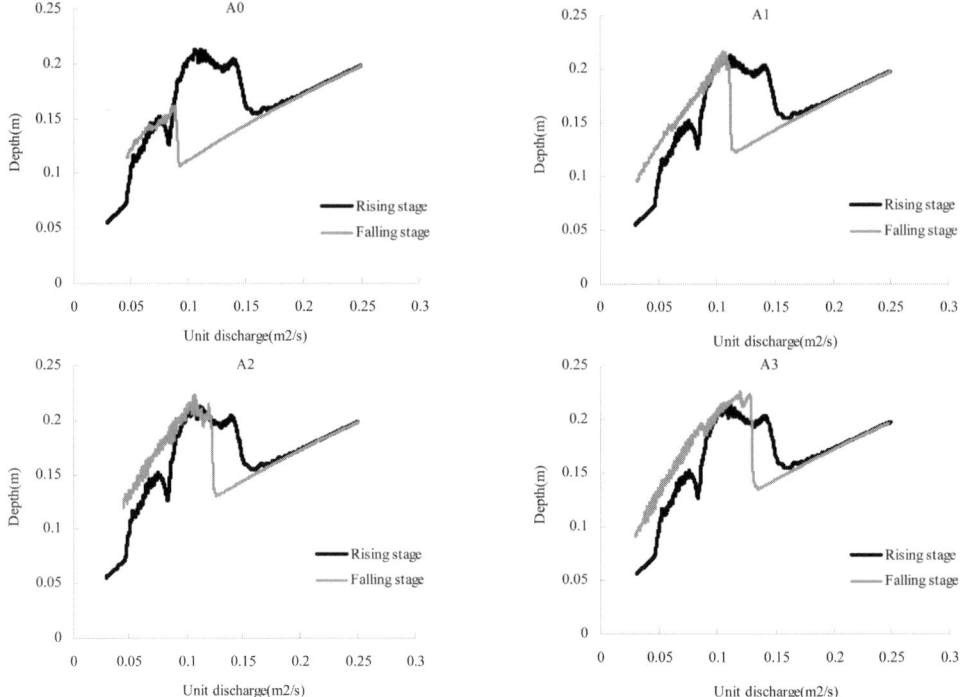

Figure 5. Stage-discharge relations for various scenarios of unit hydrograph: Constant peak discharge but different falling limb.

the wide variation of flow resistance. They described that the break in stage-discharge relation is caused by the change in bed resistance induced by bedforms evolution. The lag usually occurs between change of flow-depth and discharge, that is, the change of bed roughness lags the change of discharge. At the same discharge the flow-depth is found to be different in falling and rising stages resulting in a loop of stage-discharge curve that resembles a hysteresis curve. The break occurs on the rising stage at a larger discharge than on the falling stage; however this may alter depending on the pattern of discharge variation with time. We have conducted some numerical experiments to evaluate the form of the curve for the different scenarios of discharge variation with time. Figure 1 shows the different set of temporally varying flow condition adopted for the numerical experiments.

Firstly, we tested four different scenarios of unit hydrograph with different pattern of falling stage. From the results depicted in Figure 5, it can be seen that for gentler slope of the falling stage, the size of the loop becomes smaller. In other words, the reappearance of bedform appears to occur with greater discharge in the case of gentle slope of falling stage, namely 0.093, 0.115 and 0.126 $m^2/s$ for cases A0, A1 and A2 respectively. However, it is evident that the reappearance of

bedforms occurred almost in similar Froude number within the range of 0.8–0.89.

Secondly, stage-discharge relationship was assessed for temporally varying flows with different value of peak discharge, ranging from 0.1 $m^2/s$ to 0.3 $m^2/s$. The result is depicted in Figure 6. It can be seen that the loop tends to disappear as peak discharge decreases. This fact indicates the absence of transition due to the low peak discharge resulting in absence of lag in stage to discharge relationship in falling and rising stage. For the most lower value of peak discharge (case A10), even drop in flow-depth is not observed as bedforms almost keep the same characteristics due to the low flow intensity. However, some lag is apparent during lower discharge. In other words, existence of bedforms can be observed under the initial value of unit discharge in falling stage (Fig.6).

Finally, we evaluated the characteristic of stage-discharge relationship for different duration of discharge variation (2hrs, 4hrs, 6hrs and 8hrs) with the same peak discharge, keeping the symmetrical shape of unit hydrograph, i.e. rising and falling limbs have similar slopes. The result depicted in Figure 7 shows different characteristics of stage-discharge relationship and hysteresis loop. Analyzing simulation result, it was found that transition to flat-bed during rising

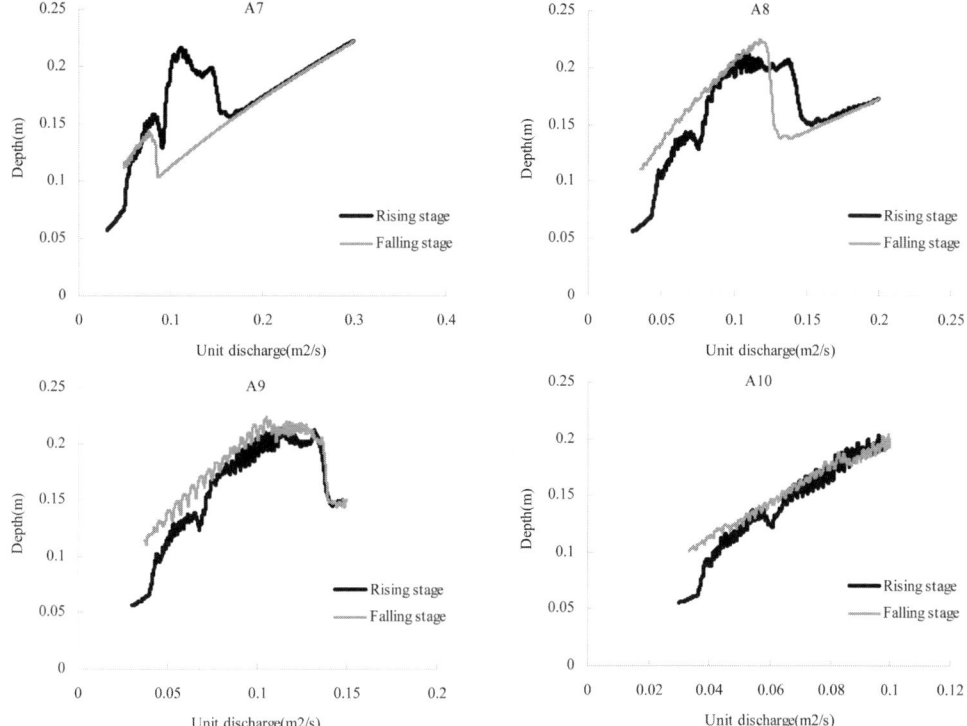

Figure 6.  Stage-discharge relations for various scenarios of unit hydrograph with different pick discharge.

stage and reappearance of bedforms during falling stage occur under identical Froude number (0.82–0.89) in all scenarios. However, bedform transition does not appear in case A6; the low Froude number is maintained due to the presence of bedforms throughout and thus no hysteresis loop is noticed.

## 5  CONCLUSION

A previously proposed morphodynamic model has been enhanced to simulate bedform evolution, transition and reappearance under varying flow with falling and rising limbs. Model is capable to replicate some significant physical processes associated with bedform evolution under temporally varying flows, such as hysteresis of stage-discharge relationship, evolution of total shear stress due to form drag exerted by bedforms. Some numerical experiments were carried out for different scenarios of discharge variation with time. Following are some conclusion that can be drawn from this study:

(1) Computational result supplements previous observations regarding the bedform evolution under varying flows, boundary shear stress variation

(simulation of form drag hydrodynamically) and hysteresis of stage-discharge relationship, which has been treated in the past only using empirical methods.

(2) Numerical experiments reveal that stage-discharge relationship significantly depends on the pattern of discharge variation with time, though can be generalized in terms of Froude number.

(3) For the same peak discharge, the hysteresis loop is appeared to be more pronounced for strongly temporally varying flows. On the other hand, comparing the computational result for four different patterns of falling limb (case A0 to A3), it was found that the reappearance of bedforms occurred almost in the same Froude number in all calculation scenarios.

(4) For the case with low peak discharge, the transitional bedforms (standing waves or/and flat-bed) do not appear; consequently a hysteresis loop is absent. However, lag in stage to discharge relation can be observed in lower part of rising and falling limb.

(5) For the scenarios of different duration of discharge variation with the same peak unit discharge (rising and falling limbs have similar slopes), result shows different characteristics of stage-discharge

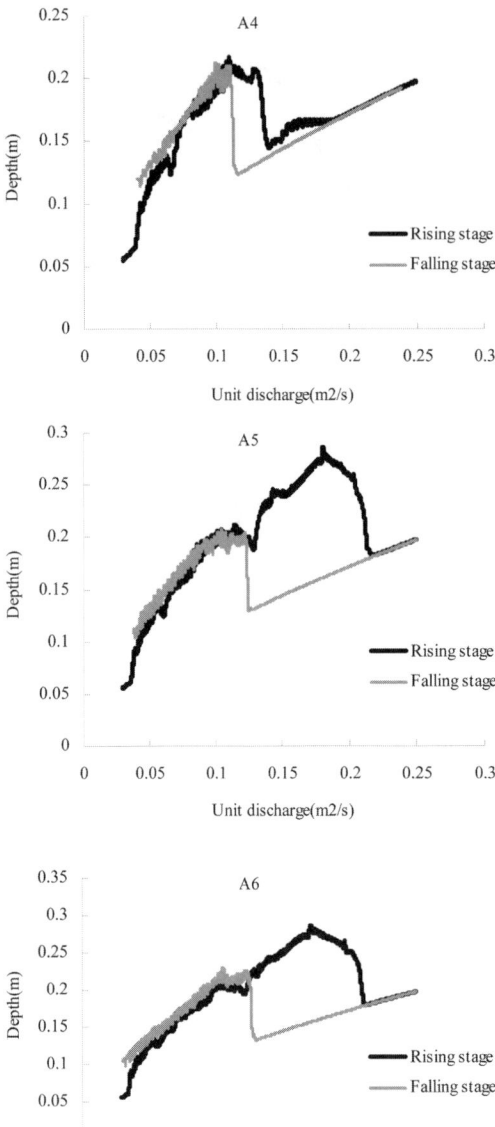

Figure 7. Stage-discharge relations for various scenarios of unit hydrograph: Symmetrical shape with constant peak discharge but different flood duration.

relationship and hysteresis loop. It is revealed that transition to flat-bed during rising stage and reappearance of bedforms in falling stage occur under identical Froude number, namely 0.82–0.89, for all scenarios except for a case, in which bedform transition does not appear and thus low Froude

number is maintained due to the presence of bedforms throughout and thus no hysteresis loop can be seen.

Results of numerical experiments corroborated previous understandings on dune-flat bed transition and associated hysteresis loop in the stage-discharge relationship. Proposed computational model is capable to quantify this important physical phenomenon of river engineering practice, which is rather promising. One subtle point, however, is that we have to assume a hypothetical relationship between mean step-length and boundary shear stress. Conventionally, mean step-length is postulated as a function of grain diameter with an empirical constant. The knowledge gained from our analysis has offered an incentive to investigate the influence of local flow variability and bed configuration on mean step-length, thereby explore and elucidate existing understanding of non-equilibrium sediment transport more comprehensively and physically.

REFERENCES

Ashida, K. & Michiue, M. 1972. Study on hydraulic resistance and bedload transport rate in alluvial streams. *Transactions of JSCE* 206: 59–69.
De Sutter, R., Huygens, M. & Verhoeven, R. 2001. Sediment transport experiments in unsteady flows. *Int. J. of Sed. Res.* 16 (1): 19–35.
Einstein, H. A. 1950. The bed-load function for sediment transportation in open channel flow. *U.S. Dep. Agric. Tech. Bull.*: 1029.
Engelund, F. & Fredsoe, J. 1982. Sediment ripples and dunes. *Ann. Rev. Fluid Mech.* 14: 13–37.
Fedele, J. J. & Garcia, M.H. 2000. Shear-stress Partition and Alluvial Roughness for Flow over Dunes. *EM2000 Proceedings*, ASCE Eng. Mech. Div. Conf., Minnesota.
Giri, S. & Shimizu, Y. 2006. Numerical computation of sand dune migration with free surface flow. *Water Resour. Res.* 42, W10422, doi:10.1029/2005WR004588.
Giri, S. & Shimizu, Y. 2007. Validation of a numerical model for flow and bedform dynamics. *Ann. J. of Hydraul. Eng.*, JSCE 51: 139–144.
Itakura, T. & Kishi, T. 1980. Open channel flow with suspended sediment, *Proceedings of ASCE* 106 (8): 1325–1343.
Itakura, T., Yamaguchi, H. & Shmizu, Y. 1986. Observations of bed topography during the 1981-flood in the Ishikari river. *J. Hydrosci. Hydraul. Eng.* 4 (2): 11–19.
Izumi, N., Kawamura, S. & Igarashi, A. 2003. Experiments on the transition between dune and flat bed regimes. *Proc. of Int. Conf. on Riv. Cosast. & Est. Morph.*, RCEM: 643–651.
Julien, P.Y., Klaassen, G.J., Ten Brinke, W.B.M. & Wilbers, A.W.E.2002. Case study: Bed resistance of Rhine river during 1998 flood. *J. Hydraul. Eng.* ASCE 128 (12): 1042–1050.
Kimura, I. & Hosoda, T. 2003. A nonlinear k-e model with realizability for prediction of flows around bluff bodies. *Int. J. Num. Meth. Fluids* 42: 813–837.

Kishi, T. & Kuroki, M. 1973. Bed forms and resistance to flow in erodible-bed channels (I).*Bulletin of Faculty of Engineering*, Hokkaido University 67: 1–23.

Kuhnle, R. A. 1992. Bed load transport during rising and falling stages on two small streams. *Earth Surface Processes and Landforms* 17: 191–197.

Lee, K. T., Liu,Y. & Cheng, K.H. 1997. Experimental investigation of bedload transport processes under unsteady flow condition. *Hydrol.Proces.*18: 2439–2454.

Nakagawa, H. & Tsujimoto, T. 1980. Sand bed instability due to bed-load motion. *J. Hyd.Div.*, ASCE 106: 2029–2051.

Nelson, J., McLean, S. & Wolfe, S., 1993. Mean flow and turbulence fields over two-dimensional bed forms", *Water Resour. Res.* 29 (12): 3935–3953.

Shimizu, Y., M. Schmeeckle, W. & Nelson, J. M. 2001. Direct numerical simulation of turbulence over two-dimensional dunes using CIP methods, *J. Hydrosci. Hydraul. Eng.* 19(2): 85–92.

Simons, D.B. & Richardson, E.V.1961. Forms of bed roughness in alluvial channels. *J. Hydraul. Div.*, Proceedings of ASCE 87 (3): 87–105.

Toyama, A., Shimizu, Y., Yamaguchi, S. & Giri, S. 2007. A study of the sediment transport rate on dune-covered bed.

Yamaguchi, S. & Izumi, N. 2003. Weakly nonlinear analysis of dunes including suspended load. *Proc. of Int. Conf. on Riv. Cosast. & Est. Morph.*, RCEM: 172–183.

Yamaguchi, S. & Izumi, N. 2003. Weakly nonlinear stability analysis of dune formation. *Proc. of Int. Conf. on Fluv. Hydraul.*, River Flow: 843–850.

*River, Coastal and Estuarine Morphodynamics: RCEM 2007 – Dohmen-Janssen & Hulscher (eds)*
*© 2008 Taylor & Francis Group, London, ISBN 978-0-415-45363-9*

# Bifurcation patterns in dune and antidune instability

M. Colombini & A. Stocchino

*DICAT – Dipartimento di Ingegneria delle Costruzioni, dell'Ambiente e del Territorio Università degli Studi di Genova, Genova, Italy*

ABSTRACT: The conditions by which a uniform flow in an infinitely wide erodible channel loses stability towards a perturbed configuration are investigated. A fully coupled differential system based on a rotational flow model plus the Exner equation is adopted, so that no simplifying assumptions are made on the characteristic flow and bed timescales. At a linear level, the analysis shows the existence of three unstable eigenvalues in the parameter space, which can be associated to dune (subcritical flow), antidune and roll-wave (supercritical flow) instabilities, respectively. In particular, for values of the Froude number larger than two, antidunes and roll-waves are both unstable. A weakly nonlinear analysis is performed to investigate the bifurcation process associated to dune, antidune and roll-wave marginal instability. Two regions of interests in the parameter space are found, one related to the dune-plane bed-antidune transition, the second to the antidune-roll-wave competition. The analysis shows that both dune-plane and plane-antidune bifurcations can be either supercritical (forward) or subcritical (backward) depending on the values of the parameters. The presence of subcritical bifurcation can theoretically justify the hysteresis often observed in dune-plane-antidune transition, where different bed patterns are detected for the same values of the flow and the sediment parameters. As for the antidune-roll-wave competition, critical points exist where the marginal curves of both modes intersect. A weakly nonlinear expansion in the neighbourhood of these points leads to a system of coupled amplitude equations that describes their growth and mutual interactions. A variety of bifurcation patterns ultimately arises from the solution of the above system depending on the equation coefficients, which include secondary bifurcations of one or both solutions.

## 1 INTRODUCTION

The idea that bedform formation in rivers can be interpreted in terms of an instability process of the system composed by the flow and the erodible bed that contains it dates back to the sixties, when the first seminal studies on this subject were published (Kennedy 1963; Kennedy 1969). This research field is still quite active and several morphodynamics patterns have been investigated making use of techniques imported from the field of hydrodynamic stability.

In a recent work, Colombini and Stocchino (2005) have revisited the linear theory of dune and antidune formation (Kennedy 1963; Fredsøe 1974; Richards 1980) showing how the coupling between bed and flow dynamics can lead to the appearance of an additional unstable mode associated to the formation of roll-waves at high Froude numbers. The present contribution is devoted to an extension of the above analysis to the weakly nonlinear regime with the purpose of tracing the bifurcation patterns that arise as dune, antidune and roll-waves become unstable and mutually interact.

## 2 FORMULATION OF THE PROBLEM

Let us consider uniform turbulent free surface flow of an incompressible fluid of density $\rho$ in a infinitely wide straight channel. In the following, variables with a star superscript are to be intended as dimensional variables. A generic quantity has been made non dimensional using the friction velocity $u_f*$ and the depth $D_0^*$ of the unperturbed uniform flow and the fluid density $\rho$. Moreover, we define the friction coefficient $C_0$ assuming that, for uniform flow condition, the friction velocity $u_f^*$ and the depth-averaged velocity $U^*$ are related to the slope $S$ and the Froude number $F$ by the following law:

$$C_0 = \frac{u_f^*}{U^*} = \frac{S^{\frac{1}{2}}}{F}. \tag{1}$$

The unsteady Reynolds and continuity equations are written in dimensionless form as:

$$U_{,t} + UU_{,x} + VU_{,y} + P_{,x} - 1 - T_{xx,x} - T_{xy,y} = 0, \tag{2}$$

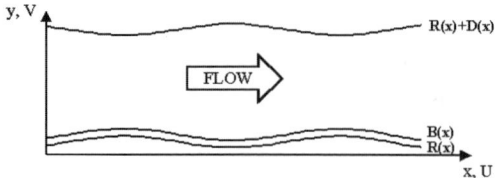

**Figure 1.** Sketch of flow configuration.

$$V_{,t} + UV_{,x} + VV_{,y} + P_{,y} - T_{xy,x} - T_{yy,y} = 0, \qquad (3)$$

$$U_{,x} + V_{,y} = 0, \qquad (4)$$

where $\mathbf{U} = (U, V)$ is the local velocity vector averaged over turbulence, $P$ is the dynamic pressure and $\mathbf{T} = \{T_{ij}\}$ is the 2-D Reynolds stress tensor.

Figure 1 shows a sketch of the flow domain referred to a Cartesian coordinate system $(x, y)$. The flow is bounded by the curves $y = R(x, t)$ and $y = R(x, t) + D(x, t)$, in a way that $D$ represents the local flow depth. Having denoted with $n$ and $t$ the directions normal and tangential to each boundary, the kinematic and dynamic boundary conditions to be associated with (2–4) read:

$$-R_{,t} + U_n = 0, \qquad U_t = 0 \qquad (5)$$

at the lower boundary, while at the upper boundary we have

$$-(R + D)_{,t} + U_n = 0, \qquad (6)$$

$$T_t = 0, \qquad T_n = -P = S^{-1}(R + D) \qquad (7)$$

where the normal stress $T_n$ includes the dynamic pressure $P$. Furthermore, we set the lower boundary of the flow domain at a predefined reference level, which conventionally represents the zero-velocity plane for the logarithmic profile. The distance $y_r$ of the latter plane from the average bed level is proportional to the nondimensional sediment diameter $d_s$.

A transformation of variables of the kind

$$\eta = \frac{y - R(\xi, \tau)}{D(\xi, \tau)}, \qquad \xi = x, \qquad \tau = t \qquad (8)$$

is then employed to map the domain shown in Figure 1 into a rectangular domain.

In order to close the above formulation a Boussinesq closure is used, which reads

$$T_{ij} = \nu_T (U_{i,j} + V_{j,i}), \qquad (9)$$

where the eddy viscosity $\nu_T$ has been evaluated by means of the mixing length approach

$$\nu_T = l^2 U_{,y} \qquad l = DL(\eta) \qquad (10)$$

$$L(\eta) = \kappa (\eta + \eta_r)(1 - \eta)^{\frac{1}{2}} \qquad (11)$$

The algebraic function $L(\eta)$ produces, for a uniform flow, a parabolic profile of eddy viscosity and, consequently, the logarithmic law of the wall. Accordingly to the above considerations, the quantity $\eta_r$ in (11) has been set to $d_s/12$.

It is also useful to recast the differential problem (2–4) in terms of the tangential $(T_t)$ and normal $(T_n)$ components of the stress acting on surfaces at constant $\eta$.

As far as bed dynamics is concerned, only bedload transport is considered and the perturbation approach adopted by Colombini (2004) to study dune and antidune formation in the decoupled framework is followed. In particular, sediment transport is assumed to be confined in a thin saltation layer close to the bottom so that the flow above it is the same as if the bed was coherent and the intensity of the sediment transport is determined by the shear stress acting at the interface between the saltation layer and the clear water flowing above it.

Based on the above considerations, the Exner equation of sediment mass conservation takes the form

$$FR_{,t} - Q_0 \Phi_{,x} = 0 \qquad Q_0 = \frac{d_s \sqrt{(s-1)d_s}}{C_0(1 - p_s)} \qquad (12)$$

where $\Phi$ is the dimensionless sediment discharge per unit width and $s$ and $p_s$ are relative density and porosity of the sediment, respectively.

The sediment transport capacity $\Phi$ is known to depend on some power of a dimensionless form of the bed shear stress, namely the Shields stress $\theta_b$. Results are only moderately affected by the choice of a particular form of the function $\Phi$. In the following, the Meyer-Peter & Müller formula

$$\Phi = 8(\theta_b - \theta_c)^{\frac{3}{2}} \qquad \theta_b \geq \theta_c \qquad (13)$$

has been employed, where $\theta_c$ is the critical Shields stress for incipient motion. Moreover, when only bedload transport is considered, as in the present case, gravity favors the downhill motion of the grains and conversely opposes to the uphill motion. This effect has been accounted for setting the critical Shields stress $\theta_c$ to

$$\theta_c = \theta_{ch} - \mu(S - R_{,x}) \qquad (14)$$

where $\theta_{ch}$ is equal to 0.047 and $\mu$ is a dimensionless constant that set to 0.1 after Fredsøe (1974).

Finally, the thickness of the saltation layer scales with the sediment grain size $d_s$ and is estimated through the relationship

$$h_b = \left[ 1 + A_b \left( \frac{\theta_r - \theta_c}{\theta_c} \right)^m \right] d_s \qquad \theta_r \geq \theta_c \qquad (15)$$

where the value of the constant $A_b$ and of the exponent $m$ have been set equal to 1.3 and to 0.55 respectively, on the basis of a regression on experimental data (Colombini 2004). The level at which the Shields stress has to be evaluated is therefore

$$B = R + \left(h_b + \frac{d_s}{12}\right) D \qquad \eta_b = \left(h_b + \frac{d_s}{12}\right) \qquad (16)$$

where the term $d_s/12$ accounts for the distance between the reference level and the top of the grains composing the bed.

## 3 LINEAR THEORY

The problem formulated in the previous section is then solved in terms of normal modes, so that a generic perturbed variable $G$ can be written as:

$$G(\xi, \eta, \tau) = G_0(\eta) + \epsilon G_1(\xi, \eta, \tau) \qquad (17)$$

$$G_1(\xi, \eta, \tau) = A G_{11}(\eta) \exp[i\lambda(\xi - \Omega\tau)] + c.c. \qquad (18)$$

where $\epsilon$ is a small parameter while $A$, $\lambda$ and $\Omega$ are the amplitude, wavenumber and complex growth rate of the bed perturbation. Note that, by this choice, the quantity $R_{11}$ is set equal to unity.

By substituting the above expansion into the governing equations, boundary conditions and turbulence closure, we are left with a sequence of problems at various orders of approximation in $\epsilon$.

### 3.1 $O(\epsilon^0)$

At leading order, integration of the system of differential equations

$$T'_{t0} = -1, \qquad T'_{n0} = 0 \qquad (19)$$

$$T_{t0} = \nu_{T0} U'_0 \qquad T_{n0} = -P_0 \qquad (20)$$

where primes stands for $\partial/\partial\eta$ and

$$\nu_{T0} = \kappa(\eta_r + \eta)(1 - \eta) \qquad (21)$$

together with the following boundary conditions

$$U_0|_{\eta=0} = 0, \qquad (22)$$

$$T_{t0}|_{\eta=1} = 0, \qquad T_{n0}|_{\eta=1} = -S_0^{-1}, \qquad (23)$$

yields the classic rough logarithmic law for the velocity

$$U_0 = \frac{1}{\kappa} \ln\left(\frac{\eta + \eta_r}{\eta_r}\right) \qquad (24)$$

Integrating once more the vertical profile of velocity, we obtain

$$\bar{U}_0 = \frac{\bar{U}^*}{u_f^*} = \frac{1}{C_0} = \frac{1}{\kappa}\left[\ln\left(\frac{1 + \eta_r}{\eta_r}\right) - 1\right] \qquad (25)$$

which relates the friction coefficient $C_0$ to $\eta_r$.

Exner equation (12) does not produce any additional information, since under uniform flow conditions the bed neither experience aggradation nor degradation. The following relationships hold:

$$\theta_{r0} = \frac{S_0}{(s-1)d_s} \qquad (26)$$

$$\theta_{b0} = \theta_{r0}(1 - \eta_b) \qquad \theta_{c0} = \theta_{ch} - \mu S_0 \qquad (27)$$

$$\Phi_0 = 8(\theta_{b0} - \theta_{c0})^{\frac{3}{2}} \qquad Q = \frac{3}{2}\frac{Q_0\Phi_0}{\theta_{b0} - \theta_{c0}} \qquad (28)$$

Note that the level $\eta_b$ has not been perturbed so that it only depends on the basic flow quantities $\theta_{r0}$ and $\theta_{c0}$.

### 3.2 $O(\epsilon^1)$

At the linear level, after some manipulations, a system of ordinary differential equations is eventually obtained that can be written in the general form

$$\mathbf{L}_{11}\mathbf{Z}_{11} - D_{11}\mathbf{D}_{11} - R_{11}\mathbf{R}_{11} = \{0\} \qquad (29)$$

where $D_{11}$ is treated as a parameter to be determined. The vector $\mathbf{Z}_{11}$ of the unknowns is:

$$\mathbf{Z}_{11} = (U_{11}, V_{11}, T_{t11}, T_{n11})^T \qquad (30)$$

The linear differential operator $\mathbf{L}_{11}$ in (29) reads

$$\mathbf{L}_{11} = \begin{pmatrix} d/d\eta & i\lambda/2 & -1/(2\nu_{T0}) & 0 \\ i\lambda & d/d\eta & 0 & 0 \\ U_0^\Omega - 4\lambda^2\nu_{T0} & -U'_0 & d/d\eta & i\lambda \\ 0 & U_0^\Omega & i\lambda & d/d\eta \end{pmatrix} \qquad (31)$$

while the vector $\mathbf{D}$ and $\mathbf{R}$ are, respectively

$$\mathbf{D}_{11} = \begin{pmatrix} 0 \\ i\lambda U'_0\eta \\ U_0^\Omega U'_0\eta - 2\lambda^2\eta(1-\eta) - 1 \\ i\lambda\eta - 2i\lambda(1-\eta) \end{pmatrix} \qquad (32)$$

$$\mathbf{R}_{11} = \begin{pmatrix} 0 \\ i\lambda U'_0 \\ U_0^\Omega U'_0 - 2\lambda^2(1-\eta) \\ i\lambda \end{pmatrix} \qquad (33)$$

and $U_0^\Omega = -i\lambda(U_0 - \Omega)$.

Linearizing the boundary conditions (5–7) we obtain, at the reference level ($\eta = 0$):

$$U_{11} = 0 \qquad V_{11} = -i\lambda\Omega \qquad (34)$$

while at the free surface ($\eta = 1$):

$$V_{11} + U_0^\Omega D_{11} = -U_0^\Omega \qquad (35)$$

$$T_{t11} = 0 \qquad T_{n11} + S_0^{-1} D_{11} = -S_0^{-1} \qquad (36)$$

The solution of the linear differential system (29) can be written in the form

$$\mathbf{Z}_{11} = c_{11}^{(1)} \mathbf{Z}_{11}^{(1)} + c_{11}^{(2)} \mathbf{Z}_{11}^{(2)} + D_{11} \mathbf{Z}_{11}^{(D)} + R_{11} \mathbf{Z}_{11}^{(R)} \qquad (37)$$

Thus $\mathbf{Z}$ is expressed as a linear combination of two linearly independent solutions of the homogeneous initial value problem, which satisfy the boundary conditions at the lower boundary

$$\mathbf{L}_{11} \mathbf{Z}_{11}^{(1,2)} = 0 \qquad (38)$$

plus particular solutions of the non-homogeneous differential systems

$$\mathbf{L}_{11} \mathbf{Z}_{11}^{(D)} = \mathbf{D}_{11}, \qquad \mathbf{L}_{11} \mathbf{Z}_{11}^{(R)} = \mathbf{R}_{11} \qquad (39)$$

again satisfying the lower boundary conditions.

Making use of the relationships (13–16), linearization of the sediment continuity equation (12) yields to

$$\Omega F_0 R_{11} - Q(\theta_{r0} T_{t11b} - i\lambda\mu R_{11}) = 0 \qquad (40)$$

where $T_{t11b}$ is the perturbation of the bed shear stress evaluated at the level $\eta_b$.

Equation (40) shows how instability is the result of a balance between destabilizing (the shear stress $T_{t11b}$ evaluated at the top of the saltation layer) and stabilizing (the 'gravity' term related to the local slope) effects.

Using the splitting (37) on the boundary conditions at the free surface (36) and on the Exner equation (40), the following algebraic homogeneous system is eventually obtained

$$\mathbf{U}_{11} \cdot \mathbf{C}_{11} = \{0\} \qquad (41)$$

where the array $\mathbf{U}_{11}$ is equal to

$$\left( \begin{bmatrix} V_{11}^{(1)} & V_{11}^{(2)} & V_{11}^{(D)} + U_0^\Omega & V_{11}^{(R)} + U_0^\Omega \\ T_{t11}^{(1)} & T_{t11}^{(2)} & T_{t11}^{(D)} & T_{t11}^{(R)} \\ T_{n11}^{(1)} & T_{n11}^{(2)} & T_{n11}^{(D)} + S_0^{-1} & T_{n11}^{(R)} + S_0^{-1} \\ T_{t11}^{(1)} & T_{t11}^{(2)} & T_{t11}^{(D)} & T_{t11}^{(R)} - \frac{\Omega F_0}{Q\theta_{r0}} - \frac{i\lambda\mu}{\theta_{r0}} \end{bmatrix}_{\eta_b} \right) \qquad (42)$$

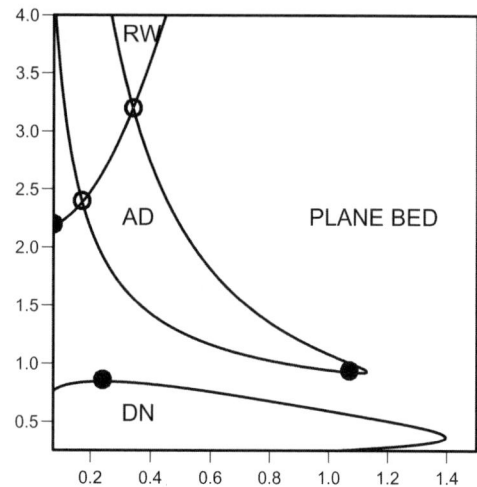

Figure 2. Regions of instability for dunes, antidunes and roll-waves.

while the vector $\mathbf{C}_{11}$ is

$$\mathbf{C}_{11} = (c_{11}^{(1)}, c_{11}^{(2)}, D_{11}, R_{11})^T \qquad (43)$$

In obtaining (41), the two linearly independent solutions of the homogeneous problem (38) have been chosen so that the unknowns $c_{11}^{(1)}$ and $c_{11}^{(2)}$ are the values of the perturbation of the tangential and normal stress at the reference level, respectively.

A nontrivial solution ($A \neq 0$) of the problem can be obtained only for the particular values of $\Omega$ (the eigenvalues) that make the determinant of $\mathbf{U}_{11}$ vanish. Note that, due to the coupling between flow and sediment transport, all the coefficients of $\mathbf{U}_{11}$ implicitly depends on the complex growth rate $\Omega$.

An extensive analysis of the behaviour of the eigenvalues is contained in Colombini and Stocchino (2005). The three most unstable eigenvalues (see Figure 2) display unstable regions in the parameter space; two of them can be readily associated with the formation of dunes (DN) and antidunes (AD), while the third (RW) describes the instability of fast sediment waves that appear at high Froude numbers associated with the presence of roll-waves. While the region of dune instability is separate from the others, the antidune and roll-wave regions partially overlap.

It may be useful at this point to briefly analyse the differences between the antidune and roll-wave modes of instability. Antidune mode is characterized by a small negative celerity (upstream propagation), with a wavelength of maximum amplification that ranges from 5 to 20 times the averaged flow depth as the Froude number increases. Roll-wave mode propagates downstream with an $O(1)$ celerity and is characterized by wavelengths larger than 30 times the uniform flow depth. Regarding the phase shift between the free surface and the bed oscillations, both antidunes and roll-waves are approximately in phase, with a positive lag for antidunes, coherently with their upstream migration, and a negative lag for roll-waves, corresponding to downstream migration. Another distinctive feature of the two modes regards free-surface and bed oscillation amplitudes. Roll-wave disturbances are characterised by small ratio of bed over free surface amplitude (e.g. $O(10^{-2})$), while the same ratio for antidunes is of order one.

# 4  WEAKLY NONLINEAR THEORY

We intend to investigate the weakly nonlinear evolution of the perturbations of the flow-bed system in a neighbourhood of a point $(F_m, \lambda_m)$ in the parameter space belonging to one or more of the marginal curves shown in figure 2. We then define

$$F = F_m(1 + \epsilon^2 F_2) \qquad \lambda = \lambda_m(1 + \epsilon^2 \lambda_2) \qquad (44)$$

It may be useful to clearly point out the kind of experiment implied by (44). Since the friction coefficient $C_0$ is assumed to be held constant, perturbing the Froude number corresponds, due to (1), to a perturbation of the average slope S

$$S = S_m(1 + 2\epsilon^2 F_2) \qquad (45)$$

We are then considering a uniform flow that is slightly perturbed with respect to the marginal situation, keeping constant the uniform flow depth and varying the discharge per unit width and the slope of the channel.

The purpose of the present analysis is twofold: (i) to look for finite amplitude equilibrium solutions in the neighbourhood of the 'critical' points for dunes, antidunes and roll-waves (shown as black circles in Figure 2); (ii) to examine the mutual nonlinear interactions between antidune and roll-wave modes in the neighbourhood of the points where the corresponding marginal curves intersect (shown as empty circles in Figure 2).

In order to investigate the nonlinear behaviour of the system we employ a multiscale (two-timing) perturbation technique and define a 'slow' timescale $T$ such that

$$T = \epsilon^2 \tau, \qquad \frac{\partial}{\partial \tau} \rightarrow \frac{\partial}{\partial \tau} + \epsilon^2 \frac{\partial}{\partial T} \qquad (46)$$

Nonlinearity gives rise to interactions between the fundamental and itself that lead to the generation of higher harmonics. Following the above cascade process, the fundamental is reproduced at third order, which leads the generation of secular terms. In order to prevent

their occurrence, the slow time dependence of the amplitude of the fundamental must also produce a contribution at third order.

We then expand the solution in the form

$$G(\xi, \eta, \tau, T) = G_0 + \epsilon G_1 + \epsilon^2 G_2 + \epsilon^3 G_3 \qquad (47)$$

and collect terms at the various order of approximation in $\epsilon$.

## 4.1  $O(\epsilon^1)$

At the linear level, the structure of the solution is analogous to (18), i.e.

$$G_1 = A(T)G_{11} \exp[i\lambda_m(\xi - \Omega_m \tau)] + c.c. \qquad (48)$$

where the complex function $A(T)$ is now a 'slowly varying' function of time to be determined which, in the linear regime, have to exhibit an exponential behaviour. Note that the complex growth rate $\Omega$ attains its marginal value $\Omega_m$ and so does the wavenumber $\lambda$.

The differential system (29) is recovered and its solution proceeds as sketched in the previous section. As expected no information are gathered on the amplitude $A$ at this level of approximation.

## 4.2  $O(\epsilon^2)$

The structure of the solution at the next-order follows from an analysis of the interaction of the fundamental with itself. Then

$$G_2 = F_2 G_{20F} + AA^* G_{20} + \qquad (49)$$

$$+ \{A^2 G_{22} \exp[2i\lambda_m(\xi - \Omega_m \tau)] + c.c.\}$$

where the real functions $G_{20F}$ and $G_{20}$ are nonlinear distortions of the basic uniform flow, following by the perturbations of the Froude number in (44) and by the interactions of the fundamental with its complex conjugate, respectively.

Three separate differential problems are then obtained at this order which can be written as

$$\mathbf{L}_{22}\mathbf{Z}_{22} - D_{22}\mathbf{D}_{22} - R_{22}\mathbf{R}_{22} = \mathbf{P}_{22}, \qquad (50)$$

$$\mathbf{L}_{20}\mathbf{Z}_{20} = \mathbf{P}_{20}, \qquad (51)$$

$$\mathbf{L}_{20}\mathbf{Z}_{20F} = 0 \qquad (52)$$

where the linear differential operators $\mathbf{L}_{22}$ and $\mathbf{L}_{20}$ are obtained from $\mathbf{L}_{11}$ substituting $\lambda$ with $2\lambda_m$ and zero, respectively. The vectors $\mathbf{P}_{22}$ and $\mathbf{P}_{20}$ are lengthy functions of $\eta$ involving interactions of the $O(\epsilon)$ components of the flow perturbation (and their complex conjugates) with themselves and with the basic flow.

953

The above differential systems are completed by the relative boundary conditions obtained from (5–7). In particular, the homogeneous system $20F$ displays a nonhomogeneous boundary condition only for the normal stress $T_{n20F}$ and immediatly integrates leading to

$$\mathbf{Z}_{20F} = (0, 0, 0, 2S_m^{-1})^T \tag{53}$$

The solution of the 22 and 20 systems follows the same procedure sketched in the previous section for the solution of the 11-system. In particular we expand the solutions as in (37) obtaining

$$\mathbf{Z}_{22} = c_{22}^{(1)} \mathbf{Z}_{22}^{(1)} + c_{22}^{(2)} \mathbf{Z}_{22}^{(2)} + \tag{54}$$

$$+ D_{22} \mathbf{Z}_{22}^{(D)} + R_{22} \mathbf{Z}_{22}^{(R)} + \mathbf{Z}_{22}^{(P)}$$

$$\mathbf{Z}_{20} = c_{20}^{(1)} \mathbf{Z}_{20}^{(1)} + c_{20}^{(2)} \mathbf{Z}_{20}^{(2)} + D_{20} \mathbf{Z}_{20}^{(D)} + \mathbf{Z}_{20}^{(P)} \tag{55}$$

and solve the resulting non-homogeneous differential problems.

The boundary conditions at the free surface plus the Exner equation eventually produce a non-homogeneous algebraic system

$$\mathbf{U}_{22} \cdot \mathbf{C}_{22} = \mathbf{U}_{22}^{(P)} \tag{56}$$

in the unknowns

$$\mathbf{C}_{22} = (c_{22}^{(1)}, c_{22}^{(2)}, D_{22}, R_{22})^T \tag{57}$$

the solution of which allows for the determination of the above constants. Similarly, for the 20-system we obtain

$$\mathbf{C}_{20} = (c_{20}^{(1)}, c_{20}^{(2)}, D_{20}, 0)^T \tag{58}$$

### 4.3  $O(\epsilon^3)$

At third order the spatial dependence of the fundamental is reproduced and therefore we assume

$$F_3 = F_{31} \exp[i\lambda_m(\xi - \Omega_m \tau)] + c.c. \tag{59}$$

The related differential system can be written as

$$\mathbf{L}_{11} \mathbf{Z}_{31} - D_{31} \mathbf{D}_{11} - R_{31} \mathbf{R}_{11} =$$

$$\frac{dA}{dT} \mathbf{P}_{31}^{(1)} + \lambda_2 A \mathbf{P}_{31}^{(2)} + A^2 A^* \mathbf{P}_{31}^{(3)} \tag{60}$$

where the vectors $\mathbf{P}_{31}^{(1,2,3)}$ are functions of $\eta$ expressed in terms of products of basic, leading and second order components of the flow field.

Once the particular solutions $\mathbf{Z}_{31}^{(P1,P2,P3)}$ of the non-homogeneous differential systems

$$\mathbf{L}_{11} \mathbf{Z}_{31}^{(P1,P2,P3)} = \mathbf{P}_{31}^{(1,2,3)} \tag{61}$$

are obtained, the boundary conditions at the free surface and the Exner equation can be cast in a similar way of (41) as

$$\mathbf{U}_{11} \cdot \mathbf{C}_{11} =$$

$$\frac{dA}{dT} \mathbf{U}_{31}^{(1)} + A(F_2 \mathbf{U}_{31}^{(2F)} + \lambda_2 \mathbf{U}_{31}^{(2\lambda)}) + A^2 A^* \mathbf{U}_{31}^{(3)} \tag{62}$$

where the second term on the right-hand side is generated by the boundary conditions and by the Exner equation.

The homogeneous part of the algebraic system (62) admits a non-trivial solution so that a solvability condition has to be satisfied, which can be found imposing the vanishing of the determinant of the array obtained by substituting the right-hand side of (62) into the last column of $\mathbf{U}_{11}$. Having set

$$\delta_i = \det(\mathbf{U}_{11}^{(i)}) \tag{63}$$

where the arrays $\mathbf{U}_{11}^{(i)}$ are obtained substituting the vector $\mathbf{U}_{31}^{(i)}$ into the last column of $\mathbf{U}_{11}$, the following Landau-Stuart amplitude equation is readily obtained

$$\delta_1 \frac{dA}{dT} + (\delta_{2F} F_2 + \delta_{2\lambda} \lambda_2) A + \delta_3 A^2 A^* = 0 \tag{64}$$

After some manipulations, the above equation can be rewritten as

$$\frac{dA}{dT} = A(\alpha_1 + \alpha_2 A^2) \tag{65}$$

where $A$, from now on, represents the modulus of the complex amplitude and

$$\alpha_1 = \alpha_{1F} F_2 + \alpha_{1\lambda} \lambda_2 =$$

$$-\Re\left(\frac{\delta_{2F}}{\delta_1}\right) F_2 - \Re\left(\frac{\delta_{2\lambda}}{\delta_1}\right) \lambda_2 \tag{66}$$

$$\alpha_2 = -\Re\left(\frac{\delta_3}{\delta_1}\right) \tag{67}$$

Note that if the coefficient $\alpha_2$ vanishes, the usual exponential behaviour of $A(T)$ is recovered with $\alpha_1$ representing the linear growth rate. Nonlinearity, expressed by the coefficient $\alpha_2$ can inhibit the exponential growth so that an equilibrium amplitude is reached as $T \to \infty$.

Before entering the discussion of the bifurcation processes described by (65), the analysis must be extended to the case of two different eigenvalues that are simultaneously marginal, which is relevant for the study of antidune and roll-wave interactions.

Following the same cascade process presented above, one finds that the fundamental is reproduced at third order for both disturbances. An amplitude equation similar to (65) is thus obtained for each disturbance, where, however, additional terms appear that describe the mutual nonlinear interactions.

A coupled system of equations is eventually obtained of the form

$$\begin{cases} \dfrac{\mathrm{d}A}{\mathrm{d}T} = A(\alpha_1 + \alpha_2 A^2 + \alpha_3 B^2) \\[2mm] \dfrac{\mathrm{d}B}{\mathrm{d}T} = B(\beta_1 + \beta_2 B^2 + \beta_3 A^2) \end{cases} \tag{68}$$

where $A$ and $B$ are the moduli of the complex amplitudes, $\alpha_1$ and $\beta_1$ are the linear coefficients, $\alpha_2$ and $\beta_2$ the cubic coefficients describing the interactions of each perturbation with itself and, finally, $\alpha_3$ and $\beta_3$ are the coefficients describing nonlinear interactions between $A$ and $B$.

## 5  ANALYSIS OF BIFURCATIONS

We first examine the kind of bifurcations that are described by (65) and their stability properties. In particular we analyze the case of critical conditions for dune and antidune formation, shown by filled circles in Figure 2. Under critical conditions, the linear coefficient $\alpha_{1\lambda}$ in (67) vanishes since the growth rate attains its maximum with respect to the wavenumber.

The Landau-Stuart equation (65) possesses only two steady solutions: the trivial one $A = 0$ and the solution

$$A = \sqrt{-\frac{\alpha_{1F} F_2}{\alpha_2}} \tag{69}$$

The latter exists only if the argument of the square root is positive, which implies that, for a positive value of the linear coefficient $\alpha_{1F}$, the coefficients $\alpha_2$ and $F_2$ must have different sign. As shown in Figure 3, two kind of bifurcations appear: the supercritical one (a) is recovered when $\alpha_2$ is negative, while a positive value of $\alpha_2$ leads to a subcritical (b) bifurcation. The picture is completely reversed when $\alpha_{1F}$ is negative.

In Figure 3 the trivial and equilibrium solutions are shown as solid lines when stable and as dashed lines when instable. An investigation of the stability properties of the solutions shows, in fact, that the trivial solution is always stable for negative values of $F_2$ and unstable for positive $F_2$, while the solution

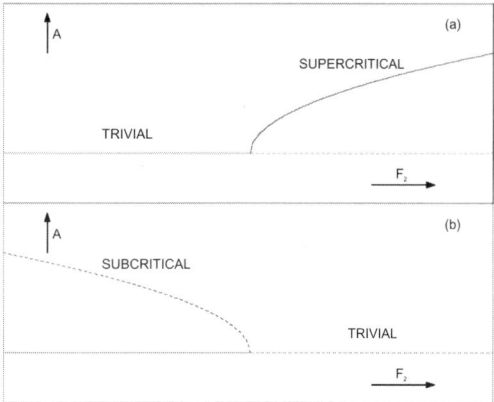

Figure 3.  Supercritical (a) and subcritical (b) bifurcations.

(69) is stable when supercritical and unstable when subcritical. Therefore, a stable steady solution can be found by means of (65) only when the bifurcation is supercritical.

In other words, if $\alpha_2$ vanishes or is positive, the cubic term in (65) is unable to balance the linear exponential growth of the disturbances and the analysis must be raised to higher order in $\epsilon$ in order to locate the damping mechanism. The point on the marginal curve in the parameter space in which the coefficient $\alpha_2$ vanishes, the so called 'tricritical' point, marks the boundary between supercritical and subcritical bifurcations.

Indeed, the present weakly nonlinear analysis shows that the critical Froude numbers for dune and antidune instability might become tricritical depending on the value of the friction coefficient $C_0$.

The plots in Figure 4 shows the regions of instability for antidunes (a) and dunes (b) in the $(Fr, C_0^{-1})$ space, together with the relative experimental data extracted from Guy, Simons, and Richardson (1966). The colour of the solid circles becomes lighter as the measured amplitude increases. An increase in the amplitude as the Froude number is raised above the critical threshold is quite visible in the antidune plot (a), but it is also detectable in the dune plot (b) as the Froude number is decreased below its critical value.

All the experimental point for antidunes fall in the subcritical range, while the opposite is true for dunes. Therefore, an equilibrium amplitude is found for dune instability while no information is produced for antidunes so far. Indeed, the fact that antidune instability is found to be subcritical suggests a possible explanation of the hysteresis processes observed during transition to plane bed.

Regarding the critical points for antidune and roll-waves, shown by empty circles in Figure 2, the bifurcation picture is complicated by the fact that

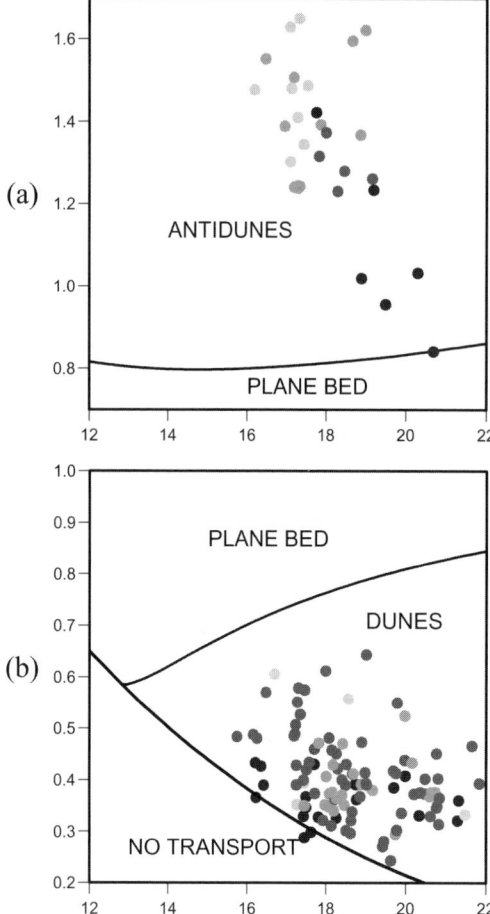

(a)

ANTIDUNES

PLANE BED

(b)

PLANE BED

DUNES

NO TRANSPORT

Figure 4. Antidune (a) and dune (b) stability regions.

the system (68) accepts three families of equilibrium solutions

$$A = \sqrt{-\frac{\alpha_1}{\alpha_2}} \qquad B = 0 \tag{70}$$

$$B = \sqrt{-\frac{\beta_1}{\beta_2}} \qquad A = 0 \tag{71}$$

$$A = \sqrt{-\frac{\alpha_1\beta_2 - \alpha_3\beta_1}{\alpha_2\beta_2 - \alpha_3\beta_3}} \qquad B = \sqrt{-\frac{\alpha_2\beta_1 - \alpha_1\beta_3}{\alpha_2\beta_2 - \alpha_3\beta_3}} \tag{72}$$

plus the trivial one. Note that the solution (72) represents a secondary bifurcation of the primary bifurcations (70) and (71) where both the disturbances might reach a finite amplitude.

The stability properties of equilibrium solutions of coupled systems of the kind (68) have been extensively investigated in the literature (e.g. Keener (1976)).

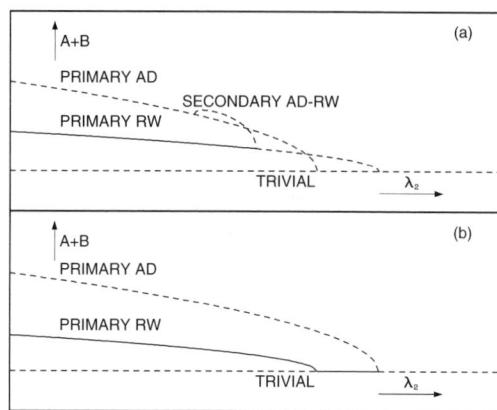

Figure 5. Primary and secondary bifurcations in AD-RW interactions; (a) $F > F_c$; (b) $F < F_c$.

As for the case of a single disturbance, the nature of the primary bifurcations depends on the sign of the coefficients $\alpha_2$, $\beta_2$ while secondary bifurcations are controlled by the signs of the coefficients $\alpha_3$, $\beta_3$. A variety of bifurcating solutions with different stability properties can in principle arise from the mutual nonlinear interactions.

For the case under investigation, the primary bifurcation for antidunes remains subcritical ($\alpha_2 > 0$) regardless the value of the friction coefficient while the roll-wave mode is consistently supercritical ($\beta_2 < 0$). Both the interaction coefficients $\alpha_3$ and $\beta_3$ turned out to be of the same sign as their cubic counterparts. This implies that the subcritical behaviour of antidunes is not affected by the interactions with the roll-wave mode. As for the case of critical conditions, higher order terms in the amplitude equation for the antidune mode are required in order to find an equilibrium solution.

In Figure 5 the bifurcation pattern in a neighbourhood of the first crossing point ($F_m = 2.4$, $\lambda_m = 0.18$), where the two marginal curves for antidunes and roll-waves intersect, is shown as a function of the wavenumber perturbation $\lambda_2$ for a value of the friction coefficient $C_0$ equal to 1/20. The uppermost and lowermost plots are relative to a Froude number above and below the threshold, respectively. Antidune primary solution is unstable in both regimes, while the opposite is true for the roll-wave primary bifurcating solution. For values of the Froude number above the threshold, a secondary bifurcation appears where both the disturbances are simultaneously of finite amplitude. Unfortunately, the latter solution is unstable so that the primary roll-wave solution (71) is recovered.

In both the plots of Figure 5 there are regions where none of the equilibrium solutions is stable, including the trivial one. It can be easily shown that the inclusion of a quintic term in the equation for the antidune amplitude is potentially able to solve this apparent inconsistency. However, the formal evaluation of the corresponding coefficient would require a huge effort in terms of the algebra involved since the weakly nonlinear analysis has to be extended to cover $O(\epsilon^4)$ and $O(\epsilon^5)$ differential problems.

REFERENCES

Colombini, M. (2004). Revisiting the linear theory of sand dune formation. *J. Fluid Mech. 502*, 1–16.

Colombini, M. and A. Stocchino (2005). Coupling or decoupling bed and flow dynamics: Fast and slow sediment waves at high froude numbers. *Phys. Fluids 17 (3)*, 9.

Fredsøe, J. (1974). On the development of dunes in erodible channels. *J. Fluid Mech. 64*, 1–16.

Guy, H. P., D. B. Simons, and E. V. Richardson (1966). Summary of alluvial channel data from flume experiments 1956–61. Prof. paper 462-I, U.S. Geol. Survay.

Keener, J. P. (1976). Secondary bifurcations in nonlinear diffusion reaction equations. *Studies in Appl. Math. 55*, 187–211.

Kennedy, J. F. (1963). The mechanism of dunes and antidunes in erodible-bed channels. *J. Fluid Mech. 16*, 521–544.

Kennedy, J. F. (1969). The formation of sediment ripples. dunes and antidunes. *Annu. Rev.. Fluid Mech.* 1, 147–168.

Richards, K. J. (1980). The formation of ripples and dunes on an erodible bed. *J. Fluid Mech. 99*, 597–618.

*Offshore bed forms*

*River, Coastal and Estuarine Morphodynamics: RCEM 2007 – Dohmen-Janssen & Hulscher (eds)*
*© 2008 Taylor & Francis Group, London, ISBN 978-0-415-45363-9*

# Spatio-temporal variability of currents and sediment fluxes over a dune field in the Dover Strait

Déborah Idier
*Bureau de Recherches Géologiques et Minières, Orléans, France*

Dominique Astruc
*Institut de Mécanique des Fluides de Toulouse, UMR 5502 CNRS-INPT-UPS, Toulouse, France*

Thierry Garlan
*Service Hydrographique et Océanographique de la Marine, Brest, France*

ABSTRACT:   A one month field campaing featuring two spring-neap tide cycles and three strong storms has been performed in a dune area located in the central part of the Dover Strait. A total of eight different meteorological regimes have been distinguished. The analysis of the currents measurements at five locations in the area shows that the eight meteorological regimes induce very different current responses at the bottom. The residual tidal currents exhibit a significant spatial variability both in direction and in intensity also observed in the total residual currents for high tidal range regimes. During strong wind periods however the wind-induced currents which are more uniform in space dominate the tidal currents. During high tidal range periods, the sediment flux are high. The residual sediment transport however was shown to depend on tidal and non-tidal currents amplitude and direction giving rise to a significant temporal variability in sediment fluxes amplitudes and directions and to different spatial sediment transport patterns over the area.

## 1 INTRODUCTION

Dunes are mobile transverse sandy bedforms observed on tidal-dominated continental shelves like the south part of the North Sea. They are several meters high and their wavelength is of the order of several hundreds of meters (Stride, 1982). Dune migration rate can reach 150 m per year but is usually of the order of a few tenths of meters.

Some insights in dune generation processes including migration rates have been gained using linear stability analysis (Hulscher 1996), (Idier et al 2004). Models dealing with finite amplitude dunes have been developed(Németh et al 2002) (de Vriend 1999). However, this approach is limited to simple hydrodynamics forcings and geometric domains so that they cannot account for short-term real dune dynamics.

Over the last decades, several field surveys have been performed in dunes areas in order to analyse their morphology and dynamics (Dyer 1970), (Terwindt 1971) and (Langhorne 1982). It has been observed that migrating dunes have an asymmetric shape, with a steep slope oriented in the same direction as the migration direction. Symetric dunes are also observed which experience a small migration rate. The finite amplitude dune migration process exhibits different times-scales

and seems to involve different hydrodynamics forcing mechanisms (Langhorne 1982). However the correlation of dune motion with forcing factors is not well established.

More recently, (Bot et al 2000) studied a dune field located in the middle of the strait in the so-called F-area (Fig. 1). The mean water depth in the F-area is about 35 m, with a weak mean transverse (i.e. north-west south-east) bedslope. This area is exposed to tidal currents and storm-generated currents. (Bot et al 2000) suggest that on the decennial time scale the F-area dune field should be divided into the North-West region, where dunes migrates towards the North Sea and the South-East region where dunes migrates towards the English Channel. In the South-East region however, on a shorter time scale (month to year), (Bot et al 2000) observed dune migration toward the North Sea, i.e. in a direction opposite to those of the decennial scale migration. They conjectured that the medium term dune dynamics was mainly influenced by storms action. However, the link between this two times scales remains unclear.

The aim of the present paper is to improve our knowledge of the relative influence of tides and storms on the dune dynamics in order to infer if the spatio-temporal variability of these forcings may explain the spacio-temporal variability of the dune motion. The analysis is based on the results of a one month field campaign performed in a dune area located in the middle of the Dover strait (F area).

## 2 PERMOD CAMPAIGN

The PERMOD field campaign has been performed over the F-area of the Dover Strait from 16th of October to 16th of November 2001 (Fig. 2), a period which was expected to offer a variety of hydrodynamics regimes.

The area is covered by a series of isolated dunes which have elongated shapes (Fig. 1). Their crests are almost perpendicular to the mean tidal current direction in the area. The dunes height reach up to 10 m, their

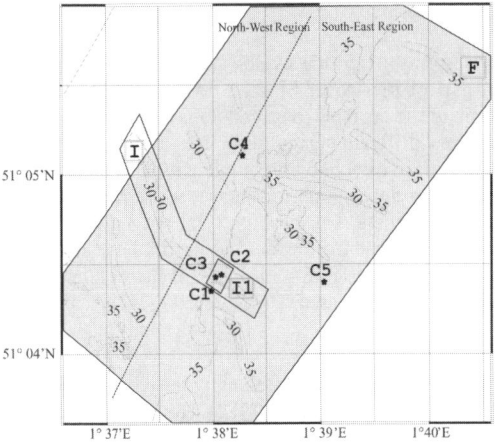

Figure 1. Bathymetry (m) of the F-area from the bathymetric survey of October 2001. Reference: hydrographic zero (lowest sea level). Projection: Lambert 93.

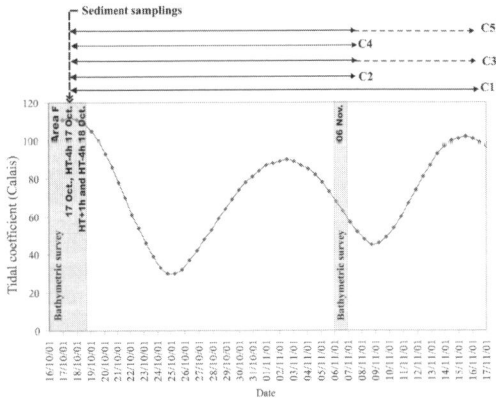

Figure 2. Field measurements of the Permod campaign. The full line arrows indicates the valid current measurements. C* is the current meter identifier.

crest length ranges from 800 m to almost 3 km. Their width ranges from 100 to 300 m, whereas the distance between two dunes ranges from 200 to 1000 m.

In this area, tidal currents are almost alternative (N.-E./S.-W.) with a typical intensity of 1 m/s for a mean spring tide and the winds climate is dominated by S.-W. (most frequents and intense) and N.-E. winds. Their direction is therefore aligned with the tidal current direction and almost normal to the dune crests.

Currents over the area have been measured at five locations (Fig. 1). To investigate the flow field over the I-dune, current-meters C1 (ADCP), C2, C3 are located almost on a line perpendicular to the I-dune (Fig. 1). From south to north, C1 is located at the foot of the steep slope, C3 at the crest and C2 at the foot of the gentle slope. Current-meter C4 is located at the boundary of the North-West and South-East F-area regions whereas C5 is located in the South-East region at the limit of the dune field. Both of them are located in dune-free areas. The current measurements are performed 1.15 m (resp. 3 m) above the seabed at C2, C3 and C4 (resp. C1). The current velocity were measured every 10 min averaging a two-minutes 1 Hz sampling rate burst. Tidal as well as storm-induced currents are properly captured whereas waves-induced flows (wind-waves and swells) are not.

The field campain almost covers two complete spring-neap tide cycles (Fig. 2). The Arpege model from Meteo France has been used to estimate wind speed (Fig. 3) and incoming direction (Fig. 4) over the Dover Strait during the campaign as no in-situ measurements where performed.

The meteorological forcing exhibits a strong temporal variability during the campain. Eight different meteorological regimes (R1 to R8) have been distinguished (Tab. 1).

R2, R4 and R7 are storm regimes with strong winds (Fig. 3). R2 and R4 experience incoming wind from the S.-W. whereas R7 features winds from the N.-E. (Fig. 4). If R2 and R4 share common wind features,

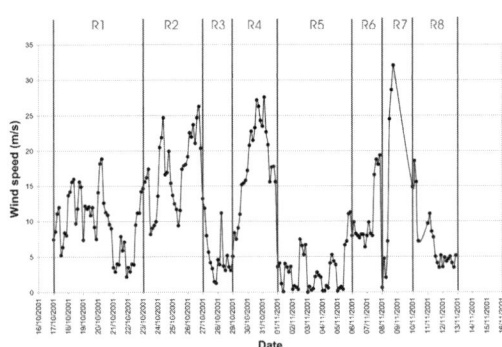

Figure 3. Wind speed (m/s) 10 m above the sea level at location: 1.7°E. – 51.1°N. (RGF93 coordinates).

the tide regime however is different with a neap tide for R2 and a mean tide for R4 (Fig. 2). R7 also experienced neap tide. R1 and R6 exhibit medium amplitude winds (Fig. 3). The R1 wind incoming direction fluctuates between S.-W. and S.-E. (Fig. 4). During R6, the direction of the wind oscillates with a one day period between N.-E. and N.-W.. These two periods also differs with respect to the tidal regime which evolves from spring to neap tide during R1 and from mean to neap during R7 (Fig. 2). These two regimes thus mainly differs in the wind direction and in the tidal amplitude. Three regimes of moderate winds (R3, R5 and R8) have been distinguished. The incoming wind direction (N. for R3 and R8, oscillating with a one day period between N.-E. and N.-W. for R5) during these regimes is probaly less significant than the tidal regime (almost neap for R3, mean for R5 and neap to mean for R8). Although not entirely similar, these three regimes bear strong similarities.

## 3  CURRENT FIELDS

First, the total currents measured by the five current meters are analysed. then the tidal and non-tidal components are investigated.

### 3.1  Total current field

Figure 5 shows that the currents in the area are dominated by tidal currents as the velocity signals exhibit a strong periodic character both at the tidal period and at the neap-spring period. Over the entire field campaign period, the maximum current velocity is observed to be in the range 1.57 to 1.80 m/s depending on the current-meter. Perturbations from pure tidal effects, due to storms, can also be depicted. On Oct. 24th, Oct. 26th (R2) and Nov. 9th (R7) the current reverse are not observed i.e. the velocity has the same direction all along the tidal cycle. It means that storm-induced currents have almost the same direction and amplitude as the tidal currents at the altitude of the measurements. The occurence of this phenomena corresponds to storm periods R2 (two times) and R7. It means that during the R4 storm regime the tidal currents reverse still exists. Current reverse inhibition phenomena is observed only during for storms occuring during a neap tide (R2 and R7) period. Storms like those observed during the PERMOD campaign cannot inhibates mean spring tide induced currents (R4) at the observed level.

The cumulated current displacements (velocity vector times the measurement interval) curves over the campain are plotted on figure 6. These curves are built by plotting one after the other, the current displacement vectors every 10 min. Two curves are related to current-meter C1, one for the velocity field at $Z = 3$ m over the bottom, and the other at $Z = 33$ m. A common feature of these curves is the S.-W./N-E. general orientation which is close to the tidal currents orientation.

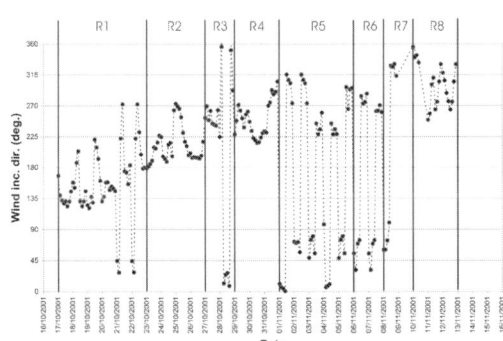

Figure 4.  Wind incoming direction 10 m above the sea level at location: 1.7°E. – 51.1°N. (RGF93 coordinates).

Figure 5.  Current velocities at C1, C2, C3, C4 and C5. Shadowed areas indicate the periods without current reverse.

Table 1.  Meteorological regimes during the PERMOD field campaign.

|  | R1 | R2 | R3 | R4 | R5 | R6 | R7 | R8 |
|---|---|---|---|---|---|---|---|---|
| Time | Oct. 17–22 | 23–26 | 27–28 | 29–31 | Nov. 1–5 | 6–7 | 8–9 | 10–12 |
| Tide | 110–40 | 40–30–50 | 50 | 50–80 | 80–82–75 | 75–45 | 45–50 | 50–80 |
| Wind | medium | strong | moder. | strong | moder. | medium | strong | moder. |
|  | S/SE | S/SW | SW/N | SW/SE | NE/NW | NE/NW | N-W | W/NW |

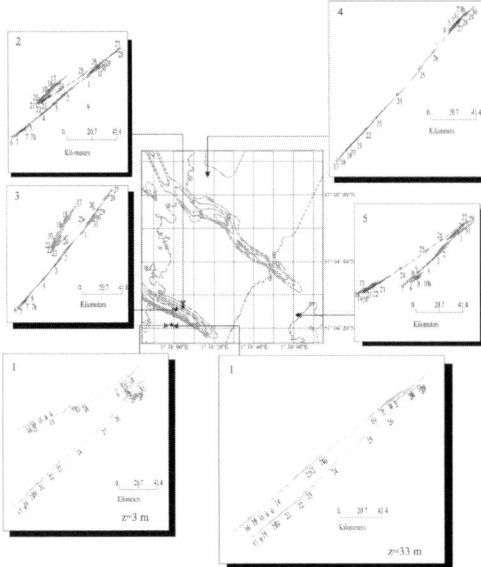

Figure 6. Cumulated current displacement curves. The scale of the cumulated vectors is not the same as this of the geographic map.

Table 2. Total residual currents characteristics (amplitude in m/s and direction) for R1–R5 meteorological regimes.

|    | R1 | R2 | R3 | R4 | R5 |
|----|-----|-----|-----|-----|-----|
| C1 | N.-E. | N.-E. | S.-W. | N.-E. | S.-W. |
|    | 0.12 | 0.34 | 0.05 | 0.06 | 0.06 |
| C2 | S.-W. | N.-E. | S.-W. | S.-W. | S.-W. |
|    | 0.04 | 0.21 | 0.03 | 0.07 | 0.18 |
| C3 | S.-W. | N.-E. | S.-W. | S.-W. | S.-W. |
|    | 0.09 | 0.22 | 0.05 | 0.11 | 0.23 |
| C4 | N.-E. | N.-E. | N.-E. | N.-E. | S.-W. |
|    | 0.11 | 0.34 | 0.05 | 0.03 | 0.05 |
| C5 | S.-W. to N.-E. | N.-E. | S.-W. | S.-E. | S.-W. |
|    | 0.03 | 0.26 | 0.05 | 0.02 | 0.05 |

We consider the period between Oct. 17th and Nov. 7th when all the current meters where operating. The mean residual current at C1 (steep side of the I- dune) is oriented N.-E. and is close to 9 cm/s at $z = 3$ m and to 10 cm/s at $z = 33$ m. The same current direction with comparable magnitude is also observed at C4 (boundary of the two regions of F-area). At C2 (top of the I- dune), the mean residual current is oriented towards the S.-W. with a magnitude of 6 cm/s. On the stoss side of the I-dune (C2) the mean current is also oriented towards the S.-W. with a smaller magnitude (3 cm/s). The current meter located on the eastern boundary of I-area (C4) records a mean current oriented towards E. with a magnitude of 2 cm/s. At this temporal scale, the magnitude of the mean residual currents is moderate with a strong spacial variability of the directions which can be opposite over the instrumented area.

However, inside this one month time scale, a strong temporal variability can be observed and deserves a detailed investigation.

Consider first the C2 (stoss face of the I-dune) curve (Fig. 6). A first period of residual South-West oriented current is observed between Oct. 17th and Oct. 22nd with noticeable current reverse. This period fits with regime R1. Then, the residual current is oriented towards the N.-E. until Oct. 26th. The amplitude of motion is larger than in the previous period and current reverses are not so well marked. This period corresponds to R2 regime. Then, the residual current is once more oriented towards the S.-E. until Nov. 6th,

but with a variable intensity and tidal reverse signature. This period covers R3, R4 and R5 regimes. The largest magnitude of the residual current is observed between Nov. 1st and Nov. 6th which can be associated to the R5 regime. The most pronounced current reverse signature occurs between Oct. 29th and Oct. 31st (R4) with limited residual current. From Nov. 7th the residuals currents are oriented North-East. The C3 (top of the I-dune) curve, is qualitatively similar the C2 curve.

From this analysis, we can conclude that the meteorological regimes classification seems to be an appropriate framework for the analysis of the bottom current fields as almost all the singularities in the curves for the C3 and C4 current meters corresponds to the boundaries of the R1 to R8 regimes.

The same analysis has been conducted for C1 (bottom and top velocity), C2 and C5. The results are sumarized in Tab. 2. From these data, some properties of the residuals currents for each meteorological regimes can be obtained. As for regimes R6 to R8 only C1 data are available, we mainly concentrate the analysis on R1 to R5 regimes.

R2 regime (neap tide amplitude, strong S.-W. winds) has several remarquable properties. The amplitude of the total residual currents are high and all the currents are oriented in the wind direction, suggesting that the currents are essentially wind-induced. The maximum is observed at C4 and the minimum at C2, on the stoss slope of the I-dune. The R2 regime leads to a residual flow pattern where the flow is everywhere N.-E. oriented and feature a weak spatial variability.

During the R4 storm period , the current intensity is significantly lower than during R2. The residual current is oriented towards the N.-E. at C1 and C4 and towards the S.-W. at C2, C3 and C5.The currents direction is no longer uniform over the area and the current reverse signature is strong except at C4. The total residual current is thus no longer dominated by the wind-induced current and the tidal-induced current plays a significant role (medium tidal range). The poor wind efficiency may resides in its Western incoming mean direction which offers a limited fetch.

During the R1 regime, wind comes from the S.-E. whereas the tidal ranges decreases from spring to neap. The observed residual currents exhibit a strong variability in space, both in magnitude and direction. The residual current is about 0.1 m/s at C1 and C4 and smaller the other locations. The magnitude of the residual currents is at least 3 times smaler than in the R2 storm regime. The highest residual currents are oriented towards the N.-E. (C1, C4) whereas the others are oriented towards the S.-E. or oscillating (C5). From these results, it is possible to conjecture that at places where the tidal residual currents are low, the wind-induced currents dominates (C1 and C4) whereas in places where the tidal is stronguer and possibly S.-W. oriented (C2 and C3), the tidal current slightly dominates the wind-induced currents. The relative effect of tide and winds is visible at C5 where the direction changes form the S.-W. for high tidal range to N.-E. for lower tidal range. At the beginning of R1 (C5 S-W oriented) the total residual current pattern is like in R4. Latter, when C5 is N-E oriented, a new pattern appears.

The R3 regime (moderate N. incoming wind, neap tide) features weak total residual currents (max. 0.11 m/s at C3), as expected from the meteorological forcings intensity. The tidal signature varies over the area. All the residual currents are oriented towards the S.-W., except C4 oriented towards the N.-E..

The R5 regime (moderate N.-E. and N.-W. winds, spring tide) leads to total residual currents with the same S.-W. direction at all locations. The intensity however varies over the domain from large (0.23 m/s at C3) to small (0.05 m/s at C4 and C5). The high values should probably be related to wind induced currents which may be increased by the contraction at the Dover Strait for N.-E. incoming winds i.e. S.-W. oriented currents coming from the North-Sea.

## 3.2 *Tidal currents*

From currents measurements, the spatial structure of real tidal current has been analysed. The maximum tidal current amplitude of about 1 m/s close to the sea bed. The tidal current directions are oriented S.-W./N.-E. as expected with a small variability in space.

In Tab. 3, the residual tidal currents amplitude for a mean spring tide (i.e. tidal coefficient of 95) are given. A strong spacial variability is observed in residual tidal current over the F-area area (Table 3). Their amplitude is very small at C1 and C5, moderate at C4 and C2 and maximum at C3 (on the top of I-dune). At C2 and C3, the residual tidal current is oriented towards the

Table 3.   Residual tidal currents.

| Name | mag. (m/s) | dir. |
|------|-----------|------|
| C1 | 0.048 | NE |
| C2 | 0.194 | SW |
| C3 | 0.216 | SW |
| C4 | 0.065 | NE |
| C5 | 0. | – |

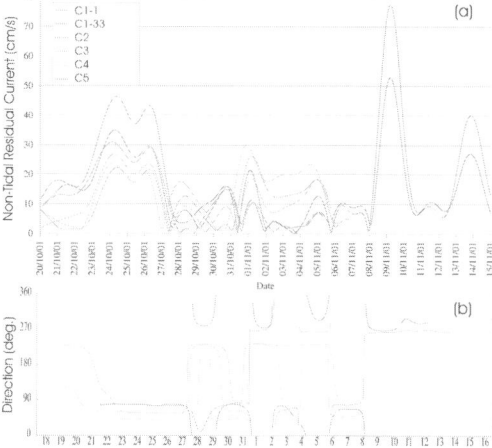

Figure 7.   Magnitude (a) and direction (b) of the non-tidal residual currents.

South-West whereas it is oriented towards the North-East at C4.

## 3.3 *Non-tidal currents*

In order to analyse the non-tidal contribution to the total residual currents, the tidal-induced currents have been removed from the total current data. The non-tidal currents magnitude and direction are plotted on Fig. 7 (a) and (b) respectively.

The direction of non-tidal current (Fig. 7 (b)) shows that the S.-W. (W. at C1) and N.-E. orientation are dominant, even if the wind direction fluctuates (Fig. 4). This result is in agreement with tidal and total residual characteristics. This shows that the flow response is not in equilibrium with the wind forcing, due to the limited size of the bassin which acts as a channel.

R2 meteorological regime corresponds to local maxima in non-tidal residual currents (Fig. 7 (a)) as expected during storm periods. During R2, the non-tidal current directions are uniformly oriented towards the N.-E. (Fig. 7 (a)), opposite to the S.-W. incoming wind direction, showing no spatial variability in direction.

During the R4 storm (incoming S.-W. wind) period however, the direction of non- tidal residual currents varies over the period and over the F-area (Fig. 7 (b)).

However, the non-tidal current amplitude for R4 is maximum on Oct. 30th whereas the direction is uniform towards the N.-E. (Fig. 7 (b)). The fluctuations in directions during R4 are probably due to fluctuations in the incoming wind direction itself (Fig. 4) and to inertia effects which tends to delay non-tidal currents inversions (Oct. 29th).

As expected, the amplitudes of the non-tidal currents during non-storm periods (R1, R3, R5) is smaller than in storm periods. However, some significant events are observed during R1 (on Oct. 20th and 21st at C1 and C5) R3 (on Oct. 28th), R5 (On Nov. 1st and Nov. 4th).

In addition, one can observe on Fig. 7 (a) that when the non-tidal residual current is oriented towards the S.-W., the maximum velocity magnitudes are measured at C3 and C2 (1st Nov.), whereas the maximum are likely to be C1 and C5 (24th Oct.) when the direction is towards N.-E.

## 4 SEDIMENT FLUXES

The purpose of the present section is to analyse the response of the system in terms of sediment fluxes. No sediment fluxes have been measured during the PERMOD campain so that all the instantaneous sediment fluxes have been estimated using the van Rijn formula (van Rijn 1989).

The instantaneous sediment fluxes at all five currents meters have been tidal-period averaged in order to compute the residual sediment fluxes at the tidal scale. In this paper we will only discuss the results for C2 (Fig. 8). The amplitudes are high during R1 and R5 periods and very low during R2 period. The direction of the tidal period averaged sediment flux (Fig. 8(b)) oscillates between two fixed directions (i.e. S.- W. and N.-E.). This evolution is very similar to those of the non-tidal currents direction (Fig. 7). One can identify several local maxima of the sediment fluxes (Fig. 8(a)) on Oct. 19th (R1), oct. 28th (R3), Nov. 1st and Nov. 5th. (R5), the main event being on Nov. 1st. Each of these maxima corresponds to a local maximum in non-tidal currents (Fig. 7). All the local maxima of the averaged sediment fluxes are oriented toward the S.-W. (Fig. 8(b)) as well as the corresponding non-tidal current (Fig. 7). However, local maxima of non-tidal currents that occurs on Oct. 24th and Oct. 26th (Fig. 7) (same intensity as the Nov. 1st event) only lead to very low residual sediment fluxes (Fig. 8 (a). The direction of the non-tidal currents on Oct. 24th and Oct. 26th (Fig. 7) is N.-E., opposite to the direction of the above mentioned maxima. The direction of the non-tidal currents is thus a critical parameter for the residual sediment flux amplitude. The maxima are observed when maxima of non-tidal currents are in the same direction (S.-W.) as the residual tidal current

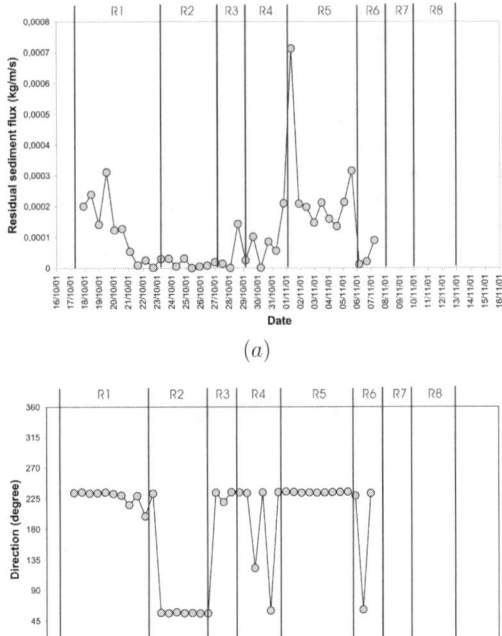

(a)

(b)

Figure 8. Tidal period averaged sediment fluxes (a) magnitude and (b) direction at C2.

(Tab. 3). In fact, when oriented towards the S.-W., non-tidal (wind induced) currents enhanced the asymmetry of the tidal currents and then increase the sediment flux (Oct. 19th and 28th, Nov. 1st and 5th) which is thus oriented towards the S.-W. (Fig. 8(a)). When oriented toward the N.-E. non-tidal currents reduce this asymmetry and thus reduce the sediment flux and even invert the residual sediment flux direction when the tidal amplitide is low. Such a residual sediment flux inversion occurs between oct. 23rd and oct. 26th (R2) as the residual sediment flux is oriented towards the N.-E. (Fig. 8(b)) like the non-tidal current (Fig. 7) and opposite the S.-W. residual tidal current direction (Tab. 3). However, due to the low tidal amplitude during this period and to the low resulting asymmetry, the sediment fluxes are low. On Oct. 30th however the residual sediment flux direction is still S.-W. as the N.-E non-tidal residual current is not strong enough to overcome the tidal asymmetry.

In order to get a more global view on the temporal variablity of the sediment fluxes, the residual over the R1-R5 regimes have been computed (Tab. 4).

The maximum of sediment transport occurs during the R5 period (Tab. 4). The R1 period experienced almost half of the R5 period averaged flux. The R4

Table 4. Residual sediment fluxes (kg/m) for R1–R5 meteorological regimes.

|    | R1 | R2 | R3 | R4 | R5 |
|----|------|------|------|------|------|
| C1 | N.-E. | N.-E. | S.-W. | N.-E. | N.-E. |
|    | 53.6 | 16.0 | 2.0 | 12.0 | 4.2 |
| C2 | S.-W. | N.-E. | S.-W. | S.-W. | S.-W. |
|    | 58.0 | 5.1 | 8.4 | 21.9 | 105.5 |
| C3 | S.-W. | N.-E. | S.-W. | S.-W. | S.-W. |
|    | 130.2 | 6.4 | 17.6 | 48.8 | 222.0 |
| C4 | E. | N.-E. | S.-W. | S.-W. | S.-W. |
|    | 3.1 | 10.9 | 4.0 | 4.8 | 94.0 |
| C5 | E./S.-E. | N.-E. | S.-W. | N.-E. | S.-W. |
|    | 1.8 | 6.4 | 2.4 | 3.0 | 59.1 |
| $\Sigma C_i$ | 246.7 | 49.3 | 34.4 | 90.5 | 484.8 |

Table 5. Mean residual sediment fluxes for the Oct.17th – Nov. 5th period.

| Location | Mean $(10^{-5}$ kg/s/m) | Direction |
|----------|------|-----------|
| C1 (1) | 4.7 | N.-E. |
| C2 | 8.4 | S.-W. |
| C3 | 23. | S.-W. |
| C4 | 3.8 | S.-W. |
| C5 | 2.7 | S.-W. |

period is the third in sediment transport intensity, followed by R3 and R2. Thus the two periods (R5 and R1) of intense sediment transport are not storm periods, whereas the period of less sediment transport (R2) is. Moreover, the periods of high sediment transport occurs during high tidal coefficients whereas, the period of neap-tide (R2) feature the lower sediment activity. However, the sediment fluxes are not fully correlated with the tidal coefficients (2) as the residual sediment flux during R2 is higher than during R3, which shows the influence of non-tidal forcings.

The spatial structure of the sediment flux also depends on the meteorological regime. During the R1 period, most of the residual sediment transport occurs at C1, C2 and C3, as the amount at C4 and C5 is very weak (Tab. 4). The spatial pattern shows a N.-E. orientation at C1 and a S.-W. orientation at C1 and C3. The orientation at C4 and C5 are not significant due to the weak amplitude.

During the R2 period, the residual sediment transport are uniformly oriented towards the N.-E. (Tab. 4). This behavior should be related to the presence of a non-tidal N.-E. current (Fig. 7). The magnitude of the sediment transport is high at C1 and C4 where the residual tidal currents are N.-E. oriented. At C2, C3 and C5 however, where the residual tidal current are S.-W. oriented, the residual sediment flux is moderate.

The R3 period feature a uniform S.-W. sediment fluxes orientation over the area (Tab. 4) which is to be related to presence of a S.-W. non tidal current (Fig. 7). The highest sediment fluxes are observed at C3 and C2 where the residual tidal current is S.-W. oriented. At C4 and C5, the sediment flux are moderate. At C1 where the residual tidal flow is oriented towards the N.-E. (i.e. opposite to the non-tidal flow) the sediment flux is weak.

During the R4 period, The sediment fluxes at C2, C3 and C4 are S.-W. oriented. At C1 and C5, the orientation is towards the N.-E.. The higher sediment flux is observed at C3 followed by C2 and C1. The sediment transport at C4 and C5 is smaller. The correlation of sediments flux and the forcings during R4 is not easy since the non-tidal currents properties at C2, C3 and C4) varies over the period (Fig. 7). However, at C1, C2 and C3, the sediment flux direction is similar the residual tidal current direction, like during R1. The amplitudes are smaller than during R1. One should observe that the non-tidal currents during R1 are moderate despite the presence of the storm (Fig. 7). At least in the vicinity of the I-dune, the sediment transport during the spring tide R1 period is controled by the tidal flow.

The R5 period is the period of maximum sediment transport. At C1, the sediment flux is oriented towards the N.-E. but is very weak. At the other locations, the sediment Transport is oriented towads the S.-W.. R5 is a period of high tidal flows (mid-spring) (Fig. 2). The non-tidal flow is of medium amplitude (Fig. 7). The sediment transport is twice during R5 than during R1, although the tidal amplitude is higher during R1. This may be due to the presence of non-tidal currents. In addition one should observe that the temporal variation of the wind forcing during the R5 period is synchonous with the tidal motion. A more detailed analysis of the high-frequency components of the non-tidal flow would be required to fully analyse this phenomena.

The sediment fluxes at all current meters locations have been time integrated from Oct. 17th to Nov. 5th. (Tab. 5). At all the points located on the dune itself and northern (C2 to C5), the sediment fluxes are directed towards the S.-W. whereas southern of the dune the sediment direction is towards the N.-E. On should notice that this sediment fluxes orientation is coherent with the observed S.-W. motion of the I-dune during this period.

5 CONCLUSION

To get a better understanding of hydrodynamics factors influencing the short term behavior of marine sand dunes in tidal-dominated continental shelves, a field campaign has been performed in an area located in the central part of the Dover Strait. In this area, a series of almost linear isolated dunes of ~10 m height and ~200 m width is observed in 35 m water depth. The distance between two successive dunes ranges from 200 to 1000 m. This campaign lasts for a month from 16th of October to 16th of November 2001. Currents have been measured 1 m above the sea-bed at five locations in the area, three along one dune profile, and the others in the middle and in the southern part of the study area. During the campaign, two spring-neap tide cycles have been observed. Three strong storms have been experienced, with opposite wind directions. Based on the three criteria of tidal amplitude, wind magnitude and wind direction, a total of eight different forcing regimes have been distinguished. The analysis of the currents measurements shows that the eight forcing regimes induce very different current responses at the bottom. The tidal currents are almost alternative in the direction of the strait (S.W.-N.E.). The residual tidal currents exhibit a significant spatial variability both in direction (regions of opposite directions) and in intensity (0 to 0.25 m/s). This variability is observed in the total residual currents during large tidal range forcing regimes. During strong wind periods however the wind-induced currents dominate the tidal currents. Periods of no current reverse are thus observed, related either to flood or ebb periods according to the wind direction. During these periods, the residual currents are almost uniform in direction over the area whereas their amplitude varies by a factor of two in space.

From the current data, sediment fluxes have been estimated. During the two storm periods corresponding to medium tidal range, extreme sediment flux events are observed, leading to an important residual sediment flux, whatever the wind direction is. However, when a storm period occurs during neap tide, no such event is observed. It has been shown that for a given tidal range, the residual sediment flux is enhenced when the non-tidal current direction coincides with residual tidal currents direction.

If the non-tidal residual current has a direction which is opposite the direction of the residual tidal current, sediments fluxes are small, unless the non-tidal current is very strong and may overcome the tidal-induced current effet. Due to spatio-temporal variability of tidal and non-tidal currents, various sediment transport patterns may arise over the area depending on the meteorological forcings. Thus according on the storm events climate and to the correlation of these events with neap-spring tidal cycle the sediment transport pattern may be strongly modified, allowing a different behavior at a different time scale as was observed by (Bot et al 2000).

REFERENCES

Bot, S. L., J. P. Herman, A. Trentesaux, T. Garlan, S. Berné, and H. Chamley (2000). Influence des tempêtes sur la mobilit des dunes tidales dans le détroit du pas-de-calais. *Oceanologica Acta 23*(2).

de Vriend, H. J. (1999). Long-term morphological prediction. In S. G. and B. P. (Eds.), *River, Coastal and Estuarine Morphodynamics*, pp. 163–190.

Dyer, K. (1970). Current velocity profiles in a tidal channel. *Geophys. J. R. Astr. Soc. 22*, 153–161.

Hulscher, S. (1996). Tidal-induced largescale regular bed form patterns in a three-dimensional shallow water model. *J. Geophys. Res. 101*(C9), 20,727–20,744.

Idier, D., D. Astruc, and S. J. M. H. Hulscher (2004). Influence of bed roughness on dune and megaripple generation. *Geophys. Res. Lett. 31*(L13214).

Langhorne, D. N. (1982). A study of the dynamics of a marine sandwave. *Sedimentology 29*(4), 574–594.

Németh, A. A., S. J. M. H. Hulscher, and H. J. de Vriend (2002). Modelling sand wave migration in shallow shelf seas. *Cont. Shelf Res. 22*.

Terwindt, J. H. J. (1971). Sand waves in the southern bight of the North Sea. *Mar. Geol. 10*, 51–67.

van Rijn, L. (1989). Handbook of sediment transport by currents and waves. Technical Report H 461, Delft Hydraulics.

*River, Coastal and Estuarine Morphodynamics: RCEM 2007 – Dohmen-Janssen & Hulscher (eds)*
*© 2008 Taylor & Francis Group, London, ISBN 978-0-415-45363-9*

# Morphodynamic evolution of sand banks: A weakly nonlinear analysis

N. Tambroni & P. Blondeaux
*DICAT University of Genoa, Genoa, Italy*

ABSTRACT: The process leading to the appearance of sand banks in shallow tidal seas is investigated studying the growth of small amplitude perturbations of the sea bottom, forced by oscillatory tidal currents. The results of the linear analysis show the existence of a critical value of the Keulegan-Carpenter number of the tide above which the flat bottom configuration is unstable and sand banks start to appear. Moreover the wavelength of the most unstable mode close to the critical conditions turns out to be finite, hence a weakly nonlinear analysis is developed to evaluate the equilibrium amplitude of the bottom forms. The configuration of the sea bottom, when the bottom forms attain their equilibrium, is characterized by the presence of long ridges, almost parallel to the tidal current but usually slightly counter-clockwise rotated, with crests sharper than the troughs and with crest-to-crests distances similar to those observed during field surveys.

## 1 INTRODUCTION

In the continental shelf, tidal currents often give rise to large scale, periodic bottom forms named sand banks. Sand banks are long ridges (the length is of the order of several tens of kilometres) with a spacing (crest to crest distance) of the order of kilometres and a height up to several tens of metres (Dyer & Huntley 1999). Moreover, their crests are almost aligned with the tidal current forming small positive or negative angles (Besio et al. 20006).

Even though it is now generally accepted that sand banks arise as free instabilities of the system describing the interaction between the cohesionless sea bottom and the water motions induced by tide propagation, and reliable models based on linear stability analyses exist to predict their main geometrical characteristics, little is known on the morphodynamics processes that shape and maintain these bottom forms in equilibrium conditions. Weakly nonlinear stability analyses are a powerful tool to investigate the equilibrium configuration attained by unstable bottom perturbations when the parameters of the problem are close to the critical values. Difficulties in the weakly nonlinear analysis of sand bank dynamics arise because the linear approaches formulated so far predict vanishing values of the wavenumber of the most unstable mode close to the critical conditions. In other words, close to the critical conditions, the bottom forms which tend to appear are characterized by an infinite wavelength.

In the present paper, first, we revisit the linear approach of Hulscher et al. (1993). In particular, we study the time development of small amplitude bottom perturbations forced by tidal currents using a different

parametrization of the bed shear stress, and a sediment transport predictor which provides vanishing values of the sediment transport rate when the bottom shear stress is smaller than a critical value. Moreover, elliptical tides are considered to describe field locations characterized by tidal currents which are rotary or have a low ellipticity. Indeed at these locations sand banks are more likely to occur (Dyer & Huntley 1999). With these improvements, close to the critical conditions, the wavelength of the most unstable mode turns out to be finite. This result opens the possibility to carry out a weakly nonlinear stability analysis. Hence, the time development of the most unstable mode is studied for values of the parameters close to the marginal conditions. The classical Landau equation for the amplitude of the most unstable bottom component of the perturbation is found. In the next section we formulate the hydrodynamic problem and we introduce the sediment transport parametrization. In Section 3 and 4, we study the time development of a bottom perturbation of small amplitude by linearizing the problem and we determine the critical conditions. Then, in Section 5, we formulate the weakly nonlinear analysis. Finally, the results of the nonlinear approach are described in Section 6 while Section 7 concludes the paper.

## 2 FORMULATION OF THE PROBLEM

The problem formulation is similar to that described in Hulscher et al. (1993). A shallow sea of small depth is considered and a Cartesian coordinate system ($x^*$, $y^*$) lying on the free surface is introduced. The seabed is supposed to be made of a cohesionless sediment of uniform size $d^*$ and density $\rho_s^*$. By using the $f$-plane approximation (LeBlond & Mysak 1978), the hydrodynamics of the problem is posed by continuity and momentum equations where the Coriolis contributions related to the Earth rotation are taken into account. Being the water depth much smaller than the horizontal scale of the problem (the wavelength of the bottom forms), the shallow water approximation is introduced and the depth averaged value of the velocity field is considered. On defining the following dimensionless variables

$$(x, y) = \frac{(x^*, y^*)\omega^*}{U_0^*} \ , \ t = t^*\omega^* \ , \ (U, V) = \frac{(U^*, V^*)}{U_0^*} \ ,$$

(1)

$$\eta = \frac{\eta^*}{a^*}, \ h = \frac{h^*}{h_0^*}, \ (\tau_x, \tau_y) = \frac{(\tau_x^*, \tau_y^*)}{\rho^*(U_0^*)^2} \ ,$$

(2)

where $t^* =$ time; $\omega^* =$ angular frequency of the tide; $(U^*, V^*) =$ depth averaged velocity components along the $x^*$ and $y^*$ axes respectively; $U_0^* =$ maximum value of the depth averaged velocity during the tidal cycle; $\eta^* =$ free surface displacement; $a^* = U_0^*\sqrt{h_0^*/g^*} =$ order of magnitude of the amplitude of the tidal wave, $h_0^* =$ average water depth, $(-h^*) =$ bottom elevation with respect of the average water level, $(\tau_x^*, \tau_y^*) = x^*$ and $y^*$ components of the bottom shear stress and $\rho^* =$ sea water density, the flow equations become:

$$\frac{\partial(h + a\eta)}{\partial t} + \frac{\partial[(h + a\eta)U]}{\partial x} + \frac{\partial[(h + a\eta)V]}{\partial y} = 0,$$

(3)

$$\frac{\partial U}{\partial t} + U\frac{\partial U}{\partial x} + V\frac{\partial U}{\partial y} = -\frac{1}{a}\frac{\partial\eta}{\partial x} - r\frac{\tau_x}{h + a\eta} + fV,$$

(4)

$$\frac{\partial V}{\partial t} + U\frac{\partial V}{\partial x} + V\frac{\partial V}{\partial y} = -\frac{1}{a}\frac{\partial\eta}{\partial y} - r\frac{\tau_y}{h + a\eta} - fU.$$

(5)

Three dimensionless parameters appear in the latter equations: $r = U_0^*/(\omega^*h_0^*)$ is the Keulegan-Carpenter number of the phenomenon, defined as the ratio between the amplitude of the fluid displacement oscillations induced by the tide in the horizontal direction and the local water depth; $a = a^*/h_0^* = U_0^*/\sqrt{g^*h_0^*}$ is a parameter proportional to the strength of the tide; finally $f = 2(\Omega^*/\omega^*)\sin\Phi$, with $\Omega^*$ angular velocity of the Earth rotation and $\Phi$ local latitude, is the Coriolis parameter.

In previous models of the phenomenon, the bed shear stress is often assumed to be proportional to the depth averaged velocity, introducing a linear friction coefficient (Zimmerman 1982). In the present work we use the following nonlinear constitutive relationship

$$(\tau_x, \tau_y) = \frac{(U\cos\varphi - V\sin\varphi, U\sin\varphi + V\cos\varphi)R}{C^2}$$

(6)

where the deviation angle $\varphi$ is introduced to account for the Coriolis effects, which deviate the bed shear stress with respect to the depth averaged velocity. Moreover in (6) $R = \sqrt{U^2 + V^2}$ is the module of the depth averaged velocity and $C$ is the friction coefficient which can be evaluated from the following standard formula

$$C = 2.5\ln[10.96(h + a\eta)/z_r]$$

(7)

where $z_r = z_r^*/h_0^*$ is the dimensionless roughness of the sea bottom. Note that, an estimate of $\varphi$ can be obtained by means of a three-dimensional model (Blondeaux & Vittori 2005, e.g.).

The morphodynamics is governed by the sediment continuity equation which simply states that convergence (or divergence) of the sediment flux must be accompanied by a rise (or fall) of the bed profile. By introducing the dimensionless sediment transport rate $(Q_x, Q_y)$ and the dimensionless morphodynamic time scale $T$:

$$(Q_x, Q_y) = \frac{(Q_x^*, Q_y^*)}{\sqrt{(\rho_s^*/\rho^* - 1)g^*d^{*3}}}, \ T = t\frac{d}{\sqrt{\psi_p(1 - p_{or})}}$$

(8)

where $(Q_x^*, Q_y^*)$ are the components of the volumetric sediment transport rate per unit width, $p_{or}$ is the sediment porosity, $d = d^*/h_0^*$ is the dimensionless sediment size and $\psi_p = (\omega^*h_0^*)^2/[(\rho_s^*/\rho^* - 1)g^*d^*]$ is the mobility number of the sediment particles, it is possible to write

$$\frac{\partial h}{\partial T} = \frac{\partial Q_x}{\partial x} + \frac{\partial Q_y}{\partial y}.$$

(9)

For values of the parameters typical of field conditions, the morphodynamic time scale is much larger than the tide period. Therefore, it follows that the time derivative of $h$ can be neglected in the continuity equation (3). Furthermore, if the small oscillations of the bottom configuration taking place during the tidal cycle around the mean position are neglected, it is possible to consider the sediment transport rate $(\overline{Q}_x, \overline{Q}_y)$ averaged over the tide period and to force the sediment balance on the net flux of sediment. The problem is closed once relationships for $(Q_x, Q_y)$ are provided. Presently, the suspended load is assumed to be negligible and the approach proposed by Fredsoe & Deigaard(1992) is used to quantify the bed load which turns out to depend on the components of the dimensionless Shields stress $(\theta_x, \theta_y) = (\tau_x^*, \tau_y^*)/[(\rho_s^* - \rho^*)g^*d^*]$. Moreover, as in Hulscher et al. (1993), a contribution is added to take into account that the sediment is transported easier downhill than uphill. For $\theta$ larger than $\theta_c$

$$(Q_x, Q_y) = \Phi_s \left[ \frac{(\theta_x, \theta_y)}{\theta} - \frac{G}{r} \left( \frac{\partial h}{\partial x}, \frac{\partial h}{\partial y} \right) \right],$$

(10)

where

$$\Phi_s = \frac{30}{\pi}(\theta - \theta_c)(\sqrt{\theta} - 0.7\sqrt{\theta_c})$$

(11)

with $\theta = \sqrt{\theta_x^2 + \theta_y^2}$ and $\theta_c$ = critical value of the Shields parameter such that for $\theta$ smaller than $\theta_c$ no sediment moves. Moreover, experimental observations of various authors (Talmon et al. 1995) provide estimates for the values of the components of the tensor $G$ (Seminara 1998). Using a reference frame $(s, n)$ such that $s$ is aligned with the bed shear stress, it turns out that

$$G_{sn} = 0, \quad G_{ss} = -\frac{\theta_c}{\mu} \frac{1}{\Phi_s} \frac{d\Phi_s}{d\theta}, \quad G_{nn} = -\frac{0.55}{\sqrt{\theta}}, \quad (12)$$

where $\mu$ is the dynamic friction coefficient of the bed material. Seminara (1998) proposed a value of $\mu$ approximatively equal to 0.5. Later Parker et al. (2003) suggested to use much smaller values. In the present investigation, the simulations have been carried out fixing $\mu$ equal to about 0.15.

## 3  THE LINEAR ANALYSIS

The linear analysis investigates the conditions required for the flow over a flat bed to loose stability to a bottom perturbation of small amplitude. Hence, the bottom profile can be assumed to be described by the superposition of different spatial components which evolve one independently from the other and the problem can be solved for the generic spatial component:

$$h = 1 - \epsilon \left[ A(T)e^{i(\alpha_x x + \alpha_y y)} + c.c. \right] + O(\epsilon^2),$$

(13)

where $A(T)$ is the amplitude of the generic component which is periodic in the $x$ and $y$ directions with wavenumbers $\alpha_x$ and $\alpha_y$ respectively and $\epsilon$ is a small quantity, strictly infinitesimal ($\epsilon \ll 1$). The small value of $\epsilon$ allows the solution to be expanded in the form

$$(U, V, \eta) = (U_0, V_0, \eta_0) +$$

(14)

$$+ \epsilon \left[ (U_1, V_1, a\eta_1) A(T)e^{i(\alpha_x x + \alpha_y y)} + c.c. \right] + O(\epsilon^2).$$

At the leading order of approximation, i.e. $O(\epsilon^0)$, the problem is reduced to the determination of both the flow and sediment transport induced by tide propagation over a flat sea bed. If the flow is supposed to be dominated by the main tide constituent, the local velocity field can be assumed to be described by

$$(U_0, V_0) = (\cos(t), e\sin(t))$$

(15)

where $e$ denotes the ratio between the minor (aligned with the $y$ axis) and major axes of the tidal ellipse. Then, momentum equations provide the surface slope while continuity equation describes the slow changes of the flow field taking place on the spatial scale $U_0^*/\omega^*$. The solution of the problem at the leading order of approximation is completed by the evaluation of the sediment transport rate which can be performed by using (10).

When (14) is substituted into equations (3–5) and terms of $O(\epsilon)$ are considered, three linear equations for $U_1$, $V_1$ and $\eta_1$ are obtained. Combining momentum equations at $O(\epsilon)$ it is possible to write the following equation for the $O(\epsilon)$ vorticity component $\zeta_1$, defined as $\zeta_1 = i(\alpha_x V_1 - \alpha_y U_1)$:

$$\frac{\partial \zeta_1}{\partial t} + F(t)\zeta_1 = G(t),$$

(16)

where

$$F(t) = i(\alpha_x U_0 + \alpha_y V_0) + r \left[ \frac{R_0}{C_0^2} + \frac{(\alpha_y U_0 - \alpha_x V_0)^2}{R_0 C_0^2 (\alpha_x^2 + \alpha_y^2)} \right]$$

$$\cos\varphi + r \frac{(\alpha_x V_0 - \alpha_y U_0)(\alpha_x U_0 + \alpha_y V_0)}{R_0 C_0^2 (\alpha_x^2 + \alpha_y^2)} \sin\varphi, \quad (17)$$

$$G(t) = -i \left( f + r \frac{R_0}{C_0^2} \sin\varphi \right)(\alpha_x U_0 + \alpha_y V_0) +$$

$$-ir(\alpha_x \tau_{y0} - \alpha_y \tau_{x0}) - ir \left( \frac{5R_0}{C_0^3} + \frac{(\alpha_x U_0 + \alpha_y V_0)^2}{C_0^2 R_0 (\alpha_x^2 + \alpha_y^2)} \right)$$

$$[\alpha_x (U_0 \sin\varphi + V_0 \cos\varphi) - \alpha_y (U_0 \cos\varphi - V_0 \sin\varphi)].$$

(18)

Moreover, the bottom shear stress components $(\tau_{x0}, \tau_{y0})$ are related to the velocity components by

$$(\tau_{x0}, \tau_{y0}) = \frac{(U_0 \cos\varphi - V_0 \sin\varphi, U_0 \sin\varphi + V_0 \cos\varphi)R_0}{C_0^2},$$

(19)

where $R_0 = \sqrt{U_0^2 + V_0^2}$ and $C_0 = 2.5\ln(10.96/z_r)$. Equation (16) can be solved by expanding the functions $F(t)$, $G(t)$ and the unknown $\zeta_1$ as Fourier series of time. Once the vorticity field at $O(\epsilon)$ is determined, the velocity components can be evaluated by using continuity equation and vorticity definition:

$$(U_1, V_1) = \frac{(\alpha_x, \alpha_y)(\alpha_x U_0 + \alpha_y V_0) + i(\alpha_y, -\alpha_x)\zeta_1}{\alpha_x^2 + \alpha_y^2}.$$

(20)

971

Finally, the equation providing the time development of the amplitude $A(T)$ of the generic bottom perturbation follows from sediment continuity equation:

$$\frac{dA(T)}{dT} = \Gamma A(T),\qquad(21)$$

where the growth rate $\Gamma$ is a complex quantity which depends on the parameters of the problem:

$$\Gamma = -i(\alpha_x \overline{Q}_{1x} + \alpha_y \overline{Q}_{1y}).\qquad(22)$$

The functions $(Q_{1x}, Q_{1y})$ can be obtained by means of the straightforward expansion of (6, 10), and once they are known, the values of $(\overline{Q}_{1x}, \overline{Q}_{1y})$ can be computed by

$$(\overline{Q}_{1x}, \overline{Q}_{1y}) = \frac{1}{2\pi}\sum_{i=1}^{2}\int_{t^{(2i-1)}}^{t^{(2i)}}(Q_{1x}, Q_{1y})\,dt,\qquad(23)$$

where $t^{(i)}$ are the phases of the cycle such that $\theta = \theta_c$. Moreover odd (even) values of $i$ are such that $d\theta/dt$ is positive (negative). Since $\theta = \theta_0 + \epsilon[A(T)\theta_1 e^{i(\alpha_x x + \alpha_y y)} + c.c.] + O(\epsilon^2)$, it follows that

$$t^{(i)} = t_0^{(i)} + \epsilon[t_1^{(i)} A(T)e^{i(\alpha_x x + \alpha_y y)} + c.c.] + O(\epsilon^2),$$

where $t_0^{(i)}$ is such that $\theta_0(t) = \theta_c$ and $t_1^{(i)} = -\theta_1(t)/(d\theta_0/dt)$ at $t = t_0^{(i)}$. Therefore (23) can be replaced by

$$(\overline{Q}_{1y}, \overline{Q}_{1y}) = \frac{1}{2\pi}\sum_{i=1}^{2}\int_{t_0^{(2i-1)}}^{t_0^{(2i)}}(Q_{1x}, Q_{1y})\,dt,\qquad(24)$$

taking into account that a contribution arising from (23) should be considered at the following order of approximation. The solution of (21) is clearly:

$$A(T) = A_0 \exp[\Gamma T].\qquad(25)$$

Hence, the growth or the decay of the perturbation is controlled by the real part $\Gamma_R$ of the growth rate $\Gamma$, while the imaginary part is related to the migration speed of the perturbations. Because of the symmetry of the forcing flow, the imaginary part of $\Gamma$ vanishes and there is no migration of the bottom forms.

## 4 THE RESULTS OF THE LINEAR ANALYSIS

The time behaviour of small amplitude perturbations of the bottom of shallow tidal seas was already analysed by Hulscher et al. (1993) and later by Hulscher (1996), showing that, for a unidirectional tide in the North Hemisphere, the most unstable modes are long ridges slightly counter-clockwise rotated with respect to the direction the main tidal current, with wavelengths of the order of a few kilometres. Therefore, in this section, we look at the qualitative and quantitative changes of the results due to the different

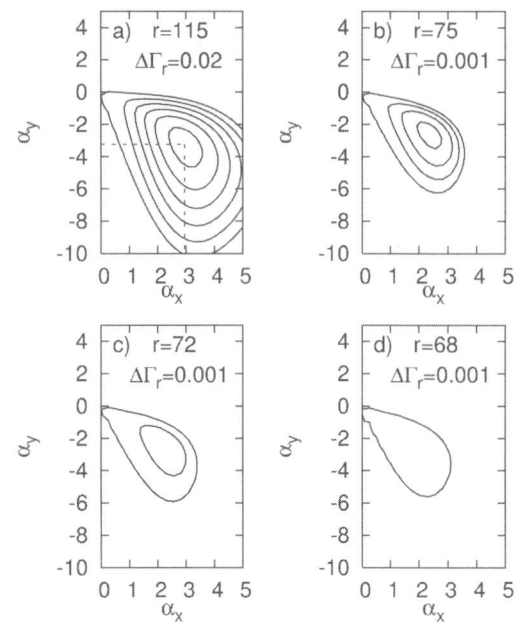

Figure 1. Amplification rate $\Gamma_R$ of the bottom perturbation as function of $\alpha_x$ and $\alpha_y$ for different values of $r$ and $e = 0$, $z_r = 1.0\ 10^{-3}$, $f = 0.8$, $\varphi = 0$, $d = 6.7\ 10^{-6}$, $\psi_d = 5.5\ 10^{-3}$. $\Delta\Gamma_R = 0.001$.

sediment transport parametrization, and to the non-linear constitutive law for the bed shear stress. We fix the values of the parameters by considering a typical site in the North Sea where the tide is dominated by the semidiurnal constituent, the water depth is relatively small and the sea bottom is made of fine sand ($h_0^* = 30$ m, $\omega^* = 1.4\ 10^{-4}$ s$^{-1}$, $U_0^* = 0.5$ m/s, $f = 0.8$, $d^* = 0.2$ mm). Finally, we assume that ripples cover the sea bottom, hence, using the empirical predictor for ripple height of Soulsby & Whitehouse (2005), we fix $z_r^*$ equal to 3 cm. The dimensionless parameters of the models turn out to be $r = 115$, $e = 0$, $z_r = 1.0\ 10^{-3}$, $f = 0.8$, $\varphi = 0$, $d = 6.7\ 10^{-6}$, $\psi_d = 5.5\ 10^{-3}$. Figure 1a shows the real part $\Gamma_R$ of the amplification rate $\Gamma$ as function of $\alpha_x$ and $\alpha_y$. A wide range of unstable modes, i.e. bottom waviness characterized by positive values of $\Gamma_R$, exists. If we assume that the mode characterized by the largest growth rate prevails on the others, the bottom forms predicted by the present linear stability analysis are ridges counter-clockwise rotated with respect to the direction of the tidal current. Indeed, the most unstable mode is characterized by wavenumbers equal to $(2.95, -3.23)$ corresponding to bottom forms with a wavelength equal to about 5 km. So far, results similar to those described by Hulscher et al. (1993) and Hulscher (1996) are found. Now, it is interesting to look at the results for different values of the Keulegan-Carpenter number of the tide. When

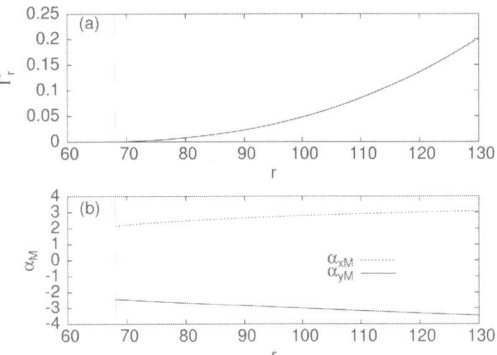

Figure 2. (a) Amplification rate $\Gamma_R$ and (b) wavenumbers $\alpha_{x,M}, \alpha_{y,M}$ of the most unstable mode as function of $r$ for $e = 0$, $z_r = 1.0 \, 10^{-3}$, $f = 0.8$, $\varphi = 0$, $d = 6.7 \, 10^{-6}$, $\psi_d = 5.5 \, 10^{-3}$.

decreasing values of $r$ are considered, even though the range of unstable modes does not change significantly, the values of $\Gamma_R$ decrease, till for $r$ equal to a critical value $r_C$, no unstable mode can be found (Fig. 1). Moreover, looking at Figure 2 where the values $(\alpha_{x,M}, \alpha_{y,M})$ of the wavenumbers characterizing the most unstable mode are plotted versus $r$, it can be observed that for $r$ tending to $r_C$, the wavenumbers $(\alpha_{x,M}, \alpha_{y,M})$ tend to finite values $(\alpha_{x,C}, \alpha_{y,C})$. The latter finding makes it impossible to perform a weakly nonlinear stability analysis similar to those developed, among other, by Colombini et al. (1987), Vittori & Blondeaux (1990), Schielen et al. (1993). However, the classical weakly nonlinear stability analysis can not be followed since, for $r$ slightly larger than $r_C$, a small but finite range of wavenumbers around $(\alpha_{x,C}, \alpha_{y,C})$ turns out to be characterized by a positive growth rate. Moreover, for $r \leq r_C$ all the modes are characterized by a vanishing growth rate. Finally, results relative to the elliptical tide case and investigations on the effects of deviation angle $\varphi$ are not reported herein since they confirm the main features emerged from the more detailed analyses of Besio et al. (2005) and Tambroni et al. (2006). In particular, the results show that if the velocity vector is counterclockwise rotating, clockwise rotated sand banks tend to be the most unstable mode when accounting for the deviation of bed shear stress from the directions of the depth averaged velocity. However, in the present context it is important to note that, because of the presence of a threshold value of the Shields parameter below which no sediment moves, in all the cases, for $r$ tending to $r_C$, the wavenumbers of the most unstable mode tend to finite values, i.e. ultra-long waves are no longer the most unstable mode close to the critical conditions. See for example the results shown in Figure 3 where the values of $\alpha_{x,M}$ and $\alpha_{y,M}$ are plotted versus $r$ for a tide characterized by a low ellipticity.

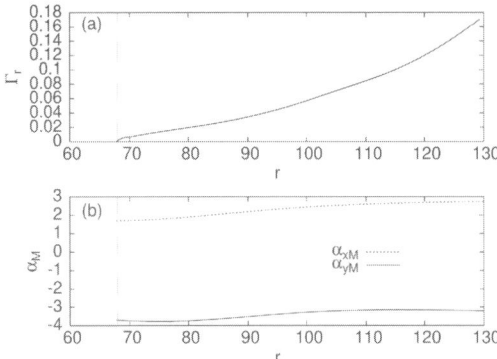

Figure 3. (a) Amplification rate $\Gamma_R$ and (b) wavenumbers $\alpha_{x,M}, \alpha_{y,M}$ of the most unstable mode as function of $r$ for $e = 0.6$, $z_r = 1.0 \, 10^{-3}$, $f = 0.8$, $\varphi = 0$, $d = 6.7 \, 10^{-6}$, $\psi_d = 5.5 \, 10^{-3}$.

## 5 THE WEAKLY NONLINEAR ANALYSIS

We seek a finite-amplitude solution, restricting our attention to the weakly nonlinear regime defined by the condition

$$r = r_C(1 + \epsilon r_1), \qquad (26)$$

where $\epsilon$ is a small quantity. Then, we consider the time development of the most unstable mode characterized by wavenumbers $(\alpha_x, \alpha_y) = (\alpha_{x,M}, \alpha_{y,M})$. The obtained results (Fig. 4) suggest that it is possible to assume that the perturbation wavenumbers $(\alpha_{x,M}, \alpha_{y,M})$ differ of a small amount from $(\alpha_{x,m}, \alpha_{y,m})$ which are the wavenumbers of a mode in marginal conditions, i.e. a mode characterized by vanishing values of the amplification rate when the Keulegan-Carpenter number is equal to $r$. In other words we assume

$$(\alpha_x, \alpha_y) = (\alpha_{x,M}, \alpha_{y,M}) = (\alpha_{x,m}, \alpha_{y,m})[1 + \epsilon(\alpha_{x,1}, \alpha_{y,1})], \qquad (27)$$

Then, because of (26), it is expected that the amplification rate of the unstable mode is of $O(\epsilon)$. Therefore, following the lead of Stuart (1971), we employ a multiple scale technique and introduce a slow time scale $\tau$ associated with the growth of the perturbation $\tau = \epsilon T$. In order to derive the order of magnitude A of the amplitude of the fastest growing perturbation, we follow the usual argument of hydrodynamic stability: nonlinearity gives rise to interactions between the fundamental mode and itself which lead to the generation of higher harmonics. Following this cascade process one finds that the fundamental is reproduced at third order which leads to the generation of secular terms. In order to prevent their occurrence the 'slow' time dependence of the amplitude of the fundamental must also be forced to produce a contribution at third order.

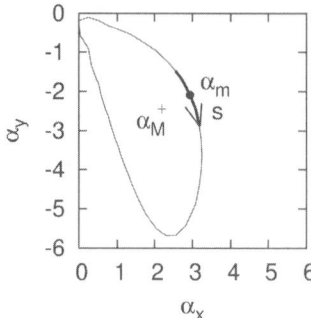

Figure 4. Sketch of the marginal stability curve for $r = 70$, $e = 0$, $z_r = 1.0\ 10^{-3}$, $f = 0.8$, $\varphi = 0$, $d = 6.7\ 10^{-6}$, $\psi_d = 5.5\ 10^{-3}$.

In other words $\epsilon \partial A / \partial \tau$ must balance $A^3$, which occurs provided $A \sim O(\epsilon^{\frac{1}{2}})$. Hence, we consider the time development of a bottom configuration in the form

$$h = 1 - \epsilon^{\frac{1}{2}} \left[ A_1(\tau) e^{i(\alpha_x x + \alpha_y y)} + c.c. \right]$$

$$- \epsilon \left[ A_2(\tau) e^{2i(\alpha_x x + \alpha_y y)} + c.c. \right] +$$

$$- \epsilon^{\frac{3}{2}} \left[ A_3(\tau) e^{i(\alpha_x x + \alpha_y y)} + c.c. + ... \right] + O(\epsilon^2). \quad (28)$$

Hereinafter $A_1(\tau)$ is the amplitude of the fastest growing mode and $A_2(\tau)$ is the amplitude of the superhamonic component which is generated by nonlinear effects. The small values of $\epsilon$ allow the solution of the problem to be expanded in the form

$$U = U_0 + \epsilon^{\frac{1}{2}} \left[ A_1(\tau) U_1 e^{i(\alpha_x x + \alpha_y y)} + c.c. \right] +$$

$$+ \epsilon \left[ A_1(\tau)^2 U_{22}^{(1)} e^{i2(\alpha_x x + \alpha_y y)} + |A_1(\tau)|^2 U_{20} + \right.$$

$$+ A_2(\tau) U_{22}^{(2)} e^{i2(\alpha_x x + \alpha_y y)} + c.c. \Big] + h.o.t. \quad (29)$$

with analogue expressions for $V$ and $\eta$.

At the leading order of approximation, i.e. $O(\epsilon^0)$, the bottom turns out to be flat and, as assumed previously, the flow is supposed to be dominated by the main tide constituent and the local velocity field is described by (15).

When (29) is plugged into the equations (3–5) and terms of $O(\epsilon^{\frac{1}{2}})$ are considered, a set of linear equations for $U_1, V_1$ and $\eta_1$ equal to the one already found in the linear analysis is derived. Hence, the solution is obtained with the approach previously outlined. Because of assumption (27), $(\alpha_x, \alpha_y)$ can be replaced by $(\alpha_{x,m}, \alpha_{y,m})$ taking into account that the terms due to $(\alpha_{x,1}, \alpha_{y,1})$ shift at $O(\epsilon^{\frac{1}{2}})$. Hence, the mode is in marginal conditions and the sediment continuity

equation simply states that the amplitude $A_1$ neither grows nor decays.

At the following order of approximation, i.e. $O(\epsilon)$, various contributions appear: the first two contributions of $O(\epsilon)$ in (29) are generated by the nonlinear interaction of the flow field at the previous order of approximation. The steady streaming associated with the second of these terms would tend to carry sediment along the bed profile causing the appearance of a bottom waviness with a wavelength equal to half of the fundamental. This new perturbation of the bottom profile produces a further component of the velocity field which is the third contribution of $O(\epsilon)$. The derivation of the equations for the unknown functions $U, V$ and $\eta$ at $O(\epsilon)$ is tedious but straightforward. The final equations are very long and for sake of space are not given herein. By means of a procedure similar to that applied at $O(\epsilon^{\frac{1}{2}})$, it is possible to solve these equations and to determine the flow fields at this order of approximation. Once the flow is known, the sediment continuity equation should be considered. If the net flux of sediment is considered, it turns out that

$$\frac{dA_2}{dT} = -2i\alpha_x \left[ P_{1x} A_1^2 + P_{2x} A_2 \right] - 2i\alpha_y \left[ P_{1y} A_1^2 + P_{2y} A_2 \right], \quad (30)$$

with

$$P_{1(x,y)} = \sum_{i=1}^{4} \frac{(-1)^i}{2\pi} \left( Q_{1(x,y)} t_0^{(i)} t_1^{(i)} + \frac{dQ_{0(x,y)}}{dt} t_0^{(i)} \frac{(t_1^{(i)})^2}{2!} \right)$$

$$+ Q_{22(x,y)}^{(1)}, \quad P_{2(x,y)} = Q_{22(x,y))}^{(2)} \quad (31)$$

where the overbar denotes the time average over the tidal cycle and the sediment transport rate vanishes when the Shields parameter $\theta_0$ is smaller than $\theta_c$. If we require that $A_2$ remains bounded, the right hand side of (30) must vanish. This condition determines the amplitude of the bottom second harmonic

$$A_2 = K A_1^2, \quad (32)$$

where $K = -(\alpha_x P_{1x} + \alpha_y P_{1y})/(\alpha_x P_{2x} + \alpha_y P_{2y})$. From a physical point of view relationship (32) forces the amplitude of the bottom second harmonic in such a way that the gravity component along the bed profile at second order exactly balances the second order component of the drag force acting on the sediment caused by the steady streaming.

At $O(\epsilon^{\frac{3}{2}})$ the flow field can be split into two components: one proportional to $|A_1|^2 A_1 e^{i(\alpha_x x + \alpha_y y)}$ and the other to $A_3 e^{i(\alpha_x x + \alpha_y y)}$. The former arises from the nonlinearity of momentum equation and the other is forced by the presence of a bottom perturbation at this order of approximation. Since the intermediate and resultant equations are very long, details are omitted. It is easy to verify that both problems are similar to that found for $(U_1, V_1)$ but for the presence of forcing terms. Hence, they can be solved with the procedure already applied at the previous orders of approximation. Let us finally come to the sediment continuity equation which reads

$$\frac{dA_3}{dT} + \frac{dA_1}{d\tau} = a_1 A_1 + a_2 |A_1|^2 A_1 + \Gamma A_3, \qquad (33)$$

where $a_1, a_2$ are quantities depending on the parameters of the problem. Their expressions are not given herein since they are very long, however coefficient $a_1$ is the leading order term of the expansion of the amplification rate of the most unstable component of the bottom perturbation in terms of $\epsilon^{1/2}$ and it is originated by the use of $(\alpha_{x,m}, \alpha_{y,m})$ instead of $(\alpha_{x,M}, \alpha_{y,M})$ into the problem at $O(\epsilon^{1/2})$. On the other hand the coefficient $a_2$ is originated by the nonlinear self-interaction of the most unstable component of the bottom perturbation. Finally, because of the assumptions already introduced, it turns $dA_3/dT = \Gamma A_3$, therefore we obtain the following ordinary differential equation for $A_1$:

$$\frac{dA_1}{d\tau} = a_1 A_1 + a_2 |A_1|^2 A_1. \qquad (34)$$

Equation (34) is of Landau-Stuart type and can be integrated in close form to obtain the time development of a small initial bottom perturbation and its asymptotic behaviour for large time

$$|A_1(\tau)| = \left[ \frac{Real(a_1)}{\exp[-2Real(a_1)\tau] - Real(a_2)} \right]^{\frac{1}{2}}. \qquad (35)$$

If the cubic term in (34) is neglected ($A_1 \ll 1$), one recovers the usual exponential behaviour of $|A_1|$ predicted by the linear theory. Then, nonlinear effects cause the perturbation to reach the equilibrium amplitude $|A_{1e}| = \sqrt{-Real(a_1)/Real(a_2)}$ when $\tau$ tends to infinity.

## 6 THE RESULTS OF THE WEAKLY NONLINEAR ANALYSIS

Let us consider the same set of parameters introduced in Section 4 ($h_0^* = 30$ m, $\omega^* = 1.4\,10^{-4}\,\mathrm{s}^{-1}$, $f = 0.8$, $d^* = 0.2$ mm, $\varphi = 0$, $z_r^* = 3$ cm). For such values of the parameters, the critical value of the Keulegan-Carpenter number turns out to be about 67, corresponding to a value of $U_0^* = 0.25$ m/s. For $r$ larger than 67, the linear analysis predicts the appearance of sand banks characterized by a finite wavelength and crests which are counter-clockwise rotated with respect to the tidal current. However, as already pointed out,

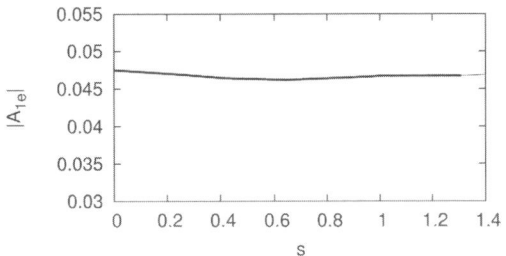

Figure 5. Equilibrium amplitude $|A_{1e}|$ as function of the curvilinear coordinate $s$ along the marginal stability curves and such that $s = 0$ corresponds to $(\alpha_{x,m}, \alpha_{y,m}) = (2.56, -1.48)$, for $r = 70$, $e = 0$, $z_r = 1.0\,10^{-3}$, $f = 0.8$, $\varphi = 0$, $d = 6.7\,10^{-6}$, $\psi_d = 5.5\,10^{-3}$ (see figure 4).

the linear analysis is unable to predict the equilibrium amplitude of the bottom features. Then, let us consider a value of $r$ slightly larger than its critical value, e.g. $r = 70$. Hence, we can apply the analysis described in the previous section by choosing a value of $(\alpha_{x,m}, \alpha_{y,m})$ along the marginal curve, sufficiently close to $(\alpha_{x,M}, \alpha_{y,M}) = (2.20, -2.44)$ (Fig. 4). For example, we can fix $(\alpha_{x,m}, \alpha_{y,m}) = (2.94, -2.09)$. From the solution of the problem at $O(\epsilon^{3/2})$ it turns $a_1 = 0.84\,10^{-3}$ and $a_2 = -0.4$. Finally, the integration of the Landau-Stuart equation provides the time behaviour of $A_1$ and its equilibrium amplitude. Note that, as shown in Figure 5 the latter value only slightly depends on the choice of $\alpha_m$ as long as $\alpha_m$ keeps close to $\alpha_M$ and it moves along the interval pointed out in Figure 4. Furthermore, as shown in Figure 6, when $r$ becomes closer to $r_C$ the equilibrium amplitude decreases while it increases when $r$ is increased. The same behavior is observed for a tide characterized by a large value of the ratio between the minor and major axes of the tidal ellipse (e.g. $e = 0.6$). Note that, we considered the same sediment parameters as those considered for the unidirectional tide, therefore the flat bottom configuration is stable only when the sediment does not move and the critical value of $r_C$ does not change. Moreover, results of Figure 6 suggest that for low ellipticities, the equilibrium amplitude of the sand banks is much larger than for large ellipticities. This finding confirms the conclusion of Dyer & Huntley (1999) which is based on the analysis of field surveys and suggests that tidal ellipses characterized by large values of $e$ are more in favour of sand bank formation. Whatever value of sediment and flow parameters, for large values of $r$, no finite value of $|A_{1e}|$ is found since weak nonlinear effects are unable to give rise to an equilibrium bottom configuration which can be attained only when nonlinear effects are strong. Moreover, the analysis fails to predict the time development of the bottom perturbations when their amplitude is so large that nonlinear effects cannot be handled by means of a perturbation approach. This clearly appears

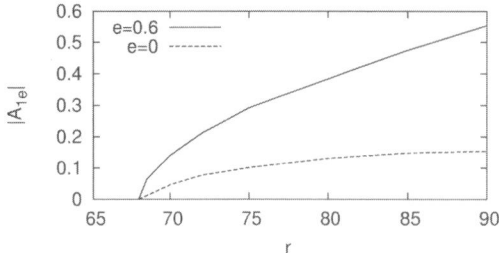

Figure 6. Equilibrium amplitude $|A_{1e}|$ as function of $r$ for $e = 0, 0.6$; $z_r = 1.0 \, 10^{-13}$; $f = 0.8$; $\varphi = 0$; $d = 6.7 \, 10^{-6}$; $\psi_d = 5.5 \, 10^{-3}$.

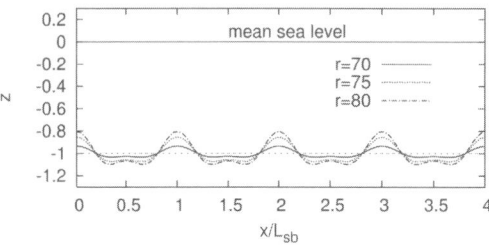

Figure 7. Equilibrium bottom profile in a plane orthogonal to the sand bank axis for different values of $r$ and $e = 0$, $z_r = 1.0 \, 10^{-3}$, $f = 0.8$, $\varphi = 0$, $d = 6.7 \, 10^{-6}$, $\psi_d = 5.5 \, 10^{-3}$.

in Figure 7 which shows the predicted bottom profile, given by the intersection of a vertical plane orthogonal to the crests of the bottom forms with the sea bottom, for the case of unidirectional tide and $r = 70$, 75, 80. The amplitude of the second harmonic of the bottom profile increases when $r$ is increased and the convergence of the series is broken for large values of $r$.

## 7 CONCLUSIONS

Previous linear stability analyses of the bottom of shallow tidal seas are revisited using a sediment transport formula which predicts a vanishing value of the sediment transport rate when the bottom shear stress drops below its critical value. Close to the critical condition, the present analysis predicts finite values of the wavenumber of the most unstable unstable mode. This finding has allowed us to develop a weakly nonlinear analysis which has been used to determine the equilibrium amplitude of the bottom forms when the parameters are not too far from the critical values. The configuration of the sea bottom, when the bottom forms attain their equilibrium, is characterized by the presence of long ridges, almost parallel to the tidal current, with crests sharper than the troughs and with crest-to-crests distances similar to those observed during field surveys. Even though the approach can be applied only for values of the parameters close to the critical conditions, the use of a fully analytical tool to

investigate the morphodynamic time development of tidal sand banks has the advantage to provide results with quite small computational costs thus allowing an exhaustive investigation of the parameter space.

This research has been supported by MIUR (Ministry of University and Research) under the contract PRIN n. 2005080197.

## REFERENCES

Besio, G., Blondeaux, P., & Vittori, G. (2005). Sand bank formation: comparison between 2D and 3D models. In G. Parker & M. H. Garcia (Eds.), *RCEM 2005*, Volume 2, pp. 973–980. Balkema.

Besio, G., Blondeaux, P., & Vittori, G. (2006). On the formation of sand waves and sand banks. *J. Fluid Mech. 557*, 1–27.

Blondeaux, P. & Vittori, G. (2005). Flow and sediment transport induced by tide propagation: 1. the flat bottom case. *J. Geophys. Res. 110(C7)*, 13 pp.

Colombini, M., Seminara, G., & Tubino, M. (1987). Finite-amplitude alternate bars. *J. Fluid Meek 181*, 213–232.

Dyer, K. R. & Huntley, D. A. (1999). The origin, classification and modelling of sand banks and ridges. *Cont. Shelf Res. 19*, 1285–1330.

Fredsoe, J. & Deigaard, R. (1992). Mechanics of coastal sediment transport. World Scientific, Singapore.

Hulscher, S. J. M. H. (1996). Tidal-induced large-scale regular bed form patterns in a three-dimensional shallow water model. *J. Geophys. Res. 101(C9)*, 20727–20744.

Hulscher, S. J. M. H., de Swart, H. E., & de Vriend, H. (1993). Generation of offshore tidal sand banks and sand waves. *Cont. Shelf Res. 13*, 1183–1204.

LeBlond, P. & Mysak, L. (1978). Waves in the ocean. Volume 20 of *Oceanography Series*. Elsevier Scientific.

Parker, G., Seminara, G., & Solari, L. (2003). Bedload transport at low shields stress on arbitrarily sloping beds: Alternative entrainment formulation. *Water Resour. Res. 39, 7*, 1183 pp.

Schielen, R., Doelman, A., & de Swart, H. E. (1993). On the nonlinear dynamics of free bars in straight channels. *J. Fluid Mech. 252*, 325–356.

Seminara, G. (1998). Stability and morfodynamics. *Meccanica 33*, 59–99.

Soulsby, R. & Whitehouse, R. (2005). Prediction of ripple properties in shelf seas. Technical Report TR155 Release 2.0, HR Wallingford Ltd, UK.

Stuart, J. (1971). Nonlinear stability theory. *Annual Review of Fluid Mechanics 3*, 347–370.

Talmon, A., Struiksma, N., & van Mierlo, M. (1995). Laboratory measurements of the direction of sediment transport on trasverse alluvial-bed slopes. *J. Hydr. Res. 33(4)*, 519–534.

Tambroni, N., Blondeaux, P., & Vittori, G. (2006). Analisi del processo di formazione delle sand banks. In *Atti del XXX Convegno di Idraulica e Costruzioni Idrauliche*.

Vittori, G. & Blondeaux, P. (1990). Sand ripples under sea waves. part 2 finite-amplitude development. *J. Fluid Mech. 218*, 19–39.

Zimmerman, J. T. F. (1982). On the lorentz linearization of a quadratically damped forced oscillator. *Phys. Lett. 89A*, 123–124.

*River, Coastal and Estuarine Morphodynamics: RCEM 2007 – Dohmen-Janssen & Hulscher (eds)*
*© 2008 Taylor & Francis Group, London, ISBN 978-0-415-45363-9*

# Prediction of sand wave migration with a non-linear spectral model

A.J. Smale
*Witteveen+Bos, Rotterdam, The Netherlands*

R. Bijker
*ACRB, Lemmer, The Netherlands*

G. Klopman
*AFR, Vollenhove, The Netherlands*

ABSTRACT: The non-linear spectral model has been developed to optimise the design of offshore submarine pipelines in dynamic sand wave areas, aiming at cost optimisation of the pre-installation dredging. The non-linear spectral model describes the sand waves on the seabed by means of a large number of sinusoidal waves and takes the non-linear shape of the individual sand waves into account, aiming at predicting seabed mobility for a period of 5 to 10 years. The model has been applied to predict the migration and seabed mobility of sand waves on the slopes of the Winterton Ridge of the Norfolk Banks on the UK continental shelf, for the design of a large diameter gas pipeline crossing this highly dynamic sand wave area. The paper describes the model and its application.

## 1 INTRODUCTION

### 1.1 Balgzand-Bacton Pipeline (BBL)

The N.V. Nederlandse Gasunie has built a pipeline to transport natural gas from The Netherlands to the United Kingdom: the Balgzand-Bacton Line (BBL), which was completed in December 2006. The pipeline route crosses a sand wave area just west of the Winterton Ridge, with sand waves up to 8 metres high. The sand waves are in average 100–200 metres long and quite dynamic: the profile indicates all possible modes of sand wave mobility (migration, oscillation and modulation).

A typical seabed profile for the project location is presented in Figure 1. The upper part of Figure 1 shows a seabed profile from pipeline km 187.5 to km 188.5, illustrating a typical singular sand wave profile. The lower part shows the seabed profile for a larger horizontal scale, illustrating the random character of the entire sand wave field.

The development of the sand wave field in time is illustrated by Figure 2 by comparing the profiles of 2002 and 2005, again showing (upper) the singular sand wave and (lower) the entire sand wave field. Figure 2 indicates that both the singular sand wave and the entire sand wave field are subject to significant changes.

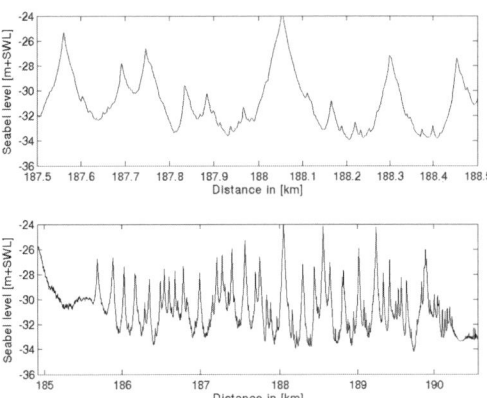

Figure 1. Typical seabed profile for project location as measured in 2002. (upper) km 187.5 to 188.5. (lower) km 185.0 to 190.5.

When a pipeline is trenched or buried in more or less stable sand waves, there is no problem: the pipeline will remain covered. If the sand waves migrate, however, the pipeline will become exposed on the 'upstream' slope of the sand wave (it becomes more buried on the 'downstream' side!). The sandwave migration had

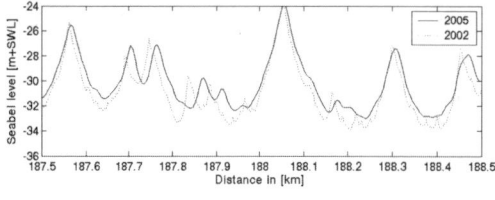

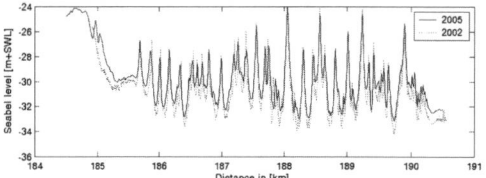

Figure 2. Typical seabed profile for project location. (upper) km 187.5 to 188.5. (lower) km 186.0 to 190.0 measured in 2002 and 2005.

to be taken into account when designing the dredging profile before installing the pipeline on the seabed.

In addition, the U.K. authorities required minimum environmental impact in the sand wave area, meaning a minimum of pre-installation and future mitigation dredging.

### 1.2 Problem definition

The amount of initial dredging to achieve adequate burial as well as reduce mitigation of future exposed pipeline sections can only be optimized if sand wave mobility can be predicted. Prediction of the sand wave mobility will indicate future pipeline exposure locations. This results in the following problem definition: when and where will the BBL become exposed again after installation due to sand wave mobility and as a function of pipeline burial profile?

### 1.3 Available methods

There are a number of methods available for the determination of the seabed mobility due to sand wave migration. Numerical, process based methods are described in Hulscher (1996) and Németh (2003). These methods require detailed information about hydrodynamic conditions, which was not available at the time of the project. Empirical models, such as Morelissen et al. (2003), are not able to predict the migration of the irregular type of sand waves present at the project location. The challenge was to develop a method able to predict the sand wave migration with use of the limited information available.

The solution was found in the development of a non-linear spectral model, describing an irregular sand wave field as a combination of sinusoidal waves: ordinary Fourier analysis. In order, however, to be able

to predict the migration of asymmetric sand waves, whilst retaining the asymmetric shape, the model was extended with bound waves. The result is a dedicated data-assimilation model, combining spectral analysis with bound waves to determine the sand wave migration and future sand wave profile development.

## 2 NON-LINEAR SPECTRAL MODEL

### 2.1 Approach

Non-linear waves are characterized by some form of asymmetry, like sharp-peaked crests and flat troughs, or a saw-toothed shape. These characteristics are often the result of local seabed material and environmental conditions, and remain constant when the waves evolve in time. This means that the elemental wave components are not sine waves, but have a different form. These different forms can still be represented by a series of sine waves, but in case of periodic waves some of these sine waves travel with the same propagation speed.

In this report we present a non-linear analysis and prediction technique. This type of analysis techniques has been developed by Bendat & Piersol for SIPM for the analysis of oscillatory wave forces on cylinders; see e.g. Bendat (1998), Vugts & Bouquet (1985), Bliek & Klopman (1988). Similar techniques have also been applied for the analysis of non-linear water waves (Laing, 1986), but in that case Stokes' wave theory can be used to compute the second-order transfer function. Linear prediction techniques for measured water waves have been presented by, among others, Zelt & Skjelbreia (1992). Here, we combine the above techniques to obtain a method for the prediction of the non-linear migration and evolution of bed forms.

### 2.2 Second order non-linear sand waves

For the moment we consider the bed elevation at a certain fixed moment in time. We assume the model has quadratic non-linearities, because of the similarity of the bed forms with sharp crested (non-breaking) or saw-toothed (breaking) water waves. For water waves we know these forms are generated by quadratic (or second-order) non-linearities.

Consider for example the following quadratic model:

$$w(x) = v(x) + \beta \, v^2(x) \qquad (1)$$

in which v(x) is the input and w(x) is the output. The input should have the form:

$$v(x) = \cos\left(\frac{2\pi x}{L}\right) \qquad (2)$$

in which L represents the characteristic sand wave length. The coefficient $\beta$ represents the quadratic transfer ($\beta=0$ results in a pure sinusoidal wave).

In order to be able to describe the saw-tooth profiles, a more general quadratic model is required. This can be achieved by describing the model presented in Equation 1 using Fourier transform:

$$v(x) = F^{-1}\{V(k)\} = \int_{-\infty}^{+\infty} V(k)e^{+ikx}\,dk \qquad (3)$$

in which $k = 2\pi/L$ and $F\{\}$ defines the Fourier transform operator applied on the function between brackets. Using the Fourier transform, Equation (1) can be described as:

$$w(x) = \int_{-\infty}^{+\infty} H^{(1)}(k)V(k)e^{ikx}\,dk + \qquad (4)$$
$$\int_{-\infty}^{+\infty}\int_{-\infty}^{+\infty} H^{(2)}(k_1,k_2)V(k_1)V(k_2)e^{ik_1x}e^{ik_2x}\,dk_1 dk_2$$

in which $H^{(1)}(k)$ and $H^{(2)}(k_1,k_2)$ are the linear (first-order) and quadratic (second-order) transfer functions, respectively. Using the transformation $k_2 = k - k_1$, we can show that the Fourier transform $W(k)$ of the output $w(k)$, Equation 4, can be written as:

$$W(x) = H^{(1)}(k)V(k) + \qquad (5)$$
$$\int_{-\infty}^{+\infty} H^{(2)}(k_1, k-k_1)V(k_1)V(k-k_1)\,dk_1$$

Large amounts of measurements and computational time are required to estimate the two-dimensional quadratic transfer function $H^{(2)}(k_1, k_2)$. This effort can be reduced, as also done by Bliek & Klopman (1988) by using the diagonal terms only:

$$J^{(2)}(k) = H^{(2)}(\tfrac{1}{2}k, \tfrac{1}{2}k) \qquad (6)$$

This reduces the model presented in Equation 5 to:

$$W(x) = H^{(1)}(k)V(k) + \qquad (7)$$
$$J^{(2)}(k)\int_{-\infty}^{+\infty} V(k_1)V(k-k_1)\,dk_1$$

This does not seem to be worthwhile at first sight, but when noting that:

$$v^2(x) = \int_{-\infty}^{+\infty}\int_{-\infty}^{+\infty} V(k_1)V(k_2)e^{ik_1x}e^{ik_2x}\,dk_1 dk_2 = \qquad (8)$$
$$\int_{-\infty}^{+\infty}\int_{-\infty}^{+\infty} V(k_1)V(k-k_1)e^{ikx}\,dk_1 dk$$

we see that the integral in Equation 7 is just the Fourier transform of $v^2(x)$. The computation time can be significantly decreased by taking the Fourier transform of $v^2(x)$:

$$W(x) = H^{(1)}(k)V(k) + J^{(2)}(k)F\{v^2(x)\} \qquad (9)$$

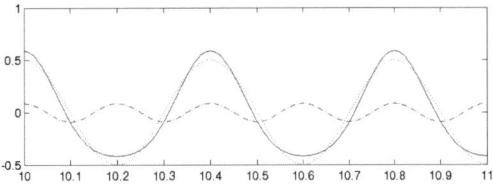

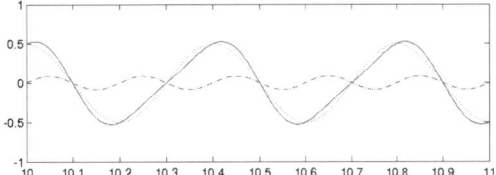

Figure 3. Non-linear waves. (upper) sharp crests and flat troughs, $\beta=0.7$ and (lower) saw-toothed waves $\beta=0.7i$. Dashed line is free wave, dash-dot line is bound wave and solid line is the resulting non-linear wave.

In case of sharp-crested waves, $J^{(2)}(k)$ will be real valued. For saw-toothed waves $J^{(2)}(k)$ will be purely imaginary valued. An example for L equal to 200 metres is plotted in Figure 3 for $J^{(2)}(k)=0.7$ (upper) and $J^{(2)}(k)=0.7i$ (lower).

### 2.3 Space-time model

When also considering the evolution in time, we have to distinguish between free-wave components and bound-wave components. We consider only those oscillatory parts of the bed elevation, which are expected to move substantially in the relevant time scale under consideration. The average part and the very long undulations are assumed to be static, and are not considered in the following. The free-wave Fourier components travel each at their own phase velocity $v(k)$, which we assume to be a function of the wave number $k$ (or equivalently of the wave length $L = 2\pi/k$). Then the linear part $a(x, t)$ of the bed evolution can be described as:

$$a(x,t) = \int_{-\infty}^{+\infty} A(k)e^{ik(x-v(k)t)}\,dk = F^{-1}\{A(k)e^{-ikv(k)t}\} \qquad (10)$$

The non-linear term can be introduced using Equation 7 with $H^{(1)}(k)=1$ and $J^{(2)}(k)=\beta(k)$:

$$b(x,t) = a(x,t) + F^{-1}\{\beta(k)\,F\{a^2(x,t)\}\} \qquad (11)$$

using forward and inverse Fourier transforms. The non-linear sand waves can now be described by the first-order waves $a(x)$ or its Fourier transform $A(k)$, phase velocity $v(k)$ of the first-order waves and the complex valued transfer function $\beta(k)$.

## 2.4 Parameter estimation

The unknown variables in Equation 11, $A(k)$, $v(k)$ and $\beta(k)$, can now be estimated with measured seabed profiles. The parameter estimation is an iterative process, which starts with the assumption (only the first iteration step) that each Fourier component of $A(k)$ is either free or bound. The free-wave components are likely to be found around the spectral peak: $k = k_p$. The corresponding bound-waves are then located near $k = 2\,k_p$. With this assumption, the first guess is that the free-wave components are located at $1/2\,k_p < k < 3/2\,k_p$. Using Equation 11 in combination with the measurements, the phase velocity and transfer function are estimated by minimizing the mean square error between $b(x,t)$ and the measured seabed profile, $z(x,t)$.

After the first iteration step, the difference between $b(x,t)$ and $z(x,t)$ is assumed to consist of free-waves. The iteration is repeated until the difference between $b(x,t)$ and $z(x,t)$ becomes sufficiently small. This requires approximately three to four iterations.

The sand wave migration can now be predicted by using the obtained free-wave components $(A(k))$ in combination with the fitted phase velocity and transfer function, $v(k)$ and $\beta(k)$ respectively.

## 2.5 Extension to higher orders (third and fourth)

The non-linear spectral model as described in Equation 11 is able to describe non-linear wave fields up to second order non-linearity. This second order model is able to describe limited asymmetry in the wave field: not too sharp crests and not too steep saw-tooth profiles. In case of strong non-linear profiles, the proposed parameter estimation might still result in an accurate representation of the measured profiles. When predicting the seabed mobility in the future, the non-bound waves result in the appearance of additional sand waves. These additional sand waves are most likely not realistic and therefore reduce the reliability of the seabed prediction.

Prediction of more asymmetry in the wave field requires that these free-waves (which are in fact bound-waves but not recognized as such!) should be included in the parameter estimation. The proposed parameter estimation, as described in Section 2.4, does not necessarily have to identify all bound-wave components. The method is therefore extended to include higher order non-linearities as well. In other words: waves with a wave length of $0.25 * k_{\text{free}}$ and $0.125 * k_{\text{free}}$ should also be recognized as bound waves.

The extension of Equation 11 to a third order model leads to:

$$b(x,t) = a(x,t) + F^{-1}\{\beta(k)\,F\{a^2(x,t)\}\}$$
$$+ F^{-1}\{\gamma(k)\,F\{a^3(x,t) - 3a(x,t)\}\} \tag{12}$$

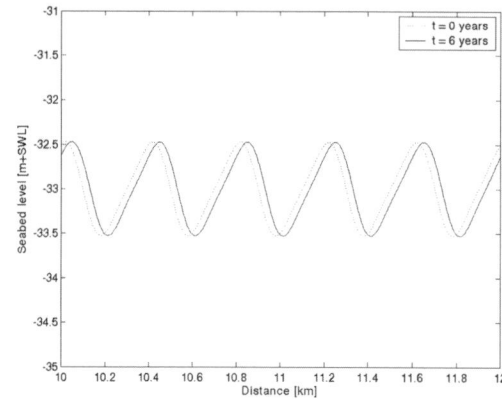

Figure 4. Theoretical saw-toothed profiles for $t=0$ and $t=6$ years.

in which $\gamma(k)$ represents the complex valued cubic transfer function. Equation 12 can be further extended to include the fourth order non-linearities:

$$b(x,t) = a(x,t)$$
$$+ F^{-1}\{\beta(k)\,F\{a^2(x,t)\}\}$$
$$+ F^{-1}\{\gamma(k)\,F\{a^3(x,t) - 3a(x,t)\}\} \tag{13}$$
$$+ F^{-1}\{\delta(k)\,F\{a^4(x,t) - a^2(x,t) + \tfrac{1}{8}\}\}$$

in which $\delta(k)$ represents the complex valued quadric transfer function. The parameter estimation is similar to the description presented in Section 2.4, except for the fact that a larger number of parameters need to be estimated.

## 2.6 Model validation on hypothetical cases

The non-linear spectral model is applied to two hypothetical seabed profiles: (i) saw-toothed profiles and (ii) sharp crests/flat troughs.

### 2.6.1 Saw-toothed profiles
Figure 4 shows the two saw-toothed profiles which serve as measured seabed profiles for respectively $t=0$ and $t=6$ years. The profiles are constructed with the following characteristics:

– quadratic transfer function $\beta(k) = 0.0 + 0.7i$;
– migration rate $v(k) = 5.3$ m/year.

Firstly, the power spectrum of each profile is determined, see Figure 5. Both power spectra are identical because the amplitude of the wave has not changed between both profiles, only the phase changed. Secondly, the Figure clearly shows a secondary peak at twice the wave number of the primary peak: this secondary peak represents the bound-wave component. Thirdly, starting at the larger wave lengths ($k_p$), the

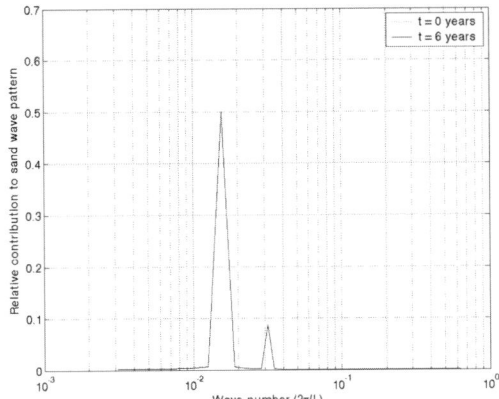

Figure 5. Power spectrum of theoretical saw-toothed profiles.

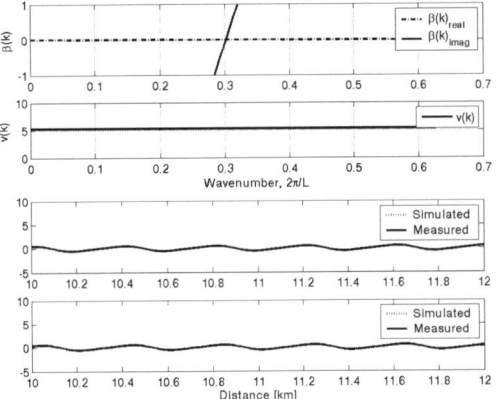

Figure 6. Parameter estimation for theoretical saw-toothed profiles as presented in Figure 4. The first subplot shows the fitted values of the transfer function $\beta(k)$, the second plot the fitted migration rate $v(k)$, the third and fourth plot show the measured and simulated sand wave profiles as a function of the distance for respectively $t = 0$ and $t = 6$ years.

transfer function and migration rate are determined for all wave lengths.

Figure 6 shows the corresponding transfer function and migration rate for the saw-toothed profiles as shown in Figure 4. The fitted transfer function shows a negligible ($\approx 0$) real part and an imaginary part of the transfer function of 0.7 for $k = 2*k_{free} = 0.314$. The transfer function for wave numbers smaller than $2*k_{free}$ are close to zero, indicating no significant non-linear behaviour. The wave numbers larger than $2*k_{free}$ show a non-realistic value for the transfer function. The effect of this non-realistic value is considered to be negligible, because the contribution of these waves to the sand wave pattern is very limited (see also the power spectrum). The deviation from the specified

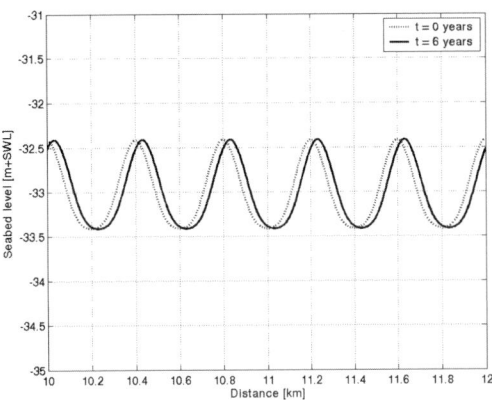

Figure 7. Theoretical sharp-crested profiles for $t = 0$ and $t = 6$ years.

transfer function is caused by the application of a Fast Fourier Transform for a monochromatic wave. It is expected that the performance of the non-linear spectral model improves with irregular sand wave fields. The migration rate is in agreement with the specified migration rate (5.3 m/year), which is as specified during the creation of the profiles.

### 2.6.2 Sharp crests/flat troughs

Figure 7 shows the two profiles which serve as measured seabed profiles for respectively $t = 0$ and $t = 6$ years. The profiles are constructed with the following characteristics:

- quadratic transfer function $\beta(k) = 0.7 + 0.0i$;
- migration rate $v(k) = 5.3$ m/year.

The transfer function and migration rate for each wave component is determined with the procedure specified in Section 2.4. The resulting transfer function and migration rate are presented in Figure 8. Again, the resulting values (for $k = 2*k_{free}$) correspond with the characteristics specified for the construction of the profiles.

## 3 APPLICATION OF THE NON-LINEAR SPECTRAL MODEL

### 3.1 Available seabed profiles

The present study used two seabed profiles taken at an interval of approximately 3 years along the proposed pipeline route.

As part of the pre-design, a multibeam survey of the entire route was performed in 2002. The resulting profile consists of an accurate and reliable seabed elevation in m + SWL every 5 metres. In November 2005, a second route survey was performed. This survey also

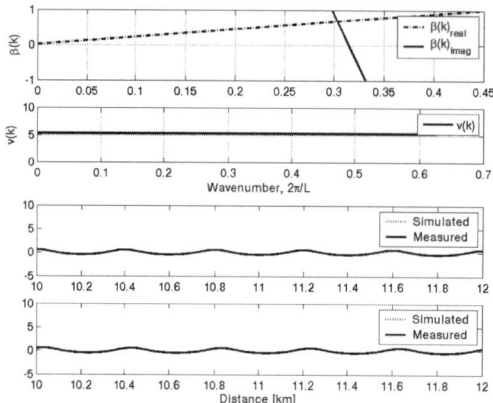

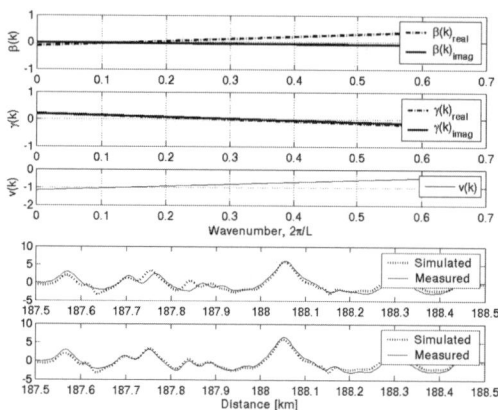

Figure 8.   Parameter estimation for theoretical sharp-crested profiles as presented in Figure 7. The first subplot shows the fitted values of the transfer function $\beta(k)$, the second plot the fitted migration rate $v(k)$, the third and fourth plot show the measured and simulated sand wave profiles as a function of the distance for respectively $t = 0$ and $t = 6$ years.

Figure 9.   Parameter estimation for pipeline section KP187.5 to KP 188.5. The first subplot shows the fitted values of the transfer function $\beta(k)$, the second plot the fitted transfer function $\gamma(k)$, the third plot the migration rate $v(k)$, the fourth and fifth plot show the measured and simulated sand wave profiles as a function of the distance for respectively $t = 2002$ and $t = 2005$.

resulted in a seabed profile with a seabed elevation in m + SWL every 5 metres.

Both the spatial resolution as well as the horizontal and vertical accuracy of the surveys is considered to be sufficient for the present analysis.

### 3.2   *Parameter estimation for the non-linear spectral model*

The parameters of the non-linear spectral model are estimated using the 2002 and 2005 surveys. The sand wave section from km 185.5 to 190.5 was split up into nine sections with a length of 1000 metres and an overlap of 500 metres, in order to reduce the effect of spatial variation of the sand wave characteristics.

Each of the sections is analysed according to the procedure specified in Section 2, with a number of settings. The measure of asymmetry (second, third or fourth order) is varied for each section. The analysis is thus performed for 3 asymmetry levels for 9 sections, resulting in 27 combinations of transfer and migration functions.

The most appropriate non-linear model (second, third or fourth order) for each section is determined using the mathematical error of the hindcast calculation and a visual inspection of the estimated parameters. An example of the estimated transfer functions and migration rate is presented in Figure 9. It should be noted that an higher order does not necessarily result in a better hindcast, because not all profiles require the highest non-linearity.

The real and imaginary parts of the transfer functions for each section are all within the expected range of $-1.0$ and $1.0$. In some cases, the values for the

transfer functions are outside the expected range. This is considered acceptable, because the wave numbers, at which these values occur, have a limited contribution to the sand wave pattern.

The estimated migration rates for each section are all between 0 and 20 metres/year, which is according to measurements. Most of the migration functions show a positive migration rate for the relevant wave numbers, indicating a migration from south to north, which also corresponds to measurements.

### 3.3   *Predicting seabed mobility*

The estimated transfer functions and migration rates are used to predict the sand wave migration from 2002 to 2032 for each section. An example is such a prediction is presented in Figure 10.

Although the example prediction extends up to 2032, it is recommended to use the results for the period up to 2022 only. After a prediction of twenty years, the prediction period becomes almost the timescale of the sand wave evolution, rendering the applied method invalid.

### 4   APPLICATION MODEL RESULTS

The model has been used to determine the burial depth of the pipeline in the sand wave field. Application of the model indicated several sections along the route in the sand wave area where additional peak-shaving was needed. Peak-shaving is removing the peak of a sand wave by dredging a trench through a sand wave to

982

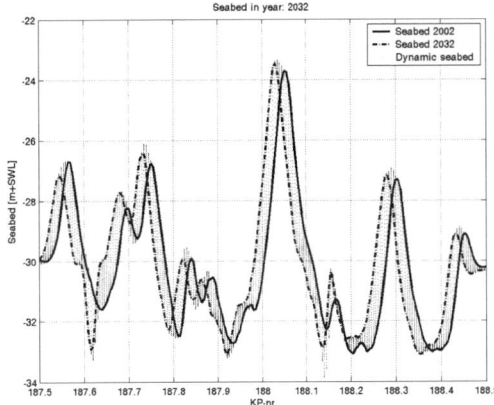

Figure 10. Example of predicted sand wave migration from 2002 to 2032.

allow pipeline installation without over-stressing; the large diameter pipeline is too stiff to follow the sand wave profile without intervention. A number of sensitivity runs of the model showed that some additional peak-shaving and slope dredging was needed to assure adequate burial to prevent unacceptable exposure of the pipe-line in the future as the result of sand wave migration and profile development. The limitation of the model has been assessed and is characterized by the statement that the model prediction is to be considered 'realistic', not 'reality'. Based on this under-standing the application of the non-linear spectral sand wave model proved a very useful tool in pipe-line-seabed interaction design.

## 5 CONCLUSIONS

We have used existing knowledge of non-linear (water) waves to create a non-linear spectral model for prediction of the migration and profile development of irregular non-linear sand waves using a limited amount of seabed measurements.

The model is validated with hypothetical cases and applied to actual measurements. The estimated parameters (migration rate and transfer functions) are used to predict the migration of sand waves and the sand wave profile development along the route of the BBL. The results allowed for an optimisation of the initial dredging and expected maintenance costs.

## 6 RECOMMENDATIONS

In order to enhance the applicability of the non-linear spectral model, we recommend to extend the model with the following items:

1 Implementation of evolutionary spectra in time and space domain, which would allow the model to predict temporal and spatial variations in the free- and bound-wave components. This enables the prediction of sand wave mobility for larger spatial scales.
2 Extension of the non-linear spectral model to a two-dimensional model. This would allow the prediction of sand wave fields in two directions, including for instance the effect of bifurcations. This is relevant because the seabed mobility is very high at these bifurcations, which is of importance for pipeline and cable design.
3 Coupling with numerical codes that are able to predict migration rates and/or asymmetry. This would further enhance the quality of the predictions made by this non-linear spectral model.

## REFERENCES

Bendat, J. S. 1998. *Nonlinear system techniques and applications*. New York: Wiley.

Bliek, A. & Klopman G. 1988. Non-linear frequency domain modelling of wave forces on large vertical and horizontal cylinders in random waves. In T. Moan, N. Janbu, and O. Faltinsen (eds.), *Proc. BOSS'88*: 821–840.

Hulscher, S. J. M. H. 1996. Tidal-induced large-scale regular bed form patterns in a three-dimensional shallow water model. *Journal of Geophysical Research*, 101(C9): 727–744.

Laing, A. K. 1986. Nonlinear properties of random gravity waves in water of finite depth. *J. Phys. Oceanography* 16 (12): 2013–2030.

Morelissen, R. & Hulscher, S. J. M. H. & Knaapen, M. A. F. & Németh, A. A. & Bijker, R. 2003. Mathematical modelling of sand wave migration and the interaction with pipelines. *Coastal Engineering*, 48(3): 197–209

Németh, A.A. 2003. *Modelling offshore sand waves*. Enschede: PrintPartners Ipskamp BV.

Vugts, J. H. & Bouquet A.G. 1985. A non-linear, frequency domain description of wave forces on an element of a vertical pile in random waves. In J. A. Battjes (ed.), *Proc. BOSS'85*: 239–253 .

Zelt, J. A. & Skjelbreia J.E. 1992. Estimating incident and reflected wave fields using an arbitrary number of wave gauges. *In Proc. 23th Int. Conf. Coastal Eng.*, Volume 1: pp. 777–789.

*River, Coastal and Estuarine Morphodynamics: RCEM 2007 – Dohmen-Janssen & Hulscher (eds)*
*© 2008 Taylor & Francis Group, London, ISBN 978-0-415-45363-9*

# Sand waves characteristics: Theoretical predictions versus field data

Giovanni Besio & Paolo Blondeaux
*Department of Civil, Environmental and Architectural Engineering, University of Genova, Genova, Italy*

Vera Van Lancker, & Els Verfaillie
*Renard Centre of Marine Geology, Ghent University, Gent, Belgium*

Giovanna Vittori
*Department of Civil, Environmental and Architectural Engineering, University of Genova, Genova, Italy*

ABSTRACT: Sand wave characteristics are predicted using the three-dimensional model proposed by Blondeaux and vittori (2005a), Blondeaux and vittori (2005b) and later developed by Besio et al. (2006). Turbulence generated by tidal currents is described by introducing an eddy viscosity which is assumed to depend on the distance from the bottom. Moreover, sediment is supposed to move both as bed load and suspended load since, at some locations, field surveys show that large amounts of sediment are put into suspension and transported by the tidal currents. The predictive capability of the model of Besio et al. (2006) is tested by comparing theoretical predictions with a large field data set. In particular the field observations carried out along the continental shelf of Belgium are used. The field data are described in Mouchet (1990) and Van Lancker et al. (2005) . The theoretical predictions are found to fairly agree with field observations, even though some of the comparisons suggest that the accuracy of the predictions depends on the accurate evaluation of the local current and sediment characteristics.

## 1 INTRODUCTION

Among the many bedforms occurring in coastal regions characterized by non-cohesive (sandy) deposits, sand waves are undoubtedly one of the most important for human activities. Sand waves typically occur in shallow seas and are induced by tidal currents. The sand waves have wavelengths of the order of hundreds of metres and heights of the order of a few metres and their crests are often almost orthogonal to the direction of the tidal current. Moreover sand waves can migrate when the symmetry of the tidal current is broken (e.g. when a residual current is present).

As pointed out by Hulscher (1996) and later discussed by Gerkema (2000), Komarova and Hulscher (2000) and Besio et al. (2006), the process which leads to the formation of these bedforms is similar to that originating sea ripples under sea gravity waves (Blondeaux 2001, Blondeaux 1990, Blondeaux 1999, Sleath 1976). In fact, the interaction of the oscillatory tidal flow with a bottom perturbation gives rise to a steady streaming in the form of recirculating cells. When the net displacement of the sediment dragged by this steady streaming is directed toward the crests of the initial bottom perturbation, the amplitude of the perturbation grows and bedforms are generated. On the other hand, the flat bottom configuration turns out to be stable when the net motion of the sediment is directed toward the troughs of the bottom perturbation.

The theoretical investigation of sand wave appearance forced by tide propagation has been mainly carried out by means of linear stability analyses which study the time development of arbitrary bottom perturbations of small amplitude superimposed to the flat bottom configuration. Employing a linear approximation, it is possible to assume that the different spatial. Fourier components, contained in a generic random perturbation, evolve each independently of the other. Hence, a normal-mode analysis can be applied and the generic component of the perturbation, characterized by an amplitude $\epsilon^* A$ and by wavenumbers $\alpha_x^*, \alpha_y^*$ in the two horizontal directions $x^*$ and $y^*$, can be considered.

The time development of the bottom configuration is provided by a sediment balance which simply states that convergence (or divergence) of the sediment flux must be accompanied by a rise (or fall) of the bed profile. To integrate such an equation, a predictor for sediment transport rate is needed and the flow must be evaluated. Different models can be used to solve the hydrodynamic problem, to evaluate the bottom shear stress and to quantify the sediment transport. However, the solution procedure invariably leads to an equation for the amplitude $A(t^*)$ of the bottom perturbation in the form $dA(t^*)/dt^* = \Gamma^*(t^*)A(t^*)$, where the complex function $\Gamma^*(t^*) = \Gamma_r^*(t^*) + i\Gamma_i^*(t^*)$ depends on the wavenumbers of the bottom waviness and on suitable flow and sediment parameters. Since the forcing tidal flow is time periodic, the quantity $\Gamma^*$ turns out to be a periodic function of time and the growth of the bottom forms is controlled by the time average $\overline{\Gamma}^*$ of $\Gamma^*$. A linear stability analysis suggests that the component of the bottom perturbation characterized by the largest value of the amplification rate (amplification rate = time averaged value of $\Gamma_r^*(t^*)$) will prevail for large times. Hence, the linear analysis allows to identify the wavelength of the appearing bedforms as well as their orientation and migration speed.

The present work aims at further supporting the validity of the model proposed by Besio et al. (2006) and at quantifying the limits of its applicability in the parameter space. In particular the model predictions are compared with different field observations carried out on the Belgian Continental Shelf. Tidal current measurements were carried out at a few locations in (Mouchet 1990), while data on sediments, morphology and bathymetry were collected over the entire shelf (Van Lancker et al. 2005).

## 2 SAND WAVES IN THE BELGIAN CONTINENTAL SHELF

The Belgian territorial waters (approximately from $51,1°$ to $51,7°$N and from $2,25°$ to $3,50°$E) occupy about $3600\,\mathrm{km}^2$ in the southern part of the North Sea. The Belgian Continental Shelf is typically very shallow, partly due to the presence of sandbanks, with water depths ranging from 5 to 55 m MLLWS (Mean Lowest Low Water at Spring, see figure 1).

The area is subject to a semi-diurnal macrotidal regime and the amplitude of the semidiurnal constituent depends on the exact location, ranging between values of 3 m for neap tides and 4.5 to 5 m for spring tides. In the present study, to estimate the tidal currents, we use the measurements carried out at 8 locations (Mouchet 1990), namely i) Kwintebank ($51°21'$N, $2°41'$E), ii) Nieuwpoortbank ($51°10'$N, $2°35'$E), iii) Trapegeer ($51°08'$N, $2°34'$E), iv) Thorntonbank ($51°34'$N, $2°59'$E), v) Wandelaar ($51°23'$N, $3°03'$E), vi) Westhinder ($51°23'$N, $2°36'$E), vii) Akkaert NE ($51°27'$N, $2°59'$E), viii) Akkaert Noord ($51°23'$N, $2°37'$E) (see figure 1). The tidal currents were measured at different water depths but they are converted to depth averaged velocities by using the velocity profile predicted by the model and using an iterative procedure. Then, the characteristics of the tidal constituents (tidal ellipse orientation and ellipticity) are computed on the basis of the available data. At these locations, the harmonic analysis shows that the semidiurnal component is the dominant one with maximum velocities ranging from 54 to 70 cm/s.

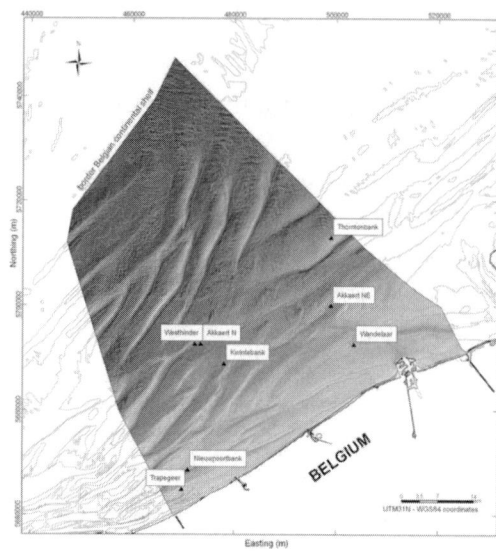

Figure 1. Bathymetry digital terrain model (DTM) of the Belgian Continental Shelf (Van Lancker et al. 2005). The current meter locations are indicated. The DTM provides a rough estimate of the presence of sandwaves.

Sediment characteristics are obtained from a digital terrain model ($250 \times 250$ m) of the median grain-size of the sand fraction (Verfaille et al. 2006). The data, originated from the sedisurf@database (Ghent University, Renard Centre of Marine Geology), contain about 7000 sampling points. The gridding is performed using the bathymetric gradient as secondary variable, allowing a more realistic modelling of the grain-size distribution (Verfaille et al. 2006).

On the Belgian Continental Shelf, the nature and differentiation of the superficial sediment are related to the unique configuration of the sandbank-swale system and to the interaction of the current with the large-scale morphological patterns which are responsible of a sorting of the sediment. The sand fraction (0.063 to 2 mm) preferentially takes part in the process of sandbank formation whilst the coarser sands ($>2$ mm) and the silt-clay fraction ($<0.063$ mm) are found in the swales. Moreover, the superficial sediment coarsens in the offshore direction. Very-fine to fine sands dominate the near coastal area whilst medium sand grains up to $400–500\,\mu\mathrm{m}$ can be found in the offshore area (figure 2). Usually, coarse sediments are associated with shell hash admixtures.

The values of the water depths are derived from a bathymetry digital terrain model ($250 \times 250$ m) compiled at Ghent University on the basis of the hydrographic data of the Flemish Hydrography, Maritime Services (Flemish Authorities, Agency for Maritime & Coastal Services, Coastal Division, Van Lancker et al.

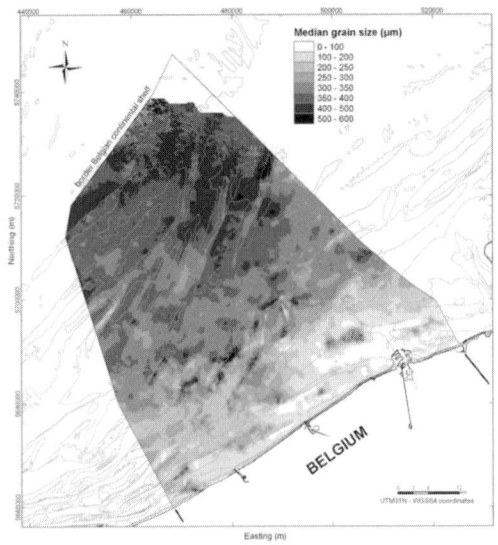

Figure 2. Median grain size of the sand fraction along the Belgian Continental Shelf (Verfaillie et al., 2006).

**Thornton Bank - Profile I**

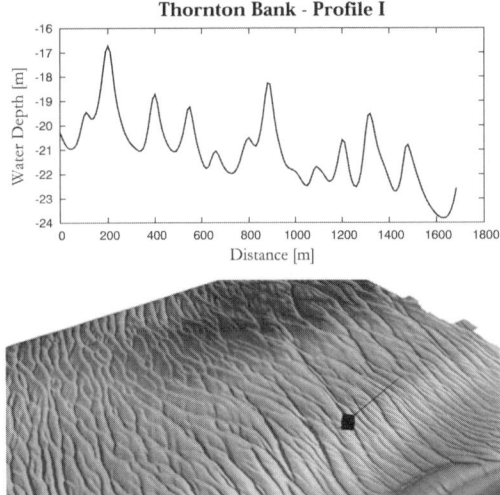

Figure 3. Sand waves map (bottom) and sand waves profile (top) on the Thornthon Bank – Profile I.

2005). The data set consists of tidally corrected single-beam measurements.

The occurrence and dimensions of sandwaves are derived from a morphological map compiled at Ghent University (Van Lancker et al. 2005). The information is gathered from high-resolution acoustic data (multibeam, side-scan sonar). Sandwaves, 2 to 4 m in height, are generally present where sandbanks occur, as shown

**Thornton Bank - Profile II**

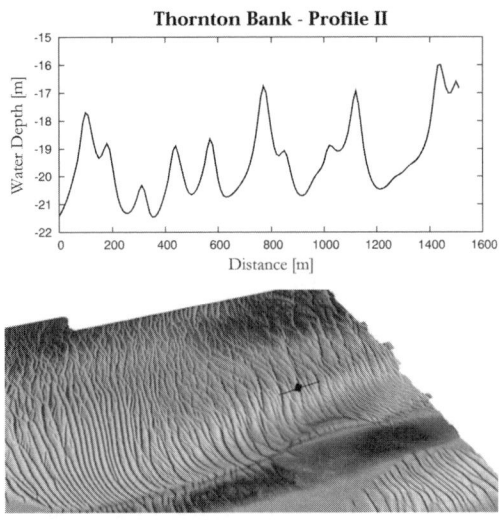

Figure 4. Sand waves map (bottom) and sand waves profile (top) on the Thornton Bank – Profile II.

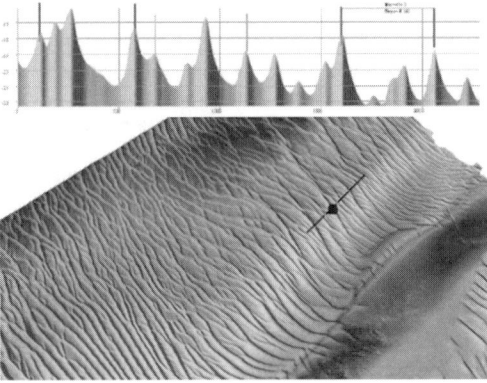

Figure 5. Sand waves map (bottom) and sand waves profile (top) on the Thornton Bank – Profile VIII.

in figures 3–5 (bottom panels) where pictorial views of sand wave fields laying on the flanks of the Thornton Bank are depicted. Similar information to those reported in figures 3–5 are obtained for all the eight locations analyzed in the present study. For each location, the geometrical characteristics of sand waves are determined analyzing different profiles of the sea bottom in order to identify the mean value of the sand wave wavelength and the minimum and maximum wavelengths observed in the field. In particular the wavelength of the bottom forms is determined along profiles perpendicular to the main strike of the sandwaves. Comparing the three different profiles obtained in proximity of the Thornton Bank, it is possible to notice that local geometrical characteristics (amplitude and wavelength of bottom forms) can vary even

Table 1. Values of dimensional wavelengths provided by field surveys.

**Observed Wavelengths**

| Site | Mean $L_o^*$ [m] | Min–Max $L_o^*$ [m] |
|---|---|---|
| Kwintebank | 316 | 240–440 |
| Trapegeer | 218 | — |
| Thorntonbank | 430 | 350–520 |
| Wandelaar | 610 | 437–800 |
| Westhinder | 197 | — |
| Akkaert NE | 617 | 400–860 |
| Akkaert N | 417 | 330–540 |

in the neighbourhood of a limited region. Therefore to compare theoretical predictions and field observations only the mean value of the sand wave wavelength is considered. The values obtained at the eight different locations of figure 1 are summarized in table 1.

Near the coast, sandwaves tend to disappear from the sandbanks flanks, as for the Coastal Banks. In the coastal zone, patches of sandwaves do occur where coarse sediment is present. To the north of the Belgian Continental Shelf, fields of sand waves characterize the sea bed with heights that are often larger than 6 m. In this area the sand waves occur both along the flanks of the sand banks and in the swales and over the flat beds. Over the Belgian Continental Shelf, sand wave wavelengths vary considerably, ranging from 100 to 800 m.

## 3 THE MODEL

The theoretical model does not differ from that of Besio et al. (2006) and the reader interested in the theoretical approach used to predict sand wave characteristics is referred to Blondeaux and Vittori (2005a), Blondeaux and Vittori (2005a) and Besio et al. (2006) for the details of the analysis. However, for the sake of clearness, hereinafter we summarize the main assumptions and characteristics of the model.

The flow generated by a tidal wave propagating over a cohesionless bed is considered and the time development of the bottom configuration, it induces, is investigated. We consider a three-dimensional turbulent flow and we employ a Boussineq-type closure with an eddy viscosity $v_T^*$ which is assumed to depend on the distance from the bottom (see Besio et al., 2006). Viscous effects are neglected. The problem of flow determination is then posed by continuity and momentum equations where the Coriolis contributions related to the Earth rotation are taken into account because they affect the tidal current. Even though inertial effects and Coriolis terms are taken into account

in the determination of the flow induced by tide propagation, they can be neglected when studying the time development of the bottom perturbations which are characterized by wavelengths of the order of a few hundreds of metres (Gerkema 2000; Besio et al. 2006). It is convenient to consider the dimensionless problem, where the mean water depth $h_0^*$ is used as length scale, the maximum value $U_0^*$ of the depth averaged fluid velocity during the tidal cycle is used as velocity scale and the inverse of the angular frequency $\omega^*$ of the tide is used as time scale. Hence the hydrodynamic problem reads

$$\nabla \cdot \mathbf{u} = 0 \tag{1}$$

$$\frac{\partial \mathbf{u}}{\partial t} + \hat{r}(\mathbf{u} \cdot \nabla)\mathbf{u} = -\nabla p + \mathbf{g} + \hat{\delta}^2 \nabla \cdot \mathbf{T} - 2\Omega \mathbf{f} \tag{2}$$

where $\mathbf{u} = (u, v, w)$ are the horizontal and vertical velocity components and $p$ is the pressure averaged over turbulence. The operator $\nabla$ is defined by $(\partial/\partial x, \partial/\partial y, \partial/\partial z)$, where $x$ and $y$ are two horizontal axes lying on the free surface and $z$ is the vertical coordinate pointing upward. $\mathbf{T}$ is a tensor defined as $\mathbf{T}_{i,j} = v_t(\frac{\partial u_i}{\partial x_j} + \frac{\partial u_j}{\partial x_i})$, while $\mathbf{f}$ represents the Coriolis terms $\mathbf{f} = (w \cos(\phi_0) - v \sin(\phi_0), u \sin(\phi_0), -u \cos(\phi_0))$ and $\mathbf{g}$ is the dimensionless gravity acceleration $(\mathbf{g} = (0, 0, -g))$.

The dimensionless hydrodynamic problem is characterized by two main dimensionless parameters, beside the values of $\Omega = \Omega^*/\omega^*$ and $\phi_0$ ($\Omega^*$ denotes the angular velocity of the Earth rotation, $\cong 0.5$ for semidiurnal and $\cong 1$ for diurnal tide, and $\phi_0$ is the local latitude):

$$\hat{r} = \frac{U_0^*}{\omega^* h_0^*}, \quad \hat{\delta} = \frac{\sqrt{v_{T0}^*/\omega^*}}{h_0^*}. \tag{3}$$

The parameter $\hat{r}$ is the ratio between the amplitude of horizontal fluid displacement oscillations and the local depth and assumes typical values of order $10^2$. The parameter $\hat{\delta}$ is the ratio between the thickness of the turbulent bottom boundary layer and the local depth. A rough estimate shows that $\hat{\delta}$ is of order one. In the definition of $\hat{\delta}$, $v_{T0}^*$ is a dimensional constant and provides the order of magnitude of the eddy viscosity. Since $v_{T0}^*$ is proportional to $U_0^*$, Besio et al. (2006) introduced the new viscous parameter $\hat{\mu} = \hat{r}/\hat{\delta}^2$ unrelated to the strength of the tidal current.

The morphodynamic problem is governed by the sediment continuity equation. The dimensionless parameters which characterize the morphodynamic problem are the sediment porosity $p_{or}$, the dimensionless sediment size $d$, the mobility number $\hat{\psi}_d$ and the particle Reynolds number $R_p$

$$d = \frac{d^*}{h_0^*} \; ; \quad \hat{\psi}_d = \frac{(\omega^* h_0^*)^2}{(\rho_s^*/\rho^* - 1)g^* d^*}; \tag{4}$$

$$R_p = \frac{\sqrt{(\rho_s^*/\rho^* - 1)g^* d^{*3}}}{\nu} \tag{5}$$

where $\rho^*$ and $\rho_s^*$ are the water and sediment density respectively.

The problem is closed once a relationship for the sediment transport rate is provided. Sediment transport is usually split into two components. The former is due to sediment moving close to the bottom (the "bed load") and the latter is due to sediment which is carried into suspension (the "suspended load"). The dimensionless bed load $(q_{Bx}, q_{By}) = (q_{Bx}^*, q_{By}^*)/\sqrt{(\rho_s^*/\rho^* - 1)g^*(d^*)^3}$ due to the tidal current is computed by means of the relationship proposed by Van Rijn (1991), which shows that $(q_{Bx}, q_{By})$ depends on the dimensionless Shields stress components $(\theta_x, \theta_y) = (\tau_x^*, \tau_y^*)/(\rho_s^* - \rho^*)g^* d^*$. The dimensional shear stress components $(\tau_x^*, \tau_y^*)$ can be easily evaluated by means of the constitutive law. Moreover, as suggested by Colombini (2004) and assumed by Cherlet et al. (2007), to compute the sediment transport rate, the shear stress is evaluated at the top of the so-called bed-load layer. In other words, it is the shear stress $\tau_t^*$, which is present at some distance from the bed, and not the shear stress $\tau_b^*$ at the bed which should be used to estimate the bedload discharge. Finally the bed is assumed to be covered by ripples whose height is computed following Soulsby and Whitehouse (2005a) and Soulsby and Whitehouse (2005b). Hence the thickness of the so-called bed load layer scales with the ripple height.

To complete the description of the bedload sediment transport, we take into account the weak effects due to a slow spatial variation of the bottom topography. If the bottom slope $\nabla h$ is small, simple dimensional arguments and linearization lead to

$$(q_{Px}, q_{Py}) = -q_B \mathbf{G} \nabla h \tag{6}$$

where $\mathbf{G}$ is a dimensionless second order 2-D tensor. (a.o. Talomon et al. 1995; Seminara 1998).

To evaluate the dimensionless suspended sediment transport $(q_{Sx}, q_{Sy})$, it is necessary to compute the concentration $c = c(x^*, y^*, z^*, t^*)$ by solving a standard convection-diffusion equation. Finally the suspended sediment transport $(q_{Sx}, q_{Sy})$ is computed as the flux of concentration over the water column.

In the analysis of the stability of the flat bottom configuration of a shallow tidal sea, small perturbations of the bottom are considered so that the bottom configuration differs from the flat one by a small (strictly infinitesimal) amount proportional to $\epsilon$. Hence, the bottom profile can be thought to be given by the superposition of different spatial components which evolve

one independently of the other. A normal mode analysis can be performed and the problem can be solved for the generic spatial component

$$1 - \frac{h^*}{h_0^*} = \epsilon A(t) e^{i(\alpha_x x + \alpha_y y)} + c.c. + O\left(\epsilon^2\right) \tag{7}$$

where $A(t)$ is the amplitude of the generic component which is periodic in the $x$- and $y$-directions with dimensionless wavenumbers $\alpha_x = \alpha_x^* h_0^*$ and $\alpha_y = \alpha_y^* h_0^*$ respectively and $\epsilon \ll 1$. The small value of $\epsilon$ allows the solution to be expanded in terms of $\epsilon$.

At the leading order of approximation, i.e. $O(\epsilon^0)$, the problem is reduced to the determination of both the flow and sediment transport induced by tide propagation over a flat seabed. Then, the hydrodynamic problem for the flow perturbations is solved by using the approach of Blondeaux and Vittori (2005b).

The equation, which provides the time development of the amplitude of the bottom perturbation, follows from the sediment continuity equation:

$$\frac{dA(T)}{dT} = \Gamma(t) A(T) \tag{8}$$

where $\Gamma$ is a periodic, complex function of $t$ which depends on the parameters of the problem. In (8), $T$ is a slow morphodynamic time scale $(T = td/[(1 - p_{or})\sqrt{\hat{\psi}_d}])$.

The solution of (8) shows that the growth or the decay of the bottom perturbations is controlled by the real part $\overline{\Gamma}_r$ of the time average $\overline{\Gamma}$ of $\Gamma$, while the imaginary part $\overline{\Gamma}_i$ is related to the migration speed of the perturbations. Because of the symmetry of the forcing flow, no migration of the bottom forms is expected and indeed $\overline{\Gamma}_i$ vanishes.

## 4  MODEL PREDICTIONS

The model has been run for values of the parameters chosen to reproduce the eight locations described in section 2. Both small (6 m) and relatively large (28 m) water depths are considered and the sediment size ranges from 0.22 mm to 0.38 mm. The dimensionless parameters, obtained for each location taking into account field observations and assuming that the sea bottom is covered by ripples whose height is obtained by means of the Soulsby and Whitehouse (2005b) predictor, are summarized in table 2.

The value of the amplification rate $\overline{\Gamma}_r$ is plotted as function of $\alpha_x$ for $\alpha_y = 0$, since preliminary runs have shown that the maximum of $\Gamma_r$ takes place for vanishing values of $\alpha_y$ (figures 6 and 7). A maximum of the amplification rate is present for all the locations presently considered and, according to the linear

Table 2. Values of dimensionless parameters used in the model runs for the different locations on the Belgian Continental Shelf.

| Site | Parameters | | | | | |
| | $z_r$ | $\hat{\mu}$ | $\hat{r}$ | $\hat{\Psi}_d$ | $R_p$ | $d$ |
| --- | --- | --- | --- | --- | --- | --- |
| Kwintebank | $8.96 \cdot 10^{-4}$ | 379.85 | 155.29 | $2.04 \cdot 10^{-3}$ | 26.16 | $1.48 \cdot 10^{-5}$ |
| Nieuwpoortbank | $18.15 \cdot 10^{-4}$ | 351.36 | 347.27 | $5.18 \cdot 10^{-4}$ | 15.03 | $2.44 \cdot 10^{-5}$ |
| Trapegeer | $27.06 \cdot 10^{-4}$ | 335.22 | 531.8 | $2.37 \cdot 10^{-4}$ | 12 | $3.35 \cdot 10^{-5}$ |
| Thorntonbank | $9.62 \cdot 10^{-4}$ | 376.99 | 151.73 | $1.79 \cdot 10^{-3}$ | 23.64 | $1.53 \cdot 10^{-5}$ |
| Wandelaar | $16.73 \cdot 10^{-4}$ | 354.64 | 306.91 | $5.97 \cdot 10^{-4}$ | 20.08 | $2.51 \cdot 10^{-5}$ |
| Westhinder | $9.74 \cdot 10^{-4}$ | 376.48 | 231.6 | $1.74 \cdot 10^{-3}$ | 24.61 | $1.57 \cdot 10^{-5}$ |
| Akkaert NE | $12.57 \cdot 10^{-4}$ | 366.2 | 279.18 | $1.05 \cdot 10^{-3}$ | 22.79 | $1.97 \cdot 10^{-5}$ |
| Akkaert N | $8.33 \cdot 10^{-4}$ | 382.8 | 182.97 | $2.40 \cdot 10^{-3}$ | 21.11 | $1.27 \cdot 10^{-5}$ |

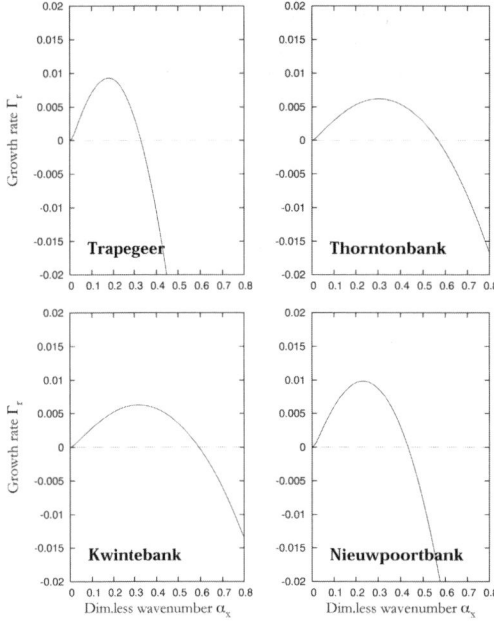
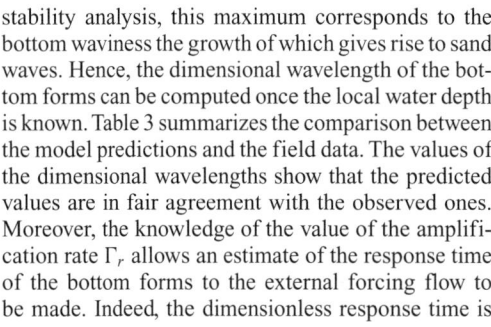

Figure 6. Growth rate $\overline{\Gamma}_r$ of bottom perturbations plotted versus the dimensionless wavenumber $\alpha_x$.

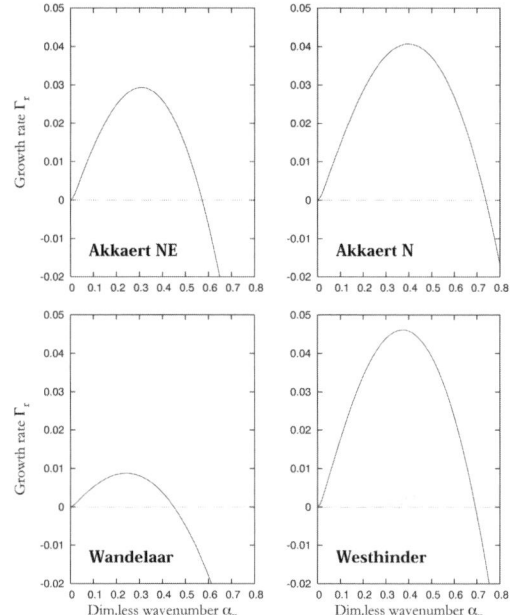

Figure 7. Growth rate $\overline{\Gamma}_r$ of bottom perturbations plotted versus the dimensionless wavenumber $\alpha_x$.

stability analysis, this maximum corresponds to the bottom waviness the growth of which gives rise to sand waves. Hence, the dimensional wavelength of the bottom forms can be computed once the local water depth is known. Table 3 summarizes the comparison between the model predictions and the field data. The values of the dimensional wavelengths show that the predicted values are in fair agreement with the observed ones. Moreover, the knowledge of the value of the amplification rate $\Gamma_r$ allows an estimate of the response time of the bottom forms to the external forcing flow to be made. Indeed, the dimensionless response time is

proportional to $\Gamma_r^{-1}$. Then, taking into account the definition of the morphodynamic dimensionless time scale $T$, it turns out that the dimensional response time $T_r^*$ is

$$T_r^* = \frac{(1 - p_{or}) \sqrt{\hat{\psi}_d}}{\Gamma_r \, d \, \omega^*} . \qquad (9)$$

Therefore, the characteristic response time $T_r^*$ of the eight different locations ranges from values of the order of $O(10^4)$ days for the sand waves observed at the Thorntonbank, to values of the order of $O(10^3)$ days for the sand waves observed at Westhinder. These

Table 3. Values of the dimensionless wavenumbers and dimensional wavelengths predicted by the model ($\alpha_p, L_p^*$) and provided by field surveys ($\alpha_o, L_o^*$).

| Site | Wavenumbers and Wavelengths | | | |
| --- | --- | --- | --- | --- |
| | $\alpha_p$ [−] | $\alpha_o$ [−] | $L_p^*$ [m] | $L_o^*$ [m] |
| Kwintebank | 0.315 | 0.474 | 475 | 316 |
| Trapegeer | 0.180 | 0.180 | 219 | 218 |
| Thorntonbank | 0.300 | 0.315 | 451 | 430 |
| Wandelaar | 0.240 | 0.121 | 309 | 610 |
| Westhinder | 0.375 | 0.687 | 361 | 197 |
| Akkaert NE | 0.310 | 0.166 | 331 | 617 |
| Akkaert N | 0.395 | 0.362 | 383 | 417 |

values are in agreement with those described by Katoh et al. (1998) and determined by Knaapen and Hulscher (2002) who showed that sand waves tend to regain their shape and amplitude after dredging and toping off. In particular Katoh et al. (1998) showed that sand waves regenerate in several years time.

## 5 CONCLUSIONS

As stated in the previous section and summarized in Table 3, the general agreement between the observed and the predicted wavenumbers/wavelengths is good for all the locations considered.

The results obtained in the framework of the present analysis confirm the validity of the morphodynamic model of Besio et al. (2006), aimed at predicting the formation of sand waves. Indeed, when evaluating the performances of the theoretical model, it should be kept in mind that in the literature, quite often, comparisons between theoretical predictions and laboratory and/or field data, concerning the geometrical characteristics of the morphological patterns observed in both fluvial and coastal environments, are made considering only the order of magnitude of the results or looking at their qualitative behaviour (e.g. Trowbridge 1995; Vittori et al. 1999; Coco et al. 2000; Komarova and Hulscher 2000; Komarova and Newell 2000; Gerkema 2000; Calvete et al. 2001) and often a relative error equal to 100% is considered to be more than acceptable.

Even though the data described in Stride et al. (1982) indicate that the whole continental shelf of Belgium is covered with sand waves, field surveys show a few locations where the seabed is devoid of sand waves. These areas are mainly located in the coastal zone and in the swales where the Quaternary cover is minimal. Also, the more detailed data summarized in Hulscher and van der Brink (2001) and van der Veen et al. (2006) show the existence of spots where sand

waves are absent. Even though no accurate measurements of tidal currents were available to us at these locations, an attempt to apply the model was made using values of the current interpolated on the basis of the data of the nearest current-meters. The model predictions indicate the presence of sand waves. On the other hand, the present model predicts a flat bed in the north part of the North Sea where Stride et al. (1982), Hulscher and van der Brink (2001) and van der Veen et al. (2006) agree to indicate the absence of sand waves. Hence, it is possible to conclude that local large scale morphological patterns can be responsible for altering the intensity and direction of the tidal current thus enhancing the growth of the sand waves or inhibiting their appearance. Therefore, only the availability of accurate data can allow the prediction of the local morphology in a peculiar site where the local characteristics differ from those of the surrounding area.

## ACKNOWLEDGEMENTS

This research has been supported by MIUR (Ministry of University and Research) under the contract PRIN n. 2005080197. The multibeam data presented in Figures 3–5 have been acquired by FPS Economy, SMEs, Self-employed and Energy, Marine Sand Fund (RV Belgica Kongsberg-Simrad EM1002)

## REFERENCES

Besio, G., P. Blondeaux, and G. Vittori (2006). On the formation of sand waves and sand banks. *J. Fluid Mech. 557*, 1–27.

Blondeaux, P. & Vittori, G. (1999). Boundary layer and sediment dynamics under sea waves. *Adv. Coastal and Ocean Eng. 4*, 133–190.

Blondeaux, P. (1990). Sand ripples under sea waves. part 1. ripple formation. *J. Fluid Mech. 218*, 1–17.

Blondeaux, P. (2001). Mechanics of coastal forms. *Annu. Rev. Fluid Mech. 33*, 339–370.

Blondeaux, P. and G. Vittori (2005a). Flow and sediment transport induced by tide propagation. part 1: the flat bottom case. *J. Geophys. Res. 110*, C07020.

Blondeaux, P. and G. Vittori (2005b). Flow and sediment transport induced by tide propagation. part 2: the wavy bottom case. *J. Geophys. Res. 110*, C08003.

Calvete, D., A. Falques, H. E. De Swart, and M. Walgreen (2001). Modelling the formation of shoreface-connected sand ridges on storm-dominated inner shelves. *J. Fluid Mech. 441*, 169–193.

Cherlet, J., G. Besio, P. Blondeaux, V. Van Lancker, E. Verfaillie, and G. Vittori (2007). Modelling sand wave characteristics on the belgian continental shelf. *J. Geophys. Res.*. Accepted for publication.

Coco, G., H. D. A., and O. T. J. (2000). Investigation of a self-organized model for beach cusp formation and development. *J. Geophys. Res. 105*(C9), 21991–22002.

Colombini, M. (2004). Revisiting the linear theory of sand dune formation. *J. Fluid Mech. 502*, 1–16.

Gerkema, T. (2000). A linear stability analysis of tidally generated sand waves. *J. Fluid Mech. 417*, 303–322.

Hulscher, S. J. M. H. (1996). Tidal-induced large-scale regular bed form patterns in a three-dimensional shallow water model. *J. Geophys. Res. 101*(C9), 20727–20744.

Hulscher, S. J. M. H. and G. M. van der Brink (2001). Comparison between predicted and observed sand waves and sand banks in the north sea. *J. Geophys. Res. 106*(C5), 9327–9338.

Katoh, K., H. Kume, K. Kuroki, and J. Hasegawa (1998). The development of sand waves and the maintenance of navigation channels in the bisanteto sea. In *Coastal Engineering 98*, pp. 3490–3502. ASCE, Reston, VA.

Knaapen, M. A. F. and S. J. M. H. Hulscher (2002). Regeneration of sand waves after dredging. *Coastal Engng 46*, 277–289.

Komarova, N. L. and S. J. M. H. Hulscher (2000). Linear instability mechanism for sand wave formation. *J. Fluid Mech. 413*, 219–246.

Komarova, N. L. and A. C. Newell (2000). Nonlinear dynamics of sand banks and sand waves. *J. Fluid Mech. 415*, 2285–312.

Mouchet, A. (1990). Analysis of tidal elevations and currents along the belgian coast. Technical Report MUMM – BH/88/29, University of Liege. 52 pp.

Seminara, G. (1998). Stability and morphodynamics. *Meccanica 33*, 59–99.

Sleath, J. F. M. (1976). On rolling grain ripples. *J. Hydraulic. Res. 14*, 69–81.

Soulsby, R. L. and R. J. S. Whitehouse (2005a). Prediction of ripples properties in shelf seas. mark 1 predictor. Technical Report TR 150, HR Wallingford.

Soulsby, R. L. and R. J. S. Whitehouse (2005b). Prediction of ripples properties in shelf seas. mark 2 predictor for time evolution. Technical Report TR 154, HR Wallingford.

Stride, A. H., R. Belderson, N. Kenyon, and M. Johnson (1982). *Offshore tidal sands*. Chapman & Hall. ed. A.H. Stride.

Talmon, A. M., N. Struiksma, and M. C. L. M. Van Mierlo (1995). Laboratory measurements of the direction of sediment transport on transverse alluvial-bed slopes. *J. Hydraulic Research 33*, 495–517.

Trowbridge, J. H. (1995). A mechanism for the formation and maintenance of shore-oblique sand ridges on stormdominated shelves. *J. Geophys. Res. 100*(C8), 16071–16086.

van der Veen, H. H., S. J. M. H. Hulscher, and M. A. F. Knaapen (2006). Grain size dependency in the occurence of sand waves. *Ocean Dynamics 56*, 228–234.

Van Lancker, V., S. Deleu, V. Bellec, E. Du Four, I Verfaillie, M. Fettweis, D. Van den Eynde, F. Francken, J. Monbaliu, A. Giardino, J. Portilla, J. Lanckneus, G. Moerkerke, and S. Degraer (2005). Management, research and budgeting of aggregates in shelf seas related to end-users (marebasse). Technical report, Belgian Science Policy. *Scientific Report Year 3* Brussels, 144pp.

Van Rijn, L. C. (1991). Sediment transport in combined waves and currents. In *Proc. Euromech 262*. Balkema.

Verfaille, E., V. Van Lancker, and M. Van Meirvenne (2006). Multivariate geostatistics for the predictive modelling of the surficial sand distribution in shelf seas. *Cont. Shelf Res. 381*, 271–303.

Vittori, G., H. E. De Swart, and P. Blondeaux (1999). Crescentic bedforms in the nearshore region. *J. Fluid Mech. 381*, 271–303.

*River, Coastal and Estuarine Morphodynamics: RCEM 2007 – Dohmen-Janssen & Hulscher (eds)*
*© 2008 Taylor & Francis Group, London, ISBN 978-0-415-45363-9*

# San Francisco Bay sand waves: Modelling and observation

Fenneke van der Meer & Suzanne J.M.H. Hulscher
*Water Engineering and Management, University of Twente, Enschede, The Netherlands*

Daniel M. Hanes
*USGS Pacific Science Center, USA*

Edwin Elias
*WL|Delft Hydraulics, Delft, The Netherlands*

ABSTRACT: Offshore sand waves are bed forms occuring in water depths of tens of meters. In these shallow areas they often influence human activities, e.g. navigation and pipelines.

In this project, field data of a sand wave area in San Francisco Bay is combined with both a full process-based model, Delft3D-online, and an idealized process-based model, the sand wave code after Van den Berg and van Damme (2006).

This comparison between these two models and field measurements increases our knowledge on sand wave areas in general and especially for the San Francisco Bay area. Next it enables calibration and validation of both models, so that we can list their strong points in relation to sand wave modelling.

## 1 INTRODUCTION

Due to tidal currents in shelf seas, wavy bed patterns can occur, varying in size from small near shore ripples up to large offshore sand banks. An intermediate pattern is the sand wave (Figure 1), which occurs abundantly in shallow seas, such as the North Sea (Van der Veen et al. 2006), the Bisanseto Sea (Knaapen and Hulscher 2002) and San Francisco Bay (Barnard et al. 2006). Sand waves can be defined as wave like bottom patterns with wavelengths between 100 to 800 m. They can migrate several meters per year and their height is a significant part (up to 1/3) of the water depth. Sand waves occur where tidal currents are strong, the water is not deep, i.e. 10–55 m (Bijker et al. 1998) and sand is in good supply. Sand waves can occur together with other bed patterns. They can coexist with large scale sand banks and can be covered with smaller scale mega ripples (Idier et al 2002; Idier et al. 2004).

Because of their migration speed and spatial dimensions, sand waves can interfere with anthropogenic structures. For example, sand wave occurrence in the region of shipping channels is one of the reasons that regularly monitoring of these channels is required, as it is important to maintain a certain navigation depth for shipping. If the nautical depth becomes shallower than the critical value required for shipping, these lanes have to be dredged.

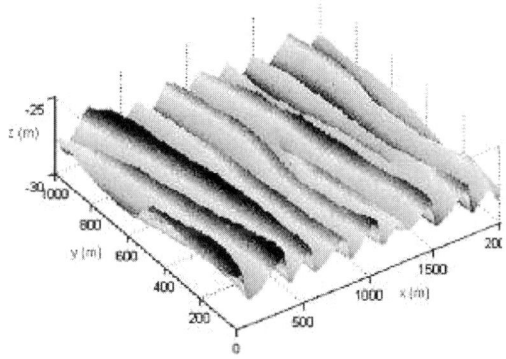

Figure 1. Example of a measured sand wave field in the North Sea.

Second, many pipelines and cables are buried in the seabed for the transportation of fossil fuels, electricity and digital information. When pipelines are buried it is not desirable for them to be exposed by a migrating sand wave. Since the burial of pipelines is expensive, excessive burial should be prevented.

More general, sand waves influence the overall sediment transport in shallow seas; directly by migration and indirectly by changing the flow. This influence can lead to erosion or deposition in (coastal) areas.

For this practical relevance, and scientific interest, research is carried out to gain knowledge regarding sand waves.

Basically two main research approaches are possible; observing the seabed and modelling the seabed. Observations provide direct information on the actual bed form shapes on the measured location and, when surveyed more than once, also about the changes over time. Information can then be destilated on sediment transport and expected sand wave behaviour. It is only since a few decades that measuring the sea floor with an appropriate precision has been possible. In order to detect sand wave migration, not only the vertical position of the sea floor should be measured accurately, but also its exact spatial (horizontal) location. Though accurate measuring is possible, surveys are often very expensive, therefore models are used to increase our knowledge on sand waves. Second reason to model sand waves is the generic knowledge a model provides. Results can be extrapolated to other locations and circumstances, and the effect of environmental factors can be studied separately.

Within modelling sand waves, one can start from different perspectives. In this project we use two process-based model approaches. The first model is fully process-based, using the known bathymetry and hydrodynamics of a location to describe the (variation in) sand waves, starting from the present form. The second model is idealized, designed specifically to describe sand wave dynamics. In this approach we model sand waves from their initial state up to their finite amplitude, starting from small disturbances of the seabed and some basic local conditions (e.g. flow velocity and mean grain size).

We combine these two model approaches, together with the detailed sand wave survey data of San Francisco Bay to increase our knowledge on sand wave areas in general and in particular for the San Francisco Bay area.

In this paper, we will start describing the case study area in San Francisco Bay (Section 2). Both numerical approaches, the full process-based and the idealised process-based approach, will be discussed in the Sections 3.1 and 3.2 respectively. Section 4 discusses the results and presents conclusions.

## 2 SAN FRANCISCO BAY SAND WAVES

San Francisco Bay is located on the west coast of the United States, covering an area of approximately 4000 km$^2$ when including the wetlands. The main river system contributing to the sediment balance in the Bay is the San Joaquin and Sacramento river system, draining around 40% of California.

San Francisco Bay is connected to the Pacific Ocean by only one outlet, the Golden Gate. Here the channel has scoured down into bedrock to a maximum depth of 113 m, where tidal currents accelerate through the narrow, erosion-resistant rocky strait. Maximum currents typically exceed 2.5 m/s .

On both sides of the Golden Gate Channel sand waves exist. Outside the Golden Gate Channel a 4 km$^2$ sand wave field occurs, with high flow velocities and sediment availability (Figure 2. The water depth ranges from 30 to 100 m. Due to the high flow velocities and a residual current there is a large sediment flux directed towards the Pacific.

Recently Barnard et al. (2006) investigated this large sand wave field (Figures 2,3 and 5). The entire mouth of San Francisco Bay was mapped in the autumn of 2004 and 2005. A region along the centerline axis of the giant sand waves was mapped four times in 2004: on 17, 18, 25, and 30 October; and three times in 2005: on 17 and 18 September and 30 October. These surveys focused on 19 distinct contiguous bed forms in water depths between approximately 35 and 80 m. Here, sand wave forms are ebb-dominated to symmetric, wavelengths are between 32 and 80 m and the mean sand wave height is 6 m. Grain size on these sand waves is coarse (0.5 to 4 mm).

Over the complete area, sand wave characteristics are diverse, due to the high variation in flow velocity and sediment. Sand waves have wavelengths up to 220 m and wave heights up to 10 meter. Net migration is approximately 7 m/yr, though the crest position can migrate approximately 3 m depending on the daily tidal current.

The flux of sand and gravel between San Francisco Bay and the coastal ocean is an important component of the littoral sediment budget. Understanding the mechanisms of transport and the quantity of material transported is integral to the proper management of sediment in the region. Key issues related to sediment management include dredging of the navigation channel, local coastal erosion, sand mining for construction and fill, and sources of sediment for beach nourishment. The sand waves are also expected to be important to the ebb and flood tidal currents at the entrance to San Francisco Bay, because their effective roughness retards the flow, and eddies shed from flow separations near the crests cause substantial generation of turbulence. The flow and turbulence influenced by the sand waves probably affect the mixing and dispersion of pollutants and contaminants both within and outside the bay.

## 3 PROCESS-BASED MODELLING

Process-based modelling aims at describing the important physical processes in terms of differential equations. In this project this mainly includes the continuity of water (Eq. 1), the hydrostatic flow equations under the shallow water approximation (for 3D Eq. 2 and

Figure 2. Oblique view of the giant sand waves and other bed forms at the mouth of San Francisco Bay. The view is from the northwest toward the Golden Gate Bridge. The city of San Francisco is in the upper right corner. The Golden Gate Bridge is approximately two kilometers long. The land was imaged using digital orthophotos draped over a U.S. Geological Survey digital elevation model. The Golden Gate Bridge model is courtesy of IVS 3D©. From Barnard et al. (2006).

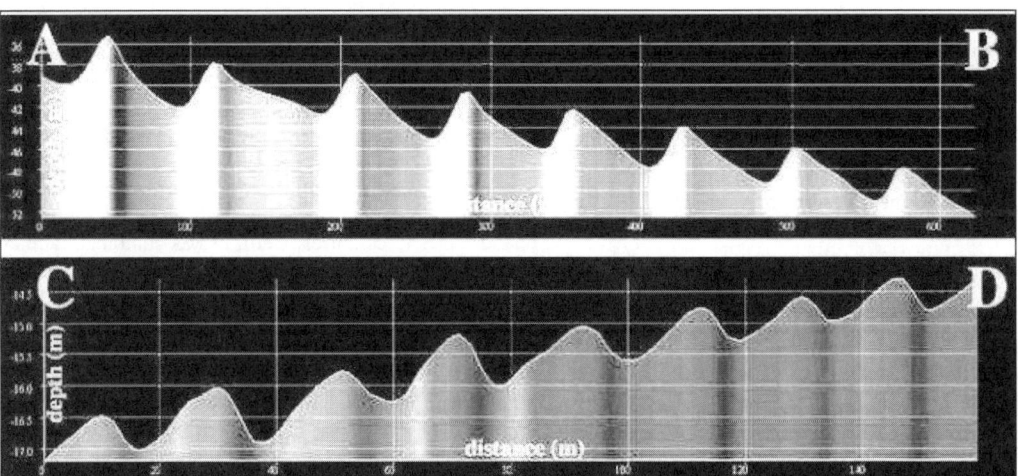

Figure 3. Two sand wave transects, one ebb dominated in the center (transect AB, see Figure 5) and one flood dominated along the periphery (transect CD, Figure 5). From Barnard et al. (2006).

3) and a sediment transport equation, for example Eq. 4, after Komarova et al. (2000). These are coupled through the continuity of sediment using sources and sinks or a coupling equation (for example Eq. 5). In both presented models the vertical momentum assumes a hydrostatic pressure gradient.

$$\frac{\partial u}{\partial x} + \frac{\partial w}{\partial z} = 0 \tag{1}$$

$$\frac{\partial u}{\partial t} + u\frac{\partial u}{\partial x} + v\frac{\partial u}{\partial x} + w\frac{\partial u}{\partial z} = -g\frac{\partial \zeta}{\partial x} + \frac{\partial}{\partial z}\left(A_v\frac{\partial u}{\partial z}\right) \tag{2}$$

995

$$\frac{\partial v}{\partial t} + u\frac{\partial v}{\partial x} + v\frac{\partial v}{\partial x} + w\frac{\partial v}{\partial z} = -g\frac{\partial \zeta}{\partial x} + \frac{\partial}{\partial z}\left(A_v\frac{\partial v}{\partial z}\right) \tag{3}$$

$$q_b = \alpha|\tau_b|^b\left[\tau_b - \lambda\frac{\partial h}{\partial x}\right] \tag{4}$$

$$\frac{\partial h}{\partial t} = -\left(\frac{\partial q_b}{\partial x} + \frac{\partial q_s}{\partial x}\right) \tag{5}$$

Full process-based modelling includes different processes (e.g. wind- and wave-driven currents, density gradients, sediment transport) within broad classes of problems over different temporal and spatial scales, and is able to deal with complex geometries. In contrast idealized modelling assumes a simple geometry, input and boundary conditions, to study morphological features at a morphodynamic time scale, i.e decades to centuries.

### 3.1 Full approach

For the full process-based approach we use the model Delft3D. The spatial discrete approach can be used in both 2DH and 3D. The model system consists of a number of integrated modules which together allow the simulation of hydrodynamic flow (under the shallow-water assumption), computation of the transport of water-borne constituents (e.g. salinity and heat), short wave generation and propagation, sediment transport and morphological changes, and the modelling of ecological processes and water quality parameters (WL|Delft Hydraulics 2006; Lesser et al. 2004).

For the sediment transport and the turbulence the user can choose between various incorporated equations. Sediment and flow are coupled by sources and sinks of sediment, leading to morphological changes.

The discretisation in Delft3D in the horizontal space is based on finite differences. The used grid can be rectangular, curvilinear or spherical. Using a staggered grid method, quantities are located on logical places within the grid cells: fluxes on boundaries and sources in the cells.

In the vertical direction (i.e. in case of 3D modelling) boundary fitting $\sigma$-coordinates are used. The number of layers is constant over the modelled area, the layer thicknesses vary to account for the changing water depth. The layer thickness is usually non-uniform distributed to encounter more resolution in the area of interest (mostly close to the bed or close to the water surface).

At the seabed and sea surface, vertical velocities are zero and friction is calculated from the shear stresses due to currents and waves. At the vertical boundaries, closed land boundaries or open boundaries are prescribed, respectively zero flow and free slip or a prescription of water level, flow velocity or discharge.

The lateral open boundaries are often described further away from the studied area to decrease the possibilty of boundary effects.

Through time Delft3D solves the equations with an ADI method (Alternating Direction Impliciet) (Leendertse 1987), which combines implicit and explicit schemes such that the result is robust for a wide range of circumstances.

To overcome a long computational time in Delft3D a morphological factor is used (Roelvink 2006). In this approach the calculations per time step within the tide are multiplied with the morphological factor, up to 100–1000, but if for example waves are present 1–50 is more likely. This lengthening of the tide increases the calculated morphological time within one tide calculation. Modelling real time for longer periods however, has the risk of amplifying small errors in the initial and boundary conditions.

With respect to the wave forcing, two methods are proposed to reduce the computational time. The first takes each wave condition into account and weighs each of the results depending on the occurrence of these wave climates. The second method uses a representative wave such that only one bed evolution computation is done. This method leads to a strong reduction in computational time. However, representative waves, wind and tidal forcing, which are commonly adopted, imply inherent predictability limits as the morphodynamic process is stochastic.

For further details we refer to (WL|Delft Hydraulics 2006; Lesser et al. 2004).

Delft3D is not yet used for long term predictions of large offshore morphological features such as sand waves. Brière (2006) investigated the effect of sand mining on sand banks using Delft3D. Sand banks have wavelengths in the order of kilometers and heights of tens of meters. Delft3D was used for the short term flow and sediment transport. For the longer term, i.e. decades to centuries, Delft3D was combined with an idealized sand bank model.

The San Francisco Bay area is described in the 2DH approach of Delft3D. Using the known bathymetry and flow characteristics, the currents and sediment transport over several tidal periods can be estimated (Figure 4 and 5). To model sand waves in detail, part of the area will be modelled as a 3D area, using the 2DH model to impose the vertical boundary conditions.

### 3.2 Idealized approach

The idealized model we use in this project (Van den Berg and van Damme 2006; Németh et al. 2007) is a 2DV sand wave code, which is developed specifically to describe sand waves from their initial state uptill their full grown state. Where Delft3D can be used for a wide range of processes, scales and environments, this sand wave code focusses on the sand wave evolution.

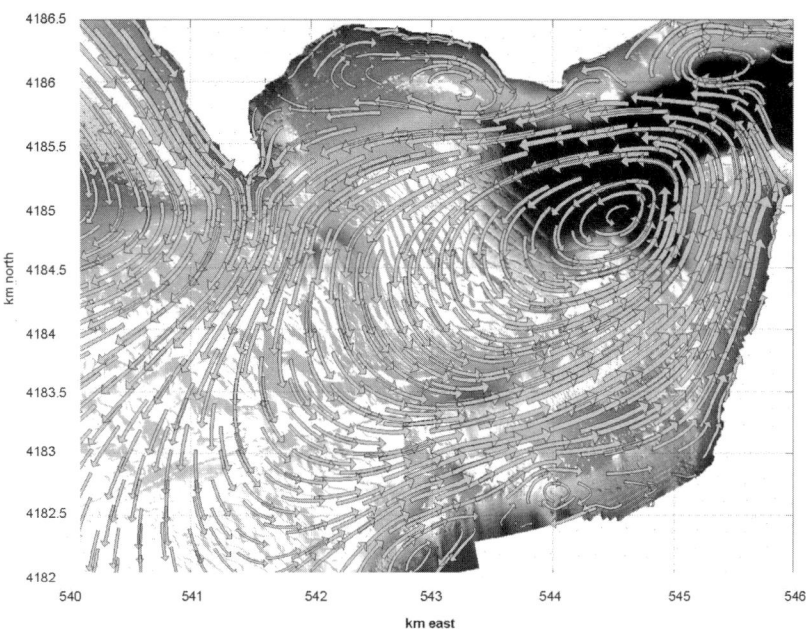

Figure 4.    Predicted mean flow pattern at the mouth of San Francisco Bay, averaged over a month. Based on the results of a 1-month non-validated flow simulation using Delft3D.

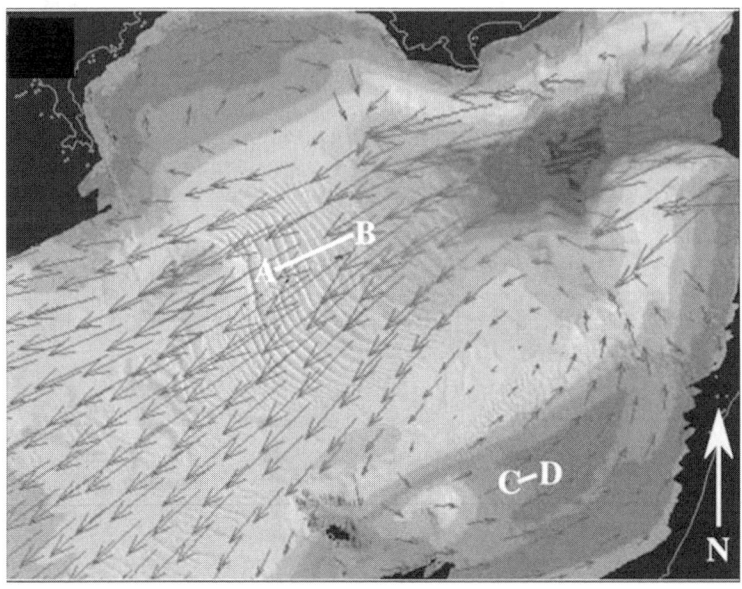

Figure 5.    Dominant transport directions at the mouth of San Francisco Bay illustrated by bed form orientation, ebb dominated in the center (transect AB, Figure 3) and flood dominated along the periphery (transect CD, Figure 3), due to persistent eddies. Predicted tidal flood currents from Delft3D are superimposed over the bathymetry, illustrating good agreement between peak flow vectors during ebbing tide and bed form morphology. From Barnard et al. (2006).

For sand wave fields in the Southern part of the North Sea the idealized model has shown good results in describing the wavelength, height, shape characteristics and migration (Van den Berg and Damme 2006; Németh et al. 2007; Németh et al. 2006).

The theory behind sand wave formation is that they are basically formed due to an interaction between a sandy seabed and a tidal flow. Sand waves occur as free instabilities in this system, i.e. there is no direct relation between the forcing (tide) scale and the morphological feature (sand wave) scale (Dodd et al. 2003). Sand wave occurrence can be understood only if the feedback mechanism between the forcing and the seabed is taken into account. Hulscher (1996) described this mechanism for sand waves, where vertical vortices play a crucial role. Small perturbations of the sea floor cause small perturbations in the flow field and vice versa. The bed can be either stable, which means that disturbances will be damped, or unstable, which means that bed perturbations will grow and the seabed is changed. If perturbations are unstable (i.e. they trigger growth) small vertical rest circulation cells occur within the tidally averaged flow field. These cells cause small net transport to the crests of the perturbation, thereby causing growth. For small amplitude perturbations, growth can be described as linear, though as sand waves grow larger, non-linear terms become important. Assuming that sand waves are only weakly non-linear the initially dominating wavelength will be dominating in the non-linear regime as well. For that wavelength we then study the finite amplitude behaviour. We have several indications to assume that sand waves are weakly non-linear: the amplitude is mostly smaller than 20% of the water depth and the predicted fastest growing mode is close to the observed mode.

The idealized approach also uses an staggered grid, rectangular in the 2DH plane. Further schematisations include that the vertical boundaries are periodic, there is no flow through the seabed and surface and at the seabed a partial slip condition compensates for the constant eddy viscosity. The grid is uniform in the horizontal and non-uniform in the vertical to obtain more resolution near the seabed.

Modelling sand waves with the idealized approach starts with a small sinusoidal disturbance of the sea floor and a characteristic tide. Since the flow changes over a timescale of hours and the morphology over a timescale of years, the bathymetry is expected to be invariant over a single tidal cycle. Once the tidal flow is known, the bed changes are calculated over this characteristic tide, using a sediment transport equation (Eq. 4). This is repeated until the bed evolution exceeds a certain value. The new bed morphology is input for the flow calculation, after which the new flow is input for the sediment transport again, and so the process is iterative. In this way, the code is able to simulate the morphological time scale accurately, while avoiding long computation times.

In case of growth, a sand wave will grow up to a certain finite amplitude and might evolve into a symmetrical or asymmetrical shape. Sand wave characteristics that can be studied using this idealized code are wavelength and height, migration and the final form characteristics, such as the crest-trough ratio.

Different from the process based approach in Section 3.1, here only idealized processes are included. Variation in the tide, storm and weather influences or surrounding coastal boundaries are not incorporated in the idealized approach. The representative tidal wave, is only influenced by changes in the bed topography.

Although large domains of sand waves can be incorporated (Van den Berg and van Damme 2006), due to the periodic boundary conditions, the sediment in the model is constant. Erosion and deposition on a larger spatial scale than within the sand wave are not taken into account and can only be estimated by the migration of the sand wave through the calculated domain.

When local characteristics, such as water depth, tidal velocity and grain sizes are known, the idealized approach computes the growth rate, which depends on the sand wavelength. The sand wavelength that grows fastest in height is called the fastest growing mode (FGM). For this sand wave the further evolution to its finite amplitude can be simulated.

The influence of environmental conditions can be tested leading to tendencies in sand wave behaviour. Typical for the San Francisco Bay are the large variation in water depth and high peak velocities. The idealized approach shows tendencies of growing sand wave amplitude for deeper water and lower velocities (Fig. 6). Sand wave length increases for increasing flow velocities and increasing water depth. In this, some basic conditions are used, not yet established for San Francisco Bay (i.e grain size 400 $\mu$m, no residual current, no suspended sediment).

## 4 DISCUSSION AND CONCLUSIONS

Sand waves can be investigated by both observation and modelling. In this paper two process-based models are introduced that can describe sand wave behaviour. The full process-based Delft3D code is expected to gain insight mostly on short time scales, in a large range of processes (sediment transport, flow, bathymetry changes), while the idealized sand wave code focusses on the change in sand wave characteristics on the morphological time scale. This combination improves our overall understanding of sand wave areas and their characteristics in various time scales.

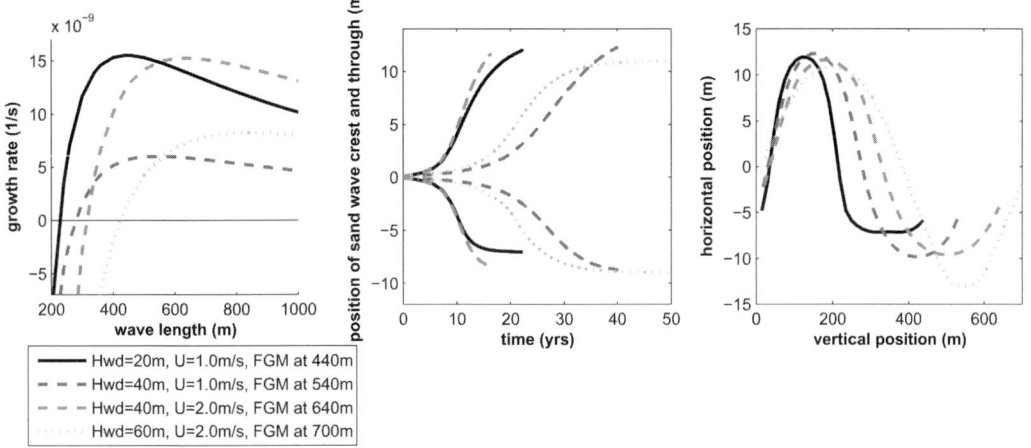

Figure 6. Idealized approach results, in the legend, Hwd = water depth, U is the maximum depth averaged tidal flow velocity. Left the growth rate for different wave lengths, showing the FGM, middle the growth to the finite amplitude and right the final sand wave shape.

Both models will be used to describe the sand wave field at the mouth of San Francisco Bay as a case study. First model results show large differences in flow velocities and sediment transport within San Francisco Bay. Sand wave heights tend to increase with decreasing flow velocities and increasing water depth. Sand wave length increases with increasing flow velocities and increasing water depth. Observations in San Francisco Bay are more ambiguous as different parameters occur together. For example when comparing section AB and CD (Fig. 3) both waterdepth and flow velocity are different. Section AB shows an decrease of sand wave height with increasing waterdepth, while section CD shows the opposite trend.

Further research will establish a parameters setting better belonging to the San Francisco Bay. With this setting, environmental influences will be studied, seperately and in combinations, to improve our understanding of sand wave characteristics and accompanying sediment transport. Besides this research enables us to calibrate and validate both model approaches, so that we can list their strong points in relation to sand wave modelling.

## ACKNOWLEDGEMENTS

This research is supported by the Technology Foundation STW, applied science division of NWO and the technology programme of the Ministry of Economic Affairs.

## REFERENCES

Barnard, P. L., D. M. Hanes, D. M. Rubin, and R. G. Kvitek (2006). Giant sand waves at the mouth of San Francisco Bay. *Eos 87*(29).

Bijker, R., J. Wilkens, and S. J. M. H. Hulscher (1998). Sand waves, where and why? In *Proceedings of the eigth international offshore and polar engineering conference*, Volume 2, Montreal, Canada, pp. 153–158.

Brière, C. (2006). EUMARSAND European Marine Sand and Gravel Resources: Evaluation and Environmental Impact of Extraction. Modelling sandbank dynamics and the effects of offshore sand extraction with Delft 3D numerical model. Technical report, University of Twente, The Netherlands.

Dodd, N., P. Blondeaux, D. Calvete, H. de Swart, A. Falqués, S. J. M. H. Hulscher, G. Rózyński, and G. Vittori (2003). Understanding coastal morphodynamics using stability methods. *Journal of Coastal Research 19*.

Hulscher, S. J. M. H. (1996). Tidal-induced largescale regular bed form patterns in a threedimensional shallow water model. *Journal of Geophysical Research 101*, 727–744.

Idier, D., D. Astruc, and S. J. M. H. Hulscher (2004). Influence of bed roughness on dune and megaripple generation. *Geophysical Research Letters 31*(L13214), 1–5.

Idier, D., A. Ehrhold, and T. Garlan (2002). Morphodynamique d'une dune sous-marine du d'etroit du pas de Calais. *Comptes Rendus Geoscience 334*, 1079–1085.

Knaapen, M. A. F. and S. J.M. H. Hulscher (2002). Regeneration of sand waves after dredging. *Coastal Engineering 46*, 277–289.

Komarova, N. L. and S. J. M. H. Hulscher (2000). Linear instability mechanisms for sand wave formation. *Journal of Fluid Mechanics 413*, 219–246.

Leendertse, J. J. (1987). A three-dimensional alternating direction implicit model with iterative fourth order dissipative non-linear advection terms. Technical Report WD-333-NETH.

Lesser, G. R., J. A. Roelvink, J. A. T. M. van Kester, and S. G. S (2004). Development and validation of a three-dimensional morphological model. *Coastal Engineering 51*, 883–915.

Németh, A. A., S. J. M. H. Hulscher, and R. M. J. Van Damme (2006). Simulating offshore sand waves. *Coastal Engineering 53*, 265–275.

Németh, A. A., S. J. M. H. Hulscher, and R. M. J. Van Damme (2007). Modelling offshore sand wave evolution. *Continental Shelf Research 27*, 713–728.

Roelvink, J. A. (2006). Coastal morphodynamic evolution techniques. *Coastal Engineering 53*, 277–287.

Van den Berg, J. and D. van Damme (2006). Sand wave simulations on large domains. In Parker and Garcia (Eds.), *River, Coastal and Estuarine Morphodynamics:RCEM2005*.

Van der Veen, H. H., S. J. M. H. Hulscher, and M. A. F. Knaapen (2006). Grain size dependency in the occurence of sand waves. *Ocean Dynamics 56*(3–4), 228–234.

WL|Delft Hydraulics (2006). Delft3D-FLOW; simulation of multi-dimensional hydrodynamic flows and transport phenomena, including sediments. User manual delft3d version 3.13, WL|Delft Hydraulics, The Netherlands.

River, Coastal and Estuarine Morphodynamics: RCEM 2007 – Dohmen-Janssen & Hulscher (eds)
© 2008 Taylor & Francis Group, London, ISBN 978-0-415-45363-9

# Computer simulations of megaripples in the nearshore

E.L. Gallagher

*Franklin and Marshall College, Lancaster, PA, USA*

ABSTRACT: Megaripples (bedforms with heights of 20–40 cm and lengths of 1–5 m) occur frequently in the shallow surf zone. However, their relationship to the flow field is unclear. It has been suggested that self-organization is responsible for the formation of many different morphological patterns (eg, river meanders, sorted-patterned ground, beach cusps, wind ripples, eolian dunes). Each of these systems is nonlinear, dissipative and open to the environment, as is the surf zone. Thus, surf zone megaripples also may be good candidates for formation via self-organization. Werner (1995) assumed self-organization and successfully predicted eolian dune morphologies using computer simulations based on simple rules governing the motion of sediment. Here, a similar model is used that describes forcing by combined flows in the nearshore. Preliminary results look remarkably like observed morphologies measured during the SandyDuck Nearshore field experiment.

## 1 INTRODUCTION

Megaripples are bedforms with heights of up to 50 cm and lengths of 1–10 m that occur frequently in the nearshore (Clifton et al. 1971, Clarke & Werner 2003). They are most common in the shallow surf zone (<2 m water depth), but are also observed as patches in deeper water (2–5 m water depth) (Gallagher et al. 2003). Megaripples are important to wave transformation, current generation and sand transport (through increased suspension by flow separation and turbulence in their lee as well as through bulk transport via megaripple migration). They act as hydraulic roughness elements, changing wave energy dissipation and water circulation patterns (Garcez-Faria et al. 1998).

Recent field observations have shown that they are ubiquitous (Clarke & Werner 2003), but that their relationship to the flow field is unclear (Gallagher et al. 2003). Because there are relatively few observations and their dynamics are not known, there are few models that predict megaripple occurrence (Clarke & Werner 2004) and thus they are not accounted for in models for sediment transport, wave transformation and nearshore circulation. A model to predict megaripples likely would improve all components of nearshore processes modeling.

### 1.1 Previous research

The limited number of observations of surf zone megaripples is due to difficulties in making measurements in the harsh nearshore environment. Early observations were made by divers, who are limited to calm conditions (Clifton et al. 1971, Davidson-Arnott & Greenwood 1976). Later measurements were made using acoustic instruments which can function during storms (Hay & Wilson 1994, Thornton et al. 1998, Gallagher et al. 1998b, Gallagher et al. 2003 and Hay & Mudge 2005). Bedform types observed in the nearshore include planar beds, regular and 3-D ripples, and megaripples. Clifton (1976) developed an empirical model for bedform regimes in oscillatory flow. He suggested that bedforms increase in size from small orbital ripples to cross-ripples to megaripples with increasing stress until the bed is planed-off during sheet flow conditions. Hay & Wilson (1994) observed similar changes in ripple types, on time scales of 1–3 hours, as flow conditions and bed stress decreased during the waning stages of a storm.

Regime-type models like Clifton's (1976) are the most common way to describe bedforms and have been developed for many different environments (eg, Middleton & Southard 1984, Raudkivi 1990). Bedform predictors have also been developed, which predict the existence, length and height of the bed ripples (Wiberg & Harris 1994, Neilsen 1981, van Rijn 1993). These models do not predict megaripples well.

Gallagher et al. (2003) made measurements of the spatial and temporal variability of bedforms in the nearshore and found that megaripples occur frequently inside the surf zone, and are common, but patchier, in deeper water (~2–5 m water depth). Bedform height was compared with mobility number and it was found that the bed became flat at high mobility numbers as is predict by regime-type models. However, for intermediate values of mobility number (between about

30–100), large ripples, small ripples and flat beds all seem to be possible. This observation is contrary to the regime-type model where different bedform types exist independently from each other.

Clarke & Werner (2003) used a video camera mounted on a high cliff overlooking the surf zone to measure bedforms. Although video techniques break down for large waves (because surface foam and suspended sediments obstruct the view), the temporal and spatial coverage of their measurements is unprecedented. Clarke & Werner (2004) made synoptic measurements of megaripples inside the surf zone and they hypothesized that megaripples will always occur inside the surf zone unless the water depth changes such that they enter the swash or approach the break point, which will destroy the bedforms. This hypothesis was based on both observations that megaripples occurred frequently and that the flow conditions in the intermediate depth between the breaker zone and the swash were right for megaripple formation. They observed megaripples forming and growing over multiple tidal cycles and they found that megaripple lengths grow continuously. This is in contrast to typical ripple predictors, which predict that the bedforms will reach an equilibrium size for given flow conditions.

These field studies have increased our understanding of megaripples in the nearshore. However, such observations have failed to elucidate the dynamics governing megaripples, thus few predictive models have been developed. Predictions of megaripple occurrence and characteristics (heights, lengths, orientation, migration, distribution) likely would be useful for and increase the accuracy of predictions of most nearshore processes. Here, megaripple occurrence and characteristics are simulated as a self-organized system using the hierarchical approach as defined by Werner (1999) and Ahl & Allen (1996).

## 1.2 Self-organization

It has been suggested that self-organization is responsible for the formation of many different types of morphological patterns, including river meanders (Stolum 1996), sorted-patterned ground (Kessler & Werner 2003), beach cusps (Coco et al. 2000), wind ripples (Nishimori & Ouchi 1993) and eolian dunes (Werner 1995). In each of these pattern-forming systems, complexity arises from nonlinear interactions between the system and the environment, from dissipative processes such as friction, turbulence and sediment transport, and from being open (both material and energy are exchanged across system boundaries) and therefore never in equilibrium (Werner 1999).

Werner (1999) uses a 'hierarchical' approach (Ahl & Allen 1996) to modeling self-organized systems, wherein processes at different temporal and

spatial scales are distinct from each other and can be separated. For example, the physics of sediment transport on the scale of the sand grains (eg, hierarchical level 1) is a system that is difficult to model, because of the large number of sand grains, their interactions, and the nonlinear relationship between the fluid and the sediment. To then simulate the individual trajectories of enough grains to predict a large-scale morphological feature like a dune (level 2) would be extremely expensive computationally. Trying to simulate height and length statistics, or spatial distribution of dunes in a dune field, or migration rates (level 3) based on level 1 physics is currently impossible. Werner (1995) created his hierarchical model by describing the transport of sediment (level 1) using a few simple parameters that sensibly represent the motion of sand being forced by a fluid, thus neglecting the complex, nonlinear physics at that scale. Instead his model emphasizes the interaction between the bedform and the gross movement of water and sand (level 2).

In the nearshore, sediment transport and feedback between morphology and forcing are highly nonlinear and dissipative processes. The system is open, with mass and energy being exchanged across boundaries and external forcing changing continually. In addition, different temporal and spatial scales are distinguishable and separable, facilitating sensible parameterization and modeling. Thus, megaripples are excellent candidates for hierarchical modeling and formation via self-organization.

## 2 MODEL

Here, a hierarchical model is used to predict megaripple occurrence, morphology, and migration. Conceptually the model follows the work of Werner (1995), who successfully predicted eolian dune morphologies using computer simulations based on simple rules governing the motion of sediment. Here, a similar model is used with rules that describe forcing by the combined flows in the nearshore.

Werner's (1995) simple rules were based on he physics of saltating sediments, where slabs of sediment on a grid were picked up randomly and moved a set jump distance, $\ell$. Whether or not the slab of sediment would land at the new location was based on a probability that depended on the conditions at the landing site (eg, bed height, presence of other grains, bed steepness, etc). If landing failed, the slab jumped another $\ell$ units and landing probability was tested again. Using this system, Werner (1995) was able to predict realistic dune morphologies (barchan, longitudinal, and star dunes) by varying the jump direction (ie, by varying the direction of the forcing winds based on the theory of Rubin & Hunter 1989). He hypothesized that the different morphological states (both predicted and observed) are attractors of the nonlinear, dissipative, complex system and thus dunes in general are self-organized features.

The present model also consists of a three-dimensional matrix of sediment slabs. However, transport is designed to represent subaqueous transport of sand. When the imposed flow is larger than a threshold value, the sediment slabs are picked up and moved in the direction of the flow a distance proportional to the magnitude of the flow. When the flow magnitude drops below the threshold value, the sediment is redeposited on the bed at the new location. During peak onshore flows, all sediment slabs at the surface are moved. This is intuitive for strong sheet flows in the nearshore where the whole sand bed moves like a carpet when the flow is strong enough, then on the return flow of the wave, sediment is transported as sheet flow in the opposite direction.

The flow field consists of an oscillatory flow with a steady current and a random fluctuation. This flow field is meant to represent that in the surf zone where oscillatory flow owing to waves (which are skewed in the onshore direction), steady currents, and turbulence are generally superimposed. In the prototype model reported here, the flows are aligned in the on-offshore direction. However, an important aspect of this modeling effort will be to investigate multi-directional flows, eg, alongshore currents normal to the on-offshore oscillatory currents and a spectrum of waves from different directions.

Feedback between the flow and the bed is also included. The primary mechanism for including feedback is to adjust the flow field over a bedform. A number of different criteria have been investigated (including variations based on bed elevation, bed slope and changes in the bed slope). The model results presented here are all run with a feedback mechanism consisting of a shadow zone in the lee of a bedform. This shadow zone is implemented by using 1/100 of the velocity if the downstream slope is greater than 0.2 (about 11°). An additional 10% of the velocity is added to simulate increased turbulence at the reattachment point. An acceleration that is proportional to the height of the bedform is also added on the upstream face of a bedform. Note that the feedback mechanisms all act to alter the flow and not the sand transport itself.

## 3 PRELIMINARY RESULTS

The results of this preliminary model (Fig 1) are encouraging and show a striking resemblance to observed megaripples (Fig 2). For the simulation shown in Fig 1, the sediment slabs are 1 cm high, 10 cm long and 10 cm wide. The fluid velocity, $u$, is modeled as a sinusoid with an oscillatory flow amplitude, $A = 50$ cm/sec and a period, $T = 10$ sec, plus a steady onshore flow, $S = 30$ cm/sec (to simulate an onshore skewed flow) and with a random fluctuation, $va = \pm 15$ cm/sec at each location on the grid (as a turbulent contribution). An angle of repose of about 20° is used (lower than normal owing to the highly dynamic bed in the nearshore).

The example in Fig 1a shows results from a model run with a crude sediment transport model where sand is moved according to simple rules:

if $u \leq 10$ cm/sec then put all sand down
if $10 < u \leq 20$ then put down half the sand that was up in the water moving
if $20 < u \leq 50$ then pick up 1 block at that location
if $50 < u \leq 90$ then pick up 2 blocks
and if $u > 90$ then pick up 3 blocks.

In Fig 1b the sediment is transported according to a typical physics-based model (Ribberink 1998),

$$Q = 11\sqrt{\frac{\rho_s - \rho_w}{\rho_w} gD_{50}{}^3} (|\theta| - \theta_c)^{1.65} (\theta/|\theta|),$$

where

$$\theta = \frac{0.5 f_w (\rho_s - \rho_w)}{gD_{50}\rho_w}|u|u$$

and, $\theta_c = \theta$ using 20 cm/sec as a threshold velocity. $Q$ is volume transport and is divided by sediment slab volume (multiplied by a packing factor, 0.6 is typical and is used here) to give a number of slabs moved. When $Q$ is below the threshold of motion, sediment slabs are deposited. The other variables in the equations are standard ($\rho_s$ and $\rho_w$ are the densities of sediment and water, $D_{50}$ is the mean grain size, (0.3 mm is used here), g is gravity, and $f_w$ is the wave friction factor (0.3).

Although the amplitudes of the two sets of bedforms in Fig 1 are different, both types of models produce similar features. This is exciting because it suggests that no matter how the sand is moved, the basic processes at work in the formation of bedforms seem to be captured with this simple model. Further, as argued by Werner (1999), the hierarchical approach, where the detailed physics at the smallest scale can be abstracted to model the processes at larger scales, seems to be a viable method for modeling megaripples.

In Fig 2, two examples of observed megaripple patches are shown in plan view (linearly interpolated from an altimeter array), one on the outer bar and one in the surf zone. The megaripples in Fig 2a are relatively long-crested, sometimes spanning the 4 m alongshore distance represented in the figure. They have measurable wavelengths of ~2 m and heights of ~30 cm. The megaripples observed in the surf zone (Fig 2b) have similar heights and lengths but are irregular in shape with short crest lengths (~2 m). This is in agreement

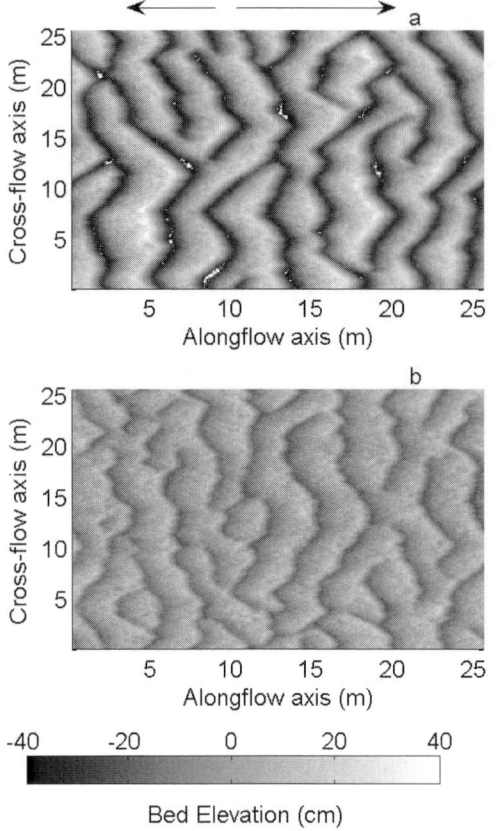

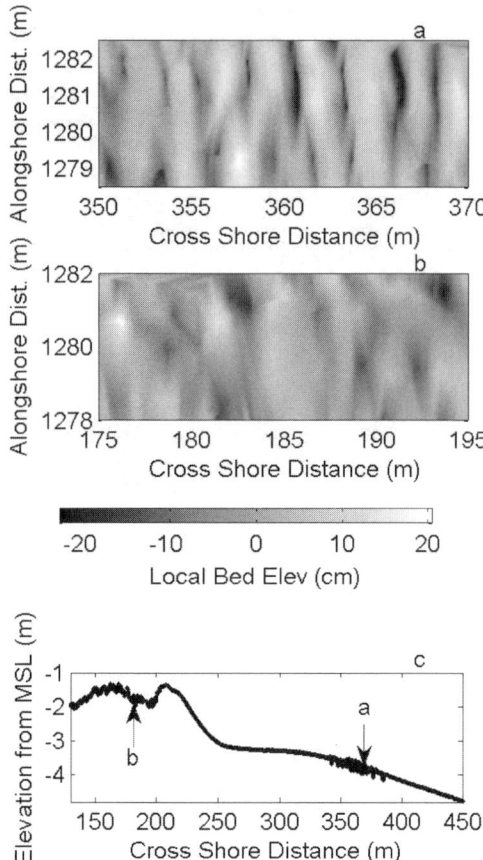

Figure 1. Results of model simulation of megaripples after 1200 secs (starting from a flat bed). Arrows illustrate skewed oscillatory flow direction (onshore direction is to the left). The result in the top panel was run with a very simple sediment transport model. The bottom panel was run with a physics-based sediment transport model (Ribberink 1998).

Figure 2. Observations from SandyDuck. Plan view of megaripples (linearly interpolated using altimeter data) from a) offshore and b) the surf zone. c) Seafloor elevation versus cross shore distance from the shoreline to 5 m water depth with patches of megaripples observed in the surf zone (a, 1–2 m water depth) and offshore (b, 4 m water depth). Note the plan views are both about 4 m × 20 m, thus they are long narrow pictures of megaripples in contrast to the 25 m × 25 m model results in Fig 1.

with observations by divers who often saw megaripples in the surf zone as irregularly spaced oval-shaped holes and with the observations of Clarke & Werner (2004, see for example their Fig 16).

Using the model to further examine megaripple dynamics in the nearshore, an interesting observation from the model is that without any turbulent component in the flow, bedforms will never form from a flat bed (Fig 3a). As has been observed in many morpho-dynamic processes, the formation of these bedforms is apparently owing to feedback between initially small perturbations on the bed and the flow field. Once the feedback is established, the orderly bedforms that we observe emerge and grow, and the feedback is reinforced. Thus the bedforms are self organized.

Similarly, without a feedback mechanism, random perturbations on the bed will not grow (Fig 3c). The examples in Fig 3 were run with the same conditions

as Fig 1 for 600 secs, except that in Figure 3a, $va = 0$, ie, there is no turbulent component and in Fig 3c the feedback mechanism was disabled. The results of the model with $va = 10$ cm/sec are shown in Fig 3b for comparison.

Most models for predicting the height and length of oscillatory ripples assume that there is an equilibrium condition that is reached and that ripples will stop growing when that condition is satisfied (Wiberg & Harris 1994, Nielsen 1982). Clarke & Werner (2004) measured megaripple wavelength using video images and found that megaripples grew continuously until they were destroyed, either in the extremely shallow water of the swash or under rapidly changing

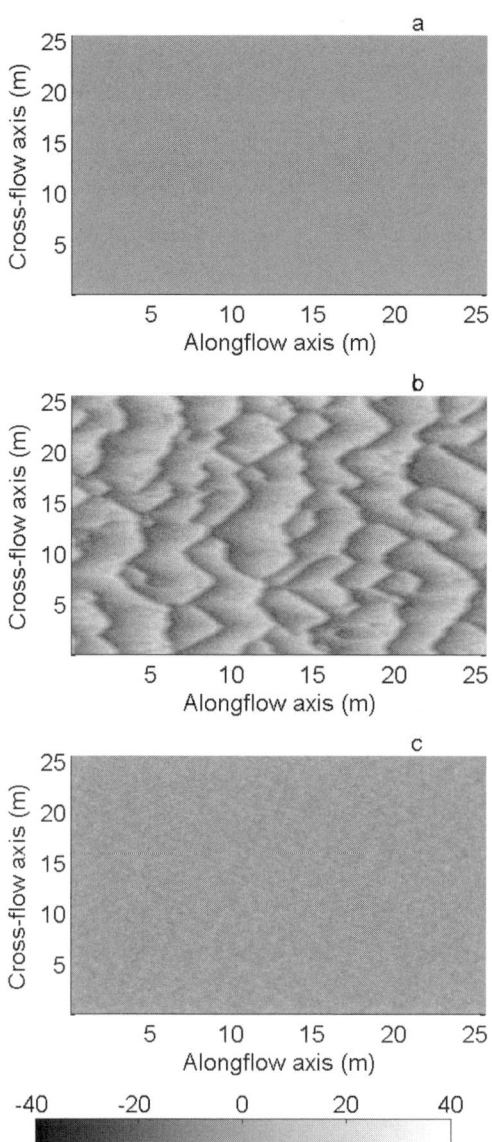

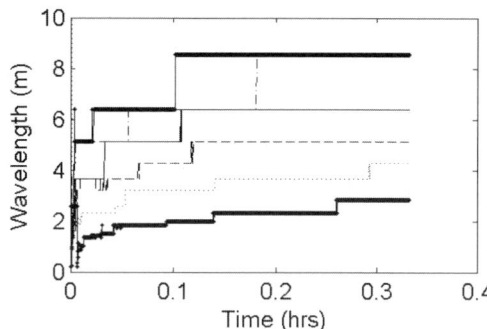

Figure 4. Growth of the predicted bed wavelength with time. The different curves were run with different oscillatory flow amplitudes, $A = 25$ cm/sec (bottom curve, solid with dots), $A = 50$ cm/sec (dotted), $A = 75$ cm/sec (dashed), $A = 100$ cm/sec (thin solid line), $A = 125$ cm/sec (dash-dot), $A = 150$ cm/sec (top curve, solid with dots) The turbulence amplitude for all runs was $va = \pm 25$ cm/sec, the steady flow magnitude was $S = 0$ cm/sec and the period, $T = 10$ sec. The 'simple rules' transport formulation was used for these model runs. Lengths were calculated using an $n = 256$ point FFT, and the wavenumber at the peak was chosen for the length presented here.

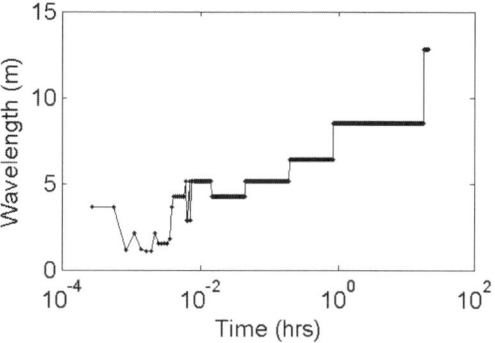

Figure 5. Growth of the predicted bed wavelength over 21 hrs. The oscillatory flow amplitude, $A = 100$ cm/sec, the steady flow amplitude, $S = 0$ cm/sec and the turbulence amplitude, $va = \pm 25$ cm/sec. The 'simple rules' transport formulation was used for this run. Lengths were calculated using an $n = 256$ point FFT, and the wavenumber at the peak was chosen for the length presented here.

Figure 3. Bed morphology resulting from the model run with the Ribberink (1998) transport model and a) with turbulence amplitude, $va = 0$ cm/sec, b) with $va = 10$ cm/sec, and c) with $va = 10$ cm/sec, but with the feedback disabled.

conditions. Here, megaripple length-scales are also observed to grow. The model was typically run for about 20 minutes (or 1200 seconds) and over that short time bedform lengths grow continuously (Fig 4). One very long run was done (21 hrs, comparable to Clarke & Werner's (2004) longer observations) and megaripple length was found to continue to grow (Fig 5). At

this time, only one such run has been completed owing to the large amount of computing time required. More runs like it will be done and presented at the RCEM meeting in Sept.

## 4 CONCLUSION

A hierarchical type morphology model has been developed to predict megaripples in the combined flows of the nearshore. Preliminary results suggest that the model is capable of predicting realistic bedforms. In addition, the dynamics of bedforms in the nearshore has been reproduced (albeit crudely). For example, feedback mechanisms do appear to be extremely important for bedform growth and development. Also, the continuous growth observed by Clarke & Werner (2004) is reproduced qualitatively.

There are many areas for continued work with the model. First of all many details are still being examined, tested and perfected. For example, the dependence of predicted bedform characteristics on the parameters in the model will be examined quantitatively. Similarly, the feedback mechanisms, the sediment transport formulations, the turbulence scales used, etc. will be tested for robust prediction skill. A primary goal of further modeling efforts is to add a flow-normal component or alongshore flow. It is not yet possible to truly simulate the complex flows in the surf zone. However, the results are exciting, suggesting that the model is capable of reproducing realistic observations with simple and sensible flow field inputs.

## REFERENCES

Ahl, V. & T.F.H. Allen 1996. *Hierarchy Theory*, Columbia Univ. Press, New York.

Clifton, H.E. 1976. Wave-formed sedimentary structures-A conceptual model. In Davis and Ethington ed: *Beach and Nearshore Sedimentation, SEPM Special Publication* 24, 126–148.

Clifton, H.E., R.E. Hunter & R.L. Phillips 1971. Depositional structures and processes in the non-barred high-energy nearshore. *Journal of Sedimentary Petrology* 41 (3): 651–670.

Clarke, L.B. & B.T. Werner 2003. Synoptic imaging of nearshore bathymetric patterns. *Journal of Geophysical Research* 108: 3005.

Clarke, L.B. & B.T. Werner 2004. Tidally modulated occurrences of megaripples in a saturated surf zone. *Journal of Geophysical Research* 109: C01012, doi:10.1029/2003JC001934.

Coco, G., D.A. Huntley & T.J. O'Hare 2000. Investigation of a self-organization model for beach cusp formation and development. *Journal of Geophysical Research* 105: 21991–22002.

Davidson-Arnott, R.G.D. & B. Greenwood 1976. Facies relationships on a barred coast, Kouchibouquac Bay, New Brunswick, Canada. *Beach and Nearshore Sedimentation, SEPM Special Pub.* 24: 149–169.

Gallagher, E.L. 2003. A note on megaripples in the surf zone: evidence for their relation to steady flow dunes, *Marine Geology* 193: 171–176.

Gallagher, E.L., S. Elgar, & E.B. Thornton 1998. Megaripple migration in a natural surf zone. *Nature* 394: 165–168.

Gallagher, E.L., E.B. Thornton & T.P. Stanton 2003. Sand bed roughness in the nearshore. *Journal of Geophysical Research* 108: 3039.

Garcez-Faria, A.F., E.B. Thornton, T.P. Stanton, C.M. Soares, & T.C. Lippmann 1998. Vertical profiles of longshore currents and related bed shear stress and bottom roughness. *Journal of Geophysical Research* 103: 3217–3232.

Hay, A.E. & D.J. Wilson 1994. Rotary side scan images of nearshore bedform evolution during a storm. *Marine Geology* 119: 57–65.

Kessler, M.A. & B.T. Werner 2003. Self-organization of sorted patterned ground. *Science* 299: 380–383.

Middleton, G.V. & J.B. Southard 1984. *Mechanics of Sediment Transport*. SEPM Short Course number 3, SEPM, 401 pp.

Nielsen, P. 1981. Dynamics and geometry of wave-generated ripples. *Journal of Geophysical Research* 86: 6467–6472.

Nishimori, H. & N. Ouchi 1993. Formation of ripple patterns and dunes by wind-blown sand. *Physical Review Letters* 71: 197–200.

Raudkivi, A.J. 1990. Loose Boundary Hydraulics, Pergamon Press, New York, 538 pp.

Stolum, H. 1996. River meandering as a self-organization process. *Science* 271: 1710–1713.

Thornton, E.B., J.L. Swayne & J.R. Dingler (1998) Small-scale morphology related to waves and currents across the surf zone. *Marine Geology* 145(3–4): 173–196.

Werner, B.T. 1995. Eolian dunes: computer simulations and attractor interpretation. *Geology* 23: 1107–1110.

Werner, B.T. 1999. Complexity in natural landform patterns. *Science* 284: 102–104.

Wiberg, P. L., & C. K. Harris 1994. Ripple geometry in wave-dominated environments, *Journal of Geophysical Research* 99(C1): 775–790.

*River dunes and antidunes*

*River, Coastal and Estuarine Morphodynamics: RCEM 2007 – Dohmen-Janssen & Hulscher (eds)*
*© 2008 Taylor & Francis Group, London, ISBN 978-0-415-45363-9*

# Froude number conditions associated with full development of 2D bedforms in flumes

G.V. Luong & M.A. Verbanck

*Department of Water Pollution Control (CP 208), Université Libre de Bruxelles (ULB), Brussels, Belgium*

ABSTRACT: Alluvial resistance and sediment transport rates are strongly dependent on bedform configuration phases. Among the various possible bedform shapes, we focus on two remarkable 2D configurations widely observed in flume studies, namely, fully-developed dunes (fdd) and stationary in-phase-waves (ipw). Combination of the characteristic bedform wavelength and the ambient stream depth and velocity at which these two flow configurations occur, allows to define a generalized Froude number $Fr_g$. The condition $Fr_g = 1$ constitutes the natural boundary between lower and upper alluvial regimes. When the material constituting the riverbed is getting coarser, progressively higher Froude number conditions are requested for the build-up of bedforms in their full developmental stage. Existence domains for fdd and ipw, respectively, are provided under the form of laws giving the necessary $Fr_g$ number conditions as a function of sediment Archimedes number. In line with what was initially suggested by Kondrat'ev (1962), the sensitivity to the value of Ar is especially marked in the case of the upper-alluvial regime bedforms. The improved knowledge on $Fr_g$ conditions associated with the formation of these canonical bedforms can be considered an asset for the future examination of the form-drag influence of transitional bedforms building in flumes. We think especially to those constituting, for a given Ar, the progressive transition between fdd and ipw (as studied experimentally, e.g., by Wang & White 1993).

## 1 INTRODUCTION

Establishing the link between the various bedform configuration phases and the resistance they oppose to flow remains an important issue in river engineering practice (Vanoni & Brooks 1957, Yalin & Ferreira da Silva 2001). When streams flow over erodible boundaries, large-scale periodic features, characterized by their wavelength $\lambda_{BF}$ along the longitudinal axis, are known to deform the bed surface, considerably affecting the effective boundary-roughness. Knowledge about equilibrium dune beds (Fig 1a) constitutes an essential prerequisite of the analysis. An ASCE Task Force recently addressed the issue by stating: *'For sediment transport engineering, a minimum contribution desired of a theory -model, understanding- for bedform development would be a reliable means of determining which equilibrium would be established, i.e., delineating stability boundaries.'* With regard to the non-equilibrium dune case (so often encountered in the field), specific research needs were highlighted by the Task Committee: *'How and at what rates bedforms change for increasing and decreasing flows remains to be quantified. The problem of transitions to a dune bed from a rippled or plane bed and from a dune bed to an upper-regime plane bed or antidune bed is also*

*of much practical interest. Has recent work shed any light on this important aspect of nonequilibrium beds?'* (ASCE Task Committee, 2002). This contribution is the preliminary step of a research effort trying to address some aspects of these two questions, focusing especially on an engineering correlation delineating stability boundaries. A methodology developed at University Brussels is also presented which, in combination with the newly developed correlations, would allow to account for the bedform transitional stages. We think especially to the progressive transition between lower-regime dunes and upper-regime bedforms, as studied experimentally, e.g., by Wang & White (1993).

## 2 WATER-SURFACE / RIVERBED INTERACTION

### 2.1 *Generalized Froude number $Fr_g$*

The present work is inspired by an early analysis conducted by Kondrat'ev et al (1962). Its detailed aim is to look at 'fully-developed' 2D bedforms as occurring in laboratory flumes (Fig 1) and to try to predict their existence domain as a function primarily of the sand size characteristics.

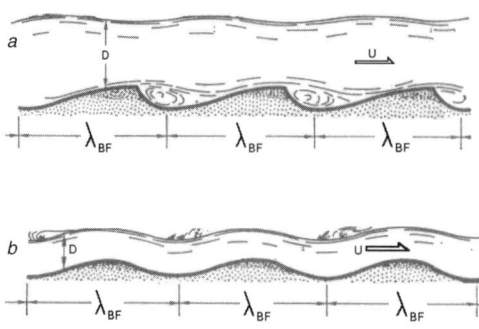

Figure 1. The 2D, mature bedforms considered in this study are the full developmental stages which can be observed in laboratory flumes. a) fully developed dunes (fdd), b) in-phase waves (ipw) *(redrawn after Yalin 1977)*.

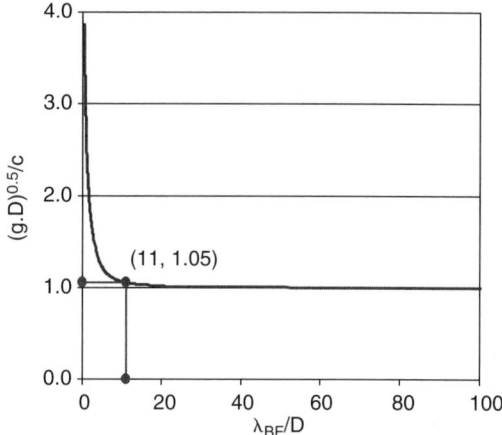

Figure 2. In the considered $\lambda_{BF}$/D ratios (typically 6 and lower), caution should be exercized regarding the gravity-wave handling. It is only for values higher than 11 that the error associated to the 'long-wave' approximation is lower than 0.05 (Stewart 2006).

In this avenue we shall evidently rely in large extent upon the Froude number concept, which is defined in the most general terms:

$$Fr^2 \equiv \frac{(stationary)\ inertial\ forces}{gravity\ forces} \qquad (1)$$

An effect of the riverbed deformation is the topographical forcing it imposes on streamlines flowing over it. The topographical forcing, which repeats itself from $\lambda_{BF}$ to $\lambda_{BF}$, generates a gravity wave tending to reconstitute a planar, free surface. At the considered disturbance scale, surface tension effects can be neglected. The speed of propagation of the gravity wave is thus simply predicted as (Airy's law):

$$c = \sqrt{\frac{g\lambda_{BF}}{2\pi} \tanh \frac{2\pi D}{\lambda_{BF}}} \qquad (2)$$

with g, acceleration of gravity and D, water depth averaged along the longitudinal axis (see Fig 1). For our purpose the relevant Froude number is thus the ratio of mean stream velocity U to celerity c. It is what we call the generalized Froude number $Fr_g$:

$$Fr_g = \frac{U}{\sqrt{\frac{g\lambda_{BF}}{2\pi} \tanh \frac{2\pi D}{\lambda_{BF}}}} \qquad (3)$$

Potential flow theory has been used extensively to treat analytically the question of the influence of the periodical deformation of the riverbed upon the amplitudes found at the water surface (Milne-Thompson 1960, Kennedy 1963). It corresponds to a linear treatment of the problem of the wave interaction, applying a 2D approach (only longitudinal axis $x$ and bed-normal axis $y$ coordinates are considered) as adopted here. Milne-Thompson (1960) was apparently the first to examine the conditions under which the profile of the surface

wave can remain in phase with the deformation of the sinuous bottom. Potential flow theory predicts that the two profiles will remain in phase only under supercritical conditions ($Fr_g > 1$). This is typically the case of the so-called 'in-phase waves' (Cheel 1996) depicted in Fig 1b. In contrast, asymmetric dune profiles, so commonly observed in natural sandbed rivers, are characteristic of subcritical conditions ($Fr_g < 1$). The flume experiments by Simons & Richardson (1966) conclusively showed that the water surface in this case (Fig. 1a) was in phase opposition with the deformation of the alluvial bed. From this we infer that the simple, linear analysis proposed by Milne-Thompson constitutes a reasonable approach to the two complex flow geometry configurations illustrated in Fig. 1. $Fr_g$, however, cannot be simplified into traditional Froude number ($U/\sqrt{gD}$) under all flow configurations, as will be shown below.

## 2.2 Relevance of gravity-wave approximations

As is well known in wave analysis, characteristic values of the ratio $\lambda_{BF}$/ D determine whether or not it is necessary to use Eq. 2 under its non-simplified form. Considering an error of 5% as acceptable in the prediction of the celerity of the gravity wave, Figure 2 recalls that we need a value $\lambda_{BF}$/ D of at least 11 to apply the long-wave approximation ($c \cong \sqrt{gD}$). The geometry of bedforms developing in laboratory flumes, generally, does not obey the rule (Raudkivi 1990): bedform wavelengths are in the order of $2\pi$ times water depth D (and, sometimes, even much shorter, such as in the case of ripples). It would therefore be strongly recommended to use Eq. 3 in its non-simplified form,

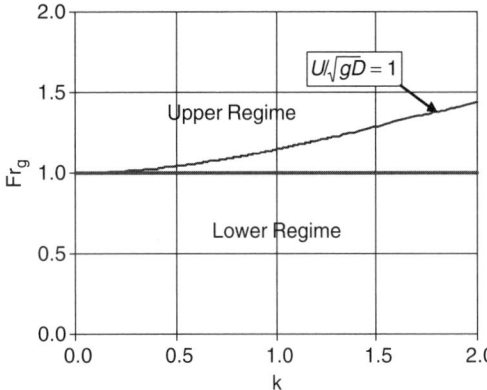

Figure 3. Consequence of the potential-flow theory in the domains separation between lower- and upper-alluvial regimes: generalized Froude number *vs.* wavenumber k.

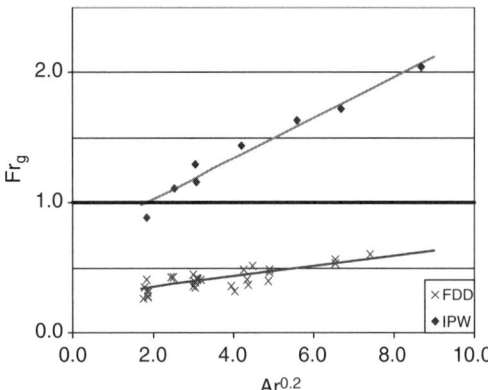

Figure 4. Generalized Froude number *vs.* Archimedes number: prediction of the typical existence domains for fdd and ipw (various flume data sources, see text).

to characterize Froude number conditions in a flume sediment-laden flow in equilibrium with its alluvial bed. Use of Eq. 3, however, supposes that $\lambda_{BF}$ can be made available as a measurable quantity with acceptable accuracy, two conditions which are obviously not easy to fulfill, even in well-designed experimental set-ups (Jain & Kennedy 1974).

# 3 LOWER- AND UPPER-ALLUVIAL REGIMES IN THE LIGHT OF FR$_G$

## 3.1 Critical flow condition and domain delineation

Figure 3, which relies on bedform wavenumber $k \equiv 2\pi D/\lambda_{BF}$, further strengthens the argument for the use of generalized Froude number Fr$_g$ in alluvial hydraulics. It capitalizes on the outcome of potential flow theory to infer what constitutes the true, natural boundary between the lower- and the upper-alluvial regimes. As highlighted by Kennedy (1963) and Parker (2004) this boundary is simply the condition Fr$_g = 1$. The interest of plotting all experimental results in a Fr$_g$-ordinate diagram, is obviously that a simple horizontal line drawn at Fr$_g = 1$, allows to delineate the two alluvial regimes consistently. As shown by Fig 3, the use of the ratio $U/\sqrt{gD}$ as Froude number could have the consequence of inappropriately designating some ipw as being subcritical features (which obviously they are not, in view of the phase considerations above). A large part of the endless arguments on the subject (see, e.g., Grant 1997 and discussion thereof by Chanson 1999, Carling & Shvidchenko 2002) could have been easily resolved by acknowledging the conceptual role an unambiguous descriptor such as Fr$_g$ could play in the analysis of sediment-laden streams in equilibrium with their alluvial bed.

For this reason, the horizontal line Fr$_g = 1$ will also constitute the convention for plotting the diagram of our own results (see Fig 4).

## 3.2 Attached-detached flow model

To account for flow resistance modulation as a function of the alluvial regime, Manning law or Chezy-derived formulations are widely used (vanRijn 1984). The attached-detached flow model is an alternative 'vortex-drag' approach recently proposed (Verbanck 2004a) to predict the value of the energy slope in alluvial streams flowing over non-cohesive particle beds. The process of vortex generation and maintenance in the lee side of dunes is described using the Strouhal frequency concept (with $m/2\pi$ as characteristic Strouhal number). (Gyr & Hoyer 2006). According to Verbanck (2004a), the resistance coefficient, in the equilibrium 2D flow configurations of Fig 1, can be predicted as

$$m = 2\pi \, Fr_g^{-1} \, S^{0.3} \tag{4}$$

Based on Fr$_g$ and its explicitness in $\lambda_{BF}$, this formulation introduces in alluvial hydraulics the concept of a Rossiter fundamental mode (m=1) which would be particularly effective in transporting both water and particle fluxes. The fundamental mode (when m in Eq. 4 is unity) would correspond to the (non breaking) in-phase waves illustrated in Fig. 1b. In their most mature stages, dune bedforms would correspond exactly to the second, higher harmonics (m = 2). These harmonics are perceived to be similar to those met in acoustics of structures (Rossiter 1962, Zima and Ackermann 2002 Gyr & Hoyer 2006). It is speculated that the attached-detached flow model is more consistent with observations in sandbed alluvial systems and actually more robust than the classical approach for these systems. Following this approach, the flow resistance formula in the alluvial configurations of Fig. 1 can be rewritten as:

$$U = \frac{2\pi}{m} S^{\frac{3}{10}} \sqrt{\frac{g\lambda_{BF}}{2\pi} \tanh \frac{2\pi D}{\lambda_{BF}}} \qquad (5)$$

According to the attached-detached flow model, in a 2D bedforms configuration the minimum possible value for control factor m is one. This, by application of Eq. 5, could explain the striking differences observed in velocity U and mean water depth D between in-phase waves and fully-developed dunes, as represented schematically in Fig. 1 (Yalin 1977). Relying on the control factor m concept, fdd bedforms would be characterized by stream velocities amounting to one half of those of their ipw counterparts (the magnitude of all other intervening factors remaining equal). Water depths are thus significantly higher for equilibrium dunes than for ipws. For a fixed water discharge to be evacuated, fdds are therefore much more prone to lead to flooding risks than are ipws (Verbanck 2004b).

## 4 DATA SOURCES

Great care has been exercized in the selection of the appropriate evidence allowing experimentally-supported analysis of the problem at stake. The focus was put on sediment-laden flows in equilibrium with their alluvial beds, as observed in laboratory flumes using uniform non-cohesive particles. The typical sediment grain size used by the experimenters was found to lie in the range 0.07–1.40 mm. Bedform wavelengths and the bedform conditions are reported in the flume studies conducted by Guy *et al* (1966), Kennedy (1961, 1963), Nordin (1976), Onishi *et al* (1972), Pratt (1970), Williams (1970) and Wang & White (1993).

The detailed comments in the description of individual runs provided in the research reports by Guy *et al* (1966) and Kennedy (1961) were found especially useful. For a given particle size, when $Fr_g$ is progressively increased for a dune bed, Guy's observations suggest that there is a maximum $Fr_g$ value beyond which the dunes are washed-out or, at least, considerably losing in geometrical steepness. The bedform condition corresponding to this maximum $Fr_g$ value (just before washing-out) is called fully-developed dune (fdd). Our understanding is that this limit bedform condition, in essence, could be considered equivalent to what is designated as 'equilibrium dunes' by the ASCE Task committee (2002) or, alternatively, as 'steady-state dunes' in the recent analysis by Coleman *et al* (2006).

The bedform phases retained here as characteristic of the upper-alluvial regime (see Fig. 1b) are the stationary, non surface-breaking, in-phase waves (ipw). In the bedform nomenclature introduced in Brownlie's (1981) sediment-transport compendium, this ipw condition receives the code BF6, as opposed to, firstly,

Table 1. Summary of flume conditions and data range considered in this study.

| Parameters (Unit) | Max–Min |
| --- | --- |
| Flume width, W (m) | 0.267–2.438 |
| Water depth, D (m) | 0.030–0.738 |
| Energy slope, S (-) | 0.00027–0.0222 |
| Discharge, Q (m³/s) | 0.008–1.352 |
| Water temperature, T (°C) | 14.0–30.5 |
| Sediment grain size, d (mm) | 0.100–1.350 |
| Sediment density, $\rho_s$ (Kg/m³) | 2640–2650 |
| Bedform wavelength, $\lambda_{BF}$ (m) | 0.180–3.410 |
| Generalized Froude number, $Fr_g$ (-) | 0.267–2.039 |

the upper-regime plane bed just before it (code BF5) and to, secondly, wave-breaking antidunes just after it (which receive bedform code BF7). The nuances introduced in this rich nomenclature (Simons & Richardson 1966) are actually supported by the interpretation analysis conducted here. In some cases, runs with upstream-migrating or downstream-migrating surface waves were retained in the analysis when insufficient material was available regarding 'genuine' ipw conditions, i.e. in-phase waves truly stationary along the longitudinal axis (Kennedy 1961).

To complete the analysis in the finest sediment-grains possible, the $100\,\mu m$ dataset by Willis *et al*. (1972) was also selected, although this study did not provide any bedform wavelength information. In that case we had to rely on the long-wave approximation ($c \cong \sqrt{gD}$), an important aspect which needs to be kept in mind when interpreting final results. A summary of experimental data ranges used for the build-up of Fig. 4 is provided in Table 1.

## 5 INFLUENCE OF ARCHIMEDES NUMBER

In this study several dimensionless parameters were considered to try to connect existence domains of ipw and fdd with their $Fr_g$ characteristics. The results indicated that the Archimedes number appears to be a good candidate. As is well-known in particle-fluid interactions (Coulson & Richardson 1955) the Archimedes number, standing for the ratio between (buoyancy-affected) gravity forces and viscous forces, is defined as:

$$Ar = \frac{g\Delta d^3}{v^2} \qquad (6)$$

with $\Delta$ relative excess density, d sediment size and $v$ kinematic viscosity. Ar is evidently directly connected to the non-dimensional sediment diameter $D_*$ widely used in sedimentation engineering.

The filter criteria for handling the experimental data are set as the ratio between the width of the flume and the water depth (W/D) being necessarily larger than 5, the control factor m = 1 and 2 (in phase waves and fully developed dunes bedforms, respectively). The correlation $Fr_g = f(Ar^{0.2})$ is successfully achieved as in Fig. 4. There is at this stage no conceptual reasoning in the selection of power $^1/_5$ in the Ar determinant: it has evidently an empirical base only.

As derived from the examination of Fig. 4 the following comments can be drawn:

– These results confirm that the border between the lower and upper regimes is well distinguished by generalized Froude number at the zone of $Fr_g = 1$.
– The in-phase waves which are formed in the upper regime tend to follow the rule:

$$Frg = 0.157Ar^{0.2} + 0.7109 \text{ with } R^2 = 0.97 \quad (7)$$

– The fully developed dunes which are formed in the lower regime tend to follow the rule:

$$Fr_g = 0.0446Ar^{0.2} + 0.2515 \text{ with } R^2 = 0.64 \quad (8)$$

The only ipw instance found in the subcritical region (Fr = 0.88) corresponds to a 100 μm run performed by Willis et al (1972). The error caused by the approximation $c \cong \sqrt{gD}$, adopted in this case, is not believed to be sufficiently high to explain this underestimate in Froude number. The consequences of assuming $c \cong \sqrt{gD}$ are indeed much more severe in the lower-alluvial regime than in the ipw case. Examination of the original data report by Joe Willis suggests that this point (at Fr = 0.88) corresponds to a wave-breaking condition (code BF7) rather than a genuine ipw (code BF6). This would have corresponded to an increased water depth comparatively to the ideal ipw configuration, in which headloss is minimized according to Equ 5 and control factor m reaching its minimal value (m = 1).

Finally, it is apparent from Fig. 4 that another trendline could have drawn to represent the ipw domain in the $Fr_g$-Ar plane. While waiting for supplementary evidence, the linear correlation proposed here is considered acceptable as a first-order analysis of the experimental evidence. In any case, as such it supports, in general terms, the concepts presented, respectively, by Kondrat'ev (1962) and Verbanck (2004a). In line with what was initially suggested by Kondrat'ev, the sensitivity to the value of Ar is especially marked in the case of the upper-alluvial regime bedforms. The improved knowledge on $Fr_g$ conditions associated with the formation of these canonical bedforms can be considered an asset for the future examination of the form-drag influence of transitional bedforms building in flumes.

## 6 CONCLUSION

The maximum limit value is shown to be strongly dependent of the Archimedes number of the sediment grain particles forming the alluvial bed. Typical occurrence conditions for fully developed dunes (fdd) and in-phase waves (ipw), respectively, appear to follow the general trendlines depicted in Fig. 4. As expected, higher Froude number conditions are requested to maintain 2D bedforms when the deposited bed material is coarser. Although these bedform prediction rules are found to stand reasonably for flume cases with single-size grain particles, they obviously cannot be exported, with the same set of regression coefficients, to actual streams as observed in the field. There is therefore merit in further research developments in this regard.

## ACKNOWLEDGEMENTS

The study contributes to the AquaTerra Project 'Integrated Modelling of the River-Sediment-Soil-Groundwater System' funded by the European 6th Framework Programme, research priority 1.1.6.3 Global Change and Ecosystems (European Commission, Contract No 505428-GOCE). It is part of Flux3 'Input/Output Mass Balances in River Basin: Dissolved and Solid Matter Load', a sub-component of the AquaTerra Integrated Project.

## NOTATION

Ar: Archimedes number (−)
c: celerity of the gravity wave (m/s)
D: total water depth (m)
d : sediment grain diameter (m)
fdd: fully developed dunes
Fr: Froude number (−)
$Fr_g$: generalized Froude number (−)
ipw: stationary in-phase waves
k: wave number (−)
g: gravitational acceleration (m/s$^2$)
m: control factor m (−)
R: hydraulic radius (m)
S: energy slope (−)
T: water temperture (C degree)
W: width of flume (m)
$\Delta = (\rho_s/\rho - 1)$ relative excess density
$\rho$ : density of the water (Kg/m$^3$)
$\rho_s$, : density of the sediment (Kg/m$^3$)
$v$: kinematic viscosity of the water at T°C (m$^2$/s)
$\lambda_{BF}$: bed form wave length (m)

## REFERENCES

ASCE Task Force on Flow and Transport over Dunes, 2002, Flow and Transport over Dunes, Forum paper, *Journal of Hydraulic Engineering*, Volume 128, No. 8, pp 726–728.
Brownlie W.R. (1981) *Prediction of flow depth and sediment discharge in open channels*, Report KH-R-43B, W.M. Keck Laboratory, Caltech (USA).

Carling P.A., Shvidchenko A.B. A consideration of the dune:antidune transition in fine gravel. *Sedimentology* 49 (6), 1269–1282.

Chanson H. 1999 Boundary shear stress measurements below free-surface standing waves – application to bed form processes. *Proc IAHR Graz Congress*, 8p.

Cheel R.J. & Udri A. (1996) *The behavior, internal stratification and fabric of in-phase waves.* Geological Society of America, Northeastern Section Annual Meeting, Buffalo, New York, Abstracts volume.

Coleman SE, Nikora VI, McLean SR, Clunie TM, Schlicke T, Melville BW (2006) Equilibrium hydrodynamics concept for developing dunes. *Physics of Fluids* 18 (10): Art. No. 105104.

Coulson JM & Richardson JF (1955) *Particle technology and separation processes, Chemical Engineering Vol 2.* Butterworth-Heinemann, Oxford (UK) 1st Edition.

Grant, G.E. (1997) Critical Flow Constrains Flow Hydraulics in Mobile-Bed Streams: a New Hypothesis. *Water Resources Res.,* Vol. 33, No. 2, pp. 349–358.

Guy, H.P., Simons, D.B., and Richardson, E.V., 1966, Summary of alluvial channel data from flume experiments, 1956–1961, *U.S. Geological Survey Professional Paper* 462-J

Gyr A., Hoyer K. 2006. *Sediment transport : a geophysical phenomenon.* Springer Dordrecht, NL, 285p.

Jain, S.C. and Kennedy, J.F., 1974, The spectral evolution of sedimentary bed forms, *Journal of Fluid Mechanics,* Vol. 63, Part 2, pp 301–314

Kennedy, J.F. (1961) *Stationary Waves and Antidunes in Alluvial Channels*, Report KH-R-2, W.M. Keck Laboratory of Hydraulics and Water Resources, California Institute of Technology, Pasadena, California.

Kennedy, J.F., 1963, The mechanics of dunes and antidunes in erodible-bed channels, *Journal of Fluid Mechanics,* Vol. 16, No. 4, pp 521–544

Kondrat'ev, N.E., Lyapin, A.N., Popov, I. V., Pinikovskii, S.I., Fedorov, N.N., Yakunin, I.I., 1962, *River flow and river channel formation*, from Russian (1959) by Israel Program for Scientific Translations, U.S. Department of the Interior

Milne-Thompson L. (1960) *Theoretical hydrodynamics.* 4th Ed, Macmillan, New York.

Parker G. (2004) *1D Sediment transport morphodynamics with application to River and Turbidity currents –* Lecture notes, e-Book, Ven Te Chow Hydrosystems Lab, Univ Illinois, November 2004.

Raudkivi, A.J., 1990, *Loose boundary hydraulics*, 3rd Edition, Pergamon Press, Oxford, England, 537 p.

Rossiter, J. E. (1962) – *The effect of cavities on the buffeting of aircraft –* Technical Memorandum 754, Royal Aircraft Establishment (UK).

Simons, D.B., and Richardson, E.V., 1966, Resistance to flow in alluvial channels, *U.S. Geological Survey Professional Paper* 422-J, 61 p.

Stewart R.H. (2006) *Introduction to Physical Oceanography.* Texas A & M University, Department of Oceanography, Electronic textbook, Fall 2006 Edition.

Van Rijn, L.C., 1984, Sediment transport, Part III: bed forms and alluvial roughness, *Journal of Hydraulic Engineering,* Volume 110, Number 12, pp 1733–1754.

Vanoni, V.A. and Brooks, N.H., 1957, *Laboratory studies of the roughness and suspended load of alluvial streams,* Sedimentation Laboratory Report No. E68, California Institute of Technology, Pasadena, California.

Verbanck M.A, 2004a. Sand transport at high stream power: towards a new generation of 1D river models? – *Proceeding of the Ninth International Symposium on River Sedimentation,* Yichang, China, 307–318.

Verbanck M.A. 2004b. Sediment-laden flows over fully-developed bedforms: first and second harmonics in a shallow, pseudo-2D turbulence environment – *Shallow Flows,* G. H. Jirka, and W. S. J. Uijttewaal, eds., Balkema, 231–236.

Yalin, M.S., 1977, *Mechanics of Sediment Transport*, 2nd Edition, Pergamon Press, Oxford, 298 p.

Yalin, M.S. and da Silva, A.M.F., 2001, *Fluvial Processes,* International Association of Hydraulic Engineering and Research Monograph, Delft, The Netherlands, 197 p.

Wang Shiqiang & White RW 1993. Alluvial Resistance in Transition Regime. *ASCE Journal of Hydraulic Engineering,* Vol. 119, No. 6, pp. 725–741.

Zima, L., & Ackermann, N. L., 2002. Wave generation in open channels by vortex shedding from channel obstructions – ASCE *Journal of Hydraulic Engineering,* 128(6), 596–603.

*River, Coastal and Estuarine Morphodynamics: RCEM 2007 – Dohmen-Janssen & Hulscher (eds)*
*© 2008 Taylor & Francis Group, London, ISBN 978-0-415-45363-9*

# Gravel dunes and antidunes in fluvial systems

P.A. Carling & R.M.D. Breakspear

*School of Geography, University of Southampton, Southampton, UK*

ABSTRACT: Dunes and antidunes may occur in water-lain gravels for given hydraulic conditions. However there is little published experimental data to assist in determining the hydraulic conditions under which these classes of bedforms might develop. In addition there are few published examples of putative gravel antidunes. A detailed review of the literature has been undertaken and unpublished data sets and illustrations have been accessed. These data together with insights obtained from flume studies undertaken by the authors provide new and additional insight into gravel bedform hydrodynamics.

## 1 INTRODUCTION

### 1.1 *Gravel transport as bedforms by wind or water*

Most gravel bedforms are associated with water flows although aeolian examples cannot be precluded. Small wind-blown ripples composed of pebbles or gravel-armoured sand ripples in deserts are common, but the existence, origin and dynamics of larger dune-scale wind blown gravel bedforms is less well attested. For example, gravel bedforms on the Bennett Platform, Shackleton Glacier area, Antarctica have a wavelength of approximately 10 to 15 meters, a wave height of 10 to 15 centimeters, and consist of gravels that range from 4 to >100 mm in maximum dimension. After allowance for grain impacts for initiating sediment transport as well as the direct frictional drag of the wind Henderson et al., (2002) calculated that clasts up to 53 mm could be moved by wind speeds of up to 60 m/s and thus concluded the bedforms were aeolian. The origin of other large bedforms in Antarctica remains enigmatic, being possible fluvial or glaciogenic (Malin, 1986), as is the origin and dynamics of large-scale gravel bedforms in the Sahara (Breed et al., 1987). Many small-scale aeolian granule ripples are armoured and it is becoming evident that armouring may play a significant role in fluvial gravel dune dynamics, such that much may be learnt from reflexive study of both aeolian and fluvial bedform morphology, sedimentology and dynamics. The remainder of this review considers only water-lain deposits.

## 2 GRAVEL DUNES

### 2.1 *Quaternary gravel dunes*

On Earth, large-scale dunes related to megafloods are probably more common than the literature would suggest but remain to be identified. Many of these floods occurred during the Quaternary period. Pardee first reported gravel ripples on Camus Prairie in the Missoula floodway in a verbal report to the American Association for the Advancement of Science in Seattle, Washington on the Tuesday afternoon, 18th June 1940. "Ripple Marks (?) in Glacial Lake Missoula" on the Camus Prairie, MT, 60 miles NW of Missoula. Other key papers on the Missoula dunes are Bretz et al, 1956, Baker, 1973, Baker and Nummendal, 1978, and Baker, (1995). Carling (1996a & b) and Carling et al. (2002) describe similar large-scale gravel dunes from Siberia. All of these dune fields are attributed to Quaternary outburst floods from former ice-dammed lakes.

Carling (1999) reviewed the literature on fossil gravel dunes. Since that time a few additional studies have been completed. These include reports of Quaternary gravel dunes by Schoeneich & Maisch (2003a & b) who describe and illustrate 40 m wavelength, 0.5 m high gravel dunes which they relate to a late glacial outbreak flood near Davos, Switzerland and a compilation of detailed unpublished data on the Camus Prairie dunes by Prof. Keenan Lee (see also http://www.mines.edu/academic/geology/faculty/klee/CamasPrairie.pdf) which dunes have also received attention from Alt (2001) and Lister (1981).

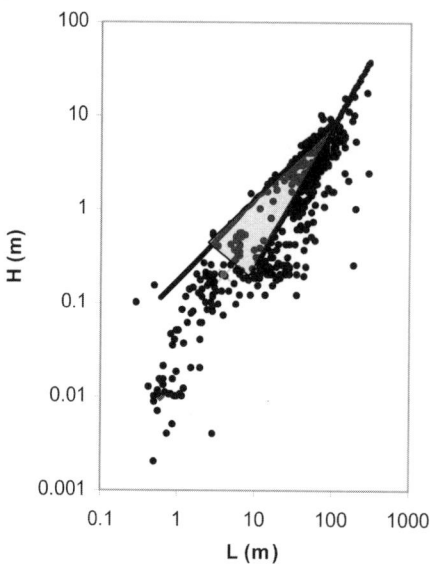

Figure 1. Height (H) and length (L) relationship for compilation of all gravel dune data. Upper limit line is curve of Ashley (1990) and the lower steeper curve is that of Carling (1996b). Data to the right of Carling's curve largely represents 2D transvers dunes whilst data in the shaded region between the curves largely represents more 3D morphologies.

Figure 2. Panton River gravel-armoured sand dunes: wavelenghs 10 to 20 m and heights <1 m. Flow top right to left. Image courtesy of Dr. P. Sandercock.

These additional data are compatible with the analyses of dune height and length data presented by Carling (1999; his fig. 6) and are included in Figure 1.

In some cases the identification as fossil gravel fluvial dunes is not verified by detailed study, or is disputed. For example, Andrea Pacifici of the International Research School of Planetary Science (IRSPS), Italy (pers com, 2006) has reported probable large-scale "fossil" fluvial dunefields in the floodplains of the Argentinian River, Patagonia, visible in satellite images but requiring ground verification. Disputed identifications in North America are considered by

Figure 3. View along the crestline of putative transverse rib at Modrudalar, Iceland. Flow right to left. Image courtesy of Dr. J. Rice.

(Munro-Stasiuk and Shaw, 1997 and Evans et al., in press).

## 2.2 "Modern" gravel dunes

Carling (1999) reviewed the literature on modern (i.e. recently active) gravel dunes and Carling et al. (2005b) used an hydraulic flume to investigate the initiation of gravel bedforms from a plane bed.

A few additional studies have been completed which, in particular, have drawn attention to the process of surface armouring by gravel on sandy dunes. Sandercock (2003) described sandy dunes armoured by gravel in the Panton River of Western Australia (Fig. 2). Carling et al. (2005a) and Williams et al. (2006 & 2007) investigated the influence of coarse-fraction armouring on initial motion of gravel dunes in an intertidal zone. Radecki-Pawlik et al. (2006) have studied small sandy-gravel dunes in the Raba River, Poland in which armouring may play a role.

## 3 GRAVEL ANTIDUNES

### 3.1 Quaternary gravel antidunes

Rice et al. (2002; see also Chapman et al. 2003, their fig. 15b) describe sub-parallel transverse bedforms with average wavelengths of 27 m and average heights of 0.84 m from Icelandic jökulhlaups as large-scale transverse ribs and use the antidune analogy to calculate palaeoflood data. Although this is a reasonable identification there are no reports of modern transverse ribs of similar scale to use as analogues and, in addition, these particular examples may be erosional remnants rather than accumulative features (Carling et al., in press).

Figure 4. Putative gravel antidunes on Camus Prarie, MT, USA. Flow left to right. Largest ridgeline spacings are c. 90 m. Image courtesy of Prof. Keenan Lee.

Figure 5. Head cuts in the Mzimvubu River (S31° 31'10" E029° 27'25") about 30 km upstream of Port St. Johns on the east coast of South Africa, a typical locality where 3D antidunes may develop. Image courtesy of Dr. J. Reddering.

Figure 6. Putative 3D gravel antidunes in the Nahoon River, South Africa (Reddering & Eserhuysen, 1987). Largest mounds are less than 0.5 m in diameter. Flow left to right. Image courtesy of Dr. J. Reddering.

According to Alt (2001), Pardee (1942) includes a section through bedforms at Camus Prarie that show sedimentary structures consistent with an antidune interpretation. At a near-by site, Lister (1976) recorded stratigraphic relationships consistent with an antidune interpretation within bedforms for which the steepest slope was facing upstream. These bedforms are on the down slope side (Fig. 4) of two passes: Markle Pass and Wills Creek Pass whereby Missoula flood water overtopped a ridge between Little Bitterroot valley to the north and Camus Prairie to the south. Sedimentary ridges transverse to the flow are up to 10.7 m high with wave lengths of up to 91 mm and contain coarse gravel up to c. 30 cm in size. In such a location, the bedforms would readily be preserved as the flood flow would stop abruptly when the water level fell below the level of the passes. On the slope below the passes the bedforms are asymmetric with steep slopes facing up flow and the sediment is very coarse (up to shoe box size), then follows a smooth bed area which Alt believes is the transitional USPB zone between the antidunes and an extensive area downslope and on the flatter prairie of lower aspect finer gravel dunes that are asymmetric with steeper slopes facing down flow.

Spectacular putative antidunes occur in the Chuja basin of southern Siberia and have been associated with the Altai megafloods (Carling et al., 2002) but these appear to be erosional features with only a veneer of deposited sediment (see Carling et al., in press). A small ($\sim$1 km$^2$) field of antidunes has been described in British Columbia, Canada (Johnsen and Brennand, 2004). These antidunes have wavelengths of 100–230 m, and heights of 3–7 m. They are 2-D bedforms, having fairly straight troughs and crestlines. Their streamwise long profiles are asymmetrical, with steeper upflow (stoss) slopes. They were created during the catastrophic drainage of a narrow, valley-filling, ice-dammed glacial lake approximately 12 kyr BP.

### 3.2 Modern gravel antidunes

Standing waves often form in both large and small rivers and antidunes may develop beneath these flow structures. Often they develop in steep channels such as at the head of alluvial fans (Zielinski, 1982), in steep, flood prone rivers (Figs. 5 & 6) and upstream and downstream of major obstacles to flow such as islands (Fig. 7).

The presence of antidunes usually indicates rapid recession of floodflow (Alexander and Fielding, 1997). For example, Reddering and Eserhuysen (1987) provide a brief description of a flood on the Nahoon River in South Africa which produced gravel antidunes (Fig. 6) but the paper contains no information on the antidunes. However the hydrograph was very abrupt lasting only a few hours and so rapid draw-down might explain preservation. The morphology of the Nahoon

Figure 7. Putative 3D antidunes formed in pebbles on the upstream side of a mid-channel island where standing waves have been observed in the Mekong River, near Luang Prabang, Lao PDR. Flow top to bottom. Note book is 25 cm × 18 cm.

antidunes are similar to those reported by Shaw and Kellerhals (1977) and additional examples from the Mekong River in Laos are shown in Figure 7.

## 4 DISCUSSION

The identification of bedforms as dunes is significant as dunes can only form in sub-critical flow conditions which limits the range of potential flow velocities and Froude numbers. A consistent sense of bedform asymmetry for several dunes within a dune train may indicate reliably the palaeoflow direction whilst the width of the dune field must indicate a minimum width of the palaeoflow field. Flow depths have to exceed the height of the dunes and average flood flow depths often scale with dune heights and/or dune wavelengths. Thus important palaeohydraulic data may be deduced from the geometry and sedimentology of palaeodunes which can be of great assistance in estimating flood discharges (e.g. Carling, 1996b).

The formation process involves both erosion and deposition but, within loose granular beds, usually only the morphology owing to accumulation of sediment is evident. The bulk hydraulics of standing waves are well known but the conditions for the preservation of antidunes is less-well understood (Carling and Schvidchenko, 2002) On falling water stages antidunes typically are erased. Antidunes usually occur as trains of self-similar ridges and intervening troughs that are roughly transverse to the main direction of flow. However, for supercritical flows the standing waves can become increasingly three-dimensional with wavy crest-lines or, for higher flow conditions, more isolated steep waves occur known as rooster-tails. Increasing three-dimensionality of the

waterwaves often produces a rhomboid water surface profile, a condition that is most prevalent in those situations where the channel margins are close-by and the rhomboid flow field is generated by waves reflected from the channel margins. The shape of the antidunes beneath these various water waves is of similar form with steep isolated mounds of sediment occurring beneath rooster-tails. The identification of large-scale antidunes is important in a number of regards. Firstly, antidunes develop in transitional and supercritical flows when the Froude number is greater than about 0.84 and may be greater than unity. Thus identification would preclude the occurrence of subcritical flows which are characterised by fairly even water surface levels during the formative phase of the bedforms; rather surface water instabilities are necessarily present for supercritical flows. Thus identification of palaeo-antidunes might indicate the presence of an irregular palaeowater surface slope. Secondly, the morphology of the antidunes, i.e. regular-transverse, wavy-transverse, or isolated mounds indicates progressively higher Froude numbers respectively. The spacing of transverse antidunes scales with the Froude number in a manner which permits an estimation of water depth or velocity; this is a powerful tool for palaeoflow reconstruction. Allen (1984) deduced from the work of Kennedy (1963) that the average wavelength ($\overline{L_w}$) of standing waves scales with the depth ($h$) of the water flow:

$$\overline{L}_w = 2\pi h \qquad (1)$$

Various studies have used this relationship to reconstruct flood hydraulic parameters from the spacing of observed standing waves. Allen (1984) showed using flume data that the average wavelength ($\overline{L_a}$) of trains of antidunes was in accord with the average wavelength of the standing waves with which they were associated. Consequently $\overline{L_a}$ can be substituted into equation [1] to estimate flow palaeodepth from "fossil" antidune wavelengths. Likewise flow velocity can also be estimated using:

$$U = \sqrt{Lg/2\pi} \qquad (2)$$

## 5 CONCLUSIONS

Gravel dunes and antidunes have been preserved in the landscape consequent upon Quaternary flooding events and recent flood flows. The geometry and stratigraphy of these bedforms may be used to infer the flow dynamics of the palaeoflows responsible for their deposition. An emerging issue, which is not germane to the more common sandy bedforms, is the importance of surface grain segregation processes, which result in armour layers being developed on the stoss sides of gravel bedforms. These armour layers may delay the entrainment process of the sub-armour and consequently the timing of the initial motion of gravel bedforms.

## REFERENCES

Allen, J.R.L. (1984) Sedimentary Structures: Their Character and Physical Basis. Elsevier, Amsterdam, 593pp.

Alt, D. (2001) Missoula and Its Humogous Flood, Montana Press Company, Missoula, Montana, 197pp.

Ashley, G.M. (1990) Classification of large-scale subaqueous bedforms: a new look at an old problem. J. Sediment Petrology, 60, 160–172.

Baker, V.R. (1973) Paleohydrology of catastrophic Pleistocene flooding in eastern Washington. Geol. Soc. Am. Spec. Pap. 144, 1–79.

Baker, V.R. (1995) Joseph Pardee and the Spokane Flood Controversy, GSA Today, 5 (9), Sept 1995.

Baker, V.R. and Nummedal, D. (1978) (Eds.) The Channeled Scabland, National Aeronautics and Space Administration, Washington, D.C.

Breed, C.S., J.F. McCauley, and Davis, P.A. (1987) Sand sheets of the eastern Sahara and ripple blankets on Mars. In Desert Sediments: Ancient and Modern, L.E. Frostick and I. Reid. (Editors) Geological Society of London Special Publica tion, no: 35, 337–359.

Bretz, J.H., Smith, H.T.U. and Neff, G.E. (1956) Channeled Scabland of Washington; new data and interpretations. Geol. Soc. America Bull., 67, 957–1049.

Carling, P.A. (1996a) A preliminary palaeohydraulic model applied to Late-Quaternary gravel dunes: Altai Mountains, Siberia. In: Global Continental Changes: The Context of Palaeohydrology. J. Branson, K.J. Gregory and A. Brown (Editors) Geol. Soc. London Spec. Publ. No. 115, 165–179.

Carling, P.A. (1996b) Morphology, sedimentology and palaeohydraulic-interpretation of large gravel dunes: Altai Mountains, Siberia. Sedimentology, 43, 647–664.

Carling, P.A. (1999) Subaqueous gravel dunes. J. Sedimentary Research, 69, 534–545. Carling, P.A., Burr, D., Johnsen, T. and Brennand, T. (in press) A review of megaflood depositional landforms on Earth and Mars, In: Megaflooding on Earth and Mars, D. Burr, V.R. Baker and P.A. Carling (Editors), Cambridge University Press, Cambridge.

Carling, P.A., Kirkbride, A.D., Parnachov, S., Borodavko, P.S. and Berger, G.W. (2002) Late Quaternary catastrophic flooding in the Altai Mountains of south-central Siberia: a synoptic overview and introduction to flood deposit sedimentology. Spec. Publ. Ass. Sediment, 32, 17–35.

Carling, P.A., Radecki-Pawlik, A., Williams, J.J., Rumble, B., Meshkova, L., Bell, P. and Breakspear, R (2005a) The morphodynamics and internal structure of intertidal fine-gravel dunes: Hills Flats, Severn Estuary, UK. Sedimentary Geology, 183, 159–179.

Carling, P.A., Richardson, K. and Ikeda, H. (2005b) A Flume Experiment of the Development of Subaqueous Fine-gravel Dunes from a Lower-stage Plane Bed. JGR Earth Surface, 110, F04S05, doi:10.1029/2004JF000205, 2005.

Carling, P.A., and Shvidchenko, A.B. (2002) The:antidune transition in fine gravel with especial consideration of downstream migrating antidunes. Sedimentology, 49, 1269–1282.

Chapman, M.G., Gudmundsson, M.T., Russell, A.J., and Hare, T. M. 2003. Possible Juventae Chasma subice volcanic eruptions and Maja Valles ice outburst floods on Mars: Implications of Mars Global Surveyor crater densities, geomorphology, and topography. J. Geophys. Res. 108, 2-1, CiteID 5113, doi: 10.1029/2002JE02009.

Evans, D.J.A., Rea, B.R., Hiemstra, J.F. and ó Cofaigh, C. (in press). A critical assessment of subglacial mega-floods: a case study of glacial sediments and landforms in south-central Alberta, Canada. Quaternary Science Reviews.

Henderson, S., Miller, M.F., Isbell, J.L. and Mabin, M.C.G. (2002) Coarse, wind blown gravel deposits, Bennett Platform Antarctica: Constraining wind velocities during transport. Abstract paper No: 7-0, GSA Joint Annual Meeting, April 3-5, 2002, Lexington, Kentucky.

Johnsen, T.F. and Brennand, T.A. (2004) Late glacial lakes in the Thompson Basin, British Columbia: paleogeography and evolution. Canadian Journal of Earth Sciences, 41, 1367–1383.

Kennedy, J.F. (1963) The mechanics of dunes and antidunes in erodible-bed channels. J. Fluid Mechanics, 16, 521–544.

Lister, J.C. (1981) The Sedimentology of Camus Prairie Basin and its Significance to the Lake Missoula Floods, Unpublished MSc thesis, University of Montana, 66pp.

Malin, M.C. (1986) Rates of geomorphic modification in ice-free areas southern Victoria Land, Antarctica. Antarctic Journal of the United States, 20(5), 18-21.

Munro-Stasiuk, M.J. and Shaw, J. (1997) Erosional origin of hummocky terrain, south-central Alberta, Canada. Geology, 25, 1027–1030.

Pardee, J.T. (1942) Unusual currents in glacial Lake Missoula, Montana. Geological Society of America, Bulletin, 53, 1569–1600.

Radecki-Pawlik A., Carling P., Słowik-Opoka E., Książek L., Breakspear R. (2006) Field investigations of sand-gravel bed forms within the Raba river, Poland. River Flow Monograph edited by Rui. M.L. Ferreira, Elsa C.T.L. Alves, Jao G.A.B. Leal and Antonio H. Cardoso, (Editors) Engineering, Water and Earth Science, BALKEMA Taylor & Francis Group, Rotterdam, vol. 1, p. 979–985.

Reddering, J. and Eserhuysen, K. (1987) The effects of river floods on sediment dispersal in small estuaries: a case study from East London. S.-Afr. Tydskr. Geol., 90, 458–470.

Rice, J.W., Parker, T.J., Russel, A.J. and Knudsen, O. (2002) Morphology of fresh outflow channel deposits on Mars. 33rd Annual Lunar and Planetary Science Conference, March 11–15, 2002, Houston, Texas, abstract no.2026.

Sandercock, P.J. (2003) Causes and nature of river channel changes in the upper Ord River catchment. Unpublished PhD thesis, University of Western Australia.

Schoeneich, P. and Maisch, M. (2003a) Témoins géomorphologiques de mégacrues et de débâcles dans les Alpes. Communication aux journées Nivologie-Glaciologie de la Société Hydrotechnique de France, 16–17 March 2003, Grenoble.

Schoeneich, P. and Maisch, M. (2003b). A lateglacial megaflood in the Swiss Alps: the outburst of the great lake of Davos. 3rd International Paleoflood Workshop, 1–8 August 2003, Hood River, Oregon.

Williams, J.J., Carling, P.A. and Bell, P.S. (2006) Dynamics of intertidal gravel dunes. J. Geophysical Research, 111, C06035, doi:10.1029/2005JC003000, 2006.

Williams J.J., Carling P.A. and Bell, P.S., (2007) Dynamics of intertidal gravel dunes. Journal of Coastal Research, SI 50 (Proceedings of the 9th International Coastal Symposium), pg – pg. Gold Coast, Australia.

Zielinski, T. (1982) Contemporary high-energy flows, their deposits and reference to the outwash depositional model. Geologia, 6, 98–108. (in Polish with extended English abstract).

*River, Coastal and Estuarine Morphodynamics: RCEM 2007 – Dohmen-Janssen & Hulscher (eds)*
© *2008 Taylor & Francis Group, London, ISBN 978-0-415-45363-9*

# Weakly nonlinear analysis of the dune-flat bed transition with the mixing length turbulent model

N. Izumi

*Hokkaido University, Sapporo, Japan*

ABSTRACT: Weakly nonlinear stability analysis of fluvial dunes is performed with the use of the growth rate expansion method incorporated with the multiple scale perturbation technique. Open channel flow is described by the mixing length turbulent model. A series of perturbation equations are solved by the use of the spectral collocation method with the Chebyshev polynomials. The analysis reveals that the dune-flat bed transition is characterized by subcritical bifurcation when the resistant coefficient $C^{-1}$ is large, or the ratio of the equivalent roughness height to the sand diameter is large.

## 1 INTRODUCTION

When the Froude number is relatively small, river beds are covered with undulations called dunes, the wave height of which is 10–50% of the flow depth. Dunes are formed in the range of relatively small Froude numbers, and they disappear in the range of large Froude numbers. In the transition between dune and flat bed regimes, it is known that the critical Froude number from dune to flat bed regimes is larger than that from flat bed to dune regimes.

Yamaguchi Izumi (2002, 2003, 2005) have proposed a mathematical model to explain this hysteresis in terms of weakly nonlinear stability analysis with the use of the constant stress layer approximation proposed by Engelund (1970) and Fredsøe(1974). In this study, a similar analysis is performed on the basis of the analysis proposed by Colombini (2004), in which the mixing length turbulent model is employed.

## 2 FORMULATION

Turbulent flow in an open channel can be expressed by the Navier-Stokes equations of the forms

$$U\frac{\partial U}{\partial x} + V\frac{\partial U}{\partial y} = -\frac{\partial P}{\partial x} - 1 + \frac{\partial T_{xx}}{\partial x} + \frac{\partial T_{xy}}{\partial y} \qquad (1)$$

$$U\frac{\partial V}{\partial x} + V\frac{\partial V}{\partial y} = -\frac{\partial P}{\partial y} + S^{-1} + \frac{\partial T_{xy}}{\partial x} + \frac{\partial T_{yy}}{\partial y} \qquad (2)$$

$$\frac{\partial U}{\partial x} + \frac{\partial V}{\partial y} = 0 \qquad (3)$$

where $x$ and $y$ are the coordinates in the streamwise and depth directions respectively, $U$ and $V$ are the $x$ and $y$ components of the flow velocity respectively, $S$ and $P$ are the average bed slope and the pressure respectively, and $T_{ij}$ $(i,j=x,y)$ is the Reynolds stress tensor. In the above equations, the time derivative terms are dropped because the time variation of flow is much faster than the time variation of bedforms. In the above equations, all the variables have been nondimensionalized as

$$(U^*, V^*) = U_{f0}^* (U, V) \qquad (4a)$$

$$(x^*, y^*) = D_0^* (x, y) \qquad (4b)$$

$$\left(P^*, T_{ij}^*\right) = \rho U_{f0}^{*2} \left(P, T_{ij}\right) \qquad (4c)$$

where $U_{f0}^*$ and $D_0^*$ are the friction velocity and the flow depth in the base state flat bed condition. With the use of the mixing length turbulent model, the Reynolds stress tensor is expressed by

$$\left(T_{xx}, T_{yy}\right) = 2\nu_T \left(\frac{\partial U}{\partial x}, \frac{\partial V}{\partial y}\right) \qquad (5a)$$

$$T_{xy} = \nu_T \left(\frac{\partial U}{\partial y} + \frac{\partial V}{\partial x}\right) \qquad (5b)$$

$$\nu_T = l^2 \left|\frac{\partial U}{\partial y}\right| \qquad (5c)$$

$$l = \kappa (y - Z)\left(\frac{D-y}{D}\right)^{1/2} \qquad (5d)$$

where $\nu_T$ is the eddy viscosity normalized by $U_{f0}^* D_0^*$, $l, Z$ and $D$ are the mixing length, the bed elevation and

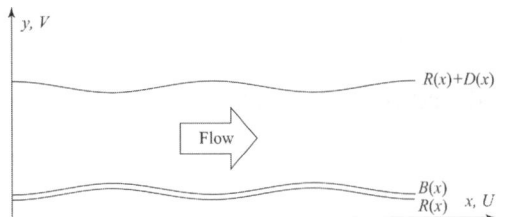

Figure 1. The conceptual diagram of flow and the coordinate system.

the flow depth normalized by $D_0^*$, and $\kappa$ is the Karman constant ($=0.4$).

We introduce the stream function defined by

$$(U, V) = \left( \frac{\partial \psi}{\partial y}, -\frac{\partial \psi}{\partial x} \right) \qquad (6)$$

Equation (1) is then rewritten as

$$\frac{\partial \psi}{\partial y} \frac{\partial^2 \psi}{\partial x \partial y} - \frac{\partial \psi}{\partial x} \frac{\partial^2 \psi}{\partial y^2} = -\frac{\partial P}{\partial x} + 1 + \frac{\partial}{\partial x} \left( 2\nu_T \frac{\partial \psi}{\partial x \partial y} \right)$$

$$+ \frac{\partial}{\partial y} \left[ \nu_T \left( \frac{\partial^2 \psi}{\partial y^2} - \frac{\partial^2 \psi}{\partial x^2} \right) \right] \quad (7)$$

Eliminating $P$ from the above equation and (2) with $(U, V)$ replaced by $(\partial \psi / \partial y, -\partial \psi / \partial x)$, we obtain

$$\frac{\partial \psi}{\partial y} \frac{\partial \nabla^2 \psi}{\partial x} - \frac{\partial \psi}{\partial x} \frac{\partial \nabla^2 \psi}{\partial y} - 4 \frac{\partial^2}{\partial x \partial y} \left( \nu_T \frac{\partial \psi}{\partial x \partial y} \right)$$

$$+ \left( \frac{\partial^2}{\partial x^2} - \frac{\partial^2}{\partial y^2} \right) \left[ \nu_T \left( \frac{\partial^2 \psi}{\partial y^2} - \frac{\partial^2 \psi}{\partial x^2} \right) \right] = 0 \quad (8)$$

In order to facilitate the application of boundary conditions at the bottom and the water surface, we introduce the variable transformation in the forms

$$(\xi, \eta) = \left( x, \frac{y - R(x)}{D(x)} \right) \qquad (9)$$

where $R$ is the reference level at which the velocity vanishes in the logarithmic velocity distribution (see Figure 1). With the use of the above variable transformation, the nondimensional mixing length $l$ is written in the form

$$l = \kappa D \left( \eta + \frac{R - Z}{D} \right) \left( 1 - \frac{R}{D} - \eta \right)^{1/2} \qquad (10)$$

The boundary conditions at the water surface and the bottom are

$$\mathbf{u} \cdot \mathbf{e}_{ns} = 0 \quad \text{at} \quad \eta = 1 \qquad (11)$$

$$\mathbf{e}_{ns} \cdot \mathbf{T} \cdot \mathbf{e}_{ns} = 0 \quad \text{at} \quad \eta = 1 \qquad (12)$$

$$\mathbf{e}_{ts} \cdot \mathbf{T} \cdot \mathbf{e}_{ns} = 0 \quad \text{at} \quad \eta = 1 \qquad (13)$$

$$\mathbf{u} \cdot \mathbf{e}_{nb} = 0 \quad \text{at} \quad \eta = 0 \qquad (14)$$

$$\mathbf{u} \cdot \mathbf{e}_{tb} = 0 \quad \text{at} \quad \eta = 0 \qquad (15)$$

where $\mathbf{u}$ is the velocity vector ($=(u, v)$), $\mathbf{e}_{ns}$ and $\mathbf{e}_{ts}$ are the unit vectors normal and tangential to the water surface respectively, $\mathbf{e}_{nb}$ and $\mathbf{e}_{tb}$ are the unit vectors normal and tangential to the bottom respectively. In the above equation, $R/D$ is assumed to be small to be neglected. In addition, $\mathbf{T}$ is the stress tensor expressed as

$$\mathbf{T} = \begin{bmatrix} -P + T_{xx} & T_{xy} \\ T_{xy} & -P + T_{yy} \end{bmatrix} \qquad (16)$$

## 3 THE ONE-DIMENSIONAL BASE STATE

The base state of the stability analysis is the flat bed normal flow condition. In the base state, the variables are written in the form

$$(U, V, D, Z, R) = (U_0, 0, 1, 0, R_0) \qquad (17)$$

The governing equations are reduced to

$$1 + \frac{dT_{xy0}}{d\eta} = 0 \qquad (18)$$

$$T_{xy0} = \nu_{T0} \frac{dU_0}{d\eta} \qquad (19)$$

$$\nu_{T0} = l_0^2 \frac{dU_0}{d\eta} \qquad (20)$$

$$l_0 = \kappa (\eta + R_0)(1 - R_0 - \eta)^{1/2} \qquad (21)$$

where $()_0$ denotes the base state solutions. To solve the above equations, the following boundary conditions are applied:

$$U = 0, \quad T_{xy0} = 1 - R_0 \quad \text{at} \quad \eta = 0 \qquad (22)$$

From (18)–(22), the following logarithmic velocity distribution is obtained:

$$U_0 = \frac{1}{\kappa} \ln \left( \frac{\eta + R_0}{R_0} \right) \qquad (23)$$

Integrating the above equation from $\eta = 0$ to 1, we obtain the friction law (friction coefficient $C$) of the form

$$C^{-1} = \frac{U_{a0}^*}{U_{f0}^*} = \frac{1}{\kappa} \left[ (1 + R_0) \ln \left( \frac{1 + R_0}{R_0} \right) - 1 \right] \qquad (24)$$

where $U_{a0}^*$ is the depth-averaged velocity in the base state.

## 4 THE TIME VARIATION OF BEDFORMS

The continuity equation of sediment on the bed is described by

$$\frac{\partial B}{\partial t} + \frac{\partial \Phi}{\partial x} = 0 \qquad (25)$$

where $B$ is the elevation of the upper surface of the bedload layer (see Figure 1), $\Phi$ is the nondimensional bedload transport rate $(= q_s^*/(R_s g d_s^{*3})^{1/2})$, $q_s^*$ is the dimensional bedload transport rate, $R_s$ is the submerged specific gravity $(=1.65)$, $t$ is time $(=[d_s^*(R_s g d_s^*)^{1/2}]/[D_0^{*2}(1-\lambda_p)]t^*)$, and $\lambda_p$ is porosity.

In order to estimate the bedload transport rate, Colombini (2004) have employed the Meyer-Peter & Müller formula with the bottom shear stress replaced by the shear stress at the upper surface of the bedload layer, that is

$$\Phi = 8 (\theta_b - \theta_c)^{3/2} \qquad (26)$$

where $\theta_b$ and $\theta_c$ are the Shields stress at the upper surface of the bedload layer, and the critical Shields stress, respectively, which are written in the forms

$$\theta_b = \frac{\tau_b^*}{\rho R_s g d_s^*} = \frac{DS}{R_s d_s^*}\tau_b \qquad (27)$$

$$\theta_c = \theta_{ch} - \mu \left( S - \frac{\partial B}{\partial x} \right) \qquad (28)$$

$$\mu = \frac{\theta_{ch}}{\tan \Psi} \qquad (29)$$

where $\theta_{ch}$ is the critical Shields stress for flat beds, and $\Psi$ is the friction angle.

From existing experimental data, Colombini (2004) estimated the bedload layer thickness $h_b$ as

$$h_b = l_b d_s \qquad (30a)$$

$$l_b = 1 + 1.3 \left( \frac{\tau_r - \tau_c}{\tau_c} \right)^{0.55} \qquad (30b)$$

where $\tau_r$ and $\tau_c$ are the shear stress at the reference level $(\eta = R)$, and the critical shear stress, respectively. The bedload layer thickness $h_b$ is related to $B$ by

$$B = h_b + d_s/6 = (l_b + 1/6) d_s \qquad (31)$$

Though there is still a room for argument whether or not this bedload layer model proposed by Colombini (2004) is physically appropriate, it is no doubt that the results of the linear stability analysis is improved by his model. In this study, Colombini's linear stability analysis is extended to a weakly nonlinear stability analysis straightforward.

## 5 LINEAR STABILITY ANALYSIS

Perturbation is imposed on the base state solution obtained in §3. Variables are expanded as

$$(\psi, P, D, Z, R, B) = (\psi_0, P_0, 1, 0, R_0, B_0)$$

$$+ A\left(\hat{\psi}_1, \hat{P}_1, \hat{D}_1, \hat{R}_1, \hat{R}_1, \hat{R}_1\right) \qquad (32)$$

where $A$ is the amplitude of perturbation, which is assumed to be infinitesimally small in the context of linear stability analysis. Substituting the above equations into (7) and (8), and equating the coefficients of like powers of $A$, we obtain the following equations at $O(A)$:

$$\hat{\mathcal{L}}^\psi \hat{\psi}_1 + \hat{\mathcal{L}}^D \hat{D}_1 + \hat{\mathcal{L}}^R \hat{R}_1 = 0 \qquad (33)$$

$$\frac{\partial \hat{P}_1}{\partial \xi} + \hat{\mathcal{R}}^\psi \hat{\psi}_1 + \hat{\mathcal{R}}^R \hat{R}_1 + \hat{\mathcal{R}}^D \hat{D}_1 = 0 \qquad (34)$$

From the boundary conditions (11)–(15), and the Exner equation (25), we obtain

$$\partial \hat{\psi}_1/\partial \xi = \hat{P}_1 = 0 \quad \text{at} \quad \eta = 1 \qquad (35)$$

$$\partial \hat{\psi}_1/\partial \xi = \partial \hat{\psi}_1/\partial \eta = 0 \quad \text{at} \quad \eta = 0 \qquad (36)$$

$$\hat{\mathcal{E}}^\psi \hat{\psi}(\eta_B) + \hat{\mathcal{E}}^R \hat{R}_1 + \hat{\mathcal{E}}^D \hat{D}_1 = 0 \quad \text{at} \quad \eta = \eta_B \qquad (37)$$

where $\eta_B = (B - R)/D$, and $\hat{\mathcal{L}}^\phi$, $\hat{\mathcal{R}}^\phi$ and $\hat{\mathcal{E}}^\phi$ $(\phi = \psi, D, R)$ are linear operators, the explicit forms of which are not shown here due to limitations of space.

The normal mode analysis is performed hereafter. The perturbation is assumed to be expressed by

$$\left(\hat{\psi}_1, \hat{P}_1, \hat{D}_1, \hat{R}_1\right) = (\psi_1, P_1, D_1, R_1) \exp\left[i\left(\alpha\xi - \Omega t\right)\right] \qquad (38)$$

where $\alpha$ and $\Omega$ are the wavenumber and the complex angular frequency of the perturbation, respectively. Substituting the above equation to (33)–(37), we obtain

$$\mathcal{L}^\psi(\eta)\psi_1(\eta) + \mathcal{L}^D(\eta) D_1 + \mathcal{L}^R(\eta) R_1 = 0 \qquad (39)$$

$$\mathcal{R}^\psi(1)\psi_1(\eta) + \mathcal{R}^D(1) D_1 + \mathcal{R}^R(1) R_1 = 0 \qquad (40)$$

$$\psi_1(1) = \psi_1(0) = \mathcal{D}\psi_1(0) = 0 \qquad (41)$$

$$\mathcal{E}^\psi \psi_1(\eta_B) + \mathcal{E}^R R_1 + \mathcal{E}^D D_1 = 0 \qquad (42)$$

where $\mathcal{D} = d/d\eta$.

With the use of the Chebyshev polynomials, $\psi_1$ can be expressed as

$$\psi_1 = \sum_{n=0}^{N} a_n T_n(\zeta) \qquad (43)$$

1023

where $T_n$ is the $n$th order Chebyshev polynomial, $\zeta$ is the independent variable of the Chebyshev polynomial defined in $[-1, 1]$ In order to improve the accuracy of the analysis, we employ the following variable transformation:

$$\zeta = 2\left\{\frac{\ln[(\eta + R_0)/R_0]}{\ln[(1 + R_0)/R_0]}\right\} - 1 \tag{44}$$

After substituting these equations, we evaluate the governing equations at the Gauss-Labatte points, written as

$$\zeta_j = \cos(j\pi/N), \quad (j = 1, \cdots, N - 2) \tag{45}$$

From the governing equations with the boundary conditions, we obtain the linear algebraic system of the form

$$\mathbf{L}a = 0 \tag{46}$$

where

$$a = [a_0, a_1, \cdots, a_N, D_1, R_1] \tag{47}$$

$$\mathbf{L} = \begin{pmatrix} \check{\mathcal{E}}^\psi T_0(\zeta_B) & \cdots & \check{\mathcal{E}}^\psi T_N(\zeta_B) & \check{\mathcal{E}}^D & \check{\mathcal{E}}^R \\ T_0(-1) & \cdots & T_N(-1) & 0 & 0 \\ \check{D}T_0(-1) & \cdots & \check{D}T_N(-1) & 0 & 0 \\ T_0(1) & \cdots & T_N(1) & 0 & 0 \\ \check{R}^\psi T_0(1) & \cdots & \check{R}^\psi T_N(1) & \check{R}^D & \check{R}^R \\ \check{\mathcal{L}}^\psi T_0(\zeta_1) & \cdots & \check{\mathcal{L}}^\psi T_N(\zeta_1) & \check{\mathcal{L}}^D & \check{\mathcal{L}}^R \\ & \cdots & & & \\ \cdot & \cdots & \cdot & \cdot & \cdot \\ \cdot & \cdots & \cdot & \cdot & \cdot \\ \check{\mathcal{L}}^\psi T_0(\zeta_{N-2}) & \cdots & \check{\mathcal{L}}^\psi T_N(\zeta_{N-2}) & \check{\mathcal{L}}^D & \check{\mathcal{L}}^R \end{pmatrix}$$

$$\tag{48}$$

In (48), $\check{}$ denotes the corresponding linear operator with $\eta$ transformed by $\zeta$. The Exner equation and all the boundary conditions are inserted in the first five rows in $\mathbf{L}$. In order for (46) to have a nontrivial solution, the solvability condition $|\mathbf{L}| = 0$ has to be satisfied. Note that $\check{\mathcal{E}}^R$ includes $\Omega$. From the solvability condition therefore, the complex angular frequency $\Omega$ is obtained in the following general functional form:

$$\Omega = f(\alpha, F; C, \mu, m) \tag{49}$$

The imaginary part of $\Omega$ corresponds to the growth rate of perturbation.

Figure 2 shows the contours of the growth rate of perturbation $\mathrm{Im}[\Omega]$ in the $\alpha - F$ plane, which is identical to the diagram that Colombini (2004) have obtained. According to Colombini (2004), experimental data of antidunes as well as dunes are well explained by this diagram.

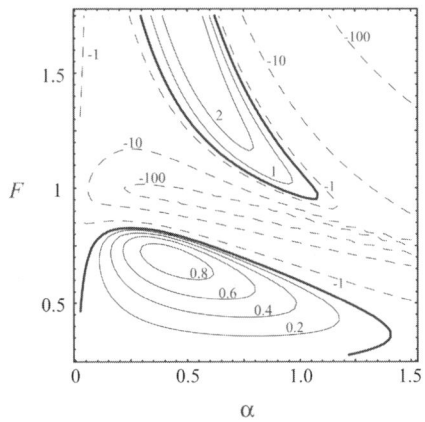

Figure 2. The contours of $\mathrm{Im}[\Omega]$. $C^{-1} = 20$, $\mu = 0.1$, $m = 2.5$.

## 6 WEAKLY NONLINEAR STABILITY ANALYSIS

### 6.1 Asymptotic expansion

Considering the vicinity slightly smaller than the critical Froude number, we introduce a small parameter $\epsilon$ defined by

$$\epsilon = \left(\frac{F_c - F}{F_c}\right)^{1/2} \tag{50}$$

where $F_c$ is the critical Froude number. With the use of the multiple-scale perturbation technique, we introduce two different time scales $T_0$ and $T_1$ defined by

$$(T_0, T_1) = \left(t, \epsilon^2 t\right) \tag{51a}$$

$$\frac{\partial}{\partial t} = \frac{\partial}{\partial T_0} + \epsilon^2 \frac{\partial}{\partial T_1} \tag{51b}$$

All the variables are expanded as

$$(\psi, D, R) = (\psi_0, 1, R_0) + \sum_{i=1}^{3} \epsilon^i (\psi_i, D_i, R_i) \tag{52}$$

If the solutions at $O(\epsilon)$ are given in the forms

$$(\psi_1, D_1, R_1) = (\psi_{11}, D_{11}, R_{11}) E \tag{53a}$$

$$E = \exp\left[i\left(\alpha\xi - \Omega t\right)\right] \tag{53b}$$

it is expected that the solutions at $O(\epsilon^2)$ and $O(\epsilon^3)$ take the forms

$$(\psi_2, D_2, R_2) = A^2 (\psi_{22}, D_{22}, R_{22}) E^2 + \text{c.c.}$$

$$+ A\bar{A} (\psi_{20}, D_{20}, R_{20}) + (\psi_{00}, 0, 0) \tag{54}$$

$(\psi_3, D_3, R_3) = A^3 (\psi_{33}, D_{33}, R_{33}) E^3 + \text{c.c.}$

$$+ (\psi_{31}, D_{31}, R_{31}) E + \text{c.c.} \quad (55)$$

Substituting (52)–(55) into (7), (8), (11)–(15) and (25) and equating the coefficients of like powers of $\epsilon$, we obtain the following results.

$O(\epsilon)$:

$$\mathcal{L}_1^{\psi} \psi_{11} + \mathcal{L}_1^{D} D_{11} + \mathcal{L}_1^{R} R_{11} = 0 \quad (56)$$

$$\mathcal{E}_1^{\psi} \psi_{11}(\eta_B) + \mathcal{E}_1^{D} D_{11} + \mathcal{E}_1^{R} R_{11} = 0 \quad (57)$$

$$\psi_1(0) = \mathcal{D}_1 \psi_{11}(0) = \psi_1(1) = 0 \quad (58)$$

$$\mathcal{R}_1^{\psi}(1)\psi_{11}(1) + \mathcal{R}_1^{D}(1)D_1 + \mathcal{R}_1^{R}(1)R_1 = 0 \quad (59)$$

$O(\epsilon^2)$:

$$\mathcal{L}_2^{\psi} \psi_{22} + \mathcal{L}_2^{D} D_{22} + \mathcal{L}_2^{R} R_{22} = \mathcal{N}_{22} \quad (60)$$

$$\mathcal{E}_2^{\psi} \psi_{22}(\eta_B) + \mathcal{E}_2^{D} D_{22} + \mathcal{E}_2^{R} R_{22} = \mathcal{S}_{22} \quad (61)$$

$$\psi_1(0) = \mathcal{D}_1 \psi_{11}(0) = \psi_1(1) = 0 \quad (62)$$

$$\mathcal{R}_1^{\psi}(1)\psi_{11}(1) + \mathcal{R}_1^{D}(1)D_1 + \mathcal{R}_1^{R}(1)R_1 = \mathcal{P}_{22} \quad (63)$$

$$\mathcal{L}_0^{\psi} \psi_{20} + \mathcal{L}_0^{D} D_{20} + \mathcal{L}_0^{R} R_{20} = \mathcal{N}_{20} \quad (64)$$

$$\mathcal{E}_0^{\psi} \psi_{20}(\eta_B) + \mathcal{E}_0^{D} D_{20} + \mathcal{E}_0^{R} R_{20} = \mathcal{S}_{20} \quad (65)$$

$$\psi_{20}(0) = \mathcal{D}\psi_{20}(0) = \psi_1(1) = 0 \quad (66)$$

$$\mathcal{R}_0^{\psi}(1)\psi_{20}(1) + \mathcal{R}_0^{D}(1)D_{20} + \mathcal{R}_0^{R}(1)R_{20} = \mathcal{P}_{20} \quad (67)$$

$O(\epsilon^3)$:

$$\mathcal{L}_1^{\psi} \psi_{31} + \mathcal{L}_1^{D} D_{31} + \mathcal{L}_1^{R} R_{31} = \mathcal{N}_{31} \quad (68)$$

$$\mathcal{E}_1^{\psi} \psi_{31}(\eta_B) + \mathcal{E}_1^{D} D_{31} + \mathcal{E}_1^{R} R_{31} = \mathcal{S}_{31} \quad (69)$$

$$\psi_{31}(0) = \mathcal{D}\psi_{31}(0) = \psi_1(1) = 0 \quad (70)$$

$$\mathcal{R}_1^{\psi}(1)\psi_{31}(1) + \mathcal{R}_1^{D}(1)D_{31} + \mathcal{R}_1^{R}(1)R_{31} = \mathcal{P}_{31} \quad (71)$$

Here $\mathcal{L}_n^{\phi}$, $\mathcal{R}_n^{\phi}$ and $\mathcal{E}_n^{\phi}$ ($\phi = \psi, D, R$; $n = 0, 1, 2$) are $\mathcal{L}^{\phi}$, $\mathcal{R}^{\phi}$ and $\mathcal{E}^{\phi}$ with $(F, k)$ replaced by $(F_c, nk)$, and $\text{Im}[\Omega]$ vanishing, and $\mathcal{N}_{ij}$, $\mathcal{S}_{ij}$ and $\mathcal{P}_{ij}$ $((i, j) = (1, 1), (2, 2), (2, 0), (3, 1))$ are inhomogeneous terms composed of lower order terms.

Solving the above equations by the use of the spectral collocation method with the Chebyshev polynomials as well as in the linear stability analysis, we obtain

$$\frac{dA}{dT_1} = \lambda_0 A + \lambda_1 |A|^2 A \quad (72)$$

where $\lambda_0$ is the linear growth rate, and $\lambda_1$ is the Landau constant. An equilibrium solution of (72), is $\sqrt{-\lambda_0/\lambda_1}$. When $\lambda_1 > 0$, the dune-flat bed transition is characterized by subcritical bifurcation, in which a real equilibrium solution exists only in the range of $\lambda_0 < 0$.

## 7 RESULTS AND DISCUSSION

Table 1 shows the critical Froude numbers and the Landau constants. The first to third rows show the case that $C^{-1} = 21$, $\mu = 0.1$ and $m = 1.7$. The critical Froude number is 0.8256, and the corresponding wavenumber is 0.227. In the vicinity of the critical Froude number, we find $\lambda_0 > 0$ and $\text{Re}(\lambda_1) < 0$. The implication is that the dune-flat bed transition is characterized by the supercritical bifurcation. The results of the case $C^{-1} = 22$ are shown in the fourth to sixth rows. When $C^{-1}$ is increased, we find $\lambda_0 > 0$ and $\text{Re}(\lambda_1) > 0$; therefore, the dune-flat bed transition is subcritical bifurcation.

The results of small $\mu (= 0.05)$ and large $m (= 2.5)$ are shown in the seventh to twelfth rows in the table. The parameters $\mu$ and $m$ represent the effect of local slope, and the ratio between the equivalent roughness height $k_s$ and the sand diameter $d_s$, respectively. With decreasing $\mu$, $\lambda_1$ is slightly increased though it is still negative. With increasing $m$, $\lambda_1$ increases to become

Table 1. The critical Froude number and the Landau constant.

| $C^{-1}$ | $\mu$ | $m$ | $\alpha$ | $F_c$ | $\Omega$ | $\lambda_0$ | $\lambda_1$ |
|---|---|---|---|---|---|---|---|
| 21 | 0.1 | 1.7 | 0.217 | 0.8254 | 12.6 | $11.2 + 51.2i$ | $-2040 + 57500i$ |
| 21 | 0.1 | 1.7 | 0.227 | 0.8256 | 11.9 | $11.6 + 54.2i$ | $-3770 + 58600i$ |
| 21 | 0.1 | 1.7 | 0.237 | 0.8254 | 13.2 | $11.9 + 57.1i$ | $-5290 + 59100i$ |
| 22 | 0.1 | 1.7 | 0.238 | 0.8491 | 26.7 | $30.4 + 139i$ | $5210 + 173000i$ |
| 22 | 0.1 | 1.7 | 0.248 | 0.8492 | 28.1 | $31.5 + 147i$ | $1890 + 176000i$ |
| 22 | 0.1 | 1.7 | 0.258 | 0.8491 | 29.4 | $32.5 + 155i$ | $-999 + 177000i$ |
| 21 | 0.05 | 1.7 | 0.219 | 0.8268 | 12.2 | $11.7 + 53.0i$ | $-1540 + 59900i$ |
| 21 | 0.05 | 1.7 | 0.229 | 0.8270 | 12.9 | $12.1 + 56.1i$ | $-3300 + 61000i$ |
| 21 | 0.05 | 1.7 | 0.239 | 0.8269 | 13.5 | $12.5 + 59.0i$ | $-4830 + 61700i$ |
| 21 | 0.1 | 2.5 | 0.258 | 0.8496 | 33.6 | $41.3 + 177i$ | $11500 + 203000i$ |
| 21 | 0.1 | 2.5 | 0.268 | 0.8496 | 35.2 | $42.8 + 187i$ | $8240 + 206000i$ |
| 21 | 0.1 | 2.5 | 0.278 | 0.8495 | 36.8 | $44.2 + 196i$ | $5360 + 208000i$ |

positive; thus, the dune-flat bed transition becomes subcritical bifurcation.

## 8 CONCLUSION

Weakly nonlinear stability analysis of fluvial dunes is performed with the use of the mixing length turbulent model. We find the following results:

- The dune-flat bed transition is characterized by subcritical bifurcation when the friction coefficient $C^{-1}$ is large.
- The dune-flat bed transition tends to become supercritical bifurcation when the local slope effect is estimated to be large.
- The dune-flat bed transition tends to become subcritical bifurcation when the ratio of the equivalent roughness height to the sand diameter is estimated to be large.

## ACKNOWLEDGEMENTS

This research was funded by the Hokkaido River Disaster Prevention Research Center. The support provided by the center, particularly Dr. Kiyoshi Hoshi, is greatly appreciated. This paper is dedicated to Dr. Kiyoshi Hoshi, who passed away in December 2006.

## REFERENCES

Colombini, M. (2004). Revisiting the linear theory of sand dune formation. *J. Fluid Mech. 502*, 1–16.

Engelund, F. (1970). Instability of erodible beds. *J. Fluid Mech. 42*, 225–244.

Fredsøe, J. (1974). On the development of dunes in erodible channels. *J. Fluid Mech. 64*, 1–16.

Yamaguchi, S. & N. Izumi (2002). Weakly nonlinear stability analysis of dune formation. *River Flow 2002 2*, 843–850.

Yamaguchi, S. & N. Izumi (2003). Weakly nonlinear analysis of dunes including suspended load. *Proceedings of 3nd IAHR Symposium on River, Coastal and Estuarine Morphodynamics I*, 172–183.

Yamaguchi, S. & N. Izumi (2005). Weakly nonlinear analysis of dunes by the use of a sediment transport formula incorporating the pressure gradient. *Proceedings of 4nd IAHR Symposium on River, Coastal and Estuarine Morphodynamics II*, 813–820.

*Morphodynamics in reservoirs*

*River, Coastal and Estuarine Morphodynamics: RCEM 2007 – Dohmen-Janssen & Hulscher (eds)*
*© 2008 Taylor & Francis Group, London, ISBN 978-0-415-45363-9*

# Effect of a reservoir release on the morphology of a gravel bar: Field observations and 2Dh modeling

M. Jodeau & A. Paquier
*Cemagref, Hydrology and Hydraulics Research Unit, Lyon Cedex, France*

A. Hauet
*LTHE, UMR 5564, Grenoble, France*

J. Le Coz, F. Thollet & T. Fournier
*Cemagref, Hydrology and Hydraulics Research Unit, Lyon Cedex, France*

ABSTRACT: In the context of a reservoir release, this paper aims at understanding the impact of a flood wave on the morphology of a gravel bar. A comprehensive field survey was conducted in the Arc River (French Alps), a highly regulated gravel-bed river. To face reservoir silting dam managers perform sediment release by flushing a series of 3 reservoirs once a year. The area of interest is located in the mid part of the river. Morphological field campaigns were performed before and after the event. Both campaigns contain topography measurements, aerial pictures and grain-size survey. Independently, during the day of the reservoir release, flow measurements were carried out using current meter and Large-Scale PIV. Observations indicate a gross trend of the reach to erosion during the flushing event. The upstream part of the main flow channel was eroded, transverse connecting channels across the bar were strengthened or were modified, and the secondary flow channel on the left side remained unchanged. Furthermore large areas of fine sediment deposits were identified and measured. Water flow in the reach was simulated with the two-dimensional model Rubar20. 2Dh modeling gives us additional information on velocity fields during the whole event. Comparison between calculation results and LS-PIV flow measurements leads to good agreement between measurements and calculations.

## 1 IMPACT OF FLUSHING OPERATIONS ON GRAVEL-BED RIVERS

Sediment flushing from mountain reservoirs is performed to preserve hydro-power performance. It is widely acknowledged that release operations often have significant effects on downstream reaches. Above all, high flow due to the sudden release of a large amount of water may deeply modify the gravel bed morphology. Mürle et al. (2003) observed scouring of alluvial fans and broad accumulation of coarse sediments in wider reaches of the River Spöl (Switzerland), next followed by scouring of these sediments. Fasolato et al. (2006) observed depositional and erosional areas of more than 4 m high. In addition, extensive amounts of flushed fine sediments may settle in specific areas. Wohl and Cenderelli (2000) showed that deposition of fine sediment occurred primarily in pools with additional deposition in lateral eddies. Brandt (1999) worked on deposition of fine sediments over a whole downstream reach. As a consequence vegetation dynamics and fish habitat may undergo specific constraints from flushing operations as observed by Wood and Armitage (1997).

This paper presents a detailed study on the morphological changes of a gravel bar following a reservoir release. Alternate bars are typical bed forms of rivers with sufficient width to depth ratio. Annual reservoir releases provide a predictable high flow event suitable to perform an extensive set of measurements. Here, flow characteristics are surveyed with specific attention to flow patterns thanks to innovating video analysis (LSPIV). Topographic and grain size surveys were conducted before and after the event, with special interest to the faith of fine sediments on the bar. This field campaign follows an exploratory work reported by Jodeau and Paquier (2006) which was a first step in studying morphological changes and flow processes over a gravel bar.

2Dh modeling was used to analyze flow measurements and observed morphological changes during the whole event.

This paper is organized in three parts. The first part is dedicated to the presentation of the study area and the flushing operation. The next part is a description of flow measurements and morphological surveys. The third part outlines field results, whereas the last part presents 2Dh numerical modeling results.

## 2 CONTEXT, STUDY AREA

### 2.1 River Arc and flushing operations

The field work was conducted in the River Arc located in the French Alps. Its hydrological regime is nival with natural discharges higher during snow melt. The River Arc channel has been rebuilt in order to allow the 1 km wide valley to contain a road, a highway and a railway in addition of villages. As a consequence almost all the reaches of the River Arc are embanked and straight. In addition flow rates are regulated by many hydraulic constructions.

Albeit the river bed is made of gravel, the River Arc is remarkable for high fine sediment transport. Most of these sediments are trapped in reservoirs. For this reason, three reservoirs in the middle part of the river are flushed once a year. On June 27th 2006, clear water was first released from the flap gates ($\approx 10\,\mathrm{m^3/s}$), then the deep gates were opened. The discharge was increased gradually to reach the maximum discharge in 6 hours, the peak discharge was maintained during 2 hours and the water discharge was decreased gradually. Dam managers estimated the maximum peak discharge to $146\,\mathrm{m^3/s}$ which is equivalent to the one year period return flood.

### 2.2 Area of interest

The study area was selected to offer a representative series of alternate bars which were easily reachable. This very straight reach with 1 m boulder embankments is located in the middle part of the river 18 km downstream the last dam. Figure 1 locates the river basin, the study area and the three flushed dams. The investigated area is restrained to one single, 400 m long gravel bar located on the left side of the river 200 m upstream a bridge. The mean slope of this reach is approximately 0.6%. The bed is made of gravel (mean diameter 10 cm) and fine sediments (sand and silt).

Owing to literature references, we distinguish typical sub-bar morphological units:

- **main channel:** right side of the stream bed where water flows at low discharge, characterized by the highest water depth in the stream reach;
- **secondary channel:** channel on the left side along the embankment of the river where about 10% of the water flows, offering a route for sediment to the bar tail;
- **bar head:** upstream part of the alternate bar with coarse material becoming finer downstream;

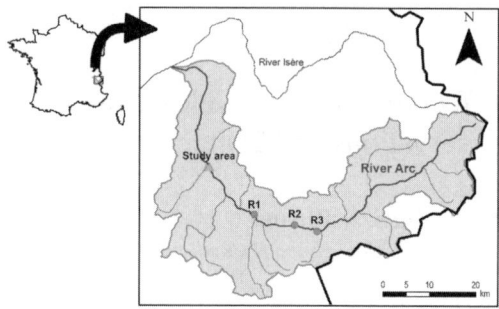

Figure 1. Location of the study area and location of the flushed reservoirs (R1, R2, R3).

- **bar tail:** downstream part of the bar made of fine sediments;
- **connecting channels:** channels connecting the secondary channel with the main flow channel. We noticed 2 permanent channels named upstream/middle ones and 1 or 2 transient connecting channels always downstream from permanent ones;
- **bar margin:** limit between the dry part of the bar and main flow channel, part of the bar oriented parallel to the flow. Along the bar edge we noticed some nooks as defined by Rempel and Church (2002).

## 3 FIELD MEASUREMENTS AND METHODS

### 3.1 Water level measurements and discharge estimates

Water levels were continuously monitored at 4 locations using bubble pipe devices. We checked the effect of water density changes due to suspended load on pressure measurements and it appeared to have no major effect on measured water depths. Measurement locations were chosen to give information about the water level at the downstream and upstream sections of the study reach, and in the middle of the reach in the main channel. The last device at the bridge section measures water level continuously over the year.

In addition to discharge assessments performed at low flow before and after the flushing event, two discharge measurements with a leaded current meter were performed from the bridge. Despite the usual uncertainty of approximately 5%, these measurements were considered as reference discharge values for the flushing event.

### 3.2 Flow velocity measurements

High flow velocities and large floating objects prevented any kind of intrusive flow measurements excepted from the bridge using leaded current-meter. In order to assess flow velocities over the bar during the flushing event, we elaborated a complete installation to create large images of the flow which can be analyzed later using Large-Scale Particle Image Velocimetry (LSPIV) an image-based measurement methods, (Hauet, 2006).

As the flow surface did not show clearly visible patterns, we developed a simple feeding device to supply the flow with Ecofoam chips, a water soluble, biodegradable, foamed filling material created from cornstarch. A 50 m-long rope was set between both banks, and a sliding container disseminated white chips over the flow. A digital video camera (Canon Mv750i) set on a 10 m-high telescopic mast was remotely controlled from the ground to record a large part of the flow. Several Ground Reference Points (GRP) were positioned to allow a geometric correction of the pictures.

Two distinct points of view were selected on the same reach of the gravel-bed river (Figure 4). The flow was recorded from the left-side embankment, with surveyed areas of approximately 60 m × 43 m. The first survey area was chosen just downstream the bar where the flow was quite simple, site D on Figure 2(a). The second survey area was the area of the upstream permanent connecting channel on the gravel bar which produced more complex flow patterns, site C on Figure 2(b). On both sites, 9 sequences of chip fed flows (and 17 sequences without artificial tracers) were recorded over a wide range of discharges (from 10 to 150 $m^3 \cdot s^{-1}$) and analyzed using LSPIV method. The LSPIV analysis was conducted on around 200 image pairs per sequence, with a 0.2 s time interval. Sequences lasted from 1'50" to 5'. Images were geometrically corrected using the GRP and changes in water surface elevation were taken into account. A classical cross-correlation algorithm was used to determine the displacement of the flow tracers. We noted that a threshold was essential to focus on chip patterns and to avoid correlation biases due to light reflection on stationary waves. LSPIV analysis gave instantaneous and time-averaged 2D surface velocity fields. Unfortunately, shadows on the right side of the stream and the absence of tracer in areas near the embankments resulted in some limited areas without velocity measurement. Here we focus on mean flow velocities. Time-averages are calculated on several minute long sequences with tracers in the camera field. Figure 2 shows examples of time-averaged velocity fields at sites D and C. Furthermore the comparison between LSPIV measurements and current meter velocities from the bridge shows acceptable agreement (same order of magnitude) and confirms the reliability of the method.

### 3.3 Suspended sediment concentration measurements

In addition to sampling stations distributed along the river channel from downstream dam (R3 on Figure 1)

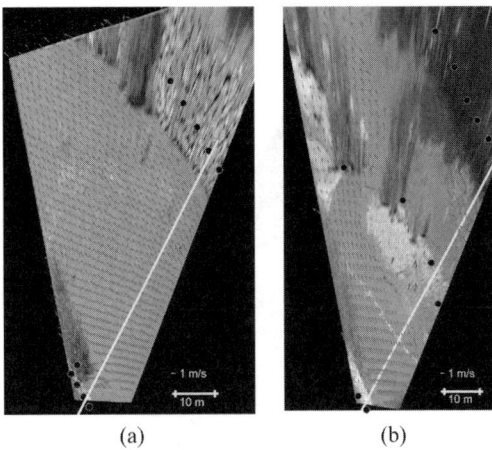

(a)                         (b)

Figure 2. Time-averaged surface velocity fields. (a) is site D and (b) is site C, white chips are clearly visible in the secondary channel. Lines are cross-sections used to plot transverse profiles. Ground Reference Points are plotted as black dots.

to the confluence with River Isère, suspended sediment concentration (SSC) measurements were performed from the bridge. Measurements along the river do not show any decrease in suspended sediment concentration except downstream a major restitution of clear water in the downstream part of the river and in areas close to the reservoir (R3) where peak concentration decreases from 80 g/l to 25 g/l over 10 km.

### 3.4 Morphological surveys and methods: topography

Digital elevation models (DEM) of the active channel were built using elevation measurements surveyed with a Trimble GPS system (5800 RTK) with a 1–3 cm vertical and horizontal precision.

Topographic surveys were performed a few days before and after the flushing event. Water level monitoring at the bridge let us assume that no geomorphologically significant flows occurred during the time period between the flushing event and each topographic survey.

A quasi systematic method was adopted in which we measured the topography of transverse sections every 20 m and we added more points where breaks of slope occurred in between. Topographic measurements of the embankments came from previous surveys and we assumed that blocks had not moved since.

First the DEM was built using linear interpolation in the main direction of the river (mean cell size 2 m). Data were linearly interpolated using guiding lines based on bar borders and embankment limits. In the connecting channel areas data were linearly interpolated using the direction of the connecting

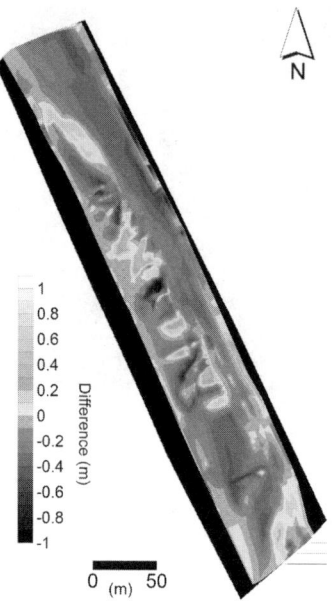

Figure 3. DEM difference ($D_2 - D_1$). Topographic changes due to the flushing event. We assumed no modification of the embankments.

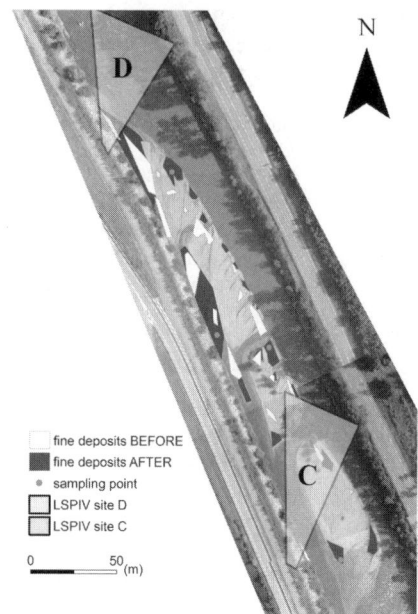

fine deposits BEFORE
fine deposits AFTER
● sampling point
☐ LSPIV site D
☐ LSPIV site C

0          50 (m)

Figure 4. Aerial picture of the study site after the flushing event, fine deposit localization and LS-PIV points of view. Water flows from bottom to top of the picture.

channels. In order to evaluate the change of topography due to the flushing event, the two DEM describing the topography before ($D_1$) and after ($D_2$) the flushing event were interpolated on the same finer lattice (cell size 0.5 m) using triangular linear interpolation. The grid step was small enough to take into account all the major topographical features within the reach including connecting channels details.

Subtraction of both DEM surfaces ($D_2 - D_1$) produced a DEM of differences between the surfaces as illustrated in Figure 3 with a minimal expected vertical precision of 10 cm. One can clearly identify the most significant changes in channel morphology within the reach for the period June–July 2006.

### 3.5 Grain size maps

Aerial pictures were taken from a drone (ABS aerolight) with a digital camera (Canon PowerShot G5 5MP or Canon PowerShot G3 4MP). Images were georeferenced using ground reference points and ArcInfo™ tools. In order to build grain size maps we used a simple and efficient sampling technique. In the field, homogeneous surfaces were defined using aerial pictures. For each zone we identified the main components of the gravel-bed. We evaluated the percentage of gravel in the intervals defined by the Wentworth scale. A reclassification was established to allow clear identification of 8 major surface grain size compositions. We then obtained spatial grain size maps to compare both states.

Using the fact that gravel areas appear whiter than fine sediment ones on aerial pictures, we identify fine sediment deposit patches on the bar corresponding to zones with clearly uniform gray color. Patch contours were manually digitized. Figure 4 reports the locations of deposits before and after the dam release. Using this method we identify that the surface covered by fine sediment deposits on the bar increased from 810 to 1590 m² between both surveys.

In addition, grain size of fine sediment deposits and grain size of suspended samples were obtained with a particle size analyzer (Malvern, 2000).

## 4 EXPERIMENTAL RESULTS

### 4.1 Flow velocity observation

Measured surface velocities are up to 4 m/s, which confirms the difficulty of using any intrusive velocity measurement device in such flow conditions.

Figure 5 reports velocity intensity (a) and (b) and velocity direction (c) and (d) in site D for discharges 94 m³/s and 113 m³/s. Velocity direction is the angle $\alpha$ between each velocity vector and the horizontal axis (East axis). The river reach has a global direction of $\alpha = 60°$. The velocity intensity distribution is classical, with the maximum intensity in the central part of the flow. This is surely due the straightness of the reach delimited by straight boulder embankments. However

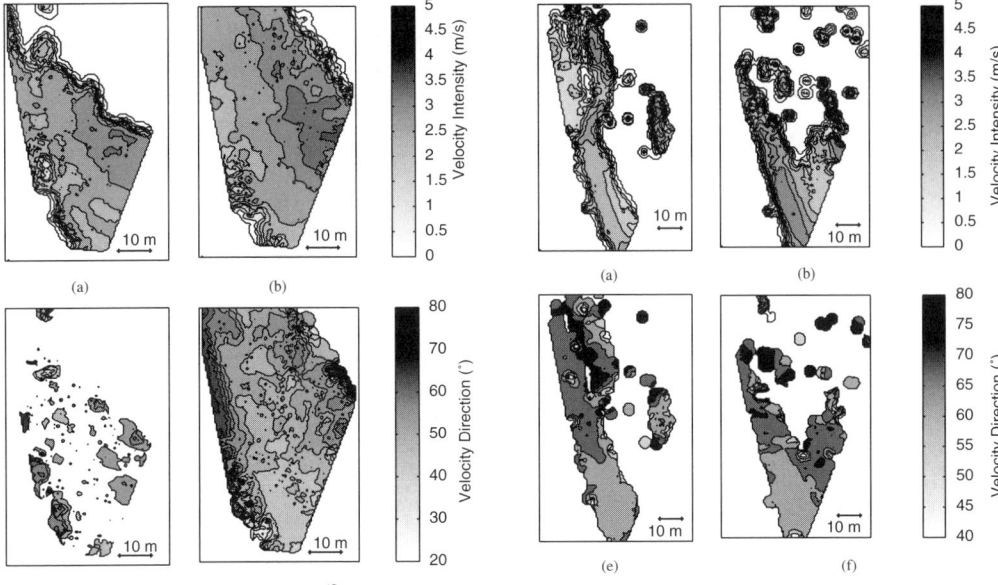

Figure 5.  Spatial distribution of velocity intensities (a, b) and orientation $\alpha$ (e, f), **site D** for discharges 94 m³/s (left) and 113 m³/s (right).

Figure 6.  Spatial distribution of velocity intensities (a, b) and orientation $\alpha$ (e, f), **site C** for discharges 77 m³/s (left) and 125 m³/s (right).

velocity intensities appear to be slightly higher on the right side of the channel. This slight asymmetry is due to the upstream presence of the bar. Asymmetry is confirmed by the analysis of spatial distribution of velocity direction.

Site C is a connecting channel zone, Figure 6 illustrates velocity features. Tracers were more visible on the left side of the river (secondary channel), in the near field which is also a slower zone. Velocities were measured for two different configurations: (a) at 10:50 ($Q = 77$ m³/s) the bar is not entirely inundated, whereas (b) at 12:38 ($Q = 125$ m³/s) the bar is under water. Despite low water depth, surface velocity intensities in the secondary channel are almost similar to intensities in the main channel (3 m/s). This can be explained by smaller bed roughness in the secondary channel due to finer bed sediments. On Figure 6(a) we clearly identify the acceleration of water flow from the secondary channel to the main channel through the connecting channel. Surface velocities over the bar are lower (2 m/s) and angle $\alpha$ is higher than the mean direction of the river reach (77° vs 60°). Water is deviated to flow directly from the secondary channel to the main channel over the bar. Previous observations highlight the complexity of flow patterns over the bar.

### 4.2  Improvement of discharge assessment

Discharge measurements performed from the bridge give peak values lower than those estimated by dam managers (120 m³/s vs. 150 m³/s). There is no loss of water between the downstream dam and the study site. Using water level monitoring performed near the bridge, low water discharge measurements and the discharge measurements from the bridge, we built a new discharge time series with a 2nd order polynomial rating curve, illustrated on Figure 7.

LSPIV measurements of surface velocities at site D are used to assess discharges over the event. Calculation is performed on a quasi-transverse stream section plotted on Figure 2(b). Bed surface from $D_1$ is used till peak discharge is reached, then $D_2$ is used. Vertical velocity distribution is assumed to be logarithmic (Smart, 1999) which is confirmed by velocity measurements at the bridge and detailed vertical velocity profiles performed at low flow. As used in many gravel-bed rivers, depth-averaged velocity is inferred from surface velocity using a straight-forward coefficient 0.85. DEM and depth-averaged projected velocities are interpolated using inverse distance weighting on 100 equally spaced points on the cross-section (0.77 m between two calculation points). In the case of no surface velocity measurement due to shadow or absence of tracers, a simple interpolation with constant value is performed. Interpolated data never form more than 30% of initial velocity measurements. Digital pictures of the flow help to deduce water level on the transverse section and a local mean longitudinal bed slope of 0.3% deduced from DEM is

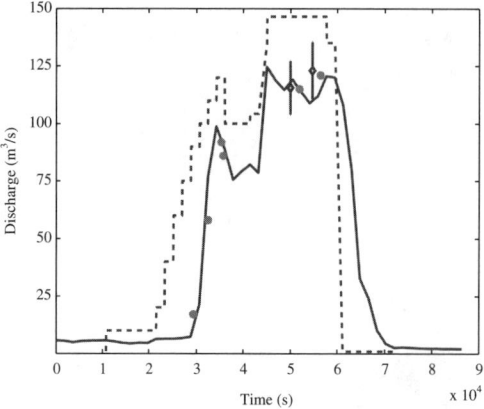

Figure 7. Discharge assessment (solid line) from water level measurements, discharge measurements (diamonds) and LS-PIV estimations (circles). Comparison with discharge assessment from dam managers (dot line).

applied. Discharges are calculated using the following formula:

$Q = \sum_{100}^{i=1} v_{Pi} h_i$, where $v_{Pi}$ is the interpolated depth-averaged projected velocity and $h_i$ is the interpolated water depth at point i on the cross section. Discharge estimates with LS-PIV measurements are reported on Figure 7. These estimations fit very well with both discharge measurements from the bridge and consequently validate the computed discharge curve.

### 4.3 Morphological changes of the gravel bar

The identification of major morphological features (main channel, secondary channel, bar margin, connecting channel...) is coherent with distinctive morphological changes. We noticed a global erosion of the study reach (Figure 3). The main flow channel was eroded especially in the downstream part of the reach which corresponds to an area of stronger velocities as noticed in paragraph 4.1. We observe erosion of the bar head, up to 0.4 m, with increase in the surface grain size, and the opposite change on the bar tail (deposition with decrease in surface grain size). The secondary channel remained at the same elevation. Major modifications of topography and grain size occurred in transient connecting channel zone: channels of the June state were filled with gravels and a new one appeared downstream. Morphological change in the bar margin are linked with changes in the equivalent transverse feature, for instance the upstream bar margin evolved like the bar head. Permanent connecting channels were strengthened and a small increase in grain size is noted.

### 4.4 Fine sediment deposits

A special attention is given to fine sediment deposits. The observation of fine sediment deposit location leads us to use and complete the specific typology developed by Wood and Armitage (1999). It distinguishes four types of deposits : Marginal deposits, Secondary channel deposits, Fine surficial laminae deposits and Vegetation and obstruction deposits. These specific places show low velocities or very low water depths that permit fine sediment deposition. Fine sediment deposits represent wider surfaces on the bar at the final state than before the flushing event. Main deposits are located at the bar margin, in back-water areas near the main flow. Secondary channel deposits are located in the downstream part of the secondary channel and comes from sediment routed by this secondary channel. Even if the grain size of fine deposits are coarser than suspended sediment, these observations highlight the contribution of flushing flow to fine sediment deposits on exposed areas of the river channel. Moreover landward and down-bar fining trends were identified in fine sediment deposit grain sizes. Fine sediment surely infiltrates in gravel bed and it could have been very interesting to quantify fine sediments in the gravel-bed interstices. However to reduce field work we chose to focus on surface characteristics only.

## 5 NUMERICAL ANALYSIS

### 5.1 Model

The water flow in the reach was modeled with the two-dimensional model Rubar20. The code Rubar20 uses a finite volume scheme that is explicit, Godunov-type and second order accurate (Paquier, 1995). It solves the 2-dimensional shallow water equations written below:

$$\frac{\partial h}{\partial t} + \frac{\partial hu}{\partial x} + \frac{\partial hv}{\partial y} = 0 \qquad (1)$$

$$\frac{\partial hu}{\partial t} + \frac{\partial (hu^2 + gh^2/2)}{\partial x} + \frac{\partial huv}{\partial y} =$$
$$- gh\frac{\partial Z}{\partial x} - g\frac{u\sqrt{u^2 + v^2}}{K_s^2 h^{1/3}} + K(\frac{\partial}{\partial x}(h\frac{\partial u}{\partial x}) + \frac{\partial}{\partial y}(h\frac{\partial u}{\partial y}) \qquad (2)$$

$$\frac{\partial hv}{\partial t} + \frac{\partial huv}{\partial x} + \frac{\partial (hv^2 + gh^2/2)}{\partial y} =$$
$$- gh\frac{\partial Z}{\partial y} - g\frac{v\sqrt{u^2 + v^2}}{K_s^2 h^{1/3}} + K(\frac{\partial}{\partial x}(h\frac{\partial v}{\partial x}) + \frac{\partial}{\partial y}(h\frac{\partial v}{\partial y})) \qquad (3)$$

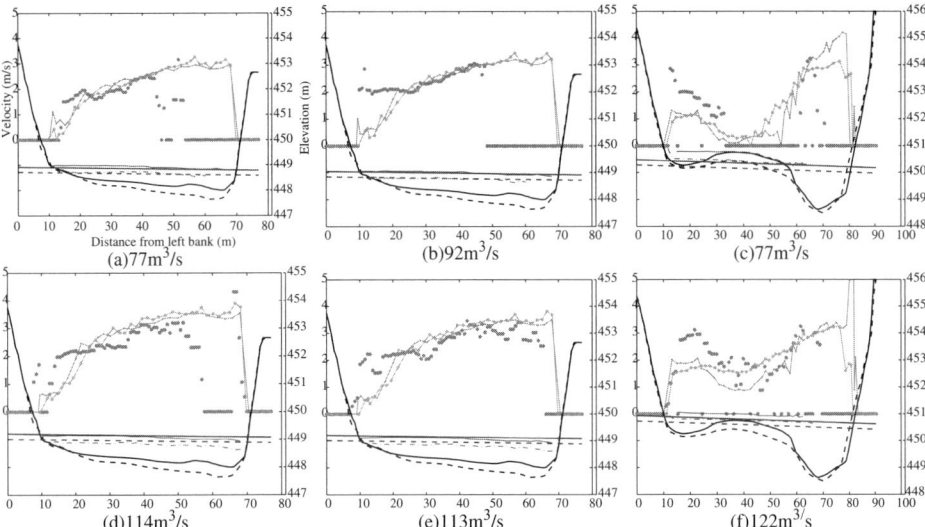

Figure 8. Velocity profiles from LS-PIV measurements (∗) and 2Dh calculations (-o- and -+-). Water lines from measurements (bold) and from calculation and gravel bed topography. Solid lines are calculated with DEM $D_1$ and dashed lines are values calculated with DEM $D_2$.

In which $u$ and $v$ are velocities along respectively $x$ and $y$ axis , $h$ water depth, $Z$ bottom level, $g$ gravity acceleration, $K_s$ Strickler coefficient, $K$ the viscosity coefficient.

## 5.2 Boundary conditions, mesh, topography and friction coefficient

The mesh is constituted of quadrangles or triangles which have 0 or 1 common edge. The mesh was built in the first step of linear interpolation in the DEM construction. Bed topographies before and after the event were meshed.

The Strickler coefficient $K_s$ is taken constant over the whole reach ($K_s = 40$) in order to obtain the mean water level along the reach of interest. At the upstream cross-section, we used an hydrograph and water level measurement as boundary conditions. The hydrograph is calculated from water level at the bridge (no time delay) and discharge measurements from current meter, using the fitted curve (Figure 7). At the downstream cross section the water level is set to critical regime.

## 5.3 Results: comparisons with field observations

Figure 8 shows comparison between LS-PIV measurements and velocity intensities calculated by the 2Dh calculations. In both cases calculated velocity vectors were interpolated on the cross sections mapped on Figure 2. Surface velocities were multiplied by coefficient 0.85 to permit comparison (paragraph 4.2).

On site D, Figures 8 (a), (b), (d) and (e) show that calculations fit fairly well with measurements except in a limited area near the left embankment. In this latter area, calculations underestimate velocity intensities. The asymmetry of transverse velocity profile is obvious on calculation results whereas measurements show an almost symmetric transverse profile. Furthermore calculations confirm the hypothesis used in discharge calculation in the case of lack of data close to the right embankment. It is acceptable to consider constant interpolation in this region. Thus the model could be a way to face lack of field measurement in space and time.

On site C, Figures 8 (c) and (f) show that calculations slightly underestimate velocity intensity in the secondary channel. Measurements provide velocities 50 cm/s higher in this region. Despite very few validated measurements in the main channel, comparison between calculations and measurements show good agreement in particular in the case of Figure 8(c). Figure 8(f) shows one strong velocity calculation close to the right embankment maybe due to a small cell, this will have to be corrected later on.

Difficulties to fit the simulated water level line to measurements were encountered for several reasons : (i) water level measurements are punctual and they must be interpolated with an assumed longitudinal slope and (ii) water level lines from calculation are deeply influenced by the topography used to build the mesh. This numerical approach does not take into account gravel bed changes. Our topography

1035

measurements show obvious changes of the bed during the flushing event. Knowing this, it is quite difficult to choose which topography corresponds to a given time. Further work will be done on this fundamental issue.

## 6   CONCLUDING REMARKS

In this paper we present field observations on morphological changes and flow processes over a gravel alternate bar. The chosen gravel bar is representative of a common morphological feature in gravel-bed rivers. Observations were performed during a reservoir release.

LS-PIV measurements and additional devices provided information on real discharge, flow patterns and fine sediment concentration. Changes in the bar morphology were assessed using topographic and grain size surveys before and after the event. Specific DEM construction and grain size mapping were developed. A typology of sub-bar morphological units and fine sediment deposits were used to analyze morphological changes. Despite a global erosion of the study reach, different trends in grain-size and in topographic evolution were observed on the bar.

To face the lack of information on flow parameters all over the bar during the whole event, a 2Dh numerical model of the study area is relevant in order to obtain values of flow parameters everywhere in the area of interest. It is useful to improve our understanding of processes. Numerical results showed good agreement with field measurements.

The major goal of this work was to develop a complete set of measurements to observe the evolution of a gravel bar during a high flow event. Despite some limitations due to measurement devices and natural field conditions, this objective was achieved. Measurements and analysis procedures developed here will be performed again to follow gravel bar evolution over years and during next reservoir releases.

Such information on on river bed morphodynamics during high flow is very useful to managers trying to understand how the river works to prevent flooding roads or bridge scouring. For instance the River Arc is often surveyed to prevent hazardous gravel deposition or aggradation and mechanical cleaning works are planned to avoid too large gravel deposition.

## ACKNOWLEDGEMENTS

Authors are grateful to people who took part in the field survey. In particular, we thank Cemagref technicians (G. Dramais, M. Lagouy) for their part in performing field measurements. Diren Rhône-Alpes performed discharge measurements from the bridge. EDF GEH Maurienne provided discharge estimates and authorized measurements during the reservoir release. UMR5600 helped on aerial topographic survey. Many thanks to Julien Hervé for his help on field measurements.

## REFERENCES

Brandt, S. (1999). Sedimentological and geomorphological effects of reservoir flushing: The Cachi reservoir, Costa Rica, 1996. *Geografiska Annaler A 81*, 391–407.

Fasolato, G., P. Ronco, and M. Tregnaghi (2006). Morphodynamics of mountain rivers following repeated sediment release from reservoirs. In *River Flow 2006*, pp. 1329–1336.

Hauet, A. (2006). *Estimation de débit et mesure de vitesse en rivière par Large-Scale Particle Image Velocimetry*. Ph. D. thesis, INP Grenoble.

Jodeau, M. and A. Paquier (2006). Analysis of water and sediment flows over an alternate bar in a gravel bed river. In *River Flow 2006*, pp. 1251–1257.

Mürle, U., J. Ortlepp, and M. Zahner (2003). Effects of experimental flooding on riverine morphology, structure and riparian vegetation : the River Spöl, Swiss National Park. *Aquatic Science 65*, 191–198.

Paquier, A. (1995). *Modélisation et simulation de la propagation de l'onde de rupture de barrage*. Ph. D. thesis, Université Jean Monnet, St Etienne, France.

Rempel, L. and M. Church (2002). Morphological and habitat classification of the lower Fraser River gravel-bed reach: Confirmation and testing. Technical report, The Fraser Basin Council.

Smart, G. M. (1999). Turbulent velocity profiles and boundary shear stress over gravel bed rivers. *Journal of Hydraulic Engineering 125*, 106–116.

Wohl, E. and D. Cenderelli (2000). Sediment deposition and transport patterns following a reservoir sediment release. *Water Resources Research 15*, 319–333.

Wood, P. and P. Armitage (1997). Biological effects of fine sediment in the lotic environment. *Environmental Management 21*, 203–217.

Wood, P. and P. Armitage (1999). Sediment deposition in a small lowland stream - management implications. *Regulated Rivers: Research and Management 15*, 199–210.

*River, Coastal and Estuarine Morphodynamics: RCEM 2007 – Dohmen-Janssen & Hulscher (eds)*
*© 2008 Taylor & Francis Group, London, ISBN 978-0-415-45363-9*

# Evolution of sediment deposition and flow patterns in a rectangular shallow reservoir under suspended sediment load

S.A. Kantoush, J.-L. Boillat & A.J. Schleiss
*Laboratory of Hydraulic Constructions (LCH), Ecole Polytechnique Fédérale de Lausanne (EPFL),*
*Lausanne, Switzerland*

ABSTRACT: The deposition behavior of fine sediment is an important phenomenon, which remains unclear to engineers concerned about reservoir sedimentation. In the present case bed level changes, suspended sediment concentrations, and flow velocities were measured in a quasi continuous way. The goal of the study was to investigate the interaction between turbulent flow structures in rectangular shallow basin, suspended particles, bed forms and other instabilities by using several measurement techniques. The results help to understand the flow mechanism and the sediment exchange process. The prediction of sediment behavior lies in the prediction of flow behavior and the results are very sensitive to the geometry and the boundary conditions. The deposition pattern is obviously strongly influenced by the inlet jet deviation and, in turn, material deposits are able to change later the pattern of the flow structure. The volume of the deposited sediments reached 50% of the total reservoir volume after 18.0 hours. The reservoir reaches to the equilibrium after 16.0 hours at which the suspended sediment release efficiency reaches 100%.

## 1 INTRODUCTION

### 1.1 *Background and objectives*

The silting-up of reservoirs is a very complex process. Shear flow over a mobile bed induces sediment transport and the generation of bed forms. The interaction between the flow and the bed usually produces different types of regular patterns characterized by a wide range of sizes and shapes (dunes, ripples, antidunes, etc.). In turn both, sediment transport and bed forms, influence the flow. The importance of studying the development and evolution of these regular patterns arises because the generated bed forms can increase the flow resistance. A mass of sediments is kept in suspension above a bed by the eddies of flow turbulence. By way of contrast, bedload grains move by rolling and saltation on the bed, with their immersed weight in dynamic equilibrium with the solid normal stress transmitted by the action of the fluid shear, Bagnold (1973). Applied to individual grains the concept of suspension is necessarily statistical, because of continuous exchange between bed load and the overlying turbulent flow. However, a steady state exists with respect to a suspended mass in a steady, uniform flow. Over a sufficiently long period of time the measured mass will itself be constant whereas the constituent grains may continuously be exchanged between bed, bedload and suspended load. A state of dynamic equilibrium thus exists.

Predicting the sediment transport rate and a better understanding of the processes of bed form generation depend on whether and how the sediment influences the flow characteristics Crowe (1993). Much debate has arisen concerning the influence of suspended sediment on the von Kármán coefficient $\kappa$, with researchers proposing either a reduction in $\kappa$ with increasing sediment suspension [e.g., Vanoni (1946, 1953); Einstein & Chien (1955)] or maintaining $\kappa$ as a constant but using an appropriate wake coefficient according to the "law of wake" [Itakura & Kishi (1980); Coleman (1981)]. Recent studies have challenged the assumption that $\kappa$ is constant and the outer wake region is most affected by suspended sediment (Lyn (1992). Furthermore, Gust & Southard (1983) believe that $\kappa$ is reduced from its clear water value, even in the case of weak bedload transport without suspended sediment. Recent work concerning two-phase flows in pipes and wind tunnels has shown that grains may enhance turbulence production when larger than the microscale of turbulence or attenuate turbulence when the grains are small enough to be enclosed within the turbulent eddies [Gore & Crowe (1989); Hetsroni (1989); Kulick (1994)]. When large grains are added to the flow, fluid turbulence may increase [e.g., Mueller (1973); Gore & Crowe (1989); Hetsroni (1989)] or remain relatively unchanged [Rayan (1980); Lyn (1992)]. Other researchers have found that adding to the flow fine-grained sediment may cause turbulence attenuation [Xingkui & Ning (1989); Kulick (1994)]. In addition, factors including the sediment concentration, grain size sorting, and sediment-to-fluid density ratio have also been shown to influence turbulence modulation Gore & Crowe (1989). However, it has also been suggested that the ratio between the response time of a particle within a flow to the scale of the turbulence may influence both turbulence attenuation and enhancement in the carrier fluid Elghobashi (1994). It is clear from the aforementioned research that the full understanding of the interaction between turbulence and sediment transport is still missing.

This study focuses on the sedimentation of shallow reservoirs by suspended load and the objective of the experiments is to confirm the existence of a morphological equilibrium in the basin by long period test (18 hour). Furthermore, examined different test durations (1.5 hr, 4.5 hr and 9 hr) are with the purpose to find the optimal one to continue with additional test configurations. Moreover, the effect of the suspended sediment mixture and sediment deposits on the flow field is illustrated. Furthermore, it is trailed to a better understand the mechanism governing the sediment exchange process between the jet entering the reservoir and the associated turbulence structures. Finally, the major physical processes responsible for the observed phenomena (asymmetric flow patterns in symmetric geometry) are analyzed.

## 2 PHYSICAL MODELING

### 2.1 Experimental setup

The experiments were performed in a closed configuration as showed in Figure 1. The experimental setup consists of a rectangular inlet channel, 0.25 m wide and 1.0 m long, a rectangular shallow basin with inner dimensions of 6.0 m length and 4.0 m width, a rectangular outlet channel 0.25 m wide and 1.0 m long. The water level in the reservoir is controlled by a flap gate 0.25 m wide and 0.30 m high at the end of the outlet. The basin is 0.30 m deep and has a flat bottom. Walls and bottom conditions are hydraulically smooth. Adjacent to the reservoir, a mixing tank is used to prepare the water-sediments mixture. A sediments supply tank is mounted above the mixing tank. The mixing tank is equipped with a propeller type mixer to create a homogenous sediments concentration. The water-sediments mixture is supplied by gravity into the water-filled rectangular basin. Along the basin side walls, a 4.0 m long, movable, aluminum frame is mounted to carry the measuring instruments.

Suspended sediments are modeled by crushed walnut shells with a median grain size $d_{50} = 50\,\mu m$ (Table 1) which is non cohesive. In Table 1, the most commonly used parameters to describe the sediment

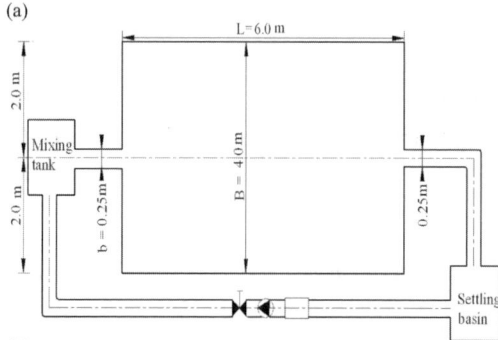

(a)

(b)

Figure 1. (a) Plan view of the reference experimental setup (L = 6 m, B = 4 m); (b) photograph, looking upstream.

Table 1. Parameters describing the physical properties and grain size distribution for the sediments material (crushed walnut shells).

| | |
|---|---|
| $\rho_s$ [kg/m$^3$] | 1500 |
| $d_{50}$ [$\mu m$] | 50 |
| $d_{84}$ [$\mu m$] | 115 |
| $d_{16}$ [$\mu m$] | 20 |
| $\sigma_g = \sqrt{d_{84}/d_{16}}$ [–] | 2.4 |

properties and grain size distribution are summarized. Several parameters were measured during every test; namely: 2D surface velocities, 3D velocities, thickness of deposited sediments, sediments concentration of inflow and outflow, water level in reservoir and discharge. Floating white polypropylene tracer particles with a 3.4 mm diameter, contrasting with dark bottom, are used to visualize the surface velocity field. Instantaneous velocity fields are obtained by a 1.3 Mpixel digital camera and the PIV algorithm of FlowManager. The bed level evolution was measured with a Miniature echo sounder (UWS). The sounder was mounted on a movable frame which allowing to scan the whole basin area.

Automated measurement of suspended sediments is crucial to study the sediment transport. The short duration, high-intensity flows that are responsible for a large fraction of sediment movement are best observed by continuous monitoring systems. For this purpose two sensors SOLITAX were installed at the inlet and outlet channels for online suspended sediment measurements. The measuring principle is based on a combined infrared absorption scattered light technique that measures the lowest turbidity values in accordance with DIN EN 27027 [Kantoush (2006)].

## 2.2 Test procedure

To account for progressive morphological evolution and to verify the final achievement of dynamic equilibrium, a long-term test has been performed with durations up to 18 hours. It has to be noticed that these long runs were performed in several time steps 1.5, 4.5, and 9.0 hour, i.e. the facility has been interrupted to allow bed morphology recording.

Six runs were conducted: run 1 with clear-water flow and no sediment feed. One hour after starting the pump and stabilizing the flow for a given discharge, LSPIV measurements were performed. Next runs 2, 3, 4, and 5 were conducted under same flow conditions but with sediment supply. The sediment concentration was kept constant over the respective test durations (1.5, 3, 4.5, 9, 18 hours). The inflow sediment mixture was controlled by turbidity meter every minute. UVP probes measured 1D vertical velocity profiles as well as local 3D velocities. Every 30 minutes, LSPIV measurements were performed. The bed morphology was measured at different cross sections and different time steps (90, 180, 270, 540, and 1080 minutes) after each run (2, 3, 4, and 5).

In all runs, the discharge is kept constant at $Q = 7.0$ l/s and the water level is controlled by a flap gate located in the downstream part of the outlet channel. The downstream water level was kept constant at $h = 0.2$ m. The inflow suspended sediments concentration was similar for all runs with 3.0 g/l.

## 3 RESULTS

### 3.1 Large coherent structures with and without suspended sediment

The flow features and large–scale structures were investigated by using LSPIV measurement technique. Figure 2 shows an overview of the velocity field and behavior of large-scale coherent structures in clear water (run 1). A plane jet issues from the narrow leading channel and enters straight ahead in the first half meter of the wider basin. Then, the main flow tends to develop in a curved way towards the right hand side over the next two meters until it touches and follows the right wall. When separating from the right wall, the main flow induces a recirculation zone. A main large stable eddy is generated in the centre of the basin,

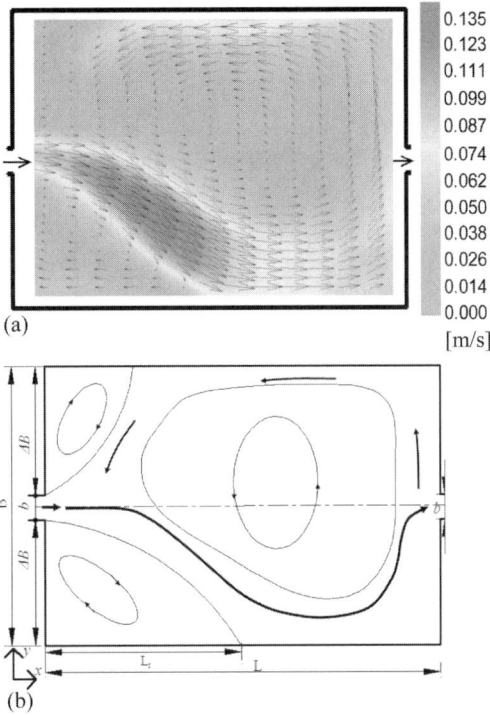

(a)

(b)

Figure 2. (a) Time averaged flow pattern and velocity magnitude (m/s) for clear water obtained by LSPIV measurements, (b) Plan view of the reference experimental setup ($L = 6$ m, $B = 4$ m) and definition of the geometrical parameter of the first recirculation cell in the basin ($L_r$). Discharge $Q = 7.0$ l/s, and water depth $h = 0.2$ m.

rotating anticlockwise. Furthermore, two small 'triangular' gyres are formed rotating clockwise in the upstream corners of the basin. The deflected jet works as a vortex shedding region between the main eddy in the centre and the triangular one in the upstream right corner. Moreover, two mixing layers left and right are observed between the main flow and both eddies in Figure 2. The jet seems to be attracted towards one side of the basin (in the test always to the right side). A second vortex shedding zone in the reverse direction is generated between the main gyre and the small triangular clockwise eddy in the upstream left corner. The reverse flow jet, which is generated by the inertia of the main gyre, pushes the incoming jet aside and forms a shedding point between four features: main flow jet, reverse flow jet, large main gyre, right and left corners gyres. The jet preference for the right side is weak, since a stable mirror image of the flow pattern can easily be established by slightly disturbing the initial conditions. The addition of sediment decreases the mixing length or eddy size, the reattachment length $L_r$ of the right corner gyre increasing with time. The flow

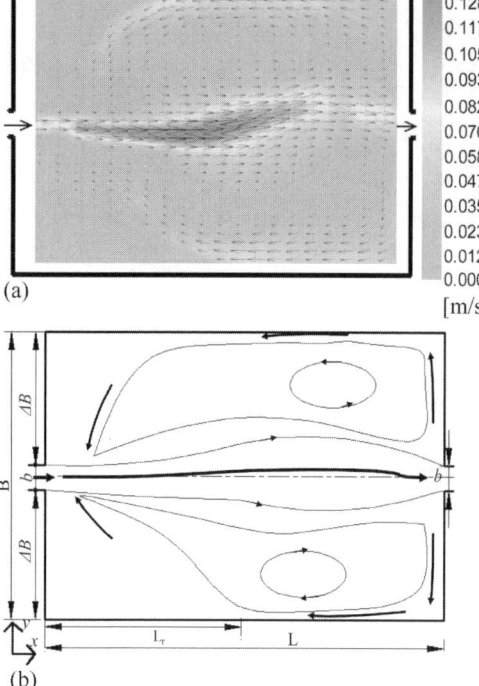

0.140
0.128
0.117
0.105
0.093
0.082
0.070
0.058
0.047
0.035
0.023
0.012
0.000
[m/s]

(a)

(b)

Figure 3. (a) Time averaged flow pattern and velocity magnitude (m/s) with sediment entrainment flow obtained by LSPIV measurements after 4.5 hours, (b) Plan view of the reference experimental setup (L = 6 m, B = 4 m). Discharge $Q = 7.0$ l/s, water depth $h = 0.2$ m, and suspended sediment concentration $C = 3.0$ g/l.

becomes also more stable and symmetric. This fact will be explained in detail hereafter. Figure 2 shows the second flow feature developed with sediment entrainment. As a result of ripple formation and suspended sediment concentrations, the flow field is completely changed. The gyres in the upstream corners disappear and a pattern emerges rather symmetric with respect to the center line. The two remaining gyres interact with the jet which shows some tendency to meander. Since the exchange with the up-stream corners of the basin is very small, it is expected that not much deposition takes place in those areas. Apparently the changes in the bed forms or effective roughness resulting from the sediment deposition can completely modify the overall flow pattern. As a conclusion, as sediment is added to the flow, the turbulence is reduced and the mixing lengths decrease which, together with increasing roughness, cause an increase in velocity gradient when compared to clearwater flow (see Figures 2 & 3). Several physical mechanisms may be invoked to cause these effects.

## 3.2 Long term morphological evolution and corresponding flow field

The average flow field and the corresponding bed morphology are shown in Figure 4 for five different runs (1.5, 3, 4.5, 9, 18 hours) allowing a comparison of the long-term bed evolution in the reservoir. For all the tested runs, two typical features were observed. The first is the development of the sediment deposition with ripples formation concentrated on the right hand side till bed thickness deposition reaches up to 15% of the water depth. The second is concentrated along the centerline with relatively steep gradients near the inlet channel and the first part of the jet.

Moreover, the deposition gradually increases generating a wider bed elevation underneath the jet centerline. The basin fills up from the center to the walls directions, starting from downstream to upstream direction. With longer period these gradient slopes regions will be eventually filled up with the finest sediment fraction. Deposits configurations in Figure 4 shows how the mixture of water and sediment is advected and diffused throughout the basin following the general flow patterns. The footprint of the flow patterns was clearly visible in the morphology. The deposition at both upstream corners is less than in other parts.

The resistance to flow is relatively small on the smooth and plane bed at the start. However, the flow resistance increases as ripples are being formed. The ripples play an important role in the interaction between the boundary layer flow structures and sediment transport. The asymmetric ripples formed after 1.5 hours (run 2) near the right side wall follow the same direction as the flow pattern. However, the increased roughness height associated with mobile sediment may contribute to increase in shear velocity and turbulence intensity. The sediment concentration and sediment deposition are higher right below the main streamlines connected to the inflow channel although of existence of high velocity.

Whereas, after 3.0 hours (run 3), as a result of ripple formation and suspended sediment concentration the flow field is completely changed. During 9.0 hours (run 5), most of the sediment deposits and suspended concentration are along the center of the basin. A symmetric ripples pattern formed on the middle of the basin is clearly visible. After 9.0 hours (run 5), the deposition on the center gradually increased generating a wider bed elevation underneath the jet centerline with a width of approximately three times the inlet channel. There is another longitudinal gradient between the upstream and downstream parts. A tongue shape deposition occurs along the centerline of the basin. The tongue average thickness is of 0.16 m and locates underneath the jet centerline with average width of approximately eight times the inlet channel.

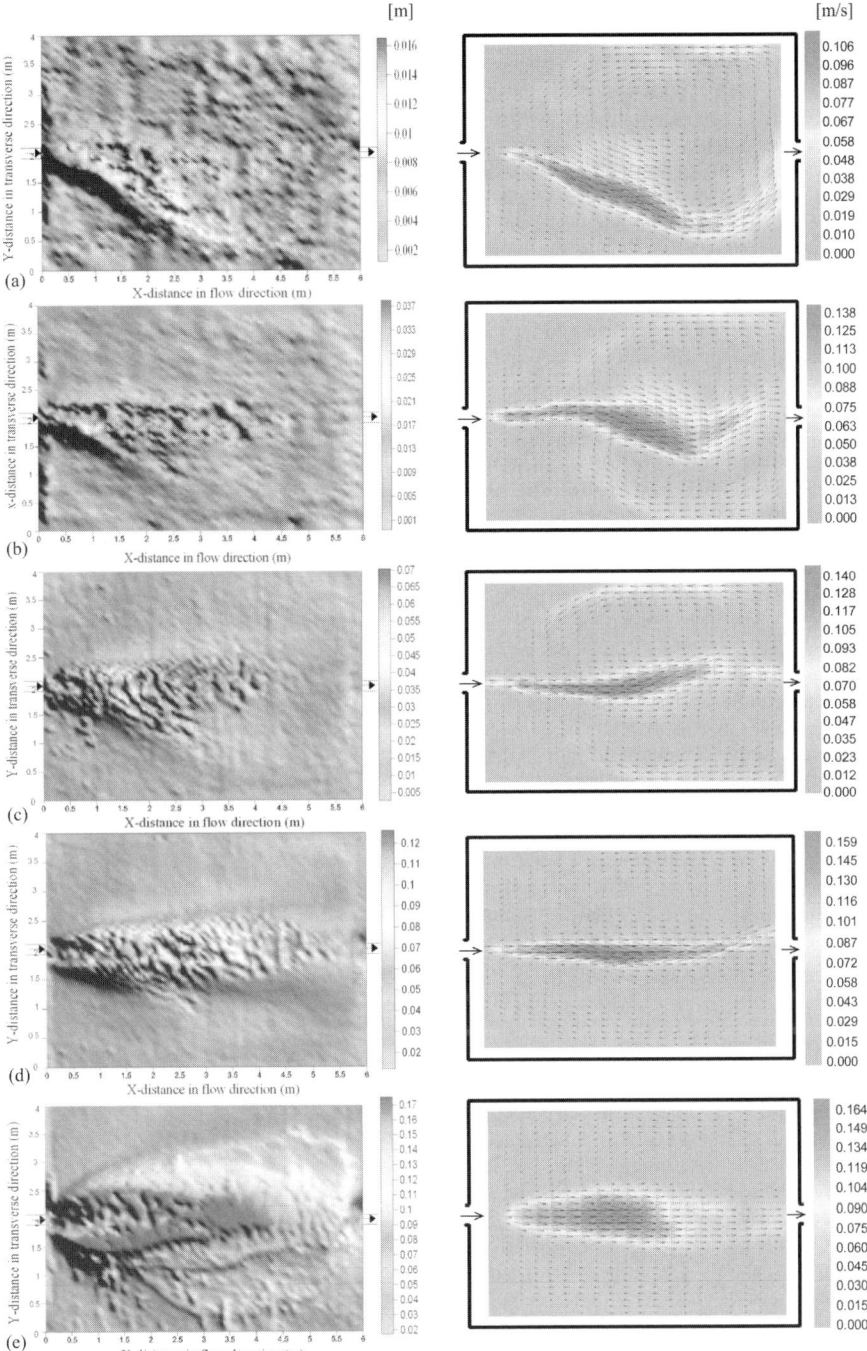

Figure 4. Long-term morphological evolution of deposition in m, (left) and flow patterns and velocities in m/s, (right) for different runs time steps (a) run 2; 1.5 hr, (b) run 3; 3 hr, (c) run 4; 4.5 hr, (d) run 5; 9.0 hr, (e) run 6; 18.0 hr. Discharge Q = 7.0 l/s, water depth $h = 0.2$ m and suspended sediment concentration C = 3.0 g/l. Caution: Full scale changes in each case.

In Figure 4(e) after 18.0 hours (run 6) a uniform deposition occurs.

During 1.5 hours (run 2) of adding sediment the observed flow pattern in Figure 4(a) does not differ much from previously explained for clearwater, except the increase in size of $L_r$ and a downstream shifted reattachment point. Apparently this is due to the suspended sediment entrainment and the associated emergence of bed forms (see Fig. 4(a)). A pattern has emerges that is rather symmetric with respect to the center line after 3.0 hours (run 3) (Fig. 4(b)).

It is also seen that the remaining two gyres interact with the jet which shows some tendency to meander. The changes in the bed forms or effective roughness resulting from the sediment deposition are able to completely chang the overall flow pattern. Through the same mechanism a further development can be expected. Velocity gradient is increased when compared to run 2 (Fig. 4(a)). Figure 4(c) still show symmetric flow patterns with two gyres coupled to the jet flow and same magnitude. Moreover, the velocity magnitude and the strength of the two gyres are increase. This pattern appears to be rather stable. The flow pattern in Figure 4(e) is similar and equal to the pattern of Figure 4(d).

Almost 50% of the basin total volume has been filled by the deposits after 18 hours. By comparing the five runs in Figure 4(a, b, c, &d), it can be concluded that asymmetric flow patterns have developed differently regarding the dimensions and strength of the circulation cells in both sides. The asymmetry leading to the subsequent pattern can already be seen. Stable bed morphology has been reached after 4.5 hours and more than 18 hours would be needed to reach full morphological equilibrium in the basin. Finally, symmetric ripple patterns, could be observed after 4.5 hours of adding sediment.

## 3.3 Cross sections comparison of deposition depth

A detailed comparison of transversal morphological development at four different sections is presented in Figure 5 (a, b, c and d). The time evolution of sediment deposits can be seen for the five runs (1.5 hr, 3 hr, 4.5 hr, 9 hr and 18 hr) at cross sections X1 = 1.5 m, X2 = 2.0 m, X3 = 3.0 m and X4 = 4.5 m, respectively. The first 2.0 meters show different bed forms (shape and height) than the last two meters downstream. Figure 5(a) shows depositions in transversal direction of the basin at distance of 1.5 m from the inlet, for the five runs. After 1.5 hr (run 2), almost a uniform depositions over the basin with average thickness of 0.015 m is observed. Due to the complete change of the flow pattern after 3 hr (run 3), sediment deposition rate is slightly increased by 0.005 m. The bed thickness observed after 4.5 hr (run 4) is almost two times

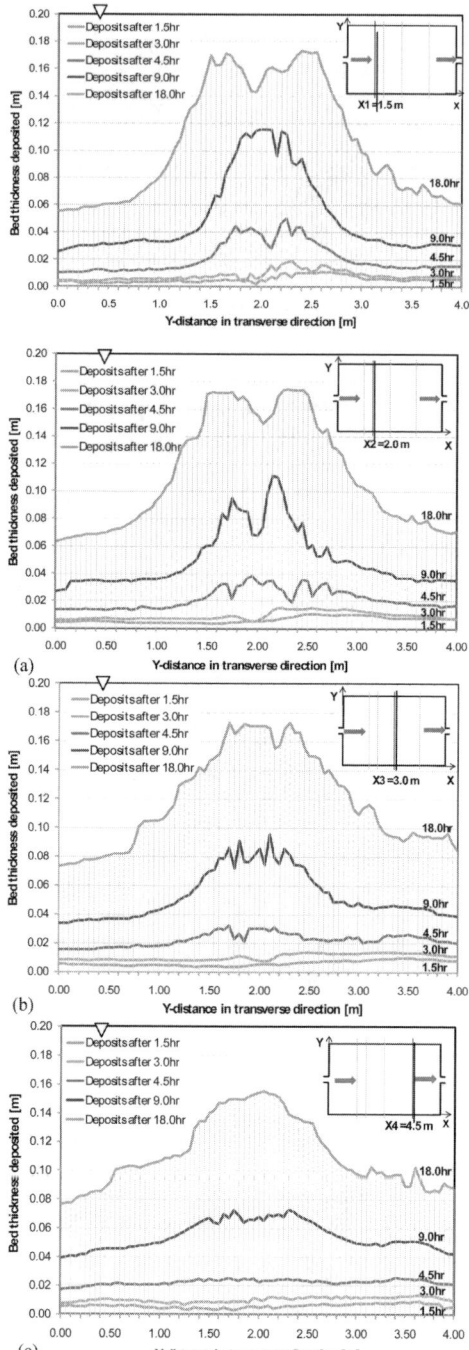

Figure 5. Comparison of bed profiles at different cross sections of the basin, (a) X1 = 1.5 m, (b) X2 = 2.0 m, (c) X3 = 3.0 m and (d) X4 = 4.5 for runs (2, 3, 4 and 5), after (1.5, 3.0, 4.5, 9.0, and 18 hours). Water depth $h = 0.2$ m.

higher than after 3.0 hr (run 3) at the center but does not differ much at left and right walls. There are two transverse mild slopes of average 2% to the center. But after 9.0 hr (run 5) steep slopes appear at both sides. A channel formed on the hill of the deposits along the centerline elevated mount channel forms at the center with a width of 0.75 m after 9.0 hr (run 5) as shown in Figure 5(a). During 9 hr of adding suspended sediment more deposits can be observed at the center and the thickness reaches to 0.17 m after 18.0 hr (run 6). A horizontal deposited reach with 2.0 m width has been formed along the centerline and still less deposits in the upstream corners.

Figure shows almost constant sediment deposits within the first hours for run 2, run 3, and run 4 but the deposits rate is increased for runs 4 and 5. It may be concluded that a stable morphology has been reached after 18.0 hours and almost morphological equilibrium in the basin has been reached for. Bed forms are almost uniform after 1.5 and 3.0 hours but 4.5 hr deposits show wavy bed forms (Fig. 5(b)). The bed thickness observed after 9.0 hr is almost three times higher than for 4.5 hr. The bed becomes thicker and even more irregular after 9.0 hours.

For the middle cross section, the influence of the flow deviating towards the centerline of the basin is clearly visible by strongly reduced bed thickness after 3.0 hr. Also, due to the recirculation eddy, the sediment deposits gradually start to increase again in the middle. The deposition progressively increases after 4.5 hr from left wall towards a peak value of 0.03 m at the middle section, followed by a small decrease at the right wall. After 18.0 hr, Figure 5(c), shows almost similar sediment deposition behavior as presented in Figure 5(b).

It is clearly seen that the deposition layers after 1.5, 3.0, and 4.5 hr are parallel with each other and almost a uniform deposition rate is reached at both sides. But after 9.0 hr the sediment deposits formed underwater-ridge at the center with mild slopes towards the sides. After 9 hours of testing, the deposition on the sides gradually increased generating a wider bed elevation underneath the jet centerline with a width of approximately 1.25 m.

### 3.4 Suspended sediment concentrations and sediment trap efficiency

The suspended sediments concentrations (SSC) at the inlet and outlet channels have been monitored in detail. Since the instruments continuously report SSC, even though these values may be spurious, it was important to scan the data to search for the periods of clogging, i.e. sudden declines in SSC, or periods when SSC remained stable despite significant changes in bed forms, or also sudden changes in flow patterns. Figure 6 shows SSC release from the reservoir every

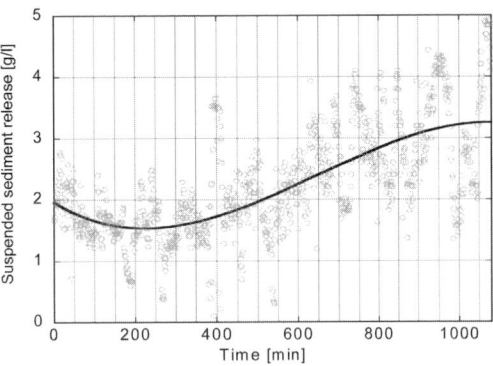

Figure 6. Suspended sediments concentration released from the reservoir during 18 hours, measured every minute.

minute. Sediments inflow was kept constant during the test around 3.0 g/l. The sediments release in the beginning of the test for runs 2, 3 & 4, is low due to the bed formation and mixing exchange between circulations. Then it gradually increases during runs 5 & 6 and relatively stable at the end of run 6. After 16 hours of long period, SSC inflow and outflow are approximately equal rates at which the basin reaches to the equilibrium.

Suspended sediments release for the long duration was used to calculate the sediments release and trap efficiencies, as shown in Figure 7. The sediments release efficiency (Erealease) of a reservoir is the mass ratio of the released sediments to the total sediments inflow over a specified time period. The sediments release efficiency (Erealease = E) of a reservoir is the mass ratio of the released sediments to the total sediments inflow over a specified time period. It is complementary to the trap efficiency (Etrap = T): E = 1 − T, where E: [−] & T: [−].

A general increasing tendency represented by parabolic curve for sediments release can be seen in Figure 7(a). The sediments release in the beginning of the test is low due to the bed formation and mixing exchange between internal circulations. Then it gradually increases. Due to the flow deflection to the right side, as shown in Figure 4(a), ripples start to form on the right side and SSC starts to decrease as shown in Figure 6. After 3.0 hr (run3), the flow pattern starts to change the direction from right to the center (see Fig. 4 (b)), and new ripples are formed at the center. The SSC decreases compared to the first run. The presented curve has the same tendency as explained in Figure 6. Continuous shallowing of a reservoir causes diminution in the area of active flow in the reservoir cross section and this process is followed by an increase in flow velocity through the length of the reservoir. Then the trap efficiency for the suspended load reaches zero as shown in Figure 7(b).

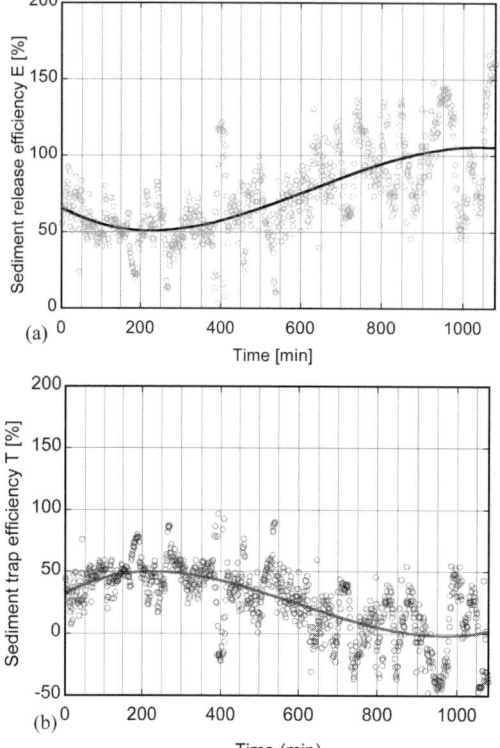

(a)

(b)

Figure 7. Evolution of sediments release E and trap efficiency T for long run duration (18 hr).

## 4 CONCLUSIONS

When suspended sediment is added to the turbulent flow over a plane bed in shallow basin and transported as bed and suspended load the study revealed:

1. The large coherent structures disappear compared to clear water flow with similar flow properties.
2. Turbulence is damped with bed deposition and suspended sediment entrainment.
3. Suspended sediment and ripples stabilize the flow and change the flow pattern from asymmetric with clear water to symmetric with sediments.
4. Ripples and bed form deposits with a thickness reaching 15% of the water height are directly responsible for changes in flow structure.
5. High sediments concentrations and deposits form along the main jet due to the formation of a large mixing layer between the primary and secondary gyres.

6. The loss of the free reservoir volume during 18 hr of testing was found to be about 50 %. The silting ratio is highly correlated with the initial reservoir capacity. Moreover, the total sedimentation volume grows linearly with time.

## REFERENCES

Bagnold, R.A. 1973. The nature of saltation and of 'bed-load'transport in water. *Proc. R. SOC. A*, (332): 473–504.
Coleman, N. 1981. Velocity profiles with suspended sediment. *J. Hydr. Res.*, 19, 211–229.
Einstein, E.A., and Chien, N. 1955. Effects of heavy sediment concentration near the bed on velocity and sediment distribution. *Univ. of California, Berkeley, and U.S. Army Corps of Engrs., Missouri River Div., Report No. 8.*
Elghobashi, S. 1994. On predicting particle-laden turbulent flows. *Appl. Scientific Res.*, (52): 309–329.
Gore, R.A., and Crowe, C.T. 1989a. Effects of particle size on modulating turbulent intensity. *Int. J. Multiphase Flow*, (15): 279–285.
Gust, G., and Southard, J.B. 1983. Effects of weak bedload on the universal law of the wall. *J. Geophys. Res.*, (88): 5939–5952.
Hetsroni, G. 1989. Particles-turbulence interaction. *Int. J. Multiphase Flow*, (5): 735–746.
Itakura, T., and Kishi, T. 1980. Open channel flow with suspended sediments. *J. Hydr Div., ASCE*, (106): 1325–1343.
Kantoush, S.A., Bollaert, E.F.R., Boillat, J.-L., Schleiss, A.J., and Uijttewaal, W.S.J. 2006. Sedimentation Processes in Shallow reservoirs with different geometries. *IAHR Proc. of the International Conference on Fluvial Hydraulics,* 1623–1631, Lisboa, Portugal, Balkema.
Kulick, J.D., Fessler, J.R., and Eaton, J.K. 1994. Particle response and turbulence modification in fully developed channel flow. *J.Mech., Cambridge*, U.K., (277): 109–134.
Lyn, D.A. 1992. Turbulence characteristics of sediment-laden flows open channels. *J. Hydr. Engrg., ASCE*, 118, 971–988.
Mueller, A. 1973. Turbulence measurements over a movable bed with sediment transport by laser-anemometry. *Proc., 15th Cong., Int. Assn. Hydr. Res.*, Vol. 1, A7-1-A7-7.
Rayan, M.A. 1980. Influence of solid particles in suspension on some turbulent characteristics. Multiphase transport: Fundamentals, reactor safety, applications, Vol. 4, Hemisphere Publ. Co., New York, N. Y., 1969–1991
Vanoni, V.A. (1946). "Transportation of sediment by water. *Trans. Am. Geophys. Union*, 3, 67 133.
Vanoni, V.A. 1953. Some effects of suspended sediment on flow characteristics, *Proc., 5th Hydr Conf., Univ. of Iowa, Iowa City, Iowa.*
Vanoni, V.A., and Nomincos, G.N. 1960. Resistance properties of sediment-laden streams. *Trans. ASCE*, (125): 140–167.
Xingkui, W., and Ning, Q. 1989. Turbulence characteristics of sediment-laden flow. *J. Hydr. Engrg., ASCE*, (115): 781–800.

*Topic E: Human interferences in morphodynamics*

KEYNOTE
# Human interferences in morphodynamics

I. Brøker
*DHI, Hørsholm, Denmark*

ABSTRACT: The present paper discusses human interferences with the natural morphological development. Examples of negative impacts such as erosion, sedimentation or degradation of water quality in rivers, estuaries and coastal zones are presented. Further, examples are given, where human activities in sensitive environments have been successfully introduced. The examples are explained and documented using numerical models; however, emphasis in this paper is on the overall description of the morphodynamic processes.

## 1 INTRODUCTION

Over the past centuries increasing human activities have taken place along rivers, estuaries and coasts. Harbours are built with ever growing requirements to the water depth, rivers have been regulated to serve navigation purposes and provide fresh water for water supply, hydro power and prevent flooding and low lying land in estuaries have been reclaimed. Many projects have been initiated at times where the morphological processes and developments were not understood and in many cases not even thought of. Today consequences of human interferences are seen many places. Many negative effects could have been prevented had understanding and tools to do the analysis been available. Today engineers and planners work closely together in an attempt to in some cases make up for historical mistakes, or to combine the various requirements to the developments in the active coastal, riverine and estuarine environments and at the same time "work with nature" to make the most of the artificially created environment. The present paper shows examples of consequences of various activities in the morphodynamic environments as well as examples where modern planning and modern analysis tools have been applied to support the design of successful projects. The morphological consequences of sand mining and regulation of rivers, reclamation of land in estuaries, fixing of tidal inlets by structures, major structures in the coastal zone, land reclamations and harbours are discussed. Further, the paper shows an overview of the analysis necessary for the successful design of a harbour on a very exposed coastline and an artificial beach park. The analysis is supported by numerical modelling. The paper therefore includes at the end a brief overview of the numerical tools applied.

## 2 MORPHODYNAMIC CONSEQUENCES OF VARIOUS ACTIVITIES

In this section effects on the morphology of various types of activities are discussed and illustrated by examples. The examples start in the "upstream" end with regulations of rivers and sand mining in rivers, continues with a brief introduction to projects in estuaries, in tidal inlets and ends with examples from the coastal zone.

### 2.1 Sand mining and regulations of rivers

Interventions in rivers take place all over the world where man need to control or use the natural resources. Interventions can be dams for collection of water for irrigation and water supply, dams for hydropower, dams for control of flooding along the lower parts of the river etc. The regulations of rivers often led to unwanted effects and in previous time also unexpected negative side effects. Many of the World's bigger rivers not only discharge water but carry loose material, stones, gravel, sand and fines from the hinterland down to the coastal zone. Some places large deltas have been built up by alluvial deposits which are eroded in the mountains or washed away by the rain on the ground and have been transported to the sea during heavy rain falls or melting of snow in the mountains. The coast lines and bed contours around a delta adjust in such a way that the wave action on the coastal zone transports the supplied material away to the adjacent beaches. Such coast lines are in a dynamic equilibrium with a yearly supply of sand and therefore adjusted to an orientation of the coastline compared to the incoming waves, which corresponds to littoral drift away from the delta of the same order of

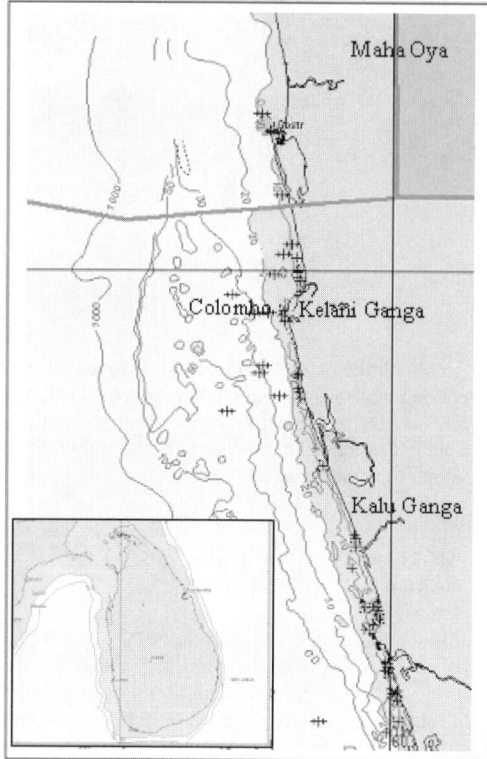

Figure 1.   Location map, three major rivers in Sri Lanka.

Table 1.   Sand mining, catchments supply and deficit in mill. m$^3$ for the period 1976–2001.

| River | Sand mining | Catchments Supply | Deficit |
|---|---|---|---|
| Maha Oya | 23.0 | 5.0 | 18.0 |
| Kelani Ganga | 32.0 | 7.5 | 24.5 |
| Kalu Ganga | 12.0 | 10.0 | 2.0 |

Figure 2.   Illustrations from Sri Lanka: upper photo: river bank erosion, lower photo: coastal erosion at the southwest coast.

magnitude as the supplied amount of material. Decrease in the supply has serious consequences. One of the most famous examples is the regulation of the Nile River, which was made to avoid floods, to improve irrigation and to provide hydropower. The regulations led to coastal erosion of up to 200 m/year at certain stretches due to lack of supply of sediment. The example presented in the following is taken from the southwest coast of Sri Lanka and is a part of the basic analysis for the development of shoreline management plans for the area.

The sand mining in and the supply from the catchments to three major rivers around Colombo have been quantified for the period 1976–2001, see Figure 1 for a location map. The result of the analysis and the deficit in supply to the coastal zone are presented in Table 1.

The supply to the coastal stretch between the three rivers (approx. 100 km) has been reduced by 44.5 mill m$^3$, which correspond to approx. 50% of the natural supply, due to sand mining.

The consequences of the sand mining have been the following:

Immediate effects:

- river bed degradation
- bank erosion

Long term effects:

- Coastal erosion
- Lowering of water table
- Increased saline intrusion

Bank erosion and coastal erosion in the area are shown in the photos in Figure 2.

2.2   *Reclamation of land in estuaries*

Through centuries estuaries and tidal lagoons have been placed where man has settled. Harbours have been built and the hinterland has been utilized more and more intensively for agriculture. Typically, estuaries include large shallow water areas, which can easily

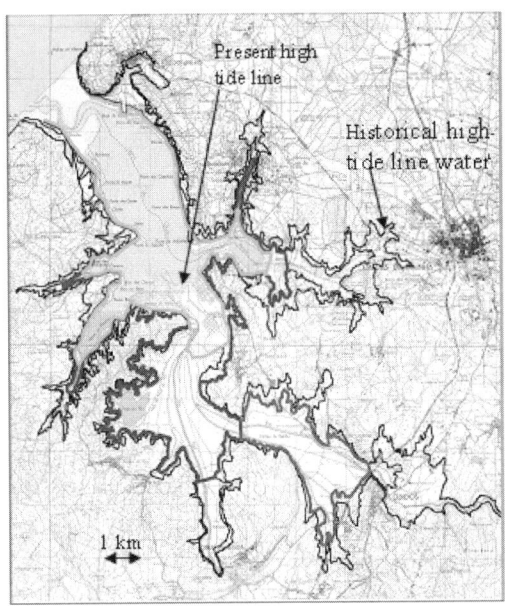

Figure 3. Obidos Lagoon, Portugal. Historical and present high tide lines.

be reclaimed and turned into land. Estuaries are at the same time often morphologically very dynamic areas. The tidal inlets are in a state of dynamic equilibrium where the tidal exchange through the inlet is sufficient to flush the inlet area and maintain sufficient cross-section area to withstand backfilling of the inlet area by littoral drift along the open coast.

The mechanisms and the effect of changing the tidal prism by reclaiming land in the estuary are illustrated by an example from Portugal. Figure 3 shows the Obidos Lagoon on the very exposed west coast of Portugal. The tidal range is 2.4 m and the gross littoral drift on the sandy Atlantic coastline is approx. 1 mill. m³/yr. In historical times the lagoon covered an area about 3 times larger than today and extended about 15 km inland compared to about 5 km today. The changes in tidal prism are due to mainly reclamations of shallow water areas. The result is a decrease in flushing capacity. The littoral drift from both sides is trapped in the inlet and is not flushed out by the tidal exchange.

After the severe winter storms the lagoon may be completely separated from the Atlantic sea with the consequences that the lagoon is not navigable and the water quality of the lagoon is deteriorated. The photo in Figure 4 shows the sand bar built up across the entrance during a series of northwesterly storms.

### 2.3 Regulation of tidal inlets

Various interventions to prevent sedimentation in the entrance to Obidos lagoon were evaluated in

Figure 4. The sand bar built up across the entrance to Obidos lagoon during severe northwesterly storms.

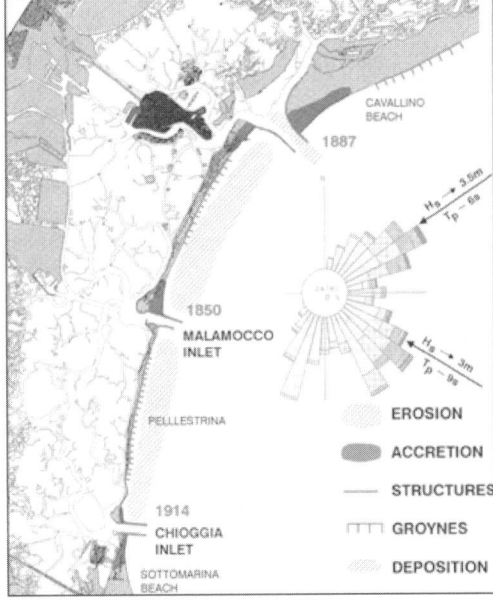

Figure 5. The Venice lagoon and the three inlets with the long jetties.

connection with the restoration. One option was the construction of jetties to fix the entrance. This is a solution which has been chosen at numerous tidal inlets worldwide and is a measure to secure navigation into the lagoon. However, the impact on the morphology of the adjacent beaches can be dramatic and cause other problems. In case of Obidos the proposed jetties were not chosen because the impact on the adjacent beaches might get out of control. The Venice lagoon is a well-known example of fixed tidal inlets, see Figure 5.

The jetties were constructed between mid eighteen hundred and beginning of nineteen hundred. The

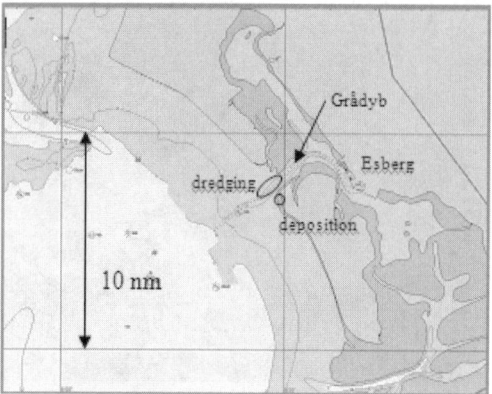

Figure 6.    Grådyb inlet, at the Wadden Sea.

2.4    *Major structures in the coastal zone, land reclamation and harbours*

Many beautiful beaches which appear to be completely natural are actually the result of human activities in the coastal zone. One example is one of the most popular swimming beaches in Denmark, Hornbæk Beach on the northeast coast of Zealand, see Figure 7. In the late eighteen hundred a small fishery port was built. The beach is sandy, with a littoral drift towards east of about 40,000 $m^3$/yr. The little port was filled with sand shortly after construction and a sand filet built up on the updrift side. The main breakwater was extended a number of times until 1991 where the main breakwater was supplemented with a symmetrical downdrift breakwater. A very attractive natural looking beach has been built up during the 100 years since the first harbour was constructed, with a wide beach plane and sand dunes. However, it is all a result of human interference. It is noted that the construction of the downdrift breakwater has actually reduced the dredging requirement to one third of what it was before the second breakwater was built and natural bypassing starts to take place, see the photo in Figure 7.

The next example of human interference with coastal morphodynamics is taken from Dubai, one of the fastest growing communities in the World. The natural coast is a straight sandy beach. The net littoral drift is northeast-going approx. 20–40,000 $m^3$/year. Before the 90'ties the two major harbours, Jebel Ali and the Dubai Dry Docks and a number of small fishery ports were the only structures along the otherwise open sandy coast. The fishery ports extended to 4–5 m of water depth and the coastline adjusted between the small harbours, which suffered from siltation. In the beginning of the 90'ties a shore parallel breakwater, a T-groyne and a curved breakwater were constructed as part of the enhancement of the recreational values of the Dubai coast to the south of the Dry Docks. Figure 9 shows the rapid response of the coastline to these structures: a tombolo developed behind the breakwater and sand was pushed to the corners of the T-groyne and the curved breakwater. The construction of Palm Jumeirah was initiated around the turn of the century. Since then The World Islands have been built and the two other palms are underway, see Figure 8. The consequences for the original coastline have been dramatic: The very big offshore structures have changed the nearshore wave climate to the extent where the littoral drift has changed direction along long stretches and the equilibrium orientations of the beaches have changed from about 10° anticlockwise turning compared to the original coastline to between 20° anticlockwise and 10° clockwise to the original coastline, see Figure 10. New coastal schemes must be developed along the entire frontage to accommodate the big changes to the nearshore wave conditions.

jetties were long, up till 4 km and with an inlet width up till 1 km. The littoral drift is northgoing south of the southern jetty and southgoing to the north of the northernmost jetty. The jetties blocked completely the littoral drift and during the years after construction the beach north of the northern jetty accreted by more than 2.5 km. At the time when the beach reached the tip of the northern jetty and sand started to bypass the jetties the offset in the coastline from updrift to downdrift of the northern inlet was about 3 km. The sand, which is not trapped inside the inlet, is deposited on ebb shoals and does not supply the downdrift beach. The central parts of the beaches between the three inlets have suffered from erosion since the construction of the jetties as sediment is either caught in the corners at the jetties or lost to offshore due to rip currents along the jetties. Huge protection schemes have been established to fix the narrow sandy barrier islands. Clearly, the jetties have provided safe navigation, but at the same time introduced coastal erosion problems, which were probably not foreseen at the time of construction.

Grådyb, Denmark is an example of a tidal inlet placed between two sandy barrier islands facing the North Sea, where navigation has been secured not by hard structures but by dredging and maintenance of a navigation channel. A sediment trap is dredged on the north side of the channel and the trapped sand is moved artificially from the trap to the active nearshore zone on the downdrift site of the channel. The site is shown in Figure 6, which also shows the location of the sediment trap and the dredged channel. The water depth in the channel is maintained at 9.3 m with a minimum width of the channel of 200 m. The southgoing littoral drift is approx. 1 mill $m^3$/yr. This inlet is an example where a management solution has proven successful, which can be seen from the fact that the downdrift coastline has not suffered from erosion.

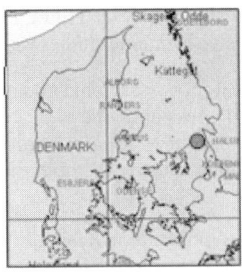

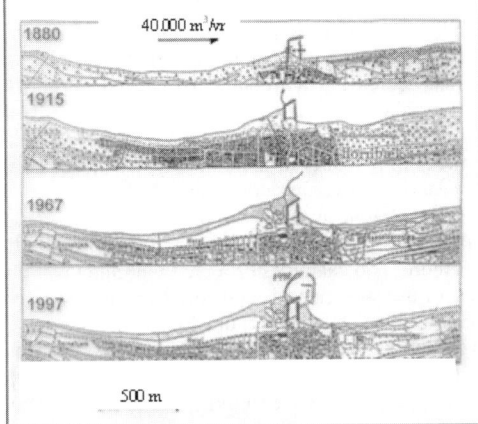

Figure 7. Location map, historical development of the coastline and the extension of the harbour. Oblique photo of the present harbour.

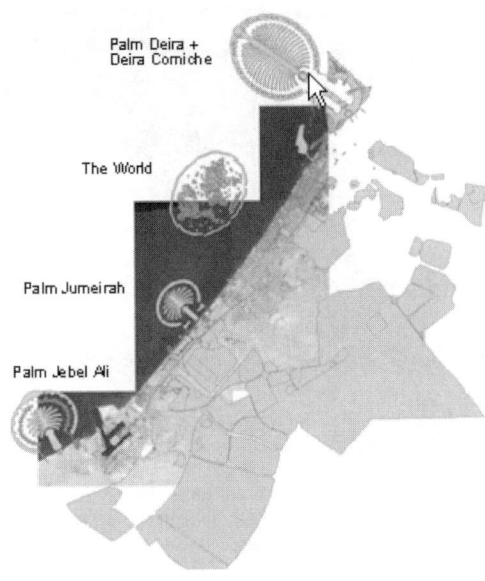

Figure 8. Sketch of the land reclamations along the Dubai coastline.

Figure 9. Shore parallel breakwaters and a small fishery port approx. 10 years after construction.

## 3 INVESTIGATIONS FOR A HARBOUR ON AN EXPOSED COAST, BAKKAFJARA, ICELAND

Westmann Islands is a group of islands about 5 nautical miles south of the south coast of Iceland. Bakkafjara is a location on Iceland immediately west of one of the outflows from melting of a glacier during spring and summer which supplies about 1 mill. $m^3$/yr fines and about 100,000 $m^3$/yr sand to the coastal zone. The offshore wave conditions are characterized by high waves that have been measured up to 16.7 m, see Figure 11. The river mouth and delta vary strongly from year to year. Storms from southeasterly directions push the river mouth to the west whereas southwesterly waves push the mouth to the east. The delta grows during periods of melting of the glacier. The tide is

1051

**Coast lines**

— Present
— + 20 years

Figure 10. New equilibrium coastlines in the lee of The World Islands between the Palm Jumeirah and Dubai Dry Docks.

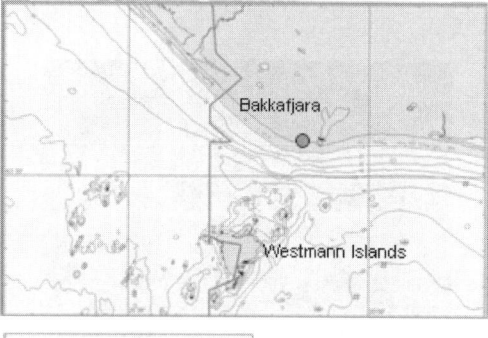

Bakkafjara

Westmann Islands

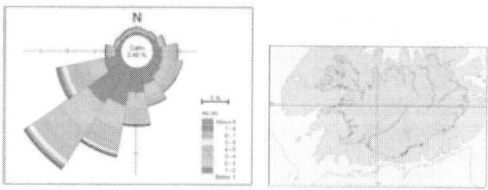

Figure 11. The location of Bakkafjara on the south coast of Iceland, illustration of the offshore wave statistics.

semidiurnal with a maximum range of 3 m. The sediment in the coastal zone and on the beach plane is medium sand with $d_{50}$ between 0.1 and 0.50 mm.

The preliminary location of the new harbour was selected based partly on analysis of the nearshore wave conditions, which showed that the Westmann Islands provide some shelter for the rough wave conditions along the coastline, and partly on the local fishermen's experience: off Bakkafjara there is almost "always" a

deep trough and a depression in the bar, see Figure 12. The purposes of comprehensive hydraulic studies were

- To understand the physical processes, which determine the location of the depression in the bar
- To optimize the harbour layout to have minimum impact on the overall morphology and minimum siltation.

The river mouth is known to be highly dynamic. A guiding wall will be established east of the harbour to avoid migration to the west of the river mouth. The stability of the river is therefore not part of the present study. Five measured bathymetries are shown in Figure 12. The depression in the bar and the trough are clearly seen. Further it appears that a well developed bar with a water depth of less than 6 m exists on the west side of the planned harbour location whereas the coast to the east of the river has got no bar.

The hydraulic studies have been initiated by transformation of the offshore wave conditions to the nearshore. Two examples of wave fields and the influence of the Westmann Islands on the nearshore wave conditions are shown in Figure 13.

Offshore wave data are available from hindcast studies starting in 1959. Littoral drift corresponding to the average wave conditions has been calculated using the assumption of long and uniform beach profiles. The overall sediment budget indicated a net eastward littoral drift west of the harbour location of approx. 300,000 m³/yr and approx. 400,000 m³/yr to the east of the harbour and the river. The river supplies in average 100,000 m³/yr of sand to the coastal zone.

The distribution along the coastal profile of the littoral drift is, however, interesting: to the west of the harbour the littoral drift is eastgoing on the bar as well as along the inner part of the profile. To the east of the river the littoral drift is concentrated along the inner part of the profile as a significant breaker bar has not developed here. At the location of the harbour the net littoral drift is eastgoing on the inner part of the profile and westgoing on the breaker bar. This sediment transport pattern is due to the effect of the Westmann Islands on the waves. The most common southwesterly waves lead to eastgoing littoral drift along the entire stretch, but the eastgoing transport capacity decreases from west to east as one moves into the zone sheltered by the Westmann Islands. The most severe waves come from southeast. These waves pass behind the Westmann Islands and push the net littoral drift on the outer bar to the west. During years of dominant waves from southwest the western breaker bar will grow towards the east. During years of severe easterly waves the tip of the western breaker bar will be pushed back towards west. The calculated average distributions of littoral drift are shown in Figure 14 for three sections. The bathymetrical measurements show years where a "spit" has developed from the delta towards

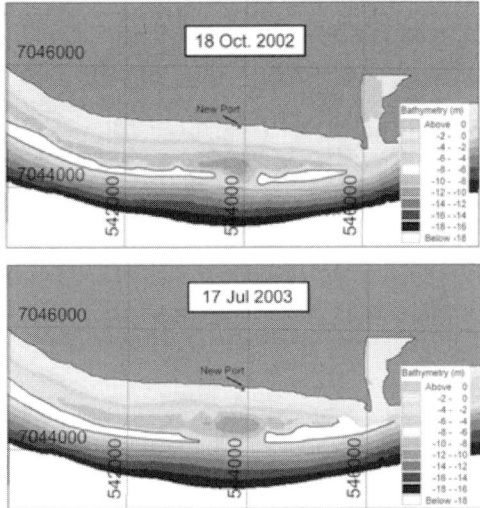

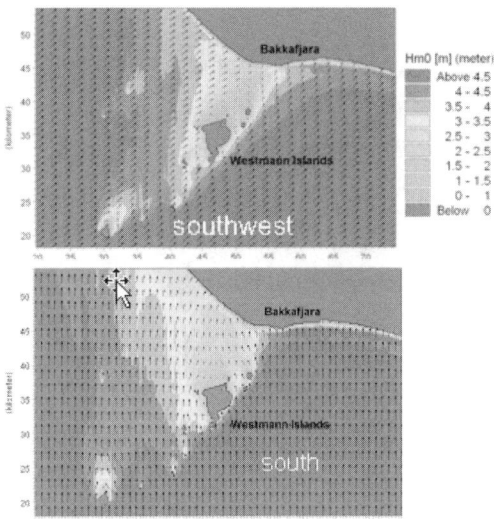

Figure 13. Results from the wave transformation study.

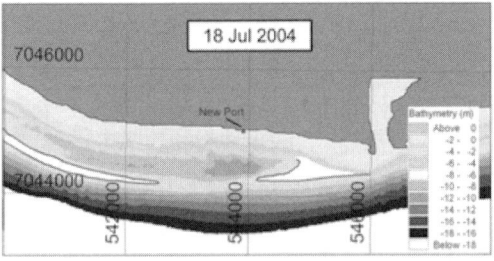

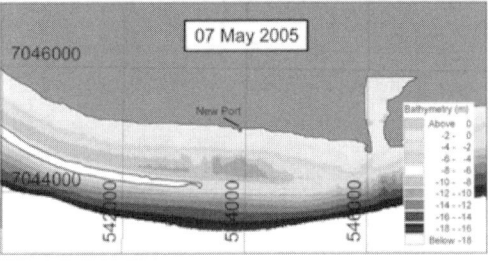

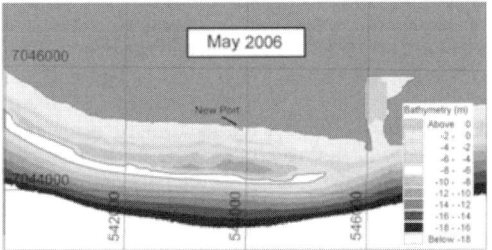

Figure 12. Bathymetries measured between October 2002 and May 2006.

west, see Figure 12, and other years where the spit is non-existing. The spit develops after the growth of the delta during spring and summer followed by severe waves from southeasterly directions. It appears that the growth of the spit is dependent both on the supply of sand from the river and a long period of persistent southeasterly waves. Analysis of littoral drift capacity during the period 1979-2006 has shown that no continuous event, or series of storms, has led to westgoing transport at the location of the delta of more than $500,000\,\mathrm{m}^3$, which corresponds to a few hundred metres of spit formation. The eastgoing littoral drift is more common and many events are registered where the littoral drift on the outer bar east of the harbour exceeded $150,000\,\mathrm{m}^3$. As seen in Figure 12 the outer bar has almost closed off the depression of the bar during the last few years. However, analysis of a long time series for average wave energy and direction of average offshore wave energy, see Figure 15 clearly indicate that this is not the most common situation. In the past years there has been a series of years with high energy from southwesterly directions which led to lengthening of the bar from west towards east. Figure 16 shows the details of the variation of the two determining parameters wave energy and average wave energy direction during the period of intensive measurements. Figure 12 and Figure 16 can be compared and it appears that the dynamic behaviour of the depression can be explained. The risk of having to dredge through the outer bar has been evaluated and found acceptable for the project. It is noted that the required navigation depth is 6 m.

The planned harbour extends approx 700 m from the shoreline out to the present 8 m depth contour and is located at the deep trough. The morphological impact of the harbour on the trough and the adjacent nearshore bathymetry as well as the expected equilibrium depth in front of the harbour have been tested and optimised

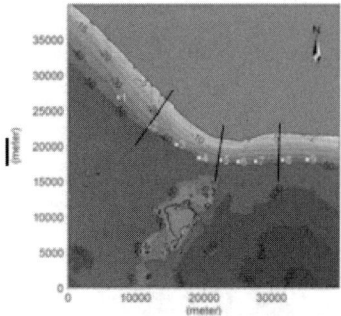

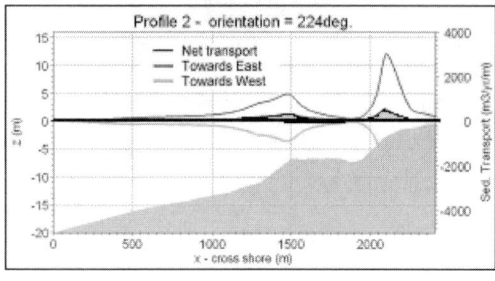

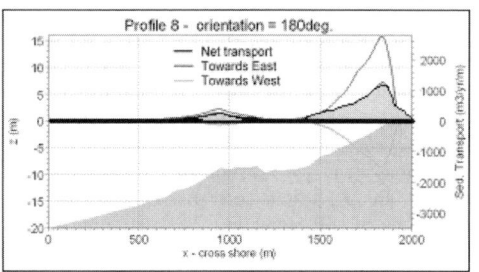

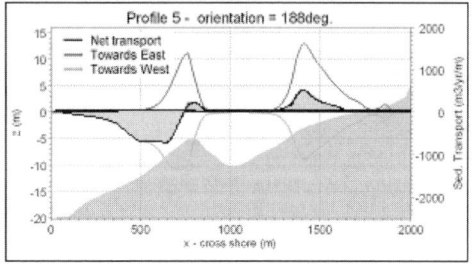

Figure 14. Average littoral drift and the distribution along the coastal profiles at locations west of, at and east of the planned harbour.

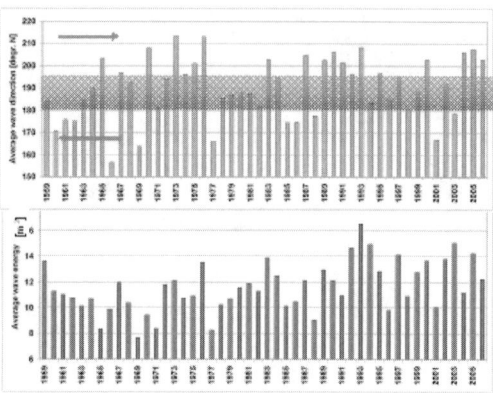

Figure 15. Average offshore wave energy direction and wave energy, 1959–2006. Eastgoing littoral drift at Bakkafjara for dir > 195° and westgoing for dir < 180°.

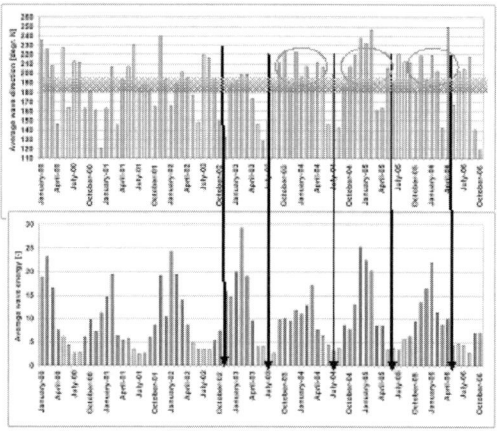

Figure 16. Average offshore wave energy direction and wave energy, 2000–2006.

by a morphological modelling complex. The model complex reproduced currents due to tide and waves, wind generated waves, and sediment transport due to depth integrated currents, wave orbital motion, turbulence due to wave breaking, see a brief description in Section 5. Examples of results are presented in Figure 17. The upper panel shows the bathymetry measured May 2006 and the harbour layout. The second panel shows the modelled bathymetry after 200 days with

constant severe waves from southwest, corresponding to a wave height of 3.7 m at 15 m of water depth off the site. It appears that the trough is not being backfilled but a sand feature will develop in front of the harbour. Due to the streamlined shape and the narrow entrance the majority of the sand will bypass the harbour and it appears that the impact on the adjacent coastline is very small. The bathymetry will adjust to a depth in front of the harbour which allows the sand migrating on the inner part of the coastal profile from the west towards the harbour to bypass. The third panel in Figure 17 shows a close-up around the entrance. The development in water depth is shown in the fourth panel in Figure 17. The water depth in front of the harbour is seen to stabilise at around 6 m. The two lower panels in Figure 17 illustrate the development around the entrance in case the angle between the two main

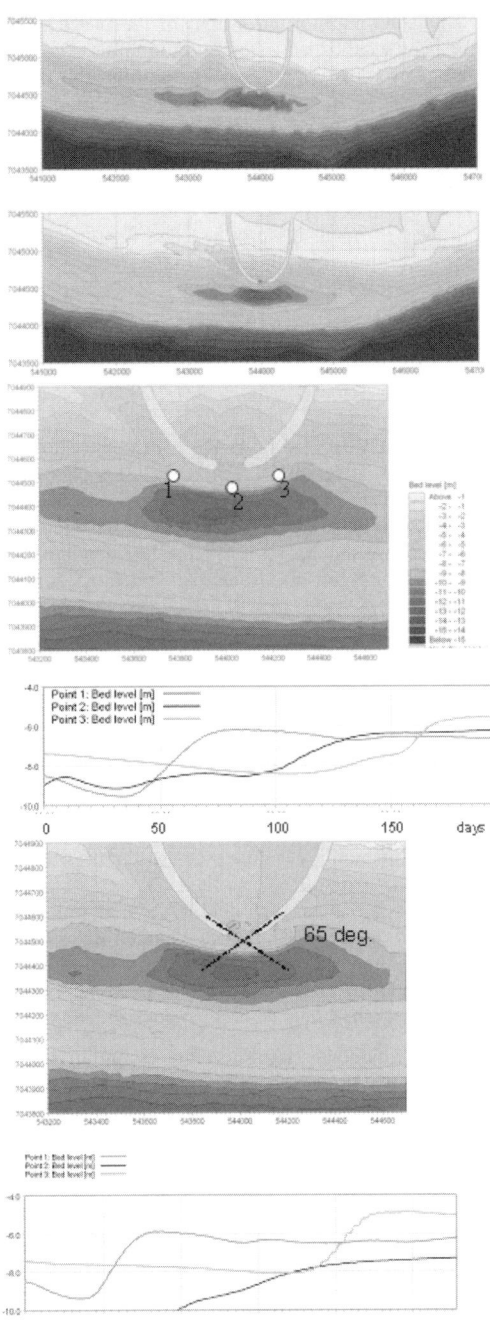

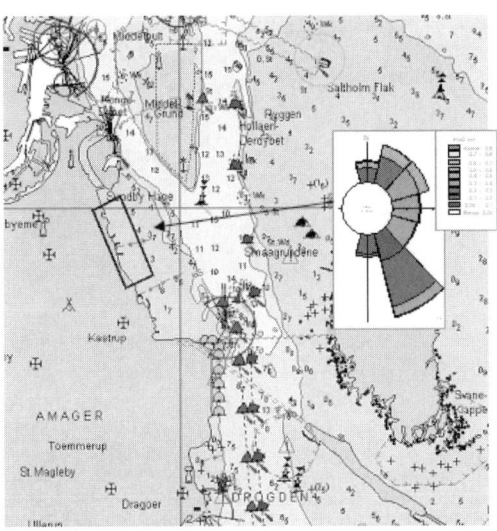

Figure 18. A mager Beach, Copenhagen, Denmark. Moderately exposed site with a protected beach due to a shallow shoreface.

breakwaters is changed from 40° to 65°. The equilibrium depth is seen to be slightly deeper, more than 7 m in this case. Several combinations of wave directions and height have been modelled to support the design of the harbour. Further, the sedimentation of sand in the fore-harbour due to eddy exchange has been estimated as well as the sedimentation of fines during the periods where the adjacent river discharges large amounts of fine sediments.

The results selected for presentation in this paper illustrate the strength of the combination of analysis of the coastal environment, quantification of littoral drift for understanding of the overall processes and the detailed analysis of the processes around the new harbour for optimising the layout with regards to impact and sedimentation.

## 4 A SUCCESSFUL BEACH PARK, AMAGER BEACH PARK, DENMARK

*Amager beach* is located out to the Sound in Copenhagen, Denmark. The problem at this site is that the existing beach is of a poor quality (muddy shoreface) due to the lack of wave exposure. The site is only moderately exposed and the existing beach is further protected due to a very shallow shoreface, see Figure 18. A new beach park was recently built at this location. The project includes 2 km of new beaches and is located less than 10 km from the Copenhagen city centre. The main concept for the new beach park has been to move the beaches seaward beyond the shallow shoreface and thereby providing the highest possible

Figure 17. Upper panel: initial bathymetry, second panel: modelled bathymetry after 200 days of constant wave action and tide, third panel: close-up around the harbour entrance, fourth panel: morphological changes at three points off the harbour (see third panel), fifth and sixth panels: similar to third and fourth, but with an angle of 65° between the breakwaters compared to 40°.

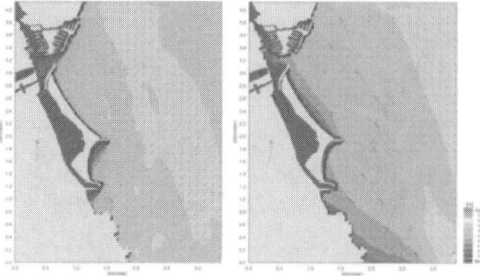

Figure 19. Modelling of wave patterns from the two main directions NE and SE.

Figure 20. Aerial photo of Amager Beach Park, which consists of the following main elements: island with terminal structures north and south and a separating headland between the northern and southern beaches and a lagoon.

wave exposure, which, however, is still moderate. The new beaches have been constructed on an island and a new lagoon (deepened) has been excavated between the island and the old shoreline. As the beach park is located near the gorge section of the Sound between Denmark and Sweden, there is always a good gradient on the water surface in the area of the beach park. This situation has been utilised to generate good flushing in the lagoon by making two openings, one at the northern and one at the southern end. The wave climate at the site is characterised by having two main directions, i.e. NE and SE, see Figure 18, which is due to the shelter provided by the island located just opposite of the site. This situation has been utilised to create two sections of beaches separated by a headland, one facing towards the NE and one facing towards the SE. The headland provides shelter at the NE facing beach for waves from the SE and shelter at the SE facing beach for waves from the NE, respectively, see Figure 19. The layout can be seen in and Figure 20.

The exact equilibrium shapes of the two beaches have been established based on modelling of a large number of wave conditions. The equilibrium orientations, i.e. orientations of zero net transport were calculated in a number of points along the coastlines. Finally, the shapes of the beaches were fitted to the series of equilibrium orientations. The length of the central headland is one of the parameters which were optimised through this process. The flushing of the lagoon was quantified by a depth-integrated hydrodynamic model. The average flushing time is about 24 hours, which will secure a good water quality. An aerial photo of the new beach park just after finalisation of the civil works is presented in Figure 20.

5 BRIEF OVERVIEW OF APPLIED NUMERICAL MODELLING TOOLS

Over the past decades, numerical models have increasingly become support tools to shoreline management and design of coastal structures. The tools applied in the present paper are all developed at DHI and comprise:

- Littoral drift and shoreline evolution models suited for relatively uniform sandy beaches, which, given proper input on variability of the wave conditions along the shore and with time, cast light over the large-scale shoreline developments.
- 2-dimensional (in plan) wave and hydrodynamic models for the detailed study of waves and flow fields on complex bathymetries and in the vicinity of coastal structures.
- Quasi 3-dimensional model for sediment transport.
- Coastal area morphological modelling tool to appraise the short- to medium-term morphological response of the coast to for example coastal structures, harbours, shoreface nourishments etc.

The littoral drift and shoreline evolution model includes wave transformation, longshore wave driven currents, longshore and cross-shore sediment transport, shoreline evolution and coastal profile evolution. The bed contours are assumed to be parallel and quasi-uniform in the longshore direction and the waves and currents are considered to be quasi-stationary. These two basic assumptions limit the use of the tool to cases of long and uniform sandy beaches and cases where the shoreline evolution is the result of the overall gradients in the longshore sediment transport capacity. On the other hand, due to these assumptions, long coastal stretches can be covered over long time spans and the complex is a fast tool for calculation of equilibrium orientations of coast lines. See for instance Fredsøe and Deigaard (1992).

The two-dimensional so-called area model consists of a series of modules capable of simulating different wave-related, current and sediment transport processes. The following modules are currently most used for coastal morphological studies:

*Wave models*: still the very advanced wave models which include all important processes in the coastal zone are not fast enough to be applied in practical cases where many different wave conditions have to be considered. Very efficient wind wave models are, however, available and are excellent tools when diffraction is not a dominant process, Holthuijsen et al. (1989). Wave models based on the parabolic approximation to the mild-slope equation, Kirby (1986) have proven to be useful for smaller areas in the order of magnitude 10 by 10 km. Both models account for the effects of shoaling, refraction, breaking, directional spreading and bed friction. The mild slope models further include partial diffraction and forward scattering, whereas the wind wave models include growth and decay of short-period waves and the effects of current may be included. In both types, the dissipation of wave energy due to breaking is parameterised. A much used approach is the model of Battjes and Janssen (1978).

*The hydrodynamic model* calculates the flow field from the solution of the depth-integrated continuity and momentum equations, Abbott (1979). In addition to wind and tide, the forcing terms may include the gradients in the radiation stress field as calculated by the wave module of the morphological modelling system. The currents and the mean water level are calculated on a bed evolving at a constant rate equal to $\partial z/\partial t$ as calculated by the sediment transport module.

*The non-cohesive sediment transport model* is used to calculate the transport rates of graded sediment and the rates of bed level change $\partial z/\partial t$ under the combined action of waves and current. The transport model for sand is an intra-wave period sediment transport model which calculates the total (bed load + suspended load) transport rates of non-cohesive sediment. The model accounts for the effects of waves propagating at an arbitrary angle to the current, breaking/unbroken waves, uniform/graded bed sediment, plane/ripple covered bed when calculating the local rates of total load transport, see e.g. Fredsøe (1984) and Deigaard et al. (1986). The model includes a quasi-3-dimensional description of the flow and the sediment transport, as described in Elfrink et al. (2000). Use of this approach allows calculation of net sediment transport rates both in the longshore and cross-shore directions.

*Morphological modelling complex.* The three above-mentioned models, waves, hydrodynamics and sediment transport are combined into a morphological modelling system, see Johnson et al. (1994) and Johnson and Zyserman (2002). The model complex is developed both in a finite difference scheme which uses rectangular grids as well as a complex which uses a fully flexible grid, triangles.

# 6 CONCLUSIONS

The paper shows examples of negative consequences of human interference with morphodynamics as well as examples of projects where understanding of the processes and the quantification of the processes by advanced numerical modelling tools have led to successful projects.

The optimal use of numerical models can briefly be characterised as follows:

- Perform basic hydraulic studies utilising existing data and regional numerical models.
- Develop concepts for schemes by utilising the basic information on processes and the concept of "work with nature".
- Utilise detailed numerical models to optimise the hydraulic and coastal performance of the schemes including optimisation through minimisation of negative impacts.

ACKNOWLEDGEMENTS

The author has drawn on experience from various studies undertaken by the coastal and estuarine experts at DHI over the past 25 years.

The studies at Bakkafjara, Iceland have been undertaken as a collaboration between DHI and Eng. Gisli Viggósson, Siglingastofnun, Iceland.

REFERENCES

Abbott M.B. 1979. *Computational hydraulics, elements of the theory of free surface flows.* Pitman, London.

Battjes, J.A. and Janssen J.P.F.M. 1978. "Energy loss and set-up due to breaking of random waves". *Procs. of the 16th Int. Conf. On Coastal Engineering,* ASCE, 569–587.

Brøker Hedegaard, Ida, Mangor, Karsten and Lintrup, Morten J. 1993: "Modelling of sediment transport in connection with tidal inlets". *Int. Colloquium and Exposition on Computer Applications in Coastal and Offshore Engineering (ICE-CACOE '93),* Kuala Lumpur, Malaysia, June 14–16, 1993.

Deigaard, R., Fredsøe, J. and Brøker Hedegaard, I. 1986. "Suspended sediment in the surf zone". *J. Waterway, Port, Coastal and Ocean Engineering,* 112 (1), ASCE, 115–128.

Deigaard, R.; Fredsøe, J., and Brøker, I. (1986). "Mathematical model for littoral drift". *Journal of Waterway, Port, Coastal and Ocean Eng., ASCE, Vol. 112, No. 3, 351–369.*

Elfrink, B., Brøker, I. and Deigaard, R. 2000. "Beach profile evolution due to oblique wave attack". *Procs. of the 27th Int. Conf. on Coastal Eng.,* ASCE, 3021–3034.

Fredsøe J. 1984. "The turbulent boundary layer in combined wave-current motion". *Journal of Hydr. Eng.,* 110(8), ASCE, 1103–1120.

Fredsøe J, Deigaard R. 1992. *Mechanics of coastal sediment transport. World Scientific.* Advanced series on ocean engineering.

Holthuijsen, L.H., Booij, N. and Herbers, T.H.C. 1989. "A prediction model for stationary, short-crested waves in shallow water with ambient currents". *Coastal Engineering,* 13, 23–54.

Johnson, H.K.; Brøker, I.; and Zyserman, J.A. 1994. "Identification of some relevant processes in coastal morphological modelling". *Proceedings of the 24th International Conference on Coastal Engineering,* ASCE, 2871–2885.

Johnson, H.K. and Zyserman, J.A. 2002. "Controlling spatial oscillations in bed level update schemes". Coastal Engineering, 46, 109–126.

Kirby, J.T. 1986. "Rational approximations in the parabolic equation method for water waves". *Coastal Engineering,* 10, 355–378.

Mangor, K. 2004. *Shoreline Management Guidelines,* DHI Water & Environment, 232p.

Sørensen, T.; Fredsøe J.; Roed Jakobsen, P.(1996) *History and heritage of coastal engineering in Denmark.* 103–141 ISBN 0784401969.

Vieira, José Rodrigues and Foster, Tom: Recovering the Óbidos Lagoon. An integrated management approach. *Medcoast, 24–27 October 199,* Tarragona, Spain.

*Human impacts offshore and near coasts*

*River, Coastal and Estuarine Morphodynamics: RCEM 2007 – Dohmen-Janssen & Hulscher (eds)*
*© 2008 Taylor & Francis Group, London, ISBN 978-0-415-45363-9*

# Seabed morphodynamics due to offshore wind farms

Henriët H. van der Veen, Suzanne J.M.H. Hulscher & Blanca Peréz Lapeña
*Water Engineering and Managment, University of Twente, Enschede, The Netherlands*

ABSTRACT: The need for sustainable energy is rising, and at the moment wind energy is one of the few forms of renewable energy that can be harvested efficiently. We investigated the influence of an offshore wind farm on the large-scale morphodynamics of the seabed. To this aim, we developed a morphodynamic model to investigate the effect of offshore wind farms on the seabed. By implementing the model in a GIS environment, the model allows us to calculate the effects of a wind farm using location specific and park design input parameters. By implementing an analytical model in a GIS, a rapid calculation of the effects of an offshore wind farm at a certain location in the North Sea can be made.

## 1 INTRODUCTION

The need for sustainable energy is rising. The members of the European Union have devoted themselves to a 21% share of renewable energy by the year 2010. And the European Parliament has adopted a resolution that stresses the need of setting a mandatory target of a share of 20% renewable energy by the year 2020. At the moment wind energy is one of the few forms of renewable energy that can be harvested efficiently. (Danish Energy Authority, 2005)

The total wind resources in the offshore area of Europe have been assessed at up to 3000 TWh, this is in theory enough energy to supply Europe's total current electricity demand. And as placing space onshore is scarce, many European countries are planning and realizing offshore wind farms to increase the share of renewable energy (e.g. figure 1).

Advantages of offshore wind energy are the strong and predictable wind speeds, the energy generating capacity offshore is approximately 40% higher than onshore. Disadvantages are the higher costs for realizing and maintaining an offshore wind farm (EWEA,Undated).

As it has only recently become technically possible to build wind farms in the truly offshore area, little research has been done at the long term morphological effects of offshore wind farms on the surrounding seabed. We want to investigate the influence of an offshore wind farm on the large-scale morphodynamics of the seabed. To this aim, we developed a morphodynamic model to investigate the effect of offshore wind farms on the seabed. In this research we focus on the large scale morphodynamics, which means that small scale processes like scour around the piles are not included.

Figure 1. North Hoyle, the first offshore wind farm in the UK (Greenpeace, 2007).

In section two the development of offshore wind farms in Europe is discussed. The morphodynamic model that is used is introduced in section three. The results are presented in section four and section five contains the discussion. In the last section the conclusions are presented.

## 2 OFFSHORE WIND FARMS IN EUROPE

The first offshore wind farms that were built consisted of a few wind turbines which were placed close to the shore in shallow waters. As the size of the offshore wind farms is increasing, the farms are located more and more in the true offshore area, with water depths going up to 40 meters and located in areas where they are exposed to full tidal forces.

Table 1. Overview of offshore energy in Europe (Greenpeace, 2005).

| Land | operational MW | planned MW | Total MW |
|------|------|------|------|
| Sweden | 67 | 2454 | 2521 |
| Denmark | 409 | 400 | 809 |
| Germany, North Sea | 5 | 30121 | 30126 |
| Germany, Baltic Sea | 0 | 4060 | 4060 |
| United Kingdom | 214 | 8699 | 8913 |
| Ireland | 25 | 1280 | 1305 |
| The Netherlands | 0 | 220 | 220 |
| Belgium | 0 | 300 | 300 |
| France | 0 | 60 | 60 |
| Spain | 0 | 2563 | 2563 |
| TOTAL | 895 | 49613 | 50508 |

Table 1 gives an overview of the capacity of offshore wind energy per country (in Europe) of operational and planned projects. Figure 2 gives an overview of the operational and planned offshore wind farms in Europe.

# 3 MODELLING

To predict the effects of offshore wind farms on the large scale bed morphodynamics we set up an analytical model that calculates the effects of offshore wind farms on the seabed.

## 3.1 The model

The morphological model describes the interaction between the depth averaged tidal flow and an erodible bed in a shallow sea, as was first described by Huthnance (1982a).

The first part of the model (eq. 1) describes the fluid motion; the equations used are the shallow water equations which can be derived from the Navier-Stokes equations. The energy loss due to the presence of the park is accounted for by the last term at the left-hand side in the first and second equation. The second part (eq. 2) describes the sediment motion, only bedload is considered; as velocities are low it is assumed that this will be the dominant mode of transport. The bedload is parameterized by an empirical relationship by Van Rijn (1993). This expression describes transport increasing with increasing velocities and that sediment is transported more easily downhill than uphill. The third part (eq. 3) of the model contains a sediment mass balance, to describe the behaviour of the seabed. For sake of convenience, the boundaries are considered to be infinitely far away, thus applying to locations well

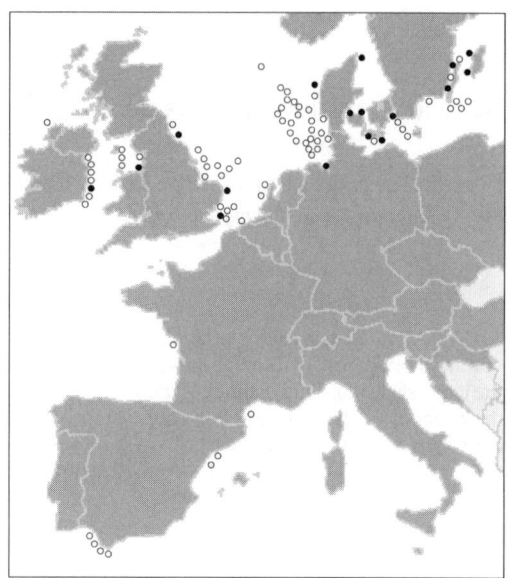

Figure 2. Overview of operational (black dots) and planned (white dots) offshore wind farms in Europe.

offshore. The effects of wind stress on the flow are neglected and the shear stress at the seabed depends linearly on the mean flow velocity.

$$
\left.\begin{aligned}
g\frac{\partial \zeta}{\partial x} + \frac{\partial u}{\partial t} + u\frac{\partial u}{\partial x} + v\frac{\partial u}{\partial y} - fv + \frac{ru}{H+\zeta-h} + e|\bar{u}|u = 0 \\
g\frac{\partial \zeta}{\partial y} + \frac{\partial v}{\partial t} + u\frac{\partial v}{\partial x} + v\frac{\partial v}{\partial y} + fu + \frac{rv}{H+\zeta-h} + e|\bar{v}|v = 0 \\
\frac{\partial \zeta}{\partial t} - \frac{\partial h}{\partial t} + \frac{\partial}{\partial x}[(H+\zeta-h)u] + \frac{\partial}{\partial y}[(H+\zeta-h)v] = 0
\end{aligned}\right\} \quad (1)
$$

$$
\bar{S} = \alpha|\bar{u}|^b\left(\frac{\bar{u}}{|\bar{u}|} - \lambda\bar{\nabla}h\right) \quad (2)
$$

$$
\frac{\partial h}{\partial t} + \bar{\nabla}\cdot\bar{S} = 0 \quad (3)
$$

Here $u$ and $v$ are the velocity components in both horizontal directions, $\zeta$ is the free surface and $h$ indicates the bed topography with respect to the undisturbed bed level $z = -H$. The sediment transport is described by $S$. The x-axis is chosen in the direction of the principal tidal current. Here $f$ denotes the Coriolis parameter ($1.16 \cdot 10 - 4\,\mathrm{s}^{-1}$, for a North Sea lattitude), $b$ is the power of sediment transport (usually 3) and $\lambda$ is the downhill constant (2).

The friction parameter of the seabed is denoted by $r$, which depends on the median grain size ($d_{50}$):

$$
r = \frac{gu}{C_h{}^2} \quad (4)
$$

1062

In which $g$ is the acceleration due to gravity ($9.81$ m/s$^2$) and $C_h$ is the grain size dependent Chezy coefficient, denoted by:

$$C_h = 18\log\left(\frac{12}{3d_{50}}\right) \text{(Van Rijn, 1989)} \qquad (5)$$

The long morphological timescale on which the seabed evolves is derived scaling eq. (3) (see for details e.g. Roos and Hulscher (2004)), denoted by:

$$T_{long} = \frac{(1-\varepsilon_p)H}{\alpha\sigma u^2}\text{(Roos and Hulscher, 2004)} \qquad (6)$$

Where $\varepsilon_p$ is the bed porosity ($\sim$0.4), and the proportionality parameter ($\alpha$) is described by:

$$\alpha = \frac{8\sqrt{g}}{sC_h^3} \text{(Van Rijn, 1989)} \qquad (7)$$

In which $s$ denotes the relative density:

$$s = \frac{\rho_s - \rho}{\rho} \qquad (8)$$

Where $\rho$ is the density of the seawater ($1000$ kg/m$^3$) and $\rho_s$ is the density of the sediment ($2560$ kg/m$^3$).

### 3.2 Inclusion of a wind farm in the morphological model

It is complicated to fully represent all details of the wind farm in the morphological model. Therefore, we aim at inclusion of the effects of wind farms though a change in flow resistance at the location of the wind farm, such that this resistance term can be included in the shallow water equations (see figure 3).

The value of the flow resistance term ($e$) is determined by spatially averaging the drag force of a single wind farm. The drag force of a single turbine is defined by:

$$F_{wt} = \frac{1}{2}\rho A C_{D\_wt}|u|u \qquad (9)$$

Where $A$ is the area of the wind turbine normal to the flow direction and $C_{D\_wt}$ is the drag force of a cylinder, determined from figure 4.

Where the Reynolds Number is defined by:

$$\text{Re} = \frac{u\,d_{wt}}{v} \qquad (10)$$

In which $v$ is the kinematic viscosity of seawater ($1.17\cdot10^6$ m$^2$/s) and $d_{wt}$ denotes the diameter of a wind turbine.

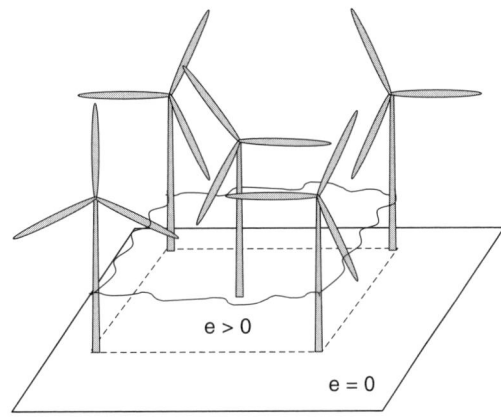

Figure 3. Representation of the wind farm in the morphological model.

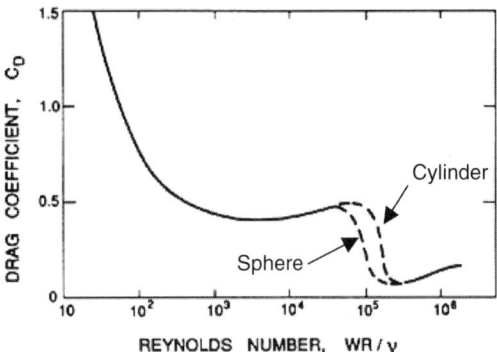

Figure 4. Drag force for smooth circular cylinders (California Institute of Technology, 2007).

To calculate the drag force of a complete wind farm ($F_{wf}$) the drag force of a single turbine is multiplied by the number of wind turbines per square meter ($N$):

$$N = \frac{1}{G^2} \qquad (11)$$

Where $G$ is the spacing between two turbines.

$$F_{wf} = \frac{1}{2}\rho\cdot A\cdot C_{D\_wt}\cdot|u|u\cdot N \qquad (12)$$

After operations at the momentum equation this leads to:

$$F_{wf} = \frac{d_{wt}\cdot C_{D\_wt}}{2G^2}|u|u \quad = e\cdot|u|u \qquad (13)$$

### 3.3 Influence of the wind farm

The influence of the wind farm is assessed by the area of influence.

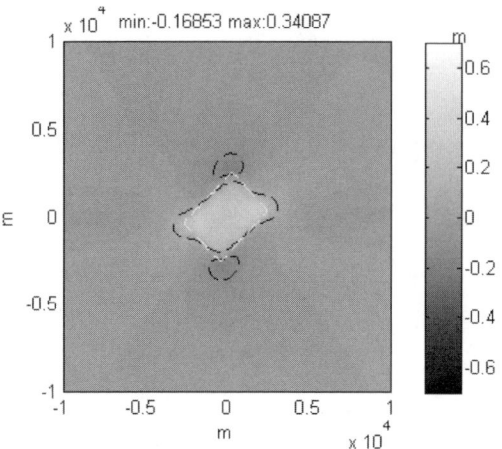

Figure 5. Morphological development of a wind farm of 4 by 3 km after 100 years. Wind turbines spaced 500 m apart (d_wt = 4.5 m). Other parameters: u = 0.7 (m/s), d50 = 200 (μm), H = 30 (m) and theta = 45°. The white dashed line marks the outline of the wind farm.

The area of influence is denoted as the area outside the wind farm, where the absolute seabed change due to the presence of the wind farm is more than 0.1 m (10 cm). In figure 5 this area is denoted by the black dashed line minus the area within the white dashed line. Note that the area of influence does not have to be a continuous area but can consist of several smaller areas. Here, $t$ is the total time in years over which the model calculates the effects of the wind farm.

## 4 RESULTS

### 4.1 Sensitivity analysis environment parameters

Figure 6a shows the dependency of the area of influence on the water depth. The area of influence decreases when the water depth increases. This means that the effect of the wind farm decreases when the water depth increases. figure 6b shows the dependency of the area of influence on the flow velocity. As can be seen the area of influence increases when the flow velocity increases, this means that the effects of a wind farm on the seabed will be stronger when the flow velocity is bigger. As can be seen from figure 6c, the area of influence increases when the median grain size increases. The last parameter that is investigated was the angle of the wind farm with respect to the flow, as is denoted in figure 6d. As can be seen, the angle of the flow with respect to the long axis has an effect on the wind farm. Two peeks occur, around 30 degrees and 210 degrees. This coincides with the tendency of sand banks to occur in an angle of 30 degrees counter-clockwise with respect to the flow in the North Sea, as

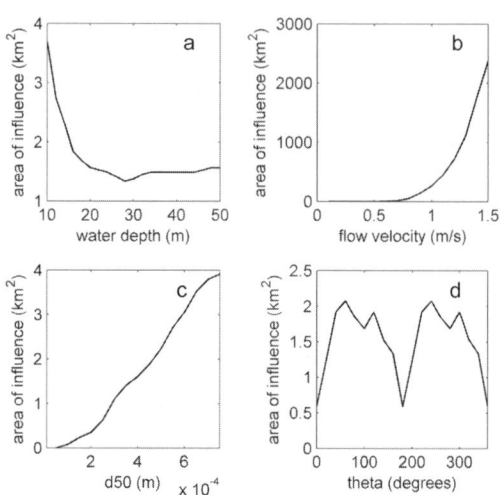

Figure 6. Dependency of the area of influence on the water depth (a), the flow velocity (b), the median grain size (c) and the angle between the flow and the axis of the wind farm (d). The wind farm that was used had the following default characteristics: L = 4.5 km, B = 3 km, G = 500 m, $d_{wt}$ = 4.5 m, d50 = 375 μm, H = 21 m, u = 0.58 m/s, angle with respect to the flow 150°.

was described by Huthnance (1982a). As a block flow is implemented in the model, the same increase occurs at 210 degrees (180 + 30).

The influence of the flow velocity on the area of influence is much bigger than the effects of the water depth, median grain size and theta.

### 4.2 Sensitivity analysis design parameters

As can be seen in figure 7, the influence of the wind farm increases linearly when the diameter of the tower of the turbines ($d_{wt}$) increases. This is due to the increase in area that blocks the flow. When the distance between the wind turbines ($G$) increases, the influence of the wind farm decreases rapidly, due to the decrease in flow resistance. When the drag coefficient of the turbines increases, the area of influence of the wind farm increases quite rapidly due to the increased friction in the wind farm. The biggest influence is caused by the spacing between the turbines, while the effect of the change in diameter of the turbines is quite small.

### 4.3 Two cases

By implementing the model in a GIS environment, see Van der Veen et al. (2006), the model allows us to calculate the effects of a wind farm using location specific input parameters.

We selected two cases to calculate the effect of a wind farm on the seabed. These cases are listed in Table 2.

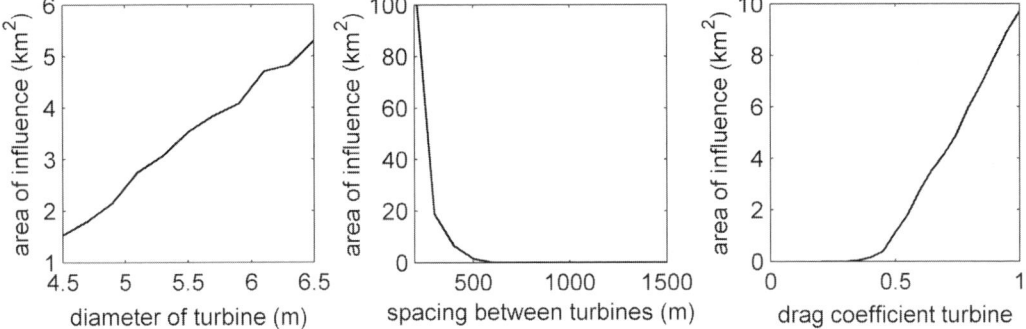

Figure 7. Dependency of the area of influence on the diameter of the turbines (G = 500 m) (upper), the spacing between the turbines ($d_{wt}$ = 4.5 m) (middle) and the drag coefficient of the turbine (lowest). The wind farm that was used had the following characteristics: L = 4.5 km, B = 3 km, d50 = 375 μm, H = 21 m, u = 0.58 m/s, angle with respect to the flow 150°.

Table 2. Overview of local parameters of the wind farms Humber Gateway and Q7.

| Name | Humber Gateway | Q7 |
|---|---|---|
| location | Mouth of the Humber (UK) | IJmuiden (NL) |
| #turbines | 70 (G = 700 m) | 60 (G = 500 m) |
| L(km) | 7.5 | 4.5 |
| B(km) | 4.5 | 3 |
| u(m/s) | 0.86 | 0.58 |
| d50 (μm) | 750 | 375 |
| H (m) | 25 | 21 |
| Theta (°) | 25 | 150 |
| $d_{wt}$ (m) | 4.5 | 4.5 |
| $C_{D\_wt}$ | 0.64 | 0.53 |

In figure 8, the morphological effects of the Humber wind farm and the Q7 wind farm are shown. The effects of the wind farm located off the coast of IJmuiden (Q7) (area of infl. 1.5 km$^2$) are much smaller than the effects of the Humber wind farm (158,1 km$^2$). The morphological timescale is shorter for the Humber farm (Tlong = 91 yr) than the Q7 wind farm (213 yr), which means that the morphological development is much faster for the Humber wind farm.

## 5 DISCUSSION

The wind farms that are implemented in the model have a rectangular shape. In reality wind farms may have a different shape, which may influence the outcome of the model. At the moment no research has been done at the effects of different shapes of the offshore wind farms.

The model starts from a flat bed in the initial situation, this means that large-scale bed forms that are present on the seabed (sand banks, sand waves)

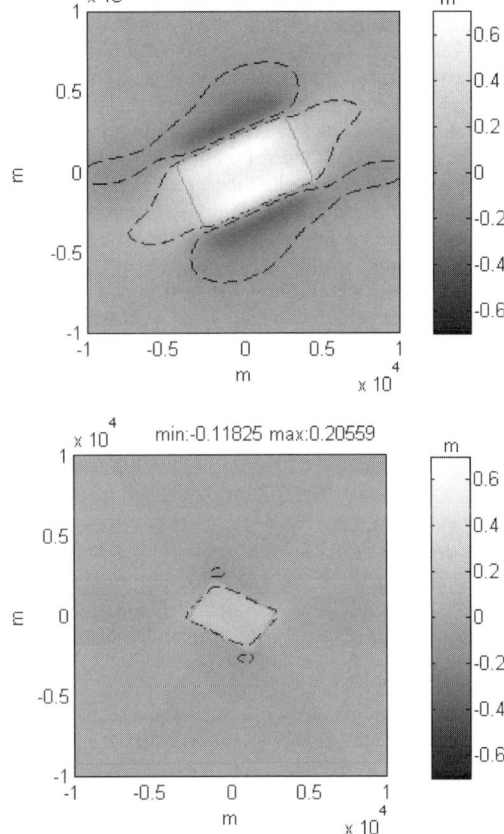

Figure 8. Seabed change in meter due to the Humber wind farm (upper) and the Q7 wind farm (lower). The values above the plot denote the lowest point of the seabed (min) and the highest bed elevation (max). The solid gray line denotes the wind farm and the black dashed line shows the area of influence.

are not taken into account. In reality this is of course not always the case. The grid of the data layer of depth in the southern part of the North Sea is sufficiently coarse so these features do not show up in the data. It is important to note however, that the morphological behaviour of a wind farm may change due to interaction with the surrounding topography as is shown by De Swart and Calvete (2003) and Roos and Hulscher (2004).

The value of the proportionality parameter ($\alpha$) is determined by the median grain size and also influences the morphological timescale ($T_{long}$). There is a lot of uncertainty about the value of $\alpha$ (for example the influence of waves on this parameter) and this can also influence the morphological timescale on which the seabed evolves due to the presence of the wind farm.

The model only studies the initial interaction between the seabed and the current flowing over it. This means that there is no damping mechanism. Therefore, the patterns that emerge at the sea bed will keep growing and are not topped off. Because it is assumed that the perturbation is very small with respect to the water depth, it is expected that this linear approximation holds for some time so that the results can be used for the timescale on which the offshore wind farms are designed. Due to the linear character of the model, it is not possible to draw conclusions about possible equilibrium states of the seabed.

The evolution of the seabed due to the presence of a wind farm takes place at the same scale as large scale bed forms like sand banks and sand waves, and therefore seems to be accurate. The model results are calculated over a time period of 100 years. However, existing wind farms are only up to 20 years old which prevents the use of field measurements for the validation of the model.

# 6 CONCLUSIONS

The share of offshore wind energy will grow and bigger offshore wind farms will be developed in the true offshore area. Offshore wind farms interact with the flow and cause changes in the large scale seabed morphology. An analytical model was used to assess the influence of an offshore wind farm on the seabed morphodynamics. The results show that the water depth, median grain size, flow direction and flow velocity influence the development of the seabed due to the

presence of a wind farm. And also, design parameters like the spacing between turbines, the diameter of the turbines and the drag force of the turbines influence the morphological development of the seabed.

By implementing an analytical model in a GIS, a rapid calculation of the effects of an offshore wind farm at a certain location in the North Sea can be made, thus providing a rapid assessment tool on the morphological effects of offshore wind farms.

ACKNOWLEDGEMENTS

This research is part of the project PhD@Sea, which is substantially funded under the BSIK-programme of the Dutch Government and supported by the consortium WE@Sea. Furthermore, we would like to thank Laura Uunk and Renske Gelderloos for their work on offshore wind farms in the Seminar Morphology course.

REFERENCES

California Institute of Technology. (2007). "http:// caltech-book.library.caltech.edu/1/04/chap5.htm"
Danish Energy Authority (2005), Copenhagen strategy on offshore wind power deployment.
De Swart, H. E. and D. Calvete (2003). Non-linear response of shoreface-connected sand ridges to interventions. Ocean Dynamics 53: 270-277, doi 10.1007/s10236-003-0044-9.
EWEA (undated), Wind Energy, the facts.
Greenpeace (2005), Offshore wind; Implementing a new powerhouse for Europe.
Greenpeace. (2007). "http://www.greenpeace.org.uk/contentlookup.cfm?ucidparam = 20031105133957&MenuPoint=D-B-C."
Huthnance, J. (1982a). On one mechanism forming linear sandbanks. Estuarine, Coastal and Shelf Science 14: 79-99.
Roos, P. C. and S. J. M. H. Hulscher (2004). Modelling the effects of different design options for offshore sand-pits. 2nd Workshop on Marine Sandwave and River Dune Dynamics (MARID), Enschede, The Netherlands.
Van der Veen, H. H., S. J. M. H. Hulscher and M. A. F. Knaapen (2006). Grain size dependency in the occurrence of sand waves. Ocean Dynamics 56: 228-234. DOI 10.1007/s10236-005-0049-7.
Van Rijn, L. C. (1989). Handbook of Sediment Transport by Currents and Waves, WL|Delft Hydraulics.
Van Rijn, L. C. (1993). Principles of sediment transport in rivers, estuaries and coastal seas. Amsterdam, Aqua publications.

*River, Coastal and Estuarine Morphodynamics: RCEM 2007 – Dohmen-Janssen & Hulscher (eds)*
© *2008 Taylor & Francis Group, London, ISBN 978-0-415-45363-9*

# Explicit formulae of shoreline change under time-varying forcing

A. Zacharioudaki & D.E. Reeve

*Centre for Coastal Dynamics and Engineering, School of Engineering, University of Plymouth, UK*

ABSTRACT:   This paper presents new analytical solutions to an extended version of the '1-line equation' for two cases of shoreline change: (1) when the evolution of an alongshore position is known in time, and (2) within a groyne compartment. Integral transform techniques are developed to obtain general analytical expressions, which can account for arbitrarily varying wave conditions, an arbitrary initial beach shape, and for sediment sources/sinks that are a known function of time and space. Example applications are described and the impact that a non-stationary forcing can have on shoreline evolution is investigated.

## 1  INTRODUCTION

One-dimensional models for shoreline evolution, known as '1-line' models, have been extensively used in applications and have shown good predictive capabilities (Hanson & Kraus 1989, Kamphuis 1993, Spivack & Reeve 2000). Because of their simplicity, i.e. the assumption of an explicit morphological equilibrium state under constant external forcing and the calculation of the shoreline positions only as a function of the alongshore sediment transport variation, they are essentially the only models to date suitable for long-term predictions of shoreline change.

1-line models, analytical and numerical, are effectively solutions to the continuity of sediment equation. However, analytical expressions can only be derived if the continuity equation, under certain assumptions, is reduced to a diffusion equation, which at its simplest is known as the '1-line equation' and takes the form (Pelnard-Considere 1956):

$$\frac{\partial y}{\partial t} = K \frac{\partial^2 y}{\partial x^2} \qquad (1)$$

where, $y$ = shoreline position from a fixed datum, $x$ = alongshore distance, $t$ = time, and $K$ = diffusion coefficient. Thus, analytical solutions involve simplifications, absent in numerical 1-line models, yet they are very useful because they pose certain advantages over them such as: (1) they enhance understanding of physical processes controlling beach response by enabling essential features to be isolated, (2) they give a quick-estimate of shoreline change, (3) numerical stability or diffusion problems are avoided, and (4) they provide an independent means of validating numerical

models for idealized conditions. Nonetheless, they may give misleading results if applied to problems of complex processes and/or complex geometries.

Amongst other assumptions, one of the least realistic typically adopted in analytical solutions is that wave conditions driving shoreline change do not very in time. Time-variation was introduced by Larson et al. (1997) and later by Dean & Dalrymple (2002). However, in the former study waves do not vary arbitrarily in time whilst both solutions lack of generality. Arbitrarily varying forcing was accounted for in a more flexible manner in Reeve's (2006) new semi-analytical solutions for the case of shoreline evolution at a single groyne.

In this paper, following Reeve's (2006) work, new semi-analytical solutions to an extended version of the one-line model are presented for two cases of shoreline change: (1) when the function of shoreline evolution is known at a location along its length, e.g. a managed beach location, and (2) between two groynes or headlands, perpendicular to the shoreline. Explicit solutions are obtained by means of integral transform techniques. In the first case, the generalized expression produces the spatial variability of the coastline explicitly with time in terms of its initial shape, the wave forcing, and a source/sink term which may vary in time and space. In the second case, an expression is derived under the assumption of an initially straight shoreline and of zero sources/sinks. The solutions hold the general advantages of analytical solutions and further provide an advanced independent means of validating time dependent numerical models and an improved straightforward tool for preliminary predictions of long-term shoreline change.

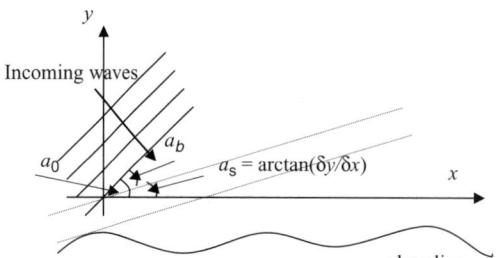

Figure 1. Angle conventions in the 1-line model.

## 2 THEORY OUTLINE

The 1-line model equation we solve here has the form (Reeve 2006):

$$\frac{\partial y}{\partial t} = K(t)\frac{\partial^2 y}{\partial x^2} + r(t)y(x,t) + s(x,t) \qquad (2)$$

where the last two terms represent sources/sinks. The diffusion coefficient, $K$, is given by:

$$K = \frac{2Q_0}{D} \qquad (3)$$

where $D$ = depth of closure and $Q_0$ = amplitude of the alongshore sand transport rate, $Q$, which is given by the CERC formula (CERC 1984):

$$Q = Q_0 \sin 2a_b \qquad (4)$$

$$Q_0 = \varepsilon \frac{\rho}{16(\rho_s - \rho)\sigma} H_b^2 C_{gb}$$

where $a$ = angle between the wave crests and the shoreline (Fig. 1), $\varepsilon$ = proportionality coefficient, $\rho$ = seawater density, $\rho_s$ = sediment density, $\sigma$ = sediment ratio, $H$ = wave height, and $C_g$ = wave group velocity. The subscript $b$ denotes these quantities at breaking. $Q, Q_0$ are in m$^3$/sec. Here, we adopt $\varepsilon = 0.41$ following Schoones & Theron (1993).

Figure 1 indicates that the angle $a_b$ may be expressed by:

$$a_b = a_0 - \arctan\left(\frac{\partial y}{\partial x}\right) \qquad (5)$$

where $a_0$ = angle between the wave crests at breaking and the x-axis, set parallel to the shoreline trend, and $\partial y/\partial x$ = local shoreline orientation. To derive Equation 2 the assumption that $a_0$ and $\partial y/\partial x$ are small is made.

Time-variation is introduced in Equation 2 through the variation of the diffusion coefficient, $K(t)$, thus the wave height variation, or through the variation of the source/sink term (which may also vary in space). The splitting of the source term in two parts, one dependent on the shoreline position, $r(t)y(x,t)$, and one independent of it, $s(x,t)$, allows the solutions to accommodate more aspects that are important in the long-term shoreline management (Reeve 2006).

Here, solutions are obtained in the region $x > 0$, $t > 0$. The initial condition is:

$$y(x,0)=g(x) \qquad\qquad x>0 \qquad (6)$$

Different boundary conditions are implemented for the two cases examined. For the case of shoreline change when the evolution of an alongshore point (located at $x = 0$) is known, the boundary condition is:

$$y(0,t) = h(t) \qquad\qquad t > 0 \qquad (7)$$

This condition may represent a situation of active beach management to maintain a particular beach width or of persistent bypassing of a groyne. For the case of shoreline change within a groyne compartment the boundary conditions are formulated at the groyne locations ($x = 0$ and $x = a$) and are given by:

$$\frac{\partial y(0,t)}{\partial x} = h(t) \qquad\qquad t > 0 \qquad (8)$$

$$\frac{\partial y(a,t)}{\partial x} = f(t) \qquad\qquad t > 0 \qquad (9)$$

The final boundary condition, required for all cases, is:

$$y \to 0 \quad\text{as}\quad x \to \infty, \quad\text{for}\quad t > 0 \qquad (10)$$

## 3 ANALYTICAL SOLUTIONS

### 3.1 Alongshore point with known time evolution

Equation 2 is solved subject to the conditions given by Equations 6, 7, and 10, with the requirement that the functions $r(t)$, $s(x,t)$, $g(x)$, and $h(t)$ are known although arbitrary. Fourier sine transforms are employed to derive the solution. The basic formulae of the Fourier sine transforms along with a few more expressions related to them are needed. For an idea on the kind of formulae required the reader is prompted to Reeve (2006). For details on the Fourier sine transforms see e.g. Sneddon (1972).

After application of the Fourier sine transform to Equations 2, 6, 7, and 10 we get:

$$\frac{d\bar{y}(\nu,t)}{dt} = -(\nu^2 K(t) - r(t))\bar{y}(\nu,t) + \nu K(t)h(t) + \bar{s}(\nu,t) \quad (11)$$

subject to the initial condition:

$$\bar{y}(\nu,0) = \bar{g}(\nu) \qquad (12)$$

where the overbar denotes the sine transform and $v$ = transform variable. Solving Equations 11 and 12 yields:

$$\bar{y}(v,t) = \bar{g}(v)e^{-\int_0^t (v^2 K(u)-r(u))du}$$

$$+ \int_0^t e^{-\int_w^t (v^2 K(u)-r(u))du} (vK(w)h(w)+\bar{s}(v,w))\, dw \tag{13}$$

Equation 13 can readily be seen as the sum of three terms. Taking the inverse Fourier sine transform of each of these terms and after some analytical manipulation, the general solution for $y(x,t)$ is obtained in terms of three integrals as follows:

$$y(x,t) = I_1 + I_2 + I_3 \tag{14}$$

$$I_1 = \frac{1}{2}\left\{\pi \int_0^t K(u)du\right\}^{-1/2} e^{\int_0^t r(u)du}$$

$$\times \int_0^\infty g(\xi)\left\{e^{-\frac{(\xi-x)^2}{4\int_0^t K(u)du}} - e^{-\frac{(\xi+x)^2}{4\int_0^t K(u)du}}\right\}d\xi \tag{15}$$

$$I_2 = \frac{x}{2\sqrt{\pi}}\int_0^t \left(\int_w^t K(u)du\right)^{-3/2} e^{-\frac{x^2}{4\int_w^t K(u)du} - \int_w^t r(u)du}\, dw \tag{16}$$

$$\times K(w)h(w)$$

$$I_3 = \frac{2}{\pi}\int_0^\infty \left(\int_0^t e^{-\int_w^t (v^2 K(u)-r(u))du}\bar{s}(v,w)\, dw\right)\sin(vx)dv \tag{17}$$

where $\xi$ = dummy variable and $w$ = dummy variable of integration running from time 0 to arbitrary time $t$.

Each integral accounts for the shoreline response to specific shaping factors. Thus, $I_1$ accounts for the initial shoreline configuration, $I_2$ for the impact of the time-varying wave conditions at the boundary, and $I_3$ for the effect of the independent source term. The dependent source function, $r$, appears in all three integrals. In all but the simplest cases, numerical evaluation of the integrals is required.

### 3.2 Groyne compartment

Equation 2 is solved subject to the conditions given by Equations 6, 8, 9, and 10. A finite Fourier cosine

transform is employed to derive the solution (for basic definitions of the finite Fourier cosine transform and related expressions see e.g. Sneddon (1972)). Applying the finite Fourier cosine transform to Equations 2, 6, 8, 9, and 10 gives:

$$\frac{d\bar{y}(v,t)}{dt} = -\left(\frac{\pi^2 v^2}{a^2} K(t)-r(t)\right)\bar{y}(v,t)+K(t)((-1)f(t)-h(t))+\bar{s}(v,t) \tag{18}$$

subject to the initial condition:

$$\bar{y}(v,0) = \bar{g}(v) \tag{19}$$

Solving Equations 18 and 19 yields:

$$\bar{y}(v,t) = \bar{g}(v)e^{-\int_0^t (\frac{\pi^2 v^2}{a^2} K(u)-r(u))du} +$$

$$\int_0^t e^{-\int_w^t (\frac{\pi^2 v^2}{a^2} K(u)-r(u))du} \times (K(w)((-1)^v f(w)-h(w))+\bar{s}(v,w))\, dw \tag{20}$$

Similarly to the previous case, Equation 20 is considered as the sum of four terms. By taking the inverse finite Fourier cosine transform of each term, the general solution for $y(x,t)$ may be obtained. Here, we solve Equation 20 only for the specific case of an initially straight shoreline with no sources/sinks of sediment and with the same boundary condition at each groyne location. In this case, $g$, $r$, $s$ equal zero, $f(x)=h(w)$, and the solution simplifies to yield:

$$y(x,t) = -\frac{2}{a}\sum_{v=1}^\infty \cos\left(\frac{v\pi x}{a}\right) \int_0^t e^{-\int_w^t (\frac{\pi^2 v^2}{a^2} K(u)-r(u))du} \times [K(w)((-1)^v f(w)-h(w))]\, dw \tag{21}$$

Numerical evaluation of the solution is required.

### 3.3 Numerical evaluation

Typically, the arbitrary functions $K(t), r(t), s(x,t), h(t)$, and $g(x)$, involved in the general solutions, will need to be specified numerically at discrete points, unless they have simple forms that could be treated analytically. Standard, readily available, and robust numerical integration routines for smoothly varying functions are sufficient to solve the integrals appearing in the new solutions. Here, we use the well-known Newton-Cotes formulae for open and semi-open integration (the latter can treat singularity points, here in $I_2$ when $w=t$), (Press et al. 1992). The error term of the specific formulae applied is of the order $O(1/n^4)$ where $n$ = number of integration steps.

The numerical evaluation of the summation in Equation 21 revealed a pattern of non-uniform convergence (i.e. the solution ultimately becomes an undamped oscillation about a mean value). This phenomenon is known as the Gibbs Phenomenon (Jerry 1998) and can be effectively eliminated with filters (see e.g. Gottlieb and Shu 1997). Here, it was satisfactorily removed with the use of an exponential filter.

The numerical evaluation of the solutions proved to be very efficient is terms of computational time. Equation 14 was solved and printed within a second or so on a Pentium IV PC for a problem with 6000 points alongshore, and for an output time of $t = 400$ hours ($dt = 1$hour). The filtered form of Equation 21 was solved and printed within less than 3 seconds.

## 4 EXAMPLE APPLICATIONS

### 4.1 Beach fill/persistent groyne bypassing

We consider the case of a rectangular beach fill on an initially straight shoreline as depicted in Figure 2. The fill is of alongshore extend $l = b - a$ and offshore extent $y_0$.

Initial and boundary conditions are formulated as:

$$y(x,0) = g(x) = \begin{cases} y_0 & a \le x \le b \\ 0 & x < a \quad x > b \end{cases} \quad x > 0 \quad (22)$$

$$y(0,t) = h(t) = 0 \qquad t > 0 \quad (23)$$

If sediment sources/sinks are taken to be zero (i.e. $s(x,t) = 0$, $r(t) = 0$) then, $I_2, I_3 = 0$ in Equation 14, which now simplifies to yield:

$$y(x,t) = I_1 = \frac{y_0}{2} \left[ \begin{array}{l} erf\left( \dfrac{a+x}{2\sqrt{\int_0^t K(u)du}} \right) - erf\left( \dfrac{a-x}{2\sqrt{\int_0^t K(u)du}} \right) \\[3ex] - erf\left( \dfrac{b+x}{2\sqrt{\int_0^t K(u)du}} \right) + erf\left( \dfrac{b-x}{2\sqrt{\int_0^t K(u)du}} \right) \end{array} \right] \quad (24)$$

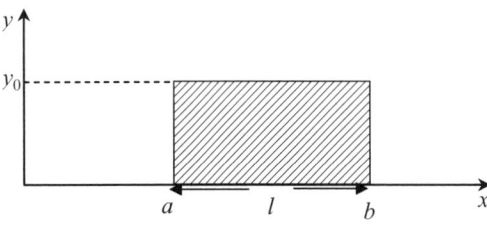

Figure 2. Rectangular beach fill.

For $K = $ constant Equation 24 is evaluated analytically.

The general case of sediment supplied to any point along the beach nourishment to preserve a certain beach width, $y_0$, at this point may also be represented by Equation 14. The boundary condition for this case is:

$$y(0,t) = h(t) = y_0 \qquad t > 0 \quad (25)$$

The initial condition is determined by the particular application. For example, if sand is supplied to the corner of a fill of $a = -l$ and $b = 0$, then the initial condition is:

$$y(x,0) = g(x) = 0 \qquad x > 0 \quad (26)$$

Now, Equation 14 gives:

$$y(x,t) = I_2 = y_0 erfc\left( \dfrac{x}{2\sqrt{\int_0^t K(u)du}} \right) \quad (27)$$

For $K = $ constant Equation 27 has a purely analytical form and is the well-known solution for this case, as given, for example by Larson et al. (1987).

The case of shoreline change when persistent bypassing of a groyne (located at $x = 0$) occurs may also be represented by Equations 25–27.

### 4.2 Beach management to maintain target plan shape

We now consider the case of Figure 2 but when a simple management policy is applied to maintain a specific shoreline configuration. The management function employed here, expressed as a source term, has the form:

$$source = a(y'_0 - y) \quad (28)$$

where $a = $ constant $= $ measure of the required supply rate and $y'_0 = $ target beach plan shape. This is a very simple, somewhat artificial situation, where beach configuration is preserved by instantaneous beach scraping or nourishment, proportional to the difference between the actual shoreline position and the target shoreline. However, it illustrates how the present solutions can be used to investigate the affect of simple beach management policies and to estimate the nourishment rates and volumes required to preserve a certain beach plan shape.

If we assume that $y'_0(x)$ is the initial shape of the fill then the initial condition is given by Equation 22 with $y(x,0)=y_0'(x)$. In Equation 14, $I_2=0$ and $y(x,t)=I_1+I_3$ with:

$$I_1 = \frac{y_0 e^{-at}}{2}\left[\left(erf\frac{a+x}{2\sqrt{\int_0^t K(u)du}} - erf\frac{a-x}{2\sqrt{\int_0^t K(u)du}}\right) - \left(erf\frac{b+x}{2\sqrt{\int_0^t K(u)du}} + erf\frac{b-x}{2\sqrt{\int_0^t K(u)du}}\right)\right] \quad (29)$$

$$I_3 = \frac{y_0 a}{2}\int_0^t e^{-a(t-w)}\left\{\left(erf\frac{a+x}{2\sqrt{\int_w^t K(u)du}} - erf\frac{a-x}{2\sqrt{\int_w^t K(u)du}}\right) - \left(erf\frac{b+x}{2\sqrt{\int_w^t K(u)du}} + erf\frac{b-x}{2\sqrt{\int_w^t K(u)du}}\right)\right\}dw \quad (30)$$

### 4.3 Groyne/headland compartment

We finally consider the case of Equation 21, i.e. of an initially straight shoreline within a groyne compartment, where there are no sources/sinks of sediment. Initial and boundary conditions are as follows:

$$y(x,0) = g(x) = 0 \qquad\qquad x > 0 \qquad (31)$$

$$\frac{\partial y(0,t)}{\partial x} = \frac{\partial y(a,t)}{\partial x} = \tan(a_0) \qquad t > 0 \qquad (32)$$

Equation 32 is obtained from the linearized form of Equation 4 (see e.g. Larson et al. 1987) for $Q=0$ at the location of the groynes. The solution after application of a 4th order exponential filter to remove the Gibbs Phenomenon is given by:

$$y(x,t) = I_3 = \frac{-4}{a}\sum_{v=1}^{v=2n+1} e^{-0.0000005v^4}\cos\left(\frac{v\pi x}{a}\right) \\ \times \int_0^t e^{-\int_w^t \frac{\pi^2 v^2}{a^2}K(u)du} K(w)\tan(a_0)dw \quad (33)$$

## 5 RESULTS

In this section we present computations of shoreline response to time-varying wave conditions for some of the applications outlined above. When possible the semi-analytical solutions are compared to analytical forms of $K=$ constant.

The time sequence of $K$ needed to perform the computations is calculated from the wave sequence described in Reeve (2006), which is expressed in terms of the significant wave height, wave period, and mean wave direction, at hourly intervals over a period of 400 hours. A water depth of 6 m, a depth of closure of 10 m, and a beach normal of $210^0$ azimuth are adopted for the calculations. The time series of significant wave height and mean wave direction are shown in Figure 3 whilst the time series of K is shown in Figure 4. The occurrence of three consecutive storms is evident in the figures, whilst waves approach mainly from the west (positive angles) relative to the shoreline normal.

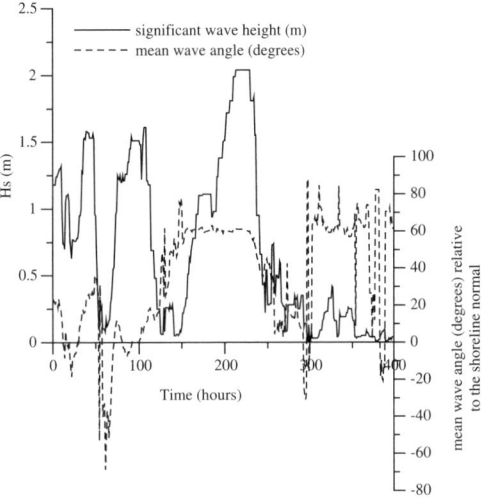

Figure 3. Sequence of wave heights and angles.

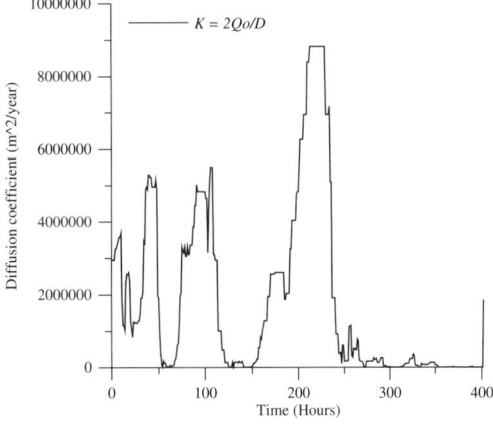

Figure 4. Sequence of the diffusion coefficient.

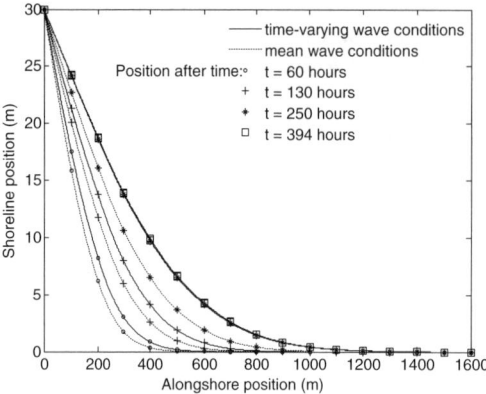

Figure 5. Computed shoreline positions when persistent bypassing of a 30 m long groyne, located at $x = 0$, occurs.

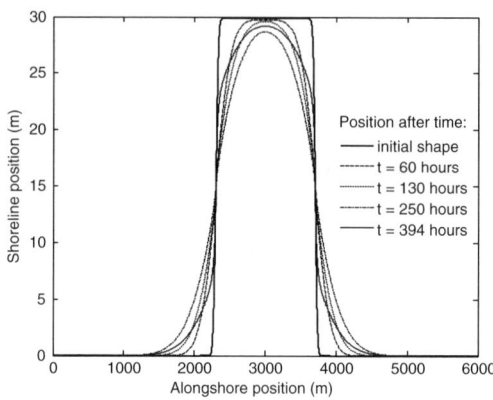

Figure 6. Computed shoreline positions when a source management function of the form $source = a(y'_0 - y)$ is applied to maintain a rectangular beach fill of initial shape $y'_0$.

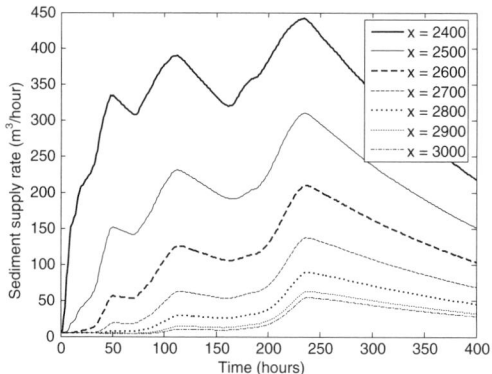

Figure 7. Time evolution of the source term of Equation 28 at different alongshore positions of the fill depicted in Figure 6. The source term is set to zero for shoreline change outside the area occupied by the initial beach fill (where $y > y'_0$).

We first examine a situation defined by Equation 27 subject to Equations 25 and 26, e.g. persistent bypassing of a groyne of length $y_0 = 30$ m. Computed shoreline positions with the semi-analytical solution, at four different times of the wave sequence, are shown in Figure 5. In addition, the results are compared with those derived from the pure analytical form of Equation 27 for constant forcing. The mean value of $K$ over the period of 400 hours ($K \approx 1.84 \times 10^6$ m²/year) is used in the latter computation. The groyne is located at $x = 0$, the alongshore extent is 2000 m and the discretisation resolution is 1 m. In the figure, the first three output times correspond to the ends of the three consecutive storms respectively whilst the last output time is close to the end of the wave sequence.

Shoreline is progressively accreting adjacent to the groyne in response to the higher waves in the time series of Figure 3. The effect of time-varying wave conditions on the shoreline better manifests itself after the end of the second, when the wave sequence deviates persistently from mean conditions. Thus, it can readily be seen that higher waves during the third storm along with a longer duration cause the shoreline to accrete faster. Moreover, the occurrence of low waves after the third storm and to the end of the time series results in trivial further shoreline response. Significantly different results are obtained when mean wave conditions are driving shoreline change. Faster shoreline advance during the third storm or slower after it, are not illustrated in the figure. In essence, purely analytical solutions do not represent the influence of the storm characteristics on the shoreline, which consequently advances only in proportion to the time elapsed.

We now examine the situation described by Equations 29 and 30 subject to Equations 22 and 23, i.e. the case of the rectangular beach fill depicted in Figure 2 but when a simple management function of the form

of Equation 28 is applied to maintain its initial shape, which here is defined by $a = 2300$ m, $b = 3700$ m, and $y_0 = 30$ m. Figure 6 shows computed shoreline positions for this case and for the same output times as described in the previous example. The alongshore extent is 6000 m with 1 m discretisation resolution. A value of $a = 40$ is adopted in Equation 28.

Shoreline advance after the end of the third storm is apparent, in contrast with minor change observed over the same period in the previous example. This advance now depends almost solely on the source terms included in the equations.

Figure 7 shows the time evolution of the rate at which sediments must be supplied to certain alongshore positions of the beach nourishment to maintain their initial offshore extent.

1072

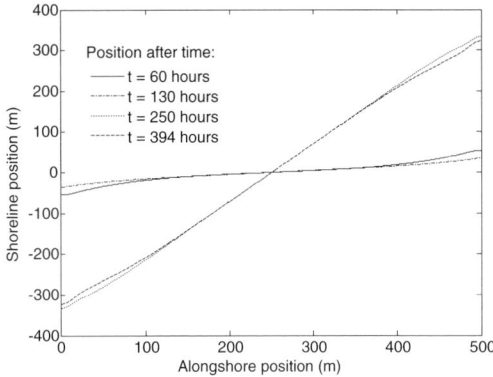

Figure 8. Computed shoreline change between two groynes which are assumed to be impermeable and of infinite length.

The general trend is one of rate increase at the beginning of the wave sequence when erosion of the fill is fast followed by rate reduction associated with a slower shoreline retreat with time which is further balanced by the supply of sediments. Higher rates near the corners of the fill are because of increasing fill erosion away from its centre. The picks in Figure 7 correspond to the largest values of the diffusion coefficient $K$ (Fig. 4) linked to periods of high wave activity (Fig. 3) and enhanced sediment transport which cause the fill to diffuse rapidly. Such an analysis could be useful to determine the optimum value of the coefficient $a$ in Equation 28.

We finally examine the case defined by Equation 36 subject to Equations 34 and 35, i.e. shoreline evolution within a groyne compartment. Figure 8 shows computed shoreline positions at the same output times as above. The groynes are assumed to be impermeable and no bypassing occurs. The spacing between them is taken to be 500 m.

Again, here, we clearly see the strong impact the third storm has on the shoreline evolution. In fact, the impact is more pronounced, with increased shoreline changing rates, erosion in this case, taking place. This is a consequence of the boundary configuration of this problem (Eq. 35), i.e. of the variation of the wave angle at the location of the groynes. The significance of this variation, absent in any other case examined in this paper, is demonstrated. However, the rates of erosion observed are somewhat exaggerated. This could be because of highly oblique wave angles frequently occurring in the wave sequence used (Fig. 3). It could also suggest that a more consistent way of incorporating wave angle variation in the solutions, than simply at the boundaries, is needed. For example, Dean and Dalrymple (2002) proposed a form of $K$ which depends both on wave height and angle variation.

## 6  CONCLUSIONS

The common assumption of simpler analytical solutions of the '1-line Equation' that a fixed average wave condition is driving shoreline change is relaxed in this paper through the derivation of new semi-analytical solutions that account for time-varying wave forcing. They new solutions are obtained for two specific but common cases of shoreline change: (1) a managed beach or persistent groyne bypassing where there is an alongshore location of known time evolution, and (2) within a groyne/headland compartment. Fourier transform techniques are used to derive the solutions. In the first case, the solution includes wave conditions as explicit but arbitrary functions of time, allows for an arbitrary initial shoreline shape, and includes source/sink terms; it is given as the sum of three integrals. In the second case, the final expression represents a situation of an initially straight shoreline with no source/sink terms, and where the same boundary condition applies at both groyne locations; it has the form of an infinite Fourier series. In all but the simplest cases the numerical evaluation of the solutions is required.

The numerical evaluation proved to be very fast. In general, less than 3 sec were needed to obtain and print the solutions for the examples examined in this paper, using a Pentium IV PC. The infinite Fourier series involved in the solution of the second case converged sufficiently fast after application of a filter to remove the Gibbs Phenomenon, originally exhibited in the results.

Somewhat exaggerated rates of shoreline change were observed when shoreline evolution within a groyne compartment was investigated. This could merely be the result of highly oblique waves frequently encountered in the wave sequence used, or could be indicative of the need for a more coherent way of incorporating wave angle variation in the solutions than simply at the boundaries, e.g. through the use of a form of the diffusion coefficient, $K$, that includes wave angle variation.

The new solutions are sufficiently flexible to allow the study of a variety of processes that may affect the shoreline, such as storminess, sea-level changes, or adaptive management policies. For example, when a wave sequence containing three consecutive storms of different nature was used to obtain shoreline positions, the impact of the storm characteristics on the results was evident. In addition, when a simple management function was applied to preserve a particular shoreline configuration, it was shown that preliminary estimates of the optimum rates of sediment supply and volumes required for maintenance can be computed.

Like all analytical solutions, the new solutions involve simplifications. However, their explicit form still permits the isolated study of essential physical processes, prevents accumulation of numerical errors inherent in time-stepping finite-difference numerical techniques, and provides a quick preliminary estimate of shoreline response, important in the conceptual design stage. Overall, the new quasi-analytical solutions give account for time varying wave conditions in estimating shoreline change and form an improved, important tool for validating numerical time-stepping models under simple geometries and non-stationary forcing.

## REFERENCES

CERC 1984. *Shore protection manual*. Coastal Engineering Research Center, U.S. Corps of Engineers, Vicksburg, Mississippi.

Dean, R.G. & Dalrymple, R.A. 2002. *Coastal processes: with engineering applications*. Cambridge: Cambridge University Press.

Gottlieb, D. & Shu, C.-W. 1997. On the Gibbs phenomenon and its resolution. *SIAM Review* 39(4): 644–668.

Hanson, H. & Kraus, N.C. 1989. *GENESIS-Generalized model for simulating shoreline change*. Technical Report No. CERC-89-19, USAE-WES. Coastal Engineering Research Center, U.S. Corps of Engineers, Vicksburg, Mississippi.

Jerri, A.J. 1998. *The Gibbs phenomenon in Fourier analysis, splines and wavelet approximations*. Dordrech: Kluwer Academic Publishers.

Kamphuis, J.W. 1993. Effective modelling of coastal morphology. *Proceedings of the 11th Australian Conference on Coastal Engineering*, Institute of Engineers of Australia, Sydney, Australia, 173–179.

Larson, M., Hanson, H., & Kraus, N.C. 1987. *Analytical solutions of the one-line model of shoreline change*. Technical Report CERC-87-15, USAE-WES, Coastal Engineering Research Center, Vicksburg, Mississippi.

Larson, M., Hanson, H., & Kraus, N.C. 1997. Analytical solutions of one-line model for shoreline change near coastal structures. *Journal of Waterway, Port, Coastal, and Ocean Engineering* 123(4): 180–191.

Pelnard-Considere, R. 1956. Essai de theorie de l'evolution des forms de rivages en plage de sable et de galets. *Societe Hydrotechnique de France, Proc. 4th Journees de l'Hydraulique, les Energies de la Mer, Question III*, Rapprot No. 1: 289–298.

Press, W.H., Teukolsky, S.A., Vettering, W.T., & Flannery, B.P. 1992. *Numerical recipes in C: the art of scientific Computing*. Cambridge: Cambridge University Press.

Reeve, D.E. 2006. Explicit expression for beach response to non-stationary forcing near a groyne. *Journal of Waterway, Port, Coastal and Ocean Engineering* 132: 125–132.

Schoones, J.S. & Theron, A.K. 1993. Review of the field-data base for longshore sediment transport. *Coastal Engineering* 19: 1–25.

Sneddon, I.H. 1972. *The use of integral transforms*. New York: McGraw-Hill.

Spivack, M. & Reeve, D.E. 2000. Source reconstruction in a coastal evolution equation. *Journal of Computational Physics* 161: 169–181.

*Human interference in estuaries*

*River, Coastal and Estuarine Morphodynamics: RCEM 2007 – Dohmen-Janssen & Hulscher (eds)*
© *2008 Taylor & Francis Group, London, ISBN 978-0-415-45363-9*

# Long term process-based morphological model of the Western Scheldt Estuary

G. Dam & A.J. Bliek
*Svašek Hydraulics, Rotterdam, The Netherlands*

R.J. Labeur
*Delft University of Technology, Delft, The Netherlands*

S.J. Ides & Y.M.G. Plancke
*Flanders Hydraulic Research, Antwerp, Belgium*

ABSTRACT: A process-based morphological model of the Western Scheldt Estuary based on the finite elements method is presented in this paper. This model is able to successfully hindcast the morphological developments of the Western Scheldt over several decades. The measured sedimentation/erosion pattern versus the pattern calculated by FINEL2d over the period 1965 – 2002 was compared. Good agreement was found in overall patterns, although many differences were still to be seen in detail. It was concluded that the model can be used to evaluate different scenarios, making it a useful tool for decision making processes e.g. future deepenings of the fairway.

## 1 INTRODUCTION

The Western Scheldt estuary lies in the south-western part of the Netherlands and is the gateway to the port of Antwerp. A large amount of dredging is necessary to maintain the required depth of the navigation channel to the port of Antwerp.

The morphological developments of this estuary are complex and are governed by natural processes and by human interventions. Since a lot of functions in the Western Scheldt are related to morphology the need for predicting future developments is high. Until now long term morphological predictions were carried out by using (semi)-empirical models, like ESTMORF (Wang et al., 1999). Process-based morphological models were not yet able to reproduce the morphological development very well over decades, partly because of the large amount of computational time involved. Hibma et al., (2003) uses a process-based morphological model with a conceptual model of an estuary like the Western Scheldt.

In this paper a process-based morphological model FINEL2d is presented which is used to reproduce the morphological developments of this estuary during the last decades. A previous successful study prepared with FINEL2d is carried out in the Haring-vliet estuary (Dam, et al., 2005).

The calibration of the model was carried out in consecutive parts. A first stage consisted of a calibration of the water motion, next the morphological module was calibrated over the period 1995 until 2002. The calibrated model was then validated from 1965 until 2002.

The content of this paper is as follows: In section 2 the Western Scheldt estuary is introduced. Section 3 describes the FINEL2d model. The model set-up of the Western Scheldt model is discussed in section 4. Section 5 presents the morphological results. Finally in section 6 some conclusions are given.

## 2 THE WESTERN SCHELDT ESTUARY

The Western Scheldt estuary is a dynamic system that has gone through may changes due to human impacts and natural developments. The morphology of the Western Scheldt is very important for all functions related to this area, e.g. navigation, ecology and sand mining.

The vertical tide in the Western Scheldt ranges from an average 3.9 m in the mouth to 5.0 m near Antwerp. Fresh water discharge from the River Scheldt is limited compared to the tidal volumes.

The Western Scheldt estuary is a multiple channel system. The tidal flats and surrounding ebb and flood channels form morphological macro cells. The entire Western Scheldt consists of such morphological cells, see Figure 1. At the locations where the cells

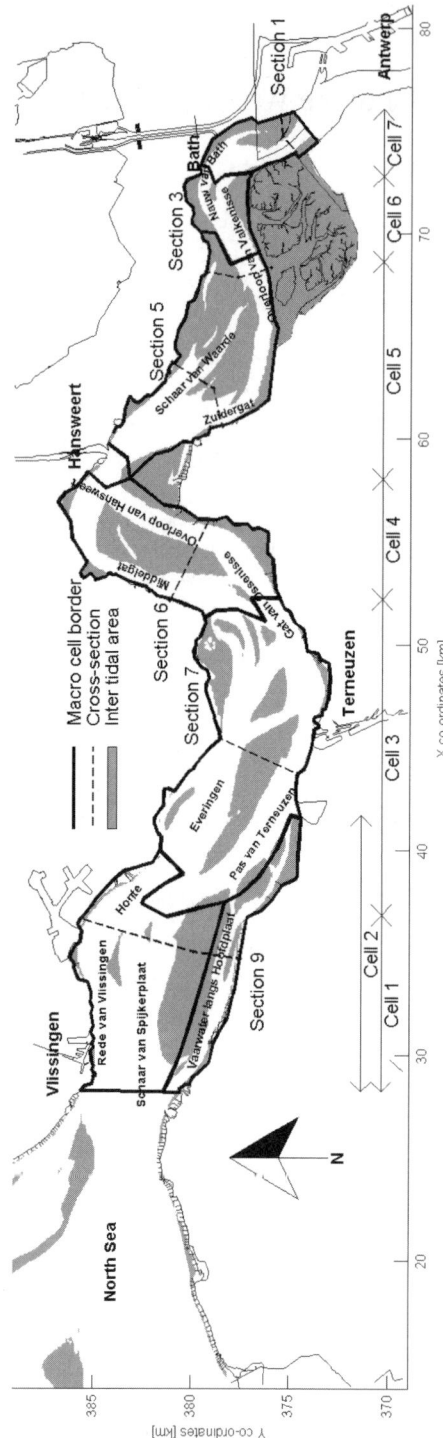

Figure 1.   Layout of the Western Scheldt Estuary.

coincide, sills develop, which block the fairway to the port of Antwerp and therefore require regular dredging (Winterwerp et al., 2001).

In the 1970's and 1990's a deepening of the navigational channel has been carried out to allow vessels with greater draught to enter the port of Antwerp. Most of the dredging had to be done at these sills. The dredging volumes (maintenance plus deepening) range from $5\,\mathrm{Mm}^3$ in 1968 to $14\,\mathrm{Mm}^3$ during the second deepening. After the second deepening the volumes seem to establish around 7 to $8\,\mathrm{Mm}^3$ per year. The dredged material is deposited into the estuary in especially designated areas, usually in the secondary branches.

Sand mining also plays a role in the Western Scheldt. Yearly approximately 2.0 to $2.5\,\mathrm{Mm}^3$ sand is mined. In contradiction to channel dredging this sand is really extracted from the estuary.

The Western Scheldt consists mainly of fine non-cohesive sediments; only at inter tidal areas silt can be found. In the model only non-cohesive sediment is taken into account.

Since the Western Scheldt is fairly sheltered from the waves of the North Sea the morphological development is mainly tidal driven and waves are neglected in this study.

## 3   THE MORPHODYNAMIC MODEL FINEL2D

### 3.1   *General*

FINEL2d is a 2DH numerical model based on the finite elements method and is developed by Svašek Hydraulics. The following sections describe the governing equations of the FINEL2d model.

### 3.2   *Hydrodynamic module*

The depth-integrated shallow water equations are the basis of the flow module. For an overview on shallow water equations see Vreugdenhil (1994).

The model equations are the continuity equation:

$$\frac{\partial h}{\partial t} + \frac{\partial uD}{\partial x} + \frac{\partial vD}{\partial y} = 0, \tag{1}$$

the x-momentum balance:

$$\frac{\partial Du}{\partial t} + \frac{\partial Du^2}{\partial x} + \frac{\partial Duv}{\partial y} + f_c Dv + gD\frac{\partial h}{\partial x} - \frac{1}{\rho}\tau_{x,b} + \frac{1}{\rho}\tau_{x,w} + \frac{1}{\rho}\tau_{x,r} = 0, \tag{2}$$

and the y-momentum balance:

$$\frac{\partial Dv}{\partial t} + \frac{\partial Duv}{\partial x} + \frac{\partial Dv^2}{\partial y} - f_c Du + gD\frac{\partial h}{\partial y} - \frac{1}{\rho}\tau_{y,b} + \frac{1}{\rho}\tau_{y,w} + \frac{1}{\rho}\tau_{y,r} = 0. \tag{3}$$

where $u =$ depth averaged velocity in x-direction [m/s]; $v =$ depth averaged velocity in y-direction [m/s];

1078

$h=$ water level [m]; $z_b=$ bottom level [m]; $D=$ water depth [m]; $f_c=$ Coriolis coefficient [1/s]; $g=$ gravitational acceleration [m/s$^2$]; $\rho=$ density of water [kg/m$^3$]; $\tau_b=$ bottom shear stress [N/m$^2$]; $\tau_w=$ wind shear stress [N/m$^2$]; and $\tau_r=$ radiation stress [N/m$^2$];

In addition to the effect of advection and pressure gradients, external forces like the Coriolis force, bottom shear stress, wind shear stress and radiation stress due to surface waves can be taken into account. It is noted that turbulent shear stresses are not taken into account: the application is therefore restricted to advection dominated flows only.

As a solution method, the discontinuous Galerkin method is adopted (Hughes, 1987) in which the flow variables are taken constant in each moment. This method has advantages in dealing with drying elements.

As the momentum equations contain first order derivatives in space, they can be written as:

$$\frac{\partial U}{\partial t} + \nabla \cdot F = H \tag{4}$$

where:

$$U = \begin{pmatrix} h \\ uD \\ vD \end{pmatrix}, \quad F = \begin{pmatrix} uD & vD \\ u^2D+\frac{1}{2}gh^2 & uvD \\ uvD & v^2D+\frac{1}{2}gh^2 \end{pmatrix}, \quad H = \begin{pmatrix} 0 \\ \frac{1}{\rho}\tau_{x,tot} - f_cvD - gDi_{b,x} \\ \frac{1}{\rho}\tau_{y,tot} + f_cuD - gDi_{b,y} \end{pmatrix} \tag{5}$$

in which $\tau_{tot,x}$ and $\tau_{tot,y}$ are summations of the external stresses in x- and y-direction respectively, while $i_{b,x}$ and $i_{b,y}$ are the bed level gradients in x- and y-direction respectively. The equation can be integrated over an element resulting in:

$$\int_{\Omega_e} \frac{\partial U}{\partial t}\,d\Omega + \int_{\Gamma_e} F\,n\,d\Gamma = \int_{\Omega_e} H\,d\Omega, \tag{6}$$

where $\Omega_e$ denotes an element, $\Gamma_e$ the associated element boundary, while $n$ is the outward pointing vector normal to $\Gamma_e$.

The problem is now reduced to the determination of the fluxes $F$ along the boundaries. As the variables are determined at the elements and not at the sides, the flux F is not known beforehand, but involves the solution of a local Riemann problem. An approximate Riemann solver according to Roe (Glaister, 1993) is applied. This method guarantees strict mass and momentum conservation, but suffers from some numerical diffusion in stream-wise direction. An explicit time integration scheme is used. As this method restricts the time step, the time step is controlled automatically for optimum performance.

A special problem in shallow waters like for example estuaries is the drying and flooding of large areas during a tidal cycle. A discontinuous discretisation is

used in combination with an explicit time-stepping. In this way this flooding and drying of the elements can be treated relatively easily. If an element tends to dry, the corresponding characteristic wave is partially reflected from this element which guarantees mass conservation.

## 3.3 Sediment transport module

FINEL2d uses the following sediment balance equation for the evolution of the bed level:

$$\frac{\partial z_b}{\partial t} + \frac{\partial q_x}{\partial x} + \frac{\partial q_y}{\partial y} = 0 \tag{7}$$

In which $z_b$ [m] is the bed level and $(q_x, q_y)$ [m$^2$/s] are the components of the sediment flux in x- and y-direction respectively.

In order to determine the non-cohesive part of the sediment fluxes, the transport formula of Engelund and Hansen formula is used (Engelund & Hansen, 1967). Since most of the sand transport in the Western Scheldt is suspended transport, a time lag effect is introduced in the model according to Gallapatti & Vreugdenhil (1985). First a dimensionless equilibrium concentration is calculated:

$$c_e = \frac{S}{D\sqrt{u^2+v^2}} \tag{8}$$

where $c_e$ is equilibrium concentration [–] and $S$ the magnitude of the equilibrium sand transport [m$^2$/s] according to Engelund and Hansen.

The concentration $c$ [–] is then calculated according from:

$$\frac{dc}{dt} = \frac{1}{T_A}[c_e(t) - c(t)] \tag{9}$$

In which $T_A$ is a characteristic timescale [s].

Equation 9 shows that if the concentration is lower than the equilibrium concentration erosion will occur ($dc/dt > 0$). If the concentration is higher than the equilibrium concentration sedimentation will occur ($dc/dt < 0$). The coefficient $T_A$ characterises the time needed for the adjustment of the concentration and is defined as $T_A = h/w_s$; where $w_s$ [m/s] is the settling velocity of the sand particles. In relative shallow areas the time scale is small and the concentration almost immediately adjusts to the equilibrium concentration.

## 4 MODEL SET UP

### 4.1 Grid schematisation and boundary conditions

FINEL2d uses unstructured triangular grids. The advantage of such meshes in comparison to for

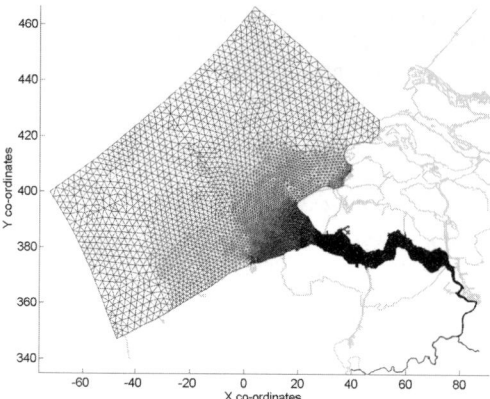

Figure 2. Computational mesh of the FINEL2d model.

example finite difference grids, is the flexible mesh generation. In this way no nesting techniques are required in regions of specific interest, where a higher degree of resolution is needed, while arbitrary coastlines and complex geometries can be resolved very well.

The seaward boundaries of the computational mesh of the Western Scheldt are chosen approximately 40 km away from the coastline and coincide with the boundaries of existing models. The latter can be used to obtain the corresponding boundary conditions. A significant part of the river system in Belgium is also included in the schematisation. In Figure 2 the overall mesh is shown. In the area of interest, the Western Scheldt, the average grid size is approximately 1.1 ha. Near the seaward boundaries the grid size is approximately 2.5 km$^2$. The total number of elements (triangles) of the mesh is 44,111.

The river discharge at the river the Scheldt and the Rupel were taken constant at respectively 43 and 65 m$^3$/s.

### 4.2 Calibration of the water motion

The first step in the calibration of a morphodynamic model is to calibrate the water motion. In this model the water motion is mainly calibrated on measured water levels in the estuary. The calibration parameter is the bottom roughness. A trial and error method was used to find the optimal settings for the bottom roughness. An 8 day calibration period was chosen. In the mouth and the western part of the estuary a Nikuradse roughness height of 0.2 cm was found. The optimal roughness gradually increases from the west to the east of the estuary. In the east a Nikuradse roughness height of 10 cm was found. The difference between measured and computed high water, low water, tidal range and mid water level at the various water level stations are shown in Table 1. See Figure 1 for an overview of the water level stations.

Table 1. Water level deviations (cm)*.

| Station | HW | LW | Tidal range | Mid |
|---|---|---|---|---|
| Vlissingen | −8 | −5 | −4 | −6 |
| Terneuzen | −7 | −9 | 2 | −8 |
| Hansweert | −3 | −12 | 9 | −7 |
| Bath | −6 | −10 | 3 | −8 |
| Antwerp | −2 | 4 | −6 | 1 |

* Averaged over a 8 day period.

Table 2. Observed and simulated tidal volumes.

| Cross-section | | Flood tidal volume | | | Ebb tidal volume | | |
|---|---|---|---|---|---|---|---|
| Nr. | Name | O | M | M/O | O | M | M/O* |
| 1b | Bath | 200 | 174 | 0.87 | 174 | 162 | 0.93 |
| 3b | Valkenisse | 288 | 270 | 0.94 | 275 | 264 | 0.96 |
| 5a | Waarde | 259 | 247 | 0.96 | 267 | 253 | 0.95 |
| 5b | Zuidergat | 160 | 159 | 1.00 | 151 | 147 | 0.97 |
| 6a | Ossenisse | 369 | 417 | 1.13 | 390 | 425 | 1.09 |
| 6b | Middelgat | 230 | 200 | 0.87 | 180 | 176 | 0.98 |
| 7a | Terneuzen | 344 | 347 | 1.01 | 417 | 383 | 0.92 |
| 7b | Everingen | 539 | 525 | 0.97 | 432 | 455 | 1.05 |
| 9a | Hoofdplaat | 112 | 112 | 1.00 | 113 | 105 | 0.93 |
| 9b | Spijkerplaat | 1051 | 1118 | 1.06 | 1063 | 1109 | 1.04 |

* O = observed volume [Mm$^3$]; M = model volume [Mm$^3$]; M/O is the model volume divided by the observed volume.

The deviations are usually within a 10 cm accuracy range. With a tidal range of approximately 3 to 5 metres in the estuary this gives a deviation of only 3%.

The next step in the calibration of the water motion concerns the tidal volumes of various cross-sections in the estuary. The cross-sections have been surveyed from 2000 to 2002 using an acoustic current profiler during 13 hours. The same periods were simulated in the model. The results are summarised in Table 2 in which the flood and ebb tidal volume of both the measurement and the simulation are shown. See figure 1 for an overview of the cross-sections.

It is concluded that all observed discharges can be simulated with an accuracy of 15% and usually 10%. Finally, a comparison was made with current data at shallow banks and tidal flats. From the results of the of all calibration steps it can be concluded that the water motion is calibrated decisively.

### 4.3 Morphological acceleration factor

A morphological acceleration factor is used to multiply the calculated bottom changes to accelerate the computational time. A small morphological acceleration factor is preferred from a numerical point of view but is not practical because of the large computational time. In order to investigate the morphological

Table 3.  Morphological acceleration factor.

| Factor | Sedimentation volume difference* | Erosion volume difference* |
|--------|----------------------------------|----------------------------|
| 1      | –                                | –                          |
| 5      | 7.6%                             | 8.4%                       |
| 24.75  | 10.3%                            | 14.6%                      |
| 49.5   | 48.1%                            | 43.2%                      |

* Difference is calculated relative to factor 1 and after 1 year.

Table 4.  Settings of the optimal calibration run.

| Factor | value | |
|--------|-------|---|
| Grain size | 150 | μm |
| Fall velocity of sand | 1.5 | cm/s |
| Morphological acceleration factor | 24.75 | [–] |
| Hydraulic roughness as described in section 4.2 | | |
| Spiral flow included as described in section 4.4 | | |
| Non erodable layers where relevant (situation of 2001) | | |
| Morphological start-up period | 1 | year |

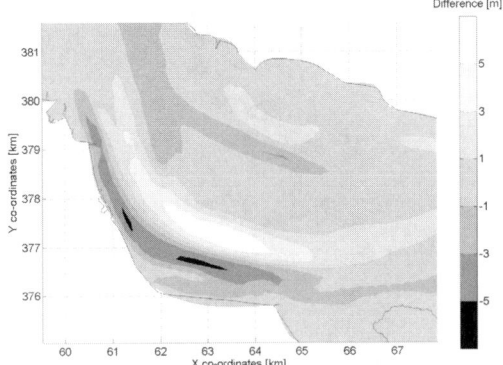

Figure 3.  Morphological effect of spiral flow in Zuidergat area.

acceleration factor several runs with different factors are carried out in the Western Scheldt model. The total amount of sedimentation and erosion volume after one year is calculated and compared to the calculation with a factor 1, since this is the run without any acceleration.

From Table 3 it is concluded that if the acceleration factor is chosen higher the difference is increased as well. The acceleration factor used in this paper is 24.75. This factor shows a reasonable difference of less than 15% and gives reasonable computational times. This factor accelerates one neap spring tidal cycle to one year.

### 4.4  Spiral flow

In the 2DH flow model the effect of spiral flow in curved channels is neglected. In FINEL2d an option is present to include a parameterisation of the spiral flow. The formulation of Booij & Pennekamp (1983) has been used. The model calculates the curvature of the current. Based on this figure the direction of the sediment transport is adjusted.

In figure 3 the morphological effect of the spiral flow is shown for the Zuidergat area over the validation period from 1965 to 2002. As expected, the outer bend shows more erosion due to the spiral flow effect, while the inner bend shows sedimentation. The effect is approximately 5 metres in this channel. From Figure 3

it is concluded that spiral flow cannot be neglected in a morphological hindcast.

### 4.5  Dredging, depositing and sand mining module

Dredging has a large impact on the morphodynamics in the Western Scheldt. Therefore this needs to be taken into account when performing a hindcast of the morphological developments of the last decades.

In reality a navigational depth is guaranteed between the navigational buoys of the fairway. If the depth becomes too shallow a dredger deepens the area to the required depth.

In FINEL2d a dredging module was developed based on the same principle. For each grid cell in the fairway a required depth is defined. If the depth in the grid cell is insufficient the sand is removed from the element and deposited according to a certain distribution key over depositing sites. The distribution key is established using historical data of the deposited material. For each historical year an input file is defined, since the buoys, the depositing areas and the maintained depth in the fairway may vary over the years.

Sand mining is simulated in the model by removing the mined volume equally spread over the specified sand mining area. Since sand mining quantities and locations differ per year an input file for each year is specified.

The morphodynamics of the estuary due to the combined effects of natural processes and dredge/depositing activities is continuously calculated, and the riverbed is accordingly updated.

### 4.6  Calibration of the morphodynamics

The next step in the set-up of the morphodynamic model is the calibration of the morphology itself. A calibration is carried out for a period of 7 years (1995–2002). Because the goal is to evaluate the long term morphodynamics the calibration results will not be shown in this paper. See for details Dam et al., (2006). Only the optimal settings of the morphodynamic model found in the calibration are given in Table 4.

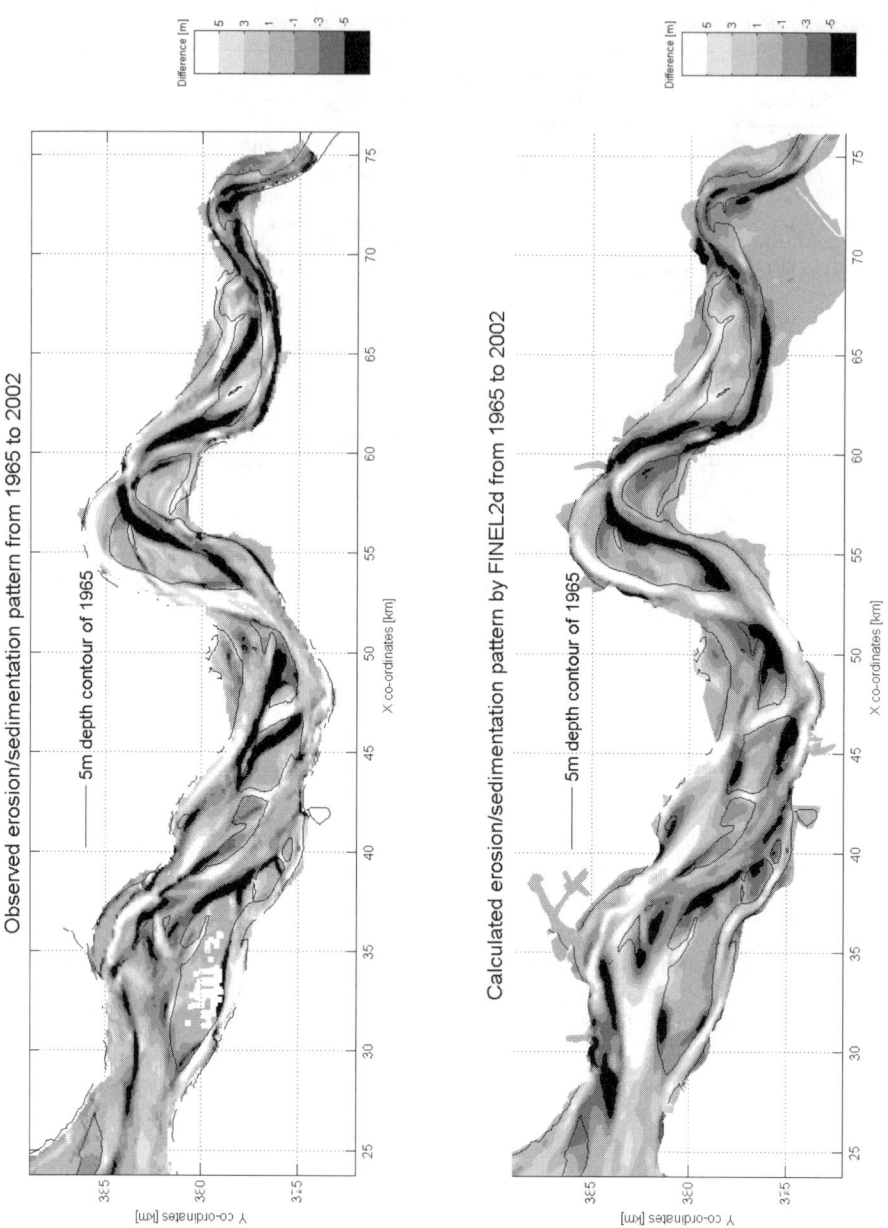

Figure 4.   Observed versus calculated erosion/sedimentation pattern from 1965 to 2002.

## 5   MORPHODYNAMIC RESULTS

In the hindcast of the years 1965 to 2002 the settings found in the calibration period are used. Starting point is the bathymetry of 1964 and a start-up period of 1 year. This means that the model needs a start-up time to enable for the initialisation of the sand transport and the adaptation of the initial bottom, which was

interpolated from GIS data. It was found that a start-up time of 1 year is sufficient to account for these factors.

The simulated erosion and sedimentation pattern of 1965 to 2002 is displayed in Figure 4 together with the observed pattern. At first sight a lot of similarities between the observed and simulated pattern occur. Macro cell 4 which is the Middelgat/ Gat van Ossenisse system shows a good resemblance.

Table 5. Total sand volume changes 1965–2002.

| Cell Nr. | Flood channel O | Flood channel M | Ebb channel O | Ebb channel M | Total O | M* |
|---|---|---|---|---|---|---|
| 1 | −14 | 31 | 1 | 1 | −13 | 32 |
| 2 | | | | | 26 | 0 |
| 3 | −20 | −12 | −8 | −43 | −29 | −55 |
| 4 | −39 | −35 | 61 | 22 | 22 | −13 |
| 5 | 12 | 26 | −49 | −79 | −37 | −53 |
| 6 | −5 | −6 | −17 | −18 | −22 | −25 |
| 7 | −5 | −1 | −16 | −16 | −22 | −18 |
| 8 | | | | | 8 | −5 |
| 9 | | | | | −2 | 9 |
| Total | | | | | −68 | −127 |
| Sand mining | | | | | −100 | −100 |
| Nett | | | | | +32 | −27 |

* O = observed volume change [Mm³]; M = modelled volume change [Mm³]; negative = erosion, positive = sedimentation.

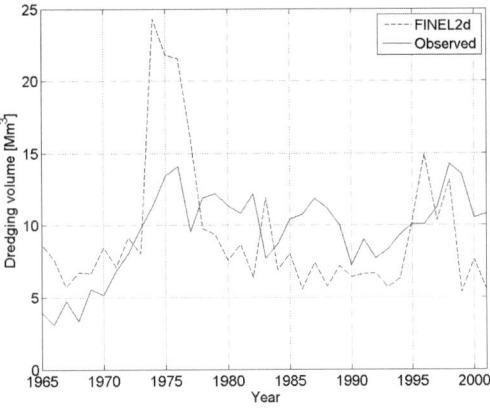

Figure 5. Observed and calculated total dredging volumes.

The Middelgat is sedimentated, while the Gat van Ossenisse is eroded. Also the Zuidergat and the Overloop van Valkenisse shows good agreement.

Inter tidal areas are eroded too much in the model, maybe because in reality cohesive sediments can be found at these shoals, which can make it harder for these areas to erode.

At Cell 1 in the Schaar van Spijkerplaat a large sedimentation can be seen in the model, which in reality does not occur. A possible reason why the Schaar van Spijkerplaat is so heavily sedimentated is the neglecting of wave effects, which might still have a stirring effect in the mouth of the estuary. The waves also cause wave driven currents which might transport the sediment further eastwards with a dominant westward wave direction.

In Table 5 the sand volume changes in the macro cells of the hindcast period are shown. From Table 5 can be seen that the model calculates volume changes that are in order of magnitude of the observed volume changes. The Macro cells 3, 5, 6 and 7 are calculated quantitatively well by the model. In Cell 4 both channels show the right direction, but the sedimentation of the Middelgat channel is too less in the model. Cell 1 and 2 are not correctly calculated.

The total sand volume change is calculated at −127 Mm³ for the model while the measurements show a change of −68 Mm³, see Table 5. Sand mining in this period extracted a total of 100 Mm³ from the Western Scheldt estuary, which leads to a net export of sand of 27 Mm³ in 37 years of the model. In reality there was an import of 32 Mm³. This problem of import versus export needs to be solved in future research, because a lot of policy questions concerning the Western Scheldt are based on a possible change in export/import of sand. Note that the last years there seems to be an export of sand out of the Western Scheldt.

The effect of dredging and depositing is large in the Western Scheldt area. In Figure 5 the actual dredging volumes in the complete Western Scheldt are plotted along with the calculated dredging volumes of FINEL2d. As pointed out in section 2.6 the navigational depth of the shipping channel is guaranteed in the model. If the depth becomes too small the sand is removed and distributed over the depositing sites.

At first notice the model can reproduce the dredging volumes in an accurate way. Around 1974 a large peak in the dredging volumes occurs in the model. This is caused by the deepening of the fairway and the transfer of the fairway from the Middelgat to the Gat van Ossenisse channel.

Around 1996 a second deepening of the channel has taken place. This also causes an increase of the dredging volumes in both the model and in reality, but the model does not show dredging volumes as large as the 1970's deepening.

In Figure 6 the simulated total dredging effort of the validation period is shown for the eastern part of the estuary, in which most of the dredging takes place. Also shown are the borders of the actual dredged areas. From Figure 6 it can be concluded that the location of the dredged material in the model gives a good agreement with reality.

6 CONCLUSION

The presented FINEL2d morphodynamic model of the Western Scheldt estuary is able to calculate the erosion/ sedimentation pattern fairly accurately over several decades. The results in the different macro cells still require optimisation, especially the import/ export

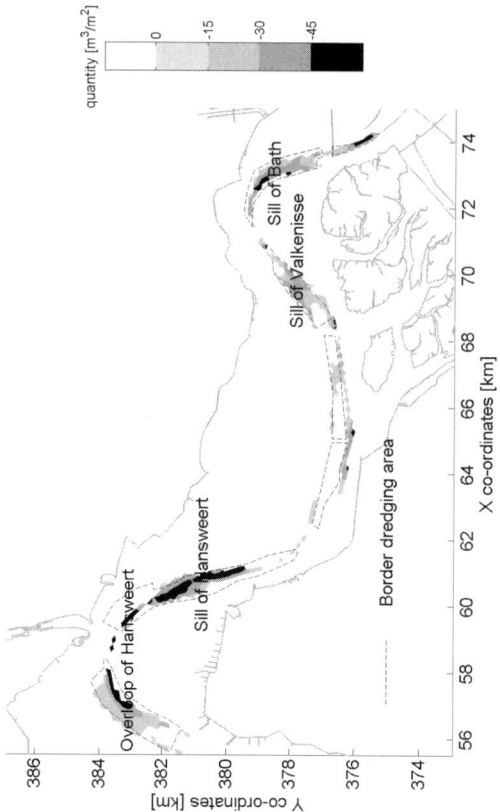

Figure 6.  Calculated total dredging quantities [m³/m²].

of sand of the complete Western Scheldt. The dredging volumes can be reproduced in a fairly accurate way, although during the deepening of the fairway in the 1970's a further optimisation of this routine of the model is required.

It is concluded that the basis for a good long term process based morphological model is present. Further improvements of the model can be obtained by additional research. However in this state the model can already be used as a decision making tool for large scale human impacts in the Western Scheldt (e.g. further deepening of the fairway).

REFERENCES

Booij, R., Pennekamp, Joh. G.S., 1983, Simulation of main flow and secondary flow in a curved open channel, Report No. 10–83, Department of Civil Engineering, Delft University of Technology.

Dam, G., Bliek, A.J., Bruens, A.W., 2005, Band Width analysis morphological predictions Haringvliet Estuary, Proceedings of 4th IAHR symposium of the River, Coastal and Estuarine Morphodynamics Conference, Urbana, Illinois, USA, Volume 2, p. 171–179.

Dam, G., Prooijen, B.C., van, Bliek. A.J., 2006, Morfodynamische berekeningen van de Westerschelde met behulp van FINEL2d (in Dutch), Svašek Hydraulics, ref. GD/06119/1339

Engelund, F., Hansen, E., 1967, A monograph on sediment transport in alluvial channels, Teknik Forlag, Copenhagen.

Gallappatti, R., Vreugdenhil, C. B., 1985, A depth-integrated model for suspended sediment transport. Journal of Hydraulic Research 23, p. 359–277.

Glaister, P. 1993, Flux difference splitting for open-channel flows, Int. J. Num. Meth. Fluids, 16, p. 629–654.

Hibma, A., Vriend, H.J., de, Stive, M.J.F., 2003, Numerical modelling of shoal pattern formation in well-mixed elongated estuaries, Estuarine, Coastal and Shelf Science, Vol. 57, 5–6, p. 981–999.

Hughes, T.J.R, 1987, The finite element method, Prentice-Hall, Englewood Cliffs, N.J.

Vreugdenhil, C.B., 1994, Numerical methods for shallow water flow, Institute for Marine and Atmospheric Research Utrecht (IMAU), Utrecht University, The Netherlands.

Wang, Z.B., Langerak, A., Fokkink, R.J., 1999, Simulation of long-term morphological development in the Western Scheldt, Symposium of the International Association for Hydraulic Research, Genova, Italy.

Winterwerp, J.C., Wang, Z.B., Stive, M.J.F. Arends, A., Jeuken, C., Kuijper, C., Thoolen, P.M.C., 2001, A new morphological schematization of the Western Scheldt estuary, the Netherlands, Proceedings of the 2nd IAHR Symposium on River, Coastal and Estuarine Morphodynamics, Obihiro, Japan, p. 525–534.

*River, Coastal and Estuarine Morphodynamics: RCEM 2007 – Dohmen-Janssen & Hulscher (eds)*
*© 2008 Taylor & Francis Group, London, ISBN 978-0-415-45363-9*

# Concept of a sustainable development of the Elbe estuary

C. Freitag, N. Ohle, T. Strotmann & H. Glindemann
*Hamburg Port Authority, Hamburg, Germany*

ABSTRACT: Apart from being an important federal waterway large parts of the Elbe estuary have been desig-nated as protected area in the sense of Natura 2000. This leads to a number of conflicts between the commercial obligation to maintain certain water depths and the preservation of these designated areas. A future action plan for the Elbe estuary was established by the Hamburg Port Authority in close cooperation with the Federal Administration for Waterways and Navigation to ensure the conservation of the invaluable estuarine system. This contribution will point out several reasons responsible for the negative developments of the past decades as well as the main problems associated with handling dredged material and further introduces the milestones for a future concept. By implementing this concept a win-win situation for economic and ecological interests can be achieved.

## 1 SUMMARY

### 1.1 *Location*

The Elbe estuary is located in Northern Germany. It extends from the estuary mouth at Cuxhaven to the tidal border at Geesthacht covering a total length of about 120 km (Fig. 1).

Apart from being an important federal waterway leading to the Port of Hamburg and being one of the biggest employees of the whole region, ecological obli-gations have become more and more relevant and can no longer be neglected.

### 1.2 *Functions*

The Tidal Elbe River has a vast variety of functions whose utilisation by the people has repeatedly led to many conflicts of interest in the past. As a federal waterway the Elbe estuary is the seaward entrance to several Elbe ports. It is one of the most important and frequented waterways within Europe. The require-ments to properly fulfil this function are dictated by the international maritime navigation.

The increasing colonisation along the Elbe estuary has led to a constant loss of areas which were formerly under permanent tidal influence. Although there is a certain safety risk accompanied with living close to the water this is a well promoted utilisation in Hamburg. Nevertheless the safety of the people must be given top priority.

Fishing has always played a major role in the Elbe estuary. Prawn fishing is typical in the mouth of the river. Thanks to the improvement of the water quality since the early 1990s the number of fish species and their abundance has significantly increased. Zones of

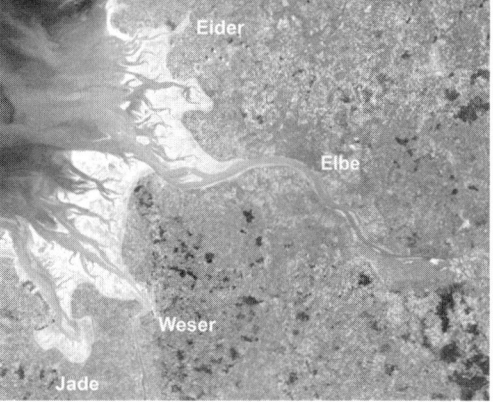

Figure 1. Satellite photo of the Eider, Elbe, Weser and Jade estuaries (Source: Brockmann Consult GmbH © 2003).

shallow water are important as hatching and spawning areas. The viability of the tributaries and side channels are of special relevance for the preservation of fishing grounds.

Agriculture is the most prevalent land utilisation in the Lower Elbe region. Pomiculture in the Elbe marsh-land enjoys an international reputation. Irrigation and drainage of the farmland play a vital role in the use of the very fertile soils in that area.

The cramped intertwining of water and land as well as the natural diversity of nature and the countryside create the special charm of the region, and is a basis for the growing sector of local recreation and tourism. In fact the Elbe is an outstanding stomping ground for recreational navigation, water sports, fishing etc.

Last but not least the dynamic estuary with its constantly changing and sometimes extreme conditions (e.g. changing water levels, fluctuating salinity, erosion and sedimentation) is a unique habitat for highly specialised flora and fauna. The estuary with its wet, intertidal and foreshore areas, with its reeds and alluvial forests, is of international importance to the avifauna, especially to the migratory birds. In fact large parts of the estuary have been designated as protected areas within the European network of Natura 2000, which result from the implementation of the Habitats Directive and the Birds Directive. The main aim concerning the quality of this complex habitat is the conservation and development of its natural dynamics and diversity.

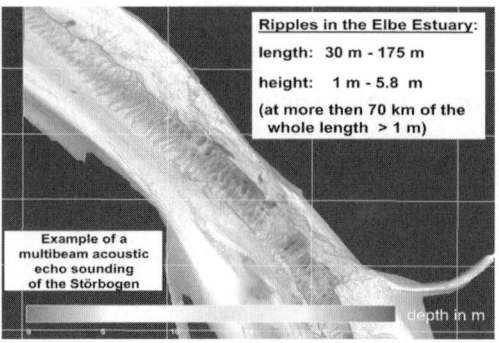

Figure 2. Examples of ripples in the Elbe estuary (Source: Federal Waterways Engineering and Research Institute BAW © 2001).

## 2 DEVELOPMENT OF THE ELBE ESTUARY

### 2.1 Morphological development

The morphological diversity of the Elbe estuary is mainly governed by tidal action and is naturally characterised by substantial sediment transport linked to the continuous remodelling within the estuary. Extensive sediment relocation takes place due to erosion and re-sedimentation enhanced by heavy storm surges. Characteristic features of natural estuaries include multiple and migrating channel systems, changing river width, scours and aggradations, intertidal areas, as well as the formation of sand banks and islands.

New channels and tidal creeks are formed continuously, serving as a habitat for different biota. Due to the periodic flooding many different freshwater and brackish water zones have developed. In the past there were only a few settlements close to the estuary so that the morphodynamic processes of a natural river system could occur without any interference by man. From the 11th century onwards the natural landscape began to change due to agricultural and pasture farming, colonisation, diking and hydraulic engineering, making it into a cultivated man-made landscape taking up more and more room which was initially governed by the river. By dewatering the hinterland accomplished with the settlement of these areas and by uncoupling the hinterland from the sedimentation processes, this area was no longer able to grow up with the continually rising water levels. This makes the dewatering more and more difficult. In total 209 km² foreshore areas along the estuary have been diked since 1955.

The construction of barriers cutting off former tributaries led to a further loss of ecological valuable inundation areas. Nowadays the sediment load within the water column is higher and the extent of intertidal areas has decreased during the last decades. This caused changes in the nature of the soil and the biota within the flood plains.

The tidal river Elbe flows through a glacial valley mostly consisting of clay covered by fine silt. Deep exploration boreholes also show the existence of peat, sand, gravel and glacial drift (see also Fig. 11). Usually the composition is coarse-grained in greater depths. The sediment budget altered due to the diking measures and the former deepening of the fairway, by dredging the existing layer of clay in some parts of the estuary. Sandy material is dominating the river bed. Alluvial material is only found in areas with low flow velocities, as it is the case in harbour basins.

Due to the marine influence coarse and fine sands can be found up to a large extent at the estuary mouth. The sandy river bed is characterized by large ripples up to 5.8 m height (Fig. 2). Once the ripples extend into the fairway they have to be dredged in order to safeguard the navigation.

Furthermore large parts of the Elbe estuary underlie a considerable loss of bed material. The amount of erosion varies depending on the different river sections. Noteworthy is the amount which has disappeared from the estuary mouth. Figure 3 shows a differential topography of the year 2002 and 1998 in this area. The loss of material leads to a widening of the river mouth and an increase of tidal energy entering the estuarine system. This is a self-energising process reinforcing the tidal pumping effect, meaning the transport of very fine sediments upstream, while the seaward transport continues. Eximinations of the Federal Waterways Engineering and Research Institute (BAW) quantify that in total more than 100 Mio · m³ have been eroded from the estuary mouth during the past 30 years.

All these described morphological changes have led to a series of hydrodynamic changes, which in turn influence the morphodynamics.

### 2.2 Hydrodynamic development

The hydrodynamic development of the tidal parameters is characterised by an increase in the high water

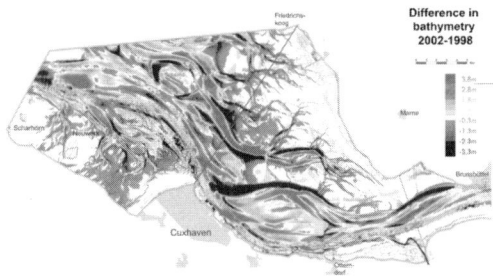

Figure 3. Differential topography 2002–1998 showing that a lot of material is being eroded from the estuary mouth.

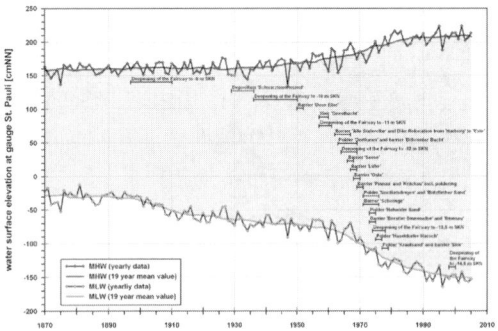

Figure 4. Development of the mean high water and mean low water as annual values and 19-year-average values at the tide gauge St. Pauli.

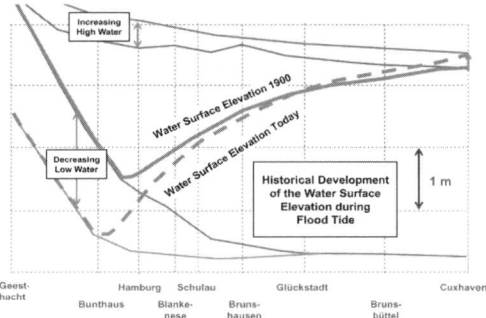

Figure 5. Steeper water level gradient between Cuxhaven at high tide and St. Pauli (Hamburg) at low tide (Source: Federal Waterways Engineering and Research Institute BAW © 2001).

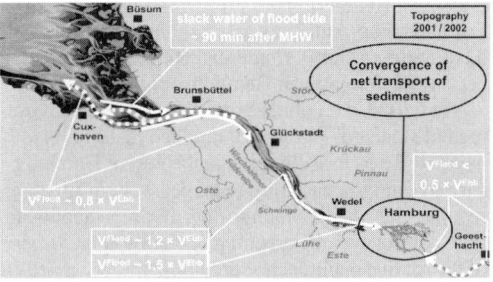

Figure 6. Flood and ebb dominance in the tidal river Elbe (Source: Federal Waterways Engineering and Research Institute BAW © 2006).

level and a decline of the low water level. This development is more significant further upstream. The maximum tidal amplitude is recorded at the tide gauge St. Pauli in Hamburg averaging 3.6 m nowadays. 150 years ago the tidal range was about 2.0 m in St. Pauli (Fig. 4).

Until 1960 there was a slight decline of the tidal amplitude visible from the mouth up to Hamburg. With the construction of the weir in Geesthacht in 1956/60 the tidal border was fixed and the tidal wave was reflected which resulted in greater tidal amplitudes in Hamburg.

The increase in tidal range is mostly due to the decline of the low water level making up about 2/3 of the variance. The most significant change of the low water level was after the deepening of the fairway in 1967 (12 m) and 1979 (13.5 m). The recent deepening in 1999 did not have such an apparent effect which leads to the suggestion that the estuarine system is somewhat stable at the moment.

But of course there are many more factors which influence the hydrology. For example the siltation in anabranches, the backfilling of port basins, the construction of barriers, the decreasing freshwater flow and of course the very recently discussed climate changes with the expected sea level rise.

Obviously the flow velocities correlate directly with the water level gradient (Fig. 5). So far an increase in flow velocities could not be determined from data, since there is only very little comparable data to evaluate, due to the constantly varying tidal characteristics and the lack of data from within the deep fairway. Logically though, the steeper gradient between Cuxhaven at high tide and Hamburg at low tide has led to increasing flow velocities during the last few decades.

These assumptions were validated by the results of the 3D model developed by the Federal Waterways Engineering and Research Institute (BAW). These show that increasing flow velocities mainly concentrate in the deep channel while the flow velocities in the anabranches continually decrease leading to sedimentation in these areas. The results of the 3D sediment transport model also show that the flood tide stream has increased far more in comparison to the ebb tide stream, which explains the observed tidal pumping effect (Fig. 6). This effect is even greater if the freshwater flow from upstream falls below 700 m³/s.

The brackwater zone is where the salt water from the North Sea mixes with the freshwater. It has moved further upstream during the last few decades. This is due to the widening cross-section at the mouth, letting more tidal energy enter the system as well as the constantly decreasing freshwater flow. The turbidity zone extends up to 20 to 40 km. The location can vary up to 25 km depending on the tidal circumstances and the flushing force from upstream. Studies of Bergemann (1995) concluded that the border of the turbidity zone has moved upstream between 5 and 20 km from 1953 until 1994.

# 3 DREDGING

## 3.1 Dredging nowadays

The Federal Administration for Waterways and Navigation is responsible for maintaining the waterway throughout Germany. Due to the historical development the Hamburg port area was delegated to the Hamburg Port Authority (HPA). Since then the HPA has a statutory and a commercial obligation to maintain certain water depths in Hamburg (Fig. 7).

There is also an obligation to meet certain environmental standards in the way dredged material is to be handled. Therefore dredging in the Port of Hamburg is carried out by a carefully managed system involving regular high quality monitoring of water depths and tight control over the dredging operation (Fig. 8). Most of the routine maintenance dredging is now carried out under contract using trailing suction hopper dredgers.

Depending on the degree of contamination and the soil composition dredged material is handled according to present regulations. Sediment too contaminated for normal disposal is treated ashore. The sediment is treated in flushing fields and in the METHA (Mechanical Separation and Dewatering of Port Sediments) plant. This process produces sand and silt fractions that can be reused; the remaining contaminated silt is separated and stored in special landfills. The METHA has a capacity up to between 1.2 and 1.4 Mio · m³/year. This material is removed permanently from the estuary.

The most efficient way to handle sediment is the disposal in the river itself. This has become possible due to the improvement in the water quality of the River Elbe. Since the mid 1990s Hamburg was allowed to relocate less contaminated sediments in the river inside the Hamburg State Boundary in agreement with the environmental authority (BSU).

The convention requires that the material is placed near the Port boundary during the ebb tide, in the belief that the material will be carried well downstream. Generally the relocation of dredged material here is not permitted during part of the summer season due to concerns about the dissolved oxygen levels and protection of sensitive organisms in the water. This can lead to problems maintaining the water depth during the summer months when the sedimentation rates are far greater due to the low freshwater flow. The

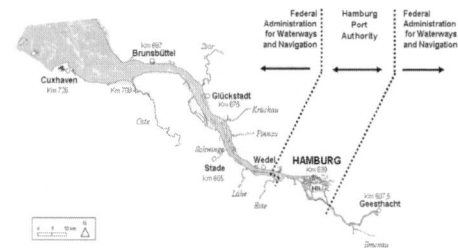

Figure 7.   Administration areas at the tidal river Elbe.

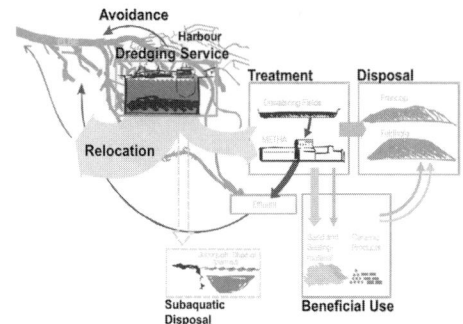

Figure 8.   Dredged material management of the Hamburg Port Authority.

practice was begun with relatively small quantities on an experimental basis in 1995. The lack of any apparent negative effects was taken to indicate that the quantity relocated could be increased. This method is cheaper than disposal on land and is also environmentally friendly because the natural sediment balance is conserved.

Due to a steady increase in the dredging rate in the port of Hamburg it became essential to place sediment further downstream. In agreement with the federal state of Schleswig-Holstein a further option to relocate dredged material from certain parts of the port area became possible in 2005. Hereupon a certain quantity of sediment limited to 1.5 Mio. m³/year is brought into the North Sea near buoy E3 close to Heligoland. There is no seasonal restriction on disposal at this location, but this strategy is only permitted till 2008 (see also Section 3.2).

Apart from maintenance dredging sand is dredged from the estuary using trailing suction hopper dredgers to supply aggregate for the construction industry. The quantity is significant enough that it should be taken into account when summarizing the dredging and disposal quantities.

## 3.2   Sediment development

With all the ongoing changes throughout the Elbe estuary affecting the morphological and hydrodynamic

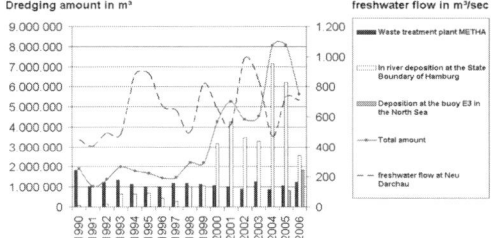

Figure 9. Development of the dredging activities in Hamburg.

development the sediment dynamics have altered too. The tidal pumping effect and the little freshwater flow have led to greater sedimentation rates within the port area. The diagram in Figure 9 shows the disposal quantities from 1990 to 2006.

The different disposal strategies are distinguished by the different bars. Low freshwater flows in 2001 and 2004 obviously affect the dredging amount. There is also an underlying trend for increased siltation reaching a maximum of over 8 Mio · m³ in 2005 due to a combination of a number of factors. The high siltation rate in 2005 in the summer period during which relocation at the State Boarder is not permitted required a new solution for the handling of dredged material instantly. This led to the permission of Schleswig-Holstein to relocate a certain amount of sediment at the buoy E3 in the North Sea. Since this permission is limited until 2008 other possibilities ought to be found soon.

In the end there are many different reasons which have led to the rapidly increasing siltation rate. They all interact with each other and it is therefore difficult to separate out their relative importance.

### 3.3 Reasons

The main factors leading to the increase in the dredging amounts are listed and explained below.

Deepening of the estuary for navigation purposes has been going on for many years. Significant changes took place in the 1970s. The most recent change was the capital dredging that took place in 1999 followed by the works in the Köhlbrand area in 2001–2002. Large amounts of material were released into suspension making it available for sedimentation in areas with low currents.

The port area of Hamburg with all its harbour basins can be referred to as a large sediment trap which is supplied with sediment from three main sources: sediment supplied by the incoming tide, sediment carried down the river and sediment resuspended from the river bed or spilt during dredging operations.

The tidal pumping mechanism identified by the 3D model results in a net transport of sediment upriver into the port area. Disposed material is continually carried back upstream so that the sediment is being recycled, the rate depending significantly on the freshwater flow. While low freshwater flows it is normal for sediment to migrate up the estuary where it can settle in the port area. Here it cumulates with the sediments from upstream leading to additional siltation. Dredging in Hamburg has been limited to that which was urgently needed. The siltation rate during the past few years was greater than the dredging rate so a backlog has been building up. Due to the increased obligation to shipping companies to ensure continuous access some of the backlog has to be cleared leading to higher dredging rates.

Other Authorities such as the WSA Hamburg and Cuxhaven also dredge and dispose material in the estuary. There are a few dumping sites close to the Hamburg State Boundary were material is deposed independent of the tidal circumstances.

Furthermore a training wall built in 1999/2000 near Glückstadt has influenced the flow velocities. Sediment which can no longer settle in this reach are transported through this zone resulting in increased concentrations and increased siltation further up or down the estuary.

Since the size and draught of vessels using the port facilities has increased and thereby reducing the under keel clearance the resuspension of sediment in the form of plumes behind the ships has also increased. Ships coming in on the rising tide may have added a component of net upriver transport, especially with regards to the tidal pumping mechanism.

By deepening and widening the cross section of the estuary in some places more erodible material is exposed. This also includes the capping of sand ripples and disturbing the natural dune formation. Obviously this could be another factor which could have increased the sediment load throughout the estuary.

Last but not least the erosion of the Medem Grund within the mouth of the estuary is an additional source of material entering the system. This also leads to a widening of the mouth letting more tidal energy enter the system and reinforcing the negative trend.

### 3.4 Sediment contamination

The sediment contamination results from the large catchment area of the river Elbe covering 148,000 km². Approximately 25 Mio. people live in the catchment area including the industrial and agricultural areas in Germany and the Czech Republic. In the past the absence of waste water treatment plants, mining sites and unconfined waste deposits have caused pollution of the waters of the Elbe and its tributaries.

In the beginning of the 1980's the river Elbe was known as one of the most polluted rivers in Europe. Since the German reunification in 1990 the pollution has considerably reduced. The mixing of silt from upstream with the non-contaminated fine sands of marine origin shows a clear gradient as the level of contamination decreases towards the North Sea.

Considering the port of Hamburg as a large sediment trap contaminated sediment from upstream is trapped in the harbour basins. Here it mingles with fine sands carried upstream by the perpetual recycling mechanism. With each sediment cycle the sediments carry a higher load of contamination so that in the end they have to be taken out of the system to be treated and separated very costly at the METHA.

In order to be able to dispose the material without any restrictions in the near future a primary aim should be to reduce the suspended heavy metal concentrations from upstream. This can only be achieved by a close collaboration with the neighbouring states.

### 3.5  Need for action

Apart from the large expenses dedicated to the high dredging amounts and their treatment it has become difficult to provide the statutory water depths during the summer season. Along with the attained expertise this required a new strategy at once.

Therefore the Hamburg Port Authority in cooperation with the Federal Administration for Waterways and Navigation has developed a "Concept for a sustainable development of the Tidal Elbe River as an artery of the metropolitan region Hamburg and beyond".

## 4  FUTURE ACTION PLAN

The concept for a sustainable development of the tidal river Elbe points out three milestones for a future program. These are:

1. Dissipation of the incoming tidal energy by hydraulic engineering constructions especially within the mouth of the estuary;
2. Establishing intertidal areas between Glückstadt and Geesthacht;
3. Optimising the sediment management considering the whole system.

Without appropriate measures the accretion of sediment within the tidal river Elbe will continue with all its ecological disadvantages as well as the increasing effort to maintain the water body and berths of the Port of Hamburg.

These milestones should be achieved by sophisticated river engineering concepts and a new integrated sediment management strategy.

### 4.1  Investigation at the mouth of the Elbe estuary

Before being able to redesign the mouth of the Elbe estuary it is necessary to investigate the boundary conditions for possible measures in this area.

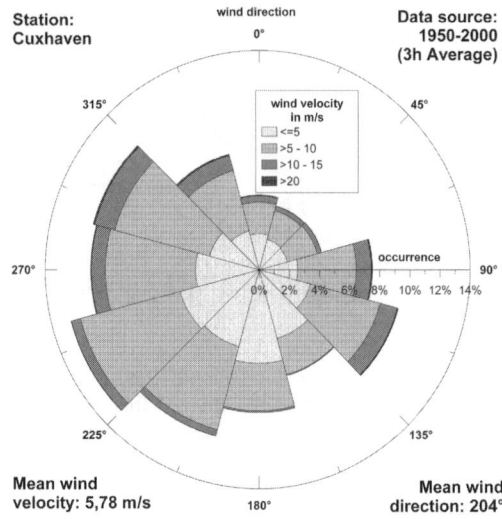

Figure 10.  Occurrence of wind velocity in Cuxhaven.

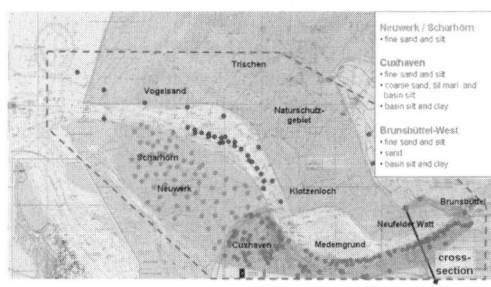

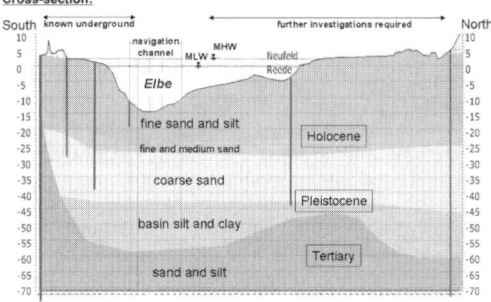

Figure 11.  Deep exploration boreholes and geological cross-section of the mouth of the Elbe estuary.

One important aspect besides water levels and flow velocities is the wave load on possible hydraulic structures. Therefore the wind conditions (Fig. 10) and the wave conditions have to be investigated.

Furthermore an important aspect in the availability of coarse sand material in that area. For this exploration a geological model of the Elbe estuary is being processed (Fig. 11).

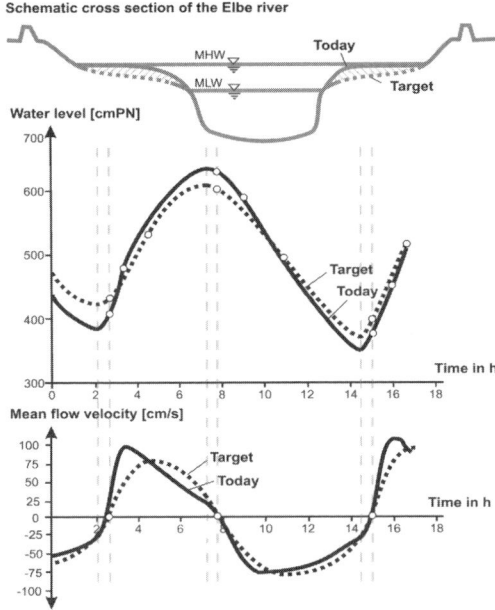

Figure 12.   Possible effects of re-profiling intertidal banks and shallow water areas.

Moreover the hydrodynamic effects of possible measures such as sand banks, sand islands, dams or sub aquatic depots have to be examined.

### 4.2   River engineering concept

The main objective is to modify the unbalanced sediment budget in a beneficial way. This can be achieved by classical river engineering methods such as training walls and sediment traps when only looking at a small reach. Considering the whole estuary and the upstream transport of sediments it is obvious that the hydrological development must be reversed. Measures which constrict the cross section at the river mouth could reduce the incoming tidal energy. Furthermore the recreation of extensive intertidal areas between Glückstadt and Geesthacht can reduce the tidal amplitude at St. Pauli (Hamburg), as estimated by the 3D sediment transport model of the BAW. Possible effects of recreating additional intertidal areas are shown in Figure 12.

These measures could include re-profiling intertidal banks and shallow water areas so that they are already governed by the tidal action at low tide and therefore increase the total tidal volume. Other measures may be reconnecting cut-off tributaries and dredging silted-up harbour basins (Fig. 13). The implementation of these measures would have an immediate effect on the hydrological parameters which also result in an alteration of the sedimentation processes.

In 2006 a project group was initiated to locate potential areas along the Elbe estuary. The total

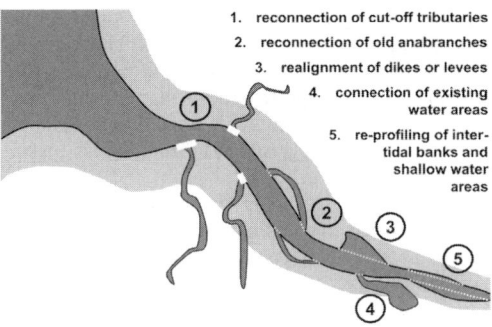

Figure 13.   Possible measures along an estuary.

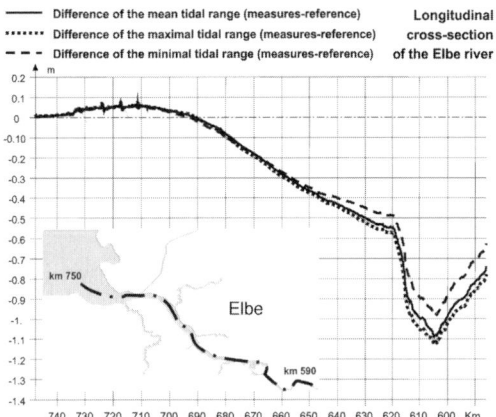

Figure 14.   Possible maximal difference of the tidal range – simulating reference situation minus situation of all possible measures along the Elbe estuary (Source: Federal Waterways Engineering and Research Institute BAW © 2007).

effect of all potential areas was then predicted by the 3D model. The outcome was astonishing. The summary results show that the tidal elevation in Hamburg St. Pauli (km 623) could be reduced by up to 50 cm (Fig. 14). This primarily leads to a reduction of the flood dominancy upstream from Glückstadt and therefore the net sediment transport upstream should be reduced as well.

### 4.3   Sediment management concept

Certainly the amount of sediment can be influenced by river engineering measures, but maintenance dredging in an anthropogenic modified water body will always be indispensable. The disposal within Hamburg's State Boundary is no longer a permanent economic solution because of the above mentioned recycling effect of sediments.

The future sediment management concept therefore intends to relocate fresh, non-contaminated sediments in areas where there is less possibility for them to return.

A good example is the relocation of "clean" sediments at the buoy E3 in the North Sea. This practice has already reduced the dredging rate in the Port of Hamburg in 2006 significantly. Although this practice seems very prosperous it is limited until 2008 by the federal state of Schleswig-Holstein, which makes a subsequent solution crucial.

Furthermore it is being examined whether sub aqueous deposits in combination with hydraulic engineering measures could be part of the future sediment management concept. Even though the sediment quality has improved the dredged material mainly from the harbour basins (backlog) is not suitable for relocation further downstream (ebb-dominant reach) due to the high screening values specified by the present regulations (HABAB, HABAK, BLABAK, etc.).

Another main objective of a future action plan should be the improvement of the sediment quality by cleaning up former sources of waste throughout the whole Elbe basin as explained in Section 3.4.

## 5 PROSPECTS & CONCLUSIONS

The sustainable development of the tidal river Elbe is a challenging and long-term task. This century-long project can only be successful if all interests are taken into account. Furthermore a close cooperation between the neighbouring states is necessary to secure the economic utilizations and maintain the valuable ecological areas of the estuary.

There is still a wide field of research necessary to be able to understand and forecast the very complex system. Close monitoring of small scale pilot projects should be used to enhance the knowledge as well as the processing of existing models in order to forecast hydrological and morphological effects of different scenarios.

The project should be understood to be under continual review and revision as experience is gained and knowledge improves.

A first draft for a future sediment management concept is being developed on a short-term basis involving the different agencies imposed to guarantee the seaward access. This will be presented in summer 2007 and will certainly identify several knowledge gaps. A next step is then to carry out specific measurements and scientific analysis in order to improve the expertise.

Particular attention should also be paid to the dissemination and education of the public in order to gain acceptance for this future concept. The natural dynamics can be experienced best by the restoration of small intertidal areas which are made accessible for recreational use. On-site information centres and an interactive communication platform via internet could help to inform the population as well.

By remodelling some specific regions along the Elbe estuary within the next couple of decades a lot of natural processes and economic aims will be influenced in a positive way on a win-win basis. There is a coming together of nature conservation and the desired hydraulic engineering changes. The new concept is an integrative approach bearing in mind the economic use, flood protection measures and the ecological objectives which can all be accomplished by the same engineering measures.

At the beginning of 2007 the Hamburg Port Authority specially established a workgroup which will promote this long-term project with all the related tasks. This group of scientists is bound to work very closely with the other federal agencies in order to achieve the designated objectives for the future.

## REFERENCES

Bakker, W.T. 1999. Effect Resonance on Morphology of Tidal Channels. In B.L. Edge, *Coastal engineering 1998: conference proceedings*: 3252–3264. Copenhagen: ASCE.

Balzano, A. 1995. On Residual Transport in Shallow Tidal Basins. In B.L. Edge, *Coastal engineering 1994: conference proceedings*: 2928–2942. Kobe: ASCE.

Bauer, B.O., Lorang, M.S. & Sherman D.J. 2002. Estimating Boat-Wake-Induced Levee Erosion using Sediment Suspension Measurements. *Journal of Waterway, Port, Coastal and Ocean Engineering* 128(4): 152–162.

Bergemann, M. 1995. Die Lage der oberen Brackwassergrenze im Elbeästuar. *Deutsche gewässerkundliche Mitteilungen – DGM* 39, Koblenz: BfG.

Bergemann, M., Blöcker, G., Harms, H.; Kerner, M., Meyer-Nehls, R., Petersen, W. & Schroeder, F. 1996. Der Sauerstoffhaushalt der Tideelbe. In KFKI, *Die Küste* 58, Heide i. Holst:Boyens & Co.

Dücker, H.P., Witte, H.-H., Glindemann, H. & Thode, K. 2006. *Concept for a sustainable development of the Tidal Elbe River as an artery of the metropolitan region Hamburg and beyond*. Hamburg: HPA.

Eichweber, G. & Lange, D. 1998. Tidal Subharmonics and Sediment Dynamics in the Elbe Estuary. In K.P. Holz, *Proceedings of the 3rd International Conference on Hydro-Science and -Engineering*. Cottbus:Brandenburg University of Technology.

Eichweber, G. 2004. Sediment Dynamics in the Elbe Estuary and the Improvement of Maintenance. *Dredging in a sensitive environment: World Dredging Congress XVII*. Hamburg: CEDA.

Erdmann, J.B., Stefan, H.G. & Brezonik, P.L. 1994. Analysis of Wind- and Ship-Induced Sediment Resuspension in Duluth-Superior Harbor. *Journal of the American Water Resources Association* 30(6): 1043–1053.

Rolinski, S. & Eichweber, G. 2000. Deformations of the Tidal Wave in the Elbe Estuary and their Effect on Suspended Particulate Matter Dynamics. *Physics and Chemistry of the Earth Part B* 25(4): 355–358.

Weilbeer, H. 2003. Zur dreidimensionalen Simulation von Strömungs- und Transportprozessen in Ästuaren. In BAW, *Mitteilungsblatt der Bundesanstalt für Wasserbau* 86. Karlsruhe: BAW.

*River, Coastal and Estuarine Morphodynamics: RCEM 2007 – Dohmen-Janssen & Hulscher (eds)*
*© 2008 Taylor & Francis Group, London, ISBN 978-0-415-45363-9*

# The Walsoorden pilot project: A first step in a morphological management of the Western Scheldt, conciliating nature preservation and port accessibility

Y.M.G. Plancke & S.J. Ides
*Flanders Hydraulics Research, Borgerhout, Belgium*

J.J. Peters
*Port of Antwerp Expert Team, Brussels, Belgium*

ABSTRACT: In the framework of the Flemish-Dutch "Long Term Vision" strategy of the Western Scheldt, the Port of Antwerp Expert Team proposed the idea of morphological dredging aiming at steering the estuarine morphology. As a pilot project, the experts proposed a new disposal strategy, where dredged material would be disposed on the eroded tip of the Walsoorden sandbar.

The feasibility of the disposal strategy was investigated at Flanders Hydraulics Research in 2002 and 2003. None of the results of this study opposed the feasibility of the proposed strategy, although final judgement would only be possible after the execution of an in situ disposal test.

At the end of 2004 500.000 m³ of sand was disposed at the tip of the Walsoorden sandbar. After a one-year extensive monitoring of the experiment, it was concluded that from morphological viewpoint the test was a success. The ecological monitoring revealed no significant negative changes in trends due to the disposal test. In 2006 the in situ disposal experiment was continued, with a new disposal of 500.000 m³ of sand.

## 1 INTRODUCTION

The morphology of an estuary is continually changing, adjusting to the forcing processes which themselves are also changing. No estuary is therefore stable and habitats and the ecological functioning of the estuary will continually change from its present status, even if man didn't intervene. This implies the need for a detailed conceptual understanding of the estuary system in question. Only such an understanding can lead to proper assessment of the effects of existing and future human activities, such as dredging and disposal, but also the construction of flow regulating structures and dikes. For any estuary there should be a holistic management plan, which takes into account the interests and effects of all uses and users of the estuary in an integrated way.

This paper focuses on the case of the Scheldt estuary, where morphological management is used to conciliate nature preservation and port accessibility. The Scheldt is the aorta to the port of Antwerp, while it is one of the few remaining European estuaries covering the entire gradient from fresh to salt water tidal areas.

## 2 OVERVIEW OF HISTORICAL EVOLUTIONS

At the end of the Pleistocene, the last ice age, rivers in North-West Europe discharged in the Atlantic Ocean in the vicinity of the Doggersbank, far away from the present shores. With the warming up of the climate, the sea level rose very quickly over more than 100 meters from 20 Kyr BP to about 7 Kyr BP, then slower to become (comparatively) rather stable over past two millenaries. The past rising sea level reshaped strongly the coastal areas and estuaries at the end of the Holocene. Many of these morphological changes are still ongoing.

In front of the actual Belgian, Dutch and German coast an almost continuous series of sandy bars and islands existed (Figure 1). An inner sea was formed, a kind of an extensive lagoon of which remains only the Wadden Sea. The sand barrier between lagoon and open sea was regularly breached during storms, scouring large channels deep into the inner sea. River sediments filled those parts of the lagoon receiving streams with large sand discharges, like the Rhine. In other parts receiving little and more silty sediment loads,

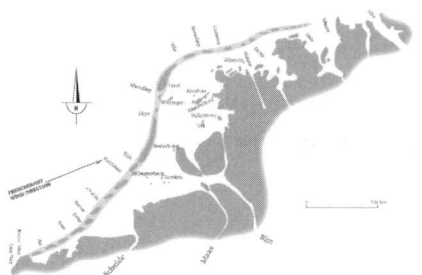

Figure 1.   Situation during the Roman Times.

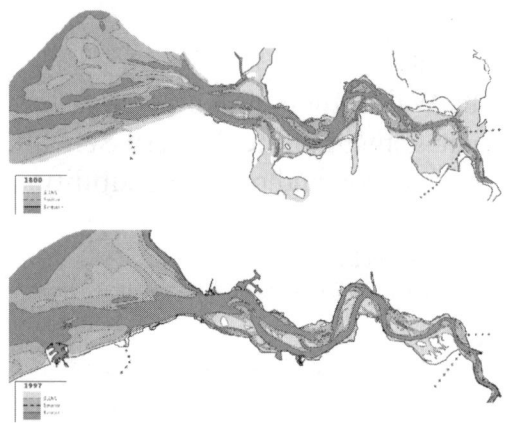

Figure 2.   Morphological evolution of the Western Scheldt estuary 1800–1997.

like from the Scheldt river, tidal action penetrated progressively, developing further the sea branches. Import of marine sediments by the tidal currents formed large shoals in the lagoon. Around 1000 A.D., Zeeland had become a patchwork of islands, surrounded by a network of tidal channels. At that time, the river Scheldt discharged in the lagoon near Bergen op Zoom and both the Honte (present Western Scheldt) and the Eastern Scheldt were conducting the Scheldt river water to the North Sea. Till the 11th century, morphological evolutions were significant but fully natural, with almost no human impact.

First signs of human impact on the estuary's environment become visible in the 11th century: locals reclaimed land that had silted up high enough and started to protect it against flooding. However, inundations due to levee breaching during storm events returned repeatedly portions of land to the river. From the 16th century on, the poldering techniques had become more sophisticated and larger areas were permanently poldered (e.g. for eastern Zeeuws-Vlaanderen 50% of the total poldering occurred during the 17th century).

Poldering was less intensive during the 19th and 20th century because a large percentage of salt marshes had been reclaimed already. However, hydraulic works and storms continued to reshape the area. In 1867 and 1871, the two remaining links (Kreekrak and Sloe) between the Honte (Western Scheldt) and the Eastern Scheldt were cut-off, modifying drastically the tidal channels network. A catastrophic storm with extensive inundation, in 1953, made the Netherlands decide about executing an extensive flood protection plan "Delta". From historical data can be concluded that these human impacts such as closure of secondary channels and poldering have strongly influenced the tidal regime of the Western Scheldt. Stronger tidal penetration enlarged the main navigation channel.

Sediment mining for providing building material started at the end of the 19th century. Since 1958, about 1 to 2 million cubic meters of sediment was mined per year, on average.

At the end of the 19th century, dredging activities were required to improve the accessibility of the port of Antwerp. Until the 1920's, these activities were concentrated on the Belgian territory ($2 \, \text{Mm}^3$/year). From 1920 till 1960 the quantities on Belgian and Dutch territory were comparable ($2 + 2 \, \text{Mm}^3$/year). The first large deepening campaign happened in the early 1970's, the main part of dredging works on Dutch territory ($3 + 10 \, \text{Mm}^3$/year). Nonetheless, the increased dredging in the Dutch part did not apparently result in significant changes of the trend in morphology or tidal action. During the late 1990's, a second dredging campaign for improving the navigation conditions was conducted. The impact of the deepening by 4 feet is monitored (MOVE programme), but no significant negative impact was noticed yet.

3   MAINTAINING ACCESSIBILITY IN A MORPHOLOGICAL DYNAMIC ESTUARY

The morphological evolution of the estuary between 1800 and 2000 (Figure 2) is one of further shoal aggradation and enlargement of the main channels. The estuary is described as a typical multiple flood and ebb channel network. The main and deeper ebb channels have usually sills at the seaward end where they join together with the flood channels. These ones are shallower and have a sill at the landward side, where they join the main ebb channel. There are also many minor channels, the "chute" channels, sometimes called "short-circuit" channels connecting the major ebb and flood channels. The reducing mobility of the channels and shoals is for a large part due to the hard bordering of the estuary (levees, bank protections, groynes, jetties and harbours); sandbars are rising too high, channels deepen, shallow water areas diminish.

Till 1970, dredging was restricted to maintaining depths on crossings in the navigation channel, formed by the main ebb channels. Traditionally, the sediments were disposed in the flood channels with the idea that it would take a rather long time before coming back into the main ebb channel.

With the demand for increased navigation depth, a first deepening started in 1970 and the dredged sediments were still disposed in flood channels. The disposal sites were decided in common by the Dutch and the Belgium administrations on the basis of the assessment of the ongoing morphological changes. The procedures were adjusted due to the increasing concern about environmental aspects and with the regionalisation making the Flanders region responsible in Belgium for public works and infrastructure. In 1995, Flanders and the Netherlands reached an agreement to deepen further the Western Scheldt shipping route. Works were executed in 1997 and 1998. However, the amount of sediment disposed in the eastern part of the Western Scheldt was reduced when aggradation was observed in some flood channels, supposedly because too much sediment had been disposed there. This siltation could eventually jeopardise the existence of the multi-channel system in that reach. Therefore, from 1997 on, more material was moved to disposal sites in the western reach of the estuary.

In 1999, the Dutch and Flemish governments decided to set up a Long-Term Vision (LTV) project with 3 objectives: to ensure maximum safety against flooding, optimal accessibility of the ports within the estuary and optimal nature development. These 3 subjects are all related to the morphology of the estuary. Directly concerned by these issues, the autonomous Port of Antwerp, independent from the Flemish administration, requested a group of experts (called Port of Antwerp Expert Team, or "PAET") to give an opinion about the prospects for a further deepening and widening of the navigation route, mainly needed for the larger container ships. One of the main questions considered in LTV was where to dispose the large volumes needed for such an enlargement? Dutch researchers had claimed that flood channels would disappear if too large quantities of sediment were to be disposed there. Their conclusions were based on some assumptions and calculations with modelling tools, of which one is based on the so-called "cell-theory" [Wang et al., 1995 – Winterwerp et al., 2001]. PAET consider their schematisation as too simplistic. Based on their analysis of past morphological changes in general and of the (temporary?) decay of some flood channels, they stated that not only disposal of sediments was to be blamed, rather the always more stringent immobilisation of the main channels and shoals. To revert the reduction in dynamic morphological behaviour of the estuary, it was proposed to steer the development of channels and shoals. Recent studies show

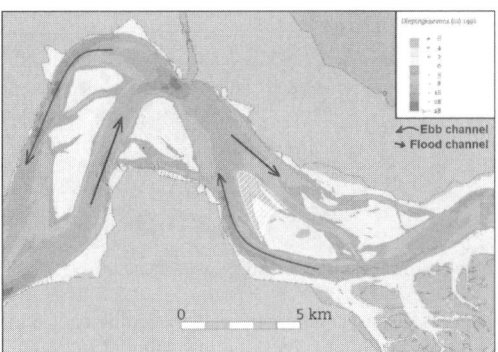

Figure 3. Walsoorden area.

that the disposal of dredging materials has a much larger impact on the estuarine morphology than the deepening of the channels [ProSes, 2004]. The main attention should therefore go to new strategies for disposal, although PAET believes that dredging may also be beneficial for morphology, e.g. rectifying the shape of sandbars.

In 2002, the Dutch and Flemish governments signed a memorandum of understanding to implement together the Long-Term Vision programme. They set up jointly an organisation called ProSes (Project Direction for the Development Scheme of the Western Scheldt Estuary) funded by both regions and which main task was to establish the development scheme with the objectives to be reached in 2010.

## 4 MORPHOLOGICAL MANAGEMENT OF THE WESTERN SCHELDT

### 4.1 *Morphological dredging*

During a meeting with the LTV's working group on morphology, in the year 2000, PAET suggested "morphological dredging" as an alternative to the present dredging strategies. It is based on the principles developed for the maintenance and the capital dredging in the navigation route in the Congo inner delta, for example by redistributing the sediment transport and using dredging and disposal to change the plan form of the river.

Disposal is a way to redistribute the sediment in the Western Scheldt, so as to feed, as an example, areas eroding too much. Not only in the flood channels, also on some parts of shoals. PAET worked out a proposal to restore the western tip of the Walsoorden sandbar that erodes since several decades. Several millions of cubic meters of sediment could be stored at that place (Figure 3 – white hatching). As a result of this disposal this eroded sandbar would be reshaped so that the flood and ebb flows would continue to maintain the multiple

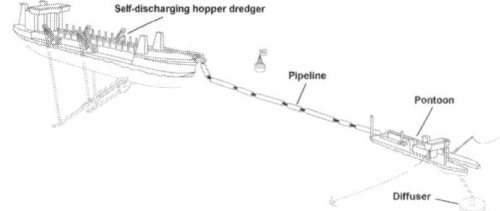

Figure 4. Schematisation of disposal technique.

channels. Besides making the estuary ecologically and morphologically healthier, the reshaping of the sandbar would also improve the self-erosive capacity of the flow on the crossing and possibly reduce the quantity of material to be dredged. The technique could also be applied in other places along the estuary.

## 4.2 *Technology for morphological dredging*

The dredging companies contacted for advice about the disposal of material in controlled way close to the riverbed have developed a system by which the sediment is disposed quietly with a diffuser in shallow water (Figure 4). This technique has already successfully been applied in coastal areas [Goossens & Bosschem, 2002].

## 4.3 *Potential benefits for the environment*

A careful choice of disposal sites, based on good field data and possibly completed with modelling, may produce a selective spatial dispersion of the sediments along the sandbar. Some particle fractions will preferentially move in the deeper areas, other moving towards the shallower ones, possibly up to the top of the bar. During the process, the change in morphology by aggrading up some parts of the bar will change the flow patterns and modify consequently the local sediment transport capacities. This will obviously also change the sedimentation pattern, also of the finest particles moving in suspension in the water column. The segregation of sediment fractions of both disposed and natural sediments will result in the formation of different substrata, some more silty than other, creating a variety of ecotopes.

## 5 RESEARCH ON THE WALSOORDEN PILOT PROJECT

### 5.1 *Study tools*

PAET stated from the beginning of the Walsoorden project proposal [Peters & Parker, 2001] that field measurements and physical and numerical models needed to be combined, as each of these study tools has advantages and limitations. They must be seen as complementary tools for the assessment of the

alternative disposal strategy. The research programme included a field measurement campaign (floats, sediment transport), physical fixed bed scale model tests for both the flow and the bed sediment movement and hydrodynamic numerical model simulations. Flanders Hydraulics Research (Flemish Community) executed this programme with the support of the Port of Antwerp and its expert team.

### 5.2 *Conclusion of the feasibility study*

The results derived from the studies concerning hydrodynamics and sediment transport [Flanders Hydraulics Research, 2003] indicated that the placement of material as proposed for the morphological dredging strategy can likely be used to influence the estuarial morphology [PAET, 2003]. Degraded areas and their associated biotopes could be regenerated. PAET insisted on having a small scale in situ disposal test to gain final evidence that the proposed strategy is feasible.

The analysis of the data has also shown that all investigative tools were needed to reach this conclusion and that morphological assessment of the Western Scheldt should not be based on modelling alone. One should realise that our knowledge about and understanding of the physical processes governing morphological changes is still not sufficient to set up trustworthy models. Combining different tools is the only way to reduce the uncertainties.

Where most of the research occurred within the scope of ProSes, a second opinion team was asked to give their comments on the methodology used for and the results gathered from the research. They confirmed that the idea to use dredged material to restore sandbars is very valuable and that an in situ disposal test is necessary to remove the remaining uncertainties about the proposed strategy.

## 6 THE 2004 WALSOORDEN IN SITU TEST

### 6.1 *Execution of the disposal test*

The execution of an in situ disposal test had to bring final proof of the feasibility of the alternative disposal strategy. The idea of the in situ test was to dispose quietly and precisely 500.000 m³ of sand with a diffuser on the bottom. The dredging vessel (self-discharging hopper dredger) was connected to a floating pipeline through which the sand is transported to a pontoon. On this pontoon the sand is pumped to a diffuser (Figure 5) that disposes the sediment in a precise way on to the bottom.

The amount of 500.000 m³ was chosen because it is on one hand large enough to affect significantly the bottom morphology, however on the other hand small enough to be reversible if something would go wrong. The choice of the disposal location was based on the

Figure 5.   Detail of the diffuser.

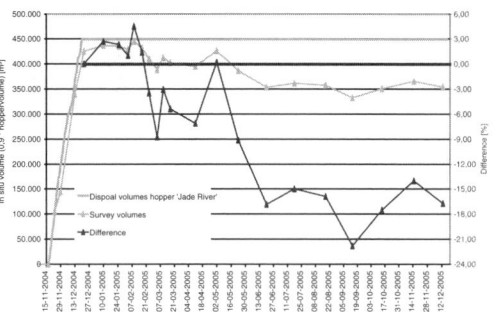

Figure 7.   Evolution of measured volumes (orange = disposed, green = measured, blue = difference).

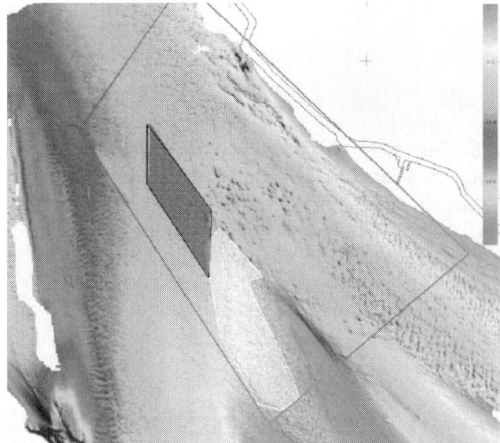

Figure 6.   Licensed disposal area (red) and disposal test areas (2004 = white, 2006 = black).

results of the feasibility study. The float measurements, the results of the numerical simulations and the physical scale model tests with moveable material on fixed bed indicated that an area between the northern sand spit and the tip of the plate was most suitable for an in situ disposal test (Figure 6).

From November 17th to December 20th 2004 500.000 m³ of sand was almost continuously disposed in the proposed area.

### 6.2   Monitoring of the disposal test

To evaluate the success of the test an extensive monitoring programme was set up. This programme, which was executed over a period of one year, included bathymetric surveys, ecological monitoring, sediment tracing tests and sediment transport measurements. Several criteria were defined before the test to evaluate its success. One of them stated that 2 weeks after finishing the disposal execution of the test, at least 80% of the disposed sediment should stay within the control

area (this was defined as the disposal area, extended slightly towards the sandbar of Walsoorden). Also the ecological parameters should not indicate a change in ongoing natural trends.

### 6.3   Bathymetric surveys

The bathymetric surveys were executed using the multibeam technique, producing high resolution bathymetric charts. Where a frequency of weekly surveys was achieved during the first months, this was reduced to once every month 6 months after disposal. Beside this possible impact area, a larger zone was measured every 2 months, to capture possible larger scale influence of the in situ test. These surveys allowed volume computations for the control area. The evolution of the sediment volume in the control area is shown in Figure 7. The amount of disposed sediment should be corrected due to the differences in density in the hopper and in situ. Therefore a correction factor 0.9 was applied to the hopper volumes. As can be seen in Figure 7 the first survey after the execution of the disposal test shows a smaller volume measured in situ than what was disposed. This small difference (25.000 m³) represents the sediment losses during the disposal of the sand, where a fraction (finer sands) was transported by the currents.

During the first 2 months the volume within the control area was even higher than after execution of the test, probably due to natural processes. Afterwards a decrease of volume was measured, a loss of approximately 10% after 6 months, almost 20% after one year. The main part of the eroded sand is transported during flood towards the Walsoorden sandbar (Figure 8). This evolution is in agreement with the predictions of the feasibility study. It may be concluded that the disposed sediments stay well in place, and the imposed criterion was successfully fulfilled.

### 6.4   Ecological monitoring

The ecological monitoring programme included both intertidal as subtidal measurements. Ecologists feared

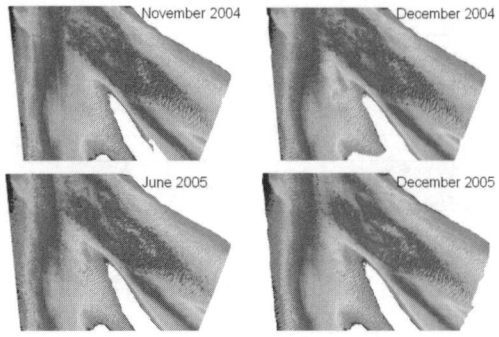

Figure 8. Evolution of bathymetry (November 2004–December 2005).

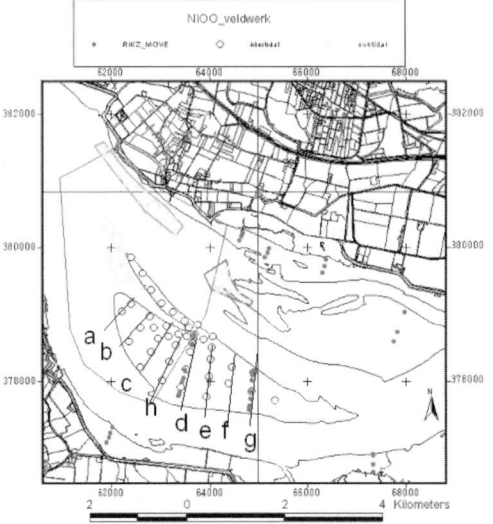

Figure 9. Sampling station ecological monitoring.

increased sedimentation, especially of coarser sediment on the sandbar, which could have a negative impact on its biotopes.

The intertidal monitoring comprised of several stations on the Walsoorden sandbar (see Figure 9) where erosion-sedimentation, sediment composition and macrobenthos were measured. None of the results from this monitoring indicated that the in situ disposal test was responsible for a significant change in ongoing trends.

The subtidal monitoring was focussed on sediment composition and macrobenthos samples, using the BACI-technique (Before-After-Control-Impact). Beside the disposal area (yellow area on Figure 11), 2 control areas were chosen: one at the traditional disposal site "Schaar van Waarde" (green area), the other (red area) where no influence from disposal activities should be expected. For the subtidal samples an initial

decrease in mud-percentage was found for the impact area. This is explained by the absence of finer mud material in the dredged sediments that were disposed. The macrobenthos samples did not show deterioration (biomass, diversity and density) for the impact area compared to the 2 other control areas.

### 6.5   Conclusions in situ disposal test

From morphological point of view, it can be concluded that the experiment using a diffuser for modifying the morphology of the sandbar by disposing precisely dredged material was very successful. The ecological monitoring did not reveal any significant negative impact, neither in the intertidal areas, nor in the subtidal areas. This in situ test confirmed the feasibility of the proposed disposal strategy.

## 7   THE 2006 WALSOORDEN IN SITU TEST

### 7.1   Execution of the disposal test

In the beginning of 2006 a second disposal test was executed, using the traditional clapping technique with hopper dredgers, instead of the diffuser. Because the disposal test was fitted in the continuous maintenance dredging works of the Scheldt estuary, the disposal of $500.000 \, m^3$ was spread over a 3 months period. The disposal location was just downstream the disposal location of 2004 (Figure 6).

### 7.2   Ongoing monitoring of the disposal test

To evaluate the success of the test an extensive monitoring programme was set up. This programme, which will is executed over a period of one year, includes bathymetric surveys and ecological monitoring, similar to those executed after the 2004 disposal test. Again morphological and ecological criteria were defined to evaluate the success of the in situ test.

The preliminary results of the bathymetric monitoring show that the second in situ test can be described as a success from morphological viewpoint. After 6 months only 32% of the dumped material has been moved out of the control volume (Figure 10). Preliminary analysis of the surveys indicate a transport of the material towards the Walsoorden sandbar. Material dumped in 2006 has reached the 2004 control area, leading to a nourishment of this area.

The ecological results will become available during around June 2007, with the preliminary results not showing any negative impact.

### 7.3   Preliminary conclusions in situ disposal test

Although the results of the monitoring campaign are only preliminary, the second disposal test seems to be a success as well. Although further investigations are necessary, these results are very satisfactory towards a further implementation of the continuation

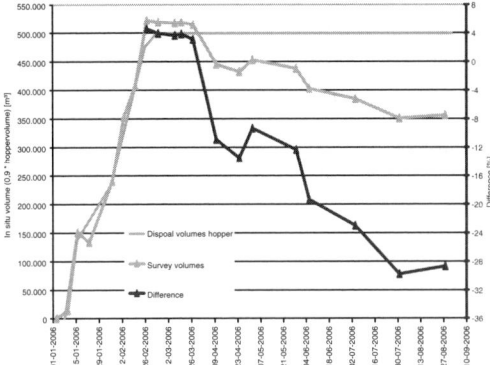

Figure 10. Evolution of measured volumes (orange = disposed, green = measured, blue = difference).

of the disposal strategy as was proposed by PAET. In total an estimated volume of 3 to 5 million m$^3$ could be disposed near the Walsoorden sandbar to reach the proposed objectives. This amount covers 50% of the dredging quantities (7 Mm$^3$ in total) necessary for a further deepening of the navigation channel in the Western Scheldt.

## 8 CONCLUSIONS AND RECOMMENDATIONS

For a long time dredging operations have been considered as producing only negative impacts on the environment. The Western Scheldt is one of the last relatively natural estuaries with a dynamic multi-channel system and exceptionally valuable eco-systems. A management with broader objectives that includes accessibility, safety and nature preservation progressively replaces the past management of the maritime access route to the Port of Antwerp, which was based almost exclusively on an engineering approach. In 2001 an international expert team appointed by the Port of Antwerp authorities, set forward new ideas about the morphological management of the estuary by using dredging and disposal of dredged material to steer the morphological behaviour of the estuarine multi-channel system. As a pilot project to demonstrate this new disposal strategy the location at the sandbar of Walsoorden was selected by the Port of Antwerp Expert Team on the basis of expertise. Reshaping the tip of this sandbar by morphological dredging might improve the self-dredging capacity of the crossing of Hansweert, reducing finally the dredging effort.

The feasibility of this project was studied by Flanders Hydraulics Research, combining desk studies, scale modelling, numerical modelling and field surveys. None of the results of this extensive study

opposed the feasibility of the proposed disposal strategy at the Walsoorden sandbar. To finally prove the proposed disposal strategy, an in situ disposal test was conducted. At the end of 2004 500.000 m$^3$ of sand was disposed at the seaward tip of the Walsoorden sandbar. In 2006 the situ disposal test was repeated, using a different disposal technique. Both experiments were intensively monitored, morphological as well as ecological.

Taking into account the results of both disposal tests, it can be concluded that a new morphological dredging and disposal strategy could be successfully embedded in the future morphological management of the Western Scheldt. However, as stated by the Port of Antwerp Expert Team, the new ways of dredging and disposing sediments should be combined with other measures, such as adapting the hard bordering of the estuary and finding alternatives to the traditional protection works of banks and shoals.

The Walsoorden experiment also confirmed the need for building the capacity of the professionals in morphological assessment techniques, giving sufficient room to expertise and visual analysis of charts, maps and remote sensing observations. A further collaboration between engineers, biologists and ecologists is needed to develop further the idea of morphological dredging and the strategies to manage the morphology of estuarine systems.

## ACKNOWLEDGEMENT

The research presented was conducted through a close collaboration between Flanders Hydraulics Research and the Port of Antwerp Authority, partly as a project under the Dutch-Flemish ProSes direction. It was funded by ProSes and by the Flemish Government. Surveys were organised with the collaboration of the Flemish Administration of Maritime Access, the Dutch Rijkswaterstaat and the contractors. The authors acknowledge the effective support of the Port of Antwerp Authority, which funded the expert team, also the participation of the other members of the Port of Antwerp expert team, Jean Cunge from France, Reginald Parker from UK, Bob Meade and Michael Stevens from USA.

## REFERENCES

Flanders Hydraulics Research, 2003. Alternative Dumping Strategy Walsoorden. Results Physical and Numerical Modelling.

Flanders Hydraulics Research, 2006. Alternatieve stortstrategie voor de Westerschelde – Proefstorting te Walsoorden – Verslag 13u meetcampagnes (In Dutch – "Alternative Dumping Strategy – In Situ Disposal Test – Results from the 13h Measurement Campaigns").

Foster R., Rossi F., Bonnie K., Heip C.H.R. and Herman P.M.J., 2006. Alternatieve stortstrategie voor de Westerschelde – Monitoringsprogramma proefstorting Walsoorden – Rapport 11/11 (In Dutch – "Alternative Dumping Strategy – Ecological Monitoring In Situ Disposal Test – Report 11/11").

Goossens M. and E. Bosschem, 2002. The innovative D.P.G.P BAYARD II at the Le Havre Port 2000 harbour extension project.

Institut für Wasserbau, 2003. Current Measurements in the Westerschelde – September and October 2002.

Peters J.J. and F. Wens, 1991. Maintenance Dredging in the Navigation Channels in the Zaire Inner Delta, COPEDEC III Conference, Mombassa.

Wang Z.B., R.J. Fokking, M. de Vries and A. Langerak, 1995. Stability of river bifurcations in 1D morphodynamic models, Journal of Hydraulic Research, vol. 33, no 6 pp. 739–750, 1995.

Peters J.J., 1994. Manejo de los Ríos en la Cuenca del Río Piraí (Management of the rivers in the Pirai river basin), editor J.J. Peters with the financial support of the European Commission.

Peters J.J., 1998. Amélioration du transport fluvial en Amazonie bolivienne, Bulletin of the Belgian Royal Academy of Overseas Sciences, vol. 44, no 3, 1998.

Peters J.J., R.H. Meade, W.R. Parker and M.A. Stevens, 2001a. Improving Navigation Conditions in the Westerschelde and Managing its Estuarine Environment. How to Harmonize Accessibility, Safetyness and Naturalness?

Peters J.J. and W.R. Parker, 2001b. A Strategy for Managing the Westerschelde's Morphology. An Addendum to the Final Report.

Port of Antwerp Expert Team, 2003. Alternative Dumping Strategy. The Feasibility of Morphological Dredging as a Tool for Managing the Westerschelde.

ProSes, 2004. Strategisch Milieueffectenrapport Ontwikkelingsschets 2010 Schelde-Estuarium – Hoofdrapport (In Dutch – "Strategic Environmental Impact Assessment Development Scheme 2010 Scheldt Estuary – Main Report).

Winterwerp J.C., Z.B. Wang, M.J.F. Stive, A. Arends, C. Jeuken, C. Kuijper and P.M.C. Thoolen, 2001. A new morphological schematization of the Western Scheldt Estuary, The Netherlands, 2nd IAHR Symposium on River, Coastal and Estuarine Morphodynamics, Obihiro, Japan, September 2001.

*River, Coastal and Estuarine Morphodynamics: RCEM 2007 – Dohmen-Janssen & Hulscher (eds)*
*© 2008 Taylor & Francis Group, London, ISBN 978-0-415-45363-9*

# Assessing the sustainability of estuarine barrages

L. Beevers
*Jacobs UK, Glasgow, UK*

G. Pender
*School of the Built Environment, Heriot Watt University, Edinburgh, UK*

ABSTRACT:  The increasing economic importance of waterside development in many cities has been the driver for major investment in barrage construction in the U.K. To ensure the sustainability of these structures it is crucial to understand their impact on sedimentation distribution and re-distribution patterns. As sedimentation is essentially a long-term phenomenon, forecasting of impoundment behavior using a computer model is necessary. This paper presents the results of a computer simulation of long-term (50 year) sediment distribution predictions for the Tees barrage impoundment, a total exclusion barrage built in 1994 and situated in the north-east of England, U.K. The simulations were undertaken using the one-dimensional "ISIS Sediment" modeling software. The upstream boundary conditions, in terms of both flow and sediment, were estimated from data collected on the river at Low Moor, and take account of possible sediment source changes during the 50 year simulation period. The predictions from the model show that after 30 years the impoundment reaches a state of dynamic equilibrium irrespective of sediment supply.

## 1 INTRODUCTION

The River Tees is situated in the North East of England, UK, and drains part of the North York moors and the Pennines (Figure 1). The source of the river is found in the Pennine hills at an altitude of around 600 m. From here it flows through a valley for approximately 160 km to the sea at Middlesbrough. Its total catchment size is 1264 km² (NRFA, 2003). The River Tees marks the boundary between the counties of Durham and Yorkshire, and has historically marked the border between various political and tribal regions.

The barrage was constructed in 1994 for amenity and urban re-generation purposes in the Teeside area (Hall, 1996). As a consequence, thirteen years on, the area around Stockton-on-Tees has attracted many new developments including the Stockton campus of Durham University.

The structure is a total exclusion barrage, which means that no saline intrusion is permitted upstream. It consists of four fishbelly flap gates, the levels of which are controlled hydraulically over the barrage's 70 m length. These gates maintain the upstream water level between +2.35 m and +2.85 m OD at all times. The impoundment affects water levels over a 25 km length of the river, from the barrage to Low Moor, which was the tidal limit of the river prior to barrage closure. A canoe slalom, a fishpass and a navigational lock are included in the barrage

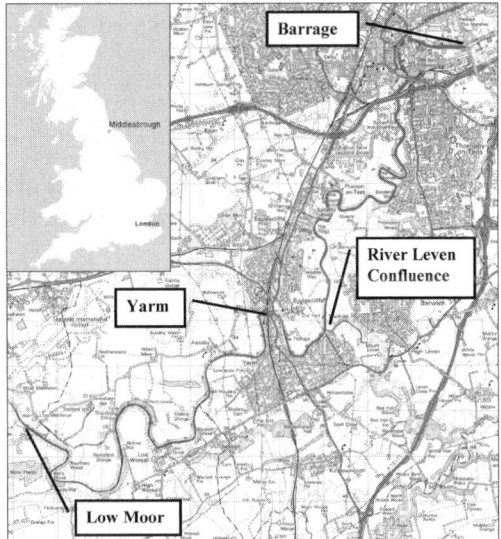

Figure 1.  Map of the Tees barrage from Low Moor to the barrage.

structure to improve the amenity value of the river upstream.

Sedimentation distribution patterns take many years to establish, and since barrage construction fundamentally alters the dynamics of estuaries in terms of sedimentation rates, velocity patterns and water quality, the closing of the barrage represents the beginning of a new sediment regime in the Tees. It is important that this new regime is sustainable as if the barrage were to silt-up it would significantly decrease it's amenity value. Additionally, because the prime purpose of impounding the river is improving amenity value to encourage urban regeneration, it is imperative that future siltation does not cause unsightly mudflats upstream of the barrage. These can bring unsavory smells and sights to an otherwise pleasant riparian environment.

Some predictive modeling of the morphological behavior of the impoundment was undertaken during the barrage design, (HR Wallingford, 1992), however little post-impoundment monitoring has been undertaken since its construction in 1994. If the morphological sustainability of the Tees barrage is to be ensured, it is essential that the pre-construction predictions are revisited periodically using updated field observations. In particular, with the increasing evidence for climate change, it is necessary to investigate the influence a changed flow regime will have on sediment build-up within the impoundment.

With this in mind the River Tees was used as a case study in a project investigating the sustainability of estuarine impoundments in general (the Sustainability In Managed Barrages project, SIMBa). As part of this study the River Tees, from the barrage to Low Moor, was modeled one-dimensionally to predict sediment behavior in the impoundment over a 50 year period. Field data consisting of flow and sediment inputs to the impoundment were collected during the year 2000. Some of the data recorded included that for a relatively extreme flood (in excess of 1 in 25 years). This data was then used as a basis for creating the long-term upstream flow and sediment boundaries.

## 2 MODEL CONSTRUCTION

### 2.1 Data availability

A survey of the river form was undertaken in 2000, which recorded topographic data of the river six years post impoundment. Data, in cross-sectional form, was recorded for a 25 km length from the barrage at the downstream extent to Low Moor, the old tidal limit.

Stockton Borough Council and the Highways Agency provided technical drawings for the majority of the structures through the surveyed reach, with the exception of the rail bridges. This data supplemented the new topographic data recorded around these structures. Details of the railway bridges within the stretch were provided by Network Rail, the national rail authority. At the downstream extent, information

on the barrage structure and operation, including canoe slalom, was obtained from British Waterways. From discussions with the then barrage manager it was understood that the barrage was operated to keep the water level upstream between +2.35 m and +2.85 m above ordnance datum at all times, apart from times of water and sediment flushing.

To complement the topographic data, flow data for the upstream extent of the model was available from the Environment Agency (EA) gauging station on the River Tees. The gauging station at Low Moor held long-term records of flows back to 1969 and is a velocity area station with good calibration (NRFA, 2002). At its position as the lowest gauging station on the River Tees, it monitors flow from the upstream headwaters of the river into the downstream section.

Information on the sediment load of the river was available from a number of sources. Monitoring of the suspended solid load, and characterization of the bed sediment through the surveyed reach, was undertaken as part of the SIMBa project. Grab samples of the riverbed were taken and analysed from Low Moor to the barrage, while suspended load was recorded at Low Moor. Continuous monitoring commenced in 1999 using a self-cleaning nephelometric turbidity sensor to monitor turbidity. The monitoring recorded the unusually high rainfall that was experienced by the Tees catchment in the early autumn and winter of 2000, following which a change in sediment supply was evident from the measured sediment data (White, 2001). The data indicated that the series of extreme flood events, which occurred in October/November 2000, resulted in a significant change to the previous sediment transport response to lower flow events. This information was then available to the modeling study.

In addition to the grab samples, sediment cores were taken and analysed from the riverbed directly upstream of the barrage. Particle size analysis using a Coulter system granulometer was completed at sections through the core. A distinct change from sand to silt was identified in the cores at a depth of 50–55 cm. This change was identified as the boundary between the river pre and post impoundment.

### 2.2 Modeling technique

For the purposes of the study, ISIS, a commercially available one dimensional, finite difference modelling software, was chosen, for reasons of computation efficiency. The model solves the St Venant shallow water equations using the Preissmann four-point implicit scheme. Calculation of the sediment transport component of the software is fully coupled to the hydrodynamic component. The Exner equation controls the sediment continuity through the model and at each cross-section. The transport of sediment is calculated using both the Ackers and White (revised)

(Ackers 1993) equation and the Westrich Jurashrek (1985) equation. The latter was developed from laboratory experiments using mainly fine sediments, consequently it was used to describe the fine component of the suspended solid load in the River Tees, while the Ackers and White equation was used to describe the coarse sediment movement.

### 2.3 Model construction

To assess the morphological sustainability of the impoundment following the construction of the barrage across the estuary, the section of the river between the barrage and the pre-impoundment tidal limit at Low Moor was modeled.

A one-dimensional computer model was constructed from topographical data collected during the summer of 2000. The principle structures were included in the model, including the main bridges through the reach. The barrage was modeled as a gated weir, which was controlled by the impoundment water level. A set of logical rules were written to maintain the water level just upstream of the barrage between +2.35 and +2.85 m AOD. The upstream extent of the model was positioned at the Low Moor gauge. Bedload sediments are mostly prevented from progressing downstream beyond the Low Moor weir, leaving suspended sediment as the principle component for sedimentation in the impoundment.

The upstream flow boundary was represented using time series flow data. For calibration purposes this data came directly from the gauging station at Low Moor. The model upstream of the barrage was controlled entirely by the gated weir function; the downstream boundary was therefore kept at a fixed water level. Measured suspended solid concentrations were used to derive an upstream boundary for suspended solids. To assess the impact in changes in sediment sources observed following the 2000 flood, data was divided into two sediment regimes collected before and after the observed change in sediment supply. Thus, two relationships exist: one which models the lower sediment supply regime pre-October 2000; and one which models the higher regime post-October 2000. Upper and lower confidence limits were calculated for each regression relationship.

In the following, the post-October 2000 regression relationship (2) has been used to model a medium level of sediment supply to the river with low and high levels of sediment supply being modeled using the pre-October 2000 (1) (upper 95% confidence limit) and post-October 2000 (3) (upper 95% confidence limit).

$$Q_s = 3.6939 Q^{0.6936} \qquad (1)$$

$$Q_s = 1.7849 Q^{0.8680} \qquad (2)$$

$$Q_s = 2.7844 Q^{0.9908} \qquad (3)$$

Predicted river levels were calibrated to two different events recoded on the 31st August 2000 and the 6th of December 2000. The resulting values for Manning's 'n' for the model were 0.03 for most of the main channel apart from areas where high energy losses were indicated during the calibration process. In these areas an 'n' value of 0.045 was used. Settling velocities within the model were calibrated to previous modeling studies (HR Wallingford, 1992), which resulted in settling velocities of 0.3–0.5 mm/s. These relatively low settling velocities are thought to result from the mix of organic, silt and clay particles found in the River Tees. Sediment input into the impoundment was validated over a period of six years using sediment cores taken at the lower end of the impoundment, just upstream from the barrage. These showed deposition of approximately 0.5 m in total or 0.84 mm/year.

## 3 PREDICTION OF LONG-TERM CHANGES ASSUMING A STABLE CLIMATE

### 3.1 Methodology

Given that the order in which flows reach a watercourse have an intrinsic link to sediment behavior within that river a Markov Chain method for creating the upstream flow boundary based on Lawrence and Kottegoda (1976) and, Issacson & Madsen (1976) was proposed. Calculations were undertaken assuming a stable climate. Markov Chains are a generic modeling technique that can be used to forward predict existing data sets based on their statistical properties, and is a short-term memory method that uses the memory of a series as a starting point for forecasting future data (Jimoh & Webster 1996; 1999). The technique uses limited historical data to establish Markov transition probabilities. These are the probabilities of moving from one state (i) at time t, to another state (j) at time $t + 1$. A transition matrix is constructed using transition probabilities, defined by the real number $p_{ij}$ such that the properties:

$$p_{ij} \geq 0$$
$$\sum_{j=0}^{\infty} p_{ij} = 1 \qquad (4)$$

are satisfied. Therefore the transition matrix P is defined by

$$\mathbf{P} = \begin{bmatrix} p_{11} & p_{12} & \cdots & p_{1k} \\ p_{21} & p_{22} & \cdots & p_{2k} \\ \vdots & & \ddots & \vdots \\ p_{k1} & p_{k2} & \cdots & p_{kk} \end{bmatrix} \qquad (5)$$

Available flow data from the National River Flow Archive at Low Moor was used to create the Markov

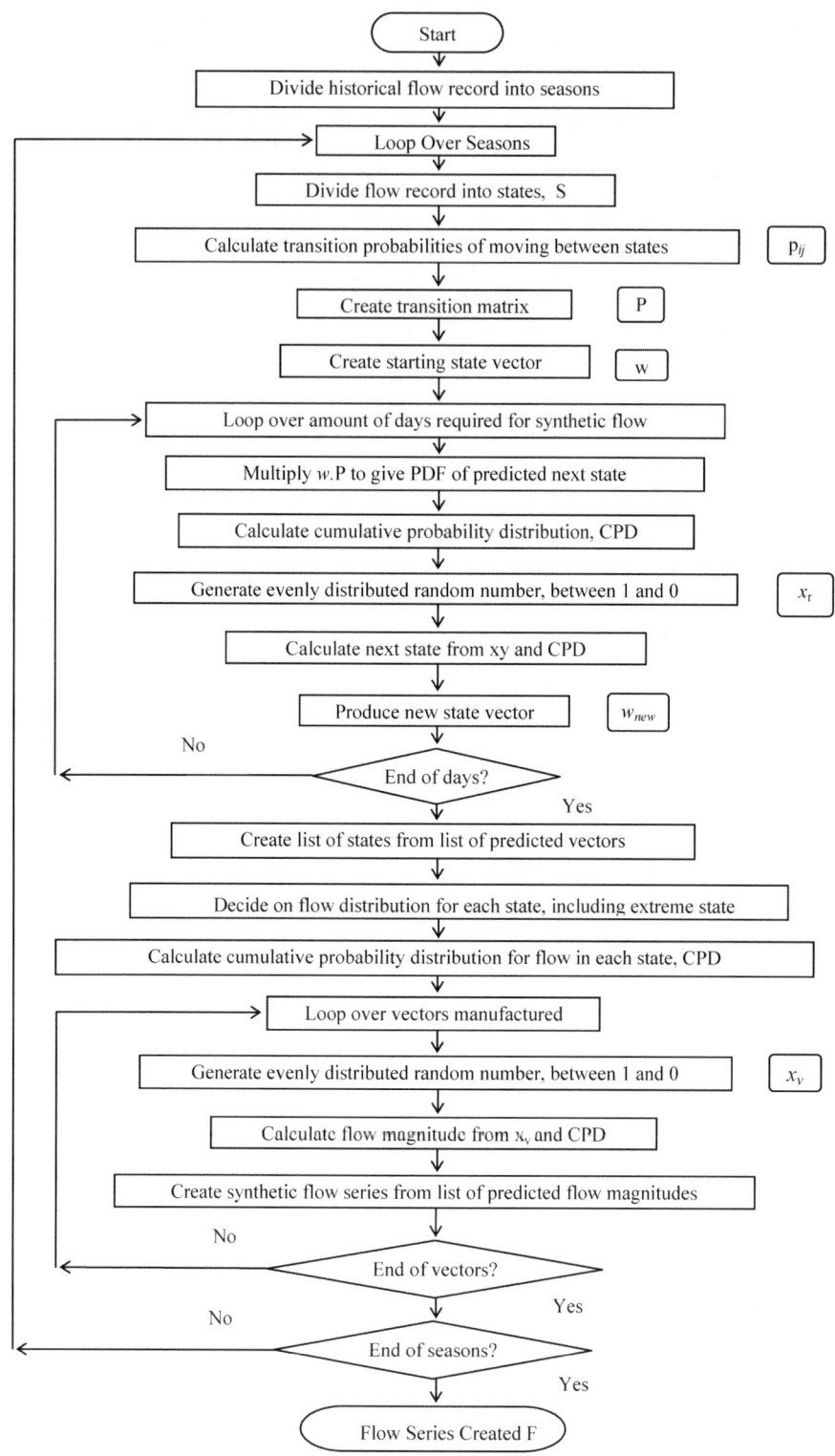

Figure 2.   Flow chart detailing the Markov Chain method for generating flows.

Chain transition matrix. The flow data recorded at Low Moor dated back to 1969 and contained one interruption of 9 months during the period 1974–1975. Data recorded over the period 1981–2000 was used to generate the Markov Chain transition matrix whilst the data recorded between 1970–1980 was used for model validation purposes.

To undertake the Markov Chain method the observed data must be binned in some meaningful way into states to allow the generation of new simulated series. The method has been used in the field of stream flow prediction previously; however the research has mainly concentrated on streams with intermittent flow where the Markov chain is used to predict the occurrence of a wet day and the height of the pulse (Aksoy & Bayazit, 2000; Xu et al. 2001). Some advice on dividing the data into states is given in Sahin & Sen (2001) where the divisions were determined using the mean and standard deviations of the series. While other papers use an analogue method to divide data based on some threshold i.e. wet or dry. However, the flow record for the Tees was such that categorization in either manner was unsuitable. The standard deviation of the series was prohibitively large which would have resulted in the lower flows being lumped into one state. This would have resulted in a less accurate portrayal of the statistics of the data due to the fact that a large proportion of the flow record is comprised of low flows. Similarly an analogue method would not have allowed reasonable discretisation of the flow series. Consequently, a compromise method was devised that combined these techniques. The analogue system was used to separate out 'high' flows from 'low' flows using Peaks over Threshold (POT) data collected at the gauge (Robson & Reed, 1999). Then to categorize the 'low' flows state divisions were devised, based on analysis of the available historical data. This method aimed to discretise flows into a series of states that balanced the requirement for the reasonable representation of the flow series, and the issues with data sparseness towards the upper end of the 'low' flow scale.

The methodology is detailed in the flow diagram Figure 2, and a five year synthetic flow series is shown in Figure 3.

### 3.2 Simulation results

The creation of different upstream boundaries for both sediment input and flows meant that several different combinations of the boundaries are required to investigate the morphological sustainability of the Tees impoundment over the next 50 years. The model was run for 50 years, using a timestep of 1 day, to predict the sedimentation build-up and distribution within the impoundment over this period. The upstream boundary was supplied from the Markov Chain model assuming a stable climate. The results (Figure 4) indicate that

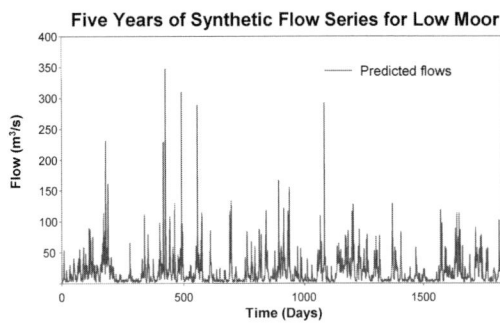

Figure 3. Five year synthetic flow series for Low Moor, created using Markov Chains.

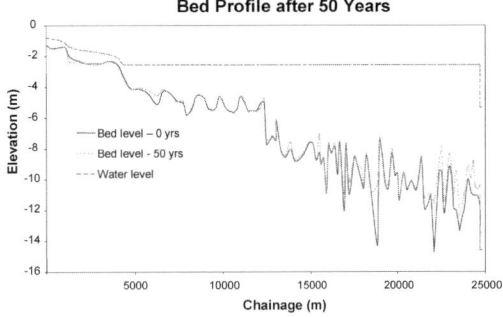

Figure 4. Bed profile of the River Tees (in long section) after 50 years.

deposition is not uniform along the impoundment but is controlled by the channel shape at particular cross-sections. The first area of deposition around chainage 6000 m is the product of the larger grain sediments >63 μm settling out due to the effect of the barrage backwater. The deposition predicted further downstream tends to be finer material as the small particles are carried further by the flow. In periods of high flow sediment is swept downstream but is still trapped by the barrage, hence the greater build-up of deposits just upstream of the barrage.

During the simulated period the model indicates that the impoundment is likely to reach a dynamic equilibrium. This can be seen from Figure 5 and may be interpreted as the impoundment reaching a new regime following the construction of the barrage, longer simulations of 80 years were undertaken and corroborate this finding. Figure 5 shows the evolution of the bed at node s4 in the model, which is situated 0.4 km upstream of the barrage. The computed bed level change under each sediment supply scenario is depicted along with the predicted water surface slope at that node for the duration of the run.

For the first 30 years of simulation the sediment gradually deposits at a rate that is dependent on the

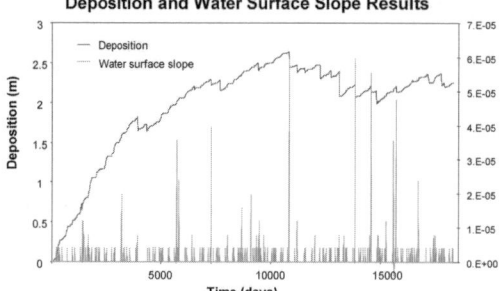

**Deposition and Water Surface Slope Results**

Figure 5.  Bed Evolution over 50 years with water surface slope – node s4 0.4 km from barrage (stable climate – high sediment supply scenarios).

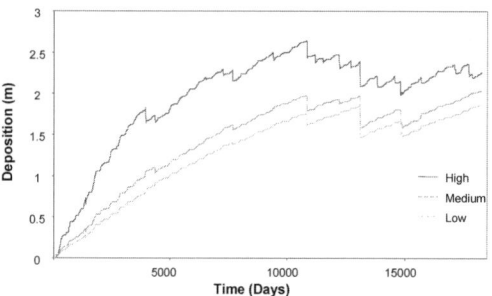

**Comparison of Deposition Rates over 50 Years (s4)**

Figure 6.  Comparison of deposition rate over 50 years at node s4 with three different sediment scenarios (low, medium and high) under existing climate conditions.

amount of sediment reaching the impoundment. After this point steeper water surface slopes become more frequent due to the change in bed levels. Steeper water surface slopes result in deposited sediment being re-entrained. This behaviour suggests that, for this particular scenario of flow patterns, and despite the continual deposition, the system is reaching a dynamic equilibrium. The bed level for the new regime is a function of the volume of sediment delivered into the impoundment, and varies slightly for each sediment supply scenario. Consequently, during the period simulated there is evidence to suggest that the river may be reaching a new regime following the construction of the barrage.

Figure 6 and Table 1 present the results from varying sediment input scenarios. This indicates that the intuitive assumption that an increase in sediment supplied results in an increase to deposited sediment within the impoundment, is true for the River Tees impoundment. However it should be noted that a linear relationship between input and deposition does not exist for this particular watercourse.

Table 1.  Comparison of deposition predictions using differing sediment supply scenarios.

| Supply Scenario | Sediment (tonnes) | | | |
|---|---|---|---|---|
| | Inflow | Outflow | Retained | (%) |
| Low | 1675636 | 1533057 | 142579 | 8.51 |
| Medium | 2225787 | 2065253 | 160534 | 7.21 |
| High | 4618494 | 4377618 | 240876 | 5.22 |

## 4  CONCLUSIONS

The overall morphological sustainability of the impoundment has been assessed assuming stable climate conditions for a fifty year period. From the discussion of the results presented the following conclusions can be drawn.

1. Markov Chain modeling can provide a simple method for extending known flow boundaries for the purposes of long-term flow modeling.
2. According to the computer modeling the impoundment is predicted to reach a dynamic equilibrium over the period of simulation, which is irrespective of the sediment supply. The results indicate that the river becomes partially self-cleaning during the course of simulation with an equilibrium being reached after approximately 30 years. The point at which the equilibrium is reached is dependent on the timing of high flows reaching the impoundment.
3. The predictions suggest that the impoundment is morphologically sustainable over the modeled 50 year period irrespective of the sediment supply scenario used. These results, which are based on the assumption that the climate will remain stable and thus flows will not change their behavior dramatically, show sedimentation occurring through the impoundment. However, the predictions suggest that the impoundment is reaching a type of dynamic equilibrium after 30 years with each cross-section becoming self stabilizing, and this equilibrium does not appear to cause the impoundment to silt up. No one cross-section reports sufficient deposition to create unsightly mudflats. Therefore it is sensible to say that the impoundment is sustainable in terms of sediments for the next 50 years and seems to reach regime after 30 years. This regime is a function of the upstream sediment concentration reaching the impoundment.
4. These findings are limited by their failure to account for climate change upon river flows. Consequently, it would be useful to incorporate climate change predictions into this study to investigate its impacts.
5. A note of caution is required however as the model used does not include time-dependent

consolidation of the deposited sediment. Consequently, the current model may over estimate the extent of flushing. However the model can be used to indicate possible trends of sediment transport within the impoundment.

ACKNOWLEDGEMENTS

The work was funded through the Engineering and Physical Sciences Research Council grants GR/M42299 and GR/M42305. The support of Wallingford Software Ltd through free access to ISIS sediment, the assistance of the National River Flow Archive for the provision of flow data, the support of Environment Agency, the British Atmospheric Data Centre and the Met. Office for the provision of catchment climate data, Stockton Borough Council, the Highways Agency and Network Rail for bridge geometry data are gratefully acknowledged.

REFERENCES

Ackers, P & White, W, 1973, Sediment Transport: New Approach and Analysis, *Journal of the Hydraulics Division. American Society Civil Engineers*, 99, pp 2041–2060.

Ackers, P, 1993, Sediment transport in open channels: Ackers and White update, *Proc. Instn Civ. Engnrs Wat. Marit. & Energy*, 101 Dec 247–249.

Aksoy, H. & Bayazit, M, 2000, A model for daily flows of intermittent streams. *Hydrological Processe,.* 14, pp.1725–1744.

Chapman, T, 1997, Stochastic models for daily rainfall in the Western Pacific. Mathematics and Computers in Simulation. 43. pp. 351–358 Davis, J. 1986. Statistics and Data Analysis in Geology. John Wiley & Sons, New York. ISBN 0-471-83743-1

Hall, D, 1996, The Tees Barrage – a success story. In: Burt, N. & Watts, J, editors, *Barrages: Engineering Design and Environmental Impact,*. Chichester John Wiley & Sons Ltd, pp 335–344

Wallingford, H.R, 1992 (a). Tees Barrage Study: Upriver Siltation, Report EX 2432. (February 1992)

Issacsen, D & Madsen, R, 1976, *Markov Chains: Theory and Applications*. John Wiley & Sons, USA, ISBN 471-42862-0

Jimoh, O & Webster, P, 1996, The optimum order of a Markov chain model for daily rainfall in Nigeria, *Journal of Hydrology*, 185, pp. 45–69.

Jimoh, O & Webster, P, 1999, Stochastic modeling of daily rainfall in Nigeria: intra-annual variations of model parameters, *Journal of Hydrology,* 222, pp. 1–17.

Lawrance, A & Kottegoda, N, 1977, Stochastic Modeling of Riverflow Time Series. J. R, *Statistical Society*, A, 140, Part 1, pp1–47.

NRFA, 2003, National River Flow Archive. http://www.ncl.ac.uk/ih/nfra/index.htm (17/05/03).

Robson, A & Reed, D, 1999, The Flood Estimation Handbook: Vol.3. Statistical procedures for flood frequency estimation. *Wallingford. Institute of Hydrology*. ISBN 0948540915. m

Sahin, A & Sen, Z, 2001, First-order Markov Chain approach to wind speed modelling. *Journal of Wind Engineering and Industrial Aerodynamics*, 89, pp. 263–269.

Westrich, B & Jurashek, M, 1985, Flow transport capacity for suspended sediment. *Presented at the 21st Congress IAHR, Melbourne, Australia.*

White, S (ed.), 2001, SIMBA Sustainability in Managed Barrages Year Two Reports, April 2001

White, S (ed.), 2002, SIMBA Sustainability in Managed Barrages Year Three Reports, April 2002

Xu, Z. Schumann, A, Brass, C. Li, J, & Ito, K, 2001, Chain-dependent Markov correlation pulse model for daily streamflow generation. *Advances in Water Resources*, 24. pp. 551–564.

*River, Coastal and Estuarine Morphodynamics: RCEM 2007 – Dohmen-Janssen & Hulscher (eds)*
*© 2008 Taylor & Francis Group, London, ISBN 978-0-415-45363-9*

# Predicting morphodynamic response of a coastal plain estuary using a Boolean model

H. Karunarathna & D.E. Reeve

*Centre for Coastal Dynamics and Engineering, School of Engineering, University of Plymouth, Plymouth, UK*

ABSTRACT: The long term morphodynamic response of coastal plain estuarine systems has been studied using a Boolean model. The method involves definition of Boolean variables to represent estuary system elements, development of Boolean networks to represent inter-relationships between system elements and external forcing and development of Boolean functions to transform Boolean networks to a formal mathematical language. The analysis of the coastal plain estuary system shows that the nature of long term morphodynamical response depends on the availability of external sediment and the Type of external forcing. If the estuary is sediment sufficient, then it is able to maintain its geomorphology and reach a stable state. If the estuary is sediment deficient, some features such as salt marshes and spits are likely to become weak or disappear in the long term.

## 1 INTRODUCTION

Coastal plain estuaries are characterised by narrow channels through fronting barrier beaches or sand spits and extensive inter-tidal flats. Hydrodynamic behaviour of a coastal plain estuary is primarily controlled by waves, tides and river flow. They can either be tide or wave dominated. River flow is low compared to the tidal prism but can be significantly large during rainy seasons. Morphodynamic behaviour depends on the hydrodynamic regime and also the nature and magnitude of sediment supply. Coastal plain estuary systems are highly complex morphological features and continuously react to changes in marine and fluvial environment. Likely effects of changes in environmental forcing such as rise in sea level, increase in storminess, etc. may cause drastic changes in sediment- and morphodynamics of these estuaries. In addition, they are subjected to morphological changes following human interference.

Understanding and predicting the morphodynamic behaviour of a coastal plain estuary is still limited because of its complexity. A central task faced by estuarine managers is to predict and manage the constantly moving coastal and estuarine environment in the medium to long term. A basic requisite for making positive management decisions would be to have a sound understanding of estuary behaviour in both short and long time scales.

Estuarine systems present a complex challenge to the modeller as they consist of multiple feedbacks that act on different time scales. The exact physical nature of many of these processes and the relevant values of the parameters involved are mostly unknown. It is therefore difficult to construct models purely based on physical concepts (Spearman, et al., 1997). Since the application of data driven models or process based models is limited, systems methods provide an alternative modelling tool (de Vriend et al., 1993). In this approach the physical system, such as the estuary morphology, is partitioned into its elements together with the linkages between them. The dependence of particular elements on other elements or external factors can then be investigated in a qualitative manner, without the need for detailed quantitative calculations.

In this paper we predict and describe the long term morphodynamic response of a coastal plain estuary to global climate change and human interference using a systems based modelling framework based on a Boolean approach. A Boolean approach was described by Nicolis (1982) in a pioneering application of the technique to climate dynamics. It is particularly suited to the mathematical formulation of conceptual models of systems that exhibit threshold behaviour, feedbacks and time delays. The approach is perhaps best considered to be a heuristic first step towards understanding problems currently too complex to model using systems of partial differential equations.

By incorporating a Boolean description of the linkages between the different elements in a coastal plain estuary system, we provide a qualitative description of the estuary morphodynamics with the advantage of logical transparency. In the context of morphological evolution of the estuary, a Boolean variable represents the volume of geomorphological elements of the estuary system. The corresponding Boolean function describes the specific feedbacks inherent in the dynamics of the system element. The set of external parameters which affect the dynamics of the system element includes waves, tides, sediment exchange between elements, and so on.

Assume that a Boolean function is quite small when the corresponding Boolean variable is less than a certain threshold value, but becomes quite appreciable when the corresponding Boolean variable is greater than the same threshold and saturates to the threshold shortly afterwards. If we idealise this situation by considering that both Boolean variable and Boolean function are zero when the variable is less than its threshold value and equal to one when the variable is greater than its threshold then, they can be considered as a discontinuous Boolean variable and Boolean function respectively. We can then say that the Boolean variable is 'low' or 'off' when it is less than the threshold value and 'high' or 'on' when it is greater. Similar considerations apply for the Boolean functions.

To indicate the state of a system element, it is assigned an associated value 1 for 'high' and 0 for 'low'. The future state of one element in the network depends on the states of the other elements in the network which are designated as that element's inputs. The element may feedback its own state as a self-input. The state of an element in a Boolean network at a future time is governed by a logical rule or Boolean function, which operates on the element's inputs. A detailed description of the Boolean approach for estuarine systems is given in Karunarathna & Reeve (2007).

Section 2 of the paper presents the Boolean model and analysis of morphodynamic evolution of a coastal plain estuary. In Section 3, the results are discussed. The application of the method to two UK estuaries is described in Section 4, with conclusions presented in Section 5.

## 2 BOOLEAN MODEL

The modelling and analysis of morphodynamic evolution of a coastal plain estuary is described in this section. The long term morphodynamic behaviour of this type of estuary is investigated under different forcing situations arising from global climate change (e.g. sea level rise, change in fluvial discharge) and human interference (dredging, construction of estuarial and flood defence structures). All forcing situations were analysed under two different scenarios of external sediment influx into the estuary: abundant flow of external sediment and constrained flow of external sediment.

Figure 1 gives system representation of a coastal plain estuary. The estuary mainly consists of five geomorphological elements: inter tidal flats, salt marshes, tidal channels, a delta and a spit. The peripheral

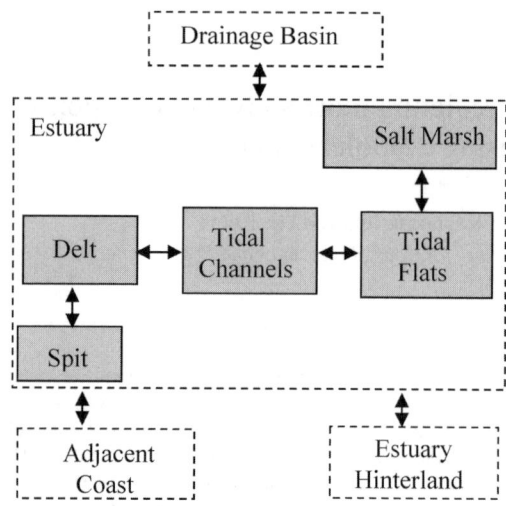

Figure 1. System representation of coastal plain estuary.

elements are adjacent beach (possibly a barrier beach), flood plain and estuary hinterland (ABP Mer, 2005).

The system is modelled and analysed with changing external forcing due to sea level rise for both deficient and abundant sediment influx. The analysis was then extended to investigate the effects of elevated fluvial flows. Finally, human interference is considered in the form of dredging and structural construction.

Figure 2 shows the Boolean networks for the coastal plain estuary shown in Figure 1 at sea level rise. (a) and (b) refer to deficient and abundant sediment inflow scenarios. Notations used in the networks are shown in Table 1.

Rising sea level increases tidal and wave forcing in the estuary. In the absence of external sediment to replenish the estuary, increased wave forcing enforce a negative feedback on salt marshes and inter-tidal flats by inundating them and disturbing loose sediment deposits thereby forcing them to erode. Elevated tidal forcing has similar effects on salt marshes and inter-tidal flats (Lee and Mehta, 1997). The spit too has a negative feedback from elevated wave and tidal forcing which could be explained on the same grounds. The feedbacks from the salt marsh, inter-tidal flats and the spit on waves and tides are negative as they tend to dissipate wave and tidal energy. Channels become wider and deeper due to the passage of strong tidal currents and large waves thereby receiving a positive feedback. The feedback on waves and tides on the channel is positive as channels allow penetration of large waves and strong tidal currents into the estuary (Lanzoni and Seminara, 2002).

Salt marshes and inter tidal flats enforce positive feedback on each other by exchanging sediment among them. Tidal flats and channels enforce positive

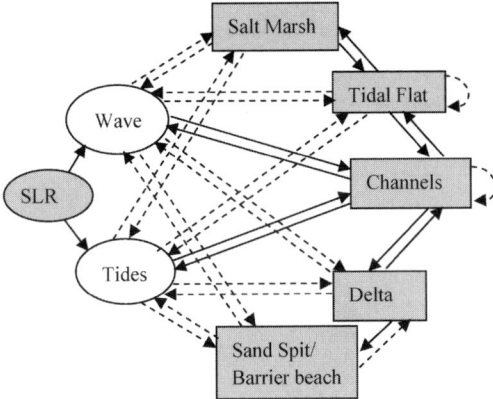

a) Constrained sediment supply

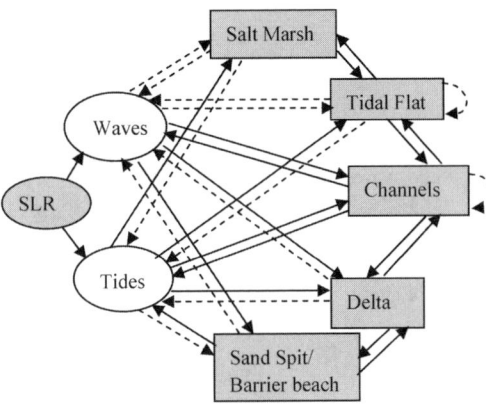

b) Abundant sediment supply

Figure 2. Boolean networks for drowned coastal plain estuary responding to sea level rise (SLR – Sea Level Rise).

Table 1. Notations for Boolean variables and Boolean functions.

| Network element | Boolean variable | Boolean function |
|---|---|---|
| wave forcing | $w$ | $W$ |
| tidal forcing | $t$ | $T$ |
| river discharge | $rd$ | $RD$ |
| Inter-tidal flat | $tf$ | $TF$ |
| salt marsh | $sm$ | $SM$ |
| channel | $cc$ | $CC$ |
| spit/barrier beach | $ss$ | $SS$ |
| delta | $dd$ | $DD$ |
| river flow | $rf$ | $RF$ |
| structure | $st$ | $ST$ |

feedback on each other. According to Pethick (1994), deep channels result in potential net import of sediment and accumulation in inter-tidal flats thereby allowing co-existence of deep channels and shallow

flats. As the changes progress, the system evolves towards having deep channels. Sediment export and lowering the tidal flats takes place simultaneously. Both tidal flats and channels have negative feedback on themselves due to the fact that shallow flats tend to be more erosive and deep channels attract more sediment and tend to become shallower.

When the estuary has an abundant influx of external sediment, the nature of the feedback between natural forcing and some estuary elements may change. For example, when the estuary is supplied with external sediment, tides and waves act as sediment carriers to inter-tidal flats, salt marshes and the spit hence enforcing positive feedbacks on them. The Boolean functions that describe the interactions and feedback between the system elements for constrained and abundant sediment cases are given in Equations 1 and 2. Some evolutionary sequences deduced by solving Equations 1 and 2 are given in Table 2.

$$
\left.
\begin{aligned}
W &= sm' \vee tf' \vee cc \vee dd' \vee ss' \\
T &= sm' \vee tf' \vee cc \vee dd' \vee ss' \\
SM &= (w' \wedge t \wedge tf) \\
TF &= (sm \vee cc) \wedge t \vee (tf' \wedge w) \\
CC &= (w \vee t) \wedge (tf \vee dd) \vee cc' \\
DD &= (w' \vee t') \wedge ss' \vee (t \wedge cc) \\
SS &= (w' \vee t') \wedge dd
\end{aligned}
\right\} \quad (1)
$$

$$
\left.
\begin{aligned}
W &= sm' \vee tf' \vee cc \vee dd' \vee ss' \\
T &= sm' \vee tf' \vee cc \vee dd' \vee ss' \\
SM &= (w' \vee t) \wedge tf \\
TF &= (w' \vee t) \vee (sm \vee cc) \wedge t \vee tf' \\
CC &= (w \vee t) \wedge (tf \vee dd) \vee (cc' \vee ss') \\
DD &= (w \vee t) \wedge (ss \vee cc) \\
SS &= (w \vee t) \wedge (dd \vee cc)
\end{aligned}
\right\} \quad (2)
$$

The Boolean matrix corresponding to the constrained sediment influx scenario indicates one stable end state given by {1,1,0,1,1,1,0} in Table 2. In this state, the estuary has little or no salt marshes and spit. The estuary is unable to retain these geomorphological features due to the lack of external sediment.

Now, consider the initial state where the estuary is wave dominated and spit-enclosed. Rising sea level increases wave and tidal forcing within the estuary system. Accordingly, the estuary evolves into a state where there are little or no salt marshes and fragmented or weak spit and has shallow channels. Evolution continues by the estuary then reaching a state with deep channels and little or no inter-tidal flats. Once the estuary reaches this state, the Boolean matrix indicates a reversal of its state into the previous state thereby following a cyclic sequence of evolution between two

Table 2. Boolean matrix for drowned coastal plain estuary subjected to change in environmental forcing (S–Stable State, WD–Wave Dominated, TD–Tide Dominated, [c]–constrained sediment supply, [a]–abundant sediment supply).

| Description | Initial State | | | | | | | End State | | | | | | |
|---|---|---|---|---|---|---|---|---|---|---|---|---|---|---|
| | w | t | sm | tf | cc | dd | ss | W | T | SM | TF | CC | DD | SS |
| S[c] | 1 | 1 | 0 | 1 | 1 | 1 | 0 | 1 | 1 | 0 | 1 | 1 | 1 | 0 |
| TD[c] | 0 | 1 | 1 | 1 | 1 | 1 | 1 | 1 | 1 | 0 | 1 | 1 | 1 | 0 |
| WD[c] | 1 | 0 | 1 | 1 | 1 | 1 | 1 | 1 | 1 | 0 | 1 | 0 | 1 | 0 |
| | | | | | | | | | | | ↕ | | | |
| | | | | | | | | 1 | 1 | 0 | 0 | 1 | 0 | 0 |
| S[a] | 1 | 1 | 1 | 1 | 1 | 1 | 1 | 1 | 1 | 1 | 1 | 1 | 1 | 1 |
| TD[a] | 0 | 1 | 1 | 1 | 1 | 1 | 1 | 1 | 1 | 1 | 1 | 1 | 1 | 1 |
| WD[a] | 1 | 0 | 1 | 1 | 1 | 1 | 1 | 1 | 1 | 1 | 1 | 1 | 1 | 1 |

states as shown in Table 2. When the estuary is tide dominated, lack of external sediment drives the estuary into the stable end state {1,1,0,1,1,1,0}.

Let us now consider that the estuary has an abundant supply of external sediment. The Boolean matrix shows one stable end state with healthy salt marshes, inter-tidal flats, spit and delta. The estuary eventually reaches this end state irrespective of the type of dominant forcing. When the estuary has a sufficient supply of external sediment, the estuary is able to maintain and up-keep its original geomorphological characteristics against the rising sea level.

In most of the estuaries in the UK, the fluvial flow is extremely small compared to the tidal flow. Therefore, under normal circumstances, the effect of change in fluvial flows is not assumed as creating a significant impact on the estuary system. However, a significantly high river flows can be experienced as a result of intense storms. Therefore the system was analysed with increased river flow as an external environmental forcing.

The effect of river flow on waves and tides is minor compared to the forcing themselves. Therefore we assume that those are not directly affected by the river flow. Elevated river flow may contribute to erosion of salt marshes and inter-tidal flats. Both delta and spit are outer estuarine features and mostly wave and tide driven. Therefore, any effect from the river flow is considered insignificant compared to the continual or cyclical changes associated with waves and tides. Tidal channels allow transmission of river flow as well as tidal flows. Therefore, increased river flows tend to erode tidal channels. Figure 3 shows the Boolean network for sediment deficient drowned coastal plain estuary with increased river flow.

The analysis of a sediment deficient estuary system with increased river flow gave identical outputs to no-river-flow case when the estuary is tide dominated. When the estuary is wave dominated, it finally

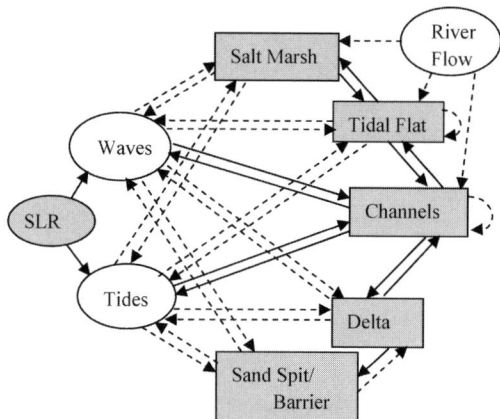

Figure 3. Boolean network for sediment deficient coastal plain estuary with increased river flow.

reached the cyclical state identical to the one without river discharge but the intermediate states encountered during the evolution were slightly different. Similar results were predicted for an estuarine system with an abundant supply of external sediment.

Human interference is considered in the form of estuarial defences and flood defence structures and channel and/or basin dredging. The effects of human interference on the evolution of estuary in the long-term mainly depend on the type and extent of interference and the size and nature of the estuary. In this study, we choose two forms of human interference which could most likely alter the evolution and stability of the estuary:

i. Construction of a training wall at the mouth of the estuary to maintain a navigation channel
ii. Dredging navigation channels and turning basin.

### i. Construction of a training wall at the mouth of the estuary

Training walls are sometimes constructed at the mouth of an estuary to control channel migration at an estuary mouth. They in general have an impact on the lower estuary. It is considered that inner estuarine elements such as inter-tidal flats and salt marshes are not affected by the structure. Tidal channels have a positive feedback from the training wall as it controls channel migration and sedimentation. Delta and spit have negative feedback from the training wall due to curtailed sediment circulation and exchange. Infiltration of tidal and wave energy into the estuary is reduced by the structure thereby enforcing a negative feedback on both forcing. These feedbacks are incorporated into the network shown in Figure 2, in a similar manner to that for river flow.

When the estuary does not have a supply of external sediment, a single stable state is predicted, and that is identical to that without the structure. The estuary stabilises with little or no salt marshes and fragmented or decayed spit. In both tide and wave dominated situations, the evolution of the estuary is identical to that without the structure. This clearly shows that when the estuary does not have an external sediment influx, construction of a training wall would not have a significant impact on the long term geomorphology of the estuary.

In a system with abundant supply of external sediment, interference at the mouth of the estuary by a construction of a structure could change external sediment influx and sediment dynamics within the system. However, the Boolean matrix in this case predicted a stable end state identical to that without the training wall for both tide and wave dominated cases. Importantly, the estuary went through different intermediate states before reaching the stable end state. This result suggests that even though the structure applies a significant impact on the system in the short to medium term time scale, it does not change the long term morphodynamic response of the system to natural forcing.

### ii. Dredging navigation channels and basin

Dredging is needed to create artificial navigation channels to enhance existing channels for navigation needs and/or to deepen ship turning basins. The two cases, abundant and constrained supply of external sediment are considered separately. In both cases, dredging was given an individual node in the network as an outlier. Dredging enforce a positive feedback to tidal channels and a negative feedback on inter-tidal flats. Any effect on delta and spit is considered negligible. Wave and tidal forcing is taken to have a positive feedback from dredging considering the fact that deep channels allow more wave and tidal forcing into the system.

When the system is sediment deficient, both wave and tide dominated estuaries evolved identically to those without dredging. Similar observations were found for the system with an abundant supply of external sediment. But, the intermediate states through which the estuary evolves before reaching the stable end state are different to those without dredging. These results indicate that localised dredging is unlikely to make an irreversible impact on the long term geomorphology of the estuary but may change the short to medium term geomorphological response.

According to the Boolean matrix given in Table 2, the estuary switches between two end states when it is wave dominated and has constrained supply of external sediment $\{1,1,0,1,0,1,0\} \leftrightarrow \{1,1,0,0,1,0,0\}$. At the first end state, the estuary has shallow channels $\{1,1,0,1,0,1,0\}$. If dredging is done at this stage to deepen shallow channels then, the state of the estuary transforms into $\{1,1,0,1,1,1,0\}$, which is the stable end state predicted by the Boolean matrix. This suggests that dredging not only provide navigation needs but may contribute to morphological evolution towards a stable state in some specific cases.

## 3 APPLICATION

The model is then applied to the Deben estuary, UK to predict historic morphological evolution of the estuary during the last two centuries. Figure 4 shows the Deben estuary and its location.

The Deben estuary located in Suffolk, UK is a mesotidal drowned coastal plain estuary with a spring tidal discharge around $2000 \, \text{m}^3/\text{s}$. The estuary has a tidal length of about 18 km and a mean tidal prism approximately $17 \times 10^{-3} \, \text{m}^3$. The mean spring tidal range spatially varies from 3.2 m to 3.6 m. Mean flow of the river Deben is about $0.74 \, \text{m}^3/\text{s}$. Little wave propagation occurs through the mouth of the estuary and the inner estuary experiences only locally generated fetch-limited waves. Coastal sediment circulation is dominated by the southerly direction littoral drift of coarse material.

The morphology of the estuary is characterised by a single meandering channel through mud flats and salt marshes. The main inlet is approximately 180 m wide. The inlet region has a landward flood tidal delta (Horse Sand) and a seaward ebb tidal delta (The Knoll). A single bar/spit extends from time to time from the Bawdsey foreland.

An extensive study of morphodynamic behaviour of Deben estuary has been reported by Burningham & French (2006). Ordnance Survey maps, Admiralty charts and Trinity House/Harwich Haven Authority Surveys have been used. Burningham & French (2006) have identified comparable states of estuary morphology such as extended up-drift shoal and fragmented delta. The survey frequency before the latter part of the 19th century was rather patchy, but as the survey frequency increased in the late 19th century, the data

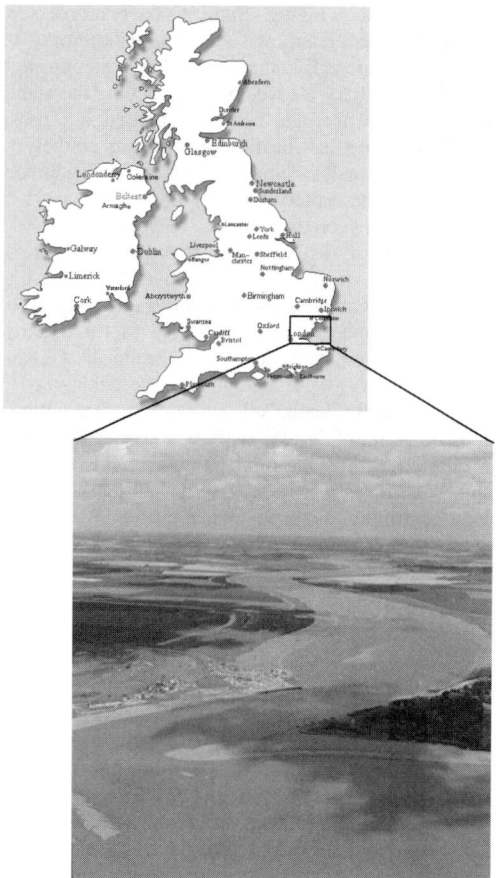

Figure 4.  Deben estuary and its location (Suffork County Council).

Therefore, feedback between them is negative. Similarly, salt marshes and tidal flats force tidal currents to slow down thereby enforcing a negative feedback on tidal forcing. Channel deepening and widening take place due to the passage of high tidal currents. On the other hand, deep and wide channels allow tidal forcing to be carried into the inner estuary. Therefore, they exert positive feedback on each other.

The spit protruding from Bawdsey foreland and the delta are fed by littoral drift. But, the spit and delta exert negative mutual feedbacks by behaving as sediment sinks to each other, obstructing sediment influx through littoral drift. The spit and delta both enforce negative feedback on wave and tidal forcing. Feedbacks between salt marshes, tidal flats and channels are similar those explained in Section 3.2.

Equation 3 gives the Boolean functions derived for Deben estuary.

$$
\left.
\begin{aligned}
W &= sm' \vee tf' \vee cc \vee dd' \vee ss' \\
T &= sm' \vee tf' \vee cc \vee dd' \vee ss' \\
SM &= w' \wedge t \wedge tf \\
TF &= (sm \vee cc) \wedge t \vee (tf' \wedge w) \\
CC &= (w \vee t) \wedge ( tf \vee dd) \vee cc' \\
DD &= w \wedge t \wedge ss' \\
SS &= w \wedge t \wedge dd'
\end{aligned}
\right\}
\qquad (3)
$$

Solution of Equation 3 gives the results shown in Table 3.

Three different initial states of the estuary are examined:

i. Both spit and delta exist
ii. Delta exists but no spit
iii. Spit exists but no delta.

For all three initial states, the estuary evolves in a cyclic sequence with two end states. In one of these states, both the delta and the spit exist. In the other state, both delta and spit are fragmented or decayed. In both end states, the estuary has only little or no salt marshes.

The Boolean model captured the cyclic sequence of fragmentation and reformation of the delta and the spit found in the historical records presented by Burningham & French (2006). This form of morphological evolution provides a severe test of any morphological model, involving a quasi-equilibrium consisting of a continual switching between a set of different states. The Boolean model has captured the essential essence of the morphological behaviour, providing very good qualitative agreement with the historic evolution of the Deben estuary.

provided greater insight into the morphological evolution of the estuary. Burningham & French (2006) found evidence of repeated morphological configuration of the estuary. They defined three characteristic system states of the estuary:

i. shortened, breached and degraded bar
ii. dominant up-drift long-shore bar
iii. over extended prograded bar

The Boolean model is applied to the Deben estuary to predict its historic morphological behaviour. The estuary is modelled as a drowned coastal plain as explained in Section 3.2. The sediment influx into the inner estuary is constrained. But the morphological behaviour of the delta and the spit in the Deben estuary is largely governed by wave and tidal forcing and littoral drift.

Salt marshes and tidal flats tend to erode due to increased wave energy associated with sea level rise.

Table 3. Boolean matrix for morphological evolution of Deben estuary, UK.

| Estuary State | w | t | sm | tf | cc | dd | ss |
|---|---|---|---|---|---|---|---|
| Both Spit and delta exist at initial state | 0 | 1 | 1 | 1 | 1 | 1 | 1 |
| | | | | ↓ | | | |
| | 1 | 1 | 1 | 1 | 1 | 0 | 0 |
| | | | | ↓ | | | |
| | 1 | 1 | 0 | 1 | 1 | 1 | 1 |
| | | | | ↕ | | | |
| | 1 | 1 | 0 | 1 | 1 | 0 | 0 |
| Spit does not exist at initial state | 0 | 1 | 1 | 1 | 1 | 1 | 0 |
| | | | | ↓ | | | |
| | 1 | 1 | 0 | 1 | 1 | 1 | 1 |
| | | | | ↕ | | | |
| | 1 | 1 | 0 | 1 | 1 | 0 | 0 |
| Delta does not exist at initial state | 0 | 1 | 1 | 1 | 1 | 0 | 1 |
| | | | | ↓ | | | |
| | 1 | 1 | 1 | 1 | 1 | 0 | 1 |
| | | | | ↓ | | | |
| | 1 | 1 | 0 | 1 | 1 | 1 | 1 |
| | | | | ↕ | | | |
| | 1 | 1 | 0 | 1 | 1 | 0 | 0 |

## 4 DISCUSSION AND CONCLUSIONS

The morphological response of a coastal plain estuary to rising sea level depends on the type of external forcing and external sediment influx. If the estuary is sediment deficient, a tide dominated coastal plain estuary reaches a stable end state with little or no salt marshes and fragmented or decayed spit. Hence the estuary is no longer spit-enclosed. If the estuary is wave dominated, it eventually evolves into a cyclic sequence between two states: In one state the estuary has little or no salt marshes and spit and has shallow tidal channels. In the other state the channels remain deep but other geomorphological elements are in recession. However, the state of the elements depends on the rate of decay. In both end states the spit is decayed or fragmented but the delta is bound to change between fragmentation and reformation.

When the estuary has an abundant supply of external sediment, the stable end state predicted by the Boolean matrix has all geomorphological elements in place. At rising sea level, both wave dominated and tide dominated coastal plain estuary reaches this stable state after going through few intermediate states. Coastal plain estuaries have well defined deep channels and extensive inter-tidal flats. If the estuary has a sufficient sediment influx, then all geomorphological elements receive enough sediment to replenish

against changing environmental forcing. Therefore, the estuary reaches the stable end state predicted by the Boolean matrix irrespective of the type of dominant forcing.

The effects of change in river flow due to global climate change on estuary morphology are found to be minimal when compared to the effects of sea level rise. This is explained by the fact that river flows in general are a few orders of magnitudes smaller than the tidal flows.

Investigation of human interference on a coastal plain estuary shows that localised structural construction or dredging do not have a significant impact on the long term morphological behaviour of the estuary. The effects of human interference are mostly short to medium term. However, in reality the time scales of these effects depend on the scale of the interference and dimensions of the estuary.

The application of the method to predict historic morphological evolution of a coastal plain estuary in nature (Deben estuary, UK) have shown that the modelling approach is very capable of capturing key attributes of long term evolution of estuary morphology.

REFERENCES

ABP Marine Environmental Research Limited, UK, 2005, ESTSIM Behavioural Statements Report FD2117, Project Report 2 (PR2).

Burningham, H. and French, J., 2006, Morphodynamic behaviour of a mixed sand-gravel ebb-tidal delta: Deben estuary, Suffolk, UK, Marine geology, No.225, 23–44.

De Vriend, H.J., Capobianco, M., Chesher, T., de Swart, H.E., Latteux, B. and Stive, M.J.F., 1993, Approaches to long term modelling of coastal morphology: A Review. Coastal Engineering, 21, 225–269.

Karunarathna, H. and Reeve, D., A Boolean approach to prediction of long-term evolution of estuary morphology, J. Coastal Res. (2007).

Lanzoni, S. and Seminara, G, 2002, Long term evolution and morphodynamic evolution of tidal channels, J. Geophysical Research, 107 (C1), 1–12.

Lee, S.C. and Mehta, A.J, 1997, Problems in characterising dynamics of mud shore profiles, J. of Hydraulic Engineering, 123(4), 1–11.

Nicolis, C., 1982, A Boolean approach to climate dynamics. Quarterly Journal of Royal Meteorological Society, No.108, 707–715.

Pethick, J.S., 1994, Estuaries and wetlands: function and form., In: Wetland Management, Thomas Telford, London, 75–87.

Spearman, J.R., Dearnaley, M.P. and Dennis, J.M., 1997, A simulation of estuary response to training wall construction using a regime approach, Coastal Engineering, 33, 71–89.

*River, Coastal and Estuarine Morphodynamics: RCEM 2007 – Dohmen-Janssen & Hulscher (eds)*
*© 2008 Taylor & Francis Group, London, ISBN 978-0-415-45363-9*

# Numerical modeling of the bed evolution downstream of a dike in the Gironde estuary

Nicolas Chini & Catherine Villaret
*EDF/LNHE, Chatou, France*

ABSTRACT: Forecasting morphological impacts of engineering constructions such as dykes is an important issue to prevent accretions or erosions of the sea bed. In this paper, a process-based finite element numerical model is applied to the central part of the Gironde estuary. Its capability to represent the medium-term (1 year) bed evolution after the construction of a submerged dike is assessed. An efficient technique for morphological updating based on the mean (cycle-averaged) solid transport is presented.

## 1 INTRODUCTION

Sediment transport rates and resulting bed evolutions can be problematic for end users with conflicting interests. We illustrate here how a process-based morphological model can be used as an operational tool:

– to reproduce the observed bed evolution, with a good accuracy and an efficient extrapolation method for long-term computation,
– to differentiate natural bed evolutions from human interactions,
– to test solutions to prevent the unwanted deposit.

The zone of interest, presented in the second part of this paper, is the central part of the Gironde estuary, which is characterized by complex estuarine hydrodynamics, strong tidal currents, large sediment transport rates and the presence of sand-mud mixture.

A large deposit formation, associated with sand bank migration, is observed since 1994 on bathymetric surveys downstream of a submerged dike. The objective of this construction is to arise the hydraulic power of the navigation channel and reduce the dredging operations. The formation of this deposit however inhibits the dilution of the outflow of a nuclear power plant, located 5 km downstream of the dike.

In this paper, the 2D morphological model Sisyphe which is part of the Telemac finite element system, is applied to represent the bed evolution in the vicinity of the dyke. Sisyphe is internally coupled to the hydrodynamic tidal model Telemac-2d. The model implementation is discussed in Part 3.

Model calibration and comparison with existing data (velocity measurements, differential bathymetry) is presented in Part 4. Different methods are compared

in order to reduce the computational time for long term evolution. The model is then used as an operational tool to study the effect of the dike on the bed evolution, and further propose some solutions in order to prevent the deposit.

## 2 CONTEXT PRESENTATION

The macrotidal Gironde estuary (southwest of France) originates from the confluence of the Dordogne and Garonne rivers, and extends to the Bay of Biscay (see location map, on Figure 1).

We focus our attention to the 40 km long central portion of the estuary, which is about 5 km wide and subdivided into three main channels: the deepest navigation channel is located on the left hand side, with a mean water depth of about 15 m. These channels are separated by a complex pattern of shoals and islands.

In 1994, rocky dredged materials were disposed between the islands of Patiras and Trompeloup (see Fig. 4) creating a porous dike, submerged during high tides.

Bathymetric surveys performed between 1995 and 2000 reveal a drastic deposit downstream of this dyke, as seen on the differential bathymetry (Fig. 2). A shoal has been formed downstream of the Patiras Island, which migrates downstream with a celerity of 360 m/yr.

The Gironde estuary is a well mixed estuary with a bed composed of a mixture of cohesive sediment and fine sand (Allen, 1970).

Sediment sample were collected in March 2006: fine sands with a mean diameter of 210 μm are predominant over shoals, whereas cohesive sediment are

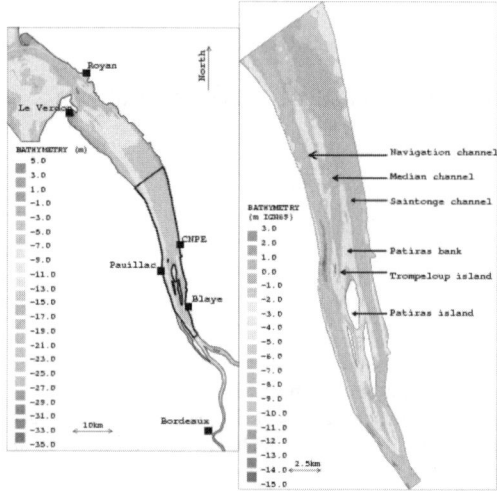

Figure 1. Location map. The left figure shows the entire estuary and the right one, the central part.

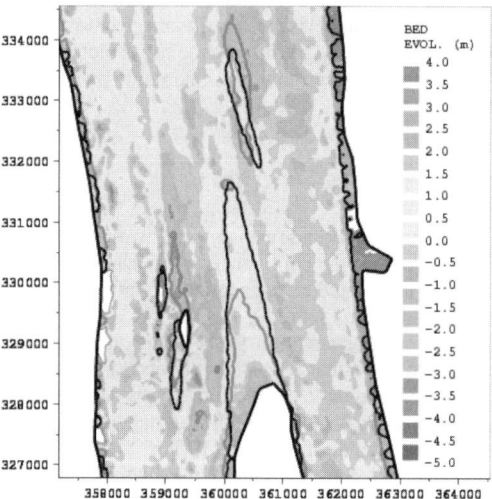

Figure 2. Bed evolution between 1995 and 2000 from bathymetric survey, the red line show the bathymetric iso-line zero in 1995, and the black one is in 2000.

found in the deeper channels (Chini & Villaret, 2007). The position of a turbidity maximum oscillates seasonally and the presence of fluid mud, with a concentration of about 200 g/l, were reported by Migniot (1985).

## 3 HYDRO-SEDIMENTARY MODEL

The finite element Telemac software, developed at EDF-LNHE, is here applied. The hydrodynamic module can be either internally coupled or chained to the

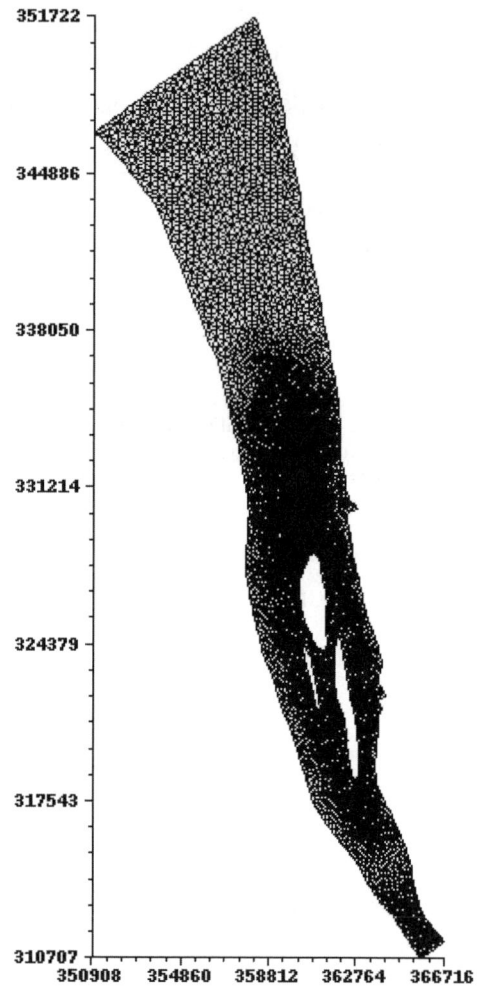

Figure 3. Domain mesh grid, with Npoin = 11500.

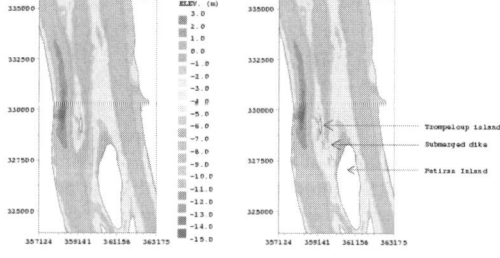

Figure 4. Bathymetries considered (left after excavation of the dike, right with the presence of dike).

morphodynamic module, Sisyphe. In the internal coupling method, the hydrodynamic flow, solid discharge and bed elevation are computed at each time step. In the chaining method, bed evolutions are assumed to be sufficiently small such that the hydrodynamic variables (flow rate and free surface) can be considered to remain constant and are obtained from a previous uncoupled Telemac-2d simulation. This simplified chaining method allows to save computer time, but with a loss of accuracy, sand transport rates and evolutions being generally overestimated: this method is only applied here in the sensitivity analysis.

### 3.1 Grid presentation

The finite element method allows a fine representation of the zone of interest, with a mesh size ranging from 20 m up to 500 m near the open boundary. The total number of nodes, $N_{poin}$, is 11 500.

Two bathymetries are considered: the first one corresponds to the actual situation observed in 1995, with the presence of the dike. The second one is an idealized bathymetry based on the situation of 1995 without the dike. The excavation of the dike with the dredging of the median channel to an elevation of $-6$ m IGN69 represents a volume of 700 000 m³.

### 3.2 Hydrodynamic module

The Telemac-2d hydrodynamic model solves the depth-averaged shallow water equations, and accounts for the presence of tidal flats.

An efficient solver for the wave equation allows to increase the time step and reduce computational time, such that the Courant number can be greater than 1 (Hervouet, 2007). The hydrodynamic model is run here with a time step of 10 s.

The model is forced on the open boundary with water levels and velocities provided from a large scale model of the entire estuary, extending from the Bay of Biscay where fourteen tidal components are imposed. Thus, the model takes account for the most important astronomical tidal waves, which are responsible for residual sand transport (Hoitink, 2003). At the downstream boundary, an average river flow discharge (1100 m³/s) is imposed (Chini, 2007).

For the large scale model, the phases of the different tidal components are calculated in order to reproduce the tidal range of 07/31/2006.

Water level simulated are compared with recomposed water level from field observation, at Richard harbour (Fig. 5).

The velocities calculated by the model are in very good agreement with ADCP velocity measurements collected during one neap-spring tidal cycle, as shown on Figure 6 (Chini & Villaret, 2007).

### 3.3 Morphodynamic module

The morphodynamic model Sisyphe accounts for both bed-load and suspended load: the bedload is calculated based on a classical sand transport formula, whereas the suspended load is determined by solving

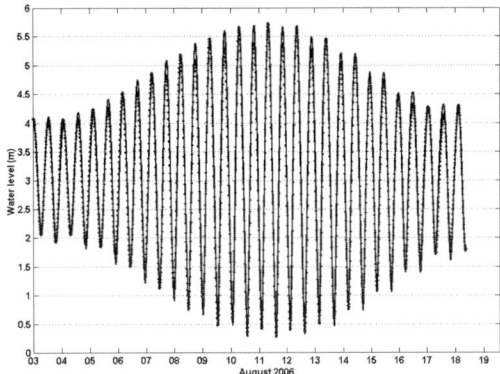

Figure 5. Comparison between water level and model results.

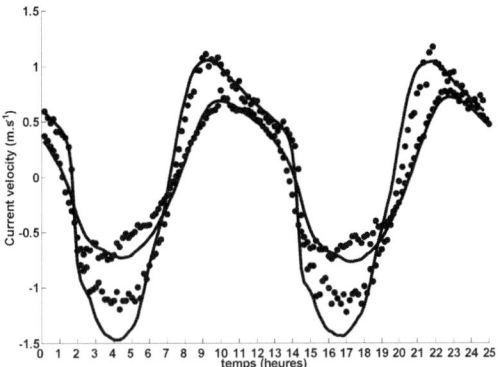

Figure 6. Comparison between velocity measurements and model results in the mid-channel (for a spring tide and for a neap tide). Positive velocities are towards the north. Time $t = 0$ corresponds to low tide.

an additional transport equation for the depth-averaged suspended sediment concentration.

A correction method is applied to the convection velocity in order to account for vertical gradients in velocity and concentration profiles (Villaret, 2005; Villaret & Hervouet, 2006).

The bed evolution is obtained by solving the continuity equation:

$$(1-n)\frac{\partial z_b}{\partial t} + \nabla \cdot \vec{Q}_b = D - E \tag{1}$$

where $n$ is the bed porosity, $z_b$ is the bottom elevation, $Q_b$ is the volumetric sediment bed-load transport per unit width. The term $D$-$E$ represents the net fluxes at the bed of deposition minus erosion.

For non-cohesive sediments, the source and sink terms in eq. (1) are calculated by:

$$E - D = W_s\left(C_a - C_{ref}\right) \tag{2}$$

Table 1. Physical parameters and numerical options.

| Parameters | Values |
|---|---|
| Mean diameter ($\mu$m) | 210 |
| Bed load formula | Bijker |
| Reference concentration | Bijker |
| Correction on convection velocity | Yes |
| Reference elevation (mm) | 0.42 |
| Strickler coefficient ($m^{1/3}$ $s^{-1}$) | 50 |
| Bed porosity | 0.4 |

where $W_s$ is the settling velocity, $C_{ref}$ the reference concentration, and $C_a$, the near bed concentration which can be extrapolated from the depth-averaged concentration, by assuming a Rouse profile for the concentration.

Equation (1) is solved with a finite volume method. The finite volume formulation enables a correction of the sediment transport rates when rigid bed is present, ensuring that the rigid bed is not eroded (Gonzales de Linares, 2003). This feature is particularly important in the present application, when rigid constructions such as dikes are present in the domain of interest.

### 3.4 Choice of sediment transport parameters and formula

Although the bed composition is highly variable in space, fine sandy sediments are predominant on the shoals, which is the key issue here.

The following table gives the values of the important sand transport parameters, which are determined after a set of sensitivity tests (Chini and Villaret, 2007). The bed-load and suspended load are calculated using the Bijker's formula (1968). According to this method, the reference elevation is set to

$$a = \max{(k_s, 2D_{50})}$$

where $k_s$ is the equivalent bed roughness, and the reference concentration is expressed as a function of the bed load transport rate.

### 3.5 Long term modeling

#### 3.5.1 Extrapolation (method 1)

The most trivial method consists of extrapolating the evolution calculated after one neap-spring tidal cycle, assuming that the bed evolution is linear. For instance to calculate the 12-month bed evolution, one should multiply the evolution obtained after one neap-spring tidal cycle (fortnight period 14 days) by a factor 24. This method presumably overestimates the bed evolution, since the natural bed evolution progressively slows down and tends towards a dynamic equilibrium.

#### 3.5.2 Morphological factor (method 2)

In the case of internal coupling, Roelvink (2006) proposes to accelerate the morphological evolution by multiplying by a 'morphological factor' the hydrodynamic time step when solving equation (1). This method is equivalent to lengthening the tide, as proposed by Latteux (1995). If the hydrodynamic model simulates one neap-spring tidal cycle, one should multiply by 24 the hydrodynamic time step in order to reproduce a 12-month evolution.

According to Roelvink (2006); the choice of the morphological factor should be limited by the criterion:

$$N < \frac{h\Delta x}{pQ_{max}Dt} \qquad (3)$$

where $h$ is the water depth, $\Delta x$ the size of the mesh, $p$ the power of solid transport law, $Q_{max}$ the maximum solid transport value and $Dt$ the time step. In the presence of tidal flats, this relation gives a relatively small value for N.

In our case, the maximum value of N should be less than 12. We use here $N = 12$ to represent a 6-month evolution, thanks to a hydrodynamic computation during one neap-spring tidal cycle, and repeat the operation twice in order to compute the 12-month bed evolution.

#### 3.5.3 Morphological tide (method 3)

Another way to compute the long-term bathymetric evolution consists of determining the tidal range which gives the mean solid transport, time-averaged during one neap-spring tidal cycle. According to Latteux (1995), this so-called morphological tide does not correspond necessarily to the mean tide and has generally a tidal range 5 to 10% higher than the mean tidal range. This is due to the fact that the solid transport is linked to the flow by a non-linear equation.

The mean solid transport, during one neap-spring tidal cycle $q_C$ (period $T_c$ of 14 days) is:

$$q_C(x, y) = \frac{1}{T_C}\int_0^{T_C} Q_b(t, x, y)dt \qquad (4)$$

while the mean solid transport during one tide $q_T$ is the summation of the solid transport during one tidal period $T$ of approximately 12 hours.

To determine this morphological tide, one needs to calculate $q_T$ for all the different tidal coefficients of a given spring-neap cycle, and then compare each value of $q_T$ to $q_C$. According to the French Hydrographic Service, the 'tidal coefficient' is defined as the ratio of the semi-diurnal tidal amplitude over the mean spring equinox tidal amplitude in Brest.

In order to do these comparisons, we introduce an error parameter ERR, which is defined as the root

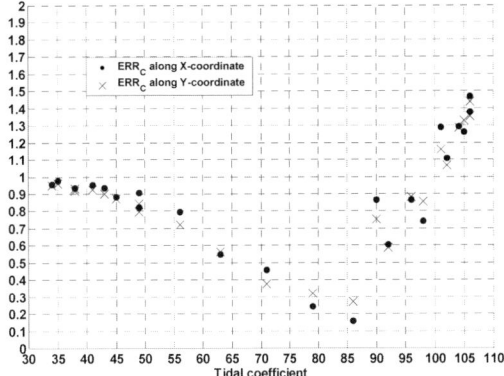

Figure 7. Residual transport rates; error critera versus tidal coefficient.

mean square error normalized by the root mean square value of $q_C$ over the whole domain:

$$ERR(T) = \left( \frac{\sum_{i=1}^{N_{poin}} (q_T(i) - q_C(i))^2}{\sum_{i=1}^{N_{poin}} q_C(i)^2} \right)^{1/2} \tag{5}$$

where $N_{poin}$ is the total number of grid points.

The mean error $ERR$ is plotted versus tidal coefficient on Figure 7, for each components of the solid transport. The morphological tide corresponds to the minimum value of $ERR$. This minimum is reached for a tidal coefficient of 86, which corresponds to a tidal range about 10% higher than the mean tide, in agreement with Latteux (1995).

The morphological tide is supposed to have a magnitude of $q_T$ close to $q_C$.

$$\frac{1}{T_c} \int_0^{T_c} Q_b(t,x,y)dt \approx \frac{1}{T} \int_0^{T} Q_b(t,x,y)dt$$

where $T$ is the period of the morphological tide.

Considering only bed-load, equation (1) can be discretised with time step $Dt$. After a summation during one neap-spring tidal cycle, the final bed evolution can be related to the divergence of the mean time-averaged residual transport rate during one cycle.

The summation over all time steps leads to:

$$z_b^{N_c} - z_b^0 = -\frac{Dt}{(1-n)} \sum_{i=1}^{N_c} \nabla \cdot \vec{Q}_b^i$$

where $N_c$ is the total number of iterations to compute one neap-spring tidal cycle.

$$z_b^{N_c} - z_b^0 = -\frac{Dt}{(1-n)} \cdot \nabla \cdot \sum_{i=1}^{N_c} \vec{Q}_b^i \approx -\frac{T_c}{(1-n)} \cdot \nabla \cdot q_C$$

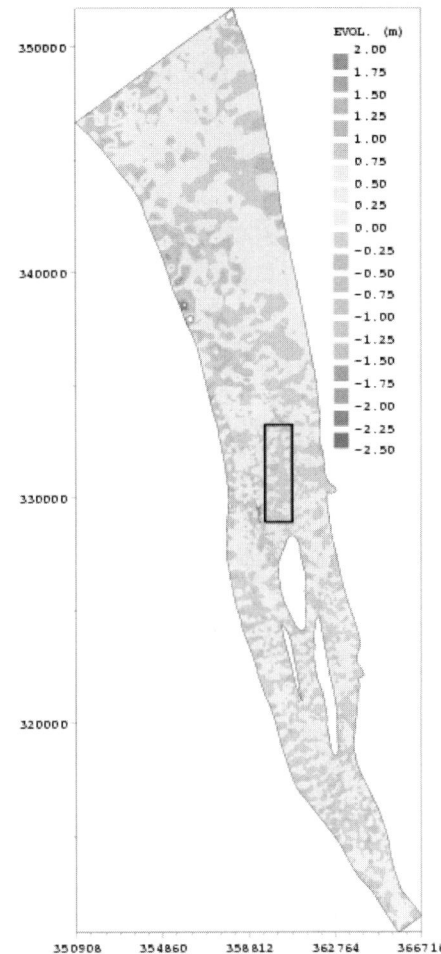

Figure 8. Bed evolution after one year, for the reference case. Zone B represented by black rectangle.

According to the previous definition of the morphological tide, it comes:

$$\frac{1}{N_c} \sum_{i=1}^{N_c} \vec{Q}_b^i \approx \frac{1}{N_T} \sum_{i=1}^{N_T} \vec{Q}_b^i$$

where $N_T$ is the number of iterations to compute one morphological tide. Then:

$$z_b^{N_c} - z_b^0 = -\frac{N_c}{N_T} \frac{Dt}{(1-n)} \cdot \nabla \cdot \sum_{i=1}^{N_T} \vec{Q}_b^i \tag{6}$$

The last expression shows that the bed evolution after one neap-spring tidal cycle should be computed thanks to only one run performed during one morphological tide. The bed evolutions computed after

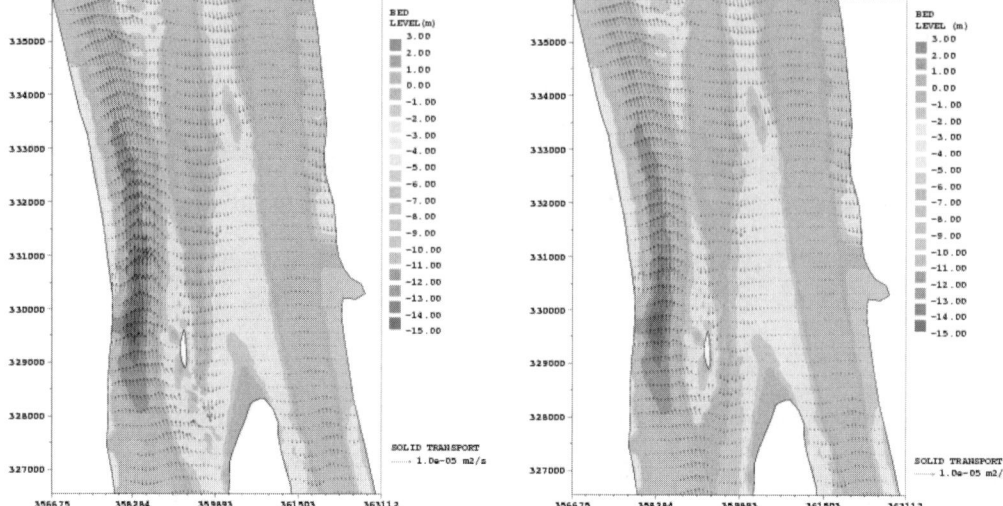

Figure 9. Residual sand transport rates (left: in presence of dike, right: after excavation of the dike.

one morphological tide can be multiplied by the factor $N_c/N_T = 28$ (since there are approximately 28 tides during one neap-spring tidal cycle) in order to represent one neap-spring tidal cycle.

The objective is to represent one year of bed evolution, which can be achieved by computing 24 morphological tides and multiplying the final bed evolution after each morphological tide by a factor $N_c/N_T = 28$.

## 4 RESULTS AND INTERPRETATION

### 4.1 Intercomparison of long term methods

The three different methods discussed above are implemented to calculate a one-year bed evolution. The results are compared to a complete reference run computed by internal coupling between both the hydrodynamic and morphodynamic modules. Figure 9 represents the final bed evolution obtained for the reference run.

We introduce the following criterion:

$$\varepsilon = \left( \frac{\displaystyle\sum_{i=1}^{N_{poin}} (B(i) - B_{ref}(i))^2}{\displaystyle\sum_{i=1}^{N_{poin}} B_{ref}(i)^2} \right)^{1/2} \quad (7)$$

where $N_{poin}$ is the number of points in the domain, $B$ is the final bottom elevation and $B_{ref}$ is the final bottom elevation of the reference run. This error parameter

Table 2. Intercomparison of the long-term methods for 6-month evolution.

| Method | CPU time | Criterion $\varepsilon$ | |
| --- | --- | --- | --- |
| | | Zone A | Zone B |
| Reference run | 10,4 days | | |
| (1) Linear Extrapolation Factor 24 | 10 hours | 16.7 | $3.10^{-2}$ |
| (2) Lengthening of the tide Factor 12 | 21 hours | 1.46 | $1.2.10^{-2}$ |
| (3) Morphological tide and linear extrapolation Factor 28 | 7 hours | 3.65 | $6.10^{-3}$ |

gives the relative magnitude of the errors for the final bed evolution.

In order to evaluate the accuracy of the different long-term methods, we calculate the error criterion (7) in two zones:

– (a) in the whole domain, including the tidal flats and rigid bed where most problems occur,
– (b) in a restricted domain located in the mid-channel, downstream of the dike (Fig. 8), which is concerned with an important accretion (see Fig. 2).

The performance of the different methods are summarized in Table 2. The use of a morphological tide (method 3) gives overall the best results and appears to be less time-consuming than the morphological factor method (2). In agreement with Latteux (1995), the linear extrapolation (method 1) overestimates the bed evolution.

## 4.2 Impact of the dike

The influence of the dyke is assessed by comparing the results obtained with and without the dike (cf §3.1). Deposits and eroded zones should not have the same locations for the different classes. First attempts have been performed using a chained method during one neap – spring tidal cycle.

Figure 8 shows the residual solid transport rates obtained with or without the dike for non cohesive sediment.

The transport of sediment is directed downstream in the lateral channels, and upstream in the median channel, with flood dominated transport. The presence of the dike modifies the intensity of sediment transport, in the median channel and in the navigation channel. The sediment transport is accelerated in the navigation channel thanks to the dike. The simulation shows that the dike is responsible to increase the hydraulic power of the navigation channel. On the other hand, the dike creates a convergence zone, in the median channel, increasing deposits.

## 5 CONCLUSION

The morphological process-based model Sisyphe, internally coupled to the hydrodynamic model Telemac-2d has been applied to simulate the medium-term bed evolution (1 year) in the central part of the Gironde macro-tidal estuary.

A new methodology for long term simulation is found to give reasonably accurate results for a large gain in computer time (a factor 34), in comparison to a complete run. The performance of this new method is overall better than previous long-term methods, based on linear extrapolation or tidal lengthening.

The model could then be used as an operational tool in order to represent the effect of the construction of a submerged dike.

In order to improve the model comparison with measured bathymetric evolutions, the effect of the mud content needs to be accounted for.

REFERENCES

Allen, G. 1972. Etude des processus sédimentaires dans l'estuaire de la Gironde. *PhD Thesis. Universite de Bordeaux 1.*

Chini, N. 2007. Modélisation morphodynamique de la partie centrale de l'estuaire de la Gironde. *Rapport de DRT.* Université de Grenoble 1.

Chini, N. & Villaret, C. 2007. Numerical modelling of the morphodynamic evolution of sand banks in the Gironde Estuary. *Submitted to the 32nd IAHR Congress.*

Gonzales de Linares, M. 2003. Finite volumes and associated rigid bed method in Sisyphe release 5.3. *EDF-LNHE report HP75/03/021.*

Hervouet, J.M. 2007. *Hydrodynamics of free surface flow, finite elements system.* Wiley.

Hoitink, A.J.F., Hoekstra, P. & van Maren, D.S. 2003. Flow asymmetry associated with astronomical tides : Implication for the residual transport of sediment. *Journal of Geophysical Research,* 108, 1–8.

Latteux, B. 1995. Techniques for long-term morphological simulation under tidal action. *Marine Geology,* 126, 129–141.

Mignot, C. 1985. Etude des vases de la Gironde. *Rapport de synthèse.* LCHF.

Roelvink, J.A. 2006. Coastal morphodynamic evolution techniques. *Coastal Engineering,* 53, 277–287.

Villaret, C. 2005. Sisyphe release 5.5 – User manual, *EDF-LNHE report HP76/05/009.*

Villaret, C. & Hervouet, J.M. 2006. Comparaison croisée de différentes approches pour le transport sédimentaire par charriage et suspension, *Proceedings of the National conference on Génie Civil – Génie Côtier, Brest.*

*River, Coastal and Estuarine Morphodynamics: RCEM 2007 – Dohmen-Janssen & Hulscher (eds)*
*© 2008 Taylor & Francis Group, London, ISBN 978-0-415-45363-9*

# Impact of setbacks on the estuarine morphology

M.C.J.L. Jeuken
*WL\Delft Hydraulics, Delft, Netherlands*

Z.B. Wang
*WL\Delft Hydraulics & Delft University of Technology, Delft, Netherlands*

D. Keiller
*Black & Veatch, London, UK*

ABSTRACT: Setting back of defences is considered as a measure in the development of flood defence strategies and to support the creation of new intertidal habitat both in the Humber estuary and the Schelde estuary. The impact of setbacks on the large-scale estuarine morphology was analyzed using the ESTMORF model. Generally, setbacks induce long-term morphological changes landward and seaward of the setback that differ from the initial response. Seaward of the setback the estuary erodes. Landward of the setback sediments are deposited that erode again at larger timescales. At the scale of the entire estuary set backs always cause a loss of sediments. For seaward situated set backs this erosion is smaller and the long-term gain of intertidal area is larger than for landward located set backs. These results can be used in the site selection and design of sustainable setbacks.

## 1 INTRODUCTION

Safety against flooding and a sustainable development of ecological values are prime management objectives in many European coastal plain estuaries. Both sea level rise (*slr*) and human interferences may threaten the safety levels and ecological values of estuaries. In particular, sea level rise may affect the ecological values of estuaries where the 'room for water' is limited. In these estuaries a future decrease of intertidal area is likely because of the so-called coastal squeeze effect related to sea level rise: the available water surface area at high water can only increase a little with increasing mean sea level (limited space), whereas the water surface area at low water can increase much more, thus reducing the intertidal area. This effect of limited storage at high water may overrule the effect of changes in intertidal range and sedimentation in the estuary related to sea level rise. In addition, man interferes in the morphologic and hydrodynamic evolution of estuaries. These interferences often include dredging and dumping of sediments to maintain shipping lanes and harbours as well as land reclamation and sand excavation projects (e.g. harbour development). These activities often affect the safety level and intertidal area in an indirect way and over longer timescales.

Setting back of defences, i.e. creating room for water, is considered as a measure in the development

of new flood defence strategies and to support the sustainable creation of new intertidal habitat, both in the Humber estuary (United Kingdom) and the Schelde estuary (Netherlands and Belgium). Setbacks provide a direct initial gain in intertidal area, thereby potentially increasing the ecological value of the estuary. Depending on the design (size and averaged height) and the exact location, the set backs dissipate the tidal energy resulting in (initially) reduced water levels, thereby increasing the safety against flooding. However, the increase in intertidal basin storage, related to the setback itself, the overall changes in tidal range and the associated changes in tidal volume may affect the large-scale morphology at longer timescales (decades) as well. This paper aims to identify and explain these short-term and long-term impacts by analyzing typical setback scenario's for the Humber estuary, using the hybrid (semi-empirical) ESTMORF model. The general effects and the use of these results in the site selection and design of sustainable set backs will be discussed in section 5. Herein the experiences for the Schelde estuary will also be briefly addressed.

## 2 STUDY AREA

The Humber estuary is a tide-dominated estuary that is situated along the east coast of England (Figure 1).

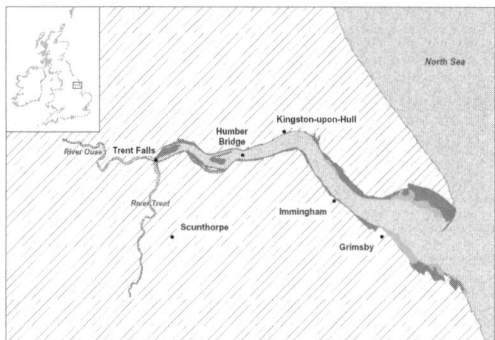

Figure 1. The Humber estuary.

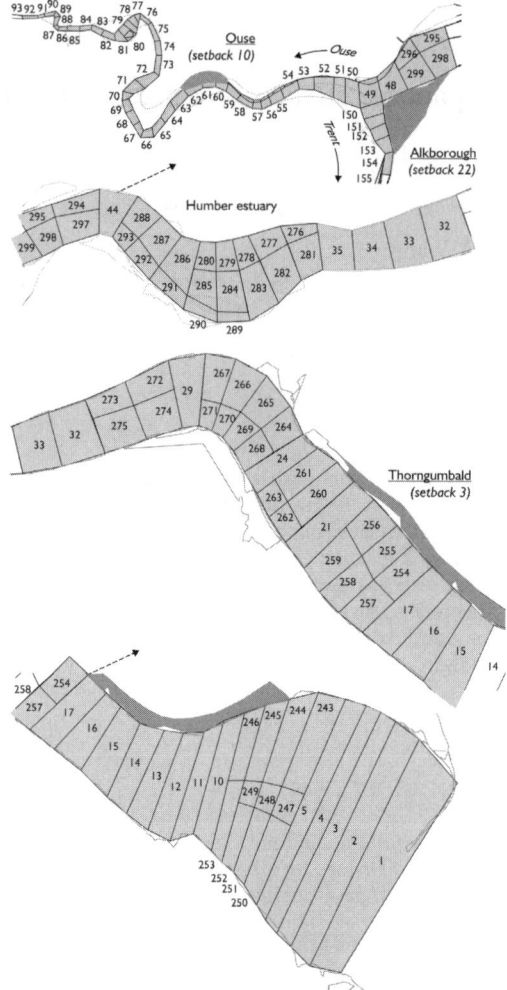

A macro-tidal regime, with a mean tidal range of 4.4 m near the mouth and average fresh water flows of 250 m³/s, characterize the estuary. The surface area of the estuary approximates 300 km². The tidal prism at averaged tide is in the order of 1300 million m³, about 115 times the averaged river inflow during a tidal period, confirming the tide-dominated character of the flow. The Humber estuary exhibits a morphologically dynamic system of parallel-aligned tidal channels that are separated and bordered by about 9700 ha of muddy inter-tidal shoals. For recent overviews of the large-scale morphologic evolution of the estuary, see e.g. Townend and Whitehead (2003) and Townend et al. (2007).

## 3   METHODS

### 3.1   The ESTMORF model

The numerical ESTMORF model was used to investigate the morphological impact of setbacks at the decadal timescale. This model was developed to predict the impact of changes such as sea-level rise and human intervention (e.g. dredging and dumping, land reclamation). It is often referred to as a hybrid (semi-empirical) model, combining the description of physical processes (hydrodynamics and sediment transport) and empirical relations for morphological equilibrium, to obtain a description of the long-term morphological development. The model uses a one-dimensional hydrodynamic network model to simulate the tidal flow which is, in turn, used to determine the morphological equilibrium of both channels and inter-tidal areas. The deviation of the actual morphological state with respect to the morphological equilibrium together with the tidal flow characteristics determines the sediment transport and the morphological changes. A more detailed description of the ESTMORF model can be found in Wang et al. (1998, 1999).

Figure 2.   Network schematization of the ESTMORF model. The locations of the considered setbacks are also indicated. For the areas where two or more channels are present (the section numbers larger than 200) the hydrodynamic results are shown for the sections that border the northern bank of the estuary. In the figures the numbers at the x-axis are subsequently numbered (e.g. section number 243 is displayed as section number 6).

The model was used to predict the likely changes in tidal wave propagation and the associated large-scale changes in sediment budget and intertidal area over the next 50 years. Within the estuary, an existing network schematization (Wang and Roelfzema, 2001) was extended by identifying the individual large channels (Figure 2) and including the river reaches of the Ouse and Trent. The definition of the estuary bed was taken from the 2000 bathymetric charts. At the downstream boundary near Spurn Head spring tidal water levels are prescribed. At the landward end long-term mean discharges are prescribed ($110\,\text{m}^3/\text{s}$ for the Ouse and $136\,\text{m}^3/\text{s}$ for the Trent).

Detailed information of long-term erosion and sedimentation at the spatial scale of individual channels was not available for the present study. Therefore the calibration was focused at larger spatial scales, i.e. the estuary as a whole and certain regions. The model was calibrated in such a way that at the start of the simulation with a sea level rise of $0.18\,\text{mm/yr}$ the overall sedimentation in the estuary up to Trent Falls approximates the observed value of about $0.1\,\text{Mm}^3/\text{yr}$ (Townend and Whitehead, 2003). This small value indicates that the sediment budget of the estuary is finely balanced. This net sediment deposition rate is smaller than the water volume increase induced by sea level rise. This is due to the additional storage at high water and an increase in tidal range resulting in a larger tidal volume and hence a larger equilibrium cross-sectional area (Townend et al., 2007).

## 3.2 Model simulations

Various setback scenario's were analyzed to investigate the short-term and long-term (50 yrs) impact of setbacks on the large-scale morphologic and hydrodynamic evolution. The results of three representative scenario's are presented in this paper, viz. (Figure 2):

1 A landward located setback (35 ha) along the river Ouse.
2 The landward located setback at Alkborough (311 ha) near the estuary head at Trent Falls. This setback was implemented in 2006.
3 The seaward located setback near Sunk Island (808 ha).

In all three model simulations a future sea level rise of $0.18\,\text{mm/yr}$ was applied. The model results were used in a relative sense before more detailed hydrodynamic modeling of individual sites. As an example, the effect of a certain setback was analyzed by comparing the results of the simulation with the setback and the baseline simulation without the setback, determined during the calibration. The following aspects were considered in the analysis:

– The temporal changes of the high water level and low water level and the resulting changes in tidal range. The changes in the high water level gives a first indication of the effect of the set back on flood control. For complete elaboration of these effect the normative storm (surge) conditions should be considered.
– In addition to the vertical tide, the changes in ebb volume (tidal prism) were determined.
– The volume change with respect to a fixed reference level (OD+5 m) to determine the large-scale erosion and sedimentation rates.

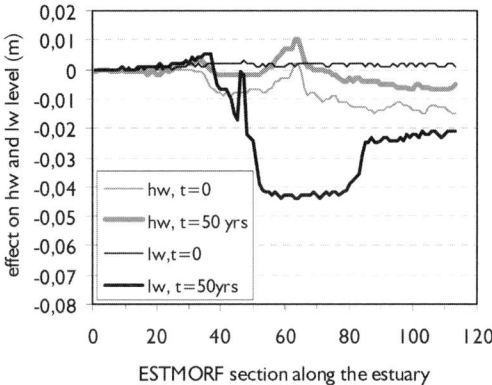

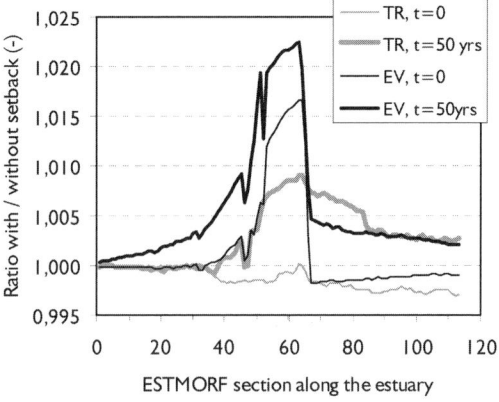

Figure 3.   Initial and long-term (after 50 yrs) effect of the setback on $hw$ and $lw$ (upper panel) and tidal range ($TR$) and ebb volume ($EV$). The setback is located along sections 60 to 63.

– The intertidal surface area, defined as the area between (instantaneous on morphological time scale) low water ($lw$) and high water ($hw$). Basically three mechanisms may cause changes of the intertidal area: (a) sealevel rise and the associated coastal squeeze effect, (b) changes of the tidal range, and (c) sedimentation and erosion processes in the intertidal region.

For presentation purposes the estuary is divided into three regions: the area seaward of the setback, the setback region and area landward of the setback.

## 4   RESULTS

### 4.1 A setback along the river Ouse

Figures 3 to 5 display the effects of the landward located setback, along the river Ouse, on the development of the large-scale hydrodynamics and morphology.

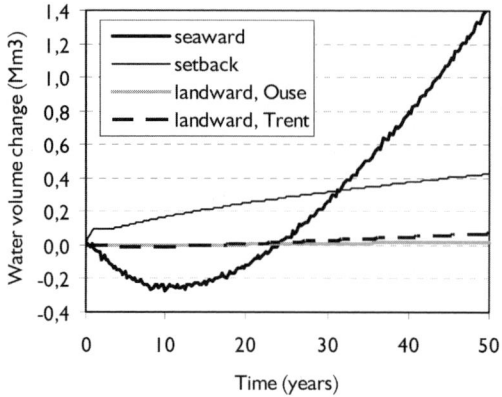

Figure 4. Effect on the change of water volume seaward, in the setback area and landward of the setback in the river Ouse (O) and Trent (T). A negative value implies sedimentation, positive value means erosion.

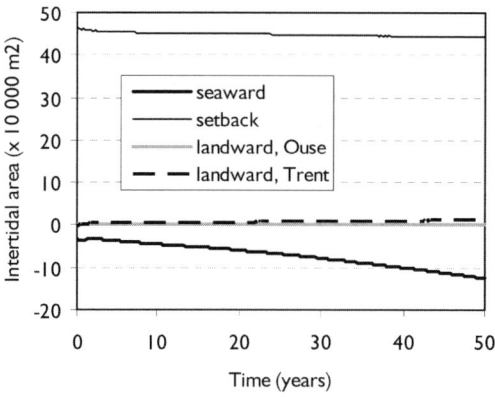

Figure 5. Effect of the setback on the evolution of the intertidal shoal area seaward, in the setback region and landward of the setback.

The initial and long-term effect of the setback on the tidal water levels and volumes differ (Figure 3). Initially the setback causes a lowering of the $hw$ level in the area upstream of section 40 including the setback (Figure 3). Initially, the $lw$ levels rise slightly thus diminishing the tidal range seaward and landward of the setback. Despite the reduction of the tidal range the ebb volume increases initially between sections 50 and 63 as a result of the increase in intertidal basin storage related to the setback. The enlarged tidal volumes induces erosion of this river reach of the Ouse ('setback' in Figure 4). This erosion in turn causes a gradual enhancement of the tidal propagation with higher $hw$ and lower $lw$ levels, i.e. an increase of the tidal range, and an increase of the tidal volumes extending landward and seaward of the setback with time. These

changes in the hydrodynamics explain the large-scale morphological changes (Figures 4 and 5): During the first ten years the reduction of tidal ranges and volumes dominate the evolution of the seaward located estuary part. The area experiences a relative sedimentation and the intertidal area decreases because of the reduced tidal range. After ten years the erosion of the setback region and associated increase of tidal ranges and volumes start to influence this part of the estuary. The estuary erodes and even after 50 years this erosion process has not stabilized. This erosion of the estuary also explains the increased loss of intertidal area seaward of the setback (Figure 5). The area landward of the setback experiences an initial sedimentation that changes into a relative erosion within a couple of years. Despite this erosion the intertidal area hardly changes. This is due to the increased tidal range.

Summarizing, the overall, long-term morphological effect is a relative erosion (loss of sediments), i.e. less sedimentation compared to the situation without setback, and an initial net gain of intertidal surface area that decreases with time (increase of more brackish intertidal area to landward and a loss of saline intertidal area to seaward).

### 4.2 Alkborough setback at the estuary head

The Alkborough setback was commissioned in 2006 (see Figure 2 for its location). The ESTMORF predicted future effects of this setback on the large-scale hydrodynamics and morphology are shown in Figures 6 to 8.

The Alkborough set back induces qualitatively similar changes as the previously discussed setback in the river Ouse. The magnitude of the changes is however larger. Initially the $hw$ level reduces with almost 5 cm landward and just seaward of the setback (Figure 6). The $lw$ level slightly increases. After 50 years the $hw$ level has increased again, whereas the $lw$ level has declined substantially, yielding a large increase in tidal range. The ebb volumes increase throughout the estuary both initially (except landward) and at the long term. This is due to the large initial increase in intertidal basin storage related to the setback and because of the large changes in tidal range. These larger hydrodynamic effects are reflected in the large-scale morphologic evolution (Figure 7): the entire estuary experiences a relative erosion that still continues after 50 years. This means that the morphological timescale of the estuary to adjust to the setback exceeds 50 years. The erosion determines the reduction of the intertidal area seaward of the setback (Figure 8). Landward of the setback, in the rivers Trent and Ouse, the intertidal area increases despite the erosion as a result of the increase in tidal range.

The initial reduction in local $hw$ level was considered likely to adversely effect navigation and

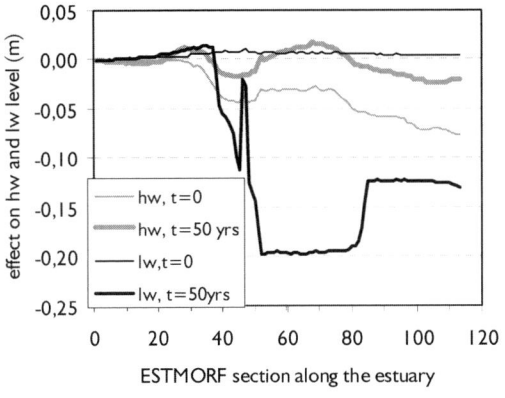

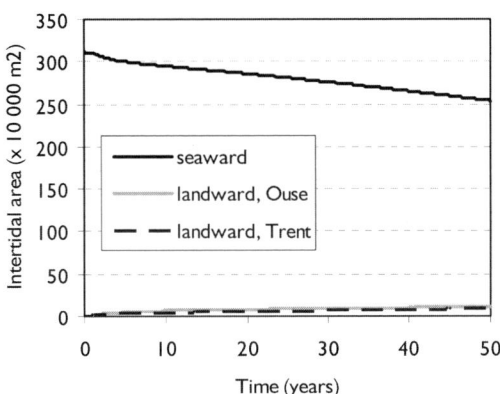

Figure 8.   Effect of the Alkborough setback on the evolution of the intertidal shoal area seaward (including the setback) and landward of the setback in the rivers Ouse and Trent.

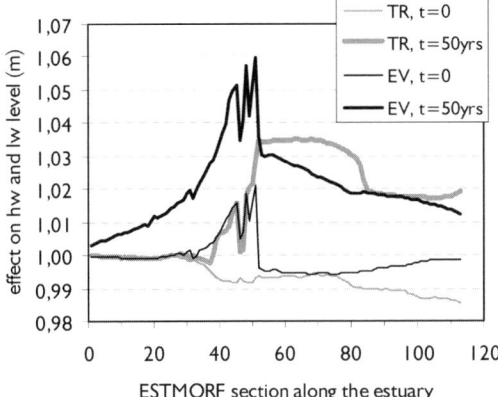

Figure 6.   Initial and long-term effect of the setback on the *hw* and *lw* levels (upper panel), tidal ranges (*TR*) and ebb volume (*EV*) along the estuary. The setback is located near sections 48 and 49 (see also Figure 2).

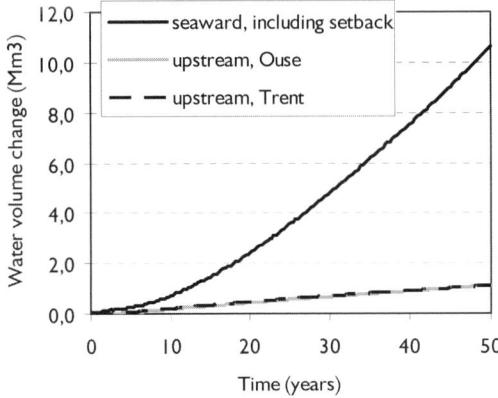

Figure 7.   Effect of the setback on the large-scale change of the water volume. A negative value implies sedimentation, positive value means erosion.

ecological interests in the inner Humber, Ouse and Trent. As a result the final design was modified to limit the area of the site that could be inundated on normal tides. This ensured there was no detectable reduction in hw level and as a result the long term morphologic changes predicted by this modeling are not expected to arise.

### 4.3   *Sunk Island setback near the estuary mouth*

The Sunk Island setback is a large elongated setback bordering the northern bank of the Humber estuary (see Figure 2). Figures 9 through 11 depict the hydrodynamic and morphologic impacts of this setback.

The influence of the setback on the *hw* levels is small and qualitatively comparable to the effect of the Alkborough setback. The initial and long-term effect is a small reduction, compared to the Alkborough setback, of the *hw* level, especially landward of the setback. Landward of the setback the *lw* levels slightly increase. In other words the tidal range diminishes. In the seaward reaches and in the setback region the tidal range slightly increases with time due to the increased tidal volumes and associated erosion of the area seaward of the setback (Figure 10). Landward of the setback the tidal ranges and ebb volumes decrease explaining the predicted sedimentation (Figure 10). The initial gain in intertidal area is preserved and continues to increase in the seaward part and setback region because of the gradual increase in tidal range. This enhancement of the initial gain stabilizes after 50 years ('setback' at right y-axis in Figure 11). In contrast, the intertidal area in the river diminishes slightly as a result of the reduction in tidal range. The overall effect is a substantial increase in intertidal area, despite a relative erosion (loss) of sediments from the estuary.

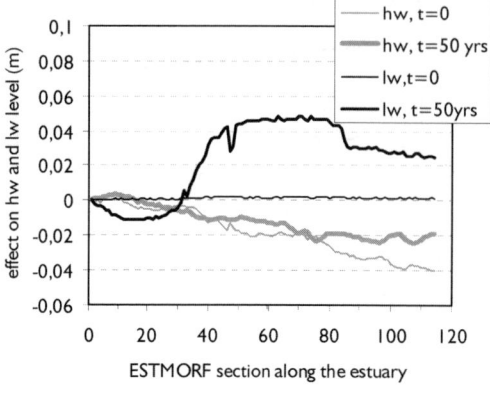

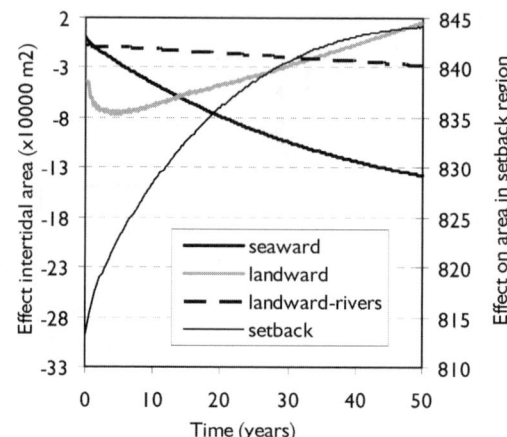

Figure 11. Effect of the setback on the evolution of the inter-tidal shoal area seaward, in the setback region and landward of the setback. The 'setback' region is displayed at the right y-axis because of the large differences in scale.

## 5 DISCUSSION

### 5.1 General responses

The response of the system to the setbacks has some common features. Immediately after the implementation of a setback the tidal range diminishes both landward and seaward of the setback. In the setback region and seaward of the setback the ebb volumes increase initially. This increase in ebb volume, despite the reduction of the tidal range, is due to the increased intertidal basin storage related to the setback. Landward of the setback region the ebb volumes initially decrease as a result of the initially reduced tidal range. This initial reduction of the ebb volumes landward of the setback region imply a reduction of the equilibrium cross-sectional area and hence an initial sedimentation (compared to the situation without setback). In contrast, the setback region and the seaward area of the estuary experience a relative erosion as a result of the increased ebb volumes. Because of this erosion the long-term response of the area landward of the setback region differs from the initial response: The erosion seaward and in the setback region enhances the tidal intrusion resulting in a general increase of the tidal ranges again. As a result the ebb volumes landward of the setback region increase again with time causing erosion after a period of initial sedimentation. The net effect of this sedimentation and erosion process depends on the location of the setback in the estuary and the considered timescale. For a seaward situated setback, in the present study the Sunk Island case, the sedimentation landward of the setback region still dominates after 50 years. For more landward located setbacks, e.g. the Alkborough and Ouse setback, the

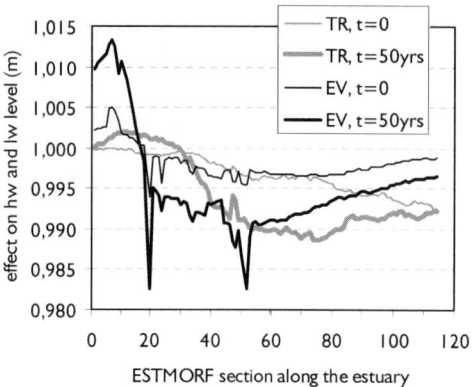

Figure 9. Initial and long-term effect of the Sunk Island setback on the hw and lw levels (upper panel), tidal ranges (TR) and ebb volume (EV) along the estuary. The setback is located between sections 10 and 21 (see also Figure 2).

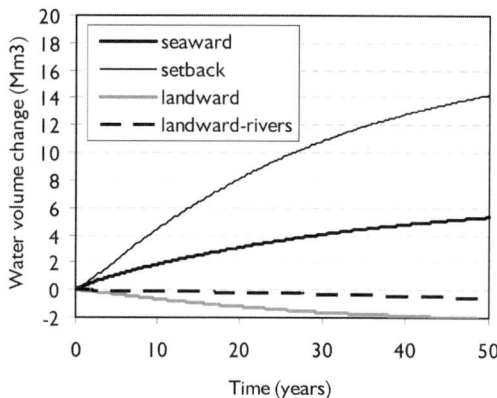

Figure 10. Effect of the setback on the large-scale change of the water volume. A negative value implies sedimentation, positive value means erosion.

initial period of sedimentation in the area landward of the setback is short and the long-term effect is erosion. For the estuary as a whole, a setback always causes erosion compared to the situation without setback. The magnitude of this erosion depends on the size of the setback in relation to its location.

Another important aspect of the morphologic evolution is the change of the intertidal area that is associated with the evolution described above. Apart from the setback region, where the intertidal area increases initially as a result of the setback itself, the interidal area decreases initially both landward and seaward of the setback. This reduction of the intertidal area is due to the initial decrease in tidal range. In the long term the intertidal area seaward of the setback continues to decrease as a result of the initial and long-term erosion described above. Landward of the setback region the long-term change of the intertidal area depends on the location of the setback and the associated changes in tidal range and the net effect of the initial sedimentation and the erosion that follows in time. For seaward located setback the initial sedimentation and reduction of tidal ranges lasts longer than for more landward located setbacks. Consequently, the intertidal area landward of the setback region decreases, as a result of (initially) reduced tidal ranges, for seaward located setback. For landward located setbacks the intertidal area increases because of the enhanced tidal ranges, i.e. the long-term hydrodynamic response manifests itself earlier.

## 5.2   *Comparison with the Schelde estuary*

Impacts of setbacks have also been investigated for the Schelde estuary, a tide-dominated meso-tidal coastal plain estuary in the Netherlands. For this estuary two setback scenario were investigated, viz.: a large seaward situated setback and a set of some smaller setback scenario's located more landward. The timescale considered in the study was 30 years, which is small compared to the morphological timescale of the estuary. For instance, the adjustment of the estuary to an accelerated sea level rise amounts 100 years (Kemerink, 2004), which is considerably larger than the timescale of the Humber estuary (30–40 years). This difference in morphological timescale manifests itself not only in, for instance, the response to the 18.6 year nodal cycle (Jeuken et al., 2003) but also in the response to setbacks. For the two setback scenario's only the initial response, as described in the previous section, was observed at the considered timescale of 30 years. The timescale at which the long-term effects become apparent has not been determined yet, but is estimated to be 50 to 100 years depending on the location and size of the setback.

The Schelde estuary has a long history of embankments (from 1600 until 1958, see e.g. Van der Spek, 1997), human interventions that will have had impacts opposite to the effect of setbacks described in this paper. Given the large morphological timescale of the Schelde estuary it is likely that the large-scale evolution of the estuary still experiences effects of these human interventions. The results of this study can be used to explain and further explore the impacts of these historic human interventions. In addition, the results may be used to assess the future impacts of the setbacks that are planned in the Flemish part of the estuary.

## 5.3   *Sustainable development of set backs*

The general hydrodynamic and morphologic responses of the estuary to setback, as derived in this study, can be used in the selection and design of setbacks. Herein the various management objectives, regarding the realization of setbacks, should be well considered.

Initially all setbacks reduce the high water level, suggesting an increased safety against flooding. This positive effect is preserved longer in time for more seaward located setbacks than for landward situated setback. In the long term this positive effect will tend to disappear and may change into a negative effect. It is noted that the model results are only indicative for flood control purposes since no extreme conditions have been simulated. It is also noted that the effect of the setbacks on river floods (extreme high river discharge) may different from storm surges from the sea.

Setbacks induce an initial and long-term erosion in the estuary seaward of the setback. Landward of the setback sediments are initially deposited, which will erode again at longer timescales. The contribution of setbacks to the objective of long-term restoration and conservation of intertidal habitats primarily depends on the location of the setback within the estuary in relation to the amount of intertidal areas seaward and landward of the setback. For seaward located setbacks the intertidal area seaward and in the setback region increases with time, despite the erosion, as a result of the increase in tidal range. Landward of the setback the initial reduction of the tidal range lasts longer (than for a landward situated setback) and causes a reduction of the intertidal area despite the initial sedimentation. An increase of the total initial gain in intertidal area is likely for seaward located setbacks. For landward situated setbacks this initial gain will decrease in time. This reduction is due to erosion of the intertidal area seaward of the setbacks that exceeds the increase in intertidal area landward of the setback. These differences in the evolution of the intertdial area landward and seaward of the setback may have implications for the distribution of different types of intertidal habitats (e.g. salt, brackish or fresh water).

Finally it is noted that primarily the timescale of the effects of setbacks, derived in this study, will also be influenced by the small-scale processes and evolutions within the setback. For example the deposition of silt within the setback will reduce its tidal volume and is expected to reduce the predicted longer term morphological changes described above.

## 6 CONCLUSIONS

Setting back of defences can be considered as a measure in the development of flood defence strategies and to support the creation of new inter-tidal habitat to maintain the estuary's conservation status. The semi-empirical model ESTMORF was used to analyze and explain the impact of set backs on the large-scale estuarine hydrodynamics and morphology at a timescale of years to decades. Generally, set backs induce long-term morphological changes that differ from the initial response. Seaward of the set back the estuary erodes. Landward of the set back initial sedimentation occurs. The deposited sediments erode again at longer time scales, with the net effect depending on the location and size of the set backs and the considered timescale. At the scale of the entire estuary set backs always cause a loss of sediments. For seaward situated set backs this erosion is smaller than for landward located set backs. The reduction of high waters and the initial gain in intertidal area is however preserved longer for seaward located setbacks than for landward situated setbacks. Moreover, the initial gain tends to increase for seaward located setbacks, whereas this gain is likely to reduce for more landward located setbacks. Finally, the above-described effects increase with the size of the setback. These results can be applied in the site selection and design of sustainable set backs in estuaries. In addition, the results can be used to assess the (reverse) impacts of estuaries to embankments.

## REFERENCES

Jeuken M.C.J.L., Wang Z.B., Keiller D., Townend I.H. & Liek G.A. 2003. Morphological response of estuaries to nodal tide variation. *Proc. Intern. Conf. on Estuaries & Coasts (ICEC-2003), 9–11 November 2003*. Hangzhou: Zhejiang University press.

Kemerink, J.S. 2004. Ontpolderen: wel of niet? (in Dutch). Msc-thesis, Delft University of Technology, Delft.

Van der Spek, A.J.F. 1997. Tidal asymmetry and long-term evolution of Holocene tidal basins in the Netherlands: simulation of paleo-tides in the Schelde estuary. *Marine Geology*. 71–90.

Townend, I.H. & Whitehead, P.A. 2003. A preliminary net sediment budget for the Humber Estuary. *Sci. Total Environ.* 314–316: 755–767

Townend, I.H., Wang, Z.B. & Rees, J.G. 2007. Millenial to annual volume changes in the Humber estuary. *Proc. R. Soc.* 463: 837–854. Published online.

Wang ZB, Karssen B, Fokkink RJ, & Langerak A, 1998, A dynamic-empirical model for estuarine morphology, In Dronkers J. & Scheffers M.B.A.M. (eds), *Physics of Estuaries and Coastal Seas; Proc. of the inter. conf., September 1996*. Rotterdam: Balkema.

Wang Z.B., Langerak A. & Fokkink R.J. 1999. Simulation of long-term morphological development in the Western Scheldt. *Proc. IAHR Symp. on River, Coastal and Estuarine Morphodynamics*. September 1999. Genova: University of Genova.

Wang Z.B. & Roelfzema A. 2001. Long-term morphological modelling for Humber Estuary with ESTMORF. *Proc. XXIX IAHR Congress, 16–21 September 2001*. Beijing: Tsinghua University Press.

*Human interference in rivers*

*River, Coastal and Estuarine Morphodynamics: RCEM 2007 – Dohmen-Janssen & Hulscher (eds)*
*© 2008 Taylor & Francis Group, London, ISBN 978-0-415-45363-9*

# Channel stability assessment in a meander cut-off

N. Kuspilic, D. Bekic & D. Carevic
*University of Zagreb, Faculty of Civil Engineering, Water Research Department, Zagreb, Croatia*

ABSTRACT: Observed during a longer period, natural watercourses are mostly in the state of dynamic equilibrium. Under the influence of external or internal changes, this morphological equilibrium of the river channel is disturbed. In new circumstances, watercourses strive to establish new dynamic equilibrium. Morphological changes are reflected by changes of the natural water and sediment regime. During works on the watercourses or in their respective catchment areas, it is necessary to monitor the new situation, in order to be able to react in time in case of adverse effects. This paper demonstrates the methodology of monitoring of hydromorphological parameters on 1.6 km long cut-off of meander in the channel of the Drava river in Croatia. During one year, changes in the river channel geometry and hydrological-hydraulic parameters were measured by sounder and by ADCP. The collected data were the basis for assessment of the global stability of the river channel in the monitored reach. The assessment of the river channel stability was made by comparison of measured bed shear stress and Shields criterion of critical stress, and on the basis of the new dimensionless hydraulic water depth.

## 1 INTRODUCTION

When the short-term changes of the river system occur without altering the general state, the river behaviour is considered in terms of a dynamic but steady system. Some river reaches maintain form and location features over long periods of time, and are considered to be in equilibrium with the inputs of water and sediment, displaying "dynamic equilibrium". Still other river reaches may switch rapidly from one set of dimensions and form to another in response to either external drivers (a recent flood or modification) or as a result of internal change (the meander cut-off). These latter river types display threshold behaviour.

This paper gives the analysis of the river channel stability after creating the meander cut-off of the main channel on the river Drava. The river Drava basin is the second largest in Croatia and is of great importance for water management of countries on its watershed. Nearly the half of its total path (305 km of total 750 km) flows through Croatia. On the entrance into Croatia the average annual river flow is 380 m³/s, and the average outflow is 620 m³/s at the mouth.

Some twenty years ago, a cut-off was made on the meander near the Osijek river port (Fig .1). The meander cut-off Nemetin was made in 1985, in the length of 1.6 km between river cross-sections PO6 and P16. Until 2005, the old river channel and the cut-off channel participated in sediment and flow transport. The channel of the cut-off in the 20-year period assumed the form of dynamic equilibrium. Due to construction

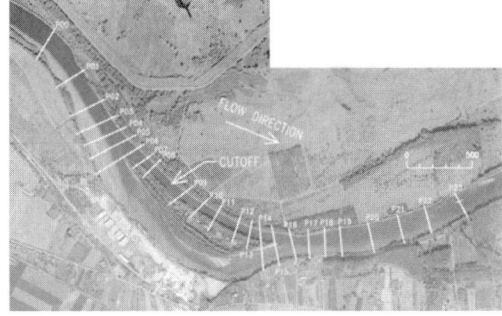

Figure 1. Site location map with position of cross-sections.

of the weir on the old river channel at the cross-section P06, the old channel has become almost inactive, and thus creating the conditions for further development of the cut-off channel.

In the new conditions, hydromorphological parameters were measured in the cut-off zone for the period of one year. On designated cross-sections (Fig. 1), measurement of channel geometry, flow and velocity profiles was carried out.

The flow and the velocity profiles were measured by boat-mounted ADCP, and geometry was measured by sounder and ADCP. Measurements were made in 2006, in 10 outings to the river in the period from February 10, to October 11.

Based on the collected data, the assessment was made of the global channel stability in the cut-off zone. The assessment was made by comparing bed shear stresses and the Shields criterion of critical stress. Further, the condition of stable channel was given on the basis of dimensionless hydraulic water depth $h\#$. The subject reach of the Drava is within the zone of influence of the Danube backwater, and the analysis must also pay attention to the effect of backwater on hydromorphologic parameters.

## 2 THEORETICAL ANALYSIS OF CHANNEL STABILITY

Disturbing of global channel stability, as a precondition for continual morphologic changes, begins when bed shear stress $\tau_0$ exceed the value of permissible stress $\tau_d$. The bed shear stress $\tau_0$, transferred from water to the channel bed, is expressed by the equation:

$$\tau_0 = \rho g h I \tag{1}$$

where $\rho$ = density of the water mass [kg/m$^3$], $g$ = acceleration by gravity [m/s$^2$], $h$ = water depth [m] and $I$ = slope of the energy grade line [m/m]. Dividing of equation (1) by density of water mass results in:

$$\frac{\tau_0}{\rho} = g h I = u_*^2 \tag{2}$$

where parameter $u_*$ = shear velocity [m/s]. Depending on the regime, hydraulically smooth, hydraulically rough or transition zone, it is possible to determine the relations of velocity $u(y)$ at a given distance from the bottom $y$ and velocity of shear stress $u(y)/u_*$ as follows:

For hydraulically smooth bottom $Re_* < 3.32$, for transition zone $3.32 < Re_* < 70$, and for hydraulically rough $Re_* > 70$. In the above expressions, $\kappa$ = von Karman's constant (0.4 for clear water and flat channel bed), $Re_*$ = Reynolds number of roughness

Table 1  Relation between velocity $u(y)$ at given distance from bottom $y$ and velocity of shear stress $(u(y)/y)$ depending on hydraulic regime (Zanke 1982).

| | |
|---|---|
| Hydraulically smooth channel bed | $\frac{1}{\kappa} \ln\left(9.0\frac{u_* y}{\nu}\right)$ |
| Transition zone | $\frac{1}{\kappa} \ln\left(\frac{y}{k_s}\right) + \frac{3.23}{\kappa Re_*} \ln(Re_*) - \frac{9.96}{Re_*} + 8.5$ |
| Hydraulically rough channel bed | $\frac{1}{\kappa} \ln\left(30\frac{y}{k_s}\right)$ |

($Re_* = u_* k_s/\nu$), $\nu$ = kinematic viscosity [m$^2$/s] and $k_s$ = absolute roughness of channel bed surface [m].

From the above expressions it is possible to determine, for straight immobile bed and clear water, the value of bed shear stress $\tau_0$ for a known value of the horizontal velocity component at a given distance from the bottom $y$, for known absolute surface roughness $k_s$ and for known conditions of hydraulic roughness. Natural conditions are considerably more complex, because the entire picture is affected by the non-prismatic nature of the channel, form of channel bottom, suspended sediment load in water, and mobility of particles at the bottom. These parameters influence the distribution of flow velocities in depth, and thus also the distribution of tangential stress.

### 2.1 Effect of channel bed form

Total shear stress on the channel bed is, in fact, the superposition of shear stress due to roughness of the bed surface $\tau_0$, shear stress due to bed form $\tau_1$ and shear stress due to presence of suspended particles in water $\tau_2$. Aware of the above effects, as well as of the fact that there are no reliable methods of calculation to determine the values $\tau_1$ and $\tau_2$, it was assumed that the predominant effect on particle movement is that of the component $\tau_0$. Therefore, the results in quantitative sense should be taken with a certain reserve. Further research is directed exactly towards determining the effect of the channel bed form and suspended sediment load on the value of total shear stress at the bed $\tau$.

Theoretically, the influence of the bed form (dunes with folds) with regard to the influence of bed surface roughness on changes of the overall roughness may be determined by the following equation (Yalin 1992):

$$\frac{1}{c^2} = \frac{1}{c_f^2} + \frac{1}{2}\sum_{i=1}^{2}\left(\frac{\Delta}{\Lambda}\right)_i^2 \frac{\Lambda_i}{h} \tag{3}$$

where $c_f$ = dimensionless Chezy coefficient ($c_f = \bar{u}/u_*$) referring to bed surface roughness, $\bar{u}$ = flow velocity (mean water velocity) [m/s], c = dimensionless Chezy coefficient referring to overall channel roughness, $\Delta_1$ = dune height [m], $\Lambda_1$ = dune length [m], $\Delta_2$ = fold height [m], $\Lambda_2$ = fold length [m] and $h$ = mean water depth (from water surface to the medium height of the dune) [m].

Further, it is possible to derive that:

$$\tau = \frac{\bar{u}^2 \rho}{c} \tag{4}$$

Following the survey of velocity profiles in the cut-off, the effect of bed form on shear stress was assessed. With the assumed geometry of the channel bed form, the effect of the bed form on the overall shear stress is approximately 15 percent. This has been assessed as an acceptable error, and the effect of the bed form was omitted from the assessment of channel stability.

## 3 SEDIMENT CHARACTERISTICS

In comprehensive analysis of morphological changes, monitoring of the bed load is important, which includes sediment discharge $Q_{sb}$ and its physical characteristics. This study did not include measuring of the sediment discharge, nor sampling of the material from the channel bed. The data on the sediment were taken over from the study Institut za elektroprivredu i energetiku d.d. (1998), which gives the grain size curve of channel bottom at the mouth into the Danube. The sediment is narrow-graded sand. It may be seen from the granulometric curve that the diameter of the 50-percent particle is $D_{50} = 0.35$ mm. In further analyses this value is used as the typical diameter of bed load $D$.

## 4 RESULTS

Measurement of discharge and velocity profile was made by using Broadband Acoustic Doppler Current Profiler (BB-ADCP) with frequency of 1200 kHz. ADCP measures vertical-velocity profile and channel-depth data, along the boat path, and over time. Velocity accuracy is declared to be ±0.25% of the water and boat, or absolute value of ±2.5 mm/s. Velocity resolution declared is 1.0 mm/s. Time between pings was set to 0.05 s and boat speed on average was 1.5 m/s.

ADCP was mounted on the boat side. At designated cross-sections, the boat location was determined by GPS. The source of data on water levels were the water levels recorded at two gauging stations, Osijek (upstream), and Aljmas (downstream). The gauging station Osijek is situated 6 km upstream from the cut-off, and there are no significant tributaries between the station and the cut-off. The gauging station Aljmas is situated 15 km downstream from the cut-off, at the mouth of the Drava into the Danube.

To reduce discharge measurement errors, it is necessary to make 4 to 6 passes on a cross-section. Surveying of all cross-sections in accordance with recommendations would require a period of 24 hours, thus surveying was done with one pass per cross-section.

### 4.1 Discharges

So far, 10 measurements were carried out in 2006, as follows: February 10, March 1, April 7, May 6, June 12, July 15, August 19, October 13, and November 10. The discharges measured at cross-sections were averaged and presented in relation to water levels measured in Osijek (Fig. 2). Figure 3 shows the discharges in relation to differences of water levels $WSE_{diff}$ between gauging stations Osijek and Aljmas. Numeric markings of measurements (m.1–m.10) are given in chronological order.

The mean long-term discharge of the Drava in the surveyed reach is $Q_{avg} = 550$ m³/s. According to Figure 2, three groups of measurements may be singled out. The first group of measurements 1, 9 and 10 was made during low discharges ($Q < 350$ m³/s). The second group, 2, 3, 4, 7, 8 was made during mean discharges ($450 < Q < 600$ m³/s), and the third group 5 & 6 during higher discharges of the Drava river ($Q > 750$ m³/s).

The subject reach is situated in the zone affected by the Danube backwater, and in morphological analysis attention must be paid to the effect of backwater. Generally, in natural watercourses (without backwater influence) the relation between water levels and discharges follows a certain rating curve. Schematic rating curve is shown in Figure 2. It may be noticed, that in measurements 4 and 6 there was a considerable influence of the backwater. These measurements considerably deviate from the rating curve (Fig. 2), with simultaneous comparatively small drop of the water table $WSE_{diff} \approx 0.4$ m (Fig. 3).

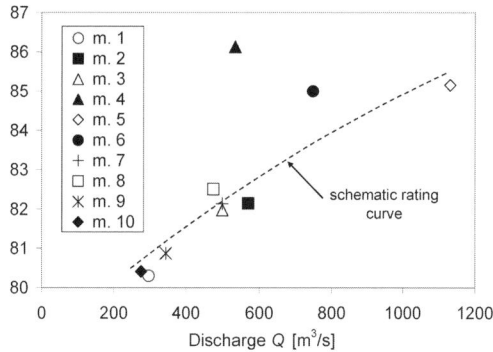

Figure 2. Measured discharges and corresponding water levels in Osijek.

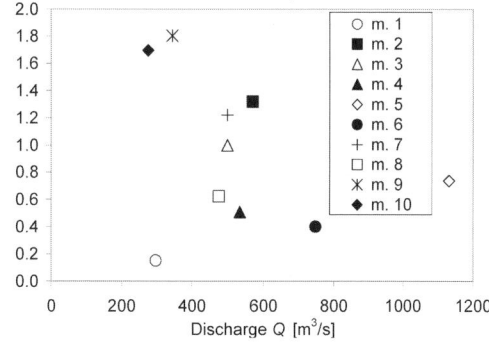

Figure 3. Measured discharges and differences of water levels in Osijek and Aljmaš.

### 4.2 Velocity profiles

By moving the ADCP across the channel vertical-velocity profile and channel-depth data are collected in the set of "ensemble" verticals. Velocity data along vertical are collected in equally sized depth layers (cells). After data collection it was found that distribution of vertical-velocity from a single ensemble is to scatter for analysis. In turbulent flows, the ADCP velocity fluctuations characterise the combined effects of the Doppler noise, signal aliasing, velocity fluctuations and other disturbances. Experiences demonstrated recurrent problems with"raw" velocity data in all velocity components evidenced by high levels of noise and spikes (Chanson et al. 2005). Therefore, it was decided to post-process velocimetry data.

For velocity recordings in a point, temporal averaging of velocity components is usual. For recordings in a cross-section this is not possible, and spatial averaging was applied. A similar method of velocity averaging, but for stationary cross-sections, was described by Mueller & Rehmel (2005).

Spatial averaging was made by averaging several vertical ensembles in a single vertical profile. The number of verticals included in averaging was chosen according to several criteria. The minimum number chosen was 3, and the maximum 10 verticals. Time between pings was set to 0.05 s and boat speed on average was 1.5 m/s, which gives the average distance of averaged verticals between 0.2 m and 0.75 m. Further, as the ADCP cell size was 0.25 m, each depth change larger than 0.25 m represents a new averaging. In this way, spatially averaged components $u$, $v$, $w$ of raw velocities $U$, $V$, $W$ were gained, which are now suitable for further analysis.

Figure 4 shows the longitudinal projection of the horizontal velocity $u_{h.long}$ (spatially averaged) for the cross-section P10 in the cut-off. The velocity profile is shown for measurements 2 and 4 during medium discharges. Considerable differences of velocity values between measurements are noticed. During measurement 2, there was no influence of the Danube backwater, while during measurement 4 this influence was significant. Considerable reduction of velocity under the influence of backwater is evident, as expected. However, it is questionable whether bed load movement takes place also at reduced values of profile velocities, or the bed is stable.

## 5 GLOBAL CHANNEL STABILITY

Spatial averaging of raw ADCP velocities $U_i$ results in averaged velocity profiles $u_i$ in control cross-sections, as described in chapter 4.2. In the analysis of channel stability, longitudinal component of horizontal velocity $u_{h.long}$ was used. The shear force of water, as a rule, is the strongest in the deepest part of the cross-section,

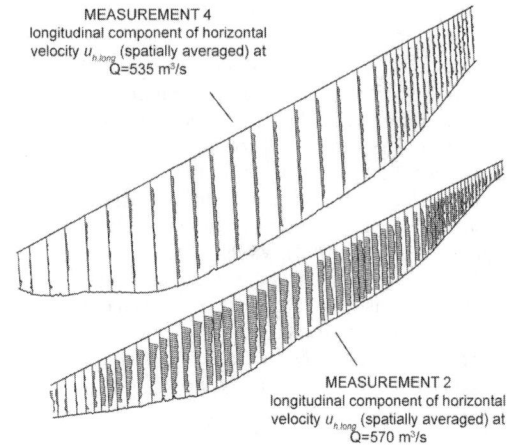

MEASUREMENT 4
longitudinal component of horizontal velocity $u_{h.long}$ (spatially averaged) at
Q=535 m³/s

MEASUREMENT 2
longitudinal component of horizontal velocity $u_{h.long}$ (spatially averaged) at
Q=570 m³/s

Figure 4. Longitudinal component of horizontal velocity $u_{h.long}$ at cross-section 10 for measurements 2 & 4. ADCP raw velocities are spatially averaged across cross-section.

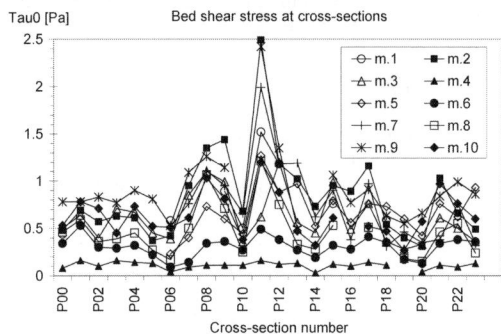

Figure 5. Average bed shear stress $\tau_0$ at highest depth of cross-section.

and the hydraulic load of water flow was calculated for the deepest part of the channel.

### 5.1 Assessment according to bed shear stress

Disturbance of the channel stability begins when bed shear stress $\tau_0$ exceeds the value of permissible stress $\tau_d$. The value of shear stress $\tau_0$ is determined on the basis of equation (2) for straight immobile bottom and clear water. Shear velocity $u_*$ is determined through the values of bottom values $u(y)$, depending on the hydraulic regime, according to equations from Table 1. Computational stress $\tau_0$ on the cross-section is determined for the deepest part of the section, as the mean average value of shear stresses for the deepest part of the channel.

Figure 5 shows the values of bed shear stress $\tau_0$ on cross-sections for measurements made in 2006. The values range between $0.1 < \tau_0 < 2.5$ Pa. Generally, the $\tau_0$ values are largest in measurements 2, 7, 9, which is in accordance with lowering of the water table

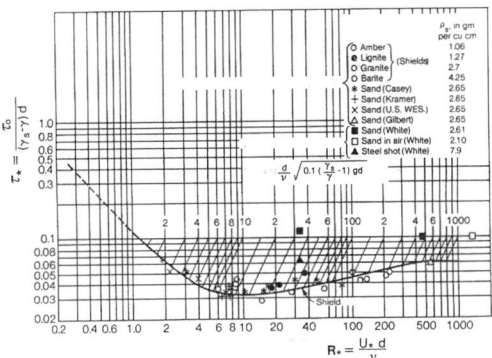

Figure 6.   Modified Shields curve.

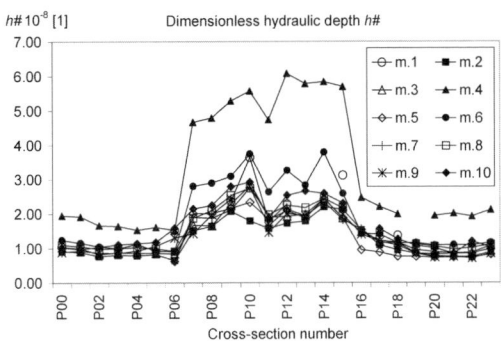

Figure 7.   Dimensionless hydraulic depth $h\#$ at cross-sections for entire range of discharges.

according to Figure 3. In the group of smaller discharges, $(Q < 350\,\mathrm{m^3/s})$ measurement 9 has the largest fall of the water table $WSE_{diff} = 1.8\,\mathrm{m}$, and in the group of mean discharges $(Q = 450–600\,\mathrm{m^3/s})$ the fall is the largest in measurements 2 and 7, $WSE_{diff} \approx 1.3\,\mathrm{m}$. The lowest values of $\tau_0$ occur in measurements 4 and 6 during backwater. Looking at changing of $\tau_0$ along the section (Fig. 5) it may be seen that the largest stress is found in the cut-off zone P06–P16.

The measured values of $\tau_0$ are compared with limit values $\tau_c$ obtained by using the Shields formula (Fig. 5). For the typical diameter of sediment load of $D_{50} = 0.35\,\mathrm{mm}$ the limit value of $\tau_c = 0.20\,\mathrm{Pa}$ is obtained.

According to Figure 2, it is noticed that the measured values $\tau_0$ are considerably higher than critical $(\tau_c = 0.20\,\mathrm{Pa})$ along the entire section, and for almost all flow conditions. The exception is measurement 4 under the influence of backwater with the mean stress value of $\tau_0 \approx 0.11\,\mathrm{Pa}$. This indicates considerable instability of the channel in the cut-off at this stage of its development.

### 5.2   Dimensionless hydraulic water depth $h\#$

Different discharges result in different transfers of hydraulic load from water flow to the bed surface. According to equation (2), shear stress $\tau_0$ increases with increased water depth. To exclude the effect of discharge magnitude in hydraulic load, the parameter $h\#$ is introduced, as follows.

Hydraulic water depth $\bar{h}$ is given as

$$\bar{h} = \frac{A}{B} \qquad (5)$$

where $A$ = flow area, $[\mathrm{m^2}]$, and $B$ = flow width at the free surface $[\mathrm{m}]$. Dividing equation 5 by the discharge $Q$ and multiplying by kinematic viscosity $v$ results in:

$$h\# = \frac{A}{B} \cdot \frac{v}{Q} = \frac{A}{B} \cdot \frac{v}{Av} = \frac{v}{Bv} \qquad (6)$$

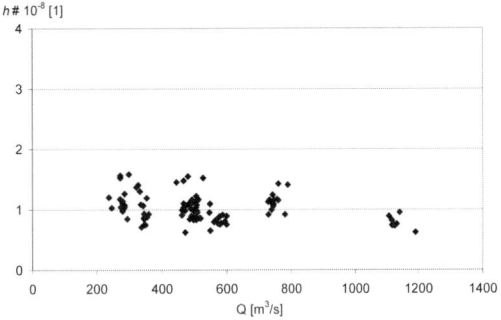

Figure 8.   Relation of parameter $h\#$ and discharge $Q$ for flow conditions without backwater in reaches beyond the cut-off.

where $v$ = kinematic viscosity $[\mathrm{m^2/s}]$, and $\bar{v}$ = average flow velocity $[\mathrm{m/s}]$. So the final expression for dimensionless hydraulic depth is obtained as:

$$h\# = \frac{v}{B\bar{v}} \qquad (7)$$

where $h\#$ = dimensionless hydraulic depth [1]. In the following analysis, the value of $v = 1.14 \times 10^{-6}\,[\mathrm{m^2/s}]$ was taken, which corresponds to water temperature of $T = 15°\mathrm{C}$.

Figure 7 shows changes of the parameter $h\#$ at cross-sections for all flow conditions. A considerable deviation of $h\#$ is noticed in measurements 4 and 6, when the influence of backwater was strong. Further, in the upstream reach from the cut-off (P00–P06), and downstream reach (P015–P23), uniform distribution of the parameter $h\#$ is noticed around the value of 1. Due to the strong backwater effect, measurements 4 & 6 were excluded in the following analysis.

The parameter $h\#$ is shown in relation to discharge, bed shear stress and water depth for reaches beyond the cut-off (Figs. 8, 9, 10), and for the cut-off (Figs. 11, 12, 13). It can be seen that in reaches beyond the cut-off distribution of parameter $h\#$ is almost uniform around

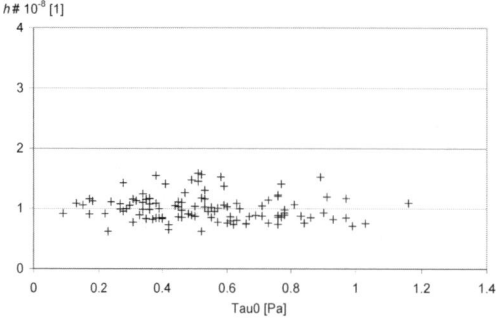

Figure 9. Relation of parameter $h\#$ and shear stress $\tau_0$ for flow conditions without backwater in reaches beyond the cut-off.

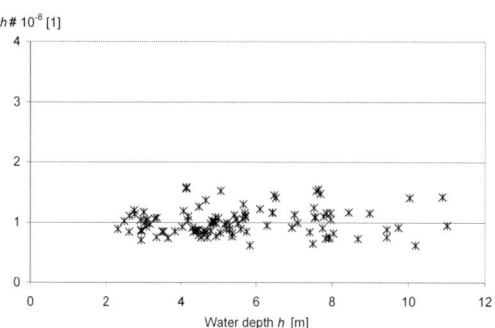

Figure 10. Relation of parameter $h\#$ and water depth $h$ for flow conditions without backwater in reaches beyond the cut-off.

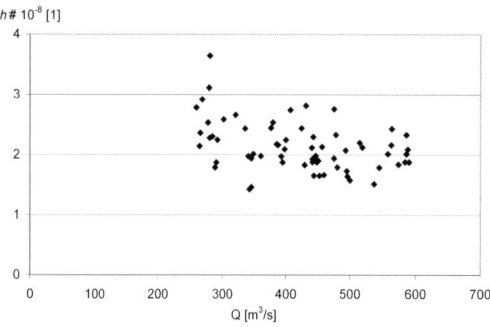
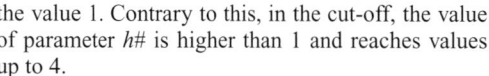

Figure 11. Relation of parameter $h\#$ and discharge $Q$ for flow conditions without backwater in the cut-off.

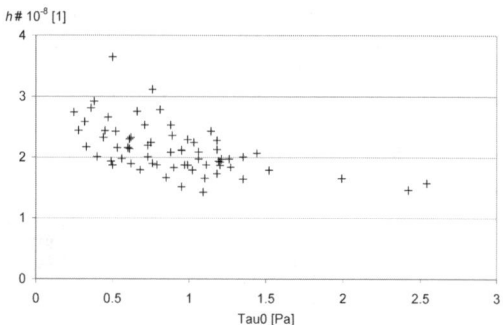

Figure 12. Relation of parameter $h\#$ and shear stress $\tau_0$ for flow conditions without backwater in the cutoff.

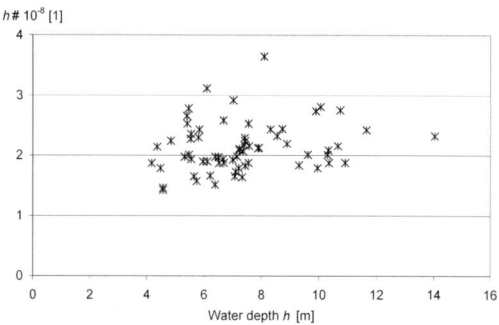

Figure 13. Relation of parameter $h\#$ and water depth $h$ for flow conditions without backwater in the cut-off.

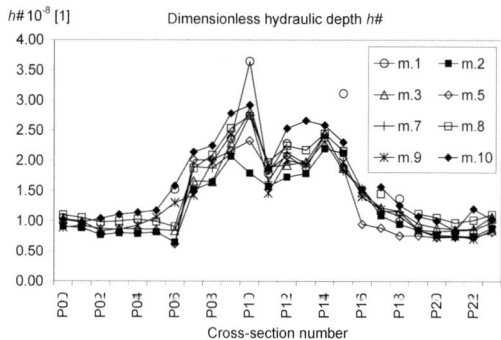

Figure 14. Dimensionless hydraulic depth $h\#$ at cross-sections for flow range without backwater effect.

the value 1. Contrary to this, in the cut-off, the value of parameter $h\#$ is higher than 1 and reaches values up to 4.

Consequently, if we exclude hydraulic conditions under backwater, we can, through value of parameter $h\#$, make the assessment of morphological stability of the reach. Figure 14 shows the change of parameter $h\#$

in cross-sections for all flow conditions except measurements 4 and 6. For all flow conditions the values beyond the cut-off are distributed around 1, and in the cut-off are higher than 1.5. Therefore, instable reach of the river channel may be determined by the value of $h\#$ greater then 1.

## 6 CONCLUSIONS

The paper presents the methodology of stability assessment of the Drava river channel in the Nemetin cut-off. The channel stability was assessed on the basis of measured shear stresses and dimensionless hydraulic depth $h\#$. Discharge and velocity data were collected with bout-mounted ADCP. It was demonstrated that "raw" ADCP velocity data in all velocity components should be post-processed before any further analysis. Spatial averaging of vertical-velocity profiles was applied by averaging several vertical ensembles in a single vertical profile. Due to the lack of sampling of all morphological parameters, the channel stability was assessed on the basis of shear stress due to roughness of the channel bed surface. Shear stress at the channel bed was calculated on the basis of spatially averaged velocity profiles, and compared to critical values. The new way of assessment is shown, using dimensionless hydraulic depth $h\#$. As the obtained bed shear stresses exceed by several times the critical stress, and as the value of parameter $h\#$ is greater then 1 in all flow conditions (with and without backwater), the conclusion is that the present channel in the cut-off is pronouncedly unstable. With regard to present hydromorphological conditions, further deepening of the channel may be expected. The general objective of the study is to forecast, or estimate, the limits of new dynamic equilibrium which will occur in the cut-off after closing of the old river channel. It is estimated that a period of three years of consecutive surveying is the minimum time required for collecting of representative sample to define the limits of new dynamic equilibrium.

## REFERENCES

Chanson, H., Trevethan, M., Aoki, S. 2005. Acoustic Doppler Velocimetry (ADV) in a Small Estuarine System. Field Experience and "Despiking". In Jun, B. H. and Lee, S. I. and Seo, I. W. and Choi, G. W. (eds.), *31st IAHR Biennial Congress*, 11–17 September, 2005, 3954–3966, Seoul, Korea.

Institut za elektroprivredu i energetiku d.d. 1998. Prognoze morfološko-psamološkh procesa u rijeci Dravi nakon izgradnje HE Novo Virje.

Mueller, D.S. & Rehmel, M. 2005. Determining Mean Velocity for Stationary Profiles using WinRiver, Office of Surface Water and Indiana WSC.

Simpson M. 2001. Discharge Measurements Using a Broad-Band Acoustic Doppler Current Profiler, *U.S. Geological Survey, Open-File Report 01–01*, Sacramento, California.

Yalin, M.S. 1992. *River Mechanics*, Pergamon Press.

Zanke, U. 1982. *Grundlagen der Sedimentbewegung*, Springer-Verlag.

*River, Coastal and Estuarine Morphodynamics: RCEM 2007 – Dohmen-Janssen & Hulscher (eds)*
*© 2008 Taylor & Francis Group, London, ISBN 978-0-415-45363-9*

# Morphologic and sediment transport adjustments around an artificial point constriction in a large meandering river (Bermejo River, Argentina)

C.G. Ramonell, M. Montagnini & M. Perez
*Facultad de Ingeniería y Ciencias Hídricas, Universidad Nacional del Litoral (UNL), Santa Fe, Argentina*

M.L. Amsler
*Instituto Nacional de Limnología, UNL-Consejo Nacional de Investigaciones Científicas y Técnicas (CONICET), Santa Fe, Argentina*

O. Orfeo
*Centro de Ecología Aplicada del Litoral, CONICET, Corrientes, Argentina*

ABSTRACT: The Bermejo River is one of the lowland rivers with more specific solid discharges in the world. For the last 60 years, mean maximum rates of meander channel shift of 600 m/year were recorded around Lavalle Bridge. The bridge was built in a cut-off formed straight reach of the river. The reach downstream the bridge remained straight to nowadays, lengthening from 4 km to 8 km during the eighties, when large floods occurred. The meander growth at the upstream reach prompted the breaching of the bridge. We suggest that the opposite river behavior around the bridge could be related with a disequilibrium in the sediment transport. The bridge acts like a dam during the high water levels promoting sedimentation in the upstream reach, and channelized bed-lowering downstream. During low water stages, the upper reach slope would exceed the limit necessary to convey the water and sediment input and meandering became the mechanism to decrease the energy excess.

## 1 INTRODUCTION

The Bermejo River is a subtropical stream flowing from the Andean Mountains to the Paraguay River through the large plains of the South American Chaco (Fig. 1). It is the more unstable large river of Argentina: channel shiftings up to 2500 m were recorded in the last ten years.

Moreover, the Bermejo River is the main hydrologic corridor supplying fine sediments to the Paraná and Río de la Plata fluvial systems: nearly 90% of the Paraná River washload, i.e., ca. $95 \times 10^6$ tn/year, was delivered by the Bermejo during the last decade of the XXth Century. Due to such large amount of sediment transport, this stream is one of the lowland rivers with more specific solid discharges in the world: the suspended sediment concentration for flood stages at Lavalle Port ($25°40'S$; $60°10'W$) averages 15,000 ppm.

In this paper a geomorphic and sedimentologic fluvial assessment was performed by the authors nearby Port Lavalle. The study involved field work (sediment sampling, surveying of geomorphic marks,

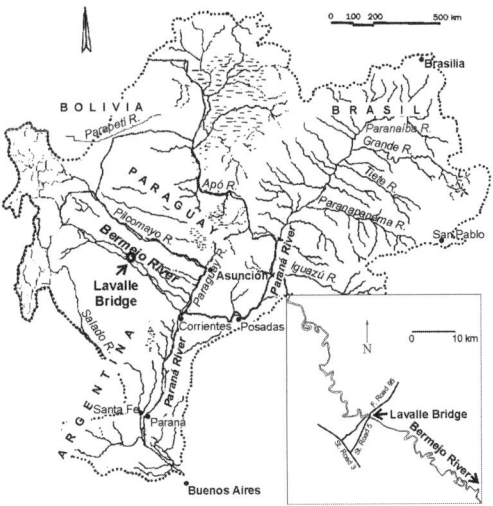

Figure 1. The Bermejo River location in the context of the Paraná River fluvial basin; the circle is on Lavalle Bridge, near to Lavalle Port village.

Table 1. Physical characteristics of the Bermejo River at Lavalle Port.

| Channel pattern | meandering |
|---|---|
| Channel Slope (m/km) | ca. 0.2 |
| Bed material | fine-to-very fine sand |
| Bank material | loose silt & very fine sand |
| Mean bankfull width (m) | 240 |
| Bankfull depth (m) | ca. 6 |
| Maximum/Minimum bankfull widths | 4.6 |
| Mean discharge (m³/sec) | ca. 400 |
| Mean flood discharge (m³/sec) | ca. 1200 |
| Mean low-water discharge (m³/sec) | ca. 60 |
| Suspended sediment concentration (ppm, average for floods) | 15,000 |

hydrometric/hydrologic measurements) and a historical data review (analysis of maps and air photographs starting in 1945 complemented with water discharge and solid gauging data from 1968 to nowadays).

It was observed that the morphologic behavior of the meandering river channel was affected by a bridge building, i.e. the Lavalle Bridge built between 1968 and 1976, in such a way do not reported in literature concerning true meandering rivers (see, for example, the extensive review performed by Briaud et al., 2001). The general picture of the change is the coexistence of a high sinuosity meandering reach upstream the bridge, and a straight, incised channel segment with a smaller slope downstream.

The related data and a conceptual qualitative explanation for such changes, as part of a research in progress which includes more sophisticated tools of analysis, are presented herein.

## 2 GENERAL DESCRIPTION OF THE BERMEJO RIVER MORPHODYNAMICS

Some physical features of the Bermejo River in the area are listed in Table 1. It is noteworthy that vegetal debris, such as trunks and trees of *Tessaria intergrifolia*, a common riverside tree in the region, are abundant during floods due to the high erodability of riverbank sediments (according to J.J. Neiff, pers. comm., 10,000 trees per hectare of *T. intergrifolia* could grow in five years, with a mean monthly growth rate of 1.4 m-height during the first five months).

Some characteristics listed in Table 1, such as the erodability of channel boundary sediments, the seasonal variability in discharge, and the huge rate of sediment transport accounts for the intense morphologic dynamics of the Bermejo River. It is highlighted below.

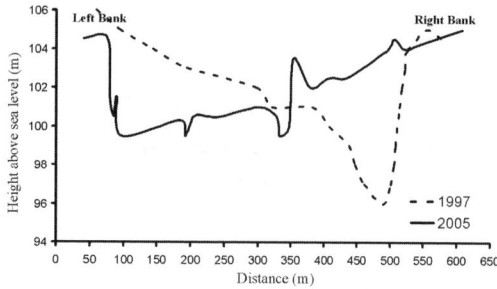

Figure 2. Example of channel change in a cross section located near Lavalle Bridge.

### 2.1 Channel changes

A typical sketch of the river thalweg wanderings and the related cross section modifications, are showed in Figure 2.

The river channel stability in a reach 65 km-long centered in Lavalle Bridge, was studied by means of historical maps. The successive tracks of the Bermejo channel axis were sketched for 8 years starting from 1945 (Fig. 3). Five additional records of a 10 km-long reach mapped from geo-corrected air photographs were also included.

All the meander shift types were identified along the reaches, e.g., meander expansion and translation, bend rotation, changes to compound meanders, and neck and chute cut-offs. Moreover, reactivation of abandoned meander was also observed (Fig. 4).

It was concluded that the channel migration processes are triggered by bend erosion and floodplain crevassing. Furthermore, bar channeling along rills formed by sand fluidization – during low-water stages – could be a significant mechanism of the main channel relocation during the period of rising water stages.

Localized old clay deposits along river banks act like point controls; these deposits part of the changes to compound meanders in the area are related to these deposits.

All these channel movements built a recent belt of meander-wanderings of 1900 m-width (in average), and provoked sinuosity variations between 1.7 and 2.0 during the last 60 years. The mean maximum rate of channel displacements around the Lavalle Bridge was appraised in 600 m/year.

### 2.2 Sediment transport main features

The sediment transport in the Bermejo River at Lavalle Port, has been measured only during short periods of time. APA (1997) compiles only ten measurements, four in 1995 and six in 1997, respectively. Due to the limited quantity of hydraulic information available at this point, the longer series of data obtained

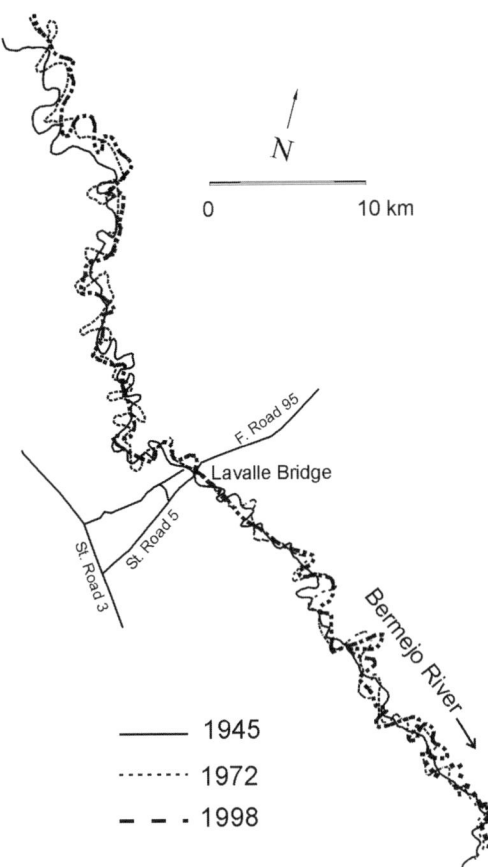

Figure 3. Channel axis tracks of the Bermejo River in 1945, 1972 and 1998 (similar records of 1989, 1993, 2000, 2002 and 2005 obtained from satellite images were included).

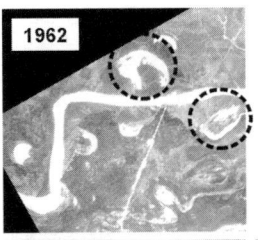

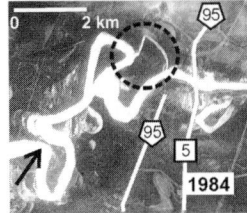
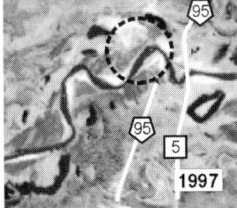

Figure 4. Air photographs of Lavalle Port surrounding area obtained between 1962 and 1993. The circles in the photo of 1962 indicate abandoned meanders since the fifties, with one of them renewed as active channel later on.

$$Q > 660 \text{ m}^3/\text{sec}$$
$$G_w = 71.055 \ Q^{1.3352} \qquad (r^2 = 0.7467) \qquad (2)$$

Suspended sands:
$$G_{ss} = 0.0278 \ Q^{2.22} \qquad (r^2 = 0.8399) \qquad (3)$$

where:

Q: water discharge; in $m^3/\text{sec}$.
$G_{ss}$: suspended sand discharge (d > 0.0625 mm); in ton/day.
$G_w$: washload discharge (d < 0,0625 mm); in ton/day.
$r^2$: determination coefficient.

It is observed that a fairly good fitness was obtained with both types of sediments. The previous relations were then used to compute values of mean concentrations for several typical discharge values between low and high waters. The results are gathered in the table below:

The predominance of washload transport in the middle and lower Bermejo River is clearly depicted. A comparison made in FICH (2006) between the limited concentration data at Lavalle Port and El Colorado for the same discharges, revealed that during the high waters the suspended sediment concentrations values have the same order. At low water stages apparently the concentrations are higher at Lavalle Port, especially those of the suspended sands.

Finally with eqs. (1) and (2) and the daily discharge data, the transport of washload during the maximum supplying period of each year was computed for the last decades. Along that period (between November–December and May–June of each hydrologic year)

at El Colorado (200 km downstream), may be used to gain a deeper insight of the sediment transport features. This station is part of the National Hydrometric System of Argentina (SNRH, 2004), and is being attended continuously since 1968 with the important detail that in 1993 began the separation between the suspended finest particles (silts and clays, the so called washload) from the suspended coarser ones (very fine and fine sands) in the concentration results from the measurements.

Based on this data Alarcón et al. (2003) and Amsler et al. (2006, in FICH, 2006), fitted the classic sediment rating curves for the wash load and the suspended sands at El Colorado station. They yielded the following results.

Wash load (silts and clays):

$$Q < 660 \text{ m}^3/\text{sec}$$
$$G_w = 0.0655 \ Q^{2.406} \qquad (r^2 = 0.9361) \qquad (1)$$

Table 2. Mean concentrations of sand, silt-clay and total in El Colorado gauging station.

| Concentration (mg/l) | Q (m³/sec) | | | | | |
|---|---|---|---|---|---|---|
| | 50 | 100 | 400 | 800 | 1000 | 1500 |
| $C_{ss}$ | 38 | 89 | 480 | 1120 | 1470 | 2410 |
| $C_w$ | 185 | 490 | 3450 | 7730 | 8330 | 9540 |
| $C_T$ | 223 | 579 | 3930 | 8850 | 9800 | 11,950 |

where:
$C_{ss}$: mean concentration of sand suspended.
$C_w$: mean concentration of washload.
$C_T$: mean concentration of total suspended sediment.

Table 3. Wash load transport ($G_w$) at El Colorado. Period: 1970–2004.

| Period | $G_w$ (ton × 10⁶) | | | |
|---|---|---|---|---|
| | Minimum | Maximum | Average | Accumulation |
| 1970–'79 | 19.9 | 115.4 | 70.8 | 636.9 |
| 1980–'89 | 36.7 | 169.7 | 112.2 | 1009.9 |
| 1990–'99 | 49.2 | 150.2 | 94.8 | 853.3 |
| 2000–'04[1] | 76.5 | 126.3 | 99.8 | 499.0 |

[1] Only 5 years considered.

nearly the total annual sediment is transported by the Bermejo River (Drago and Amsler 1988). The results are summarized in Table 3.

It is clearly seen that the eighties was a significant decade for the sediment transport in the Bermejo River. Apparently the tendency maintained after those years. Note during the first 5 years of the present decade that nearly half of the sediment discharge of the eighties was already transported.

## 3 CHANNEL ADJUSTMENTS AROUND THE LAVALLE BRIDGE

The Lavalle Bridge was finished in 1976 with lateral embankments built in a (cut-off formed) straight reach of the river (see the air view of 1962 in Fig. 4). The original bridge span of 285 m was recently widened to 400 m, because of the damages caused by the meander shifts occurred in the upstream reach. In spite of the last span enlargement, the bridge and their embankments involve a severe constriction respect to the width of the meander belt in the area from the seventies to nowadays.

The evolution of channel planform around the bridge is showed in Figure 5 (see Fig. 6 also): the straight reach downstream the bridge remained stable after its construction with a length of ca. 4 km; this length increased suddenly to 8 km during the eighties (Fig. 7), when the maximum recorded floods (and

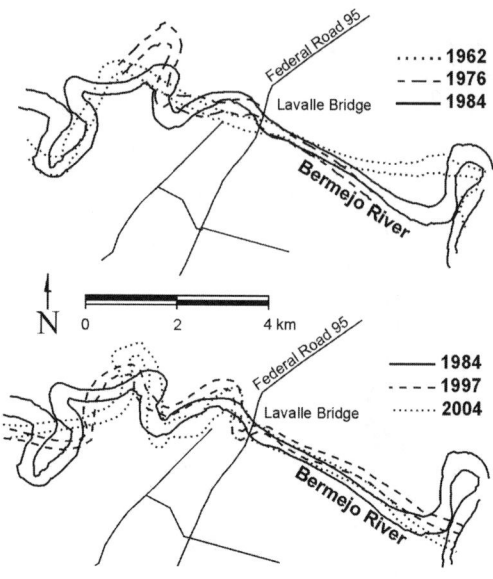

Figure 5. Changes of the channel planform around the Lavalle Bridge between 1962–2004.

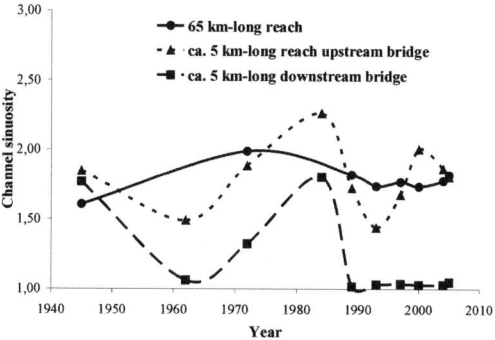

Figure 6. Channel sinuosity variations of the Bermejo River in the short and large reaches considered in the study.

also the mean rates of sediment transport; Table 3) occurred. Since that period the reach remains without major changes (Fig. 8).

It is noticeable that the upstream sinuous reach has a mean bed channel slope, as measured during low water levels, larger than the straight segment downstream (Fig. 9).

Additionally, a significant difference in the suspended sediment concentrations was recorded at either side of the bridge during a flood: in two nearly simultaneous measurements, values of 7200 ppm and 5300 ppm were obtained upstream and downstream the bridge, respectively. These values involved a decrement due to siltation of nearly 26% of the transported sediment (from ca. 9.4 tn/sec to ca. 6.9 tn/sec).

Figure 7. Changes of the straight reach length downstream of the Lavalle Bridge.

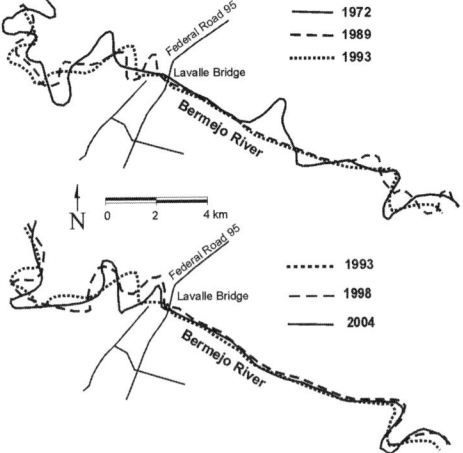

Figure 8. Channel axis tracks around the Lavalle Bridge between 1972 and 2004.

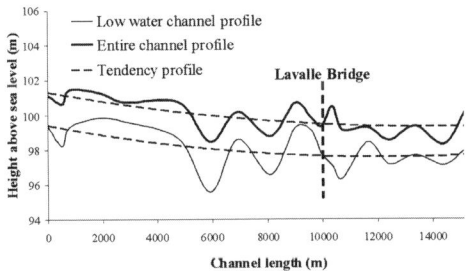

Figure 9. Average bottom profiles along the low water channel and adding to it the emerged lateral sand bars.

## 4 A CONCEPTUAL MODEL FOR THE CHANNEL ADJUSTMENTS

The channel evolution upstream the bridge showed in Figure 8 agrees fairly well with the results of the experiments conducted by Jin & Schumm (1986). According to these authors, the sinuosity and the meander amplitude of free meanders increase with time at the upstream segment of a local constriction. This "piling up" of meanders upstream of Lavalle Bridge damaged it seriously, and the bridge span had to be enlarged due to the erosive action of the river.

Unlike the above process, the permanence, lengthening and stabilization of the straight reach downstream of the bridge (Figs. 5 and 8), or the differences recorded in sediment transport and bed slopes at either side of the structure (e.g., Fig. 9), are not reported in literature concerning to meandering rivers.

A tentative explanation for the phenomena could be advanced assuming that the bridge and their embankments pose a disequilibrium condition in the interaction between the sediment transport and the local channel morphology during flood and low water stages.

By this reasoning, the manmade constriction would act like a dam during the high water levels promoting sedimentation upstream of the bridge, and channelized bed-lowering (e.g. channel incision) downstream. During low water stages, the upper reach slope would exceed the critical value necessary to convey the water and sediment input, and meandering becomes the mechanism to decrease the energy excess. The upstream dammed effect during floods possibly is increased due to the vegetation trapped by the bridge piers.

## REFERENCES

Alarcón, J.J.; Szupiany, R.; Montagnini, M.D.; Gaudin, H.; Prendes, H.H. & Amsler, M.L. 2003. Evaluación del transporte de sedimentos en el tramo medio del río Paraná. *Nuevas tendencias en Hidráulica de Ríos. Primer Simposio Regional sobre Hidráulica de Ríos* (CD ROM). Ezeiza, Buenos Aires, Argentina.

Amsler, M.L.; Montagnini, M.D.; Alarcón, J.J. & Ramonell, C.G. 2006. Washload influence in the alluvial plain building of the Paraná River (Argentina) in its middle reach: a quantitative approach. Unpublished.

APA. 1997. Acueducto Centro-Oeste Chaqueño. Estudios y recopilación de antecedentes. Administración Provincial del Agua (APA), internal report.

Briaud, J.L.; Hamn-Ching, Ch.; Edge, B.; Park, S. & Shah, A. 2001. Guidelines for bridges over degrading and migrating streams. Part 1: Synthesis of existing knowledge. Texas Transportation Institute, Report 2105–2 10, 194 pp.

FICH. 2006. Mediciones de campo y estudios hidromorfológicos de alternativas de ubicación y proyecto hidráulico del Acueducto Centro Oeste Chaqueño. Convenio: Estudio Guitelman S.A. e Inglese Consultores S.A.-UNL. Facultad de Ingeniería y Ciencias Hídricas (FICH), unpublished report.

Jin, D. & Schumm, S.A. 1986. A new technique for modelling river morphology. In: K.S. Richards, Ed., *Proc. First Internat. Geomorphology Conf.* Wiley, Chichester.

SNRH. 2004. Sistema Nacional de Información Hídrica. (www.obraspublicas.gov.ar/hidricos). Subsecretaría de Recursos Hídricos de la Nación (SNRH).

*River, Coastal and Estuarine Morphodynamics: RCEM 2007 – Dohmen-Janssen & Hulscher (eds)*
*© 2008 Taylor & Francis Group, London, ISBN 978-0-415-45363-9*

# Behavior of a river bifurcation

A. Mendoza-Resendiz & M. Berezowsky-Verduzco

*Instituto de Ingeniería, Universidad Nacional Autónoma de México, Coyoacán D.F. Mexico*

ABSTRACT: A mathematical model for fully 2D flow and bottom movement is presented. The model is used to study the Mezcalapa-Samaria-Carrizal bifurcation. Several flow conditions are tested. With the available data, as a result of the model, the water discharge distribution at the bifurcation does not change significantively for the daily discharges and during floods. However, there is a slight increment of the sediment transport capacity at the Samaria River during floods.

## 1 INTRODUCTION

The water and sediment behavior in river bifurcations is a complex phenomenon. The discharge distribution for each branch of the bifurcation is a function of several variables, among which one of the most important is the flow resistance. Of course, backwater effects can modify the flow conditions. Besides, the sediment transport can also affect the discharge distribution if there is deposition or erosion for some discharges.

In Figure 1 the Mezcalapa River bifurcation is shown; the two branches downstream are known as Samaria and Carrizal River, respectively. The river is in the Grijalva-Usumacinta basin, in the South-east of Mexico. About 70 km upstream the bifurcation, Peñitas dam is located. This is the last of a four hydropower flood control system of dams. From the dam to the sea the river drains in a very complex drainage system, see Berezowsky et al. (2003). The city of Villahermosa (520 000 inhabitants) is located 40 km downstream the bifurcation; the Carrizal river joins the Rios de la Sierra River just upstream the city.

Peñitas dam is operated in a daily basis following the energy demand in the electric system, so the flow hydrograph is rectangular and varies between 200 and 800 m³/s. At the bifurcation, due to the river regulation, the hydrograph has nearly a sinusoidal shape with discharges between 400 and 740 m³/s. The relatively fast water level variation due to these discharge hydrograph has a strong influence in the river morphology because there are problems of bank failures, and the river is widening in some regions, endangering the flood protection levees, Berezowsky and Jiménez (1993). As the region has had a strong economic development, a policy for flood control and energy optimization was designed such that the spillway of Peñitas dam is operated more or less every ten years. This is due to

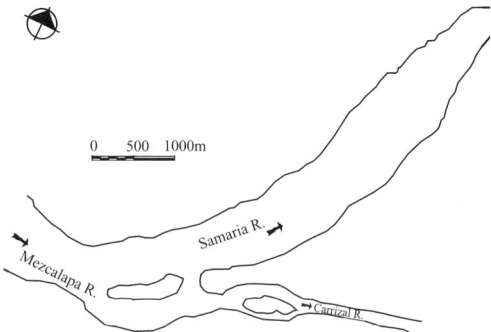

Figure 1. Bifurcation zone.

the fact that the reservoir has a relatively small storage capacity even for the local floods in the basin.

For the every day discharges, the Carrizal Samaria River carries about 40 percent of the Mezcalapa discharge. Nevertheless, during floods, the discharge distribution changes and the percentage of the discharge deviated to the Samaria River increases. During 1999 there was an extraordinary flood in the region with a peak discharge of 3660 m³/s. The local authorities estimated that about 1200 m³/s were through the Carrizal River, that is, nearly 33 percent of the total discharge. Some parts of Villahermosa City were flooded because the levees for flood protection are designed for a discharge of 850 m³/s. It has been argued that a strong sedimentation process is taking place at the bifurcation (mainly at the Samaria branch) and that the discharge at the Carrizal is increasing.

A 2D fully coupled hydrodynamic and mobile bed numerical model to study the bifurcation was implemented. With this model different conditions were modeled trying to find the reason of the change in the discharge distribution.

## 2 MATHEMATICAL MODEL

### 2.1 Equations

The model solves the shallow water equations (SWE) in 2D with a mobile bed. For the water movement, two equations for momentum conservation and one equation for mass conservation are used as follows:

$$\frac{\partial hu}{\partial t} + \frac{\partial hu^2}{\partial x} + \frac{\partial huv}{\partial x} + gh\frac{\partial H}{\partial x} + ghs_{bx} = 0 \quad (1)$$

$$\frac{\partial hv}{\partial t} + \frac{\partial hv^2}{\partial y} + \frac{\partial huv}{\partial y} + gh\frac{\partial H}{\partial y} + ghs_{by} = 0 \quad (2)$$

$$\frac{\partial h}{\partial t} + \frac{\partial hu}{\partial y} + \frac{\partial hv}{\partial y} = 0 \quad (3)$$

where $u$ and $v$ are vertical average velocities in $x$ and $y$ respectively, $h$ = water depth, $H$ = water level above datum, $g$ = acceleration due to gravity, $S_{fx}$ and $S_{fx}$ are the bottom friction slope. In this stage of the model, eddy viscosity was not considered.

The bottom movement is computed with the sediment continuity equation

$$\frac{\partial z}{\partial t} + \frac{1}{1-p}\left(\frac{\partial q_{sx}}{\partial x} + \frac{\partial q_{sy}}{\partial y}\right) = 0 \quad (4)$$

where $z$ = bottom level (above datum), $p$ = porosity of bottom material, $q_{sx}$ and $q_{sy}$ are unit total sediment transport, in m$^2$/s. The sediment transport is calculated with Engelund-Hansen Formula, Maza (1996).

$$q_s = \frac{0.05U^5}{\Delta^2\sqrt{gd_{50}}C^3} \quad (5)$$

where $U$ = vertical average velocity in m/s, $d_{50}$ = grain size for the 50 percent of the bottom material, in m, $\Delta$ = relative density of bottom material, $C$ = Chezy coefficient, in m$^{1/2}$/s. This coefficient is computed as $C = h^{1/6}/n$, where $n$ is the Manning roughness coefficient.

### 2.2 Solution algorithm

Due to the discharge variations along the day, equations (1) to (4) are fully coupled and solved with a MacCormack scheme, see for instance Chaudry (1993). The scheme has predictor and corrector steps.

Equations (1) to (4) are written as a vectorial system of equations

$$\frac{\partial}{\partial t}U + \frac{\partial}{\partial x}E_1 + gh\frac{\partial}{\partial x}E_2 + \frac{\partial}{\partial y}F_1 + gh\frac{\partial}{\partial y}F_2 + S = 0 \quad (6)$$

Where, $U, E_1, E_2, F_1, F_2$ and $S$ are defined as follows

$$U = \begin{pmatrix} h & hu & hv & z \end{pmatrix}^T \quad (7)$$

$$E_1 = \begin{pmatrix} hu & hu^2 & huv & q_{sx} \end{pmatrix}^T \quad (8)$$

$$E_2 = \begin{pmatrix} 0 & H & 0 & 0 \end{pmatrix}^T \quad (9)$$

$$F_1 = \begin{pmatrix} hv & huv & hv^2 & q_{sy} \end{pmatrix}^T \quad (10)$$

$$F_2 = \begin{pmatrix} 0 & 0 & H & 0 \end{pmatrix}^T \quad (11)$$

$$S = \begin{pmatrix} 0 & ghs_{bx} & ghs_{by} & 0 \end{pmatrix}^T \quad (12)$$

Backward and forward difference operators are defined; in $x$ direction we have

$$\Delta_x = G_{i,j} - G_{i,j-1} \quad (13)$$

$$\nabla_x = G_{i,j+1} - G_{i,j} \quad (14)$$

in $y$ direction

$$\Delta_y = G_{i,j} - G_{i-1,j} \quad (15)$$

$$\nabla_y = G_{i+1,j} - G_{i,j} \quad (16)$$

In the difference operators (13) to (16), subscript $i$ indicates moving on $y$ and $j$ on $x$ direction, respectively.

The predictor stage applied to (6) consists in the following operation

$$U_{i,j}^P = U_{i,j}^n + \begin{bmatrix} \text{change of derivatives by} \\ \text{difference operators} \\ \text{with variables in time } n \end{bmatrix} \quad (17)$$

Meanwhile the corrector step results in the following operation

$$U_{i,j}^c = U_{i,j}^n + \begin{bmatrix} \text{change derivatives by} \\ \text{difference operators} \\ \text{with variables obtained in} \\ \text{prediction} \end{bmatrix} \quad (18)$$

Finally, the solution of (6) for time $n+1$ is obtained by

$$U_{i,j}^{n+1} = \frac{1}{2}\left(U_{i,j}^P + U_{i,j}^c\right) \quad (19)$$

## 2.3 Boundary conditions

The flow is subcritical at the bifurcation. Two boundary condition upstream and two downstream are required for the water phase. Upstream a hydrograph with a given velocity distribution is imposed; flow normal to the boundary line is null. Downstream a rating curve is imposed at each branch; again, normal flow is null. For the mobile bed stage, the total sediment transport upstream is computed as a function of the hydraulic variables. The model works in a grid that can be of squares or rectangles.

The details of the model can be seen in Mendoza (2004).

## 3 ANALYSES

The measured bed topography at 1997 is used for the modeling; rating curves as reported in Mendoza (2001) were routed to the boundaries downstream of each branch. The discharge distribution at the upstream boundary was adjusted to measured data.

Different conditions were simulated with the model. Here we discuss just some of them. The calibration of flow resistance coefficient a sediment transport equation can be seen in Mendoza (2004) First a constant discharge of $1200\,\mathrm{m^3/s}$ was modeled for the entire zone shown in Figure 1. This discharge is approximately two times the average discharge in daily conditions. It was observed that the bottom movement takes place mainly in the region near the bifurcation. For that reason, a new grid was employed in a smaller region, as shown in Figure 2. A $50 \times 50\,\mathrm{m}$ grid and a time increment of one second were employed. As an average, for the higher discharge tested, the Courant number is about 0.3. A constant Manning Coefficient of 0.030 was calibrated for the bifurcation considering the imposed boundary conditions.

The following condition tested was designed to observe the dynamic behavior of the bifurcation for the daily hydrograph above described. This condition was maintained during one year. The velocity field for a discharge of $538\,\mathrm{m^3/s}$ is shown in Figure 3. A maximum velocity of $0.7\,\mathrm{m/s}$ was obtained for a region upstream the island at the right bank. For discharges between 400 and $740\,\mathrm{m^3/s}$, as an average, the discharge at Samaria branch is 70 percent of the upstream total discharge. As far as the sediment transport, the 93 percent of the total sediment transport goes through the Samaria River. The bottom has small variations, with a tendency to sedimentation at the left of the island, see Figure 2.

The next condition tested correspond to the 1999 flood. In Figure 4 the corresponding hydrograph is shown. The peak discharge is of $3660\,\mathrm{m^3/s}$. Note that there a sustained discharge of about $2000\,\mathrm{m^3/s}$ for about 13 days. In Figures 5 and 6 the flow velocity

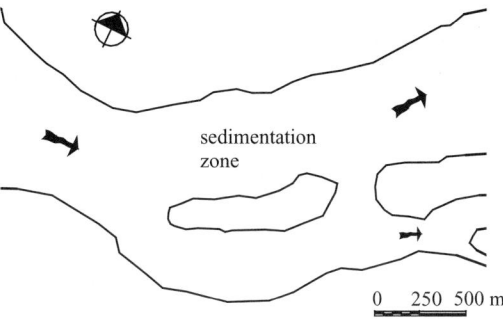

Figure 2.   Zone where is placed the movement of bottom.

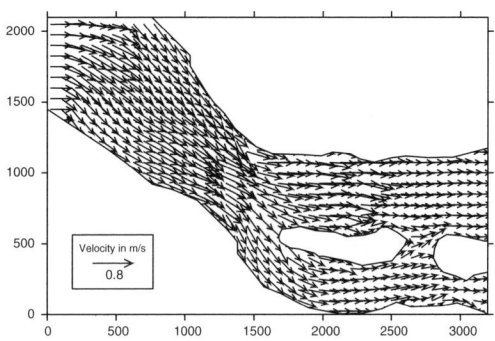

Figure 3.   Velocity distribution for a discharge of $538\,\mathrm{m^3/s}$.

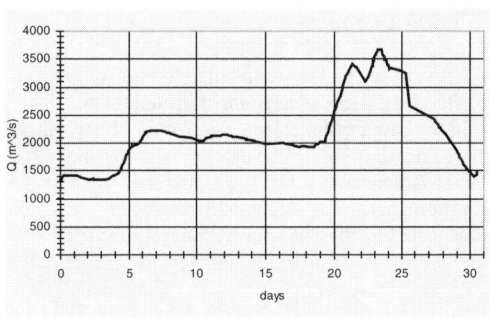

Figure 4.   Hidrograph of 1999 flood.

fields for 2000 and $3660\,\mathrm{m^3/s}$ discharges are shown. For the peak discharge, the maximum velocity is of $1.5\,\mathrm{m/s}$ near the island. As an average, 72 percent of the discharge goes through the Samaria River. The peak discharge at Carrizal River is of $1017\,\mathrm{m^3/s}$, so there is slight difference with the reported discharge of $1200\,\mathrm{m^3/s}$. As far as the sediment transport, 97 percent goes through the Samaria River; there is a tendency to deposition at the left margin of the island.

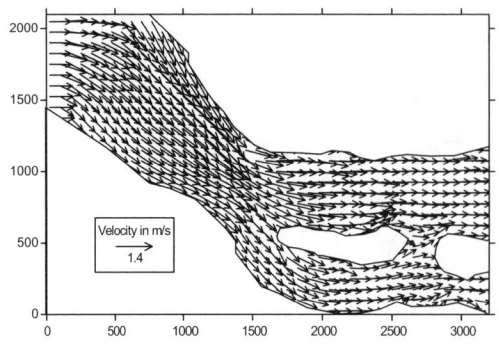

Figure 5.    Velocity distribution for a discharge of 2000 m³/s.

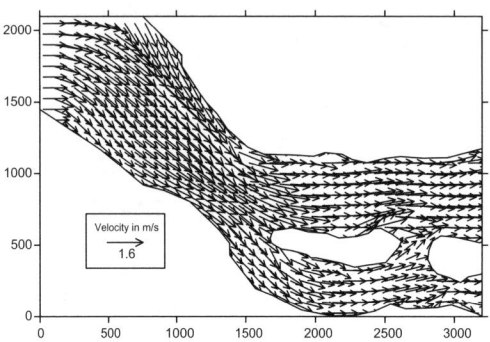

Figure 6.    Velocity distribution for a discharge of 3660 m³/s.

## 4    CONCLUSIONS

Mathematical models are a suitable tool for studying flow and sediment movement at bifurcations.

In the case of Mezcalapa bifurcation, the water discharge distribution for the daily conditions and during floods is conserved. For the sediment transport distribution there is a slight increase of the sediment transported by the Samaria river during floods.

If the sedimentation pattern at the left margin of the island continues during long time, there could be a tendency to diminish the water discharge at the Samaria River.

## REFERENCES

Berezowsky, M. & Jiménez, A.A. 1993. Effect of water level on cohesive river bank stability, *Advances in Hydro-Science and Engineering, Vol. I, Invited paper, Proceedings of Inter. Conf. on Hydroscience and Engineering*, pp 1382–1389, Washington.

Berezowsky, M., Jiménez, A.A. & Gracia-Sánchez, J. 2003. Morphodynamics of the Lower Grijalva River, Mexico, *Procc of the 3rd IAHR Symp on River, Coastal and Estuarine Morphodynamics*, pp 604–610, Vol I, Barcelona, Spain.

Chaudry M.H. 1993. *Open channel flow*. New Jersey, Prentince Hall.

Maza-Alvarez J. 1996. *Transporte de sedimentos*. Instituto de Ingeniería, UNAM, 504, Mexico.

Mendoza A. 2001. *Comparación de un modelo numérico y otro físico de la bifurcación Mezcalapa Samaria Carrizal*. Ba. Thesis. UNAM, Mexico.

Mendoza A. 2004. *Modelo de Flujo bidimensional de fondo móvil de la bifurcación Mezcalapa Samaria Carrizal*, Master Thesis, UNAM, Mexico.

*River, Coastal and Estuarine Morphodynamics: RCEM 2007 – Dohmen-Janssen & Hulscher (eds)*
*© 2008 Taylor & Francis Group, London, ISBN 978-0-415-45363-9*

# The failure of the Fonte Santa mine tailing dam (Northeast Portugal)

Mário J. Franca*, Luis Gézero & Rui M.L. Ferreira
*Instituto Superior Técnico, Av. Rovisco Pais, Lisbon, Portugal*

Sílvia Amaral
*Laboratório Nacional de Engenharia Civil, Av. do Brasil, Lisbon, Portugal*

Hugo D.B. Montenegro
*Instituto Superior Técnico, Av. Rovisco Pais, Lisbon, Portugal*

ABSTRACT: The Fonte Santa mine tailing dam situated in the Northeast Portugal failed on the 27th November 2006 due to a combination of hazards, an extraordinary rainfall and an eventual clogging of the spillway. The dam was an earthfill embankment about 25 m high and with a crest length of roughly 35 m. After the overtopping of the crest, which originated the breaching process and consequent failure, the dam was completely washed away with a fraction of the mud retained in the reservoir. The present paper constitutes a preliminary report of the accident describing the dam breaching process and the morphodynamic changes on the downstream valley. The breach geometry is characterized and estimates of the water and mud releases are presented. The geomorphic changes in the valley are described, including deposition volumes of the dam material and eroded volumes from the riverbed.

## 1 INTRODUCTION

The consequences of such flood events as the ones resulting from the collapse of a dam include economic losses related to lost project benefits and potential damage to property in the inundated area, loss of confidence in the dam owner and operators, alteration of the habitat and environment, social impacts on the local community and, most important of all, loss of lives. These consequences make dam break accidents amongst the most feared flood hazards. Within the dam safety context, tailing dams from abandoned mines present added risk given the lack of surveying and maintenance. The contamination levels of the by-products resulting from mining processes make this specific type of dams highly hazardous to the environment, which is demonstrated by the impact of the tailing dam failure of the Los Frailes lead-zinc mine at Aznalcóllar in the Doñana National Park (van Geen & Chase 1998 and Achterberg et al. 1999).

Being a rare event, the documentation of dam break accidents, including the breaching process and the downstream flood propagation, is of extreme importance to help assessing the phenomena and providing clues to further development of predicting

models. In the past, several authors have compiled important information related to accidents with natural and manmade dams, namely Babb & Mermel (1968), Johnson & Illes (1976), Combelles (1979), Serafim (1981), Ponce (1982), MacDonald & Monopolis (1984), Serafim & Coutinho-Rodrigues (1989), Lempérière (1993), Santos (1995) and Singh (1996), Franca & Almeida (2004), among others. Important contributions to the research on dam break floods were made with the detailed descriptions of such specific accidents as the Ha! Ha! lake flood (Lapointe et al. 1998) and the Tous dam break (Alcrudo & Mulet 2005). Both served as benchmark cases within the IMPACT Project (IMPACT 2005). Nevertheless, the phenomenology on the dam breach evolution in earthen dams is not sufficiently understood, as emphasized by Wahl (2001). When possible, field data provide variables describing the breach evolution, the extension of inundation areas, the magnitude of morphologic impacts and the celerity with which the dam-break wave progresses.

This paper documents preliminarily a tailing dam break occurred in November 2006 in a remote area of Northeast Portugal. We intend to document the causes associated to the accident as well as the breaching process and the effect of the passage of the flood wave through the downstream valley. The results are based on field visits, adequate topographic surveys and

---

* *Presently at: Dep. Civil Eng. and IMAR – Inst. Marine Res., Univ. Coimbra, Coimbra, Portugal*

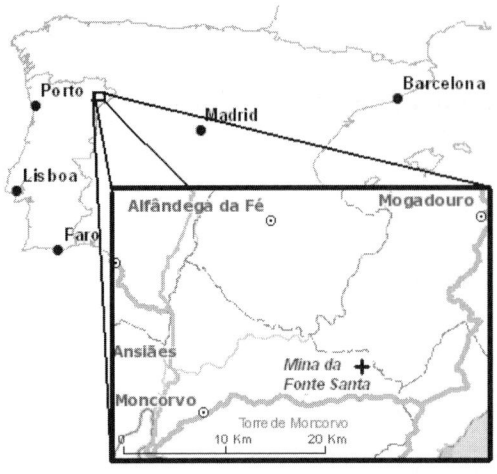

Figure 1. General location of the mining complex of the Fonte Santa. It is situated at the Northeast Portugal, north from the Douro river.

Figure 2. General view of the Fonte Santa tailing dam after the dam break. Almost the totality of the dam body was eroded. The dam is located in a rocky and very stable section of the valley. The mud deposited in the reservoir was released leaving a scar which is visible in this picture from downstream.

local meteorological records. Herein we describe the meteorological event, the dam breaching and the effect on the downstream valley. The breach geometry is characterized and estimates of the water and mud releases are presented. The topography changes in the valley are referred to, including alterations due to deposition of the dam material and due to riverbed erosion. Estimates of erosion and accretion rates in the downstream valley were assessed locally. A detailed field survey of the valley was made within the reach where geomorphic changes occurred.

## 2 CHARACTERIZATION OF THE DAM AND RESERVOIR

The Fonte Santa tailing dam is situated on the municipality of Freixo de Espada à Cinta (Bragança), on the creek Ribeiro da Ponte (Fig. 1; actually the creek changes its name in he dam section, upstream the dam it is called Ribeiro das Caravelas). It belonged to a mining complex abandoned for more than 30 years ago and its main function is to retain the mud resulting from the washing process of the extracted minerals. Fonte Santa tailing dam was an earthfill embankment about 25 m high and with a crest length of roughly 35 m. The crest elevation was at 505.0 m asl. It was not object of a special design; the construction was made progressively with coarse and fine material from the mining works. The material used to construct the dam was fine gravel ("tout-venant"). The shape of the dam was quite irregular, with a large amount of the fine gravel spreading downstream along the valley banks. The total volume of the dam embankment is estimated to be of 4 500 $m^3$. The mud deposited in the reservoir,

Figure 3. Reservoir where 1.5 $hm^3$ of the volume is filled with waste mud from the mining processes. After the accident, a temporary solution constituted by an embankment (visible in the photograph) was made to stop the downstream mud release. The mud deposited in the reservoir was released leaving a scar which is visible in this picture from upstream.

with $D_{50}$ of 0.0186 mm, contributed to keep the dam body impermeable along the years.

The dam is located in a section of the valley where a rock formation exists (Fig. 2). The flood discharge was made through an uncontrolled spillway excavated in the rock in an adjacent valley, with the upstream section at the level 504.4 m asl and a 3.5 m diameter roughly circular section. Eventually the upstream section of the spillway was clogged with rock material and its discharge was drastically reduced. The reservoir has approximately a maximum volume of 22.9 $hm^3$ for a water elevation of 505.0 m asl (crest elevation) and volume of 21.3 $hm^3$ for a water elevation of 504.4 m asl

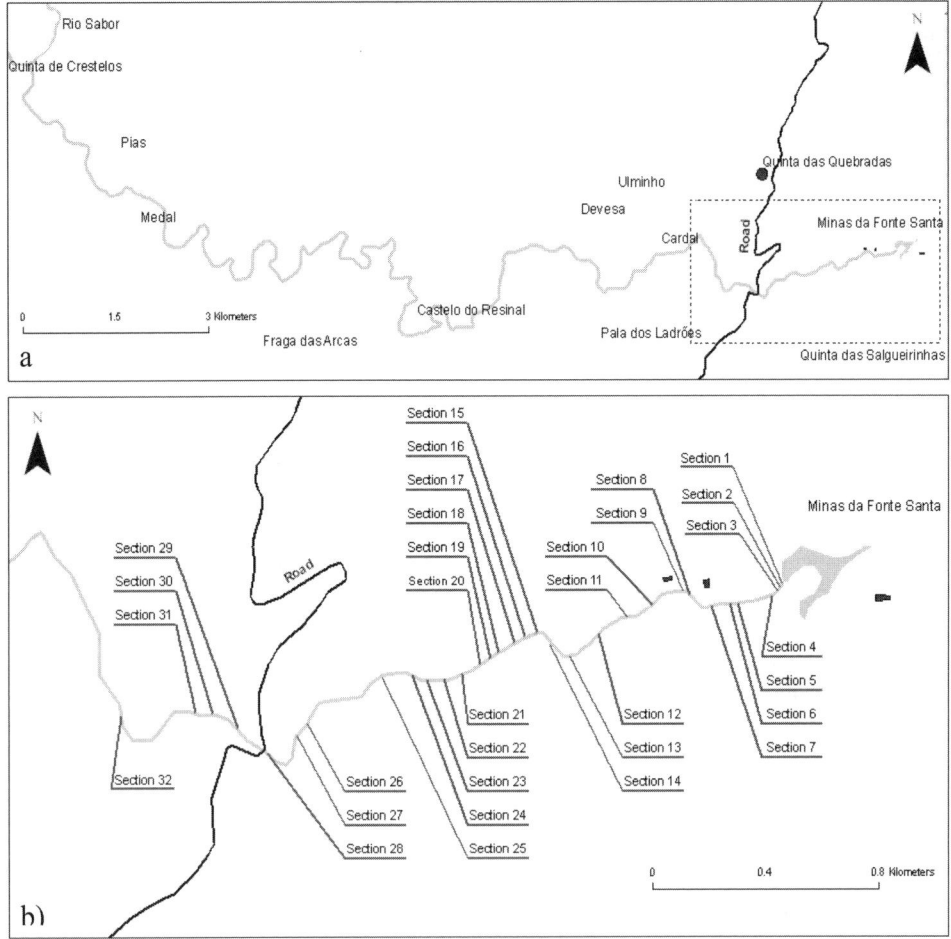

Figure 4. a) The valley of the creek Ribeiro da Ponte, between the dam section and the confluence with Sabor river (a major Portuguese river, one of the main tributaries of Douro river). Adjacent roads and towns are indicated; b) detail of the downstream valley with the sections where a detailed survey of the morphodynamic impact of the dam break event was made.

(spillway entrance). About 12.5 hm$^3$ of the reservoir is filled with the waste mud (Fig. 3).

On the 24th November 2006, an extraordinary rainfall occurred in the region. The continuous feeding of the reservoir for three days, combined with the clogging of the spillway, lead to the overtopping of the Fonte Santa dam crest originating breaching and subsequent total failure. The dam was completely washed with a portion of the mud retained in the reservoir (Fig. 2). Besides the observable morphological impacts, other immediate consequences of the accident were mainly loss of low density crops and eventual soil contamination.

Local contractors exploited illegitimately the embankment material (fine gravel used as a construction material) excavating directly from the downstream dam toe. These actions may have contributed to a

destabilization of the dam body and may be in the origin of the complete washout of the dam.

## 3 CHARACTERIZATION OF THE DOWNSTREAM VALLEY

The creek Ribeiro da Ponte flows into Sabor river, a major tributary of Douro river (Fig. 4). From the dam section until the downstream section of the creek, no major infrastructures or habitation exists within the floodplain. At a distance of 1800 m downstream the dam, a bridge exists from a secondary road linking two small towns. The main activity developed in the margins of the creek is low density agriculture for local subsistence; some cattle are breed in the floodplain of the creek. Abandoned structures such as mills and abandoned peasant villages exist.

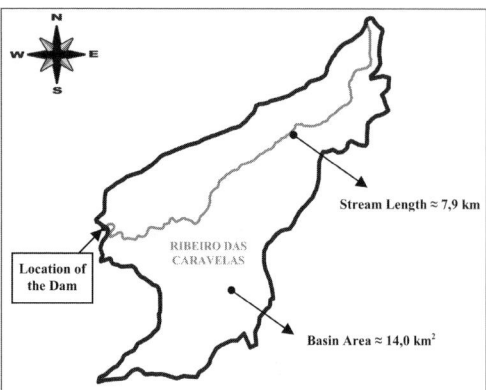

Figure 5.   River basin determined at the section of the Fonte Santa dam.

Immediately downstream the dam, roughly within the first 350 m after the dam, the creek section is very narrow and its bottom and banks are rocky thus stable. The access to the informal gravel exploitation referred previously was made through the downstream valley; for this purpose an embankment road was built in the valley providing access to the dam toe. This road was made with dam material and, after the break was washed away providing a good method to infer the erosion potential of the flood. This road was present in the first 400 m of the downstream valley and it still possible to see its remains. Downstream, the valley presents both, alluvial and fixed bed sections. A field survey allowed the identification of these different areas and an evaluation of the deposition and erosion within the valley.

## 4   HYDROLOGY OF THE EVENT

### 4.1   River basin

The dam of Fonte Santa creates a river basin of approximately 14,0 km$^2$ and its main watercourse is the Ribeiro das Caravelas (Fig. 5) which has about 7.9 km and can be qualified as torrential, attending to its longitudinal profile. The concentration time of this creek is about 3 h and the basin area soil occupation is characterized by a curve number of 90, considering the most humid antecedent conditions (AMCIII). This parameter was obtained from the CN (AMCII) which was estimated recurring to the commercial software ARC-GIS based on the Soil Hydrologic Characteristics Map available online (www.snirh.pt).

### 4.2   Extreme flood on the day of the event

On 20th November, 2006 a large low-pressure system was centered in the Atlantic North near Iceland. From the 20th till the 24th of November this system moved

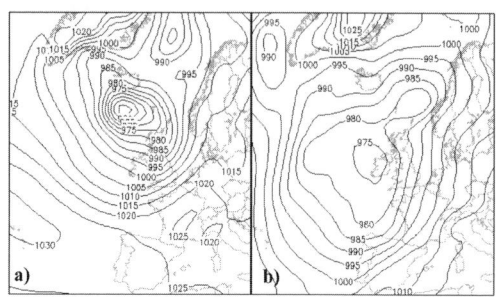

Figure 6.   Low-pressure system formed in the Atlantic North on the days a) 20th of November and b) 24th of November.

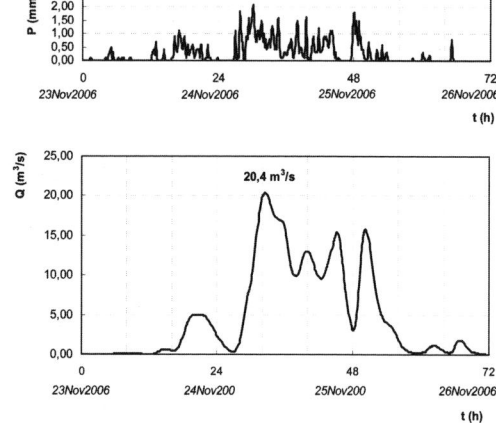

Figure 7.   Reconstructed hydrograph at the section of the Fonte Santa dam and Mogadouro precipitation hyetograph with a acquisition period of 10 m, for the 23, 24 and 25th of November.

towards the west coast of Ireland affecting the Portuguese territory as well. In the area of the Fonte Santa dam basin, a sea level pressure drop of approximately 25 mbar was observed between the days 20th and 24th of November (Fig. 6). After the 24th the system moved away from the Ireland west coast, causing a pressure ascent in all Portuguese territory. The 25 mbar pressure drop observed along those four days on the basin's area was the cause of the extreme precipitation occurred on the 24th of November.

The flood hydrograph presented in Fig. 7 was obtained with HEC-HMS 3.1.0, a software developed by the U.S. Army Corps of Engineers adequate to simulate precipitation-runoff processes of watersheds systems using, as input data, the above mentioned river basin characteristics and a pre-defined meteorological model. In the present case, observed hyetographs were used. These were measured at a meteorological station located in Mogadouro, a village distancing 12.5 km from Fonte Santa dam. This station was the only located on the interest area with precipitation registers useful in the characterization of the hydrology of the event. The meteorological data was kindly provided by the Portuguese Instituto de Meteorologia I.P.

According to the meteorological information, between the 23rd and the 25th of November 2006, a total rainfall of about 115 mm occurred on the basin area. 63% of this precipitation took place on the 24th, having the remaining precipitation occurred equally distributed between the 23 and 25th November. The peak flow of the hydrograph ($20.4 \, m^3/s$) at the Fonte Santa dam was obtained on the 24th of November at 8:30 am (Fig. 7). The total volume of water inflowing into the reservoir for the three days of rain was of $1\,300 \, m^3$.

## 5 DAM BREACHING

### 5.1 *Breaching process*

The understanding of the rupture process and the subsequent evaluation of the outflow hydrograph resulting from a dam failure is of primary importance as it constitutes an internal boundary condition for dam break flood models used for the risk management in valleys (Almeida et al. 2003). Several approaches are possible to estimate the breaching outflow hydrograph: (i) using historical dam failures data and regression approximations (cf. Wahl 2001); (ii) using semi-analytical methods established from the physical laws of breach progress and of reservoir depletion (cf. Singh 1996); and (iii) stochastic models (cf. Kast & Bieberstein 1997). However, uncertainties of about 50% in the estimate of the maximum discharge are still dominant on the results from existent models (CADAM 2000).

In the present case it is not clear how the breaking process occurred. As we may observe from the precipitation data, an extraordinary rainfall event took place in the region in the days 23rd and 24th November 2006. The meteorological information shows us that an 18 hours of continuous precipitation took place on the 24th of November. However, local accounts confirm the dam breaking event on the 27th November 2006; the description of a loud sound heard at a distance of more than 1000 m suggests a rather sudden wave travelling the valley. Two situations may have occurred: 1) the dam body may have been overtopped for a period of three days and a sudden destabilization of the body occurred finally on the 27th; 2) the spillway discharged the incoming flood efficiently and eventually got clogged on the 27th which induced the overflowing of the reservoir and consequent overtopping of the dam.

### 5.2 *Breach geometry*

Knowledge on the breach final configuration provides information for the calibration and validation breaching models (Singh 1996). Wahl (1997) affirmed that

Figure 8. Final configuration of the dam breach. In the figure, a reconstruction of what could be initially the geometry of the dam crest is represented in dashed; the crest was not horizontal. The mud deposited in the reservoir was released leaving a scar visible in the picture. The figure is not scaled and the dimensions are not proportional due to perspective. In the picture one may see works undertaken by the dam owner to stop the mud and water release.

Table 1. Breach final geometry.

| | |
|---|---|
| Top width (m) | 35 m |
| Bottom width | 11 m |
| Height | 25 m |
| Right riverbank slope (angle with the horizontal) | 50° |
| Left riverbank slope (angle with the horizontal) | 45° |

the characteristic parameters of the dam breach in embankment dams are the width, the depth the lateral bank slope and the formation time – see the overview by Singh (1996). The breach geometry was inferred in the field by adequate topography instrumentation. Locally we observed that the total volume of what used to be the informal dam body was washed away (80 to 90% of the $4\,500 \, m^3$, cf.Fig. 2) contributing to the large gravel accretion verified in the downstream valley. Fonte Santa breach has nearly a trapezoidal shape (Fig. 8). The dam breach reached the valley bottom, thus the erosion depth was about 25 m. Fonte Santa dam breach geometric parameters are presented in Table 1.

The presence of fine mud elements within the gravel composing the dam provided enough cohesion to hold such high average lateral slopes of the breach banks; locally the breach walls are nearly vertical. The deposition of the breach material was made along approximately 2500 m of the valley.

### 5.3 *Estimate of the water and mud releases*

By the configuration of the valley and the final breach configuration, we assume that the water released from the reservoir during the dam break was stored above the mud accumulated behind the dam body ($\approx$500 m asl). Taking into account the reservoir accumulation curve and the hydrograph in the dam section due to the extreme precipitation verified upon the accident, we estimate that the total amount of water released to the valley was roughly 230 000 m$^3$. According to Froehlich (1987) the maximum discharge may be estimated from the expression:

$$Q_M = 0.607 V_w^{0.295} h_w^{1.24} \qquad (1)$$

where $V_w$ = initial water volume above the final breach bottom position (m$^3$); and $h_w$ = the initial water height above the final breach bottom position (m). With expression (1), we obtain a estimation for the maximum discharge issued from the Fonte Santa breach of 3700 m$^3$/s, which seems rather an overestimation when regarding the inflow hydrograph and the reservoir volume. Froehlich's formula was empirically deduced using 22 documented dam accidents and it is widely used as a first approach. Subsequently, a breach model, calibrated with field data and the accounts of witnesses, will be used by the authors to simulate the reservoir routing during the event and to estimate a hydrograph resulting from the dam breaching.

During the dam break event the mud deposited immediately upstream the dam lost its stability and was eroded downstream, as it is evidenced by a scar in the reservoir bottom (Figs. 2 and 3). Locally we could infer that a volume of 1600 m$^3$ was released and disseminated in the valley.

## 6 MORPHODYNAMIC IMPACT ON THE DOWNSTREAM VALLEY

Dam break flood waves have a high erosive potential and are responsible for major geomorphic changes in the downstream valley as was demonstrated in the surveys made by Lapointe et al. (1998) and Alcrudo & Mulet (2005). The documentation of prototype dam break accidents are of extreme importance to test existent models as was made previously in INRS-Eau (1997) and Ferreira et al. (2005), among others.

An extensive field survey, supported by adequate topographic GPS-based surveying methods, was made along the downstream valley. GPS data was collected, in Fast Static Surveying, by a rover unit Leica-GS20. The minimal occupation time for each point was 3 minutes, with time intervals of 5 seconds. For the post-processing we used the Leica GISDataPro with a reference station based in Mirandela (coordinates in WGS84: Lat 41°31′00.41592″N; Long 07°11′10.19545″O; Ellipsoidal height 332.019 m; Mean Sea Level height 275.954 m). The reference station belongs to the GPS/GNSS network, managed by

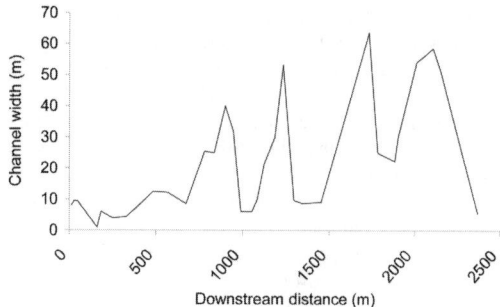

Figure 9. Lateral extension of the morphological impacts of the flood passage.

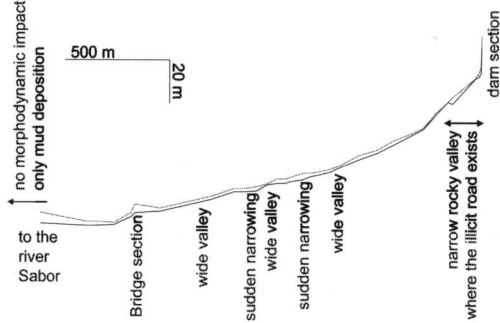

Figure 10. Valley profile with the maximum flood levels observed in the field.

the Portuguese Geographic Institute. The results of the post Processing reveal a mean position and height quality of approximately 20 cm.

The detailed survey was realized between the dam position and the last section where morphodynamic changes were visible (erosion or deposition of the riverbed and banks), see Figure 4. The dispersion of the mud released from the reservoir however was observed until the creek outlet at the Sabor river (cf. Fig. 4), though with no significant impact on the river morphology.

On Figure 9, we define the width across the channel where morphodynamic impacts from the passage of the dam break flood are visible, erosion or deposition of material.

When the field visit was made, flood marks corresponding to the maximum water levels occurred during the dam break flood were visible. This allowed the mapping of the flooded areas along the downstream valley. Figure 10 shows the maximum flood levels observed throughout the valley, caused by the dam break flood event.

The maximum water depth verified with the passage of the flood is consequent with the channel geometry (cf. Fig. 9). Just after the dam, in the narrow reach of the valley, the water depths were as high

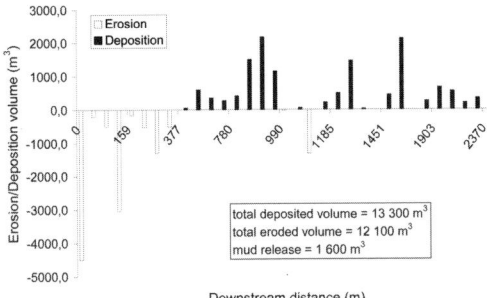

Figure 11. Profile of the estimated erosion and deposition volumes of the river bed, throughout the downstream valley and based on the field survey.

Figure 12. Deposition occurred on the riverbed, roughly 500 m downstream the dam section.

as 5.5 m. Downstream, major elevation of the water depths occurred occasionally before river constrictions and in the bridge section which was overtopped over 0.5 m. Upstream the narrowing sections, the destruction was more evident on the vegetation, riverbed and occasional natural or manmade structures, due to the apparent formation of a hydraulic jump.

Figure 11 shows the estimates of deposition and erosion volumes on the riverbed, obtained from local assessment.

The erosion peak observed at the upstream section corresponds to the erosion of the dam body (estimated in 4 500 m$^3$). The main erosion occurs immediately downstream the dam until roughly 380 m after the dam, where the valley was filled with the illicit road referred previously. Actually, for the sake of the morphodynamic impacts analysis, the existence of this structure was positive allowing a good estimation of the erosion power of the passage of the flood.

Downstream, mainly deposition of the dam material is verified due to the immense volume of fine gravel

Figure 13. Erosion occurred on the riverbank, roughly 1300 m downstream the dam section.

that composed the informal dam body (Fig. 12). Furthermore, part of the deposited material throughout the valley is originated from the gravel road embankment which is made from the same material of the dam body. The deposition occurred essentially in the river sections where the available flow area increased, allowing a reduction in the flow velocity and consequently the erosion capacity of the flow diminished (cf. Figs. 8 and 10).

Local episodes of bed erosion in narrower sections or due to local singularities in the channel geometry are observed after the section 380 m. Bank erosion (Fig. 13), though not assessed on Figure 11, was observed as well in sudden reductions of the available flow area and where the bank material allowed higher erosion rates.

The remaining downstream valley, beyond roughly 2 500 m after the dam, does not have important geometry changes; only a fine layer of mud is visible where the flood went through until the confluence with the Sabor river.

## 7 CONCLUDING REMARKS

As emphasized in the text, this paper constitutes a preliminary report on the dam break accident of the tailing dam belonging to the Fonte Santa mining complex.

The hydrologic extreme event conducting to an extraordinary inflow hydrograph in the reservoir is described. First estimates of the water and mud releases are presented. The maximum flood water depths as well as a calculation of the deposited and eroded volumes throughout the downstream valley are given. The eroded and deposited volumes throughout the area of the valley surveyed by the authors are in equilibrium, indicating that the main morphodynamic impacts produced by the flood wave are contained within this river reach.

The impact of such a flood wave as the one resulting from this event on the geomorphic characteristics of the valley was large. The quantification of this impact is of extreme importance calibration and validation of dam break flood propagation models. The first values herein presented give already useful information to be used by researchers and engineers working on the field. The work in progress on the documentation of this dam break event will produce a detailed report of the breaching process and geomorphic alterations within the downstream valley to be, hopefully, used as a benchmarking case.

At the moment the Fonte Santa dam is being reconstructed by the dam owner in order to sustain the remaining mud in the reservoir and to avoid future disasters and the spreading of the miming mud in the valley.

## ACKNOWLEDGEMENTS

The authors acknowledge the cooperation by Empresa de Desenvolvimento Mineiro, Instituto de Meteorologia I.P. and António Pedro Costa in the data collection and treatment as well as the financial support provided by the EU Project e-EcoRisk (EVG1–CT–2002–00068). M.J. Franca acknowledges the financial support by FCT through the grant BPD 21712/2005 and by Fundação Calouste Gulbenkian (88051).

## REFERENCES

Achterberg, E.P., Braungardt, C., Morley, N.H., Elbaz-Poulichet, F. & Leblanc, M. 1999. Impact of Los Frailes mine spill on riverine, estuarine and coastal waters in southern Spain. *Water Research* 33(16): 3387–3394.

Alcrudo, F. & Mulet, J. 2005. The Tous dam break case study. *Final Technical Report of the IMPACT project, Annex II, Part B WP3: Flood Propagation*, item XIX.

Almeida, A.B., Ramos, C.M., Santos M.A. & Viseu T. 2003. *Dam break flood risk management in Portugal*. Lisbon: Laboratório Nacional de Engenharia Civil.

Babb, A.O. & Mermel. T.W. 1968. *Catalog of Dam Disaster, Failures and Accidents*. Washington DC: Bureau of Reclamation.

CADAM 2000. *CADAM: Concerted Action on Dambreak Modelling*. CADAM Final Report.

Combelles, P. 1979. *Internal report*. Service de la Production Hydraulique – Electricité de France.

Ferreira, R.M.L., Leal, J.G.A.B. & Cardoso, A.H. 2005. Mathematical modelling of the morphodynamic aspects of the 1996 flood in the Ha! Ha! River. *XXXI IAHR Congress* Theme D: 3434–3445. Seoul.

Franca, M.J. & Almeida, A.B. 2004. A Computational Model of Rockfill Dam Breaching Caused by Overtopping (RoDaB). *Journal of Hydraulic Research* 42(2): 197–206.

Froehlich, D.C. 1987. Embankment-Dam Breach Parameters. *Proc. ASCE Conf. Hydr. Eng.*, Williamsburg (Virginia): 570–575.

van Geen, A. & Chase, Z. 1998. Recent Mine Spill Adds to Contamination of Southern Spain. *Eos Transactions (AGU)* 79(38): 449–455.

IMPACT 2005. *IMPACT: Investigation of Extreme Flood Processes and Uncertainty*. IMPACT Final Technical Report.

INRS-Eau 1997. *Simulation hydrodynamique et bilan sédimentaire des rivières Chicoutimi et des Ha! Ha! suite aux crues exceptionnelles de julliet 1996 – Rapport INRS-Eau No. R487*. Quebec: Institut National de la Recherche Scientifique.

Johnson, F.A. & Illes, P. 1976. A Classification of Dam Failures. *Water Power and Dam Construction* 28(12): 43–45.

Kast, K. & Bieberstein, A. 1997. Detection and Assessment of Dambreak-Scenarios. In A.B. Almeida & T. Viseu (eds), *Dams and Safety Management at Downstream Valleys*, Rotterdam: A. A. Balkema.

Lapointe, M.F., Secretan, Y., Driscoll, S.N., Bergeron, N. & Leclerc, M. 1998. Response of the Ha! Ha! River to the flood of July 1996 in the Saguenay Region of Quebec: Large-scale avulsion in a glaciated valley. *Water Resources Research* 34(9): 2383–2392.

Lempérière, F. 1993. Dams that have failed by flooding: an analysis of 70 failures. *Water Power & Dam Construction* September/October (5): 19–24.

MacDonald, T.C. & Monopolis, J.L. 1984. Breaching Characteristics of Dam Failures. *Journal of Hydraulic Engineering* 110 (5): 567–586.

Ponce, V.M. 1982. *Documented cases of earth dam breaches: Civil Eng. Series No. 82149*. San Diego: San Diego State University.

Santos, S. 1995. Dam Break Events – Accidents and Wave Propagatin, In J. Marques & J. Lima (eds), *Hydroelectric Power Plants*, Universidade de Coimbra: Coimbra.

Serafim, J.L. 1981. Safety of Dams Judged from Failures. *Water Power and Dams Construction* December: 32–35.

Serafim, J.L., Coutinho-Rodrigues, J.M. 1989. Statistics of Dams Failures: A Preliminary Report. *Water Power and Dams Construction* 41(4): 30–33.

Singh, V.P. 1996. *Dam Breaching Modeling Technology*. Dordrecht: Kluwer Academic Publishers.

Wahl, T.L. 1997. Predicting Embankment Dam Breach Parameters – A Needs Assessment, *Proceedings of Energy and Water: Sustainable Development*, São Francisco.

Wahl, T.L. 2001. The Uncertainty of Embankment Dam Breach Parameter Predictions Based on Dam Failure Case Studies, *Proceedings of USDA/FEMA Workshop on Issues – Resolutions and Research Needs Related to Dam Failure Analysis*. Oklahoma City (Ok).

*Beach and river nourishments*

*River, Coastal and Estuarine Morphodynamics: RCEM 2007 – Dohmen-Janssen & Hulscher (eds)*
© *2008 Taylor & Francis Group, London, ISBN 978-0-415-45363-9*

# Long-term simulation with 2DH and 3D models for nourishment on Mediterranean beaches

Philippe Larroudé

*Laboratoire des Écoulements Géophysiques et Industriels, UJF, Grenoble, France*

ABSTRACT: A modified 2DH morphodynamic and a 3D model was employed to simulate the evolution of large-scale features with major implications for beach nourishment. The study is focused on modeling the evolution of material artificially placed in different parts of the profile, extracting or adding material to the natural bars, and quantifying how the profile responds to different wave climates and nourishment placements. The simulated results were compared with field data from a Mediterranean beach.

## 1 INTRODUCTION

Examples of nourishment tests carried out on the near shore zone are few and far between in the relevant literature, compared to the many ones undertaken directly on the beach. SAFE, latest European project within the MAST program, acknowledges the absence of reference documents on this question, although such a technique could presumably constitute a less costly alternative (Hamm *et al.*, 2002).

The use of offshore bars to fight beach erosion, dating back to the 90th, was based on the fact that they represented a substantial reservoir of sediments. That theory turned out to be irrelevant, as beach nourishment requires coarser grain sizes. However, the essential role these bars can play in wave mitigation was evidenced by recent studies. Hence, working on reinforcing existing bars or even adding extra bars is a convincing approach, for they constitute a line of defence with no visual impact and are therefore environment-friendly. The method offers the added benefit of tapping abundant fine sands, easily available offshore, to build up the bars.

In addition, adequate depths in the inner shelf area would facilitate dredging and discharge operations and, the material reclaimed being usually clean, it can therefore be used directly without any processing. The core of the additional bar too could be made from marine mud, also easily available. All these assumptions should, of course, be systematically checked, the purpose of the exercise being to assess, through mid-term bathymetric evolution simulation, the consequences of the implementation of offshore bar nourishment and define the best location.

The understanding of these processes needs at this time the in situ data but also the development of models mathematics and numerical codes. Hence, following the work of De Vriend (1987) and De Vriend & Stive (1987), we try to improve the classic quasi-steady procedure. Objectives of this work will be therefore to model and to simulate processes of sedimentary transport on sandy beaches with varied weather conditions in the medium term time scale (from few days to few months).

## 2 PROCEDURE AND DESCRIPTION OF THE BEACH

Certain & Barusseau (2006) show that morphodynamic evolution of offshore bars in a microtidal environment and bimodal moderate wave regime follows two different conceptual models, the main one being a seasonal pattern in line with the observed cycle of hydrodynamic conditions.

The morphological evolution in the near shore region, including its large-scale features, was first investigated using a combination of a commercial 2DH model and a Multi1DH model (Camenen and Larroudé, 2003, 2003b). Simulation of the wave-driven currents was carried out with Telemac, a finite-volume elements model, and the Sisyphe sand transport module served to compute sediment transport rates and bed evolution. Since the sediment transport in the surf zone is mainly controlled by undertow, an undertow model (based on Svendsen, 1984) was added to account for that process.

These models were used in the framework of a simulated meteorological cycle describing the seasonal evolution of hydrodynamic factors. Results from monthly 2DH evolution simulations show a perfect fit with field data obtained on the "plage

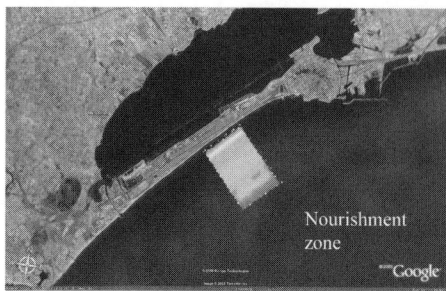

Figure 1. Localization of the "plage de la Corniche" in Mediterranean sea and zoom on the study area with the creation of an artificial offshore bar.

de la Corniche" in Sète (Certain, 2002). Morpho-hydrodynamic feedback of a bar having undergone reinforcing is also examined (see Figure 1).

## 3 THE CODES

The sedimentary evolution is modeling under the action of the oblique incident waves and is coupling with different numerical tools dedicated to the other process involved in the near shore zone. We can mention the following modules:

- module of wave with hold in account of the energy dissipation by surge (hyperbolic equation of extended Berkhoff), (LNHE, Artemis, 2002). The Artemis code (Agitation and Refraction with Telemac2d on a MIld Slope) solves Berkhoff equation taken from Navier-Stokes equations with some other hypothesis (little camber of the surface wave, little slope…).

  Main results are, for every node of the mesh, the height, the phase and the incidence of the waves. Artemis can take into account the reflection and the refraction of waves on an obstacle, the bottom friction and the breakers. One of the difficulties due to Artemis is that a fine mesh must be used to have good results when Telemac2d do not need such a fine mesh.

- module that calculates currents induced means by the surge of the waves, from the concept of radiation constraints gotten according to the module of waves, (LNHE, Telemac2d, 2002). Telemac2d is designed to simulate the free surface flow of water in coastal areas or in rivers. This code solves Barré Saint-Venant equations taken from Navier-Stokes equations vertically averaged. Then, main results are, for every node of the mesh, the water depth and the velocity averaged over the depth. Telemac2d is able to represent the following physical phenomena : propagation of long periodic waves, including non-linear effects, wetting and drying of intertidal zone, bed friction, turbulence, …

- sedimentary module integrating the combined actions of the waves and the current of waves (2D or 3D) on the transport of sediment, (LNHE, Sisyphe, 2002),

  The Sisyphe code solves the bottom evolution equation which expresses the conservation of matter using directly a current field result file given by Telemac2d. Four of the most currently empirical or semi-empirical formulas are already integrated in Sisyphe (Peter-Meyer, Einstein-Brown, Engelund-Hansen and Bijker formulas). We integrate two other ones which seem more appropriate to coastal sediment transport (Bailard, 1981 and Dibajnia-Watanabe, 1992). Main results are, for every node of the mesh, the bottom evolution and the solid transport.

- an hydrodynamic simplified model (called Multi1DH) use the following assumptions: a random wave approach, in a 1DH (cross-shore) direction. A offshore wave model (shoaling + bottom friction + wave asymmetry) is used with the break point estimation. The waves in the surf zone are modeled with the classic model of Svendsen (1984) with an undertow model (roller effect, Svendsen, 1984, Dally et al. 1984).The long shore current model is the Longuet- Higgins's model (1970).

## 4 RESULTS

### 4.1 Comparison for validation

Firstly we set up a procedure to use the coupled codes Artemis-Telemac2d-Sisyphe and especially we improved the treatment of the boundary conditions in order to be able to work on fields of calculations close to the coastal zone and equivalents in dimension for the three codes. We also used the Multi1DH code for the medium term simulations. These models were used for monthly simulations taking of account the weather conditions. These weather conditions are drawn from the data of ground for the period of November 2000 and are simplified in terms of height of swell, period of swell and direction by dividing the month into 9 significant periods (see Table 1). One can notice that the average height of the swells to broad during each period attenuated the weather events this November.

We obtain a good adequacy between numerical bathymetries after one month and those raised on the ground (see Figure 2).

We began simulations with a fattening of the zone of study at the beginning of November 2000 and we can compare the results obtained with model 2DH (waves, hydrodynamics and transport) and the model simplified Multi-1DH.

Secondly, we regarded as basic state a profile of the bathymetry of November 16, 2000, the P5 profile with X = 200 m (distance longshore compared to the

Table 1. Simplified weather data: November 2000 (Θ angle in degree in the trigonometrically direction reverses compared to the normal with the beach).

| Temps (s) | Hs (m) | Tp (s) | θ |
|---|---|---|---|
| 0j à 1j 21h | 0.244 | 7.45 | 25.475 |
| 1j 21h à 3j 12h | 1.703 | 7.92 | 27.861 |
| 3j 12h à 7j 21h | 0.351 | 7.13 | 28.094 |
| 7j 21h à 10j 9h | 1.787 | 6.76 | 6.065 |
| 10j 9h à 18j 12h | 0.222 | 6.2 | 3.97 |
| 18j 12h à 20j 3h | 1.358 | 6.78 | 14.9 |
| 20j 3h à 24j 15h | 0.251 | 7.03 | 14.33 |
| 24j 15h à 30j | 1.259 | 6.27 | −5 |

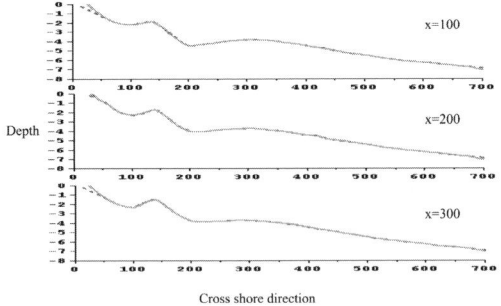

Figure 2. Sea bed for three location on the beach at the 25 November 2000, comparison between in situ data and numerical simulation after one month using the 2DH model with the meteorological model.

Table 2. Weather data simplified for the three cases of storm (Θ angle in degree in the trigonometrically direction reverses compared to the normal with the beach).

| | Temps | Hs (m) | Tp (s) | θ |
|---|---|---|---|---|
| TS | 24 h | 1 m | 6,5 s | 0° et 20° |
| FS | 24 h | 2,5 m | 7 s | 0° et 20° |
| ES | 24 h | 4 m | 10 s | 0° et 20° |

beginning of the zone of study). This approach will enable us to more easily compare the models of calculation used by the various partners of the program. For these simulations we agreed to consider three cases of climatic conditions (see Table 2): Traditional Storm (TS), falling from Storm (FS) and Exceptional Storm (ES). The whole of simulations was carried out with the model multi1DH and the results will be presented in the continuation. We also made some calculations with the chain of Artemis-Telemac2d-Sisyphis code.

For case FS, the swells do not have effects on the internal and external bars. For simulations TS the swells erode the internal bar but does not seem to attack

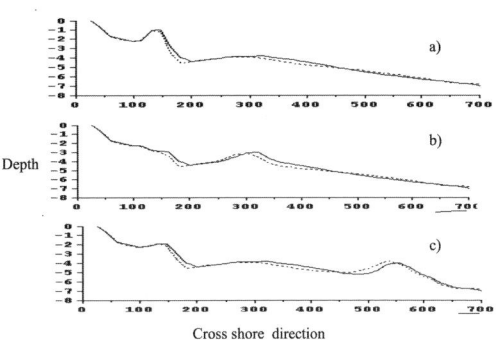

Figure 3. Sea bed evolution on 24 hours calculated with the multi-1DH model for three location of nourishment: a) on the inner bar, b) on the offshore bar and c) with the creation of new bar offshore.

to a significant degree the beach and the external bar. Only a longer-term erosion of the internal bar can be prejudicial with maintains of beach.

The case of the exceptional storms will be used to us as a basis to present the differences obtained with the model multi1DH between different the option from recharging. The case, on the basic profile of November 16, 2000, shows us an erosion of the bars internal and external with a transport of these bars towards the broad one and thus one can consider a weakening of the protection of the beach (see Figure 3). The internal bar is eroded of approximately 10 m and the external bar of 40 m, the deposit with broad with a maximum of 25 m but is very spread out.

These values are indicative to be possibly compared with other simulations but cannot be used as quantitative values for real estimates of the quantities of sands put moving. We will further see they is values are still strongly dependent on the models and in particular on the formulas of sedimentary transport.

### 4.2 Long term simulation

The good accuracy of this first result allows us to create a methodology of simulation for longer time scale. We are looking now with the coupled codes Artemis-Telemac2d-Sisyphe, the morpho-evolution of the beach with and without nourishment. The aim is to find the best way of create the simplified meteorological model from the in-situ data. In the study presented in the paper, we show the first step of this methodology which is to compare seasonal simulation with different script of meteorological event during each season. The principal question is do we have to cut out each hour, day ... Each meteorological event has time duration issued from the data. The Figures 4 and 5 show the importance of the storm in the destruction of the offshore nourishment bar. The major modification of the sea bed is numerically obtained because the time loop

1165

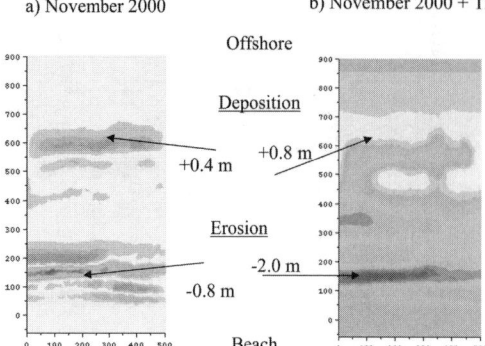

a) November 2000    b) November 2000 + T2

Offshore

Deposition

Initial    Final

+0.8 m

+0.4 m

Erosion

-2.0 m

-0.8 m

Beach

T2 : Hs=4m, Tp = 10 s for a period of 39h

Figure 4.  a) Sea bed evolution for a monthly simulation from the 04 11 to the 25 11 2000 with nourishment and b) the same with a storm of 39 h (initial on left, final on right).

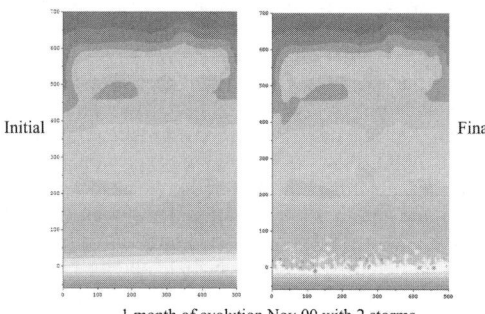

Initial    Final

1 month of evolution Nov 00 with 2 storms

Figure 5.  Iso-bathymetry of a monthly simulation from the 04 11 to the 25 11 2000 with nourishment and with a 2 consecutives storms (initial on left, final on right).

of the coupled codes Artemis-Telemac2d-Sisyphe is shorter than the time duration of the storm. The principal criteria to cut out could be the current velocity du to the waves. Indeed, when we have calm weather condition (e.g. small waves height) the feedback between the hydrodynamics and the sea bed evolution could be longer.

Some results show the numerical difference du to number of hydro-sedimentary loop we simulate for the same monthly or seasonal modeling. The differences are very important in the very near shore region and close the beach. The Figure 6 shows a complete year of simulation (the 4 seasons) with nourishment and with a cut out of 48 events. The simulation took 24 hours on a 2.36 GHz processor. The aim is to find in the same time the best accuracy of the yearly sea bed evolution and the lowest computational time. The other main goal is to be able to predict the best placement of the nourishment bar to protect the beach over five to ten years.

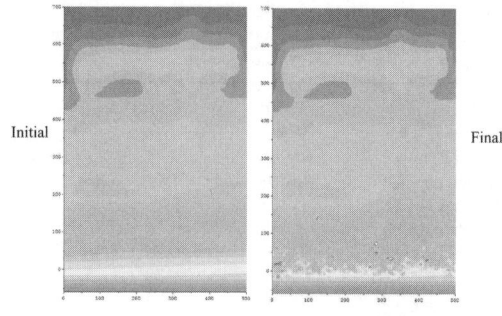

Initial    Final

1 Year of evolution Nov 99 to Oct 00 with nourishment

Figure 6.  Iso-bathymetry of a yearly simulation from November 1999 to October 2000 with nourishment (initial on left, final on right).

## 5 CONCLUSION

The interesting thing of this study is that we can compare our numerical results to the in-situ data. Then, it could be easier to modulate time steps to have the most realistic results. A modified 2DH morphodynamic model was employed to simulate the evolution of large-scale features with major implications for beach nourishment. The study is focused on modelling the evolution of sea bed and artificial material in the near shore region, extracting or adding material to the natural bars, and quantifying how the profile responds to different wave climates and nourishment placements. The simulated results were compared with field data from a Mediterranean beach over storm, monthly and yearly time scale. The results are good in term of quality and also in term of quantity for the velocity field du to waves. The first 3D simulations are in good agreement with the current data. The next step of the study is to simulate a large amount of seasons from the year 1994 to 2005 to be able to elaborate a criterion for the meteorological cut out. The second goal is to predict the next five years with different nourishment placement to have a good strategy for the beach protection.

## ACKNOWLEDGEMENTS

This paper is largely based on research work carried out in the task group Coastal Morphology of the LITEAU program supported by the French Government and the HUMOR program supported by the European Community.

## REFERENCES

Bailard J.A.,1981, An energetic total load sediment transport model for a plane sloping beach, Journal of geophysical research, vol.86 C11 pp. 10938–10954.

Camenen B. and Larroudé Ph., 2003, *Comparison of sediment transport formulae for a coastal environment*, Journal of Coastal Engineering, 48, pp. 111–132.

Camenen B. and Larroudé Ph., 2003b, Un modèle morphologique côtier pour la création de barres rythmiques, *Revue française de génie civil*, Génie côtier, vol. 7, pp. 1099–1116, 2003.

Certain, R., 2002, Morphodynamique d'une côte sableuse microtidale à barres : le golfe du Lion (Languedoc-Roussillon). PhD Thesis, University of Perpignan, 199 pp.

*Certain, R. and Barusseau J.P., 2006, Conceptual modelling of straight sand bars morphodynamics for a microtidal beach (Gulf of Lions, France), ICCE 2006, San Diego.*

Dally W.R., Dean R.G., and Dalrymple R.A. , 1984. A model for breaker decay on beaches. *In 19th Coastal Eng. Conf. Proc.*, pages 82–88. ASCE.

De Vriend H.J., 1987, 2DH Mathematical Modelling of Morphological Evolutions in Shallow Water, Coastal Engineering, 11, pp. 1–27.

De Vriend H.J. and Stive M.J.F., 1987, Quasi-3D Modelling of Nearshore Currents, Coastal Engineering, 11, pp. 565–601.

Dibajnia M. and Watanabe A., 1992, Sheet flow under nonlinear waves and currents, Coastal Engineering, pp. 2015–2029.

Hamm, L., Capobianco, M., Dette, H.H., Lechuga, A., Spanhoff, R., Stive, M.J.F, 2002, Asummary of European experience with shore nourishment, Coastal Engineering, 47, 237–264.

LNHE-Chatou, 2002, Telemac2d – modelisation system of Telemac, version 5.2 – user-validation manual, Technical report, edf-gdf

LNHE-Chatou, 2002, Sisyphe – modelisation system of Telemac, version 5.2 – user-validation manual, Technical report, edf-gdf.

LNHE-Chatou, 2002, Artemis – modelisation system of Telemac, version 5.2 – user-validation manual, Technical report, edf-gdf.

Longuet Higgins M.S., 1970, Longshore currents generated by obliquely incidentsea waves, Journal of geophysical research, vol 75, no 33, pp. 60778–60801.

Svendsen I.A., 1984, Mass flux and undertow in the surf zone, *Coastal . Eng.*, 8, pp. 347–365.

*River, Coastal and Estuarine Morphodynamics: RCEM 2007 – Dohmen-Janssen & Hulscher (eds)*
*© 2008 Taylor & Francis Group, London, ISBN 978-0-415-45363-9*

# Model assessment of sediment nourishment in rivers with graded sediment

C.J. Sloff
*WL\Delft Hydraulics & Delft University of Technology, Delft, The Netherlands*

J. Sieben
*Ministry of Transport, Public Works and Water Management, Institute of Inland Water Management and Water Treatment, Arnhem, The Netherlands*

ABSTRACT: Sediment nourishment is considered as an effective and flexible measure to reduce river-bed degradation. However, from an operational and economical perspective it is important that the efficiency of these operations is maximal. This requires a good understanding of the physical processes and morphological responses of the nourishment operations. As part of a planning study for management of the Dutch-German Rhine River a research study has been initiated to improve the knowledge on this topic. Firstly, the physics and impacts are studied using state-of-the-art mathematical models. Secondly, a trial-nourishment will be carried out in the upper Dutch Rhine River. The outcomes of the first part of this study are presented in this paper. Based on 1D and quasi-3D morphological simulations with a multi-fraction sediment model, different approaches for nourishment in the upper reach of the Dutch Rhine River have been analyzed. Not only to judge the impacts of different strategies, but also to evaluate the ability of the modeling tool to reproduce the physics. Tracers have been used in the simulations to determine the destination of nourished material. It can be concluded from these experiments that the thickness of the mixing layer is a key parameter in the spreading of the sediment (horizontal and vertical). Careful discharge dependent selection of this thickness is important with respect to time-scales of temporal storage of nourished sediment in the bed. Furthermore it has been shown that sediment must be dumped over a reasonable distance rather than at one point, and that nourishment of coarse material both enhances the efficiency to reduce the degradation, and influences the pool-bar pattern of the meandering river reach.

## 1 INTRODUCTION

Sediment nourishment or sediment feeding is considered to have great potential to effectively reduce river-bed degradation, while providing sufficient flexibility for adapting to changing conditions and anticipating on the high uncertainties in future morphology. It is therefore that sediment feeding has already successfully been applied in the German Rhine branches, and plans are drawn to extend the nourishments to the Dutch Rhine branches. Also for other rivers this approach is considered.

Sediment feeding operations are carried out by unloading sediment from bottom-dump barges. In figure 1 the unloading of the barge in the Rhine River near Rees in March 2007 is shown. Sediment with predefined composition is positioned precisely on the feeding site. The German feeding activities are ongoing at several locations along the German Rhine, as shown in figure 2, and plans are drawn to extend these activities to the lower Dutch reaches of the Rhine.

River-bed degradation is generally caused by disturbances in the sediment balance, for instance by

Figure 1. Sediment feeding in the German Rhine near Rees in March 2007.

blockage of sediment supply by an upstream dam or by the effect of river training works and dredging operations downstream. In the Rhine River a collection of mostly anthropogenic influences are causing a gradual

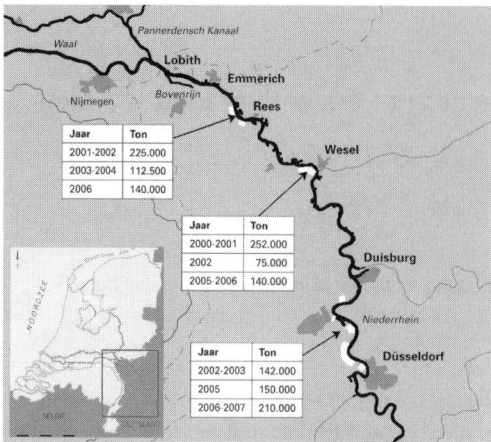

Figure 2. Sediment nourishment measures in German Rhine (Niederrhein) between 2000 and 2006.

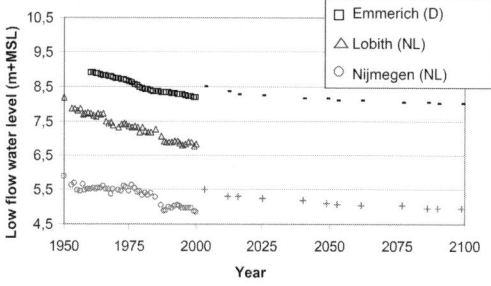

Figure 3. Observed decrease of low-flow levels in the Rhine branches in the German-Netherlands border area. Degradation rates are in the order of 1.5 to 2 cm/year.

bed degradation in the lower German and upper Dutch Rhine Branches.

The Rhine branches are an important navigation route, and it is very important from an economical perspective to maintain a safe and durable fairway. However, some fixed/non-erodible parts in the river do not go down with the degrading alluvial bed, and therefore will start to act as sills obstructing navigation. Especially the natural non-erodible layer at Emmerich gradually develops into a major bottle neck for navigation as low-flow levels continue to decrease (see figure 3). Therefore navigability of the Rhine is an important objective to stop further degradation.

For this reason Rijkswaterstaat (Ministry of Transport, Public Works and Water Management) has initiated a cross-boundary project "Duurzame Vaardiepte Rijndelta (DVR)" in which river managers in the Netherlands and Germany join forces to guarantee the navigability of the Rhine in the future. The study focusses on the fairway between Rotterdam (Netherlands) and Duisburg (Germany). At this moment

strategies are drawn in which combinations of normalization (river engineering works) and sediment management (dredging and nourishment) are studied. Most relevant tool for the studies is a two-dimensional (2D) morphological model for the entire river reach in which all relevant measures can be simulated.

For sediment nourishment or sediment feeding different approaches are being considered: feeding at one isolated location or over a certain distance, feeding with material with equal composition as the river bed or with coarser/finer composition, feeding at intervals or continuously, etc. From an operational and economical perspective it is important that the efficiency of these operations is maximal. That means that maximal counter effect is achieved with minimal time and volume of nourished sediment, and with minimal disturbance to navigation. This requires a good understanding of the physical processes and morphological responses of the nourishment operations. In this perspective Rijkswaterstaat has initiated a modeling study, in which one-dimensional (1D) and 2D models are applied to study the processes and impacts.

## 2 PHYSICS OF SEDIMENT FEEDING

River-bed degradation is a response to a positive gradient in sediment transport rate along the river (gradual increase of transport in downstream direction). Effectively sediment nourishment is used to level this gradient, for instance by replenishment of an upstream deficit of sediment supply (e.g., by feeding) or by decreasing the downstream transport capacity (e.g., by coarsening the bed).

The principle large-scale morphological response of a degrading river can be analyzed analytically by considering a parabolic model (Jansen et al., 1979). Based on this parabolic model de Vries (1975) defined a morphological time scale for degradation. For length scales larger than $x > 3hi^{-1}$, in which $h$ is average water depth and $i$ is the average bed slope, the flow and Exner equations reduce to the following model:

$$\frac{\partial z_b}{\partial t} - K(t)\frac{\partial^2 z_b}{\partial x} = 0 \qquad (1)$$

with

$$K(t) = \frac{1}{3}\frac{df(u)}{du}\frac{C^2 h}{u} \qquad (2)$$

where $C$ is the Chézy value, $K$ is a diffusion coefficient, $t$ is the time, $u$ is the average flow velocity, $x$ denotes the streamwise co-ordinate, and $z_b$ is the average bed level. Furthermore it is assumed that sediment transport $s$ (per unit of width) can be represented by a transport formula $s = f(u,$ other parameters). Using

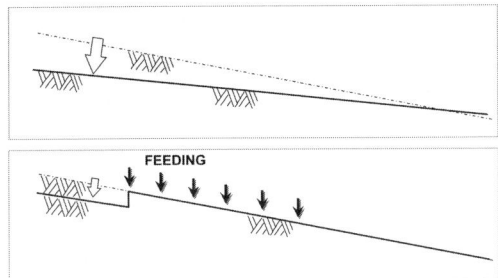

Figure 4. Schematic reproduction of equilibrium bed-level without (upper frame) and with sediment feeding (lower frame).

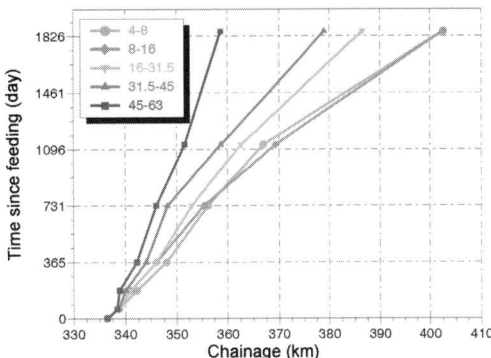

Figure 5. Longitudinal sorting effects of feeded sediment observed in the German Rhine by Gölz et al. (2006): propagation of tracer front of size fractions from 4–8 mm (finest) up to 45–63 mm (coarsest).

the analytical solution of de Vries (1975) it can be shown that considering a standard length $L_m$, a time scale $T_m$ can be defined which expresses the time after which the river bed is lowered by 50% at distance $x = L_m$ from a disturbed condition. If $T_m$ is expressed by the number of years $N_m$ then

$$N_m = \frac{L_m^2}{Y} \quad \text{with} \quad Y = \int_0^{1\,year} K(t)\,\mathrm{d}t \qquad (3)$$

Assuming a length scale of about 100 km (the lower Rhine River in the Netherlands), the value of $N_m$ is computed to be about 500 years. This implies that the observed degradation, which is assumed to be generated by anthropogenic influences in the last century, will continue in the future (although slightly decaying in time).

In figure 4, upper frame, is schematically shown how the average bed in a degrading river tends to reach eventually a new equilibrium bed with milder slope and largest bed-level lowering in the upper reaches. The lower frame shows how by appropriate sediment feeding the river bed at the nourishment section and downstream can be stabilized. It also shows that the section upstream without feeding still erodes, but final bed degradation is less than the original situation.

After feeding the created 'shoal' or 'sediment wave' migrates down. In the Rhine branches the celerity of this wave is in the order of 1 to 2 km per year. The influence of feeding therefore only slowly expands in downstream direction. Schwerdtfeger (2004) has shown that consequently degradation in the Dutch Rhine can only be treated effectively when sediment nourishment is distributed over a sufficiently long section. After all, the sediment-transport gradient causing the degradation acts upon the river bed over a long distance. According to Schwerdtfeger the amount of feeding needed to fully compensate the degradation is in the order of almost 50% of the yearly sediment transport rate, i.e. about 200,000 m³/yr or 7000 ton/week.

Due to the effect of selective sorting the downstream migration is not equal for all size fractions of a poorly sorted sediment feeding mixture. Tracer test in the German Rhine near Iffezheim in Germany (Gölz et al., 2006) reveal clearly how for such a mixture the front of the fines is migrating with a speed of 11 km/yr, while the front of the coarse fractions moves with a speed of about 4 km/yr. See figure 5.

The tracer tests in Germany also showed that the front speed is about twice the speed of the centre of the downstream migrating wave. This lead to a gradual dispersion of the wave in time.

It is supposed that the feeding of non-uniform or graded sediment can be beneficially used to modify the morphological response to sediment feeding. By feeding relatively coarse sediment the top layer of the bed will become gradually coarse, and transport capacity will reduce. Due to this effect stabilization of the bed can be reached with smaller volumes of feeding. Furthermore, it is expected that coarsening the bed will lead to a decrease of transverse slope in the river bends. The consequent lowering of point bars positively influences the navigability.

3 MODELLING CONCEPTS

The physical processes presented in the previous section require a modeling tool which is capable of routing the feeded sediment, and includes the effect of sorting of graded sediment.

In this study simulations studying the effect of sediment feeding have been carried out using both 1D cross-sectional averaged in the SOBEK modeling system, and a 2D depth-averaged morphological schematization in the Delft 3D modeling system. The flow and morphology in 2D are computed according

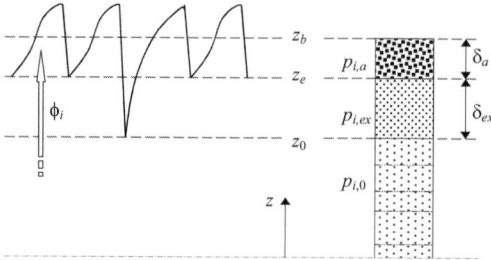

Figure 6.  Bed-layer schematization.

to Lesser et al. (2004) and Sloff et al. (2001). For simulation of sorting effects this model applies the basic bed-layer concept similar to that of Hirano (1971) as shown in figure 6, in which the bed is subdivided in transport layer with thickness $\delta_a$, an exchange layer with thickness $\delta_{ex}$ (not used in this study) and a non-moving substratum that is schematized by a number of sub-layers for which a book-keeping system takes into account the history of substrate composition.

In this figure sub-script $i$ is associated to sediment fraction $i$, $p_i$ is the probability of a size fraction, $z_b$ is the average bed level, and $\phi_i$ is the vertical sediment flux through the interface, with $\Sigma p_{i,a} = 1$, $\Sigma p_{i,ex} = 1$, $\Sigma p_{i,0} = 1$. The active layer represents the upper layer containing the material which is taking part in the actual sediment-transport process, and its thickness is generally related to the average height of bed forms (dunes, ripples). Changes in substrate during a simulation occur for instance when due to sedimentation processes material transported from upstream is deposited and added to the substratum.

The local bed-load transport rate per fraction is described using a standard transport formula. For the Rhine River simulations the Meyer-Peter and Müller formula (1948), prepared for graded sediment simulations with hiding and exposure following Egiazaroff (1965), adjusted by Ashida & Michiue (1972, 1973).

Due to the downhill gravitational transport component and the deflection of near bed shear stress by helical flow in bends, the transport direction in the 2D model does not necessarily coincide with the direction of the depth-average flow velocity. The effect is introduced by a parameterization of spiral flow and a slope effect according to Talmon et al. (1995). For graded sediment this formula has been adapted to the following relation:

$$\tan(\beta_i) = \frac{\sin(\alpha) - \dfrac{1}{f_{si}}\dfrac{\partial z_b}{\partial x}}{\cos(\alpha) - \dfrac{1}{f_{si}}\dfrac{\partial z_b}{\partial y}} \qquad (4)$$

$$\text{with} \quad f_{si} = A_{sh}\left(\frac{\tau_b}{\rho_w g \Delta D_i}\right)^{B_{sh}}\left(\frac{D_i}{h}\right)^{C_{sh}}\left(\frac{D_m}{D_i}\right)^{D_{sh}}$$

in which $A_{sh}$, $B_{sh}$, $C_{sh}$, and $D_{sh}$ are calibration coefficients, $\tau_b$ is the bed-shear stress and $\Delta$ is the relative density of the sediment under water, defined by $\Delta = (\rho_s - \rho_w)/\rho_w$, $\rho_w$ and $\rho_s$ are mass densities of water and sediment, respectively. However, as very little is known about the proper formulation for graded sediments, the distinction between different values for different size fractions is switched off by setting $C_{sh} = 0$ and $D_{sh} = 0$. The usual value for $B_{sh}$ is 0.5, and for the Rhine the value of $A_{sh}$ is 0.9. The direction of sediment transport is found to be of major importance for the development of typical 2D morphological features.

The computational models (both 1D and 2D) are using a finite-difference approach for which the river and reservoir are schematized on a numerical grid. In 2D approach a curvi-linear grid is applied.

The simulations must include the effect of the discharge hydrograph because the main erosion of the displaced sediments occurs during flood events (when the deposits are submerged).

## 4 MODEL SETTINGS AND BOUNDARY CONDITIONS

The area of interest for applying the feeding measures is the Boven-Rijn, which is the 10 km long river section between the German border and the main river bifurcation Pannerdense Kop (km 858–868). At this location a pilot study (following this modelling study) is defined in which sediment will be dumped in two outer bends, and will be followed by intensive monitoring.

This reach of the river is characterized by a main channel of about 350 m width, groyne fields on both sides, and wide flood plains. The characteristic discharge hydrograph for the simulations is based on either actual recorder discharge series or statistically averaged discharge series, for which discharges ranges between approximately 1,000 m³/s and 10,000 m³/s. Bed material in this reach is poorly sorted and non-uniformly distributed in space. Median grain sizes in samples taken in this reach range from 1 to 15 mm (computed $D_{50}$ is in the order of 5 mm). Complicating is the presence of a stable gravel/sand transition exactly at this location. In the models the sediment mixture is represented by 8 size classes.

Yearly transport of bed material is about 300,000 m³ in real river and in models (excluding wash load). It increases gradually in downstream direction (gradient of about 100,000 m³/yr), resulting in a bed-degradation of about 3 cm/year in this reach. Furthermore, the used sediment load results in a migration speed of large-scale bed features in the models of about 1 to 1.5 km/yr, which corresponds to the observed celerity.

It is important to mention that the calibration of Meyer-Peter and Müller transport formula provided

1172

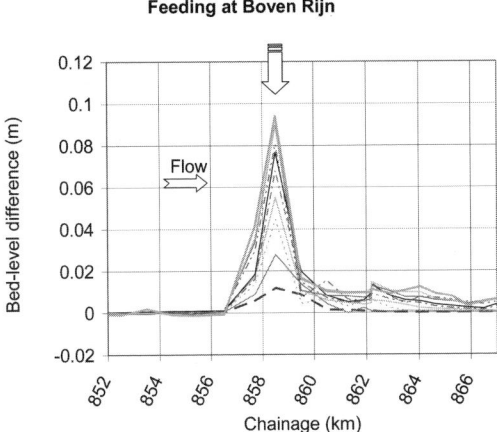

**Feeding at Boven Rijn**

Figure 7. Difference in computed bed level between a simulation with and without sediment feeding at km 858.5: yearly difference for period of 10 years.

best results if the critical Shields value in this relation is reduced from the standard value 0.047 to 0.025, or if the hiding and exposure relation (which acts upon this critical Shields value) is based on the median grain size $D_{50}$ instead of the arithmic mean diameter $D_m$. ($D_{50}$ is smaller than Dm, such that hiding of fines is reduced and exposure of coarse sediment is reduced). Analysis of these results indicated that coarse size fractions in the active layer do have an disproportional effect on transport rates. Most likely this is a deficiency related to the fact that the coarsest fractions are collected in the dune throughs where they contribute less to transport processes than following from the uniform distribution that is assumed in the active layer (Blom, 2003).

Another important conclusion is that the active layer thickness ranges from about 0.5 m (at low discharges and small bed forms) to about 3 m (at high discharges, large bed forms). An average active layer thickness of 1 m is applicable. These relatively large thicknesses are required to reproduce the bed composition and morphology, and to stabilize the sand/gravel transition.

## 5 SIMULATIONS

The 1D model has been applied to analyze the feeding of 10,000 m³/yr at km 858.5 in the Rhine River with a sieving curve comparable to the average composition of the river bed.

The results for a continuous feeding over a period of 10 years are shown in figure 7. Much sediment seems to build op at the feeding location, whereas only a small layer is found in the downstream reach. The built op sediment can be prevented by distribution the dumping sites over some distance along the river. The growth of this shoal coincides with a strong local increase in grain size, preventing erosion of the shoal.

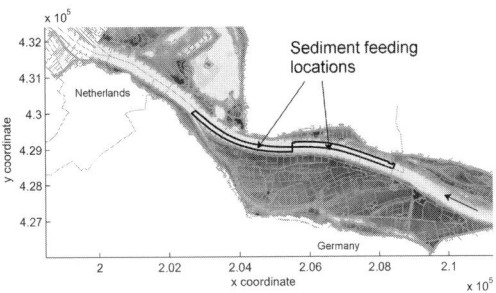

Figure 8. Location of sediment feeding.

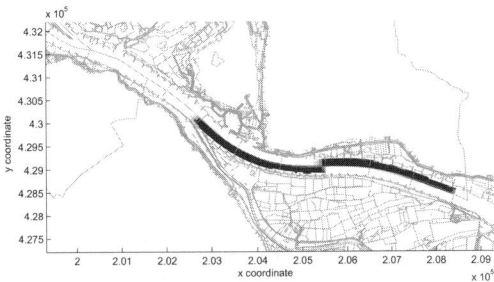

Figure 9. Amount of tracer in bed (dark color is highest concentration) directly after feeding.

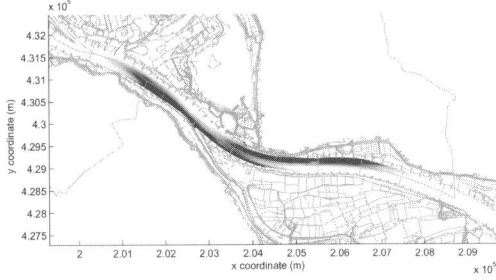

Figure 10. Amount of coarse tracer ($8 < D < 16$ mm) in bed (dark color is highest concentration) after 5 years.

The 2D model has been used to analyze the feeding of sediment at two locations. To route the sediment we have applied tracer fractions. This is done by adding separate size fractions to the sieving curve (overlapping the existing ones in the bed) with initially zero availability in the active layer. After feeding only these new size fractions it is possible to track these fractions in the model.

The model results presented below refer to a single feeding experiment of 150.000 m³. The sediment is dumped in two outer bends of the river as shown in figure 8. The composition of the sediment roughly corresponds to that of the river bed in that reach.

In figures 9 to 11 the portion of tracer is shown for the situation directly after feeding, and the situation

1173

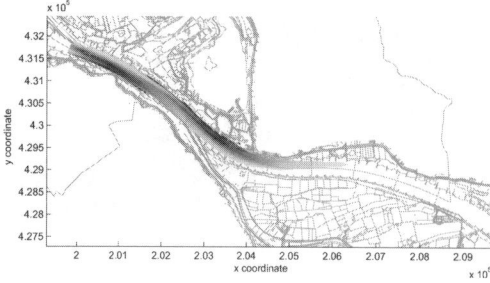

Figure 11. Amount of fine tracer ($0.5 < D < 1$ mm) in bed (dark color is highest concentration) after 5 years.

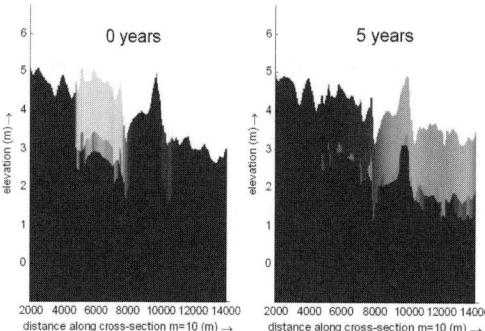

Figure 12. Side view of simulated river bed with under-layers direct after feeding, and after 5 years. Light colors represent high concentration of tracer 5 ($2.8 < D < 4$ mm).

after 10 years. Figures 10 and 11 show the difference in migration of the coarse and fines.

Figure 12 shows the side view of a section of the river along the river axis directly after feeding, and after 5 years. It shows that tracers penetrate partially into deeper layers, and remain temporary stored at lower levels.

The resulting bed-level difference after 5 years is shown in figure 13. The reduction of degradation is especially concentrated at the deeper parts of the cross-sections (such as pools in bends and near groynes).

## 6 DISCUSSION

The simulation results do reproduce the time-dependent behavior of feeded material as found from tracer experiments in prototype at Iffezheim (Gölz et al., 2006). The dispersion of sediment waves and different migrations speeds for different size fractions is simulated by the model as well.

In figures 14 and 15 is shown for two fractions how the feeded sediment tracers move down within the active layer as a damped wave with a speed of about

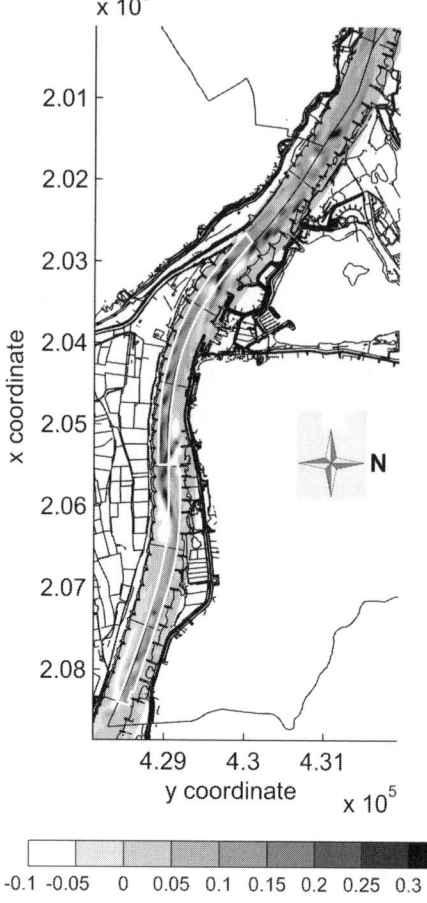

Figure 13. Bed level difference (simulated bed with feeding minus simulated bed without feeding) after 5 years.

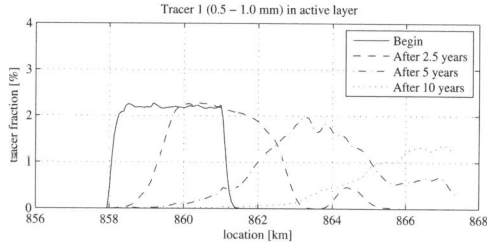

Figure 14. Migration of coarse tracer along the river as function of time (active layer, right-hand side of river).

0.3 km/year for the coarse fraction and about 1 km/year for the finer fractions. For the coarse fractions there are some interesting accelerated waves running ahead which seem to have an origin in quasi-3D effects, and need more analysis.

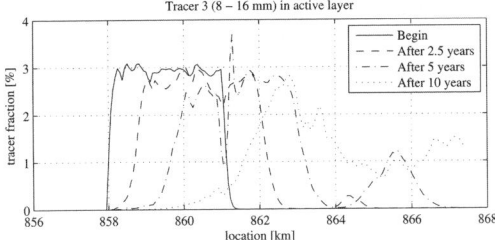

Figure 15. Migration of fine tracer along the river as function of time (active layer, right-hand side of river).

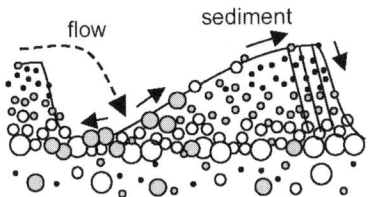

Figure 16. Vertical sorting of sediment size in a dune by Blom (2003).

The migration speed of the waves in these figures is highly dependent on the active-layer thickness. A thin layer leads to a more rapid adjustment of bed composition, and therefore leads to a higher speed. Predictability of the models is therefore highly dependent on the chosen value or model for this layer thickness. The dependence on the layer thickness has already been shown by means of analysis by means of the method of characteristics, e.g. Ribberink (1987) and Sieben (1997). Sometimes the migration of disturbances of bed level can be separated from those in the sediment mixture (two types of waves with different speeds). However, in general these disturbances in level and composition are closely interacting and can therefore not be decoupled. For instance, analysis of some of the feeding results by using the analysis of characteristics revealed that the migration speed also depends on the ration between the grain size of transported sediment and grain size of the sediment in the active layer. This ration particularly depends on the transport formula and the used hiding and exposure function, in relation with the active-layer approach. In the Rhine River bed forms generate a vertical sorting process which causes coarse sediment to collect at deeper levels, such that there is less coarse sediment available for transport than what is available averaged over the active layer (Blom, 2003), see figure 16. Correcting for this effect by using more appropriate vertical sorting models, e.g. as proposed by Blom, might therefore also be considered for improving the predictability.

It should be remarked that simulations show that a very small active-layer thickness (e.g., order of 0,1 m)

may lead to a rapid decrease of migration speed in time (and possibly even stagnation) of the feeded material. This can be explained by a rapid response of the bed composition that is developing to eliminate the local transport gradients invoked by the sediment wave. Generally such a gradient leads to a simultaneous response of bed level (erosion or sedimentation) and of bed composition (coarsening or fining). A final equilibrium is then reached after all gradients in transport have disappeared, either by smoothened bed, or by grain size gradients (such as armoring). In a model with a very thin active layer this equilibrium is reached by bed composition changes before noticeable morphological changes have occurred. This phenomenon has been explained by Mosselman and Sloff (2006). In general this effect can already be observed at the feeding location in situations with a portion of very coarse (immobile) sediment which rapidly armors the dumped sediments if the active layer is thin. For a thicker active layer more time is needed to develop a stable armor layer, such that bed has more time to erode.

In line with this comment it is important to realize that the active-layer thicknesses, as well as the transport rates are coupled to the discharge rates. During high discharges the celerity should increase because transport rates increase. However, as the dunes (and hence the layer thickness) increase this will lead to a decrease of speed. And vice-versa holds for the low discharges. Both the actual physics and the impacts on simulations of this complex system-behavior are not well studied yet for the Rhine River reach (Mosselman and Sloff, 2006).

When looking at the effects of feeding on transverse sorting and bed-level changes it has been observed that feeded sediment is convected downstream without rapid transverse sorting. The quasi-3D model shows that the tracers migrate downstream for full length of the model on the side (left or right part of the channel) where they are feeded. There is only an insignificant influence of spiral flow and transverse-slope effects that leads to a transverse sorting of the tracers in the bends downstream of the feeding site. Only when very large quantities of sediment are feeded (by a large instantaneous feeding event, or by continuous feeding) the feeded sediments are more rapidly distributed over the full width.

Furthermore, there is a slight tendency of coarsest tracers to be stored temporary on the point bars in inner bends (and slow down), whereas the fines continue migrating downstream without being stalled.

The typical quasi-3D behavior seems to have an important effect on the The computed (and observed) longitudinal sorting of size fractions also leads to a coarsening of the bed downstream of the feeding location. Especially if the single feeding event is relatively large (e.g., 300,000 m$^3$) large amounts of coarse sediments from the nourishment lag behind, while fines migrate down at a much higher rate. Consequently the slowly migrating coarse wave effectively leads to coarsening of the bed.

# 7 CONCLUSIONS

From the application of mathematical models to study the morphological effects of sediment feeding operations in the Rhine River can be concluded that they can be used to assess the faith of feeded sediment and the impact on bed level and bed composition. The simulation also reveals that the predictability is governed by the following aspects:

- the vertical sorting processes within the active layer, and the impact on composition of the active layer and of transported sediment (relevant for routing of feeded sediment)
- the thickness of the active layer, in combination with the transport capacity and its variations with discharge (relevant for routing of feeded sediment)

Furthermore for feeding of sediment it is found from the analysis that

- the efficiency is dependent on location and rate of feeded sediment, i.e. preferably the sediment is nourished evenly along a reach, with special attention to the side of the river.
- Feeding of a sediment mixture with a reasonable amount of coarse sediment will lead to (temporary) coarsening of the active layer, such that degradation is slowed down.

Uncertainties with respect to the model concepts, notably for layer thickness, can be reduced for the considered reach by carrying out a feeding experiment in the prototype. Plans are presently drawn to carry out this experiment, followed by an intensive monitoring campaign. The field experiment is the next phase of this project, and to be executed in the coming years.

# ACKNOWLEDGEMENTS

The modeling study has been carried out as component of the project "Duurzame Vaardiepte Rijndelta" (Sustainable fairway maintenance and improvement in the Rhine), and was financed by Rijkswaterstaat. A major part of the modeling study has been carried out by Saskia van Vuren, for which she is greatly acknowledged. Also the contributions to discussions and analysis by Erik Mosselman of WL | Delft Hydraulics have been essential for the outcomes of this study.

# REFERENCES

Ashida, K., Michiue, M., 1972. Study on hydraulic resistance and bed load transport rate in alluvial streams. Transactions, JSCE, 206, pp.59–69.

Ashida, K., Michiue, M., 1973. Studies on bed-load transport in open channel flows. Proc. Int. Symp. on River Mechanics, IAHR, Bangkok, Thailand, paper A36, pp.407–418.

Blom, A. (2003) A vertical sorting model for rivers with non-uniform sediment and dunes. PhD thesis. University of Twente, The Netherlands.

Gölz, E, H. Thesis, U. Trompeter (2006) Tracerversuch Iffezheim. Bericht BfG-1530, BfG-JAP-Nr. 2348, Bundesanstalt für Gewässerkunde. Germany.

Hirano, M. (1971) River bed degradation with armouring. Trans. of JSCE, Vol. 3, Part 2.

Egiazaroff, I.V., 1965. Calculation of non-uniform sediment concentrations. Journal of Hydraulic Division, ASCE, Vol.91, No.HY4, pp.225–247.

Lesser, G.R., Roelvink, J.A., Van Kester, J.A.T.M., Stelling, G.S., 2004. Development and validation of a three-dimensional morphological model. Coastal Engrg., Vol.51, Nos.8–9, pp.883–915.

Meyer-Peter, E., Müller, R., 1948. Formulas for bed-load transport. Proc. 2nd Congress IAHR, Stockholm, Paper No.2, pp.39–64.

Mosselman, E. and C.J. Sloff (2006) The importance of floods for bed topography and bed sediment composition: numerical modelling of Rhine bifurcation at Pannerden. In: Gravel-Bed Rivers 6 – From process understanding to river restoration. Edited by H.Habersack, H.Piegay, T.Hoey, M.Rinaldi P.Ergenzinger, Elsevier, in press.

Sieben, J. (1997) Modelling of hydraulics and morphology in mountain rivers. PhD. Thesis. Delft University of Technology, The Netherlands.

Sloff, C.J., H.R.A. Jagers, Y. Kitamura, P. Kitamura (2001) 2D morphodynamic modelling with graded sediment. Proc. of 2nd IAHR Symp. on River, Coastal and Estuarine Morphodynamics 10–14 sept. 2001, Obihiro, Japan.

Schwerdtfeger, C.M. (2004) Sediment feeding to reduce bed degradation of the Rhine River near the German/Dutch border. MSc thesis HE177, Unesco-IHE, The Netherlands.

Talmon, A. C., Struiksma, N. & Van Mierlo, M. C. L. M. 1995. Laboratory measurements of the direction of sediment transport on transverse alluvial-bed slopes. *J. Hydraul. Res.* 33, 495–518.

*Sedimentation and effect of structures*

River, Coastal and Estuarine Morphodynamics: RCEM 2007 – Dohmen-Janssen & Hulscher (eds)
© 2008 Taylor & Francis Group, London, ISBN 978-0-415-45363-9

# Mathematical modelling of silting in the Kugart River, Kyrgyzstan

G. Zolezzi
*Dept. of Civil and Environmental Engineering, University of Trento, Trento, Italy*

R. Repetto
*Dept. of Structures, Water and Soil, University of L'Aquila, Italy*
*on sabbatical at Dept. of Bioengineering, Imperial College of London, UK*

A. Siviglia & M. Tubino
*Dept. of Civil and Environmental Engineering, University of Trento, Trento, Italy*

M. Toropov
*Kyrgyz Russian Slavic University, Bishkek, Kyrgyzstan*

M. Serafini
*Dept. of Civil and Environmental Engineering, University of Trento, Trento, Italy*

ABSTRACT: The Kugart River, Kyrgyzstan, has been partly channelised in the past decades with the aim of reducing flooding risk for the villages of the surrounding, fairly densely populated floodplain. The reduction of the channel width was expected to improve the channel conveyance of the high sediment load produced by the river basin. The resulting longitudinal sequence of relatively sharp channel expansions and contractions has, however, triggered rapid siltation rates in the narrower reaches. Such a relatively unexpected dynamics motivated the application of a numerical model to understand the causes of the silting processes. Migrating sediment waves related to channel expansions and contractions are found to be the main driving forces of aggrading and degrading processes. The outcomes of the analysis allow to suggest possible mitigation measures to reduce the high silting rates and disclose the main parameters which govern the process.

## 1 INTRODUCTION

The Kugart River is a snow and rainfall fed piedmont river flowing in the central west part of Kyrgyzstan, Central Asia (Figure 1). It flows from the mountains of the Fergana Range to the boundary with Uzbekistan, where it joins the Kara Darya River. Its upper catchment produces high sediment loads, which potentially increase hydraulic risk for the lowland, fairly densely populated floodplain. The discharge regime is dominated by snowmelt in springtime while the largest floods originate when intense rainfall events sum to meltwater discharge.

Several regulation measures have been implemented in the last decades in order to cope with hydraulic risk and have focused on increasing sediment and flow conveyance capacity in the neighbourhood of human settlements. This has resulted in local artificial reduction of the natural river width, from original values of around 200–300 m, through

Figure 1. Location map of the Kugart River reach under investigation.

series of bank protection levees built in the '70s and in the '80s. Reducing the channel width is, indeed, generally expected to increase the sediment transport capacity and this criterion is broadly adopted in the river engineering practise to avoid in-channel sedimentation.

Channelisation has actually been one of the most ubiquitous measures used to modify river geometry during the last two centuries with the aim of navigation improvement, land reclamation and flood control (e.g. Petts 1989), especially in the most industrialised countries during the 20th century (e.g. Brookes 1988). Almost invariably channelisation works imply a reduction of channel width at local and regional scales. In the last two decades many undesired consequences of channelisation works, related to negative impacts on the riverine-floodplain ecosystems and with repeated and unforeseen failures in preventing flood-related damages, have posed the need for new approaches in the perspective of river restoration, which has been developing in the last decades as a new alternative able to comply with both human safety and ecological needs (Stanford et al. 1996; Brookes and Shields 1996). Under this umbrella widening of target river reaches is increasingly being implemented to initiate self-forming morphodynamics, a condition that has proved to be effective in hosting riparian species and improving in-stream habitat and plant diversity (Rohde et al. 2005).

A common feature between restoration-oriented local widenings and channelisation is the frequent occurrence of significant channel width variations, which pose the need to predict the altimetric bed response to both gradual and sharp spatial channel widening and/or narrowing. Longitudinal spatial width variations induce changes in sediment transport capacity and, in turn, trigger morphological changes of bed profile. Temporal variations of channel width are known to be related to river bed aggradation and degradation: narrowing is commonly associated with bed incision (e.g. Surian and Rinaldi 2003; Wyzga 1993), while widening often leads to aggradation (Simon and Thorne 1996), even though opposite correlations have been documented (see the ASCE Task Committee 1998 for a comprehensive review). Among the few available studies on spatial width variations, Repetto et al. (2002) have studied the altimetric equilibrium bed response to periodic longitudinal width variations, trying to identify the conditions which lead to the generation of a channel bifurcation.

The lower Kugart River represents a striking example of how abrupt spatial width variations may induce significant bed deformation over long distances. The most upstream and downstream parts of the considered reach (denoted as $UU$ and $DU$ in Figure 2) are in nearly natural conditions while the middle reaches ($C1$ and $C2$) are channelised. The main characteristics of such reaches are reported in Table 1.

Width reduction after river regulation works has triggered intense silting, especially in the channelised reach $C2$. Despite a relatively frequent activity of mechanical sediment cleaning along such a reach, the aggradation process has determined a higher thalweg

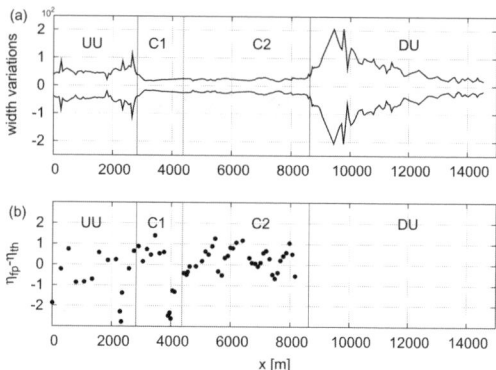

Figure 2. (a) Free surface width $b$ and (b) difference between flood plain $\eta_{fp}$ level and thalweg level $\eta_{th}$ along the Kugart River reach under consideration.

Table 1. Average characteristics of the Kugart River reaches under investigation.

| River reach | $UU$ | $C1$ | $C2$ | $DU$ s |
|---|---|---|---|---|
| Width [m] | 100 | 60 | 50 | $100 \div 400$ |
| Average slope | 0.009 | 0.007 | 0.006 | 0.005 |
| $d_{50}$ [mm] | 50 | 50 | 35 | 48 |
| Length [km] | 2.8 | 1.6 | 3.9 | 6.3 |
| % of hanging reach | 43% | 58% | 75% | – |

elevation with respect to the surrounding floodplain and has contributed to the failure of the right protection levee in the area of Suzak village (reach $C2$) by extreme floods in 1998 and in 2003. The occurrence of bed aggradation is demonstrated in Figure 2b, where the difference $\eta_{fp} - \eta_{th}$ between the average floodplain level and the thalweg elevation is plotted along reaches $UU$, $C1$ and $C2$. As shown in Table 1, it appears that the percentage of reach length in which the bed level is higher than the surrounding floodplain (i.e. $\eta_{fp} - \eta_{th} > 0$) is quite large in all reaches, attains its maximum value (75% of the length) in reach $C2$ and progressively decreases in the upstream direction.

Aggradation within the channelised reaches can be associated with the modified planform configuration, consisting of a sequence of channelised narrow reaches and unconstrained wide reaches. Figure 2a shows the presence of sharp channel cross-section variations along the Kugart River. The upstream reduction of the percentage of reach length in which the riverbed is higher than the floodplain, from reach $C2$ to reaches $C1$ and $UU$, seems to indicate upstream elongation of a deposition prism generated at the outlet of reach $C2$, where, due to a width increase from approximatively 60 m to ~200–300 m, the sediment transport capacity dramatically decreases.

In the present paper we report on the activities and the outcomes of the KG-ASSP-1 "Kugart River silting" project, aimed at understanding the causes of such an unexpected behaviour. The unsteady channel bed evolution has been modelled making use of a 1D numerical model. Possible mitigation works have also been proposed on the basis of the outcomes of the numerical model.

## 2 PROJECT KG-ASSP-1 "KUGART RIVER SILTING"

The project consisted of the mathematical modelling of the downstream reach (about 40 km) of the Kugart River, located immediately upstream of the confluence with the Kara–Darya River. The mathematical model was developed to allow the assessment of the conditions of hydraulic risk due to flooding and silting of along the Kugart River. The project was aimed at providing scientific support for decision making on the most effective river protection measures that need to be adopted to improve community safety in the lower Kugart River basin, including the monitoring procedures that need to be regularly implemented.

The project was commissioned by the Department of Water Resources of the Kyrgyz Republic to the Department of Civil and Environmental Engineering of the University of Trento (Italy) and lasted from March 1st, 2005 to October 31st, 2005. Other institutions took part in the project, namely the Kyrgyzdortransproject Design Institute and the JSC Kyrgyzsuudolboor. Both these institutions were in charge of collecting field and existing data in collaboration with staff of the Kyrgyz Russian Slavic University of Bishkek ("KRSU"). The project was an activity within the broader framework of a cooperation program, in place since 2004, aimed at sharing knowledge on sediment transport and river morphodynamic modelling.

The Kugart Silting project and other projects are pushing towards a technological improvement concerning mathematical modelling of river systems in Kyrgyzstan. Mathematical modelling appears at present as the best available tool to predict river bed evolution over long periods of time (years) and large spatial scales (tens of kilometres). The knowledge sharing program moves from the need to upgrade appropriate human competences, in order to allow river management decision making and related interventions to be increasingly locally owned and managed. The eventual aim is therefore to make river management interventions more effective and sustainable in the long term. Local demand for technical capacity upgrading in the sector has evolved during the Kugart Silting Project. In fact, local specialists have become increasingly aware of the kind of expertise that

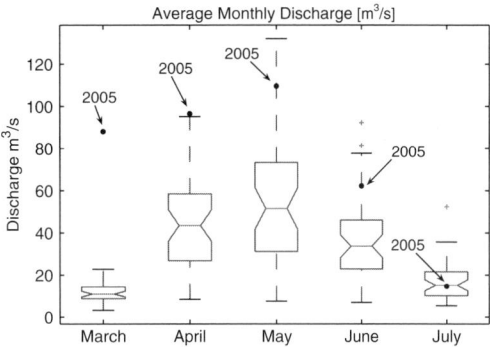

Figure 3. Box & whiskers plot relative to the monthly averaged water discharge in the Kugart River (1935–2001). Only months during which morphological activity occurs are reported. Data relative to the year 2005, for which daily data are available, are also plotted with black solid circles.

is required to be able to apply, and eventually develop, mathematical models of river morphodynamics.

## 3 FIELD DATA COLLECTION AND SITE CHARACTERISATION

The reach under investigation has been thoroughly characterised by a field activity that took place in 2005 under the framework of the Project KG-ASSP-1. We summarise the hydrology, topography and sedimentology information relevant for the present study and refer the interested reader to the Final Project Report (Tubino et al. 2005) for further details.

The Kugart river has satisfactory and reliable hydrological records. At Mikhailovka hydropost, located at the exit from the mountains, approximately 50 km from the mouth, water discharge was regularly monitored from 1935 to 2001 and monthly averaged data are available. The annual precipitation in the Kugart River basin changes from 500 mm in the downstream part up to 1500 mm in the upland reach. Most significant discharges are however due to snow melt. The box-plot reported in Figure 3 shows the monthly averaged values of the discharge computed from the available data (1935 to 2001). Morphologically relevant flows occur from March to August: 80% of the annual flow occurs at this time of the year, the highest flow being regularly observed in May.

During the year 2005, daily water discharge measures were taken from March to August. The monthly averaged values of these daily data are also reported in Figure 3 with black solid circles, showing that in 2005 water discharge has been quite large.

Channel geometry has been exhaustively characterised in 2005 through a field survey which included 177 cross-sections, the longitudinal thalweg profile,

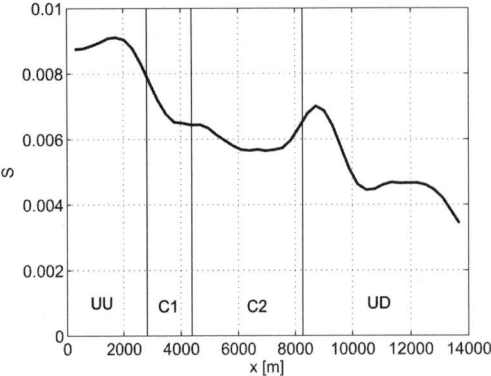

Figure 4. Average bed slope $S$ along the study reach of the Kugart River.

river banks and the surrounding floodplain together with the structures that interact with river dynamics (motor and foot bridges, levees). A GPS System 500 "Leica" with two receivers and three total stations have been employed; the whole survey has been geo-referenced with respect to the local plane zone coordinate system. The present paper focuses on the 15 km long lowest part of the study reach, depicted in Figure 1.

The longitudinal bed slope, calculated with reference to the lowest point of each cross section, decreases downstream (Figure 4 and Table 1) ranging from 0.008–0.01 in the unchannelised upstream reach $UU$ to an average value of $0.007 - 0.006$ in the channelised part ($C1$ and $C2$).

Grain size sampling has been carried out collecting 128 separate samples in 21 cross-sections, with a minimum of four samples in each cross section, characterising the surface and the sub-surface of the active riverbed and the adjacent floodplain. The longitudinal distribution of the characteristic diameters $d_{50}$ and $d_{90}$ indicates that sediment size is approximately constant ($d_{50} \approx 50$ mm) except for a weak fining in the channelised reach $C2$, suggesting a finer sediment composition of the depositional prism with respect to the average.

## 4 BED EVOLUTION IN CORRESPONDENCE OF CHANNEL EXPANSIONS: SIMPLE TEST CASES

The one-dimensional mathematical model employed solves simultaneously the De Saint-Venant and Exner equations. The numerical solution is obtained employing the MacCormack's TVD method (Garcia-Navarro and Alcrudo 1992), which is widely used for solving open-channel problems, because it can be easily implemented to work with a variable spatial step, as typically

required in modelling natural rivers, and because it captures shock waves with second order accuracy both in time and in space. We do not report any detail of the mathematical model here for the sake of space, the interested reader is referred to Siviglia et al. (2006).

We first apply the mathematical model to simple test cases with the aim of understanding the main features of river bed evolution in correspondence of a channel expansion and identifying the main dimensionless controlling parameters of the problem. In particular we consider a simple geometry consisting of an upstream channel with rectangular cross-section and width $b_u$ expanding to a downstream channel, again with rectangular cross-section, and width $b_d$. The expansion reach has length $L_w$ and, since we want to focus on sharp and abrupt width variations, we have assumed $L_w = 5b_u$. Numerical tests have shown that changing $L_w$ from 1 to 10 does not lead to qualitatively different results. We define a dimensionless longitudinal spatial coordinate $X = x/b_u$, with $x$ the dimensional coordinate. The origin of $X$ is in correspondence of the starting point of the expansion reach.

In order to isolate the effect of the width disturbance on the system, upstream and downstream boundary conditions are imposed far enough from the expansion reach to avoid their come into play. Obviously, for very long simulations an equilibrium configuration would be eventually reached, which is entirely set by the boundary conditions, and is such that sediment discharge is constant throughout the domain.

The upstream reach flow conditions are completely determined, in dimensionless form, by three parameters. We adopt in the following the Shields parameter $\vartheta$, the width to depth ratio $\beta$ and the Froude number $Fr$. Moreover we fix the value of the ratio between upstream and downstream widths $b_u/b_d$. In all the numerical runs the transport capacity is larger in the narrower (upstream) reach and it is such a difference in the sediment transport capacity between upstream and downstream reaches which triggers morphological activity.

Various values of the controlling parameters have been tested. Space does not allow here a complete exposition and discussion of the obtained results. As representative examples, we just report in Figures 5a and b instantaneous dimensionless bed profiles for two different values of the Froude number (with $\beta$ and $\vartheta$ of the upstream channel kept fixed). Figure 5a is relative to a subcritical flow, while Figure 5b to a supercritical flow. Bed profiles are relative to the initial bed configuration, which consisted of a constant channel slope in equilibrium with the upstream conditions. Each profile corresponds to a different value of the dimensionless time $T$, where time has been scaled with the morphological scale $[T = t/(Db/Q_s)]$. Such a choice for the scaling of time allows a direct comparison of plots relative to different flow conditions.

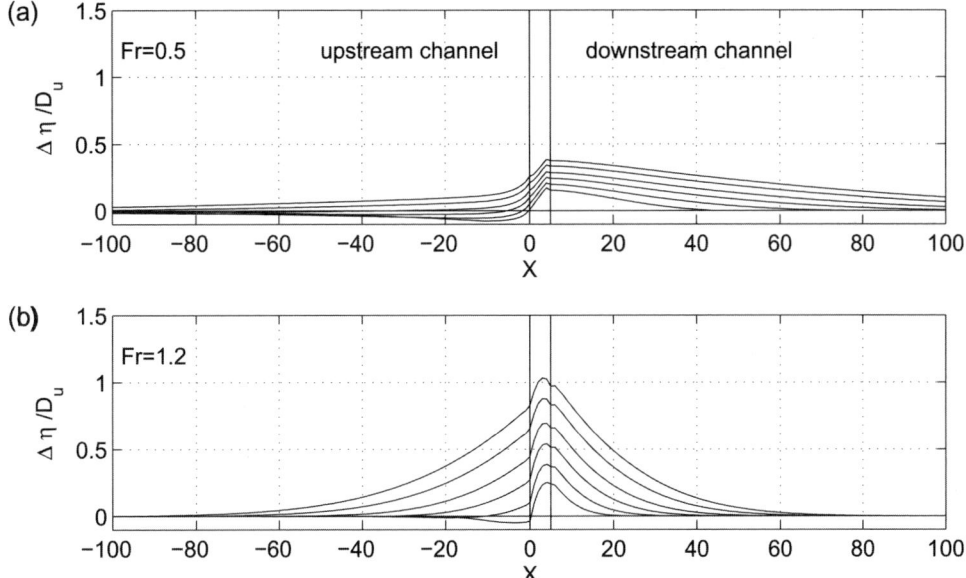

Figure 5.   Morphodynamics of an expansion: time evolution of the dimensionless bed profiles for two different values of the upstream uniform Froude number. $\Delta\eta(X,T) = \eta(X,T) - \eta(X,0)$ ($T = 5, 15, 30, 50, 110$; $\beta = 10$, $\vartheta = 0.1$, $b_u/b_d = 0.5$, $L_w/b_u = 5$).

In the initial stage of the evolution process a deposition occurs in the downstream reach while just upstream of the expansion a scour is generated. This very initial behaviour of the system is obviously strongly influenced by the initial condition adopted. After this initial stage deposition occurs both in the upstream and downstream reaches, regardless of the value of $Fr$. The initial scour in the upstream reach is progressively filled by upstream propagation of the deposition prism generated at the expansion. This prism increases its maximum height, which is invariably located within the expanding reach, and lengthens in time. Both the upstream and downstream reaches tend to be subject to strong aggradation. This shows that higher uniform sediment transport rates are not necessarily associated with river bed degradation as one may think on a intuitive basis. The prism elongation is higher in sub-critical than in supercritical flow conditions, implying that the dimensionless propagation speed of the deposit front is larger in subcritical conditions; moreover the upstream propagation of the deposit starts later (in dimensionless terms) at low values of Froude, due to the fact that scouring in the upstream reach lasts longer. The deposition height also strongly depends on the Froude number and is much higher for large $Fr$. These results suggest that the shape of the deposition prism is strongly dependent on the Froude number, the prism being much thicker at large $Fr$. Moreover, the aggradation process is more intense and fast as the ratio $b_u/b_d$ is decreased.

The above results have a great practical importance for river management. Indeed, it is found that a channel expansion invariably induces aggradation, not only in the downstream reach (where the uniform transport capacity of the flow is smaller) but also in the upstream channel (where the uniform transport capacity can be very large). The aggradation process is much more intense in supercritical flow conditions triggering very high depositional prisms.

## 5   RESULTS FROM THE KUGART RIVER CASE

We now apply the numerical model described in section 4 to simulate bed evolution in the lower Kugart River. Results presented in the previous section may help to understand the behaviour of river bed within the Kugart River since a very strong channel expansion exists at the end of the channelised reach C2.

We have performed simulations employing the daily hydrograph measured during the 2005 field campaign in the Kugart River. As shown in Figure 3, we have thus simulated a year characterised by an intense water discharge. Also notice that we have only run the code to simulate months from March to July, during which morphological activity occurs.

Field observations suggest that the upstream end of reach $UU$ has not been subject to significant aggradation/degradation processes after the banking works along reaches C1 and C2 were completed. We have then imposed the constancy of bed level in this section. The downstream boundary condition, in correspondence of the confluence of the Kugart River with the Kara-Daria River, actually depends on the morphological activity of the latter river, whose investigation was beyond the scope of the project. However, sediment transport along reach $DU$ is fairly weak, even at high stages, because river width is very large there. We therefore do not expect the downstream boundary condition to play a very important role on bed evolution within the channelised reaches $C1$ and $C2$, which are of greatest interest. Results presented below have been obtained imposing the constancy of bed level at the end of the domain. Sediment diameter $d_{50}$ has been assumed constant and equal to 50 mm. Finally, as an initial condition we have employed the bed topography surveyed during the 2005 field campaign.

Bed profiles, relative to the initial configuration, predicted by the model at the end of every month of simulation are reported in Figure 6a along the whole of the considered river reach. Results show the occurrence of strong bed aggradation in the channelised reaches $C1$ and $C2$. As discussed above, this is in good agreement with field observations (see discussion in section 1 and Figure 2b). Deposition within the channelised reaches is originated by two different effects. First, channel slope decreases along reaches $C1$ and $C2$ (see Figure 4) and, since channel width is fairly constant there, this implies a decrease of transport capacity of the flow and a consequent bed aggradation. Moreover and most importantly, in agreement with results presented in the previous section, the abrupt channel expansion at the end of reach $C2$ induces the generation of an upstream elongating deposition prism. The upstream migration of the deposit can be best appreciated in the last two months of simulation. Notice that the flow at the end of reach $C2$ is in critical conditions. This implies that, according to the previous discussion, we are in a situation in which the deposition prism generated at the expansion grows quickly and may reach a significant height.

The solution that is most likely to be technically successful in limiting the risk associated with silting and flooding in the channelised reach is to increase the lateral distance between embankments in reaches $C1$ and $C2$. In Figure 6b we show that bed aggradation in the channelised reach can be significantly reduced if channel width along reaches $C1$ and $C2$ is increased to the value of 100 m. As an initial condition for the numerical simulation we still have employed the 2005 topographic survey but river bed in the channelised reaches has been obtained by smoothing the actual thalweg with a moving average. It appears the deposition at the end of reach $C2$ almost vanishes in the present situation. The benefit arising from channel widening can be related to three main effects. (i) The expansion ratio $b_u/b_d$ has increased toward

unity and this limits the longitudinal variation of sediment transport capacity. This is the most important effect. (ii) The Froude number in the channelised reach has decreased to a value of approximatively 0.8: as previously discussed, as the Froude number decreases aggradation occurs over longer time scales and the deposition prism keeps smaller. (iii) The morphological time scale increases, implying a slowing down of the aggradation process. Reach $C1$ and the upstream part of reach $C2$ are still subject to aggradation, mainly due to the downstream decrease of bed slope. However, such a deposition is significantly lower than in the case reported in Figure 6a. Moreover, after the first two months of simulation, bed level seems to approximatively reach an equilibrium configuration.

## 6 DISCUSSION AND CONCLUSIONS

Managing the Kugart River is problematic due to both the natural configuration of the river basin, which produces very large sediment discharges, and to the present, human-modified, geometrical configuration, which is the outcome of past interventions that should have been carried out jointly considering all the relevant issues in the framework of a catchment scale view. In this respect the mathematical modelling of the Kugart River has allowed a satisfactory insight on the fluvial processes that lead to intensive silting in the channelised reach.

Predictions of morphological evolution of the Kugart riverbed under the present geometrical configuration confirm field observations and indicate that the most critical part of the river is the channelised reach ($C1$ and $C2$). In particular reach $C2$ is subject to a fast upstream migrating aggradation wave due to the sudden reduction of sediment transport capacity occurring in the transition from reach $C2$ to reach $DU$; this mechanism is particularly intense as the flow is in critical/supercritical conditions.

We have proposed, on the basis of purely morphodynamical considerations, an enlargement of the channelised reach to a width of 100 m. This leads to a significant improvement as far as silting is concerned. Deposition in certain reaches is still expected to occur and will require periodic mechanical cleaning of sediments, although with a greatly reduced frequency with respect to the present configuration. We have also tested several other possible solutions, alternative to the widening of the channelised reach, which, however, did not lead to satisfactory results in facing the problem of safe operation of the Kugart River.

In the perspective of building a general coordinated plan for the Kugart River management, there are relevant issues that could not be properly investigated in the framework of the KG-ASSP-1 project and that will necessarily deserve future assessment. In

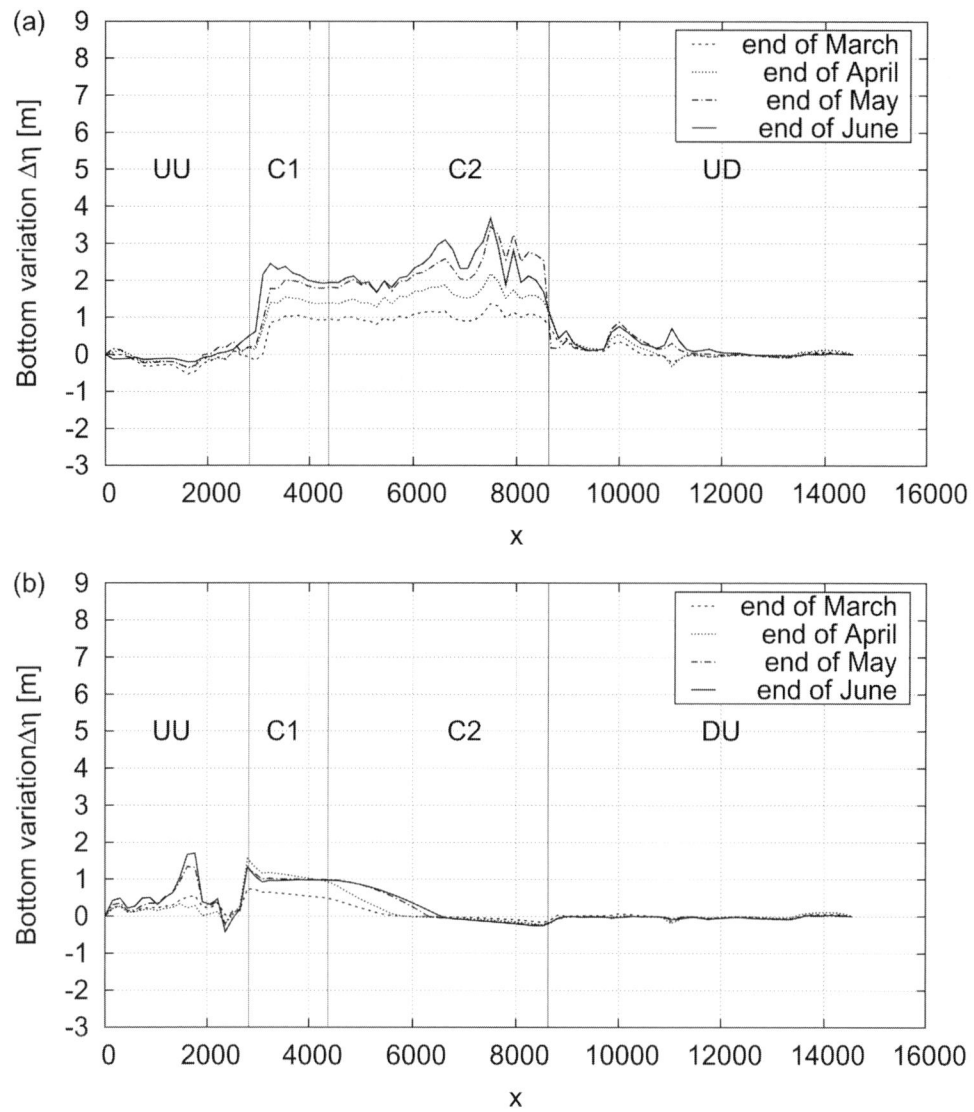

Figure 6. Predicted bed evolution corresponding to the present (a) and to the proposed (b) channel configuration in the low Kugart River.

particular, it is crucial to regularly monitor the river response to flood events and to engineering works that will be implemented in the future. The monitoring activities will have to include at least survey of bed topography after each flood season and continuous water level measurements at gauging sections. Such regular monitoring would allow a direct comparison between predictions of mathematical models and the real river response. In the perspective of a catchment-scale approach to river management, a more detailed data record on the management strategies of flow

regulation in the Kara Darya River will also need to be acquired in order to quantify the possible effect of the downstream boundary condition.

ACKNOWLEDGEMENTS

We are grateful to all the partners of Project KGASSP-1 "Mathematical modelling of silting in the Kugart River", implemented by the Kyrgyz Department of Water Resources – Project Implementation Unit – with funds from the World Bank. We also acknowledge a fruitful collaboration with the Kyrgyz Slavik University, and namely with Prof. N.Lavrov and Prof. V. Bilenko. The research activity has further been supported by the University of Trento through the grant IMAIPO – PROGETTI CRS 2006 and by the Italian Ministry for University and Scientific Research within the Project PRIN 2006. "Morphodynamical processes in river and riparian ecosystems". Rodolfo Repetto's research activity has also been supported by Cariverona within the project "MODITE".

## REFERENCES

Brookes, A. (1988). *Channelized rivers: Perspectives for environmental management*. J. Wiley, Chichester.

Brookes, A. and F. Shields (1996). *River channel restoration: guiding principles for sustainable projects*. J. Wiley.

Garcia-Navarro, P. and F. Alcrudo (1992). 1D Open Channel Flow Simulation Using TVD McCormack Scheme. *J. Hydraulic Engineering, ASCE 118*, 1359–1373.

Petts, G. (1989). Hystorical analysis of fluvial hydrosystems. *In: Historical change of large alluvial rivers: Western Europe*, 1–18. G.E. Petts (Ed.)

Repetto, R., M. Tubino, and C. Paola (2002). Planimetric instability of channels with variable width. *Journal of Fluid Mechanics 457*, 79–109.

Rohde, S., M. Schutz, F. Kienast, and P. Englmaier (2005). River widening: an approach to restoring riparian habitat and plant species. *River Research and Applications 21*, 1075–1094.

Simon, A. and C. Thorne (1996). Channel adjustment of an unstable coarse-grained stream: opposing trend of boundary and critical shear stress, and the applicability of extremal hypotesis. *Earth Surface Processes and Landforms 21*, 155–180. doi:10.1002/esp.3290140103.

Siviglia, A., G. Nobile, and M. Colombini (2006). The role of quasi-conservative form for morphodynamic modelling in river flow computations.In *Riverflow International Conference on Fluvial Hydraulics*, pp. 1475–1484.

Stanford, J., J. Ward, W. Liss, C. Frissell, R. Williams, J. Lichatowich, and C. Coutant (1996). A general protocol for restoration of regulated rivers. *Regulated Rivers: Research and Management 12*, 391413.

Surian, N. and M. Rinaldi (2003). Morphological response to river engineering and management in alluvial channels in Italy. *Geomorphology 50*, 307–326.

Task Committee, A. S. C. E. (1998, September). River width adjustment. i: processes and mechanisms. *Journal of Hydraulic Engineering 124*(9), 881–902. Prepared by the ASCE Task Committee on Hydraulics Bank mechanics River width adjustment.

Tubino, M., R. Repetto, M. Serafini, A. Siviglia, and G. Zolezzi (2005, November). Mathematical modelling of silting in the Kugart River. Project KG- FEP/ASSP – Final report, University of Trento – Dept. of Civil and Environmental Engineering.

Wyzga, B. (1993, September). River response to channel regulation: Case study of the Raba River, Carpathians, Poland. *Earth Surface Processes and Landforms 18*(6), 541–556. doi:10.1002/esp.3290180607.

*River, Coastal and Estuarine Morphodynamics: RCEM 2007 – Dohmen-Janssen & Hulscher (eds)*
*© 2008 Taylor & Francis Group, London, ISBN 978-0-415-45363-9*

# Feasibility study for implementation of sedimentation reduction measures in river harbours

H.J. Barneveld
*HKV consultants, Lelystad, The Netherlands*

J. Hugtenburg
*Ministry of Transport, Public Works and Water Management, The Hague, The Netherlands*

ABSTRACT: Sedimentation in river harbours is a serious problem in the Netherlands. Costs for removal, transport, processing and storage of this (often contaminated) sediment are high. In addition sedimentation and dredging activities hinder inland navigation.

Harbours are often designed based on requirements imposed by safe and efficient navigation. In the past decades many (mainly desk and/or laboratory) studies have been carried out aiming at the reduction of harbour siltation rates. From these studies, it follows that reductions in siltation rates up to 50% may be reached. However, only few of such sedimentation reduction measures have been implemented. This study aims at exploring the possibilities to apply simple and effective siltation reduction measures in the harbours along the rivers in the Netherlands. From about 200 harbours along the River Rhine and River Meuse three harbours were selected. This selection was made in close consultation with experts and stakeholders, based on among others the siltation rate, harbour geometry from remote sensing images, flow velocities and representativeness for Dutch river harbours. One of the three harbours was further elaborated in a first design. As distinct from preceding studies the design for this pilot project has been made using field data and expertise of river managers and specialists, but without physical scale model studies and numerical experiments. In addition a simple cost benefit analysis has been carried out.

It is expected that this study will be continued with a detailed design of the selected measure. Ultimately this will lead to a prototype implementation in 2008. Such a pilot project combined with a sound monitoring program will show the feasibility of the followed approach. In addition it will provide valuable information on morphological impacts of a structural measure in prototype situations.

## 1 INTRODUCTION

### 1.1 Background and social relevance of study

Approximately nine million cubic meters of silt and sand is annually deposited in and along the main rivers in the Netherlands (River Rhine and River Meuse). Harbours and its entrances are notorious locations for such undesirable sedimentation. Removal, transport, processing and storage of this (often contaminated) material costs about M€120 every year, for which the Ministry of Transport, Public Works and Water Management is partly responsible.

The main consideration for harbour entrance design is nautical navigation. For an optimal nautical navigation, harbour entrances should be as wide as possible. This however also maximizes the volume of sedimentation in the harbour and its entrance. Intelligent design of harbour entrances, using state of the art hydraulic en morphologic knowledge can however lead to harbours that are both well accessible and less

prone to sedimentation. This study aims to explore the possibilities to apply simple, robust and effective sedimentation reduction measures in existing and new harbours along rivers in the Netherlands.

### 1.2 Previous studies, pilot projects and barriers for implementation

In the past decades many (mainly desk and/or laboratory) studies have been carried out aiming at the reduction of harbour siltation rates. PIANC (2006) presents an overview. A prototype example is the application of a Current Deflecting Wall (CDW) in the tidal harbour Köhlfleethafen in Hamburg, Germany (Winterwerp, 2005). Despite of this successful example (a reduction of ca. 40% of the siltation rates), structures to reduce harbour sedimentation are not/hardly applied in the main river branches in the Netherlands. This is thought to be largely due to the inertia of the responsible water managers, thus preventing a possible paradigm shift to take place. This inertia is mainly based on a lack of experience with design, construction and efficacy of sedimentation reduction measures. This study (which is part of the programme Water INNovation source, WINN, supported by the Ministry of Transport, Public Works and Water Management) therefore aims at preparing and implementing a pilot project in order to gain experience with the effectiveness of sedimentation reduction measures in harbours along the rivers Rhine and Meuse.

## 2 METHOD OF STUDY

### 2.1 Process followed

First step in the project was to make an inventory of the relevant physical processes. First a literature survey on fluid dynamics, sediment load and morphology around harbour entrances was carried out. This included an assessment of the design parameters relevant for sediment exchange through the harbour entrance and sedimentation rates. In addition national and international experts were asked to fill in a questionnaire in this field. The results of the literature survey and expert consultation are found in Section 3.

In a second step river base data for harbours in the Netherlands were collected and a short list of possible pilot-locations was established.

Since the chances for a successful pilot project do not only depend on technical aspects, but also on practical and political aspects, two workshop sessions with water and harbour managers were organized as a third step. In these sessions a confrontation was made between technical data and practical aspects with regard to a possible implementation of measures. The first session, which was held in November 2006, led to the selection of three harbours of which a sedimentation reduction can be reached through an innovative harbour entrance design (adaptation of the entrance or sedimentation reduction measures). These harbours were: a recreational harbour in Roermond (along the river Meuse), the harbour of Haaften (river Rhine) and the harbour of the Amer electricity generating station (river Meuse). These harbours are described in Section 4.

Practical aspects eventually led to the harbour of the Amer electricity generating station to be the object of further study. In a second workshop session with experts and water managers, a sustainable measure for this harbour was selected to be elaborated further including a rough design and cost benefit analysis, which are described in Section 5.

### 2.2 Data collection

Within the relevant regional departments and three specialist services of the Ministry of Transport, Public

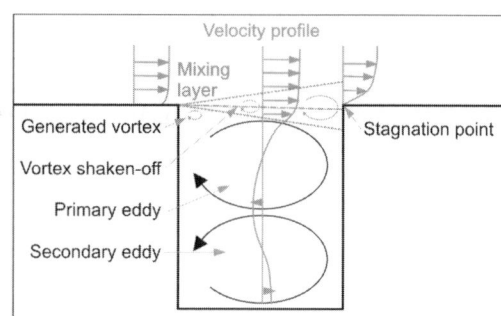

Figure 1. Horizontal entrainment mechanism.

Works and Water Management a questionnaire was distributed so as to collect data on:

1. dredging volumes in Dutch rivers, canals and harbours;
2. particular data for individual river harbours related to parties involved (ownership, stakeholders), geometry, sedimentation (rates, location, composition), causes of that, future plans for the harbour.

The collected data was combined with aerial photographs (Google Earth) to prepare a fact sheet per harbour. These fact sheets proved to be valuable in the preparation of a short list of potential harbours for the pilot.

## 3 LITERATURE SURVEY

### 3.1 Fluid dynamics, sediment load and morphology around harbour entrances

Possible mechanisms responsible for sedimentation are extensively reported in literature. Langendoen (1992), Winterwerp (2005) and PIANC (2006) provide an overview. Most important mechanisms are:

1. horizontal entrainment (turbulent mixing layer)
2. tidal filling
3. density currents

In river systems horizontal entrainment is the main mechanism, which is therefore considered in this paper (Figure 1).

According to van Schijndel and Kranenburg (1998) the sediment flux F through the harbour entrance can be described with:

$$F = ku_r(c_r - c_h)A \qquad (1)$$

with:

$c_r$ = mean sediment concentration in river
$c_h$ = mean sediment concentration in harbour
$u_r$ = mean flow velocity river
$A$ = flow area harbour entrance
$k$ = exchange coefficient

Equation (1) shows that the sediment flux will reduce linearly with smaller $c_r$, $u_r$, A and k. The first two parameters are difficult to change, but influencing A and k appears to be feasible. As the vortexes in Figure 1 prove to be important for the magnitude of the exchange coefficient $k$, additional attention was paid to the generation, growth and diminishing of these structures. It appeared that the bigger and more intense a vortex is, the larger the sediment flux will be. The sediment flux can be reduced by:

1. suppressing the intensity and dimension of the vortexes that are shaken off at the upstream point of the entrance. This can be reached by reducing the flow velocity in the river (influencing the intensity of the vortex) or influencing the shape of the primary eddy (influencing the space left over by this eddy for generation of the vortex);
2. limiting the growth of the vortex in the mixing layer. Widening of the mixing layer may realise this;
3. transition of the stagnation point so as to guide more vortexes reaching the downstream point of the harbour entrance back to the river.

Measures to realise this are presented in the next paragraph.

### 3.2 Harbour entrance designs effective for reduction of sedimentation

Based on descriptions in literature eight possible measured were identified to:

1. reduce the shaking-off of vortexes;
2. reduce the flow area of the harbour entrance;
3. move the stagnation point;
4. reduce the sediment concentration of the water flowing into the harbour;

Figure 2 shows seven of the eight identified measures. Narrowing the entrance by construction of a bottom sill is not included in the Figure.

The *current deflecting wall* (CDW) reduces the shaking-off process of vortexes, diverts the entrainment mixing layer away from the entrance and prevents high concentrations near the bed from entering the harbour. A *pile groyne* widens the mixing layer and reduction of the flow velocity in the river. This suppresses the generation and growth of vortexes. A *dividing wall with upstream sill* at the upstream side of the harbour leads relatively clear water to the harbour and prevents water with high sediment concentrations from entering the harbour. *Narrowing the entrance* and construction of a sill to *shallow the entrance* both reduce the flow area of the harbour entrance, and thus the sediment flux. Adaptation of the harbour mouth *downstream* or *upstream* may shift the stagnation point respectively suppress the process of shaking-off of vortexes. A *semi-permeable screen* diverts the upper part of the water column towards the harbour. This

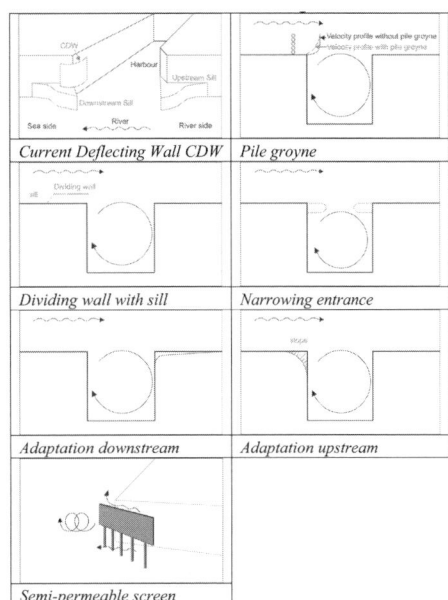

Figure 2.   Measures to reduce the sediment flux.

Table 1.   Measures and their influence.

| | Reduce the shaking-off of vortexes | Reduce flow area of harbour entrance | Move the stagnation point | Reduce incoming concentration |
|---|---|---|---|---|
| CDW | +? | | + | + |
| Pile groyne | + | | + | |
| Dividing wall & sill | + | | | + |
| Narrowing entrance | | + | | |
| Shallowing entrance | | + | | |
| Adaptation downstream | + | + | + | + |
| Adaptation upstream | | + | + | |
| Semi-permeable screen | +? | | | + |

introduces a spiral motion leading the water near the bed (with most sediment) towards the river.

Table 1 presents the potency of these measures to influence the main mechanisms of sediment flux towards river harbours.

In addition the impacts of the measures for all riverine functions (navigation, safety, nature) were assessed.

The quantitative impact of a measure depends on local circumstances. The study aimed at designing a

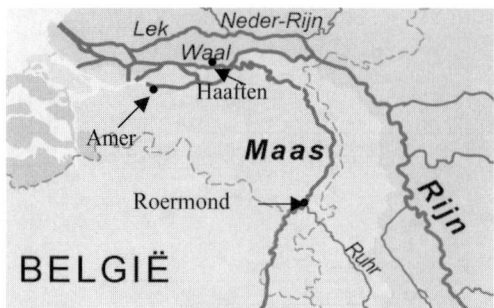

Figure 3. Location three harbours.

measure for a pilot harbour without extensive additional physical or mathematical modelling. Therefore also the possibility for quantifying the effect of the measures on the sedimentation process without additional study was assessed.

## 4 POSSIBLE LOCATIONS FOR IMPLEMENTATION OF SEDIMENTATION REDUCTION MEASURES

Figure 3 shows the location of three (out off appr. 200) selected harbours. This selection has been made in close consultation with experts and stakeholders based on among others the siltation rate, harbour geometry from remote sensing images, flow velocities and representativeness for Dutch river harbours.

### 4.1 Recreation harbour of Roermond (river Meuse)

The recreational harbour of Roermond in the river Meuse faces sedimentation problems since several years. Some studies have been executed showing that reduction of the entrance area would be a feasible solution. All stakeholders support a sustainable solution and the geometry of river and harbour appear to be representative for much more river harbours.

The municipality of Roermond recently approved a design by a consultant to realise this area reduction. Both the width and depth of the entrance will be reduced.

For this study this meant that an alternative design was not desirable. However, the implementation of the proposed measure may provide valuable information on the efficacy of it. Therefore a future activity concerning the design and implementation of a monitoring plan will be considered.

### 4.2 Harbour of Haaften (Rhine)

The harbour of Haaften serves as a resting harbour for inland navigation along the Rhine branch Waal. Apart from the sedimentation problem ($>3,000\,\mathrm{m}^3$ annually)

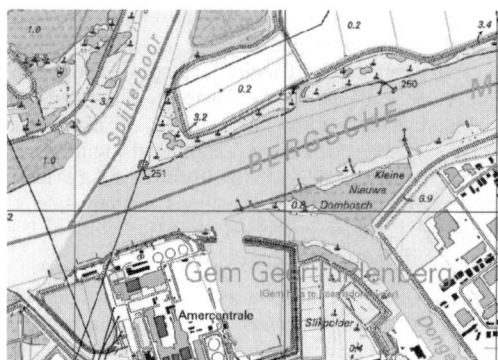

Figure 4. Harbour Amer electricity generating station (Amercentrale).

the harbour entrance causes problems for safe arrival in the harbour. This is especially the case during high water when the harbour entrance is submerged. Therefore studies are underway to adapt the harbour and entrance geometry so as to increase the capacity and make the harbour safer. Unfortunately data collection for this harbour faced delay, due to which no preliminary design for a sedimentation reduction measure could be preformed. However data collection will continue as a preparation for future studies for this harbour.

### 4.3 Harbour of the Amer electricity generating station (Bergsche Meuse)

The sedimentation in the mouth of this harbour is such, that on average a volume of $7,000–11,000\,\mathrm{m}^3$ of sand and (contaminated) silt is dredged annually. Both the management of the station and the river manager support a solution that can reduce the dredging costs and hindrance for navigation due to sedimentation and dredging works. In addition sufficient data is available to make a first design of a sedimentation reduction measure.

Although the harbour is located in a complex environment with different waterways meeting each other, and extraction of cooling-water (maximum $30\,\mathrm{m}^3/\mathrm{s}$) via the harbour, a pilot implementation for this harbour will provide valuable information on the performance of a measure and impacts on flow patterns. This pilot project is further elaborated in the next section.

## 5 TOWARDS A SUSTAINABLE REDUCTION OF SEDIMENTATION IN THE AMER POWER STATION RIVER HARBOUR

Aim of the study was to prepare a design for an adapted harbour geometry or a sedimentation reducing measure based on available field data, some exploring numerical flow simulations and engineering judgement. This means no extensive studies with physical scale models or numerical experiments. Most important result should be a pilot project and sound monitoring programme so as to learn in the field.

## 5.1 *Harbour and sedimentation characteristics*

The harbour area is about 17.5 ha large. The harbour entrance is 120 m wide and the opening between Bergsche Maas and Donge approximately 185 m wide. Main discharge comes through the Bergsche Maas with an average annual flow of about 300 m³/s and flood discharges of 1000 m³/s and higher. Discharge through the Donge is usually small (less than 3 m³/s). The influence of the tide is still noticeable in the harbour, but reverse (upstream) flow is rare.

At two locations cooling-water for the power station is withdrawn. These locations are at the Bergsche Maas and in the harbour (both maximum 30 m³/s). Especially in the winter season (September-March) full cooling-water capacity is used. Figure 5 shows a calculated flow pattern with a mean discharge Bergsche Maas of 300 m³/s and a cooling-water withdrawal of 30 m³/s.

Sedimentation in the area of the harbour, its entrance, the river and the mouth of the Donge is estimated to be 12,500–21,000 m³ per year. About 7,000–11,000 m³ of this is found in the area of the harbour entrance and Donge mouth. 50% of this material is (contaminated) silt. This means that all dredged material has to be disposed at special sites.

Most sedimentation occurs during the high water season (October-March) when river discharges and sediment concentrations are high.

## 5.2 *The proposed solution*

Based on the information available for the harbour six possible measures for a pilot project were identified. Of the 7 measures presented in Figure 2 the following measures were considered not-feasible:

- A Current Deflecting Wall was rejected because this measure certainly needs additional study in the complex situation of this harbour.
- Adaptation of the mouth/entrance upstream or downstream was considered not-feasible because of the local geometry and presence of infrastructure.

A sediment trap upstream of the harbour entrance was added as a possible measure.

In Table 2 the remaining measures are assessed on their expected efficiency for the situation with and without withdrawal of cooling-water through the harbour.

As the sedimentation process mainly takes place in the high-water season and cooling-water withdrawal is continuous then, only the diving wall with sill,

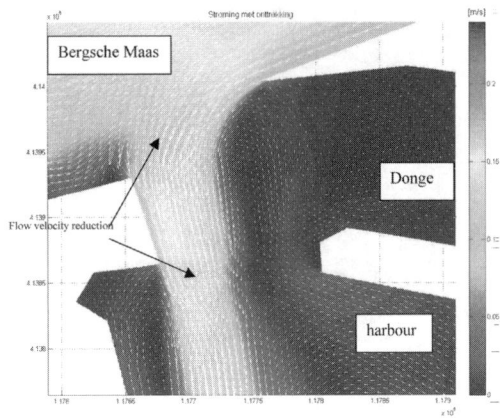

Figure 5. Flow pattern at 300 m³/s with 30 m³/s cooling-water withdrawal, calculated with FINEL.

Table 2. Assessment measures with and without cooling-water withdrawal.

| Measure | Cooling-water withdrawal withdrawal | |
|---|---|---|
| | With | Without |
| Pile groyne | ? | + |
| Dividing wall & sill | + | + |
| Narrowing entrance | ? | + |
| Secondary channel through upstream flood plain | + | + |
| Semi-permeable screen | 0 | + |
| Sediment trap upstream entrance | + | + |

secondary channel through the upstream floodplain and the sediment trap upstream of the harbour entrance appear to be feasible. The first two measures aim at leading relatively clear water from the upper part of the water column to the harbour. The sediment trap prevents the sand fraction of the sediment load from entering the harbour. Taking into consideration that the proposed measure should be applicable for other river harbours the dividing wall with sill was selected for further elaboration.

Figure 6 shows a first design for this measure.

Based on experiences from the river manager and experts from specialist services and consultants a first estimate was made of the expected impact of the measure on the sedimentation rate. It is expected that the inflow of sand may be reduced by 50%. This means that the total sedimentation rate (of sand and silt) will be reduced by 25%.

When the measure is implemented a sound monitoring programme has to be implemented. This programme should record the existing situation and follow the developments after construction.

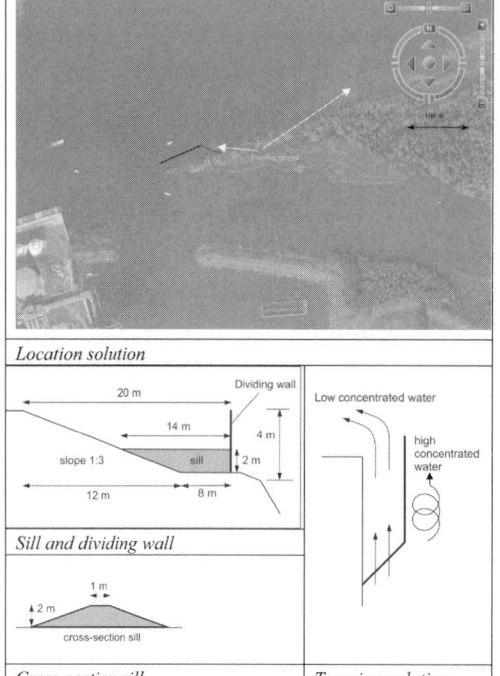

Figure 6. First design dividing wall and sill.

The monitoring programme will survey:

- Bed development. Coupled to the regular soundings some additional local measurements are foreseen after (1) floods, (2) dredging, (3) particular events (e.g. dredging in the main river);
- Flow patterns in river and harbour;
- Composition of the bed material in river, entrance and harbour.

5.3 *Rough cost benefit analysis*

Costs of the measure are related to (detailed) design, construction and maintenance. Benefits are the reduced dredging activities. Less hindrance for inland navigation is not taken into consideration for this analysis. Based on costs of € 5 per m³ dredged material, 11,250 m³ reduction in the annual dredging volume, construction costs of € 250,000, a discount-rate of 2.5% and a life-cycle period of 30 years the Net Present Value NPV (2007) for the average estimate is approximately € −9.000, meaning that the investments for the measure will not pay back completely in 30 years. For the optimistic estimate of 13,750 m³/s dredged volume reduction the NPV increases to € +45,000. However a pessimistic assumption (only 8,750 m³ reduction per year) will result in a NPV of €−62,000.

## 6 DISCUSSION

Objective of the study was to elaborate a first design of a sedimentation reduction measure in a Dutch river harbour. This should be achieved by a study based on field data and experiences of river managers and experts. Extensive physical scale modelling or numerical modelling did not form part of the study. When implementation of a measure designed in this way proves to be effective, this approach can be adopted for other locations. Intensive and specific monitoring is an important pillar of the approach that can be described with learning in the field.

It is proposed that parallel to the preparation of the studies some numerical experiments and perhaps a few simple physical scale model tests (e.g. flume study) will be carried out to further refine the design. In this period also the reference situation (existing situation) should be recorded as a first phase of the monitoring programme.

ACKNOWLEDGEMENTS

The study was carried out by three consultancy firms in close co-operation with experts from specialist services and regional divisions of the Ministry of Transport, Public Works and Water Management. The authors wish to thank the other members of the project team: Bram van Prooijen (Svašek hydraulics), Nico Struiksma (Struiksma River Engineering), Paul Termes (HKV CONSULTANTS), Maarten van der Wal (Road and Hydraulic Engineering Institute, Ministry of Transport, Public Works and Water Management) and Arjan Hijdra (Civil Engineering Division, Ministry of Transport, Public Works and Water Management).

REFERENCES

Langendoen E.J. (1992). 'Flow patterns and transport of dissolved matter in tidal harbors'. PhD thesis, Delft University of Technology.
PIANC (2006): Minimising Harbour Siltation. Report of International Navigation Association. PIANC Working Group 43 (draft).
Prooijen van B.C. (2004). 'Shallow mixing layers'. Thesis Delft Technical University.
Schijndel van & Kranenburg (1998). Reducing the siltation of a river harbour. Journal of Hydraulic Research. Vol. 36, 1998, No.5.
Winterwerp J.C. (2005): Reducing Harbor Siltation. I: Methodology. Journal of Waterway, Port, Coastal, and Ocean Engineering, Volume 131, Issue 6, pp. 258–266 (November/December 2005).

*River, Coastal and Estuarine Morphodynamics: RCEM 2007 – Dohmen-Janssen & Hulscher (eds)*
*© 2008 Taylor & Francis Group, London, ISBN 978-0-415-45363-9*

# A method to predict sediment transport process in drainage basin with dams

S. Egashira & T. Itoh
*One-time College of Department of Civil and Environmental Systems Engineering, Ritsumeikan University, Japan*

K. Horie & N. Nishimoto
*IDEA Consultants Inc., Japan*

ABSTRACT: A method, which is constituted of sub-models such as a unit slope and a unit channel for a drainage model, distributed parameter model for rainfall runoff and sediment-related channel model, is proposed for evaluating sediment transport process in river basins and is applied to investigate the sediment runoff from mountain areas and the associated bed evolution of alluvial reach in Takatoki river basin where a dam is planning to construct for flood control, supposing an occurrence of a plan size rainfall event with numerous landslides. Numerical computations were conducted for three cases: no impact of the dam on sediment transport process (Case 1), performances of flood control to reduce flow discharge with the dam (Case 2) and no performances of flood control but with the dam (Case 3). In cases 2 and 3, it is supposed that all sediment inflowing into the reservoir is caught in the storage area. Predicted results suggest that, in Case1, a fine part of sediment produced by landslides in mountain area can be transported during a period of the flood to the alluvial reach and causes the bed variation although most of the coarse sediment remains in mountain reaches. After the flood, a very active sediment runoff from the mountain area to the flat region continues for long years, and causes river bed aggradation. The dam, which drains about 45 percent of the total basin area, catches a lot of sediment and suppresses the bed variation of alluvial reach during the flood as well as for long years after the event. In addition, sediment transport activities in mountain regions are impacted by reservoir sedimentation and decreases to suppress remarkably river bed aggradation of downstream flat area.

## 1 INTRODUCTION

Sediment transportation and associated river changes have been studied for long from practical and scientific views. Specially in the period from 1960's to 1980's, the problems such as the kinematics of grain motion, the dynamics of sediment transportation, the nonlinear problems among the flow, the sediment transportation and the flow boundary were investigated actively based on the theoretical and experimental tools.

Recently there is a growing need to study the micro-mechanics of sediment transportation and related sediment phenomena as well as to investigate the sediment transport process in complex river systems. In the study of the micro-mechanics it is important how to formulate the micro-structure of relative particle-particle motion and the interaction of solid-fluid phase within the principles of the dynamics. While in the study on a basin size sediment transportation, it is a key how to integrate and to simplify the

governing equations for the flow and sediment without losing the important terms.

The present study introduces a method to predict sediment transport process in river basin and shows results obtained from the method in Takatoki river basin during a plan size flood and for several years after the flood. In Takatoki river basin, the dam is planning to construct for flood control. Numerical computations were conducted for three cases; a present drainage condition, the condition with the dam for flood control, and the condition without flood control but with the dam.

## 2 METHOD TO PREDICT SEDIMENT TRANSPORT PROCESS

### 2.1 *Brief review*

It has long been expected to develop a method for evaluating sediment discharge at any sites of complex

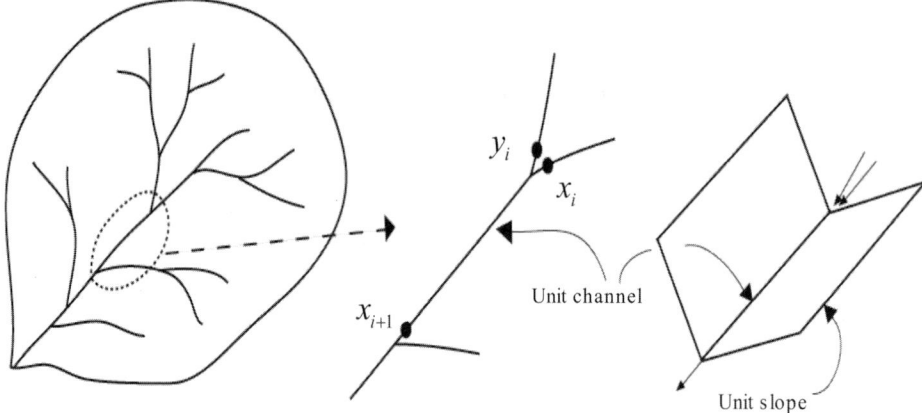

Figure 1. Drainage model with a unit channel and a unit slope.

river systems as was commonly conducted in predictions of rainfall runoffs. Muramoto, Michiue & Shimozima (1973) proposed a method to calculate a wash load connecting a simple erosion model for wash load and a distributed model for rainfall runoff. Kaneyashiki, Ashida & Egashira (1980) modified Muramoto et al.'s method introducing mechanism of wash load processes such as slope erosion, side bank erosion and bed sediment erosion. These methods, still now, are useful for evaluating discharge of wash load during floods.

Importance to evaluate a sediment runoff process from wash load to bed loads has been increasing with response to need of rehabilitation of sediment process in river systems, and has been stimulating the studies on these topics. In fact, several useful methods have been proposed (Sunada et al. 1994, 2000, Egashira et al. 1998, 2000, 2001, Takara et al. 1998, Ichikawa et al. 1999, Takahashi et al. 2000, Murakami et al. 2001, etc.). These proposed methods are composed of a rainfall runoff model and a sediment transport model. However, there are differences between them in treatments of sediment transport mechanism as well as of rainfall runoff process, which may depend largely on the aim of each study.

In order to evaluate a sediment process in river system, the method should involve a subsystem to evaluate the change of stored sediment volume in upstream reaches, sediment sorting and armoring process and sediment transport rate ranging from wash load to bed loads. It is well known that a large amount of sediment continues to run out during a specified period when the drainage basin is deserted due to a severe rainfall and an earthquake, and gradually reduces to a normal stage for long years. Such changes of sediment runoff can be caused by a gradual decrease of stored sediment volume, development of armor coats and change

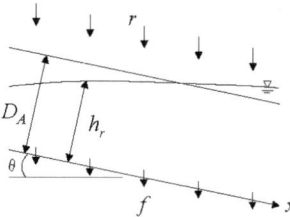

Figure 2. Schematics of surface and subsurface flows.

of surface coverage of mountain slope as well as by artificial stabilization works.

### 2.2 Present method

#### 2.2.1 Outline of the model
One of the methods, as a possible way to explain the aforementioned characteristics of sediment transport process, is introduced. Egashira (1998, 2001) and Egashira & Matsuki (2000) proposed a method that is composed of a drainage model, a distributed rainfall model and a sediment transport process model.

In Figure 1, the drainage model is illustrated. They considered that a complex channel network of a river basin could be resolved into a single unit channel and two unit slopes connecting to the both side of the channel. The unit channel has two inflow points, $x_i$ and $y_i$, and one out-flow point, $x_{i+1}$. Using the unit channel and the unit slope, we can reproduce the drainage basin. In addition, a flood hydrograph will be obtained at every confluence point of the channel network if lateral inflow discharge is computed to the unit channel from the unit slope. Lateral inflow discharge is predicted by applying a kinematic wave model for surface runoff and the Darcy law for seepage flow to each unit slope, which is shown schematically in Figure 2.

In modeling sediment transport process, Egashira (1998, 2001) focused on sediment transport modes such as landslides, debris flows, bed loads, suspended loads and wash load, and aimed to predict sediment transport rate ranging from bed loads to wash load which are a continuous function of flow discharge, sediment size distribution, bed slope, etc. On the other hand, it is well known that both of landslides and debris flows take place indeterministically and it is difficult to predict these occurrences. Therefore, he excluded to evaluate the sediment transport rate due to landslides and debris flows, but included these into a submodel to evaluate the storage volume of sediment in the unit channel, because such mass movements causes major parts of stored sediment. The stored sediment will be transported as bed loads, suspended loads and wash load.

### 2.2.2 Governing equations

The rainfall runoff is evaluated in terms of the kinematic wave runoff model and Darcy law for the unit slopes and the kinematic wave model for the unit channel.

Referring to Figure 2, the governing equations of surface and subsurface flows are as follows.

$$\alpha \frac{\partial h_r}{\partial t} + \frac{\partial q_r}{\partial x} = (r - f)\cos\theta, \quad \alpha = \begin{cases} \lambda_e & (h_r < D_A) \\ 1 & (h_r \geq D_A) \end{cases} \quad (1)$$

$$q_r = k h_r \sin\theta, \, (h_r < D_A, \, \alpha = \lambda_e) \quad (2)$$

$$q_r = k_A D_A \sin\theta + \frac{1}{N}\sqrt{\sin\theta}\,(h_r - D_A)^{5/3},$$
$$(h_r \geq D_A, \, \alpha = 1) \quad (3)$$

in which $h_r$ = depth of sum of surface flow and subsurface flows, $q_r$ = flow discharge corresponding to $h_r$ in unit width, $D_A$ = depth of surface soil layer, $\lambda_e$ = effective porosity, $r$ = rainfall intensity, $f$ = infiltration rate, $k$ = permeability coefficient, $\theta$ = inclination angle of the unit slope, $n$ = equivalent Manning's roughness; and $\alpha$ = correction factor that takes $\alpha = \lambda_e$ if $h_r < D_A$ and $\alpha = 1$ if $h_r \geq D_A$.

The equations of the kinematic wave in the unit channel shown in figure 1 can be simplified by means of an integrating and averaging methods. The continuity equation and the equation of motion for the $i$-th unit channel are:

$$\frac{\partial h}{\partial t} = \frac{1}{BL}\{Q(x_i) + Q(y_i) - Q(x_{i+1})\} + \frac{1}{B}q \quad (4)$$

$$Q(x_{i+1}) = \frac{1}{n} B I^{1/2} h^{5/3} \quad (5)$$

in which $h$ = flow depth, $Q(x_{i+1})$ = out flow discharge at $x_{i+1}$, $Q(x_i)$ = inflow discharge at $x_i$, $Q(y_i)$ = inflow

discharge at $y_i$, $B$ = flow width, $L$ = length of the $i$-th unit channel, $I$ = bed slope, $n$ = Manning's roughness; and $q$ = lateral inflow discharge from both sides of the unit channel which is predicted by Eqs. (1)-(3).

The governing equations for the sediment transport process are simplified by the same way as the equations for the flood flow. The sediment continuity equation of the unit channel can be formulated as follows.

$$\frac{\partial z}{\partial t} = \frac{1}{(1-\lambda)BL}\{Q_b(x_i) + Q_b(y_i)$$
$$-Q_b(x_{i+1}) + I_n\} + D_s - E_s + D_w - E_w \quad (6)$$

in which $z$ = bed elevation, $\lambda$ = porosity of bed sediment, $I_n$ is the special term to explain sediment supply due to landslides and debris flow which are included only in abnormal events, $Q_b$ = bed load discharge; and $Q_b(x_i)$ = bed load discharge at the inflow point $x_i$. Because of non-uniform sediment bed, $Q_b$ sums up the transport rate of each size class.

$$Q_b = \sum_j Q_{bj} \quad (7)$$

in which $Q_{bj}$ = transport rate of size class $d_j$ and can be evaluated using a bed load formula for non-uniform sediment, i.e. Ashida & Michiue's formula (1972).

$D_s$ and $E_s$ in Eq. (6) are the deposition rate and the erosion rate for suspended sediment except wash load. These are described by

$$D_s = \sum_j D_{sj}, \quad D_{sj} = \gamma_j c_{sj} w_{0j} \quad (8)$$

$$E_s = \sum_j E_{sj}, \quad E_{sj} = p_j c_{ej} w_{0j}. \quad (9)$$

In addition, the suspended sediment concentration is controlled by a simplified convection equation;

$$\frac{\partial c_{sj} h}{\partial t} = \frac{1}{BL}\{\gamma_j c_{sj}(x_i)Q(x_i)$$
$$+ \gamma_j c_{sj}(y_i)Q(y_i) - \gamma_j c_{sj}(x_{i+1})Q(x_{i+1})\} + E_{sj} - D_{sj} \quad (10)$$

In Eqs. (8)–(10), $D_{sj}$ and $E_{sj}$ are deposition rate and the erosion rate for suspended sediment of the size class $d_j$, respectively. $c_{sj}$ = average sediment concentration of the size class $d_j$, $w_{0j}$ = falling velocity of the size class $d_j$, $\gamma_j$ = correction factor to evaluate the deposition rate $D_j$ using the average sediment concentration and takes a value larger than unity. $c_{ej}$ = equilibrium concentration of size class $d_j$ at the reference level, which can be obtained by several formulas; i.e. Itakura & Kishi (1980), Ashida & Fujita (1986) and Garcia & Paker (1991). $p_j$ = fraction of size class $d_j$ in bed sediment.

$D_w$ and $E_w$ are the deposition rate and the erosion rate of wash load, respectively. In case of wash load, the sediment concentration is distributed homogeneously. Correspondingly, the deposition rate can be formulated without the correction factor;

$$D_w = c_w w_{0w} \tag{11}$$

in which $c_w$ = sediment concentration of wash load; and $w_{0w}$ = falling velocity of a reference sediment size. The fine sediment particles of the channel bed can be entrained easily into flow water when the erosion takes place. Therefore, $E_w$ will be formulated when the bed is eroded,

$$E_w = -(1-\lambda)p_w(\partial z/\partial t), \quad (\partial z/\partial t \le 0) \tag{12}$$

and otherwise,

$$E_w = 0, \quad (\partial z/\partial t > 0) \tag{13}$$

in which $p_w$ = fraction of fine sediment in the bed surface layer. $p_j$ and $p_w$ should satisfy a following condition.

$$\sum_{j}^{N} p_j + p_w = 1 \tag{14}$$

in which $N$ = number of classification for $d_j$.

Using $D_w$ and $E_w$, the sediment concentration of wash load is described by the simplified convection equation.

$$\frac{\partial c_w h}{\partial t} = \frac{1}{BL} \{ c_w(x_i)Q(x_i)$$
$$+ c_w(y_i)Q(y_i) - c_w(x_{i+1})Q(x_{i+1}) \} + E_w - D_w \tag{15}$$

in which $c_w(x_i)$ = sediment concentration of wash load at the inflow point $x_i$.

A structure of sediment size distribution in bed sediment changes continuously corresponding to the deposition and erosion of the bed loads, suspended loads and wash load. In addition, it will change discontinuously when a large amount of sediment is supplied by landslide and debris flows, which is not taken into consideration herein. In the former, the fraction, $p_j$, of sediment size $d_j$ of the bed surface layer is described by

$$\frac{\partial p_j}{\partial t} = \frac{1}{(1-\lambda)\delta BL} \{ Q_{bj}(x_i) + Q_{bj}(y_i)$$
$$- Q_{bj}(x_{i+1}) \} + D_{sj} - E_{sj} + D_w - E_w - \frac{\partial z}{\partial t}\frac{f_j}{\delta} \tag{16}$$

in which $\delta$ = depth of the surface layer (exchange layer thickness) and $f_j$ is described as follows.

$$f_j = \begin{cases} p_{j2}, (\partial z/\partial t \le 0) \\ p_j, (\partial z/\partial t > 0) \end{cases} \tag{17}$$

in which $p_{j2}$ = fraction of size class $d_j$ in 2-nd layer beneath the exchange layer. The equation to evaluate $p_{j2}$ can be obtained easily, and similarly obtained for $p_{j3}, p_{j4}$ etc.

If the rainfall condition is given, and if we specify the initial, geometric conditions such as $z$, $p_j$, $d_j$, $p_w$, $B$, $L$ and so on for every unit channel and every unit slope after reproducing the river basin in terms of channel- and slope-units, we will obtain the solutions at any place of the basin for the flow discharge, the sediment discharge including wash load, the bed elevation (the storage volume of sediment) and the structure of sediment size distribution.

## 3 COMPUTATIONS OF SEDIMENT RUNOFF AND BED VARIATION IN TAKATOKI RIVER

### 3.1 Modeling of the basin

Takatoki river basin is illustrated in Figure 3. It is located on the north of the lake Biwa and drains the area of $210\,\mathrm{km}^2$ at the confluence point of the Ane river. Niu dam whose location is shown in Figure 3 covers the area of $93.1\,\mathrm{km}^2$. The river in the downstream of the Takatoki weir spreads in flat area and has an alluvial nature, while in the upstream side it is characterized by the bed configurations formed commonly in mountain streams.

Referring to descriptions of the unit channel and the unit slope shown in Figure 1, the drainage basin is divided into 131 of the unit channel, 262 of the unit slope (two times of the unit channel) and, in addition, 19 of the unit slope which connects to the upper boundary of the unit channel located at the highest area. The length and initial bed slope of each unit channel, and the geometric factors of each unit slope such as the shape of a parallelogram were specified in terms of topographic map. In addition, the flow width of each unit channel was determined referring to a modified regime theory (Egashira et al. 2001).

$$B = 5\sqrt{Q_0 A_i / A_0} \tag{18}$$

in which $A_0$ = drainage area at the reference station, $A_i$ = drainage area at each unit channel; and $Q_0$ = reference flow discharge. In the reach located around at the Takatoki weir, it is supposed that the sediment particle of average size starts to move when the flow discharge increase to $35\,\mathrm{m}^3/\mathrm{s}$. Therefore, the reference flow discharge, $Q_0$, is specified as $35\,\mathrm{m}^3/\mathrm{s}$ in Eq. (18).

### 3.2 Computation conditions

Hydrological and hydraulic parameters necessary for the computations are specified by several sensitivity

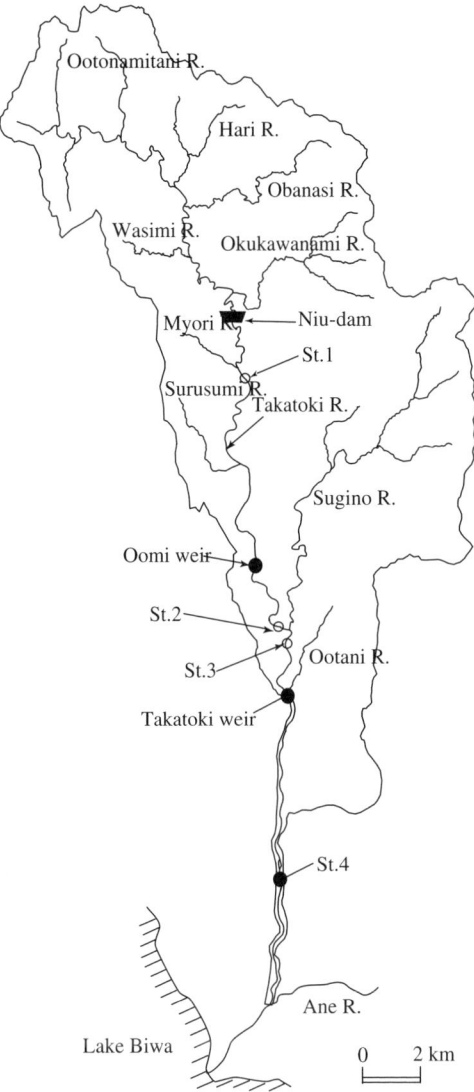

Figure 3. Takatoki river basin.

the 2nd layer, and in addition the value of $f$ means that $f_A$ = initial value (final value). $N$ is the equivalent Manning's roughness.

For the unit channels: $n = 0.04$ to $0.045$ [m-s], $d_{max} = 300$ mm (for all the unit channels), $d_{60} = 100$ to $200$ mm (for the surface bed layer), $d_{60} = 20$ to $60$ mm (for the sub-surface bed layer) and $\lambda = 0.4$. $n$ is the Manning's roughness, $d_{max}$ is the maximum sediment size, $d_{60}$ is the sediment size of 60% finer.

For the alluvial reach (downstream of Takatoki weir): $n = 0.038$ [m-s], $d_{max} = 80$ mm, $d_{60} = 50$ mm, $d_{60} = 20$ mm (sub-surface). In the alluvial reach, cross sections are surveyed at 200 m interval, which are employed for the computations.

For sediment released by landslides and debris flows: Sediment volume was estimated by a following formula (Egashira et al., 2003)

$$V = N_* b l D A \qquad (19)$$

in which $N_*$ = density of the landslides depending on slope gradient, geology and rainfall intensity, $b$ and $D$ = average width and the average depth of a single landslide, $A$ = area of a unit slope; and $l$ = average distance from a specific center to a unit channel defined by

$$l = \sum_{i=1}^{m} l_i / m \qquad (20)$$

in which $l_i$ = distance from the location of landslide to unit channel; and $m$ = number of the landslides occurring in the unit slope. Referring to many empirical data (JSCE, 1999), $N_*$ is specified and illustrated in Figure 7, and $b$ and $D$ are specified as 30 m and 3 m, respectively. In addition, $l$ is specified as a distance from a center of the unit slope to the unit channel.

Sediment which is released in the unit slope corresponding to the Eq. (19) is stored smoothly in the associated unit channel. The size distribution of the released sediment is shown in Figure 5. A fine part of the stored sediment that is considered to behave as wash load is included significantly; i.e. 20%, because a sediment sorting by flowing water does not take place.

A time series of rainfall data collected from August of the year 1993 to 1999 is employed as input data. However, the plan size rainfall was reproduced referring to the event of 1975. The reproduced plan size data was attached to the beginning of August in 1993, instead of the observed rainfall data in this period.

In Figure 6, the flood discharge predicted at station 4 (See Figure 3) in terms of the present method and the deformed flood discharge due to the flood control are illustrated. The flood control is performed during only the plan size flood. In addition, in the alluvial reach downstream of Takatoki weir, the river bed variation can be computed using general 1-D governing equations with the upper boundary conditions predicted in terms of the present method.

analyses as well as by field surveys, which are as follows.

For all the unit slope: $D_A = 0.4$ m, $\lambda_e = 0.4$, $D_B = 0.8$ m, $k_A = 3 \times 10^{-5}$ m/s, $k_B = 2.5 \times 10^{-6}$ m/s, $f_A = 150$ mm/hr (50 mm/hr), $f_B = 15$ mm/hr (10 mm/hr), $f_C = 0.05$ mm/hr (0.03 mm/hr), $N = 0.7$ [m-s]. $D_A$ is the depth of surface soil layer, $D_B$ is the depth of 2nd- soil layer underneath the surface layer, $k_A$ and $k_B$ are the coefficient of permeability of the surface and 2-nd soil layers, $f_A$, $f_B$ and $f_C$ are the infiltration coefficients of upper boundary of the surface layer, the boundary of surface to 2nd, and the lower boundary of

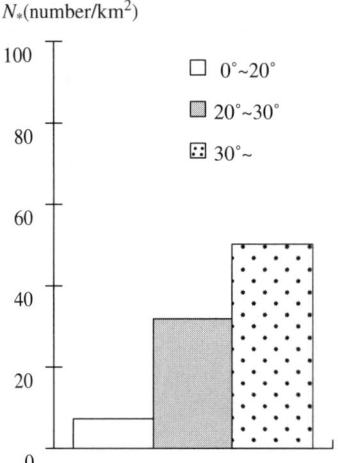

Figure 4. Density of the landslides in Takatoki River basin (Number of occurrence/Area).

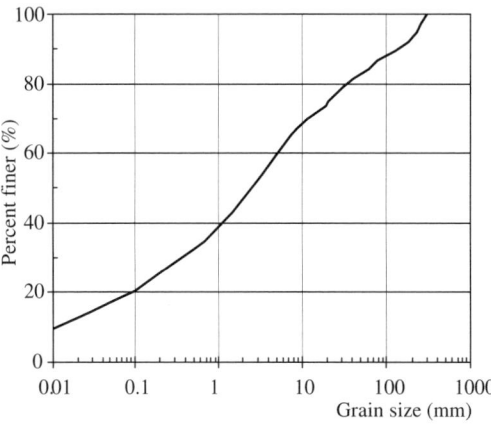

Figure 5. Particle size distribution of sediment released by landslides and debris flows.

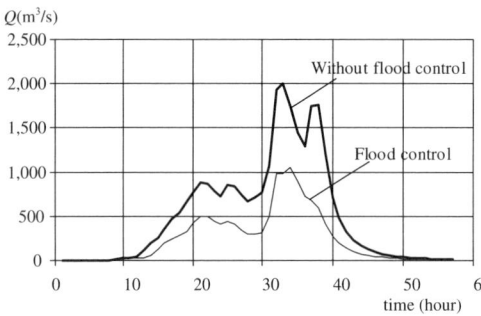

Figure 6. Flood discharge predicted by the plan size rainfall at St. 4 (See Fig. 4) and a deformed flow discharge by the flood control.

Predictions are conducted for three cases to investigate the sediment transport process during about seven years, supposing that the plan size rainfall takes place and a large amount of sediment is released at the beginning of the first year.

In Case 1, the computation is conducted for the present drainage condition, without the dam, to investigate the sediment transport process in the mountain regions as well as in the alluvial reach. In Case 2, the impact of the dam on the sediment transport process is investigated supposing the flood control is performed and all sediment transported from the upstream area of the dam is caught into the reservoir. In Case 3, the sediment transport process is investigated supposing that there are no flood control, but all sediment transported to the dam is caught in the reservoir.

## 4 RESULTS AND DISCUSSIONS

Figure 7 shows a temporal change of an accumulated sediment runoff volume computed for three cases at four sections located in mountain area.

In Case 1 denoted by the line-(1), the sediment volume increases sharply at the beginning of August 1993 when the plan size event took place, and since then it has increased discontinuously like an irregularly spaced ladder corresponding to the time series of rainfall. Such a sediment runoff characteristics is similar to each other. Influences of the dam on the sediment runoff are clearly seen in the results. Specially, in station 3 where Sugino River is confluent at immediate upstream, the sediment runoff volume in Cases 2 and 3 is much lager than the results at the stations 1 and 2.

Figure 8 illustrates the sediment runoff volume focusing on the particle size. The results shown in the left hand are the runoff volume after the event of the plan size rainfall and those in the right side are of after 7 years computation. The results show that fine material defined as wash load is dominant at all stations, and suggest that it tends to decrease temporally. In fact, bed loads and suspended loads coarser than 0.1 mm are about 30% of the total after the event and then increase to 50% after 7 years computation at the station of Niu dam.

Figure 9 shows the predicted bed evolutions of the alluvial reach which are of the immediately after the event, three and seven years computations, respectively. It is well predicted that the bed level changes corresponding with the channel width, especially with the width of low flow water stage (See the upper block of Figure 9), and in addition the bed is severely eroded in the narrow sections and aggrades largely in the wide sections in the result immediately after the event. Such severe bed variation is caused not only by active reforming of bed sediment within the channel during the plan size flood, but also by deposition

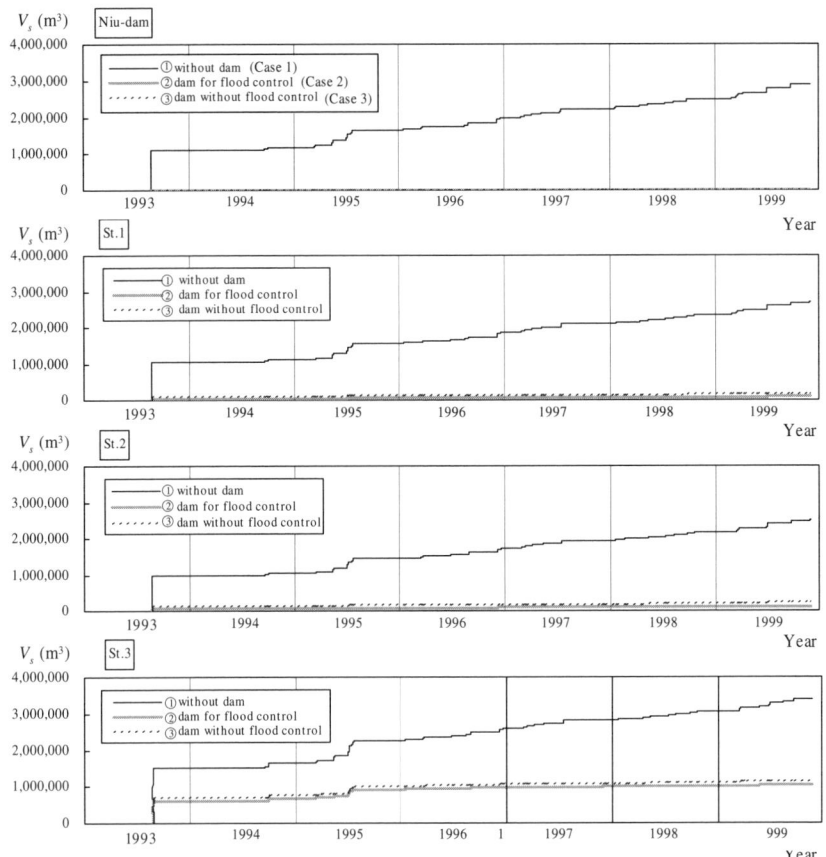

Figure 7. Temporal change of an accumulated sediment runoff volume computed for 3 cases at four sections located in mountain area.

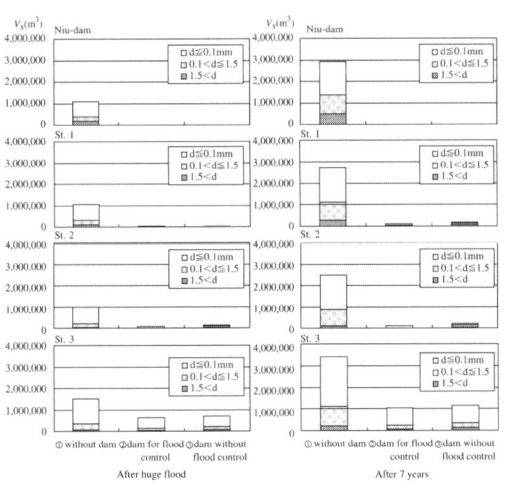

Figure 8. Sediment runoff volume focusing on the sediment particle size in each section.

of fine sediment transported from mountain reaches. Additionally, as shown in the results after three- and seven- years computations, it is found that the bed level aggrades temporally in the whole region because a lot of sediment delivered from mountain streams to the alluvial reach.

## 5 CONCLUSIONS

It is important to develop a method to evaluate sediment transport process in whole river basin size from mountain area to flood plain. A method proposed by Egashira (1998, 2001) and Egashira et al. (2000) is described and applied to Takatoki river basin to evaluate a sediment transport process in a complex river system as well as to investigate the impact of the dam on it, supposing a plan size rainfall takes place.

Part of the results suggest that river bed variation will take place unacceptably for flood hazards mitigation works if a plan size rainfall occurs and a bed

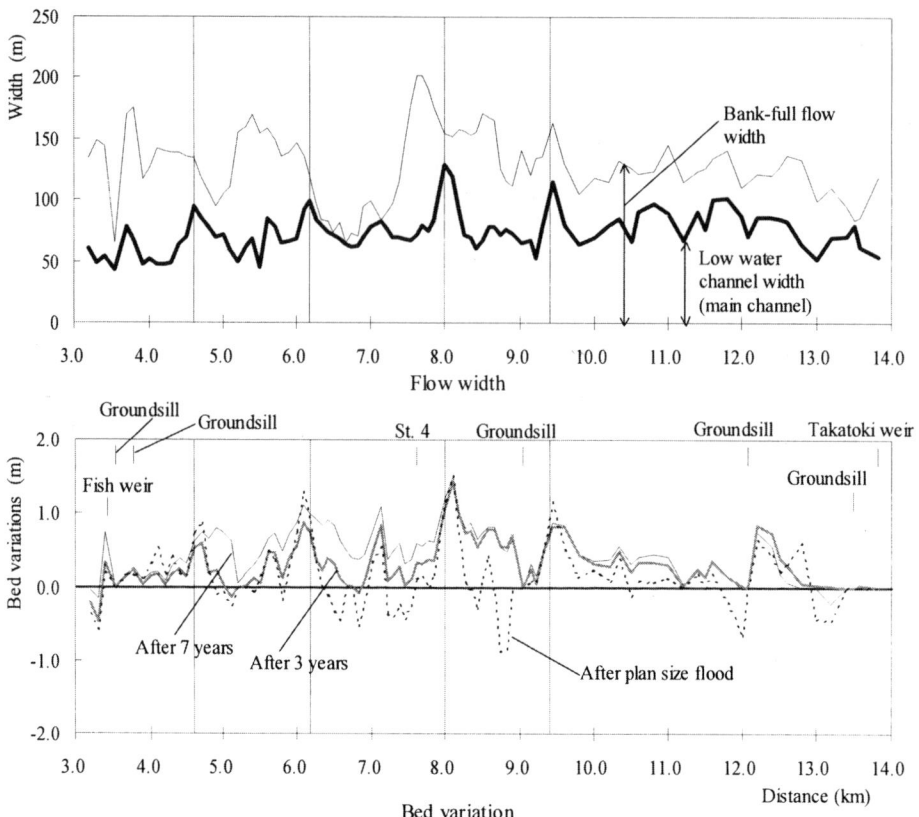

Figure 9. The results the immediately after the plan size event, 3-years computation and 7-years computation in case 1 (without the dam).

aggradation will continue for long after the event. This provides important information for countermeasures against flood- and sediment-hazards.

## ACKNOWLEDGEMENTS

The present study is financially supported by Grant-in-Aid for Scientific Research (B) (Representative: Shinji EGASHIRA) from the Japan Ministry of Education, Culture, Sports, Science and Technology (MEXT). In addition, Japan Water Agency supports the present study in intelligent aspects. These supports are greatly appreciated.

## REFERENCES

Ashida, K. & M. Michiue 1972. Study on hydraulic resistance and bed-load transport rate in alluvial streams. *Proc. of JSCE*, No. 206: 59–69 (in Japanese).

Ashida, K. & M. Fujita 1987. Simulation of reservoir sedimentation. *Annuals of Disaster Prevention Research Institute of Kyoto University*, No. 30 B-2: 457–474 (in Japanese).

Egashira, S. 1998. Research Related to Prediction of Sediment Yield and Runoff. *Symposium on Japan-Indonesia IDNDR Project*, September 21–23, 1998, Bandung, Indonesia: 373–384.

Egashira, S. & K. Matsuki 2000. A method for predicting sediment runoff caused by erosion of stream channel bed. *Annual Journal of Hydraulic Engineering*, Vol. 44: 735–740 (in Japanese).

Egashira, S. 2001. Method for predicting sediment yield and runoff. *Lecture Note of the 37th Summer Seminar on Hydraulic Engineering, 2001, Course A*, Committee on Hydraulics and Coastal Engineering Committee, JSCE: A-2-1-A-2-14 (in Japanese).

Egashira, S. & T. Itoh 2003. Sediment runoff method taking into account yielding mechanism of sediment in channels. *Proceedings of Sabo and Landslide technical reports*: 33–46 (in Japanese).

Garcia, M. & G. Parker 1991. Entrainment of bed sediment into suspension. *Jour. of Hydraulic Engineering*, ASCE, Vol. 117, No. 4: 414–435.

Japan Society of Civil Engineering 1999. *Handbook for Hydraulic Engineering*, Maruzen: p. 143 (in Japanese).

Ichikawa, Y., Y. Satoh, M. Shiiba, Y. Tachikawa & K. Takara 1999. Development of a water and sediment flow model for a mountainous area. *Annuals of Disaster Prevention Research Institute of Kyoto University*, No. 42 B-2: 211–224 (in Japanese).

Itakura, T. & T. Kishi 1980. Open channel flow with suspended sediment. *Jour. of the Hydraulics Division*, Vol. 106, No. HY-8: 1325–1343.

Kaneyashiki, T., K. Ashida & S. Egashira 1980. A hydraulic model of the yield of wash load in mountainous areas. *Proc. of the 24th Japanese Conference on Hydraulics*: 143–151 (in Japanese).

Murakami, S., S. Hayashi, S. Kameyama & M. Watanabe 2001. Fundamental study on modeling of sediment routing through forest and agricultural area in watershed. *Annual Journal of Hydraulic Engineering*, Vol. 45: 799–804 (in Japanese).

Muramoto, Y., M. Michiue & E. Shimozima 1973. On the transport process of wash load in Daido river. *Annuals of Disaster Prevention Research Institute of Kyoto University*, No. 16 B: 433–447 (in Japanese).

Sunada, K. & N. Hasagawa 1994. Study of a synthetic model for sediment routing in a mountainous river system. *Jour. of Hydraulic, Coastal and Environmental Engineering*, No. 485/II-26: 37–44 (in Japanese).

Sunada, K., K. Komatsu & H. Kobayashi 2000. Fundamental examination on the synthetic sediment routing model in a whole river system. *Annual Journal of Hydraulic Engineering*, Vol. 44: 729–734 (in Japanese).

Takahashi, T., M. Inoue, H. Nakagawa & Y. Satofuka 2000. Prediction of sediment runoff from a mountain watershed. *Annual Journal of Hydraulic Engineering*, Vol. 44: 717–722 (in Japanese).

Takara, K., K. Notsumata & R. Uesaka 1998. A Distributed Model for Flood Runoff and Sediment Yield Based on Remote Sensing and GIS. *Symposium on Japan-Indonesia IDNDR Project*, September 21–23, 1998, Bandung, Indonesia: 373–384.

*River, Coastal and Estuarine Morphodynamics: RCEM 2007 – Dohmen-Janssen & Hulscher (eds)*
*© 2008 Taylor & Francis Group, London, ISBN 978-0-415-45363-9*

# Fundamental properties of suspended sediment transport in open channel flows with a side cavity

Ichiro Kimura
*Matsue National College of Technology, Matsue, Japan*

Takashi Hosoda & Shinichiro Onda
*Kyoto University, Kyoto, Japan*

ABSTRACT: The characteristics of the suspended sediment transport in the open channel flows with a rectangular side-cavity are investigated experimentally and numerically. The properties of the deposition in the side-cavity are classified into four patterns from the experimental results. The influences of the hydraulic parameters on the deposition patterns in the dead zone are examined through the laboratory test under different hydraulic conditions. The deposition processes are numerically simulated by the plane depth-averaged 2D shallow flow equations with simple 0-quation type turbulence model. The computations could successfully capture the fundamental aspects of the flow and sedimentation. The numerical results indicate that the large scale vortex shedding at the interface promotes the sediment transport into the dead zone.

## 1 INTRODUCTION

Recently, river environment becomes one of most important research topics in civil engineering fields. It is preferable to provide various flow velocities in rivers from the viewpoints of habitat for wide variety of fauna and flora. For instance, if an inlet exists along the shore of a river, a recirculation flow with calm velocity is generated and is suitable for small fishes and insects, which cannot live in main flow part of high velocity. Figure 1 shows inlets in series formed around spur dikes in Kiso river in Japan. These inlets are called "Wando" in Japan (Kimura et al, 2009). The topography of the inlet becomes sometimes very complex like this case by the sediment transport due to complex flow pattern around the inlet.

The flow around a rectangular side-cavity in a horizontal plane is characterized by steady recirculation inside the side-cavity and unsteady vortex shedding along the interface between the main flow and the side-cavity. The latter flow is generated by shear layer instability (K–H instability) at the interface. The periodic free surface oscillations like sloshing are also notable phenomena. Kimura & Hosoda (1997) performed laboratory tests and 2D depth-averaged computations of open channel flows with rectangular side cavity and pointed out that the period of the free surface oscillation in the cavity ($=T_m$) coincides with the characteristic period of closed water basin as follows.

$$T_m = \frac{2L}{m\sqrt{gh}}, \quad m = 1,2,3\ldots \tag{1}$$

where, L: length of the closed water basin, h: mean depth, m: number of nodes and g: gravity acceleration ($=9.8\,\text{m/s}^2$). They also examined time-series of photographs of a vortex visualized by aluminum powder and found that a vortex caused by the shear instability is amplified to a large-scale vortex selectively by the interaction between the free surface oscillation and the vortex induced by the instability. The selective amplification process could be captured by a plane 2D numerical model with 0-equation formula for eddy viscosity when a numerical grid with enough spatial resolution is used. Kimura & Hosoda (2000) examined properties of amplitude of free surface oscillation which closely depend on the Froude number and found that several

Figure 1. Spur dikes and "Wando" in Kiso River.

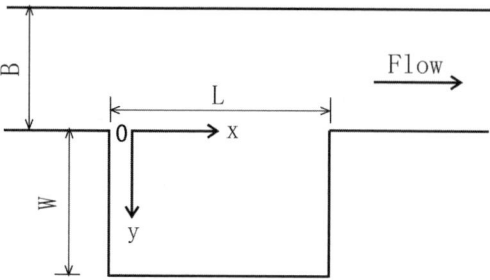

Figure 2.   Plan view of the experimental flow domain.

Table 1.   Three types of side-cavity.

| Type | L(cm) | W(cm) | B(cm) | Aspect Ratio (L/W) |
|------|-------|-------|-------|--------------------|
| Type 1 | 22.5 | 15.0 | 25.0 | 1.5 |
| Type 2 | 37.5 | 25.0 | 25.0 | 1.5 |
| Type 3 | 50.0 | 25.0 | 25.0 | 2.0 |

Table 2.   Hydraulic conditions of the laboratory tests.

| Exp. | Q(cm³/s) | h(cm) | T(°C) | Cavity type | sin θ |
|------|----------|-------|-------|-------------|-------|
| F-1 | 600 | 1.33 | 26.9 | Type 1 | 1/1000 |
| F-2 | 1354 | 3.30 | 15.0 | Type 1 | 1/300 |
| F-3 | 1354 | 1.84 | 15.0 | Type 1 | 1/300 |
| F-4 | 1354 | 1.69 | 15.0 | Type 1 | 1/300 |
| F-5 | 1354 | 5.70 | 15.0 | Type 1 | 1/300 |
| F-6 | 1177 | 9.10 | 22.2 | Type 1 | 1/300 |
| F-7 | 1177 | 10.2 | 22.2 | Type 2 | 1/300 |
| F-8 | 1177 | 1.50 | 22.2 | Type 2 | 1/300 |
| F-9 | 1177 | 3.60 | 22.2 | Type 3 | 1/300 |
| F-10 | 1307 | 2.39 | 13.0 | Type 1 | 1/1000 |
| F-11 | 758 | 1.89 | 13.0 | Type 1 | 1/1000 |
| F-12 | 719 | 6.00 | 13.0 | Type 1 | 1/1000 |
| F-13 | 324 | 1.24 | 13.0 | Type 1 | 1/1000 |
| F-14 | 591 | 1.32 | 14.0 | Type 1 | 1/1000 |

Q: discharge in main channel, h: mean depth, T: temperature, Cavity type: shape of cavity (see Table 1), sin θ: bed slope.

peaks exist in the relation between the amplitude and Froude number. The reason could be clearly explained using simple model considering the motion of large scale vortices along the interface. They also examined the mass-exchange rate between the side cavity and main flow and found that, in cases of larger Froude number (Fr ≫ 0.3), the coherent fluid oscillation is the dominant factor for the mass-exchange through the interface. On the other hand, there is no correlation between the mass-exchange and the fluid oscillation in cases of smaller Froude number (Fr ≪ 0.3).

The steady recirculation and unsteady vortex shedding in and around a side-cavity have also important rolls for sediment transport phenomena. From the environmental point of view, it is particularly important to elucidate the relation between the flow characteristics and sedimentation in open channel flows with side-cavities.

In this study, sediment transport around a rectangular side-cavity in shallow open channel flows is examined by both experimental and numerical approaches. First, the laboratory tests were performed under various hydraulic conditions and deposition patterns in the side-cavity were classified into four different cases. Then, the plane 2D computations were performed using depth-averaged shallow flow equations under the same conditions with the laboratory test. The computation was performed under both dynamic and steady flow conditions in order to consider the effects of unsteady flow characteristics. The computational results were discussed through the comparison with the experimental results.

## 2   LABORATORY TESTS

### 2.1   Experimental setup

The laboratory tests were performed using an experimental flume made of acrylic resin with a rectangular cross section. The width and length of the flume is 40 cm and 4 m, respectively. 3 types of rectangular side-cavities with different scales are made by dividing the flume using thin acrylic plates. The plan view of the side-cavity is schematically shown in Figure 1 and parameters in the 3 cavities are listed in Table 1.

We used PVC (polyvinyl chloride) powder (mean diameter, $d = 81\,\mu\mathrm{m}$ and relative density, $s = 1.21$) as sediment model instead of real sand considering the ratio of falling velocity against depth. The falling velocity $w_0$ can be estimated as 0.42 cm/s using Rubey's equation. The PVC powder was mixed with water in advance and poured uniformly in channel-width direction at upstream section of the flume. The discharge of PVC-mixed water was adjusted so as to make the concentration of the powder 0.05% as volume percentage in the main flow.

Table 2 shows the hydraulic conditions in the laboratory tests. The experiments were performed under 14 conditions with different discharge Q, mean depth h, bottom slope sinθ and shape of the side-cavity. Table 3 shows the dimensionless parameters in each case. The table shows together with the 4 deposition patterns, which will be discussed later.

### 2.2   Results and discussion of the laboratory tests

#### 2.2.1   Classification of deposition patterns

Figure 3 (a)–(d) shows the deposition patterns in and around the side-cavity in cases of Exp.F-4, Epx.F-2, Exp.F-14 and Exp.F-12. The deposition patterns in these 4 cases are qualitatively different altogether and each result in other all cases was similar to one of

Table 3. Dimensionless parameters and deposition patterns.

| Exp. | Fr | Re | $u_{*m}/w_0$ | Pattern |
|---|---|---|---|---|
| F-1 | 0.50 | 2737 | 2.71 | C |
| F-2 | 0.29 | 4747 | 2.29 | B |
| F-3 | 0.69 | 4747 | 4.11 | A |
| F-4 | 0.79 | 4747 | 4.47 | A |
| F-5 | 0.13 | 4747 | 1.33 | C |
| F-6 | 0.06 | 4925 | 0.72 | D |
| F-7 | 0.05 | 4925 | 0.64 | D |
| F-8 | 0.82 | 4925 | 4.36 | A |
| F-9 | 0.22 | 4925 | 1.81 | C |
| F-10 | 0.45 | 4343 | 3.09 | B |
| F-11 | 0.37 | 2519 | 2.43 | C |
| F-12 | 0.06 | 2389 | 0.73 | D |
| F-13 | 0.30 | 1078 | 1.80 | C |
| F-14 | 0.50 | 2018 | 2.81 | C |

Fr: Froude number, Re: Reynolds number $u_{*m}$: friction velocity in main flow, $w_0$: falling velocity of sediment.

these 4 patterns. Namely, the deposition patterns can be classified into these 4 basic patterns, which are schematically illustrated in Figure 4. The characteristics of each deposition patterns are explained briefly as follows.

### 2.2.1.1 Pattern A
The deposition occurs in the center part of the side-cavity and the bed wave can be seen in the deposition area. The deposition does not occur in main flow and around the interface. Slight deposition is also found around two corners of the side-cavity, where weak vortices are generated. Since this pattern appears in case of higher Froude numbers, the deposition and flush occur alternately due to the free surface oscillation.

### 2.2.1.2 Pattern B
The deposition occur in a doughnut like shape, namely, the thickness of the sand in the center part of the deposition is thinner than that in the surrounding part. The total deposition area is larger than that in Pattern A. In the main flow, the deposition does not occur, but streak type concentration of sand can be seen as shown in Figure 5, which is caused by secondary currents of second kind. Slight deposition can be seen in the two corners of the side-cavity like Pattern A.

### 2.2.1.3 Pattern C
The deposition area becomes larger than Patterns A and B. A distinctive sandbar along the interface between the side-cavity and main channel is generated. The deposition at two corners in the side-cavity becomes clearer. The streak pattern can be seen again in the main channel, where slight deposition was detected.

### 2.2.1.4 Pattern D
The deposition area becomes larger than other 3 patterns. The deposition occurs also in the almost whole region of the main channel. A spiral pattern, which is

Flow ⟶

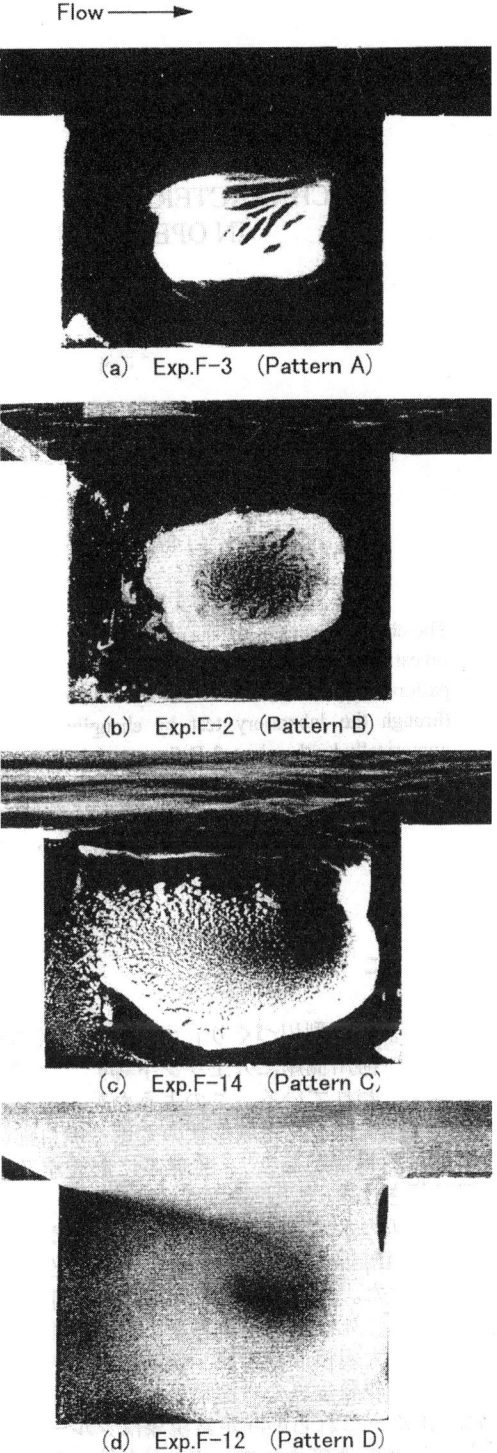

(a) Exp.F-3 (Pattern A)

(b) Exp.F-2 (Pattern B)

(c) Exp.F-14 (Pattern C)

(d) Exp.F-12 (Pattern D)

Figure 3. Deposition patterns in the laboratory tests.

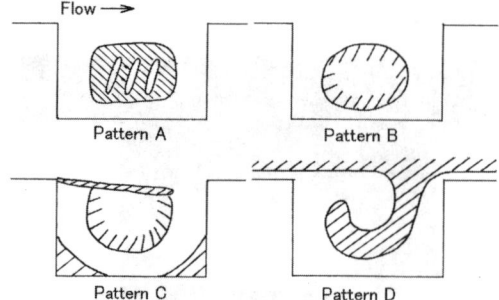

Flow →

Pattern A    Pattern B

Pattern C    Pattern D

Figure 4.   Schematic diagram of 4 deposition patterns.

Figure 5.   Streak lines observed in Exp.F-14 (Pattern C).

caused by a steady recirculation flow in the clockwise direction, can be found in the side-cavity. Only the area where the deposition can not be seen is the downstream end of the interface, where the large velocity is generated due to the impinging of unsteady vortices.

### 2.2.2   Relation between hydraulic parameters and deposition patterns

We consider two dimensionless parameters, which may closely concern with sedimentation phenomena, i.e., one is the Froude number $U/(gh)^{0.5}$ and the other is $u_{*m}/w_0$ ($u_{*m}$: mean friction velocity in the main channel, $w_0$: falling velocity). Parameter $u_{*m}/w_0$ is commonly used as the rate of suspension of sediment. The Froude number is employed as the dominant parameter because the free surface oscillation in the side-cavity considerably depends on the Froude number.

Figure 6 shows the relation between the 4 deposition patterns and the 2 dimensionless parameters. In this Figure, the 14 points of experiments align in almost single straight line and two-dimensional distribution is not obtained. The reason is that the experiments were performed under a fixed density condition in the bed material and the falling velocity $w_0$ is uniform in all 14 cases. This figure indicates that the deposition pattern changes A→B→C→D as the Froude number or

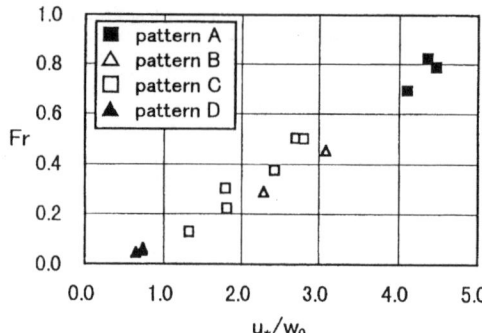

Figure 6.   Relation between dimensionless hydraulic parameters and deposition patterns.

$u_{*m}/w_0$ becomes smaller. The feature does not depend on 3 types of side-cavity in Table 1.

## 3   COMPUTATIONS

### 3.1   Basic equations

The basic equations employed in the present computations are depth-averaged plane 2D shallow flow equations composed a continuity equation, momentum equations, a suspended sediment transport equation and a bed continuity equation in a Cartesian coordinate. These equations are described as follows (Abbott, 1979, Tan, .1992).

[Continuity equation]

$$\frac{\partial h}{\partial t} + \frac{\partial M}{\partial x} + \frac{\partial N}{\partial y} = 0 \qquad (2)$$

[Momentum equation in x-direction]

$$\frac{\partial M}{\partial t} + \frac{\partial \beta u M}{\partial x} + \frac{\partial \beta v M}{\partial y} + gh\frac{\partial (h + z_b)}{\partial x} =$$

$$gh\sin\theta - \frac{\tau_{bx}}{\rho} + \frac{\partial -\overline{u'^2}h}{\partial x} + \frac{\partial -\overline{u'v'}h}{\partial y}$$

$$+ \nu\left\{\frac{\partial}{\partial x}\left(h\frac{\partial u}{\partial x}\right) + \frac{\partial}{\partial y}\left(h\frac{\partial u}{\partial y}\right)\right\} \qquad (3)$$

[Momentum equation in y-direction]

$$\frac{\partial N}{\partial t} + \frac{\partial \beta u N}{\partial x} + \frac{\partial \beta v N}{\partial y} + gh\frac{\partial (h + z_b)}{\partial y} =$$

$$-\frac{\tau_{by}}{\rho} + \frac{\partial -\overline{v'u'}h}{\partial x} + \frac{\partial -\overline{v'^2}h}{\partial y}$$

$$+ \nu\left\{\frac{\partial}{\partial x}\left(h\frac{\partial v}{\partial x}\right) + \frac{\partial}{\partial y}\left(h\frac{\partial v}{\partial y}\right)\right\} \qquad (4)$$

[Transport of suspended sediment]

$$\frac{\partial}{\partial t}(ch) + \frac{\partial}{\partial x}(cuh) + \frac{\partial}{\partial y}(cvh) =$$

$$\frac{\partial}{\partial x}\left\{ h\left(D_L \cos\theta_f + D_x\right)\frac{\partial c}{\partial x}\right\} + \qquad (5)$$

$$\frac{\partial}{\partial y}\left\{ h\left(D_L \sin\theta_f + D_y\right)\frac{\partial c}{\partial y}\right\} + q_{su} - w_0 c_b$$

[Continuity equation of bed]

$$\frac{\partial z_b}{\partial t} + \frac{1}{1-\lambda}\left(q_{su} - w_0 c_b\right) = 0 \qquad (6)$$

where, (x, y) : spatial coordinate (see Figure 1), t: time, h: water depth, $(u, v)$: depth-averaged velocity components in $(x, y)$ directions, $(M, N)$: fluxes in $(x, y)$ direction defined as $(hu, hv)$, $(u', v')$: turbulence velocities in (x, y) directions, $-\overline{u'_i u'_j}$: depth-averaged Reynolds stress tensor, $v$: dynamic viscosity coefficient, $\sin\theta$ : bed slope, $f$: friction coefficient (function of Reynolds number), $(\tau_{bx}, \tau_{by})$: bed friction stress vector, $\beta$ : momentum coefficient, c: depth-averaged concentration of suspended sediment, $z_b$: bed height□$w_0$: falling velocity of suspended sediment, $\lambda$: void rate of bed material ($\lambda = 0.5$ is used), $D_L$: convective dispersion coefficient, $\theta_f$ : angle between stream line and x-axis, $(D_x, D_y)$: eddy difusivity coefficients in $(x, y)$ directions, $(\tau_{bx}, \tau_{by})$ = bottom shear-stresss in (x, y) directions; $q_{su}$: amount of sediment suspended from bed and $c_b$: concentration of suspended sediment near bed, which is calculated from depth-averaged concentration c assuming exponential type profile.

Components of the bottom shear-stress vector are evaluated by

$$\tau_{bx} = \frac{f\rho u}{2}\sqrt{u^2 + v^2} \; ; \quad \tau_{by} = \frac{f\rho v}{2}\sqrt{u^2 + v^2} \qquad (7)$$

where $f =$ friction factor related to local Reynolds number $Re' \equiv uh/v$, evaluated as follows:

$$f = \frac{6}{R_e'} \quad (R_e' \le 430) \qquad (8a)$$

$$\sqrt{\frac{2}{f}} = A_S - \frac{1}{\kappa}\left[1 - \ln\left(R_e' \sqrt{\frac{f}{2}}\right)\right] \quad (R_e' > 430) \qquad (8b)$$

where $\kappa = 0.4$, $A_S = 5.5$. A simple 0-equation model presented in eq.(9) was used to evaluate the depth-averaged Reynolds stress tensors (Kimura & Hosoda, 1997).

$$-\overline{u_i' u_j'} = D_h\left(\frac{\partial u_i}{\partial x_j} + \frac{\partial u_j}{\partial x_i}\right) - \frac{2}{3}k\delta_{ij} \; , \quad D_h = \alpha h u_* \qquad (9)$$

where $\alpha =$ empirical constant ($\alpha = 0.3$ is used in this study); $u_* =$ local friction velocity ($\equiv \sqrt{f(u^2 + v^2)/2}$); and $k =$ depth-averaged turbulent kinetic energy evaluated by the empirical formula proposed by Neze & Nakagawa (1993), who proposed the universal expression in equation .(10) for turbulent kinetic-energy distribution

$$\frac{k}{u_*^2}/u_*^2 = 4.78\exp\left(-2\frac{z}{h}\right) \qquad (10)$$

where $z =$ direction perpendicular to the bottom bed. The depth-averaged turbulent kinetic energy becoms the following formula is $2.07u_*^2$ when equation (10) is integrated from the bottom to the surface.

$$k = 2.07u_*^2 \qquad (11)$$

$(q_{su} - w_0 c_b)$ in the equations (5) and (6) means suspend and deposition of sediment. $\sin\theta_f$ and $\cos\theta_f$ can be calculated by

$$\sin\theta_f = \frac{|v|}{\sqrt{u^2 + v^2}}, \quad \cos\theta_f = \frac{|u|}{\sqrt{u^2 + v^2}} \qquad (12)$$

Convective dispersion coefficient and eddy diffusivity coefficients are given as

$$D_L = 5.9hu_* \qquad (13a)$$

$$D_x = D_y = 5.9hu_* \qquad (13b)$$

Suspension term $q_{su}$ and concentration at the bed $c_b$ are given as

$$q_{su} = w_0 c_{be}, \quad c_b = \frac{\beta c}{1 - e^{-\beta}}, \quad \beta = \frac{w_0 h}{\varepsilon_s} \qquad (14)$$

where $\varepsilon_s$ is eddy diffusivity in the vertical direction and is supposed to be equivalent to the eddy viscosity in the horizontal direction:

$$\varepsilon_s = \frac{\gamma\kappa u_* h}{6} \qquad (15)$$

where $\gamma$ and $\kappa$ and constants (k = 0.4 (Karman constant) and $\gamma = 1.0$ arc used). From equations (14) and (15), $\beta$ becomes

$$\beta = \frac{w_0 h}{\varepsilon_s} = \frac{6w_0 h}{\gamma\kappa u_* h} = \frac{15w_0}{u_*} \qquad (16)$$

$c_{be}$ in eqution (8) means concentration at the standard section and is evaluated by the following Ashida-Michiue formula (Nagata et al, 2005).

$$c_{be} = 0.025\left\{\frac{g(\zeta_0)}{\zeta_0} - G(\zeta_0)\right\} \qquad (17)$$

1207

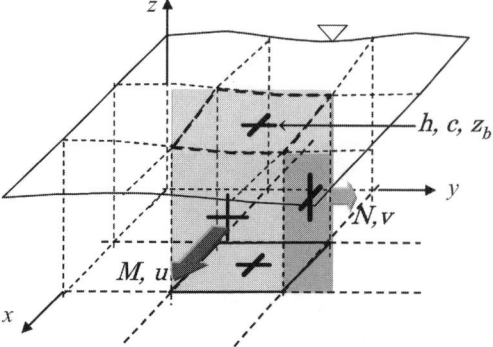

Figure 7. Arrangement of hydraulic ariables in a full-staggered grid.

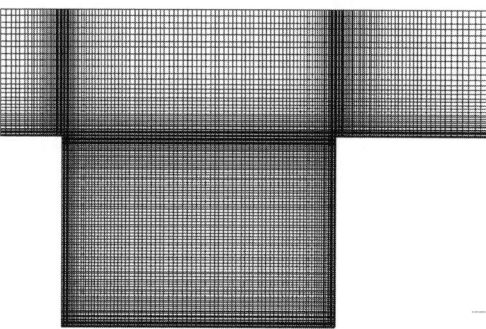

Figure 8. Computational grid around the side-cavity.

where

$$\zeta_0 = \frac{4w_0}{3u_*}, \quad g(\zeta_0) = \frac{1}{\sqrt{2\pi}}\exp\left(-0.5\zeta_0^2\right)$$

$$G(\zeta_0) = \frac{1}{\sqrt{2\pi}}\int_{\zeta_0}^{\infty}\exp(-0.5\zeta^2)d\zeta \tag{18}$$

This model has been derived assuming the sediment flux upwelling by the turbulence velocity is same as falling flux and the distribution of velocity fluctuation is Gaussian distribution.

When the $z_b$ becomes less than $10^{-2}$ mm, $q_{su}$ in equations (5) and (6) is set to be 0.

### 3.2 Computational methods

Computations were performed using finite volume method on a full-staggered. The arrangement of hydraulic variables on the grid is shown in Figure 7. The 2nd order QUICK scheme is used for convective inertia terms in momentum equations and a suspended sediment transport equation. The time integration is done fully explicitly using the 2nd order Adams-Bashforth method. Figure 8 shows the computational grid around the side-cavity. The number of grid points is about 20,000.

In order to compute the dynamic behavior of sedimentation, both the flow (equations (2) and (3) and the sedimentation (equations (4) and (5) should be solved at each time step. However, some preliminary computations reveal that the time increment $\Delta t$ for the sediment transport equation (4) is should be set much smaller than $\Delta t$ for the momentum equations for stability of computations. Therefore, the following 2 kind of computations are carried out to consider computational efficiency and effects of unsteady flow behavior on sedimentation.

#### 3.2.1 Run 1
The sedimentation is computed under dynamic flow conditions. Namely, all equations (equations (2) – (5) solved at every time step. This method can strictly consider the dynamic behavior of the sedimentation under unsteady flow behavior though it takes a lot of CPU time. The CPU time in Run 1 is about 3 times larger than that in Run 2.

#### 3.2.2 Run 2
The sedimentation is computed under time-mean steady flow conditions. First, only the flow (equations (2) and (3) is calculated from initial conditions to steady state. In the current computation, the computational result contains periodic oscillation. Second, the time-mean flow pattern is obtained by averaging flow patterns in 10 cycles of oscillations. Then only the sedimentations (equations (4) and (5) are computed under the computed time-mean flow field in a steady flow condition. This computation yields the sedimentation without effects of vortex shedding, free surface oscillation or other unsteady flow features.

### 3.3 Computational results

#### 3.3.1 Deposition pattern
Figure 9 (a) shows the computational velocity vectors, water surface, concentration and deposition pattern at $t = 2$ min using Run 1 under conditions of Exp.F-14. The velocity vectors and surface profile indicate that vortex shedding occurs along the interface. Deposition can be seen in the center port of the side-cavity similarly to the experimental result. The outer shape of deposition area is also similar to the experimental result, though, in the computational result, a lot of deposition can be seen in the center part of the cavity, where the almost no deposition occurs in the laboratory test. A sandbar at the interface, which is a unique feature in Pattern C (see figure 2), is reproduced numerically though the width of the bar is larger than the experimental result.

The comparison between experimental and numerical results indicated that the present computational model could capture the fundamental aspect of

1208

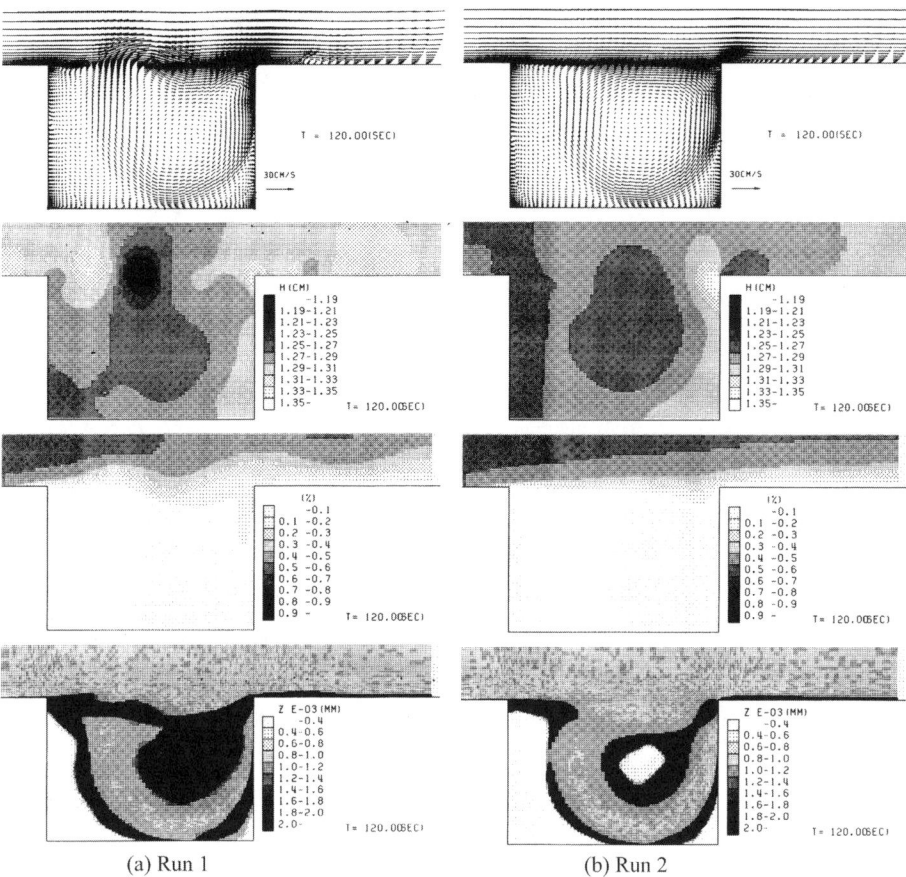

| (a) Run 1 | (b) Run 2 |

Figure 9. Computational results by the two methods at t = 2 min (velocity vectors, water surface height, suspended sediment concentration, bed height).

deposition pattern around a side-cavity to some extent though some discrepancies such as the deposition at the center part of the cavity are pointed out. The sedimentation in the flow with recirculation is affected by secondary currents of first kind caused by curvature of streamline. In the present 2D plane computations, the effects of secondary currents are completely neglected. Therefore, in the next step, we would like to apply the 2D model which takes into account the effects of the secondary currents (Hosoda et al, 2001), or 3D model with adequate turbulence models (Nagata et al, 2005, Kimura & Hosoda, 2003).

Figure 9 (b) shows the computational result using the method of Run 2 under the conditions of Epx.F-14 at 2 min computation from the initial conditions. In this computation, the effects of vortex shedding at the interface disappear through the averaging process as shown in velocity vectors and water surface pattern, and only the recirculation in the side-cavity remains. It should be noted that the sandbar at the interface is

not generated in the computation of Run 2. Therefore, it is concluded that the sandbar is generated by the effects of unsteady vortex shedding at the interface. The deposition pattern in the cavity shows a torus-like shape, namely, deposition rarely occurs at the center of the deposition area. This rare deposition area roughly coincides with the center of the recirculation zone, where the velocity is too small to carry sediment to this area. On the other hand, in the computation of Run 1, the center of the recirculation moves periodically due to the free surface oscillation. Therefore, the center of the torus becomes unclear.

Note that the Froude number in Exp.F-14 is about $Fr = 0.5$. Kimura & Hosoda (2000) showed through the consideration of a simple model that the amplitude of the free surface oscillation around $Fr = 0.5$ becomes relatively small. Therefore, it is likely that the difference of two kinds of computations becomes larger if the two methods were applied to the case with larger amplitude of free surface oscillations.

Table 4. Comparison of volume of deposition and maximum height of bed by Run 1 and Run 2.

|  | Run 1 | Run 2 |
|---|---|---|
| Volume of deposition (cm$^3$) | 1.59 | 0.73 |
| Maximum height of bed (cm) | 0.524 | 0.568 |

### 3.3.2 Comparison of volume of deposition and maximum bed height

Table 4 shows the comparison volume of the deposition and maximum height of the bed in the side-cavity at t = 2 min by the computations of Run 1 and Run 2 under the conditions of Exp.F-14. It should be noted that the volume by Run 1 is about twice larger than that by Run 2. It is therefore pointed out that the unsteady flow behavior such as vortex shedding along the interface and free surface oscillations promote the inflow of sediment into the side-cavity.

The maximum point of bed height occurs near the interface around the upstream corner (near the original point of the coordinate, see Figure 1) in both Run 1 and Run 2. The sediment transported from upstream region by the main channel flow is rapidly decelerated at this location and then deposits. The heights of the maximum bed in both Runs are almost identical.

## 4 CONCLUDING REMARKS

The suspended sediment transport in open channel flows with a rectangular side-cavity was examined experimentally and numerically in this paper. Main findings of our research are listed as follows.

1. The deposition pattern in and around the side-cavity can be classified 4 patterns.
2. The deposition pattern is affected by two dimensionless hydraulic parameters: the Froude number and $u_{*m}/w_0$.
3. The sedimentation was computed under both time-mean steady flow conditions and dynamic flow conditions. Only latter computation can yield formation of sandbar at the interface between the side-cavity and the main channel.
4. The numerical results showed that the large scale vortex shedding at the interface promote the transport of the sediment into the side-cavity.
5. The sedimentation in the side-cavity is affected by the recirculation flow. So, the computational method with effects of secondary kind of 1st kind

should be developed to improve the accuracy of computation.

ACKNOWLEDGEMENT

Professor Yoshio Muramoto is gratefully acknowledged for his kind and important scientific advices as well as the financial support.

REFERENCES

Abbott, M. B. 1979. Computational hydraulics, Pitman Publishing, Ltd., London, England.
Altai, W & Chu, V. H. 1997. Retention time in a recirculating flow. Proc. 27th Congress of IAHR, San Francisco, USA, Theme B, Vol.1, pp.9–14.
Hosoda, T. & Kimura, I. 1993. Vortex formations with free surface variation in the shear layer of plane-2D open channel flows. Proc, 9th Symp. on Turbulent Shear Flows, Kyoto, Japan, Vol. 1, P112, pp.1–4.
Hosoda, T., Nagata, N., Kimura, I., Michibata, K. & Iwata, M. 2001. A depth averaged model of open channel flows with lag between main flows and secondary currents in a generalized curvilinear coordinate system. Advances in Fluid Modeling & Turbulence Measurements, (eds. H. Ninokata, A. Wada and N. Tanaka), World Scientific, pp.63–70.
Kimura, I. & Hosoda, T. 1993. Unsteady behaviors of open channel flows with rectangular dead zone area. J. Japan Society of Fluid Mech., Vol.12, pp.399–408 (in Japanese).
Kimura, I. & Hosoda, T. 1997. Fundamental properties of flows in open channels with dead zone. J. Hydraulic. Engineering, ASCE, Vol.123 (2), pp.98–107.
Kimura, I. & Hosoda, T. 2000. Free surface oscillations and mass exchange in open channel flows with a cavity. Proc. 3rd Int. Symposium on Turbulence, Heat and Mass Transfer, Nagoya, Japan, Aichi Shuppan, pp.233–240.
Kimura, I. & Hosoda, T. 2003. A non-linear k-ε model with realizability for prediction of flows around bluff bodies. International Journal for Numerical Methods in Fluids, Vol.42, pp.813–837.
Kimura, I., Hosoda, T., Onda, S. & Tominaga, A. 2004. Computations of /3D turbulent flow structures around submerged spur dikes under various hydraulic conditions. pp.–.
Nagata, N., Hosoda, T., Nakato, T. & Muramoto, Y. 2005. Threedimensional numerical model for flow and bed deformation around river hydraulic structures. J. of Hydraulic Eng., ASCE, Vol.131, No.12, pp.1074–1087.
Nezu, I. & Nakagawa, H. 1993. Turbulence in Open-Channel Flows, IAHR Monograph, Belkema, Rotterdam, Netherlands.
Tan, W. 1992. Shallow water hydrodynamics, Elsevier Oceanography Series, 55, Elsevier Science Publishers BV, Amsterdam, The Netherlands.

*River, Coastal and Estuarine Morphodynamics: RCEM 2007 – Dohmen-Janssen & Hulscher (eds)*
*© 2008 Taylor & Francis Group, London, ISBN 978-0-415-45363-9*

# One-dimensional numerical scheme to model bed evolution in presence of a side overflow

M. Catella, G. Bechi & E. Paris
*Department of Civil Engineering, University of Florence, Florence, Italy*

B. Rosier & A.J. Schleiss
*Laboratory of Hydraulic Constructions (LCH), Swiss Fed. Inst. Of Technology, Lausanne, Switzerland*

ABSTRACT: The present study concerns a one-dimensional numerical model to analyze side weir efficiency by modeling the morphodynamic evolution of river bed during major floods and evaluating its influence on the design discharge to be diverted. The St. Venant-Exner equations are solved by employing a finite volume method, adopting a decoupled procedure. The water depth profiles in the main channel along the side weir are computed by modifying the De Marchi (1934) hypothesis and by evaluating the overflow discharge by means of two different equations in the case of free or submerged overflow. In order to verify the side weir morphodynamic modeling, the present numerical scheme is applied to reproduce the experimental results obtained by Rosier et al. (2006). Finally the model is employed to design a side weir in a real case study.

## 1 INTRODUCTION

The design of side weir is an important feature in hydraulic engineering analysis, river management and civil protection in order to control flood wave propagation in natural channels.

In particular storage areas are more and more used in river restoration and flood control management in order to prevent flooding to lower lying lands. These structures allocate a part of liquid discharge at times of unseasonably high rainfall usually by means of side flow weirs. As a part of the liquid discharge is diverted from the main channel, the downstream sediment transport capacity decreases. Thus an aggradation process is established along the side weir modifying the design efficiency. Notwithstanding many contribution to river morphodynamics a deep insight about the interaction between river bed evolution and side weir overflow is still lacking. Besides, despite the high task played by these structures to prevent flood disasters, many aspects of their performance still remains unclear because of no field data are available.

In recent years several studies have been performed on sharp-crested side weirs in rectangular channel without taking into account sediment transport and mobile bed (Nadersmoothy & Thomson, 1972; Ranga Raju et al., 1979; Hager, 1987; Milano et al., 1998; Borghei et al., 1999).

The present paper concerns a one-dimensional numerical model to analyze side weir efficiency by modeling the deposit formation during major floods and evaluating its influence on the design discharge to be diverted. The St. Venant- Exner equations are solved by employing a finite volume method, adopting a decoupled procedure (Catella & Solari, 2005). The water depth profiles in the main channel along the side weir are estimated by energy balance obtained by modifying the De Marchi (1934) hypothesis, and by evaluating the overflow discharge by means of two different equations in the case of free or submerged overflow.

To verify the side weir morphodynamic modeling, the numerical scheme is applied to reproduce the experimental results obtained by Rosier et al. (2006) at the Laboratory of Hydraulic Constructions (LCH), Swiss Federal Institute of Technology, Lausanne, Switzerland.

Finally the present model is employed to design a side weir in a real case study in a sub-catchment of the Ombrone Pistoiese River in Tuscany (Italy). Results point out the significant importance of the river bed morphodynamic evolution that has to be taken into account in the side weir design activities.

## 2 THE NUMERICAL MODEL

### 2.1 The governing equations

Let us consider a control volume $V$ at $j_{th}$ cell defined by the channel reach included between two vertical cross sections $j - 1/2$ and $j + 1/2$ distant $\Delta x_j$.

The momentum equation applied to the control volume $V$, projected on the horizontal $x$-axis, and being the banks impermeable, following Catella & Solari (2005), reads:

$$\frac{\partial}{\partial t}\int_{x_{j-1/2}}^{x_{j+1/2}}Qdx+\left[\beta\frac{Q|Q|}{S}+gI_1\right]_{x_{j-1/2}}^{x_{j+1/2}}+$$

$$-g\left[S_{j+1/2}(h_{mj}-z_{gj+1/2})-S_{j-1/2}(h_{mj}-z_{gj-1/2})\right]+ \quad (1)$$

$$+\frac{\Delta x_j}{2}\left[B_{j+1/2}\left(\frac{Q|Q|}{(CS)^2}\right)_{j+1/2}+B_{j-1/2}\left(\frac{Q|Q|}{(CS)^2}\right)_{j-1/2}\right]=0$$

where $Q=$ liquid discharge, $\beta=$ Boussinesq velocity distribution coefficient, $S=$ wetted surface, $I_1=$ first moment of the wetted cross section, $g=$ gravitational acceleration, $h_m=$ average elevation of the water surface within the cell, $z_g=$ center mass elevation of the wetted cross section, $B_{j\pm1/2}=$ wetted perimeter, and $C=$ dimensionless Chézy coefficient.

The integral form of the continuity equation under the assumption of no lateral inflow or outflow is:

$$\frac{\partial}{\partial t}\int_{x_{j-1/2}}^{x_{j+1/2}}Sdx+\left[Q\right]_{x_{j-1/2}}^{x_{j+1/2}}=0 \quad (2)$$

Finally, the integral form of the sediment continuity equation under the assumption of no lateral sediment supply or shut-off may be written as:

$$\frac{\partial}{\partial t}\int_{x_{j-1/2}}^{x_{j+1/2}}z_bdx+\frac{1}{(1-\lambda)b_{mj}}\left[Q_s\right]_{x_{j-1/2}}^{x_{j+1/2}}=0 \quad (3)$$

where $z_b=$ elevation of bed channel; $\lambda=$ porosity of bed material; $b_{mj}=$ average channel width of the water surface within the $j_{th}$ cell; and $Q_s=$ solid discharge.

To close the governing equations (1)–(2)–(3) the sediment discharge must be specified. In the present study the equilibrium transport hypothesis is applied, i.e. the solid load is able to adapt instantaneously to mild spatial or temporal variations of flow. The sediment transport capacity is evaluated by applying an arbitrary bed-load discharge formula. In particular, the Parker formula (1990) is employed herein.

## 2.2 The numerical scheme

The solution of the present equations system is conducted by following a decoupled procedure.

Firstly, by assuming a fixed-bed, the solution of water continuity and momentum equations is conducted by applying an explicit two steps predictor-corrector conservative scheme based on a finite volume method (Catella & Solari, 2005), in which

the spatial derivatives are always taken in the same directions:

$$\mathbf{U}_j^p=\mathbf{U}_j^n-\frac{\Delta t^n}{\Delta x_j}[\mathbf{F}_{j+1/2}^n-\mathbf{F}_{j-1/2}^n]+\Delta t^n\mathbf{S}_j(\mathbf{U}^n,x)(4)$$

$$\mathbf{U}_j^p=\mathbf{U}_j^n-\frac{\Delta t^n}{\Delta x_j}[\mathbf{F}_{j+1/2}^n-\mathbf{F}_{j-1/2}^n]+\Delta t^n\mathbf{S}_j(\mathbf{U}^n,x)(5)$$

where the superscripts $p$ and $c$ indicate the predictor and the corrector steps, while $\mathbf{U}$, $\mathbf{F}$ and $\mathbf{S}$ are the vectors of conserved variables, fluxes and source terms, respectively.

The solution at the new time level $n+1$ is then evaluated as an average between the two sub-steps:

$$\mathbf{U}_j^{n+1}=0.5\cdot\left(\mathbf{U}_j^p+\mathbf{U}_j^c\right) \quad (6)$$

The criterion to assign the variables $S_{j+1/2}$ and $Q_{j+1/2}$ to the $N$ interfaces from the $2(N-1)$ values $S_j$ and $Q_j$ computed in each cell and the two boundary conditions is simple and appears to be particular suitable for natural river flow modeling (Catella & Solari, 2005).

Secondly, the sediment continuity equation is solved by an explicit conservative scheme based on a finite volume method and by employing the flow variables previously computed:

$$\Delta z_{bmj}^{n+1}=\Delta z_{bmj}^n-\frac{\Delta t^n}{\Delta x_j}\frac{1}{(1-\lambda)b_m}\left[Q_{sj+1/2}^{n+1}-Q_{sj-1/2}^{n+1}\right](7)$$

To assign $\Delta z_b$, a criterion based on the Froude number of the average flow of the upstream and downstream cells across the interface to assign is applied again in addition to a proper boundary condition (Catella et al., 2005).

For the stability of the present explicit scheme, the Courant-Friederich-Lewy number $N_{CFL}$ must satisfy the following condition:

$$N_{CFL}=\Delta t\cdot\frac{\max_j\left[\left|\frac{Q}{S}\right|_{j+1/2}\right]+\max_j\left[\sqrt{\left|\frac{g\cdot S}{\alpha\cdot b}\right|_{j+1/2}}\right]}{\min_j[\Delta x_j]}<1(8)$$

$$j=1,...,N$$

## 2.3 The side overflow hydraulic analysis

Let us consider a lateral weir of fixed height $d$ (see Fig. 1), connecting the main channel to a storage area.

As soon as the water surface elevation $h$ in the channel rises above the weir crest $d$, a part of the liquid discharge spills over the side, flooding the storage

1212

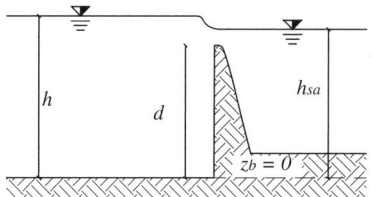

Figure 1. Sketch of a side weir connecting a natural channel and a storage area.

area. In the raising phase of the flood hydrograph, the water level inside the storage area $h_{sa}$ increases. If $h_{sa}$ results lower than the weir height ($h_{sa} < d$), the liquid discharge $dq$ spilled over a unit weir element can be computed by the same type of equation as in the frontal weir (Montes, 1998):

$$dq = \mu(h-d)\sqrt{2g(h-d)} \qquad (9)$$

where $\mu = 0.428$.

If the water level inside the storage area becomes higher than the weir height, a submerged flow over the lateral weir could occur. In particular, in the present study, we assume a submerged lateral flow whenever the water level inside the storage area overcomes the critical depth on the side weir $y_{cr}$ ($h_{sa} > d + y_{cr}$). In this case, the discharge may enter the basin or return to the channel if the storage area water elevation is lower or higher than the free water surface in the channel, respectively.

In the first case ($h > h_{sa} > d + y_{cr}$), the liquid discharge $dq$ is evaluated as follows:

$$dq = \left[\mu_1(h-h_{sa}) + \mu_2(h_{sa}-d)\right]\sqrt{2g(h-h_{sa})} \qquad (10)$$

where $\mu_1 = 0.428$, $\mu_2 = 0.65$.

In the second case ($h_{sa} > h > d + y_{cr}$), generally occurring during the decreasing phase of the flood hydrograph, the liquid discharge $dq$ is returned to the channel as follows:

$$dq = \left[\mu_1(h_{sa}-h) + \mu_2(h-d)\right]\sqrt{2g(h_{sa}-h)} \qquad (11)$$

Finally, if the free water surface in the channel becomes lower than the critical depth on the side weir ($h_{sa} > d + y_{cr} > h$), the returned liquid discharge $dq$ is:

$$dq = \mu(h_{sa}-d)\sqrt{2g(h_{sa}-d)} \qquad (12)$$

In the numerical scheme side weir flow is modeled following a two step computational procedure: firstly, before analyzing the flow over the side weir, the solution of water continuity and momentum equations is conducted by applying the explicit two steps

predictor-corrector conservative scheme based on a finite volume method (see Chapter 2.2). Secondly, the interaction between the flow and the lateral structure is considered. For this purpose, in the present model, the hydraulic analysis requires the definition of a minimum of two cross sections set immediately upstream and downstream the structure. Thus in the following the side weir is set at the $j_{th}$ cell delimited by the cross sections $j \pm 1/2_{th}$. The liquid discharge, spilled over the lateral weir, can be evaluated once the flow characteristic is known at an end cross section (i.e. the downstream cross section in case of subcritical flow or the upstream cross section in case of supercritical flow). In particular the water depth profile along the side weir is estimated by applying an energy balance between the entrance and the exit cross sections considering the head loss due to friction. The calculation of the discharge over the side weir is conducted by means of eqs. (9)–(12), by estimating the free water surface profile over the crest equal to the average values of the water depth at the cross section $j - 1/2_{th}$ and $j + 1/2_{th}$. Next the liquid discharge at the $j + 1/2_{th}$ cross section is updated as follows:

$$Q_{j+1/2}^{n+1} = Q_{j-1/2}^{n+1} - Q_{sa}^{n+1} \qquad (13)$$

where $Q_{sa}$ is the spilled liquid discharge.

Finally, if the flow regime at the side weir is subcritical, the wetted cross sectional areas at the $j - 1/2_{th}$ and $j + 1/2_{th}$ cross sections are updated respectively with the one previously computed by means of energy balance and continuity equation. If the flow regime at the side weir is supercritical, at the $j + 1/2_{th}$ cross section the wetted cross sectional area is updated with the one determined through the energy balance. If the flow regime at the $j + 1/2_{th}$ cross section is critical, the critical area is here evaluated with the updated discharge. If a hydraulic jump occurs in front of the side weir the criterion based on the momentum function is applied by using the updated values of the conserved variables.

The relation between liquid discharge and volume inside the storage area is:

$$dV^{n+1} = Q_{sa}^{n+1} \Delta t^{n+1} \qquad (14)$$

The volume inside the storage area can be expressed as a function of the water depth as follows:

$$V = A(h_{sa} - h_{sa0})^m \qquad (15)$$

where the coefficients $A$, $h_{sa0}$ and $m$ are functions of the storage area geometry.

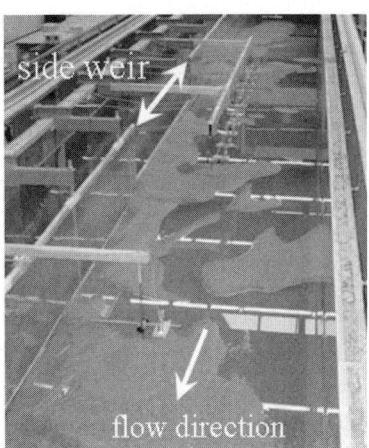

Figure 2. Laboratory setup with main-channel, side weir, mobile bed and evacuation channel.

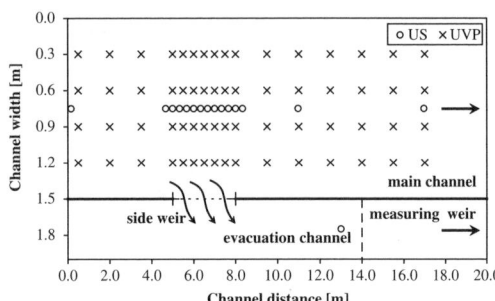

Figure 3. Experimental setup and disposition of water level (US) and velocity (UVP) recordings.

Table 1. Overview of test series and initial parameters studied.

| No of test | No of weirs $n_D$ | Length of weir crest $L$ (m) | Sill height $d$ (m) | Flume slope $S_0$ (%) | Discharge $Q_1$ (l/s) | Duration $t$ (min) |
|---|---|---|---|---|---|---|
| B01 | 1 | 3.0 | 0.10 | 0.2 | 131 | 188 |
| B02 | 1 | 3.0 | 0.10 | 0.2 | 181 | 183 |
| B05 | 1 | 3.0 | 0.10 | 0.2 | 144 | 128 |
| B06 | 1 | 3.0 | 0.10 | 0.3 | 148 | 138 |

## 3 APPLICATION TO EXPERIMENTAL TEST CASE

### 3.1 Experimental setup

The tests were conducted in a rectangular 2.0 m wide, 30 m long and 1.2 m deep glass-sided flume. The flume was subdivided longitudinally into two separate channels by a vertical 0.9 m high wall. The first channel, being 1.5 m wide, represents the actual testing facility including the mobile bed ($d_m = 1.05$ mm) and the side weir on the right bank. The second one, 0.47 m wide, constitutes a lateral channel permitting to evacuate the diverted discharge. A general layout of the experimental setup is shown in Figure 2.

In order to determine the influence of a lateral overflow on bed morphology, the main-channel discharge $Q_1$ and the channel slope $S_0$ were considered as test parameters. The crest length $L$ as well as the crest height $d$ were kept constant. Table 1 gives an overview of the tested parameters.

Sediment was fed at the channel entrance via a conveyer belt. The quantities to be supplied were estimated according to the formula of Smart & Jäggi (1983)

and adjusted during the test in order to maintain both uniform flow and equilibrium sediment transport conditions in the approach channel up-stream of the side weir.

The lateral overflow discharge was determined using a sharp-crested weir measuring located in the evacuation channel (Fig. 3). The weir was equipped with an ultrasonic gauge allowing to determine the transient evolution of the spilled discharge. The discharge coefficient for the measuring weir was computed according to the Rehbock-formula (Naudascher 1992).

Water depth was recorded continuously in the centre line of the main-channel by the use of ultrasonic gauges (Fig. 3). The 2D velocity field was measured by the help of an Ultrasonic Doppler Velocity Profiler (UVP) allowing to obtain instantaneously a 1D-velocity profile over the entire water depth. As far as the monitoring of the final bed morphology is concerned, digital photogrammetry has been applied to obtain a digital elevation model (DEM).

### 3.2 Comparison between experimental and numerical results

The numerical scheme is here applied to model the morphological evolution of the deposit that occurs in the weir alignment in experimental test cases characterized by a steady inflow liquid and solid discharges.

The present model is verified with the experimental data described in the previous chapter.

Tests were performed reproducing four experimental runs: B01, B02, B05 and B06. Inflow liquid and solid discharges, characteristic diameter of bed material, channel and side weir geometries, and duration of each numerical test were set equal to the experimental ones (see Table 1).

In the computations, the sediment rate is evaluated by means of the Meyer-Peter and Müller formula (1948), the channel is outlined by 60 cells, $\Delta x = 0.5$ m, $N_{CFL}$ is set equal to 0.9 and the bed evolution does not start until an eqilibrium steady state for the initial liquid discharge is gained.

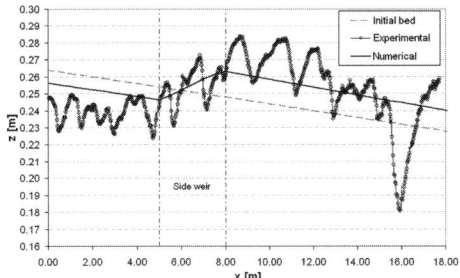

Figure 4. Comparison between experimental and numerical results for bed levels at $t = 188\,\text{min}$ (Exp. B01).

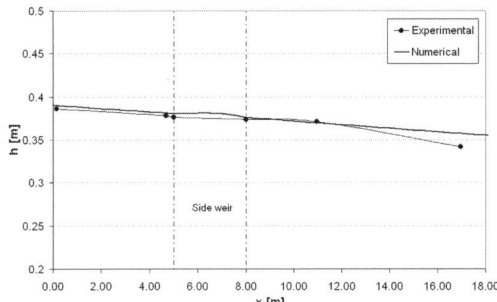

Figure 5. Comparison between experimental and numerical results for free surface elevation at $t = 188\,\text{min}$ (Exp. B01).

In Figures 4–5 a comparison between numerical and experimental results are plotted at the end of each run. The proposed numerical model shows a quite good agreement between computed and measured bed levels and water free surface elevations. In particular, Figure 5 shows a very good agreement between computed and measured free surface elevations, and Figure 4 shows that the deposit front along the side weir is well reproduced. The evolution mechanism of the bed level is reasonably catched untill the end of the experimental runs, even if the discrepancies, specifically due to the bed-form formation, can not be ignored. However, the discrepancies between numerical results and experimental data, are due to limitations of the mathematical model rather than to limitations of the numerical scheme.

In Figure 6 a comparison between experimental and numerical diverted discharge is plotted for the four test cases, showing a very good agreement. The diverted discharge is very well reproduced by the present numerical model as reported in Table 2.

Finally a comparison between fixed and mobile bed numerical results are presented (Tables 3–5), showing in all the runs considerable discrepancies in the values of the diverted discharge. Fixed bed numerical simulations could lead to underestimate the overflow

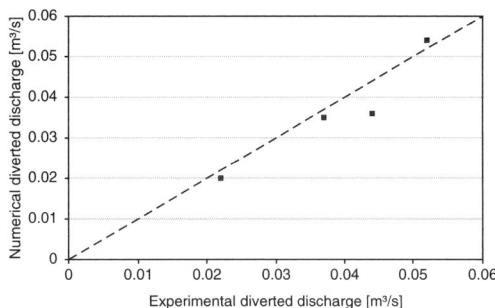

Figure 6. Comparison between experimental and numerical results for the diverted discharge.

Table 2. Comparison between experimental and numerical diverted discharge ($Q_{dEX}$ and $Q_{dNU}$ respectively; $\varepsilon_{Qen} = |1 - Q_{dEX}/Q_{dNU}|$.

| Run | $Q_{dEX}$ (m$^3$/s) | $Q_{dNU}$ (m$^3$/s) | $\varepsilon_{Qen}$ |
|---|---|---|---|
| B01 | 0.022 | 0.020 | 0.09 |
| B02 | 0.052 | 0.054 | 0.04 |
| B05 | 0.037 | 0.035 | 0.05 |
| B06 | 0.044 | 0.036 | 0.18 |

Table 3. Numerical results for liquid discharges in mobile bed case ($Q_{up}$ = liquid discharge upstream the side weir, $Q_{down}$ = liquid discharge downstream the side weir, $Q_{diverted}$ = liquid discharge diverted by the side weir).

| Run | $Q_{up}$ (m$^3$/s) | $Q_{down}$ (m$^3$/s) | $Q_{diverted}$ (m$^3$/s) |
|---|---|---|---|
| B01 | 0.131 | 0.111 | 0.020 |
| B02 | 0.181 | 0.127 | 0.054 |
| B05 | 0.144 | 0.109 | 0.035 |
| B06 | 0.148 | 0.112 | 0.036 |

Table 4. Numerical results for liquid discharges in fixed bed case.

| Run | $Q_{up}$ (m$^3$/s) | $Q_{down}$ (m$^3$/s) | $Q_{diverted}$ (m$^3$/s) |
|---|---|---|---|
| B01 | 0.131 | 0.119 | 0.012 |
| B02 | 0.181 | 0.145 | 0.036 |
| B05 | 0.144 | 0.129 | 0.015 |
| B06 | 0.148 | 0.143 | 0.005 |

discharge diverted by the side weir. As a consequence the side weir efficiency will be underestimated and the designed structures could bring about an early filling of the storage area.

Table 5. Comparison between diverted discharges in fixed and mobile bed cases ($Q_{dMB}$ and $Q_{dFB}$ respectively; $\Delta Q_r = 100\,(Q_{dMB} - Q_{dFB})/Q_{dMB}$).

| Run | $Q_{dMB}$ (m³/s) | $Q_{dFB}$ (m³/s) | $\Delta Q_r$ (%) |
|---|---|---|---|
| B01 | 0.020 | 0.012 | 40 |
| B02 | 0.054 | 0.036 | 33 |
| B05 | 0.035 | 0.015 | 57 |
| B06 | 0.036 | 0.005 | 86 |

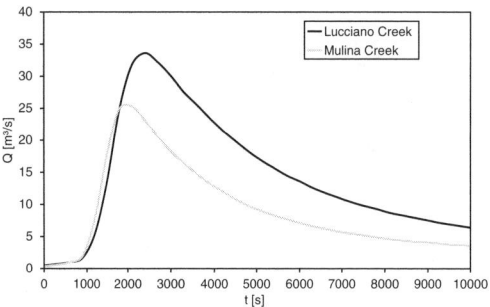

Figure 7. Flood hydrographs for return period of 200 years of Lucciano and Mulina Creeks.

## 4 THE LUCCIANO CREEK CASE STUDY

### 4.1 The study area

The Lucciano Creek is a little tributary of the Ombrone Pistoiese River, right affluent of the Arno River (Tuscany). The Lucciano sub-catchment basin (4.56 km² wide) extends in direction Sud-West Nord-East for about 4.7 km from Montalbano to the confluence of the Fermulla Creek.

The Lucciano Creek has long dry periods during summer and significant floods during autumn and spring.

The stream reach under investigation is located between Lucciano site and the confluence with the Fermulla Creek. In order to characterize the stream geometry, a detailed topographic survey has been conducted by the territorial government of Pistoia and Prato. The 1.1 km channel reach is discretized by 51 cross sections, that are characterized by a strong non-uniform grid with cell size ranging from 1.0 m to 50.8 m. No interpolation of topographic survey data is employed.

In Figure 7 the 200 years return period flood hydrographs both of the Lucciano Creek and of its major tributary are plotted.

### 4.2 Design criterion, fixed bed case

The storage area here investigated is shown in Figure 8 and it is linked to the main channel by a side weir and a derivation channel.

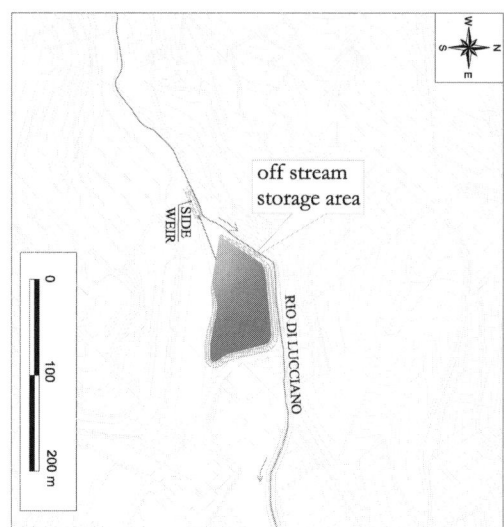

Figure 8. Plan of the study area.

Table 6. Off-stream storage area geometric characteristics.

| Storage area | Perimeter (km) | Surface (km²) | Elevation (m) | Banks height (m) |
|---|---|---|---|---|
| | 1.2 | 6.5 | 66 | 3.5 |

Table 7. Side weir geometric characteristics.

| Side weir | Average elevation (m) | Length (m) |
|---|---|---|
| | 71.65 | 25 |

The main characteristics of both the storage area and the design side weir are reported in Tables 6 and 7, respectively. In particular the design and the location of the side weir has been conducted following a criterion based on the storage area efficiency maximization in case of fixed bed leading to a reduction of the peak discharge of about 21.2%.

In Figure 9 flow hydrographs upstream and downstream the side weir, and the water elevation hydrograph inside the storage area are reported. As we can observe in this case the overflow above the side weir is always free, because the storage area is set considerably downstream the lateral structure. In Table 8 are briefly summerized results obtained performing numerical simulations with fixed bed. In the Chapter 4.3 such efficiency is verified by modeling the same flood event in case of mobile bed.

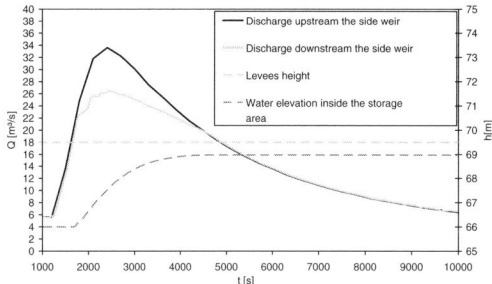

Figure 9. Flow hydrographs just upstream and downstream the side weir and water elevation inside the storage area (fixed bed case).

Table 8. Main results obtained performing numerical simulations with fixed bed (maximum volume diverted in the storage area $V_{sa}$; maximum liquid discharge upstream the side weir $Q_{max\ up}$, maximum liquid discharge downstream the side weir $Q_{max\ down}$, and side weir efficiency: $1 - Q_{max\ down}/Q_{max\ up}$).

| $V_{sa}$ (m$^3$) | $Q_{max\ up}$ (m$^3$/s) | $Q_{max\ down}$ (m$^3$/s) | $1 - Q_{max\ down}/Q_{max\ up}$ |
|---|---|---|---|
| 10418 | 33.6 | 26.5 | 21.2 |

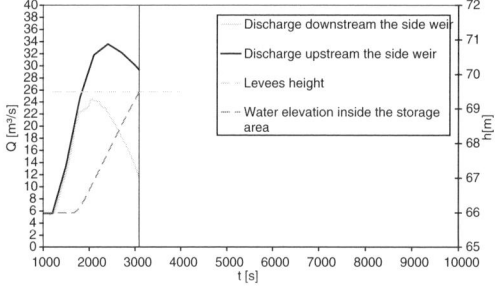

Figure 10. Flow hydrographs just upstream and downstream the side weir and water elevation inside the storage area (mobile bed case).

## 4.3 The side overflow efficiency in mobile bed case

The side weir efficiency is now tested for the 200 years return period flood in case of mobile bed. The inflow sediment discharge is set equal to the sediment rate evaluated in the first cross section.

From a sedimentologic survey the median diameter $D_{50}$ of bed material is 0.05 m and sediment porosity is 0.3.

Results obtained performing numerical simulations with mobile bed are reported in Figures 10 and 11, pointing out the significant importance to not ignore the morphodynamic evolution of the river bed during

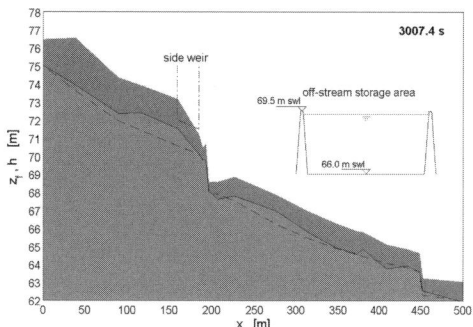

Figure 11. Longitudinal profiles near to the side weir (dashed line) in case of mobile bed; the dashed black line indicates the initial bed profile.

Table 9. Results obtained performing numerical simulations with mobile bed and the new side weir average elevation.

| $V_{sa}$ (m$^3$) | $Q_{max\ up}$ (m$^3$/s) | $Q_{max\ down}$ (m$^3$/s) | $1 - Q_{max\ down}/Q_{max\ up}$ |
|---|---|---|---|
| 9422 | 33.6 | 27.4 | 18.6 |

the side weir design activities. As a part of the liquid discharge is diverted from the main channel, the sediment transport capacity decreases. The lateral loss of water induces a reduction of the sediment transport capacity in the main channel yielding an aggradation process and the formation of a local sediment deposit in the weir alignment. As a consequence the upstream water level rises as well as the total head over the side weir. Therefore the discharge diverted over the side weir is increased in an unforeseen way. This behaviour causes an unexpected replenishment of the storage area, which is already full when the flood peak in the main channel reaches the side weir.

A comparison between Figure 9 and Figure 10 shows that in mobile bed case after about 3000 seconds the storage area results filled; besides, as the storage area bottom elevation is notably lower than the channel bed (about 5 m), simulations show inundation phenomena and therefore high risk conditions.

In order to restore the side weir efficiency in mobile bed case, the side weir average elevation is set equal to 76.8 m. In particular as we can see from Table 9 and Figure 12, the diverting efficiency is just partially restored due to the complexity of the morphodynamic effects.

## 5 CONCLUSIONS

A one-dimensional numerical model to study the interaction between river bed evolution and side weir flow is presented.

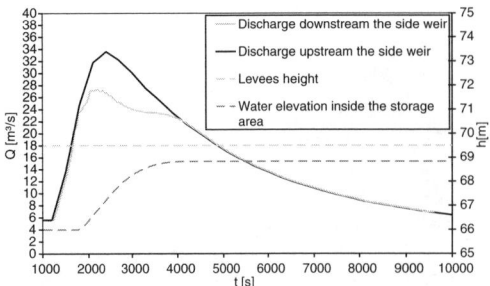

Figure 12. Flow hydrographs just upstream and downstream the side weir and water elevation inside the storage area (mobile bed case, and new side weir elevation).

The St. Venant-Exner equations are solved by employing a finite volume scheme following a decoupled procedure. The present numerical scheme is successfully applied to model river morphodynamics in real cases including subcritical, supercritical and transcritical flows (Catella & Solari, 2005).

The water depth profiles in the main channel along the side weir are computed by modifying the De Marchi (1934) hypothesis and by evaluating the overflow discharge by means of two different equations in the case of free or submerged overflow. Besides the model is able to evaluate the discharge entering the storage area or returning to the channel.

In order to verify the morphodynamic modeling in the side weir alignment, the numerical scheme is applied to reproduce the experimental results obtained by Rosier et al. (2006). A comparison between numerical and experimental results shows that bed profiles are very well reproduced as well as free surface profiles along the side weir. Moreover the numerical scheme is able to accurately reproduce the diverted discharge above the side weir. Considerable discrepancies in the diverted discharge between fixed and mobile bed numerical simulations occur.

Finally the model is employed to design a side weir in a real case study in a sub-catchment of the Ombrone Pistoiese River in Tuscany (Italy). Results point out the significant importance to not ignore the morphodynamic effects in the side weir design activities, which should concern the sediment transport effects. In the Lucciano Creek case study, the bed morphodynamics lead to an unexpected replenishment of the storage area in respect to the clear water results. In order to reestablish the side weir efficiency, the side weir average elevation is raised to an appropriate level. Particularly, fixed bed design appears to be not adequate in the considered case study, and attention has to be paid when such structures must be employed for mitigation of flood risk.

Since the extraordinarily conceptual and algorithm simplicity, the proposed model provides a great versatility, stability and robustness.

The numerical model will be implemented taking into account the momentum variation along the lateral weir.

## REFERENCES

Borghei, S.M., Jalili, M.R. & Ghodsian, M. (1999), Discharge coefficient for sharp-crested side weir in subcritical flow, *Journal of Hydraulic Engineering*, ASCE, 125(10), 1051–1056.

Catella, M. & Solari, L. (2005), Conservative scheme for numerical modeling of flow in natural geometry, *in Proceedings of XXXI Int. Congress, Int. Association for Hydraulic Research, Seoul, Korea*.

Catella, M., Paris, E. & Solari, L. (2005), 1D Morphodynamic model for natural rivers, *in Proceedings of 4th Conf. on River, Coastal and Estuarine Morphodynamics, Urbana, Illinois, USA*.

De Marchi, G. (1934), Saggio di teoria sul funzionamento degli stramazzi laterali, *L'energia elettrica*, Milano, Italia, 849–854 (in italian).

Hager, W.H. (1987), Lateral outflow over side weirs, *Journal of Hydraulic Engineering*, ASCE, 113(4), 491–504.

Meyer-Peter, E. & Müller, R. (1948), Formulas for bed load transport. *Proc., 2nd Int. Congress, Int. Association for Hydraulic Research, Stockholm, Sweden*.

Milano, V., Pagliara, S. & Venutelli, M. (1998), Sul funzionamento delle casse di espansione alimentate da soglie sfioranti rigurgitate, *in Proceedings of XXVI Convegno di Idraulica e Costruzioni Idrauliche, Catania* (in Italian).

Montes, S. (1998). *Hydraulics of open channel flow*. Reston: ASCE Press.

Nadesamoorthy, T. & Thomson, A. (1972), Discussion of 'Spatially varied flow over side weirs', *Journal of the Hydraulic Division*, ASCE, 98(HY12), 2234–2235.

Naudascher, E. (1992), *Hydraulik der Gerinne und Gerinnebauwerke*, Wien New York, Springer Verlag.

Parker, G. (1990), Surface-based bedload transport relation for gravel rivers, *Journal of Hydraulic Engineering*, ASCE, 28(4), 417–436.

Ranga Raju, K.J., Prasad, B. & Gupta, S.K. (1979), Side weir in rectangular channel, *Journal of the Hydraulic Division*, ASCE, 105(HY5), 547–554.

Rosier, B., Boillat, J.-L., Schleiss, A. J., (2005), Influence of side overflow induced local sedimentary deposit on bed form related roughness and intensity of diverted discharge, *in Proceedings of XXXI Int. Congress, Int. Association for Hydraulic Research, Seoul, Korea*.

Rosier, B., Boillat, J.-L., A. J., Schleiss, (2006), Semi-empirical model to predict mobile bed evolution in presence of a side weir overflow, *in Proceedings of River Flow 2006, Lisbon, Portugal*.

Smart, G.M. & Jäggi, M.N.R. (1983), Sedimenttransport in steilen Gerinnen, in Vischer, D. (ed.), *Mitteilungen der Versuchsanstalt für Wasserbau, Hydrologie und Glaziologie (VAW)* 64, Zürich, Eidgenössische Technische Hochschule Zürich (ETHZ).

*Groynes and scour*

*River, Coastal and Estuarine Morphodynamics: RCEM 2007 – Dohmen-Janssen & Hulscher (eds)*
*© 2008 Taylor & Francis Group, London, ISBN 978-0-415-45363-9*

# Application of intelligent system of Artificial Neural Network to predict the scour process in coastal engineering

Ali Khosronejad
*Water Eng. Dept., University of Guilan, Rasht, Iran*

Colin D. Rennie
*Civil Eng. Dept., Ottawa University, Ottawa, Canada*

S. Moghimi
*Civil Eng. Dept., University of Arak, Iran*

ABSTRACT:  Prediction of the scour hole properties around a group of pile in the field exposed to oscillatory waves is very important for many offshore structure and coastal engineering projects. Conventional predictive formulas for the geometric properties of scour hole, however, are not able to provide sufficiently accurate results. Artificial Neural Networks (ANNs) are simplified mathematical representation of the human brain. Three-layer normal feed-forward ANNs are a powerful tool for input-output mapping and have been widely used in civil engineering problems. In this article the ANNs approach is used to predict the geometric properties of the scour around a pile group using dimensionless groups of individual parameters, namely, Reynolds number, Keulegan and Carpenter number, Shields parameter, and Sediment number. Two different ANNs including multilayer Perceptron (with four different learning rules) and radial basis functions neural networks are used for this purpose. The results show that a three-layer normal feed-forward multilayer Perceptron with quick propagation (QP) learning rule can predict the scour hole properties successfully.

## 1 INTRODUCTION

Piles are structures widely used in coastal and ocean engineering. Various types of pile group arrangements may be used to protect the coastal structures. Since many of these structures are located on movable beds, the estimation of scour hole properties is an essential task, because failure of the bed or the toe could cause collapse of the entire structure. In order to predict the scour around piles, the effects of waves, currents, and tide on the structure and bed have to be considered.

The last field study on scour pile group has been done by Bayram and Larson (2000). It is extremely difficult to formulate mathematical models that accurately represent the scour process and geometry of scour hole, which develops under the influence of wave and current. Thus it is a common practice to apply empirical relationships based on laboratory data for estimation of the scour around piles. Since there are numerous effective parameters, and the interaction of these parameters is highly complicated, therefore,

the accuracy of the empirical relationships is very subjective and highly depends on the user's ability and knowledge.

An ANN, on the other hand, is an applicable and powerful tool to solve this type of problems.

In addition, its ability to learn from examples and to generalize its learning makes it well suited to situations where the problem complexity precludes the development of empirical relationships. This technique was used to estimate the scour properties around a configuration of piles (Kambekar and Deo, 2002). Also we in (Khosronejad et al, 2003) studied the scour properties around vertical pile using ANNs, In which the dimensional parameters such as wave length, water depth, wave period, maximum flow velocity and maximum shear velocity have been used as input parameters of network. Kambekar and Deo (2003) studied the scour hole properties around pile group using ANN's technique and they used combination of dimensional and non-dimensional input data as their network training set. Also they used two different learning rules (back propagation and cascade correlation algorithms) for their feed forward artificial neural network architecture.

In this paper non-dimensional parameters such as pile Regnolds number, Shields parameter, Keulegan-Carpenter number, and sediment number have been used as input parameters of networks. Using this type of network (which is trained by non-dimensional inputs) a wider range of inputs can be covered and therefore it make the trained neural network more applicable.

Moreover, in order to investigate the applicability of various artificial neural network architectures and their learning rules in hydraulic engineering area, in this study two different normal feed forward ANN's such as Radial Basis Function (*RBF*) and Multilayer Perceptron (*MLP*) and four different learning rule such as back propagation (BPN), cascade correlation (CC), quick propagation (QP), and conjugate gradient (CG) algorithms have been used.

## 2 OVERVIEW OF ANNs

The ANN is a simplified mathematical representation of the biological neural network. It has the ability to learn from examples, recognize the various pattern of input data and to process information rapidly. A neural network is characterized by its architecture that represents the pattern of connections among nodes, its method of determining the connection weights and activation function.

A typical ANN consists of number of nodes that are organized according to a particular arrangement. These nodes are generally arranged in layers, starting from the first input layer and ending at the final output layer. There can be several hidden layer, each hidden layer having one or more nodes (Jain, 2001).

Three types of the most commonly used ANNs are normal feed-forward neural network, recurrent neural network, and competitive neural network (Islam and Kothari, 2000). In this study the normal feed-forward neural network is used.

Normal feed forward neural networks are the most common among other ANNs and are widely used in function approximation and pattern classification (Islam and Kothari, 2000). The most commonly used types of normal feed-forward are the so-called multi-layered Perceptron *(MPL)* network and the radial basis function *(RBF)* network. In either of these two networks, the neurons are arranged in layered structure. Information passes from the input to the output side. The neurons in one layer are connected to those in the next layer. Thus, the output of a neuron in a layer is only dependent on the inputs it receives from pervious layer and the corresponding weights.

Consider a multilayered Perceptron network with $n$ inputs, an output layer with $o$ neurons, and a hidden

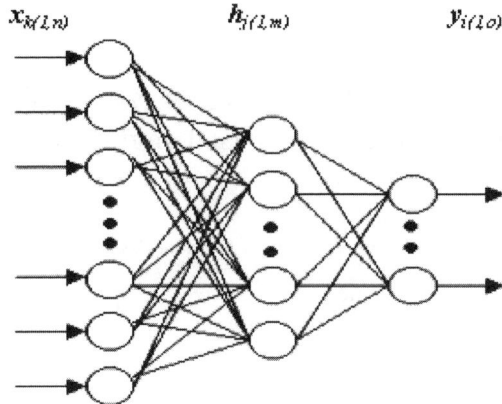

Figure 1. Schematic of multilayer Perceptron network.

layer with $m$ neurons as shown in Figure. 1. Index $i$ is referred to the individual output layer neurons, the index $j$ and $k$ refer to the hidden layer neurons and the input neurons, respectively. Inputs, feed to the hidden layer neurons through weights $W_{jk}$ and the outputs of hidden layer neurons feed to output layer neurons through weights $W_{ij}$. A hidden layer neuron produces as output:

$$h_j = f'(s_j) = f(\sum_{k=1}^{n} W_{jk} x_k) \qquad (1)$$

and an output layer neuron produces as output:

$$y_i = f'(s_i^*) = f(\sum_{j=1}^{m} W_{ij} h_j) \qquad (2)$$

where $h_j$ is the output of $j$th neuron in hidden layer, $s_j$ is the weighted sum of $j$th neuron in hidden layer, $x_k$ is the input of $k$th neuron in input layer, $y_i$ is the output of $i$th neuron in output layer and $s_i^*$ is the weighted sum of $i$th neuron in the output layer.

In this study the activation function $f'$ for hidden layer is taken to be the arctangent [$f'(x) = \arctan(x)$]. This non-linearity makes the mapping produced by the network nonlinear. Since the outputs $s$ are greater than one, the linear function is selected for output layer.

A *RBF* network is similar to *MLP* network, but the computation done by the hidden layer neurons is different, and the hidden layer neurons produce as output:

$$h_j = \exp(-\frac{\|X - V_J\|^2}{\sigma^2_J}) \qquad (3)$$

where $\sigma_J^2$ is the variance of Gaussian kernel; and $V_J = [V_{1J}, V_{2J}, \ldots, V_{nJ}]^T$ plays the role of the center of Gaussian kernel (ASCE Task Committee, 2000-a).

## 3 NETWORK TRAINING ALGORITHMS

There are two types of network training, supervised & unsupervised (ASCE Task Committee, 2000-a). In supervised training algorithm, an external teacher (supervisor) is needed to guide the training process and an unsupervised training algorithm doesn't need.

In this study, supervised training algorithm has been used to update the weight matrix of ANN. The training patterns or training set (and also test set) used in this study, were the field measurements reported by Bayram and Larson (2000). Four different training techniques are used to make sure that the training procedure is done well. These are back-propagation *(BP)*, conjugate gradient *(CG)*, cascade correlation *(CC)* and quick-propagation *(QP)*. In each of the above training algorithm, the aim is to reduce the global error E:

$$E = \frac{1}{p}\sum_{p=1}^{p} E_p \qquad (4)$$

where $p$ is total number of training patterns and $E_p$ is error for training pattern $p$ given by:

$$E_p = \frac{1}{2}\sum_{k=0}^{N}(y_k - t_k)^2 \qquad (5)$$

where $N$ is total number of output neurons; $y_k$ is network output at the $k$th output neuron and $t_k$ is target output at the $k$th output neuron.

## 4 EFFECTIVE PARAMETERS

The significant parameters which affect the depth and width of scour hole around a pile exposed to oscillatory waves are: pile diameter $D$, wave height H, water depth $h$, wave period $T$, maximum flow velocity $U_m$, maximum shear velocity $U_{fm}$, specific gravity of sediment s, mean diameter of sediment d, and acceleration due to gravity g. Therefore the maximum scour depth $S$ and the width of scour hole $L_x$ could be written as follows (see Figure. 2) (Sumer et al., 1992 and Herbich, 1991):

$$S \text{ and } L_x = f(D, H, h, T, U_m, U_{fm}, s, d, g) \qquad (6)$$

The maximum shear velocity $U_{fm}$ is defined as:

$$U_{fm} = (0.5f)^{1/2}U_m \qquad (7)$$

where $U_m$ is amplitude of the oscillatory flow velocity; $f$ is wave friction factor. As mentioned before, in this study the non-dimensional parameters controlling scour depth have been used. By applying dimensional analysis, the non-dimensional parameters affecting the dimensions of scour hole around a pile exposed to

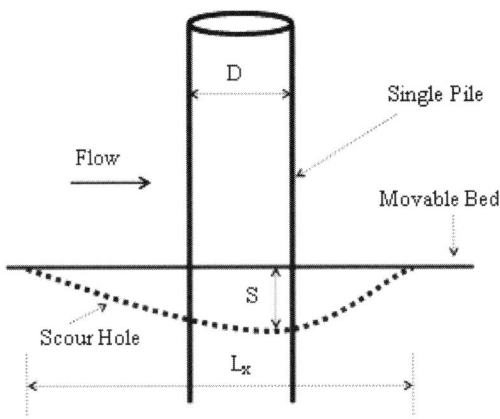

Figure 2. Definition sketch of scour depth and width around a pile.

oscillatory waves may be identified. Thus the equilibrium maximum scour depth S and the width of scour hole $L_x$ were normalized by the pile diameter D expressed as follows (Herbich 1991; Sume, et al 1992 b):

$$\frac{S}{D}, \frac{L_x}{D} = f(N_{Re}, N_s, \theta, KC) \qquad (8)$$

Where $N_{Re}$ is pile Reynolds number, $N_s$ is sediment number, $\theta$ is shields parameter and KC is Keulegan-Carpenter number. These non-dimensional numbers are defined as follow:

$$N_{Re} = \frac{U_m D}{v} \qquad (9)$$

$$N_s = \frac{U_m}{\sqrt{g(s-1)d}} \qquad (10)$$

$$\theta = \frac{U_{fm}^2}{g(s-1)d} \qquad (11)$$

$$KC = \frac{U_m T}{D} \qquad (12)$$

where $U_m$ is amplitude of the oscillatory flow velocity and $v$ is kinematical viscosity. All of these numbers depended on wave and sediment characteristics and as scour hole properties depend on these numbers, they were used in this study as input data for training and testing designed neural networks.

As mentioned above, the data set that were used for training and testing the networks were the field data reported by Bayram and Larson (2000). The range of variables is summarized in Table 1. These non-dimensional parameters have been employed in the present study as input vectors to train the designed neural network and describe the scour hole properties around vertical pile. Therefore, the number of input layer neurons are equal to four i.e. equal to the number of effective parameters affecting the scour depth and width, and the output neurons are equal to two, i.e. one for $S/D$ and one for $L_x/D$. The number of data is fifty eight and forty five of these data were used for network training and ten for testing the performance of trained networks.

## 5 DESIGNING AND TRAINING THE ANNs

This important step involves the determination of the ANN architecture and selection of a training algorithm. An optimal architecture may be considered the one yielding the best performance in term of error minimization, while retraining a simple and compact structure. No unified theory exists for determination of such an optimal ANN architecture. A trial-and-error procedure is generally applied to decide on the optimal architecture. The number of input and output neurons is problem dependent.

Table 1. Range of data set used for training and testing the Networks (Bayram and Larson, 2000).

| Variables | Range |
| --- | --- |
| Shields parameter $\theta$ | 0.08–0.64 |
| Pile Reynolds Number($N_{Re}$) | $3.4*10^5 - 1.85*10^6$ |
| Sediment Number ($N_s$) | 1.2–14.5 |
| Keulegan-Carpenter Number | 0.66–22.5 |
| Scour depth S(m) | 0.42–2.1 |
| Width of scour hole $L_x$(m) | 6.7–33.6 |

In the current study, first we used two neurons at output layer and four neurons at input layer and eight or more neurons in hidden layer. In this case the network was trained with different architectures and results showed that the network can not learn accurately.

Therefore, two separate networks with one neuron in their output layers were used, i.e. two ANNs with one output neuron were designed, one for estimation of $S/D$ and the other one for estimation of $L_x/D$.

For each ANNs, different architecture and learning rules were used and results show that these two networks together can solve the current problem. As shown in Tables 2 and 3 the normal feed-forward architecture with quick-propagation learning rule and one hidden layer is the best choice for this case, because the network learning was obtained with the least epochs and with minimum *rms* error.

## 6 ANALYZING THE OUTPUT RESULTS

Similar to other modeling approaches in engineering works, the performance of the trained ANN can be fairly evaluated by subjecting it to the new patterns that have not been seen during training process. The performance of the network can be determined by computing the error between predicted and observed values. Thus in order to assess the networks ability, the outputs of networks for new patterns is shown in Table 4 and Figures. 3(a) and (b). As shown in Table 4 and Figures. 3(a) and (b), the errors are acceptable and the trained networks could learn desired mapping successfully. The mean absolute error and correlation coefficient of non-dimensional scour depth (Figure 3 (a)) are 1.93 and 0.99, respectively. Also mean absolute error and correlation coefficient of non-dimensional

Table 2. Results of designed neural networks with one neuron in output layer for non-dimensional scour depth ($S/D$).

| Network | Learning Rule | No. of neurons (1st hidden layer) | No. of neurons (2nd hidden layer) | No. of epoachs | Mean training error | Mean testing error |
| --- | --- | --- | --- | --- | --- | --- |
| Multilayer Perceptron networks | Quick-propagation (QP) | 10 | – | 20000 | 0.015 | 0.51 |
| | | 12 | – | 32000 | 0.0011 | 0.014 |
| | | 15 | – | 34000 | 0.00012 | 0.008 |
| | Back-propagation (BPN) | 10 | – | 40000 | 0.28 | 0.28 |
| | | 12 | – | 560000 | 0.019 | 0.12 |
| | | 15 | – | 610000 | 0.008 | 0.013 |
| | Cascade correlation (CC) | 12 | – | 340000 | 0.012 | 1.4 |
| | Conjugate gradients (CG) | 12 | – | 15000 | 0.014 | 1.02 |
| Radial Basis functions networks | Quick-propagation (QP) | – | – | – | – | – |
| | Back-propagation (BPN) | 12 | – | 360000 | 0.17 | 1.44 |
| | Cascade correlation (CC) | – | – | – | – | – |
| | Conjugate gradients (CG) | – | – | – | – | – |

scour width (Figure 3 (b)) are 4.34 and 0.99, respectively. A comparison of the errors in this work and the last work of Kambekar and Deo (2003) and Khosronejad et al, 2003, show that these errors are little than above mentioned last works, and therefore the results are fairly good.

## 7 CONCLUSION

The ANN is a powerful tool to provide reasonably good solution for circumstances having complex systems that may be poorly define or understood using mathematical equations. The ANNs were applied to approximate geometric properties of scour hole around vertical piles. It was shown that use of non-dimensional as input pattern produce more accurate results and higher range of inputs parameters can be used. A network with two output neurons in output layer could not learn desired mapping. But using two networks, each with one output neuron can approximate $S/D$ and $L_x/D$, separately. These artificial neural networks could learn successfully and their *rms* error was very small. Finally, the ANN model with normal feed-forward architecture and quick-propagation learning rule and a single hidden layer with ten neurons provided higher training testing accuracy compared to those of other network architectures and learning rules.

It was demonstrated that the designed ANNs could intelligently estimate nonlinear relationships between input parameters (pile Regnolds number, Shields parameter, Keulegan-Carpenter number, and sediment

Table 3. Results of designed neural networks with one neuron in output layer for non-dimensional extended of scour hole ($L_x/D$).

| Network | Learning rule | No. of neurons (1st hidden layer) | No. of neurons (2nd hidden layer) | No. of epoachs | Mean training error | Mean testing error |
|---|---|---|---|---|---|---|
| Multilayer Perceptron networks | Quick-propagation (QP) | 10 | 5 | 100000 | 0.0021 | 0.081 |
| | | 12 | – | 230000 | 0.0018 | 0.095 |
| | | 10 | – | 18000 | 0.0024 | 0.11 |
| | Back-propagation (BPN) | 10 | 5 | 150000 | 0.002 | 0.10 |
| | | 12 | – | 890000 | 0.004 | 0.12 |
| | | 10 | – | 69000 | 0.009 | 0.51 |
| | Cascade correlation (CC) | 12 | – | 23000 | 0.001 | 0.91 |
| | Conjugate gradients (CG) | 12 | – | 36000 | 0.014 | 2.20 |
| Radial basis functions networks | Quick-propagation (QP) | – | – | – | – | – |
| | Back-propagation (BPN) | 12 | – | 51000 | 0.18 | 3.10 |
| | Cascade correlation (CC) | – | – | – | – | – |
| | Conjugate gradients (CG) | – | – | – | – | – |

Table 4. Comparisons of observed (column 5 and 7) and computed non-dimensional scour hole properties (column 6 and 8) and network errors for test data (I refers to input data and O refers to field measured output data).

| No. | $I_1$ $(N_{Re})*10^{-6}$ (1) | $I_2$ $(N_s)$ (2) | $I_3$ $(\theta)$ (3) | $I_4$ $(KC)$ (4) | $O_1$ $\left(\frac{S}{D}\right)$ (5) | Network Output for $\left(\frac{S}{D}\right)$ (6) | $O_2$ $\left(\frac{L_x}{D}\right)$ (7) | Network Output for $\left(\frac{L_x}{D}\right)$ (6) | Error Percentage for $\left(\frac{S}{D}\right)$ (9) | Error Percentage for $\left(\frac{L_x}{D}\right)$ (10) |
|---|---|---|---|---|---|---|---|---|---|---|
| 1 | 1.29 | 10.16 | 0.55 | 3.28 | 1.51 | 1.48 | 18.6 | 17.92 | 2.0 | 3.6 |
| 2 | 1.21 | 9.53 | 0.54 | 2.98 | 1.62 | 1.58 | 20.4 | 19.85 | 2.5 | 2.7 |
| 3 | 1.73 | 13.62 | 0.13 | 4.17 | 0.95 | 0.97 | 21.6 | 20.52 | 2.1 | 5.0 |
| 4 | 1.49 | 11.73 | 0.46 | 3.65 | 1.31 | 1.34 | 9.0 | 8.44 | 2.3 | 6.2 |
| 5 | 1.52 | 11.96 | 0.48 | 3.82 | 1.30 | 1.28 | 8.4 | 8.94 | 1.5 | 6.4 |
| 6 | 1.46 | 11.50 | 0.37 | 3.46 | 1.05 | 1.025 | 33.0 | 33.84 | 2.4 | 2.5 |
| 7 | 0.65 | 5.12 | 0.11 | 1.33 | 1.00 | 1.02 | 17.4 | 16.92 | 2.0 | 2.8 |
| 8 | 0.75 | 5.91 | 0.15 | 1.82 | 0.95 | 0.94 | 19.8 | 19.26 | 1.1 | 2.7 |
| 9 | 1.52 | 11.96 | 0.42 | 4.10 | 1.37 | 1.39 | 7.8 | 7.24 | 1.5 | 7.2 |
| 10 | 0.81 | 6.38 | 0.19 | 1.85 | 1.05 | 1.03 | 11.4 | 11.89 | 1.9 | 4.3 |

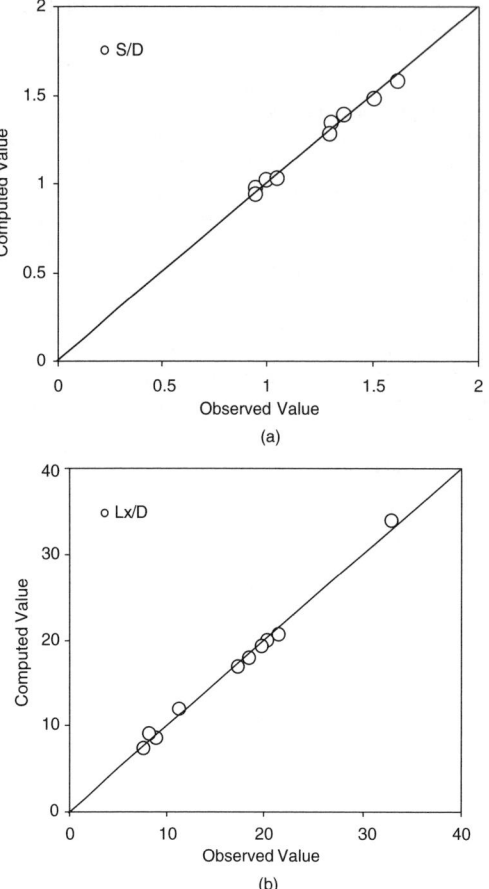

Figure 3. Comparison of observed and computed results non-dimensional scour properties: (a) Scour depth with correlation coefficient of 0.99 and mean absolute error of 1.93 percent, (b) scour width with correlation coefficient of 0.99 and mean absolute error of 4.34 percent

number) and output parameters (non-dimensional depth and width of scour hole) and mean absolute errors of estimated non-dimensional scour depth and non-dimensional scour width, relative to other last works, are negligible.

REFERENCES

A.Khosronejad, G.A.Montazer and M.Ghodsian, 2003, Estimation of scour hole properties around vertical pile using ANNs, Int.J. of Sediment Research, Vol.18, No.4,pp.290–300.

ASCE Task Committee on Application of Neural Networks in Hydrology, 2000-a, Artificial Neural Networks in hydrology, I: Preliminary concepts, J. Hydrologic Engrg. ASCE, Vol. 5, No. 2, pp. 115–123.

ASCE Task Committee on Application of Neural Networks in Hydrology, 2000-b, Artificial Neural Networks in hydrology, II: Hydrologic Applications, J. Hydrologic Engrg. ASCE, Vol.5, No. 2, pp. 124–127.

Bayram, A., and Larson, M., 2000, Analysis of scour around a group of vertical piles in the field, J. Waterway, Port, Coastal, and Ocean Engrg., Vol. 126, No. 4, pp. 215–220.

Chow, W.Y., and Herbich J. B., 1978, Scour around a group of piles, Proc. Offshore Techno. Conf.

Eadie, R.W., and Herbich, S.B., 1986, Scour around a single cylindrical pile due to combined random waves and current, Proc. 20th Coast. Engrg. Conf., ASCE, New York, pp. 1958–1870.

Herbich, J.B., 1991, Scour around pipelines, piles and seawalls, Handbook of coastal and ocean engineering, J. B. Herbich. Ed, Vol. 2, Gulf publishing Co. Book div., Houston.

Islam, S., and Kothari, R., 2000, Artificial Neural Networks in remote sensing of hydrologic processes, J. Hydrologic Engrg. ASCE, Vol. 5, No. 2, pp. 138–144.

Jain, S.K., 2001, Development of integrated sediment rating curves using ANNs, J. Hydraulic Engrg. ASCE, Vol. 127, No. 1, pp. 30–37.

Kobayashi, T., and Oda, K., 1994, Experimental study on developing process of local scour around a vertical cylinder, Proc. 24th Coast. Engrg. Conf. ASCE, Newyork, pp. 1284–1297.

Kambekar, A.R., and Deo, M.C., 2002, Neural networks to estimate oceanic pile group scour, Conference on Hydraulic, Water Resources and Ocean Engineering, HYDRO 2002, IIT Mumbai, India.

Kambekar A R and M C Deo (2003): "Estimation of pile group scour using neural networks", Applied Ocean Research, Elsevier, Oxford, UK, 25(4), 225–234.

Palmer, H.D., 1969, Wave induced scour on the sea floor, Proc. Civ. Engrg. In the Oc., ASCE, New York, pp. 703–716.

Sumer, B.M., and Fredsøe, J., 1998, Wave scour around group of vertical piles, J. Wtrwy., Port, Coast. And Oc. Engrg., ASCE , Vol. 124, No. 5, pp. 318 256.

Sumer, B.M., Fredsøe, J., and Christiansen, N., 1992, Scour around vertical pile in waves, J. Wtrwy., Port, Coast. And Oc. Engrg., ASCE , Vol. 118, No. 1, pp. 15–31.

*River, Coastal and Estuarine Morphodynamics: RCEM 2007 – Dohmen-Janssen & Hulscher (eds)*
*© 2008 Taylor & Francis Group, London, ISBN 978-0-415-45363-9*

# Local scour downstream of stilling basins

G. Oliveto, V. Comuniello & B. Onorati
*University of Basilicata, Dipartimento di Ingegneria e Fisica dell'Ambiente, Potenza, Italy*

ABSTRACT:   In this study some results on local scour downstream of stilling basins for low-head structures are presented. A number of laboratory experiments were conducted in a 1 m wide and 20 m long rectangular straight channel in which an ogee-crest spillway followed by a stilling basin type USBR IV (without chute blocks) was positioned. Two nearly-uniform bed sediments were tested. The runs were of long durations, typically 3 days, to achieve conditions of quasi-equilibrium scour hole. Based on the data collected, a new relationship was conjectured relating the quasi-equilibrium scour depth to the tail water densimetric Froude number.

## 1  INTRODUCTION

Water levels in canals and rivers are frequently controlled by weirs and low dams. The flow over such structures has a considerable potential for scour even for comparatively low heads (Breusers & Raudkivi 1991). Thus, stilling basins are frequently used to keep energy flows from scouring the streambed. Nevertheless, the flow turbulence level within them is often high, so that local bed scour processes develop.

Figure 1 provides an example of local (and general) scour phenomena observed downstream of a diversion weir on the Sauro Creek, Agri River, Southern Italy.

Although a number of studies on local scour below low-head spillways is available in literature, the related results would still appear of limited general significance. A literature review of empirical relations is provided by, for example, Breusers & Raudkivi (1991) and Hoffmans & Verheij (1997).

Breusers (1967) conducted a wide-ranging study by using various bed materials with different densities (sand, bakelite, and polystyrene) and various geometric configurations. He found that for a given geometry the scour profiles were quite similar at all values of time. Furthermore, on the basis of the data collected, a formula for the temporal evolution of the maximum scour depth was also suggested.

Catakly et al. (1973) proposed an empirical formula on the basis of several small-scale experiments in which, however, the stilling basin had always a length of 5 times the downstream flow depth.

A rather extensive set of tests was also performed by Farhoudi & Smith (1985). In particular, they used six different materials four of which were sand while the remainder bakelite. Their results were in substantial agreement with the Breusers' study.

Figure 1.   Local and general scour downstream of a stilling basin for a diversion weir on the Sauro Creek, southern Italy. The photograph on the top refers to 1998 while the other one to 2000.

Nola & Rasulo (1989) and Gisonni & Rasulo (1997) presented an approach for estimating the quasi-equilibrium scour depth as a function of a turbulence intensity coefficient and a (reduced) critical velocity for the initiation of sediment transport.

More recently, Dargahi (2003) presented an experimental study to examine the similarity development of scour profiles and the scour geometry. No experimental evidences were found in support of the above

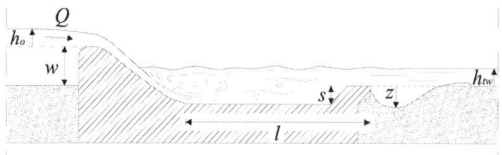

Figure 2. Definition sketch for model tests.

similarity while power type equations were introduced to predict the scour geometry, mainly in terms of the head above the spillway crest and the grain size of the sediment bed.

The impulsion for the present study was provided by a project finalised to the theoretical and experimental investigation on the hydraulic behaviour of the diversion weir on the Sauro Creek. In particular, a number of laboratory experiments, on which the findings of the present paper are based, was addressed to the analysis of local scour processes just downstream of the stilling basin. Despite the specificity of the project, the results would appear of general significance.

## 2 EXPERIMENTS

Experiments were carried out in a 1 m wide and 20 m long rectangular channel at the Hydraulic Engineering Laboratory, University of Basilicata, Italy.

An ogee-crest spillway followed by a stilling basin was mounted in the channel. The stilling basin was of type USBR IV, without chute blocks, with a sloped end sill. The model reproduced a typical section of the diversion weir on the Sauro Creek, in a scale 1:15. Figure 2 provides a definition sketch with $w$ = weir height, $l$ = basin length, $s$ = end sill height, $Q$ = discharge, $h_o$ = head over the weir crest, $h_{tw}$ = tail water depth, and $z$ = scour depth.

The weir height $w$ was 0.20 m, the basin length $l$ was 0.93 m, and the end sill height $s$ was 0.092 m. An alternate (elongated) stilling basin was also tested with length $l$ = 1.43 m.

Two different nearly uniform bed sediments were used, one with grain size $d_{50}$ = 1.7 mm and sediment gradation $\sigma = (d_{84}/d_{16})^{1/2} = 1.5$ and the other one with $d_{50}$ = 2.5 mm and $\sigma$ = 1.2. Both sediments had a density $\rho_s$ = 2.65 t/m³.

All the experiments were carried out under steady flow conditions. The runs were of long durations, typically 3 days, mainly to achieve conditions of quasi-equilibrium. To avoid the overlapping of local and general scour phenomena, the tail water depths were set to getting clear-water regime (i.e. the tail water flow intensities were less or close to 1). As a consequence and owing to the stilling basin geometry, for all runs a submerged hydraulic jump was formed. Photographs in Figure 3 show the flow features within the stilling

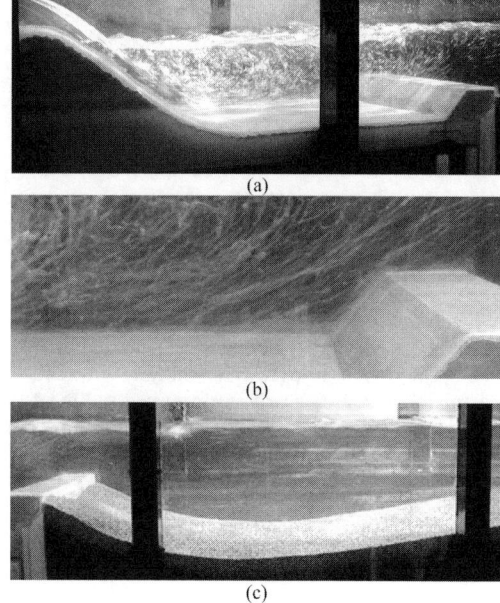

(a)

(b)

(c)

Figure 3. Photographs for the run A3 with discharge $Q$ = 83 l/s and tail water depth $h_{tw}$ = 20.4 cm: (a) flow features within the stilling basin, (b) streamlines upstream of the endsill, and (c) bed configuration at the end of the experiment. The maximum observed scour depth was 25.6 cm.

basin and the quasi-equilibrium scour profile for the run A3.

Discharges were measured with an orifice plate of ±3% accuracy. Water surface was surveyed with a conventional point gage, typically ±0.5 to 1 mm, whereas the bed morphology was surveyed with a so called shoe gage having 4 mm by 2 mm wide horizontal plate at its base.

Flow depths and bed levels were normally recorded at 2, 24, 48 and 72 hours from the start. More than 300 bed levels readings were normally recorded for each survey, so that more than 1200 readings were normally collected for each run.

Tail water depths were measured at about 3 meters downstream of the end sill. They were controlled by an adjustable sharp-crested weir located at the channel downstream end. Once the bed was accurately leveled the experiment initiated submerging the working section by setting the sharp-crested weir up. This was lowered within 10 to 20 seconds to the pre-selected flow depth and the temporal start of the experiment was set at scour inception.

The test conditions and the observed scour depths at the end of each run are given in Table 1 in which $F_d$ is the densimetric particle Froude number, $F_{di}$ the inception densimetric particle Froude number, $z_e$ the quasi-equilibrium scour depth, and $t_d$ the test duration.

Table 1. Test conditions and observed scour depths.

| Run (–) | $Q$ (l/s) | $h_{tw}$ (cm) | $F_d$ (–) | $F_{di}$ (–) | $z_e$ (cm) | $t_d$ (h) |
|---|---|---|---|---|---|---|
| A1 | 30.0 | 8.4 | 2.15 | 3.11 | 9.3 | 80.0 |
| A2 | 50.0 | 14.2 | 2.12 | 3.34 | 13.4 | 66.0 |
| A3 | 83.0 | 20.4 | 2.45 | 3.50 | 25.6 | 72.0 |
| A4 | 62.5 | 11.1 | 3.39 | 3.24 | 18.3 | 72.0 |
| A5 | 32.0 | 5.7 | 3.38 | 2.94 | 15.6 | 72.0 |
| A6 | 90.5 | 15.6 | 3.50 | 3.38 | 24.4 | 105.0 |
| A7 | 100.0 | 21.6 | 2.79 | 3.52 | 24.3 | 32.0 |
| B1 | 90.0 | 20.5 | 2.65 | 3.50 | 19.4 | 84.2 |
| B2 | 60.0 | 13.4 | 2.70 | 3.32 | 16.1 | 70.0 |
| B3 | 30.0 | 10.9 | 1.66 | 3.23 | 6.5 | 64.9 |
| B4 | 30.0 | 13.6 | 1.33 | 3.32 | 4.1 | 48.3 |
| B5 | 60.0 | 20.3 | 1.78 | 3.50 | 13.2 | 73.3 |
| B6 | 90.0 | 24.7 | 2.19 | 3.58 | 18.5 | 91.3 |
| B7 | 90.0 | 22.0 | 2.03 | 3.16 | 15.9 | 50.0 |
| B8 | 60.0 | 14.9 | 1.92 | 3.03 | 12.6 | 72.0 |
| B9 | 30.0 | 7.2 | 1.75 | 2.79 | 8.1 | 73.0 |

Both $F_d$ and $F_{di}$ refer to the tail water section. Their meaning will be better explained in the following.

Runs from A1 to A7 refer to the stilling basin with length $l = 0.93$ m while runs B1 to B9 refer to $l = 1.43$ m. Moreover, for runs from A1 to B6 the sediment bed had $d_{50} = 1.7$ mm and $\sigma = 1.5$ while for the remainder runs B7 to B9 the sediment bed had $d_{50} = 2.5$ mm and $\sigma = 1.2$.

## 3 QUASI-EQUILIBRIUM SCOUR EQUATION

In this section, an equation for the prediction of the (quasi-equilibrium) maximum scour depth, $z_e$, just downstream of a stilling basin is proposed by using a similar approach as suggested by Oliveto et al. (2006) for contraction scour.

Sediment transport formulas for uniform sediment beds are typically given in monomial form as

$$\phi = \frac{q_b}{\sqrt{g'd_s^3}} \propto \theta^n \tag{1}$$

where $q_b$ = volume of bed material in motion per unit width and time; g' = $g(\rho_s - \rho)/\rho$ with g as gravitational acceleration; $\rho$ and $\rho_s$ fluid and sediment densities; $d_s$ = grain size; $\theta = v_*^2/(g'd_s)$ = the Shields' parameter with $v_*$ shear velocity; $n$ = exponent. One should expect that $q_b$ and $z_e$ are intimately connected so that Equation 1 could be rearranged as

$$\frac{z_e}{d_s} \propto \theta^n \tag{2}$$

For uniform sediment beds, Hager & Oliveto (2002) demonstrated that

$$F_d \propto \left(\frac{R_h}{d_{50}}\right)^{1/6} \theta^{1/2} \tag{3}$$

with $R_h$ = hydraulic radius; $d_{50}$ = particle size for which 50% are finer by weight; $F_d$ = densimetric particle Froude number. In particular, $F_d = V/(g'd_{50})^{1/2}$ with $V$ = cross-sectional velocity. Thus, assuming $d_{50}$ as the characteristic grain size, Equations 2 and 3 yield

$$\frac{z_e}{d_{50}} \propto \left(\frac{R_h}{d_{50}}\right)^{-\frac{n}{3}} F_d^{2n} \tag{4}$$

For straightforwardness, $F_d$ could refer to the tail water section. Moreover, Equation 4 should be written more congruently as

$$\frac{z_e}{d_{50}} \propto \left(\frac{R_h}{d_{50}}\right)^p (F_d - F'_{di})^m \tag{5}$$

where $F'_{di}$ is the inception densimetric particle Froude number that would account for the additional turbulence induced by the stilling basin. More specifically, $F'_{di}$ is the densimetric particle Froude number at which the incipient motion, just downstream of the end sill, would occur. As for $F_d$, also $F'_{di}$ should be referred to the tail water section. Equation 5 was thus explored.

## 4 RESULTS

All the experiments made in the present study were characterized by submerged hydraulic jumps. Whereas, the most of literature formulas refers to conditions of free hydraulic jumps. Nevertheless, by accepting some approximations, the experimental data collected were preliminary compared with the approach suggested by Nola & Rasulo (1989).

Figure 4 shows the results of this comparison. Practically all the data were overestimated and such a tendency might be explained by the less scour potential of submerged flows than those with free hydraulic jumps. In support to this, one can note as the three data points closer to the line of perfect agreement refer to the runs A3 to A5 in which the tail water depths were particularly lower and flow intensities were greater (although only slightly) unity.

Figure 5 shows the ratio $\Phi = F'_{di}/F_{di}$ as a function of $F_{di}$. As above said, $F_{di}$ is the inception densimetric particle Froude number for nearly uniform flows while $F'_{di}$ is the densimetric particle Froude number at which the incipient motion just downstream of the end sill would occur. In particular, for each approaching discharge, $F_{di}$ was computed through the equations

1229

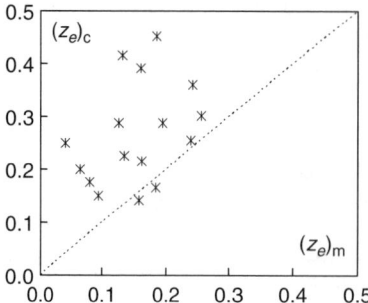

Figure 4. Comparison between measured, $(z_e)_m$, and computed, $(z_e)_c$, scour depths by applying the approach suggested by Nola & Rasulo (1989). (-----) Line of perfect agreement.

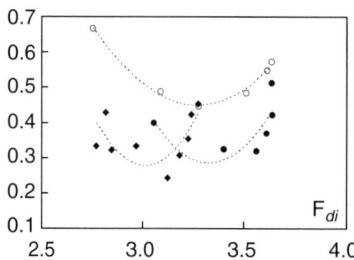

Figure 5. Ratio $\Phi = F'_{di}/F_{di}$ as a function of $F_{di}$. (○) Basin length $l = 0.93$ m and $d_{50} = 1.7$ mm, (•) $l = 1.43$ m and $d_{50} = 1.7$ mm, and (◆) $l = 1.43$ m and $d_{50} = 2.5$ mm. (-----) Quadratic regression curves.

suggested by Hager & Oliveto (2002) while $F'_{di}$ was computed on the basis of the observed tail water depth. Incidentally, the experimental data of preliminary tests on the incipient motion under normal turbulence conditions were found in satisfactory agreement with the computed values through the equations by Hager & Oliveto (2002).

For all the tests, $\Phi$ was found less than unity owing to the high level of turbulence induced by the stilling basin. Moreover, $\Phi$ exhibited a nearly parabolic dependence on $F_{di}$. This might be explained by the fact that for smaller discharges the end sill was sufficiently high to efficiently damp the energy of the outgoing flow while for higher discharges the tail water depths induced a significant submergence. Surprisingly, the $\Phi$ values for the longer apron were found significantly smaller than those for the shorter one. This, probably because by increasing the apron length the longitudinal velocity components of the outgoing flows were amplified. However, according to literature findings, for a given discharge and sediment bed, the quasi-equilibrium scour depths downstream of the longer stilling basin were found significantly smaller than those below the shorter one.

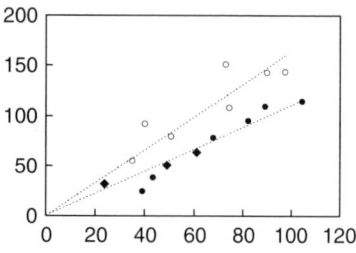

Figure 6. Ratio $\zeta = z_e/d_{50}$ as a function of $\Psi = (R_h/d_{50})^{0.5}$ $(F_{di} - F'_{di})$. (○) Basin length $l = 0.93$ m and $d_{50} = 1.7$ mm, (•) $l = 1.43$ m and $d_{50} = 1.7$ mm, and (◆) $l = 1.43$ m and $d_{50} = 2.5$ mm. (-----) Regression lines crossing the axis origin.

Figure 6 shows the dimensionless parameter $\zeta = z_e/d_{50}$ as a function of the dimensionless group $\Psi = (R_h/d_{50})^p (F_{di} - F'_{di})^m$, in conformity with Equation 5. In particular, on the basis of a sensitivity analysis the two exponents $p$ and $m$ were found 0.5 and 1, respectively. By taking into account the definition of densimetric Froude number, this would imply that the dependence of the scour depth $z_e$ on $d_{50}$ is only included in the (reduced) inception velocity.

for the shorter stilling basin the coefficient of determination was $r^2 = 0.74$ while for the longer one $r^2$ was even greater and equal to 0.93, although in this later case were used two different sediment beds. Moreover, according to literature findings for a given stilling basin configuration and sediment bed the (quasi-equilibrium) maximum scour depth would reduce by increasing the apron length.

## 5 CONCLUSIONS

Laboratory experiments on local scour below low-head spillways were carried out at the Hydraulic Engineering Laboratory, University of Basilicata, Italy.

The runs lasted typically 3 days to achieving quasi-equilibrium conditions and acquiring a suitable measurement accuracy. Tail water depths were set to avoid the overlapping of local and general scour mechanisms.

The main findings could be summarized as follows: (i) the experimental data were first interpreted on the basis of the approach suggested by Nola & Rasulo (1989). Overall, the comparison between computed and measured data revealed a significant overestimation. This could be explained by the fact that for all runs a submerged hydraulic jump occurred; (ii) starting from sediment transport formulas, a novel scour equation was conjectured in the light of the particle densimetric Froude number. Based on the experimental data collected, the proposed approach appeared promising, although further tests would be needed for its corroboration and generalization; (iii) additional tests were conducted to investigate the inception of sediment transport just downstream of the end sill. Experimental evidences revealed as scour started at a much reduced densimetric Froude number $F'_{di}$ as compared to conditions of normal turbulence (i.e. Shields' conditions). Surprisingly, for a given approaching discharge, $F'_{di}$ was found decreasing with increasing the stilling basin length. Nevertheless and according to literature findings, for a given discharge and sediment bed, the (quasi-equilibrium) maximum scour depths were found decreasing by increasing the apron length.

## REFERENCES

Breusers, H.N.C. 1967. Time scale of two-dimensional local scour. *Proc. of XII IAHR Congress, Ft. Collins* 3: 275–282.

Breusers, H.N.C. & Raudkivi, A.J. 1991. *Scouring.* Rotterdam: Balkema.

Catakli, O., Ozal, K. & Tandogan, R. 1973. A study of scours at the end of stilling basin and use of horizontal beams as energy dissipators. *Proc. of XI Int. Congress on Large Dams, Madrid* Q41 R2: 23–37.

Dargahi, B. 2003. Scour development downstream of a spillway. *Journal of Hydraulic Research* 41 (4): 417–426.

Farhoudi, J. & K.V.H. Smith. 1985. Local scour profiles downstream of hydraulic jump. *Journal of Hydraulic Research* 23 (4): 343–358.

Gisonni, C. & Rasulo, G. 1997. Local scouring downstream of positive step stilling basins. In J. S. Gulliver & P. L. Viollet (eds), *Proc. of XXVII IAHR Congress, San Francisco*, Theme D: 423–428.

Hager, W.H. & Oliveto, G. 2002. Shields' entrainment criterion in bridge hydraulics. *Journal of Hydraulic Engineering* 128 (5): 538–542.

Hoffmans, G.J.C.M. & Verheij, H.J. 1997. *Scour Manual.* Rotterdam: Balkema.

Nola, F. & Rasulo, G. 1989. Escavazioni a valle di dissipatori a risalto. *Idrotecnica* 16 (2): 63–74.

Oliveto, G., Onorati, B. & Comuniello, V. 2006. Riverbed scour induced by a long constriction. In Ferreira, Alves, Leal & Cardoso (eds), *River Flow 2006*, Vol. 2: 1737–1741.

*River, Coastal and Estuarine Morphodynamics: RCEM 2007 – Dohmen-Janssen & Hulscher (eds)*
© *2008 Taylor & Francis Group, London, ISBN 978-0-415-45363-9*

# A Large Eddy Simulation study of the bed shear stress distributions around isolated and multiple groynes

G. Constantinescu, M. Koken & A. McCoy
*Civil and Environmental Engineering Department, The University of Iowa, Iowa City, IA, US*

ABSTRACT: Large Eddy Simulation is used to investigate the main coherent structures playing a role in the erosion process (e.g., horseshoe vortex system forming around the base of the groynes, eddies shed in the separated shear layer) and the associated bed shear stress distributions around isolated and multiple groynes placed in straight channels. For isolated groynes we investigate the flow at conditions corresponding to the start of the scouring process (flat bed) and to its end (equilibrium scour bathymetry). Also, we consider the flow past two vertical groynes and we study the effect of groyne submergence (fully emerged vs. 40% relative submergence depth) on the horseshoe vortex system forming at the base of the upstream groyne and bed shear stress distribution in the groynes region. Large amplifications of the turbulence (e.g., resolved kinetic energy) inside the horseshoe vortex system and of the bed shear stress below are observed in all the cases.

## 1 INTRODUCTION

Groynes are one of the most popular remedies for controlling the scouring process along river banks, for maintaining channel navigability and restoring fish habitat (e.g., see Uijttewaal, 2005). However, scour phenomena can endanger the stability of these structures. For example, in the case of isolated groynes scour is due, to a large extent, to the presence of a system of horseshoe vortices that wrap around the base of the groyne. The flow inside the horseshoe vortex system is characterized by high levels of amplification of the turbulent kinetic energy (TKE) and pressure fluctuations (e.g., see Devenport & Simpson 1990, Simpson 2001). These vortices are thought to largely control the evolution of the scour process. These coherent structures are present over the whole duration of the scouring process from its initiation (e.g., flat bed conditions) to its end (equilibrium bathymetry is reached).

Scour and deposition take also place due to the eddies convected close to the bed in the separated shear layer forming at the tip of the groyne. In the case of groyne fields, depending on the way the groynes are disposed in the channel and of the channel geometry, a similar horseshoe vortex system may form, especially at the base of the most upstream groyne in the series. However, the structure and intensity of the horseshoe vortex system is expected to be somewhat unlike the one observed for an isolated groyne. This is mainly because the presence of the embayment modifies the nature of the recirculating flow downstream of the groyne. Additionally the lateral extent of the horizontal separated shear layer and the intensity of the turbulence in the shear layer at the embayment-channel interface may vary substantially depending on the embayment size, geometry, etc. In the case of submerged groynes, the flow becomes even more complex and three-dimensional because of the overflow and associated vertical separated shear layer forming on top of the embayment. As a result, the intensity of the horseshoe vortex system and the bed shear stress distribution in the groynes region are expected to be affected by the relative submergence depth.

The scouring process ends when the bed shear stress induced by the presence and/or convection of the coherent structures in the groyne region becomes smaller than the local critical value for sediment entrainment. This quantity that incorporates gravitational slope effects depends on several factors such as the geometry of the scour hole, the angle of repose of the sediment, the angle between the vector corresponding to the direction of maximum slope and the bed shear stress vector, etc. Thus, knowledge of the mean and instantaneous bed shear stress distributions is important. These distributions are very difficult to measure experimentally. Most up to date experimental studies focused mainly of the temporal development of the scour hole around isolated emerged groynes or bridge abutments. Models have been proposed to estimate the maximum scour depth function of the flow conditions and the form of the abutment (e.g., see Kwan & Melville 1993, Melville 1997, Melville &

Coleman 2000). Practically no such studies were conducted for groyne fields. It is not clear to what extent the models and findings from the scour studies conducted for isolated groynes apply for the case of multiple groynes in a series. No detailed investigations were conducted to study the effect of submergence for either isolated or multiple groynes.

The present study tries to use Large Eddy Simulation (LES) to describe the structure of the horseshoe vortex system at the base of the groynes and the associated bed shear stress distributions. Additionally, we try to understand how the structure and intensity of the horseshoe vortex system are affected by the evolution of the scour hole, by presence of a second groyne downstream of the first one, or by the groynes becoming submerged. Though we conduct only a limited number of simulations we think they should highlight some important general features of the flow past groynes in a channel.

## 2  NUMERICAL METHOD AND SIMULATIONS SET UP

The collocated, non-dissipative, finite volume scheme in the LES code (Mahesh et al., 2004) employed in the present work solves the filtered Navier-Stokes equations on unstructured meshes using a dynamic Smagorinsky model to account for sub-grid scale effects. No wall functions are used. The code has the capability to use unstructured hybrid meshes. All the operators in the code including the convective terms are discretized using central schemes. The length and velocity scales are selected to be the groyne depth (D) and the mean velocity (U) in the incoming channel flow, respectively.

The channel Reynolds number is equal to 18,000 in the flat bed and deformed bed simulations with an isolated groyne (Figs. 1a and 1b). The equilibrium scour bathymetry (see Fig. 1b) in the deformed bed simulation was determined from an experiment (clear water scour conditions) in which a layer of sand (mean particle diameter of 0.68 mm corresponding to a ratio of the mean friction velocity to the critical entrainment velocity of 0.42 in the region away from the groyne) was placed on the bottom of the channel containing the vertical groyne.

The other simulations were conducted in a flat bed channel containing two groynes forming an embayment (Fig. 1c). The interest in this configuration is motivated by its similarity to flow around groyne fields in natural channels. Two cases were studied. In the first one the groynes are fully emerged corresponding to normal flow conditions in the channel. In the second one, the groynes are fully submerged and the ratio between the channel depth and the embayment depth is h/D = 1.4, corresponding to flooding conditions

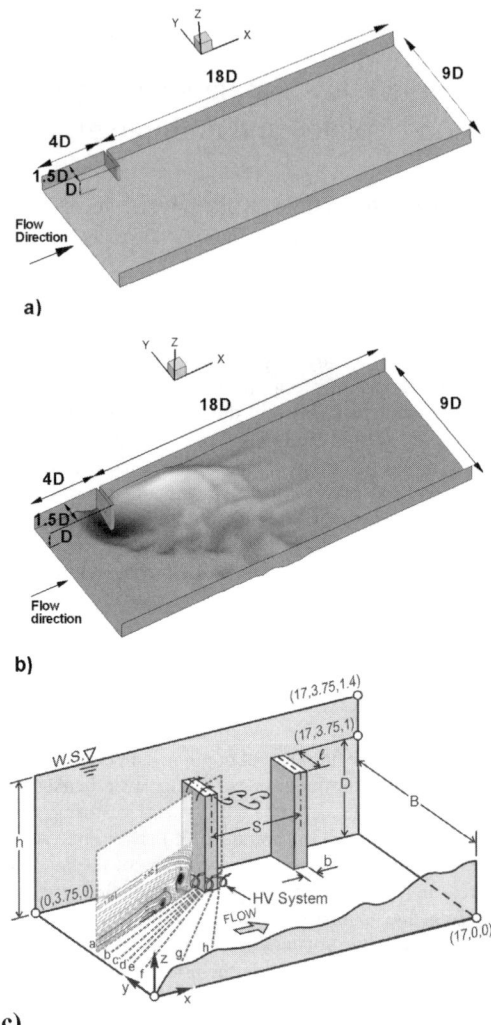

c)

Figure 1.  General sketch showing channel flow with one emerged groyne on flat bed (a) and scoured bed (b) and with two (submerged) groynes on flat bed (c).

(large relative submergence depth). The computational domains in the two simulations with emerged (see also, McCoy et al. 2006 for a more detailed description of this case) and submerged groynes are identical with the exception of the channel depth. The Reynolds number is Re = 13,600. The width of the channel is B = 3.75D. The length, l, and width, b, of each groyne are 0.625D, and 0.25D, respectively. The distance between the midplanes of the two groynes is S = 1.5D such that the embayment length is 1.25D. The ratio of the embayment length to the embayment width is close to two which is typical of many river groyne fields.

The flow in all the simulations is assumed to be fully turbulent to mimic realistic conditions. Inflow conditions are obtained from preliminary LES simulations of the flow in a periodic straight channel of identical section to the one used in the simulation containing the groynes. The turbulent velocity fields collected in a spanwise section of the straight channel simulation are then fed in a time accurate fashion through the inflow section in the simulations containing the groynes. A convective boundary condition is utilized at the outflow section. The free surface is treated as a rigid stress-free horizontal lid. The walls are considered smooth.

The domain is meshed with approximately 4 million hexahedral cells in all simulations. The mesh was generated using a paving technique which allows rapid variation in the characteristic size of the computational cells while maintaining high overall mesh quality and low stretching ratios. The mesh was refined near all solid surfaces where the minimum mesh size in wall units in the wall-normal direction was $\Delta y^+ = yu_\tau/\nu = 0.5\text{-}2$ ($\nu$ is the molecular viscosity), assuming a non-dimensional friction velocity of $u_\tau/U = 0.052$. This avoided the need to use wall functions.

## 3 RESULTS

### 3.1 *Isolated emerged groyne in a channel*

Figure 2 visualizes the main coherent structures around the groyne in the mean flow for the flat bed case using the Q criterion. In this figure the vortical structures in the detached shear layer were hidden. Besides the main necklace eddy corresponding to the primary vortex HV1, a secondary necklace vortex HV2 is also present upstream of HV1 in the region around the tip of the groyne where the intensity of the horseshoe vortex system peaks. The necklace vortices are situated near the bed and change gradually their orientation from being parallel to the groyne face until close to the tip of the groyne, to being parallel to the mean shape of the detached shear layer. Analysis of the instantaneous flowfields showed that secondary necklace vortices are shed randomly and are then convected toward the primary vortex HV1. As these smaller vortices approach HV1, they start interacting with it. These interactions are one of the main causes for the strong variations in the structure and level of coherence of the horseshoe vortex system (a similar phenomenon is observed in the deformed bed case). The primary vortex HV1 is observed in practically all of the instantaneous flowfields. This is not always the case with HV2. The main tornado like vortex in the upstream recirculation region is denoted CV1. Its role, in both the flat-bed and deformed-bed cases, is to convect flow and momentum from the free surface into the core of

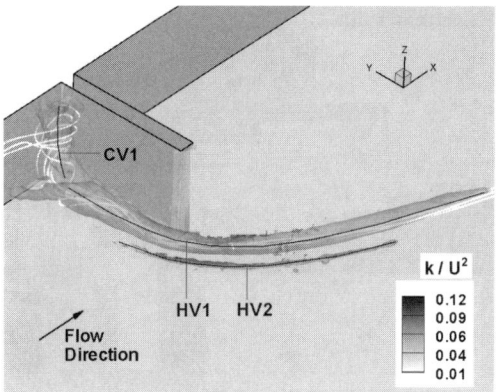

Figure 2. Visualization of the mean–flow horseshoe vortex system around an isolated groyne using Q criterion and 3D streamlines. Contour levels represent the TKE.

the main necklace vortex HV1. This shows that the dynamics of the flow in the horseshoe vortex region is affected by the large-scale content of the flow in the upstream recirculation region. The presence of the scour hole in the deformed bed case plays a stabilizing role for the overall dynamics of the horseshoe vortex system. The horseshoe vortex system in the deformed bed case contains a primary necklace eddy that corresponds to HV1 in Figure 2 (see also the 2D streamlines in Figure 4 for the mean flow) and several smaller secondary necklace vortices and bed-attached vortices.

The distributions of the resolved TKE along with 2D mean-flow streamlines are plotted in several representative vertical sections in Figures 3 and 4 for the two cases. Though the 2D streamlines capture the position of the primary necklace vortex HV1, they are not a very exact way of defining its position and size due to the strong three-dimensionality of the flow. In this regard, TKE is a much more appropriate quantity. The TKE distributions clearly show the turbulence inside the HV system dominates the background turbulence in both cases. In particular, the amplification is the largest in the regions associated with the large-scale oscillations of the main necklace vortex HV1. The strong amplification of the TKE inside the horseshoe vortex system was observed experimentally in previous investigations of turbulent flows over surface mounted bluff bodies (see Simpson, 2001). As the primary horseshoe vortex is swept past the groyne, its leg starts interacting with the detached shear layer (e.g., see Figures 3f to 3h for the flat bed case) before the core diffuses.

The TKE contours show another interesting phenomenon in both cases. Two distinct TKE peaks are present inside the horseshoe vortex regions, at the sections situated close to the tip of the groyne (see Figs. 3c to 3e and 4c to 4e. Figs. 3a and 3b also show the

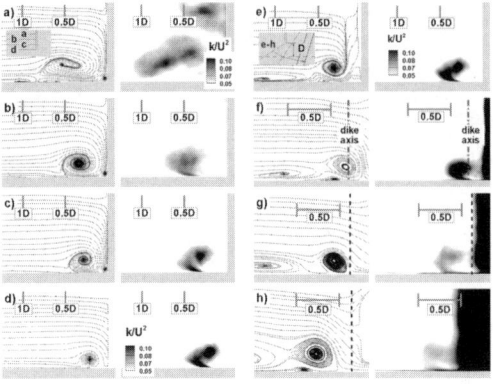

Figure 3. Streamlines and resolved TKE contours in representative vertical sections (see inset in frames a and e) for an isolated emerged groyne on flat bed.

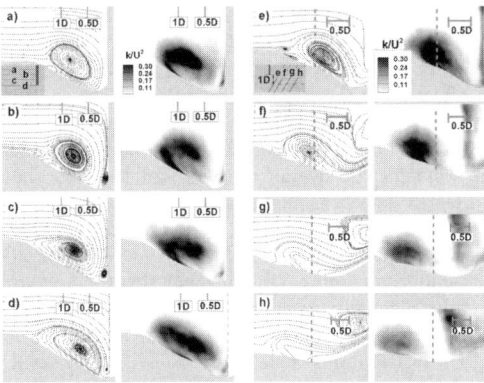

Figure 4. Streamlines and resolved TKE contours in representative vertical sections (see frames a and e) for an isolated emerged groyne on deformed bed.

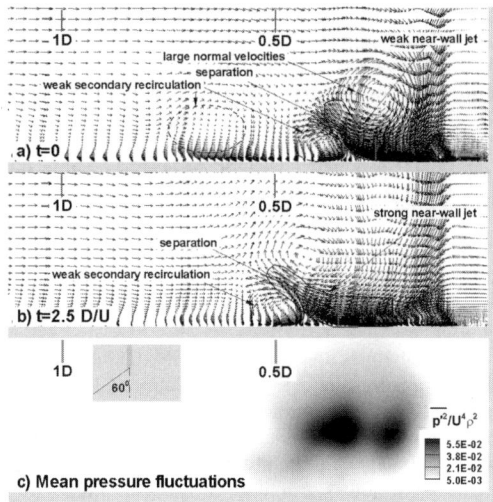

Figure 5. Instantaneous velocity vectors corresponding to the zero-flow mode and back-flow mode of the horseshoe vortex system, and mean pressure fluctuations contours in a plane inclined 60° from the groyne face (see inset in frame c).

double peak structure due to the stronger horseshoe vortex system present in the flat bed case compared to the equilibrium bathymetry case). Eventually, a one peak structure is recovered in the downstream leg of HV1 (Figs. 3f to 3h and 4f to 4h). This is mainly due to the bimodal nature of the flow inside the turbulent horseshoe vortex system. The presence of large-scale low-frequency unsteadiness of the main necklace vortices that form around the base of surface mounted bluff bodies was observed for the first time by Devenport and Simpson (1990) in their study of the flow past a wing-shaped body at high Reynolds numbers (Re > $10^5$). However, these oscillations are thought to be a general characteristic of a turbulent horseshoe vortex system.

When an irrotational patch of fluid (i.e. coming from free-surface) having a high momentum reaches the upstream face of the groyne, it is transported toward

the bottom with the downflow. There, it tries to preserve its irrotationality and forms a strong near-wall jet in the reverse flow direction which moves the center of HV1 away from the wall and induces a more elliptical shape of the core, leading to the back-flow mode. This scenario is illustrated for the flat bed case by the velocity contours in a vertical section starting at the tip of the groyne, oriented at 60° from it (Fig. 5b). The other extreme situation is when a patch of rotational fluid coming from the boundary layer region, having relatively low momentum, reaches the groyne. In this case separation is expected to occur earlier on the bottom and to produce a smaller eddy closer to the upstream groyne face. This event is known as the zero flow mode and is illustrated in the velocity contours in Figure 5a. The oscillations of the horseshoe vortex structure between these two modes are the explanation for the double-peak structure observed in the TKE plots and are the main reason for the high turbulence intensities and pressure fluctuations observed in the HV region. The phenomenon is similar in the case of a scoured bed.

The distributions of the mean friction velocity, $u_\tau$, are shown in Figure 6 for the flat-bed and deformed-bed cases. For the experimental flow conditions, the critical value for sediment entrainment is $u_\tau/U = 0.106$ if the bed is flat. Regions were the mean friction velocity is larger than the entrainment value are present beneath the mean HV1 vortex and beneath the upstream part of the separated shear layer. In the flat bed case, the largest values occur very close to the tip of the groyne due to the strong flow acceleration in that

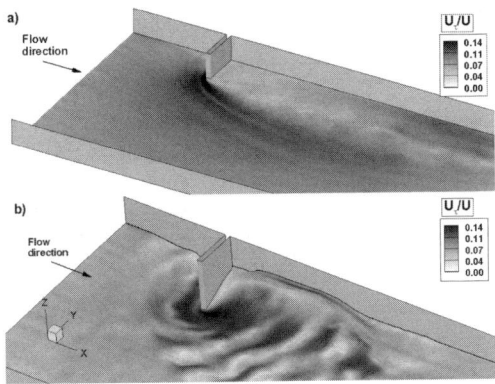

a)

Flow direction

$U_\tau/U$

0.14
0.11
0.07
0.04
0.00

b)

Flow direction

$U_\tau/U$

0.14
0.11
0.07
0.04
0.00

Figure 6. Mean-flow bed friction velocity contours for channel flow with one isolated emerged groyne (a) flat bed; (b) deformed bed (equilibrium bathymetry).

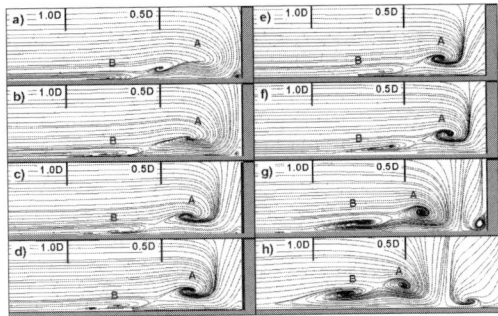

Figure 7. Structure of the mean horseshoe vortex system as it wraps around the base of the upstream groyne (two emerged groynes, one embayment case) visualized in vertical planes making an angle with the lateral channel wall of: (a) 25°; (b) 30°; (c) 35°; (d) 37.5°; (e) 40°; (f) 45°; (g) 55°; (h) 65°. See also Figure 1c for plane location.

region. Though the mean values of $u_\tau$ in Figure 6 are decaying rapidly in the downstream part of the horseshoe vortex region, high instantaneous local values of the bed shear stress ($u_\tau/U > 0.106$) are observed over long distances downstream of the groyne, especially in the detached shear layer region. These patches are induced by the presence of the vortex tubes and other coherent structures that populate the separated shear layer in the vicinity of the bed. In the equilibrium scour case (Fig. 6b) largest values are observed beneath the primary vortex HV1, beneath the upstream part of the separated shear layer and also near the crests of the bedforms present around the groyne. Observe also that the mean values of the friction velocity in some parts of the scoured region are still larger than the critical value determined for flat bed channel flow. However, if those values are adjusted for gravitational bed slope effects (e.g., using the formula proposed by Brooks & Shukry 1963) then the friction velocity becomes smaller than the local critical value over practically the entire scour hole region. This confirms indirectly the accuracy of the bed-shear stress predictions for the present LES simulation as the mean bed shear stress distribution is consistent with the equilibrium conditions present in the experiment.

## 3.2 Multiple groynes in a flat-bed channel

We focus our attention on the most upstream groyne where a horseshoe vortex system is expected to form. The following groynes are more protected and horseshoe vortices either will not form or will be much weaker compared to the horseshoe vortex system developing at the base of the first groyne in the series.

The spatial structure of the horseshoe vortex system in the emerged case (mean flow) is shown in Figure 7 using 2D streamlines in vertical planes (see also Fig. 1c for the position of these planes). The horseshoe

vortex system is composed of a primary necklace vortex A (corresponding to HV1 in the case of an isolated groyne) and an elongated secondary eddy upstream of it denoted B. Similar to the case of an isolated groyne, vortex A is initially parallel to the upstream face of the groyne. Subsequently, its core changes orientation and follows the curved shape defined by the horizontal separated shear layer near the bed. The topology of the flow in the horseshoe vortex region is relatively similar for the submerged case. Similar to the isolated groyne simulations, the flow inside the horseshoe vortex system is characterized by random interactions among the necklace eddies. Animations in several vertical planes of the instantaneous flowfields show that the primary necklace vortex A is practically always present though its shape and size can vary considerably over time. No definite frequency can be associated with the shedding of these secondary necklace vortices. The range of Strouhal numbers (fD/U, f is the frequency) associated with the advection of the secondary eddies toward the main necklace vortex and with their oscillations is between 0.45 and 2.3. Besides these frequencies, velocity and pressure power spectra show the presence of energetic frequencies between St = 0.12 and St = 0.29. These lower frequencies are associated with aperiodic oscillations (see discussion of Fig. 5) of the primary vortex. It is interesting to notice that the ratio between the mean frequency associated with the aperiodic oscillations and the main frequency associated with the shedding and advection of secondary necklace vortices (St~0.45) is close to 0.5 which is practically identical to the result of Devenport and Simpson (1990) obtained in a much higher Reynolds number flow.

Figure 8 shows the TKE distribution in the same frames as Figure 7 for the emerged case. The TKE is significantly amplified within the region occupied by the main necklace vortex. The vertical patch of

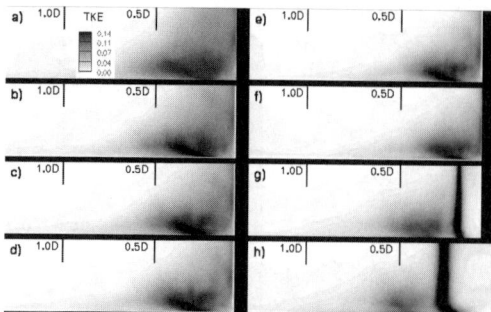

Figure 8. TKE distribution inside the horseshoe vortex system region (two emerged groynes, one embayment case) in various vertical planes making an angle with the lateral channel wall of: (a) 25°; (b) 30°; (c) 35°; (d) 37.5°; (e) 40°; (f) 45°; (g) 55°; (h) 65°. See also Figure 1c for plane location.

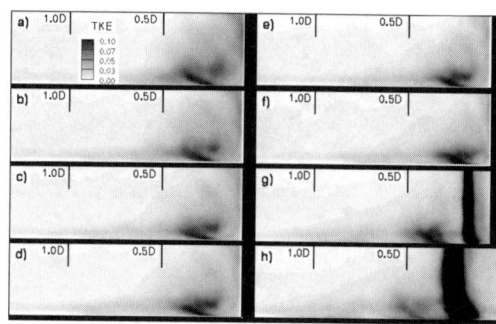

Figure 9. TKE distribution inside the horseshoe vortex system region (two submerged groynes, one embayment case) in various vertical planes making an angle with the lateral channel wall of: (a) 25°; (b) 30°; (c) 35°; (d) 37.5°; (e) 40°; (f) 45°; (g) 55°; (h) 65°. See also Figure 1c for plane location.

high TKE in Figures 8g and 8h corresponds to the cut through the horizontal separated shear layer. The large TKE values in the region associated with the separation of the attached boundary layer are due to the formation of a jet-like flow of variable strength beneath the primary necklace vortex A (see discussion of Fig. 5). This jet-like flow is the main reason why large scale oscillations of the primary vortex A are observed in the instantaneous flow fields.

By comparison, the TKE amplification in the submerged case shown in Figure 9 is smaller (by 20–30%) than the one observed in the emerged case (for the same mean channel velocity) in practically all the sections. This is expected because the flow convected into the upstream recirculation region and feeding into the main necklace vortex starts only at z/D~0.6 (z is measured from the bed) in the submerged case compared to z/D = 1 (free surface level) in the emerged case. In the submerged case, the flow above z/D~0.6 is eventually convected into the overflow on top of the embayment. However, at flooding conditions, the mean channel velocity is higher, so the dimensional values of the TKE will be higher. Consequently, the effectiveness of the horseshoe vortex system in entraining sediment will be higher in the submerged case as the sediment size remains the same at normal and flood conditions.

Figure 10 shows the contours of the non-dimensional bed shear stress relative to the mean value ($\tau_0 = \rho u_{\tau 0}^2$ where $u_{\tau 0}/U = 0.052$) corresponding to fully turbulent flow in a channel of identical section without groynes. The mean bed shear stress $\tau/\tau_o$ distributions in the emerged (McCoy et al., 2006a) and submerged cases are plotted in Figs. 16a and 16b. Qualitatively the bed shear stress distributions look similar. The maximum values of $\tau/\tau_o$ are close to 15 in the submerged case and to 16 in the emerged case in the strong acceleration region near the tip of the upstream groyne. This is consistent with previous experimental

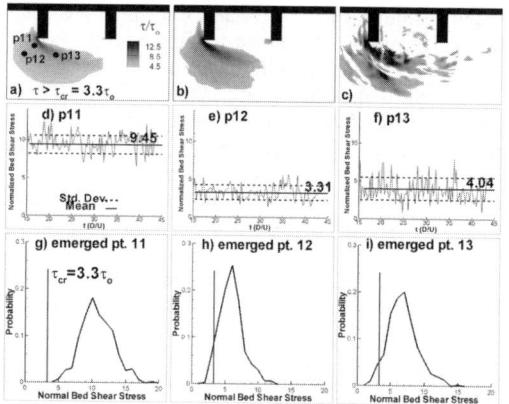

Figure 10. Non-dimensional bed shear stress distribution for simulations containing two groynes showing region where entrainment occurs. (a) mean values, emerged case; (b) mean values, submerged case; (c) instantaneous values, submerged case; (d) bed shear stress time series at point p11, submerged case; (e) bed shear stress time series at point p12, submerged case; (f) bed shear stress time series at point p13, submerged case; (g) histogram of bed shear stress at point p11, emerged case; (h) histogram of bed shear stress at point p12, emerged; (i) histogram of bed shear stress at point p13, emerged case.

observations (e.g., see Melville and Coleman, 2000) of the scour at bridge abutments (isolated emerged groynes) that show that scour is initiated in the region around the tip of the groyne. The large amplification of the bed shear stress around the tip of the upstream groyne indicates that severe scour can occur at the base of the first groyne in a series. By comparison, the amplification around the second groyne is very small.

For reference, the critical bed shear stress $\tau_{cr}$ corresponding to entrainment conditions was calculated using Shield's diagram for a sediment size of $d_{50} = 0.45\,\text{mm}$ assuming $D = 0.08\,\text{m}$ and $U = 0.171\,\text{m/s}$ ($Re_D = 13,600$). The critical shear velocity was estimated to be $0.016\,\text{m/s}$ and the particle Reynolds number was equal to 6.13. This gives a critical bed-shear stress for sediment entrainment of $\tau_{cr} = 0.035\rho\,U^2$. It corresponds to a value of $\tau_{cr}/\tau_o$ equal to 3.3. This is the reason the values beneath $3.3\tau_0$ were blanked in Figs. 10a to 16c. Observe that the entrainment area is slightly larger in the emerged case. This may look counterintuitive knowing that generally scour is more intense at flood conditions. However, the present results are consistent with this observation. The two nondimensional distributions in Figs. 10a and 10b take automatically into account the additional water depth at flood conditions while the mean incoming velocity is assumed to be identical. However, this is not the case when flooding occurs. Typically the mean velocity in the channel is higher (roughly proportional to the hydraulic radius at power 2/3) and thus same is true for the Reynolds number in the submerged case. Assuming scale effects are relatively small between the Reynolds numbers corresponding to normal and flood conditions (the incoming flow is fully turbulent in both cases), the distribution of the dimensional bed shear stress will show that higher values are predicted and the entrainment region is larger in the submerged case which corresponds to flood conditions.

The instantaneous distribution of $\tau/\tau_o$ in Fig. 10c shows the large variability of the bed shear stress distribution and the direct correspondence between the intensity of the coherent structures above the bed (e.g., vortex tubes being advected inside the separated shear layer, primary and secondary necklace vortices) and the bed shear stress value. Typical variations of the $\tau$ values around the mean value are of the order of $1.5\tau_0$ in the region of high bed shear stress amplification beneath the horseshoe vortex system and in the upstream part of the horizontal detached shear layer as also observed from the time series of the nondimensional bed shear stress in Figures 10d–10f. The positions of points p11, p12 and p13 are shown in Figure 10a. In this regard, the coherent structures responsible for most of the sediment transport phenomena around groynes are different than those playing an important role in channels containing ripples or dunes where sweep and ejection phenomena and kolk vortices play the most important role in sediment entrainment, transport and deposition. Locally, around the embayment region, the influence of these structures on local scour phenomena in the case of a loose bed is much smaller compared to that of the energetic eddies associated with the horseshoe vortex system, the horizontal separated shear layer and the recirculating motions within the embayment.

One of the quantities of real interest for sediment transport phenomena is the local probability that the instantaneous bed shear stress is larger than the critical

value for sediment entrainment ($\tau_{cr}$). This quantity is very relevant for flow around hydraulic structures in which large scale coherent structures are responsible for scour phenomena. To estimate this quantity one has to first compute the histograms of the instantaneous bed shear stress. Figures 10g–10i shows the probability plots (histograms) for the emerged case. Based on the results in Figure 10, in the emerged case the bed shear stress at point 11 has a cumulative probability (calculated by integrating the probability up to the threshold value from the histograms shown in Fig. 10) of 1.0 for being greater than the critical bed shear stress. At point 12 the cumulative probability is close to 0.90 and at point 13 is close to 0.93.

## 4 CONCLUSIONS

Large Eddy Simulation calculations were performed to investigate the horseshoe vortex system and the bed shear stress distributions around isolated (flat bed and equilibrium bathymetry) and multiple groyne configurations (one embayment with emerged and submerged groynes) present in straight channels. A wide range of coherent structures was observed inside horseshoe system in all the cases considered. Though the coherence, shape and intensity of the horseshoe system were highly variable in time, a large primary necklace vortex was always present at the base of the most upstream groyne. The main corner vortex in the upstream recirculation region was found to feed fluid and momentum from the free surface region into the primary necklace vortex.

High TKE levels were observed inside the horseshoe vortex system especially in the region around the tip of the upstream groyne. The large intensity of the turbulence inside the horseshoe vortex region was induced by the oscillations of the main necklace vortex between two modes, one in which the vortex is closer to the groyne (zero-mode flow) and one in which a strong near-bed jet flow convects the necklace vortex away from the groyne (back-flow mode), similar to the findings of Davenport and Simpson (1990). The presence of two relative maxima in the distribution of the TKE was linked to the presence of low-frequency bimodal oscillations in the horseshoe vortex region around the tip of the groyne. In contrast to that, in the regions close to the lateral wall and in the legs of the main necklace vortices past the upstream groyne, where the overall intensity of the horseshoe vortex system decayed substantially, the distribution changed gradually to a single peak shape suggesting the bimodal oscillations were absent.

High values of the bed shear stress were predicted near the tip of the abutment and beneath the primary necklace vortex. The high bed shear stress values are due, in great measure, to the presence of the bimodal large-scale oscillation, but high instantaneous values were also recorded beneath the horizontal separated shear layer. For the configurations containing two groynes, the amplification of the bed shear stress around the second groyne was much weaker due to the absence of a horseshoe vortex system. It was found that if the same mean flow channel velocity is present in the emerged and submerged cases, the intensity of the horseshoe vortex system (e.g., TKE values) and the values of the bed shear stress beneath it are slightly larger in the fully emerged case. This may look counterintuitive. However, in reality the mean flow velocity is always larger at flooding conditions (submerged case), which translates into a significant amplification of the bed shear stress levels compared to the fully emerged case.

REFERENCES

Brooks, N.H. & Shukry, A. (1963). Discussion of Boundary shear stress in curved trapezoidal channels. by A. T. Ippen and P.A. Drinker, *Proc. ASCE, Journal Hydraulics Division*, 89(3): 327–333.

Devenport, W.J. & Simpson, R.L. (1990). Time-dependent and time-averaged turbulence structure near the nose of a wing-body junction. *J. Fluid Mech.*, 210: 23–55.

Kwan, R.T. & Melville, B.W. (1994) Local scour and flow measurements at bridge abutments. *J. Hydraulic Research*, 32(5): 661–673.

Mahesh, K., Constantinescu, S.G. & Moin, P. (2004). A numerical method for large eddy simulation in complex geometries. *J. Comp. Physics*, 197(1): 215–240.

McCoy, A., Constantinescu, S.G. & Weber, L. (2006). Exchange processes in a channel with two vertical emerged obstructions. *J. of Flow, Turbulence, Combustion*, 77: 97–126.

Melville, B.W. (1997). Pier and abutment scour: integrated approach. *J. Hydr. Engrg.*, 123(2): 125–136.

Melville, B.W. & Coleman, S.E. (2000). *Bridge scour*, Water Resources, Littleton, Colorado, USA.

Simpson, R.L. (2001). Junction Flows. *Annual Rev. Fluid Mech.*, 33: 415–443.

Uijttewaal, W. (2005). Effects of groyne layout on the flow in groyne fields: laboratory experiments. *J. Hydr. Engrg.*, 131: 782–794.

*River, Coastal and Estuarine Morphodynamics: RCEM 2007 – Dohmen-Janssen & Hulscher (eds)*
© *2008 Taylor & Francis Group, London, ISBN 978-0-415-45363-9*

# Shear stress enhancement to account for increased turbulence in mixing layers along river groynes

Mohamed F.M. Yossef, H.R.A. Jagers & Erik Mosselman
*WL | Delft Hydraulics, Delft, The Netherlands*

Arjan Sieben
*Rijkswaterstaat-RIZA, Arnhem, The Netherlands*

ABSTRACT: The main channels of the Rhine branches in the Netherlands have been trained by groynes. These river training structures deepen the channel as a whole, but they also create a pattern of scour holes and a specific type of shoals called "groyne flames". The latter reduce the navigability and increase the need of dredging. However, these distinct features are not reproduced properly by the operational 2D morphological models which are based on a coarse computational grid and a Reynolds-averaged description of the flow. In this paper, we attempt to parameterise the small-scale flow characteristics deduced from a detailed simulation using a model with a fine grid and Horizontal Large-Eddy Simulation (HLES) for use in operational morphodynamic models where these details are missing. We used a well-calibrated model for a laboratory experiment of a river with groynes to analyse the differences between the time-averaged bed shear stresses from a Reynolds-averaged flow model (RANS) and those from a model with Horizontal Large-Eddy Simulation (HLES). We correlated the bed shear stress differences to the rate of deformation. Subsequently, we developed a parameterization for this relation and implemented it into a numerical model for use in operational morphodynamic models. Application of the resulting model to an idealised reach of the River Waal yielded an increased scour hole depth near the tip of the groynes in the fine grid and hardly caused any changes in the coarse grid. Apparently, coarse grids smooth the effect of the parameterised bed shear stress enhancement too much. We conclude that the enhancement is only appropriate for fine grids.

## 1 INTRODUCTION

### 1.1 *Background*

Groynes are central elements of river training in the Dutch branches of the river Rhine. Originally, they were constructed for bank protection and to prevent ice jamming. Now, they primarily serve to maintain a suitable channel for navigation. The draught of the barges and push-tow units determines the required depth. The number of required lanes and the width of the units govern the required width of the channel. For the River Waal, the most important Rhine branch in the Netherlands, a depth of 2.5 m and a width of 150 m are required during 95% of the time. Projects are ongoing to enlarge the navigation depth to 2.8 m.

Although they deepen the main channel as a whole, groynes generate complex morphological features on a smaller scale: scour holes, groyne flames and a streamwise bed undulation. The groyne flames are shoals that extend some tens of metres towards the centreline of the navigation channel. During low flow conditions, these areas often determine the critical navigation depth and corresponding need of dredging. The practical question for the river manager is how often and where dredging should take place to meet the navigation requirements. To answer this question it is important to predict the morphological development of such features so as to identify the critical locations and the associated morphological time scale in order to correctly predict their growth rate after dredging. Recent developments of morphodynamic models make this possible (see Yossef and Klaassen, 2002), albeit at the expense of computational time.

### 1.2 *Problem statement and objective*

Operational coarse-grid Reynolds-averaged (RANS) morphological models are regularly used to predict critical locations for navigation and to assess effects of interventions. The groynes-induced morphological features are not reproduced properly because the grid is too coarse and because the description of the physical processes is incomplete. The incompleteness regards 3D flow and turbulent fluctuations.

The aim of this study is to improve the prediction of groyne-induced shoals without substantially increasing the computational effort. The objective is to devise an approach that enables us to account for the missing processes in operational morphological models characterised by a rather coarse grid and Reynolds-averaged flow.

## 1.3  Approach

We analyse the contribution of neglected physical processes to the formation of groyne flames. Subsequently, we develop a suitable parameterisation of these physical processes for an operational numerical model with a coarser grid. We derive information about the dynamics of the flow by detailed numerical simulation, using Delft3D with HLES. It results in a total bed shear stress ($\tau_{bot1}$) that includes the effect of large-scale turbulent structures. Information about the time-averaged flow field can be inferred from the same simulation as well. This provides a reference bed shear stress ($\tau_{bot0}$). The excess bed shear stress is deduced as the difference between the total and the base bed shear stress. We parameterise the excess bed shear stress for representation on a coarse grid. We implement this parameterisation in a numerical model and analyse the effects by applying the model to a test case using a fine grid as well as a coarse grid.

## 2  GROYNES IN RIVERS

### 2.1  Flow near groynes

When groynes are not submerged, the flow field away from the tip of the groyne is predominantly two-dimensional with eddies in the horizontal direction (Krebs et al., 1999; Uijttewaal, 1999). They shed from the tip of the groynes and migrate in the flow direction. These eddies have horizontal dimensions in the order of tens of metres and time scales in the order of minutes. In a standard RANS flow simulation, these motions are not resolved because of using a large grid size, large time steps and, most importantly, the use of inadequate turbulence modelling. Correct reproduction of the flow pattern near groynes, using for example a k-ε model, requires substantial modifications (e.g. Tingsanchali and Maheswarn, 1990; Ouillon and Dartus, 1997; Peng et al., 1997). A better way is to use an eddy resolving approach by, for example, Horizontal Large Eddy Simulation (HLES). However, this approach requires a fine grid and a small time step that lead to a very long computational time.

### 2.2  Groynes-induced morphology

Groynes produce scour holes, groyne flames and a streamwise bed level undulation. Each has a different

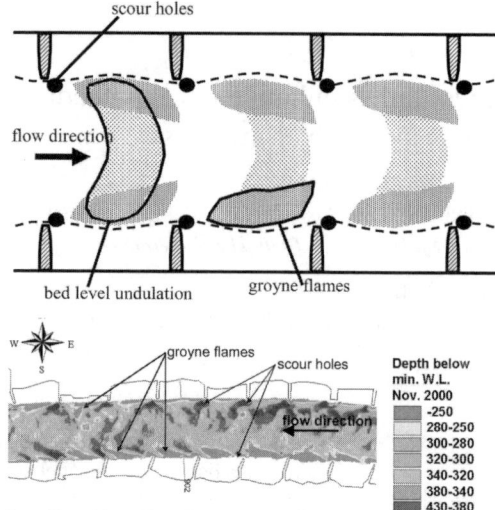

Figure 1.  Groynes-induced morphology: definition sketch (upper panal) and example from River Waal near Druten (lower panel).

spatial scale. The scour holes extend for several metres, the groyne flames extend for tens of metres and the streamwise bed level undulation extends for a maximum of the distance between groynes (see Figure 1).

The development of scour holes is mainly due to the increased turbulence intensity near the head and the formation of 3-dimensional horse-shoe vortices. The groyne flames mainly develop due to the large-scale 2-dimensional eddies. The streamwise bed undulation can be attributed to the acceleration and deceleration of the flow because of streamline convergence-divergence.

## 3  BED SHEAR STRESS ENHANCEMENT

### 3.1  Methodology

The bed shear stress enhancement is parameterised as follows:

Firstly, based on a detailed numerical simulation using HLES and a very fine grid, the bed shear stress is analysed using different approaches in order to identify an excess bed shear stress, $\Delta\tau$, due to turbulent fluctuations. The bed shear stress which can be reproduced on a coarse grid without using HLES is the base bed shear stress, $\tau_{bot0}$. The difference between $\tau_{bot1}$ and $\tau_{bot0}$ gives the excess bed shear stress due to the turbulent fluctuations.

Secondly, this excess bed shear stress is related to a hydrodynamic parameter that can be deduced from

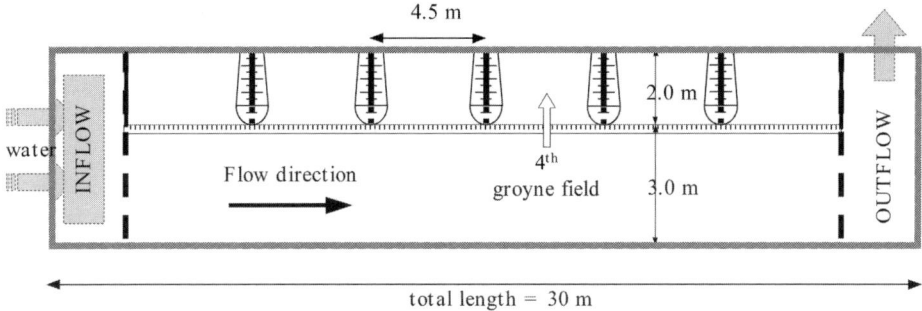

Figure 2. Experimental set-up comprising five identical groynes; Measurements were taken in the fourth groyne field.

an operational morphological model on a coarse grid without HLES. For a horizontal shear layer which develops along the interfacial line between the main channel and the groyne fields, the main driving parameter can be easily identified as the transverse velocity gradient across the mixing layer. Two parameters can be inferred as 2D generalisation of this transverse velocity gradient, viz. the vorticity, $\omega$, and the rate of deformation, $S_{xy}$, which we think is the most appropriate in this case. The effect that vorticity increases sediment entrainment by means of suction in the vortex centre rather than by bed shear stress enhancement (Yossef and De Vriend, 2004) is ignored in this paper.

Thirdly, we try to find a correlation between the rate of deformation and the excess bed shear stress.

Finally, this correlation is used as the basis of a bed shear stress enhancement term that accounts for the excess bed shear stress due to the steep velocity gradients. This enhancement term increases the sediment transport rate.

### 3.2 The analysis model

The numerical model used herein is based on a physical scale model experiment of a river with groynes on one side. The flume represents a schematised river reach based on typical dimensions of the River Waal at a scale of 1:40 (Figure 2). The measurements were carried out in the fourth groyne field (see Uijttewaal et al., 2002 for details).

The numerical computations were carried out using the depth-averaged option of Delft3D-FLOW in combination with HLES (Uittenbogaard and Van Vossen, 2003). To guarantee the highest possible accuracy, a very fine rectangular grid was applied, with typical grid sizes of 5 to 10 cm, resulting in about 68,000 active grid points. The grid was finest near the tips of the groynes.

Because of the fine grid, it was possible to model the groynes as a part of the bed topography. Figure 3 shows part of the model grid. The roughness of the main channel and the groynes is defined by a roughness height

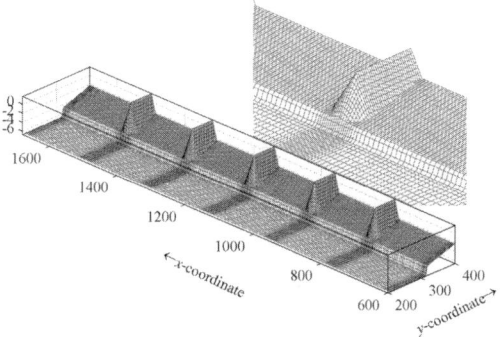

Figure 3. Model grid (2 times coarser) and coordinate system.

of 0.6 mm, based on measurements in the scale model. Friction by the lateral walls in the groyne field was accounted for using a so-called partial slip condition. The complete results of the computations and the comparison with the physical model results are given by Van Schijndel and Jagers (2003) and Uijttewaal and Van Schijndel (2004).

### 3.3 Computed excess bed shear stresses

The base bed shear stress, $\tau_{bot0}$, which is reproducible on a coarse grid without using HLES, can be represented by:

$$\tau_{bot0} = \rho_w \frac{g}{\overline{C}^2}\left(\overline{u_t}^2 + \overline{v_t}^2\right) \tag{1}$$

where $\rho_w$ = density of water, $g$ = acceleration of gravity, $C$ = Chézy coefficient, and $u_t$, $v_t$ = time-dependent velocity components in $x$- and $y$- directions.

The bed shear stress that includes time-dependent information can be expressed by:

$$\tau_{bot1} = \rho_w g \frac{1}{T}\sum_{t=1}^{T} \frac{u_t^2 + v_t^2}{\overline{C}^2} \tag{2}$$

1243

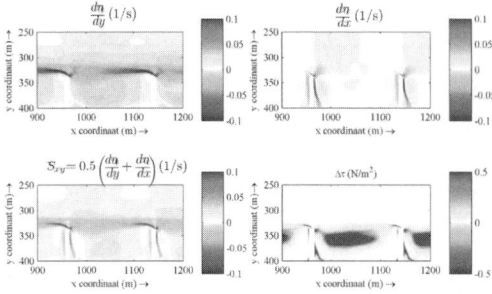

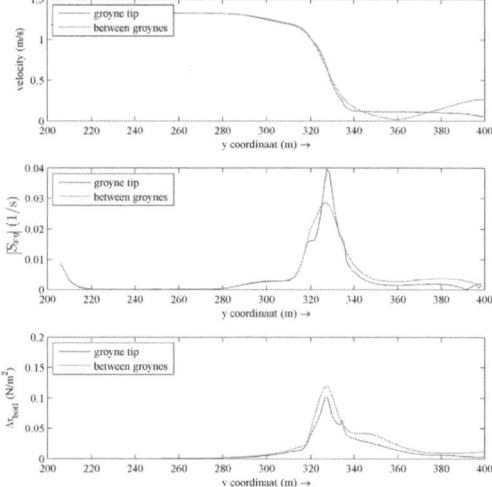

Figure 4. Two-dimensional distribution of velocity gradients, rate of deformation ($S_{xy}$) and excess bottom shear stress ($\Delta \tau$).

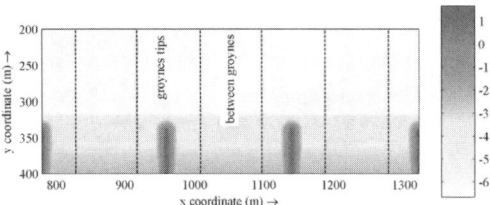

Figure 5. Bed level of the model with indication of two different zones used for spatial averaging (between groynes & near groyne tips).

The difference between $\tau_{bot1}$ and $\tau_{bot0}$ gives the excess bed shear stress due to the time-dependent velocity fluctuations:

$$\Delta \tau = \tau_{bot1} - \tau_{bot0} \qquad (3)$$

We attempt to correlate this excess bed shear stress, $\Delta \tau$, to the rate of deformation, $\overline{S}_{xy}$:

$$\overline{S}_{xy} = 0.5 \left( \frac{d\overline{u}_t}{dy} + \frac{d\overline{v}_t}{dx} \right) \qquad (4)$$

where $\overline{u}_t$ and $\overline{v}_t$ are the time-averaged velocity components. Figure 4 shows that the steep transverse velocity gradient is the dominant contribution to $S_{xy}$.

3.4  *Parameterisation of excess bed shear stress*

Now we attempt to correlate the excess bed shear stress due to time-dependent velocity fluctuations to the deformation as indicated earlier. For an aggregated view, we use two different zones to present spatially-averaged results. The first zone is the one between the groynes and the second zone is the one near the tips of the groynes (Figure 5).

Figure 6 presents the spatially-averaged transverse distribution of the time-averaged velocity, deformation

Figure 6. Spatially-averaged transverse distribution of the time-averaged velocity (upper panel), deformation $S_{xy}$ (central panel) and excess bed shear stress (lower panel).

rate and excess bed shear stress. We can see that the peak in the deformation matches the area of the steep velocity gradient.

The same peak can be observed, at the same location, in the distribution of $\Delta \tau$, which leads to expecting a good correlation between $\Delta \tau$ and $S_{xy}$.

The bed shear stress can be represented in terms of a mean and fluctuating depth-averaged flow velocity components as follows:

$$\tau = \frac{\rho g}{C^2} U^2 = \frac{\rho g}{C^2} \left( \overline{u} + u' \right)^2 = \frac{\rho g}{C^2} \left( \overline{u}^2 + 2\overline{u} u' + u'^2 \right)$$

where $U =$ instantaneous flow velocity, $\overline{u} =$ time-averaged flow velocity and $u' =$ fluctuating component of flow velocity.

Accordingly, the time-averaged bed shear stress can be written in the form:

$$\overline{\tau}_{tot} = \frac{\rho g}{C^2} \left( \overline{u}^2 + \overline{u'^2} \right) = \tau_0 \left( 1 + \left( \frac{u'}{\overline{u}} \right)^2 \right) \qquad (5)$$

Hence $\displaystyle \frac{\Delta \tau}{\tau_0} = \overline{\left( \frac{u'}{\overline{u}} \right)^2}$ $\qquad (6)$

Jagers et al. (2005) indicate that:

$$\overline{\left( \frac{u'}{\overline{u}} \right)^2} \propto \left( \frac{\tau_{hor}}{\rho \overline{u}^2} \right)^2 \qquad (7)$$

1244

where the horizontal shear stress can be expressed as:

$$\tau_{hor} = \rho \nu_{turb} \left( \frac{\partial \overline{v}}{\partial x} + \frac{\partial \overline{u}}{\partial y} \right) = \rho \nu_{turb} \left( 2 S_{xy} \right)$$

Consequently, $\dfrac{\Delta \tau}{\tau_0} \propto \left( \dfrac{\nu_{turb} S_{xy}}{\overline{u}^2} \right)^2 \qquad$ (8)

Van Prooijen (2004) reports that the eddy viscosity due to the large eddies can be represented in the form $\nu_{turb} \propto \delta^2 S_{xy}$. Hence

$$\frac{\Delta \tau}{\tau_0} \propto \left( \frac{\delta^2 S_{xy}^2}{\overline{u}^2} \right)^2 \qquad (9)$$

where $\delta$ is the mixing layer width. This mixing layer width cannot be defined easily, but several researchers indicate that there is a linear relation between the mixing layer width and the flow depth (see for example Van Prooijen, 2004). Accordingly, Eq. 9 can be written in the form:

$$\frac{\Delta \tau}{\tau_0} \propto \left( \frac{h^2}{\overline{u}^2} S_{xy}^2 \right) \qquad (10)$$

Multiplying both sides by $\tau_0 = \frac{\rho g}{C^2} \overline{u}^2$, and replacing the proportionality sign with a linear relation using a newly defined coefficient $\alpha_{shear}$, yields:

$$\Delta \tau = \alpha_{shear} \cdot \underbrace{\left( \rho \frac{g}{C^2} h^2 \right)}_{\text{for proprer scaling and dimension}} \cdot S_{xy}^2 \qquad (11)$$

An alternative scaling parameter can be deduced from the direct analogy with eddy viscosity; it takes the form:

$$\Delta \tau = \alpha_{shear2} \cdot \underbrace{\left( \rho h u \right)}_{\text{for proprer scaling and dimension}} \cdot S_{xy} \qquad (12)$$

Analysing the results of the detailed simulation in terms of Eqs. 11 and 12 helps in establishing the most appropriate scaling parameter. Figure 7 (left column) presents the excess bed shear stress versus the rate of deformation for all the points in the region between the groynes (lower left panel) and in the region near the tips of the groynes (upper left panel). In Figure 7 (right column) the excess bed shear stress is normalised by the base bed shear stress.

Figure 8 aggregates the data for the two zones of Figure 5 by spatially averaging the data from Figure 7. The plots indicate that the normalised excess bed shear stress does not correlate well with $S_{xy}$. This can be

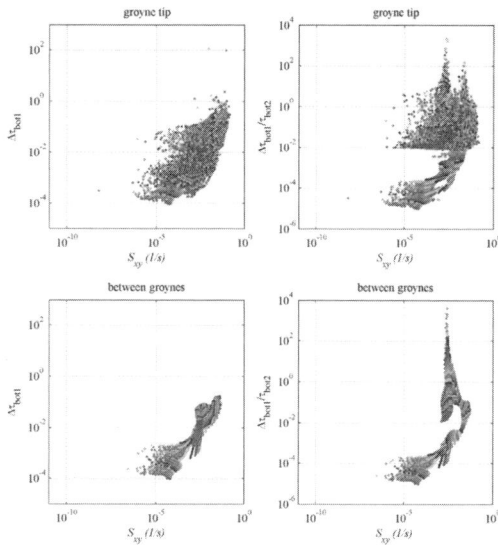

Figure 7. Relation between deformation, $S_{xy}$, and absolute excess bed shear stress, $\Delta \tau$ (left column) and normalised excess bed shear stress, $\Delta \tau / \tau_0$ (right column); all data.

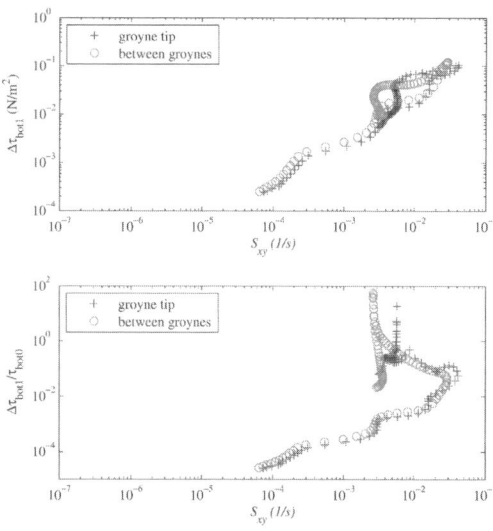

Figure 8. Relation between deformation, $S_{xy}$, and absolute excess bed shear stress, $\Delta \tau$ (upper panel) and normalised excess bed shear stress, $\Delta \tau / \tau_0$ (lower panel); spatially-averaged values.

explained from the contribution from areas with low flow velocities that produce high values of $\Delta \tau / \tau_0$ irrespective of the values of $\Delta \tau$. We conclude that scaling the excess bed shear stress with the base bed shear stress is not appropriate.

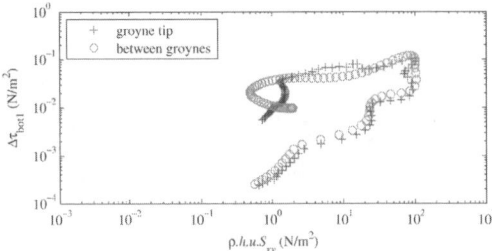

Figure 9. Relation between $(\rho.h.u.S_{xy})$ and excess bed shear stress $\Delta\tau$; spatially-averaged values.

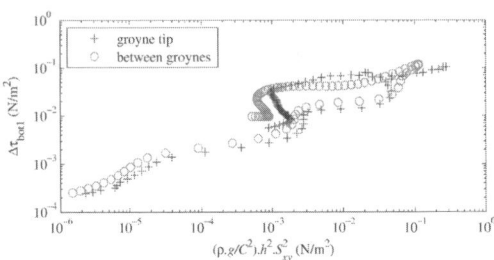

Figure 10. Relation between $(\rho.g.h^2/C^2).S_{xy}^2$ and the excess bed shear stress $\Delta\tau$ as implemented in Delft3D; spatially-averaged values.

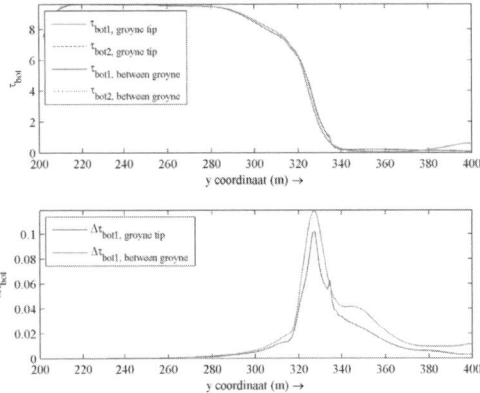

Figure 11. Spatially-averaged transverse distributions of time-averaged bed shear stress (upper panel) and excess bed shear stress (lower panel).

Figure 9 and Figure 10 indicate that a scaling based on Eq. 11 (Figure 10) is preferable over Eq. 12 (Figure 9). Accordingly, the relation given in Eq. 11 has been implemented in Delft3D. We must note, nonetheless, that the excess shear stress due to temporal fluctuations is an order of magnitude less than the time-averaged bed shear stress (see Figure 11).

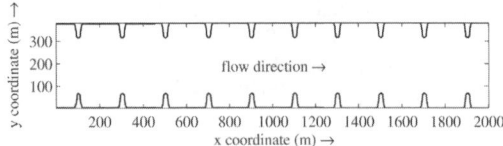

Figure 12. Model layout for prototype-scale test case.

## 4 APPLICATION TO A TEST CASE

### 4.1 The test model

To test and illustrate the effect of the newly developed bed shear stress enhancement parameter in a case on prototype scale we consider a case of river groynes subject to a constant low discharge. We apply a prototype condition of a schematised straight river reach with groynes on both sides.

The channel dimensions are chosen such that they are representative for the River Waal. Figure 12 shows the model layout. The modelled reach is 2 km long with a bed slope of 10 cm/km. The groyne length and spacing are 60 m and 200 m respectively. The width of the main channel is 260 m, the bed level difference between the main channel and the groyne fields is 2 m, and the beach slope inside the groyne fields is 1:20. The upstream boundary condition is a constant discharge of 1500 m³/s and the downstream boundary condition is a constant water level. The sediment size was chosen at 0.5 mm for the whole model. The sediment transport field is computed using the formulation of Van Rijn, utilising two separate expressions for suspended load and bed load .(Van Rijn, 1984a, b). We used a computational time step of 3.0 seconds and a morphological acceleration factor of 50 (see Roelvink, 2006, for the use of the morphological acceleration factor).

The functionality of the bed shear stress enhancement parameter was tested in a model with a fine grid and a model with a coarse grid. The fine-grid model consisted of $402 \times 78$ grid cells ($x \times y$ directions) with a uniform grid size of $5 \times 5\,\text{m}^2$ (see Figure 13, upper panel). The same model has been used earlier to investigate the possibility of reproducing the groynes-induced morphological features using Delft3D-MORSYS (see Yossef and Klaassen, 2002; Yossef and De Vriend, 2006). The coarse-grid model consisted of $82 \times 17$ grid cells (see Figure 13, lower panel).

In both cases the groynes were represented using bed topography. However, in the case of the coarse-grid model, thin dams were introduced on top of the groyne crests to ensure that no overflow is taking place. To guarantee the stability of the groynes, a fixed-layer was introduced in the groynes region. The same parameters were applied for the two models.

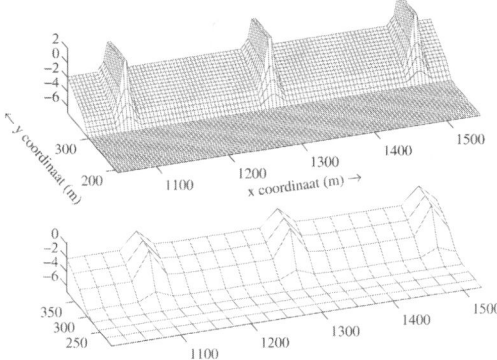

Figure 13. An excerpt from the model showing two groyne fields; comparison between the fine-grid and the coarse-grid models.

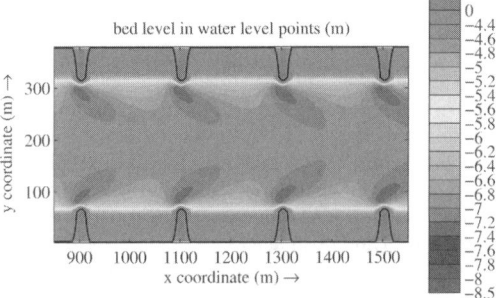

Figure 14. Bed topography after a simulation of 50 days, reference case.

## 4.2 Morphological results

Delft3D MOR was applied for a simulation of 50 days. The model was run for a reference case, i.e. without the application of the bed shear stress enhancement parameter and for several values of $\alpha_{\text{shear}}$. Herein we compare the results from the reference case with that of the case of $\alpha_{\text{shear}} = 5.0$.

The resulting morphological pattern for the reference case is presented in Figure 14, where we can observe the development of scour holes and groyne flames. The depth of the scour holes reaches nearly 1.0 m below the original depth, and the peak height of the groyne flames amounts to 0.5 m above the initial bed level. In the coarse-grid model the depth of the scour hole is less pronounced but we can still consider the result acceptable. The bed topography in Figure 14 is given as a reference. For the comparison between the results, we use the cumulative erosion-deposition pattern that represents the difference between the final and the initial bed levels.

The comparison between the reference model and the fine-grid model using the enhanced bed shear stress is given in Figure 15. We can see that enhancing

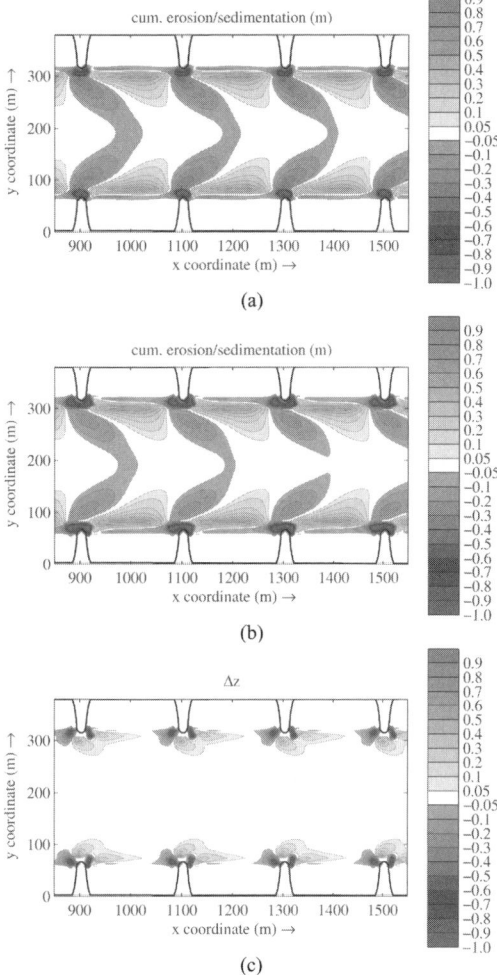

Figure 15. Results from the fine-grid model: (a) reference model, (b) model with enhanced bed shear stress, (c) difference.

the bed shear stress affects the formation of both the scour holes and the groyne flames. It increases the size and the maximum depth of scour. However, slightly away from the tip of the groyne, in the $y$-direction, the depth of the scour hole is slightly reduced (Figure 15c).

The area just upstream of the groyne experiences additional scour. Apart from a slight increase in deposition volume, the shape of the groyne flames hardly changes. Generally speaking, the effect of using shear stress enhancement is very much confined to the direct vicinity of the groynes.

The comparison between the reference model and the model using $\alpha_{\text{shear}} = 5.0$ for the coarse-grid model is given in Figure 16. We can see that the changes are

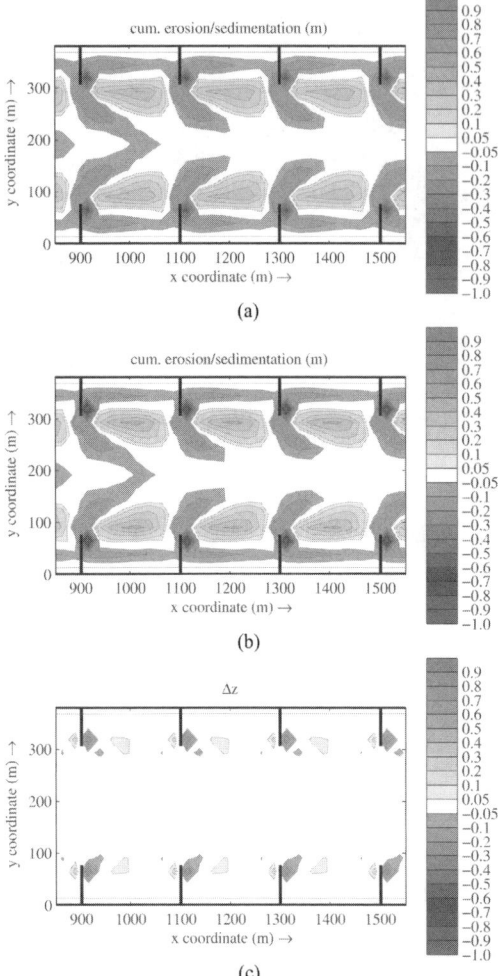

(a)

(b)

(c)

Figure 16. Results from the coarse-grid model (a) reference model, (b) model with enhanced bed shear stress, (c) difference.

rather minor in this case. Still, the scour hole is slightly deeper. The groyne flames hardly change at all.

### 4.3  *Discussion*

Figure 17 and Figure 18 show that the bed shear stress enhancement parameter clearly affects the mixing layer in both cases. The fine grid exhibits a stronger effect than the coarse grid. The fine-grid effect is concentrated near the tips of the groynes, at both upstream and downstream sides, whereas the coarse-grid effect is less significant and more evenly distributed.

The effect of the bed shear stress enhancement is an increased bed shear stress everywhere in the mixing layer, which directly increases the sediment transport

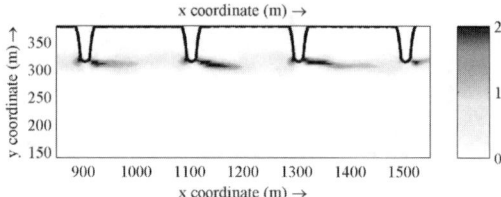

Figure 17. Computed excess bed shear stresses (N/m²) on fine grid.

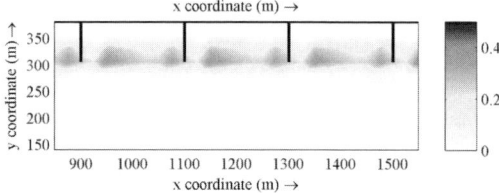

Figure 18. Computed excess bed shear stresses (N/m²) on coarse grid.

rate in this region. However, bed changes result from gradients in transport. The effect of enhancing the bed shear stress on the sediment transport gradient is most significant near the tips of the groynes, where it causes an increase in transport gradient. This effect is apparently grid-dependent as the comparison between Figure 17 and Figure 18 reveals. In the fine-grid model, the scour in the direct vicinity of the groynes significantly increases compared to hardly any change in the case of the coarse grid.

## 5  CONCLUSIONS

Using information from a detailed flow simulation near groynes, we were able to deduce a parametric relation between the rate of deformation and the excess bed shear stress due to time-dependent turbulence characteristics. Accordingly, a relation was developed to introduce an excess bed shear stress using a bed shear stress enhancement parameter, $\rho \cdot g \cdot h^2 \cdot S_{xy}^2/C^2$, that can be tuned using the proportionality coefficient ($\alpha_{shear}$). This relation has been implemented in Delft3D and tested on a model of a river with groynes on both sides. Grid resolution effects have been tested as well.

The morphological features near groynes are reproduced better by a fine-grid model, irrespective of applying an enhanced bed shear stress or not. An enhanced bed shear stress had little effect on the results of the coarse-grid model, but increased the scour depth near the tips of the groynes in the fine-grid model.

For proper reproduction of groynes-induced morphological features, the use of a fine-grid model is found to be more important than the enhancement of the bed shear stress to account for turbulent fluctuations. The results still require comparison with field observations or laboratory measurements.

ACKNOWLEDGEMENTS

This research has been funded by Rijkswaterstaat-RIZA. We thank Dr. H. van der Klis and Dr. C.J. Sloff for their contributions to the study.

REFERENCES

Jagers, B., Mosselman, E. & Van Schijndel, S. (2005), Eenvoudige modellering van bodemschuifspanningen bij kribben; Verkenning van haalbaarheid. Rapport Q4049.00, WL | Delft Hydraulics (in Dutch).

Krebs, M., Zanke, U. & Mewis, P. (1999). "Hydro-Morphodynamic modelling of groin fields" In: Proc. 28th IAHR congress, Graz, Austria.

Ouillon, S. & Dartus, D. (1997). "Three-dimensional computation of flow around groyne" Journal of Hydraulic Engineering, ASCE, 123(11), 962–970.

Peng, J., Kawahara, Y. & Tamai, N. (1997). "Numerical analysis of three-dimensional turbulent flows around submerged groynes" In: Proc. 27th IAHR congress, San Francisco, USA.

Roelvink, J. A. (2006). "Coastal morphodynamic evolution techniques" Coastal Engineering, 53(2–3), 277–287.

Tingsanchali, T. & Maheswarn, S. (1990). "2-D depth averaged flow computation near groyne" Journal of Hydraulic Engineering, ASCE, 116(1), 71–85.

Uijttewaal, W. S. J. (1999). "Groyne field velocity patterns determined with particle tracking velocimetry" In: Proc. 28th IAHR congress, Graz, Austria.

Uijttewaal, W. S. J., Berg, M. H. & Van Der Wal, M. (2002). "Experiments on physical scale models for submerged and non-submerged groynes of various types" In: River Flow 2002 – Proc. Int. Conf. on Fluvial Hydraulics, Louvain-la-Neuve, Belgium, 377–383.

Uijttewaal, W. S. J. & Van Schijndel, S. (2004). "The complex flow in groyne fields: numerical modelling compared with experiments" In: River Flow 2004 – Proc. 2nd Int. Conf. on Fluvial Hydraulics, Naples, Italy, 1331–1338.

Uittenbogaard, R. E. & Van Vossen, B. (2003). "Subgrid-scale model for quasi-2D turbulence in shallow water" In: Proc. Int. Symp. on Shallow Flows, Delft, the Netherlands, 169–176.

Van Prooijen, B. C. (2004). Shallow mixing layers, Ph.D. Thesis, Delft University of Technology, Delft, the Netherlands.

Van Rijn, L. C. (1984a). "Sediment Transport, Part I: Bed Load Transport" Journal of Hydraulic Engineering, ASCE, 110(10), 1431–1456.

Van Rijn, L. C. (1984b). "Sediment Transport, Part II: Suspended Load Transport" Journal of Hydraulic Engineering, ASCE, 110(10), 1613–1641.

Van Schijndel, S. & Jagers, B. (2003). "Complex flow around groynes, computations with Delft3D in combination with HLES" In: Proc. Int. Symp. on Shallow Flows, Delft, the Netherlands, 213–219.

Yossef, M. F. M. & Klaassen, G. J. (2002). "Reproduction of groynes-induced river bed morphology using LES in a 2-D morphological model" In: River Flow 2002 – Proc. Int. Conf. on Fluvial Hydraulics, Louvain-la-Neuve, Belgium, 1099–1108.

Yossef, M. F. M. & De Vriend, H. J. (2004). "Mobile-bed experiments on the exchange of sediment between main channel and groyne fields" In: River Flow 2004 – Proc. 2nd Int. Conf. on Fluvial Hydraulics, Naples, Italy, 127–133.

Yossef, M. F. M. & De Vriend, H. J. (2006). "Numerical modelling of groynes induced morphology" submitted for publication.

*River, Coastal and Estuarine Morphodynamics: RCEM 2007 – Dohmen-Janssen & Hulscher (eds)*
*© 2008 Taylor & Francis Group, London, ISBN 978-0-415-45363-9*

# Numerical simulation on local scouring around a spur dike using equilibrium and non-equilibrium sediment transport models

S. Onda & T. Hosoda
*Dept. of Urban Management, Kyoto University, Kyoto, Japan*

I. Kimura
*Dept. of Civil and Environmental Engineering, Matsue National College of Technology, Matsue, Japan*

M. Iwata
*CTI Engineering Co., Ltd., Osaka, Japan*

ABSTRACT:   Around a spur dike, local scouring occurs and it is important to estimate not only the maximum scour depth, but also the bed geometry around it. In this study, local scouring around the dike is simulated using various turbulence and sediment transport models. The standard and non-linear $k$-$\varepsilon$ models are employed as the turbulence model, with equilibrium and non-equilibrium sediment transport models. To evaluate the model performance, calculated results are compared with the results of experiments conducted by Michiue and Hinokidani (1992) and the applicability of all the models is assessed by examining the temporal variation of the flow near bed and the bed variation. It is shown that temporal variation of the bed elevation in the initial scouring process and the bed geometry in the equilibrium stage are different, comparing the equilibrium and non-equilibrium sediment transport models.

## 1  INTRODUCTION

Around a spur dike, local scouring occurs and the estimation of maximum scour depth is important in terms of disaster mitigation of structures. Recently, the characteristics of dikes have been found to produce aquatic habitat. Therefore, it is important to predict accurately not only the maximum scour depth, but also bed geometry around dikes, both influencing ecological issues.

Numerous experimental studies concerning the flows and bed deformation around dikes have been conducted. Ikeda et al. (2000) studied the momentum exchange rate between the main flow and the spur dike regions around the impermeable spur dikes. Ishigaki et al. (2004) and Tominaga & Matsumoto (2006) pointed out that the angle of dike affects the position of scouring and deposition.

On the other hand, a number of computations of flow and bed deformation have also been performed, such as flow around a spur dike (Ikeda et al. 2000, Muneta & Shimizu 1994, Kimura et al. 2002, 2004); flow and bed deformation around spur dike (Fukuoka et al. 1998, Michiue & Hinokidani 1992, Fukuoka et al. 2000, Zhang et al. 2006, Yossef & Rupprecht 2006). Flow models are classified as follows: (1) 2D model (Ikeda et al. 2000): (2) a quasi 3D model,

including the local transformation of velocity distribution or the force due to dikes (Muneta & Shimizu 1994, Fukuoka et al. 1998) and (3) fully 3D model (Michiue & Hinokidani 1992, Fukuoka et al. 2000, Kimura et al. 2002, 2004, Zhang et al. 2006, Yossef & Rupprecht 2006). For the bed deformation, the equilibrium sediment transport models are generally applied. In their studies, the applicability of numerical models is examined through the comparison with experimental results. Kimura et al. (2004) simulated the flow around the dikes under various hydraulic conditions, and investigated the effects of inclination angle and ratio of the interval length of two dikes to the length. It was reported that flow patterns are in close agreement with observations, by using fully 3D models. However, in many studies, the maximum scour depth is under-predicted, and the position of scouring and deposition could not be reproduced well. This is because the turbulence, bed deformation models and the calculations of sediments sliding are not employed accurately. Moreover, in some models, the assumption of hydrostatic pressure distribution is used; but it is not applicable in such cases.

Recently, Nagata et al. (2005) applied the fully 3D computational model with the nonlinear $k$-$\varepsilon$ turbulence model for simulation of flow and bed deformation around a spur dike and bridge pier. They considered the equation of motion of sediment particles in the non-equilibrium sediment transport model. They also showed that the model can predict the bed topography better than the previous models; but, since they focused on the bed geometry in the equilibrium stage, the temporal variation of scouring processes under the various turbulence and bed deformation models was not examined in details.

In this study, local scouring around the dike is simulated, using various turbulence and sediment transport models, with intention to gain a better understanding of their applicability. The numerical results are compared with the experiments conducted by Michiue & Hinokidani (1992) and the applicability of all the models is verified by examining the temporal variation of flow near bed and the bed variations.

## 2 NUMERICAL MODEL

### 2.1 Basic equations of flow model

The Reynolds averaged 3D flow model in generalized curvilinear moving coordinate system are described as follows:

[Continuity equation]

$$\frac{\partial}{\partial \xi^i}\left(\frac{U^i}{J}\right) = 0 \tag{1}$$

[Momentum equation]

$$\frac{\partial}{\partial t}\left(\frac{U^i}{J}\right) + \frac{\partial}{\partial \xi^j}\left(\frac{\left(U^j - U_G^j\right)U^i}{J}\right)$$
$$- \frac{\left(U^j - U_G^j\right)\mathbf{u}}{J}\cdot\frac{\partial}{\partial \xi^j}\left(\nabla \xi^i\right) - \frac{\mathbf{u}}{J}\cdot\frac{\partial}{\partial t}\left(\nabla \xi^i\right)$$
$$= -\frac{g^{ij}}{\rho J}\frac{\partial p}{\partial \xi^j} + \frac{\mathbf{f}}{J}\cdot\nabla \xi^i + \frac{1}{J}\frac{\partial \xi^j}{\partial x_m}\frac{\partial \xi^i}{\partial x_l}\frac{\partial}{\partial \xi^j}\left(\tau_{lm} - \overline{u_l'u_m'}\right) \tag{2}$$

[k-equation]

$$\frac{\partial}{\partial t}\left(\frac{k}{J}\right) + \frac{\partial}{\partial \xi^j}\left(\frac{\left(U^j - U_G^j\right)k}{J}\right) = \frac{-\overline{u_i'u_j'}}{J}\frac{\partial \xi^l}{\partial x_j}\frac{\partial u_i}{\partial \xi^l}$$
$$- \frac{\varepsilon}{J} + \frac{\partial}{\partial \xi^l}\left(\left(v + \frac{D_t}{\sigma_k}\right)\frac{g^{lm}}{J}\frac{\partial k}{\partial \xi^m}\right) \tag{3}$$

[ε-equation]

$$\frac{\partial}{\partial t}\left(\frac{\varepsilon}{J}\right) + \frac{\partial}{\partial \xi^j}\left(\frac{\left(U^j - U_G^j\right)\varepsilon}{J}\right) = C_{\varepsilon 1}\frac{\varepsilon}{k}\frac{\left(-\overline{u_i'u_j'}\right)}{J}\frac{\partial \xi^l}{\partial x_j}\frac{\partial u_i}{\partial \xi^l}$$
$$- \frac{C_{\varepsilon 2}}{J}\frac{\varepsilon^2}{k} + \frac{\partial}{\partial \xi^l}\left(\left(v + \frac{D_t}{\sigma_\varepsilon}\right)\frac{g^{lm}}{J}\frac{\partial \varepsilon}{\partial \xi^m}\right) \tag{4}$$

in which

$$\nabla \equiv \left(\partial/\partial x_1, \partial/\partial x_2, \partial/\partial x_3\right), \quad g^{ij} \equiv \nabla g^i \cdot \nabla g^j \tag{5a}$$

$$U^i \equiv \left(\partial \xi^i/\partial x_j\right)u_j, \quad U_G^i \equiv \left(\partial \xi^i/\partial x_j\right)u_{jG} \tag{5b}$$

$$\tau_{ij} \equiv v\left(\frac{\partial \xi^m}{\partial x_j}\frac{\partial u_i}{\partial \xi^m} + \frac{\partial \xi^m}{\partial x_i}\frac{\partial u_j}{\partial \xi^m}\right) \tag{5c}$$

where $t$ = time; $(x_1, x_2, x_3)$ = Cartesian coordinates ($x_3$ indicates the vertical coordinate); $(\xi_1, \xi_2, \xi_3)$ = generalized curvilinear coordinates; $J$ = Jacobian of transformation; $g^{ij}$ = contravariant metric tensors; $U^i$ = contravariant components of flow velocity vector; $U_G^i$ = contravariant components of grid velocity vector; $\mathbf{u}$ = velocity vector $[ = (u_1, u_2, u_3)]$; $u_i$ = velocity components in Cartesian coordinates; $u_{iG}$ = grid velocity components in Cartesian coordinate; $p$ = pressure; $\mathbf{f}$ = gravitational vector $[ = (0, 0, g)]$; $g$ = gravitational acceleration; $\tau_{ij}$ = viscous stress tensors; $-\overline{u_i'u_j'}$ = Reynolds stress tensors; $k$ = turbulent kinetic energy; $\varepsilon$ = dissipation rate; $\rho$ = fluid density; $v$ = kinematic viscosity of fluid and $i, j, k, l$ denote the value of 1, 2 and 3.

A second order nonlinear k-ε model is applied as the turbulence model, to simulate the complex turbulent flows with separation. The applicability of this model is already examined, by simulating the flow around a square cylinder (Kimura & Hosoda 1999) and a spur dike (Kimura et al. 2002, 2004, Nagata et al. 2005).

$$-\overline{u_i'u_j'} = D_t\left(\frac{\partial \xi^m}{\partial x_j}\frac{\partial u_i}{\partial \xi^m} + \frac{\partial \xi^m}{\partial x_i}\frac{\partial u_j}{\partial \xi^m}\right) - \frac{2}{3}k\delta_{ij}$$
$$- \frac{k}{\varepsilon}D_t\sum_{\beta=1}^{3}C_\beta\left(S_{\beta ij} - \frac{1}{3}S_{\beta\alpha\alpha}\delta_{ij}\right) \tag{6}$$

$$D_t = C_\mu k^2/\varepsilon \tag{7a}$$

$$S_{1ij} = \frac{\partial u_i}{\partial x_r}\frac{\partial u_j}{\partial x_r}, \quad S_{2ij} = \frac{1}{2}\left(\frac{\partial u_r}{\partial x_i}\frac{\partial u_j}{\partial x_r} + \frac{\partial u_r}{\partial x_j}\frac{\partial u_i}{\partial x_r}\right)$$
$$S_{3ij} = \frac{\partial u_r}{\partial x_i}\frac{\partial u_r}{\partial x_j} \tag{7b}$$

where $D_t$ = eddy viscosity coefficient. Equations (6) and (7) are equivalent to the expressions by Gatski & Speziale (1993). Other coefficients $C_1, C_2, C_3$ and $C_\mu$ are described as follows:

$$C_1 = \frac{0.4}{1+0.01M^2}, C_2 = 0.0, C_3 = \frac{-0.13}{1+0.01M^2} \tag{8a}$$

$$M = \max(S, \Omega) \tag{8b}$$

1252

$$S = \frac{k}{\varepsilon}\sqrt{\frac{1}{2}\left(\frac{\partial u_i}{\partial x_j}+\frac{\partial u_j}{\partial x_i}\right)^2}, \Omega = \frac{k}{\varepsilon}\sqrt{\frac{1}{2}\left(\frac{\partial u_i}{\partial x_j}-\frac{\partial u_j}{\partial x_i}\right)^2} \quad (8c)$$

$$C_\mu = \min\left[0.09, 0.3/(1+0.09M^2)\right] \quad (9)$$

Equations (8) were adjusted through the comparison of the turbulent intensity distribution in a simple shear flow (Kimura & Hosoda 2000). Equation (9) was determined in order to satisfy the realizability in a simple shear flow and the singular points in both 2D and 3D flow fields (Hosoda et al. 2001).

In this study, a standard $k$-$\varepsilon$ model is also applied, in order to compare numerical results with nonlinear $k$-$\varepsilon$ model.

### 2.2 Bed deformation model

#### 2.2.1 Non-equilibrium sediment transport model

The equilibrium and non-equilibrium sediment transport models are applied for calculation of bed variation to examine their applicability. Firstly, the non-equilibrium sediment transport model (Nagata et al. 2005) is described. This model consists of 4 steps: (1) evaluation of pick-up rate and volume of pick-up sediment; (2) calculation of the trajectory of sediment movements by solving an equation of motion of sand particles; (3) calculation of sediment deposition volume; and (4) a temporal variation of bed elevation.

The pick-up rate $p_s$ is evaluated by the following equation proposed by Nakagawa et al. (1986), in which the effect of local bed slope on sediment motion is included.

$$p_s\sqrt{\frac{d}{(\sigma/\rho-1)g}} = F_0 G_* \tau_* \left(1-\frac{k_p \Phi \tau_{*c}}{\tau_*}\right)^{m_p}$$

$$G_* = (\cos\Psi + k_L \mu_s)/(1+k_L \mu_s) \quad (10)$$

$$\Phi = \left(\frac{\mu_s \cos\theta_b - \sin\theta_b \cos\alpha}{\cos\Psi + k_L \mu_s}\right)\left(\frac{1+k_L \mu_s}{\mu_s}\right)$$

where $p_s$ = pick-up rate; $d$ = diameter of sediment particle; $\sigma$ = density of sediment; $\tau_*$ = dimensionless tractive stress[ $= u_*^2/(\sigma/\rho - 1)gd$]; $u_*$ = bed shear velocity; $\tau_{*c}$ = dimensionless critical tractive stress proposed by Iwagaki (1956); $G_*$ = coefficient of directional deviation between velocity near bed and sediment movement vectors; $\Phi$ = coefficient of side bank slope; $\mu_s$ = static friction factor ( = 0.7); $k_L$ = ratio of lift force to drag force ( = 0.85); $\theta_b$ = local bed slope angle; $\psi$ = angle between velocity vector near bed and sediment movement direction; $\alpha$ = angle between direction of maximum local bed slope and sediment movement direction; and $F_0$, $k_p$, and $m_p$ = constants (=0.03, 0.7, and 3, respectively).

Using the above pick-up rate, the volume of sediment pick-up, per unit time, ($V_p$) can be calculated as follows:

$$V_p = \frac{A_3 d}{A_2} p_s S_p, \quad S_p = \frac{\partial \xi^3/\partial x_3}{J} \quad (11)$$

where $A_2$, $A_3$ = shape characteristics of sand grain for 2 and 3 dimensional geometrical properties (=$\pi/4$, $\pi/6$); $S_p$ = area of bed surface mesh.

To calculate the sediment movement, the unit vectors, $\mathbf{p}_{b1}$ and $\mathbf{p}_{b2}$ are defined, which are parallel to a local bed surface on $(x_1 - x_3)$ and $(x_2 - x_3)$ planes, respectively.

$$\mathbf{p}_{b1} = (\cos\theta_{b1}, 0, \sin\theta_{b1}), \quad \mathbf{p}_{b2} = (0, \cos\theta_{b2}, \sin\theta_{b2}) \quad (12)$$

where $\theta_{b1}$, $\theta_{b2}$ = angles of bed inclination in the $x_1$ and $x_2$ direction, respectively. To calculate the trajectory of sediment movements, the momentum equation of sand particle in the $\mathbf{p}_{bi}$ ($i$ = 1,2) direction is solved.

$$\rho\left(\frac{\sigma}{\rho}+C_m\right)A_3 d^3 \frac{du_{sedi}}{dt} = D_i + W_i - F_i \quad (13)$$

$$D = C_D \rho (u_{bi} - u_{sedi})^2 c_e A_2 d^2 / 2 \quad (14a)$$

$$W = (\sigma - \rho)g A_3 d^3 \quad (14b)$$

$$F = \mu_k \left(W \cos\theta_{b1} \cos\theta_{b2}/\sin\theta_p - k_L D\right) \quad (14c)$$

where $C_m$ = coefficients of added mass (=0.5); $u_{sedi}$ = sediment particle velocity in the $\mathbf{p}_{bi}$ direction; $D$ = fluid drag force on a sediment particle; $W$ = submerged weight of sediment particle; $F$ = friction force between sediment particles and bed; $D_i$, $W_i$, $F_i$ = $\mathbf{p}_{bi}$ direction components of $D$, $W$ and $F$; $u_{bi}$ = fluid velocity near bed at the distance of $k_d d$ from bed; $k_d$ = constant (=0.8); $C_D$ = drag coefficients = (0.4); $\mu_k$ = kinetic friction factor (=0.35); and $\theta_p$ = angle between $\mathbf{p}_{b2}$ and $\mathbf{p}_{b1}$.

Using the vector of sediment particle velocity $u_{sedi}$, the position vector of sediment particle ($\mathbf{p}_{sed(n)}$) after $n$-th step since being picked up, and distance of sediment movement $s_{(n)}$ can be calculated.

$$\mathbf{p}_{sed(n)} = \mathbf{p}_{sed(n-1)} + \Delta t \cdot \mathbf{u}_{sed} \quad (15)$$

$$s_{(n)} = \sum \Delta t \left|\mathbf{u}_{sed(n)}\right| \quad (16)$$

The deposition volume $V_{d(j,n)}$ of sediment moving from point $j$ after $n$-th step can be obtained, using the probability density function of step length.

$$V_{d(j,n)} = V_{p(j)} f_s (s_{(n)}) \mathbf{u}_{sed} \Delta t \quad (17a)$$

$$f_s\left(s_{(n)}\right)=\frac{1}{\lambda}\exp\left(-\frac{s_{(n)}}{\lambda}\right) \tag{17b}$$

where $V_{p(j)}=$ volume of pick-up sediment at point $j$; $f_s(s_{(n)})=$ probability density function of step length; and $\lambda=$ averaged step length. To estimate $\lambda$, the equation proposed by Einstein (1950) is adopted. It should be noted that while the position of sediment pick-up is center of mesh, a point of sediment deposition does not coincide with pick-up position.

Then, a position of deposition can be found by using the position vector (Eq. (15)), and the total deposition volume is distributed to each grid point. Using the pick-up and deposition volumes calculated by Equations (11) and (17), temporal variation in bed elevation can be obtained.

$$\frac{\partial z_b}{\partial t}=\frac{A_1 A_2}{A_3}\frac{\left(V_d-V_p\right)}{S_d} \tag{18}$$

where $z_b=$ bed elevation; $A_1=$ shape coefficient of sand grain for 1 dimensional geometrical properties $(=1.0)$; and $S_d=$ area of bed surface mesh where sediment is deposited.

### 2.2.2 Equilibrium sediment transport model
In equilibrium sediment transport model, the empirical bed load transport formula, such as Meyer-Peter Muller equation is usually used. However, in order to make the critical value of sediment movement the same in the equilibrium and non-equilibrium sediment transport model, such a kind of empirical formula is not used in this study. Instead, the bed load transport flux $q_{Bs}$ is calculated by using the pick-up rate (Eq. (10)). Then, the bed load flux $q_{Bn}$ in transversal direction (Hasegawa 1983) is obtained (Eq. (19b)). The temporal variation of bed elevation is calculated by Equation (19c), after the bed load fluxes (Eqs. (19a) and (19b)) are transformed to the ones in the generalized curvilinear coordinates.

$$q_{Bs}=\frac{A_3}{A_2}p_s\Lambda d \tag{19a}$$

$$q_{Bn}=q_{Bs}\left(-\frac{\partial z_b}{\partial n}\sqrt{\frac{\tau_{*c}}{\mu_s\mu_k\tau_*}}\right) \tag{19b}$$

$$\frac{\partial}{\partial t}\left(\frac{z_b}{J}\right)+\frac{1}{1-\delta}\left\{\frac{\partial}{\partial\xi}\left(\frac{q_B^\xi}{J}\right)+\frac{\partial}{\partial\eta}\left(\frac{q_B^\eta}{J}\right)\right\}=0 \tag{19c}$$

where $\delta=$ porosity.

### 2.2.3 Sliding of bed materials
The scour hole is developed and the local bed slope in the scour hole becomes steep. It was reported that the sliding of bed materials occurs around the dike when the local bed slope exceeds a critical value $\theta_{bmax}$ (Michiue & Hinokidani 1992). In this study, the calculation of sliding is incorporated. The critical slope angle is estimated to be the angle of repose of bed materials.

The details of numerical method and procedure are described in the reference (Nagata et al. 2005) and are skipped in this paper. While $\Delta t$ is equal to 0.005 (s) in Case 1 and 3, $\Delta t$ is equal to 0.002 (s) in Case 2.

## 3 APPLICATION

The numerical models are applied to the experiments of flow and bed deformation around a spur dike conducted by Michiue & Hinokidani (1992). The experimental setup and hydraulic conditions are illustrated in Figure 1. In this study, three cases are tested by incorporating the different turbulence and sediment transport models, as shown in Table 1. Intention was to examine the appropriate combination of turbulence and bed deformation models.

## 4 RESULTS AND DISCUSSIONS

Figure 2 shows the temporal variation of maximum scour depths for three cases. It is clear that the equilibrium stages for the Case 1, 2 and 3 are reached

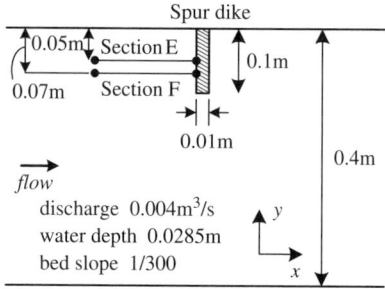

Figure 1. Experimental condition by Michiue & Hinokidani.

Table 1. Computational conditions.

| | Bed deformation model | turbulence model |
|---|---|---|
| Case 1 | non-equilibrium sediment transport model | nonlinear $k$-$\varepsilon$ model |
| Case 2 | equilibrium sediment transport model | nonlinear $k$-$\varepsilon$ model |
| Case 3 | non-equilibrium sediment transport model | standard $k$-$\varepsilon$ model |

in 6000, 1400, and 5000 (s), respectively. It is also observed that the time scale of scouring is different between the Case 1 (non-equilibrium sediment transport model) and the Case 2 (equilibrium sediment transport model), reaching the equilibrium stage faster in the latter case. By comparing the experiments with numerical results on time scale, the quantitative comparison is required. On the other hand, the scour depth for the Case 3 is shallowest in all cases. The maximum scour depth in the experiments is reported to be 0.09 m. The results for the Case 1 and the Case 2 are in close agreement with experiments, while the result for the Case 3 is under-predicted.

The flow characteristics and bed scouring in the initial processes are also examined. Figure 3 shows the temporal variation of flow patterns near bed and bed elevation contours in the initial scouring processes for the Case 1. The scouring is initiated form the tip of dike and the sediments along the stream line are picked up due to the flow converging to the tip (Fig. 3(a)), and the sediments from the tip of dike to main stream direction are scoured. While the flow is affected by the bed deformation with the development of scour hole, the water flows along the scour hole and the flow direction is not changed during the initial processes.

Figures 4 and 5 show the velocity vectors of sediment particles at $t = 60$ (s) and the representative trajectories of sediment movements (pick-up point to the final deposition point) picked up at $t = 60$(s). Some trajectories are presented in the plots for references. It is observed that the sand particles follow the flow and some particles (black) deposit behind the dike in the range of $x = 0.2$ to 0.4 (m).

Figure 6 shows the temporal variations of flow patterns near bed and bed elevation contours for the Case 2. The scouring occurs only at the tip of dike, and the characteristics of bed geometry are different

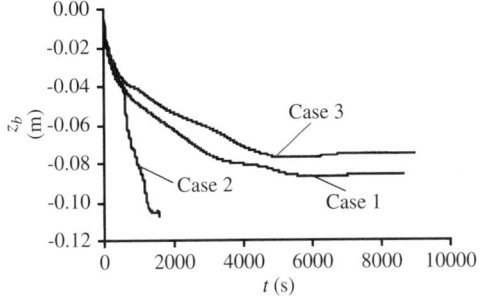

Figure 2.   Temporal variations of maximum scour depths.

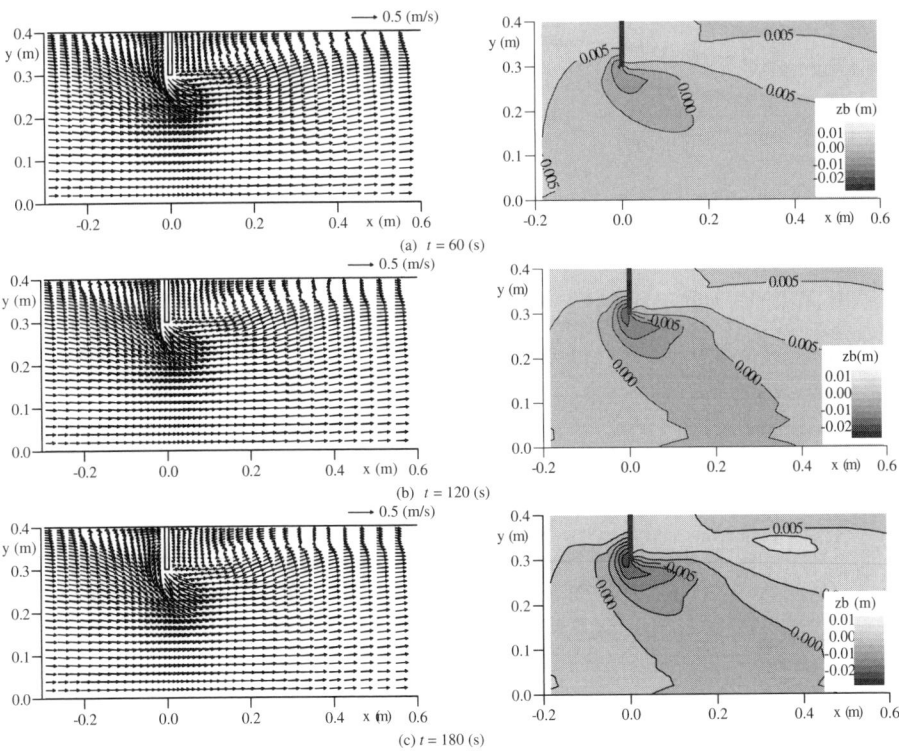

Figure 3.   Temporal variation of flows near bed and bed elevation contours (Case 1).

1255

between non-equilibrium and equilibrium sediment transport models. Then, as the scour hole develops, the flow to the behind of dike can be observed ($t = 60(s)$), and lots of sediments deposit behind the dike. The temporal change of bed-load flux vectors is examined (not shown in this paper). Directions of the flux are changed between $t = 10(s)$ and $60(s)$, as well as the flow near the bed.

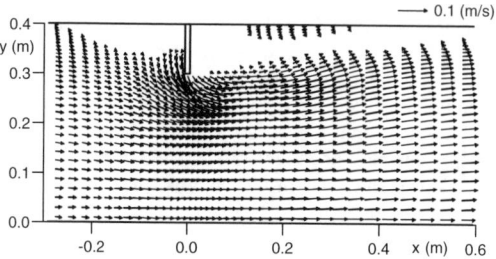

Figure 4.   Velocity vectors of sediment particles at $t = 60(s)$.

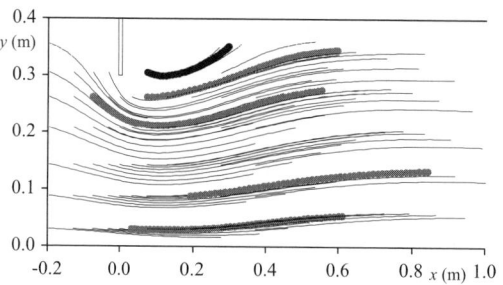

Figure 5.   Trajectories of sediment movements picked up at $t = 60(s)$.

Figure 7 shows the results of flow patterns near bed and bed elevation contours for the Case 3 and Figure 8 illustrates the velocity vectors at the longitudinal E-E sections, for the initial scouring stage ($t = 60(s)$). The flow structures for the Case 3 are almost the same as in the Case 1; but the flow separation in front of dike for the Case 3 is weaker (Fig. 7 (a)) than in Case 1. It can be also seen that the scale of vortex for the Case3 is smaller, and then the center position of vortex for the Case 3 is close to the front of dike, compared to the Case 1.

Figure 9 presents the bed contours in equilibrium stage. The maximum scour depth and the deposition region for the Case 1 are slightly smaller than the observations; however, results for the Case 1 show the maximum scour depth in front of dike uniformly, as seen in experiments, and the observed bed topography is also reproduced satisfactorily. On the other hand, the results for the Case 2 show the maximum scour depth only at the tip of dike, and the bed geometry near right bank is not reproduced, since the deposition and scouring regions are generated ($x = -0.1$ to $0.1$ (m)). For the Case 3, the maximum scour depth is under-predicted. Figure 10 represents comparison of velocity vectors at section E-E at equilibrium stage. The vertical flows, such as strong downward flow along the front face of spur dike and flow towards the upstream part of scour hole, is pronounced for the Case 1 and the Case 2. As the scour hole develops, the flow near bed directed upstream affects the pick-up of sediments. However, the vertical flow for the Case 3 is weak, compared to the Case 1 and the Case 2. This is the reason why the maximum scour depth for the Case 3 is under-predicted. Considering the results of the shape of scour hole, especially in front of dike, combination of the non-equilibrium sediment transport and

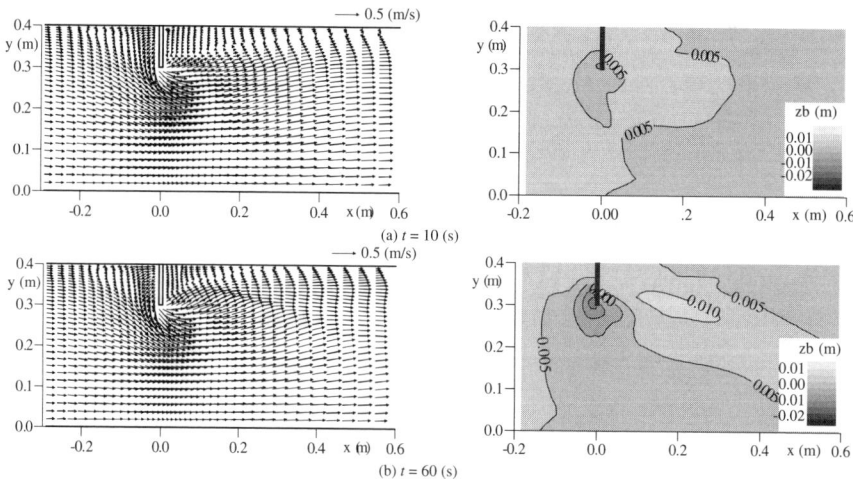

Figure 6.   Temporal variation of flows near bed and bed elevation contours (Case 2).

nonlinear $k$-$\varepsilon$ turbulence models, can predict the flow and bed geometry around the dike satisfactorily. Effects of non-equilibrium and equilibrium conditions of sediments should be further examined by changing the model scale.

## 5   CONCLUSIONS

In this study, local scouring around a spur dike is simulated by using different turbulence and bed deformation models. Applicability of models is verified by examining the temporal variation of flow near bed and bed variation. It is observed that the time scale of scouring and the bed topography, especially near the dike are different between the non-equilibrium and equilibrium sediment transport model. It is concluded that the numerical model, combining the non-equilibrium sediment transport model and nonlinear $k$-$\varepsilon$ turbulence model, can predict the flow and bed geometry accurately. By simulation of flow and local scouring under other hydraulic conditions, the applicability of model should be further examined, with eventual inclusion of suspended load into consideration.

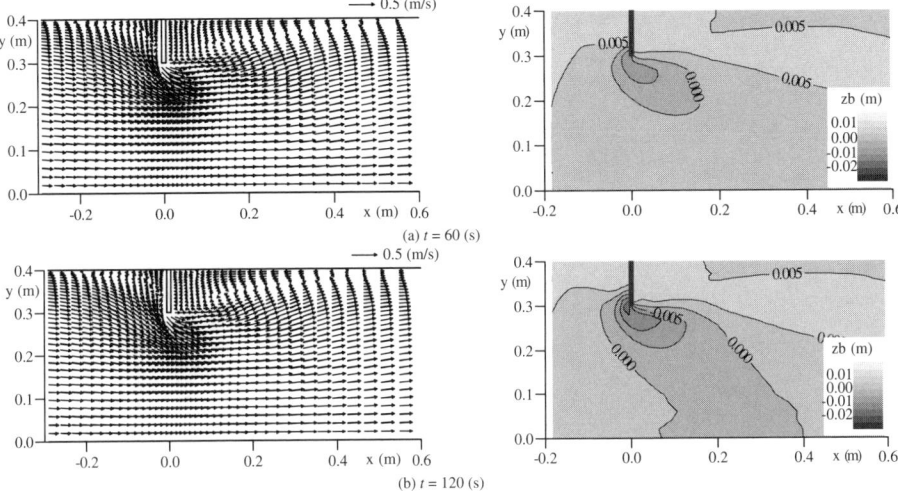

Figure 7.   Temporal variation of flows near bed and bed elevation contours (Case 3).

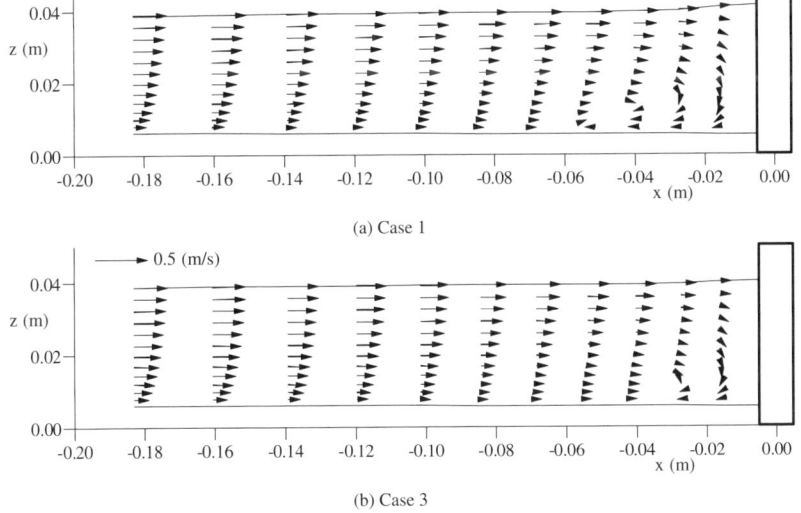

Figure 8.   Velocity vectors at section E-E at initial scouring stage ($t = 60$(s)).

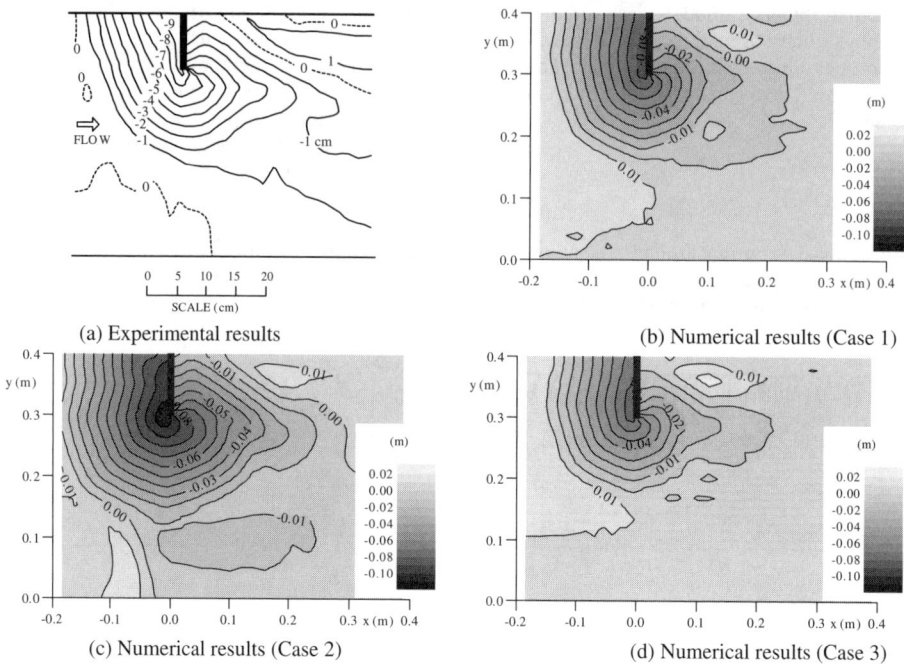

(a) Experimental results

(b) Numerical results (Case 1)

(c) Numerical results (Case 2)

(d) Numerical results (Case 3)

Figure 9. Comparison of bed contours at equilibrium stage.

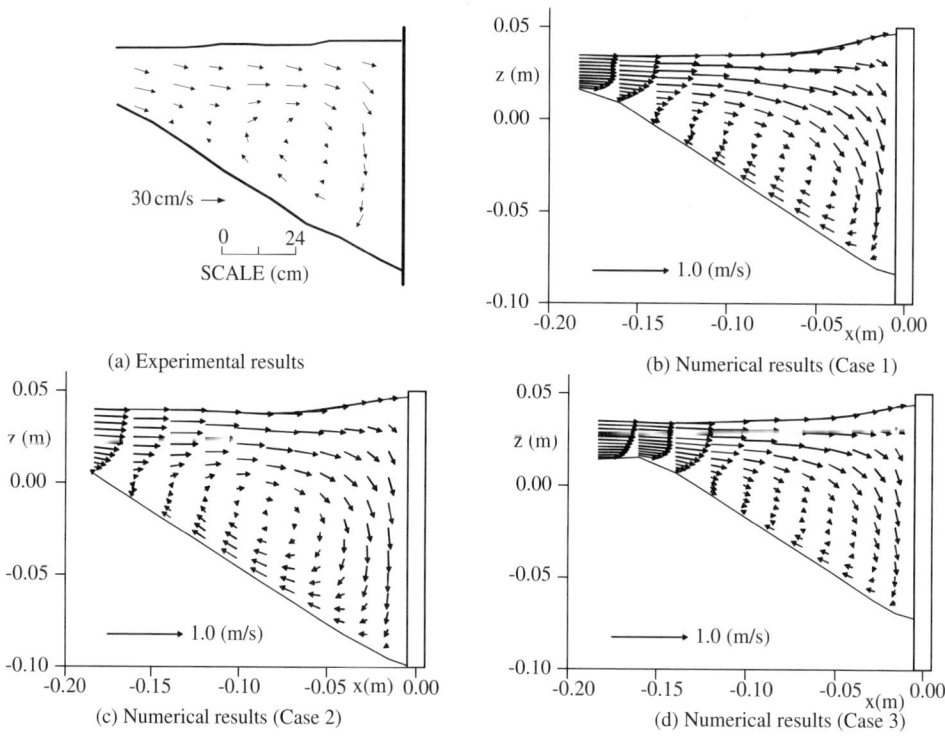

(a) Experimental results

(b) Numerical results (Case 1)

(c) Numerical results (Case 2)

(d) Numerical results (Case 3)

Figure 10. Comparison of velocity vectors at section E-E at equilibrium stage.

# REFERENCES

Bosch, G. & Rodi, W. 1998. Simulation of vortex shedding past a square cylinder with different turbulence models. *International Journal for Numerical Methods in Fluids*, 28, Wiley: 601–616.

Einstein, H.A. 1950. The bed-load function for sediment transportation in open channel flows. *Technical Bulletin*, 1026, U.S. Dept. of Agriculture, Soil Conservation Service, Washington, D.C.

Hasegawa, K. 1983. Ph.D. Thesis, Hokkaido University (in Japanese).

Fukuoka, S. Nishimura, T., Takahashi, A. Kawaguchi, A. & Okanobu, M. 1998. Design method of submerged groins. *Journal of Hydraulic, Coastal and Environmental Eng.*, 593/II–43: 51–68 (in Japanese).

Fukuoka, S. Watanabe A. Kawaguchi, H. & Yasutake, Y. 2000. A study of permeable groins in series installed in a straight channel. *Annual Journal of Hydraulic Eng.*, 44: 1047–1052 (in Japanese).

Gatski, T.B. & Speziale, C.G. 1993. On explicit algebraic stress models for complex turbulent flows. *Journal of Fluid Mechanics*, 254: 59–78.

Harlow, F.H. & Welch, J.E. 1965. Numerical calculation of time dependent viscous incompressible flow of fluid with free surface. *Physics of Fluids*, 8: 2182–2189.

Hirt, C.W., Nichols, B.D. & Romero, N.C. 1975. SOLA – a Numerical solution algorithm for transient fluid flows. *Los Alamos Scientific Report*, LA-5852.

Hosoda, T. 1990. Ph.D. Thesis, Kyoto University (in Japanese).

Hosoda, T., Kimura, I. & Onda, S. 2001. Some necessary conditions for a non-linear k-ε model in classified flow Patterns with a singular point. *Proceedings of 2nd International Symposium on Turbulence and Shear Flow Phenomena*, Stockholm, Sweden, 3: 155–160.

Ikeda, S, Sugimoto T. & Yoshiike T. 2000. Study on the characteristics of flow in channels with impermeable spur dikes. *Journal of Hydraulic, Coastal and Environmental Eng.*, 656/II–52: 145–155 (in Japanese).

Ishigaki, T., Ueno, T., Rahman, M.M. & Khaleduzzaman, A.T.M. 2004. Scouring and flow structure around an attracting groin, Greco, M., Carravetta, A. & Della Morte, R. (eds), *River Flow 2004*, 1: 521–525. Napoli: Balkema.

Iwagaki, Y. 1956. Fundamental study on critical tractive force, (I) Hydrodynamical study on critical tractive force. *Transactions of Jpn. Soc. Civ. Eng.*, 41: 1–21 (in Japanese).

Kimura, I., Hosoda, T. & Onda, S. 2002. Prediction of 3D flow structures around skewed spur dikes by means of a non-linear k-ε model. D.Bousmar and Y. Zech (eds), *River Flow 2002*, 1: 65–73. Belgium: Balkema.

Kimura, I. & Hosoda, T. 2003. A non-linear k-ε model with realizability for prediction of flows around bluff bodies. *International Journal for Numerical Methods in Fluids*, 42, Wiley: 813–837.

Kimura, I., Hosoda, T., Onda, S. & Tominaga, A. 2004. Computations of 3D turbulent flow structures around submerged spur dikes under various hydraulic conditions. Greco, M., Carravetta, A. & Della Morte, R. (eds), *River Flow 2004*, 1: 543–553. Napoli: Balkema.

Michiue, M. & Hinokidani, O. 1992. Calculation of 2-dimensional bed evolution around spur-dike. *Annual Journal of Hydraulic Eng.*, 63: 61–66 (in Japanese).

Muneta, N. & Shimizu, Y. 1994. Numerical analysis model with spur-dike concerning the vertical flow velocity distribution. *Proc. JSCE*, 497: 31–39 (in Japanese).

Nagata, N., Hosoda, T., Nakato, T. & Muramoto, Y. 2005. Three-dimensional numerical model for flow and bed deformation around river hydraulic structures, *Journal of Hydraulic Eng.*, ASCE, 131,(12): 1074–1087.

Nakagawa, H., Tsujimoto, T. & Murakami, S. 1986. Non-equilibrium bed load transport along side slope of an alluvial stream. *Proc., 3rd Int. Symp. on River Sedimentation*, Univ. of Mississippi, Miss.: 885–893.

Olsen, N.R.B. & Kjellesvig, H.M.K. 1998. Three-dimensional numerical flow modeling for estimation of maximum local scour depth. *Journal of Hydraulic Res.*, 36(4): 579–590.

Pope, S.B. 1975. A more general effective viscosity hypothesis. *Journal of Fluid Mech.*, 72: 331–340.

Sugiyama, H., Akiyama, M. & Matsubara, T. 1995. Numerical simulation of compound open channel flow on turbulence with a Reynolds stress model. *Journal of Hydraulic, Coastal and Environmental Engineering*, 515/II–31: 55–65 (in Japanese).

Tominaga, A. & Matsumoto, D. 2006. Diverse riverbed figuration by using skew spur-dike groups. Rui M.L. Ferreira, Elsa C.T.L. Alves, Joao G.A.B.Leak & Antonio H. Cardoso. (eds), *River Flow 2006*, 1: 683–691. Lisbon: Balkema.

Yoshizawa, A. 1984. Statistical analysis of the deviation of the Reynolds stress from its eddy viscosity representation. *Physics of Fluids*, 27(6): 1377–1387.

Yossef, M.F.M. & de Vriend, H.J. 2004. Mobile-bed experiments on the exchange of sediment between main channel and groyne fields. Greco, M., Carravetta, A. & Della Morte, R. (eds), *River Flow 2004*, 1: 127–133. Napoli: Balkema.

Yossef, M.F.M. & Rupprecht, R. 2006. Modeling the flow and morphology in froyne fields. Rui M.L. Ferreira, Elsa C.T.L. Alves, Joao G.A.B.Leak & Antonio H. Cardoso. (eds), *River Flow 2006*, 2: 1707–1713. Lisbon: Balkema.

Zhang, H., Nakagawa, H., Muto, Y., Baba, Y. & Ishigaki, T. 2006. Numerical simulation of flow and local scour around hydraulic structures. Rui M.L. Ferreira, Elsa C.T.L. Alves, Joao G.A.B.Leak & Antonio H. Cardoso. (eds), *River Flow 2006*, 2: 1683–1693. Lisbon: Balkema.

*River, Coastal and Estuarine Morphodynamics: RCEM 2007 – Dohmen-Janssen & Hulscher (eds)*
*© 2008 Taylor & Francis Group, London, ISBN 978-0-415-45363-9*

# Study on morphodynamics around stone-lined spur dikes in the Akashi river

K. Kanda & Y. Samoto
*Akashi National College of Technology, Akashi, Japan*

Z. Li
*Kobe University, Kobe, Japan*

ABSTRACT: In Japan, where death and injury frequently occur by flooding attributable to geographical and meteorological reasons, diversified banking and bank protection works have been provided from ancient times to protect human lives, property and agricultural crops from flood damage. The nation coexists well with rivers. Although flood control safety has been improved since the Meiji Era together with progress in modern civil engineering technology, in heavily populated urban districts where rivers are artificially straightened, fish and aquatic organisms lose their habitats, thereby accelerating destruction of natural environments. For this reason, today, a trend exists by which river construction methods are richly endowed with natural materials, numerous stones and wooden materials, to retain rich natural environments provided by rivers.

The Akashi River flowing through western part of Kobe City and Akashi City is a typical example of this sort of river. Many natural environment type constructions such as stone-lined bank protections, spur dikes and submerged wooden beds are structured from upstream to the river mouth. However, these structures are based mostly on experience and technologies gained centuries ago. Sufficient investigations to see if these river structures are adopted and functioning effectively in the highly developed modern society from flood control and environment viewpoints, have not been carried out yet. Many areas remain badly affected by flooding.

In this study, we attempted to clarify characteristics of flows and river bed deformations around spur dikes using field observations ,model experiments and 2-D numerical simulations that specifically examine stone-lined spur dikes constructed downstream from the Hirano bridge of the Akashi River. The relationship between flow and scouring characteristics was elucidated from surface flow velocity obtained by LSPIV method. For discussion, results of experiments were compared with results of observations of the actual river. Experimental results of position of scouring and dimensionless scouring depth exhibit similar characteristics to those observed with the river at the site, considering flow rate conditions and uncertainty. We also investigated a method for reducing and controlling local scouring that occurs during floods to establish a reasonable design and work execution method for stone-lined spur dikes.

## 1 INTRODUCTION

Recently, traditional methods of river construction are being reconsidered from a river-landscape and riverine ecology perspective in Japan. A spur dike is one such method of construction. However, the characteristics of flow around spur dikes are not well known. This paper describes an experimental and numerical examination of a model of flows around spur dikes. We also report observations of the riverbed level and flow velocity around spur dikes constructed at a place 8.8 km from the Akashi River mouth, along with examination of the adaptability of spur dikes to that river.

For this study of stone-lined spur dikes constructed on the downstream side of Hirano Bridge over the Akashi River, flood damage around the structure caused by a typhoon that hit this area (Typhoon No. 0423 in 2004) is identified by detailed field observation of riverbed configurations and flows around permeable-type spur dikes. Riverbed fluctuation characteristics are discussed based on numerical analyses and model experiments simulating the river channel at the site.

Furthermore, we developed a numerical analysis model for predicting local scouring configurations downstream of spur dikes. Their effectiveness was

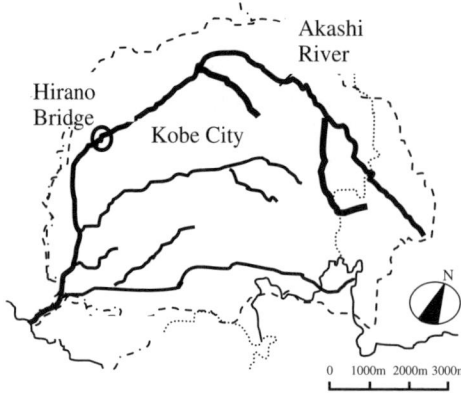

Figure 1. Outline of the Akashi River Basin.

Photograph 1. Spur dikes downstream of Hirano Bridge.

verified by field observations and results of model experiments.

## 2 OUTLINE OF SPURDIKES IN AKASHI RIVER AND FIELD OBSERVATION OF TYPHOON-FLOOD DAMAGE

### 2.1 Outline of Akashi River spur dikes

Akashi River is a class B river having river channel length of approximately 26 km and a basin area of 126.7 km². The main stream of the Akashi River line system, its source is located in Kita-ward, Kobe City. Most of its watershed is in Nishi-ward, Kobe City (Fig. 1). Photograph 1 shows the 12 sets of stone-lined impermeable overflow dikes that have been constructed on the right bank downstream from Hirano Bridge, located 8.8 km from the river mouth. These spur dikes use concrete footing protective blocks provided at the root of a high-water revetment as the

concealed bank protection. They are intended to provide greening of the berms between spur dikes and to create diversified flow conditions around dikes. Primary objectives of these spur dikes are to shift deep digging of the current foot protection towards the front edge of spur dikes while suppressing the flow velocity along the river bank.

### 2.2 Field observation of typhoon-flood damage

Akashi River was heavily damaged by flooding caused by a typhoon (No. 23; No. 0423) that hit this area in October 2004. Stones used to cover the surface of spur dikes were removed by the flood for spur dikes downstream of Hirano Bridge. Local scouring phenomena were noticed around the front edge of spur dikes. The peak flow rate at flooding, as estimated from the water level data obtained at the Fujiwara Bridge Observation Station, which is located about 4 km upstream from Hirano Bridge, was approximately 350 m³/s. The riverbed gradient was 1/180 and roughness coefficient in Manning's formula was set as 0.03. Figure 2 shows results of measurements using Total Station for riverbed configurations of surrounding parts of spur dikes after flooding. Figure 2 indicates that the riverbed around the front edge of each of spur dikes is being dug in a tongue-shape. Especially with five spur dikes downstream from $x = 0$ m, where a drainage channel merges, local scouring of as much as approximately 1.5 m was generated.

## 3 EXPERIMENTALS ON LOCAL SCOURING AROUND PREMEABLE TYPE SPUR DIKES

### 3.1 Outline of experiments

To identify local scouring characteristics around the permeable type spur dikes, moving bed experiments were performed using a water channel model (1/30 scale) for spur dikes downstream of the curved portion, where excessive local scouring was found by field observation. Figure 3 presents an outline of the experimental apparatus used.

The water channel is represented by a rectangular section having 6.3 m overall length, 0.8 m width, and 0.4 m height. For the riverbed, nearly uniform sands with mean grain size of $d = 0.088$ cm are placed with uniform 15 cm thickness. Water pumped from an underground water tank is passed through an electromagnetic flow meter, then introduced to the water channel. It flows down the water channel and is returned again to the underground water tank. A current-straightening part using filter materials is provided at the upstream end of the water channel and a water-level regulating plate, which can be adjusted to an arbitrary angle, is provided downstream.

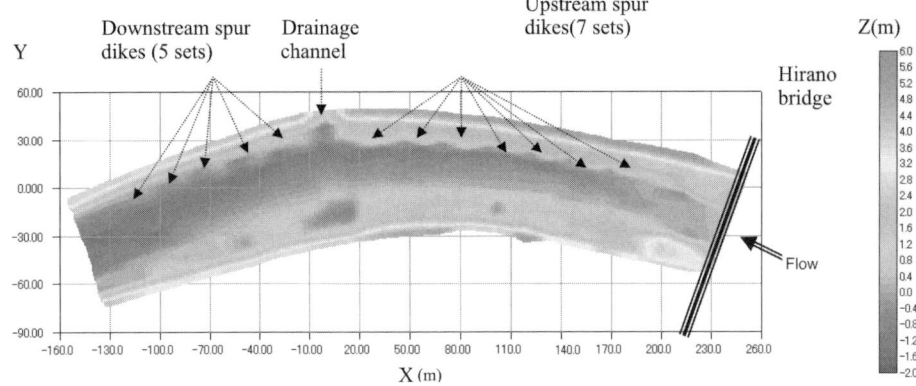

Figure 2. Riverbed profile downstream of Hirano Bridge according to field observation.

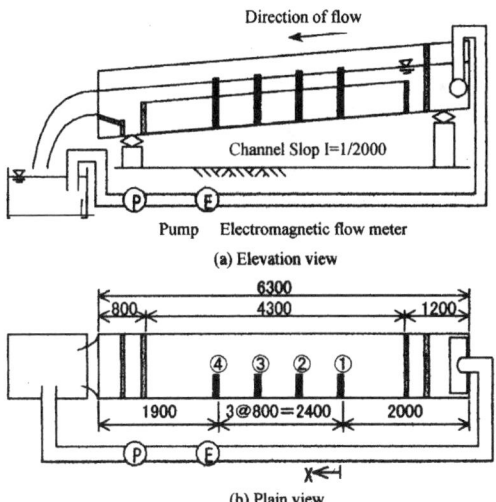

Figure 3. Outline of experimental apparatus (Unit: mm).

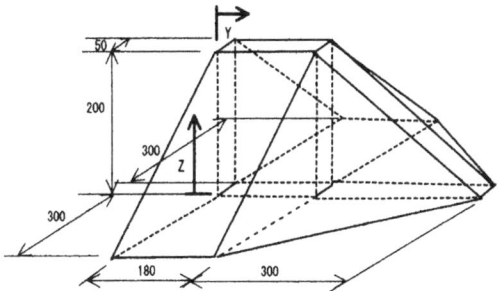

Figure 4. Profile of spur dike model (Unit: mm).

Photograph 2. Model of permeable type spur dike.

From a convenience perspective of arrangement of experimental results, the origin of the coordinate axis is set at 2 m from the upstream end of the water channel (upstream end of the first spur dike) in the downflow

direction, the $X$-axis is defined in the downstream direction, the $Y$-axis is defined in the right bank direction from the left bank wall, and the $Z$-axis is defined vertically upward from the origin at the bottom of the water channel bed.

Four spur dike models are placed on the left bank from a section 2 m from the upstream end of water channel with an 80-cm interval. To simulate spur dike profiles as closely as possible to actual ones at the site, as shown in Figure 4, a slope of 1:1.5

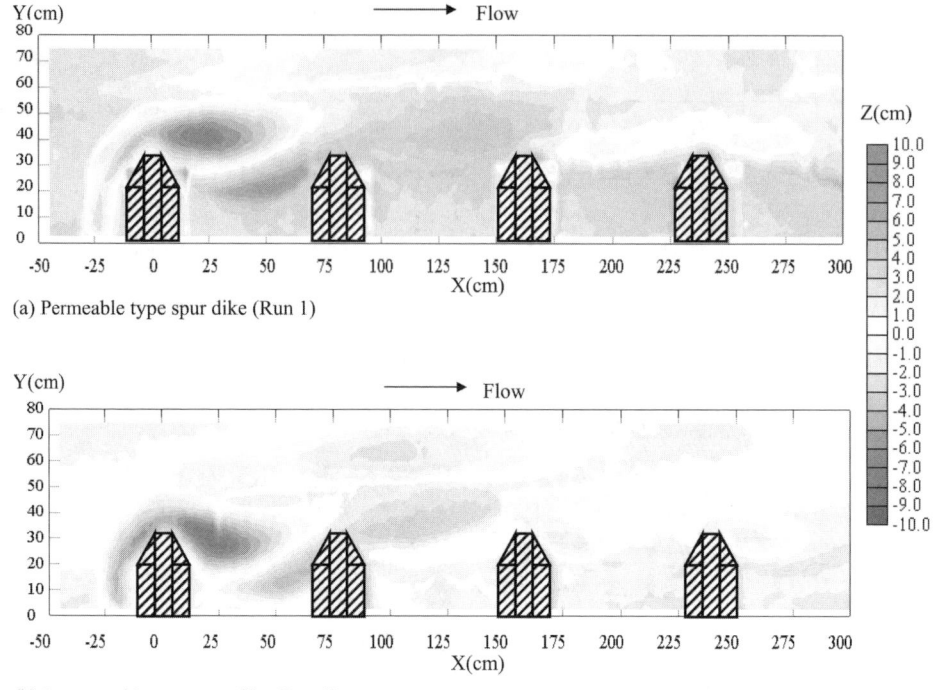

(a) Permeable type spur dike (Run 1)

(b) Impermeable type spur dike (Run 2)

Figure 5.   Comparison of amount of riverbed fluctuation around spur dikes.

Table 1.   Experimental conditions

| Run No. | Flow rate $Q$(l/s) | Uniform depth $ho$(cm) | Friction Velocity $U_*$ (cm/s) | Spur dike permeability |
|---|---|---|---|---|
| Run1 | 18.20 | 6.50 | 1.65 | Permeable |
| Run2 | | | | Impermeable |

is provided to the front and side faces of a rectangle (width:18 cm, height: 20 cm, thickness:5 cm) to generate a trapezoid. Pebbles of approximately 2 cm diameter are filled inside and aluminum angles are used as the framework. Their surface is covered with metal gauze to produce a permeable type spur dike (Photograph 2). For comparison, similar experiments (Run 2) were carried out for impermeable type spur dikes using wooden models. The flow rate used in the experiments was $Q = 18.2\,\mathrm{m}^3/\mathrm{s}$, which corresponds to the peak flow rate for the typhoon recorded at the site. Experimental conditions are shown in Table 1. The time for water introduction is 120 min for both runs.

For analyses of surface flow conditions, images taken from an oblique direction were corrected to perpendicular images and then subjected to Large Scale Particle Image Velocimetry (LSPIV) analysis

developed by (Fujita et al. 1998). In the current study, punched refuse of 5 mm diameter was used as the tracer to investigate the relationship between river bed deformations around the spur dikes and surface flow velocity. The surface flow velocity around spur dikes was measured using LSPIV.

3.2   *Scouring characteristics around spur dikes*

Local scouring around spur dikes is of two types: static scouring, by which dragging power is greater than the moving limit of the riverbed only at spur dikes and only riverbed sands around spur dikes are moved; and dynamic scouring, by which dragging power is great and running sands occur throughout the riverbed. The critical friction velocity of the riverbed materials (mean grain size $d = 0.088\,\mathrm{cm}$) evaluated using Iwagaki's formula is $U_{*cr} = 2.20\,\mathrm{cm/s}$. Meanwhile, friction Velocity $U_*$ obtained from the uniform flow depth is as shown in Table 1; this is a static scouring condition where $U_* < U_{*cr}$.

Figure 5 presents a comparison of riverbed fluctuation between permeable type spur dike 120 min after water introduction in Run 1 and an impermeable type spur dike (Run 2) (Morita et al. 2006) under the same flow rate conditions. The black region shows scouring and white represents sedimentation. The following findings were obtained.

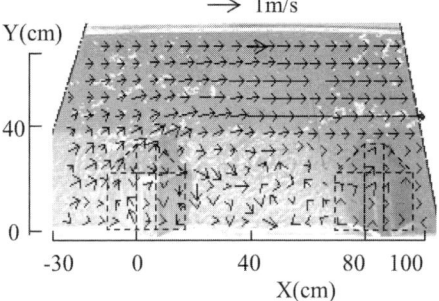

(a) Analytical result of surface velocity of Run 1

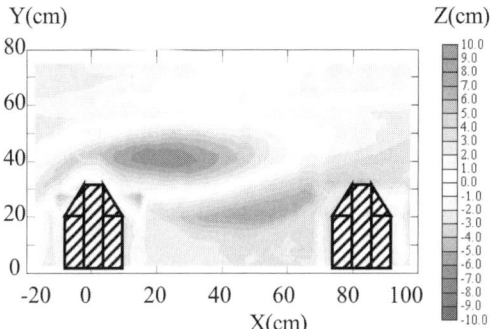

(b) Contour maps showing river bed deformation

Figure 6.   Results of LSPIV analysis and contour maps in Run 1.

For a permeable type spur dike (Run 1), the scouring profile presents an inverted cone with a cone pointing downstream of the first spur dike front and at the spur dike rear, where scouring developed from the spur dike front downstream to the water channel. The maximum scouring depth at the spur dike front is 6.2 cm, which is nearly identical to the uniform flow depth, which is approximately 40% smaller than that of the imperme-able type spur dike (Run 2). With the permeable type spur dike, scouring occurs even at the rear of the first spur dike because of the flow passing through the spur dike. In contrast, with the impermeable type, because water flow is directed toward the water channel center part because of water splash effects, scouring does not take place at the rear of spur dike. Instead, a trend of sedimentation is apparent. These features are visible even when the flow rate is changed. It is considered that, although use of permeable type spur dike can reduce the amount of local scouring around it, scouring occurs throughout a wider region.

3.3   *Surface flow velocity around spur dikes*

For flow characteristics of the surface flow in Run 1, the following findings are apparent from Figure 6. At the main stream part, the flow velocity is high and the direction of flow is nearly constant. Between spur dikes, flow velocity becomes slow and the direction of flow is not constant, thereby causing vortex flow. Upstream from the spur dikes, flow velocity is reduced because of the water splash effect and the maximum scouring depth that occurs on the extension line of fast flow generated along with the upstream slope. Accordingly, relaxation of the angle at front edge of the spur dike is useful for reducing scouring because it prevents the flow from being concentrated in one direction.

3.4   *Comparison with results of surveying in Akashi River*

Before our field observations, the Akashi River was heavily damaged by No. 23 typhoon, which hit this area in October 2004. Remarkable scouring phenom-ena resulting from flooding were apparent around spur dikes downstream from the Hirano greater bridge. The peak flow rate at flooding was calculated using water level data obtained at the Fujiwara observation station located about 4 km upstream from the Hirano greater bridge. The peak flow rate was calculated from the water level and length of the flood channel using Man-ning's formula as approximately 350 m³/s. Meanwhile, the river bed gradient was 1/180 and the roughness coefficient in Manning's formula was set as 0.03.

If Froude's similarity rule is used, the flow rate of 350 m³/s corresponds to the flow rate 18.20 l/s (actual flow rate 322 m³/s), which is a condition for overflowing for all spur dikes in model experiments. Subsequently, comparison was made with experimen-tal results (Run 1) relating to river bed deformation for spur dikes with a slope. Figure 7 shows results of the observation of the river bed profile around spur dikes in the Akashi River and experimental results (Run 1) expressed in dimensional form. For initial river bed height of spur dikes in the Akashi River, the average of the river bed height downstream was used; it was considerably less affected by spur dikes in the Akashi River.

Contour maps showing river bed deformation of the Akashi River shown in Figure 7(a) revealed that the maximum scouring depth was caused downstream from the front edge of the first spur dike. The maxi-mum scouring depth in dimensionless form is $Z/H = 1.0$ while the same obtained from experimental results is $Z/H = 1.5$. Therefore, an identical tendency as that noted with model experiments was obtained. Reasons for that position of deposition and scouring configu-rations are different are attributable to that shape and materials of the river bank differ, sands are supplied constantly from upstream in the Akashi River. Further-more, the river bed configuration is changed because of flooding that occurred in the past. Consequently, we infer that river bed deformation around the spur dikes can be predicted to a certain extent by model

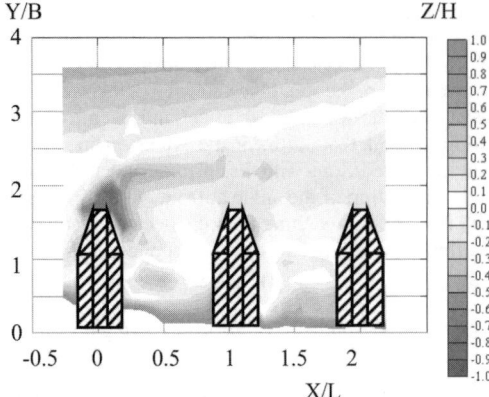

Y/B         Z/H

(a) Results of surveying in the Akashi River

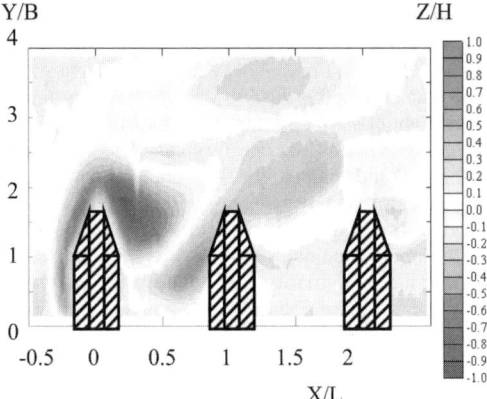

Y/B         Z/H

Figure 7. Comparison of dimensionless river bed profile (Around the 1-3rd spur dike) B:The length of the spur dike, L:Distance spur dikes, H:Height of spur dike from river bed surface.

experiments in which local hydraulic conditions are examined.

Comparison of results of field observation shows that, although a similar point is recognized in such that the maximum scouring depth is nearly identical with uniform flow depth, for scouring profiles, detailed observations are necessary, considering riverbed material characteristics at the field, geographical and hydraulic conditions.

## 4 NUMERICAL ANALYSIS ON MORPHODYNAMICS AROUND SPUR DIKES

### 4.1 Fundamental equation and analytical method

For evaluation of riverbed fluctuation around Akashi River spur dikes caused by typhoon-flooding, we carried out two-dimensional riverbed fluctuation analyses

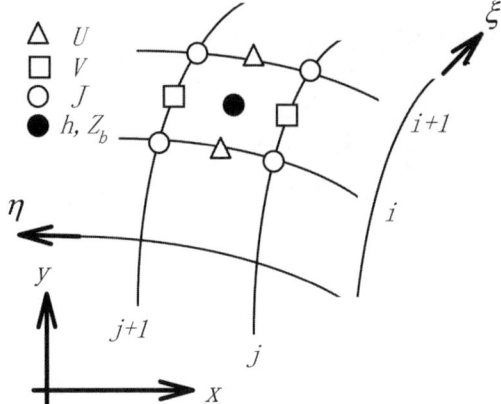

Figure 8. Boundary fitted coordinate system.

based on a general coordinate system (Figure 8). The basic equations used in the analyses are shown below (Kanda et al. 2003).

[Continuity equation of flow]

$$\frac{\partial}{\partial t}\left(\frac{h}{J}\right) + \frac{\partial}{\partial \xi}\left(\frac{Uh}{J}\right) + \frac{\partial}{\partial \eta}\left(\frac{Vh}{J}\right) = 0 \tag{1}$$

[Continuity equation for bed elevation]

$$\frac{\partial}{\partial t}\left(\frac{z_B}{J}\right) + \frac{1}{1-\lambda}\left\{\frac{\partial(q_{Bx}/J)}{\partial \xi} + \frac{\partial(q_{By}/J)}{\partial \eta}\right\} = 0 \tag{2}$$

[Momentum equation in $\xi$ direction]

$$\frac{\partial}{\partial t}\left(\frac{M}{J}\right) + \frac{\partial}{\partial \xi}\left(\frac{UM}{J}\right) + \frac{\partial}{\partial \eta}\left(\frac{VM}{J}\right) =$$

$$- gh\left(\frac{\partial \xi/\partial x}{J}\frac{\partial z_s}{\partial \xi} + \frac{\partial \eta/\partial x}{J}\frac{\partial z_s}{\partial \eta}\right) - \frac{\tau_{bx}}{\rho J}$$

$$+ \frac{\partial \xi/\partial x}{J}\frac{\partial}{\partial \xi}\left(-\overline{u'^2}h\right) + \frac{\partial \xi/\partial y}{J}\frac{\partial}{\partial \xi}\left(-\overline{u'v'}h\right)$$

$$+ \frac{\partial \eta/\partial x}{J}\frac{\partial}{\partial \eta}\left(-\overline{u'^2}h\right) + \frac{\partial \eta/\partial y}{J}\frac{\partial}{\partial \eta}\left(-\overline{u'v'}h\right) \tag{3}$$

[Momentum equation in $\eta$ direction]

$$\frac{\partial}{\partial t}\left(\frac{N}{J}\right) + \frac{\partial}{\partial \xi}\left(\frac{UN}{J}\right) + \frac{\partial}{\partial \eta}\left(\frac{VN}{J}\right) =$$

$$- gh\left(\frac{\partial \xi/\partial y}{J}\frac{\partial z_s}{\partial \xi} + \frac{\partial \eta/\partial y}{J}\frac{\partial z_s}{\partial \eta}\right) - \frac{\tau_{by}}{\rho J}$$

$$+ \frac{\partial \xi/\partial x}{J}\frac{\partial}{\partial \xi}\left(-\overline{u'v'}h\right) + \frac{\partial \xi/\partial y}{J}\frac{\partial}{\partial \xi}\left(-\overline{v'^2}h\right)$$

$$+ \frac{\partial \eta/\partial x}{J}\frac{\partial}{\partial \eta}\left(-\overline{u'v'}h\right) + \frac{\partial \eta/\partial y}{J}\frac{\partial}{\partial \eta}\left(-\overline{v'^2}h\right) \tag{4}$$

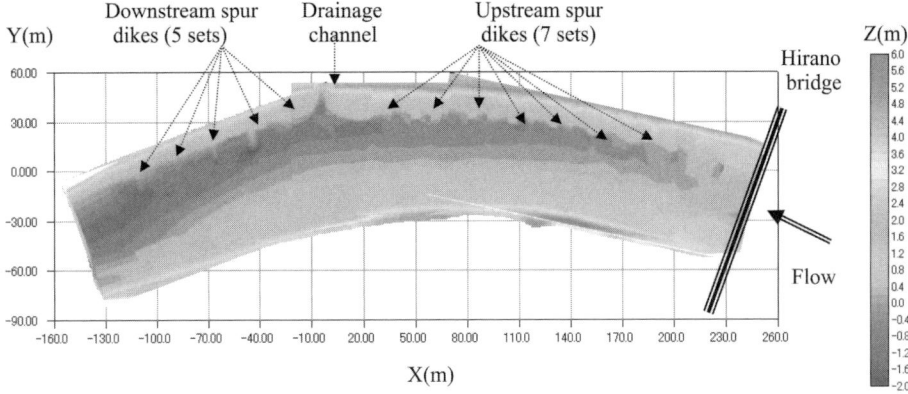

Figure 9.   Results of analysis of riverbed level ($Q = 350 \, \mathrm{m^3/s}$).

In those equations, $t$ is time, $x$ and $y$ are orthogonal space coordinate system, $M = uh$ and $N = vh$ are flux of flow in the $x$ and $y$ directions, $g$ is acceleration of gravity, $h$ is water depth, $\rho$ is the water density, $z_s$, $z_B$ are the water level and riverbed level, $\tau_{bx}$, $\tau_{by}$ are the shear stress of basal plane, $-u'^2$, $-u'v'$, $-v'^2$ are depth-averaged Reynolds stress, $q_{Bx}$, $q_{By}$ are bed load transport rates in the $x$ and $y$ directions, and $\lambda$ is the void ratio of riverbed sands. In addition, $J$ is the Jacobian of coordinate transform, and $U$, $V$ are variable components of flow velocity.

For differentiation of governing equations, the finite volume method of a staggered scheme was applied. The upwind difference was used for the special difference of the advective term and Adams-Bashforth method[1] was used for time subtraction. As for the flow rate, $350 \, \mathrm{m^3/s}$ was used from results obtained at the site, and the planned riverbed height used at the spur dike installation was used as the initial riverbed profile. The bed load transport rate is estimated by using Brown's formula, as below.

$$q_{Bs} = 10 \left( \frac{U_*^2}{\Delta g d} \right)^2 U_* d \qquad (5)$$

Here, $\Delta$ is submerged specific density of sand particle.

### 4.2   Result of numerical simulation

Results of analysis are shown in Figure 9. Although permeability of the spur dikes was not considered in the calculation, comparison with results of field observation (Figure 2) reveals that similar results are obtained for scouring profiles around the downstream spur dike. It is therefore expected that the prediction accuracy will be improved in the future if spur dike permeability is taken into account.

## 5   CONCLUSIONS

In this study, for prediction of flow around stone-lined spur dikes and river bed deformation, river bed deformation because of the flow was elucidated using experiments with scale models. The relationship between flow and scouring characteristics was elucidated from surface flow velocity obtained by LSPIV method. For discussion, results of experiments were compared with results of observations of the Akashi River. Experimental results of position of scouring and dimensionless scouring depth exhibit similar characteristics to those observed with the river at the site, considering flow rate conditions and uncertainty. We also analyzed riverbed fluctuation at flooding using a two-dimensional numerical model and clarified that riverbed fluctuation around the spur dike by typhoon-flooding is predictable to a certain degree.

## REFERENCES

Morita, A. et al. 2006. Study on Local Scour around Spur Dikes in the Akashi River, *Proceedings of 3rd International Conference on Scour and Erosion*, Amsterdam: 1197–1206.
Fujita, I. et al. 1998. Large-scale particle image velocimetry for flow analysis in hydraulic engineering applications, *Journal of Hydraulic Research Vol.36, No.3*: 397–414.
Kanda, K. et al. 2003. Experimental study on local Scour and its reduction around the Ninoatate weir in the Hyakken River. *Proceedings of 3rd IAHR Symposium on River, Coastal and Estuarine Morphodynamics,* Barcelona: 13–16.

River, Coastal and Estuarine Morphodynamics: RCEM 2007 – Dohmen-Janssen & Hulscher (eds)
© 2008 Taylor & Francis Group, London, ISBN 978-0-415-45363-9

# Author index